ADVANCED QUANTUM CONDENSED MATTER PHYSICS

Condensed matter physics has fast become the largest discipline within physics. Based on an established course, this comprehensive textbook covers one-body, many-body, and topological perspectives. It is the first textbook that presents a comprehensive coverage of topological aspects of condensed matter as a distinct, yet integrated, component. It covers topological fundamentals and their connection to physics, introduces Berry phase and Chern numbers, describes general topological features of band structures, and delineates its classification. Applications as manifest in the quantum Hall effect, topological insulators, and Weyl semimetal are presented. Modern topics of current interest are explored in-depth, helping students prepare for cutting-edge research. These include one-electron band theory, path integrals, and coherent states functional integrals as well as Green and Matsubara functions, spontaneous symmetry breaking, superfluidity, and superconductivity. Multiple chapters covering quantum magnetism are also included. With end-of-chapter exercises throughout, it is ideal for graduate students studying advanced condensed matter physics.

MICHAEL EL-BATANOUNY is a professor of physics at Boston University. He is the author of *Symmetry* and *Condensed Matter Physics: A Computational Approach* (Cambridge University Press, 2008). His research area is in surface physics, where he has an international reputation in the topics of nonlinear surface physics, surface dynamics, surface magnetism, and more recently topological aspects of surfaces, especially in electronphonon-related phenomena.

ADVANCED QUANTUM CONDENSED MATTER PHYSICS

One-Body, Many-Body, and Topological Perspectives

MICHAEL EL-BATANOUNY

Boston University

CAMBRIDGE
UNIVERSITY PRESS

CAMBRIDGE
UNIVERSITY PRESS

University Printing House, Cambridge CB2 8BS, United Kingdom

One Liberty Plaza, 20th Floor, New York, NY 10006, USA

477 Williamstown Road, Port Melbourne, VIC 3207, Australia

314–321, 3rd Floor, Plot 3, Splendor Forum, Jasola District Centre, New Delhi – 110025, India

79 Anson Road, #06–04/06, Singapore 079906

Cambridge University Press is part of the University of Cambridge.

It furthers the University's mission by disseminating knowledge in the pursuit of education, learning, and research at the highest international levels of excellence.

www.cambridge.org
Information on this title: www.cambridge.org/9781108480840
DOI: 10.1017/9781108691291

First published 2020

Printed in the United Kingdom by TJ International Ltd, Padstow Cornwall

A catalogue record for this publication is available from the British Library.

ISBN 978-1-108-48084-0 Hardback

This book is dedicated to my loving wife Gloria, whom I would like to thank for her patience, encouragement, and constant support. I would also like to dedicate the book to my sister Nagwa and to the memory of my parents Maurice and Hayat.

Contents

Preface

This book has been developed over the past 25 years, undergoing constant modifications, improvements, and attempts to include emerging developments.

The book integrates, for the first time, three aspects of condensed matter physics: one-body, many-body, and topological perspectives, and, accordingly, is organized in three parts. The presentation throughout is quite advanced in comparison with Ashcroft and Mermin's book or equivalent texts.

It is designed to accommodate an advanced two-, or possibly three-, semester graduate course, following an introductory course in solid-state physics as a prerequisite. Since the book covers a plethora of topics and subtopics, it is left to the instructor's discretion to pick and choose chapters and sections.

Part I, titled "One-Electron Physics," presents many aspects of the one-electron approach, and comprises seven chapters. It starts with an introductory chapter summarizing elementary building blocks of solid-state physics, including the Born–Oppenheimer approximation. It also reviews time-reversal symmetry and its implications. The following three chapters cover the one-electron band theory. Chapter 2 develops the one-particle formalism within Hartree–Fock and density functional frameworks and examines validity bounds. The effects of exchange and correlation are also discussed, bringing out the idea of an exchange hole for fermions. This is followed by outlines of the different methods of electronic band calculations in Chapter 3, with detailed presentation of the pseudopotential and tight-binding methods, including Harrison's matrix element scaling. Chapter 4 derives the spin–orbit interaction from the Dirac equation and discusses its manifestations in electronic structure using the $\mathbf{k} \cdot \mathbf{p}$ method. The Dresselhaus and Rashba Hamiltonians are developed within the context of bulk and structural inversion symmetry breaking, and applications in two-dimensional electron gas are presented.

The fifth chapter covers linear response from the one-electron viewpoint, including causality and the Kramers–Krönig relation. It develops the Kubo conductivity formula with special reference to the quantum Hall effect. The longitudinal and transverse dielectric functions are derived, and the ideas of intraband and interband, both direct and indirect, optical transitions are discussed.

Chapter 6 presents a detailed account of phonons, lattice dynamics, and experimental techniques for measuring phonon dispersions. It starts with deriving the electron–phonon coupling in terms of symmetry-adapted (or normal) mode coordinates. The electronic contributions to phonon energies are developed in terms of the density–density response function, and the implication of translational invariance is explained. The developed expressions are then pedagogically used to construct phenomenological models for phonons in semiconductors and insulators – "pseudocharge models." The different experimental probes used in measuring the static and dynamic structure factors of solids are introduced, with emphasis on neutron and helium atom scattering techniques that map bulk and surface dispersions, respectively.

The last chapter of Part I explains the effect of dimensionality on electronic susceptibilities, including nesting effects. It describes the onset of instabilities, as manifest in the Peierls phenomenon, and delineates their emergent order parameters. It also introduces the idea of a Kohn anomaly, and derives the giant Kohn anomaly as a consequence of the one-dimensional Peierls instability.

Part II, titled "Topological Phases," is focused on describing manifestation of topological aspects and phenomena in condensed matter systems. The concept of topological order in the electronic phases of condensed matter emerged in the early 1980s and reached maturity over a decade ago. Understanding these concepts and experimentally observed consequences has become one of the most important themes of modern condensed-matter physics. Aptly, Part II starts with a short Chapter 8, reviewing the historical development of topology in condensed matter physics. The main material of Part II is covered in four more chapters. Chapter 9 presents relevant aspects of topology, such as homeomorphism, fiber and vector bundles, connection, curvature, parallel transport, and holonomy, and ends with establishing the relevance of topology to physics. Chapter 10, titled "Berry-ology," describes Berry's phase, connection, and curvature, and derives the Chern topological number. It presents two pedagogically important but distinct examples: a two-level system, with its concomitant "magnetic monopole," and the molecular Aharonov–Bohm effect, where the interplay between the quantum electronic and ionic motions leads to fascinating topological manifestations.

Chapter 11 expounds on the topological aspects of the band structure of insulators and presents two early discoveries. It starts with developing how the ideas of *topological equivalence* and *adiabatic continuity* lead to the emergence of distinct classes of insulator Hamiltonians, and how this, in turn, leads to bulk-boundary correspondence – the connection between bulk topological invariants and edge or surface states. The classification of topologically nontrivial and trivial phases, based on fundamental discrete symmetries and dimensionality, the "*tenfold way*," is explained – an appendix outlining the procedures for the classification of insulator Hamiltonians is placed at the end of the chapter. This is followed by defining the mapping of d-dimensional Brillouin zones onto a d-dimensional Brillouin torus $\mathbb{T}^d$ and Bloch Hamiltonians, and describing the construction of Bloch bundles on the torus base manifold. Time-reversal symmetry, Kramers' band degeneracy, "time-reversal invariant momenta," and the implied vanishing of Berry's curvature are delineated.

Next, early discoveries of manifest topology in condensed matter – the integer quantum Hall effect and the modern theory of polarization – are discussed in detail. Finally, the $\mathbb{Z}_2$ topological invariant is derived using the sewing matrix, time-reversal polarization, and the non-Abelian Berry connection (the Wilczek–Zee gauge presented in the appendix).

The last chapter of Part II concerns Dirac materials and Dirac fermions. It starts with an introduction to graphene, its Dirac points and cones, and the Dirac fermion Hamiltonian in the vicinity of the K-points. This is followed by a presentation of the time-reversal symmetry-breaking Chern insulators, with special focus on Haldane's model. Next, the quantum spin Hall effect, manifest in the graphene-like model of Kane–Mele, with strong spin–orbit (SO) coupling, is described. The chapter ends with a detailed description of Weyl semimetals. An appendix describing the Dirac and Weyl equations is placed at the end of the chapter.

Part III, titled "Many-Body Physics," covers many related concepts, aspects, and phenomena. Chapter 13 teaches the techniques of many-body theory. It introduces the idea of second quantized operators in the many-particle domain, Fock spaces, field operators, and vacuum states, and outlines how canonical transformations can be applied to solve many-body problems. It also contains a section on coherent states, as eigenstates of the annihilation operator, including the development of Grassmann's algebra and calculus for fermions. Chapter 14 presents a Hartree–Fock perturbative treatment of the interacting electron gas within the jellium model and highlights its drawbacks. It also introduces the concept of the random phase approximation (RPA).

Chapter 15 develops the many-body, one-particle Green function, and explains its physical interpretation in terms of the spectral function, self-energy, and quasi-particle lifetime. Its application in angle-resolved photoemission spectroscopy is presented in detail. The time-evolution operator in the interaction picture is derived, time-ordering and adiabatic switching on are introduced as precursor tools to construct the Feynman–Dyson many-body perturbation theory. A detailed account of Wick's theorem, normal ordering, and contractions is outlined. Feynman diagrams are constructed in $(\mathbf{x}, t)$ and $(\mathbf{k}, \omega)$, and the emergence of the infinite Dyson series from irreducible diagrams is outlined. This is followed by the development of the two-particle Green function and the Particle–hole excitation spectra for noninteracting and interacting systems. The diagrammatic application of RPA in the latter case is described. Finally, the finite-temperature Matsubara Green function is introduced and developed, together with its Fourier series representation and the evaluation of Matsubara sums. Chapter 16 presents functional integral methods of quantum many-body theory. Starting with Feynman's path integral, it develops functional integrals of partition functions in imaginary time and extends these techniques to many-body systems. Finally, it expands the formulation in the coherent-state basis, and describes the application of the Hubbard–Stratononvich transformation and the saddle-point approximation.

Chapter 17 treats the Bose–Einstein condensation, and explains superfluidity from the Bogoliubov and Ginzburg–Landau perspectives. It also describes the concept of spontaneous symmetry breaking and Goldstone modes. Chapter 18 covers Landau's Fermi liquid theory, and Chapter 19 treats non-Fermi liquids and quantum critical points and describes Luttinger liquid theories. Chapter 20 develops the formalism of electron–phonon interaction within the Matsubara framework. Chapter 21 deals with superconductivity. It introduces the concept of Cooper pairing and develops a diagrammatic approach to the Cooper instability. The Bardeen–Cooper–Schrieffer (BCS) Hamiltonian is then constructed and solved with the aid of the Bogoliubov–Valatin transformation. This is followed by a presentation of the Nambu–Gorkov formalism, and the Gorkov anomalous Green function. Finally, a detailed account of the Ginzburg–Landau perspective of superconductivity is given, ending with a derivation of the Meissner effect and an explanation of the Anderson–Higgs mechanism.

The last three chapters of Part III are dedicated to several aspects of quantum magnetism. The first provides detailed analysis of mechanisms of exchange coupling: direct or potential exchange, kinetic exchange, superexchange, polarization exchange, Dzialoshinskii–Moriya, double exchange, and Ruderman–Kittel–Kasuya–Yosida (RKKY). The effects of crystal fields and the single-site anisotropy are also discussed. The second chapter on magnetism covers ferromagnetic and antiferromagnetic insulators, describing the nature of their respective ground states and deriving their spin-wave excitation spectra with the aid of the Holstein–Primakoff transformation. The final chapter deals with magnetism in itinerant systems. It starts with the Stoner mean field theory, as derived from a simple Hubbard model, and Stoner excitations and spin-waves obtained with the aid of RPA. The concept of nesting and spin-density waves is then discussed. This is followed by a detailed presentation of Anderson's magnetic impurity model, and its relation to the kondo model through the Schrieffer–Wolff transformation. Finally, a detailed account of the Kondo effect and the Kondo resonance is given.

Most chapters contain copious sets of problems, and each chapter contains a large number of helpful figures and illustrations. It is recommended pedagogically that problem solution should involve close interaction between students and instructor.

Part One
One-Electron Theory

1

Preliminaries

1.1 Periodic Lattices, Brillouin Zones, and Bloch's Theorem

We consider the single-particle Hamiltonian defined on a one-dimensional periodic lattice of period a:

$$\mathcal{H} = \frac{p^2}{2m} + V(x)$$

$$V(x) = V(x \pm na), \qquad n \text{ integer}$$

This type symmetry requires the introduction of a *discrete translation operator* $\tau(a)$. We define its action on a position ket as

$$\tau(a) \Rightarrow \tau(a) \left| x' \right\rangle = \left| x' + a \right\rangle, \quad \tau^\dagger(a) \left| x' \right\rangle = \left| x' - a \right\rangle$$

When applied to the potential energy operator $V(x)$, we get

$$\tau(a) V(x) \tau^\dagger(a) = V(x + a)$$

and we find that

$$\tau(a) \mathcal{H} \tau^\dagger(a) = \mathcal{H} \Rightarrow \mathcal{H} \tau(a) = \tau(a) \mathcal{H}$$

hence,

$$\boxed{[\mathcal{H}, \tau(a)] = 0}$$

τ is unitary, $\tau^\dagger(a) = \tau^{-1}(a)$, but not Hermitian; hence, it may have complex eigenvalues!

1.1.1 *Eigenvalues and Eigenkets of* τ

The position representation of an eigenket $|\alpha\rangle$ of τ, $\langle x' \,|\, \alpha\rangle$, is just the wavefunction $\Phi_\alpha(x')$, and we find

$$\langle x' \,|\, \alpha\rangle = \Phi_\alpha(x')$$
$$\langle x' \,|\tau(a)|\, \alpha\rangle = \langle x' - a \,|\, \alpha\rangle = \Phi_\alpha(x' - a)$$
$$\tau(a)\,\Phi_\alpha(x') = \Phi_\alpha(x' - a)$$

An eigenfunction Φ of τ, satisfies

$$\tau(a)\,\Phi_\alpha(x') = \Phi_\alpha(x' - a) = \lambda\,\Phi_\alpha(x')$$

Using the mathematical trick of fictitious periodic boundary conditions

$$\Phi_\alpha(x' - Na) = \Phi_\alpha(x')$$

we get

$$(\tau(a))^N\,\Phi_\alpha(x') = \Phi_\alpha(x' - Na) = \lambda^N\,\Phi_\alpha(x') = \Phi_\alpha(x')$$

Hence, we find that

$$\lambda^N = 1 \Rightarrow \lambda = \exp\left[i\frac{2\pi m}{N}\right] = \exp\left[i\frac{2\pi m a}{Na}\right] = e^{ik_m a},$$
$$k_m = \frac{2\pi}{a}\frac{m}{N}, \quad -\frac{N}{2} \le m \le \frac{N}{2} \tag{1.1}$$

Notice that k_m is proportional to the ratio of two integer numbers, m and N. The k_ms are uniformly distributed with intervals, or specific length $2\pi/Na = 2\pi/L$.

Now, we take the limit $N \to \infty$, so that k becomes a denumerable number, i.e., it assumes an infinite number of discrete values in the range

$$-\frac{\pi}{a} \le k \le \frac{\pi}{a} \tag{1.2}$$

which defines the one-dimensional **Brillouin zone**.

Note that the eigenvalue of Φ with respect to τ, e^{ika}, implies periodicity in the reciprocal space (k-space), because $e^{i(k+G)a} = e^{ika}$ for all reciprocal lattice vectors $G = \frac{2n\pi}{a}$, where n is an integer. We may, therefore, restrict k to the first Brillouin zone.

The notion of a Brillouin zone (BZ) can be extended to any dimension. In two- or three-dimensions, the specific area/volume is

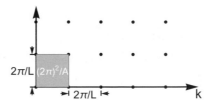

Two-dimensions (2D): specific area $\quad \left(\dfrac{2\pi}{L}\right)^2 \Rightarrow \dfrac{(2\pi)^2}{A}, \quad A$ is the system's 2D area

Three-dimensions (3D): specific volume $\quad \left(\dfrac{2\pi}{L}\right)^3 \Rightarrow \dfrac{(2\pi)^3}{V}, \quad V$ is the system's volume

Lattice Sums

A useful relation that is frequently encountered in solid-state physics involves lattice sums of the form

$$\frac{1}{N} \sum_{j=1}^{N} e^{ikja} = \begin{cases} 1, & k = 2n\pi/a = G_n, \\ \dfrac{e^{ikNa} - 1}{e^{ika} - 1} = 0, & \text{Otherwise} \end{cases}$$

The last line on the left-hand side results from substituting $k = (m/N)2\pi/a$ in the argument of the exponential in the numerator, which yields $kNa = 2m\pi$.

For two- and three-dimensions

$$\sum_{\mathbf{R}} e^{i\mathbf{k}\cdot\mathbf{R}} = N\,\delta_{\mathbf{k},\mathbf{G}}$$

$$\sum_{\mathbf{G}} e^{i\mathbf{G}\cdot\mathbf{r}} = N\,\delta_{\mathbf{r},\mathbf{R}}$$

$\mathbf{R}, \mathbf{G}$, are lattice and reciprocal lattice vectors, respectively.

Construction of the Eigenkets of τ

We consider a complete set of localized, energy-degenerate states $|n\rangle$, where $|n\rangle$ is centered at lattice site n, see Figure 1.1. Thus, n identifies the nth lattice site. Noting that

$$\tau(a)\,|n\rangle = |n+1\rangle$$

we construct the ket

$$|k_\mu\rangle = \sum_n e^{ik_\mu na}\,|n\rangle$$

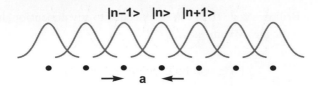

Figure 1.1 Nearest-neighbor hopping in localized orbitals.

and obtain

$$\tau(a)\,\big|k_\mu\big\rangle = \sum_n e^{ik_\mu\,na}\,\tau(a)\,|n\rangle = e^{-ik_\mu\,a}\sum_n e^{ik_\mu\,na}\,|n\rangle$$

satisfying the eigenvalue equation

$$\tau(a)\,\big|k_\mu\big\rangle = e^{-ik_\mu\,a}\,\big|k_\mu\big\rangle$$

The corresponding wavefunction is obtained as follows

$$\big\langle x'\,|\tau(a)\,|\,k_\mu\big\rangle = e^{-ik_\mu\,a}\,\big\langle x'\,|\,k_\mu\big\rangle,\quad \text{operating to the right}$$
$$\big\langle x'\,|\tau(a)\,|\,k_\mu\big\rangle = \big\langle x'-a\,|\,k_\mu\big\rangle,\qquad \text{operating to the left}$$

which means that

$$\big\langle x'-a\,|\,k_\mu\big\rangle = e^{-ik_\mu\,a}\,\big\langle x'\,|\,k_\mu\big\rangle$$

We now introduce Bloch's proposition (Bloch's theorem)

$$\left.\begin{aligned}
\Phi_\alpha(x) &= \big\langle x'\,|\,k_\mu\big\rangle = e^{ik_\mu\,x'}\,u_{k_\mu}(x')\\
u_{k_\mu}(x'\pm ja) &= u_{k_\mu}(x')
\end{aligned}\right\} \quad \text{Bloch's theorem}$$

where j is any integer. Thus,

$$\tau(a)\,\Phi_\alpha(x) = e^{ik_\mu\,(x'-a)}\,u_{k_\mu}(x'-a) = e^{ik_\mu\,x'}\,u_{k_\mu}(x')\,e^{-ik_\mu\,a} = e^{-ik_\mu\,a}\,\Phi_\alpha(x)$$

1.1.2 Eigenvalues of $\mathcal{H}$: Energy Bands (Tight-Binding)

We can envision the localized kets $|n\rangle$ as energy eigenkets of, say, an infinite parabolic potential centered at each lattice site, in which case

$$\langle n\,|\mathcal{H}|\,n\rangle = E_0$$
$$\langle n\,|\mathcal{H}|\,n\pm 1\rangle = 0$$

Infinite parabolic potential barriers

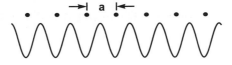

Finite periodic potential

Figure 1.2 Periodic potentials.

Now, if we replace the infinite potential barriers by a periodic potential having finite barriers between lattice sites (see Figure 1.2), a particle can tunnel from one site to its neighboring sites, and we obtain

$$\langle n |\mathcal{H}| n \rangle = E_0$$
$$\langle n |\mathcal{H}| n \pm 1 \rangle = -\Delta; \quad \text{Tight-binding approximation}$$
$$\mathcal{H} |n\rangle = E_0 |n\rangle - \Delta |n+1\rangle - \Delta |n-1\rangle$$

Since $\mathcal{H}$ and τ commute, they share the same eigenket. Thus, we examine the action of $\mathcal{H}$ on $|k_\mu\rangle$

$$\mathcal{H} |k_\mu\rangle = \sum_n e^{ik_\mu na} \mathcal{H} |n\rangle = E_0 \sum_n e^{ik_\mu na} |n\rangle - \Delta \sum_n e^{ik_\mu na} \left(|n+1\rangle + |n-1\rangle\right)$$

$$= E_0 \sum_n e^{ik_\mu na} |n\rangle - \Delta \left(e^{ik_\mu a} + e^{-ik_\mu a}\right) \sum_n e^{ik_\mu na} |n\rangle$$

$$= \left(E_0 - 2\Delta \cos\left(k_\mu a\right)\right) \sum_n e^{ik_\mu na} |n\rangle = \left(E_0 - 2\Delta \cos\left(k_\mu a\right)\right) |k_\mu\rangle$$

Hence,

$$\mathcal{H} |k_\mu\rangle = \mathcal{E}(k_\mu) |k_\mu\rangle = \left(E_0 - 2\Delta \cos\left(k_\mu a\right)\right) |k_\mu\rangle$$

The energy eigenvalue function $\left(E_0 - 2\Delta \cos\left(k_\mu a\right)\right)$ is just the electron band dispersion curve plotted in Figure 1.3 in the first BZ.

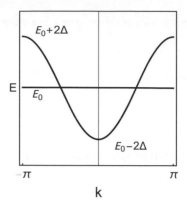

$E_0 + 2\Delta$

E E_0

$E_0 - 2\Delta$

$-\pi$ π

k

Figure 1.3 Tight-binding band dispersion, a $= 1$.

1.2 Energetics and Density Scaling for an Electron Gas

In describing most metals, we usually start from a picture of noninteracting electrons when calculating electronic structures and other properties. However, the Coulomb interaction energy among electrons is actually very large, and we might wonder why it is appropriate to assume that noninteracting electrons (an *ideal Fermi gas*) make a sensible starting point.

To answer this question, we consider an electron gas, or jellium, model, which is a simple paradigm model we shall encounter often. This system consists of N electrons together with a smooth, uniformly smeared background of positive charge density Ne/Ω, in order to ensure electrical neutrality; Ω is the volume of the system. A realization of such a system is found in high-temperature plasmas and in simplified models of metals, where the discrete ionic charges are smeared out.

We write the Hamiltonian as

$$\mathcal{H} = \sum_i \frac{\hbar^2 \nabla_i^2}{2m_e} + \frac{1}{2} \sum_{i \neq j} \frac{e^2}{\left| \mathbf{r}_i - \mathbf{r}_j \right|}$$

For an average electron separation of d and a specific volume $4\pi d^3/3 = \Omega/N$, we define $\mathbf{r}_i' = \mathbf{r}_i/d$ and $\nabla_i' = d\,\nabla_i$ and obtain

$$\mathcal{H} = \frac{1}{d^2} \sum_i \frac{\hbar^2 \nabla_i'^2}{2m_e} + \frac{1}{2} \frac{1}{d} \sum_{i \neq j} \frac{e^2}{\left| \mathbf{r}_i' - \mathbf{r}_j' \right|}$$

Setting $d = r_s a_B$, where $a_B = \hbar^2/m_e e^2$ is the Bohr radius, we get

$$\mathcal{H} = \frac{e^2}{2a_B} \left[\frac{1}{r_s^2} \sum_i \nabla_i'^2 + \frac{1}{r_s} \sum_{i \neq j} \frac{1}{\left| \mathbf{r}_i' - \mathbf{r}_j' \right|} \right]$$

We find that the relative energy scales of the physical quantities involved in this system are

$$\text{Kinetic energy} \sim \frac{\hbar^2}{m_e d^2}, \qquad \text{Coulomb energy} \sim \frac{e^2}{d}$$

so that

$$\frac{\text{Coulomb energy}}{\text{Kinetic energy}} = \frac{m_e e^2}{\hbar^2} d = \frac{d}{a_B} = r_s \qquad (1.3)$$

r_s is the *specific radius*, in units of the Bohr radius a_B.

We note that Coulomb interactions dominate when $r_s \gg 1$, while the kinetic energy dominates $r_s < 1$. It is surprising to see that interactions are, relatively speaking, weak in the high-density limit. A careful examination of (1.3) reveals that with decreasing r_s, growth in Coulomb energies is slower than kinetic energy growth. For $r_s \sim \mathcal{O}(10^2)$, Coulomb interactions prevail, and electrons are known to order into a Wigner crystal. Consequently, and contrary to conventions, we regard a high-density electron system as a gas, since interaction can be neglected, while a very low-density system becomes crystalline, as depicted in Figure 1.4.

The range of r_s for metals is Be : $1.87 \leq r_s \leq$ Cs : 5.62, which is not high enough to cause crystallization, yet the Coulomb interaction is strong. In other words, in a typical electron fluid inside metals, the Coulomb energy is comparable to the electron kinetic energy, constituting a major perturbation to the electron motions. In that sense, the electronic problem appears to be hopeless since the electron–electron interactions cannot be treated as a small perturbation. Nevertheless, the noninteracting model of the Fermi gas reproduces many qualitative features of metallic behavior, such as a well-defined Fermi surface, a linear specific heat capacity, and a temperature-independent paramagnetic susceptibility; evidence of remarkably strong robustness against perturbation. However, quantitatively, the model reveals serious discrepancies between predicted and measured values of physical properties. We shall examine this scenario in Chapter 18 when we discuss Landau's Fermi liquid theory.

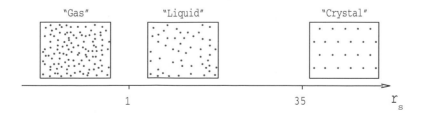

Figure 1.4 Illustration of configurations of the electron system for different densities.

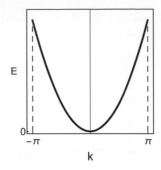

Figure 1.5 Free-electron band dispersion, $a = 1$.

1.3 Noninteracting Free Electron Gas (Sommerfeld Gas)

It is instructive at this point to rehash some properties of the free electron gas. The Hamiltonian, $\mathcal{H}_0$, for a noninteracting free electron gas system, comprised of N electrons, is given by

$$\mathcal{H}_0 = \sum_{i=1}^{N} \frac{\mathbf{p}_i^2}{2m_e}$$

The free electron gas is sometimes referred to as the *Sommerfeld gas*.

Writing the many-electron wavefunction as the product $\prod_i |\mathbf{k}_i, \sigma_i\rangle$, where electron i occupies the plane-wave ket $|\mathbf{k}_i, \sigma_i\rangle$, we obtain (see Figure 1.5)

$$\mathcal{H}_0 = \sum_{\mathbf{k}, \sigma} \frac{\hbar^2 \mathbf{k}^2}{2m_e}$$

σ represents the electron spin state. Applying the Pauli principle, the ground state is expressed as a product of single particle wavefunctions

$$|\Psi_0\rangle = \prod_{|\mathbf{k}|=0}^{k_F} |\mathbf{k} \uparrow\rangle \, |\mathbf{k} \downarrow\rangle$$

where the product runs over $|\mathbf{k}| = 0 \rightarrow k_F$. We obtain an expression for the Fermi wavevector k_F

$$N = \sum_{\mathbf{k}\sigma} \Theta(k_F - k) \rightarrow \frac{\Omega}{(2\pi)^3} \sum_{\sigma} \int d\mathbf{k} \, \Theta(k_F - k) = \frac{\Omega}{3\pi^2} k_F^3$$

$$k_F = \left(\frac{3\pi^2 N}{\Omega} \right)^{1/3} = \left(\frac{9\pi}{4a_B^3} \right)^{1/3} \frac{1}{r_s} \tag{1.4}$$

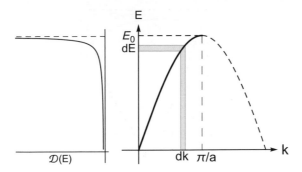

Figure 1.6 One-dimensional dispersion relation, $E(k) = E_0 \sin(ka/2)$. A singularity appears at the BZ boundary $k = \pi/a$, where $dE/dk = 0$.

where Ω is the system's volume and $\Theta(x) = 0$ for $x < 0$. The ground-state energy is

$$E_0 = \langle \Psi_0 | \mathcal{H}_0 | \Psi_0 \rangle = \left\langle \Psi_0 \left| \sum_{\mathbf{k},\sigma} \frac{\hbar^2 k^2}{2m_2} \right| \Psi_0 \right\rangle = \frac{\hbar^2}{2m_e} \sum_{\mathbf{k},\sigma} k^2 \,\Theta(k_F - k)$$

$$= \frac{\Omega}{(2\pi)^3} \frac{\hbar^2}{2m_e} \sum_{\sigma} \int d\mathbf{k}\, k^2 \,\Theta(k_F - k) = \frac{\Omega}{\pi^2} \frac{\hbar^2}{2m_e} \int_0^{k_F} dk\, k^4$$

$$= \frac{\hbar^2}{10\pi^2 m_e} \left(3\pi^2 N\right)^{5/3} \Omega^{-2/3} = \frac{e^2}{2a_B} N \frac{3}{5} \left(\frac{9\pi}{4}\right)^{2/3} \frac{1}{r_s^2} \qquad (1.5)$$

1.4 Dispersion Relations and the Density of States, $\mathcal{D}(E)$

The idea of *density of states* (DOS) provides a very useful tool in deriving physical properties of quantum systems. We start with considering the simple case shown in Figure 1.6. It represents a one-dimensional (1D) dispersion relation $E(k) \propto \sin\left(\frac{ka}{2}\right)$, where a is a lattice periodicity. From the figure, we note the following:

"Number of states in dE = Number of states in dk"

Setting the former to be $\mathcal{D}(E)\, dE$, where $\mathcal{D}(E)$ is the DOS, we write

$$\mathcal{D}(E)\, dE = \frac{L}{2\pi}\, dk \Rightarrow \mathcal{D}(E) = \frac{L/2\pi}{dE/dk}$$

Density of states in $\mathbf{k}$-space
$$\begin{cases} \text{1D} & \mathcal{D}(k) = \frac{1}{2\pi} & \text{per unit length} \\ \text{2D} & \mathcal{D}(\mathbf{k}) = \frac{1}{(2\pi)^2} & \text{per unit area} \\ \text{3D} & \mathcal{D}(\mathbf{k}) = \frac{1}{(2\pi)^3} & \text{per unit volume} \end{cases}$$

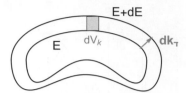

Figure 1.7 E and $E + dE$ are adjacent constant energy surfaces in reciprocal space, and $dk_n \equiv dk_\perp$ is the normal to these constant energy difference surfaces.

For 3D, we evaluate the annular volume in **k**-space enclosed between two equienergy surfaces E and $E + dE$. Expressing the annular element of volume, shown in Figure 1.7, as $dV_k = dS_k dk_\perp$, we write

$$\mathcal{D}(E) \, dE = \int \frac{1}{(2\pi)^3} \, dV_k$$

but

$$dE = \nabla_{\mathbf{k}} E(\mathbf{k}) \cdot d\mathbf{k} = |\nabla_{\mathbf{k}} E| \, dk_\perp \Rightarrow dk_\perp = \frac{dE}{|\nabla_{\mathbf{k}} E|}$$

and we obtain

$$\mathcal{D}(E) = \frac{1}{(2\pi)^3} \int \frac{d\mathbf{S}}{|\nabla_{\mathbf{k}} E|} \tag{1.6}$$

A singularity in the density of states occurs when $|\nabla_{\mathbf{k}} E| = 0$. Singularities appear when points or lines in the Brillouin zone have symmetry:

- Periodic symmetry manifests singularities at BZ boundaries.
- Rotation and reflection symmetries manifest singularities at high-symmetry points and directions (irreducible zone [IBZ] boundaries), shown for a 2D square lattice in Figure 1.8.

The existence of singularities in $\mathcal{D}(E)$ at high-symmetry points and lines in the BZ means that major contributions to the density of states occur at such points and lines. In other words, almost all states lie along high-symmetry points, directions, and planes. Thus, when we present a dispersion relation associated with a particle or a quasiparticle (electrons, phonons, magnons, etc.), we plot the dispersion curves only along such lines and at such points.

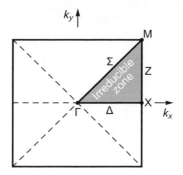

Figure 1.8 Two-dimensional square lattice BZ and its IBZ, shown in gray. There is reflection symmetry along the lines labeled Σ and Δ. The Line $X - M$ is a BZ boundary.

1.5 Born–Oppenheimer Approximation

The motion of electrons in a solid is much faster than that of the constituent nuclei, since

$$m_e/M \sim 10^{-3} - 10^{-5}$$

The mass of a single nucleon is approximately 1,800 times larger than that of the electron. This implies that at a temperature T, $\frac{E_{nuc}^{kin}}{E_{el}^{kin}} \ll 1$. Physically, this means that within the time scale that characterizes the motion of the nuclei, the much faster electrons will be able to accommodate *instantaneously* to the position of the nuclei.

These are the conditions suitable for the *adiabatic approximation*. A brief description and proof of the adiabatic theorem are given in the appendix for the interested reader.[1] The adiabatic theorem stipulates that for a time-dependent Hamiltonian $\mathcal{H}(t)$ that changes infinitely slowly (adiabatically), and for which at every time instant t the eigenvalues of $\mathcal{H}(t)$ are nondegenerate

$$E_j(t) \neq E_k(t), \quad \forall t, \text{ and all } j \neq k,$$

the system will always stay in the same eigenket j when evolving in time, as shown in Figure 1.9. We only need consider instantaneous eigenvalues and eigenvectors of $\mathcal{H}(t)$ for all t.

We write the general Hamiltonian as

$$\mathcal{H} = \mathcal{H}_i + \mathcal{H}_e + \mathcal{H}_{ei}$$

$$\mathcal{H}_i = \sum_i \frac{\mathbf{P}_i^2}{2M_i} + \sum_{i \neq j} \frac{e^2 Z_i Z_j}{\left| \mathbf{R}_i - \mathbf{R}_j \right|^2}$$

---

[1] The adiabatic theorem will be revisited later in the book, in the context of topological phases and Berry's phase.

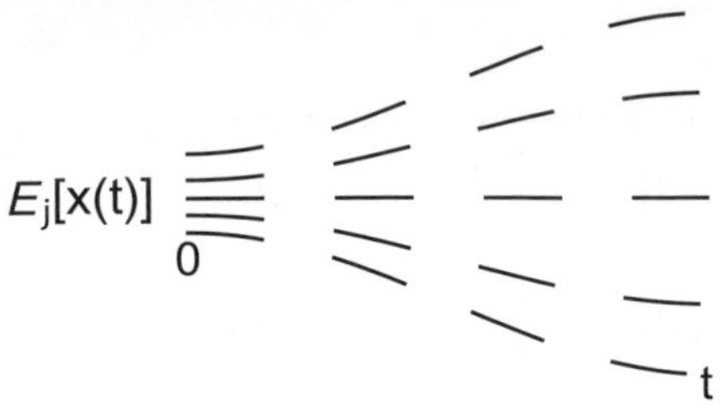

$E_j[x(t)]$

Figure 1.9 Schematic evolution of the eigenvalues in the adiabatic limit.

$$\mathcal{H}_e = \sum_i \frac{\mathbf{p}_i^2}{2m_e} + \sum_{i \neq j} \frac{e^2}{\left|\mathbf{r}_i - \mathbf{r}_j\right|^2}$$

$$\mathcal{H}_{ei} = -\sum_{i,j} \frac{e^2 Z_j}{\left|\mathbf{r}_i - \mathbf{R}_j\right|^2} \tag{1.7}$$

where the ionic positions $\mathbf{R}_i = \mathbf{R}_i(t)$ change slowly (adiabatically) in comparison to the fast movement of the electrons. In the context of molecular and condensed matter systems, the adiabatic approximation is referred to as the *Born–Oppenheimer approximation* (BOA) [36].

According to the BOA, we initially drop $\mathcal{H}_i \ll \mathcal{H}_e + \mathcal{H}_{ei}$, and we solve the Schrödinger equation corresponding to the instantaneous Hamiltonian for the electrons

$$\mathcal{H}_e + \mathcal{H}_{ei} = \sum_i \frac{\mathbf{p}_i^2}{2m_e} + \sum_{i \neq j} \frac{e^2}{\left|\mathbf{r}_i - \mathbf{r}_j\right|^2} - \sum_{i,j} \frac{e^2 Z_j}{\left|\mathbf{r}_i - \mathbf{R}_j\right|^2}$$

$$(\mathcal{H}_e + \mathcal{H}_{ei}) \left|\psi_j(\mathbf{x}, \mathbf{R})\right\rangle = E_j^{el} \left|\psi_j(\mathbf{r}, \mathbf{R})\right\rangle$$

$$\mathbf{x} = (\mathbf{r}_1, \mathbf{r}_2, \ldots, \mathbf{r}_N)$$

$$\mathbf{R} = (\mathbf{R}_1, \mathbf{R}_2, \ldots, \mathbf{R}_N)$$

freezing the positions of the nuclei $\mathbf{R}$ at every time t. Elimination of $\mathcal{H}_i$ demotes the variables $\mathbf{R}_i$ to parameter status.

Having solved for the electronic eigenkets and eigenvalues, we can now obtain the eigenfunctions $\left|\Psi(\mathbf{x}, \mathbf{R})\right\rangle$ for the whole Hamiltonian (1.7) using the product ansatz

$$\left|\Psi(\mathbf{x}, \mathbf{R})\right\rangle = \left|\psi(\mathbf{x}, \mathbf{R})\right\rangle \left|\Phi(\mathbf{R})\right\rangle$$

where $\left|\Phi(\mathbf{R})\right\rangle$ is the wavefunction associated with the motion of the nuclei. Remember that if the electrons are initially in state $\left|\psi_j(\mathbf{x}, \mathbf{R})\right\rangle$, they will remain in this state as the system evolves. Since the system was initially in its ground state, then $\left|\psi_j(\mathbf{x}, \mathbf{R})\right\rangle \Rightarrow \left|\psi_0(\mathbf{x}, \mathbf{R})\right\rangle$. First, we note that

$$\mathbf{P}_i \left|\psi(\mathbf{x}, \mathbf{R})\right\rangle \left|\Phi(\mathbf{R})\right\rangle = \left|\psi(\mathbf{x}, \mathbf{R})\right\rangle \mathbf{P}_i \left|\Phi(\mathbf{R})\right\rangle + \left|\Phi(\mathbf{R})\right\rangle \left(-i\hbar\nabla_i \left|\psi(\mathbf{x}, \mathbf{R})\right\rangle\right)$$

Thus, we obtain

$$\mathcal{H} \left| \Psi \right\rangle = \sum_i \frac{\mathbf{P}_i^2}{2M_i} \left| \Psi \right\rangle + \sum_{i \neq j} \frac{e^2 Z_i Z_j}{\left| \mathbf{R}_i - \mathbf{R}_j \right|^2} \left| \Psi \right\rangle + (\mathcal{H}_e + \mathcal{H}_{ei}) \left| \Psi \right\rangle$$

$$= \left| \psi_0(\mathbf{x}, \mathbf{R}) \right\rangle \left[-\sum_i \frac{\hbar^2}{2M_i} \nabla_i^2 + U_0(\mathbf{R}) + E_0^{\text{el}}(\mathbf{R}) \right] \left| \Phi(\mathbf{R}) \right\rangle$$

$$- \sum_i \frac{\hbar^2}{2M_i} \left[\frac{2i}{\hbar} \left(\nabla_i \left| \psi_0 \right\rangle \right) \cdot \mathbf{P}_i + \nabla_i^2 \left| \psi_0 \right\rangle \right] \left| \Phi(\mathbf{R}) \right\rangle$$

Operating by $\langle \psi_0(\mathbf{x}, \mathbf{R}) |$

$$\mathcal{H}_{\text{eff}} \left| \Phi(\mathbf{R}) \right\rangle = \left[\sum_i \frac{\mathbf{P}_i^2}{2M_i} - \frac{1}{M_i} \mathcal{A}_0 \cdot \mathbf{P}_i - U_1(\mathbf{R}) + U_0(\mathbf{R}) + E_0^{\text{el}}(\mathbf{R}) \right] \left| \Phi(\mathbf{R}) \right\rangle$$

$$= \left[\sum_i \frac{1}{2M_i} (\mathbf{P}_i - \hbar \mathcal{A}_0)^2 + U(\mathbf{R}) \right] \left| \Phi(\mathbf{R}) \right\rangle$$

where we completed the square by adding and subtracting $\mathcal{A}_0^2 \sum_i \frac{1}{2M_i}$, and where

$$\mathcal{A}_0 = i \sum_i \langle \psi_0(\mathbf{x}, \mathbf{R}) | \nabla_i | \psi_0(\mathbf{x}, \mathbf{R}) \rangle$$

$$U_1(\mathbf{R}) = \sum_i \frac{\hbar^2}{2M_i} \langle \psi_0(\mathbf{x}, \mathbf{R}) | \nabla_i^2 | \psi_0(\mathbf{x}, \mathbf{R}) \rangle$$

$$U(\mathbf{R}) = U_0(\mathbf{R}) + E_0^{\text{el}}(\mathbf{R}) - U_1(\mathbf{R}) - \mathcal{A}_0^2 \sum_i \frac{1}{2M_i}$$

We just mention here that $\mathcal{A}_0$ is known as Berry's vector potential.

We find that for the normalized ground state $| \psi_0(\mathbf{x}, \mathbf{R}) \rangle$

$$\nabla_i \langle \psi_0(\mathbf{x}, \mathbf{R}) | \psi_0(\mathbf{x}, \mathbf{R}) \rangle = 0$$

$$= \langle \nabla_i \psi_0(\mathbf{x}, \mathbf{R}) | \psi_0(\mathbf{x}, \mathbf{R}) \rangle + \langle \psi_0(\mathbf{x}, \mathbf{R}) | \nabla_i \psi_0(\mathbf{x}, \mathbf{R}) \rangle$$

$$= 2 \text{Re} \langle \psi_0(\mathbf{x}, \mathbf{R}) | \nabla_i \psi_0(\mathbf{x}, \mathbf{R}) \rangle$$

which makes $\mathcal{A}_0 = 0$, since the ground state $| \psi_0(\mathbf{x}, \mathbf{R}) \rangle$ has to be real.[2]

For the contribution, U_1, we assume the worst case where the electrons are tightly bound to the nuclei, i.e., $\Rightarrow \mathbf{x}_i \simeq \mathbf{R}_j$ for some i and j. Then a crude estimate gives

---

[2] The reality of the ground-state wavefunction is dictated by the fact that it must be nondegenerate and the Hamiltonian is time-reversal invariant, as will be shown in Section 6.6.

$$-\frac{\hbar^2}{2M} \sum_j \langle \psi_0(\mathbf{x}, \mathbf{R})| \nabla^2_{\mathbf{R}_j} |\psi_0(\mathbf{x}, \mathbf{R})\rangle \simeq$$

$$-\frac{\hbar^2}{2M} \sum_i \langle \psi_0(\mathbf{x}, \mathbf{R})| \nabla^2_{\mathbf{r}_i} |\psi_0(\mathbf{x}, \mathbf{R})\rangle = \frac{m_e}{M} E^{\text{kin}}_{\text{el}} \ll \left| E^{\text{el}}_0 \right|$$

Neglecting this term, the lattice dynamics (phonon system) is described by the following Schrödinger equation

$$\left[\sum_i \frac{\mathbf{P}_i^2}{2M_i} + U(\mathbf{R}) \right] |\Phi(\mathbf{R})\rangle = E |\Phi(\mathbf{R})\rangle$$

where now $U(\mathbf{R}) = U_0(\mathbf{R}) + E^{\text{el}}_0(\mathbf{R})$.

1.6 Time-Reversal Symmetry

The discussion of time-reversal symmetry presented in this section will follow the analysis given in the book by Sakurai and Napolitano [159].

1.6.1 Time-Reversal in Classical Mechanics

We consider a trajectory of a particle subject to some conservative force field. As shown in Figure 1.10, we stop the particle at $t = t_0$, as it moves along the trajectory, and reverse its motion:

$$\mathbf{p}(t_0) \rightarrow -\mathbf{p}(t_0)$$

The particle then moves backward along the same trajectory, such that

$$\left(\mathbf{x}(t - t_0), \mathbf{p}(t - t_0) \right) = \left(\mathbf{x}(t_0 - t), -\mathbf{p}(t_0 - t) \right)$$

It is like running a motion picture of the trajectory backward.

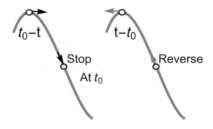

Figure 1.10 Time-reversal action at $t = t_0$.

This is a manifestation of the appearance of second-order time-derivative in the classical equation of motion

$$m\frac{d^2\mathbf{x}}{dt^2} = -\nabla V(\mathbf{x}), \tag{1.8}$$

which renders $\mathbf{x}(-t)$ a bona fide solution in the same force field derivable from the conservative potential $V(\mathbf{x})$.[3]

We also find that the Maxwell equations and the Lorentz force

$$\nabla\cdot\mathbf{E} = 4\pi\rho, \quad \nabla\times\mathbf{B} - \frac{1}{c}\frac{\partial\mathbf{E}}{\partial t} = \frac{4\pi\mathbf{j}}{c}, \quad \nabla\times\mathbf{E} = -\frac{1}{c}\frac{\partial\mathbf{B}}{\partial t}$$

$$\mathbf{F} = e\left[\mathbf{E} + \frac{1}{c}\mathbf{v}\times\mathbf{B}\right] \tag{1.9}$$

are invariant under $t \to -t$, with the proviso that all quantities that contain momenta should be reversed; namely, we have

$$\mathbf{E} \to \mathbf{E}, \quad \rho \to \rho, \quad \mathbf{j} \to -\mathbf{j}, \quad \mathbf{v} \to -\mathbf{v} \quad \mathbf{B} \to -\mathbf{B}. \tag{1.10}$$

1.6.2 Time-Reversal in Quantum Mechanics

Now, we shall explore and examine the consequences of time reversal in wave mechanics, where the basic equation is the Schrödinger wave equation

$$i\hbar\frac{\partial\Psi(\mathbf{x},t)}{\partial t} = \left(-\frac{\hbar^2}{2m}\nabla^2 + V\right)\Psi(\mathbf{x},t) = \mathcal{H}\Psi(\mathbf{x},t). \tag{1.11}$$

Here, we immediately realize that because of the first-order time derivative, $\Psi(\mathbf{x}, -t)$ is not a solution if $\Psi(\mathbf{x},t)$ is a solution. However, if we conjugate the Schrödinger equation and then do the transformation $t \to t' = -t$, we find

$$i\hbar\frac{\partial\Psi^*(\mathbf{x}, -t')}{\partial t'} = \mathcal{H}\Psi^*(\mathbf{x}, -t'). \tag{1.12}$$

Dropping the primes we find that $\Psi^*(\mathbf{x}, -t)$ is a solution, which can be obtained by the conjugation of (10.13). We therefore surmise that if Θ is a time-reversal operator, then

$$\Theta\Psi(\mathbf{x},t) = \Psi^*(\mathbf{x}, -t). \tag{1.13}$$

It is instructive to consider the case of an energy eigenstate, that is, by substituting

$$\Psi(\mathbf{x},t) = u_n(\mathbf{x})\,e^{-iE_nt/\hbar}, \qquad \Psi^*(\mathbf{x}, -t) = u_n^*(\mathbf{x})\,e^{-iE_nt/\hbar} \tag{1.14}$$

into the Schrödinger equation (10.13). Thus, we expect that complex conjugation should play a role in time reversal in quantum systems.

---

[3] It is, of course, important to note that we do not have a dissipative force here. Such a force breaks time-reversal symmetry.

1.6.3 Unitary and Antiunitary Symmetry Operations

We recall that the action of quantum operators is just a mapping of Hilbert, or ket, space onto itself. A famous theorem due to Wigner requires that a mapping of a ket space onto itself must preserve the absolute values of all scalar products, namely,

$$\left.\begin{array}{c} |\alpha\rangle \xrightarrow{\mathcal{O}} |\tilde{\alpha}\rangle, \\ |\beta\rangle \xrightarrow{\mathcal{O}} |\tilde{\beta}\rangle \end{array}\right\} \quad \left|\langle\tilde{\beta}|\tilde{\alpha}\rangle\right| = \left|\langle\beta|\alpha\rangle\right| \tag{1.15}$$

Thus, the mapping must be either a linear unitary operator or, as we will show, an antilinear, antiunitary operator. The reason that only the absolute value of scalar products need be preserved is that the only physically measurable quantities are absolute squares of scalar products – namely, probabilities. Thus, symmetry operations in quantum mechanics must preserve the probabilities of all physical outcomes – they all must be represented either by unitary or antiunitary operators.

Symmetry operations that we have encountered, such as rotations, translations, and the discrete symmetry of parity, require that the inner product be preserved; namely, we require that

$$\langle\tilde{\beta}|\tilde{\alpha}\rangle = \left\langle\beta\left|\mathcal{O}^\dagger\mathcal{O}\right|\alpha\right\rangle = \langle\beta|\alpha\rangle \tag{1.16}$$

which implies that such symmetry operators are unitary, $\mathcal{O} = U$.

Antilinear Operators

Antilinear operators have the distributive property

$$\mathcal{O}_a\big(c_1|\alpha\rangle + c_2|\beta\rangle\big) = c_1^*\mathcal{O}|\alpha\rangle + c_2^*\mathcal{O}|\beta\rangle. \tag{1.17}$$

when acting on linear combinations of kets. Actually, an antilinear operator does not commute with a constant, when the latter is regarded as a multiplicative operator in its own right, and we write

$$\mathcal{O}_a\, c = c^*\,\mathcal{O}_a \tag{1.18}$$

Thus, we find that the product of two antilinear operators is linear, and the product of a linear with an antilinear operator is antilinear. To describe antilinear operations, we need to reconsider the bra-ket picture. From the duality of bra and ket spaces, we can regard the bra as a linear operator that yields a complex value when acting on a ket, namely, the mapping

$$\text{bra}\ :\ \text{ket space}\ \rightarrow\ \mathbb{C}$$

However, if we consider the bra $\langle\beta|\,\mathcal{O}_a$, we realize that it is now an antilinear operator acting on a ket. Bras should be linear operators. Thus, we need to introduce a complex conjugation to make $\langle\beta|\,\mathcal{O}_a$ a linear operator on kets, and we set

$$(\langle\beta|\,\mathcal{O}_a)\,|\alpha\rangle = [\langle\beta|\,(\mathcal{O}_a\,|\alpha\rangle)]^* \tag{1.19}$$

Antiunitary Operators

We define an antiunitary operator A as an antilinear operator that satisfies

$$AA^\dagger = A^\dagger A = \mathbb{I}$$

We note that the products $AA^\dagger$ and $A^\dagger A$ are linear operators and thus satisfy being linear identity operators. Moreover, antiunitary operators preserve the absolute values of scalar products. To see this, consider $A\,|\alpha\rangle$ and $A\,|\beta\rangle$. Then we have

$$\left(\langle\beta|\,A^\dagger\right)(A\,|\alpha\rangle) = \left[\langle\beta|\left(A^\dagger A\,|\alpha\rangle\right)\right]^* = \langle\beta\,|\,\alpha\rangle^*$$

where we reverse the direction of $A^\dagger$ in the second equality. Antiunitary operators take scalar products into their complex conjugates, satisfying (1.15).

1.6.4 Time-Reversal Operator

We will now develop a formal theory of time-reversal symmetry. We denote the time-reversal operator by Θ, and define the time-reversed state as

$$|\alpha\rangle \;\rightarrow\; \Theta\,|\alpha\rangle. \tag{1.20}$$

We derive the fundamental property of the time-reversal operator by examining the time evolution of the time-reversed state. We start with a state ket $|\alpha\rangle$ of a physical system at $t = 0$. Then after an incremental time $t = \delta t$, the state ket becomes

$$|\alpha, \delta t\rangle = \left(1 - \frac{i\mathcal{H}}{\hbar}\delta t\right)|\alpha\rangle. \tag{1.21}$$

However, if we have applied Θ at $t = 0$, and then allowed the system to evolve to δt under the action of $\mathcal{H}$, we would obtain

$$\left(1 - \frac{i\mathcal{H}}{\hbar}\delta t\right)\Theta\,|\alpha\rangle. \tag{1.22}$$

If motion is time-reversal symmetric, then we expect the preceding state ket to coincide with

$$\Theta\,|\alpha, -\delta t\rangle, \tag{1.23}$$

where we apply Θ to the state ket at *earlier* time $t = -\delta t$ [4]. This means that

$$\left(1 - \frac{i\mathcal{H}}{\hbar}\delta t\right)\Theta\,|\alpha\rangle = \Theta\left(1 - \frac{i\mathcal{H}}{\hbar}(-\delta t)\right)|\alpha\rangle, \tag{1.24}$$

which leads to

$$-i\mathcal{H}\Theta\,|\,\rangle = \Theta i\mathcal{H}\,|\,\rangle. \tag{1.25}$$

---

[4] Recall that under Θ, both the momentum $\mathbf{p}$ and angular momentum $\mathbf{J}$ are reversed.

If we assume Θ to be unitary, then we should ignore the is in (1.25) and write the operator equation as

$$-\mathcal{H}\Theta \mid \rangle = \Theta\mathcal{H} \mid \rangle. \qquad (1.26)$$

We can show that (1.26) is physically invalid by examining the time-reversed state of an energy eigenket $|n\rangle$ with eigenvalue E_n. This time-reversed state $\Theta |n\rangle$ should then satisfy

$$\mathcal{H}\Theta |n\rangle = -\Theta\mathcal{H} |n\rangle = (-E_n)\Theta |n\rangle. \qquad (1.27)$$

This means that $\Theta |n\rangle$ must be an eigenket of the Hamiltonian with energy eigenvalue $-E_n$. We immediately find that this is nonsensical by simply considering the case of a free particle, namely,

$$\Theta^{-1}\frac{\mathbf{p}^2}{2m}\Theta = \frac{-(-\mathbf{p})^2}{2m}. \qquad (1.28)$$

All these arguments suggest that if time reversal is to be a bona fide symmetry at all, we are not allowed to ignore the i's in (1.25), hence Θ has to be antiunitary. Recognizing the conjugation action of Θ in (1.25) because of its antilinear property (1.18), we write

$$\Theta i\mathcal{H} \mid \rangle = -i\mathcal{H}\Theta \mid \rangle, \qquad (1.29)$$

which allows us to cancel the is in (1.25) leading, finally, to

$$\boxed{\Theta\mathcal{H} \mid \rangle = \mathcal{H}\Theta \mid \rangle.} \qquad (1.30)$$

Equation (1.30) expresses the fundamental property of the Hamiltonian under time reversal.

We now need to establish that an antiunitary time-reversal operator can be written as

$$\Theta = UK, \qquad (1.31)$$

where U is a unitary operator and K is the complex-conjugation operator that acts on any coefficient to its right, namely,

$$K c |\alpha\rangle = c^* K |\alpha\rangle.$$

The action of K on basis kets $\{|a\rangle\}$ can be determined by writing the expansion of a general ket $|\alpha\rangle$

$$|\alpha\rangle = \sum_a |a\rangle \langle a \mid \alpha\rangle$$

and noting that the corresponding expansion of $|a\rangle$ is

$$\langle a| = \begin{bmatrix} 0 & 0 & \dots & 1 & 0 & \dots & 0 \end{bmatrix}, \qquad (1.32)$$

we find that the action of K on the base ket does not change the base ket. Therefore, the action of K on $|\alpha\rangle$ gives

$$K |\alpha\rangle = |\tilde{\alpha}\rangle = \sum_a \langle a| \alpha\rangle^* |a\rangle. \qquad (1.33)$$

As an illustrative example, we examine how the S_y eigenkets for a spin-1/2 system change under K. We find that if the S_z eigenkets are used as base kets, the S_y eigenkets must change according to

$$K \left(\frac{1}{\sqrt{2}} |+\rangle \pm \frac{i}{\sqrt{2}} |-\rangle \right) \rightarrow \left(\frac{1}{\sqrt{2}} |+\rangle \mp \frac{i}{\sqrt{2}} |-\rangle \right). \tag{1.34}$$

Yet, if the S_y eigenkets themselves are used as base kets, they do not change under the action of K.

Thus the effect of K changes with the basis. As a result, the form of U in (23.17) depends on the particular representation used.

We note that $\Theta = UK$ satisfies (1.15),

$$\Theta\big(c_1 |\alpha\rangle + c_2 |\beta\rangle\big) = UK\big(c_1 |\alpha\rangle + c_2 |\beta\rangle\big) = c_1^* UK |\alpha\rangle + c_2^* UK |\beta\rangle$$
$$= c_1^* \Theta |\alpha\rangle + c_2^* \Theta |\beta\rangle. \tag{1.35}$$

Here, we will follow Sakurai's recommendation that it is always safer to act with Θ on kets only. We can deduce its action on bras from that on the corresponding kets, and we do not need to define $\Theta^\dagger$. Accordingly, we interpret

$$\langle \beta |\Theta| \alpha \rangle \tag{1.36}$$

as

$$\big((\langle \beta|) \cdot (\Theta |\alpha\rangle)\big), \tag{1.37}$$

and we do not attempt to define $\langle \beta| \Theta$. We write

$$|\alpha\rangle \xrightarrow{\Theta} |\tilde{\alpha}\rangle = \sum_{a'} \langle a'| \alpha\rangle^* UK |a'\rangle = \sum_{a'} \langle a'| \alpha\rangle^* U |a'\rangle = \sum_{a'} \langle \alpha| a'\rangle U |a'\rangle$$
$$|\tilde{\beta}\rangle = \sum_{a'} \langle a'| \beta\rangle^* U |a'\rangle \rightarrow \langle \tilde{\beta}| = \sum_{a'} \langle a'| \beta\rangle \langle a'| U^\dagger \tag{1.38}$$

and we obtain

$$\langle \tilde{\beta}|\tilde{\alpha}\rangle = \sum_{a',a''} \langle a''| \beta\rangle \langle a''| U^\dagger U |a'\rangle \langle \alpha| a'\rangle$$
$$= \sum_{a'} \langle \alpha| a'\rangle \langle a'| \beta\rangle = \langle \alpha| \beta\rangle = \langle \beta| \alpha\rangle^*. \tag{1.39}$$

1.6.5 Transformation of Operators under Time Reversal

We start with an important identity

$$\langle \beta| \mathcal{O} |\alpha\rangle = \Big\langle \tilde{\alpha} \Big| \Theta \mathcal{O}^\dagger \Theta^{-1} \Big| \tilde{\beta} \Big\rangle, \tag{1.40}$$

where $\mathcal{O}$ is a linear operator. This identity arises from the antiunitary nature of Θ. To prove this, we define

$$|\tilde{\alpha}\rangle = \Theta\,|\alpha\rangle, \qquad |\tilde{\beta}\rangle = \Theta\,|\beta\rangle$$
$$|\gamma\rangle = \mathcal{O}^{\dagger}\,|\beta\rangle \quad \Rightarrow \quad |\tilde{\gamma}\rangle = \Theta\,\mathcal{O}^{\dagger}\,|\beta\rangle \tag{1.41}$$

and use dual correspondence to obtain

$$\langle\beta|\,\mathcal{O} = \langle\gamma|\,. \tag{1.42}$$

We can then write

$$\langle\beta\,|\mathcal{O}|\,\alpha\rangle = \langle\gamma|\,\alpha\rangle = \langle\tilde{\alpha}|\,\tilde{\gamma}\rangle$$
$$= \langle\tilde{\alpha}|\,\Theta\mathcal{O}^{\dagger}\,|\beta\rangle = \langle\tilde{\alpha}|\,\Theta\mathcal{O}^{\dagger}\,\Theta^{-1}\Theta\,|\beta\rangle$$
$$= \langle\tilde{\alpha}\big|\,\Theta\,\mathcal{O}^{\dagger}\Theta^{-1}\big|\tilde{\beta}\rangle. \tag{1.43}$$

We are usually interested in *Hermitian* observables A, for which we get

$$\langle\beta|\,A\,|\alpha\rangle = \langle\tilde{\alpha}\big|\,\Theta A\Theta^{-1}\big|\tilde{\beta}\rangle. \tag{1.44}$$

We say that observables are even or odd under time reversal according to

$$\Theta A\Theta^{-1} = \pm A. \tag{1.45}$$

Equations (1.45) and (1.44) restrict the phase of the matrix element of A taken with respect to time-reversed states to

$$\boxed{\langle\beta|\,A\,|\alpha\rangle = \pm\langle\tilde{\alpha}\big|A\big|\tilde{\beta}\rangle = \pm\langle\tilde{\beta}\big|A\big|\tilde{\alpha}\rangle^{*}.} \tag{1.46}$$

Expectation values are obtain for $|\beta\rangle$ identical to $|\alpha\rangle$, and we have

$$\boxed{\langle\alpha|\,A\,|\alpha\rangle = \pm\langle\tilde{\alpha}|\,A\,|\tilde{\alpha}\rangle.} \tag{1.47}$$

As an example, let us consider the expectation value of $\mathbf{p}$

$$\langle\alpha|\,\mathbf{p}\,|\alpha\rangle = -\langle\tilde{\alpha}|\,\mathbf{p}\,|\tilde{\alpha}\rangle, \tag{1.48}$$

where, according to (1.44), $\mathbf{p}$ is an odd operator, namely

$$\Theta\mathbf{p}\Theta^{-1} = -\mathbf{p}, \tag{1.49}$$

which leads to

$$\mathbf{p}\Theta\,|\mathbf{p}'\rangle = -\Theta\mathbf{p}\Theta^{-1}\Theta\,|\mathbf{p}'\rangle$$
$$= (-\mathbf{p}')\Theta\,|\mathbf{p}'\rangle. \tag{1.50}$$

Similarly, we obtain

$$\Theta x \Theta^{-1} = x$$
$$\Theta \, |x'\rangle = |x'\rangle \tag{1.51}$$

from the requirement

$$\langle \alpha | \, \mathbf{x} \, | \alpha \rangle = \langle \tilde{\alpha} | \, \mathbf{x} \, | \tilde{\alpha} \rangle . \tag{1.52}$$

We can also demonstrate the invariance of the fundamental commutation relation

$$[x_i, p_j] | \, \rangle = i\hbar \delta_{ij} | \, \rangle \tag{1.53}$$

by action of Θ on both sides of (1.53)

$$\Theta [x_i, p_j] \Theta^{-1} \Theta | \, \rangle = \Theta i\hbar \delta_{ij} | \, \rangle$$
$$[x_i, (-p_j)] \Theta | \, \rangle = -i\hbar \delta_{ij} \Theta | \, \rangle . \tag{1.54}$$

We find that the antiunitary of Θ preserves the fundamental commutation relation under the action of time reversal. Similarly, we require that

$$\Theta \mathbf{J} \Theta^{-1} = -\mathbf{J} \tag{1.55}$$

to preserve

$$[J_i, J_j] | \, \rangle = i\hbar \varepsilon_{ijk} J_k \, | \, \rangle . \tag{1.56}$$

This is consistent with the case of spinless systems, where $\mathbf{J} = \mathbf{x} \times \mathbf{p}$.

1.6.6 Time-Reversal of the Wavefunction

Spinless Systems

The position representation expansion of state ket $|\alpha\rangle$ of a spinless system at $t = 0$

$$|\alpha\rangle = \int d^3x' \, |x'\rangle \langle x' \, | \, \alpha \rangle . \tag{1.57}$$

gives the corresponding wavefunction $\langle x' \, | \, \alpha \rangle = \Psi(\mathbf{x}, 0)$. Applying the time-reversal operator

$$\Theta \, |\alpha\rangle = \int d^3x' \, \Theta \, |x'\rangle \langle x' \, | \, \alpha \rangle^* = \int d^3x' \, |x'\rangle \langle x' \, | \, \alpha \rangle^* , \tag{1.58}$$

since $\Theta \, |x'\rangle = |x'\rangle$. It confirms the rule

$$\Theta \, \Psi(x') = \Psi^*(x') \tag{1.59}$$

inferred earlier from the Schrödinger equation. Thus, for a wavefunction in the position representation, Θ is just the complex conjugation operator K itself. We may note, however, that the situation is quite different when the ket $|\alpha\rangle$ is expanded in terms of the momentum eigenkets because Θ must change $|\mathbf{p}'\rangle$ into $|-\mathbf{p}'\rangle$ as follows:

$$\Theta \, |\alpha\rangle = \int d^3 p' \; |-\mathbf{p}'\rangle\langle\mathbf{p}' \mid \alpha\rangle^* = \int d^3 p' \; |\mathbf{p}'\rangle\langle-\mathbf{p}' \mid \alpha\rangle^* . \tag{1.60}$$

A particularly interesting case is that of wavefunctions of the type $\langle\mathbf{x}' \mid n,l,m\rangle = \mathcal{R}(r)Y_l^m(\theta,\phi)$. Examining the spherical harmonic component Y_l^m, we find that

$$Y_l^m(\theta,\phi) = \langle\hat{\mathbf{x}} \mid l,m\rangle \; \underset{\longrightarrow}{\Theta} \; \left(Y_l^m\right)^*(\theta,\phi) = (-1)^m \, Y_l^{-m}(\theta,\phi)$$

$$\Theta \, |l,m\rangle = (-1)^m \, |l, -m\rangle . \tag{1.61}$$

Reality of Nondegenerate Energy Eigenfunction under Time-Reversal Symmetry

For a time-reversal symmetric system, $[\mathcal{H}, \Theta] = 0$, hence

$$\mathcal{H}\Theta \, |n\rangle = \Theta\mathcal{H} \, |n\rangle = E_n \, \Theta \, |n\rangle , \tag{1.62}$$

so $|n\rangle$ and $\Theta \, |n\rangle$ have the same energy. The nondegeneracy assumption prompts us to conclude that $|n\rangle$ and $\Theta \, |n\rangle$ must represent the same state; otherwise, there would be two different states with the same energy E_n, an obvious contradiction! We recall that the wavefunctions for $|n\rangle$ and $\Theta \, |n\rangle$ are $\langle\mathbf{x}' \mid n\rangle$ and $\langle\mathbf{x}' \mid n\rangle^*$, respectively. They must satisfy

$$\langle\mathbf{x}' \mid n\rangle = \langle\mathbf{x}' \mid n\rangle^* \tag{1.63}$$

or, more precisely, they can differ at most by a phase factor independent of $\mathbf{x}$.

Thus if we have, for instance, a nondegenerate bound state, its wavefunction is always real. On the other hand, in the hydrogen atom with $l \neq 0$, $m \neq 0$, the energy eigenfunction characterized by definite (n,l,m) quantum numbers is complex because Y_l^m is complex; this does not contradict the theorem because $|n,l,m\rangle$ and $|n,l, -m\rangle$ are degenerate. Similarly, the wavefunction of a plane wave $\exp(i\mathbf{p} \cdot \mathbf{x})$ is complex, but it is degenerate with $\exp(-i\mathbf{p} \cdot \mathbf{x})$.

1.6.7 Time-Reversal of a Spin-1/2 System

Now we shall explore the action of Θ in the more interesting case of a spin-1/2 particle. We recall from quantum mechanics that the eigenket of $\mathbf{S} \cdot \hat{\mathbf{n}}$ with eigenvalue $\hbar/2$ can be written as

$$|\mathbf{n}; +\rangle = e^{-iS_z\alpha/\hbar} \, e^{-iS_y\beta/\hbar} \, |+\rangle , \tag{1.64}$$

as depicted in Figure 1.11. $\mathbf{n}$ is characterized by the polar and azimuthal angles β and α, respectively.

The action of Θ on the angular momentum operator defined in (1.55) yields

$$\Theta \, |\mathbf{n}; +\rangle = e^{-iS_z\alpha/\hbar} \, e^{-iS_y\beta/\hbar} \, \Theta \, |+\rangle = \eta \, |\mathbf{n}; -\rangle . \tag{1.65}$$

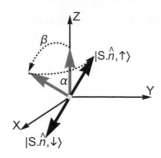

Figure 1.11 The spin eigenstates $\left|\mathbf{S} \cdot \hat{\mathbf{n}}, \uparrow\right\rangle$ and $\left|\mathbf{S} \cdot \hat{\mathbf{n}}, \downarrow\right\rangle$.

Alternatively, as depicted in Figure 1.11, and as can be easily verified, we have

$$|\mathbf{n}; -\rangle = e^{-iS_z\alpha/\hbar}\, e^{-iS_y(\pi+\beta)/\hbar}\,|+\rangle\,. \tag{1.66}$$

Comparing (1.66) with (1.65) and setting $\Theta = UK$, we obtain

$$\Theta = \eta\, e^{-iS_y\pi/\hbar}\, K = -i\eta\left(\frac{2S_y}{\hbar}\right) K, \tag{1.67}$$

where η is an arbitrary phase. In obtaining the last expression in (1.67), we used the relation

$$\exp\left(\frac{-i\mathbf{S}\cdot\hat{\mathbf{n}}\phi}{\hbar}\right) = \exp\left(\frac{-i\sigma\cdot\hat{\mathbf{n}}\phi}{2}\right) = \mathbb{1}\cos\left(\frac{\phi}{2}\right) - i\sigma\cdot\hat{\mathbf{n}}\sin\left(\frac{\phi}{2}\right).$$

It is now easy to show, with the aid of (1.67), that

$$e^{-iS_y\pi/\hbar}\,|+\rangle = +|-\rangle \qquad e^{-iS_y\pi/\hbar}\,|-\rangle = -|+\rangle\,. \tag{1.68}$$

Equation (1.68) allows us to establish the effect of Θ on the most general spin-1/2 ket

$$\Theta\left(c_+\,|+\rangle + c_-\,|-\rangle\right) = +\eta\, c_+^*\,|-\rangle - \eta\, c_-^*\,|+\rangle\,. \tag{1.69}$$

Applying Θ once more, we get

$$\Theta^2\left(c_+\,|+\rangle + c_-\,|-\rangle\right) = -|\eta|^2\, c_+\,|+\rangle - |\eta|^2\, c_-\,|-\rangle \tag{1.70}$$

or

$$\Theta^2 = -\mathbb{I} \tag{1.71}$$

for any spin orientation; $\mathbb{I}$ is the identity operator. We note that (1.71) is a fundamental relation, completely independent of the choice of the phase η. This should be contrasted with the case of a spinless particle state, which yields

$$\Theta^2 = +\mathbb{I}, \tag{1.72}$$

as is evident from (1.59).

More generally, we now prove

$$\Theta^2\,|j\text{ half integer}\rangle = -|j\text{ half integer}\rangle \tag{1.73}$$

$$\Theta^2\,|j\text{ integer}\rangle = +|j\text{ integer}\rangle\,, \tag{1.74}$$

and the eigenvalue of Θ^2 is given by $(-1)^{2j}$. We first note that (1.67) generalizes for an arbitrary j to

$$\Theta = \eta \, e^{-i J_y \pi / \hbar} \, K. \tag{1.75}$$

For a ket $|\alpha\rangle$ expanded in terms of $|j, m\rangle$ base eigenkets, we have

$$\Theta \left(\Theta \sum |j, m\rangle \langle j, m \mid \alpha\rangle \right) = \Theta \left(\eta \sum e^{-i J_y \pi / \hbar} |j, m\rangle \langle j, m \mid \alpha\rangle^* \right)$$

$$= |\eta|^2 \, e^{-2i J_y \pi / \hbar} \sum |j, m\rangle \langle j, m |\alpha\rangle . \tag{1.76}$$

But

$$e^{-2i J_y \pi / \hbar} |j, m\rangle = (-1)^{2j} |j, m\rangle , \tag{1.77}$$

as is evident from the properties of angular momentum eigenstates under 2π rotation.

We make a parenthetical remark on the phase convention η. In our earlier discussion based on the position representation, we saw that with the usual convention for spherical harmonics it is natural to choose the arbitrary phase for $|l, m\rangle$ under time reversal so that

$$\Theta \, |l, m\rangle = (-1)^m \, |l, -m\rangle . \tag{1.78}$$

Some authors find it attractive to generalize this to obtain

$$\Theta \, |j, m\rangle = (-1)^m \, |j, -m\rangle . \quad (j \text{ an integer}) \tag{1.79}$$

regardless of whether j refers to l or s (for integer spin system). We may extend this to 1/2 integer systems if we choose $\eta = +i$, so that time-reversal operation may be written as

$$\Theta \, |j, m\rangle = i^{2m} \, |j, -m\rangle \tag{1.80}$$

for any j.

Kramers' Degeneracy

We consider here the case of the Hamiltonian of a spin-1/2 particle that is time reversal invariant (TRI), then

$$[\Theta, \mathcal{H}] = 0, \qquad \left[\Theta^2, \mathcal{H} \right] = 0 \tag{1.81}$$

so that the energy eigenkets can be chosen as eigenkets of Θ^2. Since Θ commutes with $\mathcal{H}$, $\Theta \, |\psi\rangle$ has the same energy as $|\psi\rangle$. If $|\psi\rangle$ is nondegenerate, we can write

$$\Theta \, |\psi\rangle = e^{i\alpha} \, |\psi\rangle , \tag{1.82}$$

where α is an overall phase, then

$$\Theta^2 \, |\psi\rangle = \Theta \, e^{i\alpha} \, |\psi\rangle = e^{-i\alpha} \, \Theta \, |\psi\rangle = |\psi\rangle = -|\psi\rangle . \tag{1.83}$$

Hence, there must be another state with the same energy as $|\psi\rangle$. Moreover, they are orthogonal, since

$$\langle \psi \mid \Theta \psi \rangle = \left\langle \Theta \psi \mid \Theta^2 \psi \right\rangle^* = -\langle \Theta \psi \mid \psi \rangle^* = -\langle \psi \mid \Theta \psi \rangle = 0.$$

When Θ is a symmetry, eigenstates of systems with half-integer spin always come in degenerate Kramers' pairs.

Interactions with External Electric Fields

Since the appearance of electric field interactions in a system's Hamiltonian takes the form

$$V(\mathbf{x}) = e\,\Phi(\mathbf{x}),$$

where Φ is a real function of the position operator $\mathbf{x}$, and since $[\Theta, \mathbf{x}] = 0$, then

$$[\Theta, \mathcal{H}] = 0. \tag{1.84}$$

This of course leads to the reality of nondegenerate wavefunctions.

An Interesting Application

We revisit the Hamiltonian describing electron motion in a crystalline lattice encountered in §1.2. The Hamiltonian is time reversal invariant. This can be generalized to $\mathcal{H}(\mathbf{k})$, where $\hbar k$ is the electron lattice momentum. Since

$$\Theta\mathbf{p}\Theta^{-1} = -\mathbf{p},$$

we find that

$$\Theta\mathcal{H}(\mathbf{k})\Theta^{-1} = \mathcal{H}(-\mathbf{k}). \tag{1.85}$$

This result guarantees that the dispersion relation has the symmetry

$$E(\mathbf{k}) = E(-\mathbf{k}) \tag{1.86}$$

even when the system lacks space-inversion symmetry!

Interactions with External Magnetic Fields

The Hamiltonian may contain terms like

$$\mathbf{S}\cdot\mathbf{B}, \quad \mathbf{p}\cdot\mathbf{A} + \mathbf{A}\cdot\mathbf{p}.$$

Since $\mathbf{p}$ and $\mathbf{S}$ are odd under time reversal, the Hamiltonian is not invariant under time reversal, and Kramers' degeneracy is lifted.

1.7 Appendix: The Adiabatic Approximation

The concept of adiabatic motion arises when some external conditions governing the dynamics of a system change on a time scale that is much longer than the time scale characterizing the system's dynamics. A typical example is found in molecular physics: Molecules consist of electrons and nuclei (or ions). The time scale associated with the electron motion is much faster than that associated with the motion of the nucleus. Since the two systems are coupled, we find that the fast electron motion allows them to adjust to the sluggish motion of the nuclei almost "instantaneously" when viewed through the latter's time scale. This allows us to consider the time dependence of the nuclear degrees of freedom as slowly varying external parameters when dealing with the motion of the electron system of a molecule.

1.7.1 The Adiabatic Theorem

When a Hamiltonian changes <u>*gradually*</u> *and* <u>*slowly*</u> *with time from* $\mathcal{H}^i$ *to* $\mathcal{H}^f$, *and if the system is initially in the* m*th eigenket of* $\mathcal{H}^i$, *then it will be carried over into the* m*th eigenket of* $\mathcal{H}^f$

If the Hamiltonian of a system is *independent* of time, then a quantum system that starts out in eigenstate ψ_n

$$\mathcal{H} \, |\psi_n\rangle = E_n \, |\psi_n\rangle$$

remains in the nth eigenstate, simply picking up a phase factor:

$$|\Psi_n(t)\rangle = |\psi_n\rangle \, e^{-iE_n t/\hbar}.$$

If the Hamiltonian changes with time, then the eigenkets and eigenvalues are themselves time dependent:

$$\mathcal{H}(t) \, |\psi_n(t)\rangle = E_n(t) \, |\psi_n(t)\rangle. \tag{1.87}$$

The instantaneous eigenkets constitute an orthonormal and complete set:

$$\langle \psi_n(t)| \, \psi_m(t)\rangle = \delta_{nm}.$$

A general solutions satisfies

$$i\hbar \frac{\partial}{\partial t} \, |\Psi(t)\rangle = \mathcal{H}(t) \, |\Psi(t)\rangle \tag{1.88}$$

with

$$|\Psi(t)\rangle = \sum_n c_n(t) \, |\psi_n(t)\rangle \, e^{i\theta_n(t)} \tag{1.89}$$

$$\theta_n(t) = -\frac{1}{\hbar} \int_0^t dt' \, E_n(t') \qquad \textit{Dynamic phase.} \tag{1.90}$$

The definition of θ_n generalizes the "standard" phase factor to the case where E_n varies with time.

Substituting (1.89) in (1.88),

$$i\hbar \sum_n \left[\dot{c}_n \psi_n + c_n \dot{\psi}_n + i c_n \psi_n \dot{\theta}_n \right] e^{i\theta_n} = \sum_n c_n \, (\mathcal{H}\psi_n) \, e^{i\theta_n}. \tag{1.91}$$

But from (1.90), $\dot{\theta}_n = -\frac{E_n}{\hbar}$, and the last sum on the left cancels that on the right. We get

$$\sum_n \dot{c}_n \, \psi_n \, e^{i\theta_n} = -\sum_n c_n \, \dot{\psi}_n \, e^{i\theta_n}.$$

Taking the inner product with ψ_m, and invoking the orthonormality of the instantaneous eigenkets,

$$\dot{c}_n \, \delta_{nm} \, e^{i\theta_n} = -\sum_n c_n \, \langle \psi_m \mid \dot{\psi}_n \rangle \, e^{i\theta_n}$$

or

$$\dot{c}_m = -\sum_n c_n \, \langle \psi_m \mid \dot{\psi}_n \rangle \, e^{i(\theta_n - \theta_m)}. \tag{1.92}$$

In order to obtain a useful expression for $\langle \psi_m \mid \dot{\psi}_n \rangle$, we differentiate (1.87) with respect to time, take the inner product with ψ_m, and get

$$\langle \psi_m \mid \dot{\mathcal{H}} \mid \psi_n \rangle + \langle \psi_m \mid \mathcal{H} \mid \dot{\psi}_n \rangle = \dot{E}_n \, \delta_{mn} + E_n \, \langle \psi_m \mid \dot{\psi}_n \rangle \tag{1.93}$$

but

$$\langle \psi_m \mid \mathcal{H} \mid \dot{\psi}_n \rangle == E_m \, \langle \psi_m \mid \dot{\psi}_n \rangle.$$

It follows that for $n \neq m$,

$$\langle \psi_m \mid \dot{\mathcal{H}} \mid \psi_n \rangle = (E_n - E_m) \, \langle \psi_m \mid \dot{\psi}_n \rangle$$

and (1.92) becomes

$$\dot{c}_m = -c_m \, \langle \psi_m \mid \dot{\psi}_m \rangle - \sum_{n \neq m} c_n \, \frac{\langle \psi_m \mid \dot{\mathcal{H}} \mid \psi_n \rangle}{E_n - E_m} \, \exp\left(-\frac{i}{\hbar} \int_0^t dt' \, [E_n(t') - E_m(t')] \right).$$

This is an exact result.

Now comes the adiabatic approximation: *Assume that*

$$\langle \psi_m \mid \dot{\mathcal{H}} \mid \psi_n \rangle \ll (E_n - E_m)$$

is extremely small, and drop the last term. This assumption is well justified since the dynamical time scale of the system is determined by $(E_n - E_m)/\hbar$.

Thus, in the adiabatic approximation, we obtain

$$\dot{c}_m = -c_m \, \langle \psi_m \mid \dot{\psi}_m \rangle \tag{1.94}$$

$$c_m(t) = c_m(0) \, e^{i\gamma_m(t)} \tag{1.95}$$

$$\gamma_m(t) = i \int_0^t dt' \, \left\langle \psi_m(t') \mid \frac{\partial}{\partial t'} \psi_m(t') \right\rangle \qquad \textit{Geometric phase.} \tag{1.96}$$

The solution (1.95) is obtained because $\langle \psi_m \mid \dot{\psi}_m \rangle$ is purely imaginary, since

$$\frac{d}{dt} \langle \psi_m(t) \mid \psi_m(t) \rangle = \langle \psi_m \mid \dot{\psi}_m \rangle + \langle \dot{\psi}_m \mid \psi_m \rangle = 2\mathrm{Re} \, \langle \psi_m \mid \dot{\psi}_n \rangle = 0$$

If the particle starts in eigenket m such that $c_m(0) = 1$, $c_n(0) = 0 \; \forall n \neq m$, then

$$\Psi_m = e^{i\theta_m(t)} \, e^{i\gamma_m(t)} \, \psi_m(t). \tag{1.97}$$

The particle remains in the mth eigenket of the *evolving Hamiltonian*, picking up only a couple of phase factors.

Exercises

1.1   Derive an expression for the electronic pressure P_e exerted on the wall of the vessel containing an electron gas, given that

$$P_e = -\frac{\partial E_0}{\partial \Omega}.$$

Hence, obtain an expression for the bulk modulus

$$B = \frac{1}{\kappa} = -\Omega \frac{\partial P_e}{\partial \Omega},$$

where κ is the compressibility. Provide some values for Cu, where the electron density is about 8.5×10^{28} electrons/m^3.

1.2   Consider cohesive energy of free electron Fermi gas.

(a) Show that the average kinetic energy per electron in a free electron Fermi gas at 0 K is $2.21/r_s^2$, where the energy is expressed in Rydbergs, with $1 \, \mathrm{Ry} = me^4/2h^2$.

(b) Show that the coulomb energy of a positive point charge e interacting with the uniform electron distribution of one electron in the volume of radius $r_0 = r_s a_0$ is $-3e^2/r_0$, or $-3/r_s$ in Rydbergs.

(c) Show that the coulomb self-energy of the electron distribution in the sphere is $3e^2/5r_0$, or $6/5r_s$ in Rydberg.

(d) The sum of (b) and (c) gives $-1.80/r_s$ for the total coulomb energy per electron. Show that the equilibrium value of r_s is 2.45. Will such a material be stable with respect to separated H atoms?

1.3   Consider a Sommerfeld gas in two dimensions.

(a) Express k_F and E_F in terms of the electron density n.

(b) Derive an expression for the density of states.

1.4   Given the dispersion relations

$$E(k) = c\,k; \quad E(k) = c\,k^2,$$

calculate the corresponding $\mathcal{D}(E)$ in one, two, and three dimensions.

1.5   Identify critical points and singularities in the density of states.

The density of states for a single band with dispersion $E(\mathbf{k})$ is given by (1.6). A singularity in the density of states (a critical point) arises when $\nabla E(\mathbf{k}) = 0$. We may expand the band dispersion in the vicinity of a critical point $E(\mathbf{k}_c) = E_c$ as

$$E(\Delta\mathbf{k}) \sim E_c \pm \left[\frac{\hbar^2 \Delta k_x^2}{2m_x} + \frac{\hbar^2 \Delta k_y^2}{2m_y} + \frac{\hbar^2 \Delta k_z^2}{2m_z} \right], \quad \Delta\mathbf{k} = \mathbf{k} - \mathbf{k}_c,$$

$$m_\alpha = \hbar^2 \left(\frac{\partial^2 E(\Delta\mathbf{k})}{\partial \Delta k_\alpha^2} \right)^{-1}$$

(a) Identify the plus/minus signs with the nature of the critical point – a minimum, a maximum, or a saddle point – of $E(\mathbf{k})$ in 1D, 2D, and 3D.
(b) Derive the density of states $\mathcal{D}(E)$ in the vicinity of band minima and maxima (not saddle points!) in 1D and 2D. Use polar coordinates in the 2D case to simplify the derivation.
Hint: Writing $\mathcal{D}(E)$ as

$$\mathcal{D}(E) = 2 \sum_{\mathbf{k}} \delta(E(\mathbf{k}) - E),$$

use the relation for the delta function

$$\delta[f(x)] = \sum_\ell \frac{\delta(x - x_\ell)}{|f'(x_\ell)|}$$

where x_ℓ are the zeros of the function $f(x)$.

1.6 Consider classical phonons in one dimension.

The simplest model that captures the physics of phonons is a one-dimensional periodic chain of particles of mass m separated with lattice spacing a, and connected by springs with a spring constant κ. The particle at site ℓ can displace from its equilibrium position, ℓa, by u_ℓ¡ giving rise to a corresponding potential energy of $\frac{\kappa}{2}(u_{\ell+1} - u_\ell)^2$. The Hamiltonian of the system of displacing particles then reads

$$\mathcal{H} = \sum_\ell \left(\frac{p_\ell^2}{2m} + \frac{\kappa}{2}(u_{\ell+1} - u_\ell)^2 \right).$$

1. Treating this system classically, derive the infinite set of coupled equations of motion either using Hamilton's canonical equations, or writing down the Lagrangian and using the Euler–Lagrange equations.
2. Decouple these equations of motion by using an eigenket of the lattice translation operator and assuming a time dependence of the form $e^{i\omega t}$.
3. Show that the decoupling process provides the dispersion curve for the phonons, namely $\omega \equiv \omega(k)$.

1.7   Perform a Born–Oppenheimer approximation: a toy model.
      This problem is meant to convince you of the validity of the Born–Oppenheimer
      approximation through a toy model of coupled one-dimensional harmonic oscillators.
      The Hamiltonian is given by

$$\mathcal{H} = \frac{P_1^2}{2M} + \frac{P_2^2}{2M} + \frac{p_3^2}{2m} + \frac{\kappa}{2}\left(x_3 - X_1 - d\right)^2 + \frac{\kappa}{2}\left(X_2 - x_3 - d\right)^2$$

where $m \ll M$. Since we know how to solve the system exactly, we can compare the
exact solution to the BOA.

(a) Apply the BOA by neglecting the first two kinetic energy terms:

   1. Combine the potential energy into one harmonic oscillator involving x_3 plus
       a residual harmonic term independent of x_3.
   2. Solve the quantum harmonic oscillator problem of the light particle 3.
   3. Substitute the effective potential you obtain for the motion of particle 3 into
       the Hamiltonian, and solve the residual quantum eigenvalue problem for the
       two heavy particles. (Hint: use a center of mass transformation.)

(b) Obtain an exact solution for the problem. In light of this solution, how good is
     the BOA?

1.8   Consider unitary and antiunitary operators.
      A unitary operator is an operator that is linear and norm-preserving. An antiunitary
      operator is an operator that is antilinear and norm-preserving.

(a) Show that unitary operators leave scalar products unchanged

$$\langle U\psi \,|U\phi\rangle = \langle \psi \,|\phi\rangle$$

     for arbitrary kets $|\psi\rangle$ and $|\phi\rangle$.
(b) Show that antiunitary operators map a scalar product to its complex conjugate

$$\langle U\psi \,|U\phi\rangle = \langle \psi \,|\phi\rangle^*$$

     for arbitrary kets $|\psi\rangle$ and $|\phi\rangle$.

1.9   Consider time-reversal symmetry.
      In the momentum representation, the state of a spinless particle is described using a
      wavefunction $\psi(\mathbf{p})$.

(a) Give the explicit form of the time-reversal operator Θ for a spinless particle in
     the momentum representation.
(b) Give the explicit form of the Θ for a spin-1/2 particle in the momentum
     representation.

2

Electrons and Band Theory: Formalism in the One-Electron Approximation

2.1 The Many-Body Problem and One-Electron Approximation

Our objective here is to determine the properties of an interacting many-electron system subjected to an external potential (usually the ion potential $v(\mathbf{r})$) satisfying the Schrödinger equation

$$\left\{ -\frac{\hbar^2}{2m_e} \sum_i \nabla_i^2 + \sum_i v(\mathbf{x}_i) + \frac{1}{2} \sum \frac{e^2}{|\mathbf{x}_i - \mathbf{x}_j|} - E \right\} \Psi(\mathbf{x}_1, \ldots, \mathbf{x}_N) = 0. \qquad (2.1)$$

We have omitted terms involving electron spin, spin–orbit coupling, and other relativistic effects, as well as external magnetic fields, etc.

> We should note that when dealing with the electron system in a solid, the interaction potential with the ions is treated as an external potential to which the electrons respond.

Dirac is reported to have said that the Schrödinger equation (2.1) has, with one blow, taken care of all chemistry (and, by implication, of the electronic structure of solids). In a formal sense, this is, of course, true. Yet, due to the Coulomb repulsion between the electrons, this Hamiltonian is obviously not the sum of single particle Hamiltonians $\mathcal{H}_i$. In principle, this means that the total wavefunction depends on the positions of all the electrons, the coordinates of which are all correlated, so that a literal solution of (2.1) for a solid with $\simeq 10^{23}$ variables is, practically speaking, meaningless since there is no way of recording the wavefunction. Fortunately, our real interest lies in experimentally related quantities that are related to highly contracted variables, such as the one- and two-particle density matrices from which we can obtain, for example, the energy E, the particle density $n(\mathbf{r})$, particle–particle correlation functions, etc.

For most cases, therefore, the *one-electron approximation* is commonly used. In effect, it replaces the electron–electron interaction for some particular electron at $\mathbf{x}_i$ by an averaged interaction with all other electrons at $\mathbf{x}_j$, giving rise to an effective screened potential $V_e(\mathbf{x}_i)$.

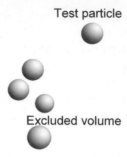

Figure 2.1 Excluded volume presented by the billiard balls to the test particle.

2.1.1 Validity of the One-Particle Approximation

Three factors mainly determine the applicability of the one-particle approximation to a given many-body system:

1. The *range* of the interparticle interaction potential:

   The importance of this factor may be demonstrated by considering the extreme case of a hard-wall-type interparticle interaction potential, such as that between billiard balls. It is impossible to use an averaged account of the interaction of a particle with the remaining ones in such a case, because of the particularity of the excluded volume, as is obviously seen in Figure 2.1.

   This severe restriction may be relaxed when the range of the potential becomes relatively long. An operational definition of a long-range potential is usually given in terms of the divergence of a particle's potential energy when the system's size tends to infinity. We consider an infinite homogeneous system, with discrete density n, surrounding a test particle that is placed at the origin. The total potential energy of our test particle is then

$$E_V = \int d\mathbf{x}\, v(\mathbf{x})n(\mathbf{x}) = 4\pi n \int v(\mathbf{x})r^2 dr, \qquad (2.2)$$

   where $r = |\mathbf{x}|$. We used the homogeneity of the system to set $n(\mathbf{x}) = n$; this integral diverges if $v(\mathbf{x}) \propto (1/r^m)$ with $m \leq 2$. Consequently, interaction potentials satisfying this condition are considered long range.

2. The particle *density*:

   The effect of the second factor is rather obvious, since for a system consisting of, say, three particles, the interaction energy of our test particle will, no doubt, depend on the specific configuration of the three particle, and an averaged potential will introduce serious errors in the corresponding energies, trajectories, and wavefunctions.

3. The degree of *localization* of the constituent particles:

   The effect of localization can be illustrated by considering a system containing a large number of particles with long-range interactions. If the system is divided into a large number of spatial cells, each accommodating a small number of particles, we find that we have to consider the dwell time of a given particle (our test particle in the present case) in each cell. Now, if the particle is completely delocalized, i.e., it is a free particle with almost zero dwell time, our test particle will only feel a potential averaged over a large number of cells and, consequently, a one-particle effective potential can be, quite justifiably, constructed. However, if our test particle has an appreciable dwell time in each cell, then its motion will be strongly correlated with the motions of the small number of particles in the cell where it is temporarily located. In this case, a simple one-particle averaged potential is inadequate.

2.1.2 Self-Consistent Formulations of the One-electron Approximation

Self-consistency means that an initial guess might be made for the one-electron potential, the eigenvalue problem is then solved, the density calculated, and a new potential found, with the aid of Poisson's equation. These steps are repeated until there is, in principle, no change in the output from one cycle to the next – self-consistency has been reached. Such a set of equations are often called self-consistent field (SCF) equations.

Two physically different self-consistent field approaches to the one-electron approximation are used to arrive at solutions to (2.1). One approach considers the electronic wavefunction and the other the electron density as the primary quantity to be determined.

2.2 The Hartree–Fock Model

D. R. Hartree pioneered the self-consistent field approach to calculating the electronic structure of many-electron systems [86]. His work was focused on the structure of atoms. However, he did not consider the exchange symmetry of the wavefunction in his work. The first use of antisymmetrized wavefunctions was reported by Fock in 1930 [64].

2.2.1 The Hartree Model

The Hartree model considers *single-particle orbitals* in a space-varying effective potential. If it were possible to write the Hamiltonian of a system of interacting electrons as a sum of single-particle Hamiltonians, it would be possible to write the many-electron wavefunction as a product of single-particle orbitals

$$\Psi(\mathbf{x}_1, \mathbf{x}_2, \ldots, \mathbf{x}_N) = \prod_{i=1}^{N} \psi_i(\mathbf{x}_i), \tag{2.3}$$

where $\{\psi_i\}$ is an orthonormal set of single-particle wavefunctions. But, since the Hamiltonian (2.1) is not a sum of single-particle Hamiltonians, the true wavefunctions cannot be written in the product form of (2.3), which, furthermore, does not have the *antisymmetry* property required for fermions.

Correlation Effects

As we have described earlier, in the case of short-range potentials the effective correlations in the motion of a many-particle system are quite complicated, and lead to the breakdown of the one-particle approximation. Such interactions depend to a large degree on the details of the instantaneous configurations of the particles, which restrict their subsequent motion.

However, long-range potentials, such as Coulomb interactions, do not exhibit similar abrupt changes. Consequently, at high particle density these correlation effects are less serious, and a test particle can feel the effective average over all particles it interacts with. In such cases, the omission of correlations may lead to tolerable errors. It is then possible to construct our wavefunctions as products of single-particle wavefunctions.

Self-Consistent Hartree Equations In the case of a system of N electrons, we have

$$|\Psi\rangle = \prod_{i=1}^{N} \psi_i(\mathbf{x}_i)$$

$$E = \sum_i \left\langle \psi_i \left| \frac{p_i^2}{2m_e} + V_{ei}(\mathbf{x}_i) \right| \psi_i \right\rangle + \frac{e^2}{2} \sum_{\substack{ij \\ i \neq j}} \left\langle \psi_i \, \psi_j \left| \frac{1}{|\mathbf{x}_i - \mathbf{x}_j|} \right| \psi_i \, \psi_j \right\rangle \tag{2.4}$$

Varying E with respect to ψ_i, and using a Lagrange multiplier ε_i to impose the normalization constraint $\langle \psi_i \,|\psi_i \rangle = 1$, we obtain

$$\left[T + v_i^{\mathrm{H}}(\mathbf{x}) \right] \psi_i(\mathbf{x}) = \varepsilon_i^{H} \, \psi_i(\mathbf{x}), \tag{2.5}$$

where the effective, or Hartree, potential $v_i^{\mathrm{H}}(\mathbf{x})$ is the mean-field expression

$$v_i^{\mathrm{H}}(\mathbf{x}) = V_{ei}(\mathbf{x}) + \frac{e^2}{2} \sum_{j \neq i} \left\langle \psi_j \left| \frac{1}{|\mathbf{x} - \mathbf{x}'|} \right| \psi_j \right\rangle \tag{2.6}$$

Equations (2.5) and (2.6) constitute the *self-consistent Hartree equations*. This is obviously also a mean-field theory in view of (2.6): each electron is regarded as moving in the external potential plus the mean potential of all the electrons. The Hartree equations are solved iteratively: An initial guess of the potential is used to solve the one-electron equations; the solutions are used to generate a new effective potential, and so on until convergence is achieved.

As we show next, an appreciable contribution of correlation effects to the interacting scenarios of electronic (fermionic) systems is accounted for through antisymmetrization of the many-electron wavefunction. It is referred to as *exchange effects*.

The Physical Origin of the Exchange Energy

To understand this effect, let us first establish the symmetry of quantum mechanical wavefunctions of systems of identical particles.

The term identical particles implies that the total energy of the system, and hence its associated Hamiltonian, remains invariant when two particles are exchanged.

The process of the two-particle exchange is effected within the framework of quantum mechanics through a permutation operator $\hat{\mathcal{P}}$. *The invariance of the system's Hamiltonian with respect to the operation of particle exchange means that $\hat{\mathcal{P}}$ commutes with the Hamiltonian $\mathcal{H}$*, and we have

$$\hat{\mathcal{P}}\mathcal{H}\Psi = \hat{\mathcal{P}}\mathcal{H}\hat{\mathcal{P}}^{-1}\hat{\mathcal{P}}\Psi$$
$$= \mathcal{H}\hat{\mathcal{P}}\Psi = E\,\hat{\mathcal{P}}\Psi \tag{2.7}$$
$$\mathcal{H}\Psi = E\,\Psi.$$

$\hat{\mathcal{P}}$ and $\mathcal{H}$ share the same eigenfunctions.

For the sake of simplicity, we consider a two-particle system. The eigenfunctions of $\hat{\mathcal{P}}$ satisfy the equation

$$\hat{\mathcal{P}}\,\Psi(\mathbf{x}_1,\mathbf{x}_2) = \Psi'(\mathbf{x}_2,\mathbf{x}_1) = p\,\Psi(\mathbf{x}_1,\mathbf{x}_2), \tag{2.8}$$

where p is the corresponding eigenvalue. Repeating the permutation operation will take us back to the original wavefunction

$$\hat{\mathcal{P}}^2\,\Psi(\mathbf{x}_1,\mathbf{x}_2) = \hat{\mathcal{P}}\,\Psi'(\mathbf{x}_2,\mathbf{x}_1)$$
$$= p^2\,\Psi(\mathbf{x}_1,\mathbf{x}_2) = \Psi(\mathbf{x}_1,\mathbf{x}_2), \tag{2.9}$$

which leads to two possible values for p, namely, $p = \pm1$. Consequently, we have two distinct classes of many-body wavefunctions under the operation of two-particle exchange, *symmetric bosons* with $p = 1$, and *antisymmetric fermions* with $p = -1$.

Now, let us explore the implication of antisymmetry of fermionic wavefunctions by considering the simple case of two spinless noninteracting particles. The condition of antisymmetry requires that

$$\Psi(\mathbf{x}_2,\mathbf{x}_1) = -\Psi(\mathbf{x}_1,\mathbf{x}_2).$$

To reveal the physical implications of this relation, we shall describe this wavefunction in terms of the relative coordinate $\Delta r = |\mathbf{x}_2 - \mathbf{x}_1|$ in the vicinity of $\Delta r = 0$, where the two particles occupy the same point in space, such that $\mathbf{x}_1 = \mathbf{x}_2 = \mathbf{x}$. The antisymmetry requirement leads to $\Psi(\mathbf{x},\mathbf{x}) = -\Psi(\mathbf{x},\mathbf{x})$ and thus to

$$\Psi(\mathbf{x},\mathbf{x}) = 0.$$

The requirement that a quantum mechanical wavefunction must have a *continuous magnitude* and a *continuous derivative* will force a gradual, rather than an abrupt, change in the magnitude of the wavefunction in the vicinity of $\Delta r = 0$, where it must vanish due to antisymmetry. This suppression in the wavefunction magnitude around $\Delta r = 0$ can be viewed as an effective *hole*, which is referred to as an *exchange hole*; see Figure 2.2. It depicts a mutual avoidance between the two particles even when they are noninteracting. This condition is obviously absent in the classical analog.

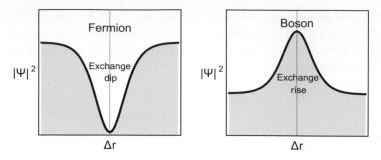

Figure 2.2 A schematic of $|\Psi|^2(\Delta r)$ for a pair of fermions and of bosons.

When a repulsive Coulomb interaction between the two particles (electrons for example) is turned on, the mutual avoidance due to the *exchange hole* in the two-particle wavefunction will lead to a reduction of the Coulomb energy in the fermionic case relative to the classical case. In contrast, the boson symmetrization will result in a repulsive energy that is higher than the classical case. The reduction due to the antisymmetrization of the total fermionic wavefunction is referred to as the *exchange energy*.

Notice that in the case of a mutually attractive interaction potential – pairing interaction in superconductivity – a symmetrized bosonic wavefunction will have a lower energy than both the classical case and the fermionic wavefunction.

Next, we consider the case of two electrons with spin. The total wavefunction is then expressed as

$$\Psi(\mathbf{1}, \mathbf{2}) = \Phi(\mathbf{x}_1, \mathbf{x}_2)\, \chi(\sigma_1, \sigma_2), \tag{2.10}$$

where σ_i is the spin degree of freedom of electron i and χ is the total spin wavefunction. Now, the overall antisymmetry of the wavefunction Ψ is satisfied by two combinations: $\Phi_s \chi_a$, singlet spin state, or, $\Phi_a \chi_s$, triplet spin state. The subscripts s, a refer to symmetric and antisymmetric, respectively. Thus, we find that in the absence of any other external interactions, the triplet state, with antisymmetrized spatial wavefunction, will have lower energy than the singlet state. The lowering in the energy of the former is still identified as due to *exchange effects*.

2.2.2 The Hartree–Fock Model

In the Hartree model, the trial many-electron wavefunction did not satisfy the required fermionic antisymmetry. Therefore, it seems more appropriate to consider a Slater determinant of single-particle wavefunctions instead, namely,

$$\Psi(\mathbf{x}_1, \ldots, \mathbf{x}_m, \ldots, \mathbf{x}_N) = \frac{1}{\sqrt{N!}} \begin{vmatrix} \psi_1(\mathbf{x}_1) & \ldots & \psi_i(\mathbf{x}_1) & \ldots & \psi_N(\mathbf{x}_1) \\ \ldots & \ldots & \ldots & \ldots & \ldots \\ \psi_1(\mathbf{x}_N) & \ldots & \psi_i(\mathbf{x}_N) & \ldots & \psi_N(\mathbf{x}_N) \end{vmatrix} \tag{2.11}$$

where the ψ_is are products of spin and space wave functions. For an arbitrary number of electrons, the wavefunction form in (2.11) can be shown to satisfy the desired antisymmetry condition. The determinant has $N!$ terms each multiplied by -1 or 1 depending on the parity of the permutation. Each term may be written as

$$(-1)^{\mathcal{P}(i_1, i_2, \ldots, i_N)} \, \psi_{i_1}(\mathbf{x}_1) \, \psi_{i_2}(\mathbf{x}_2) \ldots \psi_{i_N}(\mathbf{x}_N),$$

where the indices i_1, i_2, ... take values between 1 and N and the exponent of (-1) refers to the order of appearance of the orbital indices in the term. Terms with odd permutations assume a $-$ sign, while those with even ones have a $+$ sign. We shorten the notation and replace $\mathcal{P}(i_1, i_2, \ldots, i_N)$ by $\mathcal{P}(i)$, where i now refers to a particular sequence of the N indices. The Slater determinant may then be written as

$$\Psi = \frac{1}{(N!)^{1/2}} \sum_{i=1}^{N!} (-1)^{\mathcal{P}(i)} \left[\phi_{i_1}(\mathbf{x}_1) \ldots \phi_{i_N}(\mathbf{x}_N) \chi_1(\sigma_1) \ldots \chi_N(\sigma_N) \right]. \tag{2.12}$$

Each spatial wavefunction ϕ appears twice because of spin.

Nothing has been said so far about the form of the orbitals $\phi_i(\mathbf{x}_j)$ and they are left to be determined via a variational procedure. This is known as the Hartree–Fock (HF) approximation [64, 87].

The *best* single-particle orbitals ψ_i are obtained by varying the total energy with respect to the many-body wavefunction:

$$\delta E = \delta \int d\mathbf{x}^N \, \Psi^* \mathcal{H} \, \Psi = \int d\mathbf{x}^N \, \Psi^* \mathcal{H} \, \delta\Psi + \int d\mathbf{x}^N \, \delta\Psi^* \mathcal{H} \, \Psi. \tag{2.13}$$

The orthogonality and normalization conditions

$$\int d\mathbf{x} \, \phi_i^*(\mathbf{x}) \, \phi_j(\mathbf{x}) = \delta_{ij} \quad \Rightarrow \quad \int d\mathbf{x} \, \delta\phi_i^*(\mathbf{x}) \, \phi_j(\mathbf{x}) = 0 \tag{2.14}$$

are satisfied through the Lagrange multipliers η_{ij}.

The variation of Ψ is expressed in terms of variations of single-particle wavefunctions ψ_i, so that

$$\delta\Psi = \frac{1}{(N!)^{1/2}} \sum_{\mathcal{P}_n} (-1)^n \, \mathcal{P}_n \left[\sum_{i=1}^N \psi_1(\mathbf{x}_1) \ldots \psi_{i-1}(\mathbf{x}_{i-1}) \psi_{i+1}(\mathbf{x}_{i+1}) \ldots \psi_N(\mathbf{x}_N) \right.$$
$$\left. \times \, \delta\psi_i(\mathbf{x}_i) \chi(\sigma_1) \chi(\sigma_2) \ldots \chi(\sigma_{N-1}) \chi(\sigma_N) \right]. \tag{2.15}$$

Substituting (2.15) into (2.13), summing over spin, and using (2.14), we then find

$$\int d\mathbf{x}^N \, \delta\Psi^* \, \mathcal{H} \, \Psi = \sum_i \int d\mathbf{x}_1 \, \delta\psi_i^*(\mathbf{x}_1) \left(\left[T_1 + v(\mathbf{x}_1) + e^2 \sum_j{}' \frac{|\psi_j(\mathbf{x}_2)|^2}{r_{12}} \, d\mathbf{x}_2 \right. \right.$$

$$+ \sum_j{}' \int d\mathbf{x}_2 \, \psi_j^*(\mathbf{x}_2) \, H_2 \, \psi_j(\mathbf{x}_2) + \frac{1}{2} \sum_{j,k}{}'' e^2 \int d\mathbf{x}_2 d\mathbf{x}_3 \, \frac{|\psi_j(\mathbf{x}_2)|^2 |\psi_k(\mathbf{x}_3)|^2}{r_{23}}$$

$$\left. - \frac{1}{2} \sum_{j,k;\uparrow\uparrow}{}'' e^2 \int d\mathbf{x}_2 d\mathbf{x}_3 \, \frac{\psi_j^*(\mathbf{x}_2) \, \psi_k^*(\mathbf{x}_3) \, \psi_j(\mathbf{x}_3) \, \psi_k(\mathbf{x}_2)}{r_{23}} - \eta_{ii} \right] \psi_i(\mathbf{x}_1)$$

$$\left. - \left[\sum_{j;\uparrow\uparrow}{}' e^2 \int d\mathbf{x}_2 \, \frac{\psi_j^*(\mathbf{x}_2)\psi_i(\mathbf{x}_2)}{r_{12}} + \sum_j{}' \int d\mathbf{x}_2 \, \psi_j^*(\mathbf{x}_2) \, H_2 \, \psi_i(\mathbf{x}_2) - \eta_{ij} \right] \psi_j(\mathbf{x}_1) \right).$$

For convenience, the Lagrange multipliers η_{ij} are redefined as

$$\lambda_{ii} = \eta_{ii} - \sum_j{}' \int d\mathbf{x}_2 \, \psi_j(\mathbf{x}_2) \, H_2 \, \psi_j(\mathbf{x}_2) - \frac{1}{2} \sum_{j,k}{}' e^2 \int d\mathbf{x}_2 d\mathbf{x}_3 \, \frac{|\psi_j(\mathbf{x}_2)|^2 \, |\psi_k(\mathbf{x}_3)|^2}{r_{23}}$$

$$+ \frac{1}{2} \sum_{j,k;\uparrow\uparrow}{}'' e^2 \int d\mathbf{x}_2 d\mathbf{x}_3 \, \frac{\psi_j^*(\mathbf{x}_2) \, \psi_k^*(\mathbf{x}_3) \, \psi_j(\mathbf{x}_3) \, \psi_k(\mathbf{x}_2)}{r_{23}},$$

$$\lambda_{ij} = \eta_{ij} - \sum_j{}' \int d\mathbf{x}_2 \, \psi_j^*(\mathbf{x}_2) \, H_2 \, \psi_i(\mathbf{x}_2) \qquad (2.16)$$

Equating the coefficient of $\delta\psi_i(\mathbf{x}_1)$ to zero, we obtain

$$\left[T + v(\mathbf{x}) + \sum_j e^2 \int d\mathbf{x}' \, \frac{|\psi_j(\mathbf{x}')|^2}{|\mathbf{x} - \mathbf{x}'|} - \lambda_{ii} \right] \psi_i(\mathbf{x})$$

$$- \sum_{j;\uparrow\uparrow}{}' \left[e^2 \int d\mathbf{x}' \, \frac{\psi_j^*(\mathbf{x}') \, \psi_i(\mathbf{x}')}{|\mathbf{x} - \mathbf{x}'|} - \lambda_{ij} \right] \psi_j(\mathbf{x}) = 0. \qquad (2.17)$$

Notice that the term on the second line is an integral operator.

    There are several different solutions to (2.17), each corresponding to a different set of λ_{ij}. We choose the set of λ_{ij}, which satisfies

$$\lambda_{ij} = \delta_{ij} \, \varepsilon_i,$$

and we write

$$
\left[T + v(\mathbf{x}) + \sum_j \int d\mathbf{x}' \, \frac{e^2 |\psi_j(\mathbf{x}')|^2}{|\mathbf{x} - \mathbf{x}'|} \right] \psi_i(\mathbf{x})
$$

$$
- \sum_{j; \uparrow\uparrow}' \left[\int d\mathbf{x}' \, \frac{e^2 \psi_j^*(\mathbf{x}') \, \psi_i(\mathbf{x}')}{|\mathbf{x} - \mathbf{x}'|} \right] \psi_j(\mathbf{x}) = \varepsilon_i \, \psi_i(\mathbf{x}). \tag{2.18}
$$

where now ε_i become the set of Hartree–Fock eigenvalues, and the coupled equations (2.18) constitute the *Hartree–Fock equations*. This set is usually written in the compact form

$$
\left[T + v(\mathbf{x}) + \sum_l \left(J_l - K_l \right) \right] \psi_i(\mathbf{x}) = \varepsilon_i \, \psi_i(\mathbf{x}). \tag{2.19}
$$

The exchange operator K_l is given by

$$
K_l(1) \, \psi_i(1) = \int d\mathbf{x}_2 \, \frac{\psi_l^*(2) \, \psi_i(2)}{r_{12}} \, \psi_l(1). \tag{2.20}
$$

Since it involves an orbital exchange, it is written in the context of an orbital being operated on.

The total Hartree–Fock energy, E_{HF}, of the system of N electrons is written as

$$
E_{\mathrm{HF}} = \sum_i \langle \psi_i | \, T + v(\mathbf{x}) \, | \psi_i \rangle + \frac{1}{2} \sum_{ij} e^2 \left[\frac{|\psi_i|^2 \, |\psi_j|^2}{r_{12}} - \frac{\psi_i^*(1) \, \psi_j^*(2) \, \psi_i(2)\psi_j(1)}{r_{12}} \right],
$$

which when compared with the sum over the lowest N HF eigenenergies

$$
\sum_i \varepsilon_i = \sum_i \langle \psi_i | \, T + v(\mathbf{x}) \, | \psi_i \rangle + \sum_{ij} e^2 \left[\frac{|\psi_i|^2 \, |\psi_j|^2}{r_{12}} - \frac{\psi_i^*(1) \, \psi_j^*(2) \, \psi_i(2)\psi_j(1)}{r_{12}} \right]
$$

demonstrates that

$$
E_{\mathrm{HF}} = \sum_i \varepsilon_i - \frac{1}{2} \sum_{ij} e^2 \left[\frac{|\psi_i|^2 \, |\psi_j|^2}{r_{12}} - \frac{\psi_i^*(1) \, \psi_j^*(2) \, \psi_i(2)\psi_j(1)}{r_{12}} \right].
$$

Thus, the total energy is not just the sum of *one-particle energies*. Such sum counts each interaction twice. Finally, we stress that the resulting Hartree–Fock energy E_{HF} differs from the Hartree energy by an additional *negative exchange energy*

$$
E_{\mathrm{HF}} = E_{\mathrm{H}} + E_{\mathrm{exc}}, \tag{2.21}
$$

$$
E_{\mathrm{exc}} = -\frac{1}{2} \sum_{\substack{i \neq j \\ \uparrow\uparrow, \downarrow\downarrow}} \int d\mathbf{x}_1 \int d\mathbf{x}_2 \, \psi_i(\mathbf{x}_1)^* \, \psi_j^*(\mathbf{x}_j) \, \frac{e^2}{|\mathbf{x}_1 - \mathbf{x}_2|} \, \psi_j(\mathbf{x}_1) \psi_i(\mathbf{x}_2), \tag{2.22}
$$

where only electrons of the same spin are to be included in the sum.

Having defined the HF energy, we may now redefine the correlation energy as the difference between the expectation values of the HF and the exact wavefunctions. As we have shown, the HF wavefunction is a single determinant of one-electron orbitals, and the ground state is determined by varying the orbitals to minimize the energy. The determinant enforces the exchange component of the original correlation energy. Thus, our now standing definition of correlation refers only to effects beyond what is already contained in Hartree–Fock.

2.2.3 Koopman's Theorem: Meaning of Single-Particle ε_i

The justification of the label *one-particle energy* emerges from Koopman's theorem [111]. It provides a physical interpretation of the one-electron eigenvalues without actually solving the Hartree–Fock equations. The theorem shows that they are intimately related to the low-lying excitations of the many-body system. We consider the difference between the energies of two systems containing N and $N-1$ electrons, respectively. We assume that the one-electron orbitals ψ_i are the same in each case, but the orbital ψ_j is empty in the $N-1$ case. This approximation is quite reasonable given a large number of electrons. We obtain

$$\Delta E = \langle \psi_j | T + v(\mathbf{x}) | \psi_j \rangle + \sum_\ell e^2 \left[\frac{|\psi_\ell|^2 |\psi_j|^2}{r_{12}} - \frac{\psi_\ell^*(1) \psi_j^*(2) \psi_\ell(2) \psi_j(1)}{r_{12}} \right] = \varepsilon_j.$$

The factors of $1/2$ disappear because the removed orbital ψ_j occurs in both summations. Thus, the HF energy eigenvalue specifies the energy required to remove an electron from orbital j, leaving the remaining electrons unperturbed. Koopman's theorem states that "The eigenvalues of the HF equations ε_j correspond to total energy differences, namely, to the energies to add or subtract electrons that would result from increasing the size of the matrix by adding an empty orbital or decreasing the size by removing an orbital, if all other orbitals are frozen."

We can extend this scenario to the case of low-lying electronic excitations by the following procedure: first, remove the electron from orbital ψ_j, raising the energy by ε_j; then place it in ψ_i, resulting in an energy decrease of $-\varepsilon_i$.

2.2.4 Hartree–Fock Theory of the Jellium Model

The *jellium model* consists of an interacting free-electron gas system together with a smeared-out positive ion background of equal density to establish charge neutrality – it is translationally invariant. In spite of its simplicity, this model captures a lot of the basic physics of metals and will be considered frequently in this book. Simple systems like this play an important paradigmatic role in science. For example, the hydrogen atom is a paradigm for all of atomic physics. In the same way, the uniform electron gas is a paradigm for solid-state physics. Because of translation invariance, the single-electron wavefunctions have the familiar form,

$$\psi_{\mathbf{k}}(\mathbf{x}) = \left(\frac{e^{i\mathbf{k} \cdot \mathbf{x}}}{\Omega^{1/2}} \right) \chi(\sigma), \tag{2.23}$$

where Ω is the volume of the system. Each wavevector of magnitude less than k_F occurs twice (once for each spin orientation) in the Slater determinant, and gives a solution to the HF equation for the free electron gas. We thus obtain a uniform electron charge distribution with the same density n as the smeared positive ion background. Hence the electron–ion Coulomb energy is precisely canceled by the electron–electron one, and the only surviving term is the exchange energy term

$$E_{\text{exc}} = -\frac{1}{2} \sum_{i \neq j} \int dx_1 \int dx_2 \, \psi_i^*(x_1) \, \psi_j^*(x_2) \frac{e^2}{|x_1 - x_2|} \, \psi_j(x_1) \, \psi_i(x_2), \qquad (2.24)$$

which in the case of the jellium model becomes

$$E_{\text{exc}} = -\int_{k,k' < k_F} dk \, dk' \int dx \int dx' \, \psi_k^*(x) \, \psi_{k'}^*(x') \frac{e^2}{|x - x'|} \, \psi_{k'}(x) \, \psi_k(x')$$

$$= \int_{k < k_F} dk \, E_{\text{exc}}(k), \qquad (2.25)$$

where the fact that each wavevector k occurs twice was taken into account, and

$$E_{\text{exc}}(k) = -\int_{k' < k_F} dk' \int dx \int dx' \, \psi_k^*(x) \, \psi_{k'}^*(x') \frac{e^2}{|x - x'|} \, \psi_{k'}(x) \, \psi_k(x')$$

$$= -\frac{1}{\Omega^2} \int_{k' < k_F} dk' \, e^{-i(k-k') \cdot x} \frac{e^2}{|x - x'|} e^{i(k-k') \cdot x'} \, dx \, dx' \qquad (2.26)$$

To evaluate the preceding integral, we use the Fourier transform of the Coulomb interaction,

$$\frac{e^2}{|x - x'|} = 4\pi e^2 \int \frac{dq}{(2\pi)^3} \frac{1}{q^2} e^{iq \cdot (x - x')}. \qquad (2.27)$$

and obtain

$$E_{\text{exc}}(k) = -\frac{1}{\Omega^2} \int \frac{dq}{(2\pi)^3} \int_{k' < k_F} dk' \, e^{-i(k-k'-q) \cdot x} \frac{4\pi e^2}{|q|^2} e^{i(k-k'-q) \cdot x'} \, dx \, dx'$$

$$= -\int_{k' < k_F} \frac{dk'}{(2\pi)^3} \frac{4\pi e^2}{|k - k'|^2}$$

$$= -\frac{e^2}{\pi} \int_0^{k_F} dk' \, k'^2 \int_0^\pi \sin\theta \, d\theta \, (k^2 + k'^2 - 2kk' \cos\theta)^{-1}$$

$$= -\frac{e^2 k_F}{\pi} \left\{ 1 + \left[\frac{k^2 - k_F^2}{2kk_F} \right] \ln \left| \frac{k - k_F}{k + k_F} \right| \right\} = -\frac{2e^2 k_F}{\pi} F\left(\frac{k}{k_F} \right), \qquad (2.28)$$

$$F(x) = \frac{1}{2} + \frac{1 - x^2}{4x} \ln \left| \frac{1 + x}{1 - x} \right| \qquad (2.29)$$

$F(x)$ is plotted in Figure 2.3.

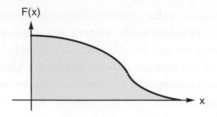

Figure 2.3 The function $F(x)$.

The following are some typical values:

$$k = 0 \quad \Rightarrow \quad E^{\text{exc}}(0) = -\frac{2}{\pi} e^2 k_F = -\frac{2}{\pi} e^2 (3\pi^2 n)^{1/3},$$

$$k = k_F \quad \Rightarrow \quad E^{\text{exc}}(k_F) = -\frac{1}{\pi} e^2 k_F = -\frac{1}{\pi} e^2 (3\pi^2 n)^{1/3}.$$

To compute the contribution of these interactions to the total energy, we must sum over all $k < k_F$, and we get[1]

$$E_{\text{exc}}^{HF} = -\frac{e^2 k_F}{\pi} \int_{k<k_F} dk \left[1 + \frac{k_F^2 - k^2}{2kk_F} \ln \left| \frac{k + k_F}{k - k_F} \right| \right] = -\frac{3}{4} \frac{e^2 k_F}{\pi} = -\frac{3e^2}{4\pi} (3\pi^2 n)^{1/3}.$$

(2.30)

2.3 The Density Functional Formalism

We find that in the previously described models the density $n(\mathbf{x})$ plays a prominent role.

Question: Could it be possible that a formally *exact* theory based on the density $n(\mathbf{x})$ may exist for the ground state of electronic systems? If such is the case, the preceding models and their variants are approximations.

This question led Pierre Hohenberg and Walter Kohn to propose and formulate the *density functional theory* (DFT) [93]. It was subsequently practically realized by Walter Kohn, and Lu Sham as a general theory of inhomogeneous electron gases in their ground state.

The central quantity in this theory is of course the *electron density*. The basic role of the density is established by a theorem that the properties of the system, in particular the ground-state energy, are *functionals* of this density. A variational principle can then be established for the energy, with the density as the varied function. The Euler equations that

---

[1] This result led John C. Slater to suggest that in nonuniform systems, and in particular in the presence of the lattice periodic potential, we can approximate and simplify the HF equations by replacing the exchange term in (2.22) with a local energy given by twice (2.30) with k_F evaluated at the local density. We effectively make the approximation $n \to n(\mathbf{x})$. Although this procedure seems to be gross and ad hoc, it is actually followed in many band structure calculations to date.

follow are formulated in two ways that are particularly convenient for the study of strongly inhomogeneous systems. One formulation predates the Hohenberg–Kohn theory and is known as the Thomas–Fermi method, the other was proposed by Kohn and Sham. In their various approximate versions, these formalisms represent classical, rather than quantum mechanical, methods – *the quantum phase is absent in the density.* DFT was proposed in the mid-1960s, but was largely ignored until the mid-1980s, when computers became good enough to solve the DFT equations accurately. It is now used by thousands of physicists, chemists, geophysicists, materials scientists, and even biochemists. In recognition of its importance, Walter Kohn (a physicist) and John Pople (a chemist who developed some of the mathematical and computational techniques used in DFT calculations) were awarded the 1998 Nobel Prize in Chemistry.

2.3.1 The Thomas–Fermi Model

This is the simplest model that adopts a density approach. It starts from the idea that a many-electron system can be broken up into cells of suitable size, such that in each cell, the electrons can be approximately regarded as a uniform electron gas having the local density $n(\mathbf{x})$, where $\mathbf{x}$ is the center of the cell.

In a uniform electron gas, we have the following connections between the Fermi momentum, k_F; Fermi energy, E_F; and the density, n (in atomic units):

$$E_F = \frac{1}{2}k_F^2, \qquad n = \frac{1}{3\pi^2} k_F^3. \tag{2.31}$$

In the case of a nonuniform electron system under the action of an external potential $v(\mathbf{x})$, we divide the space into cells, Ω_α, in such a way that we consider the electrons in a given cell as having a uniform density n_α, in a constant potential, $v_{\text{eff}}(\mathbf{x}_\alpha)$, with a Fermi energy $E_{F,\alpha}$ (see Figure 2.4).

Obviously, $E_{F,\alpha}$ *must be equal in all cells, since otherwise the energy could be lowered by transferring electrons from a cell of higher E_F to one of lower E_F.*

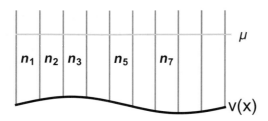

Figure 2.4 Partition of an electron system into cells.

We call this common value of E_F the *chemical potential*, μ. Then, by (2.31),

$$n(\mathbf{x}) = \frac{1}{3\pi^2} \left[k_F(\mathbf{x})\right]^2 = \frac{1}{3\pi^2} \{2[\mu - v_{\text{eff}}(\mathbf{x})]\}^{3/2}, \tag{2.32}$$

where we have dropped the subscript α on $\mathbf{x}$. The effective potential is given by

$$v_{\text{eff}} = v(\mathbf{x}) + \int \frac{n(\mathbf{x}')}{|\mathbf{x} - \mathbf{x}'|} \, d\mathbf{x}', \tag{2.33}$$

where the second term is the mean electrostatic potential at the point $\mathbf{x}$ due to the electron charge distribution. Equations (2.32) and (2.33) represent the *Thomas–Fermi model*. A posteriori, we observe that a basic assumption underlies this model:

- The system is broken up into cells in each of which $v_{\text{eff}}(\mathbf{x})$ is substantially constant (this requires sufficiently small cells).
- At the same time, each of which contain many electrons ($N_\alpha \gg 1$) so that the electron gas equations (2.31) make sense (this requires systems of many electrons and sufficiently large cells).

Notice also that, if one substitutes the expression (2.33) for $v_{\text{eff}}(\mathbf{x})$ into (2.32), one obtains a nonlinear integral equation for the density $n(\mathbf{x})$.

> Thus, under the aforementioned conditions, the Schrödinger equation (1) for Ψ, *a function of $3N$ variables, can be replaced by an equation for $n(\mathbf{x})$, a function of 3 variables.*

This is, of course, an enormous simplification and, even more important, the physical system is characterized by a quantity, $n(\mathbf{x})$, that has a simple meaning and can be easily visualized.

The same Thomas–Fermi model can be expressed in the form of a variational principle for the total energy:

$$E_{TF}[n(\mathbf{x})] = \int d\mathbf{x} \left[v(\mathbf{x}) \, n(\mathbf{x}) + \frac{3(3\pi^2)^{2/3}}{10} \, n(\mathbf{x})^{5/3} + \frac{1}{2} \int \frac{n(\mathbf{x}) \, n(\mathbf{x}')}{|\mathbf{x} - \mathbf{x}'|} \, d\mathbf{x}' \right]. \tag{2.34}$$

The first term describes (exactly) the interactions of the electrons with the external potential $v(\mathbf{x})$. The second term describes, approximately, their kinetic energy, since, for a free-electron gas of uniform density n occupying a volume Ω, the kinetic energy $\mathcal{T}$ is given by

$$\mathcal{T} = \Omega \, \frac{3}{10} \, (3\pi^2)^{2/3} \, n^{5/3}. \tag{2.35}$$

Finally, the last term in (2.34) describes the Coulomb-interaction energy in a mean-field approximation in which the correlation between the densities at two points $\mathbf{x}$ and $\mathbf{x}'$ is neglected.

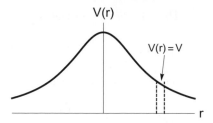

Figure 2.5 Thomas–Fermi approximation.

This energy expression must now be minimized subject to the constraint that the total number of electrons be N. This leads to

$$\delta\left[E_{TF}[n(\mathbf{x})] - \mu \int n(\mathbf{x})\, d\mathbf{x}\right] = 0, \qquad (2.36)$$

where the Lagrange multiplier defines the chemical potential. This variation is precisely equivalent to the coupled equations (2.32) and (2.33).

Thomas–Fermi Screening in the Jellium Model

We consider here a uniform electron charge density ρ_0 and an equally, uniformly distributed, positive rigid background (see Figure 2.5). We introduce a point charge Q at the origin, and we are interested in determining the emergent potential $V(\mathbf{x})$ (see Figure 2.5). We write Poisson's equation as

$$\nabla^2 V(\mathbf{x}) + 4\pi\, \rho(\mathbf{x}) = 0, \qquad (2.37)$$

where

$$\rho(\mathbf{x}) = Q\delta(\mathbf{x}) + \rho_e(\mathbf{x}) - \rho_0 = Q\delta(\mathbf{x}) + e\,[n(\mathbf{x}) - n_0].$$

The relevant electrons that will contribute to $\rho(\mathbf{x})$ are those at the Fermi energy with $k \sim k_F \Rightarrow k_F^{-1} \sim$ few angstroms. The Thomas–Fermi (TF) approximation is good if V varies on such length scale.

This allows us to write the change in electron density at point $\mathbf{x}$ as

$$\Delta n(\mathbf{x}) = n(\mathbf{x}) - n_0 = eV(\mathbf{x})\,\mathcal{N}(E_F),$$

where $\mathcal{N}(E_F)$ is the density of states at E_F. Then, (22.44) becomes

$$\nabla^2 V + 4\pi\, Q\, \delta(\mathbf{x}) - 4\pi\, e^2 \mathcal{N}(E_F)\, V(\mathbf{x}) = 0.$$

Fourier transformation gives

$$\left[k^2 + 4\pi\, e^2 \mathcal{N}(E_F)\right] V(\mathbf{k}) = \left[k^2 + \lambda^{-2}\right] V(\mathbf{k}) = 4\pi\, Q$$

$$V(\mathbf{x}) = \frac{1}{(2\pi)^3} \int d\mathbf{k} \, \frac{4\pi \, Q}{k^2 + \lambda^{-2}} e^{i\mathbf{k}\cdot\mathbf{x}} = \frac{1}{(2\pi)^2} \int_0^\infty dk \, k^2 \, \frac{4\pi \, Q}{k^2 + \lambda^{-2}} \int_{-1}^1 dx \, e^{ikrx}$$

$$= \frac{1}{(2\pi)^2} \int_{-\infty}^\infty dk \, k \, \frac{4\pi \, Q}{k^2 + \lambda^{-2}} \frac{e^{ikr}}{ir} = \frac{Q}{r} e^{-r/\lambda}.$$

Using

$$\mathcal{N}(E_F) = \frac{3}{2} \frac{n}{E_F}, \qquad E_F = \hbar v_F \, k_F, \qquad k_F^3 = 3\pi^2 n,$$

we obtain

$$\lambda^{-2} = 4\pi \, e^2 \, \mathcal{N}(E_F) = 4\pi \, \alpha \hbar c \frac{3}{2} \frac{n}{E_F} = \alpha \frac{1}{\pi} \frac{c}{v_F} k_F^2,$$

where α is the fine-structure constant. With $c/v_F \sim 100$, $\alpha = 1/137$, we find

$$\lambda \, k_F \sim 1.$$

λ is comparable to lattice spacings in normal metals

2.3.2 The Lemma of Hohenberg and Kohn

The importance of the density, $n(\mathbf{x})$, which was prominent in the approximate theories just described, is greatly highlighted by a strong lemma that demonstrates that an external potential $v(\mathbf{x})$, acting on the electron system is a *unique functional* of $n(\mathbf{x})$ (see Figure 2.6). A significant consequence of this proposition is that the ground-state Ψ is then such a functional, since $v(\mathbf{x})$ fixes $\mathcal{H}$, and $\mathcal{H}$, via the Schrödinger equation determines Ψ. The demonstration proceeds by *reductio ad absberdum*, where we assume that $v(\mathbf{x})$ is not unique: We assume that there is another potential $v'(\mathbf{x})$, differing from v by more than a constant that gives rise to the same density $n(\mathbf{x})$. The ground state associated with v' is denoted by Ψ'. It must be different from Ψ since it satisfies a different Schrödinger equation. The energy of this state is denoted by E_N'. Then we write

$$E_N = \langle \Psi | \mathcal{H} | \Psi \rangle \; < \; \langle \Psi' | \mathcal{H} | \Psi' \rangle = \langle \Psi' | \mathcal{H}' + v - v' | \Psi' \rangle. \tag{2.38}$$

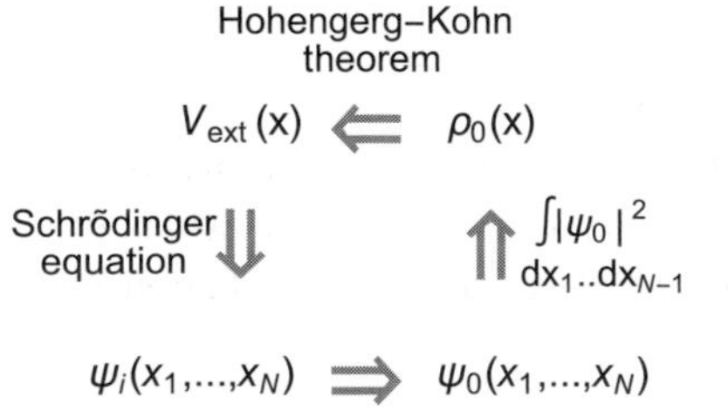

Hohengerg–Kohn
theorem

$V_{ext}(\mathbf{x}) \; \Longleftarrow \; \rho_0(\mathbf{x})$

Schrödinger
equation $\Downarrow$ $\Uparrow \; \int |\psi_0|^2 \, dx_1 .. dx_{N-1}$

$\psi_i(x_1,...,x_N) \; \Longrightarrow \; \psi_0(x_1,...,x_N)$

Figure 2.6 Density functional scheme.

Recalling the assumption that $n = n'$, we can also write

$$E_N < E'_N + \int d\mathbf{x} \left[v(\mathbf{x}) - v'(\mathbf{x}) \right] n(\mathbf{x}). \qquad (2.39)$$

Interchanging primed and unprimed quantities yields

$$E'_N < E_N + \int d\mathbf{x} \left[v'(\mathbf{x}) - v(\mathbf{x}) \right] n(\mathbf{x}). \qquad (2.40)$$

Addition of (2.39) and (2.40) then leads to the inconsistency

$$E_N + E'_N < E_N + E'_N.$$

Hence it is seen that $v(\mathbf{x})$ must be a unique functional of $n(\mathbf{x})$.

Before we formulate a density functional for the energy, we should examine the terms that appear in a Hamiltonian of a many-particle system:

- It contains a sum of the kinetic energies of individual particles of the form $T_i = -(\hbar^2/2m)\nabla_i^2$, which depends only on the particle type through the mass m.
- It contains a sum of interparticle interaction terms U, which is also determined by the type of particles comprising the system, for example, for electrons, $U = (1/2) \sum (e^2/r_{ij})$.
- It contains the energy of interaction with an external potential V that distinguishes different systems that belong to the same particle type and the same particle number N.

We this in mind, we now cast the problem in a variational approach.

Variational Principle

We define an energy functional

$$\mathcal{E}[n(\mathbf{x})] = \int d(\mathbf{x}) \, v(\mathbf{x}) n(\mathbf{x}) + F[n(\mathbf{x})],$$

$$F[n(\mathbf{x})] \equiv \langle \Psi | \mathcal{T} + U | \Psi \rangle \qquad (2.41)$$

$F[n(\mathbf{x})]$ is a universal functional of $n(\mathbf{x})$ (since Ψ is). Clearly for the correct $n(\mathbf{x})$ associated with the potential $v(\mathbf{x})$ and satisfying the integral constraint

$$\int d\mathbf{x} \, n(\mathbf{x}) = N, \qquad (2.42)$$

$\mathcal{E}[n]$ is equal to the ground-state energy E_N. Alternatively, E_N may be regarded as the minimum value of $\mathcal{E}[n]$ with respect to other density distributions corresponding to the same N. We can then consider a trial density distribution $n'(\mathbf{x}) \neq n(\mathbf{x})$ having a ground-state $\Psi' \neq \Psi$. Then

$$E_N < \langle \Psi' | \mathcal{H} | \Psi' \rangle = \int d(\mathbf{x}) \, v(\mathbf{x}) \, n'(\mathbf{x}) + \langle \Psi' | T + U | \Psi' \rangle = \mathcal{E}[n']. \qquad (2.43)$$

The vanishing of the first variation of $\mathcal{E}[n']$ about the correct density may be expressed by writing

$$\delta\mathcal{E}[n] = 0, \tag{2.44}$$

subject to the condition that all densities considered satisfy

$$\int d(\mathbf{x})\, n(\mathbf{x}) = N. \tag{2.45}$$

> Notice here that the energy of the ground state, regarded as a functional of the density, attains its minimum value with respect to *variation* of the *particle density* subject to the normalization condition when the density has the correct value.

This statement is in contrast with the quantum mechanical variational approach for the ground-state energy, where the energy attains its minimum value with respect to *variation* of the *ground-state wavefunction when the wavefunction is correct*. The density $n(\mathbf{x})$ is a real, positive function of a single vector variable, whereas the wavefunction is a complex function of N vector variables, which must have nodes if these are three or more electrons. It is remarkable that it is, in principle, sufficient to vary the density.

We also note that up to this point the Hohenberg–Kohn (HK) formulation is restricted to *spinless fermions*, which do not exist. However, we can introduce particle spin by replacing the density with a generalized four-current density within a relativistic formalism. The ground-state energy becomes a unique functional of the *four-current* density, and this quantity attains its minimum value with respect to variations that preserve the continuity equation when the density is correct. As a result of these considerations, the ground-state energy becomes a functional of the spin densities, a statement generally abbreviated as

$$E_G \equiv E_G[n_\uparrow(\mathbf{x}),\, n_\downarrow(\mathbf{x})]. \tag{2.46}$$

Separation of Direct Coulomb Terms–Hartree Terms

Because of the long range of the Coulomb interaction, compared to exchange and correlation terms, it is expedient to separate from $F[n]$ the classical Coulomb self-energy of the electrons by writing

$$F[n] = \frac{1}{2}\int d\mathbf{x}\,d\mathbf{x}'\, \frac{n(\mathbf{x})\,n(\mathbf{x}')}{|\mathbf{x}-\mathbf{x}'|} + G[n], \tag{2.47}$$

where $G[n]$ is a universal functional like $F[n]$ that contains the contributions of the kinetic, exchange, and correlation energies. The total energy functional becomes

$$\mathcal{E}[n] = \int d\mathbf{x}\, v(\mathbf{x})\, n(\mathbf{x}) + \frac{1}{2}\int d\mathbf{x}\,d\mathbf{x}'\, \frac{n(\mathbf{x})\,n(\mathbf{x}')}{|\mathbf{x}-\mathbf{x}'|} + G[n]. \tag{2.48}$$

The electrostatic potential in the system will be denoted by $\phi(\mathbf{x})$:

$$\phi(\mathbf{x}) = v(\mathbf{x}) + \int d\mathbf{x}'\, \frac{n(\mathbf{x}')}{|\mathbf{x}-\mathbf{x}'|}. \tag{2.49}$$

Introducing a Lagrange multiplier μ in conjunction with the total particle number, we write the variational equation for $n(\mathbf{x})$ as

$$\delta\left\{\mathcal{E}[n] - \mu \int d\mathbf{x}\, n(\mathbf{x})\right\} = 0, \qquad (2.50)$$

with μ being the chemical potential. Equation (2.50) implies that for the correct density,

$$\frac{\delta\mathcal{E}[n]}{\delta n(\mathbf{x})} = \mu. \qquad (2.51)$$

Using (2.49), this may be put in the form

$$\frac{\delta\phi(\mathbf{x}) + \delta G[n]}{\delta n(\mathbf{x})} = \mu, \qquad (2.52)$$

For large N, it is easily seen that μ is equal to the chemical potential $\partial E_N / \partial N$. We denote the correct densities corresponding to N and $N-1$ particles in the potential $v(\mathbf{x})$ by $n_N(\mathbf{x})$ and $n_{N-1}(\mathbf{x})$. Then

$$\left.\frac{\partial E_N}{\partial N}\right|_{v(\mathbf{x});\, T=0K} = \mathcal{E}[n_N] - \mathcal{E}[n_{N-1}]$$

$$= \int d(\mathbf{x})\, \left.\frac{\delta\mathcal{E}[n]}{\delta n(\mathbf{x})}\right|_{n=n_N} \left[n_N(\mathbf{x}) - n_{N-1}(\mathbf{x})\right] = \mu, \qquad (2.53)$$

where (2.51) is used in the last step.

As was mentioned earlier, the functional term $G[n]$ implicitly accounts for the kinetic, exchange, and correlation contributions to the energy functional. The usefulness of this form of the density functional formalism requires explicit expressions for these contribution. Although we know how to express the kinetic energy in terms of the wavefunction in the Schrödinger equation, it is not at all clear how to express it in terms of the density. In the TF method, the kinetic term was approximated by it functional form in the free particle case, namely,

$$E_{KE}[n] = \frac{3\hbar^2}{10 m_e} (3\pi^2)^{2/3} \int d\mathbf{x}\, n^{5/3}(\mathbf{x}). \qquad (2.54)$$

Since this is only valid for slowly varying potentials, it cannot be used for problems involving atomic or ionic potentials.

2.3.3 The Kohn–Sham Formulation

An approach proposed by Kohn and Sham [108] is more successful, and has become the gold standard for DFT calculations. They introduced a set of N orthonormal single-particle functions to define the particle density $n(\mathbf{x})$ as

$$n(\mathbf{x}) = \sum_{i=1}^{N} u_i^*(\mathbf{x})\, u_i(\mathbf{x}). \qquad (2.55)$$

This procedure is always possible. The variation of the density is effected by varying the functions $u_i^*(\mathbf{x})$ and $u_i(\mathbf{x})$,

$$\delta n(\mathbf{x}) = \sum_{i=1}^{N} \left\{ \delta u_i^*(\mathbf{x}) u_i(\mathbf{x}) + , u_i^*(\mathbf{x}) \, \delta u_i(\mathbf{x}) \right\}. \tag{2.56}$$

The variation need only be done with respect to $u_i^*(\mathbf{x})$. This approach is straightforward except for the kinetic energy – here we will borrow the form

$$E_{KE}(n) = \frac{\hbar^2}{2m_e} \sum_{i=1}^{N} \int d\mathbf{x} \, \nabla u_i^*(\mathbf{x}) \cdot \nabla u_i(\mathbf{x})$$

$$= \frac{\hbar^2}{2m_e} \sum_{i=1}^{N} \int d\mathbf{x} \, u_i^*(\mathbf{x})(-\nabla^2) u_i(\mathbf{x}), \tag{2.57}$$

where the N functions are the same as in (2.55). This form is an approximation; no proof has been given that (2.57) represents the exact E_{KE}. However, we can ignore any discrepancy, because it can be absorbed in $G[n]$. Thus, the variation may now be performed on the functional

$$\mathcal{E}[n] = E_{KE} + \frac{e^2}{2} \int d\mathbf{x} \, d\mathbf{x}' \, \frac{n(\mathbf{x}) \, n(\mathbf{x}')}{|\mathbf{x} - \mathbf{x}'|} + \int d\mathbf{x} \, v(\mathbf{x}) \, n(\mathbf{x}) + G[n]$$

$$= \sum_{i=1}^{N} \int d\mathbf{x} \, u_i^*(\mathbf{x}) \left[\frac{-\hbar^2}{2m_e} \nabla^2 + v(\mathbf{x}) + \frac{e^2}{2} \sum_{j=1}^{N} \int \frac{|u_j(\mathbf{x}')|^2}{|\mathbf{x} - \mathbf{x}'|} \, d\mathbf{x}' \right] u_i(\mathbf{x}) + G[n]. \tag{2.58}$$

The variation will then lead to

$$\left[\frac{-\hbar^2}{2m_e} \nabla^2 + v(\mathbf{x}) + \frac{e^2}{2} \sum_{j=1}^{N} \int \frac{|u_j(\mathbf{x}')|^2}{|\mathbf{x} - \mathbf{x}'|} \, d\mathbf{x}' + \frac{\delta G[n]}{\delta u_i^*(\mathbf{x})} \right] u_i(\mathbf{x}) = \varepsilon_i \, u_i(\mathbf{x}), \tag{2.59}$$

where we can define the exchange-correlation potential energy as

$$V_{\text{xc}} = \frac{\delta G[n]}{\delta u_i^*(\mathbf{x})}. \tag{2.60}$$

Local-Density Approximation of Exchange Correlations

Another major difficulty with the density functional approach lies in our imperfect knowledge of the exchange-correlation potential. The only system for which there is reasonably complete understanding of V_{xc} is the free electron gas, which we previously derived. In a free electron gas, however, the density is independent of position. In order to adapt these results to the case of an inhomogeneous electron gas, the *local density approximation* is introduced. In this approximation, it is assumed that V_{xc} has the same value at a point $\mathbf{x}$ that it would have in a free electron gas, namely,

$$V_{exc} \rightarrow V_{xc}[n(\mathbf{x})],$$

as the density in an inhomogeneous system is a function of position, $n \equiv n(\mathbf{x})$. In the simplest approximation in which only the exchange is considered, we have seen that the exchange energy of a free electron gas per unit volume, and not per electron, is

$$E_{xc}[n] = \frac{-3}{2} e^2 \left(\frac{3n^4}{8\pi} \right)^{1/3}. \tag{2.61}$$

Hence, the exchange potential is

$$V_{xc} = \frac{d E_{exc}[n]}{dn} = -2e^2 \left(\frac{3n(\mathbf{x})}{8\pi} \right)^{1/3}, \tag{2.62}$$

where now $n(\mathbf{x})$ is given by (2.55).

These results can be simply generalized to systems in which the densities of spin-up and spin-down electrons are not equal, namely, the case of spin-polarized systems. In this case, in accordance with (2.46), we set the energy as a functional of $n_\uparrow$ and $n_\downarrow$, and we write the exchange correlation potential as

$$V_{xc}^\sigma = \frac{\delta E[n_\uparrow, n_\downarrow]}{\delta n_\sigma}, \tag{2.63}$$

which leads to spin-dependent exchange-correlation potentials.

In the *local spin density approximation* (LSDA), the free electron gas result for E_{exc} is used. In the exchange-only case, this is

$$E_{xc} = \frac{-3}{2} e^2 \sum_\sigma n_\sigma \left(\frac{3n_\sigma(\mathbf{x})}{8\pi} \right)^{1/3}. \tag{2.64}$$

The spin densities are those obtained locally, namely,

$$n_\sigma \rightarrow n_\sigma(\mathbf{x}) = \sum_i \left| u_{i\sigma}(\mathbf{x}) \right|^2. \tag{2.65}$$

Thus the exchange potential is

$$V_{xc}^\sigma = \frac{d E_x[n]}{dn_\sigma} = -2e^2 \left(\frac{3n_\sigma(\mathbf{x})}{8\pi} \right)^{1/3}. \tag{2.66}$$

The DFT procedure is summarized in Table 2.1.

Table 2.1 *Schematic of DFT functionals and procedure.*

$$F[n] = T[n] + V_H[n] + G[n]$$

- Kinetic energy functional

Free electron gas

$$T[n] = \Omega \int \frac{d\mathbf{k}}{(2\pi)^3} \frac{\hbar^2 k^2}{2m_e} = \Omega \frac{\hbar^2 k_F^5}{10\pi^2 m_e}$$

Thomas–fermi approximation

$$k_F = \left(3\pi^2 n(\mathbf{x})\right)^{1/3}$$

(good for slow density variation)

$$T^{\mathrm{TF}}[n] = \Omega \frac{3\hbar^2 \left(3\pi^2\right)^{1/3}}{10\pi^2 m_e} n^{5/3}(\mathbf{x})$$

Kohn–Sham

$$T^{\mathrm{KS}}[n] = \frac{\hbar^2}{2m_e} \sum_{i=1}^{N} \int d\mathbf{x} u_i^*(\mathbf{x})(-\nabla^2) u_i(\mathbf{x})$$

- Hartree energy functional

$$V_H[n] = \frac{e^2}{2} \int d\mathbf{x}\, d\mathbf{x}' \frac{n(\mathbf{x})\, n(\mathbf{x}')}{|\mathbf{r} - \mathbf{r}'|}$$

- Exchange-correlation functional

Local density approximation

$$E_{\mathrm{xc}}^{\mathrm{LDA}} \simeq \int d\mathbf{x}\, n(\mathbf{x})\, V_{\mathrm{exc}}\,(n(\mathbf{x}))$$

$$V_{\mathrm{exc}}\,(n(\mathbf{x})) = -\frac{3}{4} \frac{e^2 k_F}{\pi}$$

$$= -\frac{3}{4} \frac{e^2}{\pi} \left(3\pi^2 n(\mathbf{x})\right)^{1/3}$$

Generalized gradient
approximation

$$V_{\mathrm{exc}}^{\mathrm{GGA}} \simeq \int d\mathbf{x}\, n(\mathbf{x})\, \varepsilon_{\mathrm{exc}}\,[n(\mathbf{x}), \nabla n(\mathbf{x})]$$

Kohn–Sham (KS) versus Hartree–Fock (HF)

Pedagogically, it is useful to illustrate the difference between KS and HF calculations using a simple system. Actually, the simplest system is the hydrogen molecule H_2, where we need to solve the interacting Schrödinger equation

$$\left[-\frac{1}{2} \sum_{i=1}^{2} \left(\nabla_i^2 + v_{\mathrm{ext}}(\mathbf{x}_i) \right) + \frac{1}{|\mathbf{x}_1 - \mathbf{x}_2|} \right] \Psi(\mathbf{x}_1, \mathbf{x}_2) = E\, \Psi(\mathbf{x}_1, \mathbf{x}_2), \tag{2.67}$$

where i identifies the two electrons, and $v_{\mathrm{ext}}(r)$ is the external ionic potential.

We consider the spin-singlet ground–state, with an exchange symmetric spatial wavefunction $\Psi(\mathbf{x}_2, \mathbf{x}_1) = \Psi(\mathbf{x}_1, \mathbf{x}_2)$. The presence of electron–electron repulsion complicates the solution of (2.67), as it couples the two coordinates. Obtaining an exact solution of (2.67), with six coordinates, can be quite demanding.

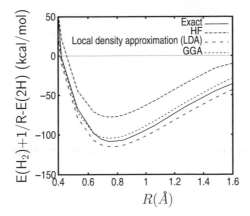

Figure 2.7 Total energy of the H_2 molecule as a function of internuclear separation.

Figure 2.7 shows results of total energy calculations of the molecule using different methods and approximations. The position of the minimum identifies the equilibrium bond length, while the depth of the minimum, minus the zero point vibrational energy, is the bond energy. The global energy minimum determines the geometry of the molecule and its vibration and rotation spectra.

(i) Hartree–Fock Approach

According to the HF method, we seek a system of two noninteracting electrons in some potential, $v_s(\mathbf{x})$, chosen somehow to *reproduce the true interacting electronic system*. Because the electrons are noninteracting, their coordinates decouple, and their wavefunction is a simple product of one-electron wavefunctions, or orbitals, satisfying

$$\left[-\frac{1}{2} \nabla^2 + v_s(\mathbf{x}) \right] \phi_i(\mathbf{x}) = \varepsilon_i \, \phi_i(\mathbf{x}), \qquad (2.68)$$

where $\Psi(\mathbf{x}_1, \mathbf{x}_2) = \phi_0(\mathbf{x}_1) \, \phi_0(\mathbf{x}_2)$. A much simpler set of equations to solve, since it has only three coordinates.[2] If we can get our noninteracting system to accurately *mimic* the true system, then we will have a computationally much more tractable problem to solve.

How do we get this mimicking? We solve the HF equations and obtain an effective potential

$$v_s^{\mathrm{HF}}(\mathbf{x}) = v_{\mathrm{ext}}(\mathbf{x}) + \frac{1}{2} \int d\mathbf{x}' \, \frac{n(\mathbf{x}')}{|\mathbf{x} - \mathbf{x}'|}.$$

The correction to the external potential represents the effect of the second electron, in particular, screening the nuclei. Insertion of $v_s^{\mathrm{HF}}(\mathbf{x})$ into (2.68) initiates a self-consistency process, since it depends on the electronic density, which in turn is calculated from the

---

[2] Even with many electrons, say N, one would still need to solve only a 3D equation, and then fill the lowest $N/2$ levels, as opposed to solving a $3N$-coordinate Schrödinger equation.

solution to the equation. An initial guess of v_s is usually made, the eigenvalue problem is then solved, the density calculated, and a new potential found. These steps are repeated until there is no change in the output from one cycle to the next – self-consistency has been reached. As we mentioned earlier, such a set of equations are often called self-consistent field (SCF) equations. As shown in Figure 2.7, we find that the HF calculations yield a quite accurate minimum energy position; however, it underbinds the molecule significantly. This has been a well-known deficiency of this method. The missing piece of energy is the *correlation energy*.

(ii) Kohn–Sham Approach

In a Kohn–Sham calculation, the basic scenario is very much the same, but the logic is entirely different. Here, we will try to find a system where the Kohn–Sham potential $v_s^{KS}(\mathbf{x})$ acting on the a pair of noninteracting electrons precisely reproduces the density $n(\mathbf{x})$ of the physical system. This is done with the aid of density functional methods, which require knowledge of how the total energy depends on the density. Thus, in order to determine $v_s^{KS}(\mathbf{x})$, we need to introduce simple approximations that allow us to obtain workable expressions for the energy dependence on the density. With this in hand, we can apply it to predict both the energy and the self-consistent potential, $v_s^{KS}(\mathbf{x})$, for fictitious noninteracting electrons that mimic all real electronic systems. However, we should underscore that the Kohn–Sham wavefunction of orbitals in this perspective is not considered an approximation to the exact wavefunction. Rather, it is a precisely defined property of any electronic system, which is determined uniquely by the density. To emphasize this point, consider the fused limit of our H_2, the He atom.

In Figure 2.8, a highly accurate many-body wavefunction for the He atom was calculated, and the density extracted. In the bottom of the figure, we plot both the bare external potential, $-2/r$, and the exact Kohn–Sham potential. Two noninteracting electrons sitting

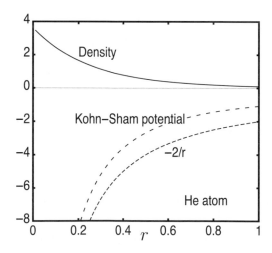

Figure 2.8 The external and Kohn–Sham potentials for the He atom.

in this potential have precisely the same density as the interacting electrons. If we can figure out some way to obtain a pretty good approximation to this potential, we have a much less demanding set of equations to solve than those of the true system. Thus we are always trying to improve a noninteracting calculation of a noninteracting wavefunction, rather than that of the full physical system. In that spirit, Figure 2.7 shows a local density approximation (LDA) and generalized gradient approximation (GGA) curves. LDA is the simplest possible density functional approximation, and it already greatly improves on HF, although it typically overbinds by about 1/20 of a Hartree (or 1 eV or 30 kcal/mol), which is too inaccurate for most quantum chemical purposes, but sufficiently reliable for many solid-state calculations. More sophisticated GGAs (and hybrids) reduce the typical error in LDA by about a factor of 5 (or more), making DFT a very useful tool in both condensed matter and quantum chemistry.

2.3.4 Reflections

The Hartree–Fock approach makes sense in the high-density limit but is obviously an approximation (see Figure 2.9). DFT provides an exact mapping from a system of interacting electrons to a system of noninteracting electrons moving through an effective potential that depends on the electron density. Solving this self-consistent noninteracting problem is easier than solving the Hartree–Fock equations and gives, in principle, the exact interacting ground-state energy E_0 and electron density for any given arrangement of the frozen (Born–Oppenheimer) nuclei. Since DFT allows us to calculate the ground-state energy quickly and reliably, we can use it to study how the total energy depends on the positions of the nuclei. This allows the application of the Hellman–Feynman theorem to determine the forces on the nuclei and hence how they move around. (Newton's laws are useless for electrons, but better for the much more massive nuclei.) We can use these forces to find the equilibrium

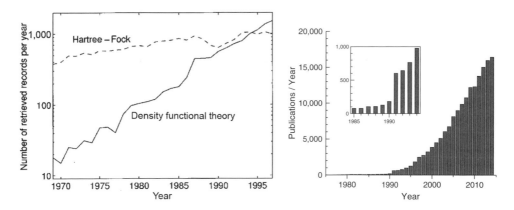

Figure 2.9 Left: Chronology of Hartree–Fock and density functional usage. Right: Histogram of the number of published papers per year using DFT as a function of time. Compiled by P. Mavropoulos (see [96]).

positions of the nuclei, watch the progress of a chemical reaction, or even follow the motion of a defect such as a dislocation.

The success of DFT, which appears to be a theory of noninteracting electrons but in fact describes a system of interacting electrons, in part explains the success of the standard model of a solid as an assembly of noninteracting electrons moving in a fixed external potential.

Exercises

2.1   Thomas–Fermi theory in two–dimensions: screening in a 2D electron gas:

Consider a system of electrons constrained to move on a plane. Use a system of coordinates where the three-dimensional vectors can be written as $\mathbf{r} = (\mathbf{R}, z)$; $\mathbf{R}$ is the vector in the plane.

(a) If an external electric potential ϕ_{ext} is applied to the system, there will be an induced charge density, $\delta\rho(\mathbf{R})\,\delta(z)$. Here $\delta\rho(\mathbf{R})$ is the induced surface charge density in the plane, and $\delta(z)$ indicates that the charge density is confined to the film at the plane $z = 0$.

Such an induced charge density produces its own induced electric potential $\delta\phi$, which is related to the induced charge density by

$$\nabla^2 \delta\phi = -4\pi\,\delta\rho(\mathbf{R})\,\delta(z).$$

By taking appropriate Fourier transforms, namely with respect to the two-dimensional wavevector $\mathbf{k}_\parallel = (k_x, k_y)$ in the xy plane, solve the previous Poisson's equations and show that the induced potential in the film is given by

$$\delta\phi(\mathbf{k}_\parallel, z = 0) = 2\pi\,\frac{\delta\rho(\mathbf{k}_\parallel)}{|\mathbf{k}_\parallel|}.$$

(Hint: You have to solve Poisson's equation in full three dimensions, even though the charge density is confined to the two-dimensional plane.)

(b) The total electric potential in the film will then be $\phi_{\text{tot}} = \phi_{\text{ext}} + \delta\phi$. If the electric susceptibility is defined by

$$\chi(\mathbf{k}_\parallel) = -\frac{\delta\rho(\mathbf{k}_\parallel)}{\phi_{\text{tot}}(\mathbf{k}_\parallel)},$$

show that

$$\phi_{\text{tot}}(\mathbf{k}_\parallel, z = 0) = \frac{\phi_{\text{ext}}(\mathbf{k}_\parallel)}{\varepsilon(\mathbf{k}_\parallel)},$$

where

$$\varepsilon(\mathbf{k}_\parallel) = 1 + 2\pi\,\frac{\chi(\mathbf{k}_\parallel)}{|\mathbf{k}_\parallel|}.$$

(c) For a point charge Q in the plane of the film, what is the screened electric potential in the plane of the film, $\phi_{\text{tot}}(\mathbf{k}_{\parallel}, z = 0)$? (Follow the steps given in the section on the Thomas–Fermi screening and relate the induced charge to the density of states.)

What is the screening length λ_{TF}?

(d) Fourier transform to get the real space potential $\phi(\mathbf{R}, z = 0)$. How does it behave in the limits $R \ll \lambda_{\text{TF}}$ and $R \gg \lambda_{\text{TF}}$?

You may need the following integrals and relations

$$\int_0^\infty dx \, \frac{x^\nu \, J_\nu(ax)}{x + b} = \frac{\pi \, b^\nu}{2 \cos(\nu\pi)} \left[H_{-\nu}(ab) - N_{-\nu}(ab) \right],$$

where $H_{-\nu}$ is Struve's function and $N_{-\nu}$ Neumann's function

$$\lim_{x \to 0} \quad x H_0(x), \; x N_0(x) = 0.$$

For large x,

$$H_0(x) - N_0(x) \simeq \frac{2}{\pi x} - \frac{1}{4\pi} \frac{8}{x^3}.$$

2.2 Kinetic energy contribution in density functional theory:

The kinetic energy contribution to the density functional formalism is difficult to obtain accurately. Here we shall attempt to derive approximate expressions in terms of the density for a noninteracting system.

(a) Express k_F in terms of the density n of a uniform electron gas. Determine the total kinetic energy of a noninteracting uniform electron gas in terms of k_F and the density n in one, two, and three-dimensions. Write the expressions in atomic units.

(b) Using the forms of total kinetic energy obtained in part (a) as local functions of position, write an expression for the kinetic energy functional.

(c) Since $k_F \equiv k_F(n)$, we can write

$$T_s[n] = A_s \int d^d x \, n^\alpha(\mathbf{x}),$$

where A_s does not depend on $\mathbf{x}$. Use dimensional analysis to determine α for one, two, and three-dimensions. A_s can be determined from parts (a) and (b).

2.3 Particle in a one-dimensional box:

Consider the elementary example of a particle in a box, with $V(x) = \infty$ everywhere, except $0 < x < L$, where $V = 0$.

(a) Since all the particle's energy is kinetic, use the approximate density functional in one dimension for the kinetic energy that you derived in problem 2.2, namely,

$$T^{\text{loc}}[n] = 1.645 \int_{-\infty}^{\infty} dx \, n^3(x)$$

to estimate the energy for single, double, and triple occupation of the box, where each energy level is singly occupied. Use the box eigenfunctions to determine the density. Compare the energies you obtain with the exact energies.

(b) For a large occupation number $N \gg 1$, obtain an expression for the exact total kinetic energy of the N particles in the box.

2.4 Exchange energy contribution in density functional theory:

Now we need to derive the density functional for the exchange energy. For the exchange energy of the uniform gas, we simply note that the Coulomb interaction has dimensions of inverse length. Use dimensional analysis to obtain an expression for the exchange energy functional, in the local approximation, in one, two, and three dimensions.

2.5 Cartoon model for the helium atom:

Consider the following one-dimensional cartoon model of a heliumlike atom, represented by the Hamiltonian

$$\mathcal{H} = -\frac{1}{2} \frac{\partial^2}{\partial x_1^2} - \frac{1}{2} \frac{\partial^2}{\partial x_2^2} - Z\delta(x_1) - Z\delta(x_2) + \delta(x_1 - x_2),$$

where x_1, x_2 are the coordinates of the two electrons along the x-axis. $-Z\delta(x)$ replaces the nucleus potential and $\delta(x_1 - x_2)$ the electron–electron Coulomb interaction.

(a) Employ the variational method with a trial wavefunction analogous to that used for the ground state of helium, namely for the single particle

$$u(x) = \sqrt{\alpha} \, \exp[-\alpha|x|],$$

where α is a variational parameter. Find the best value of $\langle \mathcal{H} \rangle$ and compare it with helium.

(b) Now take a trial wavefunction of the form

$$\Psi(x_1, x_2) = u(x_1) \, u(x_2),$$

where $u(x)$ is the variational function, à la Hartree or Hartree–Fock. Assume $u(x)$ is real; there is no loss of generality in this. Derive a Hartree-like equation for $u(x)$; it will contain a pseudoenergy eigenvalue, call it ε. Define ε so that the equation looks like a Schrödinger equation

$$\frac{1}{2} \frac{d^2 u(x)}{dx^2} + \cdots = \varepsilon \, u(x)$$

where the dots represent the remaining terms.

(c) The desired solution must be normalizable, which requires that

$$\lim_{x \to \pm\infty} u(x) = 0.$$

Use this boundary condition to solve the Schrödinger-like equation you have derived. (Hint: it's actually known as the nonlinear time-independent Schrödinger equation. Try a solution of the form $u(x) = \alpha \, \mathrm{csch}(\alpha|x| + \beta)$.)

(d) Find the normalized solution, and use it to compute an estimate for the energy E.

2.6 Derivation of the Hartree equations from the variational principle:
Consider the many-body Hamiltonian

$$\mathcal{H} = \sum_{i=1}^{N} \left(-\frac{\hbar^2}{2m_e} \nabla_i^2 + V_{\text{ext}}(\mathbf{x}) \right) + \frac{1}{2} \sum_{i \neq j} \frac{e^2}{|\mathbf{x}_i - \mathbf{x}_j|}$$

acting on the N-particle wave function $\Psi(\mathbf{x}_1, \mathbf{x}_2, \dots, \mathbf{x}_N)$. Within the Hartree approximation, the eigenstates of $\mathcal{H}$ are not antisymmetrized and determined by setting

$$\Psi(\mathbf{x}_1, \mathbf{x}_2, \dots, \mathbf{x}_N) = \prod_{i=1}^{N} \phi_i(\mathbf{x}_i \sigma_i)$$

and then minimizing the expectation value $\langle \Psi | \, calH \, | \Psi \rangle$.

(a) Show that

$$\langle \Psi | \mathcal{H} | \Psi \rangle = \sum_i \int d\mathbf{x} \, \phi_i^*(\mathbf{x}) \left(-\frac{\hbar^2}{2m_e} \nabla_i^2 + V_{\text{ext}}(\mathbf{x}) \right) \phi_i(\mathbf{x})$$

$$+ \frac{1}{2} \sum_{i \neq j} e^2 \iint d\mathbf{x}_1 \, d\mathbf{x}_2 \, \frac{|\phi_j(\mathbf{x}_2)|^2 \, |\phi_i(\mathbf{x}_1)|^2}{|\mathbf{x}_i - \mathbf{x}_j|}.$$

(a) Expressing the constraint of normalization for each ϕ_i with a Lagrange multiplier ε_i, and taking $\delta\phi_i^*$ and $\delta\phi_i$ as independent variations, show that the stationarity condition

$$\frac{\delta \langle \mathcal{H} \rangle}{\delta \phi^*}$$

leads directly to the Hartree equations

$$\left[-\frac{\hbar^2}{2m_e} \nabla_i^2 + V_{\text{ext}}(\mathbf{x}) + \sum_{i \neq j} e^2 \int d\mathbf{x}' \, \frac{|\phi_j(\mathbf{x}_2)|^2}{|\mathbf{x} - \mathbf{x}'|} \right] \phi_i(\mathbf{x}) = \varepsilon_i \, \phi_i(\mathbf{x}).$$

Compare the Hartree equations with the Hartree–Fock equations derived in class.

2.7 Sketch the shapes of the electronic and nuclear cusps as well as the Coulomb hole in the ground state of the helium atom.

2.8   Calculate the Hartree–Fock correction for the electron energy $E(\mathbf{k})$ for jellium using the screened Coulomb potential

$$V_s(\mathbf{x}) = \frac{e^2}{r} e^{-\lambda r}, \quad r = |\mathbf{x}|.$$

Does the unphysical behavior of the density of states and the electron velocity at the Fermi level still persist?

2.9   Using the Hartree–Fock exchange energy correction for the jellium, equation (2.28), we write the one electron energy as

$$E(\mathbf{k}) = \frac{\hbar^2 k^2}{2m_e} - \frac{2e^2 k_F}{\pi} F\left(\frac{k}{k_F}\right).$$

Show that near the band minimum (k=0), $E(\mathbf{k})$ can be expressed as

$$E(\mathbf{k}) \simeq \frac{\hbar^2 k^2}{2m^*},$$

where m^* is given by

$$\frac{m^*}{m_e} = \frac{1}{1 + 0.22(r_s/a_0)}.$$

3

Electrons and Band Theory: Methods of Energy-Band Calculations

3.1 The Crystal Potential

The energy levels in the free atom may be divided into *core* and *valence* levels (see Figure 3.1). The core levels comprise all states lying within the highest filled rare-gas shell, while valence levels include all occupied states outside.

(a) Core Electrons Contribution

The core levels will remain essentially unchanged in going from the free atom to the solid. We note that when atomic wavefunctions are calculated for a free ion and a free atom, no distinction is made in the tabulated core wavefunctions. Presumably in the solid, the wavefunctions of these low-lying states should be intermediate between the two extremes, and again no distinction is made.

Equipped with a knowledge of the core wavefunctions, we proceed to calculate the combined potential arising from the nucleus and these core electrons, with the proviso that the Coulomb and exchange terms coupling the valence and core states are included. The resultant potential is known as the *bare ionic potential* $V_i(\mathbf{r})$.

(b) Valence Electrons Contribution

The Coulomb and exchange contributions of the valence electrons to the one-electron potential, $V_{\text{eff}}(\mathbf{r})$, can be calculated with the aid of the free electron exchange functional discussed in the previous chapter. However, we should note the unphysical result of infinite velocity at $k = k_F$ in the HF exchange interaction for the homogeneous free electron gas, equation (2.28) and Figure 2.3:

$$E_{\text{exc}}^{\text{HF}}(\mathbf{k}) = -\frac{e^2 k_F}{\pi} \left\{ 1 + \left[\frac{k^2 - k_F^2}{2kk_F} \right] \ln \left| \frac{k - k_F}{k + k_F} \right| \right\}$$

This is really the result of the inclusion of self-interaction in the summation over $k < k_F$. Thus, in constructing the contributions of the valence electrons to the potential, $V_{\text{eff}}(\mathbf{x})$, we should avoid the inclusion of the self term.

The resulting overall potential appearing in the one-electron self-consistent equations, namely, $V_i(\mathbf{x}) + V_{\text{eff}}(\mathbf{x})$, will be called the *crystal potential* $V(\mathbf{x})$. Since all the electronic wavefunctions will assume the periodicity of the underlying lattice, the electronic

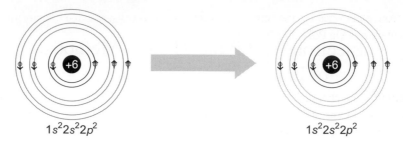

$1s^2 2s^2 2p^2$ $1s^2 2s^2 2p^2$

Figure 3.1 The carbon atom: Left – with $1s^2$ core level (black circles) and $2s^2$, $2p^2$ valence levels (gray circles). Right – defining the ionic core.

densities contributing to the Coulomb and exchange will also display the same periodicity. Consequently, we expect that $V(\mathbf{x})$ will also have the same periodic symmetry and can be simply written as

$$V(\mathbf{x}) = \sum_{\mathbf{R}, \rho} v(\mathbf{x} - \mathbf{R} - \rho), \tag{3.1}$$

where $\mathbf{R}$ are the primitive lattice vectors and ρ the atomic position vectors in the primitive cell.

3.1.1 Screened Electron–Ion Potential

A simple method for including the contribution of intravalence Coulomb interaction in the crystal potential is to treat it as an effective screening of the bare ionic potential. The physical idea associated with screening by the valence electrons is tantamount to the naive statement that if ϵ is the *dielectric constant* relevant to the valence electrons, then we expect the *bare* potential

$$V_i(\mathbf{x}) = \sum_{\mathbf{R}} v_i(|\mathbf{x} - \mathbf{R}|),$$

$$v_i(r) = -ze^2/|\mathbf{x}| \tag{3.2}$$

where ze is an effective ionic charge, to assume the modified form

$$V(\mathbf{x}) = \sum_{\mathbf{x}} v(|\mathbf{x} - \mathbf{R}|), \tag{3.3}$$

where now $v(r) = -ze^2/\epsilon|\mathbf{x}|$, as in classical electrostatics. In practice, however, one cannot treat electron screening quite so simply on a microscopic scale. A number of intermediate steps is required to see what is happening. Primarily, we have to consider that ϵ should be replaced by $\epsilon(\mathbf{x})$, a *dielectric function*.

Let us, therefore, consider the screening of the bare electron–ion potential V_i by the dielectric function of the valence electron gas in a simple metal. We suppose that we may write the *screened* potential in the form

$$V(\mathbf{x}) = V_i(\mathbf{x}) + V_{sc}(\mathbf{x}), \tag{3.4}$$

where, from electrostatics, the screening potential V_{sc} is related to the density $\rho(\mathbf{x})$ of the valence electrons via Poisson's equation

$$\nabla^2 V_{sc}(\mathbf{x}) = -4\pi e^2 \delta\rho(\mathbf{x}), \tag{3.5}$$

where $\delta\rho(\mathbf{x})$ is the change in electron density above or below its mean value in the electron gas. Such fluctuations in the local electron density are due to electron–electron interactions among the mobile valence electrons. Quantum mechanically, we know that ρ is

$$\rho(\mathbf{x}) = 2 \sum_{|\mathbf{k}|<k_F} \Psi_{\mathbf{k}}^*(\mathbf{x})\, \Psi_{\mathbf{k}}(\mathbf{x}), \tag{3.6}$$

where $\Psi_{\mathbf{k}}(\mathbf{x})$ is determined by the solution of the Schrödinger equation

$$\left\{ -\frac{\hbar^2}{2m} \nabla^2 + V(\mathbf{x}) \right\} \Psi_{\mathbf{k}}(\mathbf{x}) = E_{\mathbf{k}}\, \Psi_{\mathbf{k}}(\mathbf{x}), \tag{3.7}$$

Equations (3.3) to (3.7) suggest that in order to calculate V in (3.3), we have to solve (3.7), which itself involves V. We need a *self-consistent* solution.

The Perturbation Route!

At this point, we declare a priori that V is a weak potential, which implies that we can apply perturbation theory to the free electron gas, and obtain, to first-order,

$$\Psi_{\mathbf{k}}(\mathbf{x}) = \psi_{\mathbf{k}}^{(0)}(\mathbf{x}) + \delta\psi_{\mathbf{k}}(\mathbf{x}) = \frac{1}{\sqrt{\Omega}} e^{i\mathbf{k}\cdot\mathbf{x}} + \frac{1}{\sqrt{\Omega}} \sum_{\mathbf{q}\neq 0} \frac{V(\mathbf{q})\, e^{i(\mathbf{k}+\mathbf{q})\cdot\mathbf{x}}}{T_{\mathbf{k}} - T_{\mathbf{k}+\mathbf{q}}}, \tag{3.8}$$

$V(\mathbf{q})$ is the Fourier transform of $V(\mathbf{x})$, $T_{\mathbf{k}} = \hbar^2 k^2/2m$ and Ω is the volume occupied by the electron gas,

$$E_{\mathbf{k}} = T_{\mathbf{k}} + V(0) + \sum_{\mathbf{q}\neq 0} \frac{\left|V(\mathbf{q})\right|^2}{T_{\mathbf{k}} - T_{\mathbf{k}+\mathbf{q}}}. \tag{3.9}$$

We are now in a position to calculate the induced charge $\rho_{\mathrm{ind}}(\mathbf{x})$, but it is easier to calculate $\rho_{\mathrm{ind}}(\mathbf{q})$

$$
\begin{aligned}
\rho_{\mathrm{ind}}(\mathbf{q}) &= \frac{1}{\Omega} \int d\mathbf{x}\, \rho_{\mathrm{ind}}(\mathbf{x})\, e^{i\mathbf{q}\cdot\mathbf{x}} \\
&= \frac{2}{\Omega} \sum_{|\mathbf{k}|\leq k_F} \int d\mathbf{x} \left(\psi_{\mathbf{k}}^{(0)*}(\mathbf{x})\, \delta\psi_{\mathbf{k}}(\mathbf{x}) + \psi_{\mathbf{k}}^{(0)}(\mathbf{x})\, \delta\psi_{\mathbf{k}}^*(\mathbf{x}) \right) e^{i\mathbf{q}\cdot\mathbf{x}} \\
&= \frac{2}{\Omega} \sum_{|\mathbf{k}|\leq k_F} \left[\frac{V(\mathbf{q})}{T_{\mathbf{k}} - T_{\mathbf{k}+\mathbf{q}}} + \frac{V^*(-\mathbf{q})}{T_{\mathbf{k}} - T_{\mathbf{k}-\mathbf{q}}} \right],
\end{aligned} \tag{3.10}
$$

since the reality of V implies $V(\mathbf{q}) = V^*(-\mathbf{q})$, this reduces to

$$
\rho_{\text{ind}}(\mathbf{q}) = \frac{4}{\Omega} V(\mathbf{q}) \sum_{|\mathbf{k}| \leq k_F} \frac{1}{T_{\mathbf{k}} - T_{\mathbf{k}+\mathbf{q}}} = \frac{4}{\Omega} \frac{\Omega}{(2\pi)^3} V(\mathbf{q}) \frac{2m}{\hbar^2} \int \frac{d\mathbf{k}}{k^2 - (\mathbf{k}+\mathbf{q})^2}
$$

$$
= \frac{2m}{\pi^2 \hbar^2} V(\mathbf{q}) \int_0^{k_F} dk\, k^2 \int_{-1}^{1} \frac{-d\cos\theta}{q^2 + 2kq\cos\theta} = \frac{mk_F}{2\pi^2\hbar^2} V(\mathbf{q})\, \chi\left(\frac{q}{2k_F}\right),
$$

$$
\chi(y) = \frac{1}{2} + \frac{1-y^2}{4y} \ln\left|\frac{1+y}{1-y}\right|
$$

where $\chi(y)$ is the *Lindhard susceptibility* for a free electron gas. We substitute for $\rho_{\text{ind}}(\mathbf{q})$ in (3.5) and use the relation $k_F^3 = 3\pi^2 n$ to obtain

$$
V_{sc}(\mathbf{q}) = \frac{4\pi e^2}{q^2} \rho_{\text{ind}}(\mathbf{q}) = -\frac{\lambda^2}{q^2} V(\mathbf{q})\, \chi\left(\frac{q}{2k_F}\right),
$$

$$
\lambda^2 = \frac{4\pi e^2 m k_F}{2\pi^2 \hbar^2} = \frac{2k_F}{\pi a_0} = \frac{4\pi e^2 N}{(2/3) E_F \Omega} \tag{3.11}
$$

λ is actually the Thomas–Fermi screening length λ_{TF}.

Eliminating $V_{sc}(q)$ in (20.26), we get

$$
V(\mathbf{q}) = \frac{V_i(\mathbf{q})}{\epsilon(q)}, \tag{3.12}
$$

where

$$
\epsilon(q) = 1 + \frac{\lambda^2}{q^2} \chi\left(\frac{q}{2k_F}\right) = 1 + \frac{4\pi z e^2}{\Omega_0 q^2} \left(\frac{2}{3} E_F\right)^{-1} \chi\left(\frac{q}{2k_F}\right), \tag{3.13}
$$

$z/\Omega_0 = n$, and Ω_0 is the primitive cell volume.

Recalling that $V(q) = S(q)v(q)$ and $V_i(q) = S(q)v_i(q)$, where $S(q)$ and $v(q)$ are the structure and form factors, respectively, we get

$$
v(q) = \frac{-4\pi z e^2 / \Omega_0 q^2}{1 + \left(\frac{\lambda^2}{q^2}\right) \chi\left(\frac{q}{2k_F}\right)}. \tag{3.14}
$$

In the long wavelength limit, i.e., $\lim q \to 0$,

$$
v(q) = -\frac{4\pi z e^2 / \Omega_0}{q^2 + \lambda^2}. \tag{3.15}
$$

Figure 3.2 Friedel oscillations.

3.1.2 Screening Charge

Writing $\rho_{\text{ind}}(q) = S(q) \times \rho_a(q)$, we can derive the spatial dependence of the atomic screening charge as follows:

$$\rho_a(\mathbf{x}) = \frac{\Omega_0}{(2\pi)^3} \int \rho_a(\mathbf{q}) \, e^{i\mathbf{q}\cdot\mathbf{x}} \, d\mathbf{q}$$

$$= \frac{\Omega_0}{(2\pi)^3} \int \rho_a(\mathbf{q}) \, \frac{\sin(qr)}{qr} \times 4\pi q^2 dq$$

$$\simeq \frac{9\pi z^2}{\Omega_0 E_F} \times \frac{v(2k_F)}{\epsilon(2k_F)} \times \frac{\cos(2k_F r)}{(2k_F r)^3}. \tag{3.16}$$

The result has an oscillatory form, known as *Friedel oscillations* (see Figure 3.2).

> Despite the complicated and abstract derivation of the preceding Friedel oscillation, a simple and more physical explanation can be readily presented. By analogy to the one-dimensional slit diffraction phenomenon, where the constraint arising from the confining slit width, say a, leads to oscillatory manifestations of its reciprocal space image to
>
> $$\frac{\sin(ka)}{ka},$$
>
> we can envision that the confinement of the electronic states to $-k_F \leq k \leq k_F$ in reciprocal space will lead to similar oscillations in real space of the form
>
> $$\frac{\sin(2k_F x)}{2k_F x},$$
>
> which define similar Friedel oscillations in one dimension.

3.2 Methods of Electron Band Calculations

As we have shown in Chapter 1, the periodicity of the crystal potential leads to Bloch's theorem. It provides a general form for the electronic wavefunctions in crystal lattices, namely,

$$\Psi_{\mathbf{k}} = e^{-i\mathbf{k}\cdot\mathbf{x}} u_{\mathbf{k}}(\mathbf{x}), \tag{3.17}$$

where $\mathbf{k}$ is a wavevector and $u_{\mathbf{k}}(\mathbf{x})$ has the periodicity of the lattice,

$$u_{\mathbf{k}}(\mathbf{x} + \mathbf{R}) = u_{\mathbf{k}}(\mathbf{x}), \tag{3.18}$$

where $\mathbf{R}$ is a lattice vector.

3.2.1 Basis Sets

Two general categories of wavefunctions satisfy the periodic character of $u_{\mathbf{k}}(\mathbf{x})$:

1. Plane waves or momentum eigenfunctions:
   $u_{\mathbf{k}}$ can be expanded as a Fourier series of the periodic lattice,

$$u_{\mathbf{k}}(\mathbf{x}) = \sum_{\mathbf{G}} a_{\mathbf{k}}(\mathbf{G}) \, e^{i\mathbf{G}\cdot\mathbf{x}}, \tag{3.19}$$

where the $\mathbf{G}$s are reciprocal lattice vectors. Consequently, we find

$$\Psi_{\mathbf{k}}(\mathbf{x}) = \sum_{\mathbf{G}} a_{\mathbf{k}}(\mathbf{G}) \, e^{-i(\mathbf{k}-\mathbf{G})\cdot\mathbf{x}}. \tag{3.20}$$

The physical interpretation of this equation is that a Bloch electron with wavevector $\mathbf{k}$ can have momentum $\mathbf{k} + \mathbf{G}$; and the probability of finding momentum $\mathbf{k} - \mathbf{G}$ is simply $\left| a_{\mathbf{k}}(\mathbf{G}) \right|^2$, apart from a normalization constant.

Using the orthonormality of the functions $\exp\left[-i(\mathbf{k} - \mathbf{G}) \cdot \mathbf{x}\right]$, we can express $a_{\mathbf{k}}(\mathbf{G})$ as,

$$a_{\mathbf{k}}(\mathbf{G}) = \frac{1}{\Omega} \int d\mathbf{x} \, \Psi_{\mathbf{k}}(\mathbf{x}) \, e^{i(\mathbf{k}-\mathbf{G})\cdot\mathbf{x}}, \tag{3.21}$$

The periodicity of Bloch states in $\mathbf{k}$-space, $\Psi_{\mathbf{k}}(\mathbf{x}) = \Psi_{\mathbf{k}+\mathbf{G}}(\mathbf{x})$, yields

$$a_{\mathbf{k}}(\mathbf{G}) = \frac{1}{\Omega} \int d\mathbf{x} \, \Psi_{\mathbf{k}+\mathbf{G}}(\mathbf{x}) \, e^{i(\mathbf{k}-\mathbf{G})\cdot\mathbf{x}} = a(\mathbf{k} + \mathbf{G}). \tag{3.22}$$

2. Localized Wannier functions:
   Another implication of the periodicity of the Bloch function in reciprocal space is that it can be expanded as a Fourier series in real space

$$\Psi_{\mathbf{k}}(\mathbf{x}) = \sum_{\mathbf{R}} w_{\mathbf{R}}(\mathbf{x}) \, e^{i\mathbf{k}\cdot\mathbf{R}},$$

$$w_{\mathbf{R}}(\mathbf{x}) = \frac{1}{\Omega_B} \int d\mathbf{k} \, \Psi_{\mathbf{k}}(\mathbf{x}) \, e^{-i\mathbf{k}\cdot\mathbf{R}} \tag{3.23}$$

where Ω_B is the BZ volume. From Bloch's theorem, we can write the integrand as

$$\Psi_{\mathbf{k}}(\mathbf{x}) \, e^{-i\mathbf{k}\cdot\mathbf{R}} = \Psi_{\mathbf{k}}(\mathbf{x} - \mathbf{R}),$$

showing that

$$w_{\mathbf{R}}(\mathbf{x}) = w(\mathbf{x} - \mathbf{R}) = \frac{1}{\Omega_B} \int_{\Omega_B} d\mathbf{k}\, \Psi_{\mathbf{k}}(\mathbf{x} - \mathbf{R})$$

and that $w(\mathbf{x})$ is lattice periodic. $w(\mathbf{x})$ is known as the *Wannier function*. Consequently,

$$\Psi_{\mathbf{k}}(\mathbf{x}) = \sum_{\mathbf{R}} w(\mathbf{r} - \mathbf{R})\, e^{i\mathbf{k}\cdot\mathbf{R}}. \tag{3.24}$$

The orthonormality of the Bloch functions can be used to derive the orthonormality and completeness relations of the Wannier functions,

$$\int d\mathbf{x}\, w^*(\mathbf{x} - \mathbf{R})\, w(\mathbf{r} - \mathbf{S}) = \int_{BZ} d\mathbf{k} \int_{BZ} d\mathbf{k}'\, e^{i[\mathbf{S}\cdot\mathbf{k}' - \mathbf{R}\cdot\mathbf{k}]} \underbrace{\int d\mathbf{x}\, \Psi_{\mathbf{k}'}^*(\mathbf{x})\, \Psi_{\mathbf{k}}^*(\mathbf{x})}_{\delta_{\mathbf{k},\mathbf{k}'}}$$

$$= \int d\mathbf{k}\, e^{i\mathbf{k}\cdot(\mathbf{S}-\mathbf{R})} = \delta_{\mathbf{R},\mathbf{S}}. \tag{3.25}$$

The completeness

$$\sum_{\mathbf{R}} w^*(\mathbf{x}' - \mathbf{R})\, w(\mathbf{x} - \mathbf{R}) = \delta(\mathbf{x}' - \mathbf{x}) \tag{3.26}$$

can be similarly proven.

We should note that Wannier functions are not uniquely defined since they are expressed as the Fourier sum of the Bloch functions, which are only defined up to a phase factor. The freedom to choose different phase factor for every $\mathbf{k}$ gives us the ability to optimize the Wannier functions with respect to certain criteria, for example, that they can be maximally localized.

3.2.2 *The Orthoganalized Plane Wave (OPW) method*

In the plane wave set of Bloch wavefunctions $e^{-i(\mathbf{k}-\mathbf{G})\cdot\mathbf{x}}$, we were able to express

$$\Psi_{\mathbf{k}}(\mathbf{x}) = \sum_{\mathbf{G}} a(\mathbf{k} - \mathbf{G})\, e^{-i(\mathbf{k}-\mathbf{G})\cdot\mathbf{x}}. \tag{3.27}$$

Expanding the crystal potential

$$V(\mathbf{x}) = \sum_{\mathbf{G}} V_{\mathbf{G}}\, e^{i\mathbf{G}\cdot\mathbf{x}}, \tag{3.28}$$

and substituting (3.27) and (3.28) in the Schrödinger equation

$$\left[-\nabla^2 + V(\mathbf{x}) \right] \Psi_{\mathbf{k}}(\mathbf{x}) = E(\mathbf{k})\, \Psi_{\mathbf{k}}(\mathbf{x}),$$

we obtain the set of simultaneous equations

$$\left[\frac{|\mathbf{k} - \mathbf{G}|^2}{2} - E(k) \right] a(\mathbf{k} - \mathbf{G}) + \sum_{\mathbf{G}'} V_{\mathbf{G}'-\mathbf{G}}\, a(\mathbf{k} - \mathbf{G}') = 0. \tag{3.29}$$

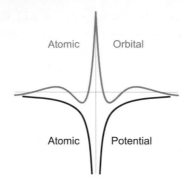

Figure 3.3 Atomic fluctuation on sub-Angstrom scales.

In principle, we choose $\mathbf{k}$ and determine all the coefficients $a(\mathbf{k} - \mathbf{G})$ by solving the coupled equations (3.29). In general, this is not convenient for practical calculations since it requires an unreasonably large number of plane waves in the expansion of the wavefunction $\Psi_{\mathbf{k}}(\mathbf{x})$, which is necessary to account for the atomic-like sub-Angstrom fluctuations in the vicinity of the nuclei (see Figure 3.3).

We now show that we can drastically reduce the number of plane waves employed by incorporating atomic core orbitals in the construction of our Bloch functions – which is the essence of the OPW method.

Construction of OPWs

We can construct Bloch functions for the tightly bound core states in a manner similar to the Wannier functions

$$\Psi_{\mathbf{k}}^{c}(\mathbf{x}) = \sum_{\mathbf{X}} e^{i\mathbf{k} \cdot \mathbf{R}} \, \phi_c(\mathbf{r} - \mathbf{R}), \tag{3.30}$$

where the ϕ_cs are atomic core orbitals. We can now write the OPWs as [40]

$$\chi_{\mathbf{k},\mathbf{G}}(\mathbf{x}) = e^{-i(\mathbf{k} - \mathbf{G}) \cdot \mathbf{x}} - \sum_{c} \mu_{\mathbf{G}}^{c}(\mathbf{k}) \, \Psi_{\mathbf{k}}^{c}(\mathbf{x}). \tag{3.31}$$

For χ to be orthogonal to all core Bloch functions, we require that

$$\mu_{\mathbf{G}}^{c*}(\mathbf{k}) = \int_{\Omega} e^{i(\mathbf{k} - \mathbf{G}) \cdot \mathbf{x}} \, \Psi_{\mathbf{k}}^{c}(\mathbf{x}) \, dr, \tag{3.32}$$

for all $\mathbf{k}$. We now expand the valence wavefunction $\Psi_{\mathbf{k}}(\mathbf{x})$ in terms of the OPWs, and write

$$\Psi_{\mathbf{k}}(\mathbf{x}) = \sum_{\mathbf{G}} b_{\mathbf{k},\mathbf{G}} \, \chi_{\mathbf{k},\mathbf{G}}(\mathbf{x}). \tag{3.33}$$

We expect this expansion to converge more rapidly than a regular plane wave expansion. Thus, we can retain a finite number in the sum, with the coefficients $b_{k,G}$

as variational parameters to be determined by minimizing the expectation value of the one-electron Hamiltonian. We then obtain the set of linear equations

$$\sum_{\mathbf{G}} b_{\mathbf{k},\mathbf{G}} \int \chi_{\mathbf{k},\mathbf{G}'}^{*} \left[\mathcal{H} - E \right] \chi_{\mathbf{k},\mathbf{G}} \, d\mathbf{x} = 0, \tag{3.34}$$

and the corresponding secular determinant

$$\left| \int \chi_{\mathbf{k},\mathbf{G}'}^{*} \left[\mathcal{H} - E \right] \chi_{\mathbf{k},\mathbf{G}} \, d\mathbf{x} \right| = 0. \tag{3.35}$$

The size of this secular equation is obviously determined by the number of OPWs we choose to retain in the expansion. For simple metals, even a single term gives a reasonable first approximation to the valence band. For more complex solids, convergence would require an appreciable number of OPWs to be included.

3.2.3 Korringa-Kohn-Rostoker (KKR) Green Function Method

This method [107, 112] solves the wave equation using the crystal Green function. It is quite general, in the sense that the form of the Green function mainly depends on the crystal structure. For simplicity, we shall consider a monatomic crystal.

However, to allow the theory to be carried through explicitly in practice, it is necessary to assume that the potential within a unit cell vanishes before the boundary of the unit cell is reached. A very important simplification then occurs. The problem can to be separated into two parts:

 (i) Determination of the scattering properties of a single potential of the *muffin-tin* form (see Figure 3.4).
(ii) Determination of the structural aspects.

This method of calculating the band structure for particular metals has been found to be highly efficient even for low symmetry points in the Brillouin zone (BZ).

1. Integral Equation
For a periodic potential $V(\mathbf{x})$, we can solve the wave equation

$$\left[\nabla^2 + E \right] \Psi = V(\mathbf{x}) \, \Psi \tag{3.36}$$

by a Green function procedure. Defining the Green function $G(\mathbf{x}, \mathbf{x}')$ through the equation

$$\left[\nabla^2 + E \right] G(\mathbf{x}, \mathbf{x}') = \delta(\mathbf{x} - \mathbf{x}'), \tag{3.37}$$

we can construct a solution for the wave function Ψ of the form

$$\Psi(\mathbf{x}) = \int_{\Omega} d\mathbf{x}' \, G(\mathbf{x}, \mathbf{x}') \, V(\mathbf{x}') \, \Psi(\mathbf{x}'). \tag{3.38}$$

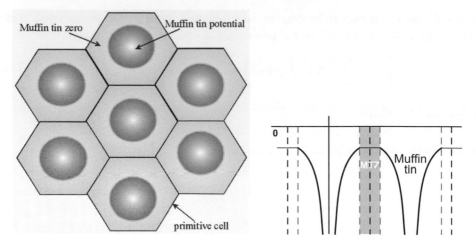

Figure 3.4 Muffin-tin configuration for a 2D hexagonal lattice. Muffin-tin configuration. Dashed vertical gray lines define the muffin-tin region, while dashed vertical black lines delineate the primitive cell boundary. Solid black horizontal lines in the interstitial region indicate the muffin-tin zero (MTZ).

2. Boundary Conditions

In order for Ψ to be a Bloch function, we must choose the Green function to satisfy the relation

$$G(\mathbf{x} + \mathbf{R}, \mathbf{x}') = e^{i\mathbf{k}\cdot\mathbf{x}}\, G(\mathbf{x}, \mathbf{x}'), \qquad (3.39)$$

where $\mathbf{R}$ is any lattice vector. It can be easily shown that the correct choice of G is

$$G(\mathbf{x}, \mathbf{x}') = -\frac{1}{\Omega} \sum_{\mathbf{G}} \frac{\exp[i(\mathbf{k} + \mathbf{G}) \cdot (\mathbf{x} - \mathbf{x}')]}{(\mathbf{k} + \mathbf{G})^2 - E}. \qquad (3.40)$$

The muffin tin assumption amounts to taking the potential to be spherical about each ion within the sphere inscribed in the unit cell and constant outside. Then we can write

$$V(\mathbf{x}) = \sum_{\mathbf{R}} v\left(|\mathbf{x} - \mathbf{R}|\right), \qquad (3.41)$$

where the vs are spherical and nonoverlapping, i.e., we can choose

$$v(r) = 0, \qquad r > r_{\mathrm{mt}}, \qquad (3.42)$$

where r_{mt} is the radius of a muffin tin sphere lying wholly within the unit cell. The wavefunction within the inscribed muffin-tin sphere may now be written as

$$\Psi(\mathbf{x}) = \sum_{L} c_L\, \mathcal{R}_l(r)\, Y_L(\theta, \phi), \qquad r < r_{\mathrm{mt}}, \qquad (3.43)$$

where $Y_L \equiv Y_{lm}$ are normalized spherical harmonics. The radial wavefunctions $\mathcal{R}_l(r)$ satisfy the equation

$$\left[\frac{-1}{r^2} \frac{d}{dr} \left(r^2 \frac{d}{dr} \right) + \frac{l(l+1)}{r^2} + v(r) - E \right] \mathcal{R}_l(r) = 0. \tag{3.44}$$

3. Secular Equation

We now explain how the coefficients c_L can be obtained from the integral equation. It will be useful to define the single-center potential by

$$v(r) = v(r), \quad r \leq r_{\mathrm{mt}} - \epsilon$$
$$v(r) = \quad 0, \quad r > r_{\mathrm{mt}} - \epsilon, \tag{3.45}$$

where ϵ is a small positive quantity, which we shall eventually allow to go to zero. The integral equation can then be written

$$\Psi(\mathbf{x}) = \int_{r' < r_{\mathrm{mt}} - \epsilon} d\mathbf{x}' \, G(\mathbf{x}, \mathbf{x}') \, v\left(\mathbf{x}'\right) \, \Psi(\mathbf{x}'). \tag{3.46}$$

We shall consider next a choice of the vector $\mathbf{x}$ such that $|\mathbf{x}| = r_{\mathrm{mt}} - 2\epsilon$. Using the fact that $\Psi(\mathbf{x}')$ and $G(\mathbf{x}, \mathbf{x}')$ satisfy

$$\left[\nabla_{\mathbf{x}'}^2 + E \right] \Psi\left(\mathbf{x}'\right) = v\left(\mathbf{x}'\right) \, \Psi\left(\mathbf{x}'\right)$$
$$\left[\nabla_{\mathbf{x}'}^2 + E \right] G(\mathbf{x}, \mathbf{x}') = \delta(\mathbf{x}' - \mathbf{x}),$$

we obtain for (3.46)

$$0 = \Psi(\mathbf{x}) - \int_{r' < r_{\mathrm{mt}} - \epsilon} d\mathbf{x}' \, G(\mathbf{x}, \mathbf{x}') \left[\nabla_{\mathbf{x}'}^2 + E \right] \Psi(\mathbf{x}')$$
$$= \int d\mathbf{x}' \, \delta\left(\mathbf{x}' - \mathbf{x}\right) \Psi(\mathbf{x}') - \int d\mathbf{x}' \, G(\mathbf{x}, \mathbf{x}') \left[\nabla_{\mathbf{x}'}^2 + E \right] \Psi(\mathbf{x}')$$
$$= \int d\mathbf{x}' \left[\nabla_{\mathbf{x}'}^2 + E \right] G(\mathbf{x}, \mathbf{x}') \Psi(\mathbf{x}') - \int d\mathbf{x}' \, G(\mathbf{x}, \mathbf{x}') \left[\nabla_{\mathbf{x}'}^2 + E \right] \Psi(\mathbf{x}')$$

Using the Green identity

$$\int_V d\mathbf{x} \left(\psi \, \nabla^2 \phi - \phi \, \nabla^2 \psi \right) = \oint_S dS \left(\psi \frac{\partial \phi}{\partial n} - \phi \frac{\partial \psi}{\partial n} \right),$$

where n is the coordinate normal to the surface S, we find

$$\int_{r' < r_{\mathrm{mt}} - \epsilon} d\mathbf{x}' \, \nabla_{\mathbf{x}'}^2 \, G(\mathbf{x}, \mathbf{x}') \, \Psi(\mathbf{x}') - \int_{r' < r_{\mathrm{mt}} - \epsilon} d\mathbf{x}' \, G(\mathbf{x}, \mathbf{x}') \, \nabla_{\mathbf{x}'}^2 \, \Psi(\mathbf{x}')$$

$$= \int_{r' < r_{\mathrm{mt}} - \epsilon} dS' \, \Psi \frac{\partial G(\mathbf{x}, \mathbf{x}')}{\partial r'} - \int_{r' < r_{\mathrm{mt}} - \epsilon} dS' \frac{\partial \Psi}{\partial r'} \, G(\mathbf{x}, \mathbf{x}'), \tag{3.47}$$

where the integrations on the second line are over the surface of the sphere of radius $r_{mt} - \epsilon$. We obtain

$$0 = \int_{r' < r_{mt} - \epsilon} dS' \, \Psi \, \frac{\partial G(\mathbf{x}, \mathbf{x}')}{\partial r'} - \int_{r' < r_{mt} - \epsilon} dS' \, \frac{\partial \Psi}{\partial r'} \, G(\mathbf{x}, \mathbf{x}'). \tag{3.48}$$

It is now clear that in order to obtain equations for the coefficients c_L, we need to express G in terms of spherical harmonics. To do this, we use the free particle Green function G_0, which satisfies the same inhomogeneous equation as G inside the atomic polyhedron, and write the difference

$$D(\mathbf{x}, \mathbf{x}') = G(\mathbf{x}, \mathbf{x}') - G_0(\mathbf{x}, \mathbf{x}'), \tag{3.49}$$

where D must satisfy the homogeneous wave equation. For a plane wave, which trivially satisfies such an equation, we have

$$e^{i\mathbf{k}\cdot\mathbf{x}} = 4\pi \sum_l (2l + 1) \, i^l \, j_l(kr) \, P_l(\cos\theta), \tag{3.50}$$

where θ is the angle between $\mathbf{k}$ and $\mathbf{x}$, and j_l is the spherical Bessel function of order l. By analogy, D can be expressed in the form

$$D(\mathbf{x}, \mathbf{x}') = \sum_{L, L'} A_{L, L'} \, j_l(\kappa r) \, j_{l'}(\kappa r') \, Y_L(\theta, \phi) \, Y_{L'}(\theta', \phi'), \tag{3.51}$$

where the inverse length scale $\kappa = \sqrt{2mE/\hbar^2}$. The As are constants to be discussed later. Moreover,

$$G_0(\mathbf{x}, \mathbf{x}') = -\frac{1}{4\pi} \frac{\cos\kappa|\mathbf{x} - \mathbf{x}'|}{|\mathbf{x} - \mathbf{x}'|}$$

$$= \kappa \sum_L j_l(\kappa r) \, n_l(\kappa r') \, Y_L(\theta, \phi) \, Y_L(\theta', \phi'), \quad r < r' \tag{3.52}$$

n_l is the spherical Neumann function. Hence we can write the full Green function in the form

$$G(\mathbf{x}, \mathbf{x}') = \sum_{L, L'} \left[A_{L, L'} \, j_l(\kappa r) \, j_{l'}(\kappa r') + \kappa \, \delta_{ll'}, \delta_{mm'} \, j_l(\kappa r) \, n_l(\kappa r') \right]$$

$$\times \, Y_L(\theta, \phi) \, Y_{L'}(\theta', \phi'), \qquad (r < r' < r_{mt}). \tag{3.53}$$

We substitute (3.53) for G and (3.43) for Ψ in (3.48), multiply through by $Y_L^*(\theta, \phi)$ and integrate over the sphere of radius $r = r_{mt} - 2\epsilon$. After letting $\epsilon \to 0$, we obtain

$$\sum_{L'} j_l \left[A_{L, L'} \left(j_{l'} L_{l'} - j'_{l'} \right) + \kappa \, \delta_{ll'} \, \delta_{mm'} \left(n_{l'} L_{l'} - n'_{l'} \right) \right] c_{L'} = 0. \tag{3.54}$$

Here

$$j'_l = \frac{d j_l(x)}{dx}, \qquad n'_l = \frac{d n_l(x)}{dx}, \tag{3.55}$$

and L is the logarithmic derivative

$$L = \frac{dR_l(\mathbf{x})/dr}{R_l(\mathbf{x})}, \tag{3.56}$$

and all functions are evaluated at $r = r_{\mathrm{mt}}$. In order that nontrivial solutions for the cs exist, the determinant of the coefficients must be zero. Since (3.54) may be written as

$$\sum_{L'} \left[A_{L,L'} + \kappa \, \delta_{ll'} \, \delta_{mm'} \, \frac{n_l \, L_l - n_l'}{j_l \, L_l - j_l'} \right] \tilde{c}_{L'} = 0, \tag{3.57}$$

where $\tilde{c}_{L'} = (j_l L_l - j_l') c_{L'}$, an equivalent condition is

$$\det \left| A_{L,L'} + \kappa \, \delta_{ll'} \, \delta_{mm'} \, \frac{n_l \, L_l - n_l'}{j_l \, L_l - j_l'} \right| = 0. \tag{3.58}$$

It is evident that the As (called *structure constants*) depend on the particular crystal lattice considered. It can be seen from (3.58) that a clear separation has been achieved between the effects of the geometry of the lattice and the effects of the potential that are manifest only in logarithmic derivatives L and through the value r_{mt} at which the functions j and n are evaluated.

4. Structure Constants $A_{lm,l'm'}$

The coefficients $A_{lm,l'm'}$ can be determined in a straightforward way. First, expand the plane waves in (3.40) in terms of spherical waves, as in (3.50); second, equate the resulting coefficients of the spherical harmonics with those of (3.53):

$$G(\mathbf{x}, \mathbf{x}') = -\frac{1}{\Omega} \sum_{\mathbf{G}} \frac{\exp[i(\mathbf{k} + \mathbf{G}) \cdot (\mathbf{x} - \mathbf{x}')]}{(\mathbf{k} + \mathbf{G})^2 - E}$$

$$= \frac{(4\pi)^2}{\Omega} \sum_{\substack{\mathbf{G} \\ lm,l'm'}} i^{l-l'} \frac{j_l(|\mathbf{k} + \mathbf{G}|r) \, j_{l'}(|\mathbf{k} + \mathbf{G}|r') \, Y_L(\mathbf{k} + \mathbf{G}) \, Y_{L'}^*(\mathbf{k} + \mathbf{G})}{(\mathbf{k} + \mathbf{G})^2 - E}$$

$$= \sum_{L} \sum_{L'} \left[A_{lm,l'm'} \, j_l(\kappa r) \, j_{l'}(\kappa r') + \kappa \, \delta_{ll'}, \delta_{mm'} \, j_l(\kappa r) \, n_l(\kappa r') \right]$$

$$\times \, Y_L(\theta, \phi) \, Y_{L'}(\theta', \phi').$$

Specifically, we find

$$A_{lm,l'm'} = -\frac{(4\pi)^2}{\Omega} \, i^{(l-l')} \left[j_l(\kappa r) \, j_{l'}(\kappa r') \right]^{-1}$$

$$\times \left(\left[\sum_{\mathbf{G}} \frac{j_l(|\mathbf{k} + \mathbf{G}|r) \, j_{l'}(|\mathbf{k} + \mathbf{G}|r') \, Y_L(\mathbf{k} + \mathbf{G}) \, Y_L^*(\mathbf{k} + \mathbf{G})}{[(\mathbf{k} + \mathbf{G})^2/2] - E} \right] \right.$$

$$\left. - \kappa \, \delta_{ll'} \, \delta_{mm'} \, \frac{n_l(\kappa r)}{j_l(\kappa r)} \right) \tag{3.59}$$

where $E = \hbar^2 \kappa^2 / 2m$.

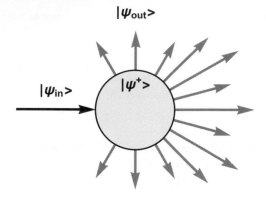

Figure 3.5 Scattering states from a spherically symmetric potential.

We can then write

$$A_{lm,l'm'} = \sum_{G} \frac{F_L(G)\, F_{L'}(G)}{[(k+G)^2/2] - E} - \kappa\, \delta_{ll'}\, \delta_{mm'}\, \frac{n_l(\kappa r_{\text{mt}})}{j_l(\kappa r_{\text{mt}})} \qquad (3.60)$$

$$F_L(G) = \frac{4\pi}{\sqrt{\Omega}}\, \frac{j_l(|k+G|r_{\text{mt}})}{j_l(\kappa r_{\text{mt}})}\, Y_L(\widehat{k+G}). \qquad (3.61)$$

5. Secular Determinant of the KKR Method in Terms of Phase Shifts: Method of Phase Shifts in Quantum Scattering: Revisited

We consider a single spherically symmetric muffin tin potential, where the scattered wavefunction has the asymptotic form shown in Figure 3.5, and given by

$$|\Psi^+\rangle = \frac{1}{(2\pi)^{3/2}} \left[\exp[i\mathbf{k}_{\text{in}} \cdot \mathbf{r}] + f(k,\theta)\, \frac{\exp[ikr]}{r} \right]$$

$$= \frac{1}{(2\pi)^{3/2}} \sum_{l} (2l+1) P_l(\cos\theta) \left[\frac{i^l \sin(kr - \frac{l\pi}{2})}{kr} + f_l(k)\, \frac{\exp[ikr]}{r} \right]$$

$$= \frac{1}{(2\pi)^{3/2}} \sum_{l} (2l+1) \frac{P_l(\cos\theta)}{2ik}$$

$$\times \left[[1 + 2ik f_l(k)] \frac{\exp[ikr]}{r} - \frac{\exp[-i(kr - l\pi)]}{r} \right].$$

In the absence of the scattering potential, the plane wave is the sum of the following:

1. A spherically outgoing wave, e^{ikr}/r
2. A spherically incoming wave, $e^{-i(kr-l\pi)}/r$, for each l

The presence of the scatterer changes the coefficient of the outgoing wave:

$$1 \;\rightarrow\; 1 + 2ik\, f_l(k).$$

The incoming wave is left completely unaffected!

Recalling the relation between the S and T operators, we obtain

$$S_l(E) = 1 + 2ik \; f_l(k).$$

The time independence of the scattering problem, $\nabla \cdot \mathbf{j} = 0$, or

$$\int_{\substack{\text{spherical} \\ \text{surface}}} \mathbf{j} \cdot d\mathbf{s} = 0,$$

stipulates that the incoming flux must equal the outgoing flux. Furthermore, because of angular-momentum conservation, this relation must hold for each partial wave separately; consequently, we obtain a **unitary relation**

$$|S_l(k)| = 1$$

for the lth partial wave, and we set

$$S_l(E) = 1 + 2ik \; f_l(k) S_l(k) = \exp[2i\delta_l(k)]$$

$$f_l(k) = \frac{\exp[2i\delta_l(k)] - 1}{2ik} = \frac{\exp[i\delta_l]\sin\delta_l}{k} = \frac{1}{k\cot\delta_l - ik}.$$

We obtain for the full scattering amplitude

$$f(k,\theta) = \frac{1}{k} \sum_{l=0} (2l+1) \; \exp[i\delta_l] \; \sin\delta_l \; P_l(\cos\theta),$$

which is the result of both rotational invariance and probability conservation.

To see how this can be done, we note that for $r \geq r_{\text{mt}}$ the radial wavefunction $R_l(r)$ (see Figure 3.6) must be a linear combination of the free particle solutions j_l and n_l, namely,

$$R_l(r) \sim \left[j_l(kr) - \tan\eta_l \; n_l(kr) \right], \tag{3.62}$$

as shown in Figure 3.6. This allows us to write down the logarithmic derivative by matching it at r_{mt} as

$$\frac{R_l'(r_{\text{mt}})}{R_l(r_{\text{mt}})} = L_l = \frac{j_l'(\kappa r_{\text{mt}}) - \tan\eta_l \; n_l'(\kappa r_{\text{mt}})}{j_l(\kappa r_{\text{mt}}) - \tan\eta_l \; n_l(\kappa r_{\text{mt}})}, \tag{3.63}$$

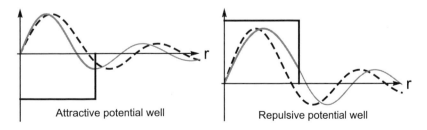

Attractive potential well   Repulsive potential well

Figure 3.6 Radial wavefunction: gray (dashed black), in the presence (absence) of the potential well.

from which we immediately obtain

$$\tan \eta_l = \frac{L_l \, j_l(\kappa r_{\text{mt}}) - j_l'(\kappa r_{\text{mt}})}{L_l \, n_l(\kappa r_{\text{mt}}) - n_l'(\kappa r_{\text{mt}})}.$$

(3.64)

Thus, (3.58) may be written

$$\det \left| A_{lm, \, l'm'} + \kappa \, \delta_{ll'} \delta_{mm'} \, \cot \eta_l \right| = 0.$$

(3.65)

Substituting from (3.60), we obtain

$$\det \left| \sum_{\mathbf{G}} \frac{F_L(\mathbf{G}) \, F_{L'}(\mathbf{G})}{[(\mathbf{k} + \mathbf{G})^2/2] - E} + \kappa \, \delta_{ll'} \delta_{mm'} \, \cot \eta_l' \right| = 0.$$

(3.66)

with

$$\cot \eta_l' = \cot \eta_l - \frac{n_l(\kappa r_{\text{mt}})}{j_l(\kappa r_{\text{mt}})}$$

where the last term comes from (3.60)

3.2.4 Augmented Plane Wave (APW) Method

As we have seen, the atomic-like behavior of the crystalline wavefunctions in the core region has prompted the formulation of the OPW method. It has also led to the proposition of the augmented plane wave formalism by Slater in 1937 [168]. This formalism makes use of the concept of muffin-tin (MT) potentials. In this scheme, one divides space up into the atomic regions near the nuclei, and the free electron regions in the interstitial space. In the atomic region, which is taken as spherical for convenience, quantities of interest are expanded in spherical harmonics. In the remaining interstitial region, which must exhibit the periodic boundary conditions, a plane wave expansion is generally chosen.

The surface of the MT sphere becomes a natural boundary enclosing an *atomic* or *core* region, where the wavefunction must be a combination of solutions of the Schrödinger equation in the spherically symmetrical $v_{MT}(r)$ and in the interstitial region. The potential in the interstitial region is practically flat, or undulates gently, and we construct the wavefunctions out of combination of plane waves. Unlike the OPW, these two constituents of the APW are never allowed to overlap, but must be matched to one another at the MT surface (see Figure 3.7).

A general APW is a function of the form

$$\phi(\mathbf{k} + \mathbf{G}, \mathbf{x}) = \begin{cases} \frac{1}{\sqrt{\Omega}} \exp(i(\mathbf{k} + \mathbf{G}) \cdot \mathbf{x}) & r > R \\ \sum_L i^l \, a_L \, \mathcal{R}_l(r, \mathcal{E}) \, Y_L(\hat{r}) & r \leq R. \end{cases}$$

(3.67)

The radial functions are chosen to satisfy the radial Schrödinger equation in the MT potential region. How should we choose the coefficients a_L? In the standard Slater scheme, the

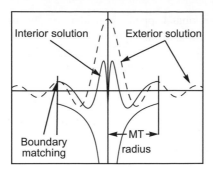

Figure 3.7 Wavefunction matching.

function is made continuous over the surface of the MT sphere. Expanding the plane wave (PW) in spherical waves

$$\frac{1}{\sqrt{\Omega}} \exp(i(\mathbf{k} + \mathbf{G}_i) \cdot \mathbf{x}) = \frac{4\pi}{\sqrt{\Omega}} \sum_L i^l \, j_l \, (|\mathbf{k} + \mathbf{G}| r) \, Y_L^*(\widehat{\mathbf{k} + \mathbf{G}}) \, Y_L(\hat{\mathbf{x}})$$

leads to

$$a_L = \frac{4\pi}{\sqrt{\Omega}} \left[\frac{j_l \, (|\mathbf{k} + \mathbf{G}| r)}{\mathcal{R}_l(r, \mathcal{E})} \right] Y_L^*(\widehat{\mathbf{k} + \mathbf{G}}). \tag{3.68}$$

A single function of this form is not, in general, a Bloch function, because it cannot be made to satisfy the Schrödinger equation in the interstitial region without a discontinuity of slope on the MT sphere. But a linear combination of such waves, of the form

$$\Psi_{\mathbf{k}}(\mathbf{x}) = \sum_{\mathbf{G}} c_{\mathbf{k}+\mathbf{G}} \, \phi(\mathbf{k} + \mathbf{G}, \mathbf{x}), \tag{3.69}$$

is acceptable as a putative solution of the Schrödinger equation. The obvious procedure is to substitute (3.69) in an expression for the expectation value of the energy, and then to vary the coefficients $c_{\mathbf{k}+\mathbf{G}}$. This yields the usual matrix, whose secular determinant must vanish.

But the discontinuity of slope at the MT surface in each APW demands a little care. In the variational approach, Slater used the standard form

$$\int_{\text{cell}} \left\{ \nabla \Psi^* \cdot \nabla \Psi + (V_{MT} - \mathcal{E}) \, \Psi^* \, \Psi \right\} d^3r$$

$$= \int_{\text{cell}} \Psi^* \left\{ -\nabla^2 + V_{MT} - \mathcal{E} \right\} \Psi \, d^3r + \int_{\text{sphere}} \left\{ \Psi_i^* \, \nabla \Psi_o - \Psi_o \, \nabla \Psi_i^* \right\} \cdot d\mathbf{S}, \tag{3.70}$$

where the surface integral vanishes only if the *inner* and *outer* wavefunctions, Ψ_i and Ψ_o, join with continuous amplitude and gradient. The algebraic steps that follow on substituting (3.69) in (3.70) are given at length in Slater's *Symmetry and Energy Bands in Crystals*, appendix 6[169]. To see what happens, consider the diagonal matrix element in $\phi_{\mathbf{k}}$. The integration over the volume of the cell is trivial. There is no contribution from within the atomic sphere, where each APW satisfies the Schrödinger equation exactly. From the interstitial region, where $V_{MT} = 0$, we get

$$\Omega^{-1} \int_{\substack{\text{interstitial} \\ \text{region}}} e^{-i\mathbf{k}\cdot\mathbf{x}} \left\{ -\nabla^2 + V_{MT}(r) - \mathcal{E} \right\} e^{i\mathbf{k}\cdot\mathbf{x}} \, d^3r = \frac{\Omega_s}{\Omega} (k^2 - \mathcal{E}), \qquad (3.71)$$

where Ω_s is the interstitial volume, and we set $\hbar^2/2m = 1$ so that the energy is measured in Rydberg. The nature of the atomic potential appears only in the surface term, which contains a factor like

$$a_L \left[j_l(kr) \frac{\partial \mathcal{R}}{\partial r} - \mathcal{R} \frac{\partial j_l(kr)}{\partial r} \right]_{r=R} = \{j_l(kR)\}^2 \left[\frac{\mathcal{R}'_l(R)}{\mathcal{R}_l(R)} - \frac{k j'_l(kR)}{j_l(kR)} \right], \qquad (3.72)$$

where the logarithmic derivative of the radial function appears explicitly.

Bringing together the various terms, and using standard identities for spherical harmonics, etc., we find that the secular determinant for the coefficients $c_{\mathbf{k}-\mathbf{G}}$ has the matrix elements

$$\Gamma_{\mathbf{G}\mathbf{G}'} = \frac{4\pi R^2}{\Omega} \left\{ -\left[|\mathbf{k} - \mathbf{G}|^2 - \mathcal{E} \right] \left[\frac{j_1(|\mathbf{G} - \mathbf{G}'|R)}{|\mathbf{G} - \mathbf{G}'|} \right] \right\}$$

$$+ \sum_{l=0}^{\infty} (2l+1) \, P_l(\cos\theta_{\mathbf{G}\mathbf{G}'}) \, j_l(|\mathbf{k} - \mathbf{G}|R) \, j_l(|\mathbf{k} - \mathbf{G}'|R)$$

$$\times \left[\frac{\mathcal{R}'(R)}{\mathcal{R}(R)} - \frac{|\mathbf{k} - \mathbf{G}| \, j'_l(|\mathbf{k} - \mathbf{G}|R)}{j_l(|\mathbf{k} - \mathbf{G}|R)} \right], \qquad (3.73)$$

where $\theta_{\mathbf{G}\mathbf{G}'}$ is the angle between $(\mathbf{k} - \mathbf{G})$ and $(\mathbf{k} - \mathbf{G}')$. For practical computation, it is convenient to sum the series in l for the term involving $j'_l(|\mathbf{k} - \mathbf{G}|R)$ by an identity that combines it with the term in $j_1(|\mathbf{G} - \mathbf{G}'|R)$. The result is a manifestly Hermitian matrix:

$$\Gamma_{\mathbf{G}\mathbf{G}'} = \frac{4\pi R^2}{\Omega} \left\{ -\left[(\mathbf{k} - \mathbf{G}) \cdot (\mathbf{k} - \mathbf{G}') - \mathcal{E} \right] \frac{j_1(|\mathbf{G} - \mathbf{G}'|R)}{|\mathbf{G} - \mathbf{G}'|} \right.$$

$$\left. + \sum_{l=0}^{\infty} (2l+1) \, P_l(\cos\theta_{\mathbf{G}\mathbf{G}'}) \, j_l(|\mathbf{k} - \mathbf{G}|R) \, j_l(|\mathbf{k} - \mathbf{G}'|R) \frac{\mathcal{R}'(R,\mathcal{E})}{\mathcal{R}(R,\mathcal{E})} \right. \qquad (3.74)$$

This is referred to as the *APW formula*.

The prescription for a band structure calculation by the APW method is fairly straightforward. The crystal structure defines a reciprocal lattice, from which we choose a finite set of Gs. The radial Schrödinger equation is solved at an energy $\mathcal{E}$, and the corresponding logarithmic derivatives evaluated on the surface of the MT sphere. For a chosen value of k, we evaluate the various matrix elements $\Gamma_{\mathbf{G}\mathbf{G}'}$, and then calculate the secular determinant of the matrix. The value of $\mathcal{E}$ is then adjusted until the determinant vanishes. This procedure demands considerable computing power, but is found to converge to physically acceptable values under almost all circumstances.

3.2.5 The Tight-Binding Method

Our physical intuition that a crystal is merely an assembly of atoms, brought close together and allowed to interact, is expressed mathematically by trying to represent the Bloch functions as *linear combination of atomic orbitals* (LCAO). Occupied valence atomic orbitals have the property that the radial function $\mathcal{R}_{nl}(r) \to 0$ as $r \to \infty$. We thus assume that our basis functions, $\phi_{nL}(\mathbf{x})$, $L \equiv (\ell, m)$, are restricted to bound states of the free atom potential $v_a(r)$.

The most general Bloch function that can be constructed out of these functions is of the form

$$\Psi_{\mathbf{k}}^{LCAO} = \sum_{nL} \alpha_{nL} \sum_{\mathbf{R}} \exp(i\mathbf{k} \cdot \mathbf{R}) \, \phi_{nL}(\mathbf{x} - \mathbf{R}), \qquad (3.75)$$

Substituting (3.75) as a trial function in the expectation value of the Hamiltonian of the crystal

$$\langle \Psi | \mathcal{H} - \mathcal{E} | \Psi \rangle = 0$$

$$\mathcal{H} = -\nabla^2 + \sum_{\mathbf{R}} v_a(|\mathbf{x} - \mathbf{R}|)$$

and minimizing with respect to the coefficients α_{nL}, we find

$$\sum_{nL'} \left\{ (\mathcal{E}_{nL} - \mathcal{E}) \, S_{nL,n'L'}(\mathbf{k}) + V_{nLn'L'}(\mathbf{k}) \right\} \alpha_{n'L'} = 0, \qquad (3.76)$$

for all values of n and L included in the sum (3.75). We used the relation

$$\left[-\nabla^2 + v_a(r) \right] \phi_{nL}(\mathbf{x}) = \mathcal{E}_{nL} \, \phi_{nL}(\mathbf{x}).$$

Because of periodic translational symmetry, the coefficients in (3.76) are expressed as lattice Fourier transforms, namely,

$$S_{nL,n'L'}(\mathbf{k}) = \sum_{\mathbf{R}} \exp(i\mathbf{k} \cdot \mathbf{R}) \, S_{nL,n'L'}(\mathbf{R}),$$

$$S_{nL,n'L'}(\mathbf{R}) = \int d\mathbf{x} \, \phi_{nL}(\mathbf{x}) \, \phi_{n'L'}(\mathbf{x} + \mathbf{R}),$$

$$V_{nL,n'L'}(\mathbf{k}) = \sum_{\mathbf{R}} \exp(i\mathbf{k} \cdot \mathbf{R}) \, V_{nL,n'L'}(\mathbf{R}),$$

$$V_{nL,n'L'}(\mathbf{R}) = \sum_{\mathbf{R}' \neq 0} \int \phi_{nL}(\mathbf{x}) \, v_a(\mathbf{x} + \mathbf{R}') \, \phi_{n'L'}(\mathbf{x} + \mathbf{R}) \, d\mathbf{x}$$

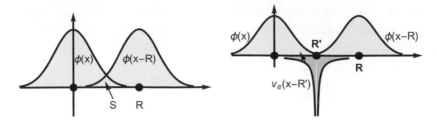

Figure 3.8 Schematic representation of $S(\mathbf{R})$ and one of the components of $V(\mathbf{R})$.

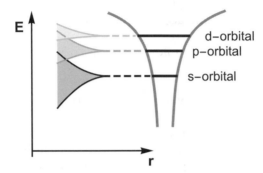

Figure 3.9 Broadening of atomic energy levels (s, p, and d orbitals) into bands with a decrease in atomic separation $\mathbf{r}$.

These integrals involving the overlap of functions centered on two or three neighboring sites, as shown in Figure 3.8, are obviously rather nasty, and can only be evaluated with great labor. Nevertheless, the preceding formulation seems quite explicit, and was regarded in the early days as capable of giving a solution to the band structure problem to an acceptable degree of accuracy.

This is the algebraic justification of one of the standard elementary principles of solid-state physics – that as the atoms come together, each of their bound states become broadened into a band of Bloch functions of appropriate symmetry, as shown in Figure 3.9. The broadening occurs, quite sharply, because bound-state functions decay exponentially outside the atomic sphere, so the overlap integrals are very small until the atoms come close together. Then, as the bands arising from distinct levels begin to overlap one another in energy, they *hybridize* into the linear combination that would be obtained by solving the prevoius equations for α_{nL}. We thus refer to the *s-p band*, the *3d band*, etc., identifying the bands by the atomic levels from which they are ultimately derived.

This confidence in the LCAO method has been, to a certain degree, misplaced. The overlap of wavefunctions eventually leads to an overlap of potentials, destroying the bound states (except for resonances) and liberating the electrons into *nearly free* bands in which the quantum numbers of the atomic levels are almost irrelevant, as illustrated in Figure 3.10.

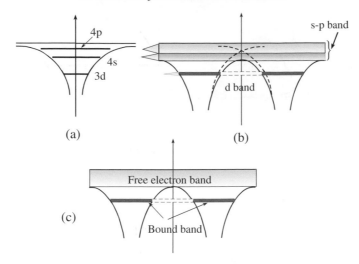

Figure 3.10 Emergence of free electron band and resonances in solids.

We should also remember that the representation (3.75) is functionally incomplete: the set of bound states of the atomic potential $v_a(r)$ does not contain the continuum of *scattered free electron* eigenfunctions of the Schrödinger equation at positive energies. It is impossible, therefore, to describe states in a metal or semiconductor in and above the valence and conduction bands, since they resemble free electron waves in the interstitial space. We, therefore, need to abandon the notion of calculating band structures by the LCAO method, yet we can still learn something from the general form of (3.76). First, we note that the nonorthogonality of the basis functions (3.75), as a consequence of nonorthogonality of atomic orbitals centered on different sites, complicates the solution of (3.76). But this is not an essential complication, since we can justifiably replace the atomic orbital basis with a Wannier-like one and set

$$S_{nL,n'L'}(\mathbf{k}) = \delta_{nL,n'L'}. \tag{3.77}$$

Moreover, we can apply symmetry constraints to define a powerful selection rule that can limit the number of the $V_{nL,n'L'}(\mathbf{k})$ coefficients for a particular lattice type. Thus, there need only be a few distinct coefficients of importance. This viewpoint leads us to one of the standard *interpolation schemes or parametric representations* of band structure theory, the *tight-binding* method. In this scheme, the energy $\mathcal{E}(\mathbf{k})$ becomes a root of the secular equation

$$\det\left|\left\{\mathcal{E}_{nL} - \mathcal{E}(\mathbf{k}))\right\} \delta_{nL,n'L'} + V_{nLn'L'}(\mathbf{k})\right| = 0, \tag{3.78}$$

where $V_{nLn'L'}(\mathbf{k})$ are then treated as parameters to be adjusted empirically to bring $\mathcal{E}(\mathbf{k})$ into agreement with the experiment. It must be emphasized, however, that the actual values of the parameters arrived at in this way cannot be interpreted physically as overlap integrals of atomic orbitals. To a large extent, this type of representation is imposed upon the band structure by the symmetry of the crystal.

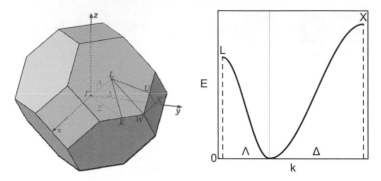

Figure 3.11 Tight-binding bands along the Δ-direction (001) and Λ-direction (111). The points $X = (0,0,2\pi)$ and $L = (\pi,\pi,\pi)$ with $a = 1$.

Example: Case of many monovalent metals

There is only a single nearly free electron conduction band to be fitted in this case. $\mathcal{E}(\mathbf{k})$ must necessarily be periodic in the reciprocal lattice, and thus it must have Fourier representation in the direct lattice, namely,

$$\mathcal{E}(\mathbf{k}) = \mathcal{E}_0 + \sum_{\mathbf{R}\neq0} \exp(i\mathbf{k}\cdot\mathbf{R})\,\mathcal{E}(\mathbf{R}).$$

But the Brillouin zone has the point group symmetries of the crystal, so that many of the coefficients $\mathcal{E}(\mathbf{R})$ must be equal to one another, or zero.

In the case of an fcc lattice, $\mathcal{E}(\mathbf{R})$ must have the same value for all nearest neighbors. Thus, the leading terms would take the form

$$\mathcal{E}(\mathbf{k}) = \mathcal{E}_0 + 4\mathcal{E}(\text{nn}) \left\{ \cos\left(\frac{k_y a}{2}\right)\cos\left(\frac{k_z a}{2}\right) + \cos\left(\frac{k_z a}{2}\right)\cos\left(\frac{k_x a}{2}\right) \right.$$
$$\left. + \cos\left(\frac{k_x a}{2}\right)\cos\left(\frac{k_y a}{2}\right) \right\}.$$

The corresponding energy bands along the Δ $(00k)$, and Λ (k,k,k) are shown in Figure 3.11, it behaves like a simple free electron energy momentum relation over a wide range of energy. Indeed, by the addition of further terms in $\mathcal{E}(\mathbf{x})$, the distorted and multiply connected Fermi surfaces of noble metals can fitted exactly in this way.

Bond Orbitals, Chemical Bonds, and Their Local Site Symmetry

This approach is quite useful in defining and determining the matrix elements between atomic-like Wannier orbitals. It also provides a pictorial approach to the construction of chemical bonds, such as the tetrahedral ones in C, Si, and Ge.

To determine the interaction matrix element between two orbitals $|\phi_L\rangle$ and $|\phi_{L'}\rangle$, we define orbital quantization with respect to the axis joining the centers of the two interacting

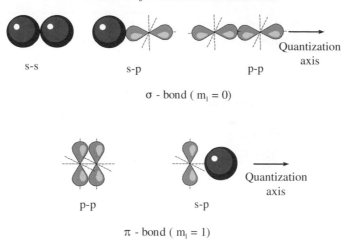

Figure 3.12 Quantization of m_ℓ along the interatomic axis uses the designation $\sigma \Rightarrow m_\ell = 0$, and $\pi \Rightarrow m_\ell = 1$.

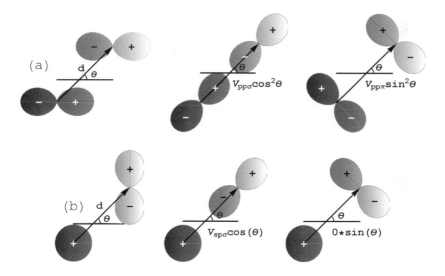

Figure 3.13 (a) pp-interactions are decomposed into a $pp\sigma$ ($m_\ell = 0$) and $pp\pi$ ($m_\ell = 1$) types; (b) $sp\sigma$ correspond to $m_\ell = 0$, while $sp\pi$ vanishes since it involves interaction between orbits of different ms.

orbitals. As we show in Figure 3.12, when the two orbitals share the same value of m_ℓ with respect to the common axis, we refer to the geometry as σ, when $m_\ell = 0$, and π, when $m_\ell = 1$. In the case when $m_\ell \neq m_{\ell'}$, such as the case of $sp\pi$, the interaction vanishes.

In the general case, shown in Figure 3.13, the orbitals and the location of their centers are defined with respect to a fixed Cartesian coordinate system. We choose the direction of the vector, **d**, joining their atomic centers as the axis of quantization. Taking the functional form of the spherical harmonics in Cartesian coordinates, namely

$$p_x = x/r, \quad p_y = y/r, \quad p_z = z/r, \tag{3.79}$$

we obtain the geometric decompositions shown in Figure 3.13 in terms of the matrix elements V_{ss}, $V_{sp\sigma}$, $V_{pp\sigma}$, and $V_{pp\pi}$.

Harrison's Scaling of Matrix Elements

By comparing electronic band structures calculated using tight-binding matrix elements V_{ss}, $V_{sp\sigma}$, $V_{pp\sigma}$, and $V_{pp\pi}$ with those based on plane-wave methods, Harrison [85] was able to establish the following scaling formula

$$V_{\ell\ell'} = \eta_{\ell\ell'} \, \frac{\hbar^2}{m_e d^2}$$

with $\hbar^2/m_e = 7.62\,\text{eV-Å}^2$. The numerical factor η depends on the angular type of the Wannier orbitals ϕ_ℓ, $\phi_{\ell'}$ and the directional cosines l_x, l_y, l_z of the vector **d** joining the nearest-neighbor sites. The values of η for s and p_x, p_y, p_z orbitals are given according to Harrison by the following expressions:

$$\eta_{s,s} = -1.32$$

$$\eta_{s,i} = 1.42\, l_i = -\eta_{i,s}, \qquad\qquad i = p_x, \, p_y, \, p_z$$

$$\eta_{i,i} = 2.22\, l_i^2 - 0.63\,(1 - l_i^2), \qquad i = p_x, \, p_y, \, p_z$$

$$\eta_{i,j} = 2.85\, l_i\, l_j, \, i \neq j, \qquad\qquad i = p_x, \, p_y, \, p_z$$

Note that reversing the order of orbitals multiplies ℓ_i by -1.

A similar argument can be given for interactions involving d-orbitals, and since these orbitals have $m_\ell = 2$, we introduce a new interaction δ that involve d-orbitals with $m_\ell = 2$ along the quantization axis, and using the d-orbital forms

$$d_1 = 3\frac{z^2}{r^2} - 1, \quad d_2 = \frac{x^2}{r^2} - \frac{y^2}{r^2}, \quad d_3 = \frac{xy}{r^2}, \quad d_4 = \frac{xz}{r^2}, \quad d_5 = \frac{yz}{r^2}. \tag{3.80}$$

The results are shown in Figure 3.14.

Harrison has also provided matrix element scaling for sd-, pd-, and dd interactions:

$$V_{ddm} = \eta_{ddm}\,\frac{\hbar^2\, r_d^3}{m_e d^5}, \quad \begin{cases} \eta_{dd\sigma} = -45/\pi \\ \eta_{dd\pi} = 30/\pi \\ \eta_{dd\delta} = -15/2\pi \end{cases} \qquad V_{ldm} = \eta_{ldm}\,\frac{\hbar^2 r_d^{3/2}}{m_e d^{7/2}}, \quad \begin{cases} \eta_{sd\sigma} = -3.16 \\ \eta_{pd\sigma} = -2.95 \\ \eta_{pd\pi} = 1.36 \end{cases}$$

r_d are atom specific effective d orbital radii, tabulated in Harrison's book.

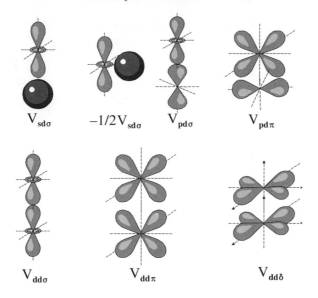

Figure 3.14 *sd*-, *pd*-, and *dd* interactions.

sp₂ and sp₃ Hybridization

Tight-binding calculations involving crystals with local $\mathcal{D}_3$ (graphite and BN) or tetrahedral site symmetry (diamond, zincblende, and wurtzite) can be simplified by the construction of symmetry-adapted orbitals, called hybrids, out of s and p orbitals.

Under planar symmetry in the xy-plane, we construct out of s, p_x, p_y, using $\mathcal{D}_3$ projection operators, three orthonormal hybrid orbitals, called sp_2 (see Figure 3.15), of the form

$$|h_1\rangle = \frac{1}{\sqrt{3}}\left(|s\rangle + |p_x\rangle + |p_y\rangle\right), \quad |h_{2,3}\rangle = \frac{1}{\sqrt{3}}\left(|s\rangle \pm |p_x\rangle \mp |p_y\rangle\right).$$

Similarly, for tetrahedral site symmetry we obtain the sp_3 orthonormal orbitals

$$|h_1\rangle = \frac{1}{2}\left(|s\rangle + |p_x\rangle + |p_y\rangle + |p_z\rangle\right), \qquad |h_2\rangle = \frac{1}{2}\left(|s\rangle + |p_x\rangle - |p_y\rangle - |p_z\rangle\right),$$

$$|h_3\rangle = \frac{1}{2}\left(|s\rangle - |p_x\rangle + |p_y\rangle - |p_z\rangle\right), \qquad |h_4\rangle = \frac{1}{2}\left(|s\rangle - |p_x\rangle - |p_y\rangle + |p_z\rangle\right).$$

3.2.6 Transition Metals: *d*-Bands and Resonances

We have been looking so far at an oversimplified picture of the electronic structure – narrow tight-bound bands below the muffin-tin zero, and nearly free electron bands above. In transition metals, these categories become blurred: a narrow band arising from the d-levels of the atoms lies within a broad band of s-electrons, as shown in Figure 3.16.

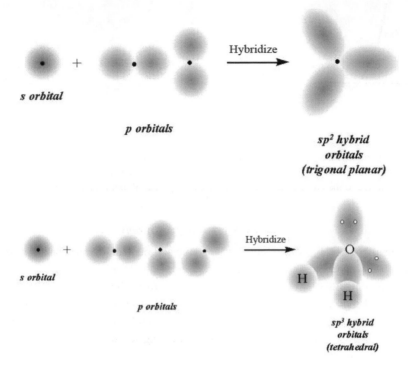

Figure 3.15 Top: *sp2* hybridization. Bottom: *sp3* hybridization.

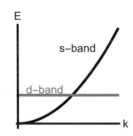

Figure 3.16 Crossing of atomic and free like bands.

Even without the additional complications of magnetic polarization in some $3d$ transition metals, this situation requires careful analysis.

In view of our previous arguments, this whole interpretation seems to be forbidden. It was maintained that all atomic bound states would disappear above the muffin-tin zero, leaving only nearly free Bloch states. What did we overlook?

The answer is that the atomic levels of high angular momentum are not completely destroyed by the overlap of potentials, but become *virtual* or *resonance* levels.

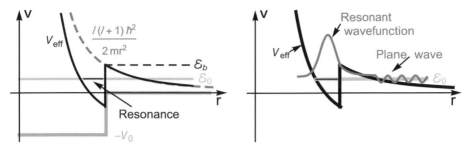

Figure 3.17 Effective potential in the presence of a centrifugal barrier (dashed gray). The light gray line represents a resonance level. For simplicity, the atomic potential is depicted as a square well. The electron is confined to the atomic annulus of the effective potential, but it can tunnel out.

Resonance Scattering from Atoms

This phenomenon, which is quite familiar in atomic and nuclear physics, arises as follows: The radial Schrödinger equation contains the term $\ell(\ell + 1)/r^2$, which behaves like the potential of a centrifugal force repelling the electron from the region of the nucleus. In a bound state of high angular momentum ($\ell \geq 2$), the electron becomes confined to the annular space between this barrier and the ordinary external Coulomb potential of the atom.

Resonances for $\ell > 0$ occur according to the following scenario:

The effective radial potential becomes

$$V_{\text{eff}}(r) = V(r) + \frac{l(l + 1)\hbar^2}{2mr^2}. \tag{3.81}$$

As shown in Figure 3.17, a particle with energy $E_0 > 0$ but less than the barrier height can tunnel through the barrier. It forms a metastable state, since the particle *trapped* inside can tunnel out.

Resonance Line Shape and Phase Shift (see Figure 3.18)

At the resonance energy E_0, the phase shift δ_ℓ goes through $\pi/2$

$$\delta_\ell(E_0) = \frac{\pi}{2}.$$

The Taylor series expansion of $\cot \delta_\ell(E)$ in the vicinity of E_0 is

$$\cot \delta_\ell(E) = \cot \delta_\ell(E_0) - \left(\frac{1}{\sin^2 \delta_\ell} \frac{d\delta_\ell}{dE}\right)_{E=E_0} \Delta E = -\frac{2}{\Gamma} \Delta E$$

and

$$f_\ell(k) = \frac{1}{k} e^{i\delta_\ell} \sin \delta_\ell = \frac{1}{k} \frac{\sin \delta_\ell}{\cos \delta_\ell - i \sin \delta_\ell} = \frac{1}{k} \frac{1}{\cot \delta_\ell - i}$$

$$= \frac{1}{k} \frac{1}{\frac{-2}{\Gamma} \Delta E - i} = -\frac{1}{k} \frac{\Gamma/2}{\Delta E + i\frac{\Gamma}{2}}.$$

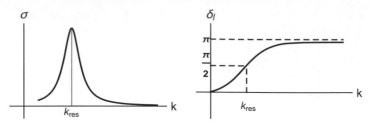

Figure 3.18 Left: cross-section resonance profile. Right: the corresponding phase shift.

The partial cross-section is given by the resonance line shape:

$$\sigma_\ell \simeq \frac{4\pi(2l+1)}{k^2} \frac{\Gamma^2/4}{(\Delta E)^2 + (\Gamma^2/4)}.$$

When atoms are brought together, this outer barrier may not be completely lost, but may still interpose a hill through which the electron in the original atomic level must tunnel if it is to escape. Thus, although we may now be above the muffin-tin zero, we find a strong tendency for the wavefunction to concentrate within the atom as we pass through the energy $\mathcal{E}_d$. This effect cannot be perfectly sharp (as it would be for a true bound state) but must spread over a width W, which would depend in detail on the characteristics of the centrifugal barrier and of the self-consistent potential of the atom (see Figure 3.19).

The simplest description is to treat the *d-electrons* and the *s-electrons* as distinct entities that occasionally interact or interchange (*s-d* scattering). The *s*-electrons are then assumed to be quite free, with energy $\propto k^2$, while the five degenerate d states, $\phi_m(\vec{r})$, of each atom are combined into a typical tight-bound band of the type previously discussed. The *d*-states of the atom have the same energy, $\mathcal{E}_d$, and in the crystal environment can be represented by a 5×5 matrix

$$\mathcal{H}_{mm'} = \mathcal{E}_d\,\delta_{mm'} + V_{mm'}(\mathbf{k}), \tag{3.82}$$

while for the *s*-states we employ the nearly free electron representation, namely, the matrix

$$\mathcal{H}_{\mathbf{GG'}} = \left|\mathbf{k} - \mathbf{G}\right|^2 \delta_{\mathbf{GG'}} + \Gamma_{\mathbf{GG'}}. \tag{3.83}$$

The matrix elements $\Gamma_{\mathbf{GG'}}$ are often treated as empirically adjustable parameters.

To combine the two systems, we construct our Bloch functions as linear combinations of d atomic orbitals and plane waves, namely,

$$\Psi_{\mathbf{k}}(\mathbf{x}) = \sum_{\mathbf{x}} e^{i\mathbf{k}\cdot\mathbf{x}} \sum_{m} \alpha_m\,\phi_m(\mathbf{x}-\mathbf{x}) + \sum_{\mathbf{G}} \alpha_{\mathbf{k}-\mathbf{G}}\,\psi_{\mathbf{k}-\mathbf{G}}, \tag{3.84}$$

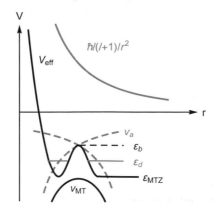

Figure 3.19 The addition of the centrifugal energy $\ell(\ell + 1)/r^2$ to an atomic potential v_a gives rise to an effective potential with a bound state at $\mathcal{E}_d$. Overlapping to produce muffin-tin potentials v_{MT} turns this into a resonance.

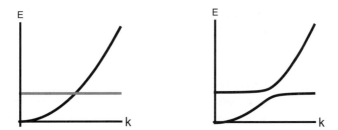

Figure 3.20 Crossing of atomic and free like bands, $\gamma_{m\mathbf{G}'} = 0$ (left). Hybridized bands, $\gamma_{m\mathbf{G}'} \neq 0$ (right).

and we construct the Hamiltonian matrix

$$\begin{pmatrix} \mathcal{H}_{\mathbf{G}\mathbf{G}'} & \gamma_{m\mathbf{G}'} \\ \gamma_{\mathbf{G}'m} & \mathcal{H}_{mm'} \end{pmatrix}$$

This Hamiltonian is called the *model Hamiltonian* [134].

The $\gamma_{m\mathbf{G}}$'s, representing the tunneling, are called hybridization coefficients. If they are zero, we are back to the simple model of noninteracting s and d bands; otherwise, these have the effect of distorting the simple bands a little, and splitting them apart at points in **k** space where they cross, as shown in Figure 3.20. A wavefunction of the form (21.25) – something like a bound state with a free electron part outside – is therefore a reasonable trial function for the Bloch function.

Figure 3.21 shows the electronic band structure of Cu along high-symmetry directions [39]. If you follow the two bands labeled Δ_1, Λ_1 and σ_1 you will notice that they result from avoided crossing.

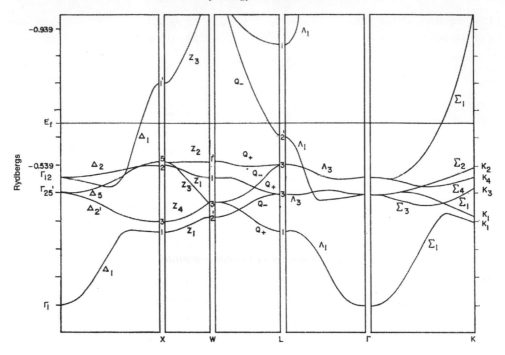

Figure 3.21 Electronic band structure of Cu, from Burdick [39].

3.3 Pseudopotentials

3.3.1 OPW Revisited

We recall that the OPW wavefunction $\chi_{\mathbf{k}-\mathbf{G}}(\mathbf{x})$ was constructed as a linear combination of a single plane wave and the set of core wavefunctions $\phi^{cL}(\mathbf{x})$:

$$\chi_{\mathbf{k}-\mathbf{G}}(\mathbf{x}) = e^{i(\mathbf{k}-\mathbf{G})\cdot\mathbf{x}} - \sum_{cL} \mu_{cL}^{\mathbf{G}} \, \Psi_{\mathbf{k}}^{cL}(\mathbf{x}), \tag{3.85}$$

$$\Psi_{\mathbf{k}}^{cL}(\mathbf{x}) = \frac{1}{\sqrt{N}} \sum_{l} e^{i\mathbf{k}\cdot\mathbf{R}_l} \, \phi^{cL}(\mathbf{x} - \mathbf{R}_l).$$

$\Psi_{\mathbf{k}}^{cL}$ is a Bloch sum of core states that satisfy the crystal Schrödinger equation

$$\mathcal{H} \, \Psi_{\mathbf{k}}^{cL}(\mathbf{x}) = \mathcal{E}_{c,L} \, \Psi_{\mathbf{k}}^{cL}(\mathbf{x}), \tag{3.86}$$

The orthogonalization condition rendered

$$\mu_{cL}^{\mathbf{G}} = \left\langle \Psi_{\mathbf{k}}^{cL}(\mathbf{x}) \left| e^{i(\mathbf{k}-\mathbf{G})\cdot\mathbf{x}} \right. \right\rangle. \tag{3.87}$$

The valence wavefunction was then constructed as a linear combination of the OPWs

$$\Psi(\mathbf{k}, \mathbf{r}) = \sum_{\mathbf{G}} \alpha_{\mathbf{k}-\mathbf{G}} \, \chi_{\mathbf{k}-\mathbf{G}}(\mathbf{x}). \tag{3.88}$$

Unfortunately, a variational calculation for the coefficients $\alpha_{\mathbf{k}-\mathbf{G}}$ runs into the complication that this basis set is *overcomplete* and its functions are not orthogonal to one another:

$$\langle \chi_{\mathbf{k}-\mathbf{G}} \mid \chi_{\mathbf{k}-\mathbf{G}'} \rangle = \delta_{\mathbf{G}\mathbf{G}'} - \sum_{cL} \langle \mathbf{k} - \mathbf{G} \mid cL \rangle \langle cL \mid \mathbf{k} - \mathbf{G}' \rangle, \qquad (3.89)$$

where $|cL\rangle \equiv \Psi_{\mathbf{k}}^{cL}(\mathbf{x})$. The matrix for the coefficients $\alpha_{\mathbf{k}-\mathbf{G}}$, whose secular determinant must vanish at the energy eigenvalue $\mathcal{E}$, is of the form

$$\langle \chi_{\mathbf{k}-\mathbf{G}} | \mathcal{H} - \mathcal{E} | \chi_{\mathbf{k}-\mathbf{G}'} \rangle = \{ |\mathbf{k} - \mathbf{G}|^2 - \mathcal{E} \} \delta_{\mathbf{G}\mathbf{G}'} + V_{\mathbf{G}-\mathbf{G}'}$$
$$+ \sum_{cL} (\mathcal{E} - \mathcal{E}_{cL}) \langle \mathbf{k} - \mathbf{G} | cL \rangle \langle cL | \mathbf{k} - \mathbf{G}' \rangle. \qquad (3.90)$$

The difficulty with nonorthogonality can be taken care of by standard algebraic procedures.

3.3.2 Concept of Pseudopotentials

The concept of a pseudopotential [89] can be developed from the idea of orthogonality between valence and core states, which has been cast above in terms of OPWs. Thus, its precise, but least *instructive* form, emerges from the structure of the secular determinant in (3.90), when we lump the last term with the crystal potential $V_{\mathbf{G}-\mathbf{G}'}$ to form a *crystal pseudopotential*

$$\langle \chi_{\mathbf{k}-\mathbf{G}} | \mathcal{H} - \mathcal{E} | \chi_{\bar{k}-\mathbf{G}'} \rangle = \left\{ |\mathbf{k} - \mathbf{G}|^2 - \mathcal{E} \right\} \delta_{\mathbf{G}\mathbf{G}'}$$
$$+ \left\langle \mathbf{k} - \mathbf{G} \left| \left[V(\mathbf{x}) + \sum_{cL} |cL\rangle (\mathcal{E} - \mathcal{E}_{cL}) \langle cL| \right] \right| \mathbf{k} - \mathbf{G}' \right\rangle$$
$$= \langle \mathbf{k} - \mathbf{G} | - \nabla^2 + \Gamma - \mathcal{E} | \mathbf{k} - \mathbf{G}' \rangle, \qquad (3.91)$$

from which a new crystal potential emerges whose Fourier components are

$$\Gamma_{\mathbf{G}\mathbf{G}'} = \left\langle \mathbf{k} - \mathbf{G} \left| V(\mathbf{x}) + \sum_{cL} |cL\rangle (\mathcal{E} - \mathcal{E}_{cL}) \langle cL| \right| \mathbf{k} - \mathbf{G}' \right\rangle. \qquad (3.92)$$

We note that the *core-orthogonality term* is always *positive* and, therefore, can go a long way toward *canceling the negative Fourier coefficients of the attractive potential V*, rendering $\Gamma_{\mathbf{G}\mathbf{G}'}$ *weak enough to be treated as a perturbation on the free electron states*.

The real mental revolution that becomes apparent is to regard (3.92) as merely a *plane-wave representation* of

$$\Gamma = V + V_R, \qquad (3.93)$$

whose repulsive part is the operator

$$V_R = \sum_{cL} |cL\rangle (\mathcal{E} - \mathcal{E}_{cL}) \langle cL|. \qquad (3.94)$$

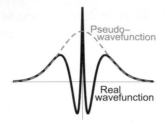

Figure 3.22 Radial (solid) and pseudo wavefunctions (dashed), showing an example of a smooth function that equals the full function outside the core region.

Consequently, the Schrödinger equation

$$\left(-\nabla^2 + V\right) \Psi = \mathcal{E}\, \Psi, \tag{3.95}$$

for the true Bloch function Ψ is thereby transformed symbolically into the equation

$$\left(-\nabla^2 + \Gamma\right) \Phi = \mathcal{E}\, \Phi, \tag{3.96}$$

for the *pseudowavefunction* Φ. Equations (3.95) and (3.96) are *isospectral* – they have identical sets of eigenvalues. A rendering of Ψ and Φ is shown in Figure 3.22.

Thus, having determined, in principle, the coefficients $\alpha_{\mathbf{k}-\mathbf{G}}$ in the conventional OPW procedure, we find Φ to be just the plane-wave part of Ψ, namely,

$$\Phi(\mathbf{k}, \mathbf{x}) = \sum_{\mathbf{G}} \alpha_{\mathbf{k}-\mathbf{G}}\, e^{i(\mathbf{k}-\mathbf{G})\cdot\mathbf{x}}. \tag{3.97}$$

Muffin-Tin Representation of Γ

The crystal potential V can be expressed as a superposition of muffin-tin potentials centered on lattice sites (j). It is then obvious that Γ can be written as

$$\Gamma = \sum_{j}\left[v_a(\mathbf{x} - \mathbf{R}_j) + \sum_{cL} \phi^*_{cL}(\mathbf{x} - \mathbf{R}_j)\left(\mathcal{E} - \mathcal{E}_{cL}\right) \phi_{cL}(\mathbf{x}' - \mathbf{R}_j)\right]$$

$$= \sum_{j} \gamma_j^a, \tag{3.98}$$

where γ_j^a is the atomic pseudopotential centered on site R_j, and takes the form

$$\gamma_j^a = v_a(\mathbf{x} - \mathbf{R}_j) + \sum_{cL} \big|cL(j)\big\rangle \left(\mathcal{E} - \mathcal{E}_{cL}\right) \big\langle cL(j)\big|, \tag{3.99}$$

with $\big|cL(j)\big\rangle \equiv \phi^{cL}(\mathbf{x} - \mathbf{R}_j)$. The worst feature of this form is that the atomic pseudopotential is not *local*. This can be seen if we go back to the basic conventions of the Dirac notation:

$$\gamma^a\, \Phi(\mathbf{x}) = v_a(\mathbf{x})\, \Phi(\mathbf{x}) + \sum_{cL} \left(\mathcal{E} - \mathcal{E}_{cL}\right) \int d\mathbf{x}'\, \psi^*_{cL}(\mathbf{x}')\, \Phi(\mathbf{x}')\, \psi_{cL}(\mathbf{x})$$

$$= v_a(\mathbf{x})\, \Phi(\mathbf{x}) + \int v(\mathbf{x}, \mathbf{x}')\, \Phi(\mathbf{x}')\, d\mathbf{x}'. \tag{3.100}$$

Thus, the repulsive part of the atomic pseudopotential can be expressed only as an operator when acting on any spatially varying function that it encounters in the algebra. As an elementary algebraic consequence of this *nonlocality*, a matrix element of the pseudopotential between two plane waves is not just a function of the difference of the two wavevectors. In general,

$$\gamma_{\mathbf{G}\mathbf{G}'}^{a} \equiv \langle \mathbf{k} - \mathbf{G} | \gamma^{a} | \mathbf{k} - \mathbf{G}' \rangle \neq \gamma^{a}(\mathbf{G} - \mathbf{G}'), \tag{3.101}$$

for all different values of $\mathbf{G}$ and $\mathbf{G}'$, so that an *atomic form factor* cannot be defined uniquely. On the other hand, it is worth noting that the pseudopotential (3.99) is *diagonal in the angular momentum quantum numbers of the core states* $|\phi^{cL}\rangle$. One may, therefore, separate γ_j^a into a sum of terms

$$\gamma_j^a = \gamma_s + \gamma_p + \gamma_d + \cdots, \tag{3.102}$$

where γ_s acts only on the part of the wavefunction transforming with s-like symmetry, etc. This is of course a consequence of the assumption of spherical symmetry of the muffin-tin. From these considerations and from the obvious dependence of Γ on the energy $\mathcal{E}$ of the state under study, it was viewed, up to 1980, that *the real value of the pseudopotential lies in the semiquantitative justification it provides for nearly free electron interpolation schemes with adjustable parameters.*[1] A serious problem with this pseudoatom representation, however, is that the pseudopotential is not unique. This follows essentially from the overcompleteness of the OPWs.

3.3.3 The Scattering Approach

A correct pseudopotential was defined as one that gave the same energy levels as the real potential, at least for a range of states of interest, with the emphasis on transforming the Schrödinger equation from (3.95) to (3.96). In this section, the final end will be the same, but the approach will lie in scattering theory.

Consider a single muffin-tin scattering center, where the potential $v_{\mathrm{mt}}(\mathbf{x})$ is taken as zero beyond the muffin-tin radius R_c. The radial wavefunction $\mathcal{R}(r, \mathcal{E})$, which is a solution of the radial equation

$$-\frac{1}{2r^2} \frac{d}{dr} \left[r^2 \frac{d\mathcal{R}_l}{dr} \right] + \left[\frac{l(l+1)}{2r^2} + v_{\mathrm{mt}}(\mathbf{x}) \right] \mathcal{R}_l^{\mathrm{in}} = \mathcal{E}\, \mathcal{R}_l^{\mathrm{in}}, \tag{3.103}$$

inside the muffin-tin radius R_c, is matched at R_c onto the solution for $r > R_c$,

$$\mathcal{R}_l^{\mathrm{out}} \propto \left[j_l(\kappa r) - \tan \eta_l\, n_l(\kappa r) \right], \tag{3.104}$$

where $\mathcal{E} = \frac{1}{2}\kappa^2$. Dropping the superscripts, we write the condition that both function and first derivative are continuous, as

---

[1] The amount of computation required for an accurate first-principle calculation of the band structure of a perfect metal or semiconductor is not reduced below that of the corresponding OPW equations to which it is exactly equivalent.

$$L_l \equiv \frac{\mathcal{R}'_l(R_c, \mathcal{E})}{\mathcal{R}_l(R_c, \mathcal{E})}$$

$$= \frac{\kappa \, j'_l(\kappa R_c) - \kappa \, \tan \eta_l \, n'(\kappa R_c)}{j_l(\kappa R_c) - \tan \eta_l \, n(\kappa R_c)}, \qquad (3.105)$$

which yields

$$\tan \eta_l(\mathcal{E}) = \frac{j_l(\kappa R_c) L_l - \kappa j'_l(\kappa R_c)}{n_l(\kappa R_c) L_l(\mathcal{E}) - \kappa n'_l(\kappa R_c)}. \qquad (3.106)$$

The phase shifts $\eta_l(\mathcal{E})$ suffice to determine the scattering amplitude

$$f(\theta) = (2i\kappa)^{-1} \sum_l (2l+1) \left[\exp(2i\eta_l) - 1 \right] P_l(\cos \theta), \qquad (3.107)$$

for an incident plane wave of energy $\mathcal{E}$ being scattered through an angle θ. The crucial point here is that η_ls may be written in the form

$$\eta_l = n_l \pi + \delta_l, \qquad (3.108)$$

where n_l is an integer chosen so that the reduced phase shift δ_l lies in the range $|\delta_l| \leq \pi/2$, or $0 \leq \delta < \pi$. We note that the integer n_l *counts the number of radial nodes in the core wiggles of* $\mathcal{R}_l(r)$.

The simplified example shown in Figure 3.23 shows that as we increase the magnitude of the square potential from zero to $-V_0$, the attractive potential pulls in more and more of the wavefunction so that the *inner core oscillations may be considered in this sense to have been pulled in by phase shifting the free wave outside.* It follows that n_l *is also equal to the number of core states of angular momentum* l.

We now define a pseudopotential as one *whose complete phase shifts are equal to* $\delta_l(\mathcal{E})$ *without the* $n_l\pi$, so that it has no core states. Equation (3.107) shows that it gives the same amplitude $f(\theta)$ as the original potential since a factor $\exp(2in_l\pi) = 1$ has no effect.

The pseudowavefunction has no radial nodes, i.e., no rapid oscillations, so that we expect a rapidly convergent series of plane waves to give a good representation of it. Figure 3.24 shows the case of the $3s$ valance state of Al. It is also important to note that

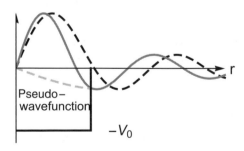

Figure 3.23 Attractive potential well. Radial wavefunction: gray (dashed black), in the presence (absence) of the potential well. The pseudo wavefunction is shown in dashed gray.

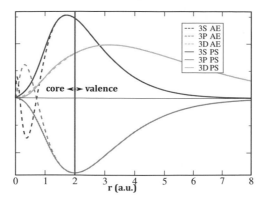

Figure 3.24 Radial wavefunctions for the real potential (dashed) and the pseudowavefunction (solid), from Paulatto, *Pseudopotential Methods for DFT Calculations*.

the *pseudo radial wavefunction* derived from the pseudopotential has the same *logarithmic derivative* $L_l(\mathcal{E})$ as the real wavefunction, the integer multiples of π again having no effect.

3.3.4 Construction of Pseudopotentials

We surmise that the pseudopotential V_{ps} can be constructed to satisfy our needs, for example we can require that it reproduces some measured quantities (*empirical* approaches), or to be the smoothest and weakest possible, while maintaining the same scattering properties of the core potential on the valence electrons (*ab initio* approaches). We shall start with describing empirical approaches [45].

The Direct Empirical Method

1. On-the-Fermi-Sphere Approximation and the Local Pseudopotential
As we have found previously, pseudopotentials are **nonlocal operators** so that they should be strictly written

$$V(\mathbf{x}, \mathbf{x}'; \mathcal{E}) = \sum_{\mathbf{R}} v(\mathbf{x} - \mathbf{R}, \mathbf{x}' - \mathbf{R}; \mathcal{E}), \qquad (3.109)$$

which is a function of seven variables $\mathbf{x}, \mathbf{x}'$, and $\mathcal{E}$, namely, two vectors and $\mathcal{E}$. In the secular determinant, we require the matrix element of (3.109) between plane waves,

$$\langle \mathbf{k} - \mathbf{G} | V(\mathbf{x}, \mathbf{x}'; \mathcal{E}) | \mathbf{k} - \mathbf{G}' \rangle. \qquad (3.110)$$

It is useful to use the abbreviation

$$\mathbf{K} = \mathbf{k} - \mathbf{G}, \quad \mathbf{K}' = \mathbf{k} - \mathbf{G}'; \quad \text{and} \quad \mathbf{K} - \mathbf{K}' = \mathbf{q} \equiv \mathbf{G}''.$$

We shall use $\mathbf{q}$ as the general scattering vector and $\mathbf{G}$ when it is necessarily equal to a reciprocal lattice vector. We shall also often take the dependence on $\mathcal{E}$ as implicit.

We now introduce the simplifying assumption of nonoverlapping spherical muffin-tin potentials, so that the matrix element becomes

$$\langle \mathbf{K} | V | \mathbf{K}' \rangle = S(\mathbf{q})\, v(\mathbf{K}, \mathbf{K}', \mathcal{E}), \qquad (3.111)$$

where

$$S(\mathbf{q}) = \frac{1}{N} \sum_{\mathbf{R}} \exp(-i\mathbf{q} \cdot \mathbf{R}), \qquad (3.112)$$

is just the *structure factor* encountered in diffraction theory; while $v(\mathbf{K}, \mathbf{K}', \mathcal{E})$ is the atomic *pseudopotential form factor*,

$$
\begin{aligned}
v(\mathbf{K}, \mathbf{K}', \mathcal{E}) &= \langle \mathbf{K} | v(\mathbf{x}, \mathbf{x}', \mathcal{E}) | \mathbf{K}' \rangle \\
&= \Omega_c^{-1} \int_{\Omega_c} d\mathbf{x}\, d\mathbf{x}'\, \exp(-i\mathbf{K} \cdot \mathbf{x})\, v(\mathbf{x}, \mathbf{x}', \mathcal{E})\, \exp(i\mathbf{K}' \cdot \mathbf{x}'), \qquad (3.113)
\end{aligned}
$$

where Ω_c is the atomic volume. The assumption of nonoverlapping spherical muffin-tin potentials $v_{\mathrm{mt}}(r, r', \mathcal{E})$, reduces the number of variables in v in (3.111) from seven to four $(r, r', \theta_{\mathbf{x}, \mathbf{x}'},$ and $\mathcal{E})$. With

$$e^{i\mathbf{k} \cdot \mathbf{x}} = 4\pi \sum_{l} \sum_{m} i^l\, j_l(kr)\, Y_{lm}^*(\hat{\mathbf{k}})\, Y_{lm}(\hat{\mathbf{x}})$$

$$P_l(\cos \gamma) = \frac{4\pi}{2l + 1} \sum_{m} Y_{lm}^*(\hat{\mathbf{x}})\, Y_{lm}(\hat{\mathbf{x}}'),$$

we obtain

$$
\begin{aligned}
v(\mathbf{K}, \mathbf{K}', \mathcal{E}) &= \frac{(4\pi)^2}{\Omega_c} \sum_{\substack{lm \\ l'm'}} \iint_{\Omega_c} d\mathbf{x}\, d\mathbf{x}'\, i^{l+l'}\, j_l(Kr)\, Y_{lm}^*(\hat{\mathbf{K}})\, Y_{lm}(\hat{\mathbf{x}}) \\
&\qquad\qquad v(r, r', \mathcal{E})\, j_{l'}(Kr')\, Y_{l'm'}^*(\hat{\mathbf{K}}')\, Y_{l'm'}(\hat{\mathbf{x}}') \\
&\propto \sum_l P_l(\theta_{\mathbf{K}, \mathbf{K}'}) \iint dr\, dr'\, j_l(Kr)\, v(r, r', \mathcal{E})\, j_l(Kr')\, P_l(\theta_{\mathbf{x}, \mathbf{x}'})\, d\theta_{\mathbf{x}, \mathbf{x}'}.
\end{aligned}
$$

We come now to the most common approximation made in the fitting and use of pseudopotentials. The assumption of *on-the-Fermi-sphere approximation* reduces this further to one variable q by setting

$$|\mathbf{K}| = |\mathbf{K}'| = k_F; \qquad \mathcal{E} = \mathcal{E}_F. \qquad (3.114)$$

Equation (3.114) says that the electron is being scattered on the Fermi sphere.[2] Figure 3.25 shows, as an example, a Fermi surface with gaps due to a pair of Brillouin zone planes. It is clear that the shape of the surface near the zone planes is determined by the mixing of waves $|\mathbf{K}\rangle$ and $|\mathbf{K}'\rangle$ where $\mathbf{K}$ and $\mathbf{K}'$ vary closely around the particular geometry

---

[2] This is obviously exactly true for the scattering of electrons by phonons or impurities in the electrical resistivity, but is also approximately so for band structures.

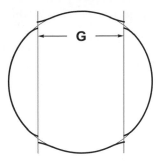

Figure 3.25 Free electron Fermi sphere (light black circle) and actual Fermi surface (thick black) due to a single pair of Brillouin zone planes separated by a reciprocal lattice vector **G**.

specified by (3.114). Similarly, when fitting a whole $\mathcal{E}(\mathbf{K})$ and not just the Fermi surface, the amplitude of the plane wave $|\mathbf{K}\rangle$ mixed into the wave $|\mathbf{K}'\rangle$ is given by

$$\frac{\langle \mathbf{K} | V | \mathbf{K}' \rangle}{\frac{1}{2}(K')^2 - \frac{1}{2}(K)^2}, \tag{3.115}$$

which is maximum at $K = K'$. Since the dependence of the pseudopotential on $\mathbf{K}, \mathbf{K}'$, and $\mathcal{E}$ is, of course, an expression of the fact that pseudopotentials are nonlocal operators, neglecting such dependence has sometimes been referred to as a *local* approximation.

2. Local Pseudopotentials for Zincblende Systems
We can generalize the form of the matrix element, in the local approximation, for a crystal with sublattices α as

$$\langle \mathbf{K} | V | \mathbf{K}' \rangle = \sum_\alpha S_\alpha(\mathbf{q}) \, v_\alpha(\mathbf{q}),$$

$$S_\alpha(\mathbf{q}) = N^{-1} \sum_{\mathbf{R}} \exp(-i\mathbf{q} \cdot (\mathbf{R} + \boldsymbol{\tau}_\alpha)),$$

$$v_\alpha(\mathbf{q}) = \Omega_c^{-1} \int v_\alpha(\mathbf{x}) \, \exp(i\mathbf{q} \cdot \mathbf{x}) \, d\mathbf{x}. \tag{3.116}$$

where $\boldsymbol{\tau}_\alpha$ is the basis vector in the unit cell for sublattice α. The lattice periodicity limits the Fourier components to reciprocal lattices. We consider the example of GaAs and other III–V as well as II–VI compounds with zincblende structure and basis vectors $\pm(1/8, 1/8, 1/8)$. We obtain

$$V(\mathbf{x}) = \sum_{\mathbf{G}} e^{-i\mathbf{G}\cdot\mathbf{x}} \left[\cos(\mathbf{G}\cdot\boldsymbol{\tau}) \underbrace{[V_{Ga}(\mathbf{G}) + V_{As}(\mathbf{G})]}_{V_S(\mathbf{G})} + i \sin(\mathbf{G}\cdot\boldsymbol{\tau}) \underbrace{[V_{Ga}(\mathbf{G}) - V_{As}(\mathbf{G})]}_{V_A(\mathbf{G})} \right]$$

where the form factors have been separated into symmetric and antisymmetric terms. The form factors V_α are treated as adjustable parameters. Only those corresponding to few **G**

Table 3.1 $V(\mathbf{G})$ *components to be included for zincblende crystals.*

G group $(2\pi/a)$	# permutations	Total number of elements	$\frac{|\mathbf{G}|^2}{(2\pi/a)^2}$
$(0,0,0)$	1	1	0
$(1,1,1)$	8	9	3
$(2,0,0)$	6	15	4
$(2,2,0)$	12	27	8
$(3,1,1)$	24	51	11
$(2,2,2)$	8	59	12
$(4,0,0)$	6	65	16
$(3,3,1)$	24	89	19

Table 3.2 *Pseudopotential form factors, in Rydbergs, derived from the experimental energy band splittings.*[a]

	V_3^S	V_8^S	V_{11}^S	V_3^A	V_8^A	V_{11}^A
Si	−0.21	+0.04	+0.08	0	0	0
Ge	−0.23	+0.01	+0.06	0	0	0
Sn	−0.20	0.00	+0.04	0	0	0
GaP	−0.22	+0.03	+0.07	+0.12	+0.07	+0.02
GaAs	−0.23	+0.01	+0.06	+0.07	+0.05	+0.01
AlSb	−0.21	+0.02	+0.06	+0.06	+0.04	+0.02
InP	−0.23	+0.01	+0.06	+0.07	+0.05	+0.01
GaSh	−0.22	0.00	+0.05	+0.06	+0.05	+0.01
InAs	−0.22	0.00	+0.05	+0.08	+0.05	+0.03
InSb	−0.20	0.00	+0.04	+0.06	+0.05	+0.01
ZnS	−0.22	+0.03	+0.07	+0.24	+0.14	+0.04
ZnSe	−0.23	+0.01	+0.06	+0.18	+0.12	+0.03
ZnTe	−0.22	0.00	+0.05	+0.13	+0.10	+0.01
CdTe	−0.20	0.00	+0.04	+0.15	+0.09	+0.04

[a] † [44]

vectors are needed. Table 3.1 shows that only $V(\mathbf{G})$ components up to $\mathbf{G} = (3,1,1)$ need be considered.

We solve the secular equations

$$\left[|\mathbf{k} - \mathbf{G}|^2 - \mathcal{E}(\mathbf{k})\right]\alpha_{\mathbf{k}-\mathbf{G}} + \sum_{\mathbf{G}'} V\left(\mathbf{G} - \mathbf{G}'\right)\alpha_{\mathbf{k}-\mathbf{G}'} = 0$$

Table 3.2 lists V_G^S and V_G^A for several crystalline systems with diamond and zincblende structures.

Model Pseudopotentials

The scattering properties of the potential depend purely on *the logarithmic derivative at the radius R_c and not on any specific details of the potential or the wavefunction inside.* Consequently, the pseudopotential may be chosen to have any convenient shape, for example a square well, provided it has some parameters adjusted to give the correct scattering. Since the adjustment has to be made separately for each l, a convenient form often is

$$v_{ps} = \sum_l f_l(r)\,\mathcal{P}_l, \tag{3.117}$$

where $f_l(r)$ is a spherically symmetric potential and $\mathcal{P}_l$ a projection operator which picks out the lth angular momentum component of the wavefunction so that $f_l(r)$ acts only on that component. This approach gained traction in the mid-1960s and the 1970s and was coined *model pseudopotentials.* A few representative model pseudopotentials are discussed here.

1. The KKRZ Pseudopotential of Ziman [207]

Our first application of these ideas is to derive the so-called KKRZ pseudopotential of Ziman, in which v_{ps} for each atom is taken as zero everywhere except for a delta function of strength B_l at radius R_c:

$$v_{ps}^{KKRZ} = \sum_l B_l\,\delta(r - R_c)\,\mathcal{P}_l. \tag{3.118}$$

This leads to a set of radial equation of the form

$$-\frac{1}{2r^2}\frac{d}{dr}\left[r^2\frac{d\mathcal{R}_l}{dr}\right] + \left[\frac{\hbar^2\ell(\ell+1)}{2mr^2} + v_{ps}^{KKRZ}\right]\mathcal{R}_l = \mathcal{E}\,\mathcal{R}_l. \tag{3.119}$$

Inside the sphere the radial pseudowavefunction satisfies the radial equation

$$-\frac{1}{2r^2}\frac{d}{dr}\left[r^2\frac{d\mathcal{R}_{in}}{dr} + \frac{\hbar^2\ell(\ell+1)}{2mr^2}\mathcal{R}_{in}\right] = \mathcal{E}\,\mathcal{R}_{in}, \tag{3.120}$$

so that $\mathcal{R}_{in}$ is just $j_l(\kappa r)$, and the effect of the delta function is to give it a kink of the right amount to make the derivative outside equal to $L_l j_l(\kappa r)$ as required, where L_l is specified by the original potential. To determine B_l, we integrate (3.119) from $r = R_c - \epsilon$ to $r = R_c + \epsilon$ and obtain

$$\mathcal{R}'_{out} - \mathcal{R}'_{in} = 2B_l\,\mathcal{R}, \tag{3.121}$$

giving

$$L_l(\mathcal{E}) - \kappa\,\frac{j'_l(\kappa R_c)}{j_l(\kappa R_c)} = 2B_l. \tag{3.122}$$

2. Heine–Abarenkov Pseudopotential [1]

Our second application will be the pseudopotential of Heine and Abarenkov, v_{ps}^{HA} (see Figure 3.26). The aim is to calculate a pseudopotential for a single bare closed-shell ion

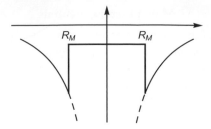

Figure 3.26 The model pseudopotential of Heine and Abarenkov for a bare ion.

such as Al^{+3}, which has the configuration $1s^2 2s^2 2p^6$. The model radius R_M is chosen at some convenient value greater than that of the core radius R_c, and outside R_M the pseudopotential is

$$v_{ps}^{\text{HA}} = -z/r, \qquad r > R_M, \tag{3.123}$$

the bare Coulomb potential of the ion of charge z. Inside R_M, it is taken as a constant

$$v_{ps}^{\text{HA}} = -\sum_l A_l(\mathcal{E}) \, \mathcal{P}_l, \qquad r < R_M \tag{3.124}$$

adjusted to give the observed energy levels $\mathcal{E}_{3s}$, $\mathcal{E}_{3p}$, $\mathcal{E}_{4s}$, etc., of an extra electron in the field of the ion. The latter are taken from spectroscopic measurements on free ions, and the procedure ensures that the pseudopotential gives the right logarithmic derivative, at least at these energies. In practice, for a given l, the value of $A_l(\mathcal{E})$ obtained from different energies, for example, $\mathcal{E}_{3s}$, $\mathcal{E}_{4s}$, $\mathcal{E}_{5s}$, differ slightly so that $A_l(\mathcal{E})$ has to be considered a weak function of energy. In accordance with the principle of pseudizing, the smallest possible value is always chosen of course.

3. Ashcroft's Empty-Core Pseudopotential [20]

Several other model-potential schemes are to be found in the literature. For example, Ashcroft has suggested that the atomic form factor of a bare ion may be adequately represented by the Fourier transform of an effective *empty-core* potential

$$
\begin{aligned}
v^{\text{ion}}(\mathbf{x}) &= 0; & r &< R_c \\
&= -\frac{z}{r}; & r &> R_c.
\end{aligned}
\tag{3.125}
$$

R_c is an adjustable parameter that has approximately the radius of the physical atomic core.

First Principles Pseudopotential

The first-principles approach to the pseudopotential construction was initially proposed by Hamann and coworkers at Bell Labs in 1979 [84]. As shown in Figure 3.27, the procedure starts with calculating the orbital eigenfunctions and eigenenergies of all occupied states of

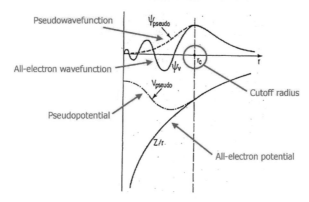

Figure 3.27 Ab initio calculations.

a given atom with the aid of DFT atomic computer codes; namely, solving the Kohn–Sham equations after invoking spherical symmetry. The solutions will have the form $\mathcal{R}_l \, Y_{lm}$, where $\mathcal{R}_l$ is the solution to

$$\left[\frac{1}{r}\frac{d^2}{dr^2} - \frac{l(l+1)}{r^2}\right]\mathcal{R}_l + \left[v_{\mathrm{H}}(r) - \frac{Ze^2}{r} + \frac{\delta G}{\delta n} - \mathcal{E}_{nl}\right]\mathcal{R}_l = 0.$$

This process is sometimes referred to as *all-electron* calculations. Next, the appropriate atomic valence configuration is identified, for example $3s^2 3p^2$ for Si, and a core radius R_c is set. The construction of an atomic pseudopotential is then carried out for the valence electrons only, such that

- The real (or all electron) and pseudo states have the same eigenvalues.
- The real and pseudowavefunctions agree beyond R_c.
- The following Normconserving condition exists:

$$\int_0^{R_c} dr\, r^2 \, |\mathcal{R}_{\mathrm{real}}(r)|^2 = \int_0^{R_c} dr\, r^2 \, |\mathcal{R}_{\mathrm{PS}}(r)|^2.$$

- The logarithmic derivatives and their first energy derivative of real and pseudowavefunctions match at the cutoff radius.

1. Matching logarithmic derivatives guarantee that the real and pseudowavefunctions are the same outside the cutoff radius.
2. Matching the energy derivative of the logarithmic derivative guarantees the above property holds for a larger range of energies. The energy range in which the logarithmic derivatives coincide give an estimate of the pseudopotential quality.

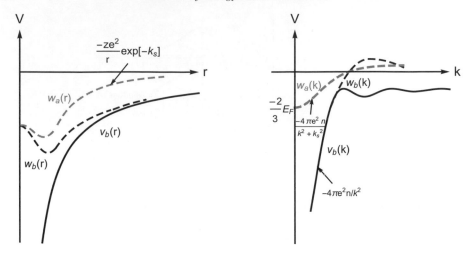

Figure 3.28 Schematic representation, (a) in real space, and (b) in reciprocal space, of transformation of bare atomic potential v_b into bare pseudopotential w_b, followed by screening to make *pseudoatom* potential w_a.

Screening

Once the pseudopotential of the bare ion is obtained, the bare ions can be planted into an electron gas that then screens them to give the total pseudopotential of the whole solid (see Figure 3.28). The point is that the screening by the electron gas can be treated by perturbation theory, the plane waves of the free electron gas being the unperturbed pseudowavefunctions. While pseudizing is the formal justification for attempting perturbation theory at all, how good is it in practice?

We start by considering the pseudopotential of the bare ion, e.g., Al^{+3}, with only tightly bound closed-shell core electrons. We plant the ions at sites x_j in a uniform electron gas of density appropriate to the number of the outer electrons, assuming the electron density to remain constrained as uniform for the moment. The potential in the system is then

$$(const) + \left(\sum_{\alpha} [v_{\alpha}^{ion}(q) S_{\alpha}(\mathbf{q})] \right) \exp(i\mathbf{q} \cdot \mathbf{x}). \qquad (3.126)$$

In reality, the electron gas relaxes to screen the ionic pseudopotential. As we have seen in Section 2.2., in lowest-order perturbation theory the screening is linearly proportional to the bare potential, so that the final potential is

$$V(\mathbf{x}) = \left(\sum_{\alpha} \frac{v_{\alpha}^{ion}(q) S_{\alpha}(\mathbf{q})}{\epsilon(q)} \right) \exp(i\mathbf{q} \cdot \mathbf{x}), \qquad (3.127)$$

where

$$\epsilon(q) = 1 - \left(\frac{8\pi e^2}{\Omega q^2}\right) \left[1 - \frac{q^2}{2(q^2 + k_F^2 + k_s^2)} \right] \chi(q),$$

$$\chi(q) = -\frac{z}{2}\frac{3}{2E_{F0}} \left(\frac{1}{2} + \frac{4k_F^2 - q^2}{8qk_F}\right) \ln\left|\frac{2k_F + q}{2k_F - q}\right| \tag{3.128}$$

is the Lindhard function, with z the number of valance electrons per atom, and $E_{F0} = k_F^2/2m_e$. The factor in square brackets in (3.128) comes from screened exchange, and k_s is the screening parameter $(2k_F/\pi)^{1/2}$ in atomic units.

In the limit as $q \to 0$, the Fourier transform of an ionic potential is always dominated by the Coulombic part outside the pseudizing radius R_c. We therefore obtain

$$v^{\text{ion}}(q \to 0) = -\frac{4\pi z}{\Omega q^2},$$

$$v(q \to 0) = -\frac{2}{3}E_{F0} + O(q^2). \tag{3.129}$$

Using the *empty core* model for the bare pseudopotential, we get

$$v(q) = -\frac{4\pi z}{q^2\epsilon(q)\Omega} \cos(qR_c), \tag{3.130}$$

and $\epsilon(q)$ has the form derived in Section 2.1. Notice that the limit of $q \to 0$

$$v(q \to 0) = -\frac{2}{3} E_{F0} + O(q^2). \tag{3.131}$$

Summary Points

 (i) What have we gained by transferring to a pseudopotential is a *secular equation* that is user-friendly. In an exact sense, this is nearly all we have gained, but then that is what is basically used in fitting band structures.

 (ii) By removing the $n_l\pi$ from the phase shifts, *all rows of the periodic table are treated on equal footing*. This makes a discussion possible of systematic trends as one goes down one group of the table. It also allows one to estimate the perturbation involved in substituting one atom for another.

(iii) In practice, one has gained far more. There are a number of situations where the pseudopotential is sufficiently weak for *perturbation theory* to be rapidly convergent, although there are also others where it may not be so useful: for example, one may fit the magnitude of a pseudopotential entirely empirically, but it may be adequate to estimate from a perturbation formulation how it changes with pressure!

In any case, perturbation theory allows one to calculate the self-consistent screening of the ions by the outer electrons, including exchange and correlation effects, for an arbitrary disposition of atomic sites and for mixtures of atomic species, which is quite out of the question in any other way.

(iv) The price that one has to pay for these gains is that pseudopotentials are *nonlocal operators*(!), but there are no major difficulties in handling them.

Exercises

3.1 Show that for the simple case of a monatomic crystal, the Fourier transform of the crystal potential

$$V(\mathbf{x}) = \sum_{\mathbf{R}} v(\mathbf{x} - \mathbf{R})$$

is of the form $F(\mathbf{q}) = S(\mathbf{q}) f(\mathbf{q})$, where

$$S(\mathbf{q}) = \frac{1}{N} \sum_{\mathbf{R}} e^{i\mathbf{q}\cdot\mathbf{R}}, \qquad \text{Structure factor,}$$

$$v(\mathbf{q}) = \frac{1}{\Omega_c} \int d\mathbf{x}\, v(\mathbf{x})\, e^{i\mathbf{q}\cdot\mathbf{x}}, \qquad \text{Form factor,}$$

N being the number of primitive calls in the crystal, and Ω_c being the primitive cell volume.

3.2 Consider a one-dimensional electronic system subject to an external potential with periodicity a, and represented by the Hamiltonian

$$\mathcal{H} = -\frac{\hbar^2}{2m_e} \nabla^2 + V(x)$$

$$V(x) = V(x + na).$$

Using Bloch's theorem, an eigenfunction of $\mathcal{H}$ can be written as

$$\Psi_{nk}(x) = \sum_G c_n(k - G)\, e^{i(k-G)x},$$

where n is the band index, $G = \dfrac{2m\pi}{a}$, and

$$u_{nk}(x) = \sum_G c_n(k - G)\, e^{-iGx}$$

has periodicity a.

(a) Determine the set of coupled equations you obtain when you vary

$$\langle \Psi_k | \mathcal{H} | \Psi_k \rangle - E \, \langle \Psi_k | \Psi_k \rangle.$$

(b) If $V(x) = 0$, we have the case of the *empty lattice*:

  1. What are the eigenvalues and the corresponding normalized eigenfunctions?

2. Draw the lowest three free electron energy bands (empty lattice) folded in the first BZ. (Set $a = 1$ and $\hbar^2/2m_e = 1$ for convenience.)

(c) If

$$V(x) = V_1 \cos\left(\frac{2\pi x}{a}\right) + V_2 \cos\left(\frac{4\pi x}{a}\right),$$

where V_1 and V_2 are given in units of $\hbar^2/2m_e$, construct the secular equation using the three plane waves with the lowest energy:

$$e^{ikx}, \quad e^{i(k-G_1)x}, \quad e^{i(k+G_1)x}.$$

(d) Determine and draw the corresponding dispersions of the lowest three free electron bands, given $V_1 = 2$, $V_2 = 3$.

3.3 If the one-dimensional crystalline potential is given as

$$V(x) = aV_0 \sum_n \delta(x - na),$$

determine the energy gaps between the bands, assuming that the nearly free electron approximation applies.

3.4 Draw the lowest two empty-lattice free electron bands for the two-dimensional square lattice along the symmetry directions Γ-X, Γ-M, and X-M shown in the BZ of Figure 3.29.

3.5 Estimate the form of the $2s$ electronic band of Li using a single OPW orthogonalized to the $1s$ core state:

$$\phi_{1s}^c(\mathbf{r}) = \sqrt{\alpha^3/\pi} \, \exp(-\alpha r).$$

Given that

$$E_{1s} = -1.883 \text{ au}, \quad V_0 = -0.5 \text{ au},$$

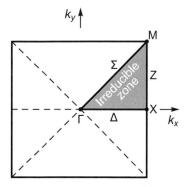

Figure 3.29 Two-dimensional square lattice Brillouin zone and its irreducible zone.

the volume of the unit cell $\Omega_c = 153$ au, the bottom of the band energy $E(0) = -0.343$ au, and $\alpha = 2.7$ au.

(Assume $\left\langle \phi^c_{1s}(\mathbf{r} - \mathbf{R}) \middle| \phi^c_{1s}(\mathbf{r} - \mathbf{R}') \right\rangle = \delta_{\mathbf{R},\mathbf{R}'}$.)

(For numerical evaluation, it is convenient to use atomic units by setting $\hbar = m = e = 1$:

- Distances are measured in Bohr radii ($a_B = 0.529$ Å).
- Masses in units of the electron mass.
- Energies are in atomic units [1 atomic unit = 2 Rydbergs = 27.2 eV].)

3.6   One-dimensional model of ionic solids:

A   B   A   B   A   B   A

n−2   n−2   n−1   n−1   n   n   n+1

⊢ d ⊣⊢ d ⊣

Figure 3.30   A composite one-dimensional crystal of period $a = 2d$ consisting of two types of atoms.

Consider the periodic arrangement of two types of atoms shown in Figure 3.30, where each atom has one orbital and one electron, with atom A being the cation and B the anion ($\epsilon_A > \epsilon_B$). Setting

$$\langle \psi_A | \mathcal{H} | \psi_A \rangle = \epsilon_A, \quad \langle \psi_B | \mathcal{H} | \psi_B \rangle = \epsilon_B$$
$$\left\langle \psi_{A,n} \middle| \mathcal{H} \middle| \psi_{B,n} \right\rangle = \left\langle \psi_{B,n} \middle| \mathcal{H} \middle| \psi_{A,n+1} \right\rangle = V_2,$$

derive an expression for the energy dispersion of its electronic bands. Plot the corresponding band dispersions.

3.7   One-dimensional solid with two electrons per primitive cell:

Consider a one-dimensional solid, of period a, consisting of a single atom type, but now each atom has a single s and a single p orbital, with two electrons per atom (see Figure 3.31).

Setting

$$\left\langle \psi_{np} \middle| \mathcal{H} \middle| \psi_{np} \right\rangle = \epsilon_p, \quad \langle \psi_{ns} | \mathcal{H} | \psi_{ns} \rangle = \epsilon_s$$
$$\left\langle \psi_{np} \middle| \mathcal{H} \middle| \psi_{n\pm1,p} \right\rangle = V_{pp\sigma} > 0, \quad \langle \psi_{ns} | \mathcal{H} | \psi_{n\pm1,s} \rangle = V_{ss\sigma} < 0$$
$$\langle \psi_{ns} | \mathcal{H} | \psi_{n+1,p} \rangle = -\langle \psi_{ns} | \mathcal{H} | \psi_{n-1,p} \rangle = V_{sp\sigma} > 0.$$

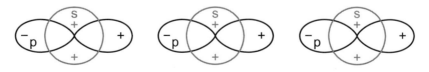

Figure 3.31   one-dimensional solid with sp hybrids.

(a) Construct the Bloch functions associated with the s and p orbitals.
(b) Determine the set of coupled equations obtained by minimizing the energy expectation value, assuming that orbitals on different sites are orthogonal.
(c) Derive an expression for the dispersion of the electronic bands, and plot the dispersion curves for the two sets of parameters:

1. Parameter set I:

$$\epsilon_s = -10.725 \text{ eV}, \ \epsilon_p = -3.525 \text{ eV}, \ V_{ss\sigma} = -2.08 \text{ eV},$$
$$V_{pp\sigma} = 3.49 \text{ eV}, \ V_{sp\sigma} = 2.24 \text{ eV}$$

Identify the maximum energy of occupied states.
2. Parameter set II:

$$\epsilon_s = -9.725 \text{ eV}, \ \epsilon_p = -4.525 \text{ eV}, \ V_{ss\sigma} = -4.294 \text{ eV},$$
$$V_{pp\sigma} = 5.2 \text{ eV}, \ V_{sp\sigma} = 2.24 \text{ eV}$$

Identify the valence band maximum and conduction band minimum in set II.

3.8 δ-function potential:

Each atom in a one-dimensional chain of lattice spacing a is represented by a potential

$$V(x) = a V_0 \, \delta(x), \quad V_0 > 0.$$

(a) Determine the ground-state wavefunction and energy of the single-atom problem, taking, in this case, $V_0 < 0$.
(b) If the atoms are located at positions na, n integer, write down the wavefunction for $0 < x < a$ in terms of $K = \sqrt{2mE/\hbar^2}$.
(c) Write down the Bloch function for this system, using the fact that the wavefunction of part (b) is periodic.
(d) Write and apply the boundary conditions at $x = na$ to the Bloch function.
(e) Show that the electron energy E and the wavevector k satisfy the relation

$$\cos(ka) = \frac{2m a V_0}{\hbar^2} \frac{1}{2K} \sin(Ka) + \cos(Ka).$$

3.9 Wannier functions:

Consider the 1D Bloch state $\psi_{\mathbf{k}}(x) = \frac{1}{\sqrt{L}} e^{ikx}$, where the system length L contains N lattice sites with periodicity a. Determine the corresponding Wannier functions $\phi(x - R)$ and check their orthonormality.

3.10 Ashcroft's empty-core model pseudopotential:

Consider Ashcroft's screened empty-core pseudopotential

$$
v(r) = \begin{cases} 0 & r < r_c \\ \dfrac{ze^2}{r}\, e^{-\lambda_{\mathrm{TF}} r} & r > r_c \end{cases}
$$

(a) Derive its reciprocal space representation.

(b) Obtain its form in the limit $q \to 0$.

3.11 KKRZ model pseudopotential:

(a) Show that the matrix element of v_{ps}^{KKRZ} between plane waves is

$$
V_{ps,GG'}^{\mathrm{KKRZ}} = \frac{4\pi R_c^2}{\Omega} \sum_l (2l+1) \left[L_l - \kappa\, \frac{j_l'(\kappa R_c)}{j_l(\kappa R_c)} \right]
$$
$$
\times\, j_l\big(|\mathbf{k}-\mathbf{G}|R_c\big)\, j_l\big(|\mathbf{k}-\mathbf{G}'|R_c\big)\, P_l(\cos\theta_{\mathbf{GG'}}), \tag{3.132}
$$

(b) Show that the term in square brackets in (3.132) can be written as

$$
-(1/\kappa)[R_c\, j_l(\kappa R_c)]^{-2}\, \tan\eta_l',
$$

where the modified phase shift η_l' is defined by

$$
\cot\eta_l' = \cot\eta_l - n_l(\kappa R_c)/j_l(\kappa R_c),
$$

which readily follows from the relation

$$
\tan\eta_l(\mathcal{E}) = \frac{j_l(\kappa R_c)L_l - \kappa j_l'(\kappa R_c)}{n_l(\kappa R_c)L_l(\mathcal{E}) - \kappa n_l'(\kappa R_c)}
$$

and using the Wronskian identity

$$
x^2\,(jn' - j'n) = 1.
$$

(c) Finally show that

$$
V_{ps,GG'}^{\mathrm{KKRZ}} = -\frac{4\pi}{\kappa\Omega} \sum_l (2l+1)\, \frac{j_l\big(|\mathbf{k}-\mathbf{G}|R_c\big)\, j_l\big(|\mathbf{k}-\mathbf{G}'|R_c\big)}{j_l^2(\kappa R_c)}\, P_l(\cos\theta_{\mathbf{GG'}})\, \cot\eta_l'.
$$

3.12 Graphene electronic band structure:

Graphene has a honeycomb lattice structure, as shown in Figure 3.32.

There are two carbon atoms per primitive cell, and each atom has four valence orbitals: $2s$, $2p_x$, $2p_y$, and $2p_z$. In graphene, the $2s$, $2p_x$, and $2p_y$ orbitals combine or hybridize to form a new three-planar orbital with angle $120°$ between each other. This is called sp_2 hybridization in which three valence electrons from each atom contribute to covalence bonds with three nearest neighbors. The remaining electron

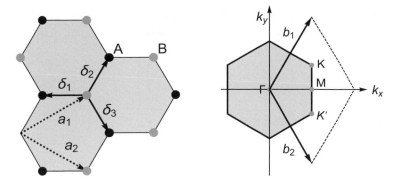

Figure 3.32 Honeycomb lattice and its Brillouin zone.

occupies the p_z orbital, and is responsible for the interesting electronic properties of graphene.

(a) Use Harrison's scaling formulas to determine numerical values for $V_{ss\sigma}$, $V_{pp\sigma}$, $V_{sp\sigma}$, and $V_{pp\pi}$, given that the nearest-neighbor distance is $d = 1.54$ Å.

(b) Each atom has three nearest neighbors at

$$\delta_1 = 1.54\,(\pm 1/2, \sqrt{3/2}), \ \delta_2 = 1.54\,(\pm 1/2, -\sqrt{3/2}), \ \delta_2 = 1.54\,(\pm 1,0).$$

Write down an expression in terms of the wavevector **k** for

$$\left\langle \phi_s^A \,\middle|\, \mathcal{H} \,\middle|\, \phi_s^B \right\rangle, \ \left\langle \phi_s^A \,\middle|\, \mathcal{H} \,\middle|\, \phi_{p_i}^B \right\rangle, \ \left\langle \phi_{p_i}^A \,\middle|\, \mathcal{H} \,\middle|\, \phi_{p_j}^B \right\rangle, \ i,j = x,y$$

and

$$\left\langle \phi_{p_z}^A \,\middle|\, \mathcal{H} \,\middle|\, \phi_{p_z}^B \right\rangle.$$

(c) Construct the 8×8 energy eigenvalue matrix and determine the dispersions of the corresponding electron bands, given

$$\epsilon_s = -17.52, \ \epsilon_p = -8.97.$$

3.13 Electron and hole pockets:
Consider a two-dimensional hexagonal lattice of lattice spacing $a = 3$Å, and one electron per unit cell. If the electrons are considered free within the two-dimensional plane,

1. What is the Fermi energy E_F? (Provide a numerical answer in eV.)
2. If there were two electrons per unit cell:

   1. Draw the free electron Fermi surface in the reduced zone scheme.
   2. What is the area of the electron and hole pockets, i.e., the density of electrons and holes?

3.14 Electron in a two-dimensional weak sinusoidal potential:

Consider electrons moving in the two-dimensional weak periodic potential:

$$V(x, y) = U \left[\cos\left(\frac{2\pi x}{a} \right) + \cos\left(\frac{2\pi y}{a} \right) \right]. \qquad U > 0$$

1. Use the variational wavefunction

$$\Psi_{\mathbf{k}} = c_{\mathbf{k}}\, e^{i\mathbf{k}\cdot\mathbf{r}} + c_{\mathbf{k}-\mathbf{b}_x}\, e^{i(\mathbf{k}-\mathbf{b}_x)\cdot\mathbf{r}} + c_{\mathbf{k}-\mathbf{b}_y}\, e^{i(\mathbf{k}-\mathbf{b}_y)\cdot\mathbf{r}} + c_{\mathbf{k}-\mathbf{b}_x-\mathbf{b}_y}\, e^{i(\mathbf{k}-\mathbf{b}_x-\mathbf{b}_y)\cdot\mathbf{r}}$$

   to calculate the electronic band structure along the high-symmetry points and lines of the irreducible Brillouin zone shown in Figure 3.29, for $U = 1$ and $U = 4$ (where $\hbar^2/2m = 1$). $\mathbf{b}_x$ and $\mathbf{b}_y$ are the basis vectors of the reciprocal lattice. Set $a = 1$.

2. Determine the four lowest-energy single-electron eigenstates at the M-point and give the corresponding wave functions.

3. Assuming that there are two electrons per unit cell, what type of material do you obtain for each value of U given in step 1?

4. In the case the system is a metal with two electrons per unit cell, make a qualitatively correct sketch of the locations of the Fermi surfaces in the first Brillouin zone.

3.15 Reading of electronic band structures (see Figure 3.33):

Consider the band structure diagram given in Figure 3.34 for tellurium, which crystallizes in a hexagonal structure with three Te atoms/unit cell. The atomic configuration for tellurium is $5s^2 5p^4$.

(a) Sketch the approximate position of the Fermi level E_F on the band diagram in Figure 3.34 and give your reasons for this placement of E_F.

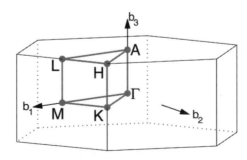

Figure 3.33 Hexagonal Brillouin zone, from [165]. Point A appears as Z in Figure 3.34.

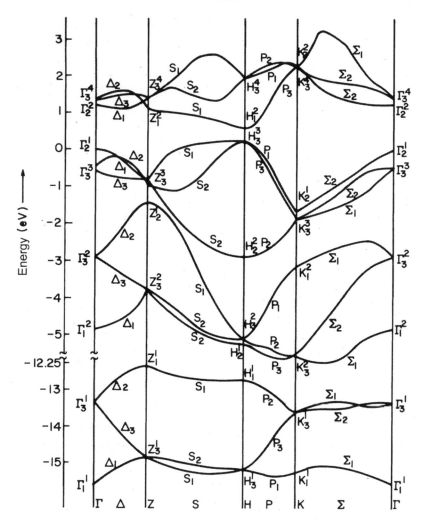

Figure 3.34 Electronic band structure of tellurium. From [46].

(b) Indicate which energy bands on the diagram correspond to the s, p, and d bands, given that

$$\Gamma_1 \to \quad x^2 + y^2 + z^2, \ z^2$$

$$\Gamma_2 \to \quad z, \ L_z$$

$$\Gamma_3 \to \quad \begin{cases} (x, y) \\ (L_x, L_y) \end{cases} \quad , \quad \begin{cases} (xz, yz) \\ (x^2 - y^2, xy), \end{cases}$$

where L_x, L_y, L_z are angular momentum components.

If the energy bands with these atomic origins are not shown, are they at higher or lower energy than those shown in the diagram?

(c) Given that z and the angular momentum L_z have the same transformation properties (they transform according to Γ_2), can you predict which symmetry is absent in the Te structure?

(d) Is tellurium transparent to visible light ($= 5,000\text{Å}$)? Explain!

4

Electrons and Band Theory: Effects of Spin–Orbit Interactions

4.1 Introduction

The first time we encounter *spin–orbit coupling* (SOC) is usually in an introductory quantum mechanics course, and in atomic physics. In solving the Schödinger equation for a hydrogen atom, we learn that in order to reproduce experimental spectra, we need to introduce additional terms in the Hamiltonian that emerge from relativistic effects. Among these effects, we discover that SOC splits the $\ell \neq 0$ orbital states; for example, the p-levels split into $p_{1/2}$ and $p_{3/2}$. This is just the fine structure of the hydrogen atom, a key evidence for the existence of electron *spin*. Further on, we learn that electron spin is an intrinsic property of the relativistic single particle theory of the Dirac equation, and that SOC is manifest in this context.

We find that spin and spin–orbit coupling play a critical role in many areas of current fundamental research and technological applications. One area is information technology, where it finds applications where the technique of encoding electron spin is used in data storage and manipulation. Another promising area is *spintronics*, which primarily relies on control of individual spins. Moreover, we note that spin is a vector quantity and, therefore, not only the magnitude but also the direction of a spin can be manipulated. In that sense, spintronics presents novel avenues for development in electronics technologies. These endeavors require understanding the underlying physical principles in the solid state, especially how SOC impacts electronic band structures – a topic that we address in this chapter.

We should also note that spin–orbit coupling has recently played an important role in emergent topological insulators, where SOC acts as an effective intrinsic magnetic field that leads to topological electronic states analogous to those produced by external magnetic fields in the quantum Hall effect, but in contrast it preserves time-reversal invariance. As we will demonstrate in Part II of this book, SOC leads to band inversions that create a state of matter characterized by new topological numbers and conjures the new phenomenon of quantum spin-Hall effect.

4.1.1 The Dirac Equation

To arrive at the correct spin-related terms in the nonrelativistic Hamiltonian, we appropriately take the Dirac equation as a starting point. The Dirac equation for an electron in an external vector potential $\mathbf{A}$, and a scalar potential V is given by

$$\mathcal{H}_D \, \Psi = +i\hbar \, \frac{\partial}{\partial t} \, \Psi = \mathcal{E}' \, \Psi$$

$$\mathcal{H}_D = \boldsymbol{\alpha} \cdot \left(c\mathbf{p} + e\,\mathbf{A}((\mathbf{x})) \right) - e\,V(\mathbf{x}) + \beta\,m_0 c^2, \tag{4.1}$$

where m_0 is the electron rest mass. $\boldsymbol{\alpha}$ and β are vector and scalar 4×4 matrices, respectively. The former is constructed in terms of Pauli spin matrices, while the latter is expressed in terms of the 2×2 unit matrix $\mathbb{I}_2$:

$$\alpha = \begin{pmatrix} \mathbf{0} & \boldsymbol{\sigma} \\ \boldsymbol{\sigma} & \mathbf{0} \end{pmatrix}, \quad \beta = \begin{pmatrix} \mathbb{I}_2 & \mathbf{0} \\ \mathbf{0} & \mathbb{I}_2 \end{pmatrix} \tag{4.2}$$

The Hamiltonian acts on a four-component (bispinor) wavefunction Ψ that describes particles with the total energy $\mathcal{E}'$, including the rest mass energy $m_0 c^2$. To obtain a spin–orbit coupling expression, we need to connect with the Schrödinger equation. To this end, we write the four-component wavefunction as two spinors dubbed *large* and *small*, and labeled ψ_A and ψ_B, respectively. We rewrite the Dirac equation as

$$\left(\mathcal{E}' - m_0 c^2 + eV(\mathbf{x}) \right) \psi_A = \boldsymbol{\sigma} \cdot \left(c\mathbf{p} + e\,\mathbf{A}((\mathbf{x})) \right) \psi_B$$

$$\left(\mathcal{E}' + m_0 c^2 + eV(\mathbf{x}) \right) \psi_B = \boldsymbol{\sigma} \cdot \left(c\mathbf{p} + e\,\mathbf{A}((\mathbf{x})) \right) \psi_A.$$

Setting $\mathcal{E} = \mathcal{E}' - m_0 c^2$, these equations can be recast as

$$\left(\mathcal{E} + eV(\mathbf{x}) \right) \psi_A = \boldsymbol{\sigma} \cdot \left(c\mathbf{p} + e\,\mathbf{A}((\mathbf{x})) \right) \psi_B \tag{4.3}$$

$$\left(\mathcal{E} + 2m_0 c^2 + eV(\mathbf{x}) \right) \psi_B = \boldsymbol{\sigma} \cdot \left(c\mathbf{p} + e\,\mathbf{A}((\mathbf{x})) \right) \psi_A. \tag{4.4}$$

In the nonrelativistic limit, we can assume that $\mathcal{E} + eV(\mathbf{x}) \ll 2m_0 c^2$, and obtain from (4.4) the approximate form for ψ_B:

$$\psi_B \approx \frac{1}{2m_0 c^2} \, \boldsymbol{\sigma} \cdot \left(c\mathbf{p} + e\,\mathbf{A}((\mathbf{x})) \right) \psi_A.$$

Inserting it into (4.3), we obtain the equation for ψ_A:

$$\left[\mathcal{E} + eV(\mathbf{x}) - \frac{1}{2m_0} \left(\mathbf{p} + \frac{e}{c}\mathbf{A}(\mathbf{x}) \right)^2 \right] \psi_A = 0. \tag{4.5}$$

We find that the large component becomes the wavefunction of the Schrödinger equation, the Pauli spin matrices are absent, and the only reminiscence of spin appears in the spinor character of ψ_A.

Figure 4.1 A rendering of an electron (red ball) moving in an electric field $\mathbf{E} = \nabla V$ near an ion core. $\mathbf{E}$ is Lorentz-transformed into a $\mathbf{B}$ field, in the electron's frame, by its orbital motion. The electron's spin, $\boldsymbol{\sigma}$, couples to the emerging $\mathbf{B}$ field.

The explicit appearance of the $\boldsymbol{\sigma}$ matrices requires applying a better approximation to the Dirac equation. We start with substituting the exact expression for ψ_B obtained from (4.4) in (4.3):

$$\left(\mathcal{E} + eV(\mathbf{x})\right)\psi_A = \boldsymbol{\sigma}\cdot\left(c\mathbf{p} + e\mathbf{A}((\mathbf{x})\right)\frac{1}{\mathcal{E} + 2m_0c^2 + eV(\mathbf{x})}\,\boldsymbol{\sigma}\cdot\left(\mathbf{p} + \frac{e}{c}\mathbf{A}(\mathbf{x})\right)\psi_A.$$

We retain terms up to order $(v/c)^2$ on the right-hand side, and obtain an equation for the large component only:[1]

$$\left[\mathcal{E} + eV(\mathbf{x}) - \frac{1}{2m}\left(\mathbf{p} + \frac{e}{c}\mathbf{A}(\mathbf{x})\right)^2 + \frac{1}{2m_0c^2}\left(\mathcal{E} + eV(\mathbf{x})\right)^2\right.$$
$$\left. +i\frac{e\hbar}{(2m_0c)^2}\mathbf{E}(\mathbf{x})\cdot\mathbf{p} - \frac{e\hbar}{2m_0c}\boldsymbol{\sigma}\cdot\mathbf{B}(\mathbf{x}) - \frac{e\hbar}{(2m_0c)^2}\boldsymbol{\sigma}\cdot\left(\mathbf{E}(\mathbf{x})\times\mathbf{p}\right)\right]\psi = 0, \quad (4.6)$$

where $\mathbf{E}(\mathbf{x}) = \nabla V$ and $\mathbf{B}(\mathbf{x}) = \nabla\times\mathbf{A}$.

- The first three terms are just the ordinary Schrödinger equation in the nonrelativistic limit.
- The fourth and fifth terms do not contain spin matrices, and are called scalar-relativistic terms. These terms are important for heavy elements:[2]
- The sixth term is just the Zeeman energy $\mu_B\,\boldsymbol{\sigma}\cdot\mathbf{B}$, where $\dfrac{e\hbar}{2m_0c} = \mu_B$, the Bohr magneton.

SOC in Atomic Physics

The last term is the spin–orbit coupling. It is a relativistic term that describes the interaction of the electron's spin with the magnetic field that appears in its local frame due to its own orbital motion through the effective atomic Coulomb field (see Figure 4.1). It emerges from a nonrelativistic approximation to the Dirac Hamiltonian. Atomic spectra are accurately described only if SOC is taken into account.

---

[1] Sometimes this equation is referred to as the Pauli equation.
[2] For example, the difference in color between silver and gold is caused by a shift of the d-band due to scalar-relativistic effects.

The spin–orbit coupling terms, sometimes referred to as the Pauli term, reads

$$\mathcal{H}_{SO} = \frac{\hbar}{(2m_0 c)^2} \, \nabla V \times \mathbf{p} \cdot \boldsymbol{\sigma} \tag{4.7}$$

For spherically symmetric potentials,

$$\begin{cases} (\nabla V \times \mathbf{p}) \cdot \boldsymbol{\sigma} = \dfrac{1}{r} \dfrac{\partial V}{\partial r} \, (\mathbf{x} \times \mathbf{p}) \cdot \boldsymbol{\sigma} = \dfrac{1}{r} \dfrac{\partial V}{\partial r} \mathbf{L} \cdot \boldsymbol{\sigma} \\[2ex] \left\langle \dfrac{1}{r} \dfrac{\partial V}{\partial r} \mathbf{L} \cdot \boldsymbol{\sigma} \right\rangle = \left\langle \dfrac{1}{r} \dfrac{\partial V}{\partial r} \right\rangle \hbar \boldsymbol{\ell} \cdot \boldsymbol{\sigma} = \xi \, \boldsymbol{\ell} \cdot \boldsymbol{\sigma} \end{cases} \tag{4.8}$$

For p-states, with $\mathbf{J} = \mathbf{L} + \mathbf{S}$, the eigenvalues are $j = 3/2, 1/2$, and we write

$$\boldsymbol{\ell} \cdot \boldsymbol{\sigma} = \frac{1}{2} \left[j(j+1) - \ell(\ell+1) - s(s+1) \right] = \begin{cases} 1/2 & \text{for } j = 3/2 \\[1ex] -1 & \text{for } j = 1/2 \end{cases} \tag{4.9}$$

and obtain the splittings

$$\Delta_{j=3/2} = \frac{1}{2} \frac{\hbar^2}{4m_0^2 c^2} \left\langle \frac{1}{r} \frac{\partial V}{\partial r} \right\rangle = \frac{1}{2} \xi \tag{4.10}$$

$$\Delta_{j=1/2} = -\frac{\hbar^2}{4m_0^2 c^2} \left\langle \frac{1}{r} \frac{\partial V}{\partial r} \right\rangle = \xi \tag{4.11}$$

so that the SO gap becomes

$$\Delta_{SO} = \frac{3}{2} \frac{\hbar^2}{4m_0^2 c^2} \left\langle \frac{1}{r} \frac{\partial V}{\partial r} \right\rangle == \frac{3}{2} \xi. \tag{4.12}$$

For a spherically symmetric atomic potential with an effective $Z^* \propto Z$ the atomic number, we have

$$\frac{1}{r} \frac{\partial V}{\partial r} \propto \frac{Z^*}{r^3} \sim \frac{Z}{r^3} |\psi(r)|^2 \, r^3,$$

where $|\psi(r)|^2 \, r^3$ is the probability of finding the electron at r. We arrive at

$$\left\langle \frac{1}{r} \frac{\partial V}{\partial r} \right\rangle \propto Z \left\langle |\psi(r)|^2 \right\rangle \sim Z^2$$

and we find that $\xi \propto Z^2$. Thus, ξ will be large for heavy atoms, and small for lighter ones.

4.2 SOC in Solid-State Physics

Now we will explore some of the consequences of spin–orbit interactions (SOI) in crystalline systems. We know that the motion of electrons in a crystal is described by energy-band dispersions $E_n(\mathbf{k})$, where n is the band index. As it turns out, SOC affects the energy band structure, especially for systems containing heavy atoms.

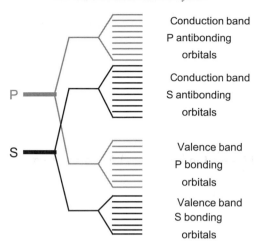

Figure 4.2 Schematic of the evolution of the atomic *s* and *p* orbitals into valence and conduction bands in a semiconductor. Note that this schematic is not general. For example, in Si the lowest conduction band is an antibonding *p*-like state.

4.2.1 SOC in Semiconductor Band Structures

In the absence of spin–orbit interaction, the bands in diamond, zincblende, and wurtzite semiconductors are derived from the outermost atomic shells of the constituent atoms, which mainly involve *s* and *p* orbitals. A schematic of the band ordering is shown in Figure 4.2. For most atomic species that are important in semiconducting materials, SOI plays a significant role. For instance, if we consider the band structure of semiconductors with zincblende structure at the Γ-point ($\mathbf{k} = 0$), Figures 4.3 and 4.4 show that the states at the bottom of the conduction band are *s*-like (Γ_1) Kramers doublet, while those at the top of the valence band are *p*-like (Γ_{15}), sixfold degenerate (orbital angular momentum $\ell = 1$). SOC cannot remove the degeneracy of the Γ_1 doublet, but the Γ_{15} states split into a doublet and a quadruplet.

Without spin		With spin
Γ_{15}	$\Leftrightarrow$	$\begin{cases} \Gamma_8 \ (j = 3/2) \ p\text{-states} \\ \Gamma_7 \ (j = 1/2) \ p\text{-states} \end{cases}$
Γ_1	$\Leftrightarrow$	$\Gamma_6 \ (j = 1/2) \ s\text{-states}$

with splitting energy $E(\Gamma_8) - E(\Gamma_7) = \Delta_0$.

It is found that the magnitude of spin–orbit splitting Δ_0 that occurs in *p*-like states of the valence band of a semiconductor follows closely that of the constituent atoms. In fact, as shown in Figure 4.5, the spin–orbit splitting energy Δ_0 of semiconductors increases as the square of the atomic number of the constituent elements. Typically, the magnitude of Δ_0 in

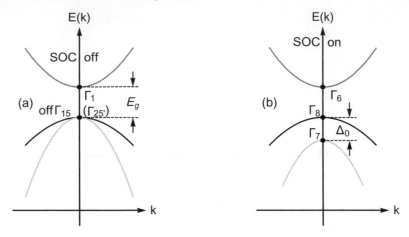

Figure 4.3 Valence and conduction band edges in zincblende semiconductors. No SOI included in (a). At the Γ-point, a sixfold degenerate p-like states (Γ_{15}) at the top of the valence band, and s-like (Γ_1) Kramers' doublet at the bottom of the conduction band. SOI is included in (b), and SOC leads to the splitting of the Γ_{15} states into a Kramers' doublet Γ_7 and a quadruplet Γ_8. The Γ_1 doublet is unaffected. It is relabeled as Γ_6.

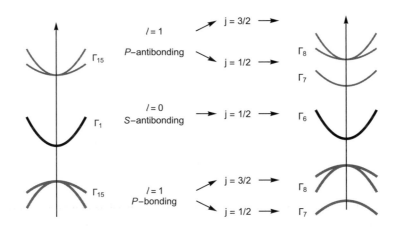

Figure 4.4 Zincblende T_d symmetry with SOC off (left) and SOC on (right).

a semiconductor is comparable to Δ_{SO} of its constituent atoms. This supports the viewpoint that these valence electrons strongly sample their respective atomic Coulomb field.

We also find that when the anion and cation in semiconductor compounds have different Δ_0, the anion contribution tends to be weighted more since the electrons are located preferentially there. Thus, the spin–orbit interaction increases with increasing atomic order number Z of the anion. Typical values of Δ_0 for semiconductor elements and compounds are given in Table 4.1.

Table 4.1 *Spin–orbit splitting of the valence band Δ_0 in eV.*

	Δ_0		Δ_0
C	0.013	InP	0.11
Si	0.044	InAs	0.38
Ge	0.295	InSb	0.81
SiC	0.014	ZnS	0.07
GaN	0.017	ZnSe	0.43
GaP	0.08	ZnTe	0.93
GaAs	0.341	CdTe	0.92
GaSb	0.75		

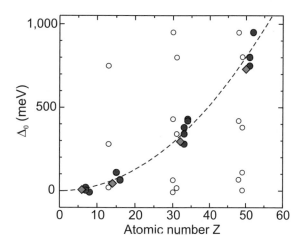

Figure 4.5 Spin–orbit splitting Δ_0 for elemental (diamonds) and various III–V and II–VI (circles) semiconductors. The data are plotted as empty (filled) circles as a function of the cation (anion) order number. Obviously, Δ_0 correlates with the anion Z. The dashed line is proportional to Z^2, from [80].

Effect of Time-Reversal and Inversion Symmetries

Under time-reversal (TR) symmetry operation, Θ, both σ and $\mathbf{p}$ change sign; consequently, $\mathcal{H}_{SO}$ remains invariant. Moreover, time-reversal symmetry implies that

$$E_n(\mathbf{k}, \uparrow) = E_n(-\mathbf{k}, \downarrow) \qquad \Rightarrow \qquad \text{Kramers' degeneracy} \qquad (4.13)$$

$$\Theta\,\psi_n(\mathbf{k}, \uparrow) = \Theta\,e^{i\mathbf{k}\cdot\mathbf{x}}\,u_{n,\mathbf{k}\uparrow}(\mathbf{x})\,|+\rangle = e^{-i\mathbf{k}\cdot\mathbf{x}}\,u^*_{n,\mathbf{k}\uparrow}(\mathbf{x})\,|-\rangle$$

$$= e^{-i\mathbf{k}\cdot\mathbf{x}}\,u_{n,-\mathbf{k}\downarrow}(\mathbf{x})\,|-\rangle = \psi_n(-\mathbf{k}, \downarrow).$$

Likewise,

$$\Theta\,\psi_n(\mathbf{k},\,\downarrow) = \Theta\,e^{i\mathbf{k}\cdot\mathbf{x}}\,u_{n,\mathbf{k}\downarrow}(\mathbf{x})\,|-\rangle = e^{-i\mathbf{k}\cdot\mathbf{x}}\left(-u_{n,\mathbf{k}\downarrow}^{*}(\mathbf{x})\right)|+\rangle$$

$$= e^{-i\mathbf{k}\cdot\mathbf{x}}\,u_{n,-\mathbf{k}\uparrow}(\mathbf{x})\,|+\rangle = \psi_n(-\mathbf{k},\,\uparrow),$$

which may lead to spin-dependent band dispersions similar to those of Figure 4.6(c).

When *space inversion symmetry* (parity) is present, it requires

$$E_n(\mathbf{k},\,\uparrow) = E_n(-\mathbf{k},\,\uparrow);\qquad E_n(\mathbf{k},\,\downarrow) = E_n(-\mathbf{k},\,\downarrow) \tag{4.14}$$

$$\pi\,\psi_n(\mathbf{k},\,\uparrow) = \pi\,e^{i\mathbf{k}\cdot\mathbf{x}}\,u_{n,\mathbf{k}\uparrow}(\mathbf{x})\,|+\rangle = e^{-i\mathbf{k}\cdot\mathbf{x}}\,u_{n,\mathbf{k}\uparrow}(-\mathbf{x})\,|+\rangle$$

$$= \psi_n(-\mathbf{k},\,\uparrow). \tag{4.15}$$

When both symmetries coexist,we find that

$$E_n(\mathbf{k},\,\uparrow) = E_n(\mathbf{k},\,\downarrow).$$

In other words, the presence of both TR and parity symmetries removes the spin-dependence band dispersion, even in the presence of SO interactions, as shown in Figure 4.6(b). Note that the dispersion of Figure 4.6(b) resembles that of Figure 4.6(a) which depicts band dispersion in the absence of SOC. The only difference between the figures is symmetry labeling.

Since in nonmagnetic semiconductors, TR is a bona fide symmetry; the electron spectrum has to satisfy the relation $E(\mathbf{k}\,\uparrow) = E(\mathbf{k}\,\downarrow)$ in the presence of parity symmetry. Thus, in semiconductors with diamond structure, such as Si and Ge, we have a global twofold degeneracy of Bloch states. Notice that at the Γ-point Γ_8 is fourfold degenerate $(\pm 3/2, \pm 1/2)$, and Γ_7 is twofold degenerate $(\pm 1/2)$.

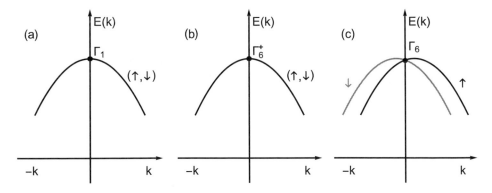

Figure 4.6 Kramers' degeneracy in the following cases: (a) No SOI is present, where each level is doubly degenerate $(\uparrow,\ \downarrow)$. (b) Both SOI and inversion symmetry are present and the levels are doubly degenerate at every $\mathbf{k}$-point. (The superscript $+$ indicates even parity.) (c) SOI is present, but inversion symmetry is absent. Γ_6 s-symmetry includes spin.

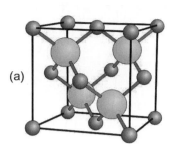
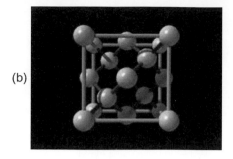

Figure 4.7 (a) Zincblende structure; (b) diamond structure. Permission from D. Wertz and North Carolina State College of Sciences.

The principal difference between zincblende and wurtzite structures, on the one hand, and diamond structures, on the other, is the lack of inversion symmetry in the former structures (see Figure 4.7). Without inversion symmetry, Kramers' theorem that $E(\mathbf{k}) = E(-\mathbf{k})$ still applies, but now the periodic part of the Bloch functions no longer satisfies the condition $u_{-\mathbf{k}}(\mathbf{x}) = u_{\mathbf{k}}(-\mathbf{x})$, and hence a twofold degeneracy throughout the Brillouin zone is not required.

4.2.2 Bulk Inversion Symmetry-Breaking and the Dresselhaus Hamiltonian

In the absence of space inversion symmetry, $E(\mathbf{k}\uparrow) \neq E(\mathbf{k}\downarrow)$. In this case, there should be some term that breaks the spin degeneracy of states at the same $\mathbf{k}$. The term should be an odd function of both $\mathbf{k}$ and σ since it breaks space inversion symmetry, but keeps time-reversal symmetry. It is referred to as the *bulk inversion asymmetry* (BIA) spin–orbit coupling term. For a spin-1/2 electron, we have

$$\sigma_i \sigma_j = \delta_{ij} + i\epsilon_{ijk}\,\sigma_k,$$

and only the linear form of σ_i can appear in any spin–orbit coupling term. Thus, any electron spin–orbit coupling that can be written in the form

$$\mathcal{H}_{SO} = \sum_{\substack{nm \\ ij}} C_{nm}^{ij}\,\sigma_i^n\,k_j^m$$

is reduced to a form linear in σ_i.

We write the BIA term as

$$\mathcal{H}_{SO} = \frac{1}{2}\,\boldsymbol{\Omega}(\mathbf{k})\cdot\boldsymbol{\sigma}, \tag{4.16}$$

where $\boldsymbol{\Omega}(\mathbf{k})$ is an odd function of $\mathbf{k}$. It acts as an effective magnetic field, where, for a given $\mathbf{k}$, $\Omega(\mathbf{k})/\hbar$ is the spin precession frequency in this field.

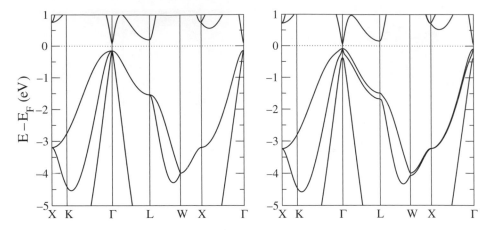

Figure 4.8 Ge band structure in the vicinity of the Fermi level. Left: without spin–orbit coupling; and right: with spin–orbit coupling. From [29].

We note that the breaking of space inversion symmetry is a prerequisite for the appearance of an electron spin–orbit coupling term in semiconductors. From another perspective, we note that the spin splitting of Bloch states in zincblende structure semiconductors must come from SOI; otherwise, the spin degree of freedom would not know whether it was in an inversion-symmetric or a non-inversion-symmetric structure.

It was G. Dresselhaus who first proposed, in 1955, that SOC may have important consequences for semiconductors lacking inversion symmetry in their bulk structure. In such semiconductors, we can have SO splitting for electron and hole states for $\mathbf{k} \neq 0$ even in the absence of magnetic fields. This behavior is called the *Dresselhaus effect*, and the previously defined SOC term is referred to as the *Dresselhaus BIA spin-orbit coupling*.

To illustrate the effect of SOI in the presence and absence of inversion symmetry, we compare the band structures of Ge (diamond) and GaSb (zincblende).

- **Ge with diamond structure**: In absence of SOC, as in Figure 4.8 (left), we have a sixfold degenerate state at the top of the valence band. Three bands emerge for $|\mathbf{k}| \neq 0$, and two of them become degenerate along some high-symmetry directions. When SOC is included, as in Figure 4.8 (right), the $\mathbf{k} = 0$ state splits into quadruply degenerate and a doubly degenerate ones. The former evolves into two bands with different dispersions, a highly dispersive one called the *light-hole* band, and the other one is termed heavy-hole band. The doubly degenerate state at Γ forms the spin–orbit split-off band.
- **GaSb, with zincblende structure**: It has the same splitting of the $\mathbf{k} = 0$ state. However, the dispersion of the bands that emerge from the quadruply degenerate state resemble that displayed in Figure 4.9, signaling the presence of BIA!

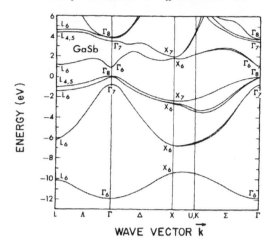

Figure 4.9 Spin–orbit splitting of the top valance bands in GaSb. From [43].

4.3 The k · p Method and the Effective Mass Tensor

Because semiconductors have characteristically small energy gaps $\lesssim$ 1 eV, the free carrier concentrations produced either by thermal excitation, where there will be an equal number of electrons and holes, or by doping, never exceed $10^{20}/\,cm^3$, and are usually much less. By contrast, the number of states in a band is on the order of $10^{23}/cm^3$. This implies that the mostly populated electron and hole states lie within a small fraction of an eV from the band edges. Thus, most physical phenomena (electronic, optical, magnetic) in semiconductors can be understood by looking at a small portion of the band structure around the band edges.

Thus, we should focus on the structure of the energy bands $E(\mathbf{k})$ in the vicinity of the conduction band minimum and the valence band maximum. The highest point of the valence bands occurs at the Γ-point; however, the conduction band minimum may occur at other points in the BZ.[3] In most compound semiconductors, the minimum of the conduction band also occurs at the Γ-point, as shown in Figure 4.10. Such semiconductors are called *direct-gap* semiconductors and form the core of most optical devices.[4]

If the wavefunctions and energies are known at the band extrema, then perturbation methods may be applied to determine the wavefunctions and energies at other $\mathbf{k}$ points in the vicinity of the extrema. Here we will present one very useful and very frequently applied perturbative approach, known as the Kohn–Luttinger $\mathbf{k} \cdot \mathbf{p}$ method.

---

[3] If both extrema occur at the Γ-point, as it is the case for GaAs and many other materials, then for small $\mathbf{k}$, $E(\mathbf{k})$ should be parabolic $E(\mathbf{k}) = \hbar^2 k^2/2m_c$ or $-\hbar^2 k^2/2m_v$, where m_c and m_v are effective masses of electrons and holes, respectively. The concepts of holes and effective masses will be discussed later. We note, however, that the effective masses may differ considerably from the free electron mass m_e, for example in GaAs $m_c = 0.067\,m_e$.

[4] If the minimum of the conduction band appears at some other point in $\mathbf{k}$-space, the semiconductor is *indirect gap*. The elemental semiconductors Si and Ge are of this type.

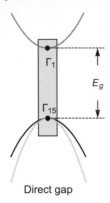

Direct gap

Figure 4.10 The case where both extrema occur at the Γ-point. The wavefunctions and energies should be determined within a small region in its vicinity.

The development of the $\mathbf{k} \cdot \mathbf{p}$ formalism starts by recasting the Schrödinger equation for Bloch electrons

$$\left(\frac{\mathbf{p}^2}{2m} + V(\mathbf{x})\right) \Psi_{\nu\mathbf{k}}(\mathbf{x}) = E_\nu(\mathbf{k}) \, \Psi_{\nu\mathbf{k}}(\mathbf{x})$$

into a form that involves their lattice-periodic part $u_{\nu\mathbf{k}}(\mathbf{x})$, where

$$\Psi_{\nu\mathbf{k}}(\mathbf{x}) = e^{i\mathbf{k}\cdot\mathbf{x}} u_{\nu\mathbf{k}}(\mathbf{x}) = e^{i\mathbf{k}\cdot\mathbf{x}} \, \langle \mathbf{x} \mid \nu\mathbf{k} \rangle ,$$

and we will use ν to denote the orbital motion type, instead of the band index n.

Since

$$\mathbf{p} \, e^{i\mathbf{k}\cdot\mathbf{x}} u_{\nu\mathbf{k}}(\mathbf{x}) = e^{i\mathbf{k}\cdot\mathbf{x}} \, (\mathbf{p} + \hbar \, \mathbf{k}) \, u_{\nu\mathbf{k}}(\mathbf{x}),$$

we obtain

$$\left[\frac{\hbar^2}{2m_e} \, (\mathbf{p} + \hbar \, \mathbf{k})^2 + V(\mathbf{x})\right] |\nu\mathbf{k}\rangle = \left[\frac{\mathbf{p}^2}{2m_e} + V(\mathbf{x}) + \frac{\hbar^2 k^2}{2m_e} + \frac{\hbar}{m_e} \, \mathbf{k} \cdot \mathbf{p}\right] |\nu\mathbf{k}\rangle = E_\nu(\mathbf{k}) \, |\nu\mathbf{k}\rangle$$

or

$$\left[\underbrace{\frac{\mathbf{p}^2}{2m_e} + V(\mathbf{x})}_{\mathcal{H}_0} + \underbrace{\frac{\hbar}{m_e} \, \mathbf{k} \cdot \mathbf{p}}_{\mathcal{H}'}\right] |\nu\mathbf{k}\rangle = \underbrace{\left(E_\nu(\mathbf{k}) - \frac{\hbar^2 k^2}{2m_e}\right)}_{\mathcal{E}_\nu(\mathbf{k})} |\nu\mathbf{k}\rangle \qquad (4.17)$$

so that

$$\left[\mathcal{H}_0 + \mathcal{H}'\right] |\nu\mathbf{k}\rangle = \mathcal{E}_\nu(\mathbf{k}) \, |\nu\mathbf{k}\rangle$$

$$u_{\nu\mathbf{k}}(\mathbf{x}) = \langle \mathbf{x} \mid \nu\mathbf{k} \rangle .$$

The advantage here is that the eigenfunctions $u_{nk_0}(\mathbf{x})$ of $\mathcal{H}_0$ have the periodicity of the lattice. Hence, the eigenvalue problem needs to be solved at the extremum $\mathbf{k}_0$ for a single primitive cell only, instead of for the whole crystal. The resulting band-edge functions $u_{nk_0}(\mathbf{x})$ form a complete and orthonormal set of functions, which can be used as a basis for perturbation theory. As we have noted earlier, for small deviations $\mathbf{k} = \mathbf{k}_0 + \Delta\mathbf{k}$, $\Delta\mathbf{k} \cdot \mathbf{p}$ perturbation theory allows us to find the $E_\nu(\mathbf{k})$ for states *near* $\mathbf{k}_o$.

For the sake of simplicity, we set the extremum $\mathbf{k}_0 = 0$. Since $\mathbf{k}_0$ is an extremum, $\nabla_{\mathbf{k}} E_{\nu\mathbf{k}}\big|_{\mathbf{k}=\mathbf{k}_0} = 0$, and the linear term vanishes. We assume that the band is nondegenerate, then to second order we get

$$E_\nu(\mathbf{k}) = E_\nu(\mathbf{0}) + \langle u_{\nu,\mathbf{0}} \,|\mathcal{H}'|\, u_{\nu,\mathbf{0}} \rangle + \sum_{\nu' \neq \nu} \frac{\langle u_{\nu,\mathbf{0}} \,|\mathcal{H}'|\, u_{\nu',\mathbf{0}} \rangle \langle u_{\nu',\mathbf{0}} \,|\mathcal{H}'|\, u_{\nu,\mathbf{0}} \rangle}{E_\nu(\mathbf{0}) - E_{\nu'}(\mathbf{0})}.$$

We also note that

$$\langle u_{\nu,\mathbf{0}} \,|\mathcal{H}'|\, u_{\nu,\mathbf{0}} \rangle = \frac{\hbar\mathbf{k}}{m_e} \cdot \langle u_{\nu,\mathbf{0}} \,|\mathbf{p}|\, u_{\nu,\mathbf{0}} \rangle,$$

vanishes for systems possessing inversion symmetry, because $\mathcal{H}'$ is odd, and the u_νs have definite parity. The matrix elements in the second-order term

$$\frac{\hbar\mathbf{k}}{m_e} \cdot \langle u_{\nu,\mathbf{0}} \,|\mathbf{p}|\, u_{\nu',\mathbf{0}} \rangle \neq 0$$

only for $u_{\nu,\mathbf{0}}$ and $u_{\nu',\mathbf{0}}$ of opposite parities. Hence, we find

$$E_\nu(\mathbf{k}) = E_\nu(\mathbf{0}) + \frac{\hbar^2 k^2}{2m_e} + \frac{\hbar^2}{m_e^2} \sum_{\substack{\nu' \neq \nu \\ \alpha\beta}} k_\alpha k_\beta \frac{\langle u_{\nu,\mathbf{0}} \,|\mathbf{p}_\alpha|\, u_{\nu',\mathbf{0}} \rangle \langle u_{\nu',\mathbf{0}} \,|\mathbf{p}_\beta|\, u_{\nu,\mathbf{0}} \rangle}{E_\nu(\mathbf{0}) - E_{\nu'}(\mathbf{0})}. \tag{4.18}$$

We define the *effective mass tensor* $m_{\alpha\beta}^*$ as

$$\frac{1}{m_{\alpha\beta}^*} = \frac{1}{\hbar^2} \frac{\partial^2 E(\mathbf{k})}{\partial k_\alpha \partial k_\beta} = \frac{1}{m_e} \delta_{\alpha\beta} + \frac{1}{m_e^2} \sum_{\nu' \neq \nu} \frac{\langle u_{\nu,\mathbf{0}} \,|\mathbf{p}_\alpha|\, u_{\nu',\mathbf{0}} \rangle \langle u_{\nu',\mathbf{0}} \,|\mathbf{p}_\beta|\, u_{\nu,\mathbf{0}} \rangle}{E_\nu(\mathbf{0}) - E_{\nu'}(\mathbf{0})}. \tag{4.19}$$

Transformation to principal axes yields

$$\frac{m_e}{m_i^*} = 1 + \frac{2}{m_e} \sum_{\nu' \neq \nu} \frac{|\langle u_{\nu,\mathbf{0}} \,|\mathbf{p}_i|\, u_{\nu',\mathbf{0}} \rangle|^2}{E_\nu(\mathbf{0}) - E_{\nu'}(\mathbf{0})}. \tag{4.20}$$

Equation (4.20) is usually applied to the calculation of the effective mass of the highest valence bands and lowest conduction bands in direct-gap semiconductors. Typical effective mass values are given in Figure 4.11. We find that the main contribution to the sum comes from the triply degenerate valence band p_x, p_y, and p_z eigenstates, and the s-like bottom of the conduction band. This yields

$$\langle u_{\mathrm{VB}x} \,|p_x|\, u_{\mathrm{CB},s} \rangle = \langle u_{\mathrm{VB}y} \,|p_y|\, u_{\mathrm{CB},s} \rangle = \langle u_{\mathrm{VB}z} \,|p_z|\, u_{\mathrm{CB},s} \rangle = p.$$

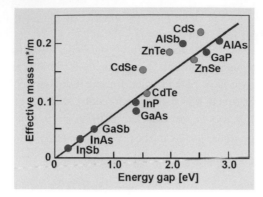

Figure 4.11 Variation of effective mass with band gap energy of different semiconductors.

Hence,

$$\frac{m_e}{m^*}\bigg|_{c,v} = 1 + \frac{2}{m_e} \frac{p^2}{\pm E_g}$$

and we can write

$$E_{_v}^{^c}(\mathbf{k}) = E_{_v}^{^c}(\mathbf{0}) \pm \frac{\hbar^2 k^2}{2|m^*|}.$$

4.3.1 Hole Concept and Valence Band Effective Masses

We find that applying the preceding definition of the effective mass tensor to states in the vicinity of semiconductor valence band maxima yields negative values! To avoid such an inconvenient conceptual outcome, we introduce the idea of a hole to describe a few empty states close to the top of an almost full band.

Consider a band containing electrons with quantum numbers $\mathbf{k}_j$, velocities $\mathbf{v}_j$, and energies $E(\mathbf{k}_j)$, where $E = 0$ is at the top of the band. For a full band, the values of $\mathbf{k}$ should all sum to zero, namely

$$\sum_j \mathbf{k}_j = 0.$$

Now we remove one electron to create an excitation, which we label a hole. If we remove the lth electron, the band acquires a net $\mathbf{k}$, which we attribute to the presence of the hole. The hole will have $\mathbf{k} = \mathbf{k}_h$, such that

$$\mathbf{k}_h = \sum_{j \neq l} \mathbf{k}_j = -\mathbf{k}_l, \tag{4.21}$$

giving the hole dispersion shown in Figure 4.12. We should note that the lower down the band the empty state, the more excitation energy the system needs. The hole's energy E_h must therefore take the form

$$E_h = -E(\mathbf{k}_l). \tag{4.22}$$

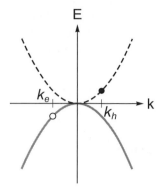

Figure 4.12 Hole dispersion (black, dashed) in relation to the electron dispersion in the valence band (solid gray).

The group velocity $\mathbf{v}_h$ associated with the hole is

$$\mathbf{v}_h = \frac{1}{\hbar} \nabla_{\mathbf{k}_h} E_h = \frac{1}{\hbar} \nabla_{-\mathbf{k}_l} (-E_{\mathbf{k}_l}) = \mathbf{v}_l, \tag{4.23}$$

where the minus signs in (4.21) and (4.22) have canceled each other. The full electron band will carry no current, namely

$$\sum_j (-e) \, \mathbf{v}_j = 0.$$

The removal of the *l*th electron produces a current

$$\sum_{j \neq l} (-e) \, \mathbf{v}_j = -(-e) \, \mathbf{v}_l = (+e) \, \mathbf{v}_h,$$

i.e., the hole appears to have an associated positive charge.

Finally, substitution of (4.21) and (4.22) into the effective mass formula yields

$$\frac{1}{m_{\alpha\beta}^{*h}} = \frac{1}{\hbar^2} \frac{\partial^2 E_h(\mathbf{k})}{\partial k_\alpha^h \, \partial k_\beta^h} = -\frac{1}{m_{\alpha\beta}^{*e}}, \tag{4.24}$$

which shows that the effective mass $m_h^* = -m_e^*$, and we restore the concept of a positive mass.

The importance of holes stems from the fact that bands are often easily well characterized only close to the band extrema. Our knowledge of the dispersion relations away from a particular extremum is not easily derived both experimentally and theoretically. It is therefore more convenient to deal with a small number of empty states close to the well-characterized maximum of an almost full band rather than attempt to treat the huge number of states lower down in the band.

Luttinger's Model of the Valence Band Structure: Light and Heavy Holes

We need to find an effective mass description of the valence band structure, taking into account its *p*-like character and the threefold degeneracy at the Γ-point. The simplest model, which will bring out the physics while avoiding complications accrued by the cubic environment, is one with isotropic symmetry. A model Hamiltonian that satisfies isotropy will involve the polar vector $\boldsymbol{\kappa} = \hbar \mathbf{k}$ and the axial angular momentum vector $\mathbf{L}$. The latter is represented by the 3×3 matrices L_x, L_y, and L_z corresponding to $l = 1$. L_z is a diagonal matrix with eigenvalues $(1, 0, -1)$. The scalar Hamiltonian to be constructed must contain a term quadratic in $\boldsymbol{\kappa}$. Rotational invariance of the Hamiltonian, imposed by isotropy, leads to the Luttinger Hamiltonian

$$\mathcal{H} = A\kappa^2 \mathbb{I} + B (\boldsymbol{\kappa} \cdot \mathbf{L})^2,$$

where A and B are arbitrary constants, and $\mathbb{I}$ is the 3×3 identity matrix. It is obvious that $\mathcal{H}$ is also a 3×3 matrix in this representation.

The energy spectrum in the valence band is obtained by diagonalizing the $\mathcal{H}$ matrix. The arbitrariness in the choice of axes allows us to drastically simplify the procedure by choosing the direction of the z-axis to be along $\boldsymbol{\kappa}$. We note that the result does not depend on the choice of axes. We find that

$$(\boldsymbol{\kappa} \cdot \mathbf{L})^2 = \kappa^2 L_z^2,$$

leaving a diagonal $\mathcal{H}$ with eigenvalues

$$E_h(\kappa) = (A + B)\kappa^2 = \frac{\kappa^2}{2m_h}, \quad L_z = \pm 1,$$

$$E_l(\kappa) = A\kappa^2 = \frac{\kappa^2}{2m_l}, \quad L_z = 0.$$

Thus the valence band energy spectrum has two parabolic branches, $E_h(\kappa)$ and $E_l(\kappa)$, the first one being twofold degenerate. The choice of the subscripts h and l stems from having two effective masses, m_h and m_l,

$$m_h = \frac{1}{2(A + B)}, \quad \text{and } m_l = \frac{1}{2A}, \quad \text{usually } B < 0, \quad \text{but } A + B > 0,$$

that define two types of holes in the valence band, the heavy and light holes. Another distinguishing feature between these particles is that the heavy hole has an orbital angular momentum $\mathbf{L}$ projection along the direction of $\boldsymbol{\kappa}$ (*helicity*) equal to ± 1, while the light hole has a projection of 0.

4.3.2 $\mathbf{k} \cdot \mathbf{p}$ *method in the presence of SO interactions*

In the presence of SO coupling, the lattice-periodic parts of the Bloch functions become spinors $|n\mathbf{k}\rangle$ with two components

$$|n\mathbf{k}\rangle = \begin{pmatrix} |\nu\mathbf{k}, \uparrow\rangle \\ |\nu\mathbf{k} \downarrow\rangle \end{pmatrix},$$

where, now, n is a common index for the orbital and spin degrees of freedom. The Pauli–Schrödinger equation

$$\left[\frac{\mathbf{p}^2}{2m_e} + V(\mathbf{x}) + \frac{\hbar}{(2m_0c^2)^2} \, \boldsymbol{\sigma} \times \nabla V \cdot \mathbf{p} \right] \psi_{n\mathbf{k}} = E_n(\mathbf{k}) \, \psi_{n\mathbf{k}}$$

becomes

$$\left[\frac{\mathbf{p}^2}{2m_e} + V(\mathbf{x}) + \frac{\hbar^2 k^2}{2m_e} + \frac{\hbar}{m_e} \mathbf{k} \cdot \boldsymbol{\pi} + \frac{\hbar}{(2m_0c^2)^2} \, \boldsymbol{\sigma} \times \nabla V \cdot \mathbf{p} \right] |n\mathbf{k}\rangle = E_n(\mathbf{k}) \, |n\mathbf{k}\rangle,$$

where

$$\boldsymbol{\pi} = \mathbf{p} + \frac{\hbar}{(2m_0c^2)^2} \, \boldsymbol{\sigma} \times \nabla V$$

since

$$\left(\boldsymbol{\sigma} \times \nabla V \cdot \mathbf{p} \right) e^{i\mathbf{k}\cdot\mathbf{x}} u_{\nu\mathbf{k}}(\mathbf{x}) = e^{i\mathbf{k}\cdot\mathbf{x}} \left(\boldsymbol{\sigma} \times \nabla V \cdot \mathbf{p} + \boldsymbol{\sigma} \times \nabla V \cdot (\hbar\mathbf{k}) \right) u_{\nu\mathbf{k}}(\mathbf{x}).$$

Note that in the presence of SO coupling, the spin quantum number is not a good quantum number.

Construction of the Eigenvalue Equations

For simplicity, we set the band extremum at $\mathbf{k} = 0$ and expand the kets $|n\mathbf{k}\rangle$ in terms of band-edge lattice-periodic functions $|\nu 0\rangle$, that provide a complete orthonormal single-particle basis in the absence of SOI. Then we have the expansion

$$|n\mathbf{k}\rangle = \sum_{\nu,\sigma=\uparrow\downarrow} C_{n\nu\sigma}(\mathbf{k}) \, |\nu\sigma\rangle$$

$$|\nu\sigma\rangle = |\nu 0\rangle \otimes |\sigma\rangle \,.$$

The Schrödinger equation involving $|n\mathbf{k}\rangle$ can be recast as

$$\sum_{\nu',\sigma'} \left[\left(E_{\nu'}(0) + \frac{\hbar^2 k^2}{2m_e} \right) \delta_{\nu,\nu'} \delta_{\sigma,\sigma'} + \frac{\hbar}{m_e} \mathbf{k} \cdot \boldsymbol{\Pi}_{\sigma,\sigma'}^{\nu,\nu'} + \Delta_{\sigma,\sigma'}^{\nu,\nu'} \right] C_{n\nu',\sigma'}(\mathbf{k})$$

$$= E_n(\mathbf{k}) \, C_{n\nu\sigma}(\mathbf{k}) \tag{4.25}$$

$$\boldsymbol{\Pi}_{\sigma,\sigma'}^{\nu,\nu'} = \langle \nu\sigma | \, \boldsymbol{\pi} \, | \nu'\sigma' \rangle$$

$$\Delta_{\sigma,\sigma'}^{\nu,\nu'} = \frac{\hbar^2}{(2m_0c^2)^2} \, \langle \nu\sigma | \, \nabla V \times \mathbf{p} \cdot \boldsymbol{\sigma} \, | \nu'\sigma' \rangle.$$

By solving the eigenvalue problem defined in (4.25), we obtain the dispersion $E_n(\mathbf{k})$ of the relevant bands. While the secular equations in (4.25) are, in principle, of infinite dimension, we are usually interested in a few bands, and in their dispersion in the vicinity of $\mathbf{k} = 0$.

Actually, it is sufficient that the $\mathbf{k} \cdot \mathbf{p}$ perturbation calculation be carried out with a basis set consisting of the triply degenerate valence band maximum Γ_{15}^v (*p*-like), the singly degenerate conduction band minimum Γ_1^c (*s*-like), and the triply degenerate conduction band Γ_{15}^c (*p*-like), shown in Figure 4.13 for GaAs.

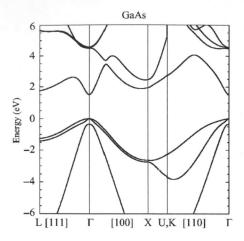

Figure 4.13 $\mathbf{k} \cdot \mathbf{p}$ calculations of GaAs band structure including spin–orbit coupling, from [156].

Notice that the mixing effected by the second term in (4.25) increases with $\mathbf{k}$. The third term is independent of $\mathbf{k}$; it arises from atomic SOI and results in Γ-point splittings similar to those defined in (4.8) through (4.12) and shown in Figure 4.3.

For $\mathbf{k} \neq 0$, the total perturbation is

$$\delta \mathcal{H} = \frac{\hbar^2 k^2}{2m_e} + \frac{\hbar}{m_e} \mathbf{k} \cdot \boldsymbol{\pi}.$$

Extension of Luttinger's Model

We note that the valence band states at $\mathbf{k} = 0$ maintain the same symmetry properties of the atomic states they are derived from. Thus, for $\mathbf{k} = 0$ we should have a fourfold degenerate state ($j = 3/2$), separated from a doubly degenerate state ($j = 1/2$) by spin–orbit splitting energy, Δ_0. The s-like bottom of the conduction band also remains doubly degenerate.

We explore changes to the $j = 3/2$ manifold at $\mathbf{k} \neq 0$ and energies $E(\mathbf{k}) \ll \Delta_0$ by constructing a Luttinger Hamiltonian in a way quite similar to the procedure outlined previously in the absence of SOI. However, we need to replace the 3×3 matrices L_x, L_y, L_z, corresponding to $l = 1$, with 4×4 matrices J_x, J_y, J_z, corresponding to $j = 3/2$

$$\mathcal{H} = A \kappa^2 \mathbb{I} + B (\boldsymbol{\kappa} \cdot \mathbf{J})^2, \tag{4.26}$$

where $\mathbb{I}$ is now a unit 4×4 matrix. Again, the matrix J_z is diagonal, but with eigenvalues $3/2, 1/2, -1/2$, and $-3/2$.

The resulting spectrum of the heavy and light holes, which is valid for energies much less than Δ_0 becomes

$$E_h(\kappa) = \left(A + \frac{9B}{4} \right) \kappa^2 = \frac{\kappa^2}{2m_h}, \qquad (J_z = \pm 3/2) \quad \text{Heavy-hole band}$$

$$E_l(\kappa) = \left(A + \frac{B}{4} \right) \kappa^2 = \frac{\kappa^2}{2m_l}, \qquad (J_z = \pm 1/2) \quad \text{Light-hole band}$$

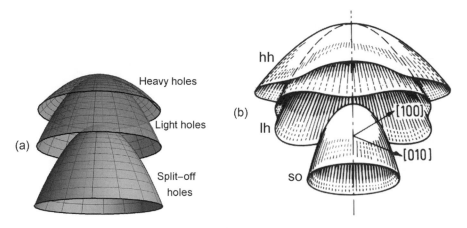

Figure 4.14 E vs. (k_x, k_y) of the top three valence bands. Left: Isotropic case. Right: Corresponding valence bands of Ge, showing warping, from [80].

depicting two doubly degenerate bands. Heavy holes have helicity of $\pm 3/2$, and the light holes' helicity is $\pm 1/2$. While $B < 0$, normally $A + 9B/4 > 0$, so that both masses are positive.[5]

Warping of Isoenergetic Surfaces

We should remember that the Luttinger Hamiltonian (4.26) presents an isotropic approximation – being rotationally invariant – as seen in Figure 4.14. In a cubic crystal, the symmetry is lower, and a more realistic Luttinger Hamiltonian takes the general form

$$\mathcal{H} = A \kappa^2 \mathbb{I} + B (\boldsymbol{\kappa} \cdot \mathbf{J})^2 + C \left(J_x^2 \kappa_x^2 + J_y^2 \kappa_y^2 + J_z^2 \kappa_z^2 \right)$$

where now the axes are not arbitrary, and coincide with the crystallographic axes. The last term introduces anisotropy in the isoenergetic dispersive surfaces of light and heavy holes, warping the simple parabolic form of the energy branches $E_h(\boldsymbol{\kappa})$ and $E_l(\boldsymbol{\kappa})$ as shown in Figure 4.14(b).

Dyakonov and Perel proposed the following Hamiltonian to describe the Dresselhaus spin–orbit interaction in the bulk of zincblende crystals

$$\mathcal{H}_D^{3D} = \gamma_D \left[\kappa_x \left(\kappa_y^2 - \kappa_z^2 \right) \sigma_x + \kappa_y \left(\kappa_z^2 - \kappa_x^2 \right) \sigma_y + \kappa_z \left(\kappa_x^2 - \kappa_y^2 \right) \sigma_z \right], \qquad (4.27)$$

$$\uparrow \qquad\qquad\qquad \uparrow \qquad\qquad\qquad \uparrow$$
$$[100] \qquad\qquad\quad [010] \qquad\qquad\quad [001]$$

where x, y, z point along the main crystallographic directions, [100], [010], and [001]. It lacks space inversion but satisfies TRI and the point group symmetry.

---

[5] In some materials, the light-hole mass becomes negative, so that this band becomes a conduction band. This is known as band inversion and will be discussed in the context of topological systems in Part II.

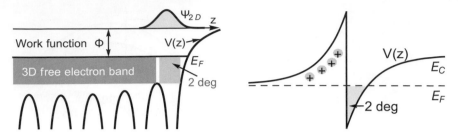

Figure 4.15 Two-dimensional electron gas confining potential at a metallic surface (left), in a heterojunction quantum well (right).

4.4 Rashba Spin–Orbit Interactions in Two-Dimensional Electron Systems

As we have shown, the global **k** spin degeneracy in semiconductor electronic bands is removed by BIA of the underlying crystal structure, as for example in zincblende and wurtzite crystals. Another structural feature that manifests inversion asymmetry and leads to the lifting of spin degeneracy is the inherent anisotropy at surfaces, interfaces, and semiconductor heterostructures: quantum wells (QWs). It is referred to as *structural inversion asymmetry* (SIA) of the confining potential $V(\mathbf{x})$ of a two-dimensional electron gas, either at high-Z metal surfaces or in QWs, shown in Figure 4.15. An sp-derived metallic surface state can be considered as a particular realization of a two dimensional electron gas. Since the surface always breaks spatial inversion symmetry, the effective potential that acts on the surface state will generally have a finite gradient along the surface normal, associated with an electric field in this direction. The physical manifestation of this field is the work function. At the surface, the potential changes from the vacuum level to the bottom of the band, a value that is approximately the work function, Φ. A rough estimate of the length scale over which the potential changes is roughly the Fermi wavelength, λ_F. We then find that $(\nabla V)_z \sim \Phi/\lambda_F$.[6]

We can elucidate the physics behind this effect with the aid of Figure 4.16. Figure 4.16(a) shows a quantum well potential possessing inversion symmetry, which guarantees spin degeneracy on a nanoscale. In contrast, the heterojunction quantum well potential of Figure 4.16(b) displays an inversion asymmetric potential, $V(z)$, in the z-direction perpendicular to the heterojunction, which is attributed to its structure, and which gives rise to the new type of SOI associated with the ensuing electric field $\mathcal{E}_{SIA} = -\dfrac{1}{e}\dfrac{\partial V}{\partial z}\hat{\mathbf{z}}$.

A Taylor expansion of the potential $V(z)$ yields

$$V(\mathbf{x}) = V_0 - e\mathcal{E}z + \cdots, \tag{4.28}$$

so that, to lowest order, the inversion asymmetry of $V(z)$ is characterized by an electric field $\mathcal{E}$. Electrons, with an effective mass m^*, will propagate with velocity

---

[6] Electrons subjected to spin–orbit interaction in magnets with SIA, generate an antisymmetric exchange between spins, known as the Dzyaloshinskii–Moriya (DM) interaction, to be discussed in Chapter 20.

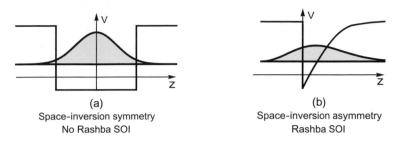

(a)

Space-inversion symmetry

No Rashba SOI

(b)

Space-inversion asymmetry

Rashba SOI

Figure 4.16 Structural inversion asymmetry.

$\mathbf{v} = \nabla_{\mathbf{k}} E(\mathbf{k}) = \mathbf{k}/m^*$ in the electric field $\mathcal{E}$, defined in the lattice frame of reference, and will experience a magnetic field

$$\mathbf{B} = \frac{1}{c} (\mathbf{v} \times \mathcal{E}) = \frac{1}{m^* c} (\mathbf{k} \times \mathcal{E})$$

in their local moving frame. The interaction of the spin with this $\mathbf{B}$ field leads to the *Rashba* or *Bychkov–Rashba* Hamiltonian.

For simplicity, we neglect the 2D periodic potential. Thus, the quasi-2D electrons, moving freely in the xy-plane with momentum $\mathbf{p}_\| = \hbar \mathbf{k}_\|$ in an electric field $\mathcal{E} = \mathcal{E} \hat{\mathbf{z}}$, will experience an SOC given by

$$\mathcal{H}_{so} = \frac{\alpha_R}{\hbar} \left(\hat{\mathbf{z}} \times \mathbf{p}_\| \right) \cdot \boldsymbol{\sigma} = \alpha_R \left(k_x \sigma_y - k_y \sigma_x \right)$$

where the Rashba parameter α_R depends on the gradient of the confining potential.

4.4.1 Spectrum of the Rashba Hamiltonian in 2D

The 2DEG Hamiltonian with Rashba SO coupling is

$$\mathcal{H}_R = \frac{\mathbf{p}_\|^2}{2m^*} + \frac{\alpha_R}{\hbar} \left(p_y \sigma_x - p_x \sigma_y \right) = \frac{\hbar^2 \mathbf{k}_\|^2}{2m^*} + \alpha_R k_\| \left(\sin \varphi \, \sigma_x - \cos \varphi \, \sigma_y \right) \tag{4.29}$$

where $\mathbf{k}_\| = k_\| (\cos \varphi, \sin \varphi)$. $\mathcal{H}_R$ commutes with the 2D momentum operator, hence, they have common 2D spatial eigenfunctions $e^{i \mathbf{k}_\| \cdot \mathbf{x}_\|}$. This allows us to write the spinor eigenvectors as

$$\psi = e^{i \mathbf{k}_\| \cdot \mathbf{x}_\|} \begin{bmatrix} a \, |\uparrow\rangle \\ b \, |\downarrow\rangle \end{bmatrix}, \qquad\qquad a^2 + b^2 = 1$$

and the eigenvalue problem becomes

$$\mathcal{H}_R \psi = \begin{pmatrix} \dfrac{\hbar^2 k_\|^2}{2m^*} & \alpha_R k_\| \left(i \cos \varphi + \sin \varphi \right) \\ \alpha_R k_\| \left(-i \cos \varphi + \sin \varphi \right) & \dfrac{\hbar^2 k_\|^2}{2m^*} \end{pmatrix} \psi = E \psi$$

with eigenvalues

$$E_\pm = \frac{\hbar^2 k_\parallel^2}{2m^*} \pm \alpha_R\, k_\parallel \tag{4.30}$$

and corresponding eigenvectors

$$\psi_+ = e^{i\mathbf{k}_\parallel \cdot \mathbf{x}_\parallel} \frac{1}{\sqrt{2}} \begin{pmatrix} -i\, e^{-i\varphi} \\ 1 \end{pmatrix} = e^{i\mathbf{k}_\parallel \cdot \mathbf{x}_\parallel} \frac{1}{\sqrt{2}} \begin{pmatrix} e^{-i(\varphi+\pi/2)} \\ 1 \end{pmatrix},$$

$$\psi_- = e^{i\mathbf{k}_\parallel \cdot \mathbf{x}_\parallel} \frac{1}{\sqrt{2}} \begin{pmatrix} i\, e^{-i\varphi} \\ 1 \end{pmatrix} - e^{i\mathbf{k}_\parallel \cdot \mathbf{x}_\parallel} \frac{1}{\sqrt{2}} \begin{pmatrix} e^{-i(\varphi-\pi/2)} \\ 1 \end{pmatrix} \tag{4.31}$$

which shows that the spins lie in the xy-plane, with orientations $\uparrow$-spin $= \pi/2$ and $\downarrow$-spin $= -\pi/2$ with respect to the wavevector $\mathbf{k}_\parallel$.

The dispersion is shown in Figure 4.17. We find the following:

- The spin splitting $E_\uparrow(\mathbf{k}_\parallel) - E_\downarrow(\mathbf{k}_\parallel) = 2\alpha_R k_\parallel$ is linear in $k_\parallel$.
- The $\pm\dfrac{\pi}{2}$-spin parabolas are shifted in opposite directions by $k_0 = m^* \alpha_R / \hbar^2$.
- The energy minimum of each parabola is at $\Delta_{SO} = -m^* \alpha_R / 2\hbar^2$.
- The spin orientation axis is independent of the magnitude $k_\parallel$ and depends only on the direction of the $\mathbf{k}_\parallel$ vector.
- We note that for $\mathbf{k}_\parallel \to -\mathbf{k}_\parallel$, the angle φ changes to $\varphi + \pi$, reversing the spin orientation axis.
- The spectrum (4.30) shows that the Kramers degeneracy $E_\uparrow(\mathbf{k}_\parallel) = E_\downarrow(-\mathbf{k}_\parallel)$ holds.

In all, the magnetic moment is zero when averaged over all states $\mathbf{k}_\parallel$. This is consistent with the absence of a **B** field.

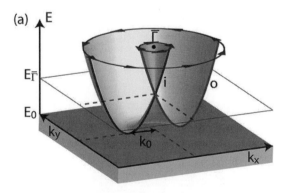

Figure 4.17 Rashba splitting.

Finally, writing the wave vector as $\mathbf{k}_\parallel = k\,(\cos\varphi,\,\sin\varphi)$, we can express the two Fermi surfaces by

$$k_\pm^{\mathrm{F}} = \mp\frac{\alpha_R m^*}{\hbar^2} + \sqrt{\left(\frac{\alpha_R m^*}{\hbar^2}\right)^2 + \frac{2m^*}{\hbar^2}\varepsilon_F}.$$

The Rashba Parameter α_R

We used a Lorentz transform to express the Rashba Hamiltonian in terms of the confining electric field. However, in general, the Rashba parameter α_R is not just proportional to it. For example, in the case of the Au(111) surface, we find that the work function $\Phi = 4.3\,\mathrm{eV}$ and $\lambda_F \sim 5\,\text{Å}$ yield a value of $\Delta E = \alpha_R k \sim 10^{-6}\,\mathrm{eV}$ – several orders of magnitude smaller than the experimentally observed splitting of $110\,\mathrm{meV}$ [117]. We note that the potential gradient in the Au atom is much larger, and leads to an atomic spin–orbit splitting of the $6p$ level of $0.47\,\mathrm{eV}$. Since the surface state is derived partially from these levels, we would suspect that the atomic SOC may contribute to the observed splitting. For pedagogical reasons, and to demonstrate how such a contribution emerges, we shall outline a tight-binding approach to determine the surface states and derive an expression for α_R that includes atomic contributions [153], leaving the actual calculations as an exercise.

Tight-Binding Model

The surface states of Au(111) are derived from s and p orbitals, but we note that the s orbital does not undergo any spin–orbit splitting, so we shall exclude it from our Wannier basis. We will consider a sheet of hexagonally arranged atoms that represents the Au(111) surface, and limit the basis per atom to the three atomiclike states p_x, p_y, and p_z. This choice is sufficient to represent the essential physics of the problem, while simplifying the calculations and making it as tractable as possible. We set the z-axis to be perpendicular to the atomic layer. The Hamiltonian, limited to nearest neighbor hopping, can then be written as

$$\mathcal{H} = \sum_{\substack{\alpha\beta \\ \langle ij\rangle}} t_{\alpha\beta}(\mathbf{d}_{ij})\,\big|p_{i\alpha}\big\rangle\big\langle p_{j\beta}\big|, \qquad \alpha,\beta = x, y, z,$$

where $\mathbf{d}_{ij}$ is the vector joining nearest neighbors i, j. The hopping parameters $t_{\alpha\beta}(\mathbf{d}_{ij})$ are expressed in terms of $V_{pp\sigma} \equiv \eta_\sigma$ and $V_{pp\pi} \equiv \eta_\pi$ with the aid of directional cosines

$$t_{\alpha\beta}(\mathbf{d}_{ij}) = \begin{cases} \eta_\sigma\,\cos^2\theta_{ij} + \eta_\pi\,\sin^2\theta_{ij}, & \alpha,\beta \to x, x \\[4pt] (\eta_\sigma + \eta_\pi)\,\cos\theta_{ij}\,\sin\theta_{ij}, & \alpha,\beta \to x, y\,; \\[4pt] \eta_\sigma\,\sin^2\theta_{ij} - \eta_\pi\,\cos^2\theta_{ij}, & \alpha,\beta \to y, y \end{cases}$$

$$t_{\alpha\beta}(\mathbf{d}_{ij}) = \begin{cases} -\eta_\pi, & \alpha,\beta \to z, z \\[4pt] \xi\,\cos\theta_{ij}, & \alpha,\beta \to x, z \\[4pt] \xi\,\sin\theta_{ij}, & \alpha,\beta \to y, z \end{cases}$$

For an isolated sheet, the transformation of p_x and p_y orbitals is even under reflection (inversion) though the sheet, while that of p_z orbitals is odd. Consequently, p_z does not mix with the other two. However, this symmetry is broken for a surface, where the surface potential $V(z)$ will induce mixing of p_z with the planar orbitals. Actually, the matrix element $\langle p_z | V | p_{x,y} \rangle = \xi$ is essentially a measure of the gradient of the surface potential, playing the role of $\partial V / \partial z$ in the free 2D free electron model.

Next, we need to consider the intratomic SOC experienced by the p states, namely

$$\mathcal{H}_{soc} = \alpha \, \mathbf{L} \cdot \mathbf{S} = \frac{\alpha}{2} \left[L^+ \sigma^- + L^- \sigma^+ + L^z \sigma^z \right].$$

In the basis $|p_x \uparrow\rangle$, $|p_x \downarrow\rangle$, $|p_y \uparrow\rangle$, $|p_y \downarrow\rangle$, $|p_z \uparrow\rangle$, $|p_z \downarrow\rangle$, we obtain

$$\mathcal{H}_{soc} = \frac{\alpha}{2} \begin{pmatrix} 0 & -i & 0 & 0 & 0 & 1 \\ i & 0 & 0 & 0 & 0 & 1 \\ 0 & 0 & 0 & -1 & i & 0 \\ 0 & 0 & -1 & 0 & i & 0 \\ 0 & 0 & -i & -i & 0 & 0 \\ 1 & 1 & 0 & 0 & 0 & 0 \end{pmatrix}$$

which shows that SOC of $\uparrow$ and $\downarrow$ spins occurs in the p_x and p_y states only – the 2×2 block of zeros in the lower-right corner of the matrix represents SOC in the p_z orbitals. Thus, if the p_x and p_y states were not included, there would be no effect of SOI on the surface states of interest, namely, p_z-derived ones.

The lowest band around the Γ-point is mainly of p_z character. We shall therefore limit our analysis to the Hilbert subspace spanned by the p_z-like bands. To focus on these bands and to incorporate effective SOI contributions, we need to include virtual transitions involving p_x and p_y bands, to second order in the coupling. We shall use the method of *projected resolvent*, described in the following, to determine the effective SOI in the p_z derived surface states.

Method of Projected Resolvent and Effective Hamiltonians

In many cases, we are interested in the physics of a low-energy Hilbert subspace that is well separated in energy from its complementary space. A useful method to integrate out the irrelevant complementary subspace states and obtain an effective Hamiltonian for the states of interest starts with the system's resolvent. The resolvent $(E - \mathcal{H})^{-1}$ is just the Laplace transform of the time-evolution operator and hence contains all relevant information. We focus on the projected resolvent $P(E - \mathcal{H})^{-1}P$, where P is the projector of the low-energy sector. To determine this projection, we start with the following partitioning of the matrix $E - \mathcal{H}$

$$E - \mathcal{H} = \begin{pmatrix} E - P\mathcal{H}P & P\mathcal{H}Q \\ Q\mathcal{H}P & E - Q\mathcal{H}Q \end{pmatrix} = \begin{pmatrix} A & B \\ C & D \end{pmatrix},$$

where $Q = \mathbb{I} - P$. After some algebraic manipulations, we find

$$\begin{pmatrix} A & B \\ C & D \end{pmatrix}^{-1} = \begin{pmatrix} (A - BD^{-1}C)^{-1} & -A^{-1}B(D - CA^{-1}B)^{-1} \\ -D^{-1}C(A - BD^{-1}C)^{-1} & (D - CA^{-1}B)^{-1} \end{pmatrix}.$$

Thus, we write

$$P\frac{1}{E - \mathcal{H}}P = P\begin{pmatrix} E - P\mathcal{H}P & P\mathcal{H}Q \\ Q\mathcal{H}P & E - Q\mathcal{H}Q \end{pmatrix}^{-1}P$$

$$= \frac{1}{E - P\mathcal{H}P - P\mathcal{H}Q\,\dfrac{1}{E - Q\mathcal{H}Q}\,Q\mathcal{H}P}, \tag{4.32}$$

which yields the effective Hamiltonian

$$\mathcal{H}_{\text{eff}} = P\mathcal{H}P + P\mathcal{H}Q\,\frac{1}{E - Q\mathcal{H}Q}\,Q\mathcal{H}P \tag{4.33}$$

This method is very useful in constructing effective Hamiltonians for low-energy Hilbert subspaces.

In the present case, we are interested in the low-lying p_z bands in the vicinity of the Γ-point. Thus, we focus on the projected resolvent $P(E - \mathcal{H})^{-1}P$, with $P = |p_z \uparrow\rangle \langle p_z \uparrow| + |p_z \downarrow\rangle \langle p_z \downarrow|$. The matrices of interest are

$$P\mathcal{H}P = \begin{pmatrix} \varepsilon_z^{(0)} - h_\pi(\mathbf{k}) & 0 \\ 0 & \varepsilon_z^{(0)} - h_\pi(\mathbf{k}) \end{pmatrix}$$

$$h_\pi(\mathbf{k}) = 2\eta_\pi \left(\cos(k_x) + \cos\left(\frac{\sqrt{3}}{2}k_y + 0.5k_x \right) + \cos\left(\frac{\sqrt{3}}{2}k_y - 0.5k_x \right) \right)$$

$$P\mathcal{H}Q = \begin{pmatrix} a(\mathbf{k}) & a(\mathbf{k}) & b(\mathbf{k}) & b(\mathbf{k}) \\ b(\mathbf{k}) & b(\mathbf{k}) & a(\mathbf{k}) & a(\mathbf{k}) \end{pmatrix}$$

$$a(\mathbf{k}) = 2i\xi \left(\sin(k_x) + \sin\left(\frac{\sqrt{3}}{2}k_y + 0.5k_x \right) + \sin\left(\frac{\sqrt{3}}{2}k_y - 0.5k_x \right) \right)$$

$$b(\mathbf{k}) = i\alpha/2 + 2i\xi \left(\sin(k_x) + \sin\left(\frac{\sqrt{3}}{2}k_y + 0.5k_x \right) + \sin\left(\frac{\sqrt{3}}{2}k_y - 0.5k_x \right) \right).$$

For the region of small wavevectors around the Γ-point, we use the approximations $\cos(x) \approx 1 - x^2/2$, $\sin(x) \approx x$. We also set $E \approx \varepsilon_z^{(0)} (\mathbf{k} = 0)$ and $Q\mathcal{H}Q \approx Q\varepsilon_{x,y}^{(0)} (\mathbf{k} = 0)$. We carry out this procedure to second order in α, ξ and $\mathbf{k}$, and arrive at

$$\mathcal{H}_{\text{eff}} = \begin{pmatrix} -6\eta_\pi + \left(\dfrac{3}{2}\eta_\pi + \dfrac{9\xi^2}{\eta_\sigma} \right) k^2 & \dfrac{-6i\alpha\xi}{\eta_\sigma}(k_x - ik_y) \\ \dfrac{6i\alpha\xi}{\eta_\sigma}(k_x - ik_y) & -6\eta_\pi + \left(\dfrac{3}{2}\eta_\pi + \dfrac{9\xi^2}{\eta_\sigma} \right) k^2 \end{pmatrix}.$$

We are now able to make contact with the free electron model discussed previously. The diagonal term is the free electron model, with an effective mass determined by η_π. The off-diagonal contribution is nothing but the Rashba term, and the parameter α_R can now be identified as $\alpha_R = 6\alpha\xi/\eta_\sigma$. We see explicitly that the spin–orbit splitting of the lowest energy band depends on the atomic spin–orbit parameter α as well as the surface potential gradient, represented by the parameter ξ.

4.4.2 Two-Dimensional Electron Gas in Heterojunctions

The ability to confine electrons in 2D is important not only for studying fundamental physics, but also for electronic applications. For example, a two-dimensional electron gas has extremely long electron mean-free paths, almost achieving ballistic electronic motion.

What is nowadays used for this purpose are interfaces of lattice-matched materials such as between GaAs and $Ga_x Al_{1-x}As$; see Figure 4.18. The Fermi energy in wide-gap $Ga_x Al_{1-x}As$ is higher than in GaAs, so that electrons will migrate toward the GaAs layer and leave behind positively charged ions (donors). This leads to a formation of an interfacial charge distribution where the ensuing electrostatic potential brings about band bending, as shown in Figure 4.19, and an electron confining potential emerges perpendicular to the interface, a 1D quantum well. The carrier concentrations in the 2D gas vary between $2 \times 10^{11} \, cm^{-2}$ to $2 \times 10^{12} \, cm^{-2}$.

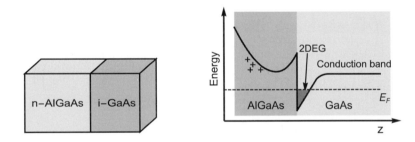

Figure 4.18 Heterojunction of $Ga_x Al_{1-x}As$-GaAs.

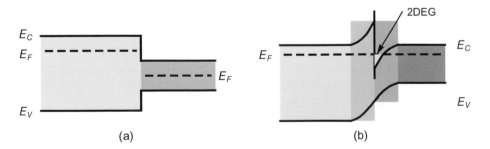

Figure 4.19 Fermi energies before, (a), and after, (b), junction formation. E_Fs lineup after charge transfer has taken place.

SOC Mechanisms in 2DEG

We can now surmise that there are two microscopic spin–orbit interaction mechanisms for a 2DEG in a heterojunction QW:

1. BIA, arising from lack of inversion symmetry in the bulk of the semiconductor crystal, and described by a Dresselhaus term
2. SIA, due to structural inversion asymmetry of the quantum well confining potential, put forward by Rashba and Bychkov

Both types of SOI result in spin splitting of conduction subbands in III-V semiconductor QW without a magnetic field.

Dresselhaus BIA Spin–Orbit Coupling

In order to gain insight in the effect of the Dresselhaus spin–orbit interaction in heterojunctions with zincblende structure, we start from the Dyakonov–Perel spin-orbit Hamiltonian for zincblende crystals in (4.27):

$$\mathcal{H}_D^{3D} = \gamma_D \left[p_x \left(p_y^2 - p_z^2 \right) \sigma_x + p_y \left(p_z^2 - p_x^2 \right) \sigma_y + p_z \left(p_x^2 - p_y^2 \right) \sigma_z \right].$$

To obtain the spin–orbit Hamiltonian in two-dimensional systems, we integrate over the growth direction. For a heterojunction grown along the (001)-direction, $\langle p_z \rangle = 0$, while $\langle p_z^2 \rangle$ is heterostructure dependent but a fixed number. The Dresselhaus Hamiltonian then reduces to

$$\mathcal{H}_D^{2D,(001)} = \gamma_D \left[-p_x \left\langle p_z^2 \right\rangle \sigma_x + p_y \left\langle p_z^2 \right\rangle \sigma_y + p_x \, p_y^2 \sigma_x - p_y \, p_x^2 \sigma_y \right].$$

The first two terms constitute the *linear Dresselhaus Hamiltonian*, and the last two form the *cubic Dresselhaus Hamiltonian*. Usually, the latter has much smaller strength, since $\langle p_z^2 \rangle \gg p_x^2, p_y^2$ due to the strong confinement along z. We then retain

$$\mathcal{H}_D^{2D,(001)} = \beta \left[-p_x \, \sigma_x + p_y \, \sigma_y \right], \tag{4.34}$$

where β depends on material properties and on $\langle p_z^2 \rangle$. It follows from (4.34) that the effective magnetic field that couples to the spin is aligned with the momentum for motion along (010), but is opposite to the momentum for motion along (100), as shown in Figure 4.20(a).

Rashba SIA Spin–Orbit Coupling

The gradient of the inversion asymmetric confining potential dominates, giving rise to an electric field along $\hat{\mathbf{z}} = (0,0,1)$. Thus, for electrons propagating in the 2DEG extended in the xy-plane, the Rashba Hamiltonian takes the form

$$\mathcal{H}_R = \frac{\hbar}{(2m_0 c^2)^2} \, \nabla V \cdot \mathbf{p} \times \boldsymbol{\sigma}$$

$$= \frac{\hbar \kappa_R}{m} \left(p_y \, \sigma_x - p_x \, \sigma_y \right).$$

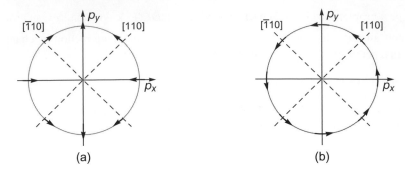

Figure 4.20 The red arrows indicate the orientation of the effective magnetic field acting on the electron spin as a result of (a) the Dresselhaus and (b) the Rashba spin–orbit interaction when the electron travels through a AlGaAs/GaAs heterojunction with momentum **p**.

Here the effective magnetic field is always orthogonal to the momentum, as shown in Figure 4.20(b). The parameter κ_R typically depends on the electric field $\mathcal{E} = \nabla V$ in the z-direction! Therefore the Rashba SO coupling can be modulated by an external field.

What the Rashba- and Dresselhaus-type SO couplings have in common is that they can cause splitting of otherwise degenerate energy levels at each **k**. From Figure 4.20, we see that the Rashba and Dresselhaus contributions add up for motion along the (110) direction and oppose each other along ($\bar{1}$10), giving rise to an anisotropic spin–orbit interaction.

Exercises

4.1 Consider the **k·p** expansion of the Hamiltonian, in the absence of spin–orbit coupling:

$$\mathcal{H} = \frac{p^2}{2m_e} + V(\mathbf{x}) + \frac{\hbar^2 k^2}{2m_e} + \frac{\hbar}{m_e}\mathbf{k} \cdot \mathbf{p}.$$

We need to construct an effective Hamiltonian for the valence band in GaAs, treating the last term perturbatively and limiting the lattice-periodic basis set to the s-like function of conduction band minimum and the triply degenerate p-like valence band maximum functions.

(a) Based on the transformation of the momentum operator **p**, the s-like and the p-like functions under inversion, which matrix elements $\langle i | \mathbf{p} | j \rangle$ are nonzero? ($i, j = s$ and p_x, p_y, p_z states)

(b) Use the perturbation expression

$$\mathcal{H}_{ij} = \left(E_p + \frac{\hbar^2 k^2}{2m_e} \right) \delta_{ij} + \frac{\hbar^2}{m_e^2} \sum_{\alpha\beta} \frac{\langle i | p_\alpha | s \rangle \langle s | p_\beta | j \rangle}{E_p - E_s}$$

to construct the effective the 3×3 Hamiltonian matrix for the valence band along a general direction **k**. Take $E_p = 0$ and $E_s = E_g$ at $k = 0$.

(c) Obtain the dispersion of the valence bands along the Δ, Σ, and Λ directions.

4.2 Repeat problem (1) in the presence of SOC, where the $\mathbf{k} \cdot \mathbf{p}$ Hamiltonian now reads

$$\mathcal{H} = \frac{p^2}{2m_e} + V(\mathbf{x}) + \frac{\hbar^2 k^2}{2m_e} + \frac{\hbar}{m_e} \mathbf{k} \cdot \boldsymbol{\pi} + \alpha \, (\boldsymbol{\sigma} \times \nabla V) \cdot \mathbf{p}$$

$$\boldsymbol{\pi} = \mathbf{p} + \alpha \, (\boldsymbol{\sigma} \times \nabla V).$$

In the presence of SOC, spin has to be included in the basis set, and the lattice-periodic basis set is expanded to eight functions. However, the s-like states remain unaffected.

(a) Starting with the $\ell = 1$ spherical harmonics, construct the spinors for $\mathbf{L} + \mathbf{S}$, $j = 3/2, 1/2$. Express your results in terms of p_x, p_y, and p_z.

(b) Based on the transformation of $\boldsymbol{\pi}$ transform under inversion, which matrix elements $\langle i | \boldsymbol{\pi} | j \rangle$ are nonzero?

(c) Ignoring SOC, construct the 6×6 Hamiltonian matrix and find the eigenenergies.

(d) Now add the spin–orbit coupling. Show that the matrix $\mathcal{H}$ is diagonalized at $\mathbf{k} = 0$, and calculate the energy gap opened by the spin–orbit coupling, using

$$\langle i | [(\nabla V) \times \mathbf{p}]_j | k \rangle = \Delta_0 \epsilon_{ijk}; \quad ijk = xyz.$$

(e) The spin–orbit coupling opens a gap between $j = 3/2$ and $j = 1/2$ terms in $\mathcal{H}'_{ij}$. Assuming $|\mathbf{k}|$ is small enough, the matrix elements that couple $j = 3/2$ and $j = 1/2$ are negligible, and the 6×6 matrix reduces to a 4×4 and a 2×2 matrix. Calculate the elements in the 4×4 matrix (remember it is hermitian), diagonalize it, and find the dispersion and the masses of the heavy hole and light hole.

4.3 Consider chiral behavior of light and heavy holes:

In the valence band, the "spin" of light and heavy holes is tightly bound to their momentum. This has many interesting consequences, particularly where external forces can mix the light- and heavy-hole states. Here, we consider reflections from an ideal flat potential wall.

(a) What happens to a heavy hole with helicity $+3/2$ ($\mathbf{J} \| \mathbf{p}$)?

(b) Now consider the case of an arbitrary angle of incidence θ. What will be the composition of the reflected state?

4.4 Consider two-dimensional electron gas:

The dimensionality of a system can be reduced by confining the electrons in certain directions. Consider an electron gas in an external potential

$$V = \begin{cases} 0 & \text{for } |z| < d/2 \\ V_0 & \text{for } |z| > d/2 \end{cases}$$

1. What is the density of states as a function of energy for $V_0 = \infty$? (Discuss what happens at low and high energies.) Assume $d = 100 \text{Å}$.

2. Show that the chemical potential of the two-dimensional electron gas is given by

$$\mu(T) = k_B T \, \ln\left(\exp\left[\frac{\pi \, n\hbar^2}{m k_B T} \right] - 1 \right)$$

for n electrons per unit area.

3. If V_0 is finite, up to what temperatures can we consider the electrons to be two-dimensional?

4. If we can produce a potential of 100 meV and reach a temperature of 20 mK, what is the range of thicknesses feasible for the study of such two-dimensional electron gas?

4.5 Derive an expression for the density of states in the presence of the Rashba interaction.

4.6 Consider the coexistence of the Rashba and Dresselhaus mechanisms in a heterojunction quantum well. Derive the dispersion relation for the 2DEG and the corresponding eigenfunctions.

4.7 Consider the case where the Dresselhaus and Rashba SOI are of equal strength, and a persistent spin helix is predicted.

(a) Express the sum of the SOI part of the Hamiltonian in terms of $k_x + k_y$ and $\sigma_x - \sigma_y$.

(b) Express the Hamiltonian in terms of $k_\pm = (k_x \pm k_y)/\sqrt{2}$.

(c) Now obtain an expression for the total Hamiltonian under the global spin rotation via the unitary transformation

$$U = \frac{1}{\sqrt{2}} \left[1 + \frac{i}{\sqrt{2}} \left(\sigma_x + \sigma_y \right) \right]$$

and determine its dispersion.

4.8 Consider the intratomic spin–orbit coupling experienced by the p states

$$\mathcal{H}_{soc} = \alpha \, \mathbf{L} \cdot \mathbf{S} = \frac{\alpha}{2} \left[L^+ \sigma^- + L^- \sigma^+ + 2 L^z \sigma^z \right].$$

Show that

$$\mathcal{H}_{soc} = \frac{\alpha}{2} \begin{pmatrix} 0 & 0 & -i & 0 & 0 & 1 \\ 0 & 0 & 0 & i & -1 & 0 \\ i & 0 & 0 & 0 & 0 & -i \\ 0 & -i & 0 & 0 & -i & 0 \\ 0 & -1 & 0 & i & 0 & 0 \\ 1 & 0 & i & 0 & 0 & 0 \end{pmatrix}$$

in the basis $|p_x \uparrow\rangle, \, |p_x \downarrow\rangle, \, |p_y \uparrow\rangle, \, |p_y \downarrow\rangle, \, |p_z \uparrow\rangle, \, |p_z \downarrow\rangle$.

4.9 Consider a tight-binding model with Rashba-type spin–orbit interaction.

Suppose there is a tight-binding Hamiltonian for electrons on a two-dimensional square lattice (lattice constant a $a = 1$) with the Rashba spin–orbit interaction ($\hbar = 1$):

$$\mathcal{H} = -t \sum_{\langle n,n'\rangle\sigma} \left(|n\sigma\rangle \langle n'\sigma| + hc \right)$$

$$+ \alpha_R \sum_{n\alpha\beta} \left(i \, |n\alpha\rangle \, \sigma_x^{\alpha\beta} \, \langle (n+y)\beta| - i \, |n\alpha\rangle \, \sigma_y^{\alpha\beta} \, \langle (n+x)\beta| + hc \right),$$

where t is the electronic hopping parameter and α_R the strength of the Rashba interaction. $\sigma_{x,y}$ are the Pauli matrices. n indexes the lattice sites, $\langle nn'\rangle$ indicates summation over nearest neighbors, while $n + x$, $n + y$ denotes the nearest neighbor of site n in the x, y directions.

(a) Show that the two electronic bands (the upper and lower Rashba bands) have dispersions given by

$$\varepsilon_{pm} = -2t \left[\cos(k_x) + \cos(k_y)\right] \pm 2\alpha_R \sqrt{\sin^2(k_x) + \sin^2(k_y)}$$

with corresponding eigenvectors

$$\psi_{\mathbf{k}}^{\pm} = \frac{1}{\sqrt{2}} \left[\phi_{\mathbf{k}\uparrow} \pm \frac{\sin(k_y) - i\sin(k_x)}{\sqrt{\sin^2(k_x) + \sin^2(k_y)}} \right].$$

(b) What is the ground-state energy ε_0 (the dispersion minima of $\varepsilon^-(\mathbf{k})$ and what is its degeneracy? What are the locations of these minimum points in the Brillouin zone? By expanding $\varepsilon^-(\mathbf{k})$ around the band minima, show that the ratio of the effective mass (in the presence of spin–orbit interaction) and the bare band mass $m_0 = 1/4t$ is given by

$$\frac{m_{SO}}{m_0} = \frac{1}{\sqrt{1 + \dfrac{\alpha_R^2}{2t^2}}}.$$

5

Linear Response and the Dielectric Function

5.1 Introduction

In order to access the physical properties of a system, one has to act on it with some external probe and observe how the system would respond. External probes include beams of photons, electrons, neutrons, atoms, and sound. In this chapter, we will present a general formalism that describes the interaction of an external probe with a material system and derive general forms for the system response function. The formalism assumes that the probe interaction is weak so that a linear approximation can be invoked. The resulting response can then be defined in terms of a *linear susceptibility*. The requirement that the response satisfies causality, that is, the response cannot precede the action producing it, is shown to lead to a relation between the real and imaginary parts of the susceptibility known as the Kramers–Krönig relation. As important applications to this formalism, we will present a treatment of electrical transport as expressed in the Kubo formula for electric conductivity, and use it to derive the Hall conductivity. We conclude this chapter by focusing on the dielectric function and the quantum mechanical treatment of the interaction of a material system with photons, which is generally known as the optical properties of the system.

5.2 Linear Response

Mathematically, the probe action is implemented by adding a time-dependent perturbation term, $\delta\mathcal{H}(t)$, to the originally unperturbed Hamiltonian $\mathcal{H}_0$ representing the isolated system, namely

$$\mathcal{H} = \mathcal{H}_0 + \delta\mathcal{H}(t). \tag{5.1}$$

The presence of the time-dependent perturbing term will induce a corresponding term in the density operator

$$\rho(t) = \rho_0 + \delta\rho(t). \tag{5.2}$$

146

We now introduce the important assumption that the action of the perturbing field on the system is *nondissipative*, which means that the production of entropy, or Joule heat, is negligible.[1] The equation of motion of the density operator gives

$$i\hbar \frac{d\rho(t)}{dt} = i\hbar \frac{d\delta\rho(t)}{dt} = [\mathcal{H}, \rho(t)]$$

$$= [\mathcal{H}_0, \delta\rho(t)] + [\delta\mathcal{H}(t), \rho_0] + [\delta\mathcal{H}(t), \delta\rho(t)], \tag{5.3}$$

where we used the fact that $[\mathcal{H}_0, \rho_0] = 0$. The approximation of *linear response theory* now is to neglect the last term, which is quadratic in the perturbation. Writing (5.3) in the interaction picture, $\delta\rho_I = \exp[i\mathcal{H}_0 t/\hbar] \, \delta\rho(t) \, \exp[-i\mathcal{H}_0 t/\hbar]$, we get

$$i\hbar \frac{d\delta\rho_I(t)}{dt} = -[\mathcal{H}_0, \delta\rho_I(t)] + [\mathcal{H}_0, \delta\rho_I(t)] + [\delta\mathcal{H}_I(t), \rho_0] = [\delta\mathcal{H}_I(t), \rho_0]. \tag{5.4}$$

Next, we adopt the scenario that the perturbation is switched on adiabatically at $t = -\infty$, and write

$$\delta\mathcal{H}_{\text{ad}} = \delta\mathcal{H} \, e^{\eta t},$$

where η is an arbitrary infinitesimal positive constant, to be set to zero after the calculation is completed. A thermal bath is introduced prior to switching the perturbation on, and the levels are populated accordingly, such that

$$\rho_0 = \frac{1}{Z_0} e^{-\beta\mathcal{H}_0} = \frac{1}{Z_0} \sum_n e^{-\beta E_n} |\phi_n\rangle \langle\phi_n|$$

$$\mathcal{H}_0 |\phi_n\rangle = E_n |\phi_n\rangle, \qquad Z_0 = \sum_n e^{-\beta E_n} \quad \textit{Partition function.} \tag{5.5}$$

The thermal bath is then removed and the perturbation switched on adiabatically slowly, so that the wavefunctions evolve without making transitions. We should note that $\delta\rho(-\infty) = 0$, so that the integration of (5.4) yields

$$\delta\rho(t) = \frac{-i}{\hbar} e^{-i\mathcal{H}_0 t/\hbar} \int_{-\infty}^{t} dt' \left[e^{i\mathcal{H}_0 t'/\hbar} \delta\mathcal{H}(t') e^{-i\mathcal{H}_0 t'/\hbar}, \rho_0 \right] e^{i\mathcal{H}_0 t/\hbar}. \tag{5.6}$$

The time-dependent part of the average value for an observable A can be expressed as

$$\langle A(t) \rangle = \text{Tr}\left[\delta\rho(t) A\right]$$

$$= \frac{-i}{\hbar} \text{Tr}\left[e^{-i\mathcal{H}_0 t/\hbar} \int_{-\infty}^{t} dt' \left[e^{i\mathcal{H}_0 t'/\hbar} \delta\mathcal{H}(t') e^{-i\mathcal{H}_0 t'/\hbar}, \rho_0 \right] e^{i\mathcal{H}_0 t/\hbar} A \right]. \tag{5.7}$$

Using the cyclic invariance of the trace, $\text{Tr}[AB \ldots C] = \text{Tr}[B \ldots CA]$, we obtain

$$\langle A(t) \rangle = \frac{-i}{\hbar} \text{Tr}\left[\int_{-\infty}^{t} dt' \left[e^{i\mathcal{H}_0 t'/\hbar} \delta\mathcal{H}(t') e^{-i\mathcal{H}_0 t'/\hbar}, \rho_0 \right] e^{i\mathcal{H}_0 t/\hbar} A \, e^{-i\mathcal{H}_0 t/\hbar} \right] \tag{5.8}$$

---

[1] The physical interpretation of this assumption is that the external probing field is turned on so slowly that the system responds *adiabatically*, that is, without making transitions to other states. More explicitly, *adiabatic response* means that, in the statistical ensemble, the internal dynamics of the system follow the imposed external variation.

Actually, a perturbation to the system is effected through the coupling of an external perturbing field to one or more of the system's operators.[2]

For simplicity, we shall assume that the coupling is through a single operator $\mathcal{O}$, which is taken to be Hermitian, and we write

$$\delta \mathcal{H}(t) = \int d\mathbf{x}\, h(\mathbf{x}, t)\, \mathcal{O}(\mathbf{x}). \tag{5.9}$$

Substituting in (5.8), we obtain

$$
\begin{aligned}
\langle A(t) \rangle &= \frac{-i}{\hbar}\, \mathrm{Tr}\left[\int_{-\infty}^{t} dt' \int d\mathbf{x}'\, h(\mathbf{x}', t) \right. \\
&\quad \left. \times \left[e^{i\mathcal{H}_0 t'/\hbar}\, \mathcal{O}(\mathbf{x}')\, e^{-i\mathcal{H}_0 t'/\hbar},\, \rho_0 \right] e^{i\mathcal{H}_0 t/\hbar}\, A\, e^{-i\mathcal{H}_0 t/\hbar} \right] \\
&= \frac{-i}{\hbar} \int_{-\infty}^{t} dt' \int d\mathbf{x}'\, h(\mathbf{x}', t)\, \mathrm{Tr}\left\{ \left[\mathcal{O}(\mathbf{x}', t'),\, \rho_0 \right] A(\mathbf{x}, t) \right\}. \tag{5.10}
\end{aligned}
$$

Using the identity

$$\mathrm{Tr}\left([A, B]\, C \right) = \mathrm{Tr}\left(ABC - BAC \right) = \mathrm{Tr}\left(B\, [C, A] \right),$$

we get

$$
\begin{aligned}
\langle A(\mathbf{x}, t) \rangle &= \frac{-i}{\hbar} \int_{-\infty}^{t} dt' \int d\mathbf{x}'\, h(\mathbf{x}', t)\, \mathrm{Tr}\left\{ \rho_0 \left[A(\mathbf{x}, t),\, \mathcal{O}(\mathbf{x}', t') \right] \right\} \\
&= \frac{-i}{\hbar} \int_{-\infty}^{t} dt' \int d\mathbf{x}'\, h(\mathbf{x}', t)\, \left\langle \left[A(\mathbf{x}, t),\, \mathcal{O}(\mathbf{x}', t') \right] \right\rangle_0, \tag{5.11}
\end{aligned}
$$

where $\langle\,\rangle_0$ denotes thermal and quantum averages taken with respect to the *unperturbed* Hamiltonian.

> In this case, the *interaction-picture operators* become *Heisenberg operators*.

Moreover, because $\mathcal{H}_0$ is time independent, or time-translationally invariant, we find from (5.8) that

$$
\begin{aligned}
\langle A(\mathbf{x}, t)\, \mathcal{O}(\mathbf{x}', t') \rangle_0 &= \frac{1}{Z_0}\, \mathrm{Tr}\left(e^{-\beta \mathcal{H}_0}\, e^{i\mathcal{H}_0 t/\hbar}\, A\, e^{-i\mathcal{H}_0 t/\hbar}\, e^{i\mathcal{H}_0 t'/\hbar}\, \mathcal{O}\, e^{-i\mathcal{H}_0 t'/\hbar} \right) \\
&= \frac{1}{Z_0}\, \mathrm{Tr}\left(e^{-\beta \mathcal{H}_0}\, e^{i\mathcal{H}_0 (t-t')/\hbar}\, A\, e^{-i\mathcal{H}_0 (t-t')/\hbar}\, \mathcal{O} \right) \\
&= \langle A(t - t')\mathcal{O}(0) \rangle_0 \\
&= \frac{1}{Z_0}\, \mathrm{Tr}\left(e^{-\beta \mathcal{H}_0}\, A\, e^{i\mathcal{H}_0 (t'-t)/\hbar}\, \mathcal{O}\, e^{-i\mathcal{H}_0 (t'-t)/\hbar} \right) \\
&= \langle A(0)\mathcal{O}(t' - t) \rangle_0. \tag{5.12}
\end{aligned}
$$

---

[2] As an example, the electric field couples to the position operator, while the vector potential couples to the momentum operator.

> *Any average of pairs of Heisenberg operators* only depends on the *time difference.*

We now define the susceptibility χ as

$$\chi(\mathbf{x},\mathbf{x}';t-t') = \frac{-i}{\hbar} \Theta\left(t-t'\right) \left\langle \left[A(\mathbf{x},t-t'), \mathcal{O}(\mathbf{x}',0) \right] \right\rangle_0. \tag{5.13}$$

The Θ function accounts for causality: Response at time t cannot precede action at time t'. If the system has spatial translation invariance, then

$$\chi(\mathbf{x},\mathbf{x}';t-t') \;\rightarrow\; \chi(\mathbf{x}-\mathbf{x}';t-t') = \chi(\mathbf{r};\tau).$$

We write

$$\langle A(\mathbf{x},t)\rangle = \int dt' \, \chi(\mathbf{x}-\mathbf{x}';t-t') \, h\left(\mathbf{x}',t'\right). \tag{5.14}$$

- Time variation of any measurable quantity is obtained through the linear response function (5.14), which is only related to averages on the unperturbed system.
- Correlations that are nonzero only for $t > t'$ are known as *retarded correlation functions*. They directly correspond to physically observable quantities.

Taking the spatial/temporal Fourier transform of (5.14), we get

$$\langle A(\mathbf{q},\omega)\rangle = \chi(\mathbf{q},\omega)\, h(\mathbf{q},\omega) \tag{5.15}$$

$$\chi(\mathbf{q},\omega) = \int d\mathbf{r} \int_0^{+\infty} d\tau \, e^{i(\mathbf{q}\cdot\mathbf{r}-\omega\tau)} \, \langle A(\mathbf{r},\tau)\mathcal{O}(0,0)\rangle_0.$$

5.2.1 Causality and the Kramers–Krönig Relation

We will now explore the analytical properties of the response function in the frequency domain. In the following, we drop the spatial dependence of the response function, which is not relevant for what we are going to demonstrate. The response of an operator A in the presence of an external probe that couples to $\mathcal{O}$ is defined by (5.14) and (5.15).

As we stated previously, because of causality the response function, or susceptibility, vanishes for $\tau < 0$. With

$$\chi(\tau) = \int_{-\infty}^{\infty} d\omega \, e^{-i\omega\tau} \, \chi(\omega), \tag{5.16}$$

we analytically continue in the complex frequency plane to $\chi(z)$. If we assume, as it is always the case, that $\chi(z)$ does not diverge exponentially for $|z| \rightarrow \infty$, we can regard (5.16) as the result of a contour integral

$$\chi(\tau) = \oint dz \, e^{-iz\tau} \, \chi(z), \tag{5.17}$$

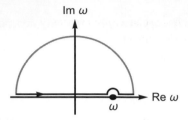

Figure 5.1 Integration contour.

where the contour is in the upper half plane for $\tau < 0$ and in the lower for $\tau > 0$. The integral captures all poles lying inside the contour.

Since $\chi(\tau) = 0$ for $\tau < 0$, it follows that *as consequence of causality, $\chi(z)$ is analytic in the upper half plane.*

We consider the contour depicted in Figure 5.1. Since there are no poles enclosed by the contour, the integral is zero:

$$\oint_C dz\, \frac{\chi(z)}{\omega - z} = 0. \tag{5.18}$$

Yet, because the integrand vanishes when $|z| \to \infty$, this integral is also equal to the line integral along the lower edge, hence

$$0 = \left\{ \int_{-\infty}^{\omega - \epsilon} + \int_{\omega + \epsilon} \right\} d\omega'\, \frac{\chi(\omega')}{\omega - \omega'} + \oint_{z = \omega + \epsilon\, \exp[i\theta],\ \theta \in [\pi, 0]} dz\, \frac{\chi(z)}{\omega - z}$$

$$= \mathcal{P} \int_{-\infty}^{\infty} d\omega'\, \frac{\chi(\omega')}{\omega - \omega'} - i \int_{\pi}^{0} d\theta\, \chi\left(\omega + \epsilon\, e^{i\theta}\right),$$

where $\mathcal{P}$ denotes the principal value of the integral. In the limit $\epsilon \to 0$, the preceding expression simplifies into

$$\mathcal{P} \int_{-\infty}^{\infty} d\omega'\, \frac{\chi(\omega')}{\omega - \omega'} - i\,\pi\,\chi(\omega) = 0. \tag{5.19}$$

Setting

$$\chi(\omega) = \chi_1(\omega) + i\,\chi_2(\omega),$$

Equation 5.19 implies that

$$\chi_2(\omega) = \mathcal{P} \int_{-\infty}^{\infty} d\omega'\, \frac{\chi_1(\omega')}{\omega - \omega'}$$

$$\chi_1(\omega) = -\mathcal{P} \int_{-\infty}^{\infty} d\omega'\, \frac{\chi_2(\omega')}{\omega - \omega'}, \tag{5.20}$$

which constitute the *Kramers–Krönig* relations.

Consequently, we find that because of causality, the real and imaginary parts of the response function are not independent.

5.2.2 Spectral Representation

In the spectral representation, the response function is expressed in terms of a complete set of eigenkets of the Hamiltonian. As we will see later, this picture actually brings to light the physical aspects of the susceptibility.

We begin with (5.5):

$$\rho_0 = \frac{1}{Z_0} \sum_n e^{-\beta E_n} |\phi_n\rangle \langle\phi_n|, \quad \mathcal{H}_0 |\phi_n\rangle = E_n |\phi_n\rangle.$$

It is simply sufficient to know that such a complete basis set exists without having to know it explicitly. Using (5.13),

$$\chi(\tau) = \frac{-i}{\hbar} \Theta(\tau) \left\langle \left[\mathcal{O}(\mathbf{x},\tau), A(\mathbf{x}',0) \right] \right\rangle_0,$$

we get

$$\chi(\tau) = \frac{-i}{\hbar} \Theta(\tau) \frac{1}{Z_0} \text{Tr}\left[e^{-\beta\mathcal{H}_0} \left(A(\tau)\mathcal{O}(0) - \mathcal{O}(0)A(\tau) \right) \right]$$

$$= \frac{-i}{\hbar} \Theta(\tau) \frac{1}{Z_0} \sum_n \left\langle \phi_n \left| e^{-\beta\mathcal{H}_0} \left(A(\tau)\mathcal{O}(0) - \mathcal{O}(0)A(\tau) \right) \right| \phi_n \right\rangle$$

$$= \frac{-i}{\hbar} \Theta(\tau) \frac{1}{Z_0} \sum_{n,m} \left[\langle\phi_n| e^{-\beta\mathcal{H}_0} A(\tau) |\phi_m\rangle \langle\phi_m| \mathcal{O}(0) |\phi_n\rangle \right.$$

$$\left. - \langle\phi_n| e^{-\beta\mathcal{H}_0} \mathcal{O}(0) |\phi_m\rangle \langle\phi_m| A(\tau) |\phi_n\rangle \right], \qquad (5.21)$$

We obtain in the Schrödinger picture

$$\chi(\tau) = \frac{-i}{\hbar} \Theta(\tau) \frac{1}{Z_0} \sum_{n,m} \left[e^{-\beta E_n} e^{i(E_n - E_m)\tau/\hbar} \langle\phi_n| A |\phi_m\rangle \langle\phi_m| \mathcal{O} |\phi_n\rangle \right.$$

$$\left. - e^{-\beta E_n} e^{i(E_m - E_n)\tau/\hbar} \langle\phi_n| \mathcal{O} |\phi_m\rangle \langle\phi_m| A |\phi_n\rangle \right], \qquad (5.22)$$

Taking the Fourier transform with respect to τ,

$$\chi(\omega) = \int_0^\infty d\tau \, e^{i(\omega+i\eta)\tau} \chi(\tau)$$

$$= \frac{-1}{\hbar Z_0} \sum_n e^{-\beta E_n} \sum_m \left[\frac{\langle\phi_n| A |\phi_m\rangle \langle\phi_m| \mathcal{O} |\phi_n\rangle}{\omega - \omega_{mn} + i\eta} - \frac{\langle\phi_n| \mathcal{O} |\phi_m\rangle \langle\phi_m| A |\phi_n\rangle}{\omega + \omega_{mn} + i\eta} \right]$$

$$= \left\langle\!\left\langle \mathcal{O} \big| A \right\rangle\!\right\rangle_\omega, \qquad (5.23)$$

where $\omega_{mn} = (E_m - E_n)/\hbar$.

Alternatively, interchanging the indices in the second term of (5.22), we get

$$\chi(\tau) = \frac{-i\,\Theta(\tau)}{\hbar\,Z_0} \sum_{n,m} e^{i(E_n - E_m)\tau/\hbar} \, \langle\phi_n|\,A\,|\phi_m\rangle \, \langle\phi_m|\,\mathcal{O}\,|\phi_n\rangle \left[e^{-\beta E_n} - e^{-\beta E_m} \right] \quad (5.24)$$

Taking the Fourier transform

$$\begin{aligned}
\chi(\omega) &= \int d\tau\, e^{i(\omega+i\eta)\tau}\, \chi(\tau) \\
&= \frac{-i}{\hbar} \int_0^{+\infty} d\tau\, e^{i(\omega+i\delta)\tau} \\
&\quad \times \frac{1}{Z_0} \sum_{n,m} e^{i(E_n - E_m)\tau/\hbar} \, \langle\phi_n|\,A\,|\phi_m\rangle \, \langle\phi_m|\,\mathcal{O}\,|\phi_n\rangle \left[e^{-\beta E_n} - e^{-\beta E_m} \right] \\
&= \frac{1}{Z_0} \sum_{n,m} \langle\phi_n|\,A\,|\phi_m\rangle \, \langle\phi_m|\,\mathcal{O}\,|\phi_n\rangle \, \frac{e^{-\beta E_n} - e^{-\beta E_m}}{\hbar\omega + E_n - E_m + i\eta}.
\end{aligned} \quad (5.25)$$

Using the identity

$$\frac{1}{x + i\eta} = \mathcal{P}\frac{1}{x} - i\pi\,\delta(x),$$

we obtain an expression for χ_2:

$$\begin{aligned}
\chi_2(\omega) &= -\frac{\pi}{Z_0} \sum_{n,m} \langle\phi_n|\,A\,|\phi_m\rangle \, \langle\phi_m|\,\mathcal{O}\,|\phi_n\rangle \left[e^{-\beta E_n} - e^{-\beta E_m} \right] \delta(\hbar\omega + E_n - E_m) \\
&= -\frac{\pi}{Z_0} \left(1 - e^{-\beta\hbar\omega} \right) \sum_{n,m} \langle\phi_n|\,A\,|\phi_m\rangle \, \langle\phi_m|\,\mathcal{O}\,|\phi_n\rangle \, e^{-\beta E_n} \delta(\hbar\omega + E_n - E_m).
\end{aligned}$$

This last form becomes transparent if we consider energy absorption due to the perturbation coupling to $\mathcal{O}$, in which case $A = \mathcal{O}$ and we write

$$\chi_2(\omega) = -\frac{\pi}{Z_0} \left(1 - e^{-\beta\hbar\omega} \right) \sum_{n,m} |\langle\phi_m|\,\mathcal{O}\,|\phi_n\rangle|^2 \, e^{-\beta E_n} \delta(\hbar\omega + E_n - E_m). \quad (5.26)$$

We may separate this expression into

$$F_n = \sum_m |\langle\phi_m|\,\mathcal{O}\,|\phi_n\rangle|^2 \, \delta(\hbar\omega + E_n - E_m),$$

which is just Fermi's golden rule for the perturbation induced transition rate from initial state $|\phi_n\rangle$ to all final states $|\phi_m\rangle$ and

$$\sum_n e^{-\beta E_n} F_n,$$

which involves the occupation probability of the initial state.

5.2.3 Kubo Formula for Electrical Conductivity

The conductivity tensor $\sigma_{\alpha\beta}(\omega)$ measures the current linearly induced by an electric field:

$$j_\alpha = \sigma_{\alpha\beta}\, E_\beta.$$

The identification of the operator $\mathcal{O}$ that couples to the external electric field $\mathcal{E}$ depends on the source of the field:

1. For a scalar potential, $\mathcal{E} = \nabla\phi$, and the position operator $\mathbf{x}$ can be identified with $\mathcal{O}$, giving rise to the perturbation

$$\delta\mathcal{H} = e\mathcal{E} \cdot \mathbf{x}.$$

The operator A with the current operator $-e\,\mathbf{v}/\Omega$, where Ω is the system's volume. An important detail must be stressed at this point. The macroscopic field inside the sample includes, by definition, screening effects due to the electronic system, while the perturbation entering via the $\mathcal{O}$ operator is the bare, or unscreened, one – for the time being, and for pedagogical reasons, we will simply ignore the screening issue. Substituting in (5.23) for $\mathcal{O}$ and A, we identify the electrical conductivity tensor as

$$\sigma_{\alpha\beta}(\omega) = \frac{-ie^2}{\Omega} \left\langle\!\!\left\langle v_\alpha \middle| x_\beta \right\rangle\!\!\right\rangle_\omega.$$

Denoting the ground state as $|\phi_0\rangle$, and using the identity

$$i\hbar \left\langle \phi_0 |\mathbf{v}| \phi_n \right\rangle = i\hbar \left\langle \phi_0 |\dot{\mathbf{x}}| \phi_n \right\rangle = \left\langle \phi_0 |[\mathbf{x},\mathcal{H}]| \phi_n \right\rangle = (E_n - E_0) \left\langle \phi_0 |\mathbf{x}| \phi_n \right\rangle, \quad (5.27)$$

we obtain the Kubo formula for the conductivity at $T = 0$, namely,

$$\sigma_{\alpha\beta}(\omega) = \frac{ie^2}{\hbar\Omega} \sum_{n\neq 0} \frac{1}{\omega_{0n}} \left[\frac{\langle\phi_0| v_\alpha |\phi_n\rangle \langle\phi_n| v_\beta |\phi_0\rangle}{\omega - \omega_{0n} + i\eta} - \frac{\langle\phi_0| v_\beta |\phi_n\rangle \langle\phi_n| v_\alpha |\phi_0\rangle}{\omega - \omega_{n0} + i\eta} \right],$$

$$(5.28)$$

so that the conductivity can be identified with the velocity–velocity correlation function.

2. In the case of a vector potential $\mathbf{A}(\mathbf{x}, t)$, we have

$$\mathcal{E} = -\frac{1}{c}\frac{\partial \mathbf{A}}{\partial t} = \frac{i\omega}{c}\mathbf{A}$$

giving rise to the perturbation

$$\delta\mathcal{H} = \frac{e}{m_e c}\mathbf{A}\cdot\mathbf{p} = -\frac{ie}{m_e\omega}\mathcal{E}\cdot\mathbf{p}.$$

Recalling that $\mathbf{j} = \operatorname{Im}\psi^*\nabla\psi/m_e$, we write $\mathbf{j} = \mathbf{p}/m_e$, and the perturbation becomes

$$\delta\mathcal{H} = -\frac{ie}{\omega}\mathcal{E}\cdot\mathbf{j} \quad\Rightarrow\quad \sigma_{\alpha\beta}(\omega) = \frac{-ie^2}{\omega\Omega}\left\langle\!\!\left\langle j_\alpha \middle| j_\beta \right\rangle\!\!\right\rangle_\omega.$$

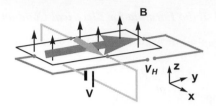

Figure 5.2 Schematic picture of the quantum Hall effect system: a magnetic field is applied to a planar electron system normal to the plane (along the z-direction), while an electric field is applied along the x-direction. This configuration leads to the development of a Hall current along the y-direction.

The Hall Conductivity

From the configuration of the Hall effect schematic shown in Figure 5.2, we write the Hall conductance as

$$\lim_{\omega \to 0} \sigma_{xy}(\omega) = \lim_{\omega \to 0} \frac{-ie^2}{\Omega \omega} \sum_{nm} f(E_n) \left[\frac{\langle n| v_x |m\rangle \langle m| v_y |n\rangle}{\hbar\omega + i\eta + E_n - E_m} + \frac{\langle n| v_y |m\rangle \langle m| v_x |n\rangle}{-\hbar\omega - i\eta + E_n - E_m} \right].$$

In the limit $\omega \to 0$,

$$\frac{1}{\pm\hbar\omega + E_n - E_m} = \frac{1}{E_n - E_m} \left(1 \mp \frac{\hbar\omega}{E_n - E_m} \right)$$

$$\sigma_{xy}(\omega \to 0) = \frac{-ie^2}{\Omega \omega} \sum_{nm} f(E_n) \left[\frac{\langle n| v_x |m\rangle \langle m| v_y |n\rangle + \langle n| v_y |m\rangle \langle m| v_x |n\rangle}{E_n - E_m} \right.$$
$$\left. + \hbar\omega \frac{- \langle n| v_x |m\rangle \langle m| v_y |n\rangle + \langle n| v_y |m\rangle \langle m| v_x |n\rangle}{(E_n - E_m)^2} \right]. \tag{5.29}$$

From the identity (5.27), we have

$$\langle n| v_x |m\rangle = \frac{(E_n - E_m)}{i\hbar} \langle n| x |m\rangle$$

and thus

$$\langle n| v_x |m\rangle \langle m| v_y |n\rangle + \langle n| v_y |m\rangle \langle m| v_x |n\rangle$$
$$= \frac{(E_n - E_m)}{i\hbar} \left(\langle n| x |m\rangle \langle m| v_y |n\rangle - \langle n| v_y |m\rangle \langle m| x |n\rangle \right)$$

substituting in the first term of (5.29), we find that

$$\sum_m \frac{\langle n| x |m\rangle \langle m| v_y |n\rangle - \langle n| v_y |m\rangle \langle m| x |n\rangle}{E_n - E_m} = \frac{1}{i\hbar} \langle n| x v_y - v_y x |n\rangle = 0$$

and we obtain

$$\sigma_{xy}(\omega \to 0) = \frac{ie^2\hbar}{\Omega} \sum_{nm} f(E_n) \left[\frac{\langle n| v_x |m\rangle \langle m| v_y |n\rangle - \langle n| v_y |m\rangle \langle m| v_x |n\rangle}{(E_n - E_m)^2} \right] \tag{5.30}$$

Using the relations

$$\mathcal{H}(\mathbf{k}) = e^{-i\mathbf{k}\cdot\mathbf{x}}\, \mathcal{H}\, e^{i\mathbf{k}\cdot\mathbf{x}}$$

$$\Rightarrow \mathcal{H}(\mathbf{k})\,|u_{n,\mathbf{k}}\rangle = \left[\frac{1}{2m}\left(\mathbf{p} + \hbar\,\mathbf{k} - \frac{e}{c}\mathbf{A}\right)^2 + V\right]|u_{n,\mathbf{k}}\rangle = E_n(\mathbf{k})\,|u_{n,\mathbf{k}}\rangle$$

$$\mathbf{v} = \dot{\mathbf{x}} = \frac{i}{\hbar}\left[\mathcal{H},\,\mathbf{x}\right]$$

$$\Rightarrow \mathbf{v}(\mathbf{k}) = e^{-i\mathbf{k}\cdot\mathbf{x}}\frac{i}{\hbar}\left[\mathcal{H},\,\mathbf{x}\right]e^{i\mathbf{k}\cdot\mathbf{x}} = \frac{i}{\hbar}\left[\mathcal{H}(\mathbf{k}),\,\mathbf{x}\right] = \nabla_{\mathbf{k}}\mathcal{H}(\mathbf{k})$$

and setting $\mathbf{k} = (k_1, k_2)$, we write

$$(v_x)_{mn} = \left\langle u_{m,\mathbf{k}} \left| \frac{\partial\mathcal{H}(\mathbf{k})}{\partial k_1} \right| u_{n,\mathbf{k}} \right\rangle$$

but

$$\nabla_{\mathbf{k}}\,\mathcal{H}(\mathbf{k})\,|u_{n,\mathbf{k}}\rangle + \mathcal{H}(\mathbf{k})\,\nabla_{\mathbf{k}}\,|u_{n,\mathbf{k}}\rangle = \nabla_{\mathbf{k}}\,E_n(\mathbf{k})\,|u_{n,\mathbf{k}}\rangle + E_n(\mathbf{k})\,\nabla_{\mathbf{k}}\,|u_{n,\mathbf{k}}\rangle$$

$$\Rightarrow \langle u_{m,\mathbf{k}}|\,\nabla_{\mathbf{k}}\,\mathcal{H}(\mathbf{k})\,|u_{n,\mathbf{k}}\rangle = (E_n - E_m)\,\langle u_{m,\mathbf{k}}|\,\nabla_{\mathbf{k}}\,u_{n,\mathbf{k}}\rangle$$

so that

$$(v_x)_{mn} = \Big(E_n(\mathbf{k}) - E_m(\mathbf{k})\Big)\left\langle u_{m,\mathbf{k}}\left|\frac{\partial u_{n,\mathbf{k}}}{\partial k_1}\right.\right\rangle = -\Big(E_n(\mathbf{k}) - E_m(\mathbf{k})\Big)\left\langle\frac{\partial u_{n,\mathbf{k}}}{\partial k_1}\left|u_{m,\mathbf{k}}\right.\right\rangle .$$

Substitution in (5.30) leads to

$$\sigma_{xy}(\omega \to 0) = \frac{e^2}{i\hbar\Omega} \sum_{E_m < E_F < E_n} \left(\left\langle\frac{\partial u_{n,\mathbf{k}}}{\partial k_1}\Big|u_{m,\mathbf{k}}\right\rangle\left\langle u_{m,\mathbf{k}}\Big|\frac{\partial u_{n,\mathbf{k}}}{\partial k_2}\right\rangle\right.$$

$$\left. - \left\langle\frac{\partial u_{n,\mathbf{k}}}{\partial k_2}\Big|u_{m,\mathbf{k}}\right\rangle\left\langle u_{m,\mathbf{k}}\Big|\frac{\partial u_{n,\mathbf{k}}}{\partial k_1}\right\rangle\right) \tag{5.31}$$

We will show in Chapter 12 that by converting the summation into an integral, David Thouless and coworkers were able to demonstrate the Hall current quantization in the phenomenon of *quantum Hall effect* (QHE).

5.3 The Dielectric Function: Linear Response to Electromagnetic Perturbations

In this section, we shall explore the various microscopic aspects of the linear response to scalar and vector electromagnetic fields, other than transport properties. Our discussion of slowly time-varying scalar potentials is mainly intended to deal with ionic perturbations in molecular and solid systems. Our analysis of fields associated with the vector potential considers mainly the interaction of electromagnetic waves (photons) with crystalline material systems.

5.3.1 Macroscopic Properties

In a regular course on electromagnetism, the electronic structure is treated in a very crude and approximate way. The assumption is that a macroscopic external electric field induces a polarization **P** counteracting the external field. The external field is given by the displacement **D**, which arises from external charges. The external field and the polarization together produce the resultant electric field inside the solid. The relations between the different electromagnetic quantities are defined by Maxwell's equations:

$$\nabla \times \mathbf{H} = \frac{4\pi \mathbf{j}}{c} + \frac{1}{c}\frac{\partial \mathbf{D}}{\partial t},$$

$$\nabla \cdot \mathbf{D} = 4\pi \rho_{\text{free}},$$

$$\nabla \times \boldsymbol{\mathcal{E}} = \frac{-1}{c}\frac{\partial \mathbf{B}}{\partial t}, \tag{5.32}$$

where

$$\mathbf{D} = \overleftrightarrow{\epsilon}\,\boldsymbol{\mathcal{E}}, \qquad \mathbf{B} = \mu \mathbf{H}. \tag{5.33}$$

$\overleftrightarrow{\epsilon}$ is a general dielectric tensor, with constant entries, and μ the magnetic permeability. We shall deal with nonmagnetic systems and set $\mu = 1$. We note that

$$\boldsymbol{\mathcal{E}} = \overleftrightarrow{\epsilon}^{-1}\mathbf{D} = \mathbf{D} - 4\pi\,\mathbf{P}$$

$$\nabla\cdot\boldsymbol{\mathcal{E}} = \nabla\cdot\mathbf{D} - 4\pi\,\nabla\cdot\mathbf{P}$$

$$= \rho_{\text{free}} + \rho_{\text{ind}}. \tag{5.34}$$

ρ_{free} gives rise to the electric displacement, while the polarization **P** results from ρ_{ind}. Linear response defines the relation

$$\mathbf{P} = \chi\,\boldsymbol{\mathcal{E}} \implies \epsilon = 1 + 4\pi\,\chi, \tag{5.35}$$

where χ is known as the electric susceptibility. This quantity is supposed to represent all the electronic structure of the solid.

In contrast, when considering microscopic details, we zoom in at the length scale of lattice spacing where the fields and susceptibilities become functions of position rather than constants.

5.3.2 The Microscopic Longitudinal Dielectric Function

For simplicity, we consider an isotropic system consisting of a large number of electrons and subject to a external scalar potential $V_{ext}(\mathbf{x}, t)$, slowly varying in time. The electric field is given by $\boldsymbol{\mathcal{E}} = -\nabla V_{ext}$, $\nabla\cdot\boldsymbol{\mathcal{E}} \neq 0$. It is a longitudinal field, in the sense that for a traveling-wave perturbation, $\boldsymbol{\mathcal{E}}$ lies along the wavevector **q**.

The applied external potential will then induce a change in the electronic density distribution. For small $V_{\text{ext}}(\mathbf{x}, t)$, the change in electronic density δn can be expressed in linear response theory as

$$\delta n(\mathbf{x}',t) = -e \int d\mathbf{x}'' \, \chi(\mathbf{x}',\mathbf{x}'';t-t') \, V_{ext}(\mathbf{x}'',t'),$$

which in turn will give rise to an induced potential

$$V_{ind}(\mathbf{x},t) = -\int d\mathbf{x}' \, \frac{e^2}{|\mathbf{x}-\mathbf{x}'|} \, \delta n(\mathbf{x}',t)$$

so that the total potential that can be measured by a probe at a point $\mathbf{x}$ is

$$
\begin{aligned}
V_{tot}(\mathbf{x},t) &= V_{ext}(\mathbf{x},t) + V_{ind}(\mathbf{x},t) \\
&= V_{ext}(\mathbf{x},t) - \int \frac{e^2}{|\mathbf{x}-\mathbf{x}'|} \, \chi(\mathbf{x}',\mathbf{x}'';t-t') \, V_{ext}(\mathbf{x}'',t') \, d\mathbf{x}' \, d\mathbf{x}'' \\
&= \int \epsilon^{-1}(\mathbf{x},\mathbf{x}'') \, V_{ext}(\mathbf{x}'') \, d\mathbf{x}'',
\end{aligned}
\tag{5.36}
$$

where both χ and ϵ^{-1} are nonlocal functions. ϵ^{-1} is the *inverse longitudinal dielectric function*. We have defined a response function ϵ^{-1} in terms of the applied and resultant potentials, and we can also define the function ϵ such that

$$\int \epsilon^{-1}(\mathbf{x},\mathbf{x}') \, \epsilon(\mathbf{x}',\mathbf{x}'') \, d\mathbf{x}' = \int \epsilon(\mathbf{x},\mathbf{x}') \, \epsilon^{-1}(\mathbf{x}',\mathbf{x}'') \, d\mathbf{x}' = \delta(\mathbf{x}-\mathbf{x}''), \tag{5.37}$$

which gives the applied potential in terms of the resultant one. According to (5.13),

$$\delta n(\mathbf{x},t) = -i \int_{-\infty}^{t} dt' \int d\mathbf{x}' \, \mathrm{Tr}\Big\{\rho_0 \big[\hat{n}(\mathbf{x},t-t'), \, \hat{n}(\mathbf{x}')\big]\Big\} \, V_{ext}(\mathbf{x}',t'), \tag{5.38}$$

where $\hat{n}$ is the density operator. Taking temporal Fourier transform

$$\delta n(\mathbf{x},\omega) = -i \int_{0}^{\infty} d\tau \, e^{i\omega\tau} \int d\mathbf{x}' \, \big\langle \big[\hat{n}(\mathbf{x},\tau), \, \hat{n}(\mathbf{x}')\big]\big\rangle_0 \, V_{ext}(\mathbf{x}',\omega). \tag{5.39}$$

For the sake of simplicity, we shall consider a spatially translation invariant system and assume

$$V_{ext}(\mathbf{x}',\omega) = \delta V(\mathbf{q},\omega) \, e^{i\mathbf{q}\cdot\mathbf{x}'}.$$

We get

$$
\begin{aligned}
\delta n(\mathbf{q},\omega) &= \int d\mathbf{x} \, e^{i\mathbf{q}\cdot\mathbf{x}} \, \delta n(\mathbf{x},\omega) = -i \int_{0}^{\infty} d\tau \, e^{i\omega\tau} \big\langle \big[\hat{n}(\mathbf{q},\tau), \, \hat{n}(-\mathbf{q})\big]\big\rangle_0 \, \delta V(\mathbf{q},\omega) \\
&= \chi(\mathbf{q},\omega) \, \delta V(\mathbf{q},\omega).
\end{aligned}
\tag{5.40}
$$

Using the spectral representation of χ given in (5.23), setting $T = 0$, where state occupation becomes $\delta_{0,n}$, and noting that $\hat{n}(-\mathbf{q}) = \hat{n}^{\dagger}(\mathbf{q})$, we get

$$\chi(\mathbf{q},\omega) = \frac{-1}{\hbar} \sum_{m} \frac{|\langle \phi_0 | n(\mathbf{q}) | \phi_m \rangle|^2}{\omega - \omega_{0m} + i\eta}, \tag{5.41}$$

where $|\phi_0\rangle$ is the ground state. $|\phi_0\rangle$ and $|\phi_m\rangle$ are many-electron wavefunctions. We may write the electron density operator as

$$\hat{n}(\mathbf{x}) = \sum_i \delta(\mathbf{x} - \mathbf{x}_i) \Rightarrow \hat{n}(\mathbf{q}) = \int d\mathbf{x}\, e^{i\mathbf{q}\cdot\mathbf{x}}\, n(\mathbf{x}) = \sum_i e^{i\mathbf{q}\cdot\mathbf{x}_i},$$

a sum of single particle operators, where $\mathbf{x}_i$ is the position of electron i, and the sum includes all electrons. We can then write

$$\chi(\mathbf{q}, \omega) = \frac{-1}{\hbar} \sum_{m,i} \frac{\left| \langle \phi_0 | e^{i\mathbf{q}\cdot\mathbf{x}_i} | \phi_m \rangle \right|^2}{\omega - \omega_{0m} + i\eta}. \tag{5.42}$$

Taking the spatial/temporal Fourier transform of (5.36), we get

$$\delta V_{\text{tot}}(\mathbf{q}, \omega) = \delta V(\mathbf{q}, \omega) + \delta V_{\text{ind}}(\mathbf{q}, \omega) = \varepsilon^{-1}(\mathbf{q}, \omega)\, \delta V(\mathbf{q}, \omega), \tag{5.43}$$

where ε^{-1} is the *inverse longitudinal dielectric function*. δV_{ind} is due to δn, satisfying the Poisson equation

$$\nabla^2 \delta V_{\text{ind}} = -4\pi\, e^2\, \delta n$$

and

$$\delta V_{\text{ind}}(\mathbf{q}, \omega) = \frac{4\pi\, e^2}{\mathbf{q}^2} \delta n(\mathbf{q}, \omega) = -\frac{4\pi\, e^2}{\mathbf{q}^2} \chi(\mathbf{q}, \omega)\, \delta V(\mathbf{q}, \omega). \tag{5.44}$$

Hence, we have

$$\varepsilon^{-1}(\mathbf{q}, \omega) = 1 - \frac{4\pi\, e^2}{\mathbf{q}^2} \chi(\mathbf{q}, \omega). \tag{5.45}$$

5.3.3 Transverse Dielectric Function and Optical Transitions

We now turn to transverse electromagnetic waves and the domain of optical spectroscopy. Here we shall interpret optical spectroscopy within the quantum mechanical theory of electronic band structures, and give a detailed description of the concepts of interband and intraband transitions. However, we should note that the wavelength of light used in optical spectroscopy[3] is $\lambda > 1,000\text{Å}$, while the BZ dimensions are of the order of an inverse Angstrom. Consequently, $|\mathbf{q}| = 2\pi/\lambda$ is very small and in some cases it will be set to zero.

We write the crystalline Bloch eigenstates in the absence of light as

$$\mathcal{H}_0 |\mathbf{k}n\rangle = E_{\mathbf{k}n} |\mathbf{k}n\rangle$$

and the external transverse vector potential $\mathbf{A}$ ($\nabla\cdot\mathbf{A} = 0$) and electric field $\mathcal{E}$ as

$$\mathbf{A}(\mathbf{x}, t) = \mathbf{A}_0\, e^{i(\mathbf{q}\cdot\mathbf{x} - \omega t)}, \quad \mathcal{E} = -\frac{1}{c}\frac{\partial \mathbf{A}}{\partial t}.$$

[3] In infrared spectroscopy, λ can be several microns.

The perturbation arising from the interaction of light with the material system is then given by

$$\delta \mathcal{H} = -\frac{ie\hbar}{m_e c} \mathbf{A} \cdot \mathbf{p} = \frac{-e\hbar}{\omega m_e} \mathcal{E} \cdot \mathbf{p} = \frac{-e\hbar \mathcal{E}_0}{\omega m_e} \boldsymbol{\varepsilon} \cdot \mathbf{p} \, e^{-i\mathbf{k} \cdot \mathbf{x}}, \tag{5.46}$$

where $\boldsymbol{\varepsilon}$ is the polarization vector.

At $T = 0$, the expression (5.26) for χ_2 becomes

$$\chi_2(\omega) = \frac{\pi \hbar^2 e^2}{m_e^2 \omega^2} \sum_{\substack{\mathbf{k}, m \\ \mathbf{k}', n}} \left| \langle \mathbf{k}'n | \, \boldsymbol{\varepsilon} \cdot \mathbf{p} \, e^{-i\mathbf{k} \cdot \mathbf{x}} \, |\mathbf{k}m \rangle \right|^2$$

$$\times \left(f(\mathbf{k}m) - f(\mathbf{k}'n) \right) \delta \left(\hbar \omega + E_{\mathbf{k}m} - E_{\mathbf{k}'n} \right)$$

$$= \frac{\pi e^2 \hbar^2}{m_e^2 \omega^2} \sum_{\substack{\mathbf{k}, m \\ \mathbf{k}', n}} \left| \boldsymbol{\varepsilon} \cdot \langle \mathbf{k}'n | \, e^{i\mathbf{q} \cdot \mathbf{x}} \, \nabla \, |\mathbf{k}m \rangle \right|^2$$

$$\times \left(f(\mathbf{k}m) - f(\mathbf{k}'n) \right) \delta \left(\hbar \omega + E_{\mathbf{k}m} - E_{\mathbf{k}'n} \right), \tag{5.47}$$

where f is the Fermi occupation number.

Now we evaluate the matrix element $\langle \mathbf{k}'n | \, e^{i\mathbf{q} \cdot \mathbf{x}} \, \nabla \, |\mathbf{k}m \rangle$. With

$$|\mathbf{k}l\rangle = e^{-i\mathbf{k} \cdot \mathbf{x}} u_{\mathbf{k}l}(\mathbf{x}) \; \Rightarrow \; u_{\mathbf{k}l}(\mathbf{x} + \mathbf{R})) = u_{\mathbf{k}l}(\mathbf{x})$$

$$\nabla \, |\mathbf{k}l\rangle = e^{-i\mathbf{k} \cdot \mathbf{x}} (-i\mathbf{k} + \nabla) u_{\mathbf{k}l}(\mathbf{x}),$$

we obtain

$$\langle \mathbf{k}'n | \, e^{i\mathbf{q} \cdot \mathbf{x}} \, \nabla \, |\mathbf{k}m \rangle = -i\mathbf{k} \, \langle \mathbf{k}'n | \, \mathbf{k}m \rangle + \int d\mathbf{x} \, e^{-i(\mathbf{k} - \mathbf{k}' - \mathbf{q}) \cdot \mathbf{x}} u_{\mathbf{k}'n}^* \, \nabla u_{\mathbf{k}m}.$$

The first term on the right vanishes since $|\mathbf{k}m\rangle$ and $|\mathbf{k}'n\rangle$ are orthogonal. In the second term, we substitute $\mathbf{x} = \mathbf{R}_j + \mathbf{x}'$, where $\mathbf{R}_j$ is a lattice vector

$$\int d\mathbf{x} \, e^{-i(\mathbf{k} - \mathbf{k}' - \mathbf{q}) \cdot \mathbf{x}} u_{\mathbf{k}'n}^* \, \nabla u_{\mathbf{k}m} = e^{-i(\mathbf{k} - \mathbf{k}' - \mathbf{q}) \cdot \mathbf{R}_j} \int d\mathbf{x}' \, e^{-i(\mathbf{k} - \mathbf{k}' - \mathbf{q}) \cdot \mathbf{x}'} u_{\mathbf{k}'n}^* \, \nabla u_{\mathbf{k}m}$$

This allows to write

$$\langle \mathbf{k}'n | \, e^{i\mathbf{q} \cdot \mathbf{x}} \, \nabla \, |\mathbf{k}m \rangle = \frac{1}{N} \sum_j e^{-i(\mathbf{k} - \mathbf{k}' - \mathbf{q}) \cdot \mathbf{R}_j} \int d\mathbf{x}' \, e^{-i(\mathbf{k} - \mathbf{k}' - \mathbf{q}) \cdot \mathbf{x}'} u_{\mathbf{k}'n}^* \, \nabla u_{\mathbf{k}m}$$

$$= \delta_{\mathbf{k}', \mathbf{k}+\mathbf{q}+\mathbf{G}} \int_{\Omega_c} d\mathbf{x} \, u_{\mathbf{k}'n}^* \, \nabla u_{\mathbf{k}m}, \tag{5.48}$$

where N is the number of primitive cells and Ω_c is the cell volume. Notice that ∇ transforms like a polar vector, hence it has odd parity; this stipulates that $u_{\mathbf{k}'n}$ and $u_{\mathbf{k}m}$ must have opposite parities. Transitions involving a reciprocal lattice vector $\mathbf{G}$ are called *Umklapp* and are usually neglected.

We then obtain

$$\chi_2(\omega) = \frac{\pi e^2 \hbar^2 \mathcal{E}_0^2}{m_e^2 \omega^2} \sum_{\mathbf{k}, m, n} f(\mathbf{k}m)\Big(1 - f((\mathbf{k}+\mathbf{q})n)\Big) \, \delta\left(\hbar\omega + E_{\mathbf{k}m} - E_{(\mathbf{k}+\mathbf{q})n}\right)$$

$$\times \left| \boldsymbol{\varepsilon} \cdot \big\langle (u_{(\mathbf{k}+\mathbf{q})n} \big| \, \nabla u_{\mathbf{k}m} \big\rangle \right|^2 . \tag{5.49}$$

Intraband Transitions

In the case of metals at $T = 0$, a band (or several bands) is partially occupied. Hence, empty and occupied states are infinitesimally close in momentum and energy. Because of the smallness of q, we are then justified to use the expansions

$$E(\mathbf{k}) - E(\mathbf{k}+\mathbf{q}) = -\mathbf{q} \cdot \nabla_{\mathbf{k}} E(\mathbf{k})$$

$$f(\mathbf{k}) - f(\mathbf{k}+\mathbf{q}) = -\frac{\partial f}{\partial E} \, \mathbf{q} \cdot \nabla_{\mathbf{k}} E(\mathbf{k}).$$

These expressions are to be used for all bands in which states straddle the Fermi energy. At $T = 0$, (5.25) for χ reduces to

$$\chi(\omega) = \frac{\pi e^2 \hbar^2 \mathcal{E}_0^2}{m_e^2 \omega^2} \sum_{\mathbf{k}} |\boldsymbol{\varepsilon} \cdot \langle (\mathbf{k}+\mathbf{q}| \, \nabla \, |\mathbf{k}\rangle|^2 \, \frac{f(\mathbf{k}) - f((\mathbf{k}+\mathbf{q}))}{\hbar\omega + E(\mathbf{k}) - E(\mathbf{k}+\mathbf{q}) + i\eta} \tag{5.50}$$

and

$$\frac{\partial f}{\partial E} = -\delta(E - E_F).$$

For small $\mathbf{q}$,

$$\frac{f(\mathbf{k}) - f((\mathbf{k}+\mathbf{q}))}{\hbar\omega + E(\mathbf{k}) - E(\mathbf{k}+\mathbf{q}) + i\eta} = \delta(E - E_F) \, \frac{\mathbf{q} \cdot \nabla_{\mathbf{k}} E(\mathbf{k})}{\hbar\omega - \mathbf{q} \cdot \nabla_{\mathbf{k}} E(\mathbf{k}) + i(\hbar\Gamma)}$$

$$= \frac{\mathbf{q} \cdot \nabla_{\mathbf{k}} E(\mathbf{k})|_{\mathbf{k}_F}/\hbar}{\omega - \mathbf{q} \cdot \nabla_{\mathbf{k}} E(\mathbf{k})|_{\mathbf{k}_F}/\hbar + i\Gamma}$$

$$= \frac{\mathbf{q} \cdot \mathbf{v}_F}{\omega - \mathbf{q} \cdot \mathbf{v}_F + i\Gamma} \simeq \frac{\mathbf{q} \cdot \mathbf{v}_F}{\omega + i\Gamma},$$

where we used $\nabla_{\mathbf{k}} E(\mathbf{k})/\hbar = \mathbf{v}$, the group velocity, and the fact that for the light wave, $\omega/q = c \gg v_F$, which translates to the following:

$$\omega \gg \mathbf{q} \cdot \mathbf{v}_F.$$

We also make use of the following approximation:

$$\frac{1}{m_e} \langle (\mathbf{k}+\mathbf{q}| \, e^{i\mathbf{q}\cdot\mathbf{r}} \, \hbar\nabla \, |\mathbf{k}\rangle \simeq \frac{1}{m_e} \langle (\mathbf{k}| \, \hbar\nabla \, |\mathbf{k}\rangle = \mathbf{v}.$$

With these approximations, we obtain

$$\chi(\omega) = \frac{\pi e^2 \hbar^2 \mathcal{E}_0^2}{\omega^2(\omega + i\Gamma)} \sum_{\mathbf{k}, \mathbf{q}} \left| \boldsymbol{\varepsilon} \cdot \left\langle \mathbf{k}_F + \mathbf{q} \left| \frac{\hbar \nabla}{m_e} \right| \mathbf{k}_F \right\rangle \right|^2 \mathbf{q} \cdot \mathbf{v}_F$$

$$= \frac{\pi e^2 \hbar^2 \mathcal{E}_0^2}{\omega^2(\omega + i\Gamma)} \sum_{|\mathbf{k}|=k_F} |\boldsymbol{\varepsilon} \cdot \mathbf{v}_F|^2 \, \mathbf{q} \cdot \mathbf{v}_F = \frac{\pi e^2 \hbar^2 \mathcal{E}_0^2}{m_e^2} \frac{\mathcal{C}_F}{\omega(\omega + i\Gamma)}, \quad (5.51)$$

which the same form as the Drude model susceptibility.

Direct Interband Transitions

We now consider excitations with finite energies, where $\mathbf{q}$ can be neglected, and we write

$$\chi_2(\omega) = \frac{\pi e^2 \hbar^2 \mathcal{E}_0^2}{m_e^2 \omega^2} \sum_{\mathbf{k}, m, n} f(\mathbf{k}m) \left(1 - f(\mathbf{k}n)\right) \delta\left(\hbar\omega - E_{nm}(\mathbf{k})\right) |\boldsymbol{\varepsilon} \cdot \langle \mathbf{k}n| \nabla |\mathbf{k}m\rangle|^2, \quad (5.52)$$

where $E_{nm}(\mathbf{k}) = E_n(\mathbf{k}) - E_m(\mathbf{k})$. Converting the summation over $\mathbf{k}$ into an integral and setting the m and n bands to be occupied and empty, respectively, we get

$$\chi_2(\omega) = \frac{\pi e^2 \hbar^2}{m_e^2} \sum_{m, n} \frac{\Omega}{(2\pi)^3} \int_{\Omega_B} d\mathbf{k} \, \delta\left(\hbar\omega - E_{nm}(\mathbf{k})\right) \frac{|\boldsymbol{\varepsilon} \cdot \langle \mathbf{k}n| \nabla |\mathbf{k}m\rangle|^2}{\omega^2}$$

$$= \frac{\pi e^2 \hbar^2}{m_e^2} \sum_{m, n} \frac{\Omega}{(2\pi)^3} \int dE_{nm} \int_S d\mathbf{S} \frac{\delta\left(\hbar\omega - E_{nm}(\mathbf{k})\right)}{\nabla_{\mathbf{k}} E_{nm}(\mathbf{k})} \frac{|\boldsymbol{\varepsilon} \cdot \langle \mathbf{k}n| \nabla |\mathbf{k}m\rangle|^2}{\omega^2}$$

$$= \frac{\pi e^2 \hbar^2}{m_e^2} \sum_{m, n} \frac{\Omega}{(2\pi)^3} \int_{\hbar\omega = E_{nm}(\mathbf{k})} \frac{d\mathbf{S}}{|\nabla_{\mathbf{k}} E_{nm}(\mathbf{k})|} \frac{|\boldsymbol{\varepsilon} \cdot \langle \mathbf{k}n| \nabla |\mathbf{k}m\rangle|^2}{\omega^2}, \quad (5.53)$$

where S is a surface of constant energy $\hbar\omega = E_{nm}(\mathbf{k})$, as shown in Figure 5.3. We used

$$\int dx \, g(x) \, \delta(f(x)) = \sum_{x_0, f(x_0)=0} \left| \frac{df(x)}{dx} \right|_{x_0}^{-1} g(x_0)$$

to define the integration in terms of energy. $1/|\nabla_{\mathbf{k}} E_{nm}(\mathbf{k})|$ represents the joint density of states. Of special interest are points in the Brillouin zone where $E_{nm}(\mathbf{k})$ is stationary and $\nabla_{\mathbf{k}} E_{nm}(\mathbf{k})$ vanishes. At such points, called joint critical points, the denominator of the integrand in (5.53) vanishes and especially large contributions can be made to χ_2.

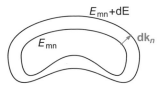

Figure 5.3 Adjacent constant energy difference surfaces in reciprocal space, E_{mn} and $E_{mn} + dE$. dk_n is the normal to these constant energy difference surfaces.

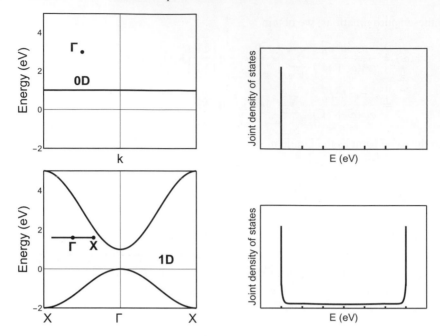

Figure 5.4 Band structures, BZs, and interband transition strengths in 0D and 1D.

This can be understood on the basis of physical considerations. Around critical points, the photon energy $\hbar\omega = E_{nm}(\mathbf{k})$ is effective in inducing electronic transitions over a relatively larger region of the Brillouin zone than would be the case for transitions about noncritical points. The relatively large contributions to the transition probability for critical points gives rise to *structure* observed in the frequency dependence of the optical properties of solids. Critical points generally occur at high symmetry points in the Brillouin zone, though there are exceptions. Typical cases for systems of different dimensions are shown in Figures 5.4 and 5.5.

Indirect Interband Transitions

Indirect interband transitions are clearly observed in materials having indirect energy gaps, as shown in the left panel of Figure 5.6. These transitions are represented by second-order time-dependent perturbation theory, and involve two transitions $|i\rangle \rightarrow |\alpha\rangle$, $|\alpha\rangle \rightarrow |f\rangle$, shown in the right panel of Figure 5.6.

A transition from $\mathbf{k}_1 m$ to $\mathbf{k}_2 n$ will be represented by

$$\sum_\alpha \left| \frac{\langle \mathbf{k}_2 n | \, \mathcal{H}_{\text{phonon}} \, |\mathbf{k}_1\alpha\rangle \, \langle \mathbf{k}_1\alpha | \, \delta\mathcal{H} \, |\mathbf{k}_1 m\rangle}{E_\alpha - E_m - \hbar\omega} \right|^2 \delta\left(E_n(\mathbf{k}_2) - E_m(\mathbf{k}_1) - \hbar\omega \mp \hbar\Omega(\mathbf{q})\right),$$

$$(5.54)$$

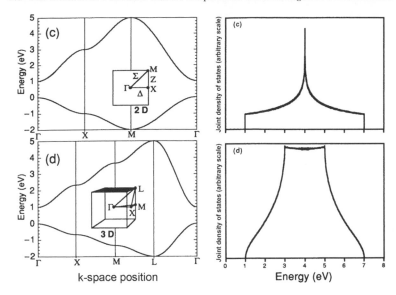

Figure 5.5 Band structures, BZs, and interband transition strengths in 2D and 3D.

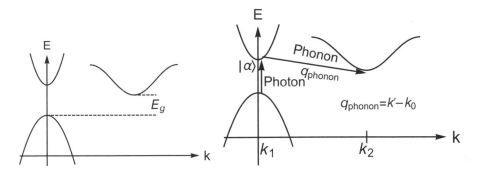

Figure 5.6 Left: indirect gap in insulators and semiconductors. Right: phonon-assisted indirect optical transition through an intermediate state $|\alpha\rangle$.

where $\Omega(\mathbf{q})$ is the phonon frequency, and $\mathbf{q} = \mathbf{k}_2 - \mathbf{k}_1$; $\mp\hbar\Omega$ correspond to phonon absorption/emission. The perturbation Hamiltonians are given by

$$\mathcal{H}_{\text{phonon}} = -\sum_n \mathbf{u}_n \cdot \nabla_{\mathbf{r}}\, V(\mathbf{r} - \mathbf{R}_n),$$

where $\mathbf{R}_n$ are lattice vectors, and $\delta\mathcal{H}$ is given in (5.46).

As a specific example of indirect absorption edge, we will consider Si. The band-structure of Si in the vicinity of the indirect gap Δ_1-$\Gamma_{25'}$ is shown schematically in Figure 5.7. Two absorption processes are shown. In process A, the electron is first excited via a virtual transition from $\Gamma_{25'}$ to an intermediate state at Γ_{15} by absorbing the

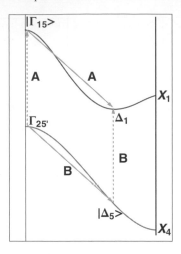

Figure 5.7 Schematic band structure of Si as an indirect-bandgap semiconductor, showing phonon-assisted transitions, labeled A and B, that contribute to the indirect absorption edge. $|\Gamma_{15}\rangle$ and $|\Delta_5\rangle$ represent intermediate states. Photon and phonon processes are shown as dashed and solid lines, respectively.

incident photon. A second virtual transition takes the electron to the Δ_1 conduction band valley state via emission of a phonon. In the final state, there is an electron in the Δ_1 conduction band, and a hole in the $\Gamma_{25'}$ valence band state, and a phonon has been created. A second possible phonon-assisted indirect optical transition is also shown as process B in Figure 5.7.

Exercises

5.1   Kramers–Kronig relation

Suppose that we model the interband transitions in Ge as a step function

$$
\epsilon_2(\omega) =
\begin{cases}
\epsilon_\ell & \text{for } E_{\min} \leq \omega \leq E_{\max} \\
0 & \text{otherwise}
\end{cases}
$$

1. Use the Kramers–Kronig relation to obtain an expression for $\epsilon_1(\omega)$ for all ω. Take $\epsilon_1 = 1$ for $\omega = \infty$ and express your answer in terms of $E_{\min}$, $E_{\max}$, and ϵ_ℓ.
2. For what photon energies does $\epsilon_1(\omega)$ exhibit structure? Is your answer physically reasonable and why?
3. Obtain an explicit expression for $\epsilon_1(0)$, and use this result to explain why narrow gap semiconductors tend to have large dielectric constants at $\omega = 0$.

5.2   The $\chi_2(\omega)$ expression (5.53) for direct-band transitions is proportional to the joint DOS and $|\boldsymbol{\varepsilon} \cdot \langle \mathbf{k}n| \mathbf{p} |\mathbf{k}m\rangle|^2$. Here, we consider a simple two-band semiconductor model, where the top valence and bottom conduction bands are far from other bands. The effective mass (4.19) takes the form

$$\left(\frac{1}{m^*}\right)_{\alpha\beta} = \frac{\delta_{\alpha\beta}}{m_e} + \frac{2}{m_e^2} \frac{|\langle u_v| p_\alpha |u_c\rangle \langle u_c| p_\beta |u_v\rangle|}{E_g}.$$

Explain how the magnitude of m^* reflect on the strength of the optical transition.

5.3 Assume that in (5.53), $|\boldsymbol{\varepsilon}\cdot \langle \mathbf{k}n| \mathbf{p} |\mathbf{k}m\rangle|^2$ does not vary strongly with angle on a surface of constant energy, and it can be taken outside the integral. Then the integral is just the joint DOS.

Now consider the two-band semiconductor model of problem (2), with dispersions

$$E_c(\mathbf{k}) = E_g + \frac{\hbar^2 k^2}{2m_c^*}, \quad E_v(\mathbf{k}) = -\frac{\hbar^2 k^2}{2m_v^*}.$$

Show that

$$E_c(\mathbf{k}) - E_v(\mathbf{k}) - \hbar\omega = E_g + \frac{\hbar^2 k^2}{2\mu} - \hbar\omega, \quad \mu = \frac{m_c^* m_v^*}{m_c^* + m_v^*}.$$

and that $\chi_2(\omega)$ becomes

$$\chi_2(\omega) = C\, |\boldsymbol{\varepsilon}\cdot \langle \mathbf{k}n| \mathbf{p} |\mathbf{k}m\rangle|^2 \left(\frac{2\mu}{\hbar\omega}\right)^{3/2} \left[1 - \frac{E_g}{\hbar\omega}\right]^{1/2}.$$

6

Phonons and Lattice Dynamics

6.1 Introduction

In the preceding chapters, we introduced the one-electron approach to obtain the electronic structure of solids and described methods for its calculation. We also presented the theory of linear response in the one-electron approximation. We will now apply some of these methods to develop the phenomenological theory of lattice vibrations, known as lattice dynamics, and describe how it can be used to calculate phonon dispersions for metals, insulators, and semiconductors. We shall also present and describe the experimental methods used to measure bulk and surface phonon dispersion curves.

In the development of the lattice dynamics method, we shall consider the lattice motion within the *Born–Oppenheimer approximation*, which was outlined in the introductory chapter. Furthermore, we shall treat the lattice displacements in the *harmonic approximation*.[1] In lattice dynamics, the coupled equations of motion of the ions are cast as a matrix eigenvalue problem, and the defining matrix is termed the *dynamical matrix*. The contributions to the dynamical matrix are separated into direct ion–ion interaction terms, plus valence electron–mediated effective ion–ion interaction terms. The latter terms arise from the scattering of valence electrons from moving ions, and are usually treated up to second-degree changes of the total electron energy in the ionic displacements.

Ionic Cores and Quantum-Mechanical Rigidity

We know that atoms contain tightly bound electrons in closed shells, together with valence electrons, which are much less bound than their closed-shell counterparts. Under the small perturbation arising from ionic displacements, an atomic electron wavefunction will distort or deform as

$$\left|\Psi_n^{(0)}\right\rangle \;\rightarrow\; \left|\Psi_n^{(0)}\right\rangle + \left|\Psi^{(1)}\right\rangle + \cdots = \left|\Psi_n^{(0)}\right\rangle + \sum_{m\neq n}\left|\Psi_m^{(0)}\right\rangle \frac{V_{nm}}{E_n^{(0)}-E_m^{(0)}} + \begin{array}{c}\text{higher-order}\\ \text{terms}\end{array}$$

---

[1] We shall expand the interatomic interaction potential up to second degree in the atomic displacements from equilibrium positions.

For the closed-shell, tightly bound electronic states,

$$\frac{V_{nm}}{E_n^{(0)} - E_m^{(0)}} \ll 1$$

and their corresponding wavefunctions hardly deform! Thus, the closed-shell electrons will move with the nucleus like a rigid body under the small perturbations produced by their respective nuclear displacements in the solid. The combined body of the nucleus and the closed-shell electrons are termed an *ionic core*. Henceforth, for the sake of brevity, we shall refer to ion cores as ions.

In general, the presence of energy gaps endow quantum-mechanical wavefunctions with some degree of rigidity.

6.2 Coupling of Phonons to Electrons

The one-electron Hamiltonian for a crystalline solid or a molecule in its equilibrium configuration is given by

$$\mathcal{H}_e = \frac{\mathbf{p}^2}{2m_e} + V(\mathbf{r}; \{\mathbf{R}_0\}),$$

where $\{\mathbf{R}_0\}$ represents the set of ionic equilibrium position vectors, and $\mathbf{r}$ is the electron position vector. Here, we shall consider how the vibronic (phonon) and electronic states couple to each other.

We start with displacing the ions by infinitesimal amounts from their equilibrium positions, and obtain the configuration $\{\mathbf{R}_0\} + \overrightarrow{d\mathbf{R}}$, where $\overrightarrow{d\mathbf{R}} = [d\mathbf{R}_1, d\mathbf{R}_2, \ldots, d\mathbf{R}_N]$ for a system with N ions. The corresponding electron Hamiltonian becomes

$$\mathcal{H}_e = \frac{\mathbf{p}^2}{2m_e} + V\left(\mathbf{r}; \{\mathbf{R}_0\} + \overrightarrow{d\mathbf{R}}\right) \simeq \frac{\mathbf{p}^2}{2m_e} + V(\mathbf{r}; \{\mathbf{R}_0\}) + \sum_j \nabla_{\mathbf{R}_j} V\bigg|_{\mathbf{R}_{0j}} \cdot d\mathbf{R}_j,$$

where j indexes the ions, and only terms linear in $d\mathbf{R}_j$ are retained. Next, we express the displacements in terms of normal, or symmetry-adapted, modes

$$\mathbf{Q}_i = \sum \alpha_{ij}\, d\mathbf{R}_j \quad \Rightarrow \quad \overrightarrow{\mathbf{Q}} = \alpha\, \overrightarrow{d\mathbf{R}}$$

as linear combinations of $d\mathbf{R}_j$ defined by the unitary transformation matrix $\boldsymbol{\alpha}$. Focusing on the last term of the Hamiltonian and dropping the arrow, we write $\nabla_{\mathbf{R}} V\big|_{\mathbf{R}_0} \cdot d\mathbf{R}$ in matrix form

$$\left(\nabla_{\mathbf{R}} V\big|_{\mathbf{R}_0}\right)^T d\mathbf{R} = \left(\nabla_{\mathbf{R}} V\big|_{\mathbf{R}_0}\right)^T \boldsymbol{\alpha}^T \boldsymbol{\alpha}\, d\mathbf{R} = \left[\boldsymbol{\alpha} \nabla_{\mathbf{R}} V\big|_{\mathbf{R}_0}\right]^T \mathbf{Q}$$

but

$$\boldsymbol{\alpha} \nabla_{\mathbf{R}} = \nabla_{\mathbf{Q}} \quad \Rightarrow \quad \left[\boldsymbol{\alpha} \nabla_{\mathbf{R}} V\big|_{\mathbf{R}_0}\right]^T = \left[\nabla_{\mathbf{Q}} V\right]^T.$$

The Hamiltonian takes the form

$$\mathcal{H}_e = \frac{\mathbf{p}^2}{2m_e} + V\left(\mathbf{r}; \{\mathbf{R}_0\}\right) + \sum_i \frac{\partial V}{\partial Q_i}\, Q_i.$$

Since the quantities $\dfrac{\partial V}{\partial Q_i}$ have the same symmetry as Q_i, the last term is an invariant scalar.

Electron–Phonon Coupling

For a crystalline solid, we write $\{\mathbf{R}_0\} \Rightarrow \{\mathbf{x}_{l\kappa}^{(0)}\}$, where l denotes the primitive cell and κ is the basis, or sublattice, index. The equilibrium position of ion $l\kappa$ is defined as

$$\mathbf{x}_{l\kappa}^{(0)} = \mathbf{R}_l + \boldsymbol{\rho}_\kappa,$$

where $\mathbf{R}_l$ is a primitive lattice vector and $\boldsymbol{\rho}_\kappa$ the basis vector of sublattice κ, while its instantaneous position is

$$\mathbf{x}_{l\kappa} = \mathbf{R}_l + \boldsymbol{\rho}_\kappa + \mathbf{u}_{l\kappa},$$

where $\mathbf{u}_{l\kappa}$ is an infinitesimal displacement of ion $l\kappa$. The underlying symmetry of interest here is translational lattice periodicity, which engenders the unitary transformation

$$\mathbf{u}_\kappa(\mathbf{q}) = \frac{1}{N} \sum_l e^{-i\mathbf{q}\cdot\mathbf{R}_l}\, \mathbf{u}_{l\kappa} \qquad \Rightarrow \qquad \text{phonons.}$$

Applying this unitary transformation to $\nabla_{\mathbf{x}} V\big|_{\mathbf{x}_{l\kappa}^{(0)}}$ we obtain $\mathbf{g}(\mathbf{q};\kappa)$, the *electron–phonon coupling* matrix element:

$$\mathbf{g}(\mathbf{q};\kappa) = \sum_l e^{i\mathbf{q}\cdot\mathbf{R}_l}\, \nabla_{\mathbf{x}_{l\kappa}} V\big|_{\mathbf{x}_{l\kappa}^{(0)}}.$$

Writing V in muffin-tin form

$$V\left(\mathbf{r}; \left\{\mathbf{x}_{l\kappa}^{(0)}\right\}\right) = \sum_{l\kappa} v_\kappa\left(\mathbf{r} - \mathbf{x}_{l\kappa}^{(0)}\right),$$

we obtain the expression

$$\mathbf{g}(\mathbf{q};\kappa) = -\sum_l e^{i\mathbf{q}\cdot\mathbf{R}_l}\, \nabla_{\mathbf{r}} v_\kappa\left(\mathbf{r} - \mathbf{x}_{l\kappa}^{(0)}\right)\Big|_{\mathbf{x}_{l\kappa}^{(0)}}.$$

where now we take the derivatives with respect to the electron position.

6.3 Ionic and Electronic Contributions to Phonon Energies

We shall regard the solid as made up of ions with equilibrium positions $\{\mathbf{x}_{l\kappa}^{(0)}\}$, mediated by valence electrons (these will be referred to simply as electrons). The choice of ion-core electrons will depend on the solid considered, and we shall leave the notion general enough for any choice. Anyhow, an ion is assumed to move rigidly. The contributions to phonon energies include the following components:

(1) Kinetic energy of the ions
(2) Direct bare ion–ion interaction in the absence of the *valence electrons*

The ion–ion interaction consists of Coulombic, v_i^c, and quantum-overlap, v_i^e, components

$$V_I = \frac{1}{2} \sum_{l\kappa, l'\kappa'}^{'} \left(v_i^c \left(|\mathbf{x}_{l\kappa} - \mathbf{x}_{l'\kappa'}| ; \kappa, \kappa' \right) + v_i^e \left(|\mathbf{x}_{l\kappa} - \mathbf{x}_{l'\kappa'}| ; \kappa, \kappa' \right) \right), \tag{6.1}$$

$\mathbf{x}_{l\kappa}$ being the instantaneous position of ion $l\kappa$. The Coulomb interaction has the form

$$v_i^c \left(|\mathbf{x}_{l\kappa} - \mathbf{x}_{l'\kappa'}| ; \kappa, \kappa' \right) = \frac{Z_\kappa Z_{\kappa'} e^2}{|\mathbf{x}_{l\kappa} - \mathbf{x}_{l'\kappa'}|}$$

with $e Z_\kappa$ the effective ionic charge of ion κ.

The quantum interaction v_i^e arises from the overlap of the tails of the electronic wave functions of the ion cores, as shown in this illustration:

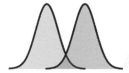

(3) Electronic contributions arise from electron–ion interactions represented by

$$V_{ei} = \sum_{l\kappa} \int d\mathbf{r} \, n(\mathbf{r}) \, v \left(\mathbf{r} - \mathbf{x}_{l\kappa}, \kappa \right), \tag{6.2}$$

where $v \left(\mathbf{r} - \mathbf{x}_{l\kappa}, \kappa \right)$ is the atomic electron–ion potential, of type κ, centered at the instantaneous position $\mathbf{x}_{l\kappa}$, and $n(\mathbf{r})$ is the corresponding electron density.

In the Born–Oppenheimer approximation, the electron–ion potential in (6.2) is taken when the ions are at rest at $\mathbf{x}_{l\kappa}$. The electronic contribution is then given by an effective potential energy equal to the lowest electron energy $E_0^{el}(\mathbf{x})$. However, in this chapter we will treat these contributions perturbatively.

In the harmonic approximation for the lattice vibration, we expand the effective ion potential energy, which includes both the direct term V_I and the electronic contribution, to second order in the displacement $\mathbf{u}_{l\kappa} = \mathbf{x}_{l\kappa} - \mathbf{x}_{l\kappa}^{(0)}$.

Direct Ion–Ion Interaction Potential in the Harmonic Approximation

Taylor expansion of V_I about equilibrium ion positions up to quadratic terms, yields the expression

$$V_I^{(0)} + V_I^{(2)},$$

$$V_I^{(2)} = \frac{1}{2} \sum_{\substack{l\kappa\alpha \\ l'\kappa'\beta}} u_{l\kappa\alpha} \, \Phi_{\alpha\beta}^I \begin{pmatrix} l & l' \\ \kappa & \kappa' \end{pmatrix} u_{l'\kappa'\beta}. \tag{6.3}$$

α and β denote Cartesian directions. Notice that the sum over terms linear in the displacements vanishes because of equilibrium.

$\Phi_{\alpha\beta} \begin{pmatrix} l & l' \\ \kappa & \kappa' \end{pmatrix}$ is the conventional notation for the matrix element of the force

constant matrix, $\Phi \begin{pmatrix} l & l' \\ \kappa & \kappa' \end{pmatrix}$; its elements are second partial derivatives of the inter

particle interaction potentials. $\Phi_{\alpha\beta} \begin{pmatrix} l & l' \\ \kappa & \kappa' \end{pmatrix}$ represents the force on the particle $l\kappa$ in

the α-direction due to the unit displacement of the $l'\kappa'$ particle in the β-direction.

$\Phi_{\alpha\beta}^I \begin{pmatrix} l & l' \\ \kappa & \kappa' \end{pmatrix}$ is the direct ion–ion *force constant* matrix element, defined as

$$
\Phi_{\alpha\beta}^I \begin{pmatrix} l & l' \\ \kappa & \kappa' \end{pmatrix} = \delta_{ll'}\delta_{\kappa\kappa'} \sum_{l''\kappa''\neq l\kappa} \frac{\partial^2}{\partial x_{l\kappa\alpha}\partial x_{l''\kappa''\beta}} \Big[v_i^c \left(|\mathbf{x}_{l\kappa} - \mathbf{x}_{l''\kappa''}| ; \kappa, \kappa'' \right)
$$

$$
+ v_i^e \left(|\mathbf{x}_{l\kappa} - \mathbf{x}_{l''\kappa''}| ; \kappa, \kappa'' \right) \Big]\Big|_{\mathbf{x}_{l\kappa}^{(0)} - \mathbf{x}_{l''\kappa''}^{(0)}}
$$

$$
- \frac{\partial^2}{\partial x_{l\kappa\alpha}\partial x_{l'\kappa'\beta}} \Big[v_i^c \left(|\mathbf{x}_{l\kappa} - \mathbf{x}_{l'\kappa'}| ; \kappa, \kappa' \right) + v_i^e \left(|\mathbf{x}_{l\kappa} - \mathbf{x}_{l'\kappa'}| ; \kappa, \kappa' \right) \Big]\Big|_{\mathbf{x}_{l\kappa}^{(0)} - \mathbf{x}_{l\kappa}^{(0)}}.
$$

The first term is known as the *self-term*; it arises from translation invariance, as explained in the following discussion.

For radial type potentials, we have

$$
\frac{\partial^2}{\partial x_\alpha \partial x_\beta} v\left(|\mathbf{x}|\right) = \frac{x_\alpha x_\beta}{x^2} \frac{d^2 v}{dx^2} - \left(\frac{x_\alpha x_\beta}{x^3} - \frac{\delta_{\alpha\beta}}{x} \right) \frac{dv}{dx}, \tag{6.4}
$$

which, for the case of Coulomb interaction, becomes

$$
\frac{\partial^2}{\partial x_\alpha \partial x_\beta} \frac{Z_\kappa Z_{\kappa'}}{x} = Z_\kappa Z_{\kappa'} \left[\frac{3x_\alpha x_\beta}{x^5} - \frac{\delta_{\alpha\beta}}{x^3} \right]. \tag{6.5}
$$

Translation Invariance Symmetry and the Self-Term

The self-term $\Phi_{\alpha\beta} \begin{pmatrix} l & l \\ \kappa & \kappa \end{pmatrix}$ is the force on particle $l\kappa$ in the α-direction due to its own displacement in the β-direction!

We determine this term with the aid of translation invariance symmetry:

We use the fact that the force $F_{l\kappa}^I$ acting on ion $(l\kappa)$ in the α-direction due to an arbitrary *uniform* displacement of the whole crystal in the β-direction

$$
u_{l\kappa\beta} = w_\beta, \ \forall \, l\kappa
$$

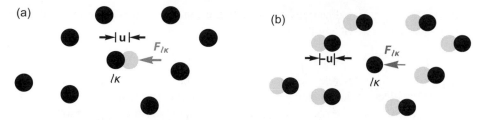

Figure 6.1 (a) The force acting on particle $\ell\kappa$ due to its own displacement by **u**, while keeping all other particles fixed. (b) The force acting on particle $\ell\kappa$ due to displacing all other particles by $-\mathbf{u}$ while keeping particle $\ell\kappa$ fixed. The two forces are identical. (Displaced particles are depicted in light gray.)

must vanish:

$$\mathbf{F}^{I}_{l\kappa\alpha} = -\left.\frac{\partial V^{(2)}_{\mathrm{I}}}{\partial x_{l\kappa\alpha}}\right|_{x_{l\kappa}=x^{(0)}_{l\kappa}} = w_{\beta}\sum_{l'\kappa'}\Phi_{\alpha\beta}\begin{pmatrix} l & l' \\ \kappa & \kappa' \end{pmatrix} = 0.$$

Since the sum over $l'\kappa'$ includes $l\kappa$, then

$$\boxed{\Phi_{\alpha\beta}\begin{pmatrix} l & l \\ \kappa & \kappa \end{pmatrix} = -\sum_{l'\kappa'\neq l\kappa}\Phi_{\alpha\beta}\begin{pmatrix} l & l' \\ \kappa & \kappa' \end{pmatrix}}$$

As illustrated in Figure 6.1, the force acting on particle $l\kappa$ due to its own displacement by **u**, while keeping all other particles fixed, is identical to the force acting on particle $l\kappa$ due to displacing all other particles by $-\mathbf{u}$ while keeping particle $l\kappa$ fixed!

6.4 Electronic Contribution to Phonon Energies

At zero temperature and in accordance with the harmonic approximation, the electronic contribution to the lattice dynamics is given by the change of the electron energy $V^{(2)}_{\mathrm{el}}$ quadratic in the ionic displacement. In the adiabatic approximation, the change in the energy of the electron system is due to the perturbative change in electron–ion interaction potential from its value at the equilibrium lattice configuration. V_{ei} in (6.2) can be expanded as

$$V_{\mathrm{ei}} = V^{(0)}_{\mathrm{ei}} + \sum_{l\kappa}\nabla_{l\kappa}V_{\mathrm{ei}}\Big|_{\mathbf{x}^{(0)}_{l\kappa}}\cdot\mathbf{u}_{l\kappa} + \sum_{l\kappa}\mathbf{u}_{l\kappa}\cdot\left[\nabla_{l\kappa}\nabla_{l\kappa}V_{\mathrm{ei}}\right]_{\mathbf{x}^{(0)}_{l\kappa}}\cdot\mathbf{u}_{l\kappa}$$

$$= \sum_{l\kappa}\int d\mathbf{r}\, n^{(0)}(\mathbf{r})\, v_{\kappa}\left(\mathbf{r}-\mathbf{x}^{(0)}_{l\kappa}\right) + \sum_{l\kappa}\int d\mathbf{r}\, n^{(1)}(\mathbf{r})\, \nabla_{l\kappa}v_{\kappa}\left(\mathbf{r}-\mathbf{x}_{l\kappa}\right)\Big|_{\mathbf{x}^{(0)}_{l\kappa}}\cdot\mathbf{u}_{l\kappa}$$

$$+ \sum_{l\kappa} \mathbf{u}_{l\kappa} \cdot \int d\mathbf{r}\, n^{(0)}(\mathbf{r}) \left[\nabla_{l\kappa} \nabla_{l\kappa} v_{\kappa}(\mathbf{r} - \mathbf{x}_{l\kappa}) \right]_{\mathbf{x}_{l\kappa}^{(0)}} \cdot \mathbf{u}_{l\kappa}$$

$$= \int d\mathbf{r}\, n^{(0)}(\mathbf{r})\, v^{[0]}(\mathbf{r}) + \int d\mathbf{r}\, n^{(1)}(\mathbf{r})\, v^{[1]}(\mathbf{r}) + \int d\mathbf{r}\, n^{(0)}(\mathbf{r})\, v^{[2]}(\mathbf{r}). \qquad (6.6)$$

$V_{ei}^{(0)}$ is just the electron–ion potential energy of the perfect crystal, $n^{(0)}(\mathbf{r})$ the corresponding electron density in the ground state, and $n^{(1)}(\mathbf{r})$ the first-order correction to the density due to ionic displacements.

Thus, the last two terms of (6.6) contribute to energy change quadratic in the ionic displacements:

(i) The first-order perturbation energy change due to the interaction term $v^{[2]}(\mathbf{r})$, which is quadratic in the ion displacements,

$$v^{[2]}(\mathbf{r}) = -\frac{1}{2} \sum_{l\kappa\alpha\beta} \frac{\partial^2 v_{\kappa}\left(\mathbf{r} - \mathbf{x}_{l\kappa}^{(0)}\right)}{\partial r_\alpha \partial r_\beta} u_{l\kappa\alpha}\, u_{l\kappa\beta}. \qquad (6.7)$$

Notice that we are now taking the derivatives with respect to the electron position $\mathbf{r}$.

The contribution to $V_{el}^{(2)}$ is simply the expectation value of $v^{[2]}$ with respect to the ground state of the electron system at the equilibrium crystal configuration:

$$E^{(22)} = \int d\mathbf{r}\, n^{(0)}(\mathbf{r})\, v^{[2]}(\mathbf{r}) = -\frac{1}{2} \sum_{\substack{l\kappa\alpha \\ l'\kappa'\beta}} \pi_{\alpha\beta}^{(2)} \begin{pmatrix} l & l' \\ \kappa & \kappa' \end{pmatrix} u_{l\kappa\alpha}\, u_{l'\kappa'\beta}, \qquad (6.8)$$

where

$$\boxed{\pi_{\alpha\beta}^{(2)} \begin{pmatrix} l & l' \\ \kappa & \kappa' \end{pmatrix} = \delta_{ll'}\, \delta_{\kappa\kappa'} \int d\mathbf{r}\, n^{(0)}(\mathbf{r}) \frac{\partial^2 v_{\kappa}\left(\mathbf{r} - \mathbf{x}_{l\kappa}^{(0)}\right)}{\partial r_\alpha \partial r_\beta}} \qquad (6.9)$$

(ii) The change in energy due to the interaction $v^{[1]}$ linear in the ionic displacements,

$$E^{(21)} = \int d\mathbf{r}\, n^{(1)}(\mathbf{r})\, v^{[1]}(\mathbf{r}) = -\sum_{l\kappa\alpha} \int d\mathbf{r}\, n^{(1)}(\mathbf{r}) \frac{\partial v_{\kappa}\left(\mathbf{r} - \mathbf{x}_{l\kappa}^{(0)}\right)}{\partial r_\alpha} u_{l\kappa\alpha}. \qquad (6.10)$$

Linear response theory gives the first-order change in the electron density as

$$n^{(1)}(\mathbf{r}) = \int d\mathbf{r}'\, \chi(\mathbf{r},\mathbf{r}')\, v^{[1]}(\mathbf{r}'), \qquad (6.11)$$

where $\chi(\mathbf{r},\mathbf{r}')$ is longitudinal susceptibility, or the static density–density response function. The Fourier transform of $\chi(\mathbf{r},\mathbf{r}')$ is given by

$$\chi(\mathbf{Q},\mathbf{Q}') = \frac{1}{\Omega} \int d\mathbf{r}\, d\mathbf{r}'\, \chi(\mathbf{r},\mathbf{r}')\, e^{-i\mathbf{Q}\cdot\mathbf{r}}\, e^{i\mathbf{Q}'\cdot\mathbf{r}'}. \qquad (6.12)$$

Periodic translation-invariance symmetry of the crystal dictates that

$$\chi(\mathbf{r} + \mathbf{R}_l, \mathbf{r}' + \mathbf{R}_l) = \chi(\mathbf{r}, \mathbf{r}'),\tag{6.13}$$

where $\mathbf{R}_l$ is a primitive lattice vector. This condition requires that

$$\begin{cases} \mathbf{Q} = \mathbf{q} + \mathbf{G} \\ \mathbf{Q}' = \mathbf{q} + \mathbf{G}' \end{cases}$$

and the Fourier transform becomes

$$\chi(\mathbf{Q}, \mathbf{Q}') \;\rightarrow\; \chi(\mathbf{q} + \mathbf{G}, \mathbf{q} + \mathbf{G}').$$

The change in the electronic energy quadratic in $v^{[1]}$ can then be cast in the form

$$E^{(21)} = \frac{1}{2} \int d\mathbf{r}\, n^{(1)}(\mathbf{r})\, v^{[1]}(\mathbf{r}) = \frac{1}{2} \iint d\mathbf{r}\, d\mathbf{r}'\, v^{[1]}(\mathbf{r})\, \chi(\mathbf{r}, \mathbf{r}')\, v^{[1]}(\mathbf{r}')$$

$$= \frac{1}{2} \sum_{\substack{l\kappa\alpha \\ l'\kappa'\beta}} \pi_{\alpha\beta}^{(1)}\begin{pmatrix} l & l' \\ \kappa & \kappa' \end{pmatrix} \mathbf{u}_{l\kappa\alpha}\, \mathbf{u}_{l'\kappa'\beta},\tag{6.14}$$

$$\pi_{\alpha\beta}^{(1)}\begin{pmatrix} l & l' \\ \kappa & \kappa' \end{pmatrix} = \iint d\mathbf{r}\, d\mathbf{r}'\, \frac{\partial v_\kappa\left(\mathbf{r} - \mathbf{x}_{l\kappa}^{(0)}\right)}{\partial r_\alpha}\, \chi(\mathbf{r}, \mathbf{r}')\, \frac{\partial v_{\kappa'}\left(\mathbf{r}' - \mathbf{x}_{l'\kappa'}^{(0)}\right)}{\partial r_\beta'}.\tag{6.15}$$

Equations (6.8) and (6.14) define the electronic contribution to the force constants $\Phi^{\text{el}}\begin{pmatrix} l & l' \\ \kappa & \kappa' \end{pmatrix}$.

Electronic Self-Term

Although the ionic part has been clearly cast in a translationally invariant form, the electronic contribution has not. It is convenient to amend this.

To establish such a form, we again make use of bodily translating the whole crystal. When the crystal undergoes a uniform and arbitrary infinitesimal displacement $\mathbf{u}$, the total change in the electronic energy should be zero. Consequently, using (6.8) and (6.14) we write

$$V_{\text{el}}^{(2)} = E^{(21)} + E^{(22)}$$

$$= \frac{1}{2} \int d\mathbf{r}\, n^{(1)}(\mathbf{r})\, v^{[1]}(\mathbf{r}) + \int d\mathbf{r}\, n^{(0)}(\mathbf{r})\, v^{[2]}(\mathbf{r}) = 0\tag{6.16}$$

under an arbitrary uniform displacement. First, we have

$$\int d\mathbf{r}\, n^{(0)}(\mathbf{r})\, v^{[2]}(\mathbf{r}) = \int d\mathbf{r}\, n^{(0)}(\mathbf{r}) \left(\sum_{l\kappa} \mathbf{u} : \nabla_\mathbf{r} \nabla_\mathbf{r}\, v_\kappa\left(\mathbf{r} - \mathbf{x}_{l\kappa}^{(0)}\right) : \mathbf{u} \right)$$

Second, we find that $v^{[1]}$ becomes

$$v^{[1]}(\mathbf{r}) = -\mathbf{u} \cdot \sum_{l\kappa} \nabla_{\mathbf{r}} v_{\kappa} \left(\mathbf{r} - \mathbf{x}_{l\kappa}^{(0)} \right), \tag{6.17}$$

and the first-order change in the electron density distribution is then given by (6.11),

$$n^{(1)}(\mathbf{r}) = \mathbf{u} \cdot \int d\mathbf{r}' \, \chi(\mathbf{r}, \mathbf{r}') \, v^{[1]}(\mathbf{r}'). \tag{6.18}$$

Substituting back in (6.16), we obtain

$$\sum_{l\kappa} \mathbf{u} : \int d\mathbf{r} \, n^{(0)}(\mathbf{r}) \left(\nabla_{\mathbf{r}} \nabla_{\mathbf{r}} v_{\kappa} \left(\mathbf{r} - \mathbf{x}_{l\kappa}^{(0)} \right) \right) : \mathbf{u}$$

$$= -\frac{1}{2} \sum_{\substack{l\kappa \\ l''\kappa''}} \mathbf{u} : \iint d\mathbf{r} \, d\mathbf{r}' \, \nabla_{\mathbf{r}} v_{\kappa} \left(\mathbf{r} - \mathbf{x}_{l\kappa}^{(0)} \right) \chi(\mathbf{r}, \mathbf{r}') \nabla_{\mathbf{r}'} v_{\kappa''} \left(\mathbf{r}' - \mathbf{x}_{l''\kappa''}^{(0)} \right) : \mathbf{u}$$

and we can write the electronic part of the force constants as

$$\Phi_{\alpha\beta}^{\mathrm{el}} \begin{pmatrix} l & l' \\ \kappa & \kappa' \end{pmatrix} = \iint d\mathbf{r} \, d\mathbf{r}' \left[\frac{\partial v_{\kappa} \left(\mathbf{r} - \mathbf{x}_{l\kappa}^{(0)} \right)}{\partial r_{\alpha}} \chi(\mathbf{r}, \mathbf{r}') \frac{\partial v_{\kappa'} \left(\mathbf{r}' - \mathbf{x}_{l'\kappa'}^{(0)} \right)}{\partial r_{\beta}'} \right. $$

$$\left. - \delta_{ll'} \delta_{\kappa\kappa'} \sum_{l''\kappa''} \frac{\partial v_{\kappa} \left(\mathbf{r} - \mathbf{x}_{l\kappa}^{(0)} \right)}{\partial r_{\alpha}} \chi(\mathbf{r}, \mathbf{r}') \frac{\partial v_{\kappa''} \left(\mathbf{r}' - \mathbf{x}_{l''\kappa''}^{(0)} \right)}{\partial r_{\beta}'} \right]. \tag{6.19}$$

This obviously satisfies the infinitesimal translational invariance relation

$$\sum_{l'\kappa'} \Phi_{\alpha\beta}^{\mathrm{el}} \begin{pmatrix} l & l' \\ \kappa & \kappa' \end{pmatrix} = 0. \tag{6.20}$$

Also, from the crystal periodic translation symmetry of the density response function, we have

$$\Phi_{\alpha\beta}^{\mathrm{el}} \begin{pmatrix} l & l' \\ \kappa & \kappa' \end{pmatrix} = \Phi_{\alpha\beta}^{\mathrm{el}} \begin{pmatrix} 0 & l' - l \\ \kappa & \kappa' \end{pmatrix} = \Phi_{\alpha\beta}^{\mathrm{el}} \begin{pmatrix} l - l' & 0 \\ \kappa & \kappa' \end{pmatrix}. \tag{6.21}$$

6.5 The Dynamical Matrix

Finally, we arrive at the total change in energy quadratic in the lattice displacements:

$$V^{(2)} = V_I^{(2)} + V_{\mathrm{el}}^{(2)} = \frac{1}{2} \sum_{l\kappa, l'\kappa'} \mathbf{u}_{l\kappa} \cdot \left[\Phi^I \begin{pmatrix} l & l' \\ \kappa & \kappa' \end{pmatrix} + \Phi^{\mathrm{el}} \begin{pmatrix} l & l' \\ \kappa & \kappa' \end{pmatrix} \right] \cdot \mathbf{u}_{l'\kappa'}. \tag{6.22}$$

With the equilibrium position of ion $l\kappa$ $\mathbf{x}_{l\kappa}^{(0)} = \mathbf{R}_l + \boldsymbol{\rho}_\kappa$, we define the Fourier transform of the ionic displacements as

$$\mathbf{U}_\kappa(\mathbf{q}) = \sqrt{\frac{M_\kappa}{N}} \sum_l e^{-i\mathbf{q}\cdot\mathbf{R}_l} \, \mathbf{u}(l\kappa)$$

$$\mathbf{u}(l\kappa) = \frac{1}{\sqrt{N M_\kappa}} \sum_\mathbf{q} e^{i\mathbf{q}\cdot\mathbf{R}_l} \, \mathbf{U}_\kappa(\mathbf{q}),$$

where N is the number of primitive cells and M_κ is the mass of the κ-type ion. Substituting for $\mathbf{u}(l\kappa)$ in (6.22), we get

$$V^{(2)} = \frac{1}{2N} \sum_{\mathbf{q};\kappa\kappa'} \frac{\mathbf{U}_\kappa(\mathbf{q})}{\sqrt{M_\kappa}} \cdot \left\{ \sum_{ll'} e^{-i\mathbf{q}\cdot\mathbf{R}_l} \left[\Phi^I \begin{pmatrix} l & l' \\ \kappa & \kappa' \end{pmatrix} \right.\right.$$

$$\left.\left. + \Phi^{\mathrm{el}} \begin{pmatrix} l & l' \\ \kappa & \kappa' \end{pmatrix} \right] e^{i\mathbf{q}\cdot\mathbf{R}_{l'}} \right\} \cdot \frac{\mathbf{U}_{\kappa'}(\mathbf{q})}{\sqrt{M_\kappa'}}$$

$$= \frac{1}{2} \sum_{\mathbf{q};\kappa\kappa'} \mathbf{U}_\kappa(\mathbf{q}) \cdot \left[\mathbf{D}^I \begin{pmatrix} \mathbf{q} \\ \kappa\kappa' \end{pmatrix} + \mathbf{D}^{\mathrm{el}} \begin{pmatrix} \mathbf{q} \\ \kappa\kappa' \end{pmatrix} \right] \cdot \mathbf{U}_{\kappa'}(\mathbf{q}). \tag{6.23}$$

We now define the *dynamical matrix* $\mathbf{D} = \mathbf{D}^I + \mathbf{D}^{\mathrm{el}}$, with components

$$\mathbf{D}^I \begin{pmatrix} \mathbf{q} \\ \kappa\kappa' \end{pmatrix} = \frac{1}{N\sqrt{M_\kappa M_{\kappa'}}} \sum_l \sum_{l'} e^{-i\mathbf{q}\cdot(\mathbf{R}_l - \mathbf{R}_{l'})} \, \Phi^I \begin{pmatrix} l & l' \\ \kappa & \kappa' \end{pmatrix}$$

$$= \frac{1}{\sqrt{M_\kappa M_{\kappa'}}} \sum_l e^{-i\mathbf{q}\cdot\mathbf{R}_l} \, \Phi^I \begin{pmatrix} 0 & l \\ \kappa & \kappa' \end{pmatrix}, \qquad \kappa \neq \kappa'$$

$$\mathbf{D}^I \begin{pmatrix} \mathbf{q} \\ \kappa\kappa \end{pmatrix} = \frac{1}{M_\kappa} \left[\sum_{l\neq 0} e^{-i\mathbf{q}\cdot\mathbf{R}_l} \, \Phi^I \begin{pmatrix} 0 & l \\ \kappa & \kappa \end{pmatrix} - \delta_{\kappa\kappa'} \sum_{l\kappa''\neq 0\kappa} \Phi^I \begin{pmatrix} 0 & l \\ \kappa & \kappa'' \end{pmatrix} \right]. \tag{6.24}$$

Similarly, we get

$$\mathbf{D}^{\mathrm{el}} \begin{pmatrix} \mathbf{q} \\ \kappa\kappa' \end{pmatrix} = \frac{1}{\sqrt{M_\kappa M_{\kappa'}}} \sum_l e^{-i\mathbf{q}\cdot\mathbf{R}_l} \, \Phi^{\mathrm{el}} \begin{pmatrix} 0 & l \\ \kappa & \kappa' \end{pmatrix}$$

$$= \frac{1}{\sqrt{M_\kappa M_{\kappa'}}} \sum_l e^{-i\mathbf{q}\cdot\mathbf{R}_l} \left(\iint d\mathbf{r}d\mathbf{r}' \left[\nabla_\mathbf{r} v(\mathbf{r} - \boldsymbol{\rho}_\kappa) \, \chi(\mathbf{r},\mathbf{r}') \, \nabla_{\mathbf{r}'} v(\mathbf{r}' - \mathbf{x}_{l\kappa'}) \right] \right.$$

$$\left. - \delta_{\kappa\kappa'} \sum_{l\kappa''\neq 0\kappa} \nabla_\mathbf{r} v(\mathbf{r} - \boldsymbol{\rho}_\kappa) \, \chi(\mathbf{r},\mathbf{r}') \, \nabla_{\mathbf{r}'} v(\mathbf{r}' - \mathbf{x}_{l\kappa''}) \right). \tag{6.25}$$

6.5.1 Simplifying the Electronic Contribution to the Dynamical Matrix

Recalling that

$$\chi(\mathbf{r}, \mathbf{r}') = \frac{1}{\Omega} \sum_{\mathbf{G}\mathbf{G}'; \mathbf{q}} e^{-i(\mathbf{q}+\mathbf{G}) \cdot \mathbf{r}} \chi\left(\mathbf{q}+\mathbf{G}, \mathbf{q}+\mathbf{G}'\right) e^{i(\mathbf{q}+\mathbf{G}') \cdot \mathbf{r}'},$$

where Ω is the volume of the system, and defining

$$v_\kappa(\mathbf{q}) = \int d\mathbf{r}\, e^{-i\mathbf{q} \cdot \mathbf{r}}\, v_\kappa(\mathbf{r}), \tag{6.26}$$

we obtain

$$
\begin{aligned}
\mathbf{D}^{\mathrm{el}}_{\alpha\beta}\begin{pmatrix} \mathbf{q} \\ \kappa\kappa' \end{pmatrix} &= \frac{1}{\Omega \sqrt{M_\kappa M_{\kappa'}}} \\[2mm]
&\quad \times \sum_{\mathbf{G}\mathbf{G}'} \left\{ \left[\chi\left(\mathbf{q}+\mathbf{G}, \mathbf{q}+\mathbf{G}'\right) \sum_l e^{-i\mathbf{q} \cdot \mathbf{R}_l} \right. \right. \\[2mm]
&\quad \times \left. \left(\int d\mathbf{r}\, \frac{\partial v(\mathbf{r} - \boldsymbol{\rho}_\kappa)}{\partial r_\alpha}\, e^{-i(\mathbf{q}+\mathbf{G}) \cdot \mathbf{r}} \right) \left(\int d\mathbf{r}'\, \frac{\partial v(\mathbf{r}' - \mathbf{x}_{l\kappa'})}{\partial r_\beta}\, e^{i(\mathbf{q}+\mathbf{G}') \cdot \mathbf{r}'} \right) \right] \\[2mm]
&\quad - \delta_{\kappa, \kappa'} \sum_{l\kappa''}' \left(\int d\mathbf{r}\, \frac{\partial v(\mathbf{r} - \boldsymbol{\rho}_\kappa)}{\partial r_\alpha}\, e^{-i\mathbf{G} \cdot \mathbf{r}} \right) \chi\left(\mathbf{G}, \mathbf{G}'\right) \\[2mm]
&\quad \times \left. \left(\int d\mathbf{r}'\, \frac{\partial v(\mathbf{r}' - \mathbf{x}_{l\kappa''})}{\partial r_\beta}\, e^{i\mathbf{G}' \cdot \mathbf{r}'} \right) \right\} \\[3mm]
&= \frac{1}{\Omega \sqrt{M_\kappa M_{\kappa'}}} \sum_{\mathbf{G}\mathbf{G}'} \left[e^{-i(\mathbf{q}+\mathbf{G}) \cdot \boldsymbol{\rho}_\kappa} (\mathbf{q}+\mathbf{G})_\alpha\, v_\kappa(-\mathbf{q}-\mathbf{G}) \right. \\[2mm]
&\quad \times \chi\left(\mathbf{q}+\mathbf{G}, \mathbf{q}+\mathbf{G}'\right) (\mathbf{q}+\mathbf{G}')_\beta\, v_{\kappa'}(\mathbf{q}+\mathbf{G})\, e^{i(\mathbf{q}+\mathbf{G}') \cdot \boldsymbol{\rho}_{\kappa'}} \\[2mm]
&\quad - \left. \delta_{\kappa\kappa'} \sum_{\kappa'' \neq \kappa} e^{i\mathbf{G} \cdot \boldsymbol{\rho}_\kappa}\, G_\alpha\, v_\kappa(-\mathbf{G})\, \chi\left(\mathbf{G}, \mathbf{G}'\right)\, v_{\kappa''}(\mathbf{G}')\, G'_\beta\, e^{-i\mathbf{G}' \cdot \boldsymbol{\rho}_{\kappa''}} \right].
\end{aligned}
$$

We define

$$X_{\alpha\beta}\begin{pmatrix} \mathbf{Q} & \mathbf{Q}' \\ \kappa & \kappa' \end{pmatrix} = \frac{1}{\Omega \sqrt{M_\kappa M_{\kappa'}}} \left[e^{i\mathbf{Q} \cdot \boldsymbol{\rho}_\kappa} Q_\alpha\, v_\kappa(-\mathbf{Q})\, \chi\left(\mathbf{Q}, \mathbf{Q}'\right) v_{\kappa'}(\mathbf{Q}')\, Q'_\beta\, e^{-i\mathbf{Q}' \cdot \boldsymbol{\rho}_{\kappa'}} \right]$$

and write the electronic contribution to the dynamical matrix as

$$D^{\mathrm{el}}_{\alpha\beta}\begin{pmatrix} \mathbf{q} \\ \kappa\kappa' \end{pmatrix} = \sum_{\mathbf{G}\mathbf{G}'} \left[X_{\alpha\beta}\begin{pmatrix} \mathbf{q}+\mathbf{G} & \mathbf{q}+\mathbf{G}' \\ \kappa & \kappa' \end{pmatrix} - \delta_{\kappa\kappa'} \sum_{\kappa'' \neq \kappa} X_{\alpha\beta}\begin{pmatrix} \mathbf{G} & \mathbf{G}' \\ \kappa & \kappa'' \end{pmatrix} \right]. \tag{6.27}$$

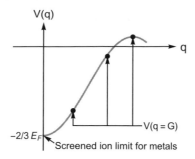

Figure 6.2 Schematic of a pseudopotential form factor in reciprocal space where $\mathbf{G}$ is a reciprocal lattice vector.

Throughout our discussion of the electronic contribution to phonon energy, we have not specified the electron–ion potential $v_\kappa \left(\mathbf{r} - \mathbf{x}_{l\kappa}^{(0)} \right)$.

In the case of metals and semiconductors, it is convenient to deal with pseudopotentials (see Figure 6.2), and thus with pseudowavefunctions, namely, of the form $e^{i(\kappa+\mathbf{G})\cdot\mathbf{r}}$ or linear combinations of such plane waves. We choose to use a pseudopotential $W(\mathbf{q}+\mathbf{G})$, and write

$$X_{\alpha\beta}\begin{pmatrix} \mathbf{q}+\mathbf{G} & \mathbf{q}+\mathbf{G}' \\ \kappa & \kappa' \end{pmatrix} = \frac{1}{\Omega\sqrt{M_\kappa M_{\kappa'}}}(\mathbf{q}+\mathbf{G})_\alpha W_\kappa(\mathbf{q}+\mathbf{G})\chi\left(\mathbf{q}+\mathbf{G},\mathbf{q}+\mathbf{G}'\right)$$

$$\times\, W_{\kappa'}(\mathbf{q}+\mathbf{G})\,(\mathbf{q}+\mathbf{G}')_\beta\, e^{-i(\mathbf{q}+\mathbf{G})\cdot\rho_\kappa}e^{i(\mathbf{q}+\mathbf{G}')\cdot\rho_{\kappa'}}. \qquad (6.28)$$

To bring out the physical content of the electronic contribution to the dynamical matrix, we will separate diagonal from nondiagonal terms in $\hat{D}^{\text{el}}$ and write

$$D_{\alpha\beta}^{\text{el}}(\mathbf{q}) = E_{\alpha\beta}(\mathbf{q}) + \overset{\leftrightarrow}{F}_{\alpha\beta}(\mathbf{q}), \qquad (6.29)$$

where $E_{\alpha\beta}(\mathbf{q})$ is given by [2]

$$E_{\alpha\beta}(\mathbf{q}) = \sum_{\mathbf{G}} (\mathbf{q}+\mathbf{G})_\alpha\,(\mathbf{q}+\mathbf{G})_\beta\,\left|W(\mathbf{q}+\mathbf{G})\right|^2\chi(\mathbf{q}+\mathbf{G})$$

$$= \sum_{\mathbf{G}} \frac{(\mathbf{q}+\mathbf{G})_\alpha\,(\mathbf{q}+\mathbf{G})_\beta}{v_c(\mathbf{q}+\mathbf{G})}\left|W(\mathbf{q}+\mathbf{G})\right|^2\left[\frac{1}{\epsilon_0(\mathbf{q}+\mathbf{G})}-1\right]. \qquad (6.30)$$

For simplicity, we have taken one atom per primitive cell. In the last line, we made the substitution

$$\frac{1}{\epsilon_0(\mathbf{q}+\mathbf{G})} = 1 + v_c(\mathbf{q}+\mathbf{G})\,\chi(\mathbf{q}+\mathbf{G}),$$

where $v_c(\mathbf{q}) = 4\pi/q^2$ is the Coulomb potential. $E_{\alpha\beta}(\mathbf{q})$ represents the free electron contribution to the lattice dynamics. It is actually the sole electronic contribution in the lattice

---

[2] The self-term contribution is implicit.

dynamics of simple metals. We also note that putting $\epsilon_0(\mathbf{q} + \mathbf{G}) = 1$ for ideal insulators leads to the vanishing of $E_{\alpha\beta}(\mathbf{q})$.

The matrix $\overleftrightarrow{F}$ in the second term of (6.29) has components

$$F_{\alpha\beta}\begin{pmatrix} \mathbf{q} \\ \kappa\kappa' \end{pmatrix} = \sum_{\mathbf{G}\neq\mathbf{G}'} \left[X_{\alpha\beta}\begin{pmatrix} \mathbf{q+G} & \mathbf{q+G}' \\ \kappa & \kappa' \end{pmatrix} - \delta_{\kappa\kappa'} \sum_{\kappa''\neq\kappa} X_{\alpha\beta}\begin{pmatrix} \mathbf{G} & \mathbf{G}' \\ \kappa & \kappa'' \end{pmatrix} \right]. \quad (6.31)$$

Actually $\overleftrightarrow{F}$ represents a kind of electronic *screened multipole* contribution to the lattice dynamics.

As we can see from the preceding analysis, the problem of finding out how the interacting electrons modify the forces between ions is reduced to the determination of the electron density response function characteristic of the perfect lattice. The scenario can be summed up as follows: When the crystal is perturbed by a lattice wave, the electrons redistribute themselves quickly to suit the new lattice potential. Their new density distribution determines the effective interaction between ions and hence the vibration frequencies.

6.5.2 The Complete Dynamical Matrix

We can now write the complete dynamical matrix as the combination of the direct ion–ion and the electronic contributions. The former is separated into Coulomb C and short-range R components. We write

$$D_{\alpha\beta}^{\text{tot}}(\mathbf{q}) = C_{\alpha\beta}(\mathbf{q}) + R_{\alpha\beta}(\mathbf{q}) + E_{\alpha\beta}(\mathbf{q}) + \overleftrightarrow{F}(\mathbf{q}). \quad (6.32)$$

The Coulomb component involves long-range interactions among the ion cores. They are absent in metals, where there is complete screening of the ion cores. The determination of Coulomb contributions in semiconductors and insulators requires special summation techniques known as *Ewald summations*.[3] The short-range interaction matrix R is treated phenomenologically and usually involves couplings to few neighbors.

In this way, the dynamical matrix of (6.32) enables us to examine the validity of various methods and models used to describe lattice vibrations in all types of solids. At this stage, we have treated the direct ion–ion contribution phenomenologically, but no approximations were invoked in the electronic contribution.

6.6 Electronic Effects on Phonons in Normal Metals

It was first reported by Woods and coworkers that a reasonable fit to measured sodium phonon dispersion curves, using a phenomenological force constant model, requires extending interactions to fairly distant neighbors [202]. By contrast, Toya found that

---

[3] The sum is broken down into two parts: long-range contributions are summed in reciprocal space, and short-rage ones are summed in real space.

treating the conduction electrons as a charged fluid, which tends to screen any charge imbalance, easily yielded a much better account[183]. Such observations support our claim that the electronic structure of simple metal systems is best treated within the pseudopotential framework, where the valence electron's pseudowavefunctions are plane waves. Consequently, we find that the basic simplification here comes from two sources:

(i) First, we note that the plane-wave form of the pseudowavefunctions reduces the matrix element for the susceptibility function $\chi\left(\mathbf{q}+\mathbf{G},\mathbf{q}+\mathbf{G}'\right)$ to a diagonal form in $\mathbf{G}$ and $\mathbf{G}'$ as in (6.30).

(ii) The pseudopotentials are considered weak enough to perturb only very slightly the electronic wavefunctions of a free electron gas. Hence, it is legitimate to set the quantity in square brackets in (6.30), the susceptibility function $\chi\left(\mathbf{q}+\mathbf{G},\mathbf{q}+\mathbf{G}'\right)$, to be that of a free electron gas,

$$\chi(\mathbf{q}) = \frac{3n_0}{4E_F} \left[1 + \frac{1-x^2}{2x} \ln \left| \frac{1+x}{1-x} \right| \right], \tag{6.33}$$

where $x = (|\mathbf{q}+\mathbf{G}|)/2k_F$, and k_F and E_F are the Fermi wavevector and energy, respectively. n_0 is the free electron density.

6.7 Electronic Effects on Phonons in Insulators and Semiconductors

In ionic crystals, the electronic charge is approximately concentrated at the ionic sites. A simple model to be considered in this case is the *rigid-ion model*[103], where the ions are treated as point charges and the short-range interactions between overlapping localized ion-core wavefunctions are introduced. The overlap arises from the rigid displacement of the wavefunctions from their equilibrium positions. Significant differences have appeared between measured phonon spectra and results obtained from this model. Moreover, such a model should dismally fail in the case of semiconductors, where covalent bonds may be manifest. However, for pedagogical reasons, we shall start with describing the simple rigid-ion model.

6.7.1 The Rigid-Ion Model

We consider the Born model of an ionic crystal, where the potential energy between two ions κ and κ' is taken to be

$$V\left(\kappa\kappa'|\, r\right) = \frac{Z_\kappa \, Z_{\kappa'} \, e^2}{r} + b_{\kappa\kappa'} \exp\left[\frac{-r}{\ell_{\kappa\kappa'}} \right], \tag{6.34}$$

where $Z_\kappa e$, $Z_{\kappa'}e$ are effective ionic charges, and $\ell_{\kappa\kappa'}$ is an effective range. The second term is known as the Born–Mayer potential; it accounts for the quantum-overlap interaction between exponentially decaying electronic wavefunctions of the respective ions.

The two potential components of (6.34) represent the rigid-ion model, and according to (6.3) give the force constants

$$\Phi \begin{pmatrix} l & l' \\ \kappa & \kappa' \end{pmatrix} = \mathbf{C} \begin{pmatrix} l & l' \\ \kappa & \kappa' \end{pmatrix} + \mathbf{R} \begin{pmatrix} l & l' \\ \kappa & \kappa' \end{pmatrix},$$

where $\mathbf{C} \begin{pmatrix} l & l' \\ \kappa & \kappa' \end{pmatrix}$ accounts for the ion–ion Coulomb interaction, and $\mathbf{R} \begin{pmatrix} l & l' \\ \kappa & \kappa' \end{pmatrix}$ are short-range forces arising from the overlap of ionic wavefunctions. The functional forms of $\mathbf{R} \begin{pmatrix} l & l' \\ \kappa & \kappa' \end{pmatrix}$ and $\mathbf{C} \begin{pmatrix} l & l' \\ \kappa & \kappa' \end{pmatrix}$ are given in (6.4) and (6.5), respectively.

The aforementioned disagreements can only be explained as arising from sizable deformations of the electronic charges from their equilibrium distributions in response to the motion of the ions. This led to the proposition and formulation of several microscopic phenomenological models aimed at explaining the origin of these deviations. However, such effects should be intrinsically incorporated in the electron density response function formalism, and a close examination of equation (6.32) should provide justification for the validity of such models.

6.7.2 Phenomenological Models of the Electronic Contributions

In order to provide a physical picture of the underlying changes in the electron density distributions, we will examine the last two terms in (6.6), $V_{\mathrm{el}}^{(2)}$:

$$V_{\mathrm{el}}^{(2)} = E^{(21)} + E^{(22)}$$

$$= \frac{1}{2} \int d\mathbf{r}\, n^{(1)}(\mathbf{r})\, v^{[1]}(\mathbf{r}) + \int d\mathbf{r}\, n^{(0)}(\mathbf{r})\, v^{[2]}(\mathbf{r}).$$

$E^{(22)}$, given in (6.8), was shown to be the $\mathbf{q}$-independent electronic self-term. It will be ignored in our following discussion.

It is possible to invert the relation

$$n^{(1)}(\mathbf{r}) = \int d\mathbf{r}\, \chi(\mathbf{r}, \mathbf{r}')\, v^{[1]}(\mathbf{r}')$$

and obtain

$$v^{[1]}(\mathbf{r}) = \int d\mathbf{r}\, K(\mathbf{r}, \mathbf{r}')\, n^{(1)}(\mathbf{r}'), \tag{6.35}$$

which allows us to express $E^{(21)}$ as

$$E^{(21)} = -\frac{1}{2} \sum_{l\kappa\alpha} \int d\mathbf{r}\, n^{(1)}(\mathbf{r})\, \frac{\partial v_\kappa \left(\mathbf{r} - \mathbf{x}_{l\kappa}^{(0)} \right)}{\partial r_\alpha}\, u_{l\kappa\alpha}$$

$$+ \frac{1}{2} \iint d\mathbf{r}\, d\mathbf{r}'\, n^{(1)}(\mathbf{r})\, [v_c(\mathbf{r} - \mathbf{r}') + K(\mathbf{r}, \mathbf{r}')]\, n^{(1)}(\mathbf{r}'), \tag{6.36}$$

where we included the Coulomb interaction between the distorted charges.

Pseudocharge Expansion

Since the electrons are fairly well localized on the ionic (or covalent) sites, we can express the first-order change in the electron density as contributions from various high-symmetry sites, known as Wyckoff positions, within a primitive cell l; they are indexed by their locations μ therein,

$$n^{(1)}(\mathbf{r}) = \sum_{l\mu} \rho^{(1)}(\mathbf{r}, l\mu), \tag{6.37}$$

where $\rho^{(1)}(\mathbf{r}, l\mu)$ vanishes for $\mathbf{r}$ more than one or two primitive cells from the site $(l\mu)$. Next, we expand the first-order change of the electron density centered at $(l\mu)$ in terms of a complete orthonormal set of symmetry-adapted *pseudocharge* functions $\phi_n(\mathbf{r})$ as

$$\rho^{(1)}(\mathbf{r}, l\mu) = \sum_n \phi_n(\mathbf{r} - \mathbf{x}_{l\mu})\, P_{nl\mu} \qquad \Leftrightarrow \qquad \int d\mathbf{r}\, \phi_n(\mathbf{r})\, \phi_{n'}(\mathbf{r}) = \delta_{nn'}, \tag{6.38}$$

where $P_{nl\mu}$ are pseudocharge expansion coefficients. We are leaving the pseudocharge basis functions unspecified for the time being, requiring only their orthonormality. We now treat the $P_{nl\mu}$s as bona fide dynamical variables alongside the ionic displacements $\mathbf{u}$. The $P_n(l\mu)$s are some kind of moments of the change of local electronic charge density,

$$P_{nl\mu} = \int d\mathbf{r}\, \phi_n(\mathbf{r} - \mathbf{x}_{l\mu})\, \rho^{(1)}(\mathbf{r}, l\mu). \tag{6.39}$$

Substituting (6.38) into (6.36), we get

$$
\begin{aligned}
E^{(21)} = &-\sum_{l\kappa\alpha, l'\mu'n} u_{l\kappa\alpha} \int d\mathbf{r}\, \phi_n(\mathbf{r} - \mathbf{x}_{l'\mu'}) \frac{\partial v_\kappa\left(\mathbf{r} - \mathbf{x}_{l\kappa}^{(0)}\right)}{\partial r_\alpha} P_{nl'\mu'} \\
&+ \frac{1}{2} \iint d\mathbf{r} d\mathbf{r}' \sum_{\substack{l\mu n \\ l'\mu'n'}} \left[\phi_n(\mathbf{r} - \mathbf{x}_{l\mu}) \left[v_c(\mathbf{r} - \mathbf{r}' + \mathbf{x}_{l\mu} - \mathbf{x}_{l'\mu'}) \right. \right. \\
&\left. \left. + K(\mathbf{r} + \mathbf{x}_{l\mu}, \mathbf{r}' + \mathbf{x}_{l'\mu'}) \right] \phi_{n'}(\mathbf{r}' - \mathbf{x}_{l'\mu'})\, P_{nl\mu}\, P_{n'l'\mu'} \right] \\
= &\ \mathbf{u} \cdot \bar{T} \cdot \mathbf{P} + \frac{1}{2} \mathbf{P} \cdot \bar{S} \cdot \mathbf{P}. \tag{6.40}
\end{aligned}
$$

This is the most general form of the electronic contribution to the phonon potential energy; it is labeled the *pseudocharge model*.

The electronic contribution to the phonon energy can then be written as

$$V_{el}^{(2)} = \sum_{\substack{l\kappa\alpha \\ l'\mu n}} T_{\alpha,n}\begin{pmatrix} l & l' \\ \kappa & \mu \end{pmatrix} u_{l\kappa\alpha}\, P_{nl'\mu} + \sum_{\substack{l\mu n \\ l'\mu'n'}} S_{nn'}\begin{pmatrix} l & l' \\ \mu & \mu' \end{pmatrix} P_{nl\mu}\, P_{n'l'\mu'}$$

and the total potential energy is given by

$$V^{(2)} = V_{\mathrm{I}}^{(2)} + V_{\mathrm{el}}^{(2)}$$

$$= \sum_{\substack{l\kappa\alpha \\ l'\kappa'\beta}} \Phi_{\alpha\beta}^{\mathrm{I}} \begin{pmatrix} l & l' \\ \kappa & \kappa' \end{pmatrix} u_{l\kappa\alpha}\, u_{l'\kappa'\beta} + \sum_{\substack{l\kappa\alpha \\ l'\mu n}} T_{\alpha,n} \begin{pmatrix} l & l' \\ \kappa & \mu \end{pmatrix} u_{l\kappa\alpha}\, P_{nl'\mu}$$

$$+ \sum_{\substack{l\kappa n \\ l'\kappa'n'}} S_{nn'} \begin{pmatrix} l & l' \\ \kappa & \kappa' \end{pmatrix} P_{nl\kappa}\, P_{n'l'\kappa'},$$

from which we obtain the equations of motion

$$M_\kappa \frac{\partial^2 u_{l\kappa\alpha}}{\partial t^2} = -\frac{\partial V^{(2)}}{\partial u_{l\kappa\alpha}} = -\sum_{l'\kappa'\beta} \Phi_{\alpha\beta}^{\mathrm{I}} \begin{pmatrix} l & l' \\ \kappa & \kappa' \end{pmatrix} u_{l'\kappa'\beta} - \sum_{l'\mu n} T_{\alpha,n} \begin{pmatrix} l & l' \\ \kappa & \mu \end{pmatrix} P_{nl'\mu}$$

$$m_n \frac{\partial^2 P_{nl\mu}}{\partial t^2} = -\frac{\partial V^{(2)}}{\partial P_{nl\mu}} = -\sum_{l'\kappa\alpha} T_{\alpha,n} \begin{pmatrix} l' & l \\ \kappa & \mu \end{pmatrix} u_{l'\kappa\alpha} - \sum_{l'\mu'n'} S_{nn'} \begin{pmatrix} l & l' \\ \mu & \mu' \end{pmatrix} P_{n'l'\mu'}.$$

Fourier transforming these equations yields

$$M_\kappa\, \omega^2\, \bar{U}_\alpha(\mathbf{q},\kappa) = \sum_{\kappa'\beta} \Phi_{\alpha\beta}^{\mathrm{I}} \begin{pmatrix} \mathbf{q} \\ \kappa,\kappa' \end{pmatrix} \bar{U}_\beta(\mathbf{q},\kappa') + \sum_{\mu n} T_{\alpha,n} \begin{pmatrix} \mathbf{q} \\ \kappa,\mu \end{pmatrix} P_n(\mathbf{q},\mu)$$

$$m_n\, \omega^2\, P_n(\mathbf{q},\kappa) = \sum_{\kappa\alpha} T_{\alpha,n} \begin{pmatrix} \mathbf{q} \\ \kappa,\mu \end{pmatrix} U_\alpha(\mathbf{q},\kappa) + \sum_{\mu'n'} S_{nn'} \begin{pmatrix} \mathbf{q} \\ \mu,\mu' \end{pmatrix} P_{n'}(\mathbf{q},\mu').$$

We can write these coupled equations in matrix form as

$$\omega^2\, M_{\mathrm{d}}\, \bar{\mathbf{U}}(\mathbf{q}) = \mathbf{\Phi}(\mathbf{q})\, \bar{\mathbf{U}}(\mathbf{q}) + \mathbf{T}(\mathbf{q})\, \mathbf{P}(\mathbf{q})$$

$$m_n\, \mathbb{I}\, \omega^2\, P(\mathbf{q}) = \mathbf{T}^\dagger(\mathbf{q})\, \mathbf{U}(\mathbf{q}) + \mathbf{S}(\mathbf{q})\, \mathbf{P}(\mathbf{q}), \qquad (6.41)$$

where M_{d} is a diagonal matrix with the masses M_κ along the diagonal. We now invoke the Born–Oppenheimer approximation by setting $m_n \propto m_e = 0$, and we obtain for **P**

$$\mathbf{P} = -\mathbf{S}^{-1}(\mathbf{q})\, \mathbf{T}^\dagger(\mathbf{q})\, \mathbf{U}(\mathbf{q}).$$

When substituted back in the first equation of (6.41), we get

$$\omega^2\, M_{\mathrm{d}}\, U(\mathbf{q}) = \left[\mathbf{\Phi}(\mathbf{q}) - \mathbf{T}(\mathbf{q})\, \mathbf{S}^{-1}(\mathbf{q})\, \mathbf{T}^\dagger(\mathbf{q}) \right] \mathbf{U}(\mathbf{q}).$$

The second term represents the effective electronic contribution to the force constant matrices. The negative sign indicates a reduction in the effective ion–ion force constant, which is expected since the electronic response tends to screen the ion displacements perturbation.

Some Typical Models

If we choose only three functions ϕ_n with *p*-like symmetry, we get the *dipolar shell* model. If we include an additional *s*-like function in the basis, we incorporate an effective

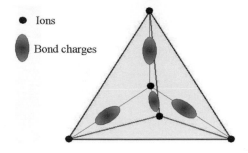

Figure 6.3 Schematic of the bond-charge model.

breathing shell model, where the shell is allowed to isotropically expand and contract. In a similar fashion, if we choose the functions to be Gaussians centered on the covalent bond-charge sites of semiconductors, we have the *bond-charge* model! shown in Figure 6.3.

The Shell Model

The shell model is very popular in treating the lattice dynamics of insulators. It allows for the inclusion of the electronic contributions as follows: Each ion is visualized as consisting of a positive *core*, with charge X, and a negative outer shell of valence electrons, with charge Y. Charge neutrality constraints require that

$$X + Y = Z, \tag{6.42}$$

where Zi is the total static charge of the ion. An electron shell is coupled to its ion core by a harmonic potential, with force constant K. At equilibrium, the shell is centered on the ion core, but, otherwise it is allowed to be displaced relative to it without deforming. This displacement gives rise to a dipolar ionic *polarizability*. The polarization may be induced by either displacements of its neighbors or an electric field.

We extend the particle labeling to the ionic shells and define the force constants for a diatomic crystal in the following table:

	κ	Charge	Displacement		Force constants	
+ve core	1	$X_1 e$	u_1^c		Core–core	D
−ve core	2	$X_2 e$	u_1^s		+ve core–shell	K_1
+ve shell	3	$Y_1 e$	u_2^c		−ve core–shell	K_2
−ve shell	4	$Y_2 e$	u_2^s		Shell–shell	S

We further define

$$W_i = u_i^s - u_i^c$$

as the relative core–shell displacement.

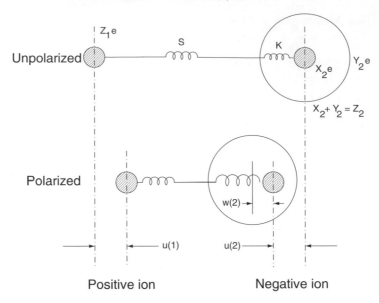

Figure 6.4 Schematic of a pair of ions where only the negative ion is polarizable.

To bring out the physical features of this model, we consider the simple case of a polarizable negative ion and a rigid positive ion, shown in Figure 6.4. Writing the equilibrium equation for the negative shell in the presence of a local electric field $\mathcal{E}$, we obtain

$$-K\,W_2 - S\,\left[\,u_2^c + W_2 - u_1^c\,\right] + Y_2 e\,\mathcal{E} = 0,$$

or

$$W_2 = \frac{S\,\left[\,u_1^c - u_2^c\,\right] + Y_2 e\,\mathcal{E}}{K + S}.$$

The dipole moment associated with the ion pair is

$$Z e\,\left[\,u_1^c - u_2^c\,\right] + Y_2 e\,W_2, \qquad Z = |Z_1| = |Z_2|.$$

After eliminating W_2, the dipole moment becomes

$$\left(Z e + \frac{Y_2 e S}{K + S}\right)\left[\,u_1^c - u_2^c\,\right] + \frac{(Y_2 e)^2}{K + S}\,\mathcal{E}.$$

We now define the negative ion polarizability as

$$\alpha_- = \frac{(Y_2 e)^2}{K + S},$$

and the additional dipole moment $\frac{Y_2 e S}{K+S}\left[u_1^c - u_2^c\right]$ defines the short-range *mechanical polarizability d* as

$$d = -\frac{Y_2 e S}{K + S}.$$

6.8 Measurement of the Structure and Dynamics of Crystals: Particle Scattering by Crystalline Solids

Experiments designed to determine the structure and dynamics of crystalline solids employ one of the following scattering probes, but in all cases the scattering potential experienced by the probe particles has the periodicity of the lattice when the crystal is in equilibrium:

- X-rays scatter from the electron charge density in the crystal.
- Electrons scatter from the crystal potential that includes all exchange and correlation effects.
- Neutrons scatter from the constituent nuclei.
- In helium atom–surface scattering, the He atoms scatter from the surface electron density, precisely, from equicharge density contours.

All have the underlying crystalline periodicity. Scattering of X-rays, neutrons, and He atoms can be treated within the Born approximation, while electrons suffer multiple scattering events, as shown in Figure 6.5, and have to be treated in a more complicated formalism, something à la the KKR method.

6.8.1 Elastic Scattering: Crystal Diffraction at 0° K

We shall confine our analysis to be within the Born approximation, so that the scattering matrix element can be written as

$$\mathcal{M}_{is} = \langle \psi_i | V(\mathbf{r}) | \psi_s \rangle,$$

where ψ_i, ψ_s are the incident and scattered wavefunctions of the applied probe particles, respectively. Taking the particle wavefunctions to be plane waves, we write

$$\mathcal{M}_{\mathbf{k},\mathbf{k}'} = \langle \mathbf{k} | V(\mathbf{r}) | \mathbf{k}' \rangle = \begin{cases} \mathcal{V}_{\mathbf{G}} & \mathbf{k}' - \mathbf{k} = \mathbf{G} \\ 0 & \text{otherwise} \end{cases}$$

since the scattering potential has the lattice periodicity. Moreover, diffraction scattering events are elastic, and we have the condition

$$|\mathbf{k}'| = |\mathbf{k}|.$$

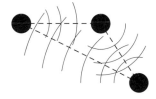

Figure 6.5 Multiple scattering of electronic waves.

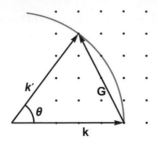

Figure 6.6 Ewald construction.

From the Ewald construction of Figure 6.6, we obtain the Bragg condition on the angle of scattering:

$$2\,|\mathbf{k}|\,\sin\frac{\theta}{2} = |\mathbf{G}|\,.$$

6.8.2 Inelastic Scattering: Measuring Phonon Dispersions

At finite temperatures, the phonon modes of the crystal are excited. We shall study how the dispersion of such modes can be measured by inelastic particle scattering. For simplicity, we consider a monatomic lattice where the scattering potential can be written as

$$V(\mathbf{x}) = \sum_l v_a(\mathbf{x} - \mathbf{x}_l).$$

$\mathbf{x}_l$ is the instantaneous position of the atom in primitive cell l.

Invoking the Born approximation, we take the scattering vector as

$$\mathbf{K} = \mathbf{k}' - \mathbf{k},$$

and write the scattering matrix element as

$$\mathcal{M}_{\mathbf{k},\mathbf{k}'} = \frac{1}{\Omega}\int d\mathbf{x}\, e^{i\mathbf{K}\cdot\mathbf{x}} \sum_l v_a(\mathbf{r} - \mathbf{x}_l) = V(\mathbf{K}) \times S(\mathbf{K})$$

$$V(\mathbf{K}) = \frac{1}{\Omega_0}\int d\mathbf{x}\, v_a(\mathbf{x})\, e^{-i\mathbf{K}\cdot\mathbf{x}}, \qquad S(\mathbf{K}) = \frac{1}{N}\sum_l e^{i\mathbf{K}\cdot\mathbf{x}_l}$$

with $N = \Omega/\Omega_0$. V and S are the form factor and structure factor, respectively. Expressing $\mathbf{x}_l$ as

$$\mathbf{x}_l = \mathbf{x}_l^{(0)} + \mathbf{u}_l = \mathbf{x}_l^{(0)} + \sum_{q>0}\left(\mathbf{U}_\mathbf{q}\, e^{i\mathbf{q}\cdot\mathbf{x}_l^{(0)}} + \mathbf{U}_\mathbf{q}^*\, e^{-i\mathbf{q}\cdot\mathbf{x}_l^{(0)}}\right),$$

we obtain

$$
\frac{1}{N}\sum_l e^{-i\mathbf{K}\cdot\mathbf{x}_l} = \frac{1}{N}\sum_l \exp\left[-i\mathbf{K}\cdot\left\{\mathbf{x}_l^{(0)} + \sum_{q>0}\left(\mathbf{U_q}e^{i\mathbf{q}\cdot\mathbf{x}_l^{(0)}} + \mathbf{U_q^*}e^{-i\mathbf{q}\cdot\mathbf{x}_l^{(0)}}\right)\right\}\right]
$$

$$
= \frac{1}{N}\sum_l e^{-i\mathbf{K}\cdot\mathbf{x}_l^{(0)}}\prod_{q>0}\exp\left[-i\mathbf{K}\cdot\left(\mathbf{U_q}e^{i\mathbf{q}\cdot\mathbf{x}_l^{(0)}} + \mathbf{U_q^*}e^{-i\mathbf{q}\cdot\mathbf{x}_l^{(0)}}\right)\right]. \qquad (6.43)
$$

This is exact! The fact that the atomic displacements are very small allows us to expand

$$
\exp\left[-i\mathbf{K}\cdot\left(\mathbf{U_q}\,e^{i\mathbf{q}\cdot\mathbf{x}_l^{(0)}} + \mathbf{U_q^*}\,e^{-i\mathbf{q}\cdot\mathbf{x}_l^{(0)}}\right)\right]
$$

$$
= 1 - i\mathbf{K}\cdot\left(\mathbf{U_q}\,e^{i\mathbf{q}\cdot\mathbf{x}_l^{(0)}} + \mathbf{U_q^*}\,e^{-i\mathbf{q}\cdot\mathbf{x}_l^{(0)}}\right) - \left|\mathbf{K}\cdot\mathbf{U_q}\right|^2\ldots \qquad (6.44)
$$

The unity term on the right of (6.44) will produce the diffraction condition in the absence of phonons, namely,

$$
\frac{1}{N}\sum_l e^{-i\mathbf{K}\cdot\mathbf{x}_l^{(0)}} = \delta_{\mathbf{K},\mathbf{G}}, \qquad \text{Static structure factor.}
$$

Single-Phonon Scattering

The second term in (6.44) will yield a sum of the form

$$
\frac{1}{N}\sum_l e^{-i\mathbf{K}\cdot\mathbf{x}_l^{(0)}}\sum_{q>0}\left(-i\mathbf{K}\cdot\mathbf{U_q}\right)e^{i\mathbf{q}\cdot\mathbf{x}_l^{(0)}} = \sum_{q>0}\left(-i\mathbf{K}\cdot\mathbf{U_q}\right)\left(\frac{1}{N}\sum_l e^{-i(\mathbf{K}-\mathbf{q})\cdot\mathbf{x}_l^{(0)}}\right)
$$

$$
= \sum_{q>}\left(-i\mathbf{K}\cdot\mathbf{U_q}\right)\delta_{\mathbf{K}-\mathbf{q},\mathbf{G}}. \qquad (6.45)
$$

Two situations arise:

(i) $\mathbf{G} = \mathbf{0}$: $\mathbf{K} = \mathbf{k}' - \mathbf{k} = \mathbf{q}$ lies in the first Brillouin zone, and the scattering matrix element becomes

$$
\mathcal{M}_{\mathbf{k},\mathbf{k}'} \quad \Rightarrow \quad \mathcal{M}_\mathbf{q} = -i\left[\mathbf{q}\cdot\mathbf{U_q}\right]V(\mathbf{q}).
$$

(ii) $\mathbf{G} \neq \mathbf{0}$: $\mathbf{K}$ does not lie in the first Brillouin zone and

$$
\mathbf{q} = \mathbf{k}' - \mathbf{k} - \mathbf{G}.
$$

The scattered particle can gain or lose an extra momentum $\hbar\mathbf{G}$ besides $\hbar\mathbf{q}$. This is known as an *Umklapp* process, first identified by Rudolph Peierls.[4]

---

[4] The name derives from the German word *umklappen* (to turn over). Rudolf Peierls, in his autobiography [152], states he was the originator of this phrase and coined it during his 1929 crystal lattice studies under the supervision of Wolfgang Pauli. Peierls wrote, "… I used the German term Umklapp (flip-over) and this rather ugly word has remained in use."

Invoking Fermi's golden rule, we write the scattering cross-section involving a phonon with energy $\hbar\omega(\mathbf{q})$ as

$$\sigma(\omega(\mathbf{q})) \propto \sum_{\mathbf{G}} |\mathcal{M}_{\mathbf{q}+\mathbf{G}}|^2 \, \delta\left(E_f - E_i \pm \hbar\omega(\mathbf{q})\right)$$

$$= \sum_{\mathbf{G}} |V(\mathbf{q}+\mathbf{G})|^2 \left|(\mathbf{q}+\mathbf{G})\cdot\mathbf{U}_{\mathbf{q}}\right|^2 \delta\left(E_f - E_i \pm \hbar\omega(\mathbf{q})\right), \qquad (6.46)$$

where E_i and E_f are the initial and final energy of the scattered probe particle. This expression shows that inelastic events associated with a phonon of energy $\hbar\omega(\mathbf{q})$ will be manifest as a multipeak spectrum in the extended BZ scheme. Usually the cross-section is significant in the first and second BZs. Equations (6.45) and (6.46) define the scattering kinematics

$$\hbar\omega(\mathbf{q}) = \frac{\hbar^2 k'^2}{2M} - \frac{\hbar^2 k^2}{2M}, \qquad \mathbf{k}' - \mathbf{k} = \mathbf{q} + \mathbf{G}. \qquad (6.47)$$

The scattering potential type will depend on the probe particles.

6.8.3 *Experimental Measurement of Phonon Dispersions*

Figure 6.7 displays all available probes for measuring the phonon and molecular vibrational excitations; they include photons, electrons, He atoms, and neutrons. The figure shows the spectral regions (energy and momentum transfer) accessible to each probe. We shall consider here methods that do not suffer from multiple scattering events and that cover inelastic scattering events throughout the Brillouin zone, namely inelastic neutron and helium scattering techniques.

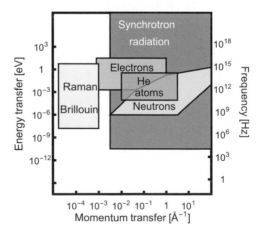

Figure 6.7 Accessible ranges in the energy-momentum space for the different probes of inelastic scattering.

Inelastic Neutron Scattering: Bulk Phonon Dispersion

The technique of inelastic neutron scattering (INS) takes advantage of the ability of neutrons to exchange energy and momentum with the atomic nuclei in a material, to measure details of its dynamics: When a neutron scatters from a crystalline solid, it can absorb or emit an amount of energy equal to a quantum of phonon energy, $\hbar\omega$. This gives rise to inelastic coherent scattering in which the neutron exchanges energy and momentum according to (6.47). In most solids, ω is within a few terahertz (THz) (1 THz = 4.18 meV).

Thermal neutrons, with energies in the meV range, are used for INS. Thus, scattering by a phonon causes an appreciable fractional change in the neutron energy, allowing accurate measurement of phonon frequencies. INS employs monochromatic neutron beams, so that their initial momentum, k_i, is well defined within a very narrow range. To determine the phonon energy and the momentum transfer vector, $\mathbf{q}$, we need to determine the neutron wavevector, k_f, after a scattering event, as shown in Figure 6.8. Several different INS spectrometers have been devised to measure phonon dispersions: time-of-flight, neutron-echo, and crystal spectrometers.

The workhorse instruments actually belong to the last type, and are known as *triple-axis spectrometers* (TAS). In a TAS setup, as shown in Figure 6.9, an incident neutron beam of well-defined wavevector $\mathbf{k}_i$ is selected from the white spectrum of the neutron source by the monochromator crystal (first axis). The monochromatic beam is then scattered from

Inelastic neutron scattering ($|k_f| \neq |k_i|$)

Neutron loses energy   **Neutron gains energy**

Figure 6.8 Scattering triangles for INS in which the neutron either loses energy ($\mathbf{k}_f < \mathbf{k}_i$) or gains energy ($\mathbf{k}_f > \mathbf{k}_i$) during the interaction with the sample. In both elastic and inelastic scattering events, the neutron is scattered through the angle 2θ, and the scattering vector is given by the vector relationship $\mathbf{q} = \mathbf{k}_i - \mathbf{k}_f$.

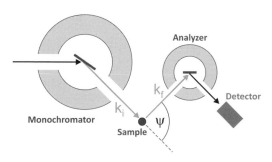

Figure 6.9 Schematic layout of a neutron triple-axis spectrometer.

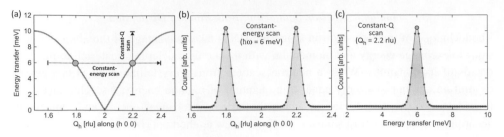

Figure 6.10 (a) Schematic view of how two points of the phonon dispersion curve can be measured using either (b) constant-energy scan or (c) constant-**q** scan. By performing multiple scans, it is possible to map out the complete dispersion.

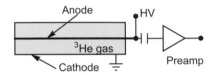

Figure 6.11 A schematic of a typical gas-filled proportional counter.

the crystalline sample (second axis). The intensity of the scattered beam with wavevector $\mathbf{k}_f$ is reflected by the analyzer crystal (third axis) onto the neutron detector (^{3}He tube), thereby defining the energy transfer $\hbar\omega$ as well. Thus, TAS spectrometers allow for controlled access to both the momentum transfer $\mathbf{q} = \mathbf{k}_i - \mathbf{k}_f$, as well as energy transfer $\hbar\omega = E_i - E_f$. The main advantage of a triple-axis spectrometer is that experimental data can be acquired at any predetermined point ($\mathbf{q}$) in reciprocal space for a selected energy transfer $\hbar\omega$. Practically, data are recorded by scanning one or both of the variables along a chosen direction. One usually chooses between *constant-*$\mathbf{q}$ scans where $\hbar\omega$ is scanned while keeping $\mathbf{q}$ fixed, or *constant-energy* scans where $\hbar\omega$ is kept at a fixed value and $\mathbf{q}$ is scanned along a selected direction in reciprocal space (see Figure 6.10). By performing one or both of these types of scans, the dispersion relation $\hbar\omega(\mathbf{q})$ for a single crystal sample can be extracted in a very controlled manner.

^{3}He Gas-Filled Proportional Counters

Neutrons can be detected using ^{3}He-filled gas proportional counters (Figure 6.11). A typical counter consists of a gas-filled tube with a high voltage applied across the anode and cathode. A neutron passing through the tube will interact with a ^{3}He atom to produce tritium (^{3}H hydrogen) and a proton. The proton ionizes the surrounding gas atoms to create charges, which in turn ionize other gas atoms in an avalanchelike multiplication process. The resulting charges are collected as measurable electrical pulses with the amplitudes proportional to the neutron energy. The pulses are compiled to form a pulse-height energy spectrum that serves as a fingerprint for the identification and quantification of the neutrons and their energies.

Inelastic Helium Scattering: Surface Phonon Dispersions

Helium atoms at thermal energies (10–80 meV) have several attributes that render them uniquely suitable for the study of surfaces in general and surface vibrations in particular: They are strictly surface sensitive, and thus provide structural and dynamical information exclusively about the outermost layer of a crystal. More precisely, the He atoms scatter from equicharge contours outside the surface (see Figure 6.12). They are chemically, electrically, magnetically, and mechanically inert. He atoms at thermal energies are particularly well matched in momentum and energy to surface phonons.

He beam monochromators, employing adiabatic nozzle expansion techniques, have especially narrow velocity distribution, giving excellent energy resolution (0.01–1 meV) and high spectral intensity.

The intense, nearly monoenergetic beam of He atoms is directed onto a target surface at a particular angle of incidence and the scattered intensity measured at a given angle of reflection (see Figure 6.13). In general, He atoms can be scattered either elastically, with no energy transfer to or from the internal degrees of freedom of the crystal surface, or

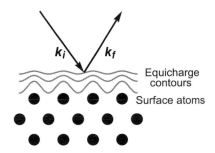

Figure 6.12 Schematic of He atom scattering from equicharge contours outside the solid surface. $\mathbf{k}_i$ and $\mathbf{k}_f$ are the incident and scattered wavevectors, respectively.

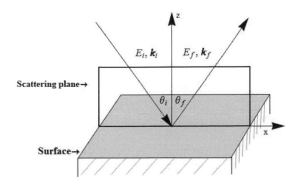

Figure 6.13 Schematic of the scattering geometry. The scattering plane is defined by $\mathbf{k}_i$ and the surface normal.

inelastically, by excitation (phonon creation process) or deexcitation (phonon annihilation process) of surface vibrational modes.

A. Elastic Scattering

For any perfect surface, whether a metal, semiconductor, or insulator, elastic scattering events for He atoms are governed by conservation of energy and of the momentum component parallel to the surface plane.[5]

$$E_f = E_i \quad \Rightarrow \quad |\mathbf{k}_f| = |\mathbf{k}_i|$$
$$\mathbf{K}_f = \mathbf{K}_i + \mathbf{G}, \quad \mathbf{G} = n_1 \mathbf{b}_1 + n_2 \mathbf{b}_2$$
$$\mathbf{a}_i \cdot \mathbf{b}_j = 2\pi \, \delta_{ij}, \quad i, j = 1, 2,$$

where $\mathbf{a}_i$ and $\mathbf{b}_i$ are lattice and reciprocal lattice basis vectors, respectively.

B. Inelastic Scattering

Schematically, an inelastic helium scattering experiment takes the form indicated in Figure 6.14. Experimental conditions are chosen to ensure that single phonon scattering dominates, in which kinematic analysis of the scattered beam yields directly the energy and momentum of the surface phonons. We consider the generally used "in-plane"[6] scattering geometry shown in Figure 6.13. Conservation of energy and momentum for a He beam

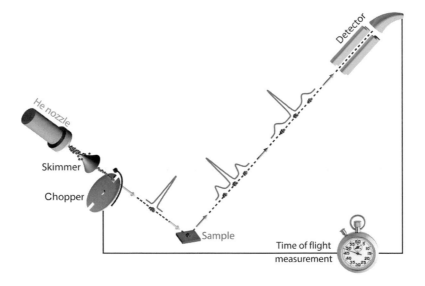

Figure 6.14 Schematic depiction of a surface phonon measurement using inelastic He scattering with time-of-flight analysis, from Anton Tamtögl's Ph.D. thesis.

---

[5] A three-dimensional momentum vector $\mathbf{k}$ is expressed as $\mathbf{k} = (\mathbf{K}, k_z)$ where $\mathbf{K}$ is parallel to the surface plane.
[6] In-plane scattering geometry of the incident and scattered wavevectors lies in the same plane containing the surface normal.

incident onto the surface at a polar angle θ_i (with respect to the surface normal) and wavevector $\mathbf{k}_i = (\mathbf{K}_i + k_{iz})$, and scatters at θ_f and wavevector $\mathbf{k}_f = (\mathbf{K}_f + k_{fz})$, yield

$$\frac{\hbar^2 k_f^2}{2M} - \frac{\hbar^2 k_i^2}{2M} = \Delta E = \hbar\omega$$

$$\Delta E > 0 \quad \Rightarrow \quad \text{Phonon annihilation and gain of energy by the beam atom}$$

$$\Delta E < 0 \quad \Rightarrow \quad \text{Phonon creation and loss of energy by the beam atom}$$

Only the momenta parallel to the surface are conserved:

$$\mathbf{K}_f - \mathbf{K}_i = \Delta \mathbf{K} = \mathbf{Q} \quad \text{Surface phonon momentum} \tag{6.48}$$

The projections onto the surface plane are $\mathbf{K}_a = \mathbf{k}_a \sin\theta_a, \quad a = i, f$.

The conservation conditions can be combined to give an expression that specifies the allowed values of energy exchange ΔE and parallel momentum exchange ΔK, for a given scattering geometry,

$$\Delta E = \frac{\hbar^2}{2M} \left(\frac{|K_i + \Delta K|^2}{\sin^2\theta_f} - k_i^2 \right) \quad \Rightarrow \quad \frac{\Delta E}{E_i} = \frac{\sin^2\theta_i}{\sin^2\theta_f} \left[1 + \frac{\Delta K}{k_i \sin\theta_i} \right]^2.$$

This expression is generally referred to as a scan curve.

To monitor the inelastic scattering events, it is necessary to energy-analyze the scattered beam. The most commonly used technique is the time-of-flight (TOF) analysis, which entails pulsing a beam with a mechanical (or electronic) *chopper* at some point in its transit from source to detector and measuring the chopper–detector flight time. The incident velocity is $v_i = \frac{x_{\text{cd}}}{t_{\text{cd}}}$, where x_{cd} is the flight path from chopper to detector, and t_{cd} is the corresponding transit time. Inelastic scattering at the target surface transfers energy to or from He atoms, changing their velocity and thereby their flight time from target to detector. So TOF spectra display peaks that are shifted relative to the elastic flight time.

6.8.4 Scattering at Finite Temperatures: Debye–Waller Factor

There are important contributions arising from terms such as $|\mathbf{K} \cdot \mathbf{U_q}|^2$ of (6.44) that do not average to a small value. Examining (6.43) and (6.44), we find that both elastic and inelastic terms ought to be multiplied by

$$\prod_{\mathbf{q}} \left(1 - |\mathbf{K} \cdot \mathbf{U_q}|^2 \right) = e^{-2W}, \tag{6.49}$$

known as the *Debye–Waller factor*.

Using the identity

$$\lim_{N\to\infty} \prod_{n=1}^{N} \left(1 - \frac{1}{N} a_n \right) = \exp\left[-\lim_{N\to\infty} \sum_{n=1}^{N} \frac{1}{N} a_n \right],$$

we obtain

$$e^{-2W} = \exp\left[-\sum_{\mathbf{q}} |\mathbf{K} \cdot \mathbf{U_q}|^2\right].$$

To evaluate $\mathbf{U_q}$, we recall that the average energy in mode $\mathbf{q}$ is

$$\bar{\mathcal{E}}_{\mathbf{q}} = \left(\bar{n}_{\mathbf{q}} + \frac{1}{2}\right) \hbar \omega_{\mathbf{q}}.$$

We can use a classical approach to relate this energy to $U_{\mathbf{q}}^2$. Treating the modes as classical harmonic oscillators, the energy is given by

$$\bar{\mathcal{E}} = \sum_{l} M |\dot{\mathbf{u}}_l|^2 = \sum_{\mathbf{q}} NM |\dot{\mathbf{u}}_{\mathbf{q}}|^2$$

$$= \sum_{\mathbf{q}} NM \omega_{\mathbf{q}}^2 |\mathbf{U_q}|^2 = \sum_{\mathbf{q}} \bar{\mathcal{E}}_{\mathbf{q}}$$

so that

$$|\mathbf{U_q}|^2 = \frac{1}{NM\omega_{\mathbf{q}}} \left(\bar{n}_{\mathbf{q}} + 1\right) \hbar.$$

If we adopt the phonon Debye model, assuming all three polarizations to be degenerate, and taking one of the polarization vectors along $\mathbf{K}$, we get

$$W = \sum_{\mathbf{q}} \frac{\hbar K^2}{6NM\omega_{\mathbf{q}}} \left(\bar{n}_{\mathbf{q}} + \frac{1}{2}\right) = \frac{\hbar K^2}{6NM} \frac{1}{(2\pi)^3} \int d\mathbf{q} \, \frac{\bar{n}_{\mathbf{q}} + \frac{1}{2}}{\omega_{\mathbf{q}}}$$

$$= \frac{\hbar^2 K^2}{6M} \int_0^{\omega_D} d\omega \, \frac{3\omega^2}{\hbar \omega \omega_D^3} \left[\frac{1}{\exp(\hbar\omega/k_B T) - 1} + \frac{1}{2}\right]$$

$$= \frac{\hbar^2 K^2 T^2}{2Mk_B \Theta^3} \int_0^{\Theta/T} dz \, z \left[\frac{1}{\exp[z] - 1} + \frac{1}{2}\right],$$

where ω_D is the Debye frequency and Θ the Debye temperature.

At high temperatures $\Theta/T \ll 1$, $\dfrac{1}{\exp(z) - 1} \sim \dfrac{1}{z}$, yielding

$$e^{-2W} = \exp\left[-\frac{\hbar^2 K^2 T}{Mk_B \Theta^2}\right],$$

while at $T = 0K$ $e^z = \infty$ and

$$e^{-2W} = \exp\left[-\frac{\hbar^2 K^2}{4Mk_B \Theta}\right],$$

which arises from quantum zero-point fluctuations.

Exercises

6.1 The one-dimensional shell model

Consider the one-dimensional shell model shown in Figure 6.15. Each atom consists of an outer shell and an inner core. Nearest-neighbor shells interact via a force-constant S, while each shell couples to its core by a force-constant K. For the sake of simplicity, we neglect the Coulomb interactions.

1. Write down the interaction potential energy in the harmonic approximation in term of the displacements u_n and v_n shown in Figure 6.15.
2. Derive the corresponding equations of motion.
3. Express the equations of motion in terms of u_n and $w_n = v_n - u_n$.
4. Obtain the Fourier transform of the equations of motion in terms of the variables $U(q)$ and $W(q)$.
5. Eliminate $W(q)$ by invoking the adiabatic approximation. Use the derived dynamical matrix to obtain the phonon dispersion curve.
6. Plot the dispersion curve of the shell model for different values of S/K. Note that the rigid atom model emerges for $K = \infty$. Comment on the form of the dispersion curves as K decreases. What happens when $K \ll S$?
7. Numerically expand the $\omega^2(q)$ as a periodic real space Fourier series for different values of S/K, and comment on your findings, especially the range of the coupling.

6.2 The one-dimensional bond-charge model (see Figure 6.16)

Consider a monatomic linear chain with bond-charges midway between the ions. The ion-BC force constant is K, and the BC–BC force constant is S. For simplicity, we neglect the Coulomb forces. Carry out the items listed in Exercise 6.1, except

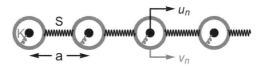

Figure 6.15 One-dimensional shell model. u_n and v_n are the ion-core and shell displacements, respectively.

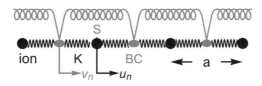

Figure 6.16 One-dimensional shell model. u_n and v_n are the ion-core and shell displacements, respectively.

for item 3. Express the Fourier transform of the equations of motion in terms of the variables $U(q)$ and $V(q)$.

6.3   Electron–phonon interaction by the method of pseudopotentials

We can express an OPW electron wavefunction as

$$\Theta_{\mathbf{k}+\mathbf{G}}(\mathbf{x}) = \frac{1}{\sqrt{N\Omega_0}} \, e^{i(\mathbf{k}+\mathbf{G})\cdot\mathbf{x}} - \sum_{c} b_c(\mathbf{k}+\mathbf{G}) \, \Psi_{\mathbf{k}}^c(\mathbf{x}),$$

where $\Psi_{\mathbf{k}}^c(\mathbf{x})$s are Bloch wavefunctions of the core electrons

$$\Psi_{\mathbf{k}}^c(\mathbf{x}) = \frac{1}{\sqrt{N}} \sum_{l} e^{i\mathbf{k}\cdot\mathbf{R}_l} \, \phi_c(\mathbf{x}-\mathbf{R}_l).$$

1. Given that the crystal potential can be written as

$$V(\mathbf{x}) = \sum_{l} v(\mathbf{x}-\mathbf{R}_l)$$

and that the overlap of the core wavefunctions is negligible, namely

$$\int d\mathbf{x} \, \phi_c^*(\mathbf{x}-\mathbf{R}_l) \, \phi_c(\mathbf{x}-\mathbf{R}_{l'}) = \delta_{ll'},$$

show that

$$b_c(\mathbf{k}+\mathbf{G}) = \frac{1}{\sqrt{\Omega_0}} \int d\mathbf{x} \, e^{i(\mathbf{k}+\mathbf{G})\cdot\mathbf{x}} \, \phi_c(\mathbf{x}).$$

2. Show that the matrix element of the repulsive potential can be written as

$$V_R(\mathbf{k}+\mathbf{G},\mathbf{k}+\mathbf{G}') = \sum_{c} (E - E_c) \, b_c^*(\mathbf{k}+\mathbf{G}) \, b_c(\mathbf{k}+\mathbf{G}').$$

3. Minimizing

$$\langle \Psi_{\mathbf{k}} | \mathcal{H} - \mathcal{E}_{\mathbf{k}} | \Psi_{\mathbf{k}} \rangle$$

with a valence electron wavefunction of the form

$$\Psi_{\mathbf{k}}(\mathbf{x}) = \sum_{\mathbf{G}} a(\mathbf{k}+\mathbf{G}) \, \Theta_{\mathbf{k}+\mathbf{G}}(\mathbf{x})$$

yields the eigenvalues $\mathcal{E}_{\mathbf{k}}$ and the eigenvectors $\{a(\mathbf{k}+\mathbf{G})\}$.

Assuming that such minimization has been carried out, and $\mathcal{E}_{\mathbf{k}}$ and $\{a(\mathbf{k}+\mathbf{G})\}$ have been determined, derive an expression for the electron–phonon interaction matrix element

$$\mathcal{I}_{\mathbf{q}\lambda}(\mathbf{k},\mathbf{k}') = -N\,\hat{\mathbf{e}}_{\mathbf{q}\lambda} \cdot \int_{N\Omega_0} d\mathbf{x} \, \Psi_{\mathbf{k}}^*(\mathbf{x}) \left(\nabla v(\mathbf{x}) \right) \Psi_{\mathbf{k}'}(\mathbf{x}).$$

4. Show that this expression is the same as that obtained in terms of the pseudopotential and the pseudowavefunction Φ, namely

$$\mathcal{I}_{\mathbf{q}\lambda}^{ps}(\mathbf{k},\mathbf{k}') = -N\,\hat{\mathbf{e}}_{\mathbf{q}\lambda} \cdot \int_{N\Omega_0} d\mathbf{x} \, \Phi_{\mathbf{k}}^*(\mathbf{x}) \left(\nabla\left[v(\mathbf{x}) + v_R(\mathbf{x})\right] \right) \Phi_{\mathbf{k}'}(\mathbf{x}),$$

if we assume a local pseudopotential that can be written as

$$V_R(\mathbf{x}) = \sum_l v_R(\mathbf{x}).$$

Hint: Make use of the relation

$$\nabla \left(-\nabla^2 + v(\mathbf{x})\right) \phi_c = \left(-\nabla^2 + v(\mathbf{x})\right) \nabla \phi_c + \nabla v(\mathbf{x}) \phi_c = \mathcal{E}_c \nabla \phi_c,$$

which is obtained from

$$\left(-\nabla^2 + v(\mathbf{x})\right) \phi_c = \mathcal{E}_c \phi_c.$$

6.4 Compare neutrons with X-ray photons

A typical velocity for thermal neutrons is about 2.20 km/sec.

1. Using the de Broglie relation, show that the wavelength of neutrons with this standard velocity is approximately 1.8Å.
2. What is the kinetic energy of these neutrons?
3. What is the energy of an X-ray photon of wavelength $\lambda = 1.8\text{Å}$?
4. Calculate the velocity of a neutron that has the same energy as this X-ray photon.

6.5 Monochromatizing neutrons

A beam of white neutrons emerges from a collimator with a divergence of $\pm 0.2°$. It is then Bragg-reflected by the (111) planes of a monochromator consisting of a single-crystal of lead.

1. Calculate the angle between the direct beam and the [111] axis of the crystal to produce a beam of wavelength $\lambda = 1.8\text{Å}$. (Unit cell edge a_0 of cubic lead is 4.94Å.)
2. What is the spread in wavelengths of the reflected beam?

6.6 Neutron scattering from crystalline solids

Since the neutron is chargeless, it only interacts with the atomic nucleus through a short-ranged nuclear interaction (Ignoring any spin–spin interaction). The range of this interaction is $\sim$ 1 Fermi (10^{-15} m). Since the neutron wavelength has to be comparable to lattice spacing

$$\lambda \sim \text{Å} \gg \text{ range of the interaction } 10^{-15} \text{ m},$$

the neutron cannot "see" the detailed structure of the nucleus, and so we may approximate the neutron–ion interaction potential as a contact interaction

$$V(\mathbf{x}) = \sum_l V_n \delta(\mathbf{x} - \mathbf{R}_l) = \frac{2\pi \hbar^2 a}{m_n} \sum_l \delta(\mathbf{x} - \mathbf{r}_l),$$

where $\mathbf{r}_l$ is the instantaneous position of the ion in primitive cell l, a the scattering length, and m_n the neutron mass.

1. Write $V(\mathbf{x})$ in terms of its Fourier transform.

2. Consider the case of an incident neutron beam of wavevector $\mathbf{k}_0$ and wavefunction

$$|\mathbf{k}_0\rangle = \frac{1}{\sqrt{\mathcal{V}}} e^{i\mathbf{k}_0 \cdot \mathbf{x}},$$

where $\mathcal{V}$ is the volume. Determine the incident neutron beam flux $\mathbf{j}_i$.

3. If part of the beam scatters with wavevector $\mathbf{k}$ within a wavevector element $d\mathbf{k}$, what is the number of scattering states involved? Express the number of states in terms of the corresponding energy element $d\mathcal{E}_n$ and the solid angle of the beam $d\Omega$.

4. Determine the matrix element

$$\langle \Psi_i | V(\mathbf{x}) | \Psi_f \rangle.$$

Remember that the initial and final wavefunctions are products of neutron and solid (phonon) ones, namely

$$|\Psi\rangle = |\mathbf{k}\rangle \, |\Phi\rangle,$$

where $|\mathbf{k}\rangle$ and $|\Phi\rangle$ are the neutron and phonon wavefunctions, respectively. Use the fact that $\mathbf{r}_l = \mathbf{R}_l + \mathbf{u}_l$.

5. With the aid of Fermi's Golden Rule, determine the probability of transition $\mathcal{P}(\mathbf{k}_0 \to \mathbf{k})$ per unit time. Show that it can be written in the form

$$\mathcal{P} = C\, S(\mathbf{q}, \omega),$$

where

$$\mathbf{q} = \mathbf{k} - \mathbf{k}_0, \qquad \hbar\omega = \frac{\hbar^2 k^2}{2m_n} - \frac{\hbar^2 k_0^2}{2m_n}.$$

6. Use the identity

$$\delta\left(\frac{E_2}{\hbar} - \frac{E_1}{\hbar} + \omega\right) = \int_{-\infty}^{\infty} \frac{dt}{2\pi} e^{i(E_2 - E_1)t/\hbar + \omega t}$$

to obtain an expression for $S(\mathbf{q}, \omega)$ of the form

$$S(\mathbf{q}, \omega) = \frac{1}{N} \int_{-\infty}^{\infty} \frac{dt}{2\pi} e^{i\omega t} \sum_{ll'} e^{i\mathbf{q}\cdot(\mathbf{R}_l - \mathbf{R}_{l'})} \left\langle \Phi_i \left| e^{i\mathbf{q}\cdot\mathbf{u}_{l'}} e^{-i\mathbf{q}\cdot\mathbf{u}_l(t)} \right| \Phi_i \right\rangle.$$

7. We can define the differential scattering cross-section as

$$j_0 \frac{d\sigma}{d\mathcal{E}\, d\Omega} d\mathcal{E}\, d\Omega = \mathcal{N}\, \mathcal{P}(\mathbf{k}_0 \to \mathbf{k}).$$

Obtain an expression for $\dfrac{d\sigma}{d\mathcal{E}\, d\Omega}$.

6.7 According to the "Lindeman" criterion, a crystal melts when the root mean square (rms) displacement of its atoms exceeds a third of the average separation of the atoms.

1. Consider a one-dimensional lattice with lattice constant a and an interparticle potential

$$V = \frac{m\omega^2}{2} \sum_j \left(\phi_j - \phi_{j-1}\right)^2.$$

Estimate the amplitude of zero point fluctuations using the uncertainty principle, to show that if

$$\frac{\hbar}{m\omega a^2} > \zeta_c,$$

where ζ_c is a dimensionless number of order one, the crystal will be unstable, even at absolute zero, and will melt due to zero-point fluctuations. (Hint: What would the answer be for a simple harmonic oscillator?)

2. Consider a three-dimensional crystal with separation a, atoms of mass m, and a nearest neighbor quadratic interaction

$$V = \frac{m\omega^2}{2} \left(\phi_{\mathbf{R}} - \phi_{\mathbf{R+a}}\right)^2.$$

Calculate ζ_c for this model. If you like, to start out, imagine that the atoms only move in one direction, so that ϕ is a scalar displacement at the site with equilibrium position R. Calculate the rms zero-point displacement of an atom $\sqrt{\langle 0 | \phi^2(x) | 0 \rangle}$. Now generalize your result to take account of the fluctuations in three orthogonal directions.

3. Suppose $\hbar\omega/k_B = 300$ K, and the atom is a helium atom. Assuming that ω is independent of atom separation a, estimate the critical atomic separation a_c at which the solid becomes unstable to quantum fluctuations. Note that in practice ω is dependent on a, and rises rapidly at short distances, with $\omega\, a^\alpha$, where $\alpha > 2$. Is the solid stable for $a < a_c$ or for $a > a_c$?

7

Dimensionality, Susceptibility, and Instabilities

In Chapter 1, we discussed in detail how the density of states (DOS) and other character-istic quantities of quasiparticle states, i.e., electrons, phonons, magnons, etc., are strongly dependent on the dimensionality of a given system. In the present chapter, we shall explore how dimensionality affects a system linear response – its susceptibilities.

7.1 Dimensionality, Susceptibility, and Nesting

7.1.1 Longitudinal Susceptibility of a Free Electron System

We start with the simple case of a jellium, where the single-electron eigenstates are plane waves

$$|\mathbf{k}\rangle = \frac{1}{\sqrt{\Omega}} e^{-i\mathbf{k}\cdot\mathbf{x}}, \qquad E_{\mathbf{k}} = \frac{\hbar^2 k^2}{2m_e}, \qquad \text{and occupation} \quad f(\mathbf{k}),$$

where Ω is the volume of the system. A many-electron wavefunction for the jellium can be written as $|\Psi\rangle = \prod_i |\mathbf{k}_i\rangle$, satisfying the Pauli principle, with the ground state given by $|\Psi_0\rangle = \prod_{|\mathbf{k}|\leq k_F} |\mathbf{k}_i\rangle$. We are then able to reduce the susceptibility expression (5.42) to the single-particle form

$$
\begin{aligned}
\chi(\mathbf{q}, \omega) &= \sum_{|\mathbf{k}|\leq k_F} \left| \langle \mathbf{k}| e^{i\mathbf{q}\cdot\mathbf{x}} |\mathbf{k}+\mathbf{q}\rangle \right|^2 \frac{f(\mathbf{k}) - f(\mathbf{k}+\mathbf{q})}{E - E_{\mathbf{k}} + E_{\mathbf{k}+\mathbf{q}} + i\hbar\eta} \\
&= \frac{1}{(2\pi)^d} \int dk^d \frac{\Theta(E_F - E_{\mathbf{k}}) - \Theta(E_F - E_{\mathbf{k}+\mathbf{q}})}{E - E_{\mathbf{k}} + E_{\mathbf{k}+\mathbf{q}} + i\hbar\eta} \\
&= \frac{1}{(2\pi)^d} \int dk^d \, \Theta(E_F - E_{\mathbf{k}}) \\
&\quad \times \left[\frac{1}{E - E_{\mathbf{k}} + E_{\mathbf{k}+\mathbf{q}} + i\hbar\eta} - \frac{1}{E - E_{\mathbf{k}-\mathbf{q}} + E_{\mathbf{k}} + i\hbar\eta} \right] \\
&= \frac{1}{(2\pi)^d} \int_{|\mathbf{k}|\leq k_F} dk^d \left[\frac{1}{E_1 + 2\gamma\mathbf{k}\cdot\mathbf{q} + i\hbar\eta} - \frac{1}{E_2 + 2\gamma\mathbf{k}\cdot\mathbf{q} + i\hbar\eta} \right]. \quad (7.1)
\end{aligned}
$$

$E_{\frac{1}{2}} = \hbar\omega \mp \gamma q^2$, and $\gamma = \hbar^2/2m_e$.

We will consider here low-frequency perturbations, such that we can set $\omega \sim 0$, so that (7.1) reduces to

$$\chi(\mathbf{q}) = \frac{-1}{(2\pi)^d} \frac{2m}{\hbar^2} \int_{|\mathbf{k}| \leq k_F} dk^d \left(\frac{1}{q^2 + 2\mathbf{k} \cdot \mathbf{q} + i(m/\hbar)\eta} + \frac{1}{q^2 - 2\mathbf{k} \cdot \mathbf{q} - i(m/\hbar)\eta} \right).$$
(7.2)

1. Susceptibility in one dimension

$$\chi(q) = \frac{m}{\pi \hbar^2 q} \int_{-k_F}^{k_F} dk \left(\frac{1}{q + 2k + i(m/\hbar)\eta} + \frac{1}{q - 2k - i(m/\hbar)\eta} \right)$$
(7.3)

We note that for $|q| < 2k_F$ there is a singular point in the integral that is removed by the presence of η, since in the limit $\eta \to 0$ we have

$$\lim_{\eta \to 0} \frac{1}{z - i\eta} = \mathcal{P}\left(\frac{1}{z}\right) + i\pi \, \delta(z).$$

We find that the integral

$$\int_{-k_F}^{k_F} dk \, \frac{1}{q + 2k}, \qquad |q| < 2k_F,$$

has a singularity at $k = -q/2$. However, the principle part

$$\mathcal{P} \int_{-k_F}^{k_F} dk \, \frac{1}{q + 2k} = \lim_{\eta \to 0} \left[\int_{-k_F}^{q/2 - \eta} dk \, \frac{1}{q + 2k} + \int_{-q/2 + \eta}^{k_F} dk \, \frac{1}{q + 2k} \right]$$

$$= \lim_{\eta \to 0} \left[\frac{1}{2} \ln|q + 2k| \Big|_{-k_F}^{-q/2 - \eta} + \frac{1}{2} \ln|q + 2k| \Big|_{-q/2 + \eta}^{k_F} \right]$$

$$= \lim_{\eta \to 0} \left[\frac{1}{2} \ln \left| \frac{q + 2k_F}{q - 2k_F} \right| + \frac{1}{2} \ln \left| \frac{-\eta}{\eta} \right| \right] = \frac{1}{2} \ln \left| \frac{q + 2k_F}{q - 2k_F} \right|.$$

Thus,

$$\Re \chi(q) = -\frac{m}{\pi \hbar^2 q} \mathcal{P} \int_{-k_F}^{k_F} dk \left(\frac{1}{q + 2k} + \frac{1}{q - 2k} \right)$$

$$= -\frac{m}{\pi \hbar^2 q} \frac{1}{2} \ln \left| \frac{q + 2k}{q - 2k} \right| \, \Big|_{-k_F}^{k_F}$$

$$= -\frac{m}{\pi \hbar^2 q} \ln \left| \frac{q + 2k_F}{q - 2k_F} \right|.$$
(7.4)

2. Susceptibility in two dimensions

$$\chi(\mathbf{q}) = \frac{2m}{(2\pi)^2 \hbar^2} \int_0^{k_F} k \, dk \int_0^{2\pi} d\varphi \left[\frac{1}{q^2 - 2qk \cos \varphi} + \frac{1}{q^2 + 2qk \cos \varphi} \right]$$

$$= -\frac{k_F^2}{(2\pi)^2 E_F} \int_0^1 dk \, k \int_0^{2\pi} \frac{d\phi}{q^2 - 4k^2 \cos^2 \phi}$$

In the last line, we set $k, q \to k, q/k_F$. With

$$\int_0^{2\pi} \frac{d\varphi}{a^2 - 4b^2 \cos^2 \varphi} = \frac{2\pi}{a\sqrt{a^2 - 4b^2}}.$$

we obtain

$$\chi(\mathbf{q}) = \begin{cases} -\dfrac{k_F^2}{2\pi E_F} \displaystyle\int_0^1 \dfrac{x\, dx}{\sqrt{1-x^2}} = \mathcal{D}(E_F)\left(1 - \sqrt{1 - \left(\dfrac{2k_F}{q}\right)^2}\right), & q \geq 2k_F \\[4mm] -\dfrac{k_F^2}{2\pi E_F} \displaystyle\int_0^{q/2} \dfrac{x\, dx}{\sqrt{1-x^2}} = \mathcal{D}(E_F), & q \leq 2k_F \end{cases}$$

3. Susceptibility in three dimensions

Set $\mathbf{q} = q\,\hat{\mathbf{e}}_z$, $\mathbf{k} = (k_z, r, \phi)$ (cylindrical coordinates)

$$\begin{aligned}
\chi(\mathbf{q}) &= \frac{-2m}{(2\pi)^3 \hbar^2 q} \int_{|\mathbf{k}| \leq k_F} d^3 k \left(\frac{1}{q + 2k_z + i(m/\hbar)\eta} + \frac{1}{q - 2k_z - i(m/\hbar)\eta}\right) \\
&= \frac{-2m}{(2\pi)^3 \hbar^2 q} \int_{-k_F}^{k_F} dk_z \int_0^{\sqrt{k_F^2 - k_z^2}} dr\, r \\
&\quad \times \int_0^{2\pi} d\phi \left(\frac{1}{q + 2k_z + i(m/\hbar)\eta} + \frac{1}{q - 2k_z - i(m/\hbar)\eta}\right) \\
&= \frac{-2m}{(2\pi)^2 \hbar^2 q} \int_{-k_F}^{k_F} dk_z \frac{k_F^2 - k_z^2}{2} \left(\frac{1}{q + 2k_z + i(m/\hbar)\eta} + \frac{1}{q - 2k_z - i(m/\hbar)\eta}\right).
\end{aligned} \tag{7.5}$$

We use the integral

$$\int_{-b}^b dx\, x^2 \left(\frac{1}{a + 2x} + \frac{1}{a - 2x}\right) = \frac{a}{2}\left[-2b + a\, \tanh^{-1}\left(\frac{2b}{a}\right)\right]$$

together with

$$\tanh \alpha = \frac{e^{2\alpha} - 1}{e^{2\alpha} + 1} = \frac{2b}{a} \Rightarrow \tanh^{-1}\left(\frac{2b}{a}\right) = \frac{1}{2} \ln\left|\frac{a + 2b}{a - 2b}\right|,$$

and obtain

$$\int_{-k_F}^{k_F} dk_z\, k_z^2 \left(\frac{1}{q + 2k_z} + \frac{1}{q - 2k_z}\right) = \frac{q}{2}\left[-2k_F + \frac{q}{2} \ln\left|\frac{q + 2k_F}{q - 2k_F}\right|\right]$$

$$k_F^2 \int_{-k_F}^{k_F} dk_z \left(\frac{1}{q + 2k_z} + \frac{1}{q - 2k_z}\right) = k_F^2 \ln\left|\frac{q + 2k_F}{q - 2k_F}\right|$$

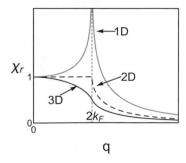

Figure 7.1 Lindhard susceptibility in one, two, and three dimensions.

$$\Re \, \chi(\mathbf{q}) = -\frac{m}{4\pi^2\hbar^2 q} \, \mathcal{P} \int_{-k_F}^{k_F} dk_z \left(k_F^2 - k_z^2\right) \left(\frac{1}{q+2k_z} + \frac{1}{q-2k_z}\right)$$

$$= -\frac{m}{4\pi^2\hbar^2 q} \left[qk_F - \left(\frac{q^2}{4} - k_F^2\right) \ln\left|\frac{q+2k_F}{q-2k_F}\right|\right]$$

$$= -\frac{mk_F}{4\pi^2\hbar^2} \left[1 - \frac{q}{4k_F}\left(1 - \frac{4k_F^2}{q^2}\right) \ln\left|\frac{q+2k_F}{q-2k_F}\right|\right]$$

$$= -\frac{mk_F}{4\pi^2\hbar^2} \left[1 + \frac{2k_F}{2q}\left(1 - \frac{q^2}{4k_F^2}\right) \ln\left|\frac{q+2k_F}{q-2k_F}\right|\right]. \tag{7.6}$$

We see that at $q = 2k_F$, $\chi(2k_F)$ diverges in 1D, while in 2D and 3D it only has a discontinuity and a divergence, respectively, in its derivative (see Figure 7.1). We note that in the case of isotropic electron systems, a wavevector $|\mathbf{q}| = 2k_F$ connects only two states.

Consequently, as can be seen from Figure 7.2, the number of participating states in the $2k_F$ transition is of the order of 1/Fermi sea length $\propto 1/k_F$ in 1D, and 1/Fermi sea area $\propto 1/k_F^2$ in 2D. It is easy to extrapolate 1/Fermi sea volume $\propto 1/k_F^3$ in 3D. Thus we can qualitatively understand why the singularity at $q = 2k_F$ weakens with increasing dimensionality for isotropic systems.

For $q > 2k_F$, $\chi(q)$ exhibits a monotonic decrease with q in all dimensions. This can be understood in terms of energy conservation: perturbations with $\hbar\omega \ll E_F$ and $q \le 2k_F$ can induce transitions on and in the vicinity of the Fermi surface that require negligible infinitesimal energies (real transitions); however, perturbations with $q > 2k_F$ induce transitions that require finite energies $\Delta E \gg \hbar\omega$ (virtual transitions).[1] In other words, electrons cannot provide adequate screening to perturbations with $q > 2k_F$.

---

[1] Recall from quantum mechanics that a virtual transition is one that violates energy conservation, hence it can only last for a time $\Delta t \le \hbar/\Delta E$, where ΔE is the energy cost of the transition. The system has to return to initial state within Δt.
This process is clear in the expression for second-order perturbation energy $\propto \langle\phi_0|V|\phi_n\rangle\langle\phi_n|V|\phi_0\rangle/\Delta E_{0n}$.

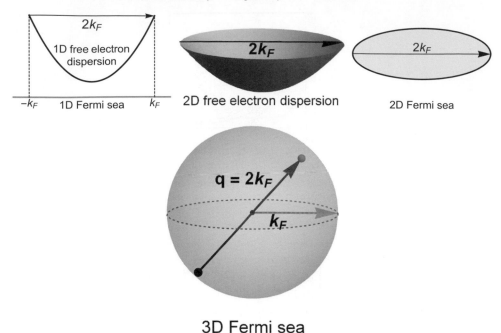

3D Fermi sea

Figure 7.2 For a transition vector $2k_F$: In 1D, the number of participating states $\propto$ 1/Fermi sea length, in 2D it is $\propto$ 1/Fermi sea area, and in 3D it is $\propto$ 1/Fermi sea volume.

This scenario implies that for a 1D free electron system, an external perturbation with a $\hbar\omega \ll E_F$ and $|\mathbf{q}| \sim 2k_F$ will induce a divergent charge redistribution

$$\rho_{\text{ind}}(2k_F) = \chi(2k_F)\, V((2k_F)).$$

Phonon perturbations present a possible mechanism that can induce such an effect.

7.1.2 Fermi Surface Nesting in Quasi-1D Systems

It is easy to build on Figure 7.3 and envision that anisotropy can change the geometry of the Fermi surface in two and three dimensions. In such situations, a vector $|\mathbf{q}| = 2k_F$ may connect many states on the Fermi surface. The process of spanning portions of a Fermi surface by a single wavevector is called nesting. If the portions of the Fermi surface that can be spanned by the single wavevector is large, a perturbation with such a wavevector will induce a large number of electron–hole excitations, all of them taking place from $-k_F$ to $+k_F$. We stress again that only in 1D the characteristic topology of the Fermi surface enforces a perfect nesting. Figure 7.3 shows a possible nesting configuration for the case of a 2D square lattice.

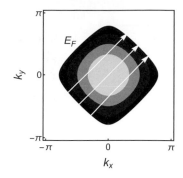

Figure 7.3 Possible Fermi surface nesting in a 2D square lattice.

Charge Density Waves

The combination of electron–phonon coupling and Fermi surface nesting induces periodic charge redistribution in some low-dimensional crystalline systems, with a wavevector $2k_F$. Such periodic charge density modulation is called a *charge density wave* (CDW). The CDW effectively screens the $2k_F$ phonon mode producing it, and thus reduces its frequency, a process known as *mode softening*. If the screening is complete, the mode frequency vanishes and the mode is said to be *frozen in*, that is, it becomes a static displacement pattern of the ions from their equilibrium positions in the normal metal state; it is usually referred to as a *periodic lattice distortion* (PLD), or *lattice modulation*. In general, the period of the modulation $\lambda = 2\pi/k_F$ is not a multiple of the lattice period, but is *incommensurate* with respect to it. The PLD phase transition introduces additional diffraction peaks due to the superposition of the modulation superstructure on the original lattice. Although the ideal case of an incommensurate CDW would suggest that it moves collectively without any resistivity, like a superconductor, pinning to impurities and defects prevents such free motion. Collective transport of the entire CDW is possible with the application of a *depinning threshold field*.

7.1.3 Finite Temperature Susceptibility

We now consider the temperature dependence of $\chi(\mathbf{q})$, which describes the system's response to perturbation energies negligible in comparison with E_F. Thus we shall confine the following derivation to an energy range of the order of the Debye energy E_D typical of phonon excitations that satisfy $E_D \ll E_F$.

At finite temperatures, the expression for $\chi(\mathbf{q})$ is dominated by the Fermi distribution functions in the numerator of (7.1), namely $f(E_{\mathbf{k}}) = (\exp(\beta E_{\mathbf{k}}) + 1)^{-1}$, where $\beta = 1/k_B T$. Examination of Figure 7.4a justifies linearizing the electronic dispersion relation about E_F, as shown in Figure 7.4b.

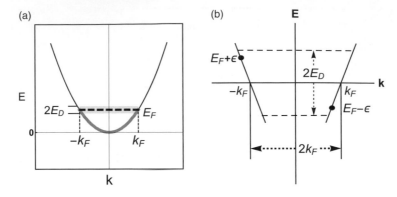

Figure 7.4 (a) Band dispersion for free electrons in one dimension. (b) Approximating the dispersion relation near the Fermi level.

If we introduce the variable ϵ such that $E_k = E_F + \epsilon$, $(\epsilon < E_D)$, then E_{k+2k_F} is given as $E_{k+2k_F} = E_F - \epsilon$, as shown in Figure 7.4b. Substituting in the expression for $\chi(2k_F)$

$$\chi(\mathbf{q}) = \frac{1}{N} \sum_{\mathbf{k} \in BZ} \frac{f(E_{\mathbf{k}}) - f(E_{\mathbf{k+q}})}{E_{\mathbf{k}} - E_{\mathbf{k+q}}} \quad \Rightarrow \quad \int \frac{dk}{2\pi}$$

with $E_{\mathbf{k}} - E_{\mathbf{k+q}} = 2\epsilon$ and $dq = \mathcal{D}(E_F)d\epsilon$, we get

$$\chi(2k_F) = \frac{\mathcal{D}(E_F)}{2N} \int_0^{E_D} \left[\frac{1}{\exp(-\beta\epsilon) + 1} - \frac{1}{\exp(\beta\epsilon) + 1} \right] \frac{d\epsilon}{\epsilon}$$

$$= \frac{\mathcal{D}(E_F)}{2N} \int_0^{E_D} \frac{\tanh(\beta\epsilon/2)}{\epsilon} d\epsilon = \frac{\mathcal{D}(E_F)}{2N} \int_0^{\beta E_D/2} \frac{\tanh x}{x} dx. \quad (7.7)$$

Using the low-temperature approximation $\beta E_D \gg 1$,

$$b \to \infty, \quad \int_0^b dx \, \frac{\tanh(x)}{x} \simeq \ln\left[\frac{4e^\gamma b}{\pi} \right], \quad \gamma = 0.5771 \text{ is Euler's constant,}$$

we obtain

$$\chi(2k_F) = \frac{\mathcal{D}(E_F)}{2N} \ln\left[\frac{1.14 E_D}{k_B T} \right], \quad (7.8)$$

which indicates that $\chi(2k_F)$ diverges logarithmically with decreasing temperature.

7.2 Peierls Instability and Peierls Transition

The instability in one-dimensional electronic systems for $q \to 2k_F$ and $T \to 0$, which gives rise to a CDW ρ_{2k_F}, is called a *Peierls instability*. The manifestation of a Peierls instability [151] is closely related to the dimensionality and extent of nesting of a system. In a free electron system without interactions, there is of course no perturbing potential present to drive the Peierls instability. However, in the presence of an underlying lattice and

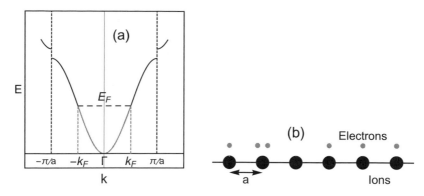

Figure 7.5 (a) Band dispersion for free electrons in 1D, in the presence of (b) a static periodic lattice with periodicity a.

its thermal phonon excitations, the Peierls instability can occur since the phonons provide the perturbing potential required to drive the electron instability.

We introduce such a lattice in the 1D electron system as shown in Figure 7.5. Figure 7.5(a) depicts a 1D band picture, and Figure 7.5(b) a classical electron picture at $T = 0$. Moreover, for the sake of simplicity, we consider the case of one free electron per lattice point. For a lattice spacing of a, we find that the electron density is

$$n = \frac{1}{a} = 2\frac{2k_F}{2\pi} \quad \Rightarrow \quad k_F = \frac{1}{4} \times \frac{2\pi}{a} = \frac{\pi}{2a}.$$

The following scenario applies to any electron density $n < 2/a$.

First we consider the case of static lattice distortions. A periodic lattice distortion with a wavevector q is described as

$$u = u_q \cos(qx), \tag{7.9}$$

where u is the ions displacement from their equilibrium positions. This lattice distortion produces a perturbing potential that acts on the electron system

$$V = V_q \cos(qx) = g u_q \cos(qx), \tag{7.10}$$

where g is the electron–lattice coupling constant. The electron system then forms a density wave

$$\rho_q = -V_q \, \chi(q),$$

and it is expected to lower its energy in the presence of this density wave.

In contrast, the periodic lattice distortion (7.9) leads to a strain energy increase of

$$\delta U = \frac{1}{2} C u_q^2, \tag{7.11}$$

due to ion–ion overlap or coulomb interactions. C is the elastic stiffness. If the decrease of the electron energy outweighs the increase of the elastic energy, the total system will stabilize itself by spontaneously forming the lattice distortion u and maintaining the electron-density wave ρ_q. As we have shown, this phenomenon occurs for $q = 2k_F$ in one dimension.

We shall now consider the problem of energy balance. A perturbation V_q creates a gap in the band dispersion of $E_g = 2\Delta = 2|V_q|$ at $k = q/2$, and modifies the dispersion in the vicinity of this gap to

$$E_k^{\pm} = \frac{E_k^0 + E_{k+q}^0}{2} \pm \sqrt{\left(\frac{E_k^0 - E_{k+q}^0}{2}\right)^2 + |V_q|^2}, \tag{7.12}$$

where E_k^0 denotes the unperturbed energies, and the $\pm$ signs denote upper and lower bands around the gap, respectively. The change in the electron energy due to the perturbation at finite T is then

$$\delta K = \sum_k E_k^{\pm} f(E_k^{\pm}) - \sum_k E_k^0 f(E_k^0). \tag{7.13}$$

Using the approximation that actually $V_q \ll E_F$, we obtain

$$\delta K \approx -|V_q|^2 \sum_k \frac{f(E_{k+q}^0) - f(E_k^0)}{E_{\mathbf{k}}^0 - E_{\mathbf{k+q}}^0} = -|V_q|^2 \, N \, \chi(q).$$

The change in the total energy of the combined electron–lattice system is then

$$(C/2)\,u_q^2 - |V_q|^2 \, N \, \chi(q) = -g^2 u_q^2 \left\{ N\,\chi(q) - \frac{C}{2g^2} \right\}$$

$$= -\Delta^2 \left\{ N\,\chi(q) - \frac{C}{2g^2} \right\}. \tag{7.14}$$

The susceptibility diverges logarithmically with decrease in T for $q = 2k_F$, while the elastic stiffness remains finite. Therefore, the total energy becomes negative below a certain temperature T_P, and the Peierls instability takes place whenever g is nonzero, giving rise to an electronic band gap at the Fermi level, and the system turns from a metal to an insulator. The periodic lattice distortion with wavevector $2k_F$ is called the *Peierls distortion*, while the metal–insulator transition arising from the Peierls instability is called the *Peierls transition*. Figure 7.6 shows schematically both the changes in the electron-band dispersion (Figure 7.6a) and the lattice distortion (Figure 7.6b) for the case of one electron per lattice point. Notice that the lattice becomes dimerized with a periodicity of $2a$.

We can derive an expression for the transition temperature T_P, by setting (7.14) to zero,

$$\frac{C}{2g^2} - N\chi(q) = 0, \tag{7.15}$$

and substituting for $\chi(q)$ from (7.8), we get

$$\frac{C}{2g^2} - N\frac{\mathcal{D}(E_F)}{2N} \ln\frac{1.14 E_D}{k_B T} = 0, \tag{7.16}$$

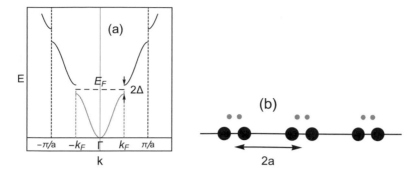

Figure 7.6 (a) Band dispersion for the insulating state caused by the Peierls transition. (b) The corresponding lattice distortion and electron localization. For the case of one electron per lattice point.

yielding

$$T_P^{MF} = 1.14 T_D \, \exp(-1/\lambda),$$

$$\lambda = \frac{g^2 \mathcal{D}(E_F)}{C},$$

(7.17)

where λ is the dimensionless parameter of the electron–lattice interactions. The superscript MF of T_P indicated that the result is derived on the basis of a mean-field approximation.

Either the neighboring atoms alternately get slightly closer and further apart, or they can get buckled (symmetrically or asymmetrically), both distortions resulting in a chain of dimers.

7.3 Electron–Phonon Coupling and the Kohn Anomaly

In the last section, we have seen that the static lattice distortion and the charge density wave with wavevector $q = 2k_F$ appear below the Peierls transition temperature T_P. As T_P is approached from above, the frequency of the $q = 2k_F$ phonon decreases, while its amplitude increases. This is the phenomenon of *phonon softening*. The softening of the $2k_F$ phonons due to interactions with an electron system is called a *Kohn anomaly*. As shown in Figure 7.7, for a 1D system, the soft phonon frequency vanishes at T_P, giving rise to the static lattice distortion, or *frozen phonon*. The Kohn anomaly can be regarded as a precursor to the Peierls transition in this case.

7.3.1 Electron–Phonon Interactions in the Jellium Model

In the jellium model, the lattice is regarded as a continuum of positive charges. The lattice vibrations then correspond to periodic distortions in the positive charge density that create electric polarizations and hence electric fields acting on the electron system. Thus, the

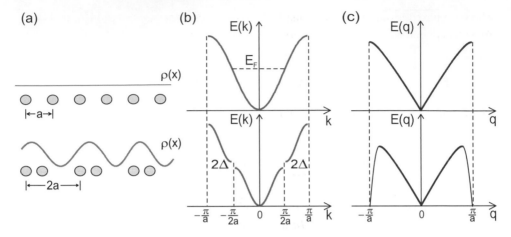

Figure 7.7 (a) Monatomic linear chain, showing the charge density, $\rho(x)$. (b) Electron dispersion curves for half band filling. (c) Acoustic phonons dispersion curve. Top and bottom in each part show before and after the Peierls transition.

electrons and the lattice waves interact with each other through these polarization fields. A lattice modulation of the elastic body of positive charges is then described by

$$u(x,t) = \frac{1}{\sqrt{\mathcal{V}}} \sum_q \eta_q(t) \exp(iqx), \tag{7.18}$$

where $\mathcal{V}$ is the volume of the system. $u(x,t)$ produces an electric polarization $P = Zen_iu$, where n_i is the "ion" density. The density of positive charge associated with the distortion is given by $\delta\rho_i = -div\ P$. Poisson's equation

$$-\nabla^2\phi = \delta\rho_i/\epsilon_0$$

then determines the potential $\phi(x)$ due to this charge density modulation. We have invoked the approximation of neglecting the Coulomb interaction among the electrons and thus used the dielectric constant of vacuum. The potential $\phi(x)$ is then given by

$$\phi(x) = -\frac{i}{\epsilon_0\sqrt{\mathcal{V}}} \sum_q \frac{Zen_i\eta_q}{q} \exp(iqx). \tag{7.19}$$

Writing the electron wavefunction as

$$\psi(x) = \frac{1}{\sqrt{\mathcal{V}}} \sum_k c_k \exp(ikx), \tag{7.20}$$

we obtain the interaction potential energy between the electron and phonons as

$$\mathcal{H}' = \int e\phi(x)\,\psi^*(x)\psi(x)\,dx = \frac{iZe^2n_i}{\sqrt{\mathcal{V}}} \sum_q \sum_k \frac{c_k^\dagger c_{k-q}\,\eta_q}{q}, \tag{7.21}$$

where we have introduced second quantization operators $c_k^\dagger$ and c_k for the electron system. We can also introduce similar operators for the phonon system through the relation

$$\eta_q = \sqrt{\frac{\hbar}{2Mn_i\omega_q}}\left(b_q + b_{-q}^\dagger\right), \tag{7.22}$$

where Mn_i is the mass density of the uniform positive charge. The electron–phonon interaction Hamiltonian is then given by

$$\mathcal{H}' = \frac{1}{\sqrt{\mathcal{V}}}\sum_q\sum_k g(q)\, c_{k+q}^\dagger c_k \left(b_q + b_{-q}^\dagger\right),$$
$$g(q) = i\sqrt{\frac{\hbar}{2Mn_i\omega_q}}\,\frac{iZe^2n_i}{\epsilon_0 q}, \tag{7.23}$$

where $g(q)$ is the electron–phonon interaction parameter.

The total Hamiltonian for the electron plus the lattice is then given by

$$\mathcal{H} = \sum_k E_k\, c_k^\dagger c_k + \sum_q \hbar\omega_q\, b_q^\dagger b_q + \frac{1}{\sqrt{\mathcal{V}}}\sum_{k,q} g(q)\, c_{k+q}^\dagger c_k \left(b_q + b_{-q}^\dagger\right). \tag{7.24}$$

This is known as the Frölich Hamiltonian for noninteracting electrons incorporating the jellium model for the lattice. The Kohn anomaly can be derived from this Hamiltonian.

7.3.2 The Kohn Anomaly

The phonon frequency $\omega(\mathbf{q})$ is determined from the magnitude of the restoring force for the corresponding lattice distortion. The origin of the restoring force in the jellium model is the Coulomb interaction among the ions. The lattice distortion associated with a phonon of wavevector $\mathbf{q}$ subjects the electron system to a potential V_q, which gives rise to a density wave $\rho_q = -\chi(q)V_q$. The electron-density wave must reduce interionic forces by screening the electric force. Thus the restoring force decreases and the phonon frequency becomes smaller than it would be in the absence of electron–phonon interactions, i.e., for the bare ions.

Using the Frölich Hamiltonian, we can write the equation of motion for the phonon operator η_q as

$$-\hbar^2 \ddot{\eta}_q = \Big[[\eta_q, \mathcal{H}], \mathcal{H}\Big]. \tag{7.25}$$

After some algebraic manipulations, we arrive at

$$\ddot{\eta}_q = -\omega_q^2 \eta_q - g_{-q}\sqrt{\frac{2\omega_q}{Mn_i\mathcal{V}\hbar}}\sum_k c_{k-q}^\dagger c_k. \tag{7.26}$$

If we have N electrons in the volume $\mathcal{V}$, the electron density can be written as

$$|\psi(x)|^2 = \frac{N}{\mathcal{V}} \sum_q \rho_q \, \exp(iqx).$$

In the language of second quantization, ρ_q becomes

$$\rho_q = (1/N) \sum_k a^\dagger_{k-q} a_k.$$

Thus we can rewrite (7.26) as

$$\ddot{\eta}_q = -\omega_q^2 \eta_q - g_{-q} \sqrt{\frac{2\omega_q}{Mn_i \mathcal{V}\hbar}} \, N \, \rho_q. \tag{7.27}$$

We can rewrite $\mathcal{H}'$ in the form

$$\mathcal{H}' = \sum_q \frac{iZe^2 n_i \, \eta_q}{q\sqrt{\mathcal{V}}} \sum_k c^\dagger_{k-q} c_k = N \sum_q \rho_q V_q = N \sum_q \chi(q) \left| V(q) \right|^2, \tag{7.28}$$

where we defined

$$V_q = g_q \eta_q \sqrt{\frac{2Mn_i \omega_q}{\hbar \mathcal{V}}}, \tag{7.29}$$

which allows us to express (7.27) in the form

$$\ddot{\eta}_q = -\Omega_q^2 \eta_q = -\omega_q^2 \, \eta_q + \frac{2\left|g_q\right|^2 N\omega_q}{\hbar \mathcal{V}} \chi(q) \, \eta_q. \tag{7.30}$$

The new phonon frequency Ω_q renormalized by the electron–phonon interaction is

$$\Omega_q^2 = \omega_q^2 - \frac{2\left|g_q\right|^2 N\omega_q}{\hbar \mathcal{V}} \chi(q,T) \leq \omega_q^2, \tag{7.31}$$

where ω_q is the bare phonon frequency.

For the case of one-dimensional electron systems, $\chi(q)$ diverges logarithmically as $q \to 2k_F$ and $T \to 0$; thus the frequency Ω_{2k_F} is always small compared with frequencies at other wavevectors, approaching zero as $T \to T_P$. In 2D and 3D, $\chi(q)$ does not diverge but decreases rapidly for $q > 2k_F$. This fact is reflected in a weak anomaly in the phonon frequency at $q = 2k_F$. Figure 7.8 depicts the form of the Kohn anomaly in each dimension. The Kohn anomaly in one dimension is sometimes called the *Giant Kohn anomaly*. The temperature at which Ω_{2k_F} becomes zero in one dimension is the *Peierls transition*

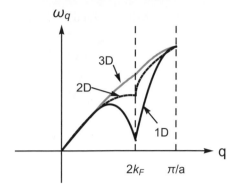

Figure 7.8 Manifestation of the Kohn anomaly in the phonon dispersion in each dimension.

temperature T_P^{MF}, or simply T_P. Setting $\Omega_{2k_F} = 0$ in (7.31) and using (7.8) for $\chi(2k_F)$, we get

$$
\begin{aligned}
\Omega_{2k_F}^2 &= \omega_{2k_F}^2 - \frac{2\left|g_{2k_F}\right|^2 N \omega_{2k_F}}{\hbar V} \chi(2k_F, T_P) \\
&= \omega_{2k_F}^2 - \frac{2\left|g_{2k_F}\right|^2 \omega_{2k_F} N}{\hbar V} \frac{\mathcal{D}(E_F)}{2N} \ln\left[1.14 E_D / k_B T_P\right] \\
&= 0.
\end{aligned}
\tag{7.32}
$$

We identify

$$
\lambda = \frac{\left|g_{2k_F}\right|^2 \mathcal{D}(E_F)}{\hbar \omega_{2k_F} V},
\tag{7.33}
$$

and write

$$
\frac{1}{\lambda} = \ln\left[1.14 T_D / T_P\right] \Rightarrow T_P = 1.14 T_D \exp(-1/\lambda),
\tag{7.34}
$$

where $T_D = E_D / k_B$. Using (7.33) and (7.34), we arrive at the following expression for the temperature dependence of $\Omega^2(2k_f)$

$$
\begin{aligned}
\Omega_{2k_F}^2(T) &= \omega_{2k_F}^2 \left(1 + \lambda \ln\left[\frac{T}{1.14 T_D}\right]\right) = \lambda \omega_{2k_F}^2 \left(\frac{1}{\lambda} + \ln\left[\frac{T}{1.14 T_D}\right]\right) \\
&= \lambda \omega_{2k_F}^2 \ln\frac{T}{T_P}.
\end{aligned}
$$

The parameter λ must be the same as in (7.17). Comparing these equations, we see that the elastic stiffness C, therefore, corresponds to $\hbar \omega_{2k_F}$. This should be expected since $C/2$ is the elastic energy per unit volume and $\hbar \omega_{2k_F}$ is the phonon energy.

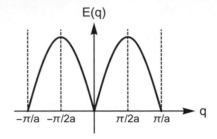

Figure 7.9 Onset of static lattice distortion of $q = 2k_F - \pi/a$ for $T \geq T_P$.

7.3.3 The Order Parameter of the Peierls Transition

Two quantities that describe the change of state that occurs in the Peierls transition are the frequency of the soft phonon, Ω_{2k_F}, and the width of the band gap, $E_g = 2\Delta$. We can employ Δ as an order parameter of the Peierls transition because $\Delta = 0$, for $T \geq T_P$, and $\Delta > 0$, for $T < T_P$.

For $T < T_P$, the restoring force for the lattice distortion with $q = 2k_F$ becomes zero, $\Omega_{2k_F} = 0$, because of the screening due to the electron-density waves. A static lattice distortion, with periodicity $2k_F = \pi/a$, appears (see Figure 7.9).

The balance of forces is represented on the right of (7.30). Using (7.30) and noting the nonvanishing expectation value $\langle \eta_{2k_F} \rangle \neq 0$ for $T < T_P$, we obtain an equation for the force balance as follows:

$$\omega_{2k_F}^2 \langle \eta_{2k_F}(T) \rangle = \frac{2\left|g_{2k_F}\right|^2 N \omega_{2k_F}}{\hbar \mathcal{V}} \chi\left(2k_F; \langle \eta_{2k_F}(T) \rangle\right) \langle \eta_{2k_F}(T) \rangle, \qquad (7.35)$$

where we used the fact the $\langle \eta_{2k_F} \rangle$ is a static distortion, and $\chi\left(2k_F; \langle \eta_{2k_F} \rangle\right)$ is the susceptibility in the presence of the gap Δ due to $\langle \eta_{2k_F} \rangle \neq 0$. This should be calculated using (7.12), which describes the electronic energy dispersion in the presence of the band gap. The energy is measured from $E = E_F$, and the calculation is done within an energy range $E_F - E_D < E < E_F + E_D$, i.e., $k \simeq k_F$, as was done in deriving (7.7). The dispersion relation near E_F is then approximated by the linear relation between E_k^0 and k,

$$E_k^0 = E_F + \epsilon_k^0, \qquad E_{k+q}^0 = E_F - \epsilon_k^0. \qquad (7.36)$$

Equation (7.12) is then written as

$$E_k^\pm = \pm\sqrt{(\epsilon_k^0)^2 + \left|V_q\right|^2} = \pm\sqrt{(\epsilon_k^0)^2 + \Delta^2}. \qquad (7.37)$$

Following the same procedure as in (7.7), we obtain

$$\chi\left(2k_F; \langle \eta_{2k_F}(T) \rangle\right) = \frac{\mathcal{D}(E_F)}{2N} \int_0^{E_D} \tanh\frac{\beta E_k^+}{2} \frac{d\epsilon_k^0}{E_k^+}. \qquad (7.38)$$

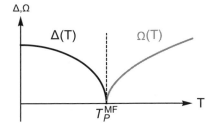

Figure 7.10 Temperature dependence of gap Δ, and the frequency Ω_{2k_F} of the soft phonon.

Substituting (7.38) and (7.33) into (7.35), we obtain

$$\frac{1}{\lambda} = \int_0^{E_D} \tanh \frac{\beta E_k^+}{2} \frac{d\epsilon_k^0}{E_k^+}. \tag{7.39}$$

This equation gives a relation between T and Δ contained in E_k^+. The form of this equation is reminiscent of the superconductivity gap. At $T = 0$, $\tanh(\beta E_+/2) = 1$, and we get

$$\frac{1}{\lambda} = \frac{1}{2} \int_{-E_D}^{E_D} \frac{d\epsilon_k^0}{\sqrt{(\epsilon_k^0)^2 + \Delta^2}} = \ln \left[\frac{\sqrt{E_D^2 + \Delta^2} + E_D}{\sqrt{E_D^2 + \Delta^2} - E_D} \right].$$

When $\Delta(0) \ll E_D$, a condition usually satisfied in real systems, we find that

$$\Delta(0) = 2E_D \exp(-1/\lambda) = 1.76 k_B T_P. \tag{7.40}$$

When the temperature is raised above zero, the numerator of the integrand of (7.39) is reduced, and in order to satisfy the left-hand side, the denominator must decrease. This implies that Δ is a monotonically decreasing function of T. The initial decrease is exponentially slow until $k_B T$ becomes of the order of $\Delta(0)$ when $\Delta(T)$ begins to drop more rapidly until it vanishes at T_P.

Figure 7.10 shows the temperature dependence of both Δ and Ω_{2k_F}. The gap Δ is zero above T_P and increases continuously with decreasing T below T_P. This suggests that the Peierls transition, like the superconducting transition, is a second-order phase transition whose order parameter is the gap Δ. Of course, we could also use η_{2k_F}, the amplitude of the lattice distortion, or ρ_{2k_F} as an order parameter.

Exercises

7.1 Consider a weak periodic potential of the form $V(x) = V_0 \cos(2k_F x)$ created in a one-dimensional electronic system, with $q = k_F$ the Fermi wavevector. We need to calculate the total energy of the system with and without this weak potential.

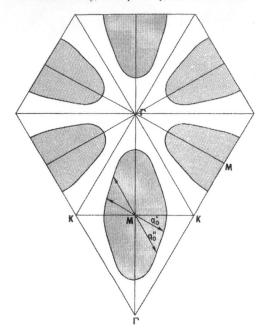

Figure 7.11 TaS$_2$ Fermi surface cross-section in the basal plane, from [137].

(a) The new perturbed bands are given by (7.12):

$$E_k^{\pm} = \frac{E_k^0 + E_{k+q}^0}{2} \pm \sqrt{\left(\frac{E_k^0 - E_{k+q}^0}{2}\right)^2 + |V_q|^2}.$$

Use this to express the change in energy in the lower band $\Delta E(k')$, where $\kappa = q/2 - k$, from the unperturbed case as

$$\Delta E(k') = \frac{\hbar^2}{2m}\left[q\kappa - \sqrt{(\kappa)^2 + \left(\frac{2mV_0}{\hbar}\right)^2}\right].$$

(b) Integrate the preceding equation over k-space and show that the total change in energy, for a weak potential, is proportional to $V_0^2 \ln V_0$.

(c) For a weak potential strength $V_0 \ll \dfrac{\hbar^2}{2m}\left(\dfrac{q}{2}\right)^2$, does the total energy increase or decrease for this gapped system compared to free electrons?

7.2 1T-TaS$_2$ is a layered compound with octahedrally coordinated S-Ta-S layers held together by van der Waals forces. Figure 7.11 shows a Fermi surface cross-sections of 1T-TaS$_2$ in the basal plane also containing an extension of one segment into the next zone. Explain qualitatively why the topology of this Fermi surface leads to strong nesting.

Part Two

Topological Phases

8

Topological Aspects of Condensed Matter Physics: A Historical Perspective

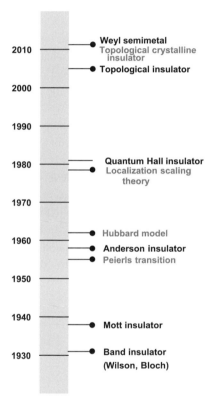

A brief history of insulators
and topological phases.

Prior to the 1980s, electronic phases have been classified as insulating, metallic (conducting), superconducting, or magnetic. Magnetic and superconducting phases were shown to be described in terms of symmetries that are spontaneously broken, which will be discussed in Part Three.

According to electronic band theory proposed by Bloch and Wilson [31, 198, 199] systems with an odd number of electron per unit cell were predicted to be metallic. However, in 1937 Jan Hendrik de Boer and Evert Johannes Willem Verwey pointed out that a variety of transition metal oxides, predicted to be conductors according to band theory, are actually insulators. Nevill Mott and Rudolf Peierls predicted that this anomaly can be explained by including interactions between electrons [133]. In 1949, Mott proposed a model where the energy gap in the metal-oxide NiO arises from a competition between Coulomb energy U and hopping of 3d electrons [132].

In 1955, Peierls proposed the idea that in a one-dimensional metal system, electron–phonon interaction leads to a Fermi surface instability that triggers a metal insulator accompanied by a lattice distortion [151], the *Peierls transition* that was presented in Chapter 7.

Shortly after, in 1958 P. W. Anderson proposed another type of insulators that results from a variety of lattice disorders [10], or randomness. This was followed in 1979 by the introduction of a scaling theory for localization [3]. In 1963, John Hubbard introduced a simple model to describe interacting electrons to describe the transition between conducting and insulating systems [94] that today bears his name. The Hubbard model can be considered an improvement on the tight-binding model, which for strong interactions can give qualitatively different behavior, correctly predicting the existence of a Mott insulator.

In 1980, the integer quantum Hall (IQH) phase was discovered by von Klitzing and coworkers [105]. At low temperature, the energy spectra of a two-dimensional electron gas subjected to a strong magnetic field, shown schematically in Figure 8.1 (left), display discrete, or flat, energy bands known as Landau levels. von Klitzing et al. observed that when the Fermi energy is in a gap between Landau levels, the system becomes insulating, yet the measured Hall conductance assumes quantized values in units of e^2/h, as manifest in the staircase plateau structure of the Hall resistivity in Figure 8.1 (right). The value of this unit of quantization brings to mind the fine-structure constant of quantum electrodynamics, $\alpha = e^2/c\hbar \sim 1/137$. The Hall conductance was found to be independent of the details of the system, such as geometry and impurities, provided the gap between the Landau

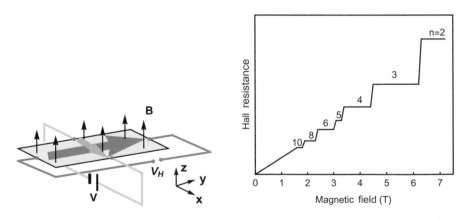

Figure 8.1 Schematic picture of the QHE system (left). Hall resistivity (right) showing the manifest plateau quantization, $R_H = h/e^2 = 25812.807557(18)\,\Omega$.

levels is sufficiently large compared to $k_B T$. Thus, the IQH state breaks no symmetry, and its quantized features are found to be robust against smooth changes in the system's constituent parameters, most obviously disorder associated with sample preparation (recall our discussion in Chapter 4 of how a 2DEG is achieved in a fabricated heterojunction).

It was then argued that novel states, confined to the edge of the system (coined *edge states*), must carry the Hall current. When an impurity is located on the edge, the robust edge current goes around it, propagating ballistically without backscatterings. Because of this *chiral* [1] character of edge states, the edge current flows in one way. Subsequent studies demonstrated that these boundary states are enforced by a topological bulk ordering, which makes the edge current robust against edge deformation, or weak perturbations: In 1982, Thouless, Kohmto, Nightingale, and den Nijs (TKNN) showed that the conductance quantization is associated with a topological invariant [176]. Calculating the Hall conductance with the aid of the Kubo formula, TKNN showed that it yields a topological invariant known as the *first Chern number*. In 1984, with the emergence of Berry's geometric phase framework [27], it became clear that the integrand that appeared in the TKNN formula was just Berry's field (or curvature), and that the integral was actually performed over the closed surface of the two-dimensional Brillouin zone.

We recall that quantum-mechanical wavefunctions are described by linear combinations of an orthonormal set of basis vectors that define a Hilbert space. We also know that in crystalline solids, the wavevector **k** becomes a good quantum number. This allows us to view the Bloch wavefunction as a mapping of a point in **k**-space, the Brillouin zone, to a subspace of Hilbert space. In the following chapter, we will identify this structure in Hilbert space as a manifold. As we will see, such identification reveals the connection of such topological structure in Hilbert space to electronic states in solids. Within this perspective, the aforementioned findings revealed that the mapping of QHE electronic states onto Hilbert space appeared as an intertwining with a nontrivial topology. These revelations led to the argument that the quantum Hall state presented a system topologically distinct from previously known electronic phases, and that a new classifying paradigm based on the notion of topology was needed.

Further progress regarding the topological characterization of material systems remained dormant until the dawn of the twenty-first century. The first half of the 2000 decade witnessed the evolution of the idea of a quantum spin Hall (QSH) system [136] and a $\mathbb{Z}_2$ topological number ν. The simplest depiction of the QSH phase involves a superposition of two quantum Hall (QH) replicas, shown in Figure 8.2, one for $\uparrow$-spins and the other for $\downarrow$-spins, thus having opposite effective magnetic fields. We discern that for the $\uparrow$-spin ($\downarrow$-spin) subsystem, the QH conductance is $\sigma_{xy}^{\uparrow} = e^2/h$ ($\sigma_{xy}^{\downarrow} = -e^2/h$), and, hence, a zero net edge current and a zero Chern number. We note, however, that the spins of the counterpropagating edge states have opposite directions, and, hence, there exists a net spin current. The combined system obeys time-reversal symmetry (TRS), since the net magnetic

---

[1] Chirality is the property that an object and its mirror image do not coincide. In the present case, it identifies the unidirectionality of electronic motion in the edge states.

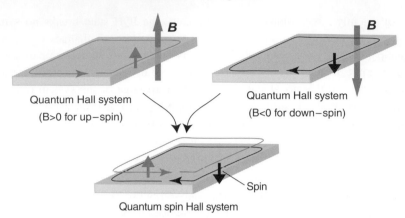

Figure 8.2 Schematic picture of the QSH system as a superposition of two QH systems; from [135].

field is zero. Actually, such a system can be realized with the aid of spin–orbit coupling, since it acts like an effective magnetic field with opposite directions for opposite spins.

In 2005, Kane and Mele (KM) proposed, in a seminal paper, a QSH mechanism for a graphene-like system [99]. Recall, that graphene is a semimetal with a gapless electronic structure. At low energies, it is represented by Dirac cones around the K-points of the 2D Brillouin zone, with Dirac points at the Fermi energy. What KM considered was a TRS model (KM model) that closely resembles graphene, but with an energy gap generated by SOI at the K-points – in other words, a graphene-like insulator because of a strong intrinsic SOC. They showed that this model exhibits *spin Hall* conductivity that is precisely quantized to $2e^2/h$. Furthermore, KM suggested that there must be an additional topological invariant, which characterizes the QSH effect. They subsequently demonstrated that such a topological invariant, $\mathbb{Z}_2$, in fact exists [98], and assumes only the two values $\nu = 0,\ 1 \bmod 2$. This is to be contrasted with integer quantum Hall effect (IQHE), which is characterized by a $\mathbb{Z}$ invariant.

In the context of the $\mathbb{Z}_2$ invariant, KM argued that it characterizes whether an insulator is of a trivial (or ordinary) or nontrivial band type. In a trivial insulator, such as diamond, electrons fully occupy energy bands that derive their ordering and character from atomic orbitals. In the nontrivial case, the energy gap between the occupied and empty states is fundamentally modified by SOI. In the KM graphene-like model, the energy gap follows the expected conventional orbital ordering around the gap away from the K-points; however, because of the strong SOC, the insulating energy gap is inverted at the K-points: the atomically derived states that should appear above and below the gap become reversed. This *twist* in the order of electronic states appears in Hilbert space like the twist in the Mobius band; it cannot be *unwound*, hence its topological nature. Such strong SOIs are manifest in materials containing heavy elements such as Bi or Sb. Further analytic and numerical studies revealed the existence of manifest edge states, and that such *nontrivial* states are robust to both weak TRS interactions and SOC terms that mix electronic states with ↑ and ↓ spins. The nontrivial insulator was later coined *topological insulator* (TI) by Moore and Balents [130].

Zhang and coworkers argued that the QSH effect was very difficult to observe in graphene [202], because of the very weak intrinsic spin–orbit coupling. Bernevig, Hughes, and Zhang (BNZ) [26] proposed, instead, that it would appear in Hg-Te quantum wells where spin–orbit coupling was unusually strong. In 2007, the QSH effect was observed experimentally by König et al. in such quantum wells [109]. In the same year, three-dimensional (3D) bulk solids of binary compounds involving bismuth were predicted to belong to the TI family. The first experimentally realized 3D TI state was observed in bismuth antimonide, and shortly afterward in pure antimony, bismuth selenide, bismuth telluride, and antimony telluride using angle-resolved photoemission spectroscopy (ARPES) [88]. The band structure and the absence of backscattering have also been reported with scanning tunneling microscopy and spectroscopy.

Three-dimensional TIs have several attributes in common with graphene, such as their low-energy electronic properties of edge (surface) states being dominated by massless Dirac fermion excitations, where the energy dispersion relations are described by a Dirac cone. As a result, we have highly conducting metallic states on the surface – a feature not seen in ordinary insulators. Moreover, the combination of strong SOI and TRS in TIs result in a spin-momentum locking property, *helicity*, whereas clean graphene exhibits a pseudospin-momentum locking not related to TRS but with the internal symmetry of the honeycomb lattice. Differences from graphene also lie in the parity of the number of Dirac cones emerging at surfaces.

Two more varieties of topological materials emerged in 2011, *topological crystalline insulators* [70] and *Weyl semimetals* (WSM) [192]. The former involves topological insulators protected by crystal point symmetries, and do not require TRS. The latter is associated with accidental twofold degeneracies of bands in a three-dimensional solid. The requirement for a WSM is that either TRS or inversion symmetry must be absent so that global band spin degeneracy can be removed and accidental double degeneracy can be realized. The dispersion in the vicinity of these band touching points is generically linear and can be described by the Weyl equation, despite the lack of Lorentz invariance.

As we will see in Chapter 12, the phenomenology of topological condensed matter phases can be understood in the framework of what has become known as *topological band theory* of solids. It is remarkable that after more than 80 years, there are still gems to be uncovered within band theory.

For further reading, the books in references [25, 66, 148, 166] are recommended.

9

Topological Preliminaries

We are used to the standard formulation of nonrelativistic quantum mechanics where pure states are described in terms of vectors and operators in a Hilbert space. However, modern approaches in mathematical physics adopt a more versatile framework [23]: the formalism of fiber bundles, in particular vector bundles, where, as we shall see, Hilbert space is repackaged [172]. To develop the fundamental aspects of topological phases, we need to introduce the basic concepts of such an approach and define the main underlying constituent components.

9.1 Defining Important Building Blocks

We first need to be familiar with some basic elements and jargon of the formalism, and we start with providing the following important but nonrigorous definitions [49, 67, 138]:

- **Manifolds** are generalizations of our familiar ideas about curves (1D) and surfaces (2D) to similar objects of arbitrary dimensions. We note an important feature of a manifold M in $\mathbb{R}^m$ space: it is an object that locally looks like $\mathbb{R}^m$ but not necessarily globally. The phrase "looks like" means that we may use, locally, a set of m coordinates just as we would in an ordinary Euclidean space, namely, a part of R^m, but we must allow for the choice of the m coordinates to vary as we move over the manifold. For example, we may describe a small part ΔS of a surface S in a three-dimensional space $\mathbb{R}^3$ around a point $\mathbf{x} \in \Delta S$, as a flat Euclidean space $\mathbb{R}^2$. Such a particular local description cannot be extended globally to other parts of the surface.

  All manifolds of dimension m can be embedded into some Euclidean space $\mathbb{R}^n, n > m$ ($M \hookrightarrow \mathbb{R}^n$). In this case, we may describe a manifold M by introducing $n-m$ constraints. For example, we may embed the two-sphere S^2 into $\mathbb{R}^3$ by representing its elements by 3-vectors $\mathbf{x}$ such that $x_1^2 + x_2^2 + x_3^2 = 1$.

- **Topology** in general deals with continuity in a given space or manifold. For example, we find that the continuity of a sphere is clearly different from that of a torus. The former has no holes while the latter has a clear hole.

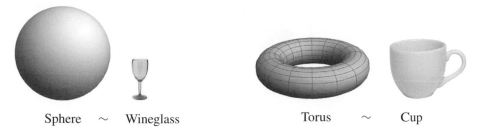

Sphere $\sim$ Wineglass   Torus $\sim$ Cup

This raises the idea of *continuous deformability*, or *homeomorphism*. Assuming our sphere is made of putty, it is then possible to continuously deform it into the form of a wineglass without punching any holes, thus we declare that the wineglass is topologically equivalent, homeomorphic, to the sphere! The same is true if we deform the torus into a cup with a handle. Hence, we introduce an *equivalence* relation ($\sim$) to classify the different types of topological continuity.

Topologically, we may say that although a sphere and a wineglass are homeomorphic manifolds, they have different *embeddings* in $\mathbb{R}^3$, but they belong to the same *homotopy* class.

- **Charts and atlases:** The idea that a manifold M may look, locally, like an $\mathbb{R}^m$ space, allows us to endow a local *open* neighborhood,[1] or patch, $U \subset M$ with an $\mathbb{R}^m$ coordinate system via a homeomorphic map, as shown in Figure 9.1.

$$\begin{cases} \varphi : U \rightarrow U' \in \mathbb{R}^m \\ \varphi : p \rightarrow \mathbf{x} \equiv \{x_1, x_2, \ldots, x_m\}, \quad p \in U, \mathbf{x} \in U' \end{cases}$$

where the mapping φ is referred to as a *coordinate function* or simply *coordinate* in $\mathbb{R}^m$. We design and collect a large number of such local, open, and overlapping U_i neighborhoods, each endowed with a homeomorphic map to $\mathbb{R}^m$, with the proviso that the collection covers the whole manifold

$$\bigcup_i U_i = M$$

Each individual pair (U_i, φ_i) is called a *chart*, and the total collection covering the complete manifold is appropriately coined an *atlas*.

- **Smoothness and differentiability:** We will be considering topological manifolds, which means we have to impose some measure of continuity on our manifolds. Since manifolds are described by overlapping charts

$$U_i \bigcap U_j \neq \emptyset \quad \longrightarrow \quad \begin{cases} \varphi_i : U_i \rightarrow U'_i \subset \mathbb{R}^m \\ \varphi_j : U_j \rightarrow U'_j \subset \mathbb{R}^m, \end{cases}$$

---

[1] An open neighborhood does not include its boundary, but a closed one does. For example, the disk $D = \left\{ (x,y) \middle| x^2 + y^2 \leq a^2 \right\}$ is a closed space with its boundary $\partial D = \left\{ (x,y) \middle| x^2 + y^2 = a^2 \right\}$, but $\tilde{D} = \left\{ (x,y) \middle| x^2 + y^2 < a^2 \right\}$ is an open space. We will refer to it as a patch.

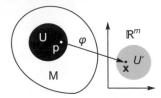

Figure 9.1   A homeomorphism φ maps U onto an open subset $U' \subset \mathbb{R}^m$, providing coordinates to a point $p \in U$.

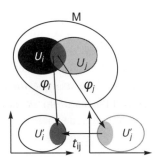

Figure 9.2   If $U_i \bigcap U_j \neq \emptyset$, the smoothness of the transition from one coordinate system to another is guaranteed via a diffeomorphic transition map t_{ij}.

we require that they are related to each other in a sufficiently *smooth* way.

We demand that the transition map t_{ij}, shown in Figure 9.2,

$$t_{ij} = \varphi_i \circ \varphi_j^{-1} \text{ from } \varphi_j \left(U_i \bigcap U_j \right) \text{ to } \varphi_i \left(U_i \bigcap U_j \right),$$

be infinitely differentiable (C^∞), so that a differentiable function in one coordinate system must be differentiable in the other system. We then have a *differentiable* manifold. t_{ij} then defines a *diffeomorphism*[2] between U_i and U_j. Homeomorphisms classify spaces according to whether it is possible to deform one space into another *continuously*. Diffeomorphisms classify spaces into equivalence classes according to whether it is possible to deform one space to another *smoothly*.

The significance of differentiable manifolds resides in the fact that we may use the usual calculus developed in $\mathbb{R}^m$.

- **Metrics and geometry:** In general, there is no naturally defined notion of *length* or *angle* on smooth manifolds; they are not coordinate invariant. Such quantities become clearer when we have an embedding of M into some $\mathbb{R}^n$, $M \hookrightarrow \mathbb{R}^n$ ($m < n$), because it then *inherits* these notions from the Euclidean structure of $\mathbb{R}^n$. However, there is a standard

---

[2] In general, a diffeomorphism is defined as an infinitely differentiable map between a manifold M in $\mathbb{R}^m$ and a manifold N in $\mathbb{R}^n$.

type of structure we can add to any manifold so that these notions are well defined: a *metric structure*, or *metric*, that ushers the starting point of geometry. We are familiar with the Riemannian metric, which defines the distance between two infinitesimally separated points as

$$ds^2 = g_{ij}\,dx_i\,dx_j,$$

where g_{ij} is the Riemannian metric tensor (Einstein summation is invoked).

9.2 Tangent and Cotangent Spaces

Tangent Space

We recall that a manifold M is a sort of a curvy surface in some $\mathbb{R}^m$ space as shown in Figure 9.3, that may not accommodate some straight arrow: a vector that joins two of its points! We may incorporate a vector in a manifold M as a tangent vector at some point $p \in M$.

We may incorporate a vector in a manifold M as a tangent vector at some point $p \in M$ (see Figure 9.4). Roughly speaking, in such a picture, a tangent vector is viewed as an infinitesimal displacement at a specific point on a manifold. However, we should caution that, in general, we have no metric to describe infinitesimal distances.

Figure 9.3 We recall that a manifold M is a sort of a curvy surface in some $\mathbb{R}^m$ space, that may not accommodate some straight arrow: a vector, that joins two of its points!

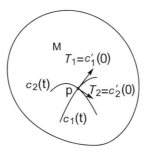

Figure 9.4 Tangent vectors at $p = c_1(0) = c_2(0)$. The prime superscripts indicate derivatives with respect to t.

We refine our definition of the tangent vector by introducing the following:

- A curve $c(t) \subset M$ that passes through p, shown in below left, with $a \leq t \leq b \in \mathbb{R}$
- A function f over an open interval U, such that $p \in c \subset U \subset M$, as shown in below right.

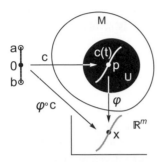   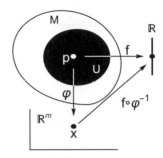

An open *curve* in an m-dimensional manifold M is a map $c : (a,b) \to M$ where $(a,b) \subset \mathbb{R}$ is an open interval with $a < 0 < b$. We assume that the curve does not intersect with itself. c is locally a map from an open interval to M. On a chart (U,φ), a curve $c(t)$ has the coordinate presentation $\mathbf{x} = \varphi \circ c : \mathbb{R} \to \mathbb{R}^m$. $\mathbf{x}(t) = \{x_\alpha(t)\}$, $\alpha = 1, \ldots, m$.

A function f on M is a smooth map from M to $\mathbb{R}$. On a chart (U, φ), the coordinate presentation of f is given by $f \circ \varphi^{-1} : \mathbb{R}^m \to \mathbb{R}$, which is a real-valued function of m variables. We denote the set of smooth functions on M by $\mathcal{F}(M)$.

We define the tangent vector at $c(0)$ as a *directional derivative*[3] of the function $f(c(t))$ along the curve $c(t)$ at $t = 0$ as shown in Figure 9.4.

The rate of change of $f(c(t))$ at $t = 0$ along the curve is

$$\left. \frac{df(c(t))}{dt} \right|_{t=0} = \sum_{\alpha=1}^{m} \frac{\partial f}{\partial x^\alpha} \left. \frac{dx^\alpha(c(t))}{dt} \right|_{t=0} \tag{9.1}$$

$$c(t) \quad \xrightarrow{\varphi} \quad \mathbf{x}(t) \equiv \{x^\alpha(t)\} \in \mathbb{R}^m \qquad \text{Chart coordinates,}$$

where $\dfrac{\partial f}{\partial x^\alpha}$ is short for $\dfrac{\partial(f \circ \varphi^{-1}(x))}{\partial x^\alpha}$.

It is more appropriate to rewrite (9.1) in the form

$$\sum_{\alpha=1}^{m} \left(\left. \frac{dx^\alpha(c(t))}{dt} \right|_{t=0} \frac{\partial}{\partial x^\alpha} \right) f = \left(\sum_{\alpha=1}^{m} X^\alpha \partial_\alpha \right) f = \hat{X}[f], \tag{9.2}$$

---

[3] As we know from vector calculus, given the gradient ∇f of a function f, the directional derivative is defined with respect to a vector $\mathbf{u} = (u^1, u^2, \ldots, u^n)$ as

$$D_\mathbf{u} f = \nabla f \cdot \mathbf{u} = \sum_i u^i \frac{\partial f}{\partial x^i}.$$

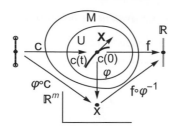

Figure 9.5 A curve c and a function f define a tangent vector along the curve in terms of the directional derivative.

where, now, $\hat{X}$ is a linear differential vector operator, which we define as the tangent vector to M at $p = c(0)$ along the direction given by the curve $c(t)$ (see Figure 9.5).

We note that

$$X^\alpha = \hat{X} x^\alpha = \frac{dx^\alpha}{dt}\bigg|_{t=0}$$

are just numbers $(X^1, X^2, \ldots, X^m)$, the components of $\hat{X}$ with respect to the coordinates $\{x^\alpha\}$.

Since $\hat{X}$ is a vector and the X^α are just numbers, then the operators $\partial_\alpha = (\partial/\partial x^\alpha)\big|_p$ must be vectors that form a basis, in some chart, for the space of all tangent vectors to M at p. Such a basis vector set defines a local *frame* over p.

For example, we consider a curve, $\mathbf{c}(t)$, in $\mathbb{R}^2$ described in polar coordinates by $r = r(t)$, $\theta = \theta(t)$. To calculate the rate of change, we use the chain rule and obtain

$$\dot{\mathbf{c}} = \dot{r}\,\frac{\partial \mathbf{c}}{\partial r} + \dot{\theta}\,\frac{\partial \mathbf{c}}{\partial \theta}$$

$$\frac{\partial \mathbf{c}}{\partial r} = \hat{\mathbf{e}}_r, \qquad \frac{\partial \mathbf{c}}{\partial \theta} = \hat{\mathbf{e}}_\theta,$$

where partial derivatives with respect to r and θ are vectors, and $\dot{r}$, $\dot{\theta}$ are scalar coefficients.

When the differential basis operates on a given function f, it yields real numbers, which together with the X^αs produce the magnitude of the desired directional derivative, so that $\hat{X}[f] \equiv \hat{X} f$ is another scalar function. Thus we have

$$f : M \to \mathbb{R} \quad \Rightarrow \quad \hat{X} f : T_p M \to \mathbb{R}.$$

$T_p M$ denotes the new vector space, called the *tangent space* to M at p. Here, it is a real vector space. This constitutes the proper *intrinsic* definition of the tangent space we seek.

9.2.1 Equivalence Classes of Curves

It is possible that several curves defined over the same open interval may have the same slope d/dt at p. We consider that any two curves giving the same value for the d/dt operation at $t = 0$ to be equivalent.[4] That is, $c_1 \sim c_2$ if

$$\frac{d(f \circ c_1)}{dt}\bigg|_{t=0} = \frac{d(f \circ c_2)}{dt}\bigg|_{t=0}, \quad \forall f \in \mathcal{F}(M).$$

Consequently, a vector X represents an equivalence class of curves $X \leftrightarrow [c]$, all passing through p at $t = 0$ and satisfying the preceding equivalence relation. This equivalence class is also interpreted as a map

$$X : \mathcal{F}(M) \to \mathbb{R} : f \mapsto \frac{d(f \circ c)}{dt}(0)$$

for any c in the equivalence class X. The tangent space at p consists of the set of all such equivalence classes.

We find from (9.2) that the dimension of the tangent space is the same as that of the manifold,

$$\dim T_p M = \dim M.$$

If we are given a chart $\varphi : U \to \mathbb{R}^n$, the tangent space at a point $p \varphi \mathbf{x}$ is the linear space spanned by the coordinate basis vectors $\hat{\mathbf{e}}_\alpha = \partial_\alpha \mathbf{x}$. This is helpful, in particular, if our manifold is defined as a parameterized surface in $\mathbb{R}^n$. For example, for $S^2 \in \mathbb{R}^3$ we consider the parameterization of $\mathbf{x}$ on S^2 by angles θ, ϕ:

$$x_1 = \cos\phi \, \sin\theta, \; x_2 = \sin\phi \, \sin\theta, \; x_3 = \cos\theta.$$

We obtain the basis vectors
$$\begin{cases} \hat{\mathbf{e}}_\theta &= (\cos\phi \, \cos\theta, \; \sin\phi \, \cos\theta, \; -\sin\theta) \\ \hat{\mathbf{e}}_\phi &= (-\sin\phi \, \sin\theta, \; \cos\phi \, \sin\theta, \; 0) \end{cases} \in \mathbb{R}^3.$$

Hence the tangent space $T_x S^2$ at $\mathbf{x} \in S^2$ can be described as the subspace of $\mathbb{R}^3$ spanned by $\hat{\mathbf{e}}_\theta$, $\hat{\mathbf{e}}_\phi$.

As we shall see, the collection of tangent spaces on a manifold forms a *vector bundle* called the *tangent bundle*.

9.2.2 Cotangent Space and One-Forms

$T_p M$ is a vector space defined by the basis set $\partial_\alpha \equiv (\partial/\partial x^\alpha)\big|_p$. We know it should have a dual vector space whose elements are *linear maps*:

$$T_p M \mapsto \mathbb{R}.$$

---

[4] The curve passing through p at $t = 0$ and p_1 at $t = \Delta t$ is free to do anything it wants when we move out of the neighborhood of p. In effect, we only need an infinitesimal segment of the curve to define the action of the d/dt operator. But we cannot talk precisely about an infinitesimal segment of a curve, so we talk about finite curve segments, that is, mappings $c : (a, b) \in M$, $a < 0 < b$, which satisfy $c(0) = p$.

The dual space is called the *cotangent* space at p, denoted by $T_p^* M$. The elements of $T_p^* M$ are called *covectors*, *cotangent vectors*, or, in the context of differential forms, *one-forms*.

We denote the basis vectors dual to the ∂_α as $dx^\alpha(\ )$, such that

$$dx^\alpha(\partial_\beta) = \delta_\beta^\alpha.$$

Thus, when operating on the tangent vector $X = X^\alpha \partial_\alpha$, the basis covectors dx_α return its components

$$dx^\alpha(\hat{X}) = dx^\alpha(X^\beta \partial_\beta) = X^\beta \, dx^\alpha(\partial_\beta) = X^\beta \delta_\beta^\alpha = X^\alpha \in \mathbb{R}.$$

In order to complete the construction of $T_p^* M$, we examine $\hat{X}[f] \equiv \hat{X} f$ from an alternative perspective.

Cotangent vectors naturally appear when we compute directional derivatives of functions. We recall that

$$X[f] = Xf = \sum_\alpha X^\alpha \, \partial_\alpha f,$$

where Xf is another scalar function. This new function gives a number $\in \mathbb{R}$ at each point $p \in M$. But this is exactly the definition of the covector as a linear map $T_p M \mapsto \mathbb{R}$: it takes in a vector at a point and returns a number. Since this operation behaves linearly also in the tangent vector, we can interpret this operation as a linear function on the tangent vector $X \in T_p M$. A common notation is to define $df_p(X) = X[f]$, so that $df_p \in T_p^* M$. The definition can be extended to a linear function on any of the tangent spaces, then we simply remove the subscript and write df.

We introduce the quantity df as a covector field, and define the following operation:

$$df(\hat{X}) = \hat{X} f = \sum_\alpha X^\alpha \frac{\partial f}{\partial x_\alpha}. \tag{9.3}$$

Considering the case $f = x^\beta$, we find

$$dx^\beta(\hat{X}) = \sum_\alpha X^\alpha \frac{\partial x^\beta}{\partial x_\alpha} = X^\beta,$$

which agrees with the result obtained previously. We can then write (9.3) as

$$df(\hat{X}) = \sum_\alpha \frac{\partial f}{\partial x_\alpha} X^\alpha = \sum_\alpha \frac{\partial f}{\partial x_\alpha} dx^\alpha(\hat{X})$$

and express the covector as

$$df = \sum_\alpha \frac{\partial f}{\partial x_\alpha} dx^\alpha.$$

It is called the differential of the scalar f at p. It is an operator instead of the infinitesimal value we are used to in calculus! It has the structure of a *one-form*.

In Lagrangian mechanics, we work with the Lagrangian $L(q; \dot{q})$, and we need to compute directional derivatives with respect to q and $\dot{q}$ respectively. In this case, dL is not a common notation, so we write

$$\frac{\partial L}{\partial q}, \quad \frac{\partial L}{\partial \dot{q}}$$

and they will both be cotangent vectors.

9.3 Fiber Bundles

Having constructed a tangent (cotangent) space at each point in the manifold, the next step is to glue together all these tangent (cotangent) spaces $T_p M$ $(T_p^* M)$ on the manifold. In mathematical terms, we take the disjoint union[5] of all $T_p M$s $(T_p^* M)$

$$TM = \bigcup_{p \in M} T_p M, \qquad T^* M = \bigcup_{p \in M} T_p^* M.$$

The result of such a gluing procedure is not a vector space, rather, it is a *tangent vector bundle*, a specific type of a *vector bundle*. TM has the natural structure of a manifold of dimension[6] $2 \dim M$, induced by the manifold structure of M.

Conversely, by "deconstruction," we define a natural map

$$\pi : TM \to M$$

that sends each vector $\mathbf{v} \in T_p M$ to the point p to which it is attached.

It is called the *projection* onto TM.

Vector bundles are specific instances of the more general notion of *fiber bundles*, where the vector space at each point in the manifold is replaced by a *fiber* over that point.[7]

General Definition of a Fiber Bundle

A fiber bundle is a differentiable manifold consisting of the following:

- The *total manifold* E, $(\dim E = m + n)$. It contains two differentiable sub-manifolds, the *base* B $(\dim B = m)$ and the *fiber* F $(\dim F = n)$.

---

[5] We should emphasize that tangent spaces at different points are, by definition, different vector spaces, which cannot have common elements. Hence, the disjoint union.

[6] A mechanical system with m degrees of freedom is described by generalized coordinates $q_i, i = 1, \ldots, m$ that parametrize its configuration space manifold M. The tangent bundle of M, TM, is the $2m$-dimensional space with $TM_q, q = (q_1, \ldots, q_m)$; if we think of the tangent vector as a velocity, the natural coordinates on TM become $(q_1, q_2, \ldots, q_m; \dot{q}_1, \dot{q}_2, \ldots, \dot{q}_m)$, and these are the variables that appear in the Lagrangian of the system.

[7] The fiber can be any mathematical object, such as a set, tensor space, or another manifold. Mathematicians picture the bundle as an assembly of fibers sprouting out of the manifold, similar to stalks of wheat emerging out of the soil.

- The gluing of the two manifolds is defined by the *projection* map from the total manifold E to the base manifold B:

$$\pi : E \rightarrow B.$$

π cannot be invertible because of the difference in dimensions.
- To every point $p \in B$, the projection associates an inverse image submanifold $F_p \subset E$

$$F_p = \pi^{-1}(p) \simeq F$$

of dimension $\dim n$ containing all points $x \in E$ with projection onto the particular point p

$$\pi(x) = p \in B, \quad \forall x \in F_p$$

called the *fiber over p*.
- Locally, E looks like the Cartesian product

$$\varphi_j : U_j \otimes F$$

of an open patch $U_j \subset B$, ($\dim U = m$), with another manifold F (the *standard fiber*) ($\dim F = n$). φ_j is a diffeomorphism.

Thus, a fiber bundle is defined according to (E, π, B, F).

Roughly speaking, a *fiber bundle* (E) is a manifold that is constructed from two submanifolds: a base manifold (B) and a fiber manifold (F), by attaching a copy of the fiber space to each point of the base space. In its simplest, or *trivial*, form, the bundle (E) is the outer, or Cartesian, product of the base and the fiber

$$E = B \otimes F.$$

In Figure 9.6, we show a simple example of a fiber bundle, where the base and fiber are closed intervals of the real line $\mathbb{R}$.

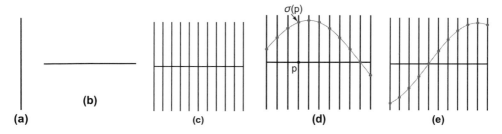

Figure 9.6 A trivial fiber bundle. Attaching a copy of the fiber (a) to each point of the base (b) gives the bundle (c). Picking a point in each fiber space defines a section (d). Another section of the same bundle (e).

If we pick a particular point on each fiber of the bundle, we define a (cross-section) *section*[8] of the bundle, also shown in Figure 9.6.

Bundle Section

A *smooth section* on a *smooth* fiber bundle $(E; \pi; B; F)$ is a *smooth* map $\sigma : B \to E$ such that

$$\sigma(p) \in F_p \equiv \pi^{-1}(p)$$

holds for all $p \in B$. The relation can also be expressed in the form

$$\sigma(p) \circ \pi = Id_B,$$

where Id_B is the identity map.

That is, σ maps each point in the base manifold into a point, or vector, in the fiber connected to that base point.

Vector Field

A smooth section on the tangent bundle TM of a base manifold M, is a mapping of each point in $p \in M$ to a vector in its tangent space $T_p M$. It is called a *smooth vector field* on M. A section of the cotangent bundle is called a *differential one-form*. The set of all smooth vector fields on E is denoted by $\mathfrak{X}(E)$, and the set of differential one-forms by $\Omega^1(E)$.

A vector field may not be defined in all of M (for instance, its domain may be the image of a curve); but when a vector field is defined in all of M, we say that it is defined globally. Otherwise, we say that it is defined only locally.

As an illustration, we consider the example of the circle (one-sphere) $S^1 \hookrightarrow \mathbb{R}^2$- a one-dimensional manifold, defined in $\mathbb{R}^2$ as $\varphi : p \to \mathbf{x} = (\cos \theta, \sin \theta)$ [59].

The tangent vectors at points $p \in S^1$ are defined as $\sigma(p) = (-\sin \theta, \cos \theta) \in \mathbb{R}^2$, forming a one-dimensional vector space $T_p(S^1)$ specified by the angle θ. All those tangent spaces, together with the base manifold S^1, again, form a manifold, as shown in Figure 9.7.

Turning all tangent spaces around by $90°$, as shown in Figure 9.8, we obtain a vector bundle $\mathbb{T}(S^1)$. The fiber in this case is a 1D vector space. We glue a replica of the fiber at each point on the base S^1. A bundle whose fiber is a one-dimensional vector space is called a *line bundle*. An open interval ΔS on S^1 in the neighborhood of a tangent vector space is indicated by a gray arc in the left figure. The ΔS neighborhood can be smoothly mapped into a line-segment $\Delta L \subset \mathbb{R}^1$, such that every point in ΔS can be represented in ΔL. It is then possible to introduce a *topology* in the entire space that is locally equivalent to the product topology of

---

[8] When one slices through a patch of wheat with a scythe, the blade exposes a cross-section of the stalks. By analogy, a choice of an element of the fiber over each point in the manifold is called a cross-section, or, more commonly, a *section* of the bundle. In this language, a tangent-vector field becomes a section of the tangent bundle, and a field of covectors becomes section of the cotangent bundle.

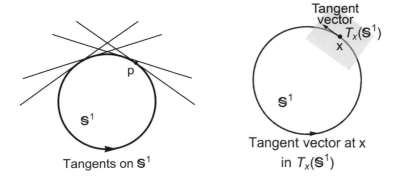

Figure 9.7 Tangents on the one-sphere $\mathbb{S}^1$ and its tangent space.

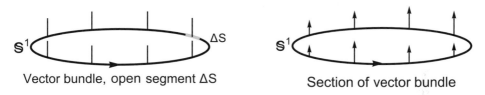

Figure 9.8 Left: vector, or line, bundle $\mathbb{T}(S^1)$ showing an open interval ΔS in gray. Right: a section on the bundle.

$$\Delta E = \Delta S \otimes \mathbb{R}^1.$$

We now set the corresponding fiber as the closed interval $F = D = [-1, 1]$. The fiber bundle $E \equiv C$ then becomes the cylinder shown in Figure 9.9. It should be intuitively clear that the cylinder is not only locally a Cartesian product, but also globally, namely $C = S^1 \otimes D$. As shown in Figure 9.9, it can be flattened into a planar strip without changing lengths and angles.

A fiber bundle will be called *trivial* if it can be described as a global Cartesian product.

The cylinder is trivial in this sense, because it is not very difficult to find a global diffeomorphism from $S^1 \otimes D$ to C.

A more interesting example is obtained when we join the edges of the open cylinder shown in Figure 9.10. But, before gluing the edges together, we perform a twist on one of the edges so that the end $-1 \in [-1; 1]$ is attached to $+1$, and vice versa. We arrive at the Möbius strip, shown on the right in Figure 9.10. The Möbius strip is not a Cartesian product, and is said to be a *twisted* bundle.

The Möbius strip is, however, locally trivial in that for each $p \in S^1$ there is an open retractable neighborhood $\Delta\mathbb{S} \subset S^1$ of p, in which E looks like a product $\Delta\mathbb{S} \otimes D$.

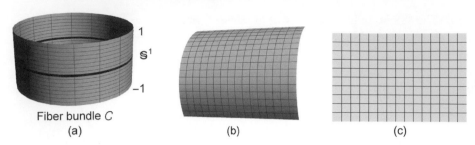

Fiber bundle C
(a)   (b)   (c)

Figure 9.9 (a) Fiber bundle C. (b), (c) A section of a cylinder can be flattened to a plane without changing any lengths or angles on the surface.

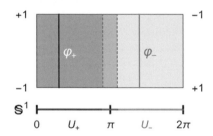

Figure 9.10 Connecting opposite corners of an open cylinder produces a Möbius strip.

Figure 9.11 The Möbius strip is shown with two open base patches U_+ and U_-. The corresponding Cartesian products are shown as overlapping red and blue areas. The fiber mappings φ_+ and φ_- are also shown.

> We declare that local triviality is a feature that *encompasses all bundles*, and use the notion of local maps, the charts, as $U \otimes F$, $\Delta \mathbb{S} \equiv U \subset B$. Again, the collection of all retractable neighborhoods U that span B, $\bigcup U_i$, is the cover of B ($B = \bigcup U_i$), or its atlas. A bundle can then be assembled out of the collection of $E = \bigcup (U_i \otimes F)$ product bundles.

The case of the Möbius strip is shown in Figure 9.11. Locally, along each open subset $U_\pm$ of $\mathbb{S}^1$, the Möbius strip still looks like a product, $U \otimes D$. Globally, however, there is no unambiguous and continuous way to write a point $p \in \mathbb{S}^1$ as a Cartesian pair $(p, x) \in$

$S^1 \otimes D$. The Möbius strip is therefore an example of a manifold that is not a global product, that is, of a nontrivial fiber bundle.

We can identify the family of pairs (U_i, φ_i), where $B = \bigcup_i U_i$, in the context of the base/fiber scenario. In that scenario, φ_i maps U_i onto the fiber, as

$$\varphi_i : \pi^{-1}(U_i) \subset E \rightarrow U_i \otimes F,$$

where the maps are required to be one-to-one, continuous, together with their inverse, and to satisfy the property that

$$\pi \, \varphi_i^{-1}(p, f) = p,$$

where $f \in F$ is a fiber point. In other words, the projection of the image in E of a base manifold point p times some fiber point f is p itself. Each pair (U_i, φ_i) is a *local trivialization*.

9.3.1 Transition Functions and Structure Groups

A very important thing to keep in mind is that even though each fiber is homeomorphic to F by some homeomorphism, there is no natural homeomorphism, and so we cannot naturally identify fibers with each other. This is somewhat unfortunate, but it is also where the interesting features of fiber bundles come from.

Since adjacent *open* local neighborhoods, U_i and U_j, may have different mapping structures onto their fibers – two different local trivializations – we need to define transition rules between them in the overlapping segment $U_i \bigcap U_j \neq \emptyset$. $U_i \bigcap U_j \neq \emptyset$ implies $\pi^{-1}(U_i) \bigcap \pi^{-1}(U_j) \neq \emptyset$, and each point $x \in \pi^{-1}(U_i \bigcap U_j)$ is mapped by φ_i and φ_j in two different pairs, $(p, f_i) \in U_i \otimes F$ and $(p, f_j) \in U_j \otimes F$, for the same $p \in U_i \bigcap U_j$. The transition rule requires the existence of a *transition function*,

$$t_{ij} \equiv \varphi_j^{-1} \circ \varphi_i : \left(U_i \bigcap U_j \right) \otimes F \rightarrow \left(U_i \bigcap U_j \right) \otimes F,$$

which acts exclusively on the fiber points in the sense that

$$t_{ij}(p, f) = (p, t_{ij}(p) \cdot f) \qquad \forall p \in \left(U_i \bigcap U_j \right), \forall f \in F$$

so that for each selected point $p \in U_i \bigcap U_j$,

$$t_{ij}(p) \quad F \mapsto F$$

is a continuous and invertible map of the standard fiber F into itself.

Structure Group, G

The set of all possible continuous invertible maps of the standard fiber F into itself

$$t : \quad F \mapsto F$$

constitutes an algebraic group, the *structure group G*. It makes sense to include G in the definition of fiber bundle, namely, (E, π, B, F, G).

Actually, G satisfies the properties of the continuous Lie groups. We shall confine our consideration of G to Lie groups of finite dimensions or, occasionally, a discrete group having a well-defined action on the standard fiber F.

Example 1: Consider two intersecting local charts (U_i, φ_i) and (U_j, φ_j) of a manifold. A tangent vector at a point $p \in M$ is written as

$$\mathbf{t}_p = X^\alpha(p) \left. \frac{\partial}{\partial x^\alpha} \right|_p.$$

Now we can consider choosing smoothly a tangent vector for each point $p \in M$, namely introducing a map

$$p \in M \mapsto \mathbf{t}_p \in T_p M.$$

Mathematically what we have obtained is a section of the tangent bundle, namely a smooth choice of a point in the fiber for each point of the base. Explicitly this just means that the components $X^\alpha(p)$ of the tangent vector are smooth functions of the base point coordinates x^α. Since we use coordinates, we need an extra label denoting in which local patch the vector components are given

$$\begin{cases} \mathbf{t} = X_i^\alpha \left. \frac{\partial}{\partial x^\alpha} \right|_p & \Rightarrow \quad \text{in chart } i \\ \mathbf{t} = Y_j^\beta \left. \frac{\partial}{\partial y^\beta} \right|_p & \Rightarrow \quad \text{in chart } j \end{cases}$$

Since the tangent vector is the same, irrespective of the coordinates used to describe it, we have

$$Y_j^\beta(y) \frac{\partial}{\partial y^\beta} = X_i^\alpha \frac{\partial y^\beta}{\partial x^\alpha} \frac{\partial}{\partial y^\beta} \qquad \Rightarrow \qquad Y_j^\beta(y) = X_i^\alpha \frac{\partial y^\beta}{\partial x^\alpha},$$

which shows that the explicit form of the transition function between two local trivializations of the tangent bundle is just the inverse Jacobian matrix associated with the transition functions between two local charts of the base manifold M. On the intersection $U_i \bigcap U_j$, we have

$$\forall p \in U_i \bigcap U_j : \quad p \to t_{ji}(p) = \frac{\partial y}{\partial x}(p) \in GL(m, \mathbb{R}),$$

where $GL(m, \mathbb{R})$ is the general linear group of real $m \times m$ matrices.

Example 2: We consider the Möbius strip with four types of intersections, shown in Figure 9.12. We find that the transition functions for the intersections V_+^{12} and V_-^{12} is a multiplication of L by $t_+^{12} = t_-^{12} = +1$, while for $V_\pm^1$, $V_\pm^2$, $t_\pm^1 = t_\pm^2 = -1$. Thus, the structure group for the Möbius bundle is just $\mathbb{Z}_2$!

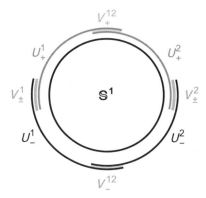

Figure 9.12 Schematic view of four open segments $(U_+^1, U_+^2, U_-^1, U_-^2)$ of $\mathbb{S}^1$, with the intersections $V_\pm^1, V_\pm^2, V_+^{12}$ and V_-^{12} of the open sets.

Principal Bundle

An important type of fiber bundles emerges when the fiber elements are just the elements of the structure group. A fiber bundle whose fibers are isomorphic to G is called a *principal bundle*. It is the most important kind of bundle for understanding the topology of gauge theories.

We note that in the preceding examples that the base and fiber are both segments of the real number line $\mathbb{R}$, a one-dimensional space with a familiar topology. However, in general, the fiber and base can both be arbitrary, multidimensional, and even complex spaces.

9.4 Covariant Derivatives and Commutators of Vector Fields

We now ask what it means to differentiate a smooth vector field with respect to another tangent vector field on the manifold, namely, how to take the corresponding directional derivative. We will show that this is effected with the aid of the *covariant derivative*.

A covariant derivative, or *derivation*, on a smooth vector bundle (E, π, B, F) is an operator

$$\nabla : \mathfrak{X}(TM) \otimes \mathfrak{X}(E) \rightarrow \mathfrak{X}(E)$$

that takes in a smooth tangent vector field X and a smooth vector bundle section σ and yields another smooth vector bundle section $\nabla_X \sigma$. The covariant derivative is required to satisfy the following properties:

1. ∇ must obey the Leibniz (or product) rule with respect to multiplication of the section by smooth functions. If $f \in C^\infty(M)$, then we require

$$\nabla_X(f\sigma) = X(f)\sigma + f\nabla_X\sigma.$$

2. ∇ must be $C^\infty(M)$-linear in the tangent vector field input. That is, for X_1, $X_2 \in \mathfrak{X}(TM)$ and $f, g \in C^\infty(M)$, we have

$$\nabla_{fX_1+gX_2}\sigma = f\nabla_{X_1}\sigma + g\nabla_{X_2}\sigma.$$

3. ∇ must be additive in the smooth section input. That is, for σ_1, $\sigma_2 \in \mathfrak{X}(E)$, we have

$$\nabla_X\left(\sigma_1 + \sigma_2\right) = \nabla_X\sigma_1 + \nabla_X\sigma_2.$$

We now examine the sequential action of two vector fields $X, Y \in \mathfrak{X}(M)$, considered as linear operators, on a function $f \in \mathcal{F}(M)$. For an arbitrary function f, we have

$$\begin{cases} X(Yf) &= X^i\,\partial_i\left(Y^j\,\partial_j f\right) = X^i\,\partial_i Y^j\,\partial_j f + X^i Y^j\,\partial^2_{ij} f \\ Y(Xf) &= Y^i\,\partial_i\left(X^j\,\partial_j f\right) = Y^i\,\partial_i X^j\,\partial_j f + Y^i X^j\,\partial^2_{ij} f \end{cases}$$

We note that the product operation (composition) XY of two vector fields X, Y as operators on functions is no longer a vector field, but is a second-order differential operator.

Our past experience with vector spaces of linear operators would suggest that the defining operation would be their commutator, namely,

$$\left[X, Y\right] = \left(XY - YX\right) f.$$

The commutator of the two vectors fields becomes

$$\left[X, Y\right] = \left(X^i\,\partial_i Y^j - Y^i\,\partial_i X^j\right) \partial_j \in \mathfrak{X},$$

which is just another binary operation on vector fields that satisfies the properties of the *Lie bracket*.[9] Hence, the *significance* of the fact that the commutator $[X, Y]$ is a vector field.

9.5 Connections and Parallel Transport

For an arbitrary manifold, tangent vectors at different points cannot be naturally compared. *There is no natural connection between different tangent spaces.* In this section, we introduce a generic geometric object that allows us to compare different fibers of the bundle and to transport elements from one fiber to another. This object, called a *connection*, plays crucial roles in many branches of physics, including general relativity [67], gauge theories [65, 200], and the theory of geometric phases [33].

---

[9] A lie bracket has the following properties:

(i) It is bilinear $\begin{cases} \left[X, aY + bZ\right] &= a\left[X, Y\right] + b\left[X, Z\right], \\ \left[aX + bY, Z\right] &= a\left[X, Z\right] + b\left[Y, Z\right] \end{cases}$ $\quad X, Y, Z \in \mathfrak{X};\ a, b \in \mathbb{R}.$

(ii) It is skew-symmetric $\left[X, Y\right] = -\left[Y, X\right].$

(iii) It satisfies the Jacobi identity $\left[X, \left[Y, Z\right]\right] + \left[Y, \left[Z, X\right]\right] + \left[Z, \left[X, Y\right]\right] = 0.$

9.5.1 Ehresmann's Definition of a Connection on a Fiber Bundle

We may represent a fiber bundle (E, B, π, F) as a family of fibers F_p whose union is just the total manifold

$$E = \bigcup_{p \in B} F_p.$$

From this perspective, the question arises as to how to connect the disjoint fibers constituting the bundle. The immediate answer is that we need to explore and classify the structure of the tangent space $T_x E = \bigcup_p T_x F_p$ at a point x in the total manifold E.

1. We identify a vector $\mathbf{v} \in T_x E$ as a vertical tangent vector if

$$\mathbf{v} \in T_x F_p \iff p = \pi(x),$$

   that is, $\mathbf{v}$ is vertical at x if it is tangent to the fiber over p.[10] We denote the space of vertical vectors at x by V_x, and call it a *vertical subspace*. $V_x E$ corresponds to motion along fibers, and is essentially fixed, its projection onto TB, the base tangent space, is the zero vector

$$V_x := \left\{ \mathbf{v} \in T_x E \,\middle|\, T_x \pi(\mathbf{v}) = \emptyset \right\}.$$

   We denote the space of vertical vector fields[11] on E by $\mathfrak{X}_{\text{ver}}(E)$, and write

$$\mathbf{v} \in \mathfrak{X}_{\text{ver}}(E) \iff \mathbf{v}(x) \in V_x$$

   for any point $x \in E$.

2. It then follows that we identify the horizontal subspace as the disjoint, transverse, and complementary subspace to V_x, and denote it by H_x. We then write

$$T_x E = V_x \oplus H_x. \tag{9.4}$$

   A vector $\mathbf{v} \in H_x$ is called a horizontal vector. Due to the decomposition (9.4), any vector $\mathbf{v} \in T_x E$ may be uniquely decomposed as

$$\mathbf{v} = \text{hor}\,\mathbf{v} + \text{ver}\,\mathbf{v}.$$

With this structure of $T_x E$ in mind, we can associate a connection of a general (Ehresmann) type with the horizontal tangent space, namely, we identify it with the smooth assignment

$$x \in E \longrightarrow H_x \subset T_x E.$$

Thus, a choice of $H_x E$ is the crucial ingredient in the definition of parallel transport. We require that vectors tangent to E, effecting parallel transport, must lie in $H_x E$.

---

[10] The label "vertical" arises from the viewpoint that neighboring fibers may be perceived as being horizontally stacked. In this picture, a tangent to a given fiber lives on that fiber and cannot link or connect to a neighboring fiber: It is vertical.

[11] Any vector field that lies in $V_x E$ is a vertical vector.

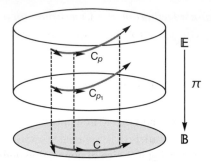

Figure 9.13   A curve $c \in B$ and its horizontal lifts on the fiber bundle.

To demonstrate how this assignment works, we introduce a curve $c \in B$

$$[0,1] \ t \ \rightarrow \ c(t) \subset B$$

and stipulate that a connection should provide us with a rule, or recipe, for parallel-transporting the fiber F along the path c from one end to the other.

Mathematically, we express this action as a map T_c[12]

$$T_c : F_{p_0} \ \rightarrow \ F_{p_1}, \quad p_0 = c(0), \ p_1 = c(1). \tag{9.5}$$

To implement this process, we need to introduce a curve $\mathcal{C}$ in E as

$$[0,1] \ t \ \longrightarrow \ \mathcal{C}(t) \subset E.$$

In particular, we can associate such a curve with c through the projection

$$\pi(\mathcal{C})(t) = c(t).$$

Such a particular curve $\mathcal{C}$ is then called a *lift* of c into E. Of course, in general, there are many lifts, as shown in Figure 9.13. But what we are interested in is a lift associated with parallel transport. To this end, we require that $\mathcal{C}$ be a horizontal lift of c. This means that its directional derivatives $d\mathcal{C}/dt$ have to be horizontal, or that the tangent spaces on $\mathcal{C}$ are horizontal.

We can now operationally define the notion of parallel transport, (9.5) as follows:

---

[12] T_c satisfies the conditions:

1. T_c depends continuously on the path c,
2. $T_{c_1 * c_2} = T_{c_1} \circ T_{c_2}$, namely, if c_1 and c_2 are two paths such that $c_1(1) = c_2(0)$, so that the end of c_1 is the beginning of c_2, then $c_1 * c_2$ is the emerging curve

$$(c_1 * c_2)(t) = \begin{cases} c_1(2t), & 0 \leq c \leq 1/2, \\ c_2(2t - 1), & 1/2 \leq t \leq 1 \end{cases}$$

3. $T_{c^{-1}} = (T_c)^{-1}$, where

$$c^{-1}(t) = c(1 - t).$$

We start with $c(t)$ as a curve in B such that $c(0) = p_0$ and $c(1) = p_1$. We then implement the map T_c by starting a horizontal lift $\mathcal{C}(t)$ of c at $x_0 \in F_{p_0}$, so that $\mathcal{C}(0) = x_0$. Then

$$T_c(x_0) = \mathcal{C}(1) \in F_{p_1}.$$

It is evident that T_c fulfills all the natural requirements of parallel transport. T_c is usually called a map (or an operator) of parallel transport determined by the connection.

Alternatively, we can consider that a choice of connection defines a choice of horizontal subspace: A connection on E is a smooth and unique separation of the tangent space $T_x E$ at each x into a vertical subspace $V_x E$ and a horizontal subspace $H_x E$, with the proviso that the choice of the horizontal subspace at x determines all the horizontal subspaces at points x' in the same fiber. This roughly means that all points above the same point $p = \pi(x) = \pi(x')$ in the base manifold will be parallel-transported in the same way.

> The connection provides a unique recipe to *lift* a base space curve to a bundle curve. More generally, it endows a bundle with a notion of parallelism. Such a map allows points on different fibers to be compared, by using the concept of parallelism provided by the connection.

9.5.2 Connection from the Derivation Perspective

In $\mathbb{R}^n$ space, translation invariance provides a convenient way to identify tangent vectors at different points. We just draw a line connecting the two points and slide a vector along the line while maintaining the relative angle between the vector and the line. However, $\mathbb{R}^n$ presents an unusually simple case of topological spaces. In general, the scenario of transporting a tangent vector from one point to another on a smooth topological manifold, a vector bundle in particular, is more complicated, since comparing vectors at different points x and y on the fibers of a manifold may not be a well-defined process. We need an extra piece of structure to *connect* these fibers in some way, at least if x and y are sufficiently close [95].

We can say that a vector is parallel-transported along an arc of a curve if the angle between the transported vector and the tangent vector to the curve remains constant throughout the entire transport process, as shown in Figure 9.14. We note, however, that in order to compare angles we need to endow the manifold with a metric.

> We introduce the derivation *connection* ∇ as an operation that delineates the rule for how to legitimately move a vector along a curve on the manifold without changing its direction – *keeping the vector parallel*.

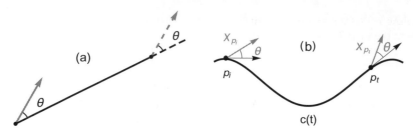

Figure 9.14 (a) Parallel transport in $\mathbb{R}^n$. (b) Parallel transport of a vector along a curve is defined through the preservation of its angle with the tangent vector. This definition is possible if we have a metric and hence the notion of scalar product of local vectors.

9.5.3 Connection on a Manifold

We recall that a tangent vector $X = X^i \, \partial_i$ defines the tangent direction along a curve $c(t)$ on the manifold, and its action on a scalar function gives the magnitude of the corresponding directional derivative of the function. What about the scenario where we replace the scalar function with a vector field $Y(c(t))$ having the diffeomorphic map $Y(c(t)) \underset{\rightarrow}{\varphi} Y(\mathbf{x}(t)) \in \mathbb{R}^m$? We have to somehow carry the vector $Y(p)$ from the tangent space $T_p M$ to the tangent space $T_{p+\epsilon X} M$, where we can subtract it from $Y(p + \epsilon X)$.

We define the connection [49], ∇_X, on M, as a rule to calculate the directional derivatives resulting from the action of X on Y:

$\nabla_X Y$ denotes the *vector field whose value at each point $p \in M$ is equal to the directional derivative of Y along X_p.*

If $\{x^i\}$ is a coordinate system in some neighborhood $U \subset M$, then the vector field $\mathbf{Y}$ can be expressed in the form $\mathbf{Y} = Y^j \, \partial_j$, and we define the connection as

$$\nabla_{\mathbf{X}} \mathbf{Y} = \nabla_{X^i \partial_i} \left(Y^j \partial_j \right) = X^i \, \nabla_{\partial_i} \left(Y^j \partial_j \right) = X^i \left[\frac{\partial Y^j}{\partial x^i} \, \partial_j + Y^j \, \nabla_{\partial_i} \partial_j \right].$$

The covariant derivatives $\nabla_{\partial_i} \partial_j$ must be differentiable vector fields. They are expressed in terms of differentiable real-valued functions Γ^k_{ji} on U, called *connection coefficients*,[a] such that

$$\nabla_{\partial_i} \partial_j = \Gamma^k_{ji} \frac{\partial}{\partial x^k} = \Gamma^k_{ji} \, \partial_k.$$

The connection coefficients specify how the basis ∂_i changes from point to point under infinitesimal displacements of the tangent space.

They can be arbitrarily chosen in any local coordinate chart, noting that different choices define different covariant derivatives. However, there are usually global compatibility constraints that appear when we assemble the charts into an atlas.

---

[a] They are referred to as Christoffel symbols in the context of Riemannian manifolds.

Once the Γ_{ji}^{k}s are defined, we write

$$\nabla_{\mathbf{X}}\mathbf{Y} = X^{i} \left[\frac{\partial Y^{j}}{\partial x^{i}} \, \partial_{j} + Y^{j} \, \Gamma_{ji}^{k} \, \partial_{k} \right] = X^{i} \left[\frac{\partial Y^{k}}{\partial x^{i}} + Y^{j} \, \Gamma_{ji}^{k} \right] \partial_{k}.$$

where we replaced the dummy index j with k in the first term of the last line. It is clear that $\nabla_{\mathbf{X}}\mathbf{Y}$ is the directional derivative of Y along X_{p}. It depends on the following:

- The components $X^{i}\big(x(p)\big)$ of $\mathbf{X}\big(x(p)\big)$.
- The rate of change $\partial_{i} Y^{k}$ of the components of Y. This requires knowledge of the values of the vector field Y in a neighborhood of x at the points of some curve to which X_{x} is tangent.
- The term $Y^{j} \, \Gamma_{ji}^{k}$ can be regarded as Jacobian-like.

When the vector field X represents the tangent vectors to a curve $c(t) \subset M$, namely

$$c(t) : \; t \; \rightarrow \; x^{i}(t)$$

$$X \; \Rightarrow \; X^{i} = \frac{\partial x^{i}}{\partial t},$$

the condition that the vector field $Y(\mathbf{x}(t))$ is parallel-transported along the curve $c(t)$ becomes

$$\nabla_{X} Y = 0, \qquad \Rightarrow \qquad \frac{\partial Y^{k}}{\partial x^{i}} + Y^{j} \, \Gamma_{ji}^{k} = 0$$

at each point $\mathbf{x}(t)$. This means that the directional derivative of the vector Y along the tangent vector fields to the curve $c(t)$ is constant.

Metric Connection on a Riemannian Manifold

So far we have left Γ arbitrary; however, when our manifold is endowed with a metric, it establishes appropriate restrictions on the structure of connections.

Riemann introduced a metric system on an m-dimensional manifold, such that it is treated as an analytic manifold in which each tangent space is equipped with an inner product defined in terms of a smooth function $g = \langle .., .. \rangle$ on the manifold. The inner product of two tangent vectors X and Y, $\langle X, Y \rangle$, gives a real number. The dot, or scalar, product is a typical example of an inner product.[13] The line element is then defined by a symmetric quadratic form in the differentials of the m coordinates

$$ds^{2} = g_{ij}(x) \, dx^{i} \, dx^{j}, \quad i,j = 1, \ldots, m,$$

---

[13] The most familiar example is that of basic high school geometry: the 2D Euclidean metric tensor, in the usual x-y coordinates, reads $g = \begin{bmatrix} 1 & 0 \\ 0 & 1 \end{bmatrix}$. The associated length of a curve is given by the familiar calculus formula: $L = \int_{a}^{b} \sqrt{(dx)^{2} + (dy)^{2}}$.

The result of parallel transport in general depends on the curve, which leads to the notion of (intrinsic) curvature (and torsion). In many applications, it is also important to be able to measure distances and angles with the parallel transport being compatible in the sense that it conserves scalar products.

The unit sphere in $\mathbb{R}^{3}$ comes equipped with a natural metric induced from the ambient Euclidean metric. In standard spherical coordinates (θ, ϕ), the metric takes the form $g = \begin{bmatrix} 1 & 0 \\ 0 & \sin^{2}\theta \end{bmatrix}$, which is usually written as $g = d\theta^{2} + \sin^{2}\theta d\phi^{2}$.

where the $g_{ij}(x)$ became known as the *Riemannian metric tensor*.

As we have previously pointed out, this allows us to define various notions such as length, specifically the length of all curves, angles, areas (volumes), curvature, gradients of functions, and divergence of vector fields.

When two vectors X and Y are parallel-transported along any curve, then their inner product remains constant under parallel transport. We therefore demand that the metric $g_{\mu\nu}$ be covariantly constant. To describe this process, we take $\mathbf{V}$ to be a tangent vector to an arbitrary curve along which the vectors are parallel-transported. Then we have

$$\nabla_{\mathbf{V}}\Big[g(\mathbf{X},\mathbf{Y})\Big] = V^k\Big[(\nabla_k\, g)\,(\mathbf{X},\mathbf{Y}) + g(\nabla_k\mathbf{X},\mathbf{Y}) + g(\mathbf{X},\nabla_k\mathbf{Y})\Big]$$
$$= V^k\, X^i\, Y^j\, (\nabla_k\, g)_{ij} = 0, \tag{9.6}$$

where we used the fact that parallel transport requires $\nabla_k\mathbf{X} = \nabla_k\mathbf{Y} = 0$. Since condition (9.6) is independent of the vector fields and curves, we require that

$$(\nabla_k\, g)_{ij} = 0.$$

We consider the simple example of a 2D Euclidean space $(\mathbb{R}^2, g)$. We define parallel transportation according to the usual sense in elementary geometry. In the Cartesian coordinate system (x, y), all the components of Γ vanish since

$$(V^i_{\text{trans}}(x + \delta x, y + \delta y) = V^i(x, y)$$

for any δx and δy. However, when we use polar coordinates (r, ϕ), with the embedding $(r, \phi) \mapsto (r\cos\phi, r\sin\phi)$, we obtain the induced metric [138],

$$g = d\mathbf{r} \otimes d\mathbf{r} + r^2\, d\boldsymbol{\phi} \otimes d\boldsymbol{\phi}.$$

Now, as shown in Figure 9.15, we parallel-transport the vector field

$$\mathbf{V} = V^r\,\frac{\partial}{\partial r} + V^\phi\,\frac{\partial}{\partial \phi}, \qquad \begin{cases} V^r &= V\cos\theta \\ V^\phi &= V(\sin\theta/r) \end{cases}$$

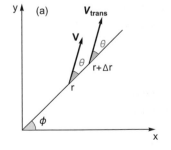
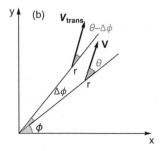

Figure 9.15 V_{trans} is a vector V parallel-transported to (a) $(r + \Delta r, \phi)$ and (b) $(r, \phi + \Delta\phi)$.

$$(r,\phi) \rightarrow (r + \Delta r, \phi): \begin{cases} V^r_{\text{trans}} &= V^r \\ V^\phi_{\text{trans}} &= \dfrac{V}{r + \Delta r} \sin\phi \simeq V^\phi - \dfrac{\Delta r}{r} V^\phi \end{cases}$$

$$(r,\phi) \rightarrow (r, \phi + \Delta\phi): \begin{cases} V^r_{\text{trans}} &= V \cos(\theta - \Delta\phi) \\ &= V \cos\theta + V \sin\theta\, \Delta\phi = V^r + V^\phi r\, \Delta\phi \\ V^\phi_{\text{trans}} &= \dfrac{V \sin(\theta - \Delta\phi)}{r} \\ &\simeq \dfrac{V \sin\theta}{r} - \dfrac{V \cos\theta}{r} \Delta\phi = V^\phi - \dfrac{\Delta\phi}{r} V^r \end{cases},$$

which yields the following connection coefficients

$$\begin{cases} \Gamma^r_{rr} &= 0; \quad \Gamma^r_{r\phi} = 0;\ \Gamma^\phi_{rr} = 0;\ \Gamma^\phi_{r\phi} = \dfrac{1}{r}; \\ \Gamma^r_{\phi\phi} &= -r;\ \Gamma^r_{\phi r} = 0;\ \Gamma^\phi_{\phi\phi} = 0;\ \Gamma^\phi_{\phi r} = \dfrac{1}{r}. \end{cases}$$

Note that Γ satisfies the symmetry $\Gamma^k_{ij} = \Gamma^k_{ji}$. It is also implicitly assumed that the norm of a vector is invariant under parallel transport. A rule of parallel transport that satisfies these two conditions is called a *Levi–Civita* connection.

9.5.4 Curvature and Torsion

The parallel transport map associated with a given connection is an important tool to "detect" the effects of curvature intrinsic in the manifold. Curvature can, in fact, be understood as a measure of the extent to which parallel transport around closed loops fails to preserve the geometrical data being transported. Anytime we have a connection, whether it be in terms of horizontal spaces or something else, we have some notion of curvature.

All fiber bundles are locally trivial, in that they are locally isomorphic to the trivial bundle over the same base space with the same fiber. But *not all connections are locally trivial*. A connection on the given bundle may or may not correspond under the local isomorphism to the trivial connection on the trivial bundle. Curvature is always some sort of mathematical entity that measures how far the given connection is from being trivial. When the curvature is zero, it means the connection, and not just the fiber bundle, is locally trivial. Such a locally trivial connection is called flat.

If $E \rightarrow M$ is a vector bundle and ∇ is a linear connection, it is natural to ask whether covariant derivative operators ∇_X and ∇_Y in different directions commute. We found that, in general, the commutative action of two vector fields on a function $f \in C^\infty(M)$ gives rise to another a third vector field that acts on the function.

Here, we consider the commutative sequential action of connection operations on a vector field V. The geometrical meaning of this relation is depicted in Figure 9.16. It shows an infinitesimally small rectangle whose two sides are given by the two vectors X and Y of

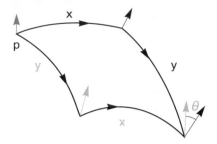

Figure 9.16 Images of the same vector parallel-transported along two different paths that converge at the same point of a manifold generically differ by a rotation angle. That angle is a measure of the intrinsic curvature of the manifold, and it is the information codified in the Riemann–Christoffel tensor.

infinitesimally short length, emanating from a point p. We consider the parallel transport of a third vector V to the diagonally opposite vertex of the rectangle. This parallel transport can be performed along two routes, both arriving at the same destination. In the first route, we displace first along δX and then δY. We follow the reverse sequence in the second route. The image vectors of these two transports are located at the terminal vertex, so they can be compared; in particular, they can be subtracted. Mathematically, we find

$$\nabla_X \nabla_Y V^k - \nabla_Y \nabla_X V^k = X^i Y^j \left[\nabla_i, \nabla_j \right] V^k + \left(\nabla_X Y^l - \nabla_Y X^l \right) \nabla_l V^k$$

$$= X^i Y^j \left(V^l R^k_{ijl} + T^l_{ji} \nabla_l V^k \right) + \left[X, Y \right]^l \nabla_l V^k.$$

T is the *torsion* tensor; it is a skew tensor (a vector) defined as

$$T(X, Y) = \nabla_X Y - \nabla_Y X - \left[X, Y \right],$$

which in a coordinate frame takes the form

$$T^l_{ij} = \Gamma^l_{ij} - \Gamma^l_{ji}.$$

The simplest, but very crude, way to visualize the action of the torsion is in the cross-vectorial product: The action of a cross-product of two vectors, a skew tensor, is a displacement normal to the plane defined by the two vectors; it prevents "the completion of the parallelogram." The torsion of a curve measures how sharply it is twisting out of its plane of curvature. **R** is the *Riemann–Christoffel tensor* that describes the intrinsic curvature of a manifold. Its components are defined in terms of commutative sequential action of connection operations on a vector field V, as

$$\left[\nabla_i, \nabla_j \right] V^k = V^l R^k_{ijl}. \tag{9.7}$$

We note that a connection leads to invariants of *curvature* and the *torsion*. Fortunately, we will be dealing with scenarios where the torsion vanishes.

9.5.5 Parallel Transport and Holonomy

Roughly speaking, a holonomy is a manifestation of a scenario where a system that undergoes a cyclic evolution ends up in a state, or a configuration, that is different from the initial one.

A crisp example is provided in the parallel transport on the unit sphere $S^2 \subset \mathbb{R}^3$, with Riemannian metric, inherited from the Euclidean metric of $\mathbb{R}^3$, and a corresponding Levi–Civita connection on $T S^2 \to S^2$. To demonstrate this process, we set $p_0 \in S^2$ to be a point on the equator, with $v_0 \in T_{p_0} S^2$ pointing along the equator. We follow the triangular path depicted in Figure 9.17, moving $90°$ along the equator, up along a longitude to the north pole, and down along another longitude to the original point p_0. As we move v_0 as a parallel vector field along this path, we see that it remains parallel to the equator on the first leg, then becoming perpendicular to the longitude as it moves toward the north pole, and parallel to the second longitude, moving back down. Parallel transport of any vector along this path has the effect of rotating the initial vector by $90°$, or acquiring a *holonomy* of $\pi/2$. Most importantly, we find that the vector at the end of the path is different from the initial vector v_0. Actually, a nonzero holonomy emerges from parallel transport along any closed path on the round sphere. Detailed analysis of the sphere holonomy will show that it has a constant positive curvature. Quantifying the holonomy for very small paths on curved manifolds gives a precise measure of the local curvature.

In contrast, we consider the case of a Möbius strip. But here we select a vector perpendicular to the surface. Transporting it along the Möbius strip, we find that after one circuit its direction will be reversed. However, we note that it will return to its initial position after a second circuit. We have again produced a *holonomy*. But the two cases are different. In the case of the sphere, the holonomy will depend on the route traversed. The holonomy in the case of curved manifolds can be traced back to the curvature properties of the underlying space, as shown in Figure 9.18. We may say that it detects the geometric properties of that

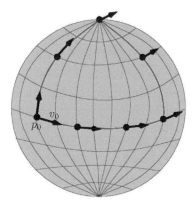

Figure 9.17 Parallel transport of a vector along a closed path in $S^2 \subset R^3$ leads to a different vector upon return.

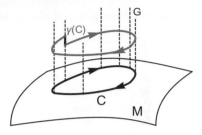

Figure 9.18 Sketch of a fiber bundle, showing a connection and holonomy. The base manifold M is generally some "curved surface." The fibers are denoted by G; the typical fiber is a Lie group, and a copy of the typical fiber is attached to each point of M. The closed curve (circuit) C is defined on M. This circuit is lifted to the bundle space by a connection; the image curve of the lifting process is generally open, as shown, but begins and ends on the same fiber. The difference between the beginning and ending points is the holonomy $\gamma(C)$.

space and can properly be called a *geometric holonomy*. By contrast, we need multiples of a full circuit for the Möbius strip, in which case we obtain two discrete holonomy outcomes: reversed (-1) or nonreversed ($+1$). The holonomy of the Möbius strip detects a topological feature of the underlying space and is therefore a *topological holonomy*. In general, parallel transport is nonintegrable, meaning that it depends on the particular path and does not only depend on initial and final points.

The difference between the sphere and the Möbius strip can be clearly expressed in the language of fiber bundles. In the first scenario, we have a tangent bundle over the sphere, while in the Möbius strip case we have a line bundle over the circle. Both bundles are nontrivial. An important question is whether the bundle has a vanishing or nonvanishing curvature tensor. In the latter case, the connection, the derivative of which gives the curvature, is flat, while in the former case the bundle has a nonflat connection. The tangent bundle over the sphere is a bundle with a nonflat connection. The Möbius strip, however, has a flat connection. Its holonomies are due to the nontrivial topology of the base space.

9.6 Relevance to the Physics of Topological Phases

An appropriate question may arise at this point: why do we need to know all these mathematical concepts? The answer lies in the structure of a quantum mechanical Hilbert space $\mathbb{H}$, which contains all possible state vectors ψ defined in terms of an eigenket basis $\{\psi_n | \mathcal{H}\psi_n = \varepsilon_n \psi_n\}$. $\mathbb{H}$ is a complex vector space $\mathbb{C}^{\mathcal{N}+1} - \{\emptyset\}$ (null vector excluded). As a vector space, it has algebraic properties, such as vector addition and subtraction. It is also endowed with the scalar product or Hermitian metric. Moreover, we find that the subset of all normalized state vectors $\mathbb{H}_N \subset \mathbb{H}$ forms the sphere

$$S^{2\mathcal{N}+1} \subset \mathbb{C}^{\mathcal{N}+1} - \{\emptyset\}, \quad S^{2\mathcal{N}+1} = \left\{ \psi \in \mathbb{C}^{\mathcal{N}+1} \middle| \langle \psi | \psi \rangle = 1 \right\}.$$

As we will show in this section, it is the starting point of constructing a fiber bundle.

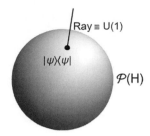

Figure 9.19 Projective space manifold $\mathcal{P}(\mathbb{H}) = S^{2\mathcal{N}+1}/U(1)$ forming a base manifold for the line bundle $E = S^{2\mathcal{N}-1}$ with line fiber $U(1)$.

9.6.1 Geometric Description and Projected Spaces

In quantum mechanics, two normalized state vectors $|\psi\rangle, |\psi'\rangle = e^{i\phi}|\psi\rangle$, $e^{i\phi} \in U(1)$[14] that differ by a phase *represent the same physical state*. Mathematically, this statement translates into an equivalence relation $|\psi'\rangle \sim |\psi\rangle$. Thus, a *physical state* is not represented by a single normalized state vector $|\psi\rangle \in \mathbb{H}$ but by a *ray*, which is an object that contains the vector $|\psi\rangle$ and all its equivalent vectors. More precisely, a ray is the one-dimensional subspace to which the vector $|\psi\rangle$ belongs; it can be understood as the space of the 1D projection operator $|\psi\rangle\langle\psi|$.

Thus, the equivalence relation induces equivalence classes on $S^{2\mathcal{N}+1}$, and, as shown in Figure 9.19, the set of all equivalence classes, rays, forms a subspace of $S^{2\mathcal{N}+1}$, the *projective* space of *physical states* that we denote by

$$\mathcal{P}(\mathbb{H}) = \frac{S^{2\mathcal{N}+1}}{U(1)} = \frac{S^{2\mathcal{N}+1}}{\mathbb{S}^1}.$$

$\mathcal{P}(\mathbb{H})$ is simultaneously a linear space and a complex analytic manifold. It is recognized as physically more basic than $\mathbb{H}$, although the information about the phase is lost. However, we can satisfy both aspects by considering $\mathcal{P}(\mathbb{H})$ as a base manifold, thus maintaining its fundamental nature, and attaching a fiber $F = U(1)$ at each point on the manifold. Such a construction is just a fiber bundle, $S^{2\mathcal{N}+1} \equiv E$ the total space (normalized vectors), the base space $\mathcal{P}(\mathbb{H})$ (physical states), and the projection $\pi \equiv |\psi\rangle\langle\psi|$. E presents a convenient tool for bookkeeping of phases.

The adoption of the bundle picture has several consequences [37]:

- We find that the physics is described within a geometric perspective, where the projective space $\mathcal{P}(\mathbb{H})$ presents a geometrical structure. No algebraic manipulations are allowed, and states cannot be added, as they are no longer elements of a vector space. The physical

---

[14] We can also identify $U(1)$ with the unit circle

$$S^1 = \left\{ z \in \mathbb{C} \,\middle|\, |z| = 1 \right\}$$

in the complex plane $\mathbb{C}$.

states are represented by *points* in $\mathcal{P}(\mathbb{H})$; they are related to each other by unitary transformations (symmetry).

- Consequently, there is no superposition principle: we cannot add rays or points. However, we still can think of the converse process: a state vector has a well-defined projection in any other state via the scalar product of any two normalized vector representatives and its modulus squared. More interestingly, any state has projections onto a complete orthonormal set of rays. In this sense, one speaks more properly of principle of decomposition of states.

- We note that wavefunctions should not be identified as complex-valued functions, but rather as sections in the line bundle. The ray structure allows for *multivaluedness* (phase multiplication). The multivaluedness allows for holonomies, which, as we will see, give rise to phenomena such as the Aharonov–Bohm effect, the Berry phase, etc.

- Despite the projective nature of physical-state space, we can apply linear or antilinear operations on vectors in the bundle, since such operations preserve the fiber integrity. This is the fundamental connection between the geometric character of the states and the algebraic nature of the operations in quantum physics.

Variations on this theme will be presented in the following chapters.

Exercises

9.1 Charts on a manifold:

Consider the unit circle $\mathbb{S}^1 \subset \mathbb{R}^2$ specified by the equation $x^2 + y^2 = 1$ in the plane, as shown in Figure 9.20.

(a) Let $P' = (u,0)$ denote the image of a point $P = (x,y) \in \mathbb{S}^1$ under the stereo-graphic projection from the north pole $N = (0,1)$. Show that $u = \dfrac{x}{1-y}$ for the projection and determine the inverse map.

(b) Express the stereographic coordinate u on $\mathbb{S}^1$ introduced previously in terms of the standard polar angle θ (where $x = r\cos\theta$, $y = r\sin\theta$ for points of $\mathbb{R}^2$).

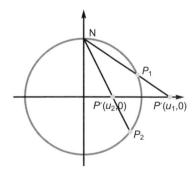

Figure 9.20 Projections from North pole of points on $\mathbb{S}^1$ onto x-axis.

(Remark: an even nicer relation with θ can be obtained if $W = (-1, 0)$ is taken as the center of projection instead of $N = (0, 1)$. You may wish to check that.)

(c) Consider the stereographic projection with center $S = (0, -1)$. Obtain the analogs of the formulas in part (a) denoting the new stereographic coordinate as u_S. (Hint: there is no need to repeat the calculations; you can just change the sign of the y-coordinate.) Rename the coordinate u introduced in (a) as u_N and find the expression of u_N via u_S and conversely.

9.2 Construction of configuration manifolds:

(a) The configuration space of a mechanical system is the set of all possible states of the system. A planar pendulum is a rod of constant length in $\mathbb{R}^2$ with one end (called the center) fixed. Show that the configuration space of a planar pendulum is a manifold and identify it with a familiar example.

(b) The double pendulum is a mechanical system consisting of two pendula linked in such a way that the center of the second pendulum is attached to the moving end of the first pendulum. Show that the configuration space of the double pendulum has a natural structure of a manifold of dimension 2. Identify it with one of the familiar examples.

9.3 Bases for tangent spaces:
Consider the bases $\{e_\theta\}$ and $\{e_u\}$ for the tangent spaces to the circle $\mathbb{S}^1$ corresponding to the polar angle θ and the stereographic coordinate u respectively.

(a) Express the vectors e_θ and e_u in terms of the ambient space $\mathbb{R}^2$. Find the transformation between these bases. (You may express the answer using either a local coordinate or the coordinates in $\mathbb{R}^2$.)

(b) Find the transformation between the bases $\{e_\theta\}$ and $\{e_u\}$ directly from the change of coordinate formula $u = \dfrac{\cos\theta}{1 - \sin\theta}$.

(c) What are the maximal domains of definitions for the bases $\{e_\theta\}$ and $\{e_u\}$?

10

Berry-ology

10.1 Introduction

In this chapter, we need to develop the formalism of *Berry's geometric quantum phase*. It is a manifestation of cyclic, adiabatic time evolution of Hamiltonians $\mathcal{H}\big[\boldsymbol{\eta}(t)\big]$ controlled by a time-dependent, multidimensional parameter $\boldsymbol{\eta}(t)$. For example, the parameter may determine variations in external magnetic and electric fields, or ionic coordinates in the Born–Oppenheimer approximation. It could also be, as we will need later, the crystal momentum $\mathbf{k}$ in a Brillouin zone. In general, $\boldsymbol{\eta}$ is defined on an n-dimensional parameter space $\boldsymbol{\eta}(t) \equiv \{\eta_1(t), \eta_2(t), \ldots, \eta_n(t)\}$. $\mathcal{H}(\boldsymbol{\eta})$ and its energy spectrum $\varepsilon_\mu(\boldsymbol{\eta})$ depend smoothly on the parameter $\boldsymbol{\eta}$.

For simplicity, we consider a quantum-mechanical system with a discrete and nondegenerate energy spectrum, with the proviso that no energy level crossing occurs during time evolution. As we have learned, for such systems the adiabatic theorem in quantum mechanics surmises that for a slowly changing Hamiltonian, the system remains in its evolving, time-dependent ground- or eigen-state. But, as revealed by Michael Berry in 1983, this is actually very incomplete [27, 159]. To understand the rationale behind this assertion, we consider the slow time variation of a Hamiltonian through its dependence on a parameter $\boldsymbol{\eta}(t)$.

We seek solutions for the Schrödinger equation

$$i \frac{\partial \psi(t)}{\partial t} = \mathcal{H}\big[\boldsymbol{\eta}(t)\big] \psi(t), \qquad \psi(t) \in \mathbb{H}. \tag{10.1}$$

For adiabatic time evolution, we write the general solution of the Schrödinger equation as

$$\psi(t) = \sum_\mu c_\mu(t) \, |\mu, \boldsymbol{\eta}(t)\rangle. \tag{10.2}$$

Substituting in (10.1), we get the following equation for the coefficients

$$\dot{c}_\mu(t) = -i \varepsilon_\mu(t) c_\mu(t) - \sum_\nu c_\nu(t) \, \langle \mu, \boldsymbol{\eta}(t)| \, \partial_t \, |\nu, \boldsymbol{\eta}(t)\rangle. \tag{10.3}$$

Thus, if we start with a certain $c_\mu(0) = 1$, $c_\nu(0) = 0$, $\forall \nu \neq \mu$, the adiabatic assumption is approximately fulfilled when $\langle \mu, \boldsymbol{\eta}(t)| \, \partial_t \, |\nu, \boldsymbol{\eta}(t)\rangle$ are small for $\nu \neq \mu$. We can show, by substituting in the eigenvalue equation

$$\mathcal{H}(t) \, |v, \boldsymbol{\eta}(t)\rangle = \varepsilon_v(t) \, |v, \boldsymbol{\eta}(t)\rangle \,,$$

that this matrix element can be expressed as

$$\langle \mu, \boldsymbol{\eta}(t)| \, \partial_t \, |v, \boldsymbol{\eta}(t)\rangle = \frac{\langle \mu, \boldsymbol{\eta}(t)| \, \partial_t \mathcal{H}(t) \, |v, \boldsymbol{\eta}(t)\rangle}{\varepsilon_v(t) - \varepsilon_\mu(t)} \ll 1 \qquad \text{adiabatic assumption.}$$

Under the adiabatic assumption, we can directly solve (10.3) for $c_\mu(t)$ by integration, and obtain

$$c_\mu(t) = e^{-i \int_0^t d\tau \, \varepsilon_\mu(\tau)} \, e^{i \int_0^t d\tau \, i \langle \mu, \boldsymbol{\eta}(\tau)| \partial_\tau |\mu, \boldsymbol{\eta}(\tau)\rangle} = e^{i\alpha_{\text{dyn}}(t)} \, e^{i\gamma(t)} \tag{10.4}$$

$$\psi(t) = e^{i\alpha_{\text{dyn}}(t)} \, e^{i\gamma(t)} \, |\mu, \boldsymbol{\eta}(t)\rangle$$

$$\gamma = \int_0^t d\tau \, i \, \langle \mu, \boldsymbol{\eta}(\tau)| \, \partial_\tau \, |\mu, \boldsymbol{\eta}(\tau)\rangle = \int_0^t d\tau \, i \, \langle \mu, \boldsymbol{\eta}(\tau)| \, \nabla_{\boldsymbol{\eta}} \, |\mu, \boldsymbol{\eta}(\tau)\rangle \cdot \frac{d\boldsymbol{\eta}}{d\tau}$$

$$= \int_{\boldsymbol{\eta}(0)}^{\boldsymbol{\eta}(t)} i \, \langle \mu, \boldsymbol{\eta}| \, \nabla_{\boldsymbol{\eta}} \, |\mu, \boldsymbol{\eta}\rangle \cdot d\boldsymbol{\eta}. \tag{10.5}$$

where the phase α arises from the usual dynamics, but γ is a manifestation of geometric aspects of the system evolution.

What Michael Berry demonstrated was that when the Hamiltonian evolves adiabatically around a closed loop C in $\boldsymbol{\eta}$-space, with $n > 1$, an irreducible geometric phase γ relative to the initial state may emerge, namely,

$$\begin{cases} \gamma & = \oint_C \boldsymbol{\mathcal{A}} \cdot d\boldsymbol{\eta} \qquad \text{mod } 2\pi, \\[2mm] \boldsymbol{\mathcal{A}} & = i\big\langle \psi(\boldsymbol{\eta}) \big| \nabla_{\boldsymbol{\eta}} \big| \psi(\boldsymbol{\eta}) \big\rangle \end{cases} \tag{10.6}$$

γ was coined *Berry's phase* by Barry Simon in 1983 [167], where he revealed the intimate connection between Berry's phase and holonomy on a line bundle.

10.1.1 Berry's Vector Potential, Field, and Flux

We find that the formulation becomes very easy to visualize when $\boldsymbol{\eta}$ is a three-dimensional parameter where ordinary vector calculus can be invoked. Here, we say that for a nondegenerate state, the only freedom in the choice of reference functions is a local phase

$$\psi(\boldsymbol{\eta}) \;\rightarrow\; e^{i\theta(\boldsymbol{\eta})} \, \psi(\boldsymbol{\eta}). \tag{10.7}$$

Under this modification, $\boldsymbol{\mathcal{A}}$ in (10.6) changes by a gradient

$$\boldsymbol{\mathcal{A}} \;\rightarrow\; \boldsymbol{\mathcal{A}} + \nabla_{\boldsymbol{\eta}}\theta,$$

a gauge transformation reminiscent of the vector potential in electrodynamics, and $\boldsymbol{\mathcal{A}}$ is sometimes referred to as *Berry's vector potential*.

So the integrals of $\boldsymbol{\mathcal{A}}$ around closed loops C

$$\oint_C \boldsymbol{\mathcal{A}} \cdot d\boldsymbol{\ell}_{\boldsymbol{\eta}}$$

will be gauge invariant. We can use Stokes, theorem and write

$$\gamma = \oint_C \mathcal{A} \cdot d\boldsymbol{\eta} \quad \mathrm{mod}\ 2\pi = \iint_{S(C)} \mathcal{F} \cdot d\mathbf{S}_{\boldsymbol{\eta}} \quad \mathrm{mod}\ 2\pi, \tag{10.8}$$

where $d\mathbf{S}_{\boldsymbol{\eta}}$ denotes the area element in $\boldsymbol{\eta}$ space, and the integral is performed over any surface bounded by the closed contour C. Again, by analogy with electrodynamics, $\mathcal{F}$ is sometimes called the *Berry field*, and is defined as

$$\mathcal{F} = \nabla_{\boldsymbol{\eta}} \times \mathcal{A}$$

$$\mathcal{F}_{\alpha\beta} = \partial_\alpha \mathcal{A}_\beta - \partial_\beta \mathcal{A}_\alpha = -2\,\Im\left\langle \partial_\alpha\,\psi(\boldsymbol{\eta}) \big| \partial_\beta\,\psi(\boldsymbol{\eta}) \right\rangle. \tag{10.9}$$

Thus, this mathematical structure has evoked analogies with classical electromagnetism. We can envision $\mathcal{A}$ as an abstract *vector potential* in parameter space, and $\mathcal{F}$ as a magnetic field. Following this analogy, we surmise from (10.8) that Berry's phase γ is akin to a flux through the surface **S**. We recall that a vector potential has considerable gauge arbitrariness, and has no physical meaning, while its loop integral is indeed gauge invariant and observable.

The set of phase factors $e^{i\theta(\boldsymbol{\eta})}$ form the group $U(1)$ of unitary 1×1 matrices. We therefore have a gauge theory with gauge (symmetry) group $U(1)$ and gauge potential $\mathcal{A}(\boldsymbol{\eta})$.

The loop integral of the Berry potential, the Berry phase γ, is nontrivial in two cases:

- Curl $\mathcal{A}$ is nonzero.
- The curl is zero but the curve C is not in a simply connected domain.

We can invoke a generalized Stokes' theorem in any dimension of $\boldsymbol{\eta}$ parameter space, when C is the boundary of a surface $\mathbf{S}(C)$ in a simply connected domain, and the curl of $\mathcal{A}$ is regular on $\mathbf{S}(C)$.

10.2 $\mathcal{A}$ as a Berry Connection

Now, we shall recast $\mathcal{A}$ as a topological geometric structure. We first note that the unitarity of time evolution preserves the scalar product, allowing the family of Hamiltonians $\mathcal{H}(\boldsymbol{\eta})$ to share the Hilbert space $\mathbb{H}$, and normalized state vectors to remain normalized and confined to the sphere $\mathbb{S}^{2\mathcal{N}+1}$. Consequently, the projected physical-state manifold $\mathbb{P}(\mathbb{H})$ continues to be the base manifold of the principal line bundle $\mathbb{E}$, with the fiber $\mathbb{F}$ being the unitary group $U(1)$. $\mathbb{P}(\mathbb{H})$ is homeomorphic with $\boldsymbol{\eta}$.

We still focus on the spectrum of *discrete, isolated,* and *nondegenerate* eigenvalues $\varepsilon_\mu(\boldsymbol{\eta})$ with corresponding normalized eigenkets $\psi_\mu(\boldsymbol{\eta})$.[1] We recall that the set of equivalent normalized eigenkets $e^{i\phi}\,\psi(\boldsymbol{\eta}) = \psi(\phi,\boldsymbol{\eta})$ form a ray, and share the projection $\boldsymbol{\pi} = |\psi(\boldsymbol{\eta})\rangle\langle\psi(\boldsymbol{\eta})| \in \mathbb{P}(\mathbb{H})$. The adiabatic time dependence of $\mathcal{H}$ amounts to traversing a

---

[1] To simplify the notation, we will drop the subscript μ.

curve $t \rightarrow \eta(t)$ in η at adiabatically slow velocity. We also note that a cyclic evolution, $\eta(0) = \eta(T)$, is uniquely associated with a loop $\mathcal{C}(\eta)$ in η, or in $\mathbb{P}(\mathbb{H})$.

From another perspective, we can envision that time evolution produces a path in the total space $\mathbb{E}$. The corresponding path in the space of physical states $\mathbb{P}(\mathbb{H}) \simeq \eta$ is obtained by a π projection of the path in $\mathbb{E}$ onto the base manifold. In the cyclic evolutions we are interested in, we find that the system terminates in the initial physical state, and the paths are represented by closed curves in η. Closed paths in η may correspond to closed or open paths in $\mathbb{E}$. In a cyclic evolution, an open path in $\mathbb{E}$ means that a state vector terminates on the same initial fiber, but it lands on another state vector that differs from the initial state vector by an overall phase.

$U(1)$ $\mathcal{A}$ Transition Functions and Gauge Transformations

We note that, in general, each wavefunction $\psi(\eta)$ can be chosen to be smooth only over patches of η, but not necessarily over the whole of η. In other words, the set of state vectors $\psi(\phi, \eta)$ form a smooth section over each patch $\mathbb{U} \subset \eta$. For two overlapping patches $\mathbb{U}_1 \cap \mathbb{U}_2 \neq \emptyset$, with sets of smooth functions $\psi^{(1)}(\eta)$ and $\psi^{(2)}(\eta)$, respectively, we know that for any $\eta \in \mathbb{U}_1 \cap \mathbb{U}_2$ the corresponding functions share the same ray above η, and can differ only by a complex phase, $\theta(\eta)$, $e^{i\theta(\eta)} \in U(1)$, namely,

$$\psi^{(2)}(\eta) = e^{i\theta(\eta)}\, \psi^{(1)}(\eta),$$

which is just a *gauge transformation*. Both $\psi^{(1)}(\eta)$ and $\psi^{(2)}(\eta)$ constitute a possible set of instantaneous solutions, and the freedom in choice of either one or another manifests the *gauge freedom*. The corresponding group that is used to formulate the condition of the gauge freedom is called the *gauge group*. In our case, the gauge group is $U(1)$.

10.2.1 Construction of the Berry Connection

Before constructing the connection [34], which is associated with the *horizontal subspace*, we shall identify the vertical subspace, or the *vertical* direction. We recall that the fibers are generated by the action of the group $U(1)$ on the physical states of the base manifold $\mathbb{P}(\eta)$

$$\psi(\phi, \eta) = e^{i\phi}\, \psi(\eta), \qquad \psi(\eta) \in \mathbb{P}(\eta).$$

This means that the normalized states of a fiber differ by a phase from each other, and, thus, point in the same direction – the vertical direction generated by the $U(1)$ action.

Since the line bundle is embedded in a complex vector space, a natural way to construct the connection (a horizontal subspace) is to use the corresponding metric, the scalar product. We can decompose the tangent vectors

$$\frac{d\psi\big(\phi(t), \eta(t)\big)}{dt} = \partial_t \psi\big(\phi(t), \eta(t)\big) \in TE$$

to the curve $\psi\big(\phi(t),\eta(t)\big) \in \mathbb{E}$, with the aid of the scalar product, into vertical and horizontal components, as

$$\frac{d\psi\big(\phi(t),\eta(t)\big)}{dt} = \big\langle\psi\big(\phi(t),\eta(t)\big)\,\big|\,\partial_t\psi\big(\phi(t),\eta(t)\big)\big\rangle\,\big|\psi\big(\phi(t),\eta(t)\big)\big\rangle$$

$$+ \big|h_\psi(t)\big\rangle \ (\text{horizontal component}) \tag{10.10}$$

$$\big\langle\psi\big(\phi(t),\eta(t)\big)\big|h_\psi(t)\big\rangle = 0, \qquad \text{parallel transport} \tag{10.11}$$

To evaluate the connection explicitly, we need to consider a local patch $\mathbb{U} \subset \eta$ and the region of $\mathbb{E}$ over $\mathbb{U}$. Since tangent vectors in TE are produced by the operator d/dt, we split it into vertical and horizontal operators as

$$\frac{d}{dt} = a_v\,\frac{\partial}{\partial\phi} + \sum_\alpha b_h^\alpha\,D_\alpha = a_v\,\frac{\partial}{\partial\phi} + \sum_\alpha b_h^\alpha\left[\frac{\partial}{\partial\eta^\alpha} + \mathcal{A}_\alpha\,\frac{\partial}{\partial\phi}\right],$$

where the first term corresponds to the vertical subspace, since it clearly depends only on variations in the phase ϕ. We therefore write

$$\big\langle\psi\big(\phi(t),\eta(t)\big)\,\big|\,\partial_t\psi\big(\phi(t),\eta(t)\big)\big\rangle\,\big|\psi\big(\phi(t),\eta(t)\big)\big\rangle = a_v\,\frac{\partial}{\partial\phi}\,\big|\psi\big(\phi(t),\eta(t)\big)\big\rangle$$

$$\big|h_\psi\big\rangle = \sum_\alpha b_h^\alpha\,D_\alpha\,\big|\psi\big(\phi(t),\eta(t)\big)\big\rangle$$

and, according to (10.11), we obtain

$$\Big\langle\psi\big(\phi(t),\eta(t)\big)\Big|b_h^\alpha\,D_\alpha\Big|\psi\big(\phi(t),\eta(t)\big)\Big\rangle = 0.$$

Since b_h^α is arbitrary, we find that

$$\Big\langle\psi\big(\phi(t),\eta(t)\big)\Big|\frac{\partial}{\partial\eta^\alpha} + \mathcal{A}_\alpha\,\frac{\partial}{\partial\phi}\Big|\psi\big(\phi(t),\eta(t)\big)\Big\rangle = 0.$$

Using the fact that $\big|\psi\big(\phi(t),\eta(t)\big)\big\rangle$ is normalized, and that

$$\frac{\partial}{\partial\phi}\,\big|\psi\big(\phi(t),\eta(t)\big)\big\rangle = \frac{\partial}{\partial\phi}\,e^{i\phi}\,\psi(\eta) = i\,\big|\psi\big(\phi(t),\eta(t)\big)\big\rangle,$$

we obtain

$$\mathcal{A}_\alpha\,\Big\langle\psi\big(\phi(t),\eta(t)\big)\Big|\frac{\partial}{\partial\phi}\Big|\psi\big(\phi(t),\eta(t)\big)\Big\rangle = i\,\mathcal{A}_\alpha\,\big\langle\psi\big(\phi(t),\eta(t)\big)\,\big|\,\psi\big(\phi(t),\eta(t)\big)\big\rangle$$

$$= -\Big\langle\psi\big(\phi(t),\eta(t)\big)\Big|\frac{\partial}{\partial\eta^\alpha}\Big|\psi\big(\phi(t),\eta(t)\big)\Big\rangle$$

or

$$\mathcal{A}_\alpha = i\,\Big\langle\psi\big(\phi(t),\eta(t)\big)\Big|\frac{\partial}{\partial\eta^\alpha}\Big|\psi\big(\phi(t),\eta(t)\big)\Big\rangle. \tag{10.12}$$

The connection $\mathcal{A}$ is then given by the one-form

$$\mathcal{A} = \sum \mathcal{A}_\alpha \, d\eta^\alpha = i \, \langle \psi\big(\phi(t), \eta(t)\big) | \, d \, |\psi\big(\phi(t), \eta(t)\big)\rangle$$

defined on the line bundle in terms of the covariant derivative ∇_η. Thus, $\mathcal{A}$ is also referred to as the *Berry connection*.[2]

With the construction of the connection in hand, we proceed to define the horizontal lift. The horizontal lift involves lifting the tangent vectors to the curve in η up along the fibers and orienting them such that they are horizontal. Consequently, for $|\psi(\phi, \eta)\rangle$ to be a horizontal lift, the vertical part of $|\partial_t \psi(\phi, \eta)\rangle$ must vanish

$$\langle \psi(\phi, \eta) \, | \partial_t \psi(\phi, \eta)\rangle = 0. \tag{10.13}$$

This condition is called the *horizontal lift* equation or the equation for *parallel transport*. We have invoked the notion of parallel transport in the line bundle $\mathbb{E}$ in a manner completely analogous to the way an ordinary covariant derivative on a smooth manifold defines parallel transport in the tangent bundle of the smooth base. As we will show, the twisting of this line bundle affects the phase of quantum mechanical wave functions.

In bundle language, $|\psi(\phi, \eta)\rangle$ can be considered as a local section of the fiber bundle, namely, a continuous mapping of a patch $\mathbb{U} \subset \eta$ into the fibers above $\mathbb{U}$. A change to a different patch $\mathbb{U}' \subset \eta$ corresponds to a change in the section

$$|\psi\big(\phi(\eta), \eta\big)\rangle \rightarrow |\psi\big(\phi'(\eta), \eta\big)\rangle$$

$$|\psi\big(\phi'(\eta), \eta\big)\rangle = e^{i\theta(\eta)} \, |\psi(\phi, \eta)\rangle$$

This means that the connection transforms as

$$\mathcal{A}'_\alpha = \mathcal{A}_\alpha - \frac{\partial \theta(\eta)}{\partial \eta^\alpha}, \tag{10.14}$$

a gauge transformation.

We note that a local section, being a section over a single patch, maps a closed path in η, with $\eta(T) = \eta(0)$, $0 \le t \le T$, into a closed path in $\mathbb{E}$. We will denote the closed path in $\mathbb{E}$ as $|\psi\big(\eta(t)\big)\rangle$, with $|\psi\big(\eta(T)\big)\rangle = |\psi\big(\eta(0)\big)\rangle$. A gauge transformation gives a different closed path in $\mathbb{E}$,

$$|\psi_\theta\big(\eta(t)\big)\rangle = e^{i\theta(t)} \, |\psi\big(\eta(t)\big)\rangle.$$

In order for $|\psi_\theta\big(\eta(t)\big)\rangle$ to be closed, the function θ must satisfy

$$\theta(T) = \theta(0) + 2n\pi \tag{10.15}$$

with n integer.

---

[2] We should note that ∇_η is not a quantum operator in Hilbert space, and thus $\mathcal{A}$ is not a physical observable.

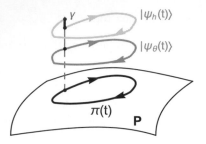

Figure 10.1 Horizontal lift of path $\pi(t)$.

10.2.2 Holonomy and Berry's Phase

We will now evaluate the holonomy produced by the horizontal lift of a closed curve in $\boldsymbol{\eta}$ with respect to the connection we previously derived (see Figure 10.1). We denote the horizontal lift by $\left|\psi_h\big(\boldsymbol{\eta}(t)\big)\right\rangle$. By definition, the tangent vectors to the curve $\left|\psi_h\big(\boldsymbol{\eta}(t)\big)\right\rangle$ must be horizontal. According to (10.10), this means

$$\left\langle\psi_h(\phi,\boldsymbol{\eta})\,\big|\partial_t\psi_h\big(\boldsymbol{\eta}(t)\big)\right\rangle = 0. \tag{10.16}$$

We can express the open path $\left|\psi_h\big(\boldsymbol{\eta}(t)\big)\right\rangle$ in $\mathbb{E}$ in terms of a closed path $\left|\psi\big(\phi(t),\boldsymbol{\eta}(t)\big)\right\rangle$ in $\mathbb{E}$ as

$$\left|\psi_h\big(\boldsymbol{\eta}(t)\big)\right\rangle = e^{i\xi(t)}\left|\psi\big(\phi(t),\boldsymbol{\eta}(t)\big)\right\rangle \tag{10.17}$$

$$\left|\psi\big(\phi(T),\boldsymbol{\eta}(T)\big)\right\rangle = \left|\psi\big(\phi(0),\boldsymbol{\eta}(0)\big)\right\rangle, \quad \phi(T) = \phi(0) + 2n\pi$$

$$\left|\psi_h\big(\boldsymbol{\eta}(T)\big)\right\rangle = e^{i\left[\xi(T)-\xi(0)\right]}\left|\psi_h\big(\boldsymbol{\eta}(0)\big)\right\rangle.$$

Setting $\gamma = \xi(T) - \xi(0)$, substituting (10.17) into (10.16) and integrating, we obtain

$$\left\langle\psi_h(\phi,\boldsymbol{\eta})\,\big|\partial_t\psi_h\big(\boldsymbol{\eta}(t)\big)\right\rangle = i\dot\xi(t)\left\langle\psi\big(\phi(t),\boldsymbol{\eta}(t)\big)\,\big|\psi\big(\phi(t),\boldsymbol{\eta}(t)\big)\right\rangle$$

$$+\left\langle\psi\big(\phi(t),\boldsymbol{\eta}(t)\big)\,\big|\dot\psi\big(\phi(t),\boldsymbol{\eta}(t)\big)\right\rangle$$

$$\dot\xi(t) = i\left\langle\psi\big(\phi(t),\boldsymbol{\eta}(t)\big)\,\big|\dot\psi\big(\phi(t),\boldsymbol{\eta}(t)\big)\right\rangle$$

$$\gamma = i\int_0^T dt\,\left\langle\psi\big(\phi,\boldsymbol{\eta}(t)\big)\big|\,\partial_t\,\psi\big(\phi,\boldsymbol{\eta}(t)\big)\right\rangle.$$

The tangent vector $\left|\dot\psi\big(\phi(t),\boldsymbol{\eta}(t)\big)\right\rangle$ is given by

$$\frac{d}{dt}\left|\psi\big(\phi(t),\boldsymbol{\eta}(t)\big)\right\rangle = \dot\phi\,\frac{\partial\left|\psi\big(\phi(t),\boldsymbol{\eta}(t)\big)\right\rangle}{\partial\phi} + \sum_\alpha\dot\eta^\alpha\,\frac{\partial\left|\psi\big(\phi,\boldsymbol{\eta}(t)\big)\right\rangle}{\partial\eta^\alpha}$$

and we write

$$
\gamma = i \int_0^T dt \, \dot{\phi} \left\langle \psi(\phi, \boldsymbol{\eta}(t)) \left| \frac{\partial}{\partial \phi} \right| \psi(\phi, \boldsymbol{\eta}(t)) \right\rangle
$$

$$
+ i \int_0^T dt \, \dot{\eta}^\mu \left\langle \psi(\phi, \boldsymbol{\eta}(t)) \left| \frac{\partial}{\partial \eta^\mu} \right| \psi(\phi, \boldsymbol{\eta}(t)) \right\rangle
$$

$$
= - \int_{\phi(0)}^{\phi(T)} d\phi + i \int_{\eta^\mu(0)}^{\eta^\mu(T)} d\eta^\mu \left\langle \psi(\boldsymbol{\eta}) \left| \frac{\partial}{\partial \eta^\mu} \right| \psi(\boldsymbol{\eta}) \right\rangle
$$

$$
= -2n\pi + \oint_C \langle \psi(\boldsymbol{\eta}) |d| \psi(\boldsymbol{\eta}) \rangle = -2n\pi + \oint_C \mathcal{A} = \gamma(\mathcal{C}), \qquad (10.18)
$$

where $\mathcal{C}$ is the closed path in $\boldsymbol{\eta}$. $\langle \psi(\boldsymbol{\eta}) |d| \psi(\boldsymbol{\eta}) \rangle$ is purely imaginary, since for normalized states $|\psi(\boldsymbol{\eta})\rangle$

$$
\frac{\partial}{\partial \eta^\mu} \langle \psi(\boldsymbol{\eta}) | \psi(\boldsymbol{\eta}) \rangle = 0 = \left\langle \psi(\boldsymbol{\eta}) \left| \frac{\partial}{\partial \eta^\mu} \, \psi(\boldsymbol{\eta}) \right\rangle + \left\langle \frac{\partial}{\partial \eta^\mu} \psi(\boldsymbol{\eta}) \right| \psi(\boldsymbol{\eta}) \right\rangle
$$

$$
= 2\mathrm{Re} \left\langle \psi(\boldsymbol{\eta}) \left| \frac{\partial}{\partial \eta^\mu} \, \psi(\boldsymbol{\eta}) \right\rangle.
$$

10.2.3 The Berry Curvature

If the base $\boldsymbol{\eta}$ is simply connected, then expression (10.18) for the Berry phase can be rewritten as a surface integral of the local curvature form. Using Stokes' theorem, we write

$$
\gamma(\mathcal{C}) = \int_{S(\mathcal{C})} \mathcal{F} \cdot d\mathbf{S}, \qquad (10.19)
$$

where $S(\mathcal{C})$ is an arbitrary submanifold, of dimension $n-1$, bounded by $\mathcal{C}$, namely, $\delta \mathbf{S} = \mathcal{C}$, and

$$
\mathcal{F} = d\mathcal{A} = -\Im \langle d\psi(\boldsymbol{\eta}) | d\psi(\boldsymbol{\eta}) \rangle \qquad \text{two-form} \qquad (10.20)
$$

is the *Berry curvature*. In local coordinates on $\boldsymbol{\eta}$ it is given by the two-form

$$
\mathcal{F} = \frac{1}{2} \mathcal{F}_{ij} \, d\eta_i \wedge d\eta_j
$$

$$
\mathcal{F}_{ij} = -\Im \left(\langle \partial_i \psi | \partial_j \psi \rangle - \langle \partial_j \psi | \partial_i \psi \rangle \right). \qquad (10.21)
$$

If the base $\boldsymbol{\eta}$ is not simply connected, then the expression for the Berry phase as surface integral (10.19) is valid only for those curves that are contractible to a point.

Berry's Curvature in Terms of Hamiltonian Derivatives: Hidden Physics

With

$$
\mathcal{F}_n(\boldsymbol{\eta}) = -\Im \langle d\psi_n(\boldsymbol{\eta}) | d \, \psi_n(\boldsymbol{\eta}) \rangle
$$

we insert the identity operator, $\mathbb{I} = \sum_m |\psi_m\rangle \langle \psi_m|$, and we obtain

$$\mathcal{F}_n(\eta) = -\Im \sum_m \langle d\psi_n(\eta) | \psi_m(\eta) \rangle \wedge \langle \psi_m(\eta) | d \psi_n(\eta) \rangle$$

$$= -\Im \sum_{m \neq n} \langle d\psi_n(\eta) | \psi_m(\eta) \rangle \wedge \langle \psi_m(\eta) | d \psi_n(\eta) \rangle, \qquad (10.22)$$

where the sum excludes $m = n$ since $\langle \psi_n(\eta) | d \psi_n(\eta) \rangle$ is purely imaginary. An important consequence of (10.22) is that the sum of Berry curvatures of all the eigenstates of the Hamiltonian is zero. If we consider Hamiltonians with discrete spectra along a closed path C, and sum over Berry's phases of all eigenstates, we obtain

$$\sum_n \mathcal{F}_n(\eta) = 0, \qquad (10.23)$$

which follows from (10.22), since

$$\mathcal{F}_n^{ij}(\eta) = -\Im \sum_{m \neq n} \left(\langle \partial_i \psi_n(\eta) | \psi_m(\eta) \rangle \langle \psi_m(\eta) | \partial_j \psi_n(\eta) \rangle - (i \rightleftharpoons j) \right)$$

$$= -\frac{1}{2} \Im \sum_{m \neq n} \left(\langle \partial_i \psi_n(\eta) | \psi_m(\eta) \rangle \langle \psi_m(\eta) | \partial_j \psi_n(\eta) \rangle \right.$$

$$\left. + \langle \partial_j \psi_m(\eta) | \psi_n(\eta) \rangle \langle \psi_n(\eta) | \partial_i \psi_m(\eta) \rangle - (i \rightleftharpoons j) \right)$$

since $\langle \partial_i \psi_n(\eta) | \psi_m(\eta) \rangle = - \langle \psi_n(\eta) | \partial_i \psi_m(\eta) \rangle$. It is clear then that

$$\sum_n \mathcal{F}_n^{ij}(\eta) = 0$$

since the sum is both symmetric and antisymmetric in i and j.

Applying the exterior derivative d to the eigenvalue equation

$$\mathcal{H}(\eta) |\psi_n(\eta)\rangle = E_n(\eta) |\psi_n(\eta)\rangle$$

yields

$$d\,\mathcal{H}(\eta) |\psi_n(\eta)\rangle + \mathcal{H}(\eta)\, d\, |\psi_n(\eta)\rangle = d\, E_n(\eta) |\psi_n(\eta)\rangle + E_n(\eta)\, d\, |\psi_n(\eta)\rangle$$

and projecting onto $\langle \psi_m(\eta)|$, we get

$$\langle \psi_m(\eta) | d\,\mathcal{H}(\eta) |\psi_n(\eta)\rangle = (E_n(\eta) - E_m(\eta)) \langle \psi_m(\eta) | d\, \psi_n(\eta) \rangle$$

or

$$\langle \psi_m(\eta) | d\, \psi_n(\eta) \rangle = \frac{\langle \psi_m(\eta) | d\,\mathcal{H}(\eta) |\psi_n(\eta)\rangle}{E_n(\eta) - E_m(\eta)}, \quad m \neq n$$

and

$$\mathcal{F}_n(\eta) = -\Im \sum_{m \neq n} \frac{\langle \psi_n(\eta) | d\,\mathcal{H}(\eta) |\psi_m(\eta)\rangle \wedge \langle \psi_m(\eta) | d\,\mathcal{H}(\eta) |\psi_n(\eta)\rangle}{\left(E_n(\eta) - E_m(\eta)\right)^2}. \qquad (10.24)$$

We note that $\mathcal{F}_n$ becomes infinite at some $\eta = \eta^*$, if $E_n(\eta^*) = E_m(\eta^*)$, a δ-function like singularity that heralds the presence of a fictitious field source. We discern that the source is static; for example, it may manifest the nature of a magnetic monopole rather than currents.

We can surmise that wherever the ground state is degenerate, the curvature becomes ill defined and singular. In such a situation, the vicinity of the singularity must be excluded, which leads to a domain that is not simply connected.

Finally, (10.24) has the advantage that it is free of wavefunction differentiation, therefore it can be evaluated under any gauge choice. This property is particularly useful for numerical calculations, in which the condition of a smooth phase choice of the eigenstates is not guaranteed in standard diagonalization procedures.

Isolated Degeneracies and Diabolical Points

We consider a quantum system parameterized by $\eta = \{\eta_1, \ldots, \eta_n\}$, and assume that for some point η^* two energy eigenvectors $\psi_1(\eta^*)$ and $\psi_2(\eta^*)$ are degenerate, with energy $E^* = E(\eta^*)$. Near the degeneracy point η^*, we may approximate the system as a two-level one, and write its Hamiltonian as

$$\mathcal{H} = \begin{pmatrix} \mathcal{H}_{11}(\eta) & \mathcal{H}_{12}(\eta) \\ \mathcal{H}_{12}^*(\eta) & \mathcal{H}_{22}(\eta) \end{pmatrix}$$

with eigenvalues

$$E_\pm = \frac{\mathcal{H}_{11} + \mathcal{H}_{22}}{2} \pm \sqrt{\left(\frac{\mathcal{H}_{11} - \mathcal{H}_{22}}{2}\right)^2 + |\mathcal{H}_{12}|^2}.$$

At η^*,

$$(\mathcal{H}_{11} - \mathcal{H}_{22})^2 + 4|\mathcal{H}_{12}|^2 = 0, \quad \mathcal{H}_{11} = \mathcal{H}_{22}, \quad \Re\,\mathcal{H}_{12} = \Im\,\mathcal{H}_{12} = 0.$$

These three constraints define an $(n-3)$-dimensional submanifold in η. If $n = 3$, then the degeneracy defines an isolated point in η.

For a system with time-reversal symmetry, the Hamiltonian $\mathcal{H}$ is real and $\Im\mathcal{H}_{12} = 0$. Hence, the subspace of degenerate points defines an $(n-2)$-dimensional submanifold of η. This behavior was recognized by von Neumann and Wigner [189]. Note that if we introduce the three parameters

$$u = \frac{1}{2}\,(\mathcal{H}_{11} - \mathcal{H}_{22}), \quad v = \Re\,\mathcal{H}_{12}, \quad w = \Im\,\mathcal{H}_{12},$$

then the eigenenergies in the vicinity of the degeneracy take a double-cone profile, as shown in Figure 10.2, with its apex at the degeneracy point, namely

$$E = E^* \pm \sqrt{u^2 + v^2 + w^2}.$$

Figure 10.2 Left: Conical intersection near a degeneracy point, showing the diabolical point. Right: A diabolo toy

The double cone is called a *diabolo* (after a spinning toy of the same shape). The apex of this cone is called a *diabolical point*.[3]

10.2.4 Berry's Quantized Flux and Topological Invariant

The flux of Berry's curvature through a surface **S** remains a bona fide physical quantity on closed surfaces, such as a sphere or a torus, in which case the closed contour C becomes the empty set. The remarkable feature in this case is that the flux is quantized. For pedagogical reasons, we will consider the simple case of a sphere in three dimensions, $\mathbf{S} \equiv \mathbb{S}^2$. For regular and divergence-free curvature on the closed surface $\mathbb{S}^2$,

$$\int_{\mathbb{S}^2} \mathcal{F}(\boldsymbol{\eta}) \cdot d\mathbf{S}$$

represents the flux of $\mathcal{F}$ across $\mathbb{S}^2$. The flux quantization is then expressed as

$$\mathcal{C}_1 = \frac{1}{2\pi} \int_{\mathbb{S}^2} \mathcal{F}(\boldsymbol{\eta}) \cdot d\mathbf{S}, \tag{10.25}$$

where $\mathcal{C}_1$ is an integer $\in \mathbb{Z}$, called the Chern number of the first class.

To demonstrate that $\mathcal{C}_1 \in \mathbb{Z}$, we shall assume that $\mathcal{F}(\boldsymbol{\eta})$ is singular at $\boldsymbol{\eta} = 0$, and that $\mathbb{S}^2$ is the spherical surface centered at the origin (see Figure 10.3).

We cut this surface at the equator, $\eta_z = 0$, and consider the flux across the two open surfaces

$$\int_{\mathbb{S}^2} \mathcal{F}(\boldsymbol{\eta}) \cdot d\mathbf{S} = \int_{S_+} \mathcal{F}(\boldsymbol{\eta}) \cdot d\mathbf{S} + \int_{S_-} \mathcal{F}(\boldsymbol{\eta}) \cdot d\mathbf{S}$$

We notice that $C_+ = C_- = C$, but the surface normals $\hat{\mathbf{n}}$ have opposite orientations. From Stokes' theorem, we get

$$\int_{S_\pm} \mathcal{F}(\boldsymbol{\eta}) \cdot d\mathbf{S} = \pm \int_C \mathcal{A}_\pm(\boldsymbol{\eta}) \cdot d\boldsymbol{\eta}$$

$$\int_{\mathbb{S}^2} \mathcal{F}(\boldsymbol{\eta}) \cdot d\mathbf{S} = \int_C \mathcal{A}_+(\boldsymbol{\eta}) \cdot d\boldsymbol{\eta} - \int_C \mathcal{A}_-(\boldsymbol{\eta}) \cdot d\boldsymbol{\eta}. \tag{10.26}$$

---

[3] A discussion of the importance of diabolical points in molecular physics is given in [28].

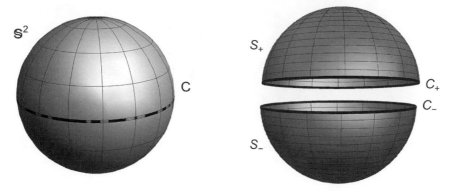

Figure 10.3 Sphere cut at the equator in two hemispheres.

The upper and lower Berry connections $\mathcal{A}_{\pm}(\eta)$ may only differ by a gauge transformation. Since the right-hand side (rhs) of (10.26) is the difference of two Berry phases on the same path, it is necessarily an integer multiple of 2π.

The Chern number turns out to be a robust topological invariant, and is associated with observable phenomena, such as the integer quantum Hall effect. More recently, other topological invariants have emerged, such as the $\mathbb{Z}_2$, which classifies time-reversal symmetric insulators into disjoint *trivial* and *topological* types.

As we have outlined, the integrand $\mathcal{F}$ is the curl of the Berry connection $\mathcal{A}$, and, in general, $\mathcal{A}$ may not be a single-valued function globally on $\mathbb{S}^2$, but only on local patches, as will be described in example 1.

10.3 Pedagogical Example 1: A Two-Level System

Pedagogically, studying this system accomplishes two goals. First, it is a simple model that demonstrates the basic concepts outlined earlier in this chapter, and reveals several important properties of the Berry phase. Second, it will emerge, in different physical guises, in the following chapters.

The generic Hamiltonian of a two-level system takes the form

$$\mathcal{H} = -\mathbf{d} \cdot \boldsymbol{\sigma} \,,$$

where $\boldsymbol{\sigma}$ represents the Pauli matrices and $\mathbf{d} \in \mathbb{R}^3$ vector space.[4]

We parametrize $\mathbf{d}$ by its polar and azimuthal angles θ and φ as

$$\mathbf{d} = d\,\hat{\mathbf{n}}, \qquad \hat{\mathbf{n}} = (\sin\theta\,\cos\varphi,\ \sin\theta\,\sin\varphi,\ \cos\theta)\,.$$

---

[4] The simplest example is the motion of a neutral spin-1/2 particle in a magnetic field $\mathbf{B}$, where we have the familiar Hamiltonian

$$\mathcal{H} = -\mu\mathbf{B}\cdot\boldsymbol{\sigma}\,,$$

where μ is the particle magnetic moment.

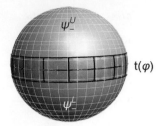

Figure 10.4 Two patches introduced by the gauge choice for the ground state $|\psi_-(\mathbf{d})\rangle$. No single patch can describe the full domain $\mathbb{S}^2$. However, the two patches can be glued together via the transition function $t(\varphi)$.

The two eigenstates, with energies $\pm d$, are

$$|\psi_+(\mathbf{d})\rangle = \begin{pmatrix} \cos(\theta/2) \\ e^{i\varphi}\sin(\theta/2) \end{pmatrix} \qquad |\psi_-(\mathbf{d})\rangle = \begin{pmatrix} -\sin(\theta/2) \\ e^{i\varphi}\cos(\theta/2) \end{pmatrix}.$$

We notice that the norm d does not affect the eigenvectors. Therefore, the parameter space is a two-sphere $\mathbb{S}^2$.

10.3.1 Patches and Transition Functions

We now choose two different gauges to write the ground state $|\psi_-(\mathbf{d})\rangle$:

$$\left|\psi_-^U\right\rangle = \begin{pmatrix} -\sin(\theta/2) \\ e^{i\varphi}\cos(\theta/2) \end{pmatrix} \quad \text{or} \quad \left|\psi_-^L\right\rangle = \begin{pmatrix} -e^{-i\varphi}\sin(\theta/2) \\ \cos(\theta/2) \end{pmatrix}.$$

Both states $\left|\psi_-^U\right\rangle$ and $\left|\psi_-^L\right\rangle$ describe the ground state. However, $\left|\psi_-^U\right\rangle$ is ill defined (singular) at $\theta = \pi$ (it displays a vortex at the south-pole $\Rightarrow$ value of φ ill defined), while $\left|\psi_-^L\right\rangle$ is singular at the north-pole. Thus, we have to introduce two patches on the sphere, shown in Figure 10.4, to obtain a smooth parameterization of the instantaneous eigenstates. Other gauges yield the singularity at a different point on the sphere, but a singularity is unavoidable. It is impossible to find a gauge that yields a smooth, nonsingular connection over the whole closed surface. Such singularity, often called *obstruction*, can be *moved* but *not removed*.[5]

However, we can glue these two patches together via a gluing phase

$$\left|\psi_-^U\right\rangle = e^{it(\varphi)}\left|\psi_-^L\right\rangle \qquad t(\varphi) = \varphi$$

along the equator.

---

[5] The algebra is the same as for Dirac's theory of the magnetic monopole: the degeneracy at the origin is the monopole, and the singularity is the *Dirac string* [52].

10.3.2 Berry Connection, Curvature, and Phase

We now calculate the Berry connection and curvature for ψ_-^U and ψ_-^L [6]

$$\nabla \left| \psi_-^U \right\rangle = -\frac{1}{d} \begin{pmatrix} \frac{1}{2} \cos \frac{\theta}{2} \\ \frac{1}{2} e^{i\phi} \sin \frac{\theta}{2} \end{pmatrix} \hat{\boldsymbol{\vartheta}} + \frac{1}{d \sin \theta} \begin{pmatrix} 0 \\ i e^{i\phi} \cos \frac{\theta}{2} \end{pmatrix} \hat{\boldsymbol{\varphi}}$$

Hence

$$\left\langle \psi_-^U \mid \nabla \psi_-^U \right\rangle = \frac{1}{2d} \left[\sin \frac{\theta}{2} \cos \frac{\theta}{2} \hat{\boldsymbol{\theta}} - \sin \frac{\theta}{2} \cos \frac{\theta}{2} \hat{\boldsymbol{\theta}} + 2i \frac{\cos^2(\theta/2)}{\sin \theta} \hat{\boldsymbol{\varphi}} \right]$$

$$= i \frac{2\cos^2(\theta/2)}{d \sin \theta} \hat{\boldsymbol{\varphi}} = i \frac{1}{d} \cot \frac{\theta}{2} \hat{\boldsymbol{\varphi}}$$

and

$$\nabla \times \left\langle \psi_-^U \mid \nabla \psi_-^U \right\rangle = \frac{1}{d \sin \theta} \frac{\partial}{\partial \theta} \left[\sin \theta \frac{i \cos^2(\theta/2)}{d \sin \theta} \right] \hat{\mathbf{n}} = -i \frac{1}{2d^2} \hat{\mathbf{n}}.$$

We obtain

$$\mathcal{A} = \frac{1}{d} \hat{\boldsymbol{\varphi}} \times \begin{cases} \cot(\vartheta/2) & \left| \psi_-^U \right\rangle \quad \text{singular at } \theta = 0, \\ \tan(\theta/2) & \left| \psi_-^L \right\rangle \quad \text{singular at } \theta = \pi, \end{cases}$$

$$\mathcal{F} = \frac{1}{2d^2} \hat{\mathbf{n}}. \tag{10.27}$$

The curvature is gauge invariant, while the connection is gauge dependent.

The curvature for the upper state $|\psi_+(\mathbf{d})\rangle$ is $\mathcal{F} = -\frac{1}{2d^2} \hat{\mathbf{n}}$. Thus, parallel transport of the eigenstates $|\psi_-(\mathbf{d})\rangle$ and $|\psi_+(\mathbf{d})\rangle$ over the same loop in parameter space *rotates* them in opposite directions (Berry phases have opposite signs). They exhibit a *twist akin to the Möbius strip*.

One- and Two-Form Approach

Alternatively, we can use the one-form expression for the connection

$$\mathcal{A} = i \left\langle \psi \mid d\psi \right\rangle$$

$$= i \left\langle \psi \mid \partial_\theta \psi \right\rangle d\theta + i \left\langle \psi \mid \partial_\varphi \psi \right\rangle d\varphi$$

and

$$\frac{\partial \psi_-}{\partial \theta} = - \begin{bmatrix} (\cos \theta)/2 \\ e^{i\varphi} (\sin \theta)/2 \end{bmatrix}, \quad \frac{\partial \psi_-}{\partial \varphi} = \begin{bmatrix} 0 \\ i e^{i\varphi} (\cos \theta)/2 \end{bmatrix}$$

$$\frac{\partial \psi_+}{\partial \theta} = \begin{bmatrix} -(\sin \theta)/2 \\ e^{i\varphi} (\cos \theta)/2 \end{bmatrix}, \quad \frac{\partial \psi_+}{\partial \varphi} = \begin{bmatrix} 0 \\ i e^{i\varphi} (\sin \theta)/2 \end{bmatrix}$$

---

[6] Recall, in spherical coordinates we have $\nabla = \frac{\partial}{\partial r} \hat{\mathbf{r}} + \frac{1}{r} \frac{\partial}{\partial \vartheta} \hat{\boldsymbol{\vartheta}} + \frac{1}{r \sin \vartheta} \frac{\partial}{\partial \varphi} \hat{\boldsymbol{\varphi}}$ (here $r \equiv |\mathbf{d}| = d$)

$$\nabla \times \mathbf{A} = \frac{1}{r \sin \vartheta} \left\{ \frac{\partial}{\partial \vartheta} \left(A_\varphi \sin \vartheta \right) - \frac{\partial A_\vartheta}{\partial \varphi} \right\} \hat{\mathbf{r}} + \frac{1}{r} \left[\frac{1}{\sin \vartheta} \frac{\partial A_r}{\partial \varphi} - \frac{\partial}{\partial r} \left(r A_\varphi \right) \right] \hat{\boldsymbol{\vartheta}}$$
$$+ \frac{1}{r} \left[\frac{\partial}{\partial r} \left(r A_\vartheta \right) - \frac{\partial A_r}{\partial \vartheta} \right] \hat{\boldsymbol{\varphi}}.$$

to obtain
$$\mathcal{A}_+ = \langle \psi_+ \,|d\psi_+\rangle = -\frac{1}{2}\,(1 - \cos\theta)\,d\varphi$$

$$\mathcal{A}_- = \langle \psi_- \,|d\psi_-\rangle = -\frac{1}{2}\,(1 + \cos\theta)\,d\varphi.$$

The corresponding Berry curvature is

$$\mathcal{F}_{\theta\varphi} = \left(\partial_\theta\,\mathcal{A}_\varphi - \partial_\varphi\,\mathcal{A}_\theta\right)\,d\theta \wedge d\varphi$$

$$\mathcal{F}^+_{\theta\varphi} = -\frac{1}{2}\,\sin\theta\,d\theta \wedge d\varphi$$

$$\mathcal{F}^-_{\theta\varphi} = \frac{1}{2}\,\sin\theta\,d\theta \wedge d\varphi$$

$$\mathcal{F}^+_{\theta\varphi} + \mathcal{F}^-_{\theta\varphi} = 0$$

and the Chern numbers are

$$\mathcal{C}_\pm = \mp\frac{1}{2\pi}\int_{\mathbb{S}^2} d\theta\,\sin\theta\,d\phi\,\frac{1}{2} = \mp 1.$$

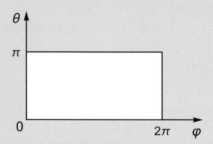

In the integration of the Berry curvature over the parameters (θ, φ), we did not include an additional $\sin\theta$ factor that is required for a sphere. This is because we were not integrating over the surface of a sphere, but over a parameter domain shown in the figure. Here, the parameter space has the topology of a sphere $\mathbb{S}^2$, but not its metric.

However, we should also note that

$$\mathcal{F}^\pm_{\theta\varphi} = \mp\frac{1}{2}\,\sin\theta\,d\theta \wedge d\varphi = \mp\frac{1}{2d^2}\,(d\,d\theta)(d\,\sin\theta d\varphi)\,\hat{\mathbf{d}}.$$

10.3.3 The Monopole: Degeneracy, Singularity, and Obstruction

If we write the curvature as $\mathcal{F} = \dfrac{g}{2d^2}\,\hat{\mathbf{n}}$, we find that it has the form of a magnetic monopole field of strength g at the origin, namely,

$$\frac{1}{2\pi}\int_{\mathbb{S}^2}\mathcal{F}\cdot d\mathbf{S} = g.$$

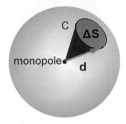

Figure 10.5 Monopole singularity.

The corresponding Berry potential $\mathcal{A}$ also shows monopole behavior at $\mathbf{d} \equiv \mathbf{B} = 0$. As expected, the vanishing of the Hamiltonian at $d = 0$ ushers a doubly degenerate state – a singularity (see Figure 10.5).

We note that integrating over the solid angle $\Delta\Omega$ subtended by $\Delta\mathbf{S}$ yields a Berry phase $\gamma = g\,\Delta\Omega/2$ proportional to half the subtended solid angle.

In fact, g cannot take arbitrary values: we calculate the Berry phase along a path C that does not contain the south pole

$$\gamma(C) = \oint_C \mathcal{A}\cdot d\boldsymbol{\ell} = \int_{\Sigma(C)} \mathcal{F}\cdot d\mathbf{S} = \frac{g}{2}\Omega(\Sigma) \quad \mathrm{mod}\ 2\pi,$$

where $\Omega(\Sigma)$ is the solid angle subtended by the surface Σ that contains the north pole. If we now choose the surface $\tilde{\Sigma} = \mathbb{S}^2 - \Sigma$

$$\gamma(C) = -\int_{\tilde{\Sigma}(C)} \mathcal{F}\cdot d\mathbf{S} = -\frac{g}{2}\left(4\pi - \Omega(\Sigma)\right) \quad \mathrm{mod}\ 2\pi = \frac{g}{2}\,\Omega(\Sigma) \quad \mathrm{mod}\ 2\pi,$$

we find that $g \in \mathbb{Z}$. This integer measures the strength of the singularity (magnetic monopole), which resides in a site inaccessible to the quantum system. With this simple yet pedagogical procedure, we can ascertain that a nonvanishing Chern number $\mathcal{C}$ is intrinsically linked to the inability to choose a smooth gauge. In other words, only if we have to choose several patches that we glue together with a gauge transformation can $\mathcal{C}$ be nonzero.

10.4 Pedagogical Example 2: Molecular Aharonov–Bohm Effect

In this example, we consider the quantum motion of the combined electronic and ionic components of a molecule. Here, the slow motion of the ions presents an adiabatic time evolution of the electronic motion and allows the application of the Born–Oppenheimer approximation (BOA).

10.4.1 The Effective Ionic Hamiltonian

We start from the complete Hamiltonian $\mathcal{H}$ of an isolated $\mathcal{N}$-atom molecular system, and explicitly separate the ionic kinetic energy

$$\mathcal{H}(\mathbf{X},[\mathbf{x}]) = \frac{1}{2} \sum_{j=1}^{\mathcal{N}} \frac{\mathbf{P}_j^2}{2M_j} + \mathcal{H}_e(\mathbf{X},[\mathbf{x}]) + E_{\text{elastic}}(\mathbf{X}), \qquad (10.28)$$

where $[\mathbf{x}]$ denotes the electronic degrees of freedom, and $\mathbf{X} = [\mathbf{X}_1, \ldots, \mathbf{X}_j, \ldots, \mathbf{X}_{\mathcal{N}}]$ represents the ionic positions, with $\mathbf{X}_j$ the position vector of ion j. $\mathbf{P}_j = -i\hbar \nabla_{\mathbf{X}_j}$ is the canonical momentum conjugate to $\mathbf{X}_j$, and M_j the mass of ion j. E_{elastic} is the ionic elastic energy for the configuration $\mathbf{X}$.

We invoke the BOA by initially ignoring the ionic kinetic energy, and applying the adiabatic approximation, where the coupling between different electronic states can be neglected. Thus, the slowly varying $3\mathcal{N}$-dimensional ionic position vector $\mathbf{X}$ in $\mathcal{H}(\mathbf{X},[\mathbf{x}])$ is demoted to a classical parameter and identified with the slow parameter η. Moreover, we use the BOA ansatz and write the eigenfunctions of (10.28), $\left| \Psi(\mathbf{X},[\mathbf{x}]) \right\rangle$, as the product $\left\langle [\mathbf{x}] \middle| \xi(\mathbf{X}) \right\rangle \chi(\mathbf{X})$, where ξ and χ denote the electronic and ionic wavefunction, respectively. We start with solving the electronic eigenvalue problem

$$\mathcal{H}_e(\mathbf{X},[\mathbf{x}]) \left| \xi_n(\mathbf{X},[\mathbf{x}]) \right\rangle = E_n(\mathbf{X}) \left| \xi_n(\mathbf{X},[\mathbf{x}]) \right\rangle$$

and determine the ground-state energy $E_0(\mathbf{X})$ and wavefunction $\left| \xi_0(\mathbf{X},[\mathbf{x}]) \right\rangle$.

Our goal is then to construct an effective Schrödinger equation for the ionic wavefunction $\chi(\mathbf{X})$ by integrating out the electronic degrees of freedom. We first consider the action of the canonical ionic momentum $\mathbf{P}$ on the product ansatz

$$\mathbf{P} \left| \xi(\mathbf{X},[\mathbf{x}]) \right\rangle \chi(\mathbf{X}) = -i\hbar \left| \xi(\mathbf{X},[\mathbf{x}]) \right\rangle \nabla_{\mathbf{X}} \chi(\mathbf{X}) - i\hbar \left| \nabla_{\mathbf{X}} \xi(\mathbf{X},[\mathbf{x}]) \right\rangle \chi(\mathbf{X}).$$

We integrate over the electronic degrees of freedom by acting on both sides with $\left\langle \xi(\mathbf{X},[\mathbf{x}]) \right|$. We obtain the effective ionic kinematic momentum Π acting on $\chi(\mathbf{X})$

$$\Pi \chi(\mathbf{X}) = \left[\mathbf{P} - i\hbar \left\langle \xi(\mathbf{X},[\mathbf{x}]) \middle| \nabla_{\mathbf{X}} \xi(\mathbf{X},[\mathbf{x}]) \right\rangle \right] \chi(\mathbf{X}), \qquad (10.29)$$

where Berry's connection is clearly recognizable.

The electronic eigenvalue $E_0(\mathbf{X})$ of the ground state plays the role of a scalar potential for ionic motion. The effective Hamiltonian acting on $\chi(\mathbf{X})$ is given by

$$\mathcal{H}_{\text{eff}} = \frac{1}{2} \sum_j \frac{1}{M_j} \Pi_j^2 + E_0(\mathbf{X}) + E_{\text{elastic}}(\mathbf{X}). \qquad (10.30)$$

10.4.2 A Simple Example: The Trimer Molecule

The smallest molecular system that exhibits *molecular Aharonov–Bohm effect* is a trimer, with ionic coordinates $\mathbf{X} = \left(\mathbf{X}_1, \mathbf{X}_2, \mathbf{X}_3 \right)$. The simplest trimers, which we consider here, are of course the homonuclear ones, where symmetry plays a major role.

Figure 10.6 A homonuclear trimer in its equilateral configuration.

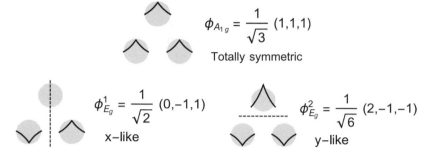

Figure 10.7 The electronic ground state of the trimer with $\mathcal{D}_{3h}$ consists of an A_{1g} totally symmetric state and a degenerate E_g doublet transforming like (x, y).

Structure of the Ground State

We focus on a trimer of monovalent atoms, such as H_3 or Na_3, where the occupied valence orbitals, $|A\rangle$, $|B\rangle$, and $|C\rangle$, are s-like.

We start with the molecule in the equilateral configuration of Figure 10.6, with $\mathcal{D}_{3h}$ symmetry.[7] Two of the valence electrons occupy a totally symmetric orbital (A_{1g} symmetry)

$$\left|\phi_{A_{1g}}\right\rangle = \frac{1}{\sqrt{3}}\left[|A\rangle + |B\rangle + |C\rangle\right].$$

The remaining unpaired electron occupies the next available orbital, which has E_g symmetry and is doubly degenerate.[8] In a simple tight-binding scheme, a possible basis is the two-dimensional manifold

$$\left|\phi_{E_g}^1\right\rangle = \frac{1}{\sqrt{2}}\left[|B\rangle - |C\rangle\right], \qquad \left|\phi_{E_g}^2\right\rangle = \frac{1}{\sqrt{6}}\left[2\,|A\rangle - |B\rangle - |C\rangle\right].$$

These orbitals are shown in Figure 10.7. The fact that the Hamiltonian is time reversal invariant and the wavefunctions vanish at infinity guarantees that the orbitals may always be chosen as real.

When we distort the molecule from its equilateral configuration, the doublet is linearly split, and one of the two E_g components becomes energetically favored. The molecule is

---

[7] The symmetry group, $\mathcal{D}_{3h}$, is the dihedral group of an equilateral triangle. It has a mirror symmetry through the plane of the triangle.

[8] The symbol A denotes a nondegenerate symmetry-adapted function, while E denotes a doubly degenerate one; g and u denote even and odd behavior, respectively, with respect to mirror reflections through the plane containing the trimer.

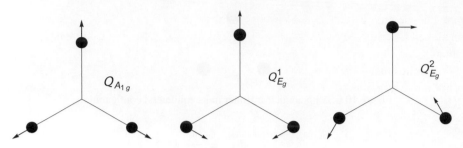

Figure 10.8 A schematic representation of the three possible normal vibration modes of a trimer, compatible with $\mathcal{D}_{3h}$ symmetry. Note that they are all planar.

said to undergo a *Jahn–Teller* distortion, where the electronic ground state in the BOA becomes nondegenerate.

The $E \otimes \varepsilon$ Jahn–Teller Problem

We shall start our study of this effect with describing the motion of the ions under $\mathcal{D}_{3h}$ symmetry. We find three symmetry-adapted (normal) modes for small oscillations of the trimer molecule, shown in Figure 10.8:

(i) A *breathing mode* of A_{1g} symmetry. It is totally symmetric and cannot remove the electronic doublet degeneracy

(ii) An E_g doubly degenerate mode. It can lift the electronic degeneracy, and is compatible with coupling to the electronic doublet and engendering invariant terms in the Hamiltonian, linear in the symmetry-adapted displacements.

The vibrational doublet, in fact, drives the dynamical Jahn–Teller effect[9].

The ionic motion lifts the electronic degeneracy. This fulfills the assertion of the *Jahn–Teller theorem* that *in the presence of electronic degeneracy, there should be distorted ionic configurations of lower energy than the symmetric one.*

Solving the One-Electron Problem

Within the Born–Oppenheimer procedure, we write

$$\Psi([\mathbf{x}], \mathbf{X}) = \chi(\mathbf{X}) \, \xi([\mathbf{x}], \mathbf{X}))$$

and we initially neglect the ionic kinetic and elastic energies in $\mathcal{H}$. The remaining component is the electronic Hamiltonian having the form

$$\mathcal{H}_e(\mathbf{X}, \mathbf{x}) = T_e + V(\mathbf{X}, \mathbf{x})$$

---

[9] The conventional notation $E \otimes \varepsilon$ given for this process means that an E vibrational mode is coupled to an E electronic state. Conventionally, the upper-case letters are used as symmetry labels for the vibrational states, and Greek lower-case ones for the electronic states.

for the one-electron case. T_e is the electronic kinetic energy. The electron–ion potential V is expressed in terms of the "parametric" ionic positions $\mathbf{X}$ and the one-electron degrees of freedom $\mathbf{x}$.

For infinitesimal vibrational displacements, $\mathbf{X} = \mathbf{X}_0 + \Delta\mathbf{X}$, where $\mathbf{X}_0$ represents the equilibrium configuration with $\mathcal{D}_{3h}$ symmetry. We expand the potential V to linear terms in $\Delta\mathbf{X}$:

$$\mathcal{H}_e(\mathbf{X}, \mathbf{x}) \simeq T_e + V(\mathbf{X}_0, \mathbf{x}) + \nabla_{\mathbf{X}}V\Big|_{\mathbf{X}=\mathbf{X}_0} \cdot \Delta\mathbf{X}.$$

As we have shown in Chapter 6, we carry out a unitary transformation to normal (or symmetry-adapted) coordinates $\mathbf{X} \rightarrow \mathbf{Q}$, and obtain

$$\mathcal{H}_e(\mathbf{Q}, \mathbf{x}) = T_e + V(\mathbf{X}_0, [\mathbf{x}]) + \sum_{\alpha} \nabla_{\mathbf{Q}_\alpha}V \cdot \mathbf{Q}_\alpha$$

$$= T_e + V(\mathbf{X}_0, \mathbf{x}) + \sum_{\alpha} \sum_{\ell=1}^{d_\alpha} K_{\alpha\ell}\, Q_\alpha^\ell = \mathcal{H}_e^{(0)} + \mathcal{H}', \tag{10.31}$$

where $\mathbf{Q}_\alpha$ is a normal mode coordinate belonging to the irreducible representation (irrep) α and d_α is the irrep dimension, or its degeneracy. In the trimer case, the relevant irrep is E_g with $d_{E_g} = 2$, yielding

$$\mathcal{H}_e = \frac{\mathbf{p}^2}{2m_e} + V(\mathbf{x}; \mathbf{X}_0) + \mathcal{V}_1(\mathbf{x})\, Q_1 + \mathcal{V}_2(\mathbf{x})\, Q_2$$

$$= \mathcal{H}_e^{(0)} + \mathcal{V}_1(\mathbf{x})\, Q_1 + \mathcal{V}_2(\mathbf{x})\, Q_2,$$

where we dropped the subscript E_g.

Diagonalization of $\mathcal{H}_e^{(0)}$ yields the solutions $\big|\xi_j(\mathbf{X}_0, \mathbf{x})\big\rangle$ of the adiabatic electronic problem for the equilateral ionic configuration. However, since $\mathcal{H}_e^{(0)}$ has the $\mathcal{D}_{3h}$ symmetry, the ground-state wavefunction is still expressed in terms of an A_{1g} singlet and an E_g doublet:

$$\big|\phi_{A_{1g}}\big\rangle, \ \big|\phi_{E_g}^1\big\rangle, \ \big|\phi_{E_g}^2\big\rangle.$$

Treating the interaction as a degenerate perturbation problem, we evaluate the matrix elements as[10]

$$\big\langle\phi^1\,|\mathcal{V}(\mathbf{x})|\,\phi^1\big\rangle = -K, \quad \big\langle\phi^2\,|\mathcal{V}(\mathbf{x})|\,\phi^2\big\rangle = K, \quad \big\langle\phi^1\,|\mathcal{V}(\mathbf{x})|\,\phi^2\big\rangle = \big\langle\phi^2\,|\mathcal{V}(\mathbf{x})|\,\phi^1\big\rangle = K,$$

where we again dropped the subscript E_g. Expressing the electron eigenket in the form

$$\xi'(\mathbf{x}, \mathbf{Q}) = A_1(\mathbf{Q})\, \phi^1(\mathbf{x}, \mathbf{Q}) + A_2(\mathbf{Q})\, \phi^2(\mathbf{x}, \mathbf{Q}),$$

we obtain

$$\begin{pmatrix} -K Q_1 & K Q_2 \\ K Q_2 & K Q_1 \end{pmatrix} \Rightarrow \mathcal{E}_\pm = \pm K\sqrt{Q_1^2 + Q_2^2}.$$

---

[10] Any real symmetric 2×2 matrix may be written as

$$\frac{a}{2}\mathbb{I} + b\begin{pmatrix} x & y \\ y & -x \end{pmatrix}.$$

In the present case, the component involving the identity matrix is unimportant.

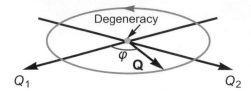

Figure 10.9 Vibrational parameter space in polar coordinates. The gray circle represents a closed loop in Q-space that encircles the origin.

Next, we transform the coordinates in Q_1-Q_2 parameter space to polar form (Q, φ):

$$Q_1 = Q \cos \varphi, \text{ and } Q_2 = Q \sin \varphi$$

as shown in Figure 10.9, and obtain

$$KQ \begin{pmatrix} -\cos\varphi & \sin\varphi \\ \sin\varphi & \cos\varphi \end{pmatrix} = KQ \left(\sin\varphi\, \sigma_x - \cos\varphi\, \sigma_z \right), \tag{10.32}$$

where σ_x and σ_z are the Pauli matrices. The eigenenergies become

$$\mathcal{E}_\pm = \pm KQ.$$

The rhs of (10.32) shows that the coupling to the vibronic modes in the electronic problem can be mapped onto a spin-1/2 problem. We can then write the eigenkets as

$$\begin{cases} |\xi_-\rangle & = \dfrac{1}{\sqrt{2}} \left[\cos(\varphi/2) \left|\phi^1\right\rangle - \sin(\varphi/2) \left|\phi^2\right\rangle \right] \\[2mm] |\xi_+\rangle & = \dfrac{1}{\sqrt{2}} \left[\sin(\varphi/2) \left|\phi^1\right\rangle + \cos(\varphi/2) \left|\phi^2\right\rangle \right] \end{cases} \tag{10.33}$$

The eigenkets $|\xi_\pm\rangle$ are real and double-valued in the angle φ. We find that although the Hamiltonian is periodic in φ, the electronic wavefunctions are *antiperiodic*:

$$|\xi_\pm(2\pi)\rangle = -|\xi_\pm(0)\rangle .$$

We can also view the sign change in the electronic eigenstates, as a result of varying the angle φ along a closed loop encircling the origin in the ionic configuration parameter space, shown in Figure 10.9, a possible holonomy. This behavior is demonstrated in Figure 10.10. As we will see, this is characteristic of a cyclic *pseudorotation*.

The sign change in the state of a two-level (fermionic) system under a 2π rotation, referred to as *spinor behavior*,[11] was known since the early days of quantum mechanics, but we shall show that Berry's formulation brings a new insight, namely, a special case of the geometric phase. However, to start with, the reality of the wavefunctions leads to the vanishing of Berry's geometric vector potential, $\langle\xi_\pm |\nabla_\mathbf{Q}| \xi_\pm\rangle = 0$.

---

[11] We recall that under the rotation operator $e^{\mathbf{J}\cdot\boldsymbol{\varphi}/\hbar}$, a spin-1/2 vector flips under spatial rotation of $\varphi = 2\pi$.

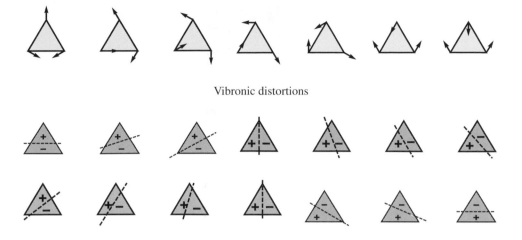

Vibronic distortions

Corresponding changes in the electronic wavefunctions ξ_+ (top) and ξ_- (bottom)

Figure 10.10 The top figure illustrates the distortions that the triangular molecule undergoes as the phase angle associated with the degenerate vibrational distortion increases from 0 to π. The bottom figure shows schematically how during this process the electronic state ξ_- (initially ϕ^1-symmetry) changes continuously into $-\phi^2$, and ξ_+ (initially ϕ^2-symmetry) into ϕ^1.

The double-valued eigenstates of (10.33) may appear disturbing and unphysical. However, this problem can be remedied in two ways. We can apply a gauge transformation that renders the wavefunctions complex, but single-valued. Or, we require that the ionic wavefunction be quantized using antiperiodic boundary conditions. Note that the eigenstate of the total molecular system, $\Psi(\mathbf{x}, \mathbf{Q})$, must be single-valued, of which the electronic term is only one factor.

10.4.3 Molecular Aharonov–Bohm effect (MABE)

Before describing the molecular Aharonov–Bohm effect, we will briefly describe the original Aharonov–Bohm effect (ABE) [6], for the sake of clarity and completeness.

The Aharonov–Bohm Effect

In the simplest description of the Aharonov–Bohm effect, an electron is placed outside a solenoid with infinite extension in the z-direction (see Figure 10.11). The region inside the solenoid is inaccessible to the electron. For simplicity, we consider the case where the electron is confined to a circular rail of radius a, in the x-y plane, as shown in figure. If the magnetic flux through the solenoid is Φ, then the vector potential at the rail will be

$$A_\varphi = \frac{\Phi}{2\pi a}.$$

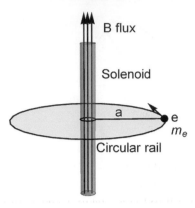

Figure 10.11 A simple model of ABE.

The corresponding Hamiltonian is

$$\mathcal{H} = \frac{\hbar^2}{2m_e} \left(\frac{i}{a} \frac{d}{d\varphi} + \frac{e}{\hbar c} A_\varphi \right)^2 = -\frac{\hbar^2}{2m_e} \left(\frac{d}{d\varphi} - i \frac{\Phi}{\Phi_0} \right)^2,$$

where $\Phi_0 = 2\pi \hbar c/e = hc/e$ is the flux quantum. The single-valued eigenfunctions satisfying periodic boundary conditions are $\psi(\varphi) \propto \exp[i\ell\varphi]$, ℓ integer, with eigenenergies

$$E_\ell = \frac{h^2}{2m_e a^2} \left(\ell + \frac{\Phi}{\Phi_0} \right)^2,$$

which clearly demonstrates that the energy quantization may have a noninteger quantum number whose value depends on the solenoid flux! Moreover, we note that for Φ not equal to integer multiples of Φ_0, energies for ℓ values that have the same magnitude but differ in sign are not degenerate. The canonical and kinematic angular momenta differ, so that clockwise and counterclockwise directions have different speeds for a given value of $|\ell|$. The reason is that the vector field $\mathbf{A}$ causes the phase of the electron wavefunction to change as the wave circulates on the ring. To satisfy the boundary condition, this phase change, in turn, results in higher or lower kinematic momentum.

The key to the Aharonov–Bohm effect is the vector potential $\mathbf{A}(\mathbf{x})$. Outside the solenoid $\mathbf{B} = \nabla \wedge \mathbf{A} = 0$, but $\mathbf{A} \neq 0$ because for any closed loop surrounding the solenoid we have a nonzero integral

$$\oint_C \mathbf{A} \cdot d\boldsymbol{\ell} = \int_{\substack{S(C) \\ \text{including} \\ \text{solenoid}}} \mathbf{B} \cdot d\mathbf{s} = \Phi.$$

Thus, locally, $\nabla \wedge \mathbf{A} = 0$ means that the vector potential is a gradient of some function, and we may gauge it away

$$\mathbf{A}(\mathbf{x}) \rightarrow \tilde{\mathbf{A}}(\mathbf{x}) = \mathbf{A}(\mathbf{x}) - \nabla \xi(\mathbf{x}) = 0,$$

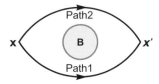

Figure 10.12 Vibrational parameter space in polar coordinates. The gray circle represents a closed loop in Q-space that encircles the origin.

but, globally, no single-valued $\xi(\mathbf{x})$ can gauge away the vector potential.[12]

To bring out the physics behind the ABE, we consider an electron beam being split at $\mathbf{x}$, and the two components travel along paths 1 and 2, of equal lengths, as shown in Figure 10.12. When they recombine at $\mathbf{x}'$, they have acquired different phases. We have two separate and different gauge transforms - $\xi_1(\mathbf{x})$ that gauges away $\mathbf{A}(\mathbf{x})$ along path 1, and $\xi_2(\mathbf{x})$ that gauges away $\mathbf{A}(\mathbf{x})$ along path 2:

$$\Delta \xi \Big|_{\text{path1}} - \Delta \xi \Big|_{\text{path2}} = \oint_{\substack{\text{closed} \\ \text{loop}}} \mathbf{A} \cdot d\boldsymbol{\ell} = \Phi.$$

Gauge Transformation of Electronic States

For pedagogical reasons, we shall use a gauge transformation to remove the double-valuedness, and we write

$$|\xi_{\pm}(\mathbf{Q}, \mathbf{x})\rangle \;\rightarrow\; \exp(i\varphi/2)\, |\xi_{\pm}(\mathbf{Q}, \mathbf{x})\rangle. \tag{10.34}$$

Such a transformation requires the addition of a gauge potential to the Hamiltonian. As we will show, the present case requires the introduction of a vector potential

$$\nabla \;\rightarrow\; \nabla - i(\hat{\boldsymbol{\varphi}}/2Q).$$

This vector potential gives rise to a *fictitious* magnetic field (a flux tube) confined to the origin. Consequently, the $E \otimes \varepsilon$ system may be regarded as an analog of the Aharonov–Bohm effect. Indeed, Mead has dubbed this the *molecular Aharonov–Bohm effect* [128]. As in BOA, the eigenstates acquire a phase γ from the vector potential.

- **Berry's connection and phase**

  We start with calculating Berry's connection for the single-valued $\xi(\mathbf{Q}, \mathbf{x})$ of (10.34). With $\nabla_Q = \left(\dfrac{\partial}{\partial Q}, \dfrac{1}{Q}\dfrac{\partial}{\partial \varphi}\right)$, we find

$$\nabla_Q |\xi_-\rangle = \frac{\hat{\boldsymbol{\varphi}}}{Q} \frac{\partial}{\partial \varphi}\left[e^{i\varphi/2}\left(-\sin\frac{\varphi}{2}\,|\phi^1\rangle + \cos\frac{\varphi}{2}\,|\phi^2\rangle\right) \right]$$

$$= \frac{\hat{\boldsymbol{\varphi}}}{2Q}\, e^{i\varphi/2}\left[i\,|\xi_-\rangle - |\xi_+\rangle \right], \tag{10.35}$$

---

[12] The obstruction presented by the impenetrable solenoid results in a domain that is not simply connected.

which yields

$$\langle \xi_- | \nabla \xi_- \rangle = \left[0, \frac{i}{2Q} \right] \quad \Rightarrow \quad \mathcal{A}_Q = i \langle \xi_- | \nabla \xi_- \rangle = -\frac{1}{2Q} \hat{\varphi}$$

and Berry's phase $\gamma_A(C)$ is given by

$$\gamma_A(C) = \oint_C \mathcal{A}_Q \cdot d\mathbf{l}_Q = \int_0^{2\pi} \left(-\frac{1}{2Q} \right) Q \, d\varphi = -\pi.$$

Quantization of Ionic Motion: Pseudorotations

- **Effective scalar potential**

Having determined the electronic energy contribution to be $\mathcal{E}_\pm = \pm KQ$, we now turn to the vibrational (slow motion) problem. The elastic energy of the vibration modes Q_1 and Q_2 is

$$E_{\text{elastic}}(\mathbf{Q}) = \frac{\kappa}{2} \left(Q_1^2 + Q_2^2 \right) = \frac{\kappa}{2} Q^2,$$

where the two modes have the same effective spring constant κ since they are degenerate. The potential energy experienced by the molecular ion system becomes

$$\frac{\kappa}{2} Q^2 - KQ,$$

with $Q \geq 0$ in polar coordinates. This is just the Born–Oppenheimer surface of the Jahn–Teller split doublet shown in Figure 10.13:

- It is a double-valued function.
- It has a degeneracy at $Q = 0$ (the point of zero distortion), together with a *conical intersection*.
- The double cone is just a diabolo, and the degeneracy point a *diabolical*.
- Near the origin, the adiabatic approximation breaks down, and the origin becomes a singularity of the Born–Oppenheimer procedure.
- The domain of Q is not simply connected, because of the singularity at the origin (an obstacle).
- The lowest sheet has a continuum of minima, a circular valley of radius
  $Q_0 = K/\kappa$, $E_- = -K^2/2\kappa$, where a classical particle travels freely.

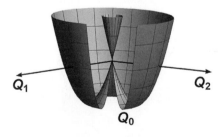

Figure 10.13 The Born–Oppenheimer surface of the Jahn–Teller split doublet: a double-valued function with a conical intersection-diabolical point. The potential minimum is a circle of radius Q_0 centered at the degeneracy point.

Nothing exotic happens if the ionic motion can be considered as classic; but when we quantize the ionic degrees of freedom, Berry's connection arising from the gauge transformation in the electronic wavefunction gives rise to half-order quantum numbers. The ionic system becomes fermionic!

- **The eigenvalue problem**

    Substituting the total eigenket

    $$|\Psi(\mathbf{r}, \mathbf{Q})\rangle = \chi(\mathbf{Q})\,|\xi_-\rangle$$

in the eigenvalue equation and setting $\kappa = \mu\omega_Q^2$, where μ is the mode effective mass, we obtain

$$\left[\frac{\mathbf{P}^2}{2\mu} + \left(\frac{\mu\omega_Q^2}{2}Q^2 - KQ\right)\right]\chi\,|\xi_-\rangle = E\,\chi\,|\xi_-\rangle$$

or

$$|\xi_-\rangle\frac{\mathbf{P}^2}{2\mu}\chi + i\hbar\nabla_Q\,|\xi_-\rangle \cdot \left(\frac{\mathbf{P}}{\mu}\chi\right) - \frac{\hbar^2}{2\mu}\chi\nabla_Q^2\,|\xi_-\rangle + \left(\frac{\mu\omega_Q^2}{2}Q^2 - KQ\right)\chi\,|\xi_-\rangle = E\chi\,|\xi_-\rangle.$$

Eliminating the electronic degrees of freedom, we get

$$\frac{\mathbf{P}^2}{2\mu}\chi + i\hbar\,\langle\xi_-|\,\nabla_Q\,|\xi_-\rangle \cdot \left(\frac{\mathbf{P}}{\mu}\chi\right) - \frac{\hbar^2}{2\mu}\,\langle\xi_-|\,\nabla_Q^2\,|\xi_-\rangle\,\chi + \left(\frac{\mu\omega_Q^2}{2}Q^2 - KQ\right)\chi = E\chi. \tag{10.36}$$

If we determine $\langle\xi_-|\,\nabla_Q^2\,|\xi_-\rangle$ by operating on (10.35) with ∇_Q, we get

$$\langle\xi_-|\,\nabla_Q^2\,|\xi_-\rangle = \langle\xi_-|\,\nabla_Q\cdot\nabla_Q\,|\xi_-\rangle = \frac{1}{2Q^2}\frac{\partial}{\partial\varphi}e^{i\varphi/2}\left[i\,|\xi_-\rangle - |\xi_+\rangle\right] = -\frac{1}{2Q^2}.$$

Using the relations

$$\mathcal{A}_Q = i\,\langle\xi_-|\,\nabla\xi_-\rangle = -\frac{1}{2Q}\hat{\varphi}, \qquad \nabla_Q = \left(\frac{\partial}{\partial Q}, \frac{1}{Q}\frac{\partial}{\partial\varphi}\right), \qquad \nabla_{\mathbf{Q}}^2 = \nabla_Q^2 + \frac{1}{Q^2}\frac{\partial^2}{\partial\varphi^2}$$

in (10.36), we obtain

$$\left[-\frac{\hbar^2}{2\mu}\nabla_{\mathbf{Q}}^2 + \frac{\hbar^2}{2\mu}\frac{1}{Q^2}\left(i\frac{\partial}{\partial\varphi}\right) + \frac{\hbar^2}{4\mu Q^2} + \left(\frac{\mu\omega_Q^2}{2}Q^2 - KQ\right)\right]\chi(\mathbf{Q}) =$$

$$\left[-\frac{\hbar^2}{2\mu}\nabla_{\mathbf{Q}}^2 + \frac{\hbar^2}{2\mu Q^2}\left(-\frac{\partial^2}{\partial\varphi^2} + i\frac{\partial}{\partial\varphi}\right) + \frac{\hbar^2}{4\mu Q^2} + \left(\frac{\mu\omega_Q^2}{2}Q^2 - KQ\right)\right]\chi(\mathbf{Q}) =$$

$$\left[-\frac{\hbar^2}{2\mu}\nabla_{\mathbf{Q}}^2 + \frac{\hbar^2}{2\mu Q^2}\left(i\frac{\partial}{\partial\varphi} + \frac{1}{2}\right)^2 + \frac{\hbar^2}{8\mu Q^2} + \left(\frac{\mu\omega_Q^2}{2}Q^2 - KQ\right)\right]\chi(\mathbf{Q}) = E\chi(\mathbf{Q}). \tag{10.37}$$

- **Quantized angular motion in Q-space: pseudorotations**

    Since χ has to be single-valued and periodic in φ, we set

    $$\chi(\mathbf{Q}) = \eta(Q)\, e^{im\varphi}$$

so that the second term in (10.37) becomes

$$\frac{\hbar^2}{2\mu Q^2}\left(m - \frac{1}{2}\right)^2 = \frac{\ell^2\hbar^2}{2\mu Q^2}, \qquad |\ell| = 1/2,\ 3/2,\ 5/2,\ \ldots \tag{10.38}$$

The pseudorotation term is obviously an ionic rotor in Q-space. But the angular energy is characterized by a *half-integer* quantum number ℓ, instead of the integer quantum number m. The lowest state with $\ell = \pm 1/2$ is an orbital doublet. This feature can be considered a manifestation of Berry's phase, which arises because we insisted that the electronic wavefunction be single-valued in Q-space!

The spectrum of (10.38) has the same structure as in ABE, if we identify the inaccessible flux Φ with *half a flux quantum* Φ_0, sometimes referred to as a π flux. There is no magnetic field in this problem; *the flux is purely topological* and can be regarded as an *obstruction*, since the ionic path cannot be contracted without crossing a degeneracy point.[13]

- **Radial motion in Q-space**

    To complete the analysis of the molecular Aharonov–Bohm phenomenon, we consider here the radial motion. With

    $$\nabla_Q^2 = \frac{1}{Q}\frac{\partial}{\partial Q}\left(Q\frac{\partial}{\partial Q}\right),$$

we write (10.37) in the form

$$\left[\frac{1}{Q}\frac{\partial}{\partial Q}\left(Q\frac{\partial}{\partial Q}\right) - \frac{A}{Q^2} - \left(\frac{\mu\omega_Q}{\hbar}\right)^2\left(Q - \frac{K}{\mu\omega_Q^2}\right)^2 + \frac{K^2}{\hbar^2\omega_Q^2} + \frac{2\mu E}{\hbar^2}\right]\eta(Q) = 0,$$
$$\tag{10.39}$$

where $A = \dfrac{\hbar^2}{2\mu Q^2}\left[\left(m - \dfrac{1}{2}\right)^2 + \dfrac{1}{4}\right]$.

If we replace $\left(Q - \frac{K}{\mu\omega^2}\right)^2$ by Q^2, we have the radial equation of the isotropic two-dimensional harmonic oscillator! Hence, (10.39) is the radial equation of a displaced isotropic oscillator, where the low-lying wavefunctions are centered on the potential

---

[13] In modern jargon, we would say that the cases $\Phi = 0$ and $\Phi = \Phi_0/2$ are topologically distinct, because of the presence of time-reversal invariance, and that other flux values are ruled out. Owing to time-reversal invariance, the electronic Berry phase in a molecule – on any closed path in configuration space – can only be either 0 or π mod 2π, hence the $\mathbb{Z}_2$ nature of this invariant is obvious. The group $\mathbb{Z}_2$ is the additive group of the integers modulo 2: it has only two elements, which can be labeled as 0 and 1, or as even and odd. This viewpoint brings out the concept of $\mathbb{Z}_2$ to be encountered later in the case of topological insulators.

minimum $Q_0 = \dfrac{K}{\mu\omega_Q^2}$, and concentrated in its trough. For these low-lying states, we can write

$$Q - \frac{K}{\mu\omega^2} = \Delta Q \;\Rightarrow\; \frac{\partial}{\partial\Delta Q} = \frac{\partial}{\partial Q}$$

with $\Delta Q \ll Q_0$. This allows us to set $Q \simeq Q_0$ in the first two terms of (10.39), so that (10.39) reduces to

$$\left[\frac{\partial^2}{\partial\Delta Q^2} - \left(\frac{\mu\omega_Q}{\hbar}\right)^2 \Delta Q^2 + \frac{2\mu E'}{\hbar^2} \right] \eta(\Delta Q) = 0,$$

which is just the equation of the one-dimensional harmonic oscillator.

To visualize the behavior of this system, we first establish the analogy between the isotropic oscillator and circularly polarized light: in the latter, we have two oscillating linear electric fields that are spatially orthogonal; in the former, we have x and y oscillators. In the present system, we have two normal modes Q^1 and Q^2 that perform circular motion in Q^1-Q^2 space. In real space, we have instantaneous linear superpositions of the two modes according to $Q^1 = Q\cos\varphi(t)$ and $Q^2 = Q\sin\varphi(t)$.

The vibrorotations, or *pseudorotations*, are depicted in Figure 10.14, with the degeneracy point represented by the equilateral triangle, and its morphing into pseudorotational distortions appears around the closed circular path.

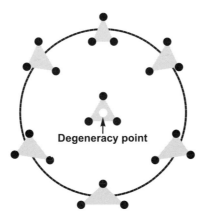

Figure 10.14 Pseudorotation in a trimer. The electronic degeneracy occurs at the symmetric configuration, shown as the equilateral triangle in the center. The distorted triangles represent low-lying vibronic states being transported along a closed loop in parameter space around the degeneracy. The transportation results in the display of pseudorotation in real space.

Exercises

10.1 Consider Dirac fermions in two dimensions described by the Hamiltonian

$$\mathcal{H}(\mathbf{k}) = \sum_\alpha d_\alpha(\mathbf{k})\,\sigma_\alpha, \qquad d_1(\mathbf{k}) = k_x,\ d_2(\mathbf{k}) = k_y,\ d_3(\mathbf{k}) = m.$$

(a) Show that the Berry connection of the lower band can be written as

$$
\begin{aligned}
\mathcal{A}_\mu &= i\,\langle \psi_-(\mathbf{k})\,|\partial_{k_\mu}\psi_-(\mathbf{k})\rangle \\
&= -\frac{1}{2d(\mathbf{k})\,[d(\mathbf{k}) - d_3(\mathbf{k})]}\Big[d_2(\mathbf{k})\partial_{k_\mu}d_1(\mathbf{k}) - d_1(\mathbf{k})\partial_{k_\mu}d_2(\mathbf{k})\Big],
\end{aligned}
$$

where $d(\mathbf{k}) = |\mathbf{d}(\mathbf{k})|$.

(b) Show that the corresponding Berry curvature is given by

$$\mathcal{F}_{\mu\nu}(\mathbf{k}) = \frac{1}{2}\,\epsilon_{\alpha\beta\gamma}\,\hat{d}_\alpha\,\partial_{k_\mu}\hat{d}_\beta\,\partial_{k_\nu}\hat{d}_\gamma, \qquad \hat{\mathbf{d}}(\mathbf{k}) = \frac{\mathbf{d}(\mathbf{k})}{|\mathbf{d}(\mathbf{k})|}.$$

10.2 Parallel transport gauge:

Consider a system subject to a weak adiabatic perturbation that is periodic in time. With the total Hamiltonian $\mathcal{H}(t + T) = \mathcal{H}(t)$, a general state $|\psi(t)\rangle$ obeys the Schrödinger equation

$$i\partial_t\,|\psi(t)\rangle = \mathcal{H}(t)\,|\psi(t)\rangle$$

and can be expressed as

$$|\psi(t)\rangle = \sum_{\ell'} \exp\left[-\frac{i}{\hbar}\int_0^t dt'\,\varepsilon_{\ell'}(t')\right] c_{\ell'}(t)\,|\ell'(t)\rangle, \tag{10.40}$$

where $|\ell'(t)\rangle$ is an eigenket of $\mathcal{H}(t)$, which are single-valued in t: $|\ell'(t + T)\rangle = |\ell'(t)\rangle$.

(a) Use the Schrödinger equation to derive an expression for $\dot{c}_\ell(t)$.

(b) Parallel transport, or horizontal lift, requires that

$$\left\langle \tilde{\ell}(t)\,\Big|\,\partial_t \tilde{\ell}(t)\right\rangle \equiv \left\langle \tilde{\ell}(\eta)\,\Big|\,\mathrm{d}\,\Big|\tilde{\ell}(t\eta)\right\rangle\frac{d\eta}{dt} = 0.$$

Show that the horizontal lift condition is satisfied by the gauge transformation

$$|\ell(t)\rangle \ \rightarrow\ \left|\tilde{\ell}(t)\right\rangle = \exp\left[i\int_0^t dt'\,\langle \ell(t')\,|\partial_{t'}\ell(t')\rangle\right]|\ell(t)\rangle.$$

(c) If the system starts in the eigenstate $\left|\tilde{\ell}(0)\right\rangle$, so that $c_\ell = 1$, $c_{\ell'} = 0$, $\ell' \neq \ell$.

(i) Determine $\dot{c}_\ell$ and $\dot{c}_{\ell'}$. Is the adiabatic theorem satisfied?

(ii) Check if the ansatz solution

$$c_{\ell'}(t) = -i\hbar \, \frac{\left\langle \tilde{\ell}'(t) \left| \partial_t \tilde{\ell}(t) \right. \right\rangle}{\varepsilon_\ell(t) - \varepsilon_{\ell'}(t)} \, \exp\left[-\frac{i}{\hbar} \int_0^t dt' \left(\varepsilon_\ell(t') - \varepsilon_{\ell'}(t') \right) \right]$$

obeys the differential equation you obtained for $\dot{c}_{\ell'}$. (Neglect terms of second order).

(iii) Write down an expression for the state function that includes the first-order approximation.

10.3 Divergence of Berry curvature:

Use expression (10.24) to show that for $E_m \neq E_n$,

$$\nabla \cdot \mathcal{F} = 0.$$

Hint: Use the Hermitian operator

$$\mathbf{O} = -i \sum_n |\nabla n\rangle \langle n|$$

to express (10.24) as

$$\mathcal{F} = \mathrm{Im} \, \langle n \, |\mathbf{O} \times \mathbf{O}| \, n \rangle .$$

10.4 In Section 10.3, we derived Berry's curvature for the two-level system using two methods: the direct geometric method, and the one- and two-form method. Alternatively, use the Hamiltonian derivative method outlined in Section 2.2.3 to derive Berry's curvature for that system.

10.5 Consider a spin-**J** particle in a slowly rotating magnetic field. The Hamiltonian is given by

$$\mathcal{H} = -\,\boldsymbol{\mu} \cdot \mathbf{J},$$

where **J** is the spin angular momentum operator, and $\boldsymbol{\mu} = \mu_B \mathbf{B}\left(\hat{\mathbf{n}}(t)\right)$. Use the Hamiltonian derivative method of Section 2.2.3 to determine the state Berry curvature and the corresponding Berry phase.

10.6 We shall consider the ABE ring of Section 10.3 from a different perspective. We shall look at the system as consisting of a one-dimensional metallic ring with circumference L, threaded by a magnetic flux Φ. The Hamiltonian is again

$$\mathcal{H} = \frac{1}{2m_e} \left(\mathbf{p} - \frac{e}{c}\mathbf{A} \right)^2, \qquad \nabla \times \mathcal{A} = 0.$$

We may recast the ring into a one-dimensional periodic structure of period L, by defining $x = L\varphi/2\pi$.

(a) If we write $\mathcal{A} = \nabla \chi(x)$, what will $\chi(x)$ be?

(b) Perform a gauge transformation on the one-dimensional Hamiltonian to eliminate $\mathcal{A}$. What boundary conditions have to be applied to the solutions of the corresponding Schrödinger equation?

(c) Determine the eigenfunctions and energy spectrum, by writing the solution in the Bloch form $A\,e^{ikx}$ and applying the boundary conditions you obtained in part (b).

(d) Calculate the current carried by an electron in a state n, $I_n(\phi) = ev_n/L = \frac{e}{\hbar L}\frac{\partial E_n}{\partial k_n}$.

(e) Let us fill the ring with and odd number N_{odd} of electrons, at zero temperature. What is the total current flowing through the system, $I(\phi)$? What happens when $\phi = \phi_0/2$?

(f) How does the current change when the number of electrons in the ring is even?

10.7 Spin–orbit coupling in an ABE-like configuration:
An electron is constrained to move on a 1D ring of radius a. The electron experiences a spin–orbit interaction with the electric field created by a charge Q placed at the center of the ring.

1. Write the Hamiltonian $\mathcal{H}(x, p, \mathbf{s})$.
2. Determine the electron eigenenergies and corresponding velocities.
3. Determine that the magnetic flux Φ through the ring would cancel the interaction of Q with a spin up electron.

10.8 We consider the case of a time-reversal symmetric system with half-integer spin, represented by a Hamiltonian $\mathcal{H}(\boldsymbol{\eta})$, with $\boldsymbol{\eta} = [\eta_1, \ldots, \eta_n]$. We label degenerate orthonormal Kramers' pair eigenkets as $|\mu, \boldsymbol{\eta}\rangle$, $\mu = \pm 1$, that are smooth functions of the parameters $\boldsymbol{\eta}$ within a given patch. We need to define a new Berry connection A_ℓ, which is a 2×2 matrix, as

$$A_\ell^{\mu\nu}(\boldsymbol{\eta}) = i\,\langle \mu, \boldsymbol{\eta} |\, \partial_\ell\, | \nu, \boldsymbol{\eta}\rangle,$$

where ∂_ℓ is the partial derivative with respect to η_ℓ.

A gauge transformation to a basis set $|\mu, \boldsymbol{\eta}\rangle'$, is obtained via a unitary transformation

$$|\mu, \boldsymbol{\eta}\rangle' = \sum_\nu |\nu, \boldsymbol{\eta}\rangle\ U_{\nu\mu}(\boldsymbol{\eta}),$$

where $U(\boldsymbol{\eta})$ is a 2×2 unitary matrix that varies smoothly with $\boldsymbol{\eta}$.[14]

The non-Abelian Berry curvatures corresponding to the Berry connections $\mathcal{A}_\ell$ are defined as

$$\mathcal{F}_{\ell\ell'} = \partial_\ell\,\mathcal{A}_{\ell'} - \partial_{\ell'}\,\mathcal{A}_\ell - i\,[\mathcal{A}_\ell, \mathcal{A}_{\ell'}].$$

[14] We note that in this case, the fiber is engendered by the unitary group $U(2)$, so that the gauge is no longer Abelian. Such non-Abelian gauges will be discussed in Chapter 11.

(a) Show that

$$\text{If } \Theta |+, \boldsymbol{\eta}\rangle = |-, \boldsymbol{\eta}\rangle \, e^{i\phi} \implies \Theta |-, \boldsymbol{\eta}\rangle = -|+, \boldsymbol{\eta}\rangle \, e^{i\phi}.$$

(b) Derive an expression for the Berry connection $\tilde{\mathcal{A}}_\ell$, corresponding to the new states $|\mu, \boldsymbol{\eta}\rangle'$, in terms of $\mathcal{A}_\ell$ and $U(\boldsymbol{\eta})$.

(c) Derive an expression for the Berry curvature $\tilde{\mathcal{F}}_{\ell\ell'}$

$$\tilde{\mathcal{F}}_{\ell\ell'} = \partial_\ell \tilde{\mathcal{A}}_{\ell'} - \partial_{\ell'} \tilde{\mathcal{A}}_\ell - i \left[\tilde{\mathcal{A}}_\ell, \tilde{\mathcal{A}}_{\ell'} \right]$$

in terms of $\mathcal{F}_{\ell\ell'}$ and $U(\boldsymbol{\eta})$.

10.9 In our consideration of the trimer problem, we only included coupling between the electronic orbitals and the E_g vibronic states linear in Q. Here, we will include a higher-order quadratic coupling term; it takes the form

$$g Q^2 \Big(-\cos(2\varphi)\, \sigma_z - \sin(2\varphi)\, \sigma_x \Big), \qquad g \ll 1.$$

This term becomes important when Q is large, since it depends on Q^2.

(a) Add the quadratic terms to the Hamiltonian matrix in (10.32), and determine the electronic energy eigenvalues.

(b) The modified Hamiltonian can still be mapped onto the spin-1/2 system Hamiltonian. Write down the eigenkets in terms of a new phase angle θ and define θ in terms of the Hamiltonian matrix components.

(c) Express the effective ion potential in terms of the new electronic eigenenergies and determine its degeneracy points.

11

Topological Aspects of Insulator Band Structure and Early Discoveries

11.1 Introduction

We start this chapter with an analogy between the way topology enters band structures of gapped one-electron Hamiltonians and the earliest topological manifestation in physics – Gauss's law in electrostatics. Gauss's law asserts that the total electric flux emanating from or flowing into a closed surface depends only on the charge enclosed inside it. The net flux does not depend on the shape of the surface or the details of the enclosed charge distribution. As we will show, the analog of the Gaussian surface is a mapping of a Brillouin zone in d-dimensions onto a closed torus surface $\mathbb{T}^d$. In this new scenario, different Gaussian surfaces correspond to different bands in the reduced zone scheme, and the electric charge is analogous to an appropriate topological invariant. The actual nature of the invariant depends on the symmetries of the system under consideration and its dimension, and it is determined by an integral over the Brillouin torus of an appropriate field, Berry's curvature, derived from the Bloch functions of the occupied bands. We can immediately see the analogy with the electric charge enclosed by a Gaussian surface being equal to an integral of the electric field over it.

11.1.1 Topological Equivalence of Insulators

We have defined topologically equivalent classes in terms of homeomorphims. The question then arises as to how we can apply this notion to topologically classify electronic phases of matter? Here, we are concerned with the electronic structure of insulators. As we have learned, the action of the Hamiltonian of an insulator on its Hilbert space produces electronic energy bands with a gap for excitations that separates the ground state from all excited states. Thus, we can envision a homeomorphism in such a case to involve continuous deformations of the electronic band structure, or, alternatively, continuous deformations of the corresponding Hamiltonian, see Figure 11.1. Moreover, to establish the principle of *topological equivalence* to insulators, the Hamiltonian deformations should be based on the principle of adiabatic continuity: insulators are changed into one another by slowly (adiabatically) and *continuously* changing the Hamiltonian. *Adiabaticity* ensures that the system remains in the ground state during the evolution process. The *scale* for how slow the adiabatic process must be is set by the *magnitude of the energy gap*.

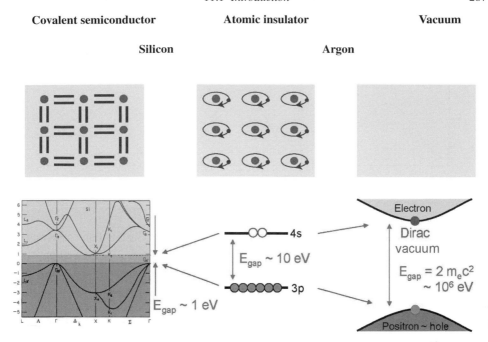

Figure 11.1 Equivalence of trivial insulators.

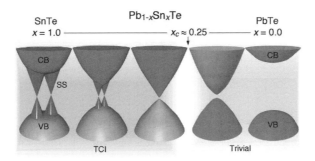

Figure 11.2 Gap closing in trivial (right) to nontrivial (left) topological quantum phase transition in $Pb_{1-x}Sn_xTe$. Figure from [175].

Based on this scenario, we can declare the following:

- Insulators are *topologically equivalent* if there exists an adiabatic path connecting them, along which the energy gap remains finite.

- By contrast, it follows that connecting *topologically inequivalent* insulators would necessarily involve a phase transition, through which the energy gap vanishes, as shown in Figure 11.2.

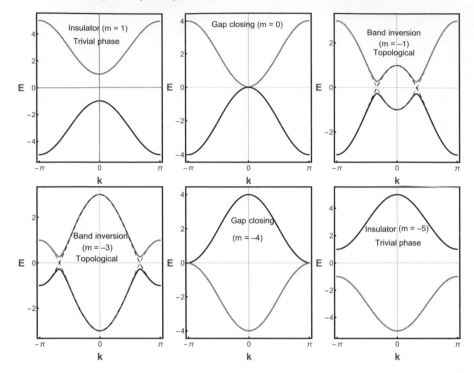

Figure 11.3 Depending on the value of the parameter m, the system transforms from trivial $\Rightarrow$ topological $\Rightarrow$ trivial. Gap closings occurs at $m = 0$, -4.

As an illustration, we consider the following simple 2D model with Hamiltonian

$$\mathcal{H} = \begin{pmatrix} d_3 & d_1 - id_2 \\ d_1 + id_2 & -d_3 \end{pmatrix} \quad \Rightarrow \quad \begin{cases} d_1 & = a\,\sin(k_x) \\ d_2 & = a\,\sin(k_y) \\ d_3 & = m + 2 - \cos(k_x) - \cos(k_y) \end{cases}$$

with eigenvalues

$$\varepsilon_{\mathbf{k}} = \sqrt{\left(m + 2 - \cos(k_x) - \cos(k_y)\right)^2 + a^2 \left(\sin^2(k_x) + \sin^2(k_y)\right)}.$$

As shown in Figure 11.3, for m spanning values from $m = 1$ to $m = -5$, the trivial phase at $m = 1$ experiences gap closure at $m = 0$, and transforms to a topological phase at $-4 < m < 0$, where a topological twist emerges due to band inversion. At $m = -4$, the gap closes again, ushering a second trivial phase for $m < -4$.

Bulk-Boundary Correspondence

The operational identification of topologically distinct Hamiltonians $\mathcal{H}(\mathbf{k})$ of insulators through gap closure allows us to contemplate the consequence of having an interface between two topologically distinct phases. We would expect that the energy gap has to vanish at such an interface, and that gapless low-energy electronic states confined to the region straddling the interface would appear. These gapless states can also be topologically classified. This picture is referred to as *bulk-boundary correspondence*.

Topological Invariants and Observables

Whenever an observable effect is endowed with a natural integer topological invariant, as we have learned, two remarkable features are known to occur:

(1) The observable can be measured, in principle, with infinite precision (10^{-9} is actually attained for the quantum Hall effect).
(2) The observable is very robust under even strong variations of the system conditions, barring an extremely disruptive perturbation that would cause a switch from one integer to another. Since we are concerned here with the topological character of insulators, such a disruptive perturbation amounts to crossing a metallic state.

11.1.2 Symmetries and Classification of Topological Phases

The preceding ideas raise the question whether a system's topological properties may be connected to its symmetries, given the interesting proposition:

Different systems that are characterized by the same fundamental symmetries may share a number of similar properties.

We noted earlier that the topological character of an insulator must be robust against deformations of its gapped Hamiltonian. This means that the topological features should survive continuous deformations of the Hamiltonian that do not close the energy gap.[1] Such deformations are implemented by adding perturbative terms that may break manifest spatial symmetries. We then realize that, ultimately, these topological features must be properties of a Hamiltonian that does not have spatial symmetries – such symmetries are represented by unitary operators $\hat{\mathcal{O}}$ that commute with the Hamiltonian

$$\left[\hat{\mathcal{O}}, \mathcal{H}\right] = 0.$$

The perturbed Hamiltonian thus becomes a gapped random Hamiltonian.[2] If the original, unperturbed phase was characterized by some spatial symmetries, and associated with some topological invariant, then the random Hamiltonian should be in the same topological phase. Alternatively, we say that a given gapped topological phase is associated with a certain class of gapped random Hamiltonians. Therefore, we find that the classification of topological phases requires that we consider the cataloging of random gapped Hamiltonians. This procedure will reveal how many different such phases a system can possess

---

[1] Actually, we observe that in this way, we map out an entire gapped phase, topological or trivial.
[2] Random in the sense of hopping parameters and site potential energies.

where in going from one phase to another a quantum phase transition (gap closing) has to be crossed.

For such a gapped random Hamiltonian, we are only left with discrete antiunitary symmetries, which, in contrast to the unitary ones, can be still preserved even in disordered systems. There exist only two such antiunitary symmetries:

(1) Time-reversal symmetry (TRS) with operator Θ:

$$\Theta^\dagger \mathcal{H}(\mathbf{p}, \mathbf{x}) \, \Theta = \mathcal{H}(\mathbf{p}, \mathbf{x}) \quad \Rightarrow \quad \left[\Theta, \mathcal{H}(\mathbf{p}, \mathbf{x})\right] = 0. \tag{11.1}$$

(2) Particle–hole conjugation symmetry (PHS) with operator $\Xi = U_{\mathrm{PH}} \, K$, which if present satisfies

$$\Xi^\dagger \mathcal{H}(\mathbf{p}, \mathbf{x}) \, \Xi = U_{\mathrm{PH}}^\dagger \, \mathcal{H}^*(-\mathbf{p}, \mathbf{x}) \, U_{\mathrm{PH}} = -\mathcal{H}(\mathbf{p}, \mathbf{x}) \quad \Rightarrow \quad \left\{\Xi, \mathcal{H}(\mathbf{p}, \mathbf{x})\right\} = 0 \tag{11.2}$$

As was demonstrated in Section 4.3.1, the operation of particle–hole conjugation reverses momentum and spin but leaves position invariant. Moreover, particle–hole symmetry requires that for every eigenstate $|\psi\rangle$ of the Hamiltonian $\mathcal{H}$ with energy ε, there is an eigenstate of the same Hamiltonian given by $\Xi |\psi\rangle$ with energy $-\varepsilon$.[3] Because of the anticommutation of Ξ with $\mathcal{H}$, $\mathcal{H}$ consists of off-diagonal blocks

$$\mathcal{H} = \begin{pmatrix} 0 & h \\ h^\dagger & 0 \end{pmatrix}.$$

where h is a square matrix.

Quantum states of matter are characterized not only by the structure of the energy spectrum but also by the nature of their wavefunctions. Thus, the action of the aforementioned symmetries on wavefunctions should be examined. We find that for the antiunitary symmetries

$$\Theta^2 |\psi\rangle = \begin{cases} +|\psi\rangle, & \text{integer spin} \\ -|\psi\rangle, & \text{1/2-integer spin} \end{cases} \qquad \Xi^2 |\psi\rangle = \begin{cases} +|\psi\rangle, & \text{triplet pairing} \\ -|\psi\rangle, & \text{singlet pairing} \end{cases} \tag{11.3}$$

The pairing states correspond to superconductor Hamiltonians.[4] We should point out that although the form of an antiunitary symmetry operation can be changed by a unitary transformation, the signs in (11.3) remain unchanged.

The presence of TRS for half-integer spin implies Kramers' degeneracy, whereas the presence of PHS implies that the energy spectrum is symmetric about zero energy.

---

[3] We find that eigenstates of the Hamiltonian with energy $\varepsilon = 0$ are special in that they may transform into themselves under this symmetry transformation.

[4] Bogoliubov–de Genne (BdG)–type Hamiltonians will be discussed in Chapter 21.

We note that an additional symmetry arises when a system possesses both TRS and PHS symmetries. It is associated with their combined action and represented by a unitary operator $\Pi = \Xi\,\Theta$, which satisfies

$$\Pi^{\dagger}\,\mathcal{H}(\mathbf{p},\mathbf{x})\,\Pi = -\mathcal{H}(\mathbf{p},\mathbf{x}) \quad \Rightarrow \quad \left\{\mathcal{H}(\mathbf{p},\mathbf{x}),\,\Pi\right\} = 0. \tag{11.4}$$

It is referred to as *chiral* or *sublattice* or *energy-reflection* symmetry.[5] However, we should caution that a Hamiltonian can have a chiral symmetry without the concurrent presence of TRS and PHS. We also note that since chiral symmetry anticommutes with the Hamiltonian, it is not a unitary symmetry in the usual sense. According to these discrete symmetries, it is easy to see that there are only *10* possible ways to describe the total symmetry of single-particle Hamiltonians $\mathcal{H}$, *the tenfold way*: TRS (PHS) can be absent (0), or Θ^2, $\Xi^2 = \pm 1$, yielding $3 \times 3 = 9$ possible combinations. The presence of only chiral symmetry gives the 10th way. The list of 10 possible types of behavior of the Hamiltonian is presented in Table 11.1. The labeling of these 10 Hamiltonian symmetries was given by Élie Cartan.

Investigation of the properties of a general Hamiltonian under such symmetries was reported in the seminal work of Altland and Zirnbauer [9], and produced the now famous 10 symmetry classes (the tenfold way). It extended and completed the "threefold way" classification scheme of Wigner and Dyson, going back to the origins of random matrix theory and the study of complex nuclei [56, 129, 195, 196].

The classification of topological phases in one, two, and three dimensions is given in Table 11.1. The analysis of classification procedures is beyond the scope of this book. However, for the sake of completeness, we shall briefly outline the approach followed by Schnyder et al. [158, 160, 161] in Appendix 1 of this chapter. It is recommended that Section 11.1.3 be read prior to Appendix 1.

11.1.3 Insulator Systems with Periodic Translational Invariance (PTI)

We introduced in the previous chapter the basic concepts of Berry-ology for generic systems described by parameter-dependent Hamiltonians, and gave two illustrative examples. We now consider how we can apply these ideas to crystalline solids. As we shall see, the band structure of crystals provides a natural platform to investigate the occurrence of the Berry phase effect.

We consider crystalline structures, with periodic translational symmetry where the crystal momentum is a good quantum number that labels the electronic states:

$$|\psi_{n\mathbf{k}}\rangle = e^{i\mathbf{k}\cdot\mathbf{x}}\,|u_{n\mathbf{k}}\rangle \quad \Rightarrow \quad \begin{cases} u_{n\mathbf{k}}(\mathbf{x}+\mathbf{R}) &= u_{n\mathbf{k}}(\mathbf{x}) \\ \psi_{n\mathbf{k}}(\mathbf{x}+\mathbf{R}) &= e^{i\mathbf{k}\cdot\mathbf{R}}\,\psi_{n\mathbf{k}}(\mathbf{x}) \end{cases}$$

---

[5] In condensed-matter systems, it is often realized as a sublattice symmetry on a bipartite lattice, namely, the symmetry operation that changes the sign of wavefunctions on all sites of one of the two sublattices of the bipartite lattice. However, in many instances, Π is not realized as a sublattice symmetry, but is simply the product of Θ and Ξ. Therefore, the term "chiral symmetry," used in quantum field theory, to describe the symmetry operation Π is sometimes more appropriate.

Table 11.1 *The 10 symmetry classes categorized using $\Theta^2 = \pm 1$, $\Xi^2 = \pm 1$ for the cases in which the respective symmetries Θ and Ξ are present. If chiral symmetry is present (not-present) we just have 1(0). The spatial-dimensionality is given by d. $-$ denotes the absence of any topologically nontrivial ground states, $\mathbb{Z}_2$ denotes two kinds of topologically distinct ground states, while $\mathbb{Z}$ indicates that the topologically distinct ground states can be labeled by the set of integers.*

| System | Cartan label | Symmetry | | | d | | |
		Θ	Ξ	Π	1	2	3
	A	0	0	0	$-$	$\mathbb{Z}$	$-$
Wigner–Dyson	AI	1	0	0	$-$	$-$	$-$
	AII	-1	0	0	$-$	$\mathbb{Z}_2$	$\mathbb{Z}_2$
	AIII	0	0	1	$\mathbb{Z}$	$-$	$\mathbb{Z}$
Chiral	BDI	$+1$	$+1$	1	$\mathbb{Z}$	$-$	$-$
	CII	-1	-1	1	$\mathbb{Z}$	$-$	$\mathbb{Z}_2$
	D	0	1	0	$\mathbb{Z}_2$	$\mathbb{Z}$	$-$
BdG	C	0	-1	0	$-$	$\mathbb{Z}$	$-$
	DIII	-1	$+1$	1	$\mathbb{Z}_2$	$\mathbb{Z}_2$	$\mathbb{Z}$
	CI	$+1$	-1	1	$-$	$-$	$\mathbb{Z}$

where n is the band index, and $\mathbf{R}$ is a primitive lattice vector. As we have shown in Chapter 3, $|\psi_{n\mathbf{k}+\mathbf{G}}\rangle$ and $|\psi_{n\mathbf{k}}\rangle$ are degenerate, and obey the same boundary conditions, since

$$\psi_{n\mathbf{k}+\mathbf{G}}(\mathbf{x} + \mathbf{R}) = e^{i\mathbf{k}\cdot\mathbf{R}} \, \psi_{n\mathbf{k}+\mathbf{G}}(\mathbf{x}),$$

suggesting that $|\psi_{n\mathbf{k}}\rangle$ and $|\psi_{n,\mathbf{k}+\mathbf{G}}\rangle$ are duplicate labels for the same state; they can differ by a phase factor, and we fix this arbitrary phase factor to be unity, so that

$$|\psi_{n,\mathbf{k}+\mathbf{G}}\rangle = |\psi_{n\mathbf{k}}\rangle. \tag{11.5}$$

It follows that $|u_{n,\mathbf{k}}\rangle$ satisfies the condition

$$|u_{n,\mathbf{k}+\mathbf{G}}\rangle = e^{-i\mathbf{G}\cdot\mathbf{x}} \, |u_{n\mathbf{k}}\rangle. \tag{11.6}$$

Thus, the u-functions have the lattice periodicity in real space, but not in reciprocal space, while for ψ-functions the reverse is true.

Periodicity of Bloch Hamiltonians

We note that the derivative $\nabla_{\mathbf{k}} |\psi_{\mathbf{k}}\rangle$ is not well behaved in the sense that

$$\nabla_{\mathbf{k}} \, \psi_{\mathbf{k}}(\mathbf{x}) = e^{i\mathbf{k}\cdot\mathbf{x}} \, \nabla_{\mathbf{k}} \, u_{\mathbf{k}}(\mathbf{x}) + i\mathbf{x} \, e^{i\mathbf{k}\cdot\mathbf{x}} \, u_{\mathbf{k}}(\mathbf{x}),$$

which shows that the second term blows up at large $|\mathbf{x}|$, since for $\psi_{\mathbf{k}}(\mathbf{x})$ the real-space argument spans the whole space. Thus, in the ensuing analysis we will prefer to deal with $|u_{\mathbf{k}}\rangle$ rather than $|\psi_{\mathbf{k}}\rangle$, where the derivatives $\nabla_{\mathbf{k}} |u_{\mathbf{k}}\rangle$ are well behaved and belong to the same Hilbert space as $|u_{\mathbf{k}}\rangle$.

We find that $|u_{\mathbf{k}}\rangle$ is actually an eigenfunction of the *Bloch Hamiltonian*

$$\mathcal{H}(\mathbf{k}) = e^{-i\mathbf{k}\cdot\mathbf{x}} \, \mathcal{H} \, e^{i\mathbf{k}\cdot\mathbf{x}} \qquad \Rightarrow \qquad \mathcal{H}(\mathbf{k}) \, |u_{n\mathbf{k}}\rangle = \varepsilon_n(\mathbf{k}) \, |u_{n\mathbf{k}}\rangle \, .$$

$\mathcal{H}(\mathbf{k})$ has the periodic property

$$\mathcal{H}(\mathbf{k} + \mathbf{G}) = \mathcal{H}(\mathbf{k})$$

in reciprocal space, since

$$\begin{aligned} \mathcal{H}(\mathbf{k} + \mathbf{G}) \, |u_{n\mathbf{k}+\mathbf{G}}\rangle &= \mathcal{H}(\mathbf{k} + \mathbf{G}) \, e^{-i\mathbf{G}\cdot\mathbf{x}} \, |u_{n\mathbf{k}}\rangle \\ &= \varepsilon_n(\mathbf{k} + \mathbf{G}) \, e^{-i\mathbf{G}\cdot\mathbf{x}} \, |u_{n\mathbf{k}}\rangle = \varepsilon_n(\mathbf{k}) \, e^{-i\mathbf{G}\cdot\mathbf{x}} \, |u_{n\mathbf{k}}\rangle \end{aligned}$$

and its Schrödinger equation takes the form

$$\left[\frac{1}{m} \left(\mathbf{p} + \hbar\mathbf{k} \right)^2 + V(\mathbf{x}) \right] |u_{n\mathbf{k}}\rangle = \varepsilon_n(\mathbf{k}) \, |u_{n\mathbf{k}}\rangle \tag{11.7}$$

In this way, we effectively reduce the $\mathcal{H}$ operator, which has an infinite-dimension matrix representation, to a one-parameter aggregate of N block matrices $\mathcal{H}(\mathbf{k})$, where N is the number of point in the Brillouin zone (BZ). Diagonalization of $\mathcal{H}(\mathbf{k})$ yields the eigenvalue spectrum $\varepsilon_n(\mathbf{k})$ and the Hilbert subspace $\mathbb{H}_{\mathbf{k}}$, with basis $u_{n\mathbf{k}} \in \mathbb{H}_{\mathbf{k}}$. Usually, we are interested in a small number of bands, which defines the dimension of $\mathbb{H}_{\mathbf{k}}$.

The periodicity of $\mathcal{H}(\mathbf{k})$ in reciprocal space defines a periodic BZ. In one dimension, this periodicity can be represented by $\mathbb{S}^1$, and in dimension d by $\underbrace{\mathbb{S}^1 \otimes \ldots \otimes \mathbb{S}^1}_{d}$.

Thus, all BZs in d-dimensions map onto a d-dimensional torus $\mathbb{T}^d$, irrespective of their microscopic geometric differences. Such a structure presents a closed manifold, which we will refer to as a Brillouin torus (BT). This is illustrated in Figures 11.4 and 11.5. Figure 11.4 shows how the one-dimensional dispersion wraps into the base torus $\mathbb{S}^1$ and a cylindrical bundle where the dispersion appears as an ellipse. Figure 11.5 indicates how the two-dimensional rectangular BZ folds into a torus.

The fact that $u_{n\mathbf{k}}$ is endowed with the lattice periodicity

$$u_{n\mathbf{k}}(\mathbf{x} + \mathbf{R}) = u_{n\mathbf{k}}(\mathbf{x})$$

allows us to regard the eigenvalue problem as that of the following:

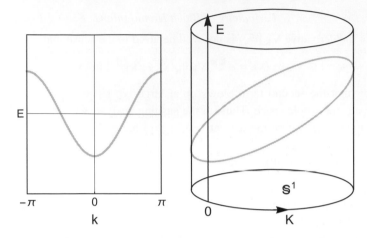

Figure 11.4 Left: conventional view of a 1D band structure in the first BZ. Right: a more topologically natural view in which the BZ is wrapped onto a circle and the band structure is plotted on a cylinder.

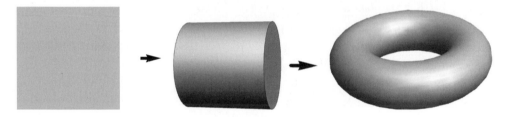

Figure 11.5 Illustrative example of mapping a two-dimensional rectangular BZ into a torus.

A **k**-dependent Hamiltonian, with **k**-independent boundary conditions.

The Hamiltonian depends on a parameter, while the eigenstates satisfy parameter-independent boundary conditions, and thus coexist in the same Hilbert space, a standard fiber. We identify the adiabatic parameter η with the wavevector **k**. Moreover, the domain where the **k** parameter varies, a BZ, is a closed surface having the geometry of a torus in 1D, 2D, and 3D. We also note that **k** appears in the Hamiltonian (11.7) as a kind of *vector potential*, although no magnetic field is present. Within this setting, we can proceed to investigate the possible occurrence of a geometric phase.

Bloch Bundles [69]

The action of a *Bloch Hamiltonian* $\mathcal{H}(\mathbf{k})$ on its $\mathfrak{N}$-dimensional Hilbert subspace, $\mathbb{H}_{\mathbf{k}} \cong \mathbb{C}^{\mathfrak{N}}$, engenders its Bloch eigenstates $u_{n\mathbf{k}}$ and eigenenergies $\varepsilon_n(\mathbf{k})$, $n = 1, \ldots, \mathfrak{N}$. $\mathbb{H}_{\mathbf{k}}$ represents the $\mathfrak{N}$ electronic degrees of freedom in the primitive lattice cell, comprising sites, orbitals, and spin, and forms the fiber over the BT at **k**. We can envision the action of $\mathcal{H}(\mathbf{k})$ as a form of reordering the fiber constituents. The union of all $\mathbb{H}_\mathbf{k}$ subspaces forms a vector bundle $\mathbb{H}$ on the base space $\mathbb{T}^d$. For periodic lattices in $d \leq 3$ dimension space, the total Berry curvature was shown to vanish [149, 167], hence the corresponding vector bundle is always trivial and isomorphic to $\mathbb{T}^d \times \mathbb{C}^{\mathfrak{N}}$.

The evolution of each eigenvalue $\varepsilon_n(\mathbf{k})$ as $\mathbf{k}$ changes on the BT defines a band n. For an insulator, the filled bands are separated from the empty ones by an energy gap. This allows us to delineate two subbundles of the complete trivial bundle: the valence bands bundle and the conduction bands bundle. The ground state consists of the aggregate of eigenstates corresponding to the filled valence bands. Such eigenstates are defined for each valence band and at each $\mathbf{k}$-point on the BT, up to a phase. The fiber bundle over the BT defined from the eigenstates of the valence bands is the object of our interest.

It is desirable to highlight the topological features of the valence bands bundle that produce the ground-state properties of topological insulators. This valence subbundle possesses a twisted topology whereas the complete bundle is trivial. To visualize the twist, the example of the normal strip and the Möbius strip provides a good starting point for a mental caricature. There, the base space manifold is a line that is joined into a circle, and the two different strips arise from different ways of identifying the fibers at the endpoint when the circle is formed. In the case of a two-dimensional insulator, for example, the base space is a two-torus, which is a square with opposite sides identified. On top of this manifold, an n-dimensional complex vector space is attached to each point as fiber. When we now perform the identification of the sides of the square to form the torus, the complex vector spaces along the edges also have to be identified. The process therefore involves gluing together the edges of a fiber bundle, with two manifold dimensions and n-complex fiber dimensions ($2 + 2n$-real dimensions). This is understandably hard to visualize, but mathematically the identification is determined by the connection, which is derived from the Hamiltonian $\mathcal{H}$.

We recall that the connection makes the covariant derivative zero through the relation

$$D_{k_\alpha} |u_{n\mathbf{k}}\rangle = \left(\partial_{k_\alpha} + i \mathcal{A}^\alpha_{n\mathbf{k}}\right) |u_{n\mathbf{k}}\rangle = 0.$$

Multiplying this from the left by $\langle u_{n\mathbf{k}}|$, we arrive at

$$\mathcal{A}^\alpha_{n\mathbf{k}} = i \langle u_{n\mathbf{k}} | \partial_{k_\alpha} u_{n\mathbf{k}} \rangle.$$

We illustrate this approach with a simple example: a two-dimensional insulator with just one occupied and one empty band, such that the Hilbert space is defined on $\mathbb{C}^2$. The eigenvalue equation

$$\mathcal{H}(\mathbf{k}) |n\mathbf{k}\rangle = \varepsilon_n(\mathbf{k}) |n\mathbf{k}\rangle$$

delineates a valence band ground state $|v\mathbf{k}\rangle$, and an empty conduction band state $|c\mathbf{k}\rangle$ with corresponding eigenvalues $\varepsilon_v(\mathbf{k}) < \varepsilon_c(\mathbf{k})$. These components can be defined over the two-torus $\mathbf{k} \in \mathbb{T}^2$. For simplicity, we focus on normalized ground states $\langle v\mathbf{k} | v\mathbf{k}\rangle = 1$.

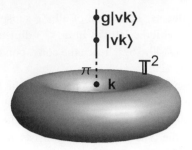

Figure 11.6 A fiber on the two-torus base.

$|v\mathbf{k}\rangle$ states are defined up to a global phase, such that $e^{i\phi(\mathbf{k})}|v\mathbf{k}\rangle \sim |v\mathbf{k}\rangle$, $\phi(\mathbf{k}) \in [0; 2\pi]$, and these states form a ray, namely,

$$\{g\,|v\mathbf{k}\rangle\,\big|\,g \in U(1)\},$$

which constitutes the fiber to be attached to the torus base manifold at $\mathbf{k}$, shown in Figure 11.6. A one-form Berry connection and a two-form Berry curvature

$$\mathcal{A}(\mathbf{k}) = i\,\langle v\mathbf{k}\,|\,d(v\mathbf{k})\rangle$$
$$\mathcal{F} = i\,\langle d(v\mathbf{k})\,|\,\wedge\,d(v\mathbf{k})\rangle = \mathcal{F}_{\mu\nu}\,dk^{\mu}\,\wedge\,dk^{\nu}$$
$$\mathcal{F}_{\mu\nu} = \text{``}\langle\partial_{\mu}(v\mathbf{k})\,|\,\wedge\,\partial_{\nu}(v\mathbf{k})\rangle \tag{11.8}$$

are defined over the bundle.

Thus we surmise that in a crystalline solid, the natural parameter space is the torus of electron crystal momentum $\mathbf{k}$. Varying $\mathbf{k}$ leads to a change in the electron Bloch eigenket $|u_{n\mathbf{k}}\rangle$ of an isolated band n within the unit cell that is described in terms of the *Abelian* Berry connection and Berry curvature as

$$\mathcal{A}_n(\mathbf{k}) = i\Big\langle u_{n\mathbf{k}}\Big|du_{n\mathbf{k}}\Big\rangle$$
$$\mathcal{F}_n(\mathbf{k}) = i\Big\langle du_{n\mathbf{k}}\Big|du_{n\mathbf{k}}\Big\rangle = \mathcal{F}_{\mu\nu}\,dk^{\mu}\,\wedge\,dk^{\nu} \Rightarrow \mathcal{F}_{\mu\nu} = \partial_{k_{\alpha}}\mathcal{A}_{\beta} - \partial_{k_{\beta}}\mathcal{A}_{\alpha}. \tag{11.9}$$

We take note here that we use $|u_{\mathbf{k}}\rangle$ and not $|\psi_{\mathbf{k}}\rangle$ in order for the $\mathbf{k}$-derivative to be well defined. We also point out that the interesting closed paths $\mathcal{C}$ on the torus are lines traversing the BZ from one face to the opposite one, as shown in Figure 11.7.

For an insulator with $\mathcal{N}$ occupied but mutually isolated bands, the Berry phase is

$$\gamma = i\sum_{n=1}^{\mathcal{N}}\oint_{\mathcal{C}}\mathcal{A}_{n\mu}(\mathbf{k})dk^{\mu} = i\sum_{n=1}^{\mathcal{N}}\oint_{\mathcal{C}}\Big\langle u_{n\mathbf{k}}\Big|\partial_{\mu}u_{n\mathbf{k}}\Big\rangle.$$

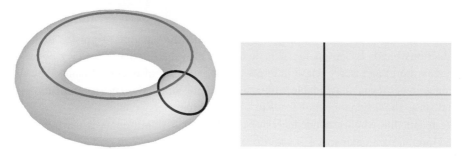

Figure 11.7 Closed paths on torus correspond to lines traversing two-dimensional rectangular BZ.

This Berry phase depends on the choice of the origin in the crystal unit cell. The only allowed values for centrosymmetric crystals with the origin at a center of inversion symmetry are $\gamma = 0$ and $\gamma = \pi$ modulo 2π, as demonstrated in the following simple example.

Berry's Phase for a One-Dimensional Insulator

We consider a single band in a 1D crystalline solid of periodicity a. We express the state $\psi_k(x)$ as

$$\psi_k(x) = e^{ikx}\, u_k(x) = \sqrt{\frac{a}{2\pi N}} \sum_n e^{ikna}\, w(x - na),$$

where $w(x)$ are Wannier functions and N the number of primitive cells. We write the corresponding Berry's phase as

$$\gamma = \frac{2\pi}{a} \int_{-\pi/a}^{\pi/a} dk \left\langle u_k \left| \frac{d}{dk} u_k \right\rangle = \frac{2\pi}{a} \int_{-\pi/a}^{\pi/a} dk \int_{-\infty}^{\infty} dx\, u_k^*(x) \frac{du_k(x)}{dk} \right.$$

$$= \frac{1}{N} \int_{-\infty}^{\infty} dx \int_{-\pi/a}^{\pi/a} dk \sum_{nm} e^{-ik(x-ma)}\, w^*(x - ma) \frac{d}{dk} e^{ik(x-na)}\, w(x - na)$$

$$= \delta_{nm} \frac{2\pi}{a} \int_{-\infty}^{\infty} dx\, x\, |w(x)|^2.$$

We identify the integral

$$\bar{x} = \int_{-\infty}^{\infty} dx\, x\, |w(x)|^2$$

with the probability center of $w(x)$ in the primitive cell. When there is no symmetry in the one-dimensional chain, $\bar{x}$ can assume any value. However, when inversion symmetry is present, it follows from the symmetry of the Wannier functions that $\bar{x}$ can assume two values only, $\bar{x} = 0, a/2$. It follows that γ can assume any value mod 2π in the absence of inversion, but can only assume the values 0 and π when inversion is present.

Degeneracy and Non-Abelian Berry Connection and Phase

In insulators, the group of occupied (valence) bands are separated from the unoccupied (conduction) bands by an energy gap. It is then useful to define the Berry connection, Berry curvature, and Berry phase in terms of the entire set of occupied bands, because in general degeneracy is unavoidable within such a group of bands. The generalization of the formalism to the degenerate case leads to a non-Abelian Berry connection.

We recall that when a nondegenerate state $|\psi_n\rangle$ of a Hamiltonian is carried around a closed loop in parameter space, it acquires Berry's phase

$$\left|\psi'_m\right\rangle - e^{i \oint_C \boldsymbol{\mathcal{A}}_m} \left|\psi_m\right\rangle = U(C) \left|\psi\right\rangle,$$

where $\left|\psi'_m\right\rangle$ is the final state vector. The geometric phase factor is just $U(C) \in U(1)$, an Abelian transformation. We show in Appendix 2 that when the eigenstate associated with an eigenvalue ε is part of an n-fold degenerate set, all n states must be considered simultaneously; and we have the matrix equation

$$\left|\psi'_i\right\rangle = \sum_{j=1}^{n} U_{ij}(C) \left|\psi_j\right\rangle, \qquad U \in U(n) \qquad \text{unitary group of order } n$$

$$U_{ij}(C) = \mathcal{P} \, e^{i \oint_C \boldsymbol{\mathcal{A}}_{ij}}$$

$$\boldsymbol{\mathcal{A}}_{ij} = i \langle \psi_i | d | \psi_j \rangle \implies \mathcal{A}_{ij}^{\alpha} = i \langle \psi_i | \partial_\alpha | \psi_j \rangle,$$

where $\alpha = 1, \ldots, d$ for a d-dimensional parameter space. The path-ordering operator $\mathcal{P}$ is necessary because $\boldsymbol{\mathcal{A}}$ is non-Abelian and does not commute with itself at different points on the circuit. $\boldsymbol{\mathcal{A}}$ is referred to as a $U(n)$ gauge potential.

In the case of $\mathcal{N}$ bands ($\mathcal{N} > 1$) separated by energy gaps from the rest of the spectrum, but not mutually isolated, the Abelian connection is replaced by its non-Abelian multiband generalization. We define the $U(\mathcal{N})$ Berry connection matrix

$$\mathbf{a}_{mn}(\mathbf{k}) = \left\langle u_{m\mathbf{k}} \middle| - i \nabla_{\mathbf{k}} \middle| u_{n\mathbf{k}} \right\rangle, \tag{11.10}$$

which shows that for a d-dimensional $\mathbf{k}$-space, $\mathbf{a}$ comprises $d \, \mathcal{N} \times \mathcal{N}$ component matrices. The non-Abelian curvature is defined as

$$\mathfrak{F}_{mn}^{\alpha\beta} = \partial_\alpha a_{mn}^{\beta} - \partial_\beta a_{mn}^{\alpha} - i \left[a^{\alpha}, a^{\beta} \right]_{mn},$$

where the $\mathbf{k}$-dependence is implicit.

11.1.4 Time-Reversal and Bloch Hamiltonian

Kramers' Degeneracy and Time-Reversal Invariant Momenta

When $\mathcal{H}$ is time-reversal invariant (TRI), namely, $[\mathcal{H}, \boldsymbol{\Theta}] = 0$, $\mathcal{H}(\mathbf{k})$ satisfies

$$\mathcal{H}(-\mathbf{k}) = \boldsymbol{\Theta} \, \mathcal{H}(\mathbf{k}) \, \boldsymbol{\Theta}^{-1} = \mathcal{H}(\mathbf{k}), \tag{11.11}$$

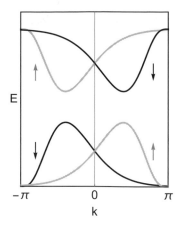

Figure 11.8 Kramers' pairs of bands. Note that each pair of bands are degenerate at the TRIM where $+k$ becomes equivalent to $-k$ due to the periodicity of the Brillouin zone. In this figure, there are two TRIMs, $k = 0$ and $k = \pi$ (which is equivalent to $k = -\pi$). The lifting of degeneracy at k values other than 0 and π comes from spin–orbit coupling.

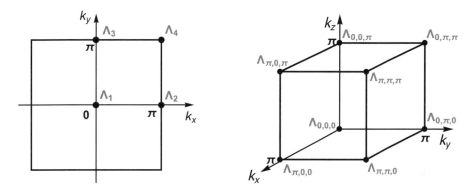

Figure 11.9 TRIMs. Left: there are four TRIMs for a 2D BZ; this figure shows the case of a square BZ. Right: there are eight TRIMs for a 3D BZ; this figure represents the case of a cubic BZ.

requiring the degeneracy of the Bloch states $\mathbf{k}$ and $-\mathbf{k}$, a Kramers' pair, as shown in Figure 11.8.

We note that there are time-reversal invariant momenta (TRIMs) in the BZ that satisfy the condition

$$\mathbf{k} = -\mathbf{k} + \mathbf{G} = \Lambda \qquad \Rightarrow \qquad \Lambda = \frac{\mathbf{G}}{2}.$$

At these points, the states $|u_\mathbf{k} \uparrow\rangle$ and $|u_\mathbf{k} \downarrow\rangle$ are degenerate, and the degeneracy is protected by TRI. TRIMs will be denoted by Λ. For a d-dimensional bulk material, there are 2^d inequivalent TRIM points, while for the boundary, or surface, the surface Brillouin zone (SBZ) is defined as a $(d-1)$-dimensional projection, in which there are 2^{d-1} number of surface TRIM. Figure 11.9 shows the 2D and 3D cases.

Another consequence of the Kramers' degeneracy is that the occupied Hilbert space of an insulator consists of an even number of bands, since the occupied states come in pairs.

Action of Θ on the Bloch Bundle

We envision that Θ will act as an anti-unitary map that relates the electronic Bloch states on the fiber at $\mathbf{k}$ to those on the fiber at $-\mathbf{k}$ of the vector bundle, over the BT base.

In a TRI system, the Bloch Hamiltonians at $\mathbf{k}$ and $-\mathbf{k}$ satisfy:

$$\mathcal{H}(-\mathbf{k}) = \Theta \mathcal{H}(\mathbf{k}) \Theta^{-1}$$

We start with defining the TR action on the BT as the map $\vartheta : \mathbb{T}^2 \to \mathbb{T}^2$, such that $\vartheta \mathbf{k} = -\mathbf{k}$ on the torus. The Θ action is then viewed as a lift to this map ϑ on the total Bloch bundle $\mathbb{T}^2 \times C^{\mathfrak{N}}$ describing the electronic states of all $\mathfrak{N}$ bands. It can be represented by a unitary matrix U_Θ that does not depend on the momentum $\mathbf{k}$ on the Brillouin torus. Hence, it is a map

$$\mathbb{T}^2 \times C^{\mathfrak{N}} \to \mathbb{T}^2 \times C^{\mathfrak{N}}$$

$$(\mathbf{k}, n) \to (\Theta \mathbf{k}, \Theta n) = (-\mathbf{k}, U_\Theta K n),$$

which sends the fiber of all bands $\mathbb{H}_\mathbf{k} \simeq C^{\mathfrak{N}}$ at $\mathbf{k}$ to the fiber $\mathbb{H}_{\Theta \mathbf{k}}$ at $\mathbf{k} = -\mathbf{k}$, or succinctly,

$$\Theta : \mathbb{H}_\mathbf{k} \to \mathbb{H}_{\Theta \mathbf{k}}.$$

We note that $\Theta^2 = -1$ maps a fiber onto itself.

Alternatively, we can say that TRI implies the existence of Kramers' pairs of eigenstates that live in different fibers, and are called Kramers' partners. $\Theta^2 = -1$ implies that these two Kramers' partners are orthogonal. Note that the orthogonality of these Kramers' partners in different fibers has only a meaning if we embed these fibers in the complete trivial bundle $\mathbb{T}^2 \times C^{\mathfrak{N}}$ corresponding to the whole state space of the Bloch Hamiltonian. Moreover, at TRIM points, the two partners of a Kramers' pair live in the same fiber. As they are orthogonal and possess the same energy, the spectrum is necessarily always degenerate at these TRIM points. We will see later in this chapter that the constraints imposed by the presence of these Kramers' partners around the valence Bloch bundle are at the origin of the $\mathbb{Z}_2$ topological order.

Berry Curvature and Time-Reversal Symmetry

We consider the wavefunctions $|u_1(\mathbf{k})\rangle$ and $|u_2(\mathbf{k})\rangle$ belonging to bands 1 and 2, respectively, which form a Kramers' pair and transform into each other under the action of the time-reversal operator Θ, namely

$$|u_2(\mathbf{k})\rangle = e^{i\chi(\mathbf{k})} \Theta |u_1(-\mathbf{k})\rangle,$$

where $\chi(\mathbf{k})$ is an arbitrary phase. The Berry connection is given by the following:

$$\mathcal{A}_1(-\mathbf{k}) = -i \langle u_1(-\mathbf{k}) | \nabla_\mathbf{k} u_1(-\mathbf{k})\rangle = -i \langle \nabla_\mathbf{k} \Theta u_1(-\mathbf{k}) | \Theta u_1(-\mathbf{k})\rangle$$
$$= i \langle \Theta u_1(-\mathbf{k}) | \nabla_\mathbf{k} \Theta u_1(-\mathbf{k})\rangle = \mathcal{A}_2(\mathbf{k}) + i \nabla \chi(\mathbf{k}).$$

Thus, the Berry curvatures $\mathcal{F}_2$ satisfy the relation

$$\mathcal{F}_2(\mathbf{k}) = \nabla_{\mathbf{k}} \times \mathcal{A}_2(\mathbf{k}) = -\mathcal{F}_1(-\mathbf{k}).$$

Therefore, the total Berry curvature

$$\mathcal{F}(\mathbf{k}) = \mathcal{F}_1(\mathbf{k}) + \mathcal{F}_2(\mathbf{k}) = -\mathcal{F}(-\mathbf{k})$$

because of time-reversal symmetry.

Symmetry Constraints on Berry Curvature

While Berry's phase depends on the path traced in the BZ, and the Berry connection depends on gauge choice, Berry's curvature $\mathcal{F}_n(\mathbf{k})$, for band n, is a uniquely defined function of $\mathbf{k}$ in the BZ. It has the following properties:

(a) Crystals with inversion (I) symmetry require $\mathfrak{F}(\mathbf{k}) = \mathfrak{F}(-\mathbf{k})$.
(b) Time-reversal symmetric crystals require $\mathfrak{F}(\mathbf{k}) = -\mathfrak{F}(-\mathbf{k})$. Thus, integrals over the BT involving $\mathfrak{F}(\mathbf{k})$ will vanish.
(c) $\mathfrak{F}(\mathbf{k})$ will vanish identically for crystal with both I and TR symmetry.
(d) The presence of other point group symmetries may impose further constraints on $\mathfrak{F}(\mathbf{k})$.

We note from (b) and (c) that the *Chern number has to vanish for time-reversal invariant systems*. Conversely, a nonzero Chern number requires breaking of time-reversal symmetry. In the quantum Hall effect, magnetic fields are used to break time-reversal symmetry. In time-reversal-invariant systems, we need another scheme for topological characterization in terms of a new topological invariant, namely, the $\mathbb{Z}_2$ topological number, which we will discuss in Section 11.4.

11.2 Integer Quantum Hall Effect

We consider a two-dimensional electron gas (2DEG) in the xy-plane, subject to a strong magnetic field $\mathbf{B}$ along the z-axis. For simplicity, we ignore the underlying lattice periodic potential. The corresponding Hamiltonian is

$$\mathcal{H} = \frac{1}{2m_e} \left(\mathbf{p} + e\,\mathbf{A} \right)^2,$$

where, for simplicity, we set $\hbar = c = 1$.

We use the Landau gauge

$$A_x = -B\,y, \qquad A_y = 0$$

and we impose periodic boundary conditions (PBCs) along the x-axis and open boundary conditions along the y-axis, while the size of the system is set to $L_x \times L_y$. The Bloch Hamiltonian for the 2DEG is then expressed as

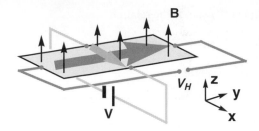

Figure 11.10 Schematic of the 2DEG system's geometry in the Hall configuration.

$$\mathcal{H}(\mathbf{k}) = e^{-ik_m x} \, \mathcal{H} \, e^{ik_m x}$$

$$= \frac{1}{2m_e} \, p_y^2 + e^{-ik_m x} \left[\frac{1}{2m_e} \, p_x^2 - \frac{e}{m_e} \, p_x \, By + e^2 \, B^2 \, y^2 \right] e^{ik_m x}$$

$$= \frac{1}{2m_e} \, p_y^2 + \frac{1}{2} m_e \omega_c^2 \left(y - k_m \ell_B^2 \right)^2, \tag{11.12}$$

where $\omega_c = \frac{eB}{m_e}$ and $\ell_B = \sqrt{\frac{\hbar}{eB}}$ is the magnetic length. The wavevector $k_m = \frac{2\pi m}{L_x}$, where m is an integer. We find that the eigenstates of this Hamiltonian are just those of a harmonic oscillator, centered at $y_m = k_m \, \ell_B^2$. Thus, we have a state that is localized in the y-direction and extended in the x-direction. The eigenvalues are the wavevector-independent Landau levels, given by

$$\varepsilon_n = \hbar\omega_c \left(\frac{1}{2} + n \right), \qquad n = 0, 1, 2 \dots \tag{11.13}$$

$$\psi_{n,k_m} = e^{-ik_m x} \, H_n(y - y_m), \tag{11.14}$$

where H_n is a Hermite polynomial. Since $y_m \leq L_y$, we have a maximum wavevector $k_{\max} = L_y / \ell_B^2$, and $m_{\max} = L_x L_y / 2\pi \ell_B^2$, which is just the number of states per Landau level. Thus, the density of states per unit area in each Landau level is

$$\mathcal{D}_{LL} = \frac{1}{2\pi \ell_B^2} = \frac{B}{\phi_0},$$

where $\phi_0 = h/e$ is the unit of quantum magnetic flux.

11.2.1 Laughlin's Gedanken Experiment

Laughlin put forward an argument, based on gauge invariance under flux insertion [118], to explain the quantization of the Hall conductivity. He made use of the PBCs to recast the experimental configuration into the cylinder of Figure 11.11, with its axis along the y-direction. Changing the wavevector $k_m \to k_{m+1}$ leaves the energy invariant, but shifts the eigenstates center by $\Delta y = \frac{2\pi \hbar}{eBL_x}$. We can express this change in terms of a gauge transformation

$$\mathbf{A} \to \mathbf{A} + \delta\mathbf{A} \implies \delta\mathbf{A} = \left(\delta A_x, \delta A_y \right) = \left(\frac{\delta\Phi}{L_x}, 0 \right),$$

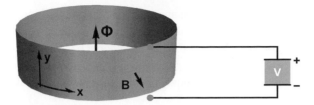

Figure 11.11 Schematic of the 2DEG system's geometry. A magnetic field **B** is applied perpendicular to the cylinder. The system is periodic along the x-axis, and the flux Φ penetrates the cylinder. An electric field is applied along the y-axis, where open boundary conditions are imposed. The Hall current is generated by the flux.

where we defined a fictitious Aharonov–Bohm type flux $\delta\Phi = \frac{2\pi\hbar}{e} = \frac{h}{e}$ that threads the cylinder along the y-axis. Within this setting, we consider the case where the Fermi energy lies in the gap between the *Landau levels* ε_n and ε_{n+1}. The change in energy due to the fictitious flux is given by

$$\delta\langle\mathcal{H}\rangle = \langle\mathcal{H}(\mathbf{A}+\delta\mathbf{A})\rangle - \langle\mathcal{H}(\mathbf{A})\rangle = e\delta A_x \left\langle \frac{\partial\mathcal{H}}{\partial A_x} \right\rangle = \frac{e\delta\Phi}{L_x} \left\langle \frac{p_x + eA_x}{m} \right\rangle$$

$$= \frac{\delta\Phi}{L_x} \int_0^{L_x} \int_0^{L_y} d\mathbf{x}\, j_x = \delta\Phi\, I_x = I_x\, \delta\Phi, \tag{11.15}$$

where we used the invariance of the states under the flux change $\delta\Phi = \frac{h}{e}$. Therefore, the Hall current is given as the change of energy by the fictitious flux as

$$I_x = \frac{\delta\langle\mathcal{H}\rangle}{\delta\Phi}.$$

The change of the energy is effected by the movement (y-shift) of the electron gas. Since the voltage $V(y)$ is applied along the y-axis, the electrons experience the electric field $\frac{V(L_y)-V(0)}{L_y}$, so that all electrons move along the y-axis by $\delta y = \frac{2\pi\ell^2}{L_x}$ by the flux. Therefore, $\delta\langle\mathcal{H}\rangle$ becomes

$$\delta\langle\mathcal{H}\rangle = -e\delta y\, \frac{V(L_y)-V(0)}{L_y}\, \nu\, m_{\max} = -e\nu\left(V(L_y)-V(0)\right) = \frac{h}{e}\, I_x, \tag{11.16}$$

where ν is the number of occupied Landau levels, and $m_{\max} = \frac{L_x L_y}{2\pi\ell^2}$ is the degeneracy per Landau level. Therefore, the Hall current becomes

$$I_x = \nu\frac{e^2}{h}\left(V(L_y)-V(0)\right),$$

showing a quantized conductivity $\nu\frac{e^2}{h}$.

Laughlin considered this process as a pump cycle, where the change in the flux drives the pump: increasing the flux creates an electromotive force (emf) around the ring, which, by the classical Hall effect, results in the transfer of charge, along the y-direction, from one edge to the other. The Aharonov–Bohm scenario stipulates that the Hamiltonian describing the system is gauge invariant under flux changes that are integer multiples of ϕ_0 [21]. Gauge invariance requires that after each cycle the pump returns to its original state. However, in general, that does not guarantee that the transported charge be the same in each cycle. How then can we explain the conductance quantization? In other words, the averaged charge transfer over many cycles are being quantized. That is where topological quantum numbers come into play.

11.2.2 Topological Invariant: The First Chern Number

The famous 1982 TKNN [176] paper, on the Hall conductance, marks the very first identification of a topological invariant integer coined the TKNN number. It was calculated from an integral representation of the Kubo formula. Later this integral was identified as the *first Chern class* of a $U(1)$ principal fiber bundle on a torus [22]: The fibers are the magnetic Bloch waves, and the torus corresponds to the magnetic Brillouin zone. The topological invariant was interpreted as the *first Chern number*, given by an integral of the Berry curvature.

To obtain the Chern number from the Kubo formula, we follow TKNN and consider a 2DEG in a periodic lattice potential, subjected to a perpendicular **B** field. We note that the Hamiltonian is not translationally invariant because of the extra magnetic length scale ℓ_B, which introduces an additional periodicity. We accommodate both periodicities by introducing a superlattice where both the lattice potential and the magnetic flux are commensurate with its periodicity. This means that an integer number of flux quanta ϕ_0 thread a *supercell*. In this case, a continuous **k** vector can be defined in the new Brillouin zone.

The Bloch Hamiltonian then takes the form

$$\mathcal{H}(\mathbf{k}) = \frac{1}{2m} \left[\mathbf{p} + \hbar\mathbf{k} + \frac{e}{c}\mathbf{A} \right]^2 + V(\mathbf{x}).$$

In Section 5.2.3 we obtained the following expression for the Hall conductivity

$$\sigma_{xy} = \frac{e^2}{i\hbar\Omega} \sum_{E_n < E_F < E_m} \left(\left\langle \frac{\partial u_{n,\mathbf{k}}}{\partial k_1} \middle| u_{m,\mathbf{k}} \right\rangle \left\langle u_{m,\mathbf{k}} \middle| \frac{\partial u_{n,\mathbf{k}}}{\partial k_2} \right\rangle - \left\langle \frac{\partial u_{n,\mathbf{k}}}{\partial k_2} \middle| u_{m,\mathbf{k}} \right\rangle \left\langle u_{m,\mathbf{k}} \middle| \frac{\partial u_{n,\mathbf{k}}}{\partial k_1} \right\rangle \right).$$

Using the completeness of the set $u_\mathbf{k}$ and the Pauli principle, and considering the case where the Fermi level lies in a gap above n filled bands (Landau levels in a flat potential), we arrive at

$$\sigma_{xy} = \frac{e^2}{i\hbar\Omega} \frac{\Omega}{(2\pi)^2} \sum_n \int_{\text{BZ}} d\mathbf{k} \left(\left\langle \frac{\partial u_{n,\mathbf{k}}}{\partial k_1} \middle| \frac{\partial u_{n,\mathbf{k}}}{\partial k_2} \right\rangle - \left\langle \frac{\partial u_{n,\mathbf{k}}}{\partial k_2} \middle| \frac{\partial u_{n,\mathbf{k}}}{\partial k_1} \right\rangle \right)$$

$$= \frac{e^2}{h} \frac{1}{2\pi i} \sum_n \int_{\mathbb{T}^2} d\mathbf{k} \left(\left\langle \frac{\partial u_{n,\mathbf{k}}}{\partial k_1} \middle| \frac{\partial u_{n,\mathbf{k}}}{\partial k_2} \right\rangle - \left\langle \frac{\partial u_{n,\mathbf{k}}}{\partial k_2} \middle| \frac{\partial u_{n,\mathbf{k}}}{\partial k_1} \right\rangle \right). \tag{11.17}$$

The last line indicates that the integral is carried out over the two-torus, and that the integrand is just the Berry curvature in **k**-space, yielding the first Chern number $\mathcal{C}$.[6]

The milestone TKNN discovery is that the Hall conductivity contains the topological invariant $\mathcal{C}$, which can be identified with the integer filling factor ν that appears in the expression given by von Klitzing:

$$\sigma_{\text{H}} = \frac{e^2}{h}\,\nu.$$

Each Landau level carries a Chern number $\mathcal{C} = 1$, and the Chern number of the system corresponds to the number of occupied Landau levels.

11.2.3 Edge States

We note that in the preceding scenario we were dealing with an insulator, where the chemical potential lies in an energy gap. This implies the absence of a net bulk current, a situation reinforced by the wavevector-independent eigenenergies. However, this picture must change near the edges of the sample, as shown in Figure 11.12, where a confining potential, $-eW(y)$, is introduced at the sample boundaries along the y-direction. Keeping the lowest-order correction to the electron dispersion, we write

$$\varepsilon_{n,k} = \hbar\omega_c\left(n + \frac{1}{2}\right) + \langle n,k| - eW(y)\,|n.k\rangle\,.$$

We assume that W varies slowly over ℓ_B, and get

$$\varepsilon_{n,k} \approx \hbar\omega_c\left(n + \frac{1}{2}\right) - eW(y_k), \qquad y_k = k\ell_B^2.$$

We are interested in the states that cross the chemical potential near either boundary, an edge state.

Since the density of states per unit area in each Landau level is $1/2\pi\ell_B^2$, the Hall current density $j_{n,x}$ for the nth Landau level is given as

$$\begin{aligned}
j_{n,x} &= -ev_x\frac{1}{2\pi\ell_B^2} = \frac{-e}{\hbar}\frac{1}{2\pi\ell_B^2}\,\Theta\left[\mu - \varepsilon_{n,k}\right]\frac{\partial\varepsilon_{n,k}}{\partial k} \\
&= \frac{e^2}{\hbar}\frac{1}{2\pi\ell_B^2}\,\Theta\left[\mu - \varepsilon_{n,k}\right]\frac{\partial W(y_k)}{\partial y_k}\frac{\partial y_k}{\partial k} \\
&= \frac{e^2}{2\pi\hbar}\,\Theta\left[\mu + eW(\mathbf{x}) - \hbar\omega_c\left(n + \frac{1}{2}\right)\right]\frac{\partial W}{\partial y},
\end{aligned} \qquad (11.18)$$

where $\Theta(x)$ is the Heaviside step function, and μ is the chemical potential.

---

[6] It is worth noticing here that the Kubo formula is a manifestation of the fluctuation-dissipation theorems, and as such it is expressed in terms of excitations of the system, whereas the Chern number is a ground-state property and the Hall conductivity is dissipationless. Moreover, we note that the topological nature of the Hall conductivity observable underscores its extreme robustness under variations of magnetic field, carrier density, substrate disorder, and more.

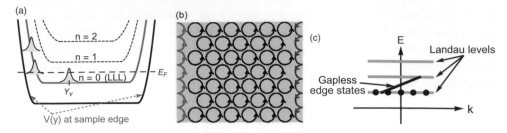

Figure 11.12 (a) Schematic of the confinement potential (black curve). The bulk Landau levels are shown as flat bands labeled by the quantum number n. The degenerate Landau harmonic oscillator states are represented by the Lorentzian on the lowest Landau level (LLL). The displaced edge states are shown as Lorentzian on the left edge of the LLL. (b) Classical chiral skipping orbits. (c) Gapless edge states dispersion (black line)

We note that the derivative $\partial W/\partial y$ has opposite signs on the opposite sample edges, and the electrons on these edges move in opposite directions! This uni-directional motion is called *chirality* of the edge state. Since the electrons of each edge are chiral, the only backscattering mechanism is via electron tunneling across the whole sample to the opposite edge. This is exponentially suppressed, therefore the Hall conductance is almost perfect.

We then write the Hall current $I_{n,x}$ at each edge as

$$I_n(\mathbf{x}) = \frac{e^2}{h} \int_0^{L_y} dy \; \Theta \left[\mu + eW(y_k) - \hbar\omega_c \left(n + \frac{1}{2} \right) \right] \frac{\partial W}{\partial y}.$$

The electric potential V applied across the sample in the y-direction becomes the difference of the confinement potential, so that the net Hall current can be expressed as

$$I_{n,x}^{\text{net}} = \frac{e^2}{h} \left[W(L_y) - W(0) \right] = \frac{e^2}{h} V.$$

When the nth Landau level with $0 \leq n \leq \nu - 1$ is filled, the Hall current I_x is given by

$$I_n(\mathbf{x}) = \nu \frac{e^2}{h} V.$$

In this case, each Landau level is shown to have the Chern number unity.

We note that the edge mode can also be visualized from a picture depicting a collection of cyclotron orbital motions. In this picture, the cyclotron orbital motions, arising from the magnetic field, cancel each other in the bulk, but not on the edges, where they form skipping orbits running parallel to the edges, as shown in Figure 11.12(b), which are just the chiral edge states that give rise to edge currents. From this intuitive picture, we see that the direction of the chiral edge mode results from that of cyclotron motion.

To sum up, we have learned that the electronic states move in momentum space by flux insertion, which together with the Hall current imply the existence of electronic states that cross the Fermi energy. Such states cannot be bulk states, since the chemical potential lies in a bulk gap. The fact that they are localized edge states is consistent with the quantized conductivity being independent of the system size. We also learned that a nonzero Chern number enforces the existence of boundary states carrying quantized conductivity. This illustrates the basic notion of bulk-boundary correspondence, which applies to broad classes of topological insulators.

11.3 Modern Theory of Crystalline Polarization

The microscopic modeling of bulk macroscopic polarization has been conventionally based on employing discrete and well-separated dipoles, à la Clausius–Mossotti. However, real dielectrics are quite different from such a simplistic model; for example, in some covalently bonded dielectrics, the valence electron charge density is continuous and may be delocalized.

If we consider a finite system, such as a molecule, we can clearly, and correctly, define its electronic contribution to the dipole moment as

$$\mathbf{d} = -e \int d\mathbf{x}\, \mathbf{x}\, n(\mathbf{x}), \tag{11.19}$$

where $n(\mathbf{x}) = \sum_i |\psi_i(\mathbf{x})|^2$ is the charge density of occupied states ψ_i. For a macroscopic solid, the quantity of interest is the macroscopic polarization, which we may naively define as the dipole of a macroscopic sample divided by its volume. However, when using integrals such as (11.19), we should be concerned about surface contributions. We recall that the standard approach to avoid difficulties arising from surface effects is to invoke periodic Born–von Kármán (BvK) boundary conditions. The BvK boundary conditions usually work in the thermodynamic limit of a large but finite sample. However, such an approach is inadequate for polarization, since surface contributions to the dipole moment do not vanish in the thermodynamic limit. Actually, the main contribution to the integral depends on what happens at the sample surface. Alternatively, we know that the macroscopic polarization must be an intensive quantity, which means that in the thermodynamic limit it must be insensitive to surface effects. Moreover, because of the unbounded nature of the position operator $\hat{\mathbf{x}}$, the integral (11.19) becomes ill defined when the wavefunctions involved satisfy the BvK boundary condition.

Because of these paradoxes, the formulation of a viable theory of macroscopic polarization, in the context of electronic structure theory, presented a major challenge for many years. The breakthrough came in 1992, when Resta proposed that for periodic systems, we should actually consider differences in polarization between an initial state and a final state of a crystal that can be connected by an adiabatic switching process [154]. Thus, within this new framework, the macroscopic polarization of an extended system is more appropriately treated as a dynamical property of currents in the adiabatic limit. Resta was able, with the aid of this approach, to define the polarization in terms of wavefunctions, rather than of charge densities, since the current essentially depends on the phase of the wavefunction, while the density depends on the square of its modulus. This definition presents a clear

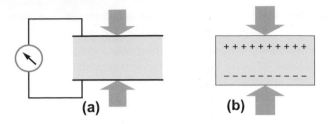

Figure 11.13 Implementations of piezoelectric effect measurement: (a) The insulator is in a shorted capacitor configuration, where the current is measured. (b) The sample is isolated, and piezoelectric strain creates a depolarizing field $\mathbf{E} - 4\pi\mathbf{P}$.

thermodynamic limit where BvK and Bloch states can be correctly introduced. Shortly afterwards King-Smith and Vanderbilt introduced a modern theory of macroscopic polarization in crystalline dielectrics, whereby the polarization was expressed in terms of a Berry phase [104].

The rationalization behind this modern viewpoint emerged from reflections on some experimental conventions: the absolute polarization of a crystal in a given state has never been measured as a bulk property, since it depends on sample termination. Instead, changes of the polarization with respect to physically relevant perturbations have been measured, producing familiar bulk properties such as permittivity, pyroelectricity, and piezoelectricity. In all cases, such derivatives or differences in the polarization are typically extracted from measurement of a macroscopic current.

We use the example of the piezoelectric effect to illustrate the main viewpoint of this modern theory. Schematics of two possible manifestations of the piezoelectric effect are shown in Figure 11.13. In both cases shown, the crystal is uniaxially strained along a piezoelectric axis. In (a), the strain is applied while the crystal is kept in a shorted capacitor configuration. In (b), the sample is kept isolated while being strained. In measuring process (a), the piezoelectricity appears as a flow of bulk current across the sample; nothing peculiar happens at the surfaces. Experiments measure such current flow. As we previously mentioned, in the modern theory of polarization, a gauge-invariant phase, Berry's phase, appears in the description of electronic polarization. This is actually consistent with the fact that quantum mechanically, currents are derived from the phase of the wavefunctions.

Accordingly, in the following analysis, we profile the polarization difference $\Delta\mathbf{P}$ between two different states of the same crystal as an integrated macroscopic current. The two states are adiabatically connected, and the time-dependent Hamiltonian is assumed to remain insulating. The Born–Oppenheimer approximation allows us to separate the ionic and electronic contributions as

$$\Delta\mathbf{P} = \Delta\mathbf{P}_{\text{ion}} + \Delta\mathbf{P}_{\text{el}},$$

where the ionic contribution comprises trivial electrostatic point charges. We shall, therefore, focus on the electronic contributions. We should caution that each of the two terms actually depends on the choice of origin in the primitive cell, whereas their sum is a macroscopic, origin-independent, bulk observable.

We consider the adiabatic change in the electronic polarization per unit volume of a crystal due to an adiabatic change in the crystal potential. The change in the potential is effected via a parameter λ, which varies in the range $0 \to 1$. For pedagogical simplicity, we shall consider the case where potential changes preserve the translational symmetry of the crystal. This means absence of macroscopic electric fields.

We compute the total change in polarization per unit volume, $\Delta \mathbf{P}$, using

$$\Delta \mathbf{P} = \int_0^1 d\lambda \, \frac{\partial \mathbf{P}}{\partial \lambda} = \int_0^{\Delta t} dt \, \frac{\partial \mathbf{P}}{\partial \lambda} \frac{d\lambda}{dt} = \int_0^{\Delta t} dt \, \mathbf{j}(t). \tag{11.20}$$

For adiabatic switching on, we require that $\frac{d\lambda}{dt} \ll 1/\Delta t E_g$, where E_g is the insulating gap energy. We note that (11.20) shows that $\Delta \mathbf{P}$ arises from the flow of polarization currents in the crystal.[7]

11.3.1 Berry Phase Theory of Polarization [155]

We now derive an expression for the current induced by an adiabatic change of the crystal Hamiltonian that connects the initial and final states. The adiabatic instantaneous eigenstates and ground state at time t are denoted by $\Psi_{n\mathbf{k}}(t)$ and $\Psi_0(t)$, respectively. To eliminate the dynamical phase, we shall work with the density operator that represents the adiabatic evolution, namely,

$$\rho(t) = |\Psi_0(t)\rangle \langle \Psi_0(t)| + \delta\rho(t),$$

where $\delta\rho(t)$ contains first-order correction to the instantaneous adiabatic $\Psi_0(t)$. We express the instantaneous velocity in this adiabatic process as

$$\mathbf{v}(t) = \text{Tr}\,[\mathbf{v}\,\rho(t)] = \langle\Psi_0(t)|\,\mathbf{v}\,|\Psi_0(t)\rangle + \sum_n \langle\Psi_0(t)|\,\delta\rho(t)\,|\Psi_n(t)\rangle\,\langle\Psi_n(t)|\,\mathbf{v}\,|\Psi_0(t)\rangle.$$

Next, we make the approximation

$$\frac{d\rho(t)}{dt} \simeq \frac{d}{dt}\,|\Psi_0(t)\rangle\,\langle\Psi_0(t)| = \big|\dot{\Psi}_0(t)\big\rangle\big\langle\Psi_0(t)\big| + \big|\Psi_0(t)\big\rangle\big\langle\dot{\Psi}_0(t)\big|,$$

where we neglect the higher-order term in the adiabatic parameter. The Heisenberg equation of motion for ρ becomes

$$i\hbar\left[\big|\dot{\Psi}_0(t)\big\rangle\big\langle\Psi_0(t)\big| + \big|\Psi_0(t)\big\rangle\big\langle\dot{\Psi}_0(t)\big|\right] = [\mathcal{H}(t), \rho(t)] = [\mathcal{H}(t), \delta\rho(t)] \tag{11.21}$$

since $|\Psi_0(t)\rangle\,\langle\Psi_0(t)|$ commutes with $\mathcal{H}(t)$. Taking the matrix elements of (11.21) between $|\Psi_0\rangle$ and $|\Psi_n\rangle$,

---

[7] We recognize here that in the adiabatic limit, $\Delta t \to \infty$, $\mathbf{j}_{\text{el}}(t) \to 0$. Thus, the mean value of the adiabatic current in the instantaneous crystalline ground state vanishes at any λ. Yet, we also note that the integral in (11.20) remains finite. This reveals the fundamental reason why the ground-state electron density does not provide the macroscopic polarization, while it provides other familiar adiabatic one-body observables.

$$\langle \Psi_0 | [\mathcal{H}(t), \delta\rho(t)] | \Psi_n \rangle = (\varepsilon_0 - \varepsilon_n) \langle \Psi_n | \delta\rho(t) | \Psi_n \rangle$$

$$= i\hbar \left[\langle \Psi_n(t) | \dot{\Psi}_0(t) \rangle + \langle \dot{\Psi}_0(t) | \Psi_n(t) \rangle \right]$$

$$\langle \Psi_0 | \delta\rho(t) | \Psi_n \rangle = \frac{i\hbar}{\varepsilon_0 - \varepsilon_n} \left[\langle \Psi_n(t) | \dot{\Psi}_0(t) \rangle + \langle \dot{\Psi}_0(t) | \Psi_n(t) \rangle \right].$$

We should note that the term with $m = n$ vanishes because of norm conservation.

Using the preceding relations, we write

$$\mathbf{v}(t) = \langle \Psi_0(t) | \mathbf{v} | \Psi_0(t) \rangle + i\hbar \sum_{n \neq 0} \left[\frac{\langle \dot{\Psi}_0(t) | \Psi_n(t) \rangle \langle \Psi_n(t) | \mathbf{v} | \Psi_0(t) \rangle}{\varepsilon_0 - \varepsilon_n} - c.c. \right]. \qquad (11.22)$$

Current Carried by a Filled Band

We consider the simple case of a crystalline insulator with a single filled band 0. First, we obtain the current carried by state $\Psi_{0\mathbf{k}}$ as

$$\mathbf{j}_{0\mathbf{k}} = -e \langle \Psi_{0\mathbf{k}}(t) | \mathbf{v} | \Psi_{0\mathbf{k}}(t) \rangle - ie\hbar \sum_{n \neq 0} \left[\frac{\langle \dot{\Psi}_{0\mathbf{k}}(t) | \Psi_{n\mathbf{k}}(t) \rangle \langle \Psi_{n\mathbf{k}}(t) | \mathbf{v} | \Psi_{0\mathbf{k}}(t) \rangle}{\varepsilon_{0\mathbf{k}}(t) - \varepsilon_{n\mathbf{k}}(t)} - c.c. \right]$$

$$= -e \langle u_{0\mathbf{k}}(t) | \mathbf{v} | u_{0\mathbf{k}}(t) \rangle - ie\hbar \sum_{n \neq 0} \left[\frac{\langle \dot{u}_{0\mathbf{k}}(t) | u_{n\mathbf{k}}(t) \rangle \langle u_{n\mathbf{k}}(t) | \mathbf{v} | u_{0\mathbf{k}}(t) \rangle}{\varepsilon_{0\mathbf{k}}(t) - \varepsilon_{n\mathbf{k}}(t)} - c.c. \right].$$

Next, we use the relations

$$\mathbf{v} = \frac{1}{\hbar} \nabla_{\mathbf{k}} \mathcal{H}(\mathbf{k})$$

$$| \nabla_{\mathbf{k}} u_{0,\mathbf{k}} \rangle = \frac{1}{\hbar} \sum_{n \neq 0} | u_{n,\mathbf{k}} \rangle \frac{\langle u_{n,\mathbf{k}} | \nabla_{\mathbf{k}} \mathcal{H}(\mathbf{k}) | u_{0,\mathbf{k}} \rangle}{\varepsilon_{0,\mathbf{k}} - \varepsilon_{n,\mathbf{k}}} = \sum_{n \neq 0} | u_{n,\mathbf{k}} \rangle \frac{\langle u_{n,\mathbf{k}} | \mathbf{v} | u_{0,\mathbf{k}} \rangle}{\varepsilon_{0,\mathbf{k}} - \varepsilon_{n,\mathbf{k}}}$$

to write

$$\mathbf{j}_{0\mathbf{k}}(t) = -e \langle u_{0\mathbf{k}}(t) | \mathbf{v} | u_{0\mathbf{k}}(t) \rangle - ie \left[\langle \dot{u}_{0\mathbf{k}}(t) | \nabla_{\mathbf{k}} u_{0\mathbf{k}}(t) \rangle - c.c. \right].$$

The current arising from the entire occupied band 0 is then

$$\mathbf{j}_0(t) = -\frac{e}{(2\pi)^3} \left[\int_{\text{BZ}} d\mathbf{k} \langle u_{0\mathbf{k}}(t) | \mathbf{v} | u_{0\mathbf{k}}(t) \rangle - i \int_{\text{BZ}} d\mathbf{k} \left[\langle \dot{u}_{0\mathbf{k}}(t) | \nabla_{\mathbf{k}} u_{0\mathbf{k}}(t) \rangle - c.c. \right] \right]. \qquad (11.23)$$

The first integral over the Brillouin zone vanishes. The integrand in the second term of (11.23) is just a Berry curvature component in the four-dimensional $(\mathbf{k}, t)$-space. Taking into account band double occupancy, the total band current is

$$\mathbf{j}_0(t) = -i \frac{2e}{(2\pi)^3} \int_{\text{BZ}} d\mathbf{k} \left[\langle \dot{u}_{n\mathbf{k}}(t) | \nabla_{\mathbf{k}} u_{n\mathbf{k}}(t) \rangle - \langle \nabla_{\mathbf{k}} u_{n\mathbf{k}}(t) | \dot{u}_{n\mathbf{k}}(t) \rangle \right]. \qquad (11.24)$$

Band Contribution to Polarization

Using (11.24), we arrive at the single-band contribution to the polarization

$$\Delta\mathbf{P}_n = \frac{-i2e}{(2\pi)^3} \int_0^{\Delta t} dt \int_{\text{BZ}} d\mathbf{k} \left[\left\langle \dot{u}_{n\mathbf{k}}(t) \middle| \nabla_{\mathbf{k}} u_{n\mathbf{k}}(t) \right\rangle - \left\langle \nabla_{\mathbf{k}} u_{n\mathbf{k}}(t) \middle| \dot{u}_{n\mathbf{k}}(t) \right\rangle \right]$$

$$= \frac{-i2e}{(2\pi)^3} \int_0^1 d\lambda \int_{\text{BZ}} d\mathbf{k} \left[\left\langle \partial_\lambda u_{n\mathbf{k}}(\lambda) \middle| \nabla_{\mathbf{k}} u_{n\mathbf{k}}(\lambda) \right\rangle - \left\langle \nabla_{\mathbf{k}} u_{n\mathbf{k}}(\lambda) \middle| \partial_\lambda u_{n\mathbf{k}}(\lambda) \right\rangle \right], \quad (11.25)$$

where we used the chain rule $\dfrac{\partial u}{\partial t} = \dfrac{\partial u}{\partial \lambda} \dfrac{d\lambda}{dt}$ in the last line. Curiously, we find that the sum over empty states disappears from this expression, underscoring the physics that the rate of change of polarization with λ should be a property of an occupied band.

We now carry out the integral over λ. To reveal the physical content of (11.25), we first adopt the gauge defined in (11.5) and (11.6):

$$\psi_{\mathbf{k}v}^{(\lambda)}(\mathbf{x}) = \psi_{\mathbf{k}+\mathbf{G}, v}^{(\lambda)}(\mathbf{x}), \quad u_{\mathbf{k}v}^{(\lambda)}(\mathbf{x}) = e^{-i\mathbf{G}\cdot\mathbf{x}} u_{\mathbf{k}+\mathbf{G}, v}^{(\lambda)}(\mathbf{x}).$$

Then we integrate (11.25) by parts, and obtain

$$\int_0^1 d\lambda \left[\left\langle \partial_\lambda u_{n\mathbf{k}}(\lambda) \middle| \nabla_{\mathbf{k}} u_{n\mathbf{k}}(\lambda) \right\rangle - \left\langle \nabla_{\mathbf{k}} u_{n\mathbf{k}}(\lambda) \middle| \partial_\lambda u_{n\mathbf{k}}(\lambda) \right\rangle \right]$$

$$= \left[\langle u_{n\mathbf{k}}(\lambda) | \nabla_{\mathbf{k}} | u_{n\mathbf{k}}(\lambda) \rangle \right]_0^1 - \int_0^1 d\lambda \, \nabla_{\mathbf{k}} \, \langle u_{n\mathbf{k}}(\lambda) | \partial/\partial\lambda | u_{n\mathbf{k}}(\lambda) \rangle. \quad (11.26)$$

This choice of gauge renders $\langle u_{n\mathbf{k}}(\lambda) | \partial/\partial\lambda | u_{n\mathbf{k}}(\lambda) \rangle$ periodic in $\mathbf{k}$. Consequently, the integral of its gradient over BZ vanishes, and the second term makes no contribution. We arrive at

$$\Delta\mathbf{P}_n = \frac{-i2e}{(2\pi)^3} \int_{\text{BZ}} d\mathbf{k} \left[\langle u_{n\mathbf{k}}(\lambda) | \nabla_{\mathbf{k}} | u_{n\mathbf{k}}(\lambda) \rangle \right]_0^1 = \mathbf{P}_n(1) - \mathbf{P}_n(0) \quad (11.27)$$

$$\mathbf{P}_n(\lambda) = \frac{-i2e}{(2\pi)^3} \int_{\text{BZ}} d\mathbf{k} \, \langle u_{n\mathbf{k}}(\lambda) | \nabla_{\mathbf{k}} | u_{n\mathbf{k}}(\lambda) \rangle. \quad (11.28)$$

The integrand is just Berry's connection of band n. Equation (11.27) is the central result of the modern theory of polarization.

If the crystal remains in its insulating state for $0 \leq \lambda \leq 1$, we sum over M occupied bands and obtain

$$\Delta\mathbf{P} = \frac{-i2e}{(2\pi)^3} \sum_{n=1}^M \int_{\text{BZ}} d\mathbf{k} \left[\langle u_{n\mathbf{k}}(\lambda) | \nabla_{\mathbf{k}} | u_{n\mathbf{k}}(\lambda) \rangle \right]_0^1 = \mathbf{P}(1) - \mathbf{P}(0). \quad (11.29)$$

To reveal the Berry phase content of the integral, we consider the special case of a simple cubic crystal with lattice constant a, and write the component of $\mathbf{P}(\lambda)$ in the z-direction as

$$P_{nz}(\lambda) = -\frac{2e}{(2\pi)^3} \int_{-\pi/a}^{\pi/a} dk_x \int_{-\pi/a}^{\pi/a} dk_y \left[i \int_{-\pi/a}^{\pi/a} dk_z \, \langle u_{n\mathbf{k}} | \partial_{k_z} u_{n\mathbf{k}} \rangle \right],$$

where the square parenthesis highlights the Berry phase.

The One-Dimensional Case

We shall carry out, in detail, the analysis for the a 1D crystal of periodicity a, and with only one doubly occupied band. In this simplified case, (11.27) takes the form

$$\Delta \mathbf{P}_{\mathrm{el}} = \frac{2e}{\pi} \, \Im \int_0^1 d\lambda \int_{-\pi/a}^{\pi/a} dk \left\langle \frac{\partial}{\partial k} u_{\mathbf{k}v}^{(\lambda)} \Big| \frac{\partial}{\partial \lambda} u_{\mathbf{k}v}^{(\lambda)} \right\rangle. \tag{11.30}$$

Equation (11.30) is a two-dimensional integral in the (k, λ) plane, over the domain shown in Figure 11.14. The integrand can also be expressed as the curl of a vector field:

$$\Im \left\langle \frac{\partial}{\partial k} u_{\mathbf{k}v}^{(\lambda)} \Big| \frac{\partial}{\partial \lambda} u_{\mathbf{k}v}^{(\lambda)} \right\rangle = -\frac{i}{2} \left[\frac{\partial}{\partial k} \left\langle u_{\mathbf{k}v}^{(\lambda)} \Big| \frac{\partial}{\partial \lambda} u_{\mathbf{k}v}^{(\lambda)} \right\rangle - \frac{\partial}{\partial \lambda} \left\langle u_{\mathbf{k}v}^{(\lambda)} \Big| \frac{\partial}{\partial k} u_{\mathbf{k}v}^{(\lambda)} \right\rangle \right].$$

We can convert this area integral into a line integral with the aid of Stokes' theorem, and define the two-dimensional Berry connection as

$$d\gamma = \mathcal{A} \cdot d\boldsymbol{\ell} = i \left[\left\langle u_{\mathbf{k}v}^{(\lambda)} \Big| \frac{\partial}{\partial k} u_{\mathbf{k}v}^{(\lambda)} \right\rangle dk + \left\langle u_{\mathbf{k}v}^{(\lambda)} \Big| \frac{\partial}{\partial \lambda} u_{\mathbf{k}v}^{(\lambda)} \right\rangle d\lambda \right],$$

which defines the phase change of $u_{\mathbf{k}v}^{(\lambda)}$ for an infinitesimal variation of k and λ. Thus, we can express (11.30) as a Berry phase by integrating this connection over the boundary of the two-dimensional domain shown in Figure 11.14, namely,

$$\Delta \mathbf{P}_{\mathrm{el}} = \frac{e}{\pi} \oint d\gamma.$$

It is easy to show that the contributions of the two vertical sides in figure cancel. Thus, the polarization difference becomes

$$\Delta \mathbf{P}_{\mathrm{el}} = \frac{ie}{\pi} \int_{-\pi/a}^{\pi/a} dk \left[\left\langle u_{\mathbf{k}v}^{(1)} \Big| \frac{\partial}{\partial k} u_{\mathbf{k}v}^{(1)} \right\rangle - \left\langle u_{\mathbf{k}v}^{(0)} \Big| \frac{\partial}{\partial k} u_{\mathbf{k}v}^{(0)} \right\rangle \right]. \tag{11.31}$$

Each of the two terms is separately gauge invariant, but we have to take into account both terms, as well as the ionic contribution, in order to obtain a *translationally invariant* form for the total polarization difference.

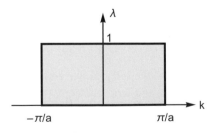

Figure 11.14 Integration domain in the (k, λ) plane. Using Stokes' theorem, the integral is transformed into a circuit integral over the boundary.

11.3.2 Reformulation in Terms of Wannier Functions

To provide a more intuitive picture of the modern theory, we reformulate the foregoing results using localized Wannier functions instead of delocalized Bloch states. In this setting, the electronic contribution to the macroscopic polarization P is expressed in terms of the dipole of the Wannier charge distribution associated with a single unit cell. P is then reformulated as a property of a localized charge distribution, apparently free of phase information. We find that the form of (11.28) is particularly simple when written in terms of the Wannier functions $W_n^{(\lambda)}(\mathbf{x})$ of the occupied bands. We recall the relation between Bloch states and Wannier function as

$$u_{n\mathbf{k}}^{(\lambda)}(\mathbf{x}) = \frac{1}{\sqrt{N}} \sum_{\mathbf{R}} e^{-i\mathbf{k}\cdot(\mathbf{x}-\mathbf{R})} W_n^{(\lambda)}(\mathbf{x}-\mathbf{R}),$$

where N is the number of primitive cells, and the sum over $\mathbf{R}$ runs over all real-space lattice vectors. Substituting in (11.28), we find the simple result that

$$\mathbf{P}(\lambda) = \frac{-i2e}{(2\pi)^3} \sum_{n=1}^{M} \int_{BZ} d\mathbf{k} \langle u_{n\mathbf{k}}(\lambda) | \nabla_{\mathbf{k}} | u_{n\mathbf{k}}(\lambda) \rangle = \frac{2e}{\Omega} \sum_{n=1}^{M} \int d\mathbf{x}\, \mathbf{x} \left| W_n^{(\lambda)}(\mathbf{x}) \right|^2. \quad (11.32)$$

Physically, (11.27) and (11.32) state that the change in polarization of the solid is proportional to the displacement of the center of charge of the Wannier functions induced by the adiabatic change in the Hamiltonian.

Wannier functions are strongly gauge dependent, since they depend on the particular choice of phases of $|u_{\mathbf{k}}\rangle$ used in the periodic gauge, which can be arbitrary. However, their centers are gauge-invariant modulo a lattice vector. To demonstrate this uncertainty, we consider the case where the Hamiltonians at $\lambda = 0$ and 1 are identical, and $u_{\mathbf{k}n}^{(0)}(\mathbf{x})$ and $u_{\mathbf{k}n}^{(1)}(\mathbf{x})$ can at most differ by a phase factor so that

$$u_{\mathbf{k}n}^{(1)}(\mathbf{x}) = e^{i\theta_{\mathbf{k}n}} u_{\mathbf{k}n}^{(0)}(\mathbf{x}).$$

In this limit, (11.29) reduces to

$$\Delta\mathbf{P} = \frac{2e}{4\pi^3} \sum_{n=1}^{M} \int_{BZ} d\mathbf{k}\, \nabla_{\mathbf{k}}\theta_{\mathbf{k}n}.$$

With the periodic choice of gauge, $e^{i\theta_{\mathbf{k}n}}$ must be periodic in $\mathbf{k}$. The most general form for the phase angle under these circumstances is $\theta_{\mathbf{k}n} = \beta_{\mathbf{k}n} + \mathbf{k}\cdot\mathbf{R}_n$, where $\beta_{\mathbf{k}n}$ is periodic in $\mathbf{k}$. We thus conclude that

$$\Delta\mathbf{P} = \frac{2e}{4\pi^3} \sum_{n=1}^{M} \mathbf{R}_n.$$

The change in polarization per unit volume for paths where the Hamiltonian returns to itself is therefore quantized in units of $(2e/\Omega)\mathbf{R}$.

The One-Dimensional Case

We return to one band in 1D: the Wannier function center in the unit cell at the origin is just

$$x_0 = \langle w_0 | \, x \, | w_0 \rangle \, .$$

Writing the Wannier function as

$$|w_0\rangle = \frac{a}{2\pi} \int dk \, e^{ikx} \, |u_k\rangle \, ,$$

we find that

$$x \, |w_0\rangle = \frac{a}{2\pi} \int dk \, \left(-i \partial_k \, e^{ikx} \right) |u_k\rangle = \frac{a}{2\pi} \int dk \, e^{ikx} \, i \, |\partial_k \, u_k\rangle \, ,$$

where an integration by parts has been used. Then

$$x_0 = -\frac{a}{2\pi} \, \Im \int_0^{2\pi/a} dk \, \langle u_k | \, \partial_k \, u_k \rangle = \frac{a}{2\pi} \, \gamma \, .$$

Consequently, the Berry phase γ is nothing other than a measure of the location of the Wannier center in the unit cell. The fact that γ was previously shown to be invariant with respect to choice of gauge implies that the same is true of the Wannier center x_0.

11.4 Time-Reversal Symmetry, Time-Reversal Polarization, and $\mathbb{Z}_2$

We have demonstrated in Section 11.1.4 that the Chern number has to vanish for systems obeying time-reversal symmetry. However, in their 2005 seminal work [98], C. Kane and G. Mele demonstrated that topological order and TRI can actually coexist in insulators. They showed that the presence of strong, TR symmetric spin–orbit interaction, which acts like an opposite magnetic field on $\uparrow$ and $\downarrow$ spins, may result in an electronic band structure with a topological order characterized by a new topological invariant coined $\mathbb{Z}_2$.[8] This invariant assumes only two integer values 0 and 1, indicating a trivial and a topological phase, respectively.

In this section, we present a method, developed by Fu and Kane [71], to derive an expression for the $\mathbb{Z}_2$ topological invariant. It employs a contrived implement, defined as a *time-reversal polarization*, to differentiate between the two classes of TRI insulator systems. We start with developing a matrix representation of Θ within Bloch state bases.

11.4.1 Matrix Representation of Θ: The Sewing Matrix

We recall that spin-1/2 systems obey the condition $\Theta^2 = -1$. Writing $\Theta = U \, K$, where U and K are the unitary and conjugation operators, respectively, yields

---

[8] Remember, in mathematics the group of integers is called $\mathbb{Z}$ and its quotient group classifying odd and even integers is called $\mathbb{Z}_2$. Hence, a $\mathbb{Z}_2$ index gives a topological classification based on parity.

$$U\,K\,U\,K = U\,U = -1 \qquad \Rightarrow \qquad U = -U^{\mathrm{T}}$$

so that U is antisymmetric and unitary.

Now we discuss a very useful matrix representation of the TR operator in the Bloch states basis. We start with defining the matrix as

$$w_{mn}(\mathbf{k}) = \big\langle u_{m,-\mathbf{k}} \big| \Theta \big| u_{n\mathbf{k}} \big\rangle, \tag{11.33}$$

with m, n band indices. It relates states at $\mathbf{k}$ with ones at $-\mathbf{k}$. It has been coined, appropriately, the *sewing matrix*. We then find that

$$\begin{aligned}
w_{mn}(\mathbf{k}) &= \big\langle u_{m,-\mathbf{k}} \big| \Theta \big| u_{n\mathbf{k}} \big\rangle \rightarrow u^*_{m,-\mathbf{k}}\, U\, K\, u_{n,\mathbf{k}} = u^*_{m,-\mathbf{k}}\, U\, u^*_{n,\mathbf{k}} \\
&= -u^*_{n,\mathbf{k}}\, U^{\mathrm{T}}\, K\, u_{m,-\mathbf{k}} \rightarrow -\big\langle u_{n,\mathbf{k}} \big| \Theta \big| u_{m,-\mathbf{k}} \big\rangle = -w_{nm}(-\mathbf{k}) = -w^*_{mn}(\mathbf{k}).
\end{aligned} \tag{11.34}$$

Using (11.33) and (11.34) we can obtain an expression relating the Bloch states $\big| u_{m,-\mathbf{k}} \big\rangle$ and $\big| u_{n\mathbf{k}} \big\rangle$, in the form

$$\big| u_{m,-\mathbf{k}} \big\rangle = \Theta \big| u_{m,\mathbf{k}} \big\rangle = \sum_n \big| u_{n,-\mathbf{k}} \big\rangle \big\langle u_{n,-\mathbf{k}} \big| \Theta \big| u_{m,\mathbf{k}} \big\rangle = \sum_n w^*_{mn}(\mathbf{k}) \Theta \big| u_{n\mathbf{k}} \big\rangle . \tag{11.35}$$

Unitarity of $w_{mn}(\mathbf{k})$

$$\begin{aligned}
\sum_m w^\dagger_{lm}(\mathbf{k})\, w_{mn}(\mathbf{k}) &= \sum_m \big\langle u_{l,\mathbf{k}} \big| \Theta \big| u_{m,-\mathbf{k}} \big\rangle \big\langle u_{m,-\mathbf{k}} \big| \Theta \big| u_{n,\mathbf{k}} \big\rangle \\
&= \big\langle u_{l,\mathbf{k}} \big| u_{n,\mathbf{k}} \big\rangle = \delta_{ln}
\end{aligned}$$

where the sum over m is taken over the complete set of occupied and empty states at $-\mathbf{k}$.

Equation (11.34) implies that at a TRIM $\mathbf{\Lambda}$, the matrix $\mathbf{w}$ becomes antisymmetric

$$w_{mn}(\mathbf{\Lambda}) = -w_{nm}(\mathbf{\Lambda}).$$

In case of only two occupied bands, $\mathbf{w}$ is a 2×2 matrix, and at $\mathbf{\Lambda}$ it takes the form

$$\mathbf{w}(\mathbf{\Lambda}) = \begin{pmatrix} 0 & w_{12}(\mathbf{\Lambda}) \\ -w_{12}(\mathbf{\Lambda}) & 0 \end{pmatrix} = w_{12}(\mathbf{\Lambda}) \begin{pmatrix} 0 & 1 \\ -1 & 0 \end{pmatrix}.$$

Non-Abelian Berry Connection

For systems satisfying TRI, we can further explore the properties of the Berry connection matrix by applying (11.35) together with the TR property

$$\langle \Theta\psi | \Theta\phi \rangle = \langle \phi | \psi \rangle$$

to find

$$
\begin{aligned}
a_{mn}(-\mathbf{k}) &= \langle u_{m,-\mathbf{k}}| -i\nabla_{-\mathbf{k}} |u_{n,-\mathbf{k}}\rangle = \langle u_{m,-\mathbf{k}}| i\nabla_{\mathbf{k}} |u_{n,-\mathbf{k}}\rangle \\
&= i\sum_{jl} w_{mj}(\mathbf{k})\langle \Theta u_{j\mathbf{k}}| \nabla_{\mathbf{k}} w_{nl}^*(\mathbf{k}) |\Theta u_{l\mathbf{k}}\rangle \\
&= \sum_{jl} w_{mj}(\mathbf{k})\langle \Theta u_{j\mathbf{k}}| i\nabla_{\mathbf{k}} |\Theta u_{l\mathbf{k}}\rangle w_{nl}^*(\mathbf{k}) + i\sum_{jl} w_{mj}(\mathbf{k})\langle \Theta u_{j\mathbf{k}} |\Theta u_{l\mathbf{k}}\rangle \nabla_{\mathbf{k}} w_{nl}^*(\mathbf{k}) \\
&= \sum_{jl} w_{mj}(\mathbf{k})\langle u_{j\mathbf{k}}| -i\nabla_{\mathbf{k}} |u_{l\mathbf{k}}\rangle^* w_{nl}^*(\mathbf{k}) + i\sum_{l} w_{ml}(\mathbf{k}) \nabla_{\mathbf{k}} w_{nl}^*(\mathbf{k})
\end{aligned}
$$

and obtain

$$
\mathbf{a}(-\mathbf{k}) = \mathbf{w}(\mathbf{k})\, \mathbf{a}^*(\mathbf{k})\, \mathbf{w}^\dagger(\mathbf{k}) + i\mathbf{w}(\mathbf{k})\, \nabla_{\mathbf{k}} \mathbf{w}^\dagger(\mathbf{k}) \tag{11.36}
$$

Taking the trace, we get

$$
\mathrm{Tr}\,[\mathbf{a}(-\mathbf{k})] = \mathrm{Tr}\left[\mathbf{a}^*(\mathbf{k})\right] + i\,\mathrm{Tr}\left[\mathbf{w}(\mathbf{k})\, \nabla_{\mathbf{k}} \mathbf{w}^\dagger(\mathbf{k})\right],
$$

where we used the cyclic property of the trace in the first term on the right. Using the following properties

$$
\mathrm{Tr}\,[\mathbf{a}(\mathbf{k})] = \mathrm{Tr}\left[\mathbf{a}^*(\mathbf{k})\right]
$$

$$
\mathbf{w}(\mathbf{k})\, \nabla_{\mathbf{k}} \mathbf{w}^\dagger(\mathbf{k}) = -\left(\nabla_{\mathbf{k}} \mathbf{w}^\dagger(\mathbf{k})\right)\mathbf{w}(\mathbf{k}) \qquad \Leftrightarrow \qquad \mathbf{w}(\mathbf{k})\, \mathbf{w}^\dagger(\mathbf{k}) = \mathbb{I},
$$

we interchange $\mathbf{k}$ and $-\mathbf{k}$ and write

$$
\mathrm{Tr}\,[\mathbf{a}(\mathbf{k})] = \mathrm{Tr}\,[\mathbf{a}(-\mathbf{k})] + i\,\mathrm{Tr}\left[\mathbf{w}^\dagger(\mathbf{k})\, \nabla_{\mathbf{k}} \mathbf{w}(\mathbf{k})\right]. \tag{11.37}
$$

This relation will be used in the calculation of the $\mathbb{Z}_2$ topological invariant.

Time-Reversal Polarization

In order to pedagogically derive the topological invariant associated with 2D electron systems preserving TRS, we follow Fu and Kane and consider a 1D system with length L and lattice constant $a = 1$. We only consider two bands that form a Kramers' pair, and denote their Bloch kets as $|u_1(k)\rangle$ and $|u_2(k)\rangle$. We write the polarization as

$$
P = \frac{1}{2\pi} \int_{-\pi}^{\pi} dk\, \mathcal{A}(k)
$$

$$
\mathcal{A}(k) = -i\,\langle u_1(k)| \partial_k |u_1(k)\rangle - i\,\langle u_2(k)| \partial_k |u_2(k)\rangle
$$

$$
= a_{11}(k) + a_{22}(k) = \mathrm{Tr}[\mathbf{a}].
$$

We note that for such a band pair, the TR operation allows us to obtain a physical quantity of one band, if we know the same quantity of its partner. We thus define the contribution from each band as a partial polarization, namely,

$$P_i = \frac{1}{2\pi} \int_{-\pi}^{\pi} dk \, a_{ii}(k), \qquad i = 1, 2$$

$$P = P_1 + P_2.$$

We note here that the 1D BZ can be mapped onto the torus $\mathbb{T}^1$. Consequently, the integral for P is just that of the corresponding Chern number, which vanishes for TRI systems. Fu and Kane defined a *time-reversed polarization* as

$$P_\theta = P_1 - P_2 = 2P_1 - P.$$

> Intuitively, P_θ gives the difference in charge polarization between spin-up and spin-down bands, since $|u_1(k)\rangle$ and $|u_2(k)\rangle$ form a Kramers' pair.

We need to highlight the physical significance of P_θ, and the role it plays in TRI insulators classification. Fu and Kane considered a family of 1D bulk-gapped Hamiltonians $\mathcal{H}(t)$, parameterized by a cyclic parameter t subject to the constraints

$$\mathcal{H}(t + T) = \mathcal{H}(t) \tag{11.38}$$

$$\frac{d\mathcal{H}}{dt} \ll \frac{\mathcal{H} \, \Delta E}{\hbar} \qquad \text{Adiabaticity condition}$$

$$\Theta \, \mathcal{H}(t) \, \Theta^{-1} = \mathcal{H}(-t) \tag{11.39}$$

where ΔE is the gap energy. This can be understood as an adiabatic pumping cycle, with t playing the role of time or pumping parameter.

From (11.38) and (11.39), one can see that the system is TR symmetric at $t = 0$ and $T/2$. At these times, the Kramers' degeneracy must be fulfilled at every k. Consequently, we require that the time-reversed version of $|u_2(k)\rangle \; \to \; \Theta \, |u_2(k)\rangle$ be equal to $|u_1(-k)\rangle$ except for a phase factor. Hence, at $t = 0$ and $T/2$,

$$\Theta \, |u_2(k)\rangle = e^{-i\chi(k)} \, |u_1(-k)\rangle . \tag{11.40}$$

$$\Theta \, |u_1(k)\rangle = -e^{-i\chi(-k)} \, |u_2(-k)\rangle. \tag{11.41}$$

Equation (11.41) follows from (11.40), because of the property $\Theta^2 = -1$. Using these relations, the **w** matrix can be shown to become

$$\mathbf{w}(k) = \begin{pmatrix} 0 & e^{-i\chi(k)} \\ -e^{-i\chi(-k)} & 0 \end{pmatrix}. \tag{11.42}$$

Now we calculate P_1 at the TR symmetric times. First, using (11.41) one may obtain

$$a_{11}(-k) = \langle u_1(-k)| \, i\partial_k \, |u_1(-k)\rangle = \left\langle \Theta \, u_2(k) \, \left| e^{-i\chi(k)} \, (i\partial_k) \, e^{i\chi(k)} \right| \Theta \, u_2(k) \right\rangle$$

$$= \langle \Theta \, u_2(k) \, |(i\partial_k)| \, \Theta \, u_2(k)\rangle - \frac{\partial}{\partial k} \chi(k) = \langle u_2(k) \, |(-i\partial_k)| \, u_2(k)\rangle - \frac{\partial}{\partial k} \chi(k)$$

$$= a_{22}(k) - \frac{\partial}{\partial k} \chi(k), \tag{11.43}$$

which leads to

$$P_1 = \frac{1}{2\pi} \left(\int_0^\pi dk \, a_{11}(k) + \int_{-\pi}^0 dk \, a_{11}(k) \right)$$

$$= \frac{1}{2\pi} \int_0^\pi dk \left(a_{11}(k) + a_{22}(k) - \frac{\partial}{\partial k} \chi(k) \right)$$

$$= \frac{1}{2\pi} \int_0^\pi dk \, \mathcal{A}(k) - \frac{1}{2\pi} [\chi(\pi) - \chi(0)]. \tag{11.44}$$

Since $w_{12}(k) = e^{-i\chi(k)}$ from (11.42), $\chi(k)$ can be written as

$$\chi(k) = i \ln w_{12}(k) \tag{11.45}$$

and (11.44) reduces to

$$P_1 = \frac{1}{2\pi} \int_0^\pi dk \, \mathcal{A}(k) - \frac{i}{2\pi} \ln \frac{w_{12}(\pi)}{w_{12}(0)}. \tag{11.46}$$

This expression leads to

$$P_\theta = 2P_1 - P = \frac{1}{2\pi} \int_0^\pi dk \left[\mathcal{A}(k) - \mathcal{A}(-k) \right] - \frac{i}{\pi} \ln \frac{w_{12}(\pi)}{w_{12}(0)}.$$

With $\mathcal{A}(k) = \mathrm{Tr}[\mathbf{a}(k)]$ and $\mathrm{Tr}[\mathbf{a}(k)]$ given by (11.37), one obtains

$$P_\theta = \frac{i}{2\pi} \int_0^\pi dk \, \mathrm{Tr}\left[\mathbf{w}^\dagger(k) \frac{\partial}{\partial k} \mathbf{w}(k) \right] - \frac{i}{\pi} \ln \frac{w_{12}(\pi)}{w_{12}(0)}.$$

But we find from (11.42) that

$$\mathrm{Tr}\left[\mathbf{w}^\dagger(k) \frac{\partial}{\partial k} \mathbf{w}(k) \right] = \mathrm{Tr}\left[\begin{pmatrix} 0 & -e^{i\chi(-k)} \\ e^{i\chi(k)} & 0 \end{pmatrix} \begin{pmatrix} 0 & \partial_k e^{-i\chi(k)} \\ -\partial_k e^{-i\chi(-k)} & 0 \end{pmatrix} \right]$$

$$= -i \left(\partial_k \chi(k) + \partial_k \chi(-k) \right) = \frac{\partial_k \det[\mathbf{w}(k)]}{\det[\mathbf{w}(k)]}.$$

This allows us to write

$$P_\theta = \frac{i}{2\pi} \int_0^\pi dk \, \frac{\partial}{\partial k} \ln (\det[\mathbf{w}(k)]) - \frac{i}{\pi} \ln \frac{w_{12}(\pi)}{w_{12}(0)}$$

$$= \frac{i}{\pi} \frac{1}{2} \ln \frac{\det[\mathbf{w}(\pi)]}{\det[\mathbf{w}(0)]} - \frac{i}{\pi} \ln \frac{w_{12}(\pi)}{w_{12}(0)}. \tag{11.47}$$

Since $\det[\mathbf{w}(k)] = w_{12}^2(k)$, we obtain

$$P_\theta = \frac{1}{i\pi} \ln \left(\frac{\sqrt{w_{12}^2(0)}}{w_{12}(0)} \cdot \frac{w_{12}(\pi)}{\sqrt{w_{12}^2(\pi)}} \right). \tag{11.48}$$

The argument of the logarithm takes the values $+1$ or -1 only. With $\ln(1) = 0$ and $\ln(-1) = i\pi$, we find that P_θ is 0 or $1 \mod 2$.

The two values of P_θ physically correspond to two different polarization states that the 1D system can accommodate in going from $t = 0$ to $T/2$. The wavefunction $|u_i(k,t)\rangle$ can be viewed as a map from $\mathbb{T}^2$, of 2D phase space $(k;t)$, to Hilbert space. Accordingly, this Hilbert space can be classified into two possible realizations, depending on the difference in P_θ from $t = 0$ to $T/2$:

$$\Delta = P_\theta[T/2] - P_\theta[0] \mod 2. \tag{11.49}$$

This Δ is specified only in *mod* 2, so it gives a $\mathbb{Z}_2$ topological invariant to characterize the Hilbert space. When P_θ changes between $t = 0$ to $T/2$, we can think of a *twisted* Hilbert space for $\Delta = 1$, and a trivial Hilbert space for ($\Delta = 0$), when there is no change in P_θ.

Using (11.48), P_θ can be given in terms of

$$(-1)^\Delta = \prod_{i=1}^{4} \frac{w_{12}(\Lambda_i)}{\sqrt{w_{12}^2(\Lambda_i)}}, \tag{11.50}$$

where $\Lambda_1 = (0,0)$, $\Lambda_2 = (\pi,0)$, $\Lambda_3 = (0,T/2)$, and $\Lambda_4 = (\pi,T/2)$, as shown in Figure 11.15. The physical consequence of a cycle with $\Delta = 1$ is spin pumping from one end of the 1D system to the other.

General Formula for the $\mathbb{Z}_2$ Invariant

Extension of the preceding argument to a multiband system is not very difficult. The Hamiltonian still satisfies (11.38) and (11.39), and in the following we take $T = 2\pi$ for simplicity. Consider that $2N$ bands are occupied, and form N Kramers' pairs. For each Kramers' pair n, at the TR symmetric times $t = 0$ and π the wavefunctions are related by

$$\Theta \left|u_2^n(k)\right\rangle = e^{-i\chi_n(k)} \left|u_1^n(-k)\right\rangle \tag{11.51}$$

$$\Theta \left|u_1^n(k)\right\rangle = -e^{-i\chi_n(-k)} \left|u_2^n(-k)\right\rangle, \tag{11.52}$$

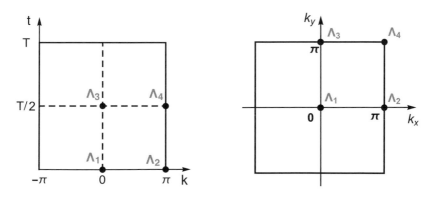

Figure 11.15 The TRIMs for a TR-symmetric 1D system. One can see that identifying $t \to ky$ maps the periodic (k,t) space to the ordinary 2D BZ.

and the w matrix is given by

$$
w(k) = \begin{pmatrix} 0 & e^{-i\chi_1(k)} & 0 & 0 & \cdots \\ -e^{-i\chi_1(-k)} & 0 & 0 & 0 & \cdots \\ 0 & 0 & 0 & e^{-i\chi_2(k)} & \cdots \\ 0 & 0 & -e^{-i\chi_2(-k)} & 0 & \cdots \\ \vdots & \vdots & \vdots & \vdots & \end{pmatrix}.
\tag{11.53}
$$

Hence, at $t = 0$ and π, $w(0)$ and $w(\pi)$ become antisymmetric, and we have

$$
w_{12}(\Lambda_i)\, w_{12}(\Lambda_i) \ldots w_{2N-1,2N}(\Lambda_i) = \exp\left[-i \sum_{n=1}^{N} \chi_n(\Lambda_i) \right]
\tag{11.54}
$$

$$
= \mathrm{Pf}[w(\Lambda_i)].
\tag{11.55}
$$

Note that in the preceding equation, w is viewed as a function of k and t, and we used the formula for the Pfaffian of a $2N \times 2N$ skew-symmetric matrix with 2×2 blocks on the diagonal. In general, Pfaffian is defined for an antisymmetric matrix and is related to the determinant by

$$
\left(\mathrm{Pf}[A] \right)^2 = \det[A].
$$

It is straightforward to extend the previous calculations for the TR symmetric times $t = 0$ and π to obtain

$$
P_1 = \frac{1}{2\pi} \int_0^\pi dk\, \mathcal{A}(k) - \frac{1}{2\pi} \sum_{n=1}^{N} [\chi_n(\pi) - \chi_n(0)]
\tag{11.56}
$$

$$
= \frac{1}{2\pi} \int_0^\pi dk\, \mathcal{A}(k) - \frac{1}{2\pi} \ln\left(\frac{\mathrm{Pf}[w(\pi)]}{\mathrm{Pf}[w(0)]} \right),
\tag{11.57}
$$

which in turn gives the TR polarization

$$
P_\theta = \frac{1}{i\pi} \ln\left(\frac{\sqrt{\det[w(0)]}}{\mathrm{Pf}[w(0)]} \cdot \frac{\mathrm{Pf}[w(\pi)]}{\sqrt{\det[w(\pi)]}} \right).
\tag{11.58}
$$

Therefore, the $\mathbb{Z}_2$ topological invariant ν is given by

$$
(-1)^\nu = \prod_{i=1}^{4} \frac{\mathrm{Pf}[w(\Lambda_i)]}{\sqrt{\det[w(\Lambda_i)]}},
\tag{11.59}
$$

which classifies the Hilbert space into *twisted* ($\nu = 1$) and trivial ($\nu = 0$) ones.

By reinterpreting the periodic 2D phase space (k, t), which forms a torus, as the 2D Brillouin zone (k_x, k_y), the preceding theory provides a $\mathbb{Z}_2$ topological classification of 2D TR-invariant insulators with $2N$ occupied bands.

11.5 Appendices

11.5.1 Classification of Topological Phases

We start with a translationally invariant topological insulator. It is a band insulator of noninteracting fermions, where there is a gap between valence and conduction bands, and the Fermi level E_F lies in this gap. Because of translational invariance, the insulator is represented in momentum space by a matrix equation for every value of momentum $\mathbf{k}$ in the Brillouin zone

$$\mathcal{H}(\mathbf{k}) \, |u_\alpha(\mathbf{k})\rangle = \varepsilon_\alpha \, |u_\alpha(\mathbf{k})\rangle \, ,$$

where α labels the different bands, with m filled and n empty bands. We define the $\mathbf{k}$-dependent projection operator onto the filled Bloch states,

$$\hat{P}(\mathbf{k}) = \sum_\alpha^m |u_\alpha(\mathbf{k})\rangle \langle u_\alpha(\mathbf{k})| \, ,$$

which, in turn, allows us to define the more convenient operator

$$\hat{Q}(\mathbf{k}) = 2\hat{P}(\mathbf{k}) - \mathbb{I}_{m+n}.$$

$\hat{Q}$ has the following properties:

$$\hat{Q}^\dagger(\mathbf{k}) = \hat{Q}(\mathbf{k}), \quad \hat{Q}^2(\mathbf{k}) = 4\hat{P}^2(\mathbf{k}) - 4\hat{P}(\mathbf{k}) + \mathbb{I}_{m+n} = \mathbb{I}_{m+n}, \quad \mathrm{Tr}\left[\hat{Q}(\mathbf{k})\right] = m - n.$$

Depending on the symmetry class, additional conditions may be imposed on $\hat{Q}(\mathbf{k})$.

In the general case, any $\hat{Q}$-matrix can be viewed as a set of $m+n$ complex eigenvectors, and these can be represented as an element of the $(m + n)$-dimensional complex unitary matrices: $U(m+n)$. However, the $\hat{Q}$ matrix does not change if the filled or empty states are rearranged among themselves, thus we have a gauge symmetry $U(m)$, $U(n) \subset U(m + n)$ for each of these state types. Each allowed projector is then described by an element of the coset

$$\hat{Q} \in \frac{U(n + m)}{U(m) \times U(n)}, \tag{11.60}$$

$$\hat{Q} = U^\dagger \Lambda \, U, \qquad \Lambda = \begin{pmatrix} \mathbb{I}_m & 0 \\ 0 & -I_n \end{pmatrix}, \qquad U \in U(m + n).$$

The Hermitian operator $\hat{Q}(\mathbf{k})$ plays the role of the Hamiltonian, carrying only the essential information about the insulator in question. It has eigenvalues ± 1. We can think of it as a "simplified Hamiltonian" that can be obtained from $\mathcal{H}(\mathbf{k})$ by assigning the energy $+1$ to all occupied bands, and the energy -1 to all empty bands, while keeping all wavefunctions unchanged. Since we are only interested in the properties of the phase described by the insulator, we may deform the actual Hamiltonian of the band insulator until it acquires the simple form $\hat{Q}(\mathbf{k})$, while remaining in the same phase – a topological homeomorphic map. Now the question as to how many inequivalent phases there are amounts to asking how many different maps there are that cannot be continuously deformed into each other. The answer to this question is given by the so-called homotopy group π_d of the topological map. It depends on the symmetry constraints and the system's dimension d.

Example

As an example, we consider a band insulator in the simplest symmetry class A, in which there are no conditions whatsoever imposed on the Hamiltonian. The Hamiltonian $\mathcal{H}$ is just a general Hermitian matrix. In this case, (11.60) applies. It turns out that for $d = 2$, the homotopy group is

$$\pi_2\Big[U(n+m)/U(m) \times U(n)\Big] = \mathbb{Z}.$$

This means that for every integer there exists a band insulator for symmetry class A and $d = 2$. We find that band insulators corresponding to different integers cannot be continuously deformed into each other without crossing a quantum phase transition. This is precisely the case of *integer quantum Hall* insulators. The integer characterizing the insulator denotes precisely the number of chiral edge states. When the number of edge states changes, a quantum phase transition necessarily has to be crossed.

In $d = 3$, the relevant homotopy group is (for sufficiently large values of n and m)

$$\pi_3\Big[U(n+m)/U(m) \times U(n)\Big] = \{\text{Identity element}\},$$

which is the trivial group of only one element. This means that for symmetry class A in $d = 3$ spatial dimensions, there can only be one insulator phase, which precludes nontrivial topological insulators.

11.5.2 Non-Abelian Berry Connection: Wilczek–Zee Gauge

In 1984, Wilczek and Zee [197] extended Berry's formulation to Hamiltonians with degenerate spectra, and showed that in such cases it leads to non-Abelian gauge fields (connections). Thus, a non-Abelian phase factor emerges as a natural generalization of the Berry phase to cover such systems.

In this appendix, we consider the case where the spectrum of a Hamiltonian $\mathcal{H}(\boldsymbol{\eta})$ contains degenerate but nonintersecting subspaces on the entire parameter base manifold $\boldsymbol{\eta}$. This means that the condition

$$\varepsilon_\ell(\boldsymbol{\eta}) \neq \varepsilon_j(\boldsymbol{\eta}) \quad \Rightarrow \quad \mathbb{H}_\ell(\boldsymbol{\eta}) \cap \mathbb{H}_j(\boldsymbol{\eta}) = \emptyset$$

applies for the entire manifold. Furthermore, this condition guarantees that in a cyclic evolution the system returns to the same eigensubspace $\mathbb{H}_\ell$ that contains the initial state vector.

Adiabatic Evolution Approach

We consider an $n$fold degenerate ℓth eigenvalue of a Hamiltonian, such that

$$\mathcal{H}(\eta)\,\psi_{\ell\alpha}(\eta) = \varepsilon_\ell(\eta)\,\psi_{\ell\alpha}(\eta), \quad \alpha = 1,\ldots n$$

$$\mathcal{H}(\eta)\,P_\ell(\eta) = P_\ell(\eta)\,\mathcal{H}(\eta) = \varepsilon_\ell(\eta)\,P_\ell(\eta), \quad P_\ell = \sum_{\alpha=1}^{n} |\psi_{\ell\alpha}(\eta)\rangle\,\langle\psi_{\ell\alpha}(\eta)|$$

with a corresponding n-dimensional eigenspace

$$\mathbb{H}_\ell(\eta) = \left\{ \sum_{\alpha=1}^{n} c_\alpha\,\psi_{\ell\alpha}(\eta_i) \middle| c_\alpha \in \mathbb{C} \right\},$$

where the orthonormal set of eigenvectors $\psi_{\ell\alpha}(\eta_i)$ locally span $\mathbb{H}_\ell$, $\eta_i \in U \subset \eta$, with

$$\langle\psi_{\ell\alpha}(\eta)|\psi_{\ell\beta}(\eta)\rangle = \delta_{\alpha\beta}.$$

The Hilbert space is a direct sum of all eigenspaces $\mathbb{H}_\ell$

$$\mathbb{H} = \bigoplus_\ell \mathbb{H}_\ell(\eta).$$

We can transform the set $\{\psi_{\ell\alpha}(\eta)\}$ to another orthonormal set with the aid of a unitary transformation as

$$|\psi_{\ell\alpha}(\eta)\rangle \longleftrightarrow |\psi'_{\ell\alpha}(\eta)\rangle = \sum_{\beta=1}^{n} |\psi_{\ell\beta}(\eta)\rangle\,U_{\beta\alpha}(\eta), \qquad |\psi'_{\ell\alpha}(\eta)\rangle \in H_\ell(\eta), \qquad (11.61)$$

where U is an $n \times n$ matrix that is an element of the unitary group $U(n)$.

Now we consider an adiabatic cyclic evolution of the state vector $\psi_\ell(t)$, effected by an adiabatic change of the external parameters

$$[0, T]\,t \to \eta_t \in \eta, \qquad \eta_0 = \eta_T$$

$$\psi_{\ell\alpha}(\eta_0) = \psi_{\ell\alpha}(\eta_T) \qquad \text{single-valuedness of basis set.}$$

This means that the state returns to the same degeneracy subspace $\mathbb{H}_\ell(\eta_0)$. Thus, in general, the final state vector $|\psi_\ell(\eta_T)\rangle$ is related to the initial state vector $|\psi_\ell(\eta_0)\rangle$ through the action of an unitary matrix

$$|\psi_\ell(\eta_T)\rangle = |\psi_\ell(\eta_0)\rangle\,U_\psi.$$

Adiabatic evolution means that $\psi(\eta_t) \in \mathbb{H}_\ell(\eta)$, and satisfies the Schrödinger equation

$$i\hbar\,\frac{\partial\psi_\ell(\eta_t)}{\partial t} = \mathcal{H}(\eta_t)\,\psi_\ell(\eta_t).$$

We choose the initial state vector as

$$\psi_\ell(\eta_0) = \sum_\alpha c_{\ell\alpha}(0)\psi_{\ell\alpha}(\eta_0),$$

and we use the ansatz

$$\psi_\ell(\eta_t) = \sum_\alpha c_{\ell\alpha}(t)\psi_{\ell\alpha}(\eta_t)$$

to solve the Schrödinger equation. We obtain a set of coupled differential equations of the form

$$\frac{dc_{\ell\beta}(t)}{dt} + \sum_\alpha \left[i\varepsilon_\ell(\eta_t)\,\delta\alpha\beta + \left\langle \psi_{\ell\beta}(\eta_t) \left| \frac{d}{dt} \right| \psi_{\ell\alpha}(\eta_t) \right\rangle \right] c_{\ell\alpha}(t) = 0,$$

and using the initial conditions we get the solution

$$c_{\ell\beta}(t) = \sum_{\alpha=1}^n \left[T \exp \int_0^t d\tau \left(-i\varepsilon_\ell(\eta_\tau)\,\mathbb{I} + i\mathcal{A}^\ell(\eta_\tau) \right) \right]_{\beta\alpha} c_{\ell\alpha}(0), \tag{11.62}$$

where we introduced the one-form Wilczek–Zee gauge potential, or the non-Abelian Berry connection

$$\mathcal{A}_{\beta\alpha}^\ell(\eta_\tau) = i \left\langle \psi_{\ell\beta}(\eta_\tau) \left| \frac{d}{d\tau} \right| \psi_{\ell\alpha}(\eta_\tau) \right\rangle d\tau$$

$$\mathcal{A}_{\beta\alpha}^\ell(\eta) = i \left\langle \psi_{\ell\beta}(\eta)\,|d|\,\psi_{\ell\alpha}(\eta) \right\rangle, \tag{11.63}$$

which is an $n \times n$ Hermitian matrix, $\mathcal{A}_{\beta\alpha}^{(\ell)*} = \mathcal{A}_{\alpha\beta}^{(\ell)}$. We also introduced the time-ordering operator T to maintain the chronological sequencing of events, since the non-Abelian character of $\mathcal{A}_{\alpha\beta}^{(\ell)}$ indicates that the commutator

$$\left[\mathcal{A}_{\alpha\beta}^{(\ell)}(\eta_t),\ \mathcal{A}_{\alpha\beta}^{(\ell)}(\eta_{t'}) \right] \neq 0.$$

We can write the state vector

$$|\psi_\ell(\eta_t)\rangle = \sum_{\alpha,\beta=1}^n |\psi_{\ell\beta}(\eta_0)\rangle\, e^{-i\int_0^t dt'\,\varepsilon_\ell(\eta_{t'})} \left[\mathcal{P} \exp \left(i\int_{\eta_0}^{\eta_t} \mathcal{A}^\ell(\eta) \right) \right]_{\beta\alpha} c_{\ell\alpha}(0),$$

where we replaced the time-ordering with a path-ordering operator $\mathcal{P}$.

Under the unitary transformation,

$$\psi_{\ell\alpha}(\eta)\ \rightarrow\ \psi_{\ell\alpha}'(\eta) = \sum_{\beta=1}^n \psi_{\ell\beta}(\eta)\, U_{\beta\alpha}^{(\ell)}(\eta). \tag{11.64}$$

The gauge potential $\mathcal{A}^\ell$ transforms according to

$$\mathcal{A}^\ell(\eta)\ \rightarrow\ \mathcal{A}'^\ell(\eta) = U^{-1}(\eta)\,\mathcal{A}^\ell(\eta)\,U(\eta) + iU^{-1}(\eta) \cdot dU(\eta). \tag{11.65}$$

Alternatively, we can use a solution to the Schrödinger equation of the form

$$\psi_\ell(t) = \exp\left(-\frac{i}{\hbar} \int_0^t dt'\,\varepsilon_\ell\left(t'\right) \right) \sum_{\beta=1}^n U_{\alpha\beta}^{(\ell)}(\eta_t)\, |\psi_{\ell\beta}(\eta_t)\rangle \tag{11.66}$$

and obtain the following equation for the time dependence of U

$$\left(\left(U^{(\ell)}\right)^{-1} \dot{U}^{(\ell)}\right)_{\alpha\beta} = -\langle\psi_{\ell\alpha}|\dot{\psi}_{\ell\beta}\rangle \tag{11.67}$$

when substituting (11.66) in the Schrödinger equation. Integrating (11.67) yields

$$U_{\alpha\beta}(t) = \mathrm{T} \exp\left[-\int_0^t d\tau \left\langle\psi_{\ell\alpha}(\tau)\left|\frac{d\psi_{\ell\beta}(\tau)}{d\tau}\right\rangle\right]\right.$$

$$= \mathcal{P} \exp\left[-\sum_\mu \int_{\eta_\mu(0)}^{\eta_\mu(t)} \left\langle\psi_{\ell\alpha}(\eta)\left|\frac{d\psi_{\ell\beta}(\eta)}{d\eta_\mu}\right\rangle d\eta_\mu\right]\right.$$

$$= \mathcal{P} \exp\left[i \int_{\eta_\mu(0)}^{\eta_\mu(t)} \mathcal{A}^{(\ell)}_{\alpha\beta}(\eta(t))\right],$$

where T, $\mathcal{P}$ are the time-ordering and path-ordering operators, respectively, since the integrand, namely, $\mathcal{A}^{(\ell)}_{\alpha\beta}$ being non-Abelian, does not commute with itself at different points on the manifold.

For cyclic evolution, $[0, T]$, $\eta_0 = \eta_T$, and we obtain

$$U_{\alpha\beta}(\mathcal{C}) = \mathcal{P} \exp\left(i \oint_\mathcal{C} \mathcal{A}^{(\ell)}\right). \tag{11.68}$$

Such a potential is called a $U(n)$ gauge potential. The phase factors $U_{\alpha\beta}(\mathcal{C})$ represent the extension of the Berry phase to the unitary group $U(n)$. It is again a pure geometrical object, in the sense that it depends only on the geometry of the degenerate space. While having most of the geometric particulars of the Abelian Berry phase, the non-Abelian phase is a matrix with elements that are not separately gauge invariant. Its most relevant quantities are the trace (also known as the *Wilson loop* in gauge theory) and its eigenvalues, which are indeed gauge-invariant quantities.

Fiber Bundle Approach

The Wilczek–Zee gauge can be cast as a holonomy element in a proper fiber bundle. Each point of a base space, $\eta_i \in \eta$, is associated with an n-dimensional Hilbert subspace $H_\ell(\eta)$, a typical fiber. We define the corresponding bundle as

$$E^{(\ell)} = \bigcup_{\eta_i \subset \eta} H_\ell(\eta_i)$$

$$F_\ell = \mathbb{C}^n \qquad \text{standard fiber.}$$

Alternatively, we can use the eigensubspace basis $\{\psi_{\ell\alpha}\}$, and define the typical fiber as

$$F^{(\ell)}(\eta) = \left\{\psi_{\ell\alpha}(\eta) = \sum_\beta U_{\alpha\beta} \phi_\beta \middle| U \in U(n)\right\} \simeq U(n) \tag{11.69}$$

with a corresponding $U(n)$ principal bundle

$$P^{(\ell)} = \bigcup_{\eta_i \in \eta} F_\eta^{(\ell)}(\eta_i)$$

having $E^{(\ell)}$ as its associated bundle.

We define a curve $t \to C(t)$ on η, and set a corresponding lift to $E^{(\ell)}$ as $(\psi_{\ell 1}(t), \ldots, \psi_{\ell n}(t))$. We identify $\psi_\alpha(t)$ as a horizontal lift with respect to the WZ connection, when

$$\langle \psi_{\ell\beta} | \dot\psi_\alpha \rangle \to \langle \psi_{\ell\beta} | d\phi_\alpha \rangle = 0, \qquad \alpha, \beta = 1, \ldots, n.$$

The Gauge Field, or Berry Curvature $\mathcal{F}^{(\ell)}$

We define the corresponding gauge field

$$\mathcal{F}^{(\ell)} = d\mathcal{A}^{(\ell)} - i\, \mathcal{A}^{(\ell)} \wedge \mathcal{A}^{(\ell)}$$

$$\left(\mathcal{F}_{jk}^{(\ell)} \right)_{\alpha\beta} = \partial_j \left(\mathcal{A}_k^{(\ell)} \right)_{\alpha\beta} - \partial_k \left(\mathcal{A}_j^{(\ell)} \right)_{\alpha\beta} - i \left[\mathcal{A}_j^{(\ell)}, \mathcal{A}_k^{(\ell)} \right]_{\alpha\beta}$$

$$\left(\mathcal{A}_k^{(\ell)} \right)_{\alpha\beta} = i\, \langle \psi_{\ell\alpha} | \partial_k | \psi_{\ell\beta} \rangle, \qquad \partial_k = \partial/\partial\eta^k, \qquad (11.70)$$

where $(\eta^1, \ldots, \eta^d)$ are local coordinates on a d-dimensional base η.

An important physical consequence of non-Abelian holonomies is that not only may each degenerate state acquire a phase change, but population transfers among the different degenerate levels are possible as well.

11.5.3 Topology and Phase Singularity in QHE

The nontrivial topology of the Berry connection $\mathcal{A}$ arises when the phase of the wavefunction cannot be determined uniquely and smoothly in the entire magnetic BZ. To show this phenomenon, we rewrite the last integral in (5.31) for a single band as [106, 173]

$$\sigma_{xy} = \int_{\mathbb{T}^2} d\mathcal{A} = \oint_C \mathcal{A}.$$

We could naively argue that since the wavefunction is expected to be periodic over the torus, the integral should vanish because the path of the integral contains the same states with opposite directions, as depicted in Figure 11.16.

However, ν can assume nonzero values when the phase of the wavefunction cannot be determined uniquely and smoothly in the whole BZ, due to the presence of a nonremovable phase singularity in the BZ – in other words, a singularity that cannot be removed by any gauge transformation.

In order to clarify this behavior, we shall use the following hypothetical scenario: we consider an arbitrary eigenstate $|u_\mathbf{k}\rangle$ in a 2D system, and try to determine its phase over the whole BZ. A simple way to accomplish this is to first fix a certain position $\mathbf{x} = (x_1, x_2)$ and set $u_\mathbf{k}(\mathbf{x}) = \langle \mathbf{x} | u_\mathbf{k} \rangle$ to be real by performing a gauge transformation,

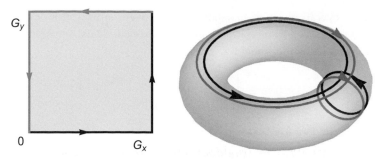

Figure 11.16 Schematic of the 2D magnetic BZ. The arrows show the direction of the path integral for the Hall conductivity. Parallel paths traversing the BZ correspond to paths of opposite directions on the torus. Such paths cancel each other when the phase of the wavefunction is defined smoothly in the whole BZ.

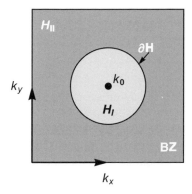

Figure 11.17 Schematic diagram of the wavefunction phase. To determine the phase on the entire BZ, it is divided into two regions. At $\mathbf{k}_0$, there is a singularity of the phase for $u_{\mathbf{k}}(\mathbf{x})$.

$$\left|u'_{\mathbf{k}}\right\rangle = e^{-ig_{\mathbf{k}}}\left|u_{\mathbf{k}}\right\rangle, \qquad g_{\mathbf{k}} = \frac{u_{\mathbf{k}}(\mathbf{x})}{|u_{\mathbf{k}}(\mathbf{x})|}.$$

We should note that we cannot apply such transformation at $\mathbf{k}$ points where the wavefunction vanishes. For pedagogy, we assume that $u_{\mathbf{k}}(\mathbf{x})$ vanishes only at $\mathbf{k} = \mathbf{k}_0$, and the phase of $u_{\mathbf{k}}(\mathbf{x})$ cannot be determined at $\mathbf{k} = \mathbf{k}_0$ because of the singularity. However, the $u_{\mathbf{k}}(\mathbf{x})$ phase can be determined at $\mathbf{k} \neq \mathbf{k}_0$. As shown in Figure 11.17, we divide the BZ into two patches, H_I and H_{II}, where H_I contains a vicinity of $\mathbf{k}_0$, while H_{II} is the complement of H_I in the BZ. We fix the phase of $u_{\mathbf{k}}(\mathbf{x})$ to be real for $\mathbf{k} \in H_{II}$ using $g_{\mathbf{k}}$.

In H_I, we choose another position $\mathbf{x}'$ such that $u_{\mathbf{k}}(\mathbf{x}')$ does not vanish in the whole region H_I *including* $\mathbf{k} = \mathbf{k}_0$.

Since $u_{\mathbf{k}}(\mathbf{x}')$ does not have a singularity in H_I, we can force it to be real by using a real function $h_{\mathbf{k}}$ as

$$\left|u''_{\mathbf{k}}(\mathbf{x}')\right\rangle = e^{-ih_{\mathbf{k}}}\left|u_{\mathbf{k}}(\mathbf{x}')\right\rangle, \qquad h_{\mathbf{k}} = \frac{u_{\mathbf{k}}(\mathbf{x}')}{|u_{\mathbf{k}}(\mathbf{x}')|}.$$

Note that the overall phase of the state vector is well defined at $\mathbf{k}_0$ even though the phase of $u_{\mathbf{k}}(\mathbf{x})$ at position $\mathbf{x} = (x_1, x_2)$ cannot be defined. Thus, we have a state vector whose phase is defined in the whole BZ in

$$|u_{\mathbf{k}}\rangle = \begin{cases} |u_{\mathbf{k} \in H_I}\rangle & = e^{-ih_{\mathbf{k}}} |u_{\mathbf{k}}\rangle \\ |u_{\mathbf{k} \in H_{II}}\rangle & = e^{-ig_{\mathbf{k}}} |u_{\mathbf{k}}\rangle \end{cases}$$

However, at the boundary ∂H between H_I and H_{II}, we have a phase mismatch

$$|u_{\mathbf{k} \in H_I}\rangle = e^{-it_{\mathbf{k}}} |u_{\mathbf{k} \in H_{II}}\rangle,$$

where $t_{\mathbf{k}} = g_{\mathbf{k}} - h_{\mathbf{k}}$. We define the Berry connections as $\mathcal{A}_I$ and $\mathcal{A}_{II}$ on H_I and H_{II}, respectively, because connection depend on the gauge. The contributions to the Hall conductivity from the two regions is expressed as

$$\sigma_{xy} = \frac{e^2}{h} \frac{1}{2\pi} \int d\mathcal{A} = \frac{e^2}{h} \frac{1}{2\pi} \left[\int_{H_I} d\mathcal{A}_I + \int_{H_{II}} d\mathcal{A}_{II} \right]$$

$$= \frac{1}{2\pi} \oint_{\partial H} d\mathbf{k} \cdot [\mathcal{A}_I - \mathcal{A}_{II}] = \frac{1}{2\pi} \oint_{\partial H} d\mathbf{k} \cdot \nabla_{\mathbf{k}} t_{\mathbf{k}}$$

$$= \frac{e^2}{h} \mathcal{C}.$$

The sign change occurs because the line integral along ∂H has opposite orientations for H_I and H_{II}. Again, $\mathcal{C}$ must be an integer because $t_{\mathbf{k}}$ arises from gauge transformations, and its change around ∂H is an integer multiple of 2π in order for each of the state vectors to fit together exactly when a full revolution around is completed. $\mathcal{C}$ is a topological invariant; it does not change by any smooth deformation of the path ∂H. In addition, since the nontrivial winding of the phase difference $t_{\mathbf{k}}$ is supported by the existence of the singularity, it never changes from the integer by small external perturbation that does not close the band gap.

Exercises

11.1 Anomalous velocity:

Consider the effect of a weak uniform electric field $\mathbf{E}$ perturbation on a crystal. In the absence of the perturbation, the Bloch Hamiltonian is given by

$$\mathcal{H}(\mathbf{k}) = \frac{1}{2m} (\mathbf{p} + \hbar\mathbf{k})^2 + V(\mathbf{x}).$$

Notice that $\mathbf{k}$ behaves like a vector potential. What are the eigenfunctions of $\mathcal{H}(\mathbf{k})$?

(a) What are the drawbacks of using an electrostatic potential $\Phi(\mathbf{x})$ that produces the uniform $\mathbf{E}$?

(b) We can avoid these obstacles by introducing a time-varying uniform vector potential $\mathbf{A}_0(t)$. Describe how such a potential is manifest in the crystal's Bloch Hamiltonian.

(c) You may define a modified time-dependent crystal momentum, $\mathbf{q} = \mathbf{k} - e\mathbf{E}t$, with its corresponding Bloch Hamiltonian $\mathcal{H}(\mathbf{q}(t))$. What is the condition for adiabatic change?

(d) Write down an expression for the velocity operator $\mathbf{v}(\mathbf{q})$ in terms of $\mathcal{H}(\mathbf{q})$.

(e) Use the expression you obtained for the first-order corrected adiabatic eigenket in Exercise 10.2 to show that the velocity $\mathbf{v}_n = \langle u_n(q,t)| v(q) |u_n(q,t)\rangle$ is given by

$$\mathbf{v}_n(\mathbf{k}) = \frac{\partial \varepsilon_n(\mathbf{k})}{\hbar \partial \mathbf{k}} - \frac{e}{\hbar} \mathbf{E} \times \mathcal{F}_n(\mathbf{k}).$$

The second term involving the Berry curvature is the so-called anomalous velocity – note that its direction is transverse to the electric field and thus will give rise to a Hall current.

(f) Comment on the effect of time-reversal symmetry on the anomalous velocity and its consequences.

11.2 Writing a general wavefunction as a superposition of the Bloch waves

$$\Psi(\mathbf{x}) = \sum_\ell \int d\mathbf{k}\, \Psi_\ell(\mathbf{k})\, \psi_{\ell\mathbf{k}}(\mathbf{x}) = \sum_\ell \int d\mathbf{k}\, \Psi_\ell(\mathbf{k})\, u_{\ell\mathbf{k}}(\mathbf{x})\, e^{i\mathbf{k}\cdot\mathbf{x}},$$

show that

$$\mathbf{x}\,\Psi(\mathbf{x}) = \sum_\ell \int d\mathbf{k}\, \sum_m (i\nabla\delta_{m\ell} - a_{m\ell})\, \Psi_m(\mathbf{k})\, \psi_{m\mathbf{k}}(\mathbf{x}).$$

11.3 Non-Abelian gauges and parallel transport: Wilson loops:

Consider an ABE-type two-component wavefunction $\Psi^T = \begin{bmatrix} \psi^1 & \psi^2 \end{bmatrix}$. A change of phase is obtained via the transformation

$$\Psi'(\mathbf{x}) = S(\mathbf{x})\,\Psi(\mathbf{x}), \tag{11.71}$$

where S is a 2×2 $SU(2)$ matrix. Local gauge invariance is now the requirement that physics is unchanged under arbitrary $SU(2)$ transformation.

    Parallel transport of the phase at a point $\mathbf{x}$ to the phase at a neighboring point $\mathbf{x} + d\mathbf{x}$ is effected by a phase change of $q\,\mathcal{A}_\alpha(\mathbf{x})\,dx^\alpha$, for a particle with charge q, where $\mathcal{A}$ is a generalized non-Abelian gauge potential 2×2 matrix.

(a) Express $\Psi(\mathbf{x} + d\mathbf{x})$ in terms of $\Psi(\mathbf{x})$ and $\mathcal{A}$.

(b) Transform $\Psi(\mathbf{x} + d\mathbf{x})$ according to (11.71), namely, to the primed frame.

(c) Use the result in (a) to express $\Psi'(\mathbf{x} + d\mathbf{x})$ in terms of $\Psi(\mathbf{x})$.

(d) In the primed frame, express parallel transport in terms of $\Psi'(\mathbf{x} + d\mathbf{x})$ and $\Psi'(\mathbf{x})$.

(e) Using the preceding results and expanding to first order in dx^α, show that

$$\mathcal{A}'_\alpha(\mathbf{x}) = S(\mathbf{x})\,\mathcal{A}_\alpha(\mathbf{x})\,S^{-1}(\mathbf{x}) - \frac{i}{q}\,[\partial_\alpha S(\mathbf{x})]\,S^{-1}(\mathbf{x}). \tag{11.72}$$

(f) Setting $S(\mathbf{x}) = 1 + iq\Lambda(\mathbf{x})$, show that, to first order in Λ,

$$A'_\alpha(\mathbf{x}) = \mathcal{A}_\alpha(\mathbf{x}) + \partial_\alpha \Lambda(\mathbf{x}) + iq\left[\Lambda(\mathbf{x}), \mathcal{A}_\alpha(\mathbf{x})\right].$$

This yields the *path-ordered phase* for a closed-loop C for the non-Abelian case as

$$\gamma(C) = P \exp\left[iq \oint_C \mathcal{A}\cdot d\mathbf{x}\right]$$

known as the Wilson loop. Since in general cases of the Wilson loops the matrices at different points do not commute, path-ordering P is used.

11.4 Charge conjugation symmetry:
Consider the Hamiltonian

$$\mathcal{H} = \sum_{\mathbf{k}\alpha\beta} h_{\alpha\beta}(\mathbf{k}) \ |\mathbf{k},\alpha\rangle \langle\mathbf{k},\beta|,$$

where α, β are basis indices. Charge conjugation, or particle–hole transformation, is defined as

$$|\mathbf{k},\alpha\rangle = \sum_\beta U_{\alpha\beta} \langle-\mathbf{k},\beta|.$$

For the Hamiltonian to satisfy particle–hole symmetry, it should be invariant under $U_{\alpha\beta}(\mathbf{k})$: $\mathcal{H}$ is the same when expressed in terms of electron bases and hole bases.

(a) Find the conditions that $h(\mathbf{k})$ must satisfy for $\mathcal{H}$ to be particle–hole symmetric (there should be an extra condition on the trace of $h(\mathbf{k})$; what is it?).

(b) Show that, for a system with particle–hole symmetry, the energy eigenvalues at each $\mathbf{k}$ come in pairs ($\varepsilon_\mathbf{k}$, $-\varepsilon_\mathbf{k}$), provided that either time-reversal or inversion symmetry is present. Hint: First show that if $|\psi_\mathbf{k}\rangle$ is an eigenstate of $h^*(\mathbf{k})$ with energy $\varepsilon_\mathbf{k}$, then $U(-\mathbf{k})|\psi_\mathbf{k}\rangle$ is an eigenstate of $h(-\mathbf{k})$, with energy $-\varepsilon_\mathbf{k}$. Next, note that the spectrum of $h^*(\mathbf{k})$ and $h(\mathbf{k})$ must be identical (because $h(\mathbf{k})$ is Hermitian). Finally, recall the consequence of time-reversal symmetry and/or inversion symmetry on the energy spectrum.

12

Dirac Materials and Dirac Fermions

12.1 Introduction

Historically, the Dirac and Weyl fields[1] were first used in high-energy physics to describe elementary particles (electrons, quarks, neutrinos, etc.). They constitute the simplest building blocks for constructing Lorentz-invariant Lagrangians that describe interacting particles within the standard model. Recently, however, a wide variety of materials, ranging from graphene to topological insulators and d-wave superconductors, were found to have a common low-energy fermionic dispersion that resembles massless Dirac particles, rather than the usual free particle parabolic dispersion of the Schrödinger type, conventionally referred to as *Schrödinger fermions*. Materials having this unifying emergent Dirac fermion spectral character are now referred to as *Dirac materials*. Particular symmetries, such as TRS in topological insulators, point group symmetries in topological crystalline insulators, and sublattice symmetry in graphene, control the appearance of Dirac cones and points in their excitation spectra.

Such systems comprise lattices where electrons are described by Bloch states indexed by the crystal momentum $\mathbf{k}$ and defined on the Brillouin torus. As we will see, the Bloch bundle topology implies, under certain conditions, that the Weyl or Dirac equations, and their corresponding dispersion relations, are not be satisfied globally over the whole manifold, but only locally.

The most studied *Dirac material* is the semimetal graphene [2, 41], where low-energy massless Weyl excitations emerge in the vicinity of the K and K' points of the 2D BZ. The relativistic-like behavior originates from the two-sublattice structure of the honeycomb lattice of graphene. This structure introduces a *momentum-pseudospin* coupling term in the Hamiltonian, namely, $\mathbf{p} \cdot \boldsymbol{\sigma}$, where $\boldsymbol{\sigma}$ is the Pauli spin matrix. The pseudospin is actually the sublattice index, and not the real electronic spin, which remains decoupled from orbital motion and is neglected in this picture because spin–orbit coupling is very weak in graphene. Theoretical predictions that graphene should host such massless Dirac–Weyl fermions that obey the 2D Weyl equations have preceded the experimental realization of the single monolayer [191].

---

[1] A detailed review of the Dirac and Weyl equations and fields is given in Appendix 2 at the end of this chapter.

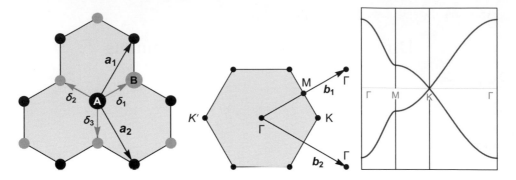

Figure 12.1 The graphene lattice (left). The $\boldsymbol{\delta}$s and $\mathbf{a}_i$s are nearest neighbor vectors and lattice basis vectors, respectively. A, B denote the two honeycomb sublattices. The honeycomb BZ (middle), showing the high-symmetry points and the reciprocal lattice basis vectors $\mathbf{b}_i$s. The π-bands along high-symmetry directions (right).

12.2 Graphene: The Gate to Dirac Fermions

As shown in Figure 12.1, the honeycomb structure of graphene has two interpenetrating triangular sublattices A and B, and primitive lattice basis

$$\mathbf{a}_1 = \frac{a}{2}\left(\sqrt{3}, 3\right), \ \mathbf{a}_2 = \frac{a}{2}\left(\sqrt{3}, -3\right).$$

Its reciprocal lattice basis is given by

$$\mathbf{b}_1 = \frac{2\pi}{3a}\left(\sqrt{3}, 1\right), \ \mathbf{b}_2 = \frac{2\pi}{3a}\left(\sqrt{3}, -1\right).$$

In Exercise 3.12, we studied the complete electronic structure of graphene. We learned that the graphene bands are separated into a set of σ-bands emerging from interacting $2s$, $2p_x$, and $2p_y$ orbitals, and a pair of π-bands in the vicinity of the Fermi energy derived from dangling p_z-orbitals. Here, we shall focus on the π-bands.

The hexagonal Brillouin zone of graphene, shown in Figure 12.1, has three high-symmetry points:

- Γ and M demarcate the positions of the Van Hove singularities of the $\pi - \pi^*$ bands, where the density of states (DOS) is logarithmically divergent
- The K points, where the $\pi - \pi^*$ bands touch and the DOS vanishes linearly

We write the corresponding Hamiltonian as

$$\mathcal{H}_\pi = -t \sum_{A,\ell} \left(|p_z, A\rangle \langle p_z, B, \delta_\ell| + \text{hermitian conjugate}\right),$$

where t is the hopping energy, and δ_ℓ are the nearest B neighbor vectors, shown in Figure 12.1 and given by

$$\delta_1 = \frac{a}{2}\left(\sqrt{3}, 1\right), \qquad \delta_2 = \frac{a}{2}\left(-\sqrt{3}, 1\right) \qquad \delta_3 = a\,(0, -1).$$

Expressing the eigenkets $|\mathbf{k}, A\rangle$ and $|\mathbf{k}, B\rangle$ as

$$|\mathbf{k}, A/B\rangle = \frac{1}{\sqrt{N}} \sum_{\mathbf{R}} e^{i\mathbf{k}\cdot\mathbf{R}} \, |p_z, A/B\rangle,$$

we obtain

$$\mathcal{H}_\pi = \sum_{\mathbf{k}} \langle \psi_{\mathbf{k}} | \, \mathcal{H}(\mathbf{k}) \, |\psi_{\mathbf{k}}\rangle,$$

where

$$|\psi_{\mathbf{k}}\rangle = \begin{bmatrix} |\mathbf{k}, A\rangle \\ |\mathbf{k}, B\rangle \end{bmatrix}$$

$$\mathcal{H}(\mathbf{k}) = \begin{pmatrix} 0 & h(\mathbf{k}) \\ h^*(\mathbf{k}) & 0 \end{pmatrix} = h_1(\mathbf{k})\,\sigma_1 + h_2(\mathbf{k})\,\sigma_2 = \mathbf{h}(\mathbf{k})\cdot\boldsymbol{\sigma} \tag{12.1}$$

$$h(\mathbf{k}) = -t \sum_\ell e^{i\mathbf{k}\cdot\boldsymbol{\delta}_\ell} = h_1(\mathbf{k}) + i h_2(\mathbf{k}),$$

where σ_1, σ_2 are the Pauli matrices. We note that

$$h(-\mathbf{k}) = -t \sum_\ell e^{-i\mathbf{k}\cdot\boldsymbol{\delta}_\ell} = h_1(-\mathbf{k}) - i h_2(-\mathbf{k}) = h(\mathbf{k}) \tag{12.2}$$

since the functions $h_1(\mathbf{k})$ and $h_2(\mathbf{k})$ are even and odd in momentum, respectively, which will be important for symmetry analysis. Furthermore, using the anticommutation properties of the Pauli matrices, we obtain

$$\sigma_3 \, \mathcal{H}(\mathbf{k}) \, \sigma_3 = -\mathcal{H}(\mathbf{k}),$$

which establishes the chiral symmetry (or particle–hole symmetry)[2] of the bipartite honeycomb lattice with nearest-neighbor hopping: if $|\psi\rangle$ is an eigenstate with an eigenvalue E, then $\sigma_3 \, |\psi\rangle$ is an eigenstate with energy $-E$.

Diagonalization of the Bloch Hamiltonian yields the spectrum of the two π bands of graphene in tight-binding approximation:

$$\varepsilon_\lambda(\mathbf{k}) = \lambda t \left| \sum_\ell e^{-i\mathbf{k}\cdot\boldsymbol{\delta}_\ell} \right|, \quad \lambda = \pm. \tag{12.3}$$

The $+$ $(-)$ sign in the spectrum corresponds to the conduction (valence) band, and establishes their particle–hole symmetry: because each carbon atom contributes one p electron and each electron may occupy either a spin-up or a spin-down state, the lower band $\lambda = -$ (the π or valence band) is completely filled, and that with $\lambda = +$ (the π^* or conduction band) completely empty.

---

[2] This symmetry operation, usually denoted by S, is sometimes called *sublattice symmetry*, hence the notation S. However, in many instances, it is not realized as a sublattice symmetry, but is simply the product of Θ and C. Therefore, the term *chiral symmetry* to describe the symmetry operation S is sometimes more appropriate.

At the BZ points K,

$$h(\mathbf{K}) = 1 + \omega + \omega^2 = 0, \qquad \omega = e^{i\pi/3}.$$

Thus, the $\pi - \pi^*$ bands are degenerate at these points with energy $\varepsilon_\pm(\mathbf{K}) = \varepsilon_F = 0$. The topology of the Fermi surface (FS) in graphene is defined by the six K *Weyl* points, where the conduction and valence bands touch. These special points form two inequivalent sets K and K', with $\mathbf{K}' = -\mathbf{K}$, $|\mathbf{K}| = 4\pi/3\sqrt{3}a$; they cannot be connected by the $\mathbf{b}_i$s.

$$K = \frac{2\pi}{3a}\left(\frac{2}{\sqrt{3}}, 0\right), \qquad K' = \frac{2\pi}{3a}\left(-\frac{2}{\sqrt{3}}, 0\right).$$

These points are referred to as Dirac points, motivated by the development described later. However, we note that the Hamiltonian (12.1) is not periodic in **k**-space. To obtain a periodic form, we need apply the following gauge transformation

$$|\mathbf{k}, B\rangle \;\rightarrow\; i e^{-i\mathbf{k}\cdot\boldsymbol{\delta}_3}\,|\mathbf{k}, B\rangle \;\rightarrow\; h(\mathbf{k}) \;\rightarrow\; i e^{i\mathbf{k}\cdot\boldsymbol{\delta}_3}\,h(\mathbf{k}) = i\left[1 + e^{-i\mathbf{k}\cdot(\boldsymbol{\delta}_1 - \boldsymbol{\delta}_3)} + e^{-i\mathbf{k}\cdot(\boldsymbol{\delta}_3 - \boldsymbol{\delta}_2)}\right],$$

where $\boldsymbol{\delta}_1 - \boldsymbol{\delta}_3 = \mathbf{a}_1$ and $\boldsymbol{\delta}_3 - \boldsymbol{\delta}_2 = \mathbf{a}_2$, rendering $h(\mathbf{k} + \mathbf{G}) = h(\mathbf{k})$, $\mathbf{G} = n_1\mathbf{b}_1 + n_2\mathbf{b}_2$.

We may write the eigenstates of the effective Hamiltonian $\mathcal{H}_\pi$ as the spinors

$$\psi_{\mathbf{k}}^\lambda = \begin{bmatrix} a_{\mathbf{k}}^\lambda \\ b_{\mathbf{k}}^\lambda \end{bmatrix},$$

the components of which are the probability amplitudes of the Bloch wavefunction on the two different sublattices A and B. They can be determined by considering the eigenvalue equation

$$\mathcal{H}_\pi\,\psi_{\mathbf{k}}^\lambda = \lambda\,|h(\mathbf{k})|\,\psi_{\mathbf{k}}^\lambda$$

$$\psi_{\mathbf{k}}^\lambda = \frac{1}{\sqrt{2}}\begin{bmatrix} e^{-i\phi_{\mathbf{k}}} \\ \lambda \end{bmatrix}, \qquad \phi_{\mathbf{k}} = \mathrm{Arctan}\left[\frac{\Im\,h(\mathbf{k})}{\mathrm{Re}\,h(\mathbf{k})}\right].$$

$\psi_{\mathbf{k}}^\lambda$ is defined up to a gauge. As expected, the spinor represents an equal probability to find an electron on the A as on the B sublattice because both sublattices are built from carbon atoms with the same onsite energy.

> We note that for 2D systems with honeycomb structure and nondegenerate onsite sublattice energies, $\varepsilon_A \neq \varepsilon_B$, we need to add the term
>
> $$h_3 = \Delta\begin{pmatrix} 1 & 0 \\ 0 & -1 \end{pmatrix} = \Delta\,\sigma_3,$$
>
> where $\varepsilon_A - \varepsilon_B = 2\Delta$. This term opens a gap of magnitude 2Δ in the energy dispersion at the points K and K'. A typical example is boron nitride, BN, where one sublattice consists of boron atoms, energy ε_A, and the other sublattice consists of nitrogen atoms, energy ε_B.

12.2.1 Dirac Fermion Hamiltonian for Graphene

In order to describe the low-energy excitations, namely, electronic excitations with energies much smaller than the band width $3t$, we focus on energies close to the Fermi level. This effectively means restricting the excitations between states to the vicinity of the K and K' points. We expand the energy dispersion around $\pm\mathbf{K}$, and consider the states at $\pm\mathbf{K} + \mathbf{q}$, with $q \ll |\mathbf{K}|$, namely

$$h^{\pm}(\mathbf{q}) = h(\mathbf{k} = \pm\mathbf{K} + \mathbf{q}) = -t \left[1 + e^{\pm i\mathbf{K}\cdot\mathbf{a}_1} e^{i\mathbf{q}\cdot\mathbf{a}_1} + e^{\pm i\mathbf{K}\cdot\mathbf{a}_2} e^{i\mathbf{q}\cdot\mathbf{a}_2} \right]$$

$$\simeq -t \left[1 + e^{\pm i2\pi/3} \left(1 + i\mathbf{q}\cdot\mathbf{a}_1 \right) + e^{\mp i2\pi/3} \left(1 + i\mathbf{q}\cdot\mathbf{a}_2 \right) \right]$$

$$= -i\frac{\sqrt{3}at}{2} \left[\left(q_x + \sqrt{3}\, q_y \right) e^{\pm i2\pi/3} + \left(-q_x + \sqrt{3}\, q_y \right) e^{\mp i2\pi/3} \right] \tag{12.4}$$

yielding

$$h^+ = \frac{3at}{2} \left(q_x - i\, q_y \right), \qquad h^- = -\frac{3at}{2} \left(q_x + i\, q_y \right). \tag{12.5}$$

Replacing the qs by ks, the linearized Hamiltonian takes the form

$$\mathcal{H}_\pi(\mathbf{k}) = \hbar\, v_F \begin{cases} k_x\,\sigma_1 + k_y\,\sigma_2, & K \\ -k_x\,\sigma_1 + k_y\,\sigma_2, & K' \end{cases} \tag{12.6}$$

which is just the 2D massless Dirac, or Weyl Hamiltonian, having conical dispersion about the K, K' points, as shown in Figure 12.2. Consequently, these points are referred to as Dirac points, and the neighborhoods of these points are called valleys. The σ matrices reflect the pseudospin character of the two sublattices. Moreover, we note that the twofold valley degeneracy survives when considering the low-energy excitations in the vicinity of the Dirac points. Introducing $\xi = \pm 1$ to denote the valleys K and K', respectively, we write the effective low-energy Hamiltonian as

$$\mathcal{H}_\xi^{\text{eff}}(\mathbf{k}) = \xi\, h\, v_F \left(k_x\,\sigma_1 + \xi\, k_y\,\sigma_2 \right) = \xi\, \hbar\, v_F\, |\mathbf{k}| \left(\cos\phi_{\mathbf{k}}\,\sigma_1 + \xi\, \sin\phi_{\mathbf{k}}\,\sigma_2 \right), \tag{12.7}$$

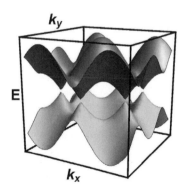

Figure 12.2 Graphene band structure in the vicinity of E_F, showing the **K**-points.

where now $\tan\phi_{\mathbf{k}} = k_y/k_x$. The energy dispersion still reads

$$\varepsilon^\lambda_\xi(\mathbf{k}) = \lambda\,\hbar\,v_F\,|\mathbf{k}|$$

independent of the valley pseudospin ξ. The pseudospin direction is associated with the momentum of the particles. This means that the wavefunctions in the vicinity of Dirac points represent chiral fermions. One consequence of this is that any backscattering, namely, scattering of particles from the wavevector $\mathbf{k}$ to $-\mathbf{k}$, is suppressed [17]. Particles have opposite chirality in the K and K' valleys.

A more convenient form of the Hamiltonian is obtained by inverting the spinor components at the K' point

$$\psi_{\mathbf{k}}^{(K)} = \begin{bmatrix} \psi_{\mathbf{k}}^A \\ \psi_{\mathbf{k}}^B \end{bmatrix}, \qquad \psi_{\mathbf{k}}^{(K')} = \begin{bmatrix} \psi_{\mathbf{k}}^B \\ \psi_{\mathbf{k}}^A \end{bmatrix},$$

which amounts to interchanging the role of the two sublattices. The Hamiltonian now takes the form

$$\mathcal{H}^{\text{eff}}_\xi(\mathbf{k}) = \xi\,\hbar\,v_F\left(k_x\,\sigma_1 + k_y\,\sigma_2\right). \tag{12.8}$$

Equation (12.8) suggests that $\mathcal{H}^{\text{eff}}(\mathbf{k})$ can be written in the more compact form if we define the double spinor

$$\langle\Psi_{\mathbf{k}}| = \begin{bmatrix} \psi_{\mathbf{k}+}^A & \psi_{\mathbf{k}+}^B & \psi_{\mathbf{k}-}^B & \psi_{\mathbf{k}-}^A \end{bmatrix}.$$

In this double-spinor representation, the first two components denote the sublattice components at the K point and the last two components those at the K' point. This allows us to write

$$\mathcal{H}^{\text{eff}}(\mathbf{k}) = \hbar\,v_F\,\tau_3\otimes\boldsymbol{\sigma}\cdot\mathbf{k}$$

$$\tau_3\otimes\boldsymbol{\sigma} = \begin{pmatrix} \boldsymbol{\sigma} & 0 \\ 0 & -\boldsymbol{\sigma} \end{pmatrix}, \tag{12.9}$$

where $\boldsymbol{\sigma} = (\sigma_1,\sigma_2)$, and the new pseudospin degree of freedom, represented by the Pauli matrix τ_3, describes the two valleys.

The eigenstates of the Hamiltonian (12.9) are the double-spinors

$$\Psi_{\mathbf{k}+}^\lambda = \frac{1}{\sqrt{2}}\begin{bmatrix} e^{-i\phi_{\mathbf{k}}/2} \\ \lambda\,e^{i\phi_{\mathbf{k}}/2} \\ 0 \\ 0 \end{bmatrix}, \qquad \Psi_{\mathbf{k}-}^\lambda = \frac{1}{\sqrt{2}}\begin{bmatrix} 0 \\ 0 \\ e^{-i\phi_{\mathbf{k}}/2} \\ -\lambda\,e^{i\phi_{\mathbf{k}}/2} \end{bmatrix}. \tag{12.10}$$

In summary, the effective low-energy Hamiltonian of graphene is just the Dirac Hamiltonian; it depicts the low-energy continuum limit of the graphene lattice Hamiltonian. The absence of a mass term qualifies the corresponding Dirac fermions as Weyl fermions; it is responsible for the high-mobility properties of graphene. We also discern that the presence of the Dirac, or Fermi, points renders graphene a semimetal: it behaves as a metal because of the vanishing gap between the valence and conducting bands, yet it has a small number of charge carriers due to the vanishing density of states at the Fermi level.

We should clearly distinguish the two types of pseudospin introduced in this section:

(a) The sublattice pseudospin, given by the Pauli matrices σ_i, has *spin-up* representing one sublattice component and *spin-down* representing the component on the second sublattice. The diagonalization of the graphene Hamiltonian can be regarded as a $SU(2)$ rotation in the sublattice pseudospin space that yields the band indices $\lambda = \pm$.

(b) The valley pseudospin Pauli matrix τ_3 defines the twofold valley degeneracy and is only indirectly related to the presence of two sublattices.

12.2.2 Symmetries of the Graphene Bloch Hamiltonian

We consider a honeycomb system with two pseudospin and one spin-1/2 degrees of freedom:

(1) Space inversion $\hat{\pi}$:

$$\hat{\pi} : (\mathbf{x}, \mathbf{p}, \mathbf{s}) \rightarrow (-\mathbf{x}, -\mathbf{p}, \mathbf{s}).$$

For a Hamiltonian on a lattice, we write

$$\mathcal{H} = \sum_{\mathbf{k}n} |\mathbf{k}, n\rangle \, \mathcal{H}(\mathbf{k}) \, \langle \mathbf{k}, n|.$$

The action of $\hat{\pi}$ yields

$$\hat{\pi} \, \mathcal{H} \, \hat{\pi} = \hat{\pi} \sum_{\mathbf{k}n} |\mathbf{k}, n\rangle \, \mathcal{H}(\mathbf{k}) \, \langle \mathbf{k}, n| \, \hat{\pi}$$

$$= \sum_{\mathbf{k}n} |-\mathbf{k}, n\rangle \, \mathcal{H}(\mathbf{k}) \, \langle -\mathbf{k}, n| = \sum_{\mathbf{k}n} |\mathbf{k}, n\rangle \, \mathcal{H}(-\mathbf{k}) \, \langle \mathbf{k}, n|.$$

Effectively, we can consider the action on the Hamiltonian as

$$\hat{\pi} : \mathcal{H}(\mathbf{k}) \rightarrow \mathcal{H}(-\mathbf{k}).$$

In two dimensions, the inversion operation is[3]

$$\hat{\pi} : (x, y) \rightarrow \begin{cases} (x, -y) \\ (-x, y) \end{cases}$$

The action of $\hat{\pi}$ on the honeycomb lattice is to interchange the sublattices. This is achieved by setting

$$\hat{\pi} = \hat{\sigma}_1.$$

---

[3] Inversion only flips one coordinate so that the determinant of the transformation will be -1. If we flipped both coordinates, the determinant would be $+1$ and therefore a rotation. This is a common subtlety seen in even spatial dimensions.

It would be instructive here to consider the $\hat{\pi}$ action when the sublattice symmetry is broken, as in BN, namely

$$\hat{\pi}\,\mathcal{H}\,\hat{\pi} = k_x\,\sigma_1\,\sigma_1\,\sigma_1 - k_y\,\sigma_1\,\sigma_2\,\sigma_1 + m\,\sigma_1\,\sigma_z\,\sigma_1$$
$$= k_x\,\sigma_1 + k_y\,\sigma_2 - m\,\sigma_z,$$

which implies that the mass term (onsite energy anisotropy) clearly breaks parity. The absence of this term in graphene guarantees the invariance of its Hamiltonian under inversion, where $\mathcal{H}(\mathbf{k}) = h(\mathbf{k})\cdot\boldsymbol{\sigma}$, and

$$\hat{\pi} : \begin{cases} h(k_x,k_y) \rightarrow h(k_x, -k_y) = \Big(h_1(k_x, -k_y), h_2(k_x, -k_y)\Big), \ h_1(\text{even}), h_2(\text{odd}) \\ (\sigma_1,\sigma_2,\sigma_z) \rightarrow (\sigma_1, -\sigma_2, -\sigma_3), \qquad \text{Sublattices } A \text{ and } B \text{ interchanged.} \end{cases}$$

Moreover, as seen in Figure 12.1, the transformation $(k_x,k_y) \rightarrow (k_x, -k_y)$ leads to valley switching $\hat{\pi} : \mathbf{K} \rightarrow \mathbf{K}'$

(2) Time-reversal Θ:

$$\Theta : (\mathbf{x}, \mathbf{p}, \mathbf{s}) \rightarrow (\mathbf{x}, -\mathbf{p}, -\mathbf{s})$$

Its action on the preceding degrees of freedom is given by

$$\Theta = \begin{cases} i\sigma_0\,\hat{s}_y\,K, & \text{spin-1/2} \\ \sigma_0\,K, & \text{pseudospin} \end{cases}$$

where σ_0 is the identity matrix. The TR operation leaves the sublattice invariant, but conjugates the wavefunctions amplitudes, and thus for pseudospin it obeys $\Theta^2 = 1$. The spinless graphene Hamiltonian $\mathcal{H}(\mathbf{k})$ is also TR invariant

$$\Theta\,\mathcal{H}(\mathbf{k})\,\Theta^{-1} = \mathcal{H}^*(-\mathbf{k}) = h_1(-\mathbf{k})\,\sigma_1 + h_2(-\mathbf{k})\,\sigma_2^* = h_1(\mathbf{k})\,\sigma_1 + h_2(\mathbf{k})\,\sigma_2 = \mathcal{H}(\mathbf{k})$$

but Θ switches valleys. We note that at TRIMs, such as the BZ center, the action of Θ on the wavefunction yields

$$\langle \psi(\mathbf{0}) | \Theta\psi(\mathbf{0}) \rangle = \int \psi^*(\mathbf{0})\,\psi^*(\mathbf{0}) \neq 0,$$

indicating that $\psi(\mathbf{0})$ and $\Theta\psi(\mathbf{0})$ can be the same state, or that $\psi(\mathbf{k})$ and $\Theta\psi(\mathbf{k})$ can belong to the same band. In contrast, for spin-1/2, $\Theta^2 = -1$, we have Kramers' degeneracy as a fundamental consequence.

(3) The combination of both symmetries requires that

$$\mathcal{H}(\mathbf{k}) = \sigma_1\,\mathcal{H}^*(-\mathbf{k})\,\sigma_1,$$

which forces mass terms proportional to σ_3 to vanish.

Finally, in the low-energy effective theory, the representation combines the valley and sublattice pseudospins. Then the valley switching is implemented by writing

$$\hat{\pi} = \sigma_1 \otimes \tau_1, \qquad \Theta = \sigma_0 \otimes \tau_1\,K.$$

Generic Two- and Four-Band Hamiltonians

Two-Band Hamiltonians

In the preceding section, we expressed a two-band Hamiltonian in terms of the Pauli spin matrices. In general, we can express any two-band Bloch Hamiltonian as

$$\mathcal{H}(\mathbf{k}) = h_0(\mathbf{k})\,\sigma_0 + \mathbf{h}(\mathbf{k})\cdot\boldsymbol{\sigma}\,. \tag{12.11}$$

The Pauli matrices represent some internal degree of freedom. This degree of freedom can be a real spin, the sublattice index (A and B sublattices of graphene), or an orbital index (s and p orbitals defined on the same site). The details of the coupling are described by the vector $\mathbf{h}(\mathbf{k}) = (h_1(\mathbf{k}), h_2(\mathbf{k}), h_3(\mathbf{k}))$ of periodic functions of $\mathbf{k}$. The structure of the Bloch Hamiltonian $\mathcal{H}(\mathbf{k})$ is constrained by the symmetries of the problem.

All the information about the topology of wavefunctions is encoded in four real and periodic functions of the momentum, $(h_0(\mathbf{k}), h_1(\mathbf{k}), h_2(\mathbf{k}), h_3(\mathbf{k}))$, all defined on the whole Brillouin torus $\mathbb{T}^2$. The function $h_0(\mathbf{k})$ simply shifts the eigenvalues without affecting the eigenstates, and therefore it has no effect on the topological properties of the material. Nevertheless, this function is very important because it enters the spectrum dispersion and thus determines the position of the Fermi level.

Four-Band Hamiltonians

Anticipating that we will consider spin–orbit (SO) interactions, we expand our two-band model to a four-band one. Generally, any 4×4 Hamiltonian can be expanded in terms of the complete basis of the Dirac Γ matrices as

$$\mathcal{H}(\mathbf{k}) = h_0(\mathbf{k})\,\mathbb{I} + \sum_i h_i(\mathbf{k})\Gamma_i + \sum_{ij} h_{ij}(\mathbf{k})\Gamma_{ij},$$

where $\mathbb{I}$ is the 4×4 identity matrix, $\Gamma_i (i = 1 \ldots 5)$ denote the five Dirac Γ matrices satisfying $\{\Gamma_i, \Gamma_j\} = 2\delta_{ij}$, and the 10 commutators of Γ matrices are given by $\Gamma_{ij} = [\Gamma_i, \Gamma_j]/2i$. The quantities h_0, h_i, h_{ij} can be constructed according to the transformation properties of the basis set used under the symmetry operations of the group of the Hamiltonian.

In Sections 12.3 and 12.4, we will be dealing with two-dimensional systems that represent different guises of graphene.

12.3 Chern Topological Insulators

Chern insulators identify two-dimensional insulators that break time-reversal symmetry and exhibit topological properties in their electronic band structure, even in the absence of a net magnetic field. Consequently, when time-reversal symmetry is broken, we would expect that they exhibit phases having Bloch bands with nonzero Chern numbers. Here, we shall explore how to construct lattice systems with Hamiltonians that engender nonzero Chern numbers. Such insulators would be termed Chern insulators. We will describe Haldane's Chern insulator in detail, and leave other models as exercises.

Our initial goal will be to transform the graphene sheet into a quantum Hall–like phase, with chiral edge states. The necessary first step is to gap the bulk of the system. We know that the Dirac points are protected by both sublattice-inversion and time-reversal symmetry. So we can break one or both of these symmetries, and open energy gaps at K and K'.

12.3.1 The Trivial Semenoff Insulator

The simplest way to break sublattice symmetry is to assign opposite onsite energies [164] $\Delta\,(-\Delta)$ to the $A\,(B)$ sites, à la boron nitride, while leaving time-reversal symmetry intact. The Hamiltonian is then given by

$$h_1(\mathbf{k})\sigma_1 + h_2(\mathbf{k})\sigma_2 + \Delta\,\sigma_3.$$

This leads to a gapped spectrum,

$$\varepsilon(\mathbf{k}) = \pm\sqrt{|h(\mathbf{k})|^2 + \Delta^2}.$$

However, we immediately realize that this leads to a rather boring situation: it preserves time-reversal symmetry. And with the time-reversal symmetry present, it is impossible to obtain a nonzero Chern insulator accompanied by chiral edge states. Moreover, taking the limit $|\Delta| \gg t$, we obtain electronic states that are localized on one of the two sublattices A or B, irrespective of the sign of Δ.

12.3.2 Haldane's Chern Insulator

In his seminal paper [82], Haldane considered a model of a honeycomb lattice subjected to a periodic magnetic field normal to its 2D plane. Most importantly, the field he proposed has zero total flux in the unit cell and respects the full symmetry of the lattice. Figure 12.3(a) depicts a rendering of a staggered flux pattern that can give rise to such a scenario. It is clear that the presence of this magnetic field does not generate Landau levels, but breaks TRS in the system, and introduces complex phases $e^{\pm i\phi}$ in the next-nearest-neighbor (NNN) hoppings.

Origin of the Complex Phase

The complex phase acquired during electron hopping can be seen as a type of an AB effect, where a vector potential $\mathbf{A}(\mathbf{x})$, due to the tailored magnetic field, augments the hopping amplitude with a multiplicative phase term of the form $e^{-i\phi} = e^{-i(e/\hbar)\int d\mathbf{x}\cdot\mathbf{A}(\mathbf{x})}$. The integral is taken along the shortest closed path that connects NNN sites, delineated by the black and gray triangular loops in Figure 12.3(b). An electron effectively hops along such

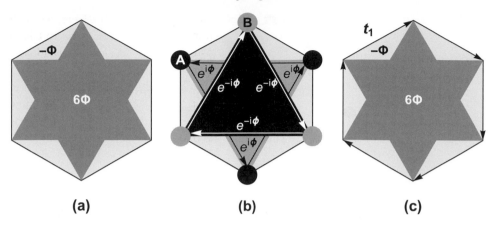

Figure 12.3 Haldane's model. (a) A possible pattern of the magnetic flux. (b) Next-nearest-neighbor hopping t_2. The direction of the arrows indicate the direction in which the hopping has a phase $e^{i\phi}$ or $e^{-i\phi}$. (c) Closed path of nearest-neighbor (NN) hoppings (blue arrows), indicates a zero net flux flowing through the loop.

a loop on one sublattice and accumulates a phase proportional to the flux of the magnetic field through the corresponding triangle. Since this phase is opposite for electron hopping on the two sublattices, the total field acting on electrons averages to zero within the unit cell. By contrast, Figure 12.3(c) shows that a closed loop of NN hoppings (black arrows) encircles the full hexagon, and hence experiences a zero net flux, indicating that an NN hopping does not acquire a phase. By Fourier transforming the lattice Hamiltonian to a **k**-space representation, Haldane showed that the Chern number of this model equals +1 for $-\pi < \phi < 0$, and −1 for $0 < \phi < \pi$. The point $\phi = 0$ can be seen to give rise to Weyl points in $(\mathbf{k}; \phi)$-space.

Another important ingredient of Haldane's model is the breaking of inversion symmetry via a Semenoff-like staggered sublattice potential. It is by breaking both TR and inversion symmetries of the honeycomb lattice, and opening gaps in the band structure at the **K** and **K′** momentum points, that Haldane could demonstrate the possibility of observing a Hall conductivity of $\sigma_{xy} = \pm e^2/h$, under certain conditions. Conceptually, this model, proposed by Haldane in 1988, marked the beginning of tremendous advances in topological condensed matter physics.

The Model Hamiltonian of Haldane

The Hamiltonian of the Haldane model can be written as

$$
\mathcal{H} = t_1 \sum_{\langle ij \rangle} \left(|A,i\rangle \langle B,j| + HC \right) + t_2 \sum_{\langle\langle ij \rangle\rangle} \left(e^{i\phi} |A,i\rangle \langle A,j| + e^{-i\phi} |B,i\rangle \langle B,j| \right)
$$

$$
+ \Delta \sum_i \left(|A,i\rangle \langle A,i| - |B,i\rangle \langle B,i| \right). \tag{12.12}
$$

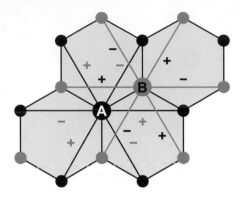

Figure 12.4 The sign convention for the acquired phases for NNN hoppings. The sign structure assumes that hopping in the counterclockwise direction has a $+\phi$, while the clockwise direction leads to a $-\phi$.

The summations $\langle ij \rangle$ and $\langle\langle ij \rangle\rangle$ are over NN and NNN, respectively. The sign structure of the phases ϕ is shown in Figure 12.4. Transforming to **k**-space, we obtain the Bloch Hamiltonian

$$\mathcal{H}(\mathbf{k}) = \sum_{i=0}^{3} h_i(\mathbf{k})\,\sigma_i = h_0(\mathbf{k})\,\sigma_0 + \mathbf{h}(\mathbf{k})\!\cdot\!\boldsymbol{\sigma}$$

$$h_0(\mathbf{k}) = 2t_2\,\cos(\phi) \sum_{i=1}^{3} \cos\!\big[\mathbf{k}\cdot\mathbf{v}_i\big], \quad h_1(\mathbf{k}) = t_1\Big[\cos(\mathbf{k}\cdot\mathbf{a}_1) + \cos(\mathbf{k}\cdot\mathbf{a}_2) + 1\Big]$$

$$h_2(\mathbf{k}) = t_1\Big[\sin(\mathbf{k}\cdot\mathbf{a}_1) + \sin(\mathbf{k}\cdot\mathbf{a}_2)\Big], \quad h_3(\mathbf{k}) = \Delta + 2t_2\,\sin(\phi) \sum_{i=1}^{3} \sin\!\big[\mathbf{k}\cdot\mathbf{v}_i\big].$$

$\mathbf{a}_1$ and $\mathbf{a}_2$ were defined for graphene, and

$$\mathbf{v}_1 = \boldsymbol{\delta}_2 - \boldsymbol{\delta}_3, \quad \mathbf{v}_2 = \boldsymbol{\delta}_3 - \boldsymbol{\delta}_1, \quad \mathbf{v}_3 = \boldsymbol{\delta}_1 - \boldsymbol{\delta}_2.$$

The eigenvalues and eigenfunctions are

$$\varepsilon(\mathbf{k}) = h_0(\mathbf{k}) \pm |\mathbf{h}(\mathbf{k})| = h_0(\mathbf{k}) \pm h(\mathbf{k})$$

$$|\mathbf{h}, \pm\rangle = \frac{1}{\sqrt{2h(h \mp h_3)}} \begin{bmatrix} h_3 \pm h \\ h_1 - ih_2 \end{bmatrix}. \tag{12.13}$$

We ignore the term $h_0(\mathbf{k})$, since it just shifts the energies and removes the electron–hole symmetry of the original NN model. Exploring the symmetries of this Hamiltonian, we find the following:

- If both Δ and $t_2 \sin\phi$ vanish, the bands touch at the BZ corners, K or K', where the symmetry group of the wavevector is the subgroup C_{3v}. It contains a reflection that interchanges the A and B sublattices. This group has a two-dimensional irreducible representation, and the degenerate states at these points belong to this representation.

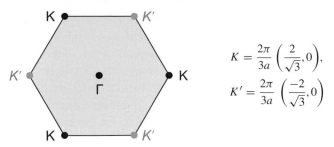

Figure 12.5 Gap closings for the Haldane Chern insulator occur at both K and K' points when both Δ and $t_2 \sin \phi$ vanish. Gap closing occurs at either K or K' when $|\Delta| = 3\sqrt{3}\, t_2 \sin \phi$.

- The degeneracy of the bands at these points is lifted either by nonzero Δ or nonzero $t_2 \sin \phi$, either of which reduce the unitary subgroup to C_3, which has only one-dimensional irreducible representations.
- The terms $h_1 \sigma_1$ and $h_2 \sigma_2$ remain invariant under Θ. However, we note that $h_3(\mathbf{k}) = h_3(-\mathbf{k})$ holds only for $\phi = 0, \pi$.

At the K and K' points (se Figure 12.5),

$$\sum_{i=1}^{3} \cos\Big[\mathbf{K} \cdot \mathbf{v}_i\Big] = -\frac{3}{2}; \quad \sum_{i=1}^{3} \sin\Big[\pm\mathbf{K} \cdot \mathbf{v}_i\Big] = \mp\frac{3\sqrt{3}}{2}$$

and the Hamiltonian reduces to

$$\begin{cases} \mathcal{H}(\mathbf{K}) = \Big[\Delta + 3\sqrt{3}\, t_2 \sin \phi\Big] \sigma_3 \\ \mathcal{H}(\mathbf{K'}) = \Big[\Delta - 3\sqrt{3}\, t_2 \sin \phi\Big] \sigma_3 \end{cases} \Rightarrow \Delta\Big(\sigma_3 \otimes \tau_0\Big) + 3\sqrt{3} t_2 \sin \phi \Big(\sigma_3 \otimes \tau_3\Big) \quad (12.14)$$

We note that the corresponding eigenkets at each K or K' point in the $\{A, B\}$ basis will always be

$$|1\rangle = \begin{bmatrix} 1 \\ 0 \end{bmatrix}, \qquad |2\rangle = \begin{bmatrix} 0 \\ 1 \end{bmatrix}$$

The eigenvalues will depend on the quantities Δ, t_2 and ϕ, and we have

1. $|\Delta| > 3\sqrt{3}\, t_2 \sin \phi$: the system is gapped, with the gaps $\Delta\varepsilon(\mathbf{K})$ and $\Delta\varepsilon(\mathbf{K'})$ having the same sign.
2. $|\Delta| = 3\sqrt{3}\, t_2 \sin \phi$: the system has a single vanishing gap either at K ($\Delta < 0$) or K' ($\Delta > 0$).
3. $|\Delta| < 3\sqrt{3}\, t_2 \sin \phi$: the system is again gapped. Taking $\Delta > 0$, we obtain

$$\begin{cases} \Delta\varepsilon(\mathbf{K}) &= 2\Big[3\sqrt{3}\, t_2 \sin \phi + \Delta\Big] > 0 \\ \Delta\varepsilon(\mathbf{K'}) &= -2\Big[3\sqrt{3}\, t_2 \sin \phi - \Delta\Big] < 0 \end{cases}$$

The opposite signs of the gaps at K and K' reflect an interchange of the two eigenstates, locations with respect to the gap! This is referred to as a *band inversion*. Such a band inversion presents an intertwining, or entanglement, of the states as we move from, say, K to K'. We cannot go continuously from this case 3 to case 1 without closing the gap. The transition from a topological insulator to a trivial insulator is marked by a semimetal state, where the gap closes at one Dirac-point type when $\Delta = \pm 3\sqrt{3}\, t_2 \sin\phi$. In the trivial insulating phase, Dirac points have identical associated mass sign, and the Chern number goes to zero.

Mapping to the Two-Level System

The Haldane Hamiltonian has the form $\mathbf{h}\cdot\boldsymbol{\sigma}$, which resembles that of the two-level system of Chapter 11. By establishing this analogy, we realize that the physics of the Haldane problem is actually governed by the Dirac monopole. To perform the mapping, we introduce polar and azimuthal angles, $(\theta_{\mathbf{k}}, \varphi_{\mathbf{k}})$, that define the direction of $\mathbf{h}(\mathbf{k})$, namely,

$$\cos\theta_{\mathbf{k}} = \frac{h_3(\mathbf{k})}{|\mathbf{h}(\mathbf{k})|}, \quad \tan\varphi_{\mathbf{k}} = \frac{h_2(\mathbf{k})}{h_1(\mathbf{k})}.$$

This allows us to express the eingenvectors as

$$|\mathbf{h}, -\rangle = \begin{pmatrix} u_-^{(1)} \\ u_-^{(2)} \end{pmatrix} = \begin{pmatrix} \sin\frac{\theta_{\mathbf{k}}}{2}\, e^{-i\varphi_{\mathbf{k}}} \\ -\cos\frac{\theta_{\mathbf{k}}}{2} \end{pmatrix}; \qquad |\mathbf{h}, +\rangle = \begin{pmatrix} u_+^{(1)} \\ u_+^{(2)} \end{pmatrix} = \begin{pmatrix} \cos\frac{\theta_{\mathbf{k}}}{2}\, e^{i\varphi_{\mathbf{k}}} \\ \sin\frac{\theta_{\mathbf{k}}}{2} \end{pmatrix}$$

and to write the occupied state projection [4] in terms of the unit vector

$$\hat{\mathbf{h}}_{\mathbf{k}} = \frac{\mathbf{h}(\mathbf{k})}{h(\mathbf{k})} = \begin{pmatrix} \cos\varphi_{\mathbf{k}} \sin\theta_{\mathbf{k}} \\ \sin\varphi_{\mathbf{k}} \sin\theta_{\mathbf{k}} \\ \cos\theta_{\mathbf{k}} \end{pmatrix}.$$

$\hat{\mathbf{h}}_{\mathbf{k}}$ resides on the unit sphere $\mathbb{S}^2$, while $\mathbf{k}$ spans the two-dimensional BT $\mathbb{T}^2$.

---

[4] We note that

$$\hat{\mathbf{h}}_{\mathbf{k}}\cdot\boldsymbol{\sigma} = \frac{h_1(\mathbf{k})}{h(\mathbf{k})}\sigma_1 + \frac{h_2(\mathbf{k})}{h(\mathbf{k})}\sigma_2 + \frac{h_3(\mathbf{k})}{h(\mathbf{k})}\sigma_3 = \frac{h_1(\mathbf{k})}{\sqrt{h_1^2(\mathbf{k})+h_2^2(\mathbf{k})}}\frac{\sqrt{h_1^2(\mathbf{k})+h_2^2(\mathbf{k})}}{h(\mathbf{k})}\sigma_1$$

$$+ \frac{h_2(\mathbf{k})}{\sqrt{h_1^2(\mathbf{k})+h_2^2(\mathbf{k})}}\frac{\sqrt{h_1^2(\mathbf{k})+h_2^2(\mathbf{k})}}{h(\mathbf{k})}\sigma_2 + \frac{h_3(\mathbf{k})}{h(\mathbf{k})}\sigma_3$$

$$= \cos\varphi_{\mathbf{k}} \sin\theta_{\mathbf{k}}\, \sigma_1 + \sin\varphi_{\mathbf{k}} \sin\theta_{\mathbf{k}}\, \sigma_2 + \cos\theta_{\mathbf{k}}\, \sigma_3 = \begin{pmatrix} \cos\theta_{\mathbf{k}} & \sin\theta_{\mathbf{k}}\, e^{i\varphi_{\mathbf{k}}} \\ \sin\theta_{\mathbf{k}}\, e^{-i\varphi_{\mathbf{k}}} & -\cos\theta_{\mathbf{k}} \end{pmatrix}$$

$$|\psi_-\rangle\langle\psi_-| = \begin{pmatrix} \sin\frac{\theta_{\mathbf{k}}}{2}\, e^{-i\varphi_{\mathbf{k}}} \\ -\cos\frac{\theta_{\mathbf{k}}}{2} \end{pmatrix} \begin{pmatrix} \sin\frac{\theta_{\mathbf{k}}}{2}\, e^{i\varphi_{\mathbf{k}}} & -\cos\frac{\theta_{\mathbf{k}}}{2} \end{pmatrix}$$

$$= \begin{pmatrix} \sin^2\frac{\theta_{\mathbf{k}}}{2} & -\cos\frac{\theta_{\mathbf{k}}}{2}\sin\frac{\theta_{\mathbf{k}}}{2}\, e^{i\varphi_{\mathbf{k}}} \\ -\cos\frac{\theta_{\mathbf{k}}}{2}\sin\frac{\theta_{\mathbf{k}}}{2}\, e^{-i\varphi_{\mathbf{k}}} & \cos^2\frac{\theta_{\mathbf{k}}}{2} \end{pmatrix} = \frac{1}{2}\left[\sigma_0 - \hat{\mathbf{h}}_{\mathbf{k}}\cdot\boldsymbol{\sigma}\right]$$

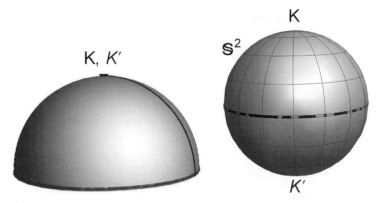

Figure 12.6 The map $\mathbf{k} \to \hat{\mathbf{h}}_{\mathbf{k}}$: for $|\Delta| > 3\sqrt{3}\,t_2 \sin\phi$, both K, K' lie at the north pole ($\Delta > 3\sqrt{3}\,t_2 \sin\phi$), according to (12.14), and the $\hat{\mathbf{h}}_{\mathbf{k}}$s essentially fluctuate around the north pole, as shown in (a), so the map can be continuously deformed to a point – the north pole. The same scenario plays around the south pole for $\Delta < -3\sqrt{3}\,t_2 \sin\phi$. Essentially, the $\hat{\mathbf{h}}_{\mathbf{k}}$-vectors can be combed straight up or down. (b) For $|\Delta| < 3\sqrt{3}\,t_2 \sin\phi$, the $\hat{\mathbf{h}}_{\mathbf{k}}$-vectors around K map to the north pole, and those around K' map to the south pole. The entire sphere is covered, and that gives a Chern number of 1.

The mapping $\mathbf{k} \to \hat{\mathbf{h}}_{\mathbf{k}}$ for the gapped system is well defined over the entire BZ, since $h(\mathbf{k}) \neq 0$ everywhere, and it captures the topological properties of the Hamiltonian. We find that for $\Delta > 3\sqrt{3}\,t_2 \sin\phi$, the image of $\mathbb{T}^2$ on $\mathbb{S}^2$ only covers a portion of the upper hemisphere, as depicted in Figure 12.6(a), hence the Chern number $\mathcal{C}_H = 0$. In contrast, for $|\Delta| < 3\sqrt{3}\,t_2 \sin\phi$ it covers the whole $\mathbb{S}^2$, and hence yields $\mathcal{C}_H = 1$; see Figure 12.6(b). In that sense, the Chern number becomes a winding number that counts how many times the surface traced by $\hat{\mathbf{h}}_{\mathbf{k}}$ wraps around the origin $(0,0,0)$ of $\mathbb{S}^2$, when $\mathbf{k}$ spans $\mathbb{T}^2$.

Actually, the Berry phase that is accumulated when going along a closed-loop $\mathcal{C}$ in $\mathbf{k}$-space can be easily related to that picked up when going along the image loop of $\mathcal{C}$ on a unit sphere $\mathbb{S}^2$.

Berry Curvature Associated with $\mathbf{h}(\mathbf{k})$

We use

$$\frac{\partial}{\partial k_\ell}\,|\mathbf{h},\pm\rangle = \sum_i \frac{\partial}{\partial h_i}\,|\mathbf{h},\pm\rangle\,\frac{\partial h_i}{\partial k_\ell}, \qquad \ell = 1,2,3$$

to write the Berry connection as

$$\mathcal{A}_\ell^{(\pm)} = \langle \mathbf{h},\pm|\,\partial_{k_\ell}\,|\mathbf{h},\pm\rangle = \sum_i \frac{\partial h_i}{\partial k_\ell}\,a_i^{(\pm)}(\mathbf{h})$$

$$a_i^{(\pm)}(\mathbf{h}) = i\,\langle \mathbf{h},\pm|\,\partial_{h_i}\,|\mathbf{h},\pm\rangle \qquad \text{The two-level Berry connection.}$$

The Berry curvature is then written as

$$
\begin{aligned}
\mathcal{F}^{(\pm)}_{xy} &= \partial_{k_x}\mathcal{A}^{(\pm)}_y - \partial_{k_y}\mathcal{A}^{(\pm)}_x \\
&= \frac{\partial}{\partial k_x}\left(\sum_i \frac{\partial h_j}{\partial k_y} a^{(\pm)}_j(\mathbf{h})\right) - \frac{\partial}{\partial k_y}\left(\sum_i \frac{\partial h_i}{\partial k_x} a^{(\pm)}_i(\mathbf{h})\right) \\
&= \sum_{ij} \frac{\partial h_i}{\partial k_x}\frac{\partial h_j}{\partial k_y}\left[\frac{\partial a^{(\pm)}_j}{\partial h_i} - \frac{\partial a^{(\pm)}_i}{\partial h_j}\right] = \sum_{ij} \frac{\partial h_i}{\partial k_x}\frac{\partial h_j}{\partial k_y} f^{(\pm)}_k \epsilon_{ijk} \\
&= \mp \frac{1}{h^3}\,\mathbf{h}\cdot\partial_{k_x}\mathbf{h}\wedge\partial_{k_y}\mathbf{h},
\end{aligned}
$$

where $f^{(\pm)}$ is just the Berry curvature of the two-level Hamiltonian.

We then obtain the following expression for the Chern number:

$$
\mathcal{C} = \frac{1}{2\pi}\int_{\mathbb{T}^2} d\mathbf{k}\,\mathcal{F}^{(\pm)}_{xy}(\mathbf{k}) = \mp\frac{1}{2\pi}\int_{\mathbb{T}^2} d\mathbf{k}\,\frac{1}{h^3}\,\mathbf{h}\cdot\partial_{k_x}\mathbf{h}\wedge\partial_{k_y}\mathbf{h}.
$$

Considering $\hat{\mathbf{h}}(\mathbf{k}): \mathbb{T}^2 \to \mathbb{S}^2$ as a mapping from the Brillouin zone to the unit sphere, the integrand $\frac{1}{h^3}\,\mathbf{h}\cdot\partial_{k_x}\mathbf{h}\wedge\partial_{k_y}\mathbf{h}$ is simply the Jacobian of this mapping. Thus the integration over it gives the total area of the image of the Brillouin zone on $\mathbb{S}^2$, which is a topological winding number with quantized value, the Chern number.

We can now construct the phase diagram in Figure 12.7.

The Hall conductance was expressed in terms of an integral of Berry's curvature over the Brillouin torus in Chapter 11, and thus can be expressed in terms of the Chern number as

$$
\sigma_{xy} = \frac{e^2}{h}\,\mathcal{C}.
$$

Interface between Topologically Distinct Insulators: Chiral Ddge States

We now describe how fermionic edge modes emerge in the gapless interfacial region between two topologically distinct insulators, and are confined between the two fully gapped bulk insulating regions. To clearly illustrate this scenario, we consider the edge states located at the interface between a parity-breaking Semenoff, phase and a Haldane

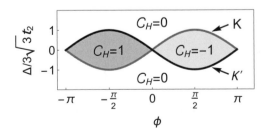

Figure 12.7   Phase diagram of the Haldane model.

phase, with additional TRS breaking, and show how they are protected. In this 2D system, the edge modes run along the 1D interface [42].

In the model, the Semenoff insulator fills the half-plane $x < 0$, while the Haldane insulator occupies the half-plane $x > 0$. The interface lies along the y-direction. We will use the low-energy effective Hamiltonian that is valid in the vicinity of K and K', since we are interested in the ensuing zero modes residing near the interface $x = 0$. Because of translational invariance along the y-direction, the wave equation becomes

$$\left(-i\hbar v_F \sigma_1 \tau_3 \partial_x + \hbar v_F \, k_y \sigma_2 + m(x)\right) \Psi = \varepsilon \, \Psi \qquad (12.15)$$

$$m(x) = \Delta \, \Theta(-x) \, \sigma_3 + m_\phi \, \Theta(x) \, \sigma_3 \tau_3,$$

where Θ is the Heaviside step function. This simple model with a sharp step at the boundary is correct as long as the decay length scale of the edge state in the interfacial region, $\hbar v_F / \max(\Delta, m_\phi)$, is much larger than the lattice length scale (lattice constant). We note that a solution at $E = 0$ and $k_y = 0$ always exists. To show that, we multiply both sides of (12.15) with $i\sigma_1 \tau_3$, and obtain

$$\hbar v_F \partial_x \Psi = -i\sigma_1 \tau_3 \, m(x) \, \Psi = \begin{cases} \hbar v_F \, \partial_x \Psi = -\sigma_2 \, m_\phi \, \Psi, & x > 0 \\[2mm] \hbar v_F \, \partial_x \Psi = -\sigma_2 \tau_3 \Delta \, \Psi, & x < 0 \end{cases} \qquad (12.16)$$

With σ_2 having eigenvalues ± 1 and τ_3 representing the valley sign ξ, we obtain bounded solutions $\text{sign}(m_\phi)$, $x > 0$ and $-\text{sign}(\xi \Delta)$, $x < 0$. There will be a zero mode at the boundary, only if the two preceding solutions correspond to the same eigenvalue of σ_2, namely if

$$\text{sign}(m_\phi) = -\text{sign}(\xi \Delta).$$

This equality is always valid in one valley determined by the relative signs of m_ϕ and Δ. Thus, a zero mode always exists in the valley $\xi = -\text{sign}(m_\phi \Delta)$, irrespective of the choice of masses.

We obtain the wavefunction and dispersion $E(k_y)$ of the edge mode (shown in Figure 12.8), by allowing $E \neq 0$ and $k_y \neq 0$ in (12.15). We note that the zero mode

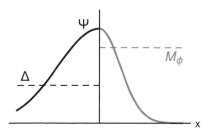
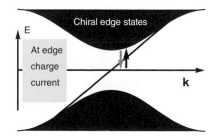

Figure 12.8 Left: edge mode wavefunction Ψ at interface between a Sememoff and Haldane insulators. Right: Chiral right-propagating dispersion along the interface.

at $k_y = 0$ is also an eigenstate of $\hbar v_F k_y \sigma_2$, which extends its validity to finite energies and k_ys. This allows us to write the edge mode dispersion as

$$\varepsilon(k_y) = -\text{sign}(\xi\Delta)\,\hbar v_F k_y = \text{sign}(m_\phi)\,\hbar v_F k_y.$$

It is obvious that the edge mode is chiral, and that it appears in the valley that undergoes a mass inversion at the interface.

We note that for large Δ, the Semenoff insulator can represent the vacuum. We then find that a 2D square of Haldane insulator, of edge length L, surrounded by a large Δ Semenoff insulator, will have a 1D edge chiral edge mode that circulates clockwise if $\text{sign}(m_\phi\Delta)$ is positive, and anticlockwise for negative $m_\phi\Delta$. We also note that if we assume that the vacuum is represented by a large positive Δ, then the sign of $m_\phi\Delta$ is simply the sign of $m_\phi = 3\sqrt{3}t2\sin(\phi)$ which is set by the chirality of the flux pattern in the microscopic Haldane model.

12.4 Quantum Spin Hall Insulator: The Kane–Mele Model

In 2004–2005, it was realized that TR-invariant 2D electronic insulator systems may exhibit new topological invariants, since the Chern number must vanish for TR symmetric systems. This topological invariant turned out to be the two-valued $\mathbb{Z}_2$ invariant derived in Chapter 11. Insulators were then classified into two categories: *ordinary* or *trivial* insulators that do not have protected edge state, and *quantum spin Hall* systems or *topological insulators* with a bulk topological invariant that protects edge states.

The idea that sparked this development was the quantum spin Hall effect introduced in Chapter 8 with the aid of a simple model. It satisfied TRS, having zero net Chern number and vanishing edge charge current. However, the counter-propagating edge electrons have opposite spins, and produce a nonzero spin current, hence the *quantum spin Hall* effect. We note that in such a model, spin-rotation invariance, as defined by $SU(2)$, is clearly broken, yet, S_z is a conserved quantum number. Such a setting can be realized by spin–orbit coupling (SOC) arising from intraatomic terms like $\mathbf{L}\cdot\mathbf{S}$, specifically, $L_z S_z$. For an electron of fixed spin, the coupling to the orbital motion L_z mimics the coupling to a constant magnetic field, recalling that the orbital motion L_z effectively generates a magnetic dipole moment. Because SOC is invariant under both TR and inversion, it opens a gap but respects Kramers' spin degeneracy of the energy bands.

In 2005, C. L. Kane and E. G. Mele proposed a variant of the Haldane model that respects TRS and includes the spin via spin–orbit interactions. Actually, their proposal launched the field of topological insulators and triggered its rapid expansion.

To convert the graphene semimetal into an insulator, we need to open a gap at the K and K' points. We have noted earlier that this can be achieved if we add a term $\propto \sigma_3$ to the sublattice pseudospin matrices. Moreover, to realize this scenario, we need a spin-dependent term that couples to the two copies of the QHE model. We consider the combination for the K point, which produces gaps of opposite signs for the two spins, namely,

$$\sigma_3 \otimes s_3 = \begin{pmatrix} \sigma_3 & 0 \\ 0 & -\sigma_3 \end{pmatrix}.$$

Next, we explore the action of Θ on the combined sub-lattice and spin spaces:

$$\Theta = \sigma_0 \otimes i s_2 \, K = \begin{pmatrix} 0 & -\sigma_0 \\ \sigma_0 & 0 \end{pmatrix}.$$

We find that the term $\propto \sigma_3 \otimes s_3$ transforms as

$$\Theta \left(\sigma_3 \otimes s_3 \right) \Theta^{-1} = \begin{pmatrix} 0 & -\sigma_0 \\ \sigma_0 & 0 \end{pmatrix} \begin{pmatrix} \sigma_3 & 0 \\ 0 & -\sigma_3 \end{pmatrix} \begin{pmatrix} 0 & \sigma_0 \\ -\sigma_0 & 0 \end{pmatrix}$$

$$= \begin{pmatrix} 0 & -\sigma_0 \\ \sigma_0 & 0 \end{pmatrix} \begin{pmatrix} 0 & \sigma_3 \\ \sigma_3 & 0 \end{pmatrix} = \begin{pmatrix} -\sigma_3 & 0 \\ 0 & \sigma_3 \end{pmatrix} = -\sigma_3 \otimes s_3.$$

Since under $\Theta : K \rightarrow K'$, to preserve TR invariance, namely, $\Theta \, \mathcal{H}(\mathbf{k}) \, \Theta^{-1} = \mathcal{H}(-\mathbf{k})$, we need the gap opening term at K' to be $\Theta \, \sigma_3 \otimes s_3 \, \Theta^{-1} = -\sigma_3 \otimes s_3$. This can be achieved if we incorporate the valley degree of freedom, τ_3, as proposed by Kane and Mele, namely,

$$\mathcal{H}_{\text{so}} = \lambda_{\text{so}} \, \sigma_3 \otimes \tau_3 \otimes s_3, \tag{12.17}$$

where λ_{so} is the SO coupling magnitude. Adding a term of this form gives a system of two decoupled Hamiltonians – one for spin-up and one for spin-down. According to (12.14), each of these subsystems corresponds to Haldane's model with $\Delta = 0$, $\lambda_{so} = 3\sqrt{3}t_2 \sin \phi$ and gives a nontrivial topology as in case 3 of the Haldane model. The spin degree of freedom made it possible to have an effective magnetic field for each spin without breaking TRS.

Before we complete the construction of the Kane–Mele Hamiltonian, we shall discuss how to properly construct terms that represent both intrinsic and Rashba-type SOI in graphene.

12.4.1 Spin–Orbit Couplings in Graphenelike Systems

Intrinsic Spin–Orbit Coupling

The intrinsic SOC (ISOC) Hamiltonian is $\mathcal{H}_{SOC} \propto \mathbf{L} \cdot \boldsymbol{\sigma}$. It induces a change in the orbital magnetic quantum number, m, together with an electron spin flip. Hence, the relevant ISOC exists either among the $2p$ orbitals or among the unoccupied $3d$ orbitals of carbon. Its magnitude ξ_p (ξ_d) is of the of order $4\,(1)$ meV, as determined from the SOC overlap integral between $2p_z$ ($3d_{xz}$) and $2p_x$ ($3d_{yz}$) orbitals.

ISOC satisfies all D_{3h} symmetries of the graphene lattice. Reflection symmetry with respect to the lattice plane restricts the intersite SOC to the normal direction, thus only $L_z \sigma_z$, a spin-conserving term, can acquire a nonzero value. However, another reflection symmetry forces this term to vanish for NN π orbitals: a mirror reflection in a vertical plane bisecting the NN bond leaves the $2p_z$ wavefunctions unchanged, but changes the sign of the angular momentum L_z. This symmetry precludes direct spin-dependent or spin-flipping

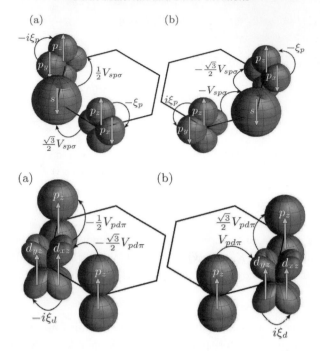

Figure 12.9 NNN hopping paths through s, p orbitals (left), and through d orbitals (right). Interorbital hoppings are shown as black arrows, and orbital spin shown by the gray arrows on the orbitals. The opposite sign for the clockwise (a) and the anticlockwise (b) effective hopping is determined by the signs of the two SOCs of the p or d orbitals. Figure and caption from [110].

NN hopping terms, yet such an effective spin-conserving SOC term may survive in NNN hopping. Moreover, we note that ISOC acts only in the vicinity of the atom core. Thus, to obtain an effective nonvanishing SOC contribution to the π-band, we need to consider hopping to NN d_{xz}, d_{yz} orbitals, or onsite SOC involving spin-flip $|p_z \uparrow\rangle \to |p_{x,y} \downarrow\rangle$ from the π-band to the σ-band; the latter extends our calculations to second order in the SO interaction. These processes lead to an effective NNN contribution to the π-Hamiltonian, represented by the virtual transition mechanisms shown in Figure 12.9 and given by [110]

$$\left|p_z^A \uparrow\right\rangle \underset{\text{SOC}}{\longrightarrow} \left|p_y^A \downarrow\right\rangle \underset{V_{sp\sigma}}{\longrightarrow} \left|s^B \downarrow\right\rangle \underset{V_{sp\sigma}}{\longrightarrow} \left|p_x^{A'} \downarrow\right\rangle \underset{\text{SOC}}{\longrightarrow} \left|p_z^{A'} \uparrow\right\rangle$$

$$\left|p_z^A \uparrow\right\rangle \underset{V_{pd\pi}}{\longrightarrow} \left|d_{xz}^B \uparrow\right\rangle \underset{\text{SOC}}{\longrightarrow} \left|d_{yz}^B \uparrow\right\rangle \underset{V_{pd\pi}}{\longrightarrow} \left|p_z^{A'} \uparrow\right\rangle.$$

The ISOC contribution of the p-orbitals channel involves two spin-flip processes, and thus leads to NNN electron hopping with no net spin-flip. Since it is $\propto \xi_p^2$, it is negligibly small – it produces a gap of about $\sim 1\,\mu$eV. The contribution via the d orbitals is linear in ξ_d, since it involves a spin-conserving onsite SOC transition between d_{xz}, d_{yz} orbitals, and gives a gap of about 23 μeV. These NNN spin-conserving hoppings introduce complex

Table 12.1 *Matrix elements of the SOC operator* $\mathbf{L} \cdot \mathbf{s}$ *in the basis of s-,
p-, and d-directed orbitals.*

Orbital			s	p_x	p_y	p_z
s			0	0	0	0
p_x			0	0	$-is_3$	is_2
p_y			0	is_3	0	$-is_1$
p_z			0	$-is_2$	is_1	0

Orbital	d_{xy}	$d_{x^2-y^2}$	d_{xz}	d_{yz}	d_{z^2}
d_{xy}	0	$2is_3$	$-is_3$	is_2	0
$d_{x^2-y^2}$	$-2is_3$	0	is_2	is_1	0
d_{xz}	is_1	$-is_2$	0	$-is_3$	$i\sqrt{3}s_2$
d_{yz}	$-is_2$	$-is_1$	is_3	0	$-i\sqrt{3}s_2$
d_{z^2}	0	0	$-i\sqrt{3}s_2$	$i\sqrt{3}s_2$	0

$$is_1 = \begin{pmatrix} 0 & i \\ i & 0 \end{pmatrix}, \quad is_2 = \begin{pmatrix} 0 & 1 \\ -1 & 0 \end{pmatrix}, \quad is_3 = \begin{pmatrix} i & 0 \\ 0 & -i \end{pmatrix}.$$

phases in the hopping process. We note from Figure 12.9 and Table 12.1 that the SOC induced transitions give rise to

$$\left\langle p_z \middle| H_{SOC} \middle| p_x \right\rangle \left\langle p_y \middle| H_{SOC} \middle| p_z \right\rangle \propto (-is_2)(is_1) = is_3,$$

$$\left\langle p_z \middle| H_{SOC} \middle| p_y \right\rangle \left\langle p_x \middle| H_{SOC} \middle| p_z \right\rangle \propto (is_1)(is_2) = -is_3$$

$$\langle d_{xz}| H_{SOC} |d_{yz}\rangle \propto -is_3, \qquad \langle d_{yz}| H_{SOC} |d_{xz}\rangle \propto is_3,$$

which are analogous to the key ingredient of the Haldane model.

Seen from an alternative perspective [29], we discern two types of hoppings in Figure 12.10: for an electron that hops on the A-sublattice with momentum $\mathbf{K}$, the nearest B-site is on the left of its path. Similarly, for hopping on the B-sublattice with momentum $\mathbf{K}'$, $(B, \mathbf{K}')$, the nearest A-site is on the left. For the other state hoppings $(A, \mathbf{K}')$ and $(B, \mathbf{K})$, the nearest-neighbor sites are on the right. If we think of the influence of the potential gradient of the nearest-neighbor atom (electric field E) on an electron moving with momentum $\mathbf{K}$ as a Rashba field $\mathbf{E} \times \mathbf{k}$, it is clear that SOC will lower the energy for one type of states, $(A, \mathbf{K})$ or $(B, \mathbf{K}')$, for one spin direction ($\uparrow$). For the other cases, $(A, \mathbf{K}')$ or $(B, \mathbf{K})$, the other spin direction ($\downarrow$) will be preferred. We can formalize this effect by introducing a factor v_{ij}, as

$$v_{ij} = \frac{2}{\sqrt{3}} \mathbf{d}_{ik} \times \mathbf{d}_{kj} = \pm 1,$$

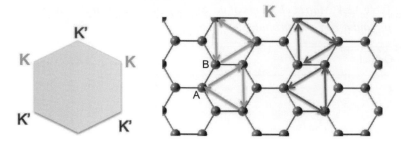

Figure 12.10 The Brillouin zone (left) indicating the position of the **K** (black) and **K**′ (gray) points. Hopping processes for electrons with momentum **K** and **K**′ are indicated by arrows (right).

where i, j are NNN, while $\mathbf{d}_{ik}$, $\mathbf{d}_{kj}$ are unit vectors along the two bonds that connect i to j through their NN k. This would modify (12.17) to read

$$\mathcal{H}_{\text{intrinsic}} = i\lambda_{\text{SO}} \sum_{\langle\langle ij \rangle\rangle} \nu_{ij}\, \sigma_3 \otimes \tau_3 \otimes s_3.$$

In the spirit of the Haldane model, we can view the role of ISOC as effectively producing some sort of intrinsic rotation of the electron via the hopping process: all electrons of one spin direction pick up one sense of rotation, while all electrons of opposite spin direction rotate in the opposite way. The factor ν_{ij} describes the *chirality* of the circulating electron, whereby the *orbital momentum* **L** in **L** · **S** coupling is associated with the chirality. Such rotations produce opposing currents in the two spin channels. In the system bulk, such currents are hard to follow, but can be cast into topological numbers (a spin Chern number). However, at the boundary it leads to a quantized spin conductivity, the *quantum spin Hall effect* (QSHE).

Rashba Spin–Orbit Coupling

It is quite unlikely that the spin–orbit coupling is only manifest through the spin-conserving intrinsic terms, which produce two decoupled copies of the Haldane insulator. Other spin–orbit coupling terms, which mix spin components, should also be present. What we need to explore is whether the edge states with spin helicity will survive the ensuing mixing between the two copies of the Haldane model. As we will see, in the Kane and Mele model the counterpropagating edge states remain quite robust as long as the bulk is gapped and TRS is obeyed. Mixing the two spin directions, together with the concomitant violation of the conservation of S_z, can actually be achieved by a Rashba term,[5] which emerges in the presence of a substrate or a perpendicular electric field. It turns out that when the Rashba spin-orbit coupling (RSOC) is increased, the bulk gap at the K, K' points decreases. However, as long as the bulk gap is finite, the helical edge states remain gapless (metallic). The bulk spectrum becomes gapped when RSOC exceeds ISOC.

[5] see Section 4.4.

A Rashba term $\propto (\mathbf{p} \times \mathbf{s}) \cdot \hat{\mathbf{z}}$ is allowed, if the mirror symmetry about the graphene plane is broken, either by a perpendicular electric field or by interaction with a substrate. For low energies, the coupling at the K-point reduces to

$$(\boldsymbol{\sigma} \times \mathbf{s}) \cdot \hat{\mathbf{z}} = \sigma_1 s_2 - \sigma_2 s_1.$$

At K', we find

$$\Theta (\sigma_1 s_2 - \sigma_2 s_1) \Theta = \mathbb{I}_\sigma \otimes i s_2 \, K \, (\sigma_1 s_2 - \sigma_2 s_1) \, \mathbb{I}_\sigma \otimes i s_2 \, K = -\sigma_1 s_2 - \sigma_2 s_1.$$

Combining the Rashba Hamiltonians at K and K', we obtain

$$\mathcal{H}_R \propto \sigma_1 \tau_3 s_2 - \sigma_2 \tau_0 s_1.$$

The inclusion of the Rashba term leads to suppression of s_3 spin conservation, since it couples up- and down-spins. Moreover, unlike in semiconductor heterostructures (discussed in Chapter 4), the coupling at low energies in graphene does not depend on the magnitude of the electron momentum, as the electrons at K (K') have a constant velocity.

To simulate the Rashba interaction, we introduce an external potential eEz, which gives rise to an onsite single-particle atomic Stark effect. This field couples the p_z orbital to either the s or the d_{z^2} orbitals, on the same atom, via the matrix elements $eE \langle s | z | p_z \rangle = eE z_{ps}$ and $eE \langle d_{z^2} | z | p_z \rangle = eE z_{pd}$.

The onsite interorbital Stark coupling allows for a spin-flip RSOC process to be manifest in an effective hopping connecting two NN p_z orbitals of opposite spins. The effective NN spin-flip hopping is achieved via the virtual processes shown in Figure 12.11 and given by [110]:

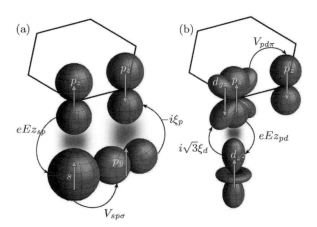

Figure 12.11 Rashba SOC mechanism hopping paths are shown as black arrows, orbital spins as gray arrows. The effective hopping is between nearest neighbors. (a) The dominant p-orbital contribution. (b) The negligible d-orbital contribution. For clarity, the orbitals of the same atoms are separated vertically, according to their contribution either to the bands (bottom) or to the bands (top). Figure and caption from [110].

$$\left| p_z^A \uparrow \right\rangle \xrightarrow{\text{Stark}} \left| s^A \uparrow \right\rangle \xrightarrow{V_{sp\sigma}} \left| p_y^B \uparrow \right\rangle \xrightarrow{SOC} \left| p_z^B \downarrow \right\rangle \tag{12.18}$$

$$\left| p_z^A \uparrow \right\rangle \xrightarrow{\text{Stark}} \left| d_{z^2}^A \uparrow \right\rangle \xrightarrow{SOC} \left| d_{yz}^A \downarrow \right\rangle \xrightarrow{V_{pd\pi}} \left| p_z^B \downarrow \right\rangle .$$

In contrast to the ISOC case, the contribution of the d orbitals channel is negligible, for the following reasons:

1. The Stark coupling of the p_z and d_{z^2} orbitals is much smaller than that of p_z to s orbitals.
2. The d orbitals SOC appears to be an order of magnitude smaller than that of the p orbitals.
3. Hopping between p and d orbitals, $V_{pd\pi}$, is also smaller than the one between p and s orbitals, $V_{sp\sigma}$.

Thus, the contribution of the sequence of the hoppipaths that include d orbitals is quite small. Figure 12.11 shows that in the p-channel, the character of the intermediate p-orbital on the B-site involved in the NN $V_{sp\sigma}$-hopping will actually depend on the direction of the vector along which the hopping occurs. This, in turn, will determine the components of the **s** operator involved. With this in mind, we can write the expression for the Rashba term in the compact form

$$\mathcal{H}_R^{\text{lattice}} = \lambda_R \sum_{\langle ij \rangle} c_i^\dagger \left(\mathbf{s} \times \widehat{\mathbf{d}}_{ij} \right) c_j,$$

where $\widehat{\mathbf{d}}_{ij}$ is the unit vector along the bond direction, and $c_i = [c_{i\uparrow}, c_{i\downarrow}]$.[6]

12.4.2 Kane–Mele Hamiltonian

The preceding presentation allows us to write the complete Kane–Mele lattice Hamiltonian as

$$\mathcal{H}_{\text{KM}}$$

$$= t \sum_{<ij>,s} (|A,i,s\rangle \langle B,j,s| + HC) + \Delta \sum_{i,s} (|A,i,s\rangle \langle A,i,s| - |B,i,s\rangle \langle B,i,s|)$$

$$+ i\lambda_{so} \sum_{\ll ij \gg} v_{ij} \left([|A,i,\uparrow\rangle \; |A,i,\downarrow\rangle] s_3 \begin{bmatrix} \langle A,j,\uparrow| \\ \langle A,j,\downarrow| \end{bmatrix} + [|B,i,\uparrow\rangle \; |B,i,\downarrow\rangle] s_3 \begin{bmatrix} \langle B,j,\uparrow| \\ \langle B,j,\downarrow| \end{bmatrix} \right)$$

$$+ i\lambda_R \sum_{<ij>} [|A,i,\uparrow\rangle \; |A,i,\downarrow\rangle] \left(\mathbf{s} \times \widehat{\mathbf{d}}_{ij} \right) \begin{bmatrix} \langle B,j,\uparrow| \\ \langle B,j,\downarrow| \end{bmatrix}. \tag{12.19}$$

The second term represents the Semenoff staggered potential.

---

[6] With a complete set of vectors $|\mu\rangle$ in Hilbert space, we can expand a potential operator as

$$V(x) = \sum_{\mu\nu} |\mu\rangle \langle \mu| \, V(x) \, |\nu\rangle \langle \nu| = \sum_{\mu\nu} \langle \mu| \, V(x) \, |\nu\rangle \, |\mu\rangle \langle \nu| = \sum_{\mu\nu} \langle \mu| \, V(x) \, |\nu\rangle \, c_\mu^\dagger \, c_\nu,$$

where $c_\mu^\dagger$, c_ν are creation and annihilation operators for states μ and ν, respectively.

We have two sites per unit cell and two spin orientations. The tensor product of a sublattice basis (A, B) and a spin basis $(\uparrow, \downarrow)$ provides the basis

$$(A, B) \otimes (\uparrow, \downarrow) = \left(\psi_{A\uparrow}, \psi_{B\uparrow}, \psi_{A\downarrow}, \psi_{B\downarrow}\right),$$

a four-spinor representation. We use the Dirac gamma matrices Γ_i, $i = 1 \ldots 5$, which are constructed as tensor products of Pauli matrices that represent the two-level systems associated with the sublattice degree of freedom and the spin of electrons. They are given by

$$\Gamma_1 = \sigma_1 \otimes s_0, \ \Gamma_2 = \sigma_3 \otimes s_0 \ \Gamma_3 = \sigma_2 \otimes s_1,$$
$$\Gamma_4 = \sigma_2 \otimes s_2, \ \Gamma_5 = \sigma_2 \otimes s_3, \ s_0 = \sigma_0 = \mathbb{I}_2$$

and the commutators $\Gamma_{ij} = \frac{1}{2i}\left[\Gamma_i, \Gamma_j\right]$.

The Bloch Hamiltonian $\mathcal{H}_{\mathrm{KM}}(\mathbf{k})$ will consist of the following:

- $-t \sum_{\langle ij \rangle} c_i^\dagger c_j \rightarrow$
$$\begin{cases} -t\left[1 + \cos(x - y) + \cos(x + y)\right]\Gamma_1 + 2t\,\cos(x)\sin(y)\,\Gamma_{12} \\ = h_1(\mathbf{k})\,\Gamma_1 + h_{12}(\mathbf{k})\,\Gamma_{12} \end{cases}$$

$$\Gamma_{12} = \frac{1}{2i}\,[\Gamma_1, \Gamma_2] = -\sigma_2 \otimes s_0$$

- $\Delta \sum_{i,s} \left(|A,i,s\rangle\langle A,i,s| - |B,i,s\rangle\langle B,i,s|\right) \rightarrow \Delta\,\Gamma_2 = h_2(\mathbf{k})\,\Gamma_2$

- $i\lambda_{so} \sum_{\ll ij \gg} \ldots\ldots \rightarrow$
$$\begin{cases} 2\lambda_{so} - \left[2\sin(2x) - 4\sin(x)\cos(y)\right]\Gamma_{15} \\ = h_{15}(\mathbf{k})\,\Gamma_{15} \end{cases}$$

$$\Gamma_{15} = \frac{1}{2i}\,[\Gamma_1, \Gamma_5] = \sigma_3 \otimes s_3,$$

where $x = \frac{k_x a}{2}$ and $y = \frac{\sqrt{3}k_y a}{2}$. All the preceding terms are invariant under mirror symmetry through the graphene plane, but the Semenoff term breaks spatial inversion within the plane. The last and most involved term to consider is the Rashba contribution, which is allowed in the absence of the mirror symmetry through the graphene plane:

$$\mathcal{H}_R = i\lambda_R \sum_{<ij>} \left[|A,i,\uparrow\rangle \quad |A,i,\downarrow\rangle\right] (\mathbf{s} \times \widehat{\mathbf{d}}_{ij}) \begin{bmatrix} \langle B,j,\uparrow| \\ \langle B,j,\downarrow| \end{bmatrix}$$

$$= i\lambda_R \sum_{<ij>,k\ell} \left[|A,i,\uparrow\rangle \quad |A,i,\downarrow\rangle\right] \epsilon_{zk\ell}\, s_k\, d_{ij}^\ell \begin{bmatrix} \langle B,j,\uparrow| \\ \langle B,j,\downarrow| \end{bmatrix}$$

$$= \lambda_R \left\{\left[1 - \cos(x)\cos(y)\right]\Gamma_3 - \sqrt{3}\,\sin(x)\sin(y)\,\Gamma_4 + \cos(x)\sin(y)\,\Gamma_{23}\right.$$
$$\left. + \sqrt{3}\sin(x)\cos(y)\,\Gamma_{24}\right\} = h_3(\mathbf{k})\,\Gamma_3 + h_4(\mathbf{k})\,\Gamma_4 + h_{23}(\mathbf{k})\,\Gamma_{23} + h_{24}(\mathbf{k})\,\Gamma_{24}$$

$$\Gamma_{23} = \frac{1}{2i}\,[\Gamma_2, \Gamma_3] = -\sigma_1 \otimes s_1$$

$$\Gamma_{24} = \frac{1}{2i}\,[\Gamma_2, \Gamma_4] = -\sigma_1 \otimes s_2.$$

The five Dirac matrices Γ_i are even under Θ, while the 10 commutators are odd,

$$\Theta\,\Gamma_i\,\Theta^{-1} = \Gamma_i$$
$$\Theta\,\Gamma_{ij}\,\Theta^{-1} = -\Gamma_{ij},$$

whereas the coefficients in the Kane–Mele Hamiltonian transform as

$$h_i(-\mathbf{k}) = h_i(\mathbf{k})$$
$$h_{ij}(-\mathbf{k}) = -h_{ij}(\mathbf{k}),$$

which preserves TRS.

The diagonalization of $\mathcal{H}_{\mathrm{KM}}(\mathbf{k})$ will yield four bands.

Two copies of the Haldane Hamiltonian: We note that in the absence of the Rashba term, $\lambda_R = 0$,

$$[\mathcal{H}_{\mathrm{KM}}, s_3] = 0,$$

and the model can be decoupled into two subsystems for spin-up ($s_3 = 1$) and spin-down ($s_3 = -1$), respectively, with low-energy valley dispersion

$$\varepsilon(k) = \pm\sqrt{(\hbar v_F k)^2 + (\Delta_{so} - \Delta)^2}$$
$$\Delta_{so} = 3\sqrt{3}\lambda_{so}$$

and an energy gap of $2\,|\Delta_{so} - \Delta|$. The Hamiltonian for spin-up (spin-down) electrons is just the Haldane Hamiltonian, with $\lambda_{so} = 3\sqrt{3}t_2$ and $\phi = +\pi/2\,(-\pi/2)$:

$$\mathcal{H} = \begin{pmatrix} \mathcal{H}_\uparrow & \\ & \mathcal{H}_\downarrow \end{pmatrix} = \begin{pmatrix} \mathcal{H}_{\mathrm{Haldane}} & \\ & \mathcal{H}^*_{\mathrm{Haldane}} \end{pmatrix}.$$

Hence many properties can be deduced from our knowledge of the Haldane model for spinless fermions. Most importantly, the gaps at K and K' have opposite signs, as shown in Figure 12.12, which establishes the topological nontriviality of the system. Moreover, taken separately, the copies for the $s_3 = \pm 1$ spins violate time-reversal symmetry, and will have Chern numbers ± 1, with corresponding Hall conductance of $\sigma_{xy} = \pm e^2/h$. Since the

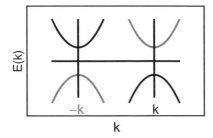
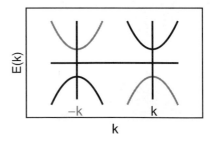

Figure 12.12 Low-energy dispersion for the Kane–Mele insulator $\lambda_R = 0$, $\lambda_{so} \neq 0$. It depicts two time-reversed copies of a Haldane insulator, one for $\uparrow$-spin (left) and $\downarrow$-spin (right).

signs of the gaps are opposite for opposite spins, an electric field will induce opposite edge currents for the opposite spins, leading to a spin current $\mathbf{J}_s = (\hbar/2e)\,(\mathbf{J}_\uparrow - \mathbf{J}_\downarrow)$ characterized by a quantized spin Hall conductivity

$$\sigma_H^s = \frac{\hbar}{2e}\left[\frac{e^2}{h} - \left(\frac{-e^2}{h}\right)\right] = \frac{e}{2\pi}.$$

Although the total Chern number cannot be nonzero, the spin projections $|\uparrow\rangle$, $|\downarrow\rangle$ are good eigenstates. We can use the Chern number C_s in each spin sector to characterize the phases and define

$$\nu = \frac{C_\uparrow - C_\downarrow}{2} \quad \mathrm{mod}\ 2 \ \in \mathbb{Z}_2$$

as a good topological index.

We find that the global Kane–Mele system has a nonchiral, *helical* edge state consisting of two spin-filtered counterpropagating gapless edge modes. Within this spin-conserving model, the emerging edge states inherit the topological character of the Haldane model edge states. We note that what creates the edge states is the presence of a topologically nontrivial band structure in the bulk, characterized by band inversion between K and K', which is induced by the ISOC. Moreover, this ISOC also introduces the helicity and maintains the transverse spin conductance, since it distinguishes between $\uparrow$ and $\downarrow$ in real space (see Figure 12.13.).

Spin nonconserving Rashba term: The inclusion of the Rashba term removes the spin conserving feature of the previous model, and lifts the quantization of the spin Hall conductance. However, as long as $\Delta_{so} > \lambda_R$, we still get a QSH phase with topological edge states. The electronic bands near K (K') are now modified to

$$\varepsilon_{\mu\nu}(k) = \mu\lambda_R + \nu\sqrt{(\hbar v_F k)^2 + (\Delta_R - \mu\Delta_{so})^2}$$

$$\Delta_R = 3\lambda_R,$$

where $\mu, \nu = \pm 1$ are band indices. For simplicity, we set $\Delta = 0$. The electron bands have $\nu = +1$, and the hole bands $\nu = -1$. At K (K'), the energy is $\mu\lambda_R + \nu\,|\Delta_{so} - \mu\lambda_R|$.

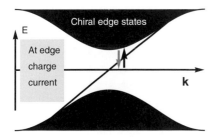 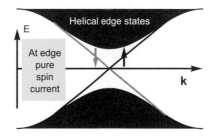

Figure 12.13 Left: edge-state dispersion for single Haldane copy, showing chiral character. Right: Kane–Mele insulator $\lambda_R = 0$, $\lambda_{so} \neq 0$, with two time-reversed copies of a Haldane insulator, showing helical spin-filtered edge-state dispersion.

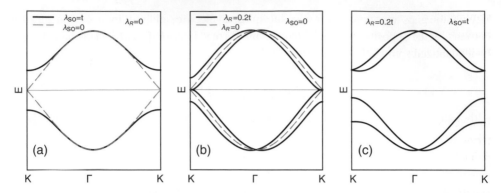

Figure 12.14 Dispersion along the Γ-K direction for the Kane–Mele model: : (a) $\lambda_R = \lambda_v = 0$, $\lambda_{so} = t/3\sqrt{3}$; (b) $\lambda_v = \lambda_{SO} = 0$, $\lambda_R = 0.2t$; (c) $\lambda_v = \lambda_{SO} = t/3\sqrt{3}$, $\lambda_R = 0.2t$.

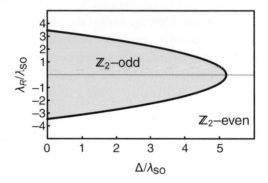

Figure 12.15 Phase diagram of the Kane–Mele model for $\Delta/\lambda_{SO} > 0$.

Figure 12.14 shows the modification to the band dispersion for $\lambda_R = 0$, $\lambda_{so} = t$ in (a), and $\lambda_R = 0.2t$, $\lambda_{so} = 0$ in (b). It clearly shows that the intrinsic SOC opens a gap at $K(K')$, while the Rashba term does not. For $\lambda_R = \lambda_{so}$, the band structure of graphene depends strongly on the interplay of the two spin–orbit coupling effects. For $\lambda_R < \Delta_{so}$, the spectral gap gets smaller, but the electron branch is still degenerate at K (K'); in contrast, the hole branch is split by $2\Delta_R$.

The phase diagram for $\mathbb{Z}_2$ in terms of λ_R/λ_{so} versus $\Delta/\lambda_{so} \geq 0$ is shown in Figure 12.15.

12.5 Weyl and Dirac Semimetals

The presence of an energy gap in topological insulator phases facilitated the description of a topological band twist in the nontrivial phases – the band inversion about the gap. Here we will attempt to describe topological band features in systems with no global bulk gap. We will close the band gap, and explore whether topological properties can survive in semimetals or metals. We will find that topological band attributes can exist, provided we maintain translation symmetry. We will encounter new and amusing nontrivial topological features in gapless Weyl[7] semimetals [192].

12.5.1 Accidental Degeneracies and Dimensionality

In electronic band theory of crystals, degeneracy at each wavevector $\mathbf{k}$ is understood in terms of symmetry. The dimension of an irreducible representation (irrep) at a given $\mathbf{k}$ point is equal to the degeneracy at that point. However, topological semimetals have band degeneracies that arises from topology. In a topological semimetal, a band gap closes at generic $\mathbf{k}$ points, and this closing of the gap originates from and is protected by topological factors, and not symmetry.

The conditions under which degeneracies occur in electronic band structures was investigated by Herring [92] in 1937.[8] As a starting point, we reconsider the basic idea of accidental or isolated degeneracies [189] that we briefly considered in Section 10.2. We follow the analysis of [184] and focus on a pair of energy bands and ask if one can bring these bands into degeneracy by tuning Hamiltonian parameters. This system is represented by the most general 2×2 Hamiltonian [184]

$$\mathcal{H}(\mathbf{k}) = h_0(\mathbf{k})\,\sigma_0 + h_1(\mathbf{k})\,\sigma_1 + h_2(\mathbf{k})\,\sigma_2 + h_3(\mathbf{k})\,\sigma_3$$

with an energy splitting between the levels

$$\varepsilon_\pm(\mathbf{k}) = h_0(\mathbf{k}) \pm \sqrt{h_1^2(\mathbf{k}) + h_2^2(\mathbf{k}) + h_3^2(\mathbf{k})}.$$

For a general $\mathbf{k}$ point and in the absence of spatial and TR symmetries, $h_j(\mathbf{k}) \neq 0$ for each j. However, the expression for $\varepsilon_\pm(\mathbf{k})$ shows that the two bands touch only if $h_j(\mathbf{k}) = 0$ for each $j > 0$ at some $\mathbf{k}_0$.

In three dimensions, we can vary each of the three components of $\mathbf{k}$ and look for simultaneous zeros of each of the three components $h_j(\mathbf{k})$, $j > 0$. To show the possibility of such occurrence without fine-tuning, we note that each one of the three equations $h_j(\mathbf{k}) = 0$ describes a two-dimensional surface in $\mathbf{k}$-space. Two such surfaces may generally intersect along lines, and such lines may then intersect the third surface at points without the need for fine-tuning. In general, such points appear in pairs, and the dispersion near each may be linearized. The effective Hamiltonian near one such point $\mathbf{k}_0 + \delta\mathbf{k}$ takes the form

---

[7] The Dirac equation in d spatial dimension and effective speed of light $c = 1$ is

$$\left(\gamma^\mu\,\partial_\mu - m\right)\Psi = 0, \qquad \mu = 0, 1, \ldots, d.$$

In 1929, Hermann Weyl noticed that this equation can be further simplified in certain cases. In odd spatial dimensions and $m = 0$, the equation decouples the two-spinor wavefunction into two independent spinors with opposite chirality. Hence, the existence of a massless fermion in the Dirac equation, which was later called the Weyl fermion [193]. For further details, see the appendix.

[8] Conyers Herring's thesis was titled "Energy Coincidences in the Theory of Brillouin Zones," and his thesis advisor was of course Eugene Wigner.

$$\mathcal{H}(\delta\mathbf{k}) = \varepsilon_{\mathbf{k}_0} + \hbar\mathbf{v}_0 \cdot \delta\mathbf{k}\,\sigma_0 + \sum_{j=1}^{3} \hbar\mathbf{v}_j \cdot \delta\mathbf{k}\,\sigma_j, \qquad (12.20)$$

where $\mathbf{v}_\mu = \nabla_{\mathbf{k}} h_\mu(\mathbf{k})\big|_{\delta\mathbf{k}=0}$, $\mu = 0,\dots,3$. If $|\mathbf{v}_0| = 0$ and the three velocity vectors $\mathbf{v}_j$ are mutually orthogonal, $\mathcal{H}(\mathbf{k})$ has the form of an anisotropic Weyl Hamiltonian. It is clear that far away from $\mathbf{k}_0$, both bands may disperse in any direction, in which case even if the Fermi level could be set to $\varepsilon_{\mathbf{k}_0}$, there would be additional FSs.

In two dimensions, there are only two $\mathbf{k}$ components that can be varied. Consequently, it is impossible to find simultaneous zeros of three functions $h_j(\mathbf{k})$ without additional fine-tuning. This means that in the absence of additional symmetries that may constrain the number of independent $h_j(\mathbf{k}) = 0$, the two bands will avoid each other.

It was, therefore, noted that even in the absence of any symmetry, it is possible to obtain accidental twofold degeneracies of bands in a three-dimensional solid. The dispersion in the vicinity of these band touching points is generically linear and resembles the Weyl equation [193], notably in the absence of Lorentz invariance. Although such gapless band-touching has long been known, its corresponding topological nature has been appreciated only recently.

We note that for $|\mathbf{v}_0| = 0$, (12.2) resembles the two-level topological system of Chapter 10. We therefore surmise that the node at $\mathbf{k}_0$ is associated with a Berry curvature of

$$\mathcal{F} = \pm\frac{\widehat{\Delta\mathbf{k}}}{2|\Delta\mathbf{k}|^2},$$

where $\Delta\mathbf{k} = \mathbf{k} - \mathbf{k}_0$. We also recall that the Berry curvature field resembled that of a magnetic monopole with positive or negative magnetic charge. Weyl points are monopoles of Berry flux, chirality, and charge.

We now consider one of the node with $+$ or $-$ Berry charge,

$$\mathcal{H} = \pm v\,\boldsymbol{\sigma}\cdot(\mathbf{k} - \mathbf{k}_0)\,.$$

Under TRS (if the pseudospin behaves like a spin),

$$\mathbf{k} \to -\mathbf{k}, \quad \boldsymbol{\sigma} \to -\boldsymbol{\sigma}, \quad \text{and} \quad \mathcal{H} \to \mathcal{H}' = \pm v\,\boldsymbol{\sigma}\cdot(\mathbf{k} + \mathbf{k}_0)\,.$$

This means that TRS implies that there must be another nodal point at $-\mathbf{k}_0$ with the same charge. Under space inversion transformation,

$$\mathbf{k} \to -\mathbf{k}, \quad \boldsymbol{\sigma} \to \boldsymbol{\sigma}, \quad \text{and} \quad \mathcal{H} \to \mathcal{H}' = \mp v\,\boldsymbol{\sigma}\cdot(\mathbf{k} + \mathbf{k}_0)\,.$$

Inversion symmetry (IS) dictates that there must be another nodal point at $-\mathbf{k}_0$ with opposite charge.

When both TRS and IS exist, each node would have two monopoles with opposite charges. The net topological charge of a nodal point (with four levels) is zero. Alternatively, we can argue that the presence of both symmetries implies global band spin degeneracy and any extra band-touching degeneracy has to involve four bands. Therefore, we need to break at least one of these two symmetries to obtain spin-nondegenerate bands.

Table 12.2 *Classification of Weyl point types.*

TRS	I	Implications	Minimum number
No	No	Weyl nodes can be at any $\mathbf{k}$ and may have different energies	2
Yes	No	Weyl node at $\mathbf{k}_0$ $\Leftrightarrow$ Weyl node of same chirality at $-\mathbf{k}_0$	4
No	Yes	Weyl node at $\mathbf{k}_0$ $\Leftrightarrow$ Weyl node of same chirality at $-\mathbf{k}_0$	2
Yes	Yes	No stable, individually separated Weyl nodes possible	None

Note that when there is IS but no TRS, the minimum number of Weyl points in a solid is two degenerate points at $\pm\mathbf{k}_0$ with opposite charge. If there is TRS but no IS, the minimum number is four, since the TRS doublet would have a partner doublet with opposite helicity. In the absence of TRS or space inversion, massless lattice fermions are required to come in pairs with opposite helicities, or Berry charges. This is known as the Nielsen–Ninomiya theorem [144, 145, 146], or fermion-doubling theorem. These classifications are given in Table 12.2. The constraint that Weyl nodes should come in opposite chirality pairs can be simply demonstrated if we examine the topology of the Brillouin zone. We first note that the net Berry flux emanating from a surface enclosing a Weyl point is given by $2\pi C$, where C is a nonvanishing Chern number $C = \pm 1$. However, if we expand this surface so that it covers the entire Brillouin zone, then by periodicity it is actually equivalent to a point and must have net Chern number zero. Therefore, the net Chern number of all Weyl points in the Brillouin zone must vanish.

Now the stability of Weyl points can be connected to Gauss's law: a Gaussian surface surrounding the Weyl point detects its charge, preventing it from disappearing surreptitiously. It can only disappear after an oppositely charged monopole goes through the surface, and later annihilates with it. Also, the net charge of all the Weyl points in the Brillouin zone has to be zero (which is seen by taking a Gauss-law surface that goes around the whole Brillouin zone). Thus, the minimum number of Weyl points (if there are some) is two, and they have to have the opposite chirality, as in the preceding model. This is the proof of the fermion doubling theorem. This also shows that Weyl nodes can be eliminated in a pairwise fashion: annihilation of a Weyl node pair of opposite chirality.

We surmise from this reasoning that all we need to do to realize a WSM is to explore 3D crystals with nondegenerate bands by breaking appropriate symmetries and looking for band crossings. In order to observe clear effects of Weyl nodes, we need to introduce an additional requirement: they should be close to the Fermi energy. This means that we find candidates where $h_0(\mathbf{k})$ is nearly zero. We recall that in the special limit of $|\mathbf{v}_0| = 0$ and $\mathbf{v}_j$ being mutually orthogonal, we obtain the Weyl equation. That is why we refer to these band crossings as *Weyl nodes*, and we makes a connection with Weyl fermions with a fixed chirality [$C = \text{sign}(\mathbf{v}_x \cdot \mathbf{v}_y \times \mathbf{v}_z)$].

12.5.2 Weyl Semimetals with Broken TRS

We consider a WSM system having IS, but with broken TRS. Breaking TRS requires the presence of an intrinsic or external magnetic field. The minimal number of Weyl nodes is then two, with opposite chirality, and IS guarantees they have the same energy. We can also show that the simple criterion based on the parity eigenvalues at the TRIM, presented in Chapter 11, can be used to determine the existence of Weyl points.

We follow the example given in [19] of a magnetically ordered system where bands have no spin degeneracy. We introduce a pair of orbitals with opposite parity, say s, p orbitals, on each site of a simple cubic lattice. The orbitals are represented by τ_z, and then inversion symmetry is given by $\mathcal{H}(\mathbf{k}) \rightarrow \tau_z \mathcal{H}(-\mathbf{k}) \tau_z$. The Hamiltonian is

$$\mathcal{H} = t_z \Big(2 - \cos(k_x a) - \cos(k_y a) + \gamma - \cos(k_z a) \Big) \tau_z + t_x \sin(k_x a)\, \tau_x + t_y \sin(k_y a)\, \tau_y$$

The example is structured such that it allows for the existence of a pair of Weyl nodes at location $\pm\mathbf{k}_0 = (0; 0; \pm k_0)$ for $-1 < \gamma < 1$, where $\cos k_0 a = \gamma$. Starting at $\gamma = -1$, Weyl nodes form at the BZ boundaries and move toward each other before annihilating at the zone center at $\gamma = 1$. The low-energy excitations for $\mathbf{k} = \pm\mathbf{k}_0 + \mathbf{q}$, $|\mathbf{q}| \ll k_0$, are given by

$$\mathcal{H}_\pm \approx \mathcal{H}(\pm\mathbf{k}_0 + \mathbf{q}) = \sum_i v_i^\pm q_i \tau_i$$

$$v_i^\pm = \Big(t_x, t_y, \pm t_z \sin(k_0 a) \Big).$$

To employ the parity criterion, we focus on the eight TRIM momenta (see Figure 11.9), where we find that only the first term in the Hamiltonian survives. For $\gamma > 1$ the parity eigenvalues of all the TRIM are the same and the bands are not inverted. However, we find that for $\gamma = 0$, the parity eigenvalue at the Γ-point changes sign signaling band inversion and the emergence of Weyl nodes, where odd number of inverted parity eigenvalues is a diagnostic of Weyl physics.

We now compare the Chern numbers $C(k_z)$ of two planes in momentum space $k_z = 0$ and $k_z = \pi/a$: the Chern number vanishes at $C(\pi/a) = 0$, but $C(0) = 1$. As $\gamma \rightarrow 1$, the entire Brillouin zone is filled with a unit Chern number along the k_z direction, and a 3D version of the integer quantum Hall state is realized. Therefore, the WSM appears as a transitional state between a trivial insulator and a TI.

When the chemical potential is at $E_F = 0$, the FS consists solely of two points $\pm k_0$. On increasing E_F, two nearly spherical FSs appear around the Weyl points and a metal exists. The FSs are closed 2D manifolds within the BZ. One can therefore define the total Berry flux penetrating each, which by general arguments is required to be an integer, and in the present case is quantized to ± 1 – a particular feature characteristic of a Weyl metal. When $E_F > E^* = t_z (1 - \gamma)$, the FSs merge through a Lifshitz transition and the net Chern number on an FS vanishes. At this point, one would cease to call this phase a Weyl metal. This discussion highlights the importance of the Weyl nodes being sufficiently close to the chemical potential as compared to E^*. Ideally, we want the chemical potential to be tuned to the location of the Weyl nodes just from stoichiometry, as occurs for ideal graphene.

12.5.3 Weyl Semimetals with Broken $\mathcal{P}$ Symmetry

If TRS is preserved, then inversion symmetry must be broken to realize a WSM. As we have shown, a key difference from broken TRS WSM is that the total number of Weyl points must now be a multiple of four. Moreover, TRS guarantees that a Weyl node at $\mathbf{k}_0$ is converted into a degenerate Weyl node at $-\mathbf{k}_0$ with the same chirality. Since the net chirality in the BZ must vanish, another pair with the opposite chirality must exist, but not necessarily degenerate with the $\pm\mathbf{k}_0$ ones.

12.5.4 Fermi Arc Surface States

In the case of topological insulators, we were able to clearly identify Dirac fermion surface states as well-defined metallic states that exist within the bulk band gap and that are exponentially localized near the surface. Here, the question arises as to how to define such states in the surface BZ (SBZ) for WSMs where the bulk does not have an absolute gap.

We consider a macroscopic 3D slab of Weyl semimetal bounded by surfaces at $z = \pm Z_0$. Translation invariance along the x and y directions defines an SBZ. For pedagogical clarity, we consider the idealized limit of a pair of Weyl nodes that lie at E_F. Hence, at E_F the projection of bulk states onto the SBZ only contains the Weyl points at momenta $\pm\bar{\mathbf{k}}_0$. Surface states can exist wherever there is a projected gap in the SBZ, because they should not resonate with bulk states. Consequently, we can define surface states at E_F at all momenta except at $\pm\bar{\mathbf{k}}_0$. As the Weyl points are approached, the surface states penetrate deeper into the bulk and are not well defined. This results in arc-shaped FSs, as depicted in Figure 12.16(a) and 12.16(b), instead of the 2D closed loops we are used to. They are defined by $\varepsilon(k_x, k_y) = E_F$.

These states are topologically defined as follows: the Berry curvature field emanates from one WP and terminates on the other. The net Berry phase accumulated in any 2D $\mathbf{k}$-plane between the WP pair induces a nonzero Chern number $C = 1$ with a quantized Hall effect that supports edge states, whereas the Berry phase is zero in other planes with $C = 0$, as shown in Figure 12.16(c). This is a pure topological effect from the band structure, because the bulk FS vanishes at the WPs. On the boundary, topological edge states exist at edges of two-dimensional planes with $C = 1$ and vanish at edges of other planes where $C = 0$.

If we consider FSs at energies that do not coincide with the WPs, we find that the momentum region occupied by bulk states grows, as shown in Figure 12.17. The presence of these bulk states allows for surface states that are impossible to realize in both strictly

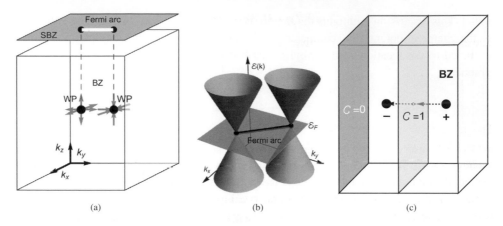

(a)   (b)   (c)

Figure 12.16 (a) Surface states of a Weyl semimetal appear as arcs in the SBZ that connect two Weyl points projections to one another. (b) A point on a Fermi arc can be regarded as an edge state of a two-dimensional insulator, where a nonzero net Berry flux leads to an integer Chern number of Hall states, which have edge modes. (c) A depiction of surface modes dispersion and how it joins to the bulk states.

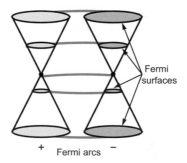

Figure 12.17 Manifestation of Fermi arcs at the FS of the surface band structure. A pair of bulk Weyl cones exists as a pair of Fermi pockets at $E_F \neq 0$. A Fermi arc (thick gray line) appears on the top or bottom surface to tangentially connect such a pair of Fermi pockets.

2D and on the surface of any 3D insulator, where there is a finite energy gap throughout the entire Brillouin zone. Haldane [83] argued that the Fermi arc surface states must be tangent to the bulk Fermi surfaces projected onto the SBZ. This follows from the fact that the surface states must convert seamlessly into the bulk states as they approach their termination points. Putting this differently, the evanescent depth of the surface state wavefunction grows, until at the point of projection onto the bulk states, the surface states merge with the bulk states. They should inherit the velocity of the bulk states, which implies they must be attached tangentially to the bulk Fermi surface projections, as shown in Figure 12.17.

12.5.5 Type II Weyl Semimental

The energy spectrum of (12.2) is given by

$$\varepsilon_{\pm}(\delta\mathbf{k}) = \varepsilon_{\mathbf{k}_0} + \sum_i v_{0i}\delta k_i \pm \sqrt{\sum_{j=1}^{3}\left(\sum_{i=1}^{3} k_i v_{ij}\right)^2}. \qquad (12.21)$$

The term involving the velocity parameter $\mathbf{v}_0$ introduces an overall tilt of the Weyl cone [170]. Such a term is forbidden by Lorentz symmetry for the Weyl Hamiltonian in vacuum. However, because Lorentz invariance does not need to be respected in condensed matter, its inclusion is important and leads to a finer classification of distinct Fermi surfaces. It can generically appear in a linearized long-wavelength theory near an isolated twofold band crossing in a crystal. Small $\mathbf{v}_0$ simply induces a crystal field anisotropy into the band dispersion near a Weyl point. However, sufficiently large $\mathbf{v}_0$ produces a qualitatively new momentum space geometry wherein the constant energy surfaces are open rather than closed and the resulting electron and hole pocket contact at a point as shown in Figure 12.18. This new semimetallic phase has been termed a "type II" WSM, in contrast to a "type I" WSM with closed constant energy surfaces. Type I respects Lorentz symmetry, while type II does not. Although type I and type II WSMs cannot be smoothly deformed into each other, they share electronic behavior that derives from the presence of an isolated band contact point in their bulk spectra. Interestingly the topological character of the Weyl point is still fully controlled by the last term in (12.2) and persists even for type II WSM. Thus type II Weyl semimetals support surface Fermi arcs that terminate on the surface projections of their band contact points, which are the signature of the topological nature of the semimetallic state.

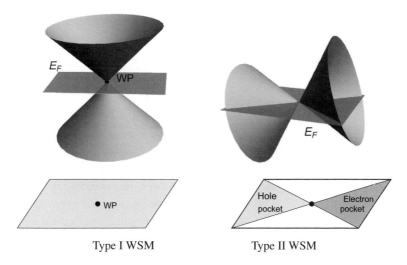

Type I WSM   Type II WSM

Figure 12.18 Type-II WSM. Due to the strong tilting of the Weyl cone, the Weyl point acts as the touching point between electron and hole pockets in the FS.

12.6 Appendix: Dirac and Weyl Equations

Historical Perspective

Dirac first introduced his famous relativistic equation for a fermion field [50] to primarily describe free electrons. It was constructed to encompass both single-particle quantum mechanics and special relativity. However, in addition to a massive electron with negative charge and positive energies, his solutions contained negative energy particles which Dirac predicted as an antiparticle, the positron, which has the same mass as the electron but with an opposite charge.

In the following year, Weyl showed that a simpler equation [193] involving two-component fields rather than the four-component Dirac field was enough to describe massless fermions. In 1930, Pauli [38] suggested the possible existence of neutrinos to account for the continuous electronic energy spectrum manifest in beta decay. Because of charge conservation, neutrinos had to be neutral, and beta-decay data indicated that they should be massless. Consequently, it was assumed that the neutrino properties should be described by Weyl's theory. Another issue with neutrinos concerned whether they were their own antiparticles since they are uncharged. A description of such dual fermion fields was then proposed by Majorana in 1937 [124]. However, at that time, the community was convinced that neutrinos are Weyl fermions. It was only in the 1960s, when evidence supporting small but nonzero neutrino masses emerged, that the idea of Weyl-like neutrinos was abandoned, and the possibility of being Dirac fermions or Majorana fermions emerged.

Today, in a broad frontier of condensed matter research, ranging from graphene to high T_c superconductors to topological insulators and beyond, systems are found to exhibit electronic properties that can be fittingly described by the Dirac and Weyl equations. Consequently, understanding Dirac fermions has become a must in modern condensed matter physics. It is no longer an exclusive domain of quantum field and high-energy theories; instead, we find Dirac particles describing the physics of real condensed matter systems in two and three dimensions. While the physics that gives rise to massless Dirac fermions in different systems may be different, the low-energy properties are governed by the same Dirac physics.

The Massive Dirac Equation

The Klein–Gordon equation suffered from negative energy solutions accompanied with negative probability densities. Dirac realized that his equation should be linear in energy in order to interpret the wavefunction as a probability amplitude. This meant that the equation must be first order in the time derivative ∂_t. Furthermore, he noted that in order to satisfy Lorentz invariance, space derivatives should appear at the same order as the time derivative. The only possible formulation for such an equation, in d spatial dimensions, is

$$\mathcal{H} = \beta \, mc^2 + c \, \boldsymbol{\alpha} \cdot \mathbf{p}$$

$$i\hbar \, \partial_t \, \Psi(\mathbf{x}, t) = \left[\beta \, mc^2 + c \, \boldsymbol{\alpha} \cdot \mathbf{p} \right] \Psi(\mathbf{x}, t), \tag{12.22}$$

where now Ψ is a four spinor. Besides the particle mass m and the speed of light c, β and $\boldsymbol{\alpha} = (\alpha_1, \ldots, \alpha_d)$ are dimensionless matrices. In order for (12.22) to satisfy the relativistic dispersion relation, we need

$$E^2 = m^2 c^4 + c^2 \mathbf{p}^2 = \beta^2 m^2 c^4 + c^2 \,(\boldsymbol{\alpha} \cdot \mathbf{p})^2 + mc^3 \,(\beta \,\boldsymbol{\alpha} \cdot \mathbf{p} + \boldsymbol{\alpha} \cdot \mathbf{p} \,\beta) \,.$$

This implies that the quantities β and $\boldsymbol{\alpha}$ must anticommute:

$$\beta^2 = \mathbb{I}, \quad \{\alpha_i, \beta\} = 0, \quad \{\alpha_i, \alpha_j\} = 2\delta_{ij}. \tag{12.23}$$

Equation (12.23) is a manifestation of a Clifford algebra, where β and $\boldsymbol{\alpha}$ are square matrices, with dimension $N > 1$.[9] They must also satisfy the following properties:

(a) They must be Hermitian,

$$\alpha_i^\dagger = \alpha_i, \qquad \beta^\dagger = \beta,$$

in order to conform with the hermiticity of the Hamiltonian (12.22).

(b) They must be unitary,

$$\alpha_i^\dagger = \alpha_i = \alpha_i^{-1}, \qquad \beta^\dagger = \beta = \beta^{-1},$$

since $\alpha_i^2 = \beta^2 = \mathbb{I}$. Their eigenvalues are, therefore, ± 1.

(c) The anticommutation, $\alpha_i \alpha_j = -\alpha_j \alpha_i$, leads to the determinantal relation

$$\det[\alpha_i \alpha_j] = \det[-\alpha_j \alpha_i] = (-1)^N \det[\alpha_j \alpha_i] = (-1)^N \det[\alpha_i \alpha_j]$$

which requires that N be even, $N = 2m$, $m = 1, 2, \ldots$ As we will see, the value of m depends on the spatial dimension d.

The 2D Massive Dirac Equation

For $d = 2$, we need three mutually anticommuting matrices, β, α_1 and α_2; they can be identified with the Pauli matrices

$$\sigma_z = \beta, \qquad \sigma_1 = \alpha_1, \qquad \sigma_2 = \alpha_2$$

since σ_1, σ_2, and σ_z are 2×2 matrices that satisfy the Clifford algebra. This implies that the wavefunction must be a two-dimensional vector: a spinor. The 2D Dirac Hamiltonian then has the form

$$\mathcal{H}_D^{2D} = c \,\boldsymbol{\alpha} \cdot \mathbf{p} + mc^2 \,\sigma_z. \tag{12.24}$$

To simplify the ensuing analysis, we set $c = \hbar = 1$, and write the 2D Dirac Hamiltonian as

$$\mathcal{H}_D^{2D} = \boldsymbol{\alpha} \cdot \mathbf{p} + m \,\sigma_z = \varepsilon \begin{pmatrix} \cos\theta & e^{-i\phi_\mathbf{k}} \sin\theta \\ e^{i\phi_\mathbf{k}} \sin\theta & -\cos\theta \end{pmatrix}, \tag{12.25}$$

---

[9] Not to be confused with the spatial dimension d.

where $\varepsilon = \sqrt{m^2 + |\mathbf{p}|^2}$, $\cos\theta = m/\varepsilon$, $\sin\theta = |\mathbf{p}|/\varepsilon$, and $\phi_\mathbf{k} = \arctan(p_y/p_x)$. The eigenvalues of the unitary and Hermitian matrix on the rhs of (12.25) are $\lambda = \pm 1$ and correspond to the positive and negative energy states, respectively, with

$$E_\lambda = \lambda \sqrt{m^2 + \mathbf{p}^2} \tag{12.26}$$

We consider plane-wave spinor eigenstates of the form

$$\Psi_{\mathbf{p}\lambda}(\mathbf{x}) = \frac{1}{\sqrt{2}} \begin{bmatrix} u_\lambda \\ v_\lambda \end{bmatrix} e^{i\mathbf{p}\cdot\mathbf{x}}.$$

Substituting in (12.25), we obtain

$$(\cos\theta - \lambda)\, u_\lambda + \sin\theta\, e^{-i\phi_\mathbf{k}}\, v_\lambda = 0.$$

We write the solution as

$$u_\lambda = \sqrt{1 + \lambda\,\cos\theta}, \qquad v_\lambda = \lambda\,\sqrt{1 - \lambda\,\cos\theta}\; e^{i\phi_\mathbf{k}},$$

which yields the spinor eigenstates

$$\Psi_{\mathbf{p}\lambda}(\mathbf{x}) = \frac{1}{\sqrt{2}} \begin{bmatrix} \sqrt{1 + \frac{\lambda m}{\varepsilon}} \\ \lambda\sqrt{1 - \frac{\lambda m}{\varepsilon}}\; e^{i\phi_\mathbf{k}} \end{bmatrix}. \tag{12.27}$$

We consider two limits:

- The ultrarelativistic limit ($m \to 0$) yields

$$\Psi_{\mathbf{p}\lambda}(\mathbf{x}) = \frac{1}{\sqrt{2}} \begin{bmatrix} 1 \\ \lambda\, e^{i\phi_\mathbf{k}} \end{bmatrix},$$

  which can be identified with those of low-energy massless electrons in graphene, equation (12.10), at the Dirac points K, k'.
- The case ($mc \gg |\mathbf{p}|$)

$$\Psi_{\mathbf{p}\lambda=+1}(\mathbf{x}) = \frac{1}{\sqrt{2}} \begin{bmatrix} 1 \\ 0 \end{bmatrix} e^{i\mathbf{p}\cdot\mathbf{x}}, \qquad \Psi_{\mathbf{p}\lambda=-1}(\mathbf{x}) = \frac{1}{\sqrt{2}} \begin{bmatrix} 0 \\ 1 \end{bmatrix} e^{i\mathbf{p}\cdot\mathbf{x}}.$$

The 3D Massive Dirac Equation

There are only three Pauli matrices, whereas four anticommuting matrices are needed for the massive Dirac Hamiltonian in $3 + 1$ space-time: one for each space dimension and one for the mass. It is then necessary to use an $N = 4$ representation with 4×4 matrices. The Clifford algebra can be satisfied by taking tensor products of Pauli matrices σ_i and τ_i, namely,

$$\alpha_i = \sigma_i \otimes \tau_1 = \begin{pmatrix} 0 & \sigma_i \\ \sigma_i & 0 \end{pmatrix}; \qquad \beta = \sigma_0 \otimes \tau_3 = \begin{pmatrix} \sigma_0 & 0 \\ 0 & -\sigma_0 \end{pmatrix},$$

which is called the *standard representation*. In this representation, we obtain a Dirac Hamiltonian of the form

$$\mathcal{H}_D^{3D} = \begin{pmatrix} mc^2\mathbb{I} & c\,\boldsymbol{\sigma}\cdot\mathbf{p} \\ c\,\boldsymbol{\sigma}\cdot\mathbf{p} & -mc^2\mathbb{I} \end{pmatrix}$$

and the wavefunction becomes a two-spinor.

The Dirac equation can also be written in the covariant form

$$\left(i\gamma^\mu\,\partial_\mu - m\right)\Psi = 0, \qquad \hbar = c = 1$$

with Dirac matrices $\boldsymbol{\gamma} = (\beta, \beta\,\boldsymbol{\alpha})$ The gamma matrices form the Dirac algebra, which is a special case of the Clifford algebra

$$\left\{\gamma^\mu, \gamma^\nu\right\} = 2g^{\mu\nu},$$

where $g^{\mu\nu}$ is the Minkowski metric tensor ($g^{00} = -g^{ii} = 1$). Therefore, the gamma matrices can be identified as

$$\gamma^i = i\sigma_i \otimes \tau_2 = \underbrace{\begin{pmatrix} 0 & \sigma_i \\ -\sigma_i & 0 \end{pmatrix}}_{\text{Anti-Hermitian}}; \qquad \gamma^0 = \underbrace{\begin{pmatrix} \sigma_0 & 0 \\ 0 & -\sigma_0 \end{pmatrix}}_{\text{Hermitian}}$$

Discrete Symmetries of the Dirac Equation

(i) Charge conjugation: In order to derive an expression for a charge conjugation operator, we write the Dirac equation for a free particle with charge q in an external electromagnetic field as

$$i\,\frac{\partial\Psi(\mathbf{x},t)}{\partial t} = \left[\alpha_i\,(-i\partial_i - q\,A_i) + \beta\,m\right]\Psi(\mathbf{x},t), \qquad \hbar = c = 1 \tag{12.28}$$

and we apply the complex conjugation operator K, to obtain

$$i\,\frac{\partial\Psi^*(\mathbf{x},t)}{\partial t} = \left[\alpha_i^*\,(-i\partial_i + q\,A_i) - \beta^*\,m\right]\Psi^*(\mathbf{x},t). \tag{12.29}$$

We find that the charge $q \to -q$. However, the new equation also contains the conjugations: $\Psi(\mathbf{x},t) \to \Psi^*(\mathbf{x},t)$, $\alpha_i \to \alpha_i^*$ and $\beta \to -\beta^*$. To restore (12.29) to the form of (12.28) but with a change in the charge sign, we apply a unitary transformation U_C such that $U_C \Psi * (\mathbf{x},t)$ exactly satisfies (12.28), but with the simple replacement $q \to -q$. The requirements on U_C are

$$U_C^{-1}\alpha_i^*\,U_C = \alpha_i, \qquad U_C^{-1}\beta^*\,U_C = -\beta,$$

which means that U_C commutes with α_1 and α_3, but anticommutes with α_2 and β. The choice $U_C = i\tau_2\sigma_2 = \gamma^2$ satisfies these conditions, and we obtain the antiunitary charge conjugation operator

$$\hat{C} = \gamma^2\,K.$$

It relates the wavefunction of the particle (charge q) to the wavefunction of its antiparticle (charge $-q$).

(ii) Inversion symmetry and spinors: We recall the action of the inversion operator on the position and momentum operators as

$$\pi^{\dagger} \mathbf{x} \, \pi = -\mathbf{x}, \qquad \pi^{\dagger} \mathbf{p} \, \pi = -\mathbf{p}.$$

However, in the present case we need to construct a parity transformation that acts on two spinors. To this end, we consider a 4×4 matrix parity operator, $\mathcal{P}$, of the form

$$\mathcal{P} - \pi \, U_P,$$

where the unitary matrix U_P must satisfy

$$U_p \, \alpha \, U_P^{\dagger} = -\alpha, \quad U_P \, \beta \, U_P^{\dagger} = \beta, \quad U_P^2 = \mathbb{I}.$$

It is obvious that $U_P = \beta$ would satisfy these three conditions.

(iii) Time-reversal: Again, we are familiar of the action of the time-reversal operator on the position and momentum operators in the nonrelativistic case, namely

$$\Theta \, \mathbf{x} \, \Theta^{-1} = \mathbf{x}, \qquad \Theta \, \mathbf{p} \, \Theta^{-1} = -\mathbf{p}$$

Here, we will use the Dirac equation in the covariant form

$$i \partial_t \, \Psi(\mathbf{x}, t) = \left(-i \boldsymbol{\gamma} \cdot \nabla + \gamma^0 m \right) \Psi(\mathbf{x}, t)$$

and we define the time-reversal operator in the form

$$\mathcal{T} = U_T \, K.$$

We consider the action of $\mathcal{T}$ on the Dirac equation

$$\mathcal{T}(i \, \partial_t) \mathcal{T}^{-1} \mathcal{T} \Psi(\mathbf{x}, t) = U_T \, K \, (i \partial_t) K \, U_T^{\dagger} \, U_T \, \Psi^*(\mathbf{x}, t)$$
$$= -i \partial_t \, U_T \, \Psi^*(\mathbf{x}, t) = i \partial_{-t} \left[U_T \, \Psi^*(\mathbf{x}, t) \right].$$

For $\left[U_T \, \Psi^*(\mathbf{x}, t) \right]$ to satisfy the time-reversed Dirac equation, we need to satisfy the conditions

$$\mathcal{T} \, (-i \boldsymbol{\gamma}) \, \mathcal{T}^{-1} = i \boldsymbol{\gamma} \quad \Rightarrow \quad U_T \, \boldsymbol{\gamma} \, U_T^{\dagger} = -\boldsymbol{\gamma}^*$$
$$\mathcal{T} \gamma^0 \, \mathcal{T}^{-1} = \gamma^0.$$

We find the choice $U_T = \gamma^1 \gamma^3$ satisfies these conditions, and we obtain

$$\mathcal{T} = \gamma^1 \gamma^3 \, K.$$

Finally, the product $\pi \, C \, \Theta = \gamma^5$ is also a symmetry operation representing *chirality*:

$$\gamma^5 = i \gamma^0 \gamma^1 \gamma^2 \gamma^3 = i \sigma_0 \otimes \tau_1 = \begin{pmatrix} 0 & \sigma_0 \\ \sigma_0 & 0 \end{pmatrix}.$$

Weyl Equation for Massless Particles: Helicity and Chirality

We now introduce the Weyl or chiral representation, which is very useful for ultra-relativistic particles, and particularly for massless particles, as we will see in this section. The Weyl representation is defined by the matrices

$$\alpha_i = \sigma_i\,\tau_3 = \begin{pmatrix} \sigma_i & 0 \\ 0 & -\sigma_i \end{pmatrix}, \quad \beta = \sigma_0\,\tau_1 = \begin{pmatrix} 0 & \sigma_0 \\ \sigma_0 & 0 \end{pmatrix}$$

with corresponding γ matrices

$$\gamma^\mu = \beta\alpha_\mu = \sigma_0\,\tau_1\sigma_\mu\,\tau_3 = \sigma_\mu\,(i\tau_2) = \begin{pmatrix} 0 & -\sigma_i \\ \sigma_i & 0 \end{pmatrix}, \quad \gamma_0 = \beta = \sigma_0\,\tau_1 = \begin{pmatrix} 0 & \sigma_0 \\ \sigma_0 & 0 \end{pmatrix}.$$

In the Weyl or chiral representation, the Dirac equation for the Weyl spinors $\Psi = (\psi_R, \psi_L)$[10] becomes

$$\begin{cases} (i\sigma_0\,\partial_t + ic\sigma_i\,\partial_i)\,\psi_R & = \dfrac{mc^2}{\hbar}\,\psi_L \\[2mm] (i\sigma_0\,\partial_t - ic\sigma_i\,\partial_i)\,\psi_L & = \dfrac{mc^2}{\hbar}\,\psi_R \end{cases} \tag{12.30}$$

We note that the mass determines the coupling between the two-component spinors ψ_R and ψ_L, in other words, it mixes ψ_R and ψ_L. We also note that an effective length $\hbar/mc$, the Compton length, naturally emerges in the equations. Alternative effective lengths are manifest in condensed matter systems. For example, a length scale, roughly $\hbar\,v_F/E_g$, determines the spatial extent of edge states confined between two insulator classes, with a characteristic energy gap E_g and Fermi velocity v_F.

For a massless particle ($m = 0$), the two-component spinors, R and L, become decoupled and the equations become scale invariant. The β matrix drops out and only the three α_i matrices associated with the three spatial directions are left in the Hamiltonian $\mathcal{H}_D$, and we obtain

$$(i\sigma_0\,\partial_t + ic\boldsymbol{\sigma}\cdot\partial_{\mathbf{x}})\,\psi_R = \left(E - \boldsymbol{\sigma}\cdot\mathbf{p}\right)\psi_R = 0 \tag{12.31}$$

$$(i\sigma_0\,\partial_t - ic\boldsymbol{\sigma}\cdot\partial_{\mathbf{x}})\,\psi_L = \left(E + \boldsymbol{\sigma}\cdot\mathbf{p}\right)\psi_L = 0. \tag{12.32}$$

The operator $\boldsymbol{\sigma}\cdot\mathbf{p}$ projects the spin onto the momentum direction. We call the sign of the projection of the spin vector onto the direction of motion, the momentum vector, the *helicity*. Alternatively, *chirality* defines the handedness of the spin. We find that the left- and right-hand states are eigenstates of $\boldsymbol{\sigma}\cdot\mathbf{p}$ with opposite helicity eigenvalue.

For a plane-wave solution, $\psi \sim e^{i(\mathbf{p}\cdot\mathbf{x} - Et)/\hbar}$, we find the following:

- ψ_L has energy $E_L = -c\,\boldsymbol{\sigma}\cdot\mathbf{p}$. For a particle solution, $E_L > 0$, the spin projection has to point opposite to the momentum $\mathbf{p}$ ($\boldsymbol{\sigma}\cdot\mathbf{p} < 0$), hence, the label *left-handed fermion* and the subscript L.

---

[10] The subscripts of the Weyl spinors indicate left- and right-handedness, as will be apparent shortly.

- Similarly, ψ_R for the right-handed particle, has energy $E = c\,\boldsymbol{\sigma} \cdot \mathbf{p}$ and a spin pointing in the direction of motion ($\boldsymbol{\sigma} \cdot \mathbf{p} > 0$).

Hence, a massless particle can be described by a two-component spinor with definite *helicity*, or *chirality*. Since the mass mixes left- and right-handed states, massive fermions are not eigenstates of helicity. This becomes obvious in the electrons rest frame: there is no direction of motion, so helicity is not well defined. In the massless limit, there is no rest frame, and a left-handed electron can never turn into a right-handed electron, and a right-handed electron can never turn into a left-handed electron, even in the presence of electromagnetic fields. In contrast to the Dirac equation, the Weyl equation breaks space inversion symmetry because it can be written separately for left-handed and right-handed Weyl fermions.

Part Three
Many-Body Physics

13

Many-Body Physics and Second Quantization

13.1 Introduction

Quantum condensed matter physics deals with the interplay of quantum mechanics with the presence of a very large number of coupled degrees of freedom $\sim 10^{24}$. This interplay leads to a rich plethora of emerging properties that find their way into our daily life: metals, insulators, superconductors, optical and carbon fibers, etc. It also gives rise to amazing phenomena, such as giant magnetoresistance, responsible for large capacity hard drives and, most recently, topological materials.

When considering such systems, one primary problem emerges: the interaction among its constituent particles. In the first part of this book, we have dealt with systems where details of such interactions were not important, and we were able to average over these interactions, such that the large number of degrees of freedom are decoupled. We solved the one-electron problem, including ground states and excitations. This does not mean that such systems are boring. We found out that particle indistinguishability still requires obeying the Pauli principle, and that this constraint led to many interesting properties.

In this part of the book, we will tackle systems where interparticle interactions cannot be neglected or averaged. This poses a formidable problem, since all the $\sim 10^{24}$ degrees of freedom in the system are coupled. Solving a Schrödinger equation with $\sim 10^{24}$ variables is intractable. We need to develop tools that are capable of tackling this type of problem. Here, we shall introduce the formalism of second quantization, and use it, in subsequent chapters, to study phenomena such as superconductivity, superfluidity, magnetism, and the Kondo effect.

The outstanding attributes of the second quantization formulation are manifest in its constructive structure. It eliminates the unwieldy and limiting process of explicit symmetry adaptation (symmetrize/antisymmetrize) of many particle states inherent in standard, first-quantized, quantum mechanics. The second-quantized structure supports ladder operators, endowed with appropriate commutators, that implicitly and automatically provide the necessary implementation of many-particle symmetry. Thus, the second-quantization formalism offers a compact package for representing many-body particle space of excitations in condensed matter systems. It is amenable to be generalized to a comprehensive and highly efficient formulation of many-body quantum mechanics in general. As a matter of fact, second quantization can be considered the first major cornerstone on which the theoretical framework of quantum field theory was built.

The development we follow here avoids full mathematical rigor for the sake of peda-gogy. We will start with outlining the first quantization approach to many-body physics and highlighting its limits. Next, we motivate the second quantization procedure with the aid of an analogy to wave fields, to be followed with a presentation of the second-quantized version of standard quantum mechanical operations, such as basis change, operator rep-resentations, matrix element manipulation, etc. Finally, we will explore the power of the formalism through several applications.

13.2 Symmetry Adaptation of Many-Particle Wavefunctions

We begin our discussion by recapitulating some fundamental notions of many body quan-tum mechanics, as formulated in the traditional language of symmetrized/antisymmetrized wavefunctions of indistinguishable particles.

To construct a many-particle wavefunction for systems with more than one electron, we start with products of orthonormalized single-particle eigenfunctions, à la Hartree,

$$|\psi(\mathbf{r}_1, \ldots, \mathbf{r}_k, \ldots)\rangle = |\phi_{v_1}(\mathbf{r}_1)\rangle \ldots |\phi_{v_k}(\mathbf{r}_k)\rangle \ldots,$$

where we use numeral subscripts to designate particles, and Greek subscripts for the eigen-states. Next we apply all possible permutation operators P_{ij}, $P_{ij,kl}$, $P_{ij,kl,mn}$, ... to the product and sum all ensuing terms, with the proviso that in the case of fermions, a minus sign will appear when the number of transpositions in the permutation is odd. For N particles, we obtain

$$\begin{cases} \left|\psi_A^{\text{boson}}(\mathbf{r}_1, \ldots, \mathbf{r}_N)\right\rangle = \dfrac{1}{\sqrt{N!}} \sum_{P(v_1, \ldots, v_N)} \phi_{v_1}(1)\phi_{v_2}(2)..\phi_{v_N}(N) \\[2em] \left|\psi_A^{\text{fermion}}(\mathbf{r}_1, \ldots, \mathbf{r}_N)\right\rangle = \dfrac{1}{\sqrt{N!}} \sum_{P(v_1, \ldots, v_N)} (-1)^{|P|}\phi_{v_1}(1)\phi_{v_2}(2)..\phi_{v_N}(N), \end{cases} \tag{13.1}$$

where $\phi_{v_1}(k) \equiv \phi_{v_1}(\mathbf{r}_k, \sigma_k)$. The summation is over all possible permutations, and $|P|$ stands for the number of two-particle transpositions performed in a given permutation. Again, as we recall, a way of constructing a fermionic wavefunction with the correct antisymmetrization properties, proposed by John C. Slater, is to write the N-particle wave-function in the form of a determinant, namely

$$\psi_A = \frac{1}{\sqrt{N!}} \begin{vmatrix} \phi_{v_1}(1) & \phi_{v_2}(1) & \ldots & \phi_{v_k}(1) & \ldots & \phi_{v_N}(1) \\ \phi_{v_1}(2) & \phi_{v_2}(2) & \ldots & \phi_{v_k}(2) & \ldots & \phi_{v_N}(2) \\ \ldots & \ldots & \ldots & \ldots & \ldots & \ldots \\ \phi_{v_1}(k) & \phi_{v_2}(k) & \ldots & \phi_{v_k}(k) & \ldots & \phi_{v_N}(k) \\ \ldots & \ldots & \ldots & \ldots & \ldots & \ldots \\ \phi_{v_1}(N) & \phi_{v_2}(N) & \ldots & \phi_{v_k}(N) & \ldots & \phi_{v_N}(N) \end{vmatrix} \tag{13.2}$$

Each row has the particle label, while each column has the same eigenfunction. Exchanging the eigenfunction label on two particles is equivalent to exchanging two columns. But exchanging two columns of a determinant changes the sign, in agreement with fermion statistics.

Often there exists a natural basis of single-particles states, for instance the states associated with the energy levels of a single atomic particle or the Bloch states for a particle in a periodic potential.

In general, we set $\{\phi_\alpha\}$ as an orthonormal basis of the single-particle Hilbert space $\mathbb{H}^{(1)}$, $\langle \phi_\alpha | \phi_\beta \rangle = \delta_{\alpha,\beta}$. A basis for fully symmetric and antisymmetric N-particle states is then given by (13.1). Thus the relevant many-particle Hilbert space of an N-particle system will comprise all such symmetry-adapted wavefunctions, or states

$$\mathbb{H}^{\otimes N}_{\text{boson}} = \mathbb{H}^{\otimes N}_+ = \bigotimes_{\text{symm}}^{N} \mathbb{H}^{(1)}$$

$$\mathbb{H}^{\otimes N}_{\text{fermion}} = \mathbb{H}^{\otimes N}_- = \bigotimes_{\text{antisymm}}^{N} \mathbb{H}^{(1)}. \tag{13.3}$$

13.3 Many-Particle Systems and Second Quantization

The preceding prescriptions for constructing symmetric and antisymmetric many-particle wavefunctions are useful for systems consisting of some "manageable" number of particles such as atoms and small enough molecules. The reason is that the standard quantum-mechanical procedure outlined in the previous section would require wavefunction expressions consisting of $N!$ terms for an N-particle system. An additional complication arises from operator representations, which are expressed as sums over single-particle operators, namely,

$$\mathcal{O}_N = \bigoplus_1^N \mathcal{O}_i$$

$$\mathcal{O}_i = \mathbb{I} \otimes \mathbb{I} \otimes \ldots \mathcal{O} \otimes \mathbb{I} \otimes \ldots \mathbb{I},$$

where $\mathbb{I}$ is the identity and $\mathcal{O}$ is inserted at the ith position. We find that the operator form depends explicitly on the number of particles. Such representations hamper the process of taking the thermodynamic limit. Consequently, standard quantum-mechanics prescriptions become intractable for systems with very large particle numbers, such as solids.

A desirable formalism would mitigate the aforementioned impediments by providing the following:

• Proper symmetry adaptation without dealing explicitly with the $N!$ terms

• An operator representation independent of particle number

This would not only facilitate taking the thermodynamic limit, but also handling situations where the particle number can change, such as in photoemission experiments.

Physicists have devised the formalism of *second quantization* to meet these requirement, and thus to handle large particle number systems. It is described in the following subsection.

13.3.1 Fock Space

Here we shall adopt an approach that will turn the idea of particle indistinguishability into an asset. We start by claiming that we do not need to specify the quantum state of each individual particle; all we simply need to do is to identify the number of particles occupying a given quantum state. Thus we replace the definition (13.1) by the more convenient procedure of indicating n_α, the number of times each single-particle state $|\phi_\alpha\rangle$ appears in the product. This number n_α is the *occupation number* of the state $|\phi_\alpha\rangle$. Then the state (13.1) can be specified as

$$\psi = |n_\alpha, n_\beta, \ldots\rangle, \tag{13.4}$$

where there are n_α particles in $|\phi_\alpha\rangle$, n_β particles in $|\phi_\beta\rangle$, and so forth. For bosons $n_\nu = 0, 1, 2, 3, \ldots$, and for fermions $n_\nu = 0, 1$, according to the Pauli principle and as implicit in the determinant structure of (13.2). For an N-particle state, we have the restriction $\sum_\nu n_\nu = N$.

Two states $|n_\alpha, n_\beta, \ldots\rangle, |n'_\alpha, n'_\beta, \ldots\rangle$ are orthogonal if they differ in at least one occupation number, namely, $n_\alpha \neq n'_\alpha$ for some index α. If all the occupation numbers coincide, we find

$$\langle\psi|\psi\rangle = \mathfrak{N}^2 \, N! \, n_\alpha! \, n_\beta! \ldots$$

Thus the normalization factor is $\mathfrak{N} = 1/\sqrt{N! \, n_\alpha! \, n_\beta! \ldots}$, and we get

$$\left\langle n'_\alpha, n'_\beta, \ldots \middle| n_\alpha, n_\beta, \ldots \right\rangle = \delta_{n_\alpha, n'_\alpha} \, \delta_{n_\beta, n'_\beta} \ldots$$

It is important to realize that the *occupation number representation* (13.4) depends on the single-particle basis. In general, one tries to make a judicious choice dictated by the physical problem to be studied.

> One has to remember that the states in the occupation number representation are symmetric for bosons and antisymmetric for fermions.

The states $|n_\alpha, n_\beta, \ldots\rangle$ form an orthonormal basis of the N-particle Hilbert space $\mathbb{H}_\pm^{\otimes N}$, where plus and minus signs refer to bosonic and fermionic systems, respectively.

Thus any state of $\mathbb{H}_\pm^{\otimes N}$ can be written as a linear combination

$$|\psi\rangle = \sum_{\substack{n_\alpha, n_\beta, \ldots \\ \sum_i n_i = N}} c(n_\alpha, n_\beta, \ldots) \, |n_\alpha, n_\beta, \ldots\rangle. \tag{13.5}$$

We will now consider the more general case where, a priori, the number of particles is unrestricted. Consequently, we have a significantly enlarged space of states in which the number of particles is allowed to fluctuate. In statistical physics jargon, we are switching from *Canonical* to *Grand Canonical Ensemble*.

When we remove the restriction $\sum_i n_i = N$, our linear combination includes states with different number of particles, namely, the summation does not specify N:

$$|\psi\rangle = \sum_{n_\alpha, n_\beta, \ldots} c(n_\alpha, n_\beta, \ldots) \, |n_\alpha, n_\beta, \ldots\rangle. \tag{13.6}$$

We should note that two Hilbert spaces with different particle numbers have no state vector in common. Thus states of the form (13.6) belong to the Hilbert space formed by the direct sum

$$\bigoplus_{N=0}^{\infty} \mathbb{H}_\pm^{\otimes N} = \mathbb{F}^\pm.$$

In this expression, $\mathbb{H}^0$ consists of the vacuum state $|\varnothing\rangle = |0, 0, 0, \ldots\rangle$, and its properties will be discussed later. The space $\mathbb{F}^\pm$ is called Fock space. It consists of symmetric (bosons) and antisymmetric (fermions) state vectors, the number of particles being unspecified. $\mathbb{F}^\pm$ is the appropriate Hilbert space for the formalism of second quantization.

13.3.2 Fields and Second Quantization

We have learned in quantum mechanics that upon quantizing the classical radiation field, it becomes endowed with particle properties: *photon quanta* with energy $\hbar\omega$ and momentum $\hbar k$ emerge!

Generalizing this scenario, we might infer that this is an attribute of all wave fields, whereby manifest particle qualities will emerge upon their quantization. Conversely, we may argue that all particles occurring in nature could be construed as quanta of some field. This would invite the question as to what wave field we may associate with particles such as electrons. We might conjecture that the field is manifest in the wavefunction $\psi(x, t)$, with the Schrödinger equation

$$i\hbar \frac{\partial \psi(\mathbf{x}, t)}{\partial t} = -\frac{\hbar^2}{2m} \nabla^2 \psi(\mathbf{x}, t) + V(\mathbf{x}) \psi(\mathbf{x}, t) = \mathcal{H} \psi(\mathbf{x}, t) \tag{13.7}$$

representing a particle of mass m, in analogy to Maxwell's equations being the field equations of the electromagnetic fields.

Thus, we are encouraged to adopt the Schrödinger equation as the field equation for the wave field $\psi(\mathbf{x}, t)$. We proceed to quantize Schrödinger's wave field following the same scenario for the quantization of the vector potential in the electromagnetic case. We define the complete orthonormal set $\{\phi_\nu(\mathbf{x})\}$ as

$$\mathcal{H} \phi_\nu(\mathbf{x}) = E_\nu \phi_\nu(\mathbf{x}) \tag{13.8}$$

and write the general normalized solution of (13.7) as

$$\psi(\mathbf{x}, t) = \sum_{\nu} b_{\nu}(t)\, \phi_{\nu}(\mathbf{x}) \tag{13.9}$$

with $\sum_n |b_{\nu}|^2 = 1$ and $|b_{\nu}|^2 \leq 1$. The b_{ν}s can be interpreted as normal coordinates of the system, in which case, $\phi_{\nu}(\mathbf{x})$ would be identified as normal states, or modes. Substituting (13.9) in (13.7), we obtain

$$\frac{db_{\nu}}{dt} = -\frac{i}{\hbar} E_{\nu} b_{\nu}, \tag{13.10}$$

which are just the equations of motion of the normal coordinates. This step is called *first,* or standard, *quantization.*

Next we shall proceed to introduce creation and annihilation operators for field quanta (which may be particles, photons, etc.) The quantum-mechanical procedure of introducing such operators is called *second quantization.*

In analogy with the radiation field quantization, we now need to construct a Hamiltonian that only depends on the normal coordinates b_{ν}, and that yields the equation of motion (13.10). We can choose

$$\mathcal{H} = \int d\mathbf{x} \psi^*(\mathbf{x}, t) \left[-\frac{\hbar^2}{2m} \nabla^2 + V(\mathbf{x}) \right] \psi(\mathbf{x}, t) = \int d\mathbf{x}\, \psi^*(\mathbf{x}, t) \mathcal{H} \psi(\mathbf{x}, t), \tag{13.11}$$

which is just the energy expectation value, with

$$\psi^*(\mathbf{x}, t) \left[-\frac{\hbar^2}{2m} \nabla^2 + V(\mathbf{x}) \right] \psi(\mathbf{x}, t)$$

being the energy density.

Using (13.8) and (13.9) and the orthonormality of $\{\phi_{\nu}\}$, we obtain

$$\mathcal{H} = \sum_{\nu} E_{\nu} b_{\nu}^* b_{\nu}. \tag{13.12}$$

Bosons

Examination of (13.12) reveals that if we promote the normal coordinates b_{ν}^*, b_{ν} to *creation* and *annihilation operators,* namely,

$$b_{\nu}^* \rightarrow \hat{b}_{\nu}^{\dagger}, \qquad b_{\nu} \rightarrow \hat{b}_{\nu}, \tag{13.13}$$

obeying the commutation relations

$$\left[b_{\nu}, b_{\nu'} \right] = \left[b_{\nu}^{\dagger}, b_{\nu'}^{\dagger} \right] = 0, \qquad \left[b_{\nu}, b_{\nu'}^{\dagger} \right] = \delta_{\nu, \nu'}, \tag{13.14}$$

we obtain a second-quantized Hamiltonian representing an infinite number of harmonic oscillators with energies E_{ν}:

$$\hat{\mathcal{H}} = \sum_{\nu} E_{\nu} b_{\nu}^{\dagger} b_{\nu}. \tag{13.15}$$

The equation of motion for b_ν is

$$
\begin{aligned}
i\hbar \frac{db_\nu}{dt} = [b_\nu, \mathcal{H}] &= \sum_{\nu'} \left[b_\nu, E_{\nu'} b_{\nu'}^\dagger b_{\nu'} \right] = \sum_{\nu'} E_{\nu'} \left[b_\nu b_{n'}^\dagger b_{\nu'} - b_{\nu'}^\dagger b_{\nu'} b_\nu \right] \\
&= \sum_{\nu'} E_{\nu'} \left[\left(b_{\nu'}^\dagger b_\nu + \delta_{\nu,\nu'} \right) b_{\nu'} - b_{\nu'}^\dagger b_{\nu'} b_\nu \right] = \sum_{\nu'} E_{\nu'} \, b_{\nu'} \, \delta_{\nu,\nu'} \\
&= E_\nu \, b_\nu,
\end{aligned}
\tag{13.16}
$$

consistent with the "classical" equations of motion.

We should stress at this point that the commutator

$$
\left[b_\nu^\dagger, b_\mu^\dagger \right] = 0 \quad \Rightarrow \quad b_\nu^\dagger b_\mu^\dagger = b_\mu^\dagger b_\nu^\dagger
\tag{13.17}
$$

ensures the wavefunction symmetrization for bosonic particles. To establish this action, we interpret the commutation as an exchange operation: We first associate particle i with the left operator entry and particle j with the right one. Thus, we interpret $b_\nu^\dagger b_\mu^\dagger$ as creating particle i in state ν and particle j in state μ, while $b_\mu^\dagger b_\nu^\dagger$ as creating particle i in state μ and particle j in state ν. Consequently, keeping in mind this operator ordering convention, (13.17) would represent an exchange of particles i and j. The fact that it has a positive sign indicates that the particles are bosons!

> Because of the commutation relations satisfied by the operators, the theory developed here describes quanta that obey *Bose–Einstein statistics*.

These commutation relations also allow us to construct states of the form

$$
\left| n_\mu \right\rangle = \frac{1}{\sqrt{n_\mu!}} \left(b_\mu^\dagger \right)^{n_\mu} |0\rangle,
\tag{13.18}
$$

in which n_μ particles appear with the same wavefunction $\phi_\mu(\mathbf{x})$, and to define a *number operator* $\hat{N}_\mu$ for particles occupying a state of type μ as

$$
\hat{N}_\mu = b_\mu^\dagger \, b_\mu.
\tag{13.19}
$$

> Thus, by second quantizing the quantum field we have ensured that the *number of particles in the field is a positive integer* !

The general second-quantized state vector is now written as a direct product of eigenkets, à la Hartree

$$
\begin{aligned}
|\psi\rangle &= |\ldots, n_\mu, \ldots, n_{\mu'}, \ldots\rangle = \ldots, \left| n_\mu \right\rangle, \ldots, \left| n_{\mu'} \right\rangle, \ldots \\
\hat{N}_\mu |\ldots, n_\mu, \ldots, n_{\mu'}, \ldots\rangle &= n_\mu |\ldots, n_\mu, \ldots, n_{\mu'}, \ldots\rangle.
\end{aligned}
\tag{13.20}
$$

The last line is just the eigenvalue equation for the number operator $\hat{N}_\mu$. Notice that here, unlike our earlier discussion of particle exchange, the locations $|\ldots, n_\kappa, \ldots, n_\mu, \ldots\rangle$ identify eigenstates κ, μ, etc., and not particles.

We note that with the aid of the creation and annihilation operators, we are to modify the number of particles in a given quantum state, and thus span the whole Fock space. However, the formalism we have just developed applies only for *bosons*, since it allows multiple occupation of eigensates.

Fermions

Now we shall expand the formalism to cover *fermionic* particles. We maintain the form of the Hamiltonian

$$\hat{\mathcal{H}} = \sum_\nu E_\nu \, c_\nu^\dagger c_\nu \tag{13.21}$$

but replace bosonic operators b with fermionic operators c that satisfy anticommutation relations

$$\left\{ c_\nu, c_{\nu'} \right\} = \left\{ c_\nu^\dagger, c_{\nu'}^\dagger \right\} = 0, \qquad \left\{ c_\nu, c_{\nu'}^\dagger \right\} = \delta_{\nu, \nu'} \tag{13.22}$$

in order to achieve the antisymmetrization constraints imposed on the exchange of two fermions. Defining operator ordering as in the bosons case, we find that the anticommutation operation

$$c_\nu^\dagger c_\mu^\dagger = -c_\mu^\dagger c_\nu^\dagger$$

provides the required antisymmetrization for fermions.

We require that equations of motion for the c operators to conform with equations like (13.10), namely,

$$
\begin{aligned}
i\hbar \frac{dc_\nu}{dt} &= [c_\nu, \mathcal{H}] = \sum_{\nu'} \left[c_\nu, E_{\nu'} c_{\nu'}^\dagger c_{\nu'} \right] = \sum_{\nu'} E_{\nu'} \left\{ c_\nu c_{\nu'}^\dagger c_{\nu'} - c_{\nu'}^\dagger c_{\nu'} c_\nu \right\} \\
&= \sum_{\nu'} E_{\nu'} \left\{ \left(\delta_{\nu\,\nu'} - c_{\nu'}^\dagger c_\nu \right) c_{\nu'} - c_{\nu'}^\dagger c_{\nu'} c_\nu \right\} = \sum_{\nu'} E_{\nu'} \, c_{\nu'} \, \delta_{\nu, \nu'} \\
&= E_\nu c_\nu,
\end{aligned} \tag{13.23}
$$

where we used *anticommutation relations*.

Now we determine the eigenvalues of $\hat{N}_\mu = c_\mu^\dagger c_\mu$. We shall use the relation

$$
\begin{aligned}
N_\mu^2 &= \left(c_\mu^\dagger c_\mu \right)\left(c_\mu^\dagger c_\mu \right) = c_\mu^\dagger \left(1 - c_\mu^\dagger c_\mu \right) c_\mu \\
&= c_\mu^\dagger c_\mu - c_\mu^\dagger c_\mu^\dagger c_\mu c_\mu = c_\mu^\dagger c_\mu = N_\mu
\end{aligned} \tag{13.24}
$$

since $c_\mu^\dagger c_\mu^\dagger c_\mu c_\mu$ vanishes because

$$
\begin{cases}
\left\{ c_\mu^\dagger, c_\mu^\dagger \right\} = 0 & \Rightarrow & c_\mu^\dagger c_\mu^\dagger = 0 \\
\left\{ c_\mu, c_\mu \right\} = 0 & \Rightarrow & c_\mu c_\mu = 0.
\end{cases}
$$

Thus, we arrive at

$$N_\mu^2 \left|n_\mu\right\rangle = n_\mu^2 \left|n_\mu\right\rangle = n_\mu \left|n_\mu\right\rangle, \qquad n_\mu^2 = n_\mu, \;\Rightarrow\; n_\mu = 1, 0, \qquad (13.25)$$

which satisfies the requirement of the *Pauli principle*, and the *Fermi–Dirac statistics*.

Next, we determine the matrix elements of $c_\mu^\dagger$ and c_μ. We start with $c_\mu^\dagger \left|n_\mu\right\rangle$

$$\hat{N}_\mu c_\mu^\dagger \left|n_\mu\right\rangle = c_\mu^\dagger c_\mu c_\mu^\dagger \left|n_\mu\right\rangle = c_\mu^\dagger \left(1 - c_\mu^\dagger c_\mu\right) \left|n_\mu\right\rangle$$
$$= c_\mu^\dagger \left(1 - \hat{N}_\mu\right) \left|n_\mu\right\rangle = \left(1 - n_\mu\right) c_\mu^\dagger \left|n_\mu\right\rangle, \qquad (13.26)$$

which reveals that $c_\mu^\dagger \left|n_\mu\right\rangle$ represents an eigenvector of $\hat{N}_\mu$ with eigenvalue $1 - n_\mu$, namely

$$c_\mu^\dagger \left|n_\mu\right\rangle = A_\mu \left|1 - n_\mu\right\rangle.$$

To evaluate A_μ, we use

$$\left(c_\mu^\dagger \left|n_\mu\right\rangle\right)^\dagger c_\mu^\dagger \left|n_\mu\right\rangle = \left\langle n_\mu \left|c_\mu c_\mu^\dagger\right| n_\mu\right\rangle = \left\langle n_\mu \left|1 - c_\mu^\dagger c_\mu\right| n_\mu\right\rangle$$
$$= 1 - n_\mu = |A|^2 \;\Rightarrow\; A = e^{i\alpha_\mu} \sqrt{1 - n_\mu}. \qquad (13.27)$$

Similarly,

$$\hat{N}_\mu c_\mu \left|n_\mu\right\rangle = c_\mu^\dagger c_\mu c_\mu \left|n_\mu\right\rangle = \left(1 - c_\mu c_\mu^\dagger\right) c_\mu \left|n_\mu\right\rangle$$
$$= \left(1 - c_\mu \hat{N}_\mu\right) \left|n_\mu\right\rangle = \left(1 - n_\mu\right) c_\mu \left|n_\mu\right\rangle, \qquad (13.28)$$

and with $c_\mu \left|n_\mu\right\rangle = D_\mu \left|1 - n_\mu\right\rangle$, we find

$$|D_\mu|^2 = \left(c_\mu \left|n_\mu\right\rangle\right)^\dagger c_\mu \left|n_\mu\right\rangle = \left\langle n_\mu \left|c_\mu^\dagger c_\mu\right| n_\mu\right\rangle = n_\mu, \;\Rightarrow\; D = e^{i\alpha_\mu'} \sqrt{n_\mu}.$$

Phase Factor Choices for Fermions and Bosons

We can now write for fermions

$$c_\mu \left|\ldots, n_\mu, \ldots\right\rangle = e^{i\alpha_\mu'} \sqrt{n_\mu} \left|\ldots, 1 - n_\mu, \ldots\right\rangle$$
$$c_\mu^\dagger \left|\ldots, n_\nu, \ldots\right\rangle = e^{i\alpha_\mu} \sqrt{1 - n_\mu} \left|\ldots, 1 - n_\mu, \ldots\right\rangle. \qquad (13.29)$$

Following a similar procedure for bosons, we find

$$b_\nu \left|\ldots, n_\nu, \ldots\right\rangle = \sqrt{n_\nu} \left|\ldots, n_\nu - 1, \ldots\right\rangle$$
$$b_\nu^\dagger \left|\ldots, n_\nu, \ldots\right\rangle = \sqrt{n_\nu + 1} \left|\ldots, n_\nu + 1, \ldots\right\rangle. \qquad (13.30)$$

The phase factors for bosons are set equal to unity, since this choice allows the derivation of the corresponding commutation relations from (13.30).

The choice of the fermion phase factors is more complicated. Recalling that the occupation number for the μth state is $n_\mu = 1, 0$, we can construct the general ket as

$$\left|n_\alpha, \ldots, n_\kappa, \ldots\right\rangle = \left(c_\alpha^\dagger\right)^{n_\alpha} \left(c_\beta^\dagger\right)^{n_\beta} \cdots \left(c_\kappa^\dagger\right)^{n_\kappa} \cdots \left|0\right\rangle.$$

Then the operation

$$c_\kappa \left|n_\alpha, \ldots, n_\kappa, \ldots\right\rangle = c_\kappa \left(c_\alpha^\dagger\right)^{n_\alpha} \left(c_\beta^\dagger\right)^{n_\beta} \ldots \left(c_\kappa^\dagger\right)^{n_\kappa} \ldots |0\rangle$$

makes the commutation of c_κ with c_λ, $\lambda < \kappa$ necessary, when $n_\lambda \neq 0$. Since

$$c_\kappa c_\lambda^\dagger = -c_\lambda^\dagger c_\kappa, \text{ for } \lambda \neq \kappa,$$

a factor -1 comes up whenever $n_\lambda \neq 0$; and we accumulate an exponent of $\sum_{\nu=1}^{\kappa-1} n_\nu$. This leads to

$$c_\kappa \left(c_\alpha^\dagger\right)^{n_\alpha} \left(c_\beta^\dagger\right)^{n_\beta} \ldots \left(c_\kappa^\dagger\right)^{n_\kappa} \ldots |0\rangle = \left(c_\alpha^\dagger\right)^{n_\alpha} \left(c_\beta^\dagger\right)^{n_\beta} \ldots (-1)^{\sum_{\nu=1}^{\kappa-1} n_\nu} c_\kappa \left(c_\kappa^\dagger\right)^{n_\kappa} \ldots |0\rangle .$$

Thus the phase $e^{i\alpha_\kappa}$ depends on the occupation numbers of the states preceding $|n_\kappa\rangle$, and we obtain

$$c_\mu \left|\ldots, n_\mu, \ldots\right\rangle = (-1)^{\sum_{\nu=1}^{\mu-1} n_\nu} \sqrt{n_\mu} \left|\ldots, 1 - n_\mu, \ldots\right\rangle$$

$$c_\mu^\dagger \left|\ldots, n_\mu, \ldots\right\rangle = (-1)^{\sum_{\nu=1}^{\mu-1} n_\nu} \sqrt{1 - n_\mu} \left|\ldots, 1 - n_\mu, \ldots\right\rangle. \tag{13.31}$$

With this choice of phase, the fermion commutatators can be derived from (13.31).

13.3.3 Field Operators

If we replace the coefficients b_ν in (13.9) by $\hat{b}_\nu$ or $\hat{c}_\nu$, the field $\psi(\mathbf{x}, t)$ becomes a field operator $\hat{\psi}(\mathbf{x}, t)$. We write for the *field operator*

$$\hat{\psi}_b(\mathbf{x}, t) = \sum_\nu \hat{b}_\nu \phi_\nu(\mathbf{x}) = \sum_\nu \hat{b}_\nu \langle \mathbf{x}| \nu\rangle, \quad \hat{\psi}_b^\dagger(\mathbf{x}, t) = \sum_\nu \hat{b}_\nu^\dagger \phi_\nu^*(\mathbf{x}) = \sum_\nu \hat{b}_\nu^\dagger \langle \nu| \mathbf{x}\rangle$$

$$\hat{\psi}_f(\mathbf{x}, t) = \sum_\nu \hat{c}_\nu \phi_\nu(\mathbf{x}) = \sum_\nu \hat{c}_\nu \langle \mathbf{x}| \nu\rangle, \quad \hat{\psi}_f^\dagger(\mathbf{x}, t) = \sum_\nu \hat{c}_\nu^\dagger \phi_\nu^*(\mathbf{x}) = \sum_\nu \hat{c}_\nu^\dagger \langle \nu| \mathbf{x}\rangle .$$

The operator $\hat{\psi}(\mathbf{x}, t)$ is a linear combination of annihilation operators, which are position dependent; it annihilates a particle at position $\mathbf{x}$ and time t. $\hat{\psi}^\dagger(\mathbf{x}, t)$ is construed as an operator that creates a particle at position $\mathbf{x}$ and time t. They obey the commutation relations

$$\left[\hat{\psi}_b(\mathbf{x}, t), \hat{\psi}_b^\dagger(\mathbf{x}', t)\right] = \sum_{\nu, \nu'} \left[\hat{b}_\nu, \hat{b}_{\nu'}^\dagger\right] \phi_\nu(\mathbf{x}) \phi_{\nu'}^*(\mathbf{x}') = \sum_{\nu, \nu'} \delta_{\nu, \nu'} \phi_\nu(\mathbf{x}) \phi_{\nu'}^*(\mathbf{x}')$$

$$= \sum_\nu \phi_\nu(\mathbf{x}) \phi_\nu^*(\mathbf{x}') = \delta(\mathbf{x} - \mathbf{x}') \tag{13.32}$$

$$\left\{\hat{\psi}_f(\mathbf{x}, t), \hat{\psi}_f^\dagger(\mathbf{x}', t)\right\} = \sum_{\nu, \nu'} \left\{\hat{c}_\nu, \hat{c}_{\nu'}^\dagger\right\} \phi_\nu(\mathbf{x}) \phi_{\nu'}^*(\mathbf{x}') = \sum_{\nu, \nu'} \delta_{\nu, \nu'} \phi_\nu(\mathbf{x}) \phi_{\nu'}^*(\mathbf{x}')$$

$$= \sum_\nu \phi_\nu(\mathbf{x}) \phi_\nu^*(\mathbf{x}') = \delta(\mathbf{x} - \mathbf{x}'). \tag{13.33}$$

We can easily demonstrate that the bosonic and fermionic field operators satisfy the remaining commutation and anticommutation relations. These relations are known as *equal-time commutation relations for field operators*.

The Hamiltonian $\mathcal{H}$ can be obtained from the expectation value of the one-particle Hamiltonian with respect to the field operators:

$$\mathcal{H} = \int d\mathbf{x}\, \hat{\psi}_b^\dagger(\mathbf{x},t) \left[-\frac{\hbar^2}{2m}\nabla^2 + V \right] \hat{\psi}_b(\mathbf{x},t)$$

$$= \sum_{\nu,\nu'} b_\nu^\dagger b_{\nu'} \int d^l\mathbf{x}\, \phi_\nu^*(\mathbf{x}) \left[-\frac{\hbar^2}{2m}\nabla^2 + V \right] \phi_{\nu'}(\mathbf{x})$$

$$= \sum_{\nu,\nu'} E_\nu\, b_\nu^\dagger b_{\nu'} \int d^l\mathbf{x}\, \phi_\nu^*(\mathbf{x})\, \phi_{\nu'}(\mathbf{x}) = \sum_\nu E_\nu\, b_\nu^\dagger b_\nu. \tag{13.34}$$

A similar expression can be obtained for the fermionic Hamiltonian.

It is noteworthy that the time-dependent Schrödinger equation can be derived from Heisenberg's equations of motion for the field operators for both bosons and fermions

$$i\hbar\, \frac{\partial}{\partial t}\, \hat{\psi}(\mathbf{x},t) = \left[\hat{\psi}(\mathbf{x},t), \mathcal{H} \right] \tag{13.35}$$

with

$$\mathcal{H} = -\frac{\hbar^2}{2m}\nabla^2 + V(\mathbf{x}),$$

which demonstrates the consistency of the theory. We dropped the subscripts on field operators, since this applies to both bosons and fermions. We find that

$$\left[\hat{\psi}(\mathbf{x},t), \mathcal{H} \right] = \left[\hat{\psi}(\mathbf{x},t), \int d\mathbf{x}'\, \hat{\psi}^\dagger(\mathbf{x}',t) \left[-\frac{\hbar^2}{2m}\nabla'^2 + V(\mathbf{x}') \right] \hat{\psi}(\mathbf{x}',t) \right]$$

$$= \int d\mathbf{x}' \left(\hat{\psi}(\mathbf{x},t)\, \hat{\psi}^\dagger(\mathbf{x}',t) \left[-\frac{\hbar^2}{2m}\nabla'^2 + V(\mathbf{x}') \right] \hat{\psi}(\mathbf{x}',t) \right.$$

$$\left. -\hat{\psi}^\dagger(\mathbf{x}',t) \left[-\frac{\hbar^2}{2m}\nabla'^2 + V(\mathbf{x}') \right] \hat{\psi}(\mathbf{x}',t)\, \hat{\psi}(\mathbf{x},t) \right), \tag{13.36}$$

But

$$\hat{\psi}(\mathbf{x},t)\, \hat{\psi}^\dagger(\mathbf{x}',t) = \delta(\mathbf{x}-\mathbf{x}') \pm \hat{\psi}^\dagger(\mathbf{x}',t)\, \hat{\psi}(\mathbf{x},t),$$

where the plus sign applies to bosons, and the minus sign to fermions, and it leads to

$$\int d\mathbf{x}'\, \hat{\psi}(\mathbf{x},t)\, \hat{\psi}^\dagger(\mathbf{x}',t) \left[-\frac{\hbar^2}{2m}\nabla'^2 + V(\mathbf{x}') \right] \hat{\psi}(\mathbf{x}',t)$$

$$= \int d\mathbf{x}'\, \delta(\mathbf{x}-\mathbf{x}') \left[-\frac{\hbar^2}{2m}\nabla'^2 + V(\mathbf{x}') \right] \hat{\psi}(\mathbf{x}',t)$$

$$\pm \int d\mathbf{x}'\, \hat{\psi}^\dagger(\mathbf{x}',t)\, \hat{\psi}(\mathbf{x},t) \left[-\frac{\hbar^2}{2m}\nabla'^2 + V(\mathbf{x}') \right] \hat{\psi}(\mathbf{x}',t)$$

$$= \left[-\frac{\hbar^2}{2m} \nabla^2 + V(\mathbf{x}) \right] \hat{\psi}(\mathbf{x}, t)$$

$$+ \int d\mathbf{x}' \, \hat{\psi}^\dagger(\mathbf{x}', t) \left[-\frac{\hbar^2}{2m} \nabla'^2 + V(\mathbf{x}') \right] \hat{\psi}(\mathbf{x}', t) \, \hat{\psi}(\mathbf{x}, t).$$

The last term cancels the last term in(13.36), and we get

$$i\hbar \frac{\partial}{\partial t} \hat{\psi}(\mathbf{x}, t) = \left[-\frac{\hbar^2}{2m} \nabla^2 + V(\mathbf{x}) \right] \hat{\psi}(\mathbf{x}, t), \tag{13.37}$$

which demonstrates that the Schrödinger equation also holds for the field operators.

We can also use the field operators to define the spatiotemporal *particle number-density operator* as

$$\hat{n}(\mathbf{x}, t) = \hat{\psi}^\dagger(\mathbf{x}, t) \, \hat{\psi}(\mathbf{x}, t) \tag{13.38}$$

as well as the *total particle number operator*

$$\hat{N}(t) = \int d\mathbf{x} \, \hat{n}(\mathbf{x}, t) = \int d\mathbf{x} \, \hat{\psi}^\dagger(\mathbf{x}, t) \, \hat{\psi}(\mathbf{x}, t)$$

$$= \int d\mathbf{x} \left(\sum_n b_n^\dagger \phi_\nu^*(\mathbf{x}) \right) \left(\sum_{n'} b_{n'} \phi_{n'}(\mathbf{x}) \right)$$

$$= \sum_{n, n'} b_n^\dagger b_{n'} \int d\mathbf{x} \, \phi_\nu^*(\mathbf{x}) \, \phi_{n'}(\mathbf{x}) = \sum_n b_n^\dagger b_\nu = \sum_n \hat{N}_n. \tag{13.39}$$

A similar derivation for fermions yields

$$\hat{N}(t) = \sum_n c_n^\dagger c_n = \sum_n \hat{N}_n.$$

Moreover, it can be shown that

$$\frac{d\hat{N}}{dt} = -\frac{i}{\hbar} \left[\hat{N}, \mathcal{H} \right] = 0. \tag{13.40}$$

13.3.4 The Vacuum State

The ket vector $|\varnothing\rangle$ represents the *vacuum state* – the state without particles – so that

$$|\varnothing\rangle = |0, 0, 0 \dots, 0, 0, \dots\rangle = \bigotimes_{\mu=1}^{\infty} |n_\mu = 0\rangle, \qquad \begin{cases} b_\mu |\varnothing\rangle = 0, \\ c_\mu |\varnothing\rangle = 0, \end{cases} \quad \forall \mu.$$

The vacuum state cannot be described by a wavefunction, in the sense that we do not expect to be able to ask the same questions about probabilities that we may ask about a state $|n\rangle$. However, we declare that $|\varnothing\rangle$ is normalized, so that

$$\langle \varnothing | \varnothing \rangle = 1 \tag{13.41}$$

and that it is orthogonal to all the single-particle kets $|\nu\rangle$, namely

$$\langle \nu \, | \varnothing \rangle = 0. \tag{13.42}$$

Now we recall the concept of transfer operators

$$|\nu\rangle \langle \mu|, \tag{13.43}$$

which removes a particle from eigenstate $|\mu\rangle$ and places it in eigenstate $|\nu\rangle$. We find that the vacuum state serves as a useful device when we insert (13.41) into (13.43), and write

$$|\nu\rangle \langle \mu| = |\nu\rangle \langle \varnothing \, | \varnothing \rangle \langle \mu| = (|\nu\rangle \langle \varnothing|) \, (|\varnothing\rangle \langle \mu|). \tag{13.44}$$

We immediately recognize $|\varnothing\rangle \langle \mu|$ as the annihilation operator c_μ, or b_μ, that annihilates a particle from state $|\mu\rangle$, and $|\nu\rangle \langle \varnothing|$ as a creation operator $c_\nu^\dagger$, or $b_\nu^\dagger$, that creates a particle in $|\nu\rangle$.

Single-Particle Interaction Potential

With the identification established in (13.44), we derive an expression for the potential acting on a single fermionic (or bosonic) particle in terms of creation and annihilation operators. We make use of the identity operator $\sum_n |n\rangle \langle n|$ to write the single-particle potential as

$$V(\mathbf{x}) = \sum_{\mu,\nu} |\nu\rangle \langle \nu| \, V(\mathbf{x}) \, |\mu\rangle \langle \mu|$$

$$= \sum_{\mu,\nu} \langle \nu| \, V(\mathbf{x}) \, |\mu\rangle \, |\nu\rangle \langle \mu| = \sum_{\mu,\nu} V_{\mu\nu} \, c_\nu^\dagger c_\mu. \tag{13.45}$$

If $V_{\mu\nu} \neq 0$, then the potential $V(\mathbf{x})$ serves to annihilate a particle from eigenstate $|\mu\rangle$ and create it in eigenstate $|\nu\rangle$, thus *conserving the number of particles*.

> That is why the creation and annihilation operators must occur in pairs!

A potential cannot remove a particle from a state without putting it back in some other state. This argument applies as well for bosons with nonzero chemical potential.

If we have a set of N noninteracting electrons, the potential is written as

$$\sum_i V(\mathbf{x}_i) \;\rightarrow\; \sum_{m,n} V_{\mu\nu} \sum_{i=1}^{N} c_\nu^{(i)\dagger} c_\mu^{(i)}.$$

We obtain a similar expression for the kinetic energy $T = -\frac{\hbar^2}{2m} \nabla^2$:

$$T = \sum_{i=1}^{N} T^{(i)} = \sum_{i=1}^{N} \sum_{\mu,\nu} T_{\mu\nu}^{(i)} \, c_\nu^{(i)\dagger} c_\mu^{(i)}. \tag{13.46}$$

Two-Particle Interaction Potential

We consider the case when we add the interaction potential of two fermionic (or bosonic) particles, namely

$$V = \sum_{i \neq j} \frac{1}{2} v \left(\mathbf{x}_i - \mathbf{x}_j \right). \tag{13.47}$$

We obtain a second-quantized expression for $v \left(\mathbf{x}_i - \mathbf{x}_j \right)$ by performing the following insertions:

$$v \left(\mathbf{x}_i - \mathbf{x}_j \right) = \left(|\kappa\rangle_i \langle\kappa| \right) \left(|\lambda\rangle_j \langle\lambda| \right) v \left(\mathbf{x}_i - \mathbf{x}_j \right) \left(|\mu\rangle_j \langle\mu| \right) \left(|\nu\rangle_i \langle\nu| \right)$$

$$= V_{\kappa\lambda,\mu\nu} \, c_\kappa^{(i)\dagger} c_\lambda^{(j)\dagger} c_\mu^{(j)} c_\nu^{(i)}. \tag{13.48}$$

In order to verify that we chose the right ordering of the creation and annihilation operators, we consider the case $v = 1$. Then for a total of N particles, (13.47) becomes

$$V = \frac{1}{2} \sum_{i \neq j} 1 = \frac{1}{2} N(N-1). \tag{13.49}$$

Alternatively, we evaluate (13.48), using the matrix element

$$v_{\kappa\lambda,\mu\nu} = \delta_{\kappa\nu} \, \delta_{\lambda\mu}.$$

which leads to

$$V = \frac{1}{2} \sum_{i \neq j} \sum_{\kappa\lambda} c_\kappa^{(i)\dagger} c_\lambda^{(j)\dagger} c_\lambda^{(j)} c_\kappa^{(i)} = \frac{1}{2} \sum_{i \neq j} \sum_{\kappa\lambda} c_\kappa^{(i)\dagger} \left(c_\kappa^{(i)} c_\lambda^{(j)\dagger} - \delta_{\kappa\lambda} \right) c_\lambda^{(j)}$$

$$= \frac{1}{2} \sum_{ij} \sum_{\kappa\lambda} \left(c_\kappa^{(i)\dagger} c_\kappa^{(i)} c_\lambda^{(j)\dagger} c_\lambda^{(j)} - c_\kappa^{(i)\dagger} c_\kappa^{(i)} \right) = \frac{1}{2} \left(N^2 - N \right) \tag{13.50}$$

in agreement with (13.49).

Field Operators and the Vacuum State

As for the field operators, we immediately realize that

$$\hat{\psi}(\mathbf{x}, t) \, |\varnothing\rangle = 0. \tag{13.51}$$

This result confirms that the *vacuum does not contain particles*. Thus, it becomes clear that the *vacuum expectation value of the field operator* vanishes:

$$\left\langle \varnothing \left| \hat{\psi}(\mathbf{x}, t) \right| \varnothing \right\rangle = 0. \tag{13.52}$$

Hence, we expect that

$$\hat{\psi}^\dagger(\mathbf{x}, t) \, |\varnothing\rangle \tag{13.53}$$

describes a state where a particle stays at position $\mathbf{x}$. To convince ourselves, we first calculate the action of the particle density operator $\hat{n}(\mathbf{x}, t)$ on this state:

$$
\begin{aligned}
\hat{n}(\mathbf{x},t)\,\hat{\psi}^{\dagger}(\mathbf{x},t)\,|\varnothing\rangle &= \hat{\psi}^{\dagger}(\mathbf{x},t)\,\hat{\psi}(\mathbf{x},t)\,\hat{\psi}^{\dagger}(\mathbf{x},t)\,|\varnothing\rangle \\
&= \hat{\psi}^{\dagger}(\mathbf{x},t)\left[\delta(\mathbf{x}-\mathbf{x}') + \hat{\psi}^{\dagger}(\mathbf{x},t)\,\hat{\psi}(\mathbf{x},t)\right]|\varnothing\rangle \\
&= \delta(\mathbf{x}-\mathbf{x}')\,\hat{\psi}^{\dagger}(\mathbf{x},t)\,|\varnothing\rangle .
\end{aligned}
\tag{13.54}
$$

We realize that $\hat{\psi}^{\dagger}(\mathbf{x}, t)\,|\varnothing\rangle$ represents an *eigenvector of the particle-number density operator* with eigenvalue $\delta(\mathbf{x} - \mathbf{x}')$. At the point $\mathbf{x}$, the particle density becomes so large that an integration over the vicinity of this position yields 1. Then the validity of the following relation becomes clear:

$$
\hat{N}\,\hat{\psi}^{\dagger}(\mathbf{x},t)\,|\varnothing\rangle = \hat{\psi}^{\dagger}(\mathbf{x},t)\,|\varnothing\rangle .
$$

It can be proven as follows:

$$
\int d\mathbf{x}'\,\hat{n}(\mathbf{x}',t)\,\hat{\psi}^{\dagger}(\mathbf{x},t)\,|\varnothing\rangle = \int d\mathbf{x}'\,\delta(\mathbf{x}-\mathbf{x}')\,\hat{\psi}^{\dagger}(\mathbf{x},t)\,|\varnothing\rangle = \hat{\psi}^{\dagger}(\mathbf{x},t)\,|\varnothing\rangle ,
\tag{13.55}
$$

which shows that $\hat{\psi}^{\dagger}(\mathbf{x}, t)\,|\varnothing\rangle$ is an *eigenvector* of $\hat{N}$ with eigenvalue 1; and the interpretation of $\hat{\psi}^{\dagger}(\mathbf{x}, t)\,|\varnothing\rangle$ as a *one-particle state* is justified!

Similarly,

$$
\hat{\psi}^{\dagger}(\mathbf{x}_1,t)\,\hat{\psi}^{\dagger}(\mathbf{x}_2,t)\,|\varnothing\rangle
\tag{13.56}
$$

represents a *two-particle state* with a particle at $\mathbf{x}_1$ and one at $\mathbf{x}_2$. *Many-particle states* can be constructed in an analogous fashion.

13.4 Canonical Transformations

We now know how to construct operators in second-quantized form, and how to incorporate them in Hamiltonians or other physical observables. Now the question arises as to how to carry out solutions to problems within that framework, where coping with wavefunctions is to be avoided.

13.4.1 Diagonal Quadratic Hamiltonians

The simplest second-quantized Hamiltonian has the general diagonal quadratic form

$$
\mathcal{H} = \sum_{\mu} \mathcal{E}_{\mu}\,c_{\mu}^{\dagger}\,c_{\mu},
\tag{13.57}
$$

where μ represents a complete orthonormal basis, and the $\mathcal{E}_{\mu}$ are arbitrary energies. Obviously, the Hamiltonian (13.57) yields an eigenvalue

$$
E = \sum_{i=1}^{n} \mathcal{E}_i
\tag{13.58}
$$

for a state vector $c_{\mu_1}^\dagger \, c_{\mu_2}^\dagger \, c_{\mu_3}^\dagger \ldots c_{\mu_n}^\dagger \, |\varnothing\rangle$. It is instructive to consider the simple case of a fermionic two-particle state $|\psi\rangle = c_\lambda^\dagger \, c_\kappa^\dagger \, |\varnothing\rangle$:[1]

$$\mathcal{H} \, |\psi\rangle = \left(\sum_\mu \mathcal{E}_\mu \, c_\mu^\dagger \, c_\mu \right) c_\lambda^\dagger \, c_\kappa^\dagger \, |\varnothing\rangle = \sum_\mu \mathcal{E}_\mu \, c_\mu^\dagger \left(\delta_{\mu,\lambda} - c_\lambda^\dagger \, c_\mu \right) c_\kappa^\dagger \, |\varnothing\rangle$$

$$= \mathcal{E}_\lambda \, |\psi\rangle - \sum_\mu \mathcal{E}_\mu \, c_\mu^\dagger \, c_\lambda^\dagger \, c_\mu \, c_\kappa^\dagger \, |\varnothing\rangle = \mathcal{E}_\lambda \, |\psi\rangle - \mathcal{E}_\kappa \, c_\kappa^\dagger \, c_\lambda^\dagger \, |\varnothing\rangle = \left(\mathcal{E}_\lambda + \mathcal{E}_\kappa \right) |\psi\rangle .$$

Hence, we realize that we can effortlessly get eigenvalues for any ket in Fock space for Hamiltonians in diagonal form.

Particle–Hole Transformation

A simple example of a canonical transformation, namely, one that preserves commutation relations is the particle–hole transformation,

$$h_\mu = c_\mu^\dagger, \qquad h_\mu^\dagger = c_\mu \tag{13.59}$$

$$\left\{ h_\mu, h_\nu^\dagger \right\} = \left\{ c_\mu^\dagger, c_\nu \right\} = \delta_{\mu,\nu} .$$

While c_μ and $c_\mu^\dagger$ annihilate and create a particle in state μ, respectively, h_α and $h_\alpha^\dagger$ remove or introduce a hole, and still obey fermionic statistics and preserve Fock space. One caveat, however, is that the two representations have different vaccum states. But it is easy to see that

$$|\varnothing_h\rangle = \prod_\mu c_\mu^\dagger \, |\varnothing_c\rangle \quad \Rightarrow \quad h_\nu \, |\varnothing_h\rangle = c_\nu^\dagger \prod_\mu c_\mu^\dagger \, |\varnothing_c\rangle = 0. \tag{13.60}$$

13.4.2 *Quadratic Nondiagonal Hamiltonians*

In general, Hamiltonians that we encounter are neither quadratic nor diagonal. Quartic Hamiltonian terms, encountered in two-body interactions, cannot be solved within the second-quantization framework. A physical mean-field approximation that reduces such terms to general quadratic form has to be employed. Thus, we need to devise methods to diagonalize general quadratic Hamiltonians. Such methods involve introducing transformations of creation and annihilation operators that lead to quadratic diagonal forms. However, we need to seek transformations that preserve Fock space, or, in other words, preserve the canonical commutation relations. It follows that these transformations are coined as *canonical* ones. In general, obtaining such transformations is a daunting undertaking. Sometimes the physics of the problem under consideration mitigates the diagonalization procedure.

---

[1] A similar procedure can be used for bosons.

Diagonalizing Quadratic Hamiltonians

A class of nondiagonal quadratic Hamiltonians of the form

$$\mathcal{H} = \sum_{\mu,\nu} \mathcal{E}_{\mu\nu} c_\mu^\dagger c_\nu \tag{13.61}$$

covers a very large number of system, including those reduced to quadratic form in mean field. Here, the matrix $\mathcal{E}_{\mu\nu}$ is diagonalized by unitary transformation involving linear combinations of the c_μ, $c_\mu^\dagger$ operators.

Example: The Tight-Binding Hamiltonian

We shall consider here a simplified 1D version of the tight-binding Hamiltonian we encountered in graphene. It has the form

$$\mathcal{H} = \mathcal{E}_0 \sum_i c_i^\dagger c_i - t \sum_{\substack{\langle ij \rangle \\ i \neq j}} c_i^\dagger c_j. \tag{13.62}$$

The t is the hopping energy, and $\langle ij \rangle$ indicates nearest neighbor. The mitigating physics of this problem is translational invariance and conservation of momentum (momentum must be a good quantum number). This then suggests the transformation

$$f_{\mathbf{k}}^\dagger = \frac{1}{\sqrt{N}} \sum_{j=0}^{N-1} e^{i\mathbf{k}\cdot\mathbf{x}_j} c_j^\dagger \tag{13.63}$$

with $\mathbf{k} \subset$ first BZ. The $f_{\mathbf{k}}$ operators satisfy the anticommutation rules

$$\left\{ f_{k_1}, f_{k_2}^\dagger \right\} = \frac{1}{N} \sum_{j,l} e^{-ik_1 x_j} e^{ik_2 x_l} \left\{ c_j, c_l^\dagger \right\} = \frac{1}{N} \sum_{j,l} e^{-ik_1 x_j} e^{ik_2 x_l} \delta_{j,l}$$

$$= \frac{1}{N} \sum_j e^{i(k_2 - k_1)x_j} = \delta_{k_1,k_2}, \tag{13.64}$$

and the Hilbert space dimension is preserved since there are exactly N pairs of operators – the transformation is canonical! Moreover, $|\varnothing_f\rangle = |\varnothing_c\rangle$. Substituting for c_js by f_ks in the Hamiltonian, we arrive at

$$\mathcal{H} = \mathcal{E}_0 \sum_k f_k^\dagger f_k - \sum_k 2t \cos(ka) f_k^\dagger f_k = \sum_k \varepsilon(k) f_k^\dagger f_k. \tag{13.65}$$

We can use the $f_k^\dagger$s that diagonalized the Hamiltonian to construct the ground state and to determine other related physical quantities.

A more interesting and instructive example is obtained by adding to (13.62) a staggered periodic potential of the form

$$\mathcal{H}' = \Delta \sum_j (-1)^j |j\rangle \langle j| = \Delta \sum_j (-1)^j c_j^\dagger c_j, \tag{13.66}$$

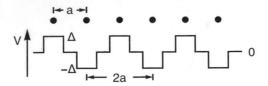

Figure 13.1 The staggered potential doubles the periodicity and halves the BZ.

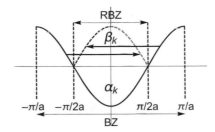

Figure 13.2 Going to the reduced zone scheme yields two species of fermions corresponding to the lower and upper bands.

shown in Figure 13.1, which effectively doubles the periodicity of the system. Using (13.63) and setting $(-1)^j = e^{i\pi r_j/a}$, we write

$$\mathcal{H}' = \Delta \sum_j e^{i\pi r_j/a} \frac{1}{N} \sum_{k_1, k_2} e^{-ik_1 r_j} e^{ik_2 r_j} f_{k_1}^\dagger f_{k_2} \qquad k_1,\ k_2 \in \quad \text{first BZ } [-\pi/a, \pi/a]$$

$$= \Delta \sum_k f_{k+\pi/a}^\dagger f_k, \tag{13.67}$$

which is nondiagonal in the f operators. To diagonalize the total Hamiltonian, we notice that the state k is only coupled to the state $k + \pi/a$, since the latter is coupled to $k + 2\pi/a = k$. Thus, the diagonalization implies a linear combination of f_k and $f_{k+\pi/a}$.

A more illuminating perspective is to work in the reduced BZ, $[-\pi/2a, \pi/2a]$, associated with the $2a$ periodicity. We recover translation invariance and render k a good quantum number. We see from Figure 13.2 that an umklapp process yields two bands, corresponding to the two degrees of freedom now in the primitive cell.

Accordingly, we define the operators

$$\begin{cases} \alpha_k = f_k & |k| \le \dfrac{\pi}{2a} \\[2mm] \beta_k = f_k & |k| \ge \dfrac{\pi}{2a} \end{cases} \tag{13.68}$$

so that α_k, β_k are defined in the reduced BZ, and they obey fermionic commutators. We now recast the tight-binding Hamiltonian (13.65) in terms of the new operators as

$$\sum_{k \in \text{RBZ}} \varepsilon(k) \left(\alpha_k^\dagger \alpha_k - \beta_k^\dagger \beta_k \right), \tag{13.69}$$

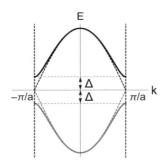

Figure 13.3 The dispersion relation $\mathcal{E}_{\pm}(k)$. Because of the staggered potential, there is now a gap 2Δ at the zone edge.

where we used $\varepsilon(k \pm \pi/a) = -\varepsilon(k)$. The mapping of (13.69) is displayed in Figure 13.2. We also express $\mathcal{H}'$ as

$$\sum_{k \in \text{BZ}} f^{\dagger}_{k+\frac{\pi}{a} (\text{mod} \frac{2\pi}{a})} f_k \rightarrow \sum_{k \in \text{RBZ}} \left(\alpha^{\dagger}_k \beta_k + \beta^{\dagger}_k \alpha_k \right), \tag{13.70}$$

giving

$$\mathcal{H} = \sum_{k \in \text{RBZ}} \left[\mathcal{E}(k) \left(\beta^{\dagger}_k \beta_k - \alpha^{\dagger}_k \alpha_k \right) + \Delta \left(\alpha^{\dagger}_k \beta_k + \beta^{\dagger}_k \alpha_k \right) \right]$$

$$= \sum_{k \in \text{RBZ}} \begin{bmatrix} \alpha^{\dagger}_k & \beta^{\dagger}_k \end{bmatrix} \begin{pmatrix} -\mathcal{E}(k) & \Delta \\ \Delta & \mathcal{E}(k) \end{pmatrix} \begin{bmatrix} \alpha_k \\ \beta_k \end{bmatrix}, \tag{13.71}$$

where we set $\mathcal{E}(k) = -\varepsilon(k)$.

Bogoliubov Transformation

The structure of (13.71) indicates that the two degrees of freedom α_k, β_k are coupled by Δ, making the Hamiltonian nondiagonal quadratic. To diagonalize (13.71), we introduce the unitary *Bogoliubov* canonical transformation: we define the new operators

$$\begin{bmatrix} \gamma_{k-} & \gamma_{k+} \end{bmatrix} = \begin{pmatrix} a & b \\ c & d \end{pmatrix} \begin{bmatrix} \alpha_k \\ \beta_k \end{bmatrix}. \tag{13.72}$$

To satisfy orthogonality, we set

$$\begin{bmatrix} \gamma_{k-} & \gamma_{k+} \end{bmatrix} = \begin{pmatrix} u_k & -v_k \\ v_k & u_k \end{pmatrix} \begin{bmatrix} \alpha_k \\ \beta_k \end{bmatrix}. \tag{13.73}$$

Moreover, to satify anticommutation relations, we require

$$\left\{ \gamma_{k-}, \gamma^{\dagger}_{k-} \right\} = \left\{ u_k \alpha_k - v_k \beta_k, u_k \alpha^{\dagger}_k - v_k \beta^{\dagger}_k \right\}$$

$$= u^2_k \left\{ \alpha_k, \alpha^{\dagger}_k \right\} + v^2_k \left\{ \beta_k, \beta^{\dagger}_k \right\} + u_k v_k \left(\left\{ \beta_k, \alpha^{\dagger}_k \right\} + \left\{ \alpha_k, \beta^{\dagger}_k \right\} \right)$$

$$= u^2_k + v^2_k = 1. \tag{13.74}$$

This last relation suggests that we set $u_k = \cos\theta_k$, $v_k = \sin\theta_k$. Inverting (13.73),

$$\begin{bmatrix} \alpha_k \\ \beta_k \end{bmatrix} = \begin{pmatrix} u_k & v_k \\ -v_k & u_k \end{pmatrix} \begin{bmatrix} \gamma_{k-} & \gamma_{k+} \end{bmatrix} \tag{13.75}$$

and substituting in (13.71), we obtain

$$\mathcal{H} = \sum_{k \in \text{RBZ}} \begin{bmatrix} \gamma^\dagger_{k-} & \gamma^\dagger_{k+} \end{bmatrix}$$

$$\times \begin{pmatrix} -\left[\mathcal{E}(k)(u_k^2 - v_k^2) + 2\Delta u_k v_k \right] & \left[\Delta(u_k^2 - v_k^2) - 2\mathcal{E}(k)u_k v_k \right] \\ \left[\Delta(u_k^2 - v_k^2) - 2\mathcal{E}(k)u_k v_k \right] & \left[\mathcal{E}(k)(u_k^2 - v_k^2) + 2\Delta u_k v_k \right] \end{pmatrix} \begin{bmatrix} \gamma_{k-} \\ \gamma_{k+} \end{bmatrix}.$$

This leads to the condition

$$\Delta(u_k^2 - v_k^2) - 2\mathcal{E}(k)u_k v_k = 0 \;\Rightarrow\; \tan(2\theta_k) = \frac{\Delta}{\mathcal{E}(k)} \tag{13.76}$$

giving the dispersion

$$\mathcal{E}_{\pm}(k) = \pm \left[\mathcal{E}(k)(u_k^2 - v_k^2) + 2\Delta u_k v_k \right] = \pm \left[\mathcal{E}^2(k) + \Delta^2 \right]^{1/2}, \tag{13.77}$$

shown in Figure 13.3, and

$$u_k = \left[\frac{1}{2} \left(1 + \frac{\mathcal{E}(k)}{\sqrt{\mathcal{E}(k)^2 + \Delta^2}} \right) \right]^{1/2}, \quad v_k = \left[\frac{1}{2} \left(1 - \frac{\mathcal{E}(k)}{\sqrt{\mathcal{E}(k)^2 + \Delta^2}} \right) \right]^{1/2}.$$

13.5 Coherent States

In field theory, we have a field. It can be described as an operator $\psi(\mathbf{x})$ at each point in space. As we will discuss later when we deal with path integral formalism, we anticipate that the path integral will be related to an integration over the field configurations $\psi(\mathbf{x}, t)$. To make sense of this, it is clear that we first need to work in a basis that diagonalizes the field operators. The states that do this are called coherent states. Moreover, coherent states comprise states with varying particle numbers, thus they provide an extremely useful basis of Fock space. Although it is not an orthonormal set, it does span the whole Fock space.

There are some crucial differences between the coherent states corresponding to boson fields and those corresponding to fermion fields, which, as we will see, arise from their characteristic commutators.

13.5.1 *Bosonic Coherent States*

For a bosonic state with an annihilation operator b, the coherent states are defined as eigenstates of b, namely,

$$b\,|z\rangle = z\,|z\rangle, \tag{13.78}$$

where $z = |z| e^{i\varphi}$ is a complex c-number. Every state in Hilbert space can be written as

$$|z\rangle = B(b^\dagger) |\emptyset\rangle,$$

where the function $B(b^\dagger)$ has to satisfy

$$\left[b, B(b^\dagger) \right] = z\, B(b^\dagger). \tag{13.79}$$

We use the identity

$$\left[b, B(b^\dagger) \right] = \frac{\partial}{\partial b^\dagger}\, B(b^\dagger)$$

to obtain

$$\frac{\partial}{\partial b^\dagger}\, B(b^\dagger) = z\, B(b^\dagger) \quad \Rightarrow \quad B(b^\dagger) = C\, e^{z\, b^\dagger}. \tag{13.80}$$

We can then write the eigenstate of b as

$$|z\rangle = C\, e^{z\, b^\dagger} |\emptyset\rangle \tag{13.81}$$

subject to the normalization

$$1 = \langle z | z \rangle = |C|^2 \left\langle \emptyset \left| e^{z^* b}\, e^{z b^\dagger} \right| \emptyset \right\rangle.$$

Using the Baker–Housdorff identity

$$e^A\, e^B = e^{A+B+[A,\, B]/2} = e^{[A,\, B]}\, e^B\, e^A,$$

we obtain

$$|C|^2 = e^{-z^* z}$$

yielding

$$|z\rangle = e^{-|z|^2/2}\, e^{z\, b^\dagger} |\emptyset\rangle = e^{-|z|^2/2} \sum_{n=0}^{\infty} \frac{z^n}{n!}\, |n\rangle. \tag{13.82}$$

We can generalize and write the coherent state for a system with N bosonic states as

$$|\mathbf{z}\rangle = \prod_{i=1}^{N} e^{-|z_i|^2/2}\, e^{z_i\, b_i^\dagger} |\emptyset\rangle = e^{-\sum_{i=1}^{N} |z_i|^2/2}\, e^{\sum_{i=1}^{N} z_i\, b_i^\dagger} |\emptyset\rangle. \tag{13.83}$$

Properties of Bosonic Coherent States

- The coherent states are not orthogonal to each other:

$$\langle z_1 | z_2 \rangle = e^{-(|z_1|^2 + |z_2|^2)/2} \left\langle \emptyset \left| e^{z_1^* b}\, e^{z_2 b^\dagger} \right| \emptyset \right\rangle$$

$$= e^{-(|z_1|^2 + |z_2|^2)/2}\, e^{z_1^* z_2}. \tag{13.84}$$

- The coherent states form an overcomplete basis

$$
\int dz^* \, dz \, |z\rangle \langle z| = \sum_{n,m} \int dz^* \, dz \, e^{-|z|^2} \frac{z^{*n} \, z^m}{n! \, m!} \, b^{\dagger n} \, |\emptyset\rangle \langle \emptyset| \, b^m
$$

$$
= \sum_{n,m} \int_0^\infty r \, dr \, e^{-r^2} \frac{r^{m+n}}{n! \, m!} \int_0^{2\pi} d\varphi \, e^{i(m-n)\varphi} \, b^{\dagger n} \, |\emptyset\rangle \langle \emptyset| \, b^m
$$

$$
= \pi \sum_n |n\rangle \langle n| \frac{1}{n!} \int_0^\infty dy \, e^{-y} \, y^n = \pi \sum_n |n\rangle \langle n| = \pi \mathbb{I}
$$

and the *resolvent identity* is

$$
\mathbb{I} = \int \frac{dz^* \, dz}{\pi} \, |z\rangle \langle z| . \tag{13.85}
$$

- Next, we consider the action of $b^\dagger$ on the coherent state $|z\rangle$:

$$
b^\dagger \, |z\rangle = b^\dagger \sum_{n=0}^\infty \frac{z^n}{n!} \, b^{\dagger n} \, |\emptyset\rangle = \sum_{n=0}^\infty \frac{z^n}{n!} \, b^{\dagger(n+1)} \, |\emptyset\rangle
$$

$$
= \sum_{n=0}^\infty (n+1) \frac{z^n}{(n+1)!} \, b^{\dagger(n+1)} \, |\emptyset\rangle = \sum_{n=1}^\infty \frac{z^{n-1}}{(n-1)!} \, b^{\dagger n} \, |\emptyset\rangle .
$$

We find that

$$
b^\dagger \, |z\rangle = \frac{\partial}{\partial z} \, |z\rangle . \tag{13.86}
$$

- Now consider matrix elements of normal-ordered operators.

  We note that

$$
\langle z_1| \, b^{\dagger n} \, b^m \, |z_2\rangle = z_1^{*n} \, z_2^m \, \langle z_1| \, z_2\rangle = z_1^{*n} \, z_2^m \, e^{z_1^* \, z_2},
$$

which allows us to write the matrix element of the operator

$$
A = \sum_{m,n} a_{mn} \, b^{\dagger n} \, b^m
$$

as

$$
\left\langle z_1 \left| \sum_{m,n} a_{mn} \, b^{\dagger n} \, b^m \right| z_2 \right\rangle = \left(\sum_{m,n} a_{mn} \, z_1^{*m} \, z_2^n \right) e^{z_1^* z_2}. \tag{13.87}
$$

- We can express the trace of operator A in terms of coherent states, namely,

$$
\mathrm{Tr} \, A = \sum_n \langle n \, |A| \, n\rangle = \int \frac{dz^* \, dz}{\pi} \, e^{-|z|^2} \sum_n \langle n \, | \, z\rangle \langle z \, |A| \, n\rangle
$$

$$
= \int \frac{dz^* \, dz}{\pi} \, e^{-|z|^2} \, \langle z| \, A \sum_n |n\rangle \langle n \, | \, z\rangle
$$

$$
= \int \frac{dz^* \, dz}{\pi} \, e^{-|z|^2} \, \langle z \, |A| \, z\rangle . \tag{13.88}
$$

- We can expand a Fock ket in terms of coherent states, as

$$|\psi\rangle = \int \frac{dz^* \, dz}{\pi} \, |z\rangle \, \langle z| \, \psi\rangle = \int \frac{dz^* \, dz}{\pi} \, \phi(z^*) \, |z\rangle. \qquad (13.89)$$

- Action of $b^\dagger$ and b in coherent state representation is as follows:

$$\left\langle z \left| b^\dagger \right| \psi \right\rangle = z^* \, \psi\left(z^*\right) \qquad \Rightarrow \qquad b^\dagger \to z^*$$

$$\langle z \, |b| \, \psi\rangle = \frac{\partial}{\partial z^*} \, \psi\left(z^*\right) \qquad \Rightarrow \qquad b \to \frac{\partial}{\partial z^*}. \qquad (13.90)$$

The coherent state variables satisfy the bosonic commutators

$$\left[z_i^*, z_j^*\right] = \left[\frac{\partial}{\partial z_i^*}, \frac{\partial}{\partial z_j^*}\right] = 0$$

$$\left[\frac{\partial}{\partial z_i^*}, z_j^*\right] = \delta_{ij}.$$

- Consider coherent state representation of the Schrödinger equation:
  The Schrödinger equation for a Hamiltonian $\mathcal{H}\left(b^\dagger, b\right)$ is given by

$$\mathcal{H}\left(b^\dagger, b\right) \, |\psi\rangle = E \, |\psi\rangle.$$

Acting from the left by $\langle z|$, we get

$$\mathcal{H}\left(z^*, \frac{\partial}{\partial Z^*}\right) \psi\left(z^*\right) = E \, \psi\left(z^*\right), \qquad (13.91)$$

which, for a typical many-body Hamiltonian, reads

$$\sum_{ij} \varepsilon_{ij} \, z_i^* \frac{\partial}{\partial z_j^*} + \frac{1}{2} \sum_{ijkl} \langle ij| \, V \, |kl\rangle \, z_i^* \, z_j^* \frac{\partial}{\partial z_l^*} \frac{\partial}{\partial z_k^*}. \qquad (13.92)$$

13.5.2 Fermionic Coherent States

Coherent states for fermions can be defined in a similar way to bosonic ones. However, when we consider fermionic systems, things get more complicated. The eigenstates of a fermionic annihilation operator should satisfy the equation

$$c \, |\xi\rangle = \xi \, |\xi\rangle.$$

Since for fermions we can have only two possible state occupations, $|\xi\rangle = \alpha \, |0\rangle + \beta \, |1\rangle$, we find that

$$c \, |\xi\rangle = \beta \, |0\rangle = \xi \, (\alpha \, |0\rangle + \beta \, |1\rangle),$$

which implies that the eigenket of c is $|0\rangle$ with eigenvalue 0. To obtain nontrivial fermionic coherent states, we should start with considering the anticommutators of annihilation operators. For two operators c_i and c_j with coherent eigenstates

$$c_i \, |\boldsymbol{\xi}\rangle = \xi_i \, |\boldsymbol{\xi}\rangle, \quad c_j \, |\boldsymbol{\xi}\rangle = \xi_j \, |\boldsymbol{\xi}\rangle,$$

we find that

$$\{c_i, c_j\} = 0 \quad \Rightarrow \quad \{\xi_i, \xi_j\} = 0, \tag{13.93}$$

which also implies that $\xi_i^2 = 0$!. Clearly, the quantities ξ are no ordinary numbers. As a matter of fact, they turn out to satisfy what is mathematically known as Grassmann algebra, and such objects are identified as Grassmann numbers.

Grassmann Algebra

Grassmann algebra and calculus can be viewed as a clever construct that can manipulate sign changes arising from the inherent anticommutations of its variables. A Grassmann algebra is defined by the set of its generating elements $\{1, \xi_1, \bar{\xi}_1, \xi_2, \bar{\xi}_2, \dots, \xi_n, \bar{\xi}_n\}$ that satisfy

$$\{\xi_i, \xi_j\} = \{\bar{\xi}_i, \bar{\xi}_j\} = \{\xi_i, \bar{\xi}_j\} = \xi_i^2 = \bar{\xi}_i^2 = 0$$

$$\{\xi_i, c_j\} = \{\bar{\xi}_i, c_j\} = \{\xi_i, c_j^\dagger\} = \{\bar{\xi}_i, c_j^\dagger\} = 0. \tag{13.94}$$

The basis set of Grassmann algebra consists of all distinct products of its generators, namely, $\{1, \xi_1, \dots, \xi_1 \xi_2, \dots, \xi_1 \xi_2 \cdots \xi_n\}$.

For a Grassmann algebra with two generators $\{\xi, \bar{\xi}\}$, we find that all analytic functions of a Grassmann variable reduce to a first-degree polynomial,

$$\psi(\xi) = \psi_0 + \psi_1 \, \xi,$$

where ψ_0, Ψ_1 are complex numbers. We can define a complex conjugation as

$$\overline{\psi(\xi)} = \bar{\psi}_0 + \bar{\psi}_1 \, \bar{\xi}.$$

We can also define functions of two Grassmann variables as

$$F(\xi, \bar{\xi}) = f_0 + f_1 \, \xi + \bar{f}_1 \, \bar{\xi} + f_{12} \bar{\xi} \, \xi,$$

where, again, a_0, a_1, a_{12} are complex numbers, but a_1 and $\bar{a}_1$ need not be mutual conjugates.

Grassmann Calculus

1. Differentiation of Grassmann variables:

   Since analytic functions of Grassmann variables have such a simple structure, differentiation is just as simple. Indeed, we define the derivative as the coefficient of the linear term

$$\partial_\xi \psi(\xi) = \psi_1, \quad \partial_{\bar\xi} \overline{\psi(\xi)} = \bar\psi_1$$

$$\partial_\xi \left(\bar\xi\,\xi\right) = -\bar\xi, \tag{13.95}$$

where in the last line we moved ξ to the left of $\bar\xi$. We also obtain

$$\partial_\xi F(\xi,\bar\xi) = f_1 - f_{12}\bar\xi$$

$$\partial_{\bar\xi} F(\xi,\bar\xi) = f_1 + f_{12}0\,\xi$$

$$\partial_{\bar\xi}\partial_\xi F(\xi,\bar\xi) = -f_{12} = -\partial_\xi\partial_{\bar\xi} F(\xi,\bar\xi), \tag{13.96}$$

and we find that

$$\left\{\partial_{\bar\xi},\, \partial_\xi\right\} = 0. \tag{13.97}$$

The chain rule for differentiation then reads

$$\partial_\xi F(\psi) = \partial_\xi \psi\, \partial_\psi F. \tag{13.98}$$

Contrary to ordinary variables, the order of the terms in the rhs matters.

2. Integration over Grassmann variables:

Integration and differentiation are identical for Grassmann variables

$$\partial_{\xi_i} F(\xi_1,\dots,\xi_n) = \int d\xi_i\, F(\xi_1,\dots,\xi_n). \tag{13.99}$$

This property ensures that two fundamental properties of ordinary integrals over functions vanishing at infinity are satisfied:

- The integral of an exact differential form is zero:

$$\int d\xi_i\, \partial_{\xi_i} F(\xi_1,\dots,\xi_n) = 0. \tag{13.100}$$

- The integral over ξ_i of $F(\xi_1,\dots,\xi_n)$ does not depend on ξ_i so that its derivative vanishes:

$$\partial_{\xi_i} \int d\xi_i\, F(\xi_1,\dots,\xi_n) = 0. \tag{13.101}$$

Both properties (13.100) and (13.101) follow from the definition (13.99) and the nilpotence of the differential operator ∂_ξ. Another consequence of (13.99) is

$$\int d\xi\, 1 = 0.$$

For a Grassmann algebra with a single generator, we have

$$F(\xi) = c_0 + c_1\xi \quad \Rightarrow \quad \int d\xi\, F(\xi) = c_1 = \partial_\xi F(\xi). \tag{13.102}$$

Properties of Fermionic Coherent States

Grassmann variables can now be used to define coherent states analogs of the ones we have delineated for bosons. We define a fermionic coherent state as

$$|\xi\rangle = \mathcal{N} e^{-\xi c^\dagger} |\emptyset\rangle, \qquad \langle\xi| = \langle\emptyset| e^{\bar{\xi} c} \mathcal{N} \tag{13.103}$$

since

$$c\,|\xi\rangle = c\,e^{-\xi c^\dagger} |\emptyset\rangle = c\left(1 - \xi c^\dagger\right)|\emptyset\rangle = \xi\,|\emptyset\rangle = \xi\left(1 - \xi c^\dagger\right)|\emptyset\rangle. \tag{13.104}$$

$\mathcal{N}$ is a normalization constant. We also find

$$c^\dagger\,|\xi\rangle = c^\dagger e^{-\xi c^\dagger} |\emptyset\rangle = c^\dagger\left(1 - \xi c^\dagger\right)|\emptyset\rangle = c^\dagger\,|\emptyset\rangle = -\partial_\xi\left(1 - \xi c^\dagger\right)|\emptyset\rangle = -\partial_\xi\,|\xi\rangle. \tag{13.105}$$

For a Grassmann variable η, we obtain the inner product

$$\langle\xi\,|\,\eta\rangle = \langle\emptyset| e^{\bar{\xi} c}\, e^{-\eta c^\dagger} |\emptyset\rangle = 1 + \bar{\xi}\eta = e^{\bar{\xi}\eta}. \tag{13.106}$$

When $\eta = \xi$, we obtain a normalization factor of $\mathcal{N} = e^{-\bar{\xi}\xi/2}$. For n fermions, we have

$$|\boldsymbol{\xi}\rangle = |\xi_1\dots\xi_n\rangle = \prod_{i=1}^{n} e^{-\bar{\xi}_i \xi_i/2}\, e^{-\xi_i c_i^\dagger} |\emptyset\rangle = e^{-\sum_i \bar{\xi}_i \xi_i/2}\, e^{-\sum_i \xi_i c_i^\dagger} |\emptyset\rangle \tag{13.107}$$

where we used the commutator

$$\left[\xi_i c_i^\dagger,\, \xi_j c_j^\dagger\right] = 0.$$

The resolvent identity is given by

$$\int d\bar{\xi}\,d\xi\,\,|\xi\rangle\,\langle\xi| = \int d\bar{\xi}\,d\xi\,e^{-\bar{\xi}\xi/2}\left(1 - \xi c^\dagger\right)|\emptyset\rangle\,\langle\emptyset|\left(1 + \bar{\xi} c\right)\,e^{-\bar{\xi}\xi/2}$$

$$= \int d\bar{\xi}\,d\xi\left(1 - \frac{1}{2}\bar{\xi}\xi\right)\left(1 - \xi c^\dagger\right)|\emptyset\rangle\,\langle\emptyset|\left(1 + \bar{\xi} c\right)\left(1 - \frac{1}{2}\bar{\xi}\xi\right)$$

$$= \int d\bar{\xi}\,d\xi\left[\left(1 - \frac{1}{2}\bar{\xi}\xi\right)|\emptyset\rangle - \xi\,|1\rangle\right]\left[\langle\emptyset|\left(1 - \frac{1}{2}\bar{\xi}\xi\right) + \langle 1|\bar{\xi}\right] = \mathbb{I}. \tag{13.108}$$

This can be generalized to

$$\mathbb{I} = \int \prod_{i=1}^{n} d\bar{\xi}_i\,d\xi_i\,\,|\boldsymbol{\xi}\rangle\,\langle\boldsymbol{\xi}|$$

$$|\boldsymbol{\xi}\rangle = |\xi_1\dots\xi_n\rangle. \tag{13.109}$$

Its proof requires the use of Grassmann variables permutations.

We can use (13.109) to expand a ket $|\psi\rangle$ in Fock space as

$$|\psi\rangle = \int \prod_{i=1}^{n} d\bar{\xi}_i\,d\xi_i\,\,\psi(\boldsymbol{\xi})\,|\boldsymbol{\xi}\rangle. \tag{13.110}$$

We can then determine the following matrix elements

$$\langle \boldsymbol{\xi} | c_i | \psi \rangle = \partial_{\bar{\xi}_i} \psi(\boldsymbol{\xi}), \quad \langle \boldsymbol{\xi} | c_i^\dagger | \psi \rangle = \bar{\xi}_i \, \psi(\boldsymbol{\xi}) \tag{13.111}$$

and the matrix element of a normal-ordered operator $A(\{c_i^\dagger\}, \{c_i\})$ becomes

$$\langle \boldsymbol{\xi} | A | \boldsymbol{\xi}' \rangle = e^{-(|\boldsymbol{\xi}|^2 + |\boldsymbol{\xi}'|^2)/2} \, e^{\sum_i \bar{\xi}_i \xi_i'} \, A(\{\bar{\xi}_i\}, \{\xi_i'\}). \tag{13.112}$$

In the case of the fermion number operator $N = \sum_i \, c_i^\dagger c_i$ gives the expectation value

$$\frac{\langle \boldsymbol{\xi} | N | \boldsymbol{\xi} \rangle}{\langle \boldsymbol{\xi} | \boldsymbol{\xi} \rangle} = \sum_i \bar{\xi}_i \xi_i.$$

13.5.3 Gaussian Integrals

When we develop the many-body path integral formalism, we will frequently encounter integrals involving exponentials of complex or Grassmann variables. For actions having quadratic forms, the integrals are just generalizations of the simple Gaussian integrals. We will derive several useful such integrals here.

1. Real variables:

   The integral has the general form

$$Z(J) = \int dx_1 \ldots dx_n \, e^{-\frac{1}{2} \sum_{i,j+1}^n x_i A_{ij} x_j + \sum_i J_i x_i} = \int d\mathbf{x} \, e^{-\frac{1}{2} \mathbf{x}^T \mathbf{A} \mathbf{x} + \mathbf{J}^T \mathbf{x}}.$$

   The n-dimensional real matrix $\mathbf{A}$ is assumed to be symmetric and positive definite. It can be diagonalized by an orthogonal transformation: $\mathbf{A} = \mathbf{O}^T \mathbf{D} \mathbf{O}$, with $\mathbf{O}$ an orthogonal matrix ($\mathbf{O}^T \mathbf{O} = \mathbf{O}\mathbf{O}^T = 1$) and $\mathbf{D}$ a diagonal matrix with nonnegative elements ($D_{ii} > 0$). With the change of variables $\mathbf{y} = \mathbf{O}(\mathbf{x} - \mathbf{A}^{-1}\mathbf{J})$ (Jacobian $|\det \mathbf{O}| = 1$), we obtain

$$-\frac{1}{2} \mathbf{x}^T \mathbf{A} \mathbf{x} + \mathbf{J}^T \mathbf{x} = -\frac{1}{2} \mathbf{y}^T \mathbf{D} \mathbf{y} + \frac{1}{2} \mathbf{J}^T \mathbf{A}^{-1} \mathbf{J}$$

   and $Z(J)$ reduces to a product of Gaussian integrals, [2]

$$Z(J) = e^{\frac{1}{2} \mathbf{J}^T \mathbf{A}^{-1} \mathbf{J}} \prod_{i=1}^n \int_{-\infty}^\infty dy \, e^{-y^2 D_{ii}/2}$$

$$= e^{\frac{1}{2} \mathbf{J}^T \mathbf{A}^{-1} \mathbf{J}} \prod_{i=1}^n \sqrt{\frac{2\pi}{D_{ii}}}$$

$$= e^{\frac{1}{2} \mathbf{J}^T \mathbf{A}^{-1} \mathbf{J}} (2\pi)^{n/2} (\det \mathbf{A})^{1/2}. \tag{13.113}$$

---

[2] Recall that $\int_{-\infty}^\infty dy \, e^{-ay^2/2} = \sqrt{2\pi/a}$.

2. Complex variables:

$$Z(\mathbf{J}^*, \mathbf{J}) = \int_{-\infty}^{\infty} \prod_i \frac{dz_i^* \, dz_i}{2\pi i} \, e^{-\sum_{ij} z_i^* A_{ij} z_j + \sum_i (J_i^* z_i + cc)}$$

$$= \int_{-\infty}^{\infty} \prod_i \frac{dz_i^* \, dz_i}{2\pi i} \, e^{-\mathbf{z}^\dagger \mathbf{A} \mathbf{z} + (\mathbf{J}^\dagger \mathbf{z} + cc)},$$

where $\mathbf{A}$ is positive definite and Hermitian, with $\mathbf{D} = U^\dagger \mathbf{A} U$. Again, we change variables

$$\mathbf{z}' = U\left(z - \mathbf{A}^{-1}\mathbf{J}\right), \quad \mathbf{z}'^\dagger = \left(z^\dagger - \mathbf{J}^\dagger \mathbf{A}^{-1}\right),$$

which yields [3]

$$Z(\mathbf{J}^*, \mathbf{J}) = e^{\mathbf{J}^\dagger \mathbf{A}^{-1} \mathbf{J}} \int_{-\infty}^{\infty} \prod_i \frac{dz_i^* \, dz_i}{2\pi i} \, e^{-\mathbf{z}'^\dagger \mathbf{D} \mathbf{z}'} = e^{\mathbf{J}^\dagger \mathbf{A}^{-1} \mathbf{J}} \, (\det A)^{-1}. \quad (13.114)$$

3. Grassmannian integrals:

A Gaussian integral involving a conjugate pair of Grassmann variables takes the form

$$\int d\bar{\xi} \, d\xi \, e^{-\bar{\xi} a \xi} = \int d\bar{\xi} \, d\xi \, \left(1 - \bar{\xi} a \xi\right) = a. \quad (13.115)$$

By analogy with the previous Gaussian integrals, we may write the Grassmann Gaussian integral as

$$Z(\bar{\zeta}, \zeta) = \int \prod_i d\bar{\xi}_i \, d\xi_i \, e^{-\sum_{ij} \bar{\xi}_i A_{ij} \xi_j + \sum_i (\zeta^* \xi_i + cc)}. \quad (13.116)$$

Thus, if we can bring this Grassmann integral into a diagonal form, we may be able to reduce the integral to det A in the numerator.

Transformation of Grassmann Variables

We consider the integral

$$\mathcal{I} = \int d\xi \, F(\xi)$$

subject to the transformation $\xi \to \alpha \xi' + \beta$, where α and β are complex numbers. Using (13.99) and the chain rule, we get

$$\mathcal{I} = \partial_\xi F(\xi) = \partial_\xi(\xi') \, \partial_{\xi'} F = \frac{1}{\alpha} \partial_{\xi'} F = \frac{1}{a} \int d\xi' \, F\left(\alpha \xi' + \beta\right)$$

---

[3] We have $\int_{-\infty}^{\infty} \prod_i \frac{dz^* \, dz}{2\pi i} \, e^{-a|z|^2/2} = a^{-1}$.

so that the Jacobian of the transformation is α^{-1} instead of α for ordinary transformations. This result can be generalized to a multidimensional integral,

$$\mathcal{I} = \int \prod_i d\xi_1 \ldots d\xi_n \, F(\xi_1, \ldots, \xi_n) = \prod_i \partial_{\xi_i} F(\xi_1, \ldots, \xi_n)$$

$$= \prod_i \left(\sum_j \left(\partial_{\xi_i} \xi_j' \right) \partial_{\xi_j'} \right) F(\xi_1, \ldots, \xi_n). \tag{13.117}$$

For a linear transformation $\xi_i' = \sum_j M_{ij} \xi_j$, we get

$$\mathcal{I} = \prod_i \left(\sum_j M_{ji} \, \partial_{\xi_j'} \right) F(\xi_1, \ldots, \xi_n).$$

Since the differential operators $\partial_{\xi_j'}$ anticommute, the differential operator acting on $F(\xi_1, \ldots, \xi_n)$ is proportional to $\partial_{\xi_1'} \ldots \partial_{\xi_n'}$, and the prefactor is easily seen to be the determinant of the matrix $\mathbf{M}$,

$$\mathcal{I} = \det \left(\partial_{\xi_j} \xi_i' \right) \prod_i \partial_{\xi_i'} F(\xi_1, \ldots, \xi_n)$$

so that

$$\prod_i d\xi_i = \det \left(\partial_{\xi_j} \xi_i' \right) \prod_i \partial_{\xi_i'}. \tag{13.118}$$

The Jacobian is the inverse of the determinant $\det \left(\partial_{\xi_j'} \xi_i \right)$.

Equation (13.118) can now be used to determine the Grassmannian integral

$$Z = \int \prod_i d\bar{\xi}_i \, d\xi_i \; e^{-\sum_{ij} \bar{\xi}_i^* A_{ij} \xi_j},$$

using the linear transformation $\xi_i' = \sum_j A_{ij} \xi_j$, and obtain

$$Z = \det \mathbf{A} \int \prod_i d\bar{\xi}_i \, d\xi_i' \; e^{-\sum_i \bar{\xi}_i \xi_i'} = \det \mathbf{A} \prod_i \int d\bar{\xi}_i \, d\xi_i' \left(1 - \bar{x}i_i \, \xi_i' \right) = \det \mathbf{A}. \tag{13.119}$$

Or, for the more general integral in (13.116), we get

$$Z(\bar{\zeta}, \zeta) = \det \mathbf{A} \, e^{\bar{\zeta} \mathbf{A}^{-1} \zeta}. \tag{13.120}$$

Exercises

13.1 In this question, $c_i^\dagger$ and c_i are fermion creation and annihilation operators and the states are fermion states. Use the convention $|11111000\ldots\rangle = c_5^\dagger\, c_4^\dagger\, c_3^\dagger\, c_2^\dagger\, c_1^\dagger\, |\varnothing\rangle$.

   (i) Evaluate $c_3^\dagger\, c_6^\dagger\, c_6\, c_4\, c_6^\dagger\, c_3\, |11111\ldots\rangle$.

   (ii) Write $|1101100100\ldots\rangle$ in terms of excitations about the filled Fermi sea $|1111100000\ldots\rangle$. Interpret your answer in terms of electron and hole excitations.

   (iii) Find $\langle\psi|\,\widehat{N}\,|\psi\rangle$, where $|\psi\rangle = A\,|100\rangle + B\,|111000\rangle$, where $\widehat{N} = \sum_i c_i^\dagger c_i$.

13.2 Derive expressions for the spin, density, and current density operators of a spin-1/2 system in the plane-wave representation.

13.3 At finite T, the occupancy of state μ_i is given by

$$\left\langle a_\mu^\dagger\, a_\mu \right\rangle = \frac{\mathrm{Tr}\left[e^{-\beta\mathcal{H}}\, a_\mu^\dagger\, a_\mu\right]}{\mathrm{Tr}\left[e^{-\beta\mathcal{H}}\right]},$$

where a_μ can either be fermionic c_μ or bosonic b_μ operators. Use the simple Hamiltonian (13.57) to show that for fermions

$$\left\langle c_\mu^\dagger\, c_\mu \right\rangle == \frac{1}{1 + e^{\beta\,\mathcal{E}_\mu}}, \qquad \text{Fermi factor.}$$

and for bosons

$$\left\langle b_\mu^\dagger\, b_\mu \right\rangle == \frac{1}{e^{\beta\,\mathcal{E}_\mu} - 1}, \qquad \text{Bose factor}$$

Hint: Since $\left[a_\mu^\dagger\, a_\mu,\, a_\nu\right] = 0$, use the Baker–Hausdorff identity

$$e^{-\beta\mathcal{H}} = \prod_{\mu=1}^{N} e^{-\beta\,\mathcal{E}_\mu\, a_\mu^\dagger\, a_\mu}.$$

13.4 Consider a problem where bosons may accumulate on a site. The Hamiltonian is given by

$$\mathcal{H} = \varepsilon\, b^\dagger\, b + t\left(b^{\dagger 2} + b^2\right).$$

The energy of a boson on the site is ε, while the nature of the problem is such that the bosons appear or disappear in pairs, with probability t. Notice that the total number of bosons in this problem is not conserved.

Employ the unitary (Bogoliubov) transformation

$$a = u\, b + v\, b^\dagger$$
$$a^\dagger = u\, b^\dagger + v\, b$$

to diagonalize the Hamiltonian.

(a) Ensure that the boson commutation relation is satisfied for the new operators $a^\dagger$, a. What condition on u and v do you obtain, and what form can they assume?

(b) Show that a Hamiltonian of the form

$$\mathcal{H} = \omega\, a^\dagger a + \Omega$$

can be obtained. Express ω and Ω in terms of ε and t.

(c) The ground state of the system corresponds to the absence of particles of type a. But that does not mean there are no b-particles there. Indeed, in the ground state $\langle a\, a^\dagger \rangle = 1$ (while of course $\langle a^\dagger a \rangle = 0$ and $\langle a\, a \rangle = 0$), and

$$\langle b^\dagger b \rangle = \sinh^2(\phi),$$

where ϕ is the hyperbolic angle. Use the expression for ϕ in terms of t and ε to find the number of particles in the ground state.

(c) What happens when $t = \varepsilon/2$? Give a physical explanation of this result (try to express the Hamiltonian in terms of x and p).

13.5 The effective Hamiltonian

$$\mathcal{H} = \varepsilon_c\, c^\dagger c + \varepsilon_d\, d^\dagger d - \Delta\, c\, d - \Delta^*\, c^\dagger d^\dagger$$

contains fermions in two kinds of states c and d. We would like to diagonalize this Hamiltonian in the form

$$\mathcal{H} = \mathcal{E}_\alpha\, \alpha^\dagger \alpha + \mathcal{E}_\beta\, \beta^\dagger \beta + \mathcal{E}_0$$

by introducing quasiparticle operators α, β through the Bogoliubov transformation

$$\begin{cases} c^\dagger &= u^*\,\alpha^\dagger + v\,\beta \\ \beta^\dagger &= u^*\,\beta^\dagger - v\,\alpha \end{cases} \qquad u,\, v \in \mathbb{C}.$$

(a) Show that the coefficients have to fulfill $|u|^2 + |v|^2 = 1$.

(b) Express $\mathcal{H}$ in terms of α and β, and determine u and v that diagonalize $\mathcal{H}$.
  Hint: Introduce new variables ϕ and θ by setting $u = \cos\theta$, $v = e^{i\phi} \sin\theta$, and determine their form.

(c) Determine the energy spectrum of the new quasiparticles in the special case $\varepsilon_c = \varepsilon_d = \varepsilon$, and u, v real.

(d) Discuss the meaning of $\mathcal{E}_\alpha$, $\mathcal{E}_\beta$, and $\mathcal{E}_0$.

13.6 The classical Lagrangian for a one-dimensional diatomic chain is given by

$$\mathcal{L} = \sum_j \left\{ \frac{1}{2} m_\phi\, \dot\phi^2 + \frac{1}{2} m_\psi\, \dot\psi^2 - \frac{\kappa}{2} \left[(\phi_j - \psi_j)^2 + (\phi_j - \psi_{j-1})^2 \right] \right\}.$$

(i) Determine and draw the dispersion curves.

(ii) What is the gap in the excitation spectrum?

(iii) Write the diagonalized Hamiltonian in second-quantized form and give a detailed account of how you arrived at your final answer. You will now need two types of creation operator. Hint: Diagonalize the first-quantized Hamiltonian before you second quantize.

13.7 Perturbational canonical transformation:

Consider a Hamiltonian $\mathcal{H}$ and the canonical transformation $\tilde{\mathcal{H}} = e^{-S} \mathcal{H} e^{S}$, with $S^{\dagger} = -S$.

(a) By expanding in a power series and collecting terms, show that

$$\tilde{\mathcal{H}} = \mathcal{H} + [\mathcal{H}, S] + \frac{1}{2} [[\mathcal{H}, S], S] + \cdots$$

(b) Now take $\mathcal{H} = \mathcal{H}_0 + \lambda \mathcal{H}'$, with $\mathcal{H}_0$ diagonal. Show that if we choose S such that

$$\lambda \mathcal{H}' + [\mathcal{H}_0, S] = 0,$$

then $S \propto \lambda$, and

$$\tilde{\mathcal{H}} = \mathcal{H}_0 + \frac{1}{2} [\lambda \mathcal{H}', S] + O(\lambda^3) + \cdots$$

(c) Show that the matrix elements of S are given by

$$\langle n | S | m \rangle = \lambda \frac{\langle n | \mathcal{H}' | m \rangle}{E_m - E_n}.$$

13.8 Phonons interacting with localized particle:

A system of phonons interacting with a localized level that can contain a single (spinless) fermionic particle is described by

$$\mathcal{H} = \varepsilon_0 c^{\dagger} c + \sum_{\mathbf{q}} \mathfrak{M}_{\mathbf{q}} \left(b_{\mathbf{q}} + b_{\mathbf{q}}^{\dagger} \right) c^{\dagger} c + \sum_{\mathbf{q}} \omega_{\mathbf{q}} b_{\mathbf{q}}^{\dagger} b_{\mathbf{q}}.$$

where $\mathfrak{M}_{\mathbf{q}}$ is the fermion–phonon coupling matrix element, with $\mathfrak{M}_{\mathbf{q}} = \mathfrak{M}_{-\mathbf{q}}$ to be real; $b_{\mathbf{q}}$ and $b_{\mathbf{q}}^{\dagger}$ destroy and create phonons in state $\mathbf{q}$, and c and $c^{\dagger}$ destroy and create fermions in the localized state with energy ε_0. Use the following steps to determine the energy spectrum of this Hamiltonian.

(a) Consider the canonical transformation

$$\mathcal{H} \to \tilde{\mathcal{H}} = e^{S} \mathcal{H} e^{-S}$$

$$S = c^{\dagger} c \sum_{\mathbf{q}} \frac{\mathfrak{M}_{\mathbf{q}}}{\omega_{\mathbf{q}}} \left(b_{\mathbf{q}}^{\dagger} - b_{-\mathbf{q}} \right).$$

(b) Use the following properties of canonical transformations:

   i. Eigenvalues of Hermitian operators are invariant.
   ii. The transform of product is the product of transforms.

(c) Try to put transformed H in the form

$$\mathcal{H} = (\varepsilon_0 + \Sigma)\, c^\dagger c + \sum_\mathbf{q} \omega_\mathbf{q}\, b_\mathbf{q}^\dagger b_\mathbf{q}$$

and determine the self-energy Σ.
(d) The Hamiltonian is now a sum of independent diagonal quadratic forms. Read off the eigenvalues

13.9   Consider a charged oscillator in an external electric field (charge e = 1):

$$\mathcal{H} = \frac{p^2}{2m} + \frac{m\omega_0^2}{2} x^2 + Ex.$$

(a) Express the Hamiltonian in second-quantized form using $x; p \rightarrow a^\dagger a$ for the simple harmonic oscillator.
(b) Consider the canonical transformation $\mathcal{H} = e^S \mathcal{H} e^{-S}$ with $S = \lambda \left(a - a^\dagger\right)$. Show that $S^\dagger = -S$ since $\mathcal{H}$ must be Hermitian.
(c) Use the expansion $\tilde{\mathcal{H}} = \mathcal{H} + [\mathcal{H}, S] + \ldots$ to find $\tilde{\mathcal{H}}$.
(d) Find the choice of λ that makes $\tilde{\mathcal{H}}$ diagonal. Write down the resulting $\tilde{\mathcal{H}}$, including the constant energy term.
(e) Express "old" creation-annihilation operators in terms of the "new" ones. Can you use that as an alternative canonical transformation?

13.10 You have shown in Exercise (6.6) that the dynamic structure factor can be expressed as

$$S(\mathbf{Q}, \omega) - \frac{1}{N} \int_{-\infty}^{\infty} \frac{dt}{2\pi}\, e^{i\omega t} \sum_{ll'} e^{i\mathbf{Q}\cdot(\mathbf{R}_l - \mathbf{R}_{l'})} \left\langle \Phi_i \left| e^{i\mathbf{Q}\cdot\mathbf{u}_{l'}}\, e^{-i\mathbf{Q}\cdot\mathbf{u}_l(t)} \right| \Phi_i \right\rangle.$$

where $\mathbf{Q}$ is the momentum transfer. At finite temperatures, one has to thermal-average over all initial states so that

$$S(\mathbf{Q}, \omega) = \frac{1}{N} \int_{-\infty}^{\infty} \frac{dt}{2\pi}\, e^{i\omega t} \sum_{ll'} e^{i\mathbf{Q}\cdot(\mathbf{R}_l - \mathbf{R}_{l'})} \left\langle e^{i\mathbf{Q}\cdot\mathbf{u}_{l'}(0)}\, e^{-i\mathbf{Q}\cdot\mathbf{u}_l(t)} \right\rangle_T,$$

where

$$\left\langle e^{i\mathbf{Q}\cdot\mathbf{u}_{l'}(0)}\, e^{-i\mathbf{Q}\cdot\mathbf{u}_l(t)} \right\rangle_T = \frac{\mathrm{Tr}\left[e^{-\beta\mathcal{H}}\, e^{i\mathbf{Q}\cdot\mathbf{u}_{l'}}\, e^{-i\mathbf{Q}\cdot\mathbf{u}_l(t)} \right]}{\mathrm{Tr}\, e^{-\beta\mathcal{H}}}.$$

(a) Use the identity

$$\left\langle e^{\hat{A}}\, e^{\hat{B}} \right\rangle_T = e^{\left\langle \hat{A}^2 + 2\hat{A}\hat{B} + \hat{B}^2 \right\rangle_T / 2},$$

where operators $\hat{A}$ and $\hat{B}$ are linear combinations of $\hat{a}$ and $\hat{a}^\dagger$, together with spatial and temporal translation invariance, to show that the dynamic structure factor $S(\mathbf{Q}, \omega)$ can be written as

$$S(\mathbf{Q}, \omega) = \frac{e^{-2W}}{N} \sum_l e^{i\mathbf{Q} \cdot \mathbf{R}_l} \int_{-\infty}^{\infty} \frac{dt}{2\pi} e^{i\omega t} e^{\langle\langle (\mathbf{Q} \cdot \mathbf{u}_0(0)) (\mathbf{Q} \cdot \mathbf{u}_l(t))\rangle\rangle_T},$$

where $W = \frac{1}{2} \langle (\mathbf{Q} \cdot \mathbf{u}_0)^2 \rangle_T$ is the Debye–Waller factor.

(b) To obtain the long-range modulations, take the term in the exponent, namely, the correlator

$$\langle\langle (\mathbf{Q} \cdot \mathbf{u}_0(0)) \; (\mathbf{Q} \cdot \mathbf{u}_l(t)) \rangle\rangle_T,$$

to be unity. Determine $S(\mathbf{Q}, \omega)$ under this high-correlation condition. This result implies that diffraction peaks are weighted by e^{-2W}.

(c) Calculate the Debye–Waller factor in d-dimensional space. (Hint: Expand $\mathbf{u}_0$ in terms of creation and annihilation operators of the phonon modes, and assume that the phonons have a linear dispersion, which is cut off by the Debye frequency.)

(d) Show that in one dimension W diverges for all temperatures. Is zero temperature different? In what way?

(e) Show that in two dimensions W diverges at finite temperatures.

(f) Do you identify any similarity between the one-dimensional case at zero temperature and the two-dimensional case at finite temperatures?

13.11 The Hamiltonian of the system of bosons (a, $a^\dagger$, b, and $b^\dagger$ are Bose operators) is

$$\mathcal{H} = \varepsilon_1 \, a^\dagger \, a + \varepsilon_2 \, b^\dagger \, b + \frac{\Delta}{2} \left(a^\dagger \, b^\dagger + b \, a \right),$$

where ε_1, ε_2, and Δ are real and positive, $\Delta < (\varepsilon_1 + \varepsilon_2)$.

Find canonical transformation diagonalizing this Hamiltonian. Write down explicit expressions for the eigenenergies and parameters of the transformation.

14

The Interacting Electron Gas

As we have previously pointed out, the jellium model, shown in Figure 14.1, is broadly considered a paradigm of condensed matter electronic systems. It is the simplest possible interacting model, and as such we will used it in this chapter to illustrate the breadth and limitations of applying conventional perturbation theory to the general treatment of interacting electronic systems.

14.1 The Jellium Model

We write the total Hamiltonian as

$$\mathcal{H} = \mathcal{H}_+ + \mathcal{H}_{+e} + \mathcal{H}_e$$

$$\mathcal{H}_+ = \frac{1}{2} \int d\mathbf{x}\, d\mathbf{x}'\, \rho_+(\mathbf{x})\rho_+(\mathbf{x}') \frac{e^{-\mu|\mathbf{x}-\mathbf{x}'|}}{|\mathbf{x}-\mathbf{x}'|} \qquad \text{(Positive background Hamiltonian)}$$

$$= \frac{1}{2} e^2 \left(\frac{N}{\Omega}\right)^2 \int d\mathbf{x}' \int d\mathbf{x}\, \frac{e^{-\mu r}}{r} = \frac{1}{2} e^2 \frac{N^2}{\Omega} \frac{4\pi}{\mu^2}$$

$$\mathcal{H}_{+e} = -e \sum_i \int d\mathbf{x}\, \rho_+(\mathbf{x}) \frac{e^{-\mu|\mathbf{x}-\mathbf{x}_i|}}{|\mathbf{x}-\mathbf{x}_i|} = -\frac{e^2 N}{\Omega} \sum_i \int d\mathbf{x}\, \frac{e^{-\mu|\mathbf{x}-\mathbf{x}_i|}}{|\mathbf{x}-\mathbf{x}_i|}$$

$$= -\frac{e^2 N^2}{\Omega} \frac{4\pi}{\mu^2} \qquad \text{(Electron–positive background Hamiltonian)}. \tag{14.1}$$

Next, we consider the electron Hamiltonian

$$\mathcal{H}_e = \sum_i \frac{\mathbf{p}_i^2}{2m_e} + \frac{1}{2} e^2 \sum_{i \neq j} \frac{e^{-\mu|\mathbf{x}_i-\mathbf{x}_j|}}{|\mathbf{x}_i - \mathbf{x}j|} \tag{14.2}$$

and recall that the single-electron wavefunctions are just plane waves. The single-particle kinetic energy is given by

$$T = \sum_i \frac{\hbar^2 k_i^2}{2m_e} c_i^\dagger c_i. \tag{14.3}$$

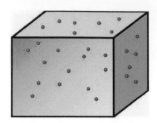

Figure 14.1 The jellium model consists of electrons, represented by dots, and a smeared positive background, depicted in gray.

and the matrix element for the two-electron repulsion is

$$
\left\langle \mathbf{k}_4\sigma_i, \mathbf{k}_3\sigma_j, \left| \frac{e^{-\mu|\mathbf{x}_i-\mathbf{x}_j|}}{|\mathbf{x}_i - \mathbf{x}j|} \right| \mathbf{k}_2\sigma_j, \mathbf{k}_1\sigma_i \right\rangle
$$

$$
= \frac{1}{\Omega^2} \int d\mathbf{x}_i\, \mathbf{x}_j\, e^{i(\mathbf{k}_1-\mathbf{k}_4)\cdot\mathbf{x}_i}\, e^{i(\mathbf{k}_2-\mathbf{k}_3)\cdot\mathbf{x}_j}\, \frac{e^{-\mu|\mathbf{x}_i-\mathbf{x}_j\mathbf{x}_i-\mathbf{x}_j|}}{|\mathbf{x}_i - \mathbf{x}j|}
$$

$$
= \frac{1}{\Omega} \delta_{\mathbf{k}_1+\mathbf{k}_2-\mathbf{k}_3-\mathbf{k}_4} \int d\mathbf{x}\, e^{-i\mathbf{q}\cdot\mathbf{x}} \frac{e^{-\mu r}}{r} = \frac{1}{\Omega} \delta_{\mathbf{k}_1+\mathbf{k}_2-\mathbf{k}_3-\mathbf{k}_4} \frac{4\pi}{\mu^2 + q^2}, \qquad (14.4)
$$

where $\mathbf{x} = \mathbf{x}_i - \mathbf{x}_j$, $r = |\mathbf{x}$. Substituting from (14.3) and (14.4) into (14.2), we get

$$
\mathcal{H}_e = \sum_{\mathbf{k},\sigma} \frac{\hbar^2 k^2}{2m_2} c^\dagger_{\mathbf{k},\sigma} c_{\mathbf{k},\sigma} + \frac{2\pi e^2}{\Omega} \sum_{\substack{\mathbf{k}\mathbf{k}',\mathbf{q} \\ \sigma\sigma'}} \frac{1}{\mu^2 + q^2} c^\dagger_{\mathbf{k}+\mathbf{q},\sigma} c^\dagger_{\mathbf{k}'-\mathbf{q},\sigma'} c_{\mathbf{k}',\sigma'} c_{\mathbf{k},\sigma}. \qquad (14.5)
$$

The term with $\mathbf{q} = \mathbf{0}$ in (14.5) will be treated separately here:

$$
\frac{2\pi e^2}{\mu^2\Omega} \sum_{\substack{\mathbf{k}\mathbf{k}' \\ \sigma\sigma'}} c^\dagger_{\mathbf{k},\sigma} c^\dagger_{\mathbf{k}',\sigma'} c_{\mathbf{k}',\sigma'} c_{\mathbf{k},\sigma} = \frac{2\pi e^2}{\mu^2\Omega} \sum_{\substack{\mathbf{k}\mathbf{k}' \\ \sigma\sigma'}} c^\dagger_{\mathbf{k},\sigma} c_{\mathbf{k},\sigma} \left(c^\dagger_{\mathbf{k}',\sigma'} c_{\mathbf{k}',\sigma'} - \delta_{\mathbf{k}\mathbf{k}'}\delta_{\sigma\sigma'} \right)
$$

$$
= \frac{2\pi e^2}{\mu^2\Omega} \left(N^2 - N \right). \qquad (14.6)
$$

From (14.1) and (14.6), we find that all the constant terms that contain $N^2 e^2/\mu$ add up to zero. The remaining term $-\frac{2\pi N e^2}{\mu^2\Omega}$ vanishes when $\Omega \to \infty$. The Hamiltonian is finally reduced to

$$
\mathcal{H} = \sum_{\mathbf{k},\sigma} \frac{\hbar^2 k^2}{2m_2} c^\dagger_{\mathbf{k},\sigma} c_{\mathbf{k},\sigma} + \frac{2\pi e^2}{\Omega} \sum_{\substack{\mathbf{k}\mathbf{k}',\mathbf{q} \\ \sigma\sigma'}}{}' \frac{1}{q^2} c^\dagger_{\mathbf{k}+\mathbf{q},\sigma} c^\dagger_{\mathbf{k}'-\mathbf{q},\sigma'} c_{\mathbf{k}',\sigma'} c_{\mathbf{k},\sigma} = \mathcal{H}_0 + \mathcal{H}', \qquad (14.7)
$$

where we set $\mu = 0$ since the singularity at $q = 0$ is now removed. $\mathcal{H}_0$ is the Hamiltonian for the *Sommerfeld gas*, discussed in detail in Section 1.3. We shall treat $\mathcal{H}'$ as a perturbation on $\mathcal{H}_0$. The prime over summation in $\mathcal{H}'$ indicates the exclusion of the $\mathbf{q} = \mathbf{0}$ term. Notice that in the second-quantization formalism, the ground state is expressed as a product of single particle wavefunctions and not as an antisymmetrized linear combination of such terms, since the antisymmetrization is now manifest in the creation and annihilation operator commutators. Thus, we write the jellium ground-state wavefunction as

$$|\Psi_0\rangle = \prod_{\mathbf{k}=0}^{k=k_F} |\mathbf{k} \uparrow\rangle \, |\mathbf{k} \downarrow\rangle .$$

14.1.1 The Wigner Crystal

In the small-density limit, the Coulomb energy becomes dominant, and therefore it is more appropriate to start from the ground state of the interaction term than from that of the kinetic energy. Wigner argued that the free electron–like Fermi sphere ground state could become unstable to an insulating lattice of localized electrons when the electron density becomes sufficiently small. We are then faced with a purely classical problem, namely to calculate the lowest-energy configuration of charged particles immersed into a homogeneous background of opposite charge. This problem was addressed by Wigner already in 1934 as that of an inverted alkali metal, and he argued that at low enough densities the electrons would form a crystal. A consistent theory has of course to take into account the kinetic energy, which leads to zero-point fluctuations of the electrons around their equilibrium positions. As the lattice constant decreases, these fluctuations become more and more important until the Wigner crystal melts. Numerical simulations indicate that this happens for $r_s \sim 100$.

A two-dimensional Wigner crystal with a triangular structure has actually been observed for a very low-density electron system ($r_s \sim 10^4$) dispersed over the surface of liquid helium.

14.1.2 First-Order Perturbation

The first-order correction to the energy is given by

$$E^{(1)} = \langle \Psi_0 \, | \mathcal{H}' | \, \Psi_0 \rangle = \left\langle \Psi_0 \left| \frac{2\pi e^2}{\Omega} \sum_{\substack{\mathbf{k}\mathbf{k}',\mathbf{q} \\ \sigma\sigma'}}' \frac{1}{q^2} c_{\mathbf{k}+\mathbf{q},\sigma}^{\dagger} \, c_{\mathbf{k}'-\mathbf{q},\sigma'}^{\dagger} \, c_{\mathbf{k}',\sigma'} c_{\mathbf{k},\sigma} \right| \Psi_0 \right\rangle . \tag{14.8}$$

Notice that the states $\mathbf{k} + \mathbf{q}$, $\mathbf{k}' - \mathbf{q}$, $\mathbf{k}'$, $\mathbf{k}$ must be contained in the ground-state wavefunction product; otherwise, the matrix element is zero. This means that the only terms that do not vanish will be those that *do not change the occupation numbers of* $|\Psi_0\rangle$, as shown in Figures 14.2 and 14.3, so we arrive at

$$\begin{cases} \mathbf{k} &= \mathbf{k} + \mathbf{q}, \qquad \mathbf{k}' = \mathbf{k}' - \mathbf{q}, \quad \mathbf{q} = 0 \\ \mathbf{k}' &= \mathbf{k} + \mathbf{q}, \qquad \mathbf{k} = \mathbf{k}' - \mathbf{q} \end{cases} \tag{14.9}$$

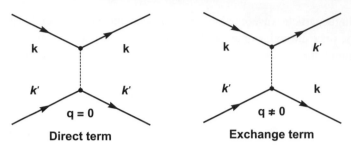

Figure 14.2 Diagrams for (14.9). The dashed line represents the interaction.

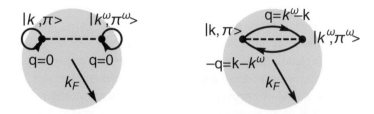

Figure 14.3 The two possible processes in first-order perturbation theory for two states $|\mathbf{k}\sigma\rangle$ and $|\mathbf{k}'\sigma'\rangle$ in the Fermi sea. The direct process having $\mathbf{q} = 0$ is already taken into account in the homogeneous part, hence only the exchange process contributes to the energy correction.

The first set implies that $\mathbf{q} = 0$, which is the case we considered in (14.6), and it does not involve any scattering. The remaining term, known as *exchange scattering*, and represented by the right diagrams in Figures 14.2 and 14.3, depicts the case where the particle that was in state $\mathbf{k}$ is scattered into the state $\mathbf{k}'$, and vice versa. The correction this gives to the energy E_0 is known as the *exchange energy*. We obtain

$$
E^{(1)} = \left\langle \Psi_0 \left| \frac{2\pi e^2}{\Omega} \sum_{\substack{\mathbf{k}\mathbf{k}' \\ \sigma}}' \frac{1}{q^2} c_{\mathbf{k}',\sigma}^{\dagger} c_{\mathbf{k},\sigma}^{\dagger} c_{\mathbf{k}',\sigma} c_{\mathbf{k},\sigma} \right| \Psi_0 \right\rangle
$$

$$
= \left\langle \Psi_0 \left| \frac{2\pi e^2}{\Omega} \sum_{\substack{\mathbf{k},\mathbf{q} \\ \sigma}}' \frac{1}{q^2} c_{\mathbf{k}+\mathbf{q},\sigma}^{\dagger} c_{\mathbf{k},\sigma}^{\dagger} c_{\mathbf{k}+\mathbf{q},\sigma} c_{\mathbf{k},\sigma} \right| \Psi_0 \right\rangle
$$

$$
= \left\langle \Psi_0 \left| \frac{2\pi e^2}{\Omega} \sum_{\substack{\mathbf{k},\mathbf{q} \\ \sigma}}' \frac{1}{q^2} \left(-N_{\mathbf{k}+\mathbf{q},\sigma} N_{\mathbf{k},\sigma} \right) \right| \Psi_0 \right\rangle. \tag{14.10}
$$

Notice that the correction to the energy of state $\mathbf{k}$ is

$$
\varepsilon^{(1)}(\mathbf{k}) = -\frac{2\pi e^2}{\Omega} \sum_{\mathbf{q}} \frac{N_{\mathbf{k}+\mathbf{q}}}{q^2}.
$$

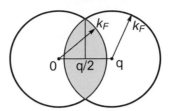

Figure 14.4 The geometry for the **k**-integration is shown for an arbitrary but fixed value of **q**. Integration domain is just the intersection of two spheres.

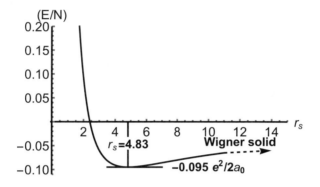

Figure 14.5 The ground-state energy per electron, including first-order correction, for the jellium model.

Next, we carry out the integration as follows:

$$
\begin{aligned}
E^{(1)} &= -\frac{4\pi e^2}{\Omega} \frac{\Omega^2}{(2\pi)^6} \int d\mathbf{k} \int d\mathbf{q} \frac{1}{q^2} \Theta\left(k_F - |\mathbf{k} + \mathbf{q}|\right) \Theta\left(k_F - k\right) \\
&= -\frac{4\pi e^2 \Omega}{(2\pi)^6} \int d\mathbf{q} \frac{1}{q^2} \int d\mathbf{p} \,\Theta\left(k_F - \left|\mathbf{p} + \frac{1}{2}\mathbf{q}\right|\right) \Theta\left(k_F - \frac{1}{2}\mathbf{q}\right).
\end{aligned} \tag{14.11}
$$

The integral over **p** is just the volume of intersection between two sphere each of radius k_F but separated by **q**, as shown in Figure 14.4. The result is

$$
E^{(1)} = -\frac{e^2}{2a_0} N \frac{3}{2\pi} \left(\frac{9\pi}{4}\right)^{1/3} \frac{1}{r_s}. \tag{14.12}
$$

The ground-state energy per electron is then given by

$$
\frac{E}{N} = \frac{e^2}{2a_0} \left(\frac{9\pi}{4}\right)^{1/3} \left(\frac{3}{5} \frac{1}{r_s^2} - \frac{3}{2\pi} \frac{1}{r_s}\right). \tag{14.13}
$$

E/N, in units of Rydberg, is plotted in Figure 14.5.

This result shows that the electron gas is stable when the repulsive Coulomb interaction is turned on. No external confinement potential is needed to hold the electron gas in the ion jellium together. There exists an optimal density n^*, or interparticle distance r_s^*, which minimizes the energy and furthermore yields an energy $E^* < 0$. The negative exchange energy overcomes the positive kinetic energy. This method of treating the electron gas is a special case of the *Hartree–Fock (HF) approximation*, and is the simplest way we can take the interactions into account. The interesting point to note is that in this approximation the energy is reduced *below* that of the Sommerfeld gas. This is just a manifestation of the exchange hole!

14.1.3 Hartree–Fock Approximation as Mean-Field Theory

We start with the electronic Hamiltonian

$$\mathcal{H} = \sum_{\mathbf{k},\sigma} \frac{\hbar^2 k^2}{2m_2} c_{\mathbf{k},\sigma}^\dagger c_{\mathbf{k},\sigma} + \frac{2\pi e^2}{\Omega} \sideset{}{'}\sum_{\substack{\mathbf{k}\mathbf{k}',\mathbf{q} \\ \sigma\sigma'}} \frac{1}{q^2} c_{\mathbf{k}+\mathbf{q},\sigma}^\dagger c_{\mathbf{k}'-\mathbf{q},\sigma'}^\dagger c_{\mathbf{k}',\sigma'} c_{\mathbf{k},\sigma}.$$

The goal is to write down an effective two-body Hamiltonian, which takes into account the average effects of the interactions. We therefore replace the four-Fermi interaction with a sum of all possible two-body terms:

$$\left\langle c_1^\dagger c_2^\dagger c_3 c_4 \right\rangle = -\left\langle c_1^\dagger c_3 \right\rangle c_2^\dagger c_4 - \left\langle c_2^\dagger c_4 \right\rangle c_1^\dagger c_3 + \left\langle c_1^\dagger c_4 \right\rangle c_2^\dagger c_3 + \left\langle c_2^\dagger c_3 \right\rangle c_1^\dagger c_4.$$

This can be thought of as mean-field terms, where

$$\left\langle c_{\mathbf{k}\sigma}^\dagger c_{\mathbf{k}'\sigma'} \right\rangle = n_{\mathbf{k}\sigma} \, \delta_{\mathbf{k}\mathbf{k}'} \, \delta_{\sigma\sigma'}$$

is the average number of particles $n_{\mathbf{k}\sigma}$ in the state $\mathbf{k}\sigma$, which will be weighted with the two-body interaction $V(\mathbf{q})$ to give the average interaction due to all other particles.

Substituting this mean-field approximation in the interaction term, we obtain

$$V_{\text{HF}} = \frac{1}{2} \sum_{\substack{\mathbf{k}\mathbf{k}'\mathbf{q} \\ \sigma\sigma'}} V(\mathbf{q}) \left[-\left\langle c_{\mathbf{k},\sigma}^\dagger c_{\mathbf{k}',\sigma'} \right\rangle c_{\mathbf{k}'-\mathbf{q},\sigma'}^\dagger c_{\mathbf{k}-\mathbf{q},\sigma} - \left\langle c_{\mathbf{k}'-\mathbf{q},\sigma'}^\dagger c_{\mathbf{k}-\mathbf{q},\sigma} \right\rangle c_{\mathbf{k},\sigma}^\dagger c_{\mathbf{k}',\sigma'} \right.$$

$$\left. + \left\langle c_{\mathbf{k},\sigma}^\dagger c_{\mathbf{k}-\mathbf{q},\sigma} \right\rangle c_{\mathbf{k}'-\mathbf{q},\sigma'}^\dagger c_{\mathbf{k}',\sigma'} + \left\langle c_{\mathbf{k}'-\mathbf{q},\sigma'}^\dagger c_{\mathbf{k}',\sigma'} \right\rangle c_{\mathbf{k},\sigma}^\dagger c_{\mathbf{k}-\mathbf{q},\sigma} \right]$$

$$= -\sum_{\mathbf{k}\mathbf{q}\sigma} V(\mathbf{q}) \left\langle c_{\mathbf{k},\sigma}^\dagger c_{\mathbf{k},\sigma} \right\rangle c_{\mathbf{k}-\mathbf{q},\sigma}^\dagger c_{\mathbf{k}-\mathbf{q},\sigma} + V(0) \sum_{\substack{\mathbf{k}\mathbf{k}' \\ \sigma\sigma'}} \left\langle c_{\mathbf{k},\sigma}^\dagger c_{\mathbf{k},\sigma} \right\rangle c_{\mathbf{k}',\sigma'}^\dagger c_{\mathbf{k}',\sigma'}$$

$$= \sum_{\mathbf{k}\sigma} \left(-\sum_{\mathbf{q}} n_{\mathbf{k}+\mathbf{q},\sigma} \, V(\mathbf{q}) + n V(0) \right) c_{\mathbf{k},\sigma}^\dagger c_{\mathbf{k},\sigma},$$

where the total density n is defined to be $n = \sum_{\mathbf{k}\sigma} n_{\mathbf{k}\sigma}$. Since this is now a one-body term of the form $\sum_{\mathbf{k}\sigma} \mathcal{E}_{\text{HF}}(\mathbf{k}) c_{\mathbf{k},\sigma}^\dagger c_{\mathbf{k},\sigma}$, it is clear the full Hartree–Fock Hamiltonian may be written in terms of a k-dependent energy shift

$$\mathcal{H}_{HF} = \sum_{\mathbf{k}\sigma} \left(\frac{\hbar^2 k^2}{2m} + \mathcal{E}_{HF}(\mathbf{k}) \right) c^\dagger_{\mathbf{k},\sigma} c_{\mathbf{k},\sigma}$$

$$\mathcal{E}_{HF}(\mathbf{k}) = - \underbrace{\sum_{\mathbf{q}} n_{\mathbf{k}+\mathbf{q},\sigma} V(\mathbf{q})}_{\text{Fock}} + \underbrace{n V(0)}_{\text{Hartree}}$$

Note the Hartree or direct Coulomb term, which represents the average interaction energy of the electron $\mathbf{k}\sigma$ with all the other electrons in the system, is a constant; when summed over all electrons, it cancels exactly with the constant arising from the sum of the self-energy of the positive background and the interaction energy of the electron gas with that background. The Fock, or exchange term, is a momentum-dependent shift.

14.1.4 Problem with Hartree–Fock Theory

Although we argued that the Hartree–Fock approximation becomes a better approximation in the limit of high density for electrons interacting via the Coulomb interaction, it never becomes exact. We recall from Section 2.2.4 that the HF energy correction to the energy $\varepsilon(\mathbf{k})$ is

$$\varepsilon^{(1)}(\mathbf{k}) = \frac{e^2 k_F}{4\pi^2} \left[\frac{k_F^2 - k^2}{k k_F} \ln \left| \frac{k_F + k}{k_F - k} \right| + 2 \right].$$

The first term in parentheses has a logarithmic divergence in slope at $k = k_F$. This means that while the energy shift might be small compared to the Fermi energy, the Fermi velocity, $v_F = \nabla_{\mathbf{k}} \varepsilon(\mathbf{k})$, contains a term that is infinite. This problem can be traced back to the long-range nature of the Coulomb force. Two electrons at large distances $\mathbf{x} - \mathbf{x}'$ do not really feel the full $1/|\mathbf{x} - \mathbf{x}'|$, but a "screened" version due to the presence of the intervening medium, namely the electron gas, rearranges itself to cancel out the long-range part of V.

Electron Interactions in Second-Order Perturbation Theory

One may try to improve on the first-order result by going to second-order perturbation theory. However, the result is disastrous. The matrix elements diverge without giving hope for a simple cure.

Here we can only reveal what goes wrong, and then later learn how to deal correctly with the infinities occurring in the calculations. According to second-order perturbation theory, $E^{(2)}$ is given by

$$\frac{E^{(2)}}{N} = \frac{1}{N} \sum_{\nu \neq |\psi_0\rangle} \frac{\langle \psi_0 | V_c | \nu \rangle \langle \nu | V_c | \psi_0 \rangle}{E^{(0)} - E_\nu}, \tag{14.14}$$

where all the intermediate states $|\nu\rangle$ must be different from $|\psi_0\rangle$. As sketched in Figure 14.6, this combined with the momentum conserving Coulomb interaction yields

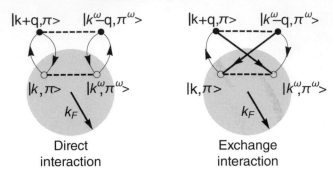

Figure 14.6 The two possible processes in second-order perturbation theory for two states $|\mathbf{k}_1\sigma_1\rangle$ and $|\mathbf{k}_2\sigma_2\rangle$ in the Fermi sea. The direct process gives a divergent contribution to E/N while the exchange process gives a finite contribution.

intermediate states where two particles are injected out of the Fermi sphere. From such an intermediate state, $|\psi_0\rangle$ is restored by putting the excited electrons back into the holes they left behind. Only two types of processes are possible: the direct and the exchange process.

We now proceed to show that the direct interaction process gives a divergent contribution $E_{\mathrm{dir}}^{(2)}$ to $E^{(2)}$ due to the singular behavior of the Coulomb interaction at small momentum transfers $\mathbf{q}$. For the direct process, the constraint $|\nu\rangle \neq |\psi_0\rangle$ leads to

$$|\nu\rangle = \Theta\left(|\mathbf{k}_1 + \mathbf{q}| - k_F\right) \Theta\left(|\mathbf{k}_2 - \mathbf{q}| - k_F\right) \Theta\left(k_F - |\mathbf{k}_2|\right) \Theta\left(k_F - |\mathbf{k}_1|\right)$$

$$\times\; c_{\mathbf{k}_1+\mathbf{q}}^\dagger\, c_{\mathbf{k}_2-\mathbf{q}}^\dagger\, c_{\mathbf{k}_2}\, c_{\mathbf{k}_1}\; |\psi_0\rangle\,. \tag{14.15}$$

To restore $|\psi_0\rangle$ the same momentum transfer, $\mathbf{q}$ must be involved in both $\langle\nu| V_c |\psi_0\rangle$ and $\langle\psi_0| V_c |\nu\rangle$, and writing $V(\mathbf{q}) = \frac{4\pi e^2}{q^2}$, we find

$$E_{\mathrm{dir}}^{(2)} = \frac{1}{\Omega^2} \sum_{\mathbf{q}} \sum_{\mathbf{k}_1\sigma_1,\mathbf{k}_2\sigma_2} \frac{(V(\mathbf{q})/2)^2}{E^{(0)} - E_\nu}\, \Theta\left(|\mathbf{k}_1 + \mathbf{q}| - k_F\right) \Theta\left(|\mathbf{k}_2 - \mathbf{q}| - k_F\right)$$

$$\times\; \Theta\left(k_F - |\mathbf{k}_2|\right) \Theta\left(k_F - |\mathbf{k}_1|\right)\,. \tag{14.16}$$

The contribution from small values of $\mathbf{q}$ to $E_{\mathrm{dir}}^{(2)}$ is found by noting that

$$V^2(\mathbf{q}) \propto \frac{1}{q^4}$$

$$\mathbf{q} \to 0 \quad E^{(0)} - E_\nu \propto \mathbf{k}_1^2 + \mathbf{k}_2^2 - (\mathbf{k}_1 + \mathbf{q})^2 - (\mathbf{k}_2 - \mathbf{q})^2 \propto q$$

$$\mathbf{q} \to 0 \quad \sum_{\mathbf{k}_1\sigma_1,\mathbf{k}_2\sigma_2} \cdots \Theta\left(k_F - |\mathbf{k}_2|\right) \Theta\left(k_F - |\mathbf{k}_1|\right) \propto q,$$

from which we obtain

$$E_{\mathrm{dir}}^{(2)} \propto \int_0 dq\, q^2\, \frac{1}{q^4}\, \frac{1}{q}\, q\, q = \int_0 dq\, \frac{1}{q} = \ln(q)\Big|_0 \propto \infty. \tag{14.17}$$

The exchange process does not lead to a divergence since in this case the momentum transfer in the excitation part is $\mathbf{q}$, but in the relaxation part it is $\mathbf{k}_2 - \mathbf{k}_1 - \mathbf{q}$. Thus $V^2(\mathbf{q})$ is replaced by $V(\mathbf{q})V((\mathbf{k}_2 - \mathbf{k}_1 - \mathbf{q}) \propto q^{-2}$ for $\mathbf{q} \to 0$, which is less singular than $V^2(\mathbf{q}) \propto q^{-4}$.

This divergent behavior of second-order perturbation theory is a nasty surprise. We know that physically the energy of the electron gas must be finite. The only hope for rescue lies in regularization of the divergent behavior by taking higher-order perturbation terms into account. In fact, it turns out that one has to consider perturbation theory to infinite order, which is possible using the full machinery of quantum field theory to be developed in the coming chapters.

One of the important early problems was to find the ground-state energy of the jellium model of a three-dimensional electron gas. It was anticipated that in the regime where $r_s < 1$, as defined in Section 1.2, this energy can be expressed in terms of a power series in r_s, as

$$E_0 = \frac{K}{r_s^2} \left[[1 + b\,r_s + c\,r_s^2 + \cdots] \right]$$

where K, a, b, c, etc. are constants. However, this turned out to be not quite right. Indeed, 1st order perturbation theory gives a term of the form $b\,r_s$ in this series.

But if one goes one step further and considers second-order perturbation theory, one finds a contribution that diverges like $\int_0 dq/q$, where q is the momentum transfer in the Fourier transform v_q of the Coulomb interaction ($v_q \propto 1/q^2$). That is, there is a logarithmic divergence from the lower limit 0 of the momentum transfers. This divergence is associated with the long range of the Coulomb interaction. Furthermore, if one examines higher-order terms in the perturbation series, one finds that they diverge even more strongly. Thus standard perturbation theory appears to be worthless.

On physical grounds, however, one does of course expect the energy of the interacting electron gas to be a finite and well-defined number, and no phase transitions occur as one "turns on" the repulsive interactions, so this failure of standard perturbation theory appears just to be a signal that the energy does not have a standard power series expansion in r_s. In 1957, Gell-Mann and Bruckner resolved this issue by using the recently developed many-body perturbation theory [75]. Essentially what they did was to sum all the most divergent terms in the series (an infinite number of them) before doing the momentum integrals, and showed that one could then arrive at a result which was well defined and finite (this is called a resummation of the perturbation series). They found that there was a term in the series for E_0 that is $\propto \ln r_s$, and thus indeed is not analytic at $r_s = 0$. This assumes at the very least that the interaction doesn't cause any drastic changes to the system, such as a phase transition.

The solution of this problem thus requires one to include an infinite number of terms in the perturbation theory. Clearly one needs to develop a new method to be able to do this in an efficient way, and this is one of the main strengths of many-body perturbation theory. We will also see other examples where one needs to include an infinite number of terms in the perturbation theory.

14.2 The Random Phase Approximation

Previously, in the discussion of second quantization, we derived expressions for the kinetic energy operator, the single-particle, and pair interaction operators. Here, we start by deriving expressions for the remaining electron, or fermion, operators.

14.2.1 The Density Operator

We can write the fermionic field operators in a plane-wave basis as

$$\Psi(\mathbf{x},t) = \frac{1}{\sqrt{\Omega}} \sum_{\mathbf{k}} e^{i\mathbf{k}\cdot\mathbf{x}} c_{\mathbf{k}}, \qquad \Psi^{\dagger}(\mathbf{x},t) - \frac{1}{\sqrt{\Omega}} \sum_{\mathbf{k}} e^{-i\mathbf{k}\cdot\mathbf{x}} c_{\mathbf{k}}^{\dagger}.$$

We then express the density of particles $\rho(\mathbf{x})$, at point $\mathbf{x}$

$$\rho(\mathbf{x}) = \sum_{i} \delta(\mathbf{x} - \mathbf{x}_i), \tag{14.18}$$

where the sum is taken over particle coordinates $\mathbf{x}_i$, in terms of the field operators as

$$\rho(\mathbf{x}) = \int d\mathbf{x}\, \Psi^{\dagger}(\mathbf{x}) \left(\sum_{i} \delta(\mathbf{x} - \mathbf{x}_i) \right) \Psi(\mathbf{x}) = \sum_{i} \Psi^{\dagger}(\mathbf{x}_i)\, \Psi(\mathbf{x}_i). \tag{14.19}$$

This is equivalent to the operator $\Psi^{\dagger}(\mathbf{x})\, \Psi(\mathbf{x})$, since if there is no particle at $\mathbf{x}$, this operator gives zero, and if there is a particle it gives the right answer, (14.19). The Fourier transform of the density operator is

$$\rho_{\mathbf{q}} = \frac{1}{\Omega} \int d\mathbf{x}\, e^{i\mathbf{q}\cdot\mathbf{x}} \rho(\mathbf{x}) = \frac{1}{\Omega^2} \int d\mathbf{x}\, e^{i\mathbf{q}\cdot\mathbf{x}} \sum_{\mathbf{k},\mathbf{k}'} e^{-i(\mathbf{k}-\mathbf{k}')\cdot\mathbf{x}}\, c_{\mathbf{k}}^{\dagger} c_{\mathbf{k}'}$$

$$= \frac{1}{\Omega} \sum_{\mathbf{k},\mathbf{k}'} c_{\mathbf{k}}^{\dagger} c_{\mathbf{k}'}\, \delta_{\mathbf{k}'-\mathbf{k},\mathbf{q}} = \sum_{\mathbf{k}} c_{\mathbf{k}+\mathbf{q}}^{\dagger} c_{\mathbf{k}}. \tag{14.20}$$

The physical meaning of the density operator is quite straightforward: it gives rise to charge fluctuations that are particle–hole pairs described by (14.25) around the Fermi surface. Since ρ is hermitian, $\rho^{\dagger} = \rho$, it follows that $\rho_{\mathbf{q}}^{\dagger} = \rho_{-\mathbf{q}}$. We also note the density operators $\rho(\mathbf{q})$ commute, namely,

$$\left[\rho_{\mathbf{q}}, \rho_{\mathbf{q}'} \right] = \left[\sum_{\mathbf{k}} c_{\mathbf{k}+\mathbf{q}}^{\dagger} c_{\mathbf{k}}, \sum_{\mathbf{k}'} c_{\mathbf{k}'+\mathbf{q}'}^{\dagger} c_{\mathbf{k}'} \right]$$

$$= \left(\sum_{\mathbf{k}} c_{\mathbf{k}+\mathbf{q}}^{\dagger} c_{\mathbf{k}} \right) \left(\sum_{\mathbf{k}'} c_{\mathbf{k}'+\mathbf{q}'}^{\dagger} c_{\mathbf{k}'} \right) - \left(\sum_{\mathbf{k}'} c_{\mathbf{k}'+\mathbf{q}'}^{\dagger} c_{\mathbf{k}'} \right) \left(\sum_{\mathbf{k}} c_{\mathbf{k}+\mathbf{q}}^{\dagger} c_{\mathbf{k}} \right) = 0, \tag{14.21}$$

which shows that they are bosonic operators!

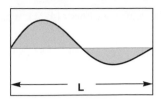

Figure 14.7 For a system of length L and $q = 2\pi/L$, ρ_q measures a quantity approximately equal to the difference in the number of particles in the two halves of the container.

As an example of the usefulness of the density operator, we show how the Hamiltonian (14.7) can be expressed in terms of the number $n_{\mathbf{k}}$ and density operators. We recast the sum over $\mathbf{k}$, $\mathbf{k}'$ in second term in the form

$$\sum_{\mathbf{k},\mathbf{k}'} c^\dagger_{\mathbf{k}-\mathbf{q}} c^\dagger_{\mathbf{k}'+\mathbf{q}} c_{\mathbf{k}'} c_{\mathbf{k}} = -\sum_{\mathbf{k},\mathbf{k}'} c^\dagger_{\mathbf{k}-\mathbf{q}} c^\dagger_{\mathbf{k}'+\mathbf{q}} c_{\mathbf{k}} c_{\mathbf{k}'} = -\sum_{\mathbf{k},\mathbf{k}',\mathbf{q}} c^\dagger_{\mathbf{k}-\mathbf{q}} \left(\delta_{\mathbf{k}'+\mathbf{q},\mathbf{k}} - c_{\mathbf{k}} c^\dagger_{\mathbf{k}'+\mathbf{q}} \right) c_{\mathbf{k}'}$$

$$= -\sum_{\mathbf{k}'} c^\dagger_{\mathbf{k}'} c_{\mathbf{k}'} + \left(\sum_{\mathbf{k}} c^\dagger_{\mathbf{k}-\mathbf{q}} c_{\mathbf{k}} \right) \left(\sum_{\mathbf{k}'} c^\dagger_{\mathbf{k}'+\mathbf{q}} c_{\mathbf{k}'} \right)$$

$$= -\left(\sum_{\mathbf{k}'} n_{\mathbf{k}'} \right) + \rho_{\mathbf{q}} \rho_{-\mathbf{q}}.$$

But since $\sum_{\mathbf{k}'} n_{\mathbf{k}'} = N$, the total number of particles, we have

$$\mathcal{H} = \sum_{\mathbf{k}} \mathcal{E}_{\mathbf{k}} n_{\mathbf{k}} + \sum_{\mathbf{q}} \frac{2\pi e^2}{q^2} \left(\rho_{\mathbf{q}} \rho_{-\mathbf{q}} - N \right). \tag{14.22}$$

14.2.2 The Random Phase Approximation

We now consider under what conditions for an interacting system, with

$$\mathcal{H} = \sum_{\mathbf{k}} \mathcal{E}_{\mathbf{k}} c^\dagger_{\mathbf{k}} c_{\mathbf{k}} + \frac{1}{2} \sum_{\mathbf{k},\mathbf{k}',\mathbf{q}} V_{\mathbf{q}} \left(\rho_{\mathbf{q}} \rho_{-\mathbf{q}} - N \right), \tag{14.23}$$

and ground-state $|\Psi_0\rangle$

$$\mathcal{H} |\Psi_0\rangle = \mathcal{E}_0 |\Psi_0\rangle,$$

can support density excitations associated with $\rho_{\mathbf{q}}$, namely,

$$\mathcal{H} \rho^\dagger_{\mathbf{q}} |\Psi_0\rangle = \left(\mathcal{E}_0 + \hbar\omega_{\mathbf{q}} \right) \rho^\dagger_{\mathbf{q}} |\Psi_0\rangle, \tag{14.24}$$

We can recast (14.24) in the commutator form

$$\left[\mathcal{H}, \rho^\dagger_{\mathbf{q}} \right] = \hbar\omega_{\mathbf{q}} \rho^\dagger_{\mathbf{q}}. \tag{14.25}$$

Using the commutation of the density operators (14.21), we obtain

$$[\mathcal{H}, \rho_{\mathbf{q}}^{\dagger}] = \sum_{\mathbf{k}} \left(\mathcal{E}_{\mathbf{k}+\mathbf{q}} - \mathcal{E}_{\mathbf{k}} \right) c_{\mathbf{k}+\mathbf{q}}^{\dagger} c_{\mathbf{k}}. \tag{14.26}$$

This does not look very much like (14.25), and we take the commutator with $\mathcal{H}$ once again. If $\rho_{\mathbf{q}}^{\dagger}$ actually creates an excitation of energy $\hbar\omega$, we should have

$$\left[\mathcal{H}, [\mathcal{H}, \rho_{\mathbf{q}}^{\dagger}] \right] = (\hbar\omega)^2 \, \rho_{\mathbf{q}}^{\dagger}.$$

After tedious algebra manipulation, we find that

$$[\mathcal{H}, c_{\mathbf{k}+\mathbf{q}}^{\dagger} c_{\mathbf{k}}] = \left(\mathcal{E}_{\mathbf{k}+\mathbf{q}} - \mathcal{E}_{\mathbf{k}} \right) c_{\mathbf{k}+\mathbf{q}}^{\dagger} c_{\mathbf{k}}$$
$$+ \sum_{\mathbf{q}'} \frac{V_{\mathbf{q}'}}{2} \left[(c_{\mathbf{k}+\mathbf{q}-\mathbf{q}'}^{\dagger} c_{\mathbf{k}} - c_{\mathbf{k}+\mathbf{q}}^{\dagger} c_{\mathbf{k}+\mathbf{q}'}) \rho_{\mathbf{q}'}^{\dagger} + \rho_{\mathbf{q}'} (c_{\mathbf{k}+\mathbf{q}+\mathbf{q}'}^{\dagger} c_{\mathbf{k}} - c_{\mathbf{k}+\mathbf{q}}^{\dagger} c_{\mathbf{k}-\mathbf{q}'}) \right].$$

and we obtain

$$(\hbar\omega)^2 \, \rho_{\mathbf{q}}^{\dagger} = \sum_{\mathbf{k}} \left(\mathcal{E}_{\mathbf{k}+\mathbf{q}} - \mathcal{E}_{\mathbf{k}} \right) [\mathcal{H}, c_{\mathbf{k}+\mathbf{q}}^{\dagger} c_{\mathbf{k}}]$$
$$= \sum_{\mathbf{k}} \left(\mathcal{E}_{\mathbf{k}+\mathbf{q}} - \mathcal{E}_{\mathbf{k}} \right)^2 c_{\mathbf{k}+\mathbf{q}}^{\dagger} c_{\mathbf{k}} + \sum_{\mathbf{k},\mathbf{q}'} \left(\mathcal{E}_{\mathbf{k}+\mathbf{q}} - \mathcal{E}_{\mathbf{k}} \right) \frac{V_{\mathbf{q}'}}{2}$$
$$\times \left[(c_{\mathbf{k}+\mathbf{q}-\mathbf{q}'}^{\dagger} c_{\mathbf{k}} - c_{\mathbf{k}+\mathbf{q}}^{\dagger} c_{\mathbf{k}+\mathbf{q}'}) \rho_{\mathbf{q}'}^{\dagger} + \rho_{\mathbf{q}'} (c_{\mathbf{k}+\mathbf{q}+\mathbf{q}'}^{\dagger} c_{\mathbf{k}} - c_{\mathbf{k}+\mathbf{q}}^{\dagger} c_{\mathbf{k}-\mathbf{q}'}) \right]. \tag{14.27}$$

With $\mathcal{E}_{\mathbf{k}+\mathbf{q}} - \mathcal{E}_{\mathbf{k}} = \hbar^2 (2\mathbf{k} \cdot \mathbf{q} + q^2)/2m$, we write

$$\sum_{\mathbf{k}} (\mathcal{E}_{\mathbf{k}+\mathbf{q}} - \mathcal{E}_{\mathbf{k}}) \left(c_{\mathbf{k}+\mathbf{q}-\mathbf{q}'}^{\dagger} c_{\mathbf{k}} - c_{\mathbf{k}+\mathbf{q}}^{\dagger} c_{\mathbf{k}+\mathbf{q}'} \right)$$
$$= \frac{\hbar^2}{2m} (2\mathbf{k} \cdot \mathbf{q} + q^2) c_{\mathbf{k}+\mathbf{q}-\mathbf{q}'}^{\dagger} c_{\mathbf{k}} - \frac{\hbar^2}{2m} \sum_{\mathbf{k}'} [2(\mathbf{k}' - \mathbf{q}') \cdot \mathbf{q} + q^2] c_{\mathbf{k}'+\mathbf{q}-\mathbf{q}'}^{\dagger} c_{\mathbf{k}'}$$
$$= \frac{\hbar^2 \mathbf{q}' \cdot \mathbf{q}}{m} \, \rho_{\mathbf{q}-\mathbf{q}'}^{\dagger}, \tag{14.28}$$

where we set $\mathbf{k}' = \mathbf{k} + \mathbf{q}'$, and we arrive at

$$(\hbar\omega)^2 \, \rho_{\mathbf{q}}^{\dagger} = \sum_{\mathbf{k}} \left[\frac{\hbar^2}{2m} (2\mathbf{k} \cdot \mathbf{q} + q^2) \right]^2 c_{\mathbf{k}+\mathbf{q}}^{\dagger} c_{\mathbf{k}}$$
$$+ \sum_{\mathbf{q}'} \frac{V_{\mathbf{q}'}}{2} \frac{\hbar^2 \mathbf{q} \cdot \mathbf{q}'}{m} \left(\rho_{\mathbf{q}'-\mathbf{q}} \rho_{\mathbf{q}'}^{\dagger} + \rho_{-\mathbf{q}'}^{\dagger} \rho_{-\mathbf{q}-\mathbf{q}'} \right). \tag{14.29}$$

We note that ρ_0 is actually much larger than all the other components. It is just the average density of particles in the system. Consider a box of length L containing a large number of electrons of average density ρ_0, as shown in Figure 14.7. The first Fourier component of ρ

$$\int d\mathbf{r}\, \rho(\mathbf{r}) \exp\left(\frac{2i\pi x}{L}\right),$$

will be approximately equal to the difference between the number of particles in the left- and right-hand sides of the box, the gray area. The number difference will be very small compared to the total number, and so ρ_0 will prevail in the summation over $\mathbf{q}'$ in (14.29). Because the term with $\mathbf{q}' = 0$ is omitted, ρ_0 will only appear when $\mathbf{q}' = \pm\mathbf{q}$. The neglect of all terms for which $\mathbf{q}' \neq \pm\mathbf{q}$ is known as the **random phase approximation** (RPA).

Alternative Interpretation of RPA

Consider now the first quantized version of the density operator $\rho_{\mathbf{q}}$ that is given by

$$\rho_{\mathbf{q}} = \frac{1}{\Omega} \sum_n e^{i\mathbf{q}\cdot\mathbf{x}_n},$$

where $\mathbf{x}_n$ is the position of the nth particle. We see that $\rho_{\mathbf{q}\pm\mathbf{q}'}$ can be written in first quantized form as

$$\rho_{\mathbf{q}\pm\mathbf{q}'} = \frac{1}{\Omega} \sum_n e^{i(\mathbf{q}\pm\mathbf{q}')\cdot\mathbf{x}_n}.$$

We note that if the electrons are (statistically) randomly distributed in the system, the preceding sum only gives a finite result if $\mathbf{q} = \mp\mathbf{q}'$, because the sum involves random phases that add destructively. In RPA, we approximate (14.29) by keeping only terms with $\mathbf{q} = \mp\mathbf{q}'$.

Plasma Oscillations

We consider the case where $\mathbf{q}$ is small enough that, for the moment, we neglect the first term of (14.29) and we apply RPA to the second term. We obtain

$$\left(\hbar\omega\right)^2 \rho_{\mathbf{q}}^\dagger = \frac{4\pi e^2}{|\mathbf{q}|^2}\, \frac{\hbar^2\, |\mathbf{q}|^2}{m}\, \left(\rho_0\rho_{\mathbf{q}}^\dagger + \rho_{\mathbf{q}}^\dagger \rho_0\right), \tag{14.30}$$

and since all the $\rho_{\mathbf{q}}$ commute, we have

$$\omega^2 = \frac{4\pi e^2 \rho_0}{m}, \tag{14.31}$$

which gives just the classical plasma frequency, ω_p. We now approximate the contribution of the first term by setting $|\mathbf{k}| \sim k_F$, and $\langle 2\mathbf{k}_F \cdot \mathbf{q}\rangle \gg \mathbf{q}^2)$. Averaging over the angles, we obtain the approximation

$$\left\langle (\mathbf{k}_F \cdot \mathbf{q})^2 \right\rangle \simeq \frac{1}{4\pi}\, 4\pi^2 k_F^2 q^2 \int_{-1}^{1} x^2\, dx = 2\pi k_F^2\, q^2,$$

which yields the plasmon dispersion relation

$$\omega^2(\mathbf{q}) = \omega_p^2 + \frac{2\pi\hbar^2 k_F^2}{3m^2} q^2 \simeq \omega_p + \frac{\pi v_F^2}{3\omega_p} q^2. \qquad (14.32)$$

Thus, the relevant long-wavelength excitations are collective motions of the electron gas. Bohm and Pines [35] suggested that electrons act as interacting particles through a bare Coulomb potential $4\pi e^2/|\mathbf{q}|^2$ only for length scales less than an effective screening length, which we can take as the Thomas–Fermi $\sim q_{TF}^{-1}$. Below q_{TF}, the interactions contribute only to the plasma oscillations.

Exercises

14.1 Second-order contribution to the ground-state energy of an electron gas from Rayleigh–Schrödinger perturbation theory:

(a) Show that the second-order contribution to the ground-state energy (per particle) of an electron gas can be expressed as

$$\frac{E^{(2)}}{N} = \frac{e^2}{2a_0} \left(\varepsilon_2^d + \varepsilon_2^{exc} \right)$$

where

$$\varepsilon_2^d = \frac{3}{8\pi^5} \int \frac{d\mathbf{q}}{q^4} \int_{|\mathbf{k}+\mathbf{q}|>1} d\mathbf{k} \int_{|\mathbf{p}+\mathbf{q}|>1} d\mathbf{p} \, \frac{\Theta\left(1-|\mathbf{k}|\right)\Theta\left(1-|\mathbf{p}|\right)}{q^2 + \mathbf{q}\cdot(\mathbf{k}+\mathbf{p})}$$

$$\varepsilon_2^{exc} = \frac{3}{16\pi^5} \int \frac{d\mathbf{q}}{q^4} \int_{|\mathbf{k}+\mathbf{q}|>1} d\mathbf{k} \int_{|\mathbf{p}+\mathbf{q}|>1} d\mathbf{p} \, \frac{\Theta\left(1-|\mathbf{k}|\right)\Theta\left(1-|\mathbf{p}|\right)}{(\mathbf{q}+\mathbf{k}+\mathbf{p})^2 \left[q^2 + \mathbf{q}\cdot(\mathbf{k}+\mathbf{p})\right]}$$

are the contributions stemming from the direct and exchange processes, respectively. All the momenta on the right-hand sides are expressed in units of the Fermi momentum k_F.

Hint: Recall that the second-order change of the energy in ordinary (Rayleigh–Schrödinger) perturbation theory for the Hamiltonian $\mathcal{H} = \mathcal{H}_0 + V$ (V is the perturbation) is

$$E^{(2)} = \sum_{n>0} \frac{|\langle 0|V|n\rangle|^2}{E_0 - E_n}.$$

Here $|0\rangle$ is the ground state of $\mathcal{H}_0$, E_0 is the corresponding ground-state energy, while $|n\rangle$ are the excited states with energies E_n.

(b) We now want to show that the term ε_2^d is actually divergent. Consider the function

$$F(|\mathbf{q}|) = \int_{|\mathbf{k}+\mathbf{q}|>1} d\mathbf{k} \int_{|\mathbf{p}+\mathbf{q}|>1} d\mathbf{p} \, \frac{\Theta\left(1-|\mathbf{k}|\right)\Theta\left(1-|\mathbf{p}|\right)}{q^2 + \mathbf{q}\cdot(\mathbf{k}+\mathbf{p})}.$$

Show that

$$F(|\mathbf{q}|) \sim \left(\frac{4\pi}{3}\right)^2 |\mathbf{q}|^{-2} \quad |\mathbf{q}| \to \infty$$

and that

$$F(|\mathbf{q}|) \sim \frac{2}{3} (2\pi)^2 (1 - \ln 2) |\mathbf{q}| \quad |\mathbf{q}| \to 0$$

Using these results, show that ε_2^d diverges logarithmically because of the contributions at small momentum transfer $\mathbf{q}$.

14.2 Delta-function pair interactions:
If we replace the Coulomb electron–electron interaction with a delta-function

$$v\left(\mathbf{x}, \mathbf{x}'\right) = g\,\delta\left(\mathbf{x} - \mathbf{x}'\right).$$

Show that

$$\mathcal{H}' = \frac{g}{2\Omega} \sum_{\substack{\mathbf{k}, \mathbf{k}', \mathbf{q} \\ \sigma \sigma'}} c^\dagger_{\mathbf{k}+\mathbf{q}, \sigma}\, c^\dagger_{\mathbf{k}'-\mathbf{q}, \sigma'}\, c_{\mathbf{k}', \sigma'}\, c_{\mathbf{k}, \sigma}$$

and that the total exchange energy is $-3gN/16\pi r_s^3$. Does $\Delta E^{(2)}$ diverge in this case?

14.3 Magnetic interactions in the homogeneous electron gas:

(a) In order to derive the ground-state energy of N free electrons in volume Ω, we assumed that every one-electron level with a wavevector less than k_F is occupied by two electrons of opposite spin. Consequently, we obtained the following ground-state energy E_{unpol} of the unpolarized electrons system

$$\frac{E_{\text{unpol}}}{N} = \frac{e^2}{2a_0} \left(\frac{9\pi}{4}\right)^{1/3} \left(\frac{3}{5}\frac{1}{r_s^2} - \frac{3}{2\pi}\frac{1}{r_s}\right),$$

where a_0 is the Bohr radius, and E/N is given in units of Rydberg $= e^2/2a_0$.
Show that it can be expressed in terms of k_F and a_0 as

$$\frac{E_{\text{unpol}}}{N} = \frac{e^2}{2a_B} \left[\frac{3}{5} (k_F a_B)^2 - \frac{3}{2\pi} (k_F a_B)\right].$$

However, a more general possibility would be to fill each one-electron level with $k < k_{F\uparrow}$ with spin-up electrons and each level with $k < k_{F\downarrow}$ with spin-down electrons.

(b) Show that the ground-state energy of a fully magnetized gas of N electrons (i.e., all the electrons have the same value of the spin) is given by

$$\frac{E_{\text{pol}}}{N} = \frac{e^2}{2a_B} \left[\frac{3}{5} 2^{2/3} (k_F a_B)^2 - \frac{3}{2\pi} 2^{1/3} (k_F a_B)\right].$$

k_F and $k_{F\uparrow}$ are obtained from the condition that the number of electrons is preserved upon the magnetization.)

(c) Derive under which condition the fully magnetized (ferromagnetic) ground state has lower energy than the unmagnetized one ($E_{\text{pol}} < E_{\text{unpol}}$). Which physical regime corresponds to a magnetized ground state?

14.4 Commutator of the density operators:
Show that

$$[\rho_{\mathbf{q}}, \rho_{\mathbf{q}'}] = 0 \quad \forall \, \mathbf{q}, \mathbf{q}'.$$

14.5 Hamiltonian of an electron in a Wigner crystal:
Comparison of the kinetic to Coulomb energy contributions to the jellium Hamiltonian showed that at low electron densities the Coulomb contribution will prevail. Electrons will localize at positions far from each other to lower their Coulomb interactions. This leads to electron crystallization into what is known as a Wigner crystal.

Consider an electron in such a configuration of volume Ω that contains N particles. Assume that it is located in a unit cell, which we approximate as a sphere of radius $r_0 \propto (\Omega/N)^{1/3}$.

(a) Determine the electron potential energy due to the underlying uniform positive charge, when the electron is at a position $\mathbf{x}$, $|\mathbf{x}| < r_0$ from the center of the sphere.
(b) Write down the Hamiltonian for the electron at $\mathbf{x}$ and $\mathbf{p}$, and show that it resembles an isotropic simple harmonic oscillator. Express its frequency in terms of an effective plasma frequency.
(c) What is its ground-state energy? Express it in terms of r_s. (Remember that the harmonic oscillator has three degenerate modes in the ground state.)

15

Green Functions for Many-Body Systems and Feynman Diagrams

15.1 Introduction

To explore the properties of a macroscopic system, we probe it. Experimental *condensed matter physics* provides a plethora of probes to investigate different aspects of such systems: Neutron scattering, electron scattering, atom scattering, photon scattering (Raman, etc.), nuclear magnetic resonance, resistivity, thermal conductivity, muon resonance; it is actually a long list. All probes interact weakly with systems, and scattering experiments employing a weak probe actually measure one of a system's equilibrium correlation functions. Consequently, we find that correlation functions are at the heart of modern condensed-matter theory. In Chapter 5, we developed the one-body correlation functions within linear response theory, which we delineated as susceptibilities. In this chapter, we will develop the mathematical tools to construct appropriate correlation functions for many-body systems, whose basic building blocks we identify as *many-body Green functions*.

As we shall demonstrate, Green functions techniques play a fundamental role in the treatment of many-body systems. We will find that various fundamental physical properties of a many-body system – ground-state energy and excitation spectra, response functions, such as electric and magnetic susceptibilities, as well as thermodynamic quantities – can be derived from many-body Green functions (GFs). The applications of nonrelativistic Green function techniques to many-body interactions and linear response of condensed matter systems will be surveyed here.

To explore the structure of quantum correlation functions, we start with the ground state $|\Phi_0\rangle$ of a system. At some time t', we act on the system with a local perturbation, represented by a local operator $O^\dagger(\mathbf{x}, t)$, creating an initial state $O^\dagger(\mathbf{x}', t') |\Phi_0\rangle$. Thus, we prepared the system in some initial state that deviates from the ground state by some local excitations, a disturbance, and we examine how this disturbance evolves in space and time. Such a scenario describes typical local measurements done on a quantum many-particle system. Since we will be dealing with time dependence, it is expedient to use the Heisenberg representation of the quantum evolution. We will be interested in determining the quantum-mechanical amplitude to find the system in state $O^\dagger(\mathbf{x}, t) |\Phi_0\rangle$ at time $t > t'$, namely,

$$\text{Amplitude} = \langle \Phi_0 | O(\mathbf{x}, t) O^\dagger(\mathbf{x}', t') |\Phi_0\rangle .$$

We are aware by now that quantum-mechanical operators can be expressed in bilinear forms of the field operators $\hat{\Psi}(\mathbf{x}, t)$. However, it turns out that correlation functions of the field $\hat{\Psi}(\mathbf{x}, t)$ itself are related to experiments such as photoemission and tunneling spectroscopies. These experimental techniques involve extraction or injection of a single electron. We can clearly identify the field correlation functions as just quantum mechanical propagators, and we know from quantum mechanics that such propagators satisfy the Green function equation! Hence, the name Green functions. We will demonstrate in this chapter that Green functions form the building blocks for correlation functions that appear in linear response.

15.2 The One-Particle Green Function of Many-Body Systems

15.2.1 Noninteracting Particle Propagator

We consider a particle in free space described by a single-particle time-independent Hamiltonian $\mathcal{H}_1$. Its eigenstates and eigenenergies are

$$\mathcal{H}_1 \, |\phi_n\rangle = \varepsilon_n \, |\phi_n\rangle .$$

In general, if we put the particle in one of its $|\phi_n\rangle$ eigenstates, it will remain in the same state forever. Instead, we imagine preparing the system in a generic state $|\psi_{\text{trial}}\rangle$ and then follow its time evolution. If the trial state is created at time t', the wavefunction at a later time t is given by

$$|\psi(t)\rangle = e^{-i\mathcal{H}_1(t-t')/\hbar} \, |\psi_{\text{trial}}(t')\rangle = \sum_n |\phi_n\rangle \, e^{-i\varepsilon_n(t-t')/\hbar} \, \langle \phi_n | \psi_{\text{trial}}(t')\rangle .$$

Eventually, at time t, we want to know the probability amplitude that a measurement would find the particle at position $\mathbf{x}$,

$$\langle \mathbf{x} \, | \psi(t)\rangle \, \Theta(t - t') = \int d\mathbf{x}' \, \langle \mathbf{x} | e^{-i\mathcal{H}_1(t-t')/\hbar} |\mathbf{x}'\rangle \langle \mathbf{x}' \, | \psi_{\text{trial}}(t')\rangle \, \Theta(t - t') \tag{15.1}$$

$$= \int d\mathbf{x}' \, \sum_n \langle \mathbf{x} \, |\phi_n\rangle \, e^{-i\varepsilon_n(t-t')/\hbar} \, \langle \phi_n \, |\mathbf{x}'\rangle \langle \mathbf{x}' \, | \psi_{\text{trial}}(t')\rangle \, \Theta(t - t'), \tag{15.2}$$

where $\Theta(t - t')$ establishes causality. Equation (15.1) may be rewritten as

$$\psi(\mathbf{x}, t) \, \Theta(t - t') = i \int d\mathbf{x}' \, G^{\text{R}}(\mathbf{x}t, \mathbf{x}'t') \, \psi_{\text{trial}}(\mathbf{x}', t') \, \Theta(t - t'), \tag{15.3}$$

where we introduce the retarded Green function, or propagator, in the position representation as

$$G^{\text{R}}(\mathbf{x}t, \mathbf{x}'t') = -i \, \Theta(t - t') \, \langle \mathbf{x} | e^{-i\mathcal{H}_1(t-t')/\hbar} |\mathbf{x}'\rangle . \tag{15.4}$$

We note that the $\tau = t - t'$ time dependence is due to time-translation invariance of the Hamiltonian.

It is clear that once $G(\mathbf{x}, \mathbf{x}'; \tau)$ is known it can be used to calculate the evolution of any initial state. However, there is more information included in the propagator:

- We find that the bracket $\langle \phi_n \, | \mathbf{x} \rangle = \langle \phi_n | \, \hat{\psi}^\dagger(\mathbf{x}) \, |0\rangle$ in (15.1) is just the probability amplitude that inserting a particle at position $\mathbf{x}$ and measuring its energy right away would force the system to collapse into the eigenstate $|\phi_n\rangle$.

- We find the resolvant. We note from (15.2) that the time evolution is a superposition of waves propagating with different energies. In principle, it could be inverted to determine the eigenspectrum. This can be achieved by transforming $G(\mathbf{x}, \mathbf{x}'; \tau)$ in (15.4) to the frequency domain

$$G(\mathbf{x}, \mathbf{x}'; \omega) = \int dt \, e^{i(\omega + i\eta)\tau} \, \langle \mathbf{x} | \, e^{-i\mathcal{H}_1 \tau/\hbar} \, |\mathbf{x}'\rangle = \left\langle \mathbf{x} \left| \frac{1}{\omega - \mathcal{H}_1 + i\eta} \right| \mathbf{x}' \right\rangle.$$

This is just the resolvent whose poles constitute the quantum system's spectrum.

Operationally, we can devise a gedanken experiment in which the particle is initially inserted at position $\mathbf{x}'$ and picked up at $\mathbf{x}$ after some time t. When performing this recipe with adequate resolution and for a sufficiently large number of different positions and elapsed times, the Fourier transform of the recorded data would provide the full eigenvalue spectrum. Such an experiment would give us complete information on our particle.

This scenario reveals that we can express all observables in terms of the Green function and thus avoid the explicit use of wavefunctions.

Differential Equation for the Green Function

Operating on the left-hand side of (15.3) with $i\partial/\partial t$, we obtain

$$i \frac{\partial}{\partial t} \left[\psi(\mathbf{x}, t) \, \Theta(t - t') \right] = i \, \delta(t - t') \, \psi(\mathbf{x}, t) + i \, \Theta(t - t') \, \frac{\partial}{\partial t} \psi(\mathbf{x}, t)$$

$$= i \, \delta(t - t') \, \psi(\mathbf{x}, t) + \Theta(t - t') \, \mathcal{H}_1(\mathbf{x}) \, \psi(\mathbf{x}, t).$$

Replacing $\psi(\mathbf{x}, t) \, \Theta(t - t')$ by the right-hand side of (15.3), and setting $\psi(\mathbf{x}, t) \, \delta(t - t') = \psi(\mathbf{x}, t')$, we get

$$i \frac{\partial}{\partial t} \left(i \int d\mathbf{x}' \, G^R(\mathbf{x}t, \mathbf{x}'t') \, \psi(\mathbf{x}', t') \right) = i \, \delta(t - t') \int d\mathbf{x}' \, \delta(\mathbf{x} - \mathbf{x}') \, \psi(\mathbf{x}', t')$$

$$+ \mathcal{H}_1(\mathbf{x}) \left[i \int d\mathbf{x}' \, G^R(\mathbf{x}t, \mathbf{x}'t') \, \psi(\mathbf{x}', t') \right].$$

Rearranging and choosing our initial wavefunction $\psi(\mathbf{x}', t') = \delta(\mathbf{x}' - \mathbf{x}_1)$, we obtain

$$\left[i \frac{\partial}{\partial t} - \mathcal{H}_1(\mathbf{x}) \right] G^R(\mathbf{x}t, \mathbf{x}_1 t') = \delta(t - t') \, \delta(\mathbf{x}' - \mathbf{x}_1), \tag{15.5}$$

which is just the definition of the Green function for the Schrödinger equation in a differential equation form.

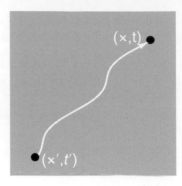

Figure 15.1 One way to understand the structure of the excitation in an interacting system is to inject a particle at $(\mathbf{x}',t')$, allow the resulting system to evolve, and then remove the particle at $(\mathbf{x},t)$. The amplitude of such a process indicates how much the propagation of the excitation between the two points is similar to the case of noninteracting particle systems.

15.2.2 Construction of the One-Particle Green Function of Many-Body Systems

Now, we want to explore how the notion of a Green function such as that just described can be adapted to the situation where we inject a particle into a system containing many similar interacting particles and subsequently remove it from the system. Conversely, we can remove a particle from the system and then inject it back. These processes can induce unexpected events, such as collective excitations of the system. Physically, we would like to have a correlation function that tells us how much excitations in interacting systems look like noninteracting particle excitations. The scenario described so for such a function is depicted in Figure 15.1.

We will consider in the present section only the zero temperature case, since it is pedagogically simpler than the general case of finite temperatures. It will serve as a conduit to develop the necessary theoretical techniques for a many-body system.

The one-particle Green function is defined in the position representation as[1]

$$i G_{\alpha\beta}\left(\mathbf{x},t;\mathbf{x}',t'\right) = \frac{\left\langle \Phi_0 \left| T\left[\hat{\psi}_{H\alpha}(\mathbf{x},t)\,\hat{\psi}^{\dagger}_{H\beta}\left(\mathbf{x}',t'\right)\right]\right| \Phi_0\right\rangle}{\langle\Phi_0 \mid \Phi_0\rangle}, \tag{15.6}$$

where $|\Phi_0\rangle$ is the ground state of the interacting system in the Heisenberg picture,[2] and $\hat{\psi}_{H\alpha}(x,t)$ is a Heisenberg field operator; it can be expressed in terms of the corresponding Schrödinger field operator as

$$\hat{\psi}_{H\alpha}(x,t) = e^{i\mathcal{H}t/\hbar}\,\hat{\psi}_{S\alpha}\,e^{-i\mathcal{H}t/\hbar},$$

---

[1] The factor i is introduced for convenience to simplify further formulas.
[2] We note that the role played by the physical vacuum in the case of a noninteracting particle is now assumed by a many-body state, its ground state.

where α represents a set of quantum numbers associated with the particle, such as spin in the case of electrons.[3]

T is a time-ordering operator, where the time-ordered product is defined as

$$T\left[\hat{\psi}_{H\alpha}(\mathbf{x},t)\,\hat{\psi}^{\dagger}_{H\beta}\left(\mathbf{x}',t'\right)\right] = \begin{cases} \hat{\psi}_{H\alpha}(\mathbf{x},t)\,\hat{\psi}^{\dagger}_{H\beta}\left(\mathbf{x}',t'\right) & t > t' \\ -\hat{\psi}^{\dagger}_{H\beta}\left(\mathbf{x}',t'\right)\,\hat{\psi}_{H\alpha}(\mathbf{x},t) & t' > t \end{cases}. \tag{15.7}$$

The minus sign arises from the anticommutation of the fermionic field operators. Equation (15.7) can also be written as

$$T\left[\hat{\psi}_{H\alpha}(\mathbf{x},t)\,\hat{\psi}^{\dagger}_{H\beta}\left(\mathbf{x}',t'\right)\right] = \Theta(t-t')\,\hat{\psi}_{H\alpha}(\mathbf{x},t)\,\hat{\psi}^{\dagger}_{H\beta}\left(\mathbf{x}',t'\right)$$
$$- \Theta(t'-t)\,\hat{\psi}^{\dagger}_{H\beta}\left(\mathbf{x}',t'\right)\,\hat{\psi}_{H\alpha}(\mathbf{x},t). \tag{15.8}$$

The *one-particle* label stems from the fact that (15.6) represents a propagator for a particle created at $\mathbf{x}',t'$ with quantum numbers β, and annihilated at $\mathbf{x},t$ with quantum numbers α, as shown in Figure 15.1. We should caution that these *single-particle Green functions* are truly many-body objects because they describe the propagation of a single particle obeying the full many-body Hamiltonian, and therefore contain all effects of interactions with the other particles in the system.

There are a number of other important variant morphs in which Green functions can be expressed. These variants have important relations between them. Quite often, it is more expedient to use a certain version for a particular application.

First, we have the so-called *greater* and *lesser* Green functions, defined as

$$G^{>}\left(\mathbf{x},t;\mathbf{x}',t'\right) = -i\,\left\langle \Phi_0 \left| \hat{\psi}_{H\alpha}(\mathbf{x},t),\,\hat{\psi}^{\dagger}_{H\beta}\left(\mathbf{x}',t'\right) \right| \Phi_0 \right\rangle$$
$$G^{<}\left(\mathbf{x},t;\mathbf{x}',t'\right) = \pm i\,\left\langle \Phi_0 \left| \hat{\psi}^{\dagger}_{H\beta}\left(\mathbf{x}',t'\right)\,\hat{\psi}_{H\alpha}(\mathbf{x},t) \right| \Phi_0 \right\rangle, \tag{15.9}$$

where the $+$ and $-$ signs are for fermions and for bosons, respectively. Next, we have the retarded and advanced Green functions

$$G^{R}\left(\mathbf{x},t;\mathbf{x}',t'\right) = \begin{cases} -i\Theta(t-t')\,\left\langle \Phi_0 \left| \left[\hat{\psi}_{H\alpha}(\mathbf{x},t),\,\hat{\psi}^{\dagger}_{H\beta}\left(\mathbf{x}',t'\right)\right]_{\pm} \right| \Phi_0 \right\rangle \\ \Theta(t-t')\left[G^{>}\left(\mathbf{x},t;\mathbf{x}',t'\right) - G^{<}\left(\mathbf{x},t;\mathbf{x}',t'\right)\right] \end{cases} \tag{15.10}$$

$$G^{A}\left(\mathbf{x},t;\mathbf{x}',t'\right) = \begin{cases} i\Theta(t'-t)\,\left\langle \Phi_0 \left| \left[\hat{\psi}^{\dagger}_{H\beta}\left(\mathbf{x}',t'\right),\,\hat{\psi}_{H\alpha}(\mathbf{x},t)\right]_{\pm} \right| \Phi_0 \right\rangle \\ \Theta(t'-t)\left[G^{<}\left(\mathbf{x},t;\mathbf{x}',t'\right) - G^{>}\left(\mathbf{x},t;\mathbf{x}',t'\right)\right], \end{cases} \tag{15.11}$$

---

[3] For the most part, we shall consider our particles to be electrons, so that the quantum numbers can only take the values $\pm 1/2$, and the field operators obey anticommutators.

where the brackets with the $+$ and $-$ subscripts denote fermion anticommutators and bosons commutators, respectively. For compactness, we set $\langle \Phi_0 \mid \Phi_0 \rangle = 1$. We note that, by definition, $G^R = 0$ for $t - t' < 0$, and $G^A = 0$ for $t - t' > 0$. The Green functions defined so far are called *space-time* Green functions, because they involve creation and annihilation of particles at definite locations in space and time. We can also define analogous Green functions in bases other than space-time. For example, for translationally invariant systems in space-time, it is usually convenient to work in the momentum-frequency domain. Depending on the problem at hand, Green functions with particle creation/annihilation represented in other types of single-particle states may also be expedient.

As a many-body correlation function, a Green function conveys only part of the full information available in the many-body wavefunctions of the systems, but it provides the relevant information for the problem at hand.

Although no direct calculations will be performed before discussing perturbation theory in the following sections, we will describe here how to extract a large variety of physical information from the Green function.

15.2.3   Equation of Motion for the One-Particle Green Function

Hierarchy of Infinite Equations and Green Functions

We consider a system with Hamiltonian

$$\mathcal{H} = \mathcal{H}_0 + \mathcal{H}',$$

where $\mathcal{H}_0$ is bilinear in the field operators, and diagonal in the basis α, such that

$$\mathcal{H}_0 = \sum_\alpha \varepsilon_\alpha \, c_\alpha^\dagger \, c_\alpha,$$

while $\mathcal{H}'$ is at least quartic in the field operators. In the α basis, G^R is given by

$$G^R(\alpha, t; \beta, t') = -i \, \Theta(t - t') \left\langle \left[c_\alpha(t), c_\beta^\dagger(t') \right]_\pm \right\rangle. \tag{15.12}$$

Differentiating (15.12) with respect to t, we get

$$
\begin{aligned}
i \frac{\partial}{\partial t} G^R(\alpha, t; \beta, t') &= i(-i) \left[\delta(t - t') \left\langle \left[c_\alpha(t), c_\beta^\dagger(t') \right]_\pm \right\rangle \right. \\
&\quad \left. + \Theta(t - t') \left\langle \left[\frac{\partial c_\alpha(t)}{\partial t}, c_\beta^\dagger(t') \right]_\pm \right\rangle \right] \\
&= \delta(t - t') \delta_{\alpha\beta} - i \, \Theta(t - t') \left\langle \left[[c_\alpha(t), \mathcal{H}], c_\beta^\dagger(t') \right]_\pm \right\rangle, \tag{15.13}
\end{aligned}
$$

where we used the equation of motion for $c_\alpha(t)$

$$i \frac{\partial c_\alpha(t)}{\partial t} = \left[c_\alpha, \mathcal{H} \right].$$

We also have

$$\left[c_\alpha, \mathcal{H}_0 \right] = -\varepsilon_\alpha\, c_\alpha$$

and we obtain

$$\left[i \frac{\partial}{\partial t} - \varepsilon_\alpha \right] G^R(\alpha, t; \beta, t') = \delta(t - t')\, \delta_{\alpha\beta} - i\, \Theta(t - t') \left\langle \left[[c_\alpha(t), \mathcal{H}'], c_\beta^\dagger(t') \right]_\pm \right\rangle.$$
(15.14)

Now, if $\mathcal{H}'$ contains only terms quartic in the field operators, then $\left[c_\alpha(t), \mathcal{H}' \right]$ will be cubic in the c operators. Writing

$$G_2^R(\alpha, t; \beta, t') = -i\, \Theta(t - t') \left\langle \left[[c_\alpha(t), \mathcal{H}], c_\beta^\dagger(t') \right]_\pm \right\rangle,$$
(15.15)

we have a two-particle Green function. Our one-particle Green function becomes coupled to a two-particle Green function! This two-particle propagator is in turn related to the three-particle propagator, etc. Due to the Heisenberg equation of motion for the field operator, every equation of motion for the Green function involves a Green function of higher order. Unless we have a particular case where the couplings terminate at a reasonable order, we will have an infinite hierarchy of equations and corresponding Green functions.

Equation of Motion and Microscopic Self-Energy Phenomenology

We consider the many-electron Hamiltonian

$$\mathcal{H} = \int d\mathbf{x} \sum_\sigma \hat{\psi}_\sigma^\dagger(\mathbf{x}) \left[\frac{-\hbar^2}{2m} \nabla^2 + U(\mathbf{x}) \right] \hat{\psi}_\sigma(\mathbf{x})$$

$$+ \frac{1}{2} \iint d\mathbf{x}\, d\mathbf{x}' \sum_{\sigma\sigma'} \hat{\psi}_\sigma^\dagger(\mathbf{x})\, \hat{\psi}_{\sigma'}^\dagger(\mathbf{x}')\, V\left(|\mathbf{x} - \mathbf{x}'|\right) \hat{\psi}_{\sigma'}(\mathbf{x}')\, \hat{\psi}_\sigma(\mathbf{x}),$$
(15.16)

where V represents pair interaction. The Hamiltonian includes a one-electron potential U, which appears in the case of electrons in solids.

Using field operators equations of motion, like

$$i\hbar \frac{\partial}{\partial t} \hat{\psi}_{H\sigma}(\mathbf{x}, t) = \left[\hat{\psi}_{H\sigma}(\mathbf{x}, t), \mathcal{H} \right],$$
(15.17)

we obtain the equation of motion for the Green function. First, operating with $i\hbar\partial/\partial t$ on (15.6), we get

$$i\hbar \frac{\partial}{\partial t} G_{\sigma\sigma'}\left(\mathbf{x}, t; \mathbf{x}', t'\right) = (-i)\, i\hbar \frac{\partial}{\partial t} \Big[\Theta(t - t') \left\langle \hat{\psi}_{H\sigma}(\mathbf{x}, t)\, \hat{\psi}_{H\sigma'}^\dagger(\mathbf{x}', t') \right\rangle$$

$$- \Theta(t' - t) \left\langle \hat{\psi}_{H\sigma'}^\dagger(\mathbf{x}', t')\, \hat{\psi}_{H\sigma}(\mathbf{x}, t) \right\rangle \Big],$$
(15.18)

where we use the notation

$$\langle \mathcal{O} \rangle = \frac{\langle \Phi_0 | \mathcal{O} | \Phi_0 \rangle}{\langle \Phi_0 | \Phi_0 \rangle}.$$

Performing the time derivative on the right-hand side, we have

$$i\hbar \frac{\partial}{\partial t} G_{\sigma\sigma'}\left(\mathbf{x},t;\mathbf{x}',t'\right) = \hbar\left[\delta(t-t')\left\langle \hat{\psi}_{H\sigma}(\mathbf{x},t)\hat{\psi}^{\dagger}_{H\sigma'}\left(\mathbf{x}',t'\right) + \hat{\psi}^{\dagger}_{H\sigma'}\left(\mathbf{x}',t'\right)\hat{\psi}_{H\sigma}(\mathbf{x},t)\right\rangle\right.$$

$$+ \Theta(t-t')\left\langle \frac{\partial}{\partial t}\hat{\psi}_{H\sigma}(\mathbf{x},t)\,\hat{\psi}^{\dagger}_{H\sigma'}\left(\mathbf{x}',t'\right)\right\rangle$$

$$\left. - \Theta(t'-t)\left\langle \hat{\psi}^{\dagger}_{H\sigma'}\left(\mathbf{x}',t'\right)\frac{\partial}{\partial t}\hat{\psi}_{H\sigma}(\mathbf{x},t)\right\rangle\right]. \tag{15.19}$$

But,

$$\delta(t-t')\left\langle \hat{\psi}_{H\sigma}(\mathbf{x},t)\,\hat{\psi}^{\dagger}_{H\sigma'}\left(\mathbf{x}',t'\right) + \hat{\psi}^{\dagger}_{H\sigma'}\left(\mathbf{x}',t'\right)\,\hat{\psi}_{H\sigma}(\mathbf{x},t)\right\rangle$$

$$= \hat{\psi}_{H\sigma}(\mathbf{x},t)\,\hat{\psi}^{\dagger}_{H\sigma'}\left(\mathbf{x}',t\right) + \hat{\psi}^{\dagger}_{H\sigma'}\left(\mathbf{x}',t\right)\,\hat{\psi}_{H\sigma}(\mathbf{x},t)$$

$$= \left\langle e^{i\mathcal{H}t/\hbar}\left[\hat{\psi}_{S\sigma}(\mathbf{x}),\hat{\psi}^{\dagger}_{S\sigma'}\left(\mathbf{x}'\right)\right]_{+} e^{-i\mathcal{H}t/\hbar}\right\rangle = \delta\left(\mathbf{x}-\mathbf{x}'\right)\delta_{\sigma\sigma'},$$

which yields

$$i\frac{\partial}{\partial t}G_{\sigma\sigma'}\left(\mathbf{x},t;\mathbf{x}',t'\right) = \delta\left(\mathbf{x}-\mathbf{x}'\right)\delta(t-t')\delta_{\sigma\sigma'}$$

$$+ \left\langle T\left\{-\frac{i}{\hbar}\left[\hat{\psi}_{H\sigma}(\mathbf{x},t),\mathcal{H}\right]\hat{\psi}^{\dagger}_{H\sigma'}\left(\mathbf{x}',t'\right)\right\}\right\rangle. \tag{15.20}$$

By transforming the commutator in (15.20) to the Schrödinger picture

$$\left[\hat{\psi}_{H\sigma}(\mathbf{x},t),\mathcal{H}\right] = e^{i\mathcal{H}t/\hbar}\left[\hat{\psi}_{\sigma}(\mathbf{x}),\mathcal{H}\right]e^{-i\mathcal{H}t/\hbar},$$

we obtain

$$\left[\hat{\psi}_{\sigma}(\mathbf{x}),\mathcal{H}\right] = \left[\frac{-\hbar^2}{2m}\nabla^2 + U(\mathbf{x})\right]\hat{\psi}_{\sigma}(\mathbf{x})$$

$$+ \int d\mathbf{x}'' \sum_{\sigma''} \hat{\psi}^{\dagger}_{\sigma''}(\mathbf{x}'')\,V\left(\left|\mathbf{x}-\mathbf{x}''\right|\right)\hat{\psi}_{\sigma''}(\mathbf{x}'')\hat{\psi}_{\sigma}(\mathbf{x}), \tag{15.21}$$

which in the Heisenberg picture takes the form

$$\left[\hat{\psi}_{H\sigma}(\mathbf{x},t),\mathcal{H}\right] = \left[\frac{-\hbar^2}{2m}\nabla^2 + U(\mathbf{x})\right]\hat{\psi}_{H\sigma}(\mathbf{x},t)$$

$$+ \int d\mathbf{x}'' \sum_{\sigma''} \hat{\psi}^{\dagger}_{H\sigma''}(\mathbf{x}'',t)V\left(\left|\mathbf{x}-\mathbf{x}''\right|\right)\hat{\psi}_{H\sigma''}(\mathbf{x}'',t)\hat{\psi}_{H\sigma}(\mathbf{x},t). \tag{15.22}$$

To generalize to time-dependent interactions, which will be useful later, we write

$$V\left(\mathbf{x},t;\mathbf{x}'',t''\right) = V\left(\left|\mathbf{x}-\mathbf{x}''\right|\right)\delta(t-t'').$$

Substituting back in (15.20), we write the Green function equation of motion as

$$
\mathbb{I} = \left\{ i \frac{\partial}{\partial t} - \frac{1}{\hbar} \left[\frac{-\hbar^2}{2m} \nabla^2 + U(\mathbf{x}) \right] \right\} G_{\sigma\sigma'}(\mathbf{x},t;\mathbf{x}',t') + \frac{i}{\hbar} \int d\mathbf{x}'' \, dt'' \, V(\mathbf{x},t;\mathbf{x}'',t'')
$$

$$
\times \left\langle T \left[\hat{\psi}_{H\sigma''}^{\dagger}(\mathbf{x}'',t'') \, \hat{\psi}_{H\sigma''}(\mathbf{x}'',t'') \, \hat{\psi}_{H\sigma}(\mathbf{x},t) \, \hat{\psi}_{H\sigma'}^{\dagger}(\mathbf{x}',t') \right] \right\rangle, \tag{15.23}
$$

where we use the notation

$$
\mathbb{I} = \delta(\mathbf{x} - \mathbf{x}') \, \delta(t - t') \, \delta_{\sigma\sigma'}.
$$

In the absence of interactions, (15.23) becomes

$$
\left\{ i \frac{\partial}{\partial t} - \frac{1}{\hbar} \left[\frac{-\hbar^2}{2m} \nabla^2 + U(\mathbf{x}) \right] \right\} G_{\sigma\sigma'}^{(0)}(\mathbf{x},t;\mathbf{x}',t') = \mathbb{I}, \tag{15.24}
$$

i.e., $G_{\sigma\sigma'}^{(0)}(\mathbf{x},t;\mathbf{x}',t')$, which is just (15.5). Formally, the solution is

$$
G_{\sigma\sigma}^{(0)}(\mathbf{x},t;\mathbf{x}',t') = \left\{ i \frac{\partial}{\partial t} - \frac{1}{\hbar} \left[\frac{-\hbar^2}{2m} \nabla^2 + U(\mathbf{x}) \right] \right\}^{-1}. \tag{15.25}
$$

For a given potential $U(\mathbf{x})$, it is possible in many cases, for example in the case of a periodic potential, to obtain an explicit solution for the propagator.

Equation (15.23) is an integro-differential equation for the Green function, which in general has no explicit solution. We can view the linear differential equation (15.23) from a matrix perspective as having an infinite number of continuous and discrete indices. In that spirit, we can express (15.23) in such a way that a formal solution can be obtained as in the noninteracting case. We define the new matrix form

$$
\frac{i}{\hbar} \int d\mathbf{x}'' dt'' V(\mathbf{x},t;\mathbf{x}'',t'') \left\langle T \left[\hat{\psi}_{H\sigma''}^{\dagger}(\mathbf{x}'',t'') \hat{\psi}_{H\sigma''}(\mathbf{x}'',t'') \, \hat{\psi}_{H\sigma}(\mathbf{x},t) \hat{\psi}_{H\sigma'}^{\dagger}(\mathbf{x}',t') \right] \right\rangle
$$

$$
\equiv - \int d\mathbf{x}'' dt'' \, \Sigma_{\sigma\sigma''}^{*}(\mathbf{x},t;\mathbf{x}'',t'') \, G_{\sigma''\sigma'}(\mathbf{x}'',t'';\mathbf{x}',t') \tag{15.26}
$$

and express (15.23) in matrix notation as

$$
\left[\left(G^{(0)} \right)^{-1} - \Sigma^* \right] G = \mathbb{I}, \tag{15.27}
$$

which yields the formal solution

$$
G = \left[\left(G^{(0)} \right)^{-1} - \Sigma^* \right]^{-1} = G^{(0)} + G^{(0)} \Sigma^* G, \tag{15.28}
$$

which is the matrix form of the Dyson equation for the one-particle Green function, or the Dyson series. As we will show in Section 15.2.6, Σ^* is the one-particle *irreducible self-energy*; its physical meaning will be explained in the next section. We will discuss the Dyson equation in the context of a perturbation theory and Feynman diagrams.

15.2.4 Physical Interpretation of the One-Particle Green Function and the Self-Energy

Before we develop the full formal structure of the Green function framework, we shall present several of its general features that illustrate its physical content. To further bring to light the physical content of these features, we will present tangible connections to experimental applications.

Lehmann Representation

We consider the Green function of (15.6), with $\langle \Phi_0 | \Phi_0 \rangle = 1$, namely

$$i G_{\sigma\sigma'} \left(\mathbf{x}, t; \mathbf{x}', t' \right) = \left\langle \Phi_0 \left| T \left[\hat{\psi}_{H\sigma} (\mathbf{x}, t) \, \hat{\psi}^{\dagger}_{H\sigma'} \left(\mathbf{x}', t' \right) \right] \right| \Phi_0 \right\rangle, \tag{15.29}$$

where, again, $|\Phi_o\rangle$ is the exact Heisenberg ground state. We introduce the complete set of eigenstates $\{|\Phi_n\rangle\}$ of the Hamiltonian, defined on the Fock space, and thus contain any number of particles, namely

$$\sum_n |\Phi_n\rangle \langle \Phi_n| = \mathbb{I} \qquad \text{Sum over Fock space}$$

$$e^{-i\mathcal{H}t/\hbar} |\Phi_n(N)\rangle = e^{-i E_n(N)t/\hbar} |\Phi_n(N)\rangle, \tag{15.30}$$

where $|\Phi_n(N)\rangle$ and $E_n(N)$ are the nth eigenstate in the sub-Hilbert space for N particles and the corresponding eigenvalue, respectively.

We insert this identity operator between the creation and annihilation field operators:

$$i G_{\sigma\sigma'} \left(\mathbf{x}, t; \mathbf{x}', t' \right) = \sum_n \left[\Theta(t - t') \, \langle \Phi_0 | \, \hat{\psi}_{H\sigma} (\mathbf{x}, t) \, |\Phi_n\rangle \, \langle \Phi_n | \, \hat{\psi}^{\dagger}_{H\sigma'} (\mathbf{x}', t') \, |\Phi_0\rangle \right.$$

$$\left. - \Theta(t' - t) \, \langle \Phi_0 | \, \hat{\psi}^{\dagger}_{H\sigma'} (\mathbf{x}', t') \, |\Phi_n\rangle \, \langle \Phi_n | \, \hat{\psi}_{H\sigma} (\mathbf{x}, t) \, |\Phi_0\rangle \right]. \tag{15.31}$$

Transforming the Heisenberg creation and annihilation operators to the Schrödinger picture and using (15.30), we obtain

$$i G_{\sigma\sigma'} \left(\mathbf{x}, t; \mathbf{x}', t' \right) = \sum_n \left[\Theta(t - t') e^{-i(E_n - E_0)(t-t')/\hbar} \, \langle \Phi_0 | \, \hat{\psi}_{\sigma} (\mathbf{x}) \, |\Phi_n\rangle \, \langle \Phi_n | \, \hat{\psi}^{\dagger}_{\sigma'} (\mathbf{x}') \, |\Phi_0\rangle \right.$$

$$\left. - \Theta(t' - t) \, e^{i(E_n - E_0)(t-t')/\hbar} \, \langle \Phi_0 | \, \hat{\psi}^{\dagger}_{\sigma'} (\mathbf{x}') \, |\Phi_n\rangle \, \langle \Phi_n | \, \hat{\psi}_{\sigma} (\mathbf{x}) \, |\Phi_0\rangle \right], \tag{15.32}$$

where E_0 is the ground-state energy. Because the Hamiltonian is time independent, namely, time-translational invariant, G depends only on the difference $\tau = t - t'$.

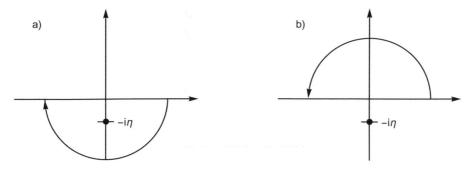

Figure 15.2 (a) Contour integral for $t > t'$. (b) Contour integral for $t < t'$.

To simplify our analysis, we use the following integral transformation for the time-ordering Θ functions

$$\Theta(\tau) = \lim_{\eta \to 0} - \int_{-\infty}^{\infty} \frac{d\omega}{2\pi i} \frac{e^{-i\omega\tau}}{\omega + i\eta}.$$

Furthermore, we Fourier-transform the Green function into the frequency domain, and obtain the desired form of the Green function via analytic continuation in the complex ω-plane, as shown in Figure 15.2. To delineate this approach, we consider the first term in (15.32), and write

$$\int d\tau \, e^{i\omega\tau} \, G_{\sigma\sigma'}\left(\mathbf{x}, \mathbf{x}'; \tau\right) \to (-i) \int d\tau \, e^{i\omega\tau} \, \Theta(\tau) \, e^{-i(E_n - E_0)\tau/\hbar}$$

$$= -(-i) \int_{-\infty}^{\infty} \frac{d\omega'}{2\pi i} \frac{1}{\omega' + i\eta} \int d\tau \, e^{i\omega\tau} \, e^{-i\omega'\tau} \, e^{-i(E_n - E_0)\tau/\hbar}$$

$$= \frac{1}{\omega - \frac{1}{\hbar}(E_n - E_0) + i\eta}, \tag{15.33}$$

where we have used the residue theorem.

Examination of the matrix element in first term in (15.32)

$$\langle \Phi_0 | \, \hat{\psi}_\sigma(\mathbf{x}) \, | \Phi_n \rangle \, \langle \Phi_n | \, \hat{\psi}_{\sigma'}^\dagger(\mathbf{x}') \, | \Phi_0 \rangle$$

shows that the state $|\Phi_n\rangle$ must correspond to an excited state with $N + 1$ particles, since $|\Phi_0\rangle$ is the ground state for N particles. Consequently, its eigenvalue E_n is the energy of an excited state of the system with $N+1$ particles. We denote it explicitly as $E_n \to E_n(N+1)$. Thus, we should explicitly identify the energy denominator in (15.33) as

$$E_n - E_0 \to E_n(N + 1) - E_0(N). \tag{15.34}$$

A similar calculation shows that for the second term in (15.32), the energies in the denominator should be modified to read $E_0 - E_n \to E_0(N) - E_n(N - 1)$. Thus, the final result is

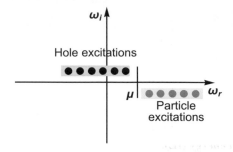

Figure 15.3 Representation of the poles of $G(\omega)$ in the complex ω plane. Black and gray discs correspond to hole excitations below the chemical potential μ, and to particle excitations above μ, respectively.

$$G_{\sigma\beta}\left(\mathbf{x},\mathbf{x}';\omega\right) = \hbar \sum_n \left[\frac{\langle \Phi_0| \hat{\psi}_\sigma(\mathbf{x}) |\Phi_n\rangle \langle \Phi_n| \hat{\psi}^\dagger_{\sigma'}(\mathbf{x}') |\Phi_0\rangle}{\hbar\omega - [E_n(N+1) - E_0(N)] + i\hbar\eta} \right.$$

$$\left. + \frac{\langle \Phi_0| \hat{\psi}^\dagger_{\sigma'}(\mathbf{x}') |\Phi_n\rangle \langle \Phi_n| \hat{\psi}_\sigma(\mathbf{x}) |\Phi_0\rangle}{\hbar\omega - [E_0(N) - E_n(N-1)] - i\hbar\eta} \right]. \tag{15.35}$$

Equation (15.35) reveals that the poles of the one-particle Green function are precisely the one-particle excitations of the interacting system. The first term describes excitations associated with a particle added to the system. Thus, it defines particle excitations with energies above the chemical potential, namely,

$$E_n(N+1) - E_0 = E_n(N+1) - E_0(N+1) + E_0(N+1) - E_0(N) = \varepsilon_n(N+1) + \mu,$$

where $E_0(N+1)$ is the ground state of the $N+1$ system. The quantity $E_0(N+1) - E_0(N)$ is identified with the chemical potential μ, since the volume of the system is kept constant. Consequently, the energy difference $E_n(N+1) - E_0(N+1) = \varepsilon_n(N+1)$ must be just the *excitation energy* of the $N+1$ system, with the proviso that $\varepsilon_n(N+1) \geq 0$. The second term describes the transition to a system with one particle less, with manifest hole excitations having energies below the chemical potential. This merger of hole and electron excitations in a single function is one of the notable advantages of the time-ordered Green function. Figure 15.3 demarcates the locations of the singularities of the Green function in the Lehmann representation.

An Alternative Identification

Let us consider the energy terms appearing in the denominators. They can be rewritten as

$$E_n(N+1) - E_0(N) = \left(E_n(N+1) - E_0(N+1)\right) + \left(E_0(N+1) - E_0(N)\right)$$

$$E_0(N) - E_n(N-1) = \left(E_0(N) - E_0(N-1)\right) + \left(E_0(N-1) - E_n(N-1)\right).$$

- The difference $E_0(N+1) - E_0(N)$ represents the minimum energy needed to add one electron to a system of N electrons.

  It is the *electron affinity* (EA):

$$EA = E_0(N+1) - E_0(N).$$

- The difference $E_0(N) - E_0(N-1)$ represents the minimum energy needed to remove one electron to a system of N electrons.

  It is the *ionization energy* (IE):

$$IE = E_0(N) - E_0(N-1).$$

Green Function for the Fermi Gas

As a simple example, we derive the Green function for the Sommerfeld gas, for which all the excited states are known. The corresponding field operators are given by

$$\hat{\psi}_\sigma(\mathbf{x}) = \frac{1}{\sqrt{\Omega}} \sum_{\mathbf{k},\sigma} e^{i\mathbf{k}\cdot\mathbf{x}}\, c_{\mathbf{k}\sigma}, \qquad \hat{\psi}_\sigma^\dagger(\mathbf{x}) = \frac{1}{\sqrt{\Omega}} \sum_{\mathbf{k},\sigma} e^{-i\mathbf{k}\cdot\mathbf{x}}\, c_{\mathbf{k}\sigma}^\dagger,$$

where Ω is the volume of the system. We write the matrix elements in (15.35) as

$$
\begin{aligned}
M &= \langle\Phi_0|\,\hat{\psi}_\sigma(\mathbf{x})\,|\Phi_n\rangle\,\langle\Phi_n|\,\hat{\psi}_{\sigma'}^\dagger(\mathbf{x}')\,|\Phi_0\rangle \\
&= \frac{1}{\Omega} \sum_{\mathbf{k},\mathbf{k}'} e^{i\mathbf{k}\cdot\mathbf{x}}\, e^{-i\mathbf{k}'\cdot\mathbf{x}'}\,\langle\Phi_0|\,c_{\mathbf{k}\sigma}\,|\Phi_n\rangle\,\langle\Phi_n|\,c_{\mathbf{k}'\sigma'}^\dagger\,|\Phi_0\rangle
\end{aligned}
\tag{15.36}
$$

and with eigenstates $|\Phi_n\rangle = |\mathbf{k}\sigma\rangle = c_{\mathbf{k}\sigma}^\dagger\,|\Phi_0\rangle$,

$$M = \frac{1}{\Omega} \sum_{\mathbf{k}} e^{i\mathbf{k}\cdot(\mathbf{x}-\mathbf{x}')}\,\langle\Phi_0|\,c_{\mathbf{k}\sigma}\,|\mathbf{k}\sigma\rangle\,\langle\mathbf{k}\sigma|\,c_{\mathbf{k}\sigma}^\dagger\,|\Phi_0\rangle. \tag{15.37}$$

The translational invariance of the electron gas leads to the $\mathbf{x} - \mathbf{x}'$ dependence. Inserting this expression into (15.35), and applying spatiotemporal Fourier transformation, we arrive at

$$G_{\sigma\sigma'}(\mathbf{k},\omega) = \delta_{\sigma\sigma'}\,\hbar \left[\frac{|\langle\Phi_0|\,c_{\mathbf{k}\sigma}\,|\mathbf{k}\sigma\rangle|^2}{\hbar\omega - \varepsilon_{\mathbf{k}} + i\hbar\eta} + \frac{\left|\langle\Phi_0|\,c_{\mathbf{k}\sigma}^\dagger\,|\overline{\mathbf{k}\sigma}\rangle\right|^2}{\hbar\omega - \varepsilon_{\mathbf{k}} - i\hbar\eta} \right], \tag{15.38}$$

where $|\overline{\mathbf{k}\sigma}\rangle = c_{\mathbf{k}\sigma}\,|\Phi_0\rangle$ represents the ground state with one particle at energy $\varepsilon_{\mathbf{k}}$ removed. We notice that the poles of the one-particle Green function are just the one-particle excitation energies for particles and holes, with the energies of particle states above the chemical potential, and the states for holes energies below the chemical potential, satisfying $\varepsilon_{\mathbf{k}} > 0$, as we stressed in the preceding discussions.

Since the ground state of the Sommerfeld gas is just the Fermi sea, $|\Phi_0\rangle = \prod_{|\mathbf{k}|<k_F,\sigma} c_{\mathbf{k}\sigma}^\dagger |0\rangle$, we can simply write

$$|\langle\Phi_0| c_{\mathbf{k}\sigma} |\mathbf{k}\sigma\rangle|^2 = \Theta(|\mathbf{k}| - k_F), \quad \left|\langle\Phi_0| c_{\mathbf{k}\sigma}^\dagger |\overline{\mathbf{k}\sigma}\rangle\right|^2 = \Theta(k_F - |\mathbf{k}|). \tag{15.39}$$

15.2.5 Spectral Functions

The spectral function, or spectral weight, $A(\nu, \omega)$ can be thought of as either the quantum state resolution of a particle with given energy ω or as the energy resolution for a particle in a given quantum number ν. It gives an indication of how suitable the excitation created by adding or removing a particle in state ν can be described by a free noninteracting particle or hole, respectively.

To be more specific, we define two spectral functions $A^+(\mathbf{k}, \omega)$ and $A^-(\mathbf{k}, \omega)$ that give the probability of a particle $(+)$ or a hole $(-)$ with momentum $\mathbf{k}$ and energy ω to be in an exact eigenstate of the system with $N + 1$ or $N - 1$ particles, respectively. They are defined as

$$A^+(\mathbf{k}, \omega) = \frac{1}{\Omega} \sum_n |\langle\Phi_0| c_{\mathbf{k}\sigma} |\Phi_n\rangle|^2 \, \delta\left(\hbar\omega - [E_n(N+1) - E_0(N)]\right)$$

$$A^-(\mathbf{k}, \omega) = \frac{1}{\Omega} \sum_n \left|\langle\Phi_0| c_{\mathbf{k}\sigma}^\dagger |\Phi_n\rangle\right|^2 \, \delta\left(\hbar\omega - [E_0(N) - E_n(N-1)]\right). \tag{15.40}$$

It is clear from this definition and (15.35) that

$$A^+(\mathbf{k}, \omega) + A^-(\mathbf{k}, \omega) = \text{Im}\left[G(\mathbf{k}, \omega)\right]. \tag{15.41}$$

The spectral functions are manifestly real and positive definite, which allows their interpretation as a probability to find a single particle excitation with energy ω and momentum $\mathbf{k}$, in the present case. We find that

$$\int_{-\infty}^{\infty} d\omega A(\mathbf{k}, \omega) = \frac{1}{\Omega} \sum_n \left|\langle\Phi_0| c_{\mathbf{k}\sigma}^\dagger |\Phi_n\rangle\right|^2 \int_{-\infty}^{\infty} d\omega \delta\left(\hbar\omega - [E_0(N) - E_n(N-1)]\right)$$

$$= \frac{1}{\Omega} \sum_n \langle\Phi_0| c_{\mathbf{k}\sigma} |\Phi_n\rangle \langle\Phi_n| c_{\mathbf{k}\sigma}^\dagger |\Phi_0\rangle = \langle\Phi_0| c_{\mathbf{k}\sigma} c_{\mathbf{k}\sigma}^\dagger |\Phi_0\rangle$$

$$= \langle\Phi_0| \Phi_0\rangle = 1.$$

Thus, the spectral function satisfies the sum rule

$$\int_{-\infty}^{\infty} d\omega \, A(\mathbf{k}, \omega) = 1 \tag{15.42}$$

being a probability, irrespective of the Hamiltonian. The matrix elements in (15.40) include only the states involving one particle *more* or *less* in the system. Consequently, they specify the one-particle density of states resolved in momentum.

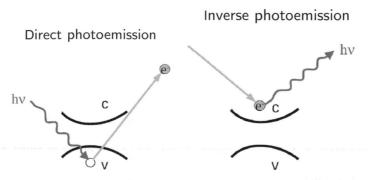

Figure 15.4 Green function contains spectral information on single-particle excitations changing the number of particles by one! The poles of the GF give the corresponding excitation energies.

We note that the spectral functions are accessible experimentally with the techniques of angle-resolved photoemission spectroscopy (ARPES) and inverse photoemission spectroscopy, as shown in Figure 15.4. However, inverse photoemission has not achieved the accuracy necessary to measure spectral functions.

Spectral Function in ARPES

In order to see how the spectral function of an electron system can be studied experimentally by ARPES, we need to introduce a few simplifying assumptions. We note from the outset that in photoemission spectroscopy a photon absorbed by an electron provides enough energy to the electron to be ejected out of the system. The collected body of data is compiled by recording the energy and momentum of the outgoing electrons. Here, we will describe how the measured spectrum of the photo-ejected electrons is a direct manifestation of the spectral function $A(\mathbf{k}, \omega)$.

A schematic of the experimental setup is shown in upper part of Figure 15.5. The upper-left of the figure depicts the different stages of photon beam collimation, as well as the beam monochromator, prior to sample photoexcitation. The ejected photoelectrons are collected, angle-resolved, and energy-resolved in the electron energy analyzer shown in the bottom-right of the figure.

The bottom of Figure 15.5 shows the geometric setup of the ARPES experiment. The direction of the incident photons together with the surface normal define the scattering plane.

The energy-resolving capability of the detector allows the determination of the photo-ejected electron energy ε, while its orientation and angle-resolving capability allows the determination of its momentum $\mathbf{k}$.

We choose the Coulomb gauge for the electromagnetic field, $\nabla \cdot \mathbf{A}(\mathbf{x}, t) = 0$, and set the electrostatic potential to zero, $\Phi(\mathbf{x}, t) = 0$. The linear coupling of the electrons to the electromagnetic field is given by

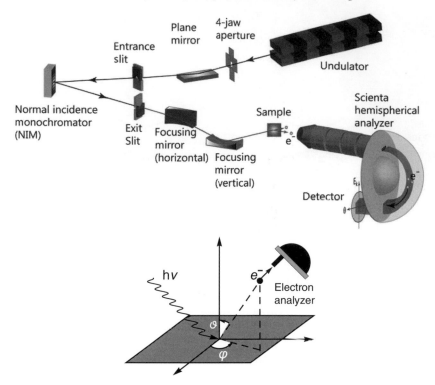

Figure 15.5 Top: a schematic of a typical ARPES end station in a synchrotron light source facility. From [119]. Bottom: geometric setup of the ARPES experiment, defining the orientation of the incident photon and the angle-resolving detector.

$$V(\mathbf{x},t) = \frac{e}{m} \sum_i \mathbf{A}(\mathbf{x}_i,t) \cdot \mathbf{p}_i,$$

where $\mathbf{x}_i$ and $\mathbf{p}_i$ are the position and momentum of electron i, respectively. We consider a periodic solid of volume Ω, where the application of Born–von Karman boundary conditions yield Bloch function eigenstates

$$\Psi_{\nu\mathbf{k}}(\mathbf{x}) = e^{i\mathbf{k}\cdot\mathbf{x}} u_{\nu\mathbf{k}}(\mathbf{x})$$

of the single-particle Hamiltonian

$$\mathcal{H}_0 = \sum_i \left[\frac{p_i^2}{2m} + U(\mathbf{x}_i) \right]$$

with band index ν and $\mathbf{k} \in$ BZ.

In second-quantized representation with respect to the Bloch function basis, we have

$$\mathcal{H}_0 = \sum_{\nu \mathbf{k} \sigma} \varepsilon_{\nu \mathbf{k}} c^\dagger_{\nu \mathbf{k} \sigma} c_{\nu \mathbf{k} \sigma}$$

$$V(\mathbf{x}, t) = \sum_{\substack{\nu \mu \\ \mathbf{k}, \mathbf{k}' \sigma}} \left\langle \nu \mathbf{k} \left| \frac{e}{m} \mathbf{A}(\mathbf{x}, t) \cdot \mathbf{p} \right| \mu \mathbf{k}' \right\rangle c^\dagger_{\nu \mathbf{k} \sigma} c_{\mu \mathbf{k}' \sigma}. \tag{15.43}$$

We consider a monochromatic electromagnetic wave

$$\mathbf{A}(\mathbf{x}, t) = \mathbf{A}_0 \cos (\mathbf{q} \cdot \mathbf{x} - \omega t)$$

so that the matrix elements become

$$\left\langle \nu \mathbf{k} \left| \mathbf{A}(\mathbf{x}) \cdot \mathbf{p} \right| \mu \mathbf{k}' \right\rangle = \left[\delta \left(\mathbf{k}' - \mathbf{k} + \mathbf{q} \right) + \delta \left(\mathbf{k}' - \mathbf{k} - \mathbf{q} \right) \right] v_{\nu \mu} \left(\mathbf{k}, \mathbf{k}' \right)$$

$$v_{\nu \mu} \left(\mathbf{k}, \mathbf{k}' \right) = \frac{e \hbar}{2m} \int d\mathbf{x} \, u_{\nu \mathbf{k}}(\mathbf{x}) \, \mathbf{A}_0 \cdot \left(\mathbf{k}' - i \nabla \right) u_{\mu \mathbf{k}'}(\mathbf{x}). \tag{15.44}$$

We obtain

$$V(\mathbf{x}, t) = V(\mathbf{q}) \, e^{-i \omega t} + V(-\mathbf{q}) \, e^{i \omega t}$$

$$V(\mathbf{q}) = \sum_{\substack{\nu \mu \\ \mathbf{k} \sigma}} v_{\nu \mu} \left(\mathbf{k}, \mathbf{k} - \mathbf{q} \right) c^\dagger_{\nu \mathbf{k} \sigma} c_{\mu (\mathbf{k} - \mathbf{q}) \sigma}. \tag{15.45}$$

The total Hamiltonian of the system is

$$\mathcal{H} = \mathcal{H}_0 + \mathcal{H}_{ee} + V(\mathbf{x}, t),$$

where $\mathcal{H}_{ee}$ represents the electron–electron interactions.

The experiment starts with the system in its N-particle ground state $|\Psi_0(N)\rangle$. In the first step of a two-step process shown in Figure 15.6, photon absorption induces an electron to transition from the single-particle state $\mu \mathbf{k}'$, with energy $\varepsilon_{\mu \mathbf{k}'}$, to the single-particle state $\nu \mathbf{k}$, with energy $\varepsilon_{\nu \mathbf{k}} = \varepsilon_{\mu \mathbf{k}'} + \hbar \omega$, above the vacuum level.

The transition rate due to absorption of energy $\hbar \omega$ and momentum $\hbar \mathbf{q}$ from the electromagnetic field is given by Fermi's Golden Rule,

$$P_{0 \to n} = \frac{2\pi}{\hbar} \sum_n |\langle \Psi_n(N) | V(\mathbf{q}) | \Psi_0(N) \rangle|^2 \, \delta \left(\hbar \omega - E_n(N) + E_0(N) \right). \tag{15.46}$$

In the second step of the photoemission process, the high-energy outgoing electron, $\nu \mathbf{k}$ remains unscattered by other electrons, and is ejected and recorded by the detector.

Effectively, we can think that the photon removes one particle with momentum $\mathbf{k}$, directly measured, from the system. The determination of $\mu \mathbf{k}'$ and $\varepsilon_{\mu \mathbf{k}'}$ allows for the mapping out of the band structure of the occupied levels.

Thus, we may write

$$|\Psi_n(N)\rangle = c^\dagger_{\nu \mathbf{k} \sigma} |\Psi_n(N - 1)\rangle$$

$$E_n(N) \simeq \varepsilon_{\nu \mathbf{k}} + E_n(N - 1), \tag{15.47}$$

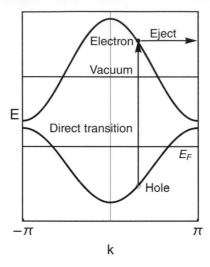

Figure 15.6 The two-step photoemission process.

where $|\Psi_n(N-1)\rangle$ are the eigenstates of $\mathcal{H}_0 + \mathcal{H}_{ee}$ for $N-1$ electrons. The transition matrix elements are then given by

$$\left\langle \Psi_n(N-1)c_{\mu,\mathbf{k}+\mathbf{q},\sigma} \left| V(\mathbf{q}) \right| \Psi_0(N) \right\rangle = \sum_{\mu\sigma} v_{\nu\mu}(\mathbf{k},\mathbf{k}+\mathbf{q}) \left\langle \Psi_n(N-1) \left| c_{\mu,\mathbf{k}+\mathbf{q},\sigma} \right| \Psi_0(N) \right\rangle.$$

For pedagogical reasons, we assume that in the ground state all bands are occupied except one. In this case, there are no contributions to (15.46) originating from two different bands, and we get

$$P_{0\to n} = \frac{2\pi}{\hbar} \sum_{n\mu\sigma} \left| v_{\nu\mu}(\mathbf{k},\mathbf{k}+\mathbf{q}) \right|^2 \left| \langle \Psi_n(N-1) | c_{\mu,\mathbf{k}+\mathbf{q},\sigma} | \Psi_0(N)\rangle \right|^2$$

$$\times \, \delta \left(\hbar\omega - \varepsilon_{\nu\mathbf{k}} - E_n(N-1) + E_0(N) \right). \tag{15.48}$$

The contribution from a single band λ is proportional to

$$A_\lambda = \sum_n \left| \langle \Psi_n(N-1) | c_{\mu,\mathbf{k}+\mathbf{q},\sigma} | \Psi_0(N)\rangle \right|^2 \, \delta \left(\omega' - \frac{E_n(N-1) + E_n(N) + \mu}{\hbar} \right),$$

where $\omega' = \omega - \dfrac{\varepsilon_{\nu\mathbf{k}} - \mu}{\hbar}$. Comparison with (15.40) confirms that this is the spectral density of the single-particle Green function.

Density of States

The density of states for one-particle excitations is obtained by summing over momenta

$$\mathcal{D}(\omega) = \sum_{\mathbf{k}} \left[A^+(\mathbf{k},\omega) + A^-(\mathbf{k},\omega) \right] \tag{15.49}$$

and covers states with energies above and below the Fermi energy. We can discern that the spectral function is related to the Green function as

$$G(\mathbf{k}, \omega) = \int_{-\infty}^{\infty} d\omega' \left[\frac{A^+(\mathbf{k}, \omega')}{\omega - \omega' + i\eta} + \frac{A^-(\mathbf{k}, \omega')}{\omega - \omega' - i\eta} \right]. \tag{15.50}$$

Using the principal value

$$\frac{1}{\omega \pm i\eta} = \mathcal{P}\frac{1}{\omega} \mp i\pi\delta(\omega),$$

we write

$$G(\mathbf{k}, \omega) = \mathcal{P} \int_{-\infty}^{\infty} d\omega' \frac{A^+(\mathbf{k}, \omega')}{\omega - \omega' + i\eta} - i\pi A^+(\mathbf{k}, \omega) \tag{15.51}$$

for $\hbar\omega > \mu = E_F$. The reality of the integral allows to write

$$A^+(\mathbf{k}, \omega) = -\frac{1}{\pi} \operatorname{Im} G(\mathbf{k}, \omega) \qquad \text{for } \omega > \frac{\mu}{\hbar}$$

$$A^-(\mathbf{k}, \omega) = \frac{1}{\pi} \operatorname{Im} G(\mathbf{k}, \omega) \qquad \text{for } \omega < \frac{\mu}{\hbar}. \tag{15.52}$$

Thus, demonstrating that the one particle Green function contains the spectral function for particles and holes measured from the chemical potential. Since the density of states is given by the sum over momentum of the spectral functions, it can be expressed in terms of the local Green function

$$\mathcal{D}(\omega) = -\frac{1}{\pi} \operatorname{Im} G(\mathbf{x}, \mathbf{x}; \omega) \qquad \text{for } \omega > \frac{\mu}{\hbar}$$

$$\mathcal{D}(\omega) = \frac{1}{\pi} \operatorname{Im} G(\mathbf{x}, \mathbf{x}; \omega) \qquad \text{for } \omega > \frac{\mu}{\hbar}. \tag{15.53}$$

We again note that the one-particle Green function is directly related to experimentally accessible quantities such as the spectral function (angular-resolved photoemission and inverse photoemission) and the density of states (angle-integrated photoemission).

15.2.6 Spectral Signatures in the One-Particle Green Function

Having demonstrated that the Green function contains useful information about the one-particle states, we will now survey the identifiable signatures of those states.

Noninteracting Fermion System

We start with the simple case of noninteracting fermions with a dispersion relation $\varepsilon_0(\mathbf{k})$, shown in Figure 15.7 for the one-dimension case.

The Green function for the noninteracting case can be written, using (15.38) and (15.39), as

$$G_0(\mathbf{k}, \omega) = \frac{\hbar}{\Omega} \left[\frac{\Theta(|\mathbf{k}| - k_F)}{\hbar\omega - \varepsilon_0(\mathbf{k}) + i\hbar\eta} + \frac{\Theta(k_F - |\mathbf{k}|)}{\hbar\omega - \varepsilon_0(\mathbf{k}) - i\hbar\eta} \right]. \tag{15.54}$$

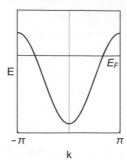

Figure 15.7 Dispersion relation for a one-dimensional free electron tight-binding model with nearest neighbor hopping t: $\varepsilon_0(\mathbf{k}) = -2t \cos ka$.

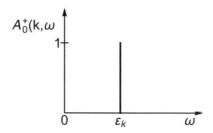

Figure 15.8 Free fermion spectral weight.

The corresponding spectral function, obtained from (15.52), is

$$A_0^+(\mathbf{k}, \omega) = \frac{\hbar}{\Omega} \delta(\hbar\omega - \varepsilon_0(\mathbf{k})) \qquad \text{for } |\mathbf{k}| > k_F$$

$$A_0^-(\mathbf{k}, \omega) = \frac{\hbar}{\Omega} \delta(\hbar\omega - \varepsilon_0(\mathbf{k})) \qquad \text{for } |\mathbf{k}| < k_F. \qquad (15.55)$$

The spectral function shows that for a noninteracting system, every single-particle state has a unity weight, aside from a normalization factor, as shown in Figure 15.8. It confirms that all the weight is on the one-particle, or hole, state. We note that an excitation with energy ω can only happen by adding an electron, or a hole, to the state $\mathbf{k}$ such that $\varepsilon_0(\mathbf{k}) = \omega$.

Interacting Fermion System: Quasiparticles' Effective Mass and Spectral Weight

Next, we turn on the interactions, so that the Green function is now given by (15.28), namely,

$$G(\mathbf{k}, \mathbf{k}'; \omega) = \frac{1}{\left[G^{(0)}(\mathbf{k}, \omega)\right]^{-1} - \Sigma^*(\mathbf{k}, \mathbf{k}'; \omega)},$$

where $\left(G^{(0)}\right)^{-1} = \hbar\omega - \varepsilon_0(\mathbf{k})$.

G and the self-energy Σ^* are, in general, nonlocal functions that depend on two spatial or momentum points. However, in the presence of translational or periodic invariance, they becomes diagonal in momentum. We can then write

$$G(\mathbf{k},\omega) = \frac{1}{\hbar\omega - \varepsilon_0(\mathbf{k}) - \Sigma^*(\mathbf{k},\omega)} = \frac{1}{\hbar\omega - \varepsilon_0(\mathbf{k}) - \mathrm{Re}\,\Sigma^*(\mathbf{k},\omega) - \mathrm{Im}\,\Sigma^*(\mathbf{k},\omega)}. \tag{15.56}$$

$\Sigma^*(\mathbf{k},\omega)$ could be a messy complex function, but it is clear that its real part modifies, or renormalizes, the noninteracting dispersion $\varepsilon_0(\mathbf{k})$, while its imaginary part replaces the delta function of the pole with a finite width profile, a Lorentzian line shape.

The imaginary part accounts for a finite lifetime in the state $(\mathbf{k},\omega)$, and arises from electron–electron, electron–phonon, or other scattering events. As will be demonstrated in the subsections to follow, only states close to the Fermi surface are relevant to our analysis, and for such states, scattering events due to the electron–electron interaction lead to a quadratic contribution to the line broadening, namely, $-\mathrm{Im}\,\Sigma^*(\mathbf{k},\omega) \propto (\hbar\omega - E_F)^2$.

For simplicity, we set $E_F = 0$ in the following, which can be achieved by shifting $\varepsilon_0(\mathbf{k})$ correspondingly. Since we are only interested in low-energy excitations close to the Fermi level, we perform a Taylor expansion of $\varepsilon(\mathbf{k}) = \varepsilon_0(\mathbf{k}) + \mathrm{Re}\,\Sigma^*(\mathbf{k},\omega)$ around $k = k_F$ and $\omega = E_F = 0$. We thus only retain terms linear in $\Delta\mathbf{k} = k - k_F$ and ω, and obtain

$$\varepsilon(\mathbf{k}) = \varepsilon_0(\mathbf{k}) + \mathrm{Re}\,\Sigma^*(\mathbf{k},\omega) = \left[\frac{k_F}{m} + \frac{\partial\,\mathrm{Re}\Sigma^*}{\partial k}\bigg|_{k=k_F}\right]\Delta\mathbf{k} + \frac{\partial\mathrm{Re}\,\Sigma^*}{\partial\omega}\bigg|_{\omega=0}\omega. \tag{15.57}$$

We caution that in (15.57), $\Delta\mathbf{k}$ and the derivatives are taken along a direction perpendicular to the Fermi surface.

After substituting (15.57) in (15.56), we can cast the interacting Green function for small ω and $\Delta\mathbf{k}$ in the form

$$G(\mathbf{k},\omega) = \frac{Z}{\omega - \varepsilon(\mathbf{k})}, \tag{15.58}$$

which resembles the expression for $G^{(0)}$, with the following modifications:

1. Spectral weight factor

$$Z = \left(1 - \frac{\partial\,\mathrm{Re}\,\Sigma^*(\mathbf{k},\omega)}{\partial\omega}\bigg|_{\omega=0}\right)^{-1}. \tag{15.59}$$

2. Mass renormalization

$$\varepsilon_0(\mathbf{k}) = (k - k_F)\frac{k_F}{m} \quad\Rightarrow\quad \varepsilon(\mathbf{k}) = (k - k_F)\frac{k_F}{m^*} \tag{15.60}$$

$$\frac{m^*}{m} = Z^{-1}\left(1 + \nabla_{\mathbf{k}}\mathrm{Re}\,\Sigma^*\bigg|_{k=k_F}\right)^{-1}, \tag{15.61}$$

which indicates that the dispersion relation has been changed by the interactions. We thus see that in keeping single-particle excitations, they will have to be modified due to interactions: a different mass from the one of independent particle emerges. This renormalization of the mass by interaction conforms with the experimental findings for fermionic systems, as for example in the measurements of electronic specific heat discussed in Chapter 18 on Fermi liquids.

A particle whose lifetime becomes finite and/or its energy gets renormalized is referred to as a *quasiparticle*, in order to differentiate it from its kin in a noninteracting environment. The quasiparticle consists of the original real, individual particle, plus a cloud of disturbed neighbors interacting with it. It behaves very much like an individual particle, except that it has an effective mass and a lifetime.[4] The idea of quasiparticles was first proposed by Lev Landau in his phenomenological theory of Fermi liquids, which was originally formulated for studying liquid ^{3}He. More will be said about quasiparticles in Chapter 18 on Fermi liquids.

Particle Lifetime in the Vicinity of the Fermi Surface

In the previous section, we introduced the concept of dispersion renormalization and particle lifetime through the new quantity of self-energy arising from interparticle interactions. Here, we examine the idea of particle lifetime and its relation to the Fermi surface in more detail. An important revelation from the following analysis is that the concept of a quasiparticle is crucially dependent on the existence of a Fermi surface.

We consider the consequences of injecting a quasiparticle into a state above but close to the Fermi surface:

$$(\varepsilon = E_F + \delta\varepsilon, \mathbf{k}), \qquad k > k_F, \quad \frac{\delta\varepsilon}{E_F} \ll 1.$$

We expect that the quasiparticle will be scattered by particles in the Fermi sea, with $\left(\varepsilon' < E_F, \; k' < k_F\right)$, as shown in Figure 15.9. We denote the final scattering states with a subscript "1." Since all the states in the Fermi sea are occupied, both particles should be scattered to states outside the Fermi sea. Applying momentum and energy conservation to the scattering process, we write

$$\varepsilon + \varepsilon' = \varepsilon_1 + \varepsilon_1'$$
$$\mathbf{k} + \mathbf{k}' = \mathbf{k}_1 + \mathbf{k}_1'.$$

We expect that the scattering rate $\gamma_\mathbf{k}$ to be proportional to the accessible phase space, namely,

$$\gamma_\mathbf{k} \propto \int d\mathbf{k}' \int d\mathbf{k}_1 \int d\mathbf{k}_1' \, \delta\left(\mathbf{k} + \mathbf{k}' - [\mathbf{k}_1 + \mathbf{k}_1']\right). \tag{15.62}$$

---

[4] There also exist other kinds of fictitious particles in many-body systems, namely, manifestations of collective excitations. These do not center around individual particles, but instead involve collective, wavelike motion of all the particles in the system simultaneously. They are typified by phonons, magnons, rotons, etc.

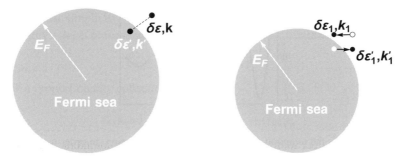

Figure 15.9 Left: initial state with a single quasiparticle excitation above a filled Fermi sea. Right: final state with two quasiparticles and a quasihole.

We use the densities of states to change the integrations over momenta in (15.62) into integrations over energies:

$$
\gamma_{\mathbf{k}} \propto \int_{-\infty}^{E_F} d\varepsilon'\, \mathcal{D}(\varepsilon') \int_{E_F}^{\infty} d\varepsilon_1\, \mathcal{D}(\varepsilon_1) \int_{E_F}^{\infty} d\varepsilon_1'\, \mathcal{D}(\varepsilon_1')\, \delta\left(\varepsilon + \varepsilon' - [\varepsilon_1 + \varepsilon_1']\right)
$$

$$
\leq \int_{E_F}^{E_F+\delta\varepsilon} d\varepsilon_1\, \mathcal{D}(\varepsilon_1) \int_{E_F}^{E_F+\delta\varepsilon} d\varepsilon_1'\, \mathcal{D}(\varepsilon_1')\, \mathcal{D}\left(\varepsilon_1 + \varepsilon_1' - \varepsilon\right)
$$

$$
\sim \mathcal{D}^3(E_F)\, (\delta\varepsilon)^2. \tag{15.63}
$$

The fact that $\gamma \propto (\varepsilon - E_F)^2$ in the vicinity of the Fermi energy leads us to conclude that the existence of a Fermi surface allows quasiparticles to become more clearly defined as the Fermi surface is approached. As a matter of fact, the lifetime becomes infinite on the Fermi surface, since γ has to change sign on crossing the Fermi surface. The same phase space argument reveals that the particle lifetime quickly approaches zero as we move away from the vicinity of the Fermi surface! The implication of this finding is that a particle injected into a state with $\varepsilon \gg E_F$ will immediately scatter into other states, leaving no trace of coherent information about its propagation; in other words, it will propagate *incoherently*. There are no poles in the Green function for such scattering events.

This reasoning reveals that a quasiparticle lifetime is determined via the constraint imposed by the Pauli principle on accessible states, and that it is independent of the interaction type, and therefore is quite general. In the absence of such constraint, it is not clear how manifest sharp quasiparticles would be discernible in the presence of interactions. Actually, we would have naively expected exactly the opposite scenario to the one derived: the strength and type of the interaction should determine the lifetime, $\tau = 1/\Gamma$, of single-particle states, while the particle's energy $\varepsilon(\mathbf{k})$ defines the periodicity of its wavefunction oscillations in time $e^{-i\varepsilon(\mathbf{k})t/\hbar}$. Thus, in order to be

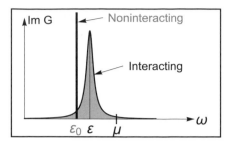

Figure 15.12 Spectral function for a particle in a noninteracting system (black delta function) and of the corresponding quasiparticle after turning on interactions (gray filled curve). Notice the shift in energy and the linewidth broadening.

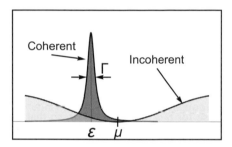

Figure 15.13 Typical spectral function for a quasiparticle displaying its coherent component and the incoherent background.

The modifications to the spectral function arising from the presence of interactions are shown in Figures 15.12 and 15.13.

We can also write the full spectral function as

$$A^+(\mathbf{k}, \omega) = A^+_{\text{coherent}}(\mathbf{k}, \omega) + A^+_{\text{incoherent}}(\mathbf{k}, \omega),$$

which is represented schematically in Figure 15.13: the contributions from the coherent part of the spectral function appears as a Lorentzian line shape, while the background arises from the incoherent part. If the quasiparticle weight vanishes, then the background is the only manifestation in the spectral function.

The coherent part of the spectral function has the following features:

- It has a Lorentzian profile peaked at $\varepsilon(\mathbf{k})$.
- The single-particle energies are renormalized, $\varepsilon_0(\mathbf{k}) \rightarrow \varepsilon(\mathbf{k})$.
- Its width is given by the inverse of the lifetime.
- Its magnitude is proportional to the quasiparticle weight Z, a positive number between 0 and 1.
- The area under the peak has decreased (from 1 to Z).

The incoherent part $A_{\text{incoherent}}(\mathbf{k}, \omega)$ presents a continuum, not a peak. It must be there if $Z \neq 1$, in order to satisfy sum rule (15.42). A similar result is obtained for the hole spectral function, but we should note that $\Gamma < 0$ in this case.

Fermi Liquids and Non-Fermi Liquids

Fermionic systems in which this picture holds are called Fermi liquids, and the theory describing them is known as Fermi liquid theory. It was first developed phenomenologically by Landau around 1957, and, in the subsequent years, was given a microscopic foundation with the aid of many-body perturbation theory. The phenomenological version will be described in Chapter 18.

We also note that the identification and investigation of interacting fermionic systems that do not obey Fermi liquid theory is an important research area in current many-body physics. Such non-Fermi liquids by definition have spectral functions that cannot be approximated by the form (15.67), and therefore they cannot be qualitatively understood in terms of a picture of noninteracting fermions. One prominent example of a non-Fermi liquid is the so-called Luttinger liquid, which occurs in one spatial dimension. It will also be covered in Chapter 18.

15.3 Time-Evolution Operator in the Interaction Picture

We have seen in Section 2.3 that for a general many-body problem, attempts to construct an equation of motion for the Green function led to the generation of an infinite hierarchy of coupled differential equations involving higher- and higher-order Green functions. However, it is often possible to extract meaningful answers using decoupling approximations, where a link in the chain of equations is broken at some order, and the higher-order Green functions, beyond the link, expressed in terms of lower ones.

In the following section, we shall develop an alternative approach, based on a powerful perturbation theory method for the Green function. It presents an iterative integral representation in time. This Feynman–Dyson method [54, 55, 61, 62], forms the basis of relativistic quantum field theory, and has been adapted quite successfully for nonrelativistic many-body physics [4, 5]. It will also be presented in the concise and systematic language of Feynman rules and diagrams.[5]

We recall that the definition of the Green function (15.6) involved operators in the Heisenberg picture, or Heisenberg operators. However, the development and practical realization of the perturbation theory hinges on transforming state vectors and operators to the interaction picture. We remember from quantum mechanics that the interaction framework can be considered as intermediate between the Schrödinger and Heisenberg pictures, where now both operators and state vectors are time dependent. Most importantly, operators in this picture acquire a simple time dependence, where time evolution is determined only by the unperturbed Hamiltonian, while the nontrivial part corresponding to the interactions is

---

[5] See [93]. This article gives an interesting historical perspective to the development of Feynman diagrams.

incorporated into the wavefunctions. Thus, we find that the time-evolution operator plays an important role in the development of quantum field theory and quantum many-body physics. We will delineate this interaction picture formulation here.

We start with the Hamiltonian

$$\mathcal{H} = \mathcal{H}_0 + \mathcal{H}'. \tag{15.68}$$

In the absence of the interaction term $\mathcal{H}'$, the problem can be solved exactly. We define the state vector in the interaction picture, $|\psi_I(t)\rangle$, in terms of its counterpart in the Schrödinger one, $|\psi_S(t)\rangle$, as

$$|\psi_I((t)\rangle = \exp\left[i\frac{\mathcal{H}_0 t}{\hbar}\right] |\psi_S(t)\rangle = \exp\left[i\frac{\mathcal{H}_0 t}{\hbar}\right] \exp\left[-i\frac{\mathcal{H}t}{\hbar}\right] |\psi_S(0)\rangle. \tag{15.69}$$

Notice that

$$\exp\left[i\frac{\mathcal{H}_0 t}{\hbar}\right] \exp\left[-i\frac{\mathcal{H}t}{\hbar}\right] \neq \exp\left[-i\frac{\mathcal{H}'t}{\hbar}\right]$$

when $\mathcal{H}_0$ and $\mathcal{H}$ do not commute. Now, consider the time evolution of $|\psi_I((t)\rangle$

$$i\hbar\frac{\partial \psi_I}{\partial t} = -\mathcal{H}_0 \exp\left[i\frac{\mathcal{H}_0 t}{\hbar}\right] |\psi_S(t)\rangle + \exp\left[i\frac{\mathcal{H}_0 t}{\hbar}\right] \left(i\hbar\frac{\partial}{\partial t}\right) |\psi_S(t)\rangle$$

$$= \left\{-\mathcal{H}_0 + \exp\left[i\frac{\mathcal{H}_0 t}{\hbar}\right] \mathcal{H} \exp\left[-i\frac{\mathcal{H}_0 t}{\hbar}\right]\right\} |\psi_I(t)\rangle = \mathcal{H}_I(t) |\psi_I(t)\rangle$$

$$\mathcal{H}_I(t) = \exp\left[i\frac{\mathcal{H}_0 t}{\hbar}\right] \mathcal{H}' \exp\left[-i\frac{\mathcal{H}_0 t}{\hbar}\right]. \tag{15.70}$$

Since in general $\mathcal{H}_0$ and $\mathcal{H}'$ do not commute, we have to respect the order in which they appear. We note that the definition of $\mathcal{H}_I(t)$ in (15.70) can be extended to other operators. To see this, we consider an arbitrary matrix element of some operator

$$\langle\psi_S(t)| \mathcal{O}_S |\psi_S(t)\rangle = \langle\psi_I(t)| e^{i\mathcal{H}_0 t/\hbar} \mathcal{O}_S e^{-i\mathcal{H}_0 t/\hbar} |\psi_I(t)\rangle = \langle\psi_I(t)| \mathcal{O}_I(t) |\psi_I(t)\rangle \tag{15.71}$$

and we find that

$$\mathcal{O}_I = e^{i\mathcal{H}_0 t/\hbar} \mathcal{O}_S e^{-i\mathcal{H}_0 t/\hbar}$$

Thus, in the interaction picture both states and operators depend on time. It is straightforward then to write the equation of motion for an interaction picture operator as

$$i\hbar\frac{\partial \mathcal{O}_I(t)}{\partial t} = \exp\left[i\frac{\mathcal{H}_0 t}{\hbar}\right] \left(\mathcal{O}_S \mathcal{H}_0 - \mathcal{H}_0 \mathcal{O}_S\right) \exp\left[-i\frac{\mathcal{H}_0 t}{\hbar}\right] = \left[\mathcal{O}_I(t), \mathcal{H}_0\right], \tag{15.72}$$

that is, it is simply determined by the noninteracting part.

Next we consider the time-evolution operator for a state vector in the interaction picture. It should be unitary and engenders the state at time t, namely,

$$|\psi_I(t)\rangle = U(t) |\psi_I(0)\rangle. \tag{15.73}$$

U should also fulfill the initial condition

$$U(0) = 1. \tag{15.74}$$

From (15.70) and (15.73), we have

$$i\hbar \frac{\partial U(t)}{\partial t} = \mathcal{H}_I\, U(t). \tag{15.75}$$

Integrating this equation from time $t = 0$ to time t, we have

$$U(t) - U(0) = -\frac{i}{\hbar} \int_0^t dt_1\, \mathcal{H}_I(t_1)\, U(t_1), \tag{15.76}$$

or, given the initial condition,

$$U(t) = 1 - \frac{i}{\hbar} \int_0^t dt_1\, \mathcal{H}_I(t_1)\, U(t_1). \tag{15.77}$$

This is an integral equation, and an iterative solution can be obtained as

$$U(t) = 1 - \frac{i}{\hbar} \int_0^t dt_1\, \mathcal{H}_I(t_1) + \left(\frac{i}{\hbar}\right)^2 \int_0^t dt_1 \int_0^{t_1} dt_2\, \mathcal{H}_I(t_1)\, \mathcal{H}_I(t_2) + \cdots \tag{15.78}$$

As we remarked earlier, in general $\mathcal{H}_0$ and $\mathcal{H}_I$ do not commute. To overcome this difficulty, Schwinger invented a device called the *time-ordering operator*.

Time-Ordering Operator

Suppose $\{O_1(t_1), O_2(t_2)\ldots O_N(t_N)\}$ is a set of operators at different times $\{t_1, t_2 \ldots t_N\}$. If P is the permutation that orders the times, so that $t_{P1} > t_{P2} \ldots > t_{PN}$, then if the operators are entirely bosonic, containing an even number of fermionic operators, the time-ordering operator is defined as

$$T\,[O_1(t_1)\, O_2(t_2)\ldots O_N(t_N)] = O_{P1}(t_{P1}),\, O_{P2}(t_{P2})\ldots O_{PN}(t_{PN}). \tag{15.79}$$

We note that if the operator set contains fermionic operators, composed of an odd number of fermionic operators, then

$$T\,[F_1(t_1)\, F_2(t_2)\ldots F_N(t_N)] = (-1)^P\, F_{P1}(t_{P1}),\, F_{P2}(t_{P2})\ldots F_{PN}(t_{PN}), \tag{15.80}$$

where P is the number of pairwise permutations of fermions involved in the time-ordering process.

We cast (15.78) in a more symmetric form, by rewriting the term with two integrals as

$$\int_0^t dt_1 \int_0^{t_1} dt_2\, \mathcal{H}_I(t_1)\, \mathcal{H}_I(t_2) = \int_0^t dt_2 \int_{t_2}^t dt_1\, \mathcal{H}_I(t_1)\, \mathcal{H}_I(t_2)$$

$$= \int_0^t dt_1 \int_{t_1}^t dt_2\, \mathcal{H}_I(t_2)\, \mathcal{H}_I(t_1), \tag{15.81}$$

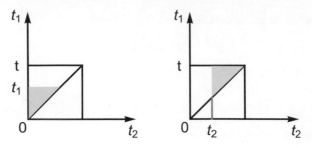

Figure 15.14 Regions of integration for the first two integrals in (15.80).

where the first equality was obtained because the integration regions are equal as shown in Figure 15.14. The second line results from a convenient change of variables. The contribution to the time-evolution operator in second order can be written, with the aid of (15.80), as

$$\int_0^t dt_1 \int_0^{t_1} dt_2 \, \mathcal{H}_I(t_1) \, \mathcal{H}_I(t_2) = \frac{1}{2} \int_0^t dt_1 \int_0^t dt_2 \, [\mathcal{H}_I(t_1) \, \mathcal{H}_I(t_2) \, \Theta(t_1 - t_2)$$

$$+ \mathcal{H}_I(t_2) \, \mathcal{H}_I(t_1)) \, \Theta(t_2 - t_1)]$$

$$= \frac{1}{2} \int_0^t dt_1 \int_0^t dt_2 \, T \, [\mathcal{H}_I(t_1) \, \mathcal{H}_I(t_2)] . \qquad (15.82)$$

where we introduced a time-ordering operator T that operates on the square bracket to its right. The bracket contains a product of operators affected by it, with the operator at later time to the left of that at earlier time. It is clear that this can be generalized to any order n, by properly taking into account the number of possible time orderings, which is just the permutations of n time variables, and, of course, we have to divide by $n!$.

$$U(t) = \sum_{n=0}^{\infty} \frac{1}{n!} \left(\frac{-i}{\hbar} \right)^n \int_0^t dt_1 \ldots \int_0^t dt_n \, T \, [\mathcal{H}_I(t_1) \ldots \mathcal{H}_I(t_n)]$$

$$= T \, \exp\left[-\frac{i}{\hbar} \int_0^t dt_1 \, \mathcal{H}_I(t_1) \right] . \qquad (15.83)$$

15.3.1 The S-Matrix (S-Operator)

We now come to one of the most important constructions in diagrammatic perturbation theory, which is known as the S-matrix, although in our context it is more convenient to think of it as an operator than a matrix. We have discussed at length the time-evolution operator $U(t)$, which evolves the wavefunction in the interaction representation $|\psi_I(t)\rangle = U(t) \, |\psi_I(0)\rangle$.

Now we would like to define a similar, but slightly more general object $S(t, t_0)$, which evolves the system from time t_0 to time t

$$|\psi_I(t)\rangle = S(t, t_0) |\psi_I(t_0)\rangle \tag{15.84}$$

$$S(t_0, t_0) = 1.$$

Writing

$$|\psi_I(t)\rangle = U(t) |\psi_I(0)\rangle$$

$$|\psi_I(t_0)\rangle = U(t_0) |\psi_I(0)\rangle \quad \Rightarrow \quad U^\dagger(t_0) |\psi_I(t_0)\rangle = |\psi_I(0)\rangle,$$

we find

$$S(t, t_0) = U(t) U^\dagger(t_0) = U(t) U^{-1}(t_0)$$

since U is unitary. It follows that

- $S(t_0, t) = S^\dagger(t, t_0)$.
- S has the composition property

$$S(t, t_1) = S(t, t_0) S(t_0, t_1), \quad t > t_0 > t_1.$$

- S satisfies the equation

$$i\hbar \frac{\partial S(t, t_0)}{\partial t} = \mathcal{H}_I S(t, t_0)$$

and, hence,

$$S(t, t_0) = \sum_{n=0}^{\infty} \frac{1}{n!} \left(\frac{-i}{\hbar} \right)^n \int_{t_0}^{t} dt_1 \ldots \int_{t_0}^{t} dt_n \, T \, [\mathcal{H}_I(t_1) \ldots \mathcal{H}_I(t_n)]$$

$$= T \exp\left[-\frac{i}{\hbar} \int_{t_0}^{t} dt_1 \, \mathcal{H}_I(t_1) \right]. \tag{15.85}$$

15.4 Perturbation Theory and Feynman Diagrams

In this section, we present the basic ingredients for constructing a perturbation expansion for the Green function, and develop a pedagogical outline of the method of Feynman diagrams, focusing on fermionic systems. The goal is not to present a complete account of the diagrams, but to lay down the groundwork to understanding them.

(15.84) plays an important role in the development of the Feynman–Dyson perturbation theory. In the definition (15.6) of the Green function, the average value was taken with respect to the Heisenberg ground-state vector, $|\psi_H\rangle$, which is time independent, but usually very complicated for interacting systems, and no exact solution is achievable. The strategy adopted in the theory is to set

$$|\psi_H\rangle = |\psi_I(0)\rangle = S^\dagger(t, 0) |\psi_I(t)\rangle$$

and use the time $t \to -\infty$ to interpolate to the noninteracting ground state with the aid of the principle of adiabatic switching on, as will be described in the following subsection.

15.4.1 The Concept of Adiabaticity in Quantum Physics

We have earlier explored the attributes and consequences of adiabaticity in the context of Berry's phase. The construct of adiabaticity can also play an important role in quantum many-body theory. It can possibly pave a path that allows us to understand a many-body problem, even when we can only arrive at an approximate solution.

In order to examine the usefulness of adiabaticity in our current endeavor, we consider the scenario where we have a many-body system with Hamiltonian

$$\mathcal{H} = \mathcal{H}_0 + \mathcal{H}'$$

for which no exact solution of its ground state $\left|\Psi_g\right\rangle$ can be obtained. But, we do know the exact solution for $\mathcal{H}_0$, with ground state $\left|\Phi_g\right\rangle$. To connect the two systems, we start with the Hamiltonian $\mathcal{H}_0$, having a ground state $\left|\Phi_g\right\rangle$, and adiabatically slowly turn on $\mathcal{H}'$ until we reach $\mathcal{H}$. Mathematically, this can be achieved by setting

$$\mathcal{H}(t) = \mathcal{H}_0 + \lambda(t)\,\mathcal{H}', \quad \Rightarrow \quad \lambda(t) = e^{-\eta|t|} \quad \eta > 0 \ \text{ is arbitrarily small.} \tag{15.86}$$

and varying $-\infty \leq t \rightarrow 0$. The efficacy of this approach hinges on whether $\left|\Phi_g\right\rangle$ has the same basic symmetries as $\left|\Psi_g\right\rangle$, namely, they share the same quantum number labels.

There are two possible paths for the process of adiabatic switching on in a many-body system:

(i) *The ground-state symmetry remains invariant throughout the adiabatic evolution*, as λ increases smoothly from 0 to 1. In such cases, the corresponding evolution of the energy spectrum, within the symmetry subspace of the ground state, should behave as the one drawn schematically in Figure 15.15(a). It shows, for simplicity, a discrete spectrum with a series of *avoided crossings* that allow us to follow adiabatically the nth excited level at $\lambda = 0$ into the nth excited level at $\lambda = 1$. Note that the crossings are avoided because the states have the same symmetry. The *avoided crossing* arises because an intersection of two levels having the same symmetry is removed because, in general, the off-diagonal matrix elements between the two states will not vanish, and thus will give rise to mutual repulsion.

(ii) *The ground-state symmetry changes during the evolution due to a crossing of an energy level with different symmetry.* As shown in Figure 15.15(b), it is possible for the energy levels of states with different symmetries to cross, because selection rules give vanishing matrix elements between the states, and thus prevent them from mixing. This scenario may lead to an adiabatic evolution where *level crossing* occurs at some $\lambda = \lambda_c$. Within such a scenario, an excited state Φ of $\mathcal{H}_0$, with a symmetry different from that of the ground state $\left|\Phi_g\right\rangle$, may cross at λ_c to a lower energy than the ground state, resulting in a symmetry change of the ground state. A simple example is when a ferromagnetic ground state becomes stabilized by interactions; however, in this case a continuous rotational symmetry of the spin is broken (spontaneous symmetry breaking).

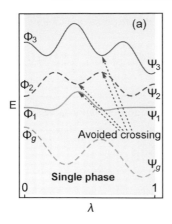 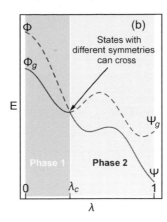

Figure 15.15 (a) Adiabatic evolution of a discrete spectrum within the symmetry subspace of the ground state. The ground state can be adiabatically evolved all the way to $\lambda = 1$, from Φ_g to Ψ_g. (b) Adiabatic evolution of Hilbert space with λ. A phase transition occurs at $\lambda = \lambda_c$, where an excited state of $\mathcal{H}_0$, with a symmetry different from that of the ground state, crosses below the ground state.

Thus the switching-on procedure becomes troublesome if it leads to instabilities at some critical value of $\lambda = \lambda_c$. The presence of a λ_c ushers a phase transition whereby the ground state changes its symmetry. If the transition is continuous, it signals a quantum critical transition initiated at a quantum critical point.

In many cases of interest, there is no symmetry-changing phase transition as the interaction $\mathcal{H}'$ is turned on, and the procedure of adiabatic evolution can be used to turn on "interactions" and to evolve the ground state from $\left|\Phi_g\right\rangle$ to $\left|\Psi_g\right\rangle$. The adiabatic concept, together with the Gell-Mann–Low theorem, are key elements in the development of perturbation theory and Feynman diagrams.

15.4.2 Adiabatic "Switching on" and the Gell-Mann–Low Theorem

In a 1951 seminal paper, Murray Gell-Mann and Francis Low [73] used the adiabatic construct to establish the connection between the interacting and noninteracting Green functions, in their now famous Gell-Mann–Low theorem.

We note that the ground state $\left|\Psi_0\right\rangle$, that appeared in the definition of the Green function of the interacting system, (15.6), must be a time-independent state in the Heisenberg picture. Since, so far, we only know how to calculate expectation values in the noninteracting ground state $\left|\Phi_0\right\rangle$, we need to connect the states of the exactly soluble Hamiltonian $\mathcal{H}_0$, where in principle everything is known, to the ones of the full Hamiltonian $\mathcal{H} = \mathcal{H}_0 + \mathcal{H}'$. For this purpose, we use the time-dependent Hamiltonian of (15.86), which allows for the adiabatic switching on:

- For $|t| \to \infty$, the Hamiltonian reduces to $\mathcal{H}_0$.
- At $t = 0$, it becomes the interacting problem we are interested in.
- In the limit $\eta \to 0$, the interaction is switched on adiabatically.

Replacing $\mathcal{H}' \rightarrow e^{-\eta|t|}\mathcal{H}'$ in the scattering operator expression (15.85), we obtain

$$S_\eta(t, t_0) = \sum_{n=0}^{\infty} \frac{1}{n!} \left(\frac{-i}{\hbar}\right)^n \int_{t_0}^{t} dt_1 \ldots \int_{t_0}^{t} dt_n \, e^{-\eta(|t_1|\ldots+|t_n|)} \, T \left[\mathcal{H}_I(t_1) \ldots \mathcal{H}_I(t_n)\right].$$

$$(15.87)$$

As $t_0 \rightarrow -\infty$, the effect of the interaction vanishes and one approaches $\mathcal{H}_0$:

$$t_0 \rightarrow -\infty \qquad \begin{cases} \mathcal{H}_0 \, |\Phi_0\rangle \;=\; E^{(0)} \, |\Phi_0\rangle \\[2mm] |\Psi_I(t_0)\rangle \;\rightarrow\; |\Phi_0\rangle . \end{cases} \qquad (15.88)$$

Recalling the relation among state vectors in the different pictures, we write

$$|\Psi_H\rangle = |\Psi_S(t = 0)\rangle = |\Psi_I(t = 0)\rangle,$$

where $t = 0$ corresponds to the fully interacting Hamiltonian. In particular, we obtain the following relation between the Heisenberg and interaction state vectors,

$$|\Psi_H\rangle = |\Psi_I(0)\rangle = S_\eta(0, -\infty) \, |\Phi_0\rangle \equiv S_\eta(0, -\infty) \, |-\infty\rangle, \qquad (15.89)$$

where introducing $|-\infty\rangle \equiv |\Phi_0\rangle$ simplifies the chronology of the evolving ground state. While we realize that (15.89) is determined through an adiabatic switching process, we recognize that it gives us the link between the eigenstates of the noninteracting system and the eigenstates of the fully interacting one, with $|\Psi_I(0)\rangle = |\Psi_H\rangle$ being the ground state of the interacting problem.

Gell-Mann–Low Theorem

"In the absence of any symmetry breaking transition, and in the limit $\eta \rightarrow 0$, $S_\eta \rightarrow S_I$ exists and provides the right solution."

$$|\Psi_H\rangle = |\Psi_I(0)\rangle = S_I(0, -\infty) \, |\Phi_0\rangle \equiv S_I(0, -\infty) \, |-\infty\rangle \qquad (15.90)$$

With this theorem in hand, we shall reexpress the Green function in a form in which the expectation values to be determined are those of the unperturbed ground state $|-\infty\rangle \equiv |\Phi_0\rangle$. The state vector defined in (15.90) can be written as

$$|\Psi_H\rangle = |\Psi_0\rangle = S_I(0, -\infty) \, |-\infty\rangle = S_I(0, \infty) \, S_I(\infty, -\infty) \, |-\infty\rangle.$$

Because of time-reversal symmetry of $e^{-\eta|t|}\mathcal{H}'$, namely, the interaction is again zero at $t = +\infty$, it is clear that

$$|+\infty\rangle = S_I(\infty, -\infty) \, |-\infty\rangle$$

is just the ground state $|\Phi_0\rangle$ of $\mathcal{H}_0$ up to a phase factor, provided that the state that develops from $|-\infty\rangle$ remains nondegenerate throughout the evolution process. We can then write

$$|+\infty\rangle = S_I(\infty, -\infty) \, |-\infty\rangle = e^{i2\sigma} \, |-\infty\rangle \quad \Rightarrow \quad e^{i2\sigma} = \langle-\infty \, |+\infty\rangle$$

and we obtain

$$\langle +\infty | = e^{-i2\alpha} \langle -\infty | = \frac{\langle -\infty |}{\langle -\infty | +\infty \rangle} = \frac{\langle -\infty |}{\langle -\infty | S_I(\infty, -\infty) | -\infty \rangle}. \tag{15.91}$$

Therefore, by using the expansion (15.85) of the S-matrix, we get a perturbation expansion in $\mathcal{H}'_I$, where our calculations now involve expectation values and time evolution expressed in terms of the unperturbed system.

15.4.3 One-Particle Green Function as Power Series in the Interaction

We have established the connection between the eigenstates of the noninteracting system with those of the fully interacting one, with the aid of the interaction picture. What remains is to relate the Heisenberg operators that appear in the definition of the Green function to the corresponding interaction picture operators.

To accomplish this, we use the relations between operators in the Heisenberg, Schrödinger, and interaction pictures to write

$$\mathcal{O}_H(t) = e^{i\mathcal{H}t/\hbar} \mathcal{O}_S e^{-i\mathcal{H}t/\hbar} = e^{i\mathcal{H}t/\hbar} e^{-i\mathcal{H}_0 t/\hbar} \mathcal{O}_I e^{i\mathcal{H}_0 t/\hbar} e^{-i\mathcal{H}t/\hbar}. \tag{15.92}$$

We recall from (15.69) that

$$|\psi_I((t)\rangle = \exp\left[i\frac{\mathcal{H}_0 t}{\hbar}\right] \exp\left[-i\frac{\mathcal{H}t}{\hbar}\right] |\psi_I(0)\rangle = S_I(t, 0) |\psi_I(0)\rangle$$

and we obtain

$$\mathcal{O}_H(t) = S_I(0, t) \mathcal{O}_I(t) S_I(t, 0). \tag{15.93}$$

For the product of two Heisenberg operators, we get

$$\mathcal{O}_H(t) \mathcal{O}'_H(t_1) = S_I(0, t) \mathcal{O}_I(t) S_I(t, 0) S_I(0, t_1) \mathcal{O}'_I(t_1) S_I(t_1, 0)$$

$$= S_I(0, t) \mathcal{O}_I(t) S_I(t, t_1) \mathcal{O}'_I(t_1) S_I(t_1, 0). \tag{15.94}$$

We now have all the tools to construct a perturbative form for the Green function

$$iG\left(x; x'\right) = \langle \Psi_H | T\left[\hat{\psi}_H(x)\, \hat{\psi}_H^\dagger\left(x'\right)\right] |\Psi_H \rangle$$

$$= \frac{\langle -\infty | S_I(\infty, 0) T\left[S_I(0, t)\, \hat{\psi}_I(x)\, S_I(t, 0)\, S_I(0, t')\hat{\psi}_I^\dagger\left(x'\right) S_I(t', 0)\right] S_I(0, -\infty) | -\infty \rangle}{\langle -\infty | S_I(\infty, 0) S_I(0, -\infty) | -\infty \rangle}, \tag{15.95}$$

where $x \equiv (\mathbf{x}, t, \sigma)$ for compactness. We replaced the Heisenberg operator $\hat{\psi}_H$ ($\hat{\psi}_H^\dagger$) by its interaction representation counterpart, which simply evolves only under the action of $\mathcal{H}_0$. We also replaced the ground state of the interaction problem $|\Psi_0\rangle$ by the noninteracting one $|\Phi_0\rangle$. We can combine terms such as $S_I(t', 0) S_I(0, -\infty)$ to $S_I(t', -\infty)$. Moreover, since the time-ordering operator is taking care of the proper time ordering anyhow, we can even combine all time propagations to a single one, $S_I(\infty, -\infty)$. Then the Green function becomes

$$
iG_{\sigma\sigma'}\left(\mathbf{x}, t; \mathbf{x}', t'\right) = \frac{\langle -\infty| \, T\left[S_I(\infty, -\infty)\hat{\psi}_{I\sigma}(\mathbf{x}, t)\,\hat{\psi}^{\dagger}_{I\sigma'}\left(\mathbf{x}', t'\right)\right]|-\infty\rangle}{\langle -\infty\,|S_I(\infty, -\infty)|-\infty\rangle}. \tag{15.96}
$$

Finally, we can replace $S_I(\infty, -\infty)$ by its Taylor expansion

$$
iG_{\sigma\sigma'}\left(\mathbf{x}, t; \mathbf{x}', t'\right) = \sum_{n=0}^{\infty} \frac{1}{n!} \left(\frac{-i}{\hbar}\right)^n \int_{-\infty}^{\infty} dt_1 \ldots \int_{-\infty}^{\infty} dt_n
$$

$$
\times \frac{\left\langle \Phi_0 \left| T\left[\mathcal{H}_I(t_1)\ldots\mathcal{H}_I(t_n)\,\hat{\psi}_{I\sigma}(\mathbf{x}, t)\,\hat{\psi}^{\dagger}_{I\sigma'}\left(\mathbf{x}', t'\right)\right]\right| \Phi_0\right\rangle}{\langle \Phi_0 \,|S_I(\infty, -\infty)|\, \Phi_0\rangle}. \tag{15.97}
$$

Now the one-particle Green function is explicitly written as an expansion in powers of the perturbation $\mathcal{H}_I$. There is just one more hurdle to overcome before actually implementing this scheme: we need to devise a method that would mitigate taking expectation values of large numbers of operators of a noninteracting system. This procedure is systematized with the application of Wick's theorem and simplified with the aid of Feynman diagrams.

15.4.4 Wick's Theorem, Normal Ordering, and Contractions

The arraying of the operators $\mathcal{H}_I$ is fixed by time ordering. However, as the interaction Hamiltonian consists of creation and annihilation operators, it would be convenient to move all annihilation operators to the right, where they can eliminate particles. This ordering process is called *normal ordering*. We note that the action of annihilation and creation operators $\hat{\psi}$ and $\hat{\psi}^{\dagger}$ on the vacuum is to annihilate it, namely,

$$
\hat{\psi}\,|\varnothing\rangle = 0, \quad \langle\varnothing|\,\hat{\psi}^{\dagger} = 0,
$$

which is true for both bosons and fermions. Moreover, since observables have null expectation values in the vacuum state, it is convenient to construct them in *normal order, with creation operators on the left of destruction operators*. Wick's theorem[6] tells us how to go from time-ordered to normal-ordered products – it is valid when the expectation values are with respect to a Hamiltonian that is quadratic in fermion operators.

Identification of Vacuum States

We recall from the case of superfluidity that the Bogoliubov canonical transformation generated new noninteracting quasiparticles from the original interacting bosons, and that the emerging ground state acted as a vacuum state for the new quasiparticles. We shall often encounter in coming chapters that a ground state $|\Psi_g\rangle$ of some incipient effective Hamiltonian will be taken as a reference vacuum state. Such a state is filled with particles or

---

[6] The general problem of bringing products of field operators into a normal form was solved in 1950 by Gian-Carlo Wick (1909–1992). He completed the work on his theorem while in Berkeley. It was aimed at providing a clear derivation of Feynman's diagrammatic rules for perturbation theory.

quasiparticles, and the effective Hamiltonian is expressed in a basis of *canonical* operators α_a and $\alpha_a^\dagger$, $a = 1, 2, \ldots$, where

$$\alpha_a \left| \Psi_g \right\rangle = 0, \qquad \left\langle \Psi_g \right| \alpha_a^\dagger = 0.$$

This action only holds for such operators, but not for other creation and destruction operators associated with the system.

To illustrate this process, we shall consider the noninteracting ground state $|\Phi_0\rangle$ describing the Fermi sea. We examine the action of the field operators $\hat{\psi}(\mathbf{x})$ and $\hat{\psi}^\dagger(\mathbf{x})$, and apply the canonical particle–hole transformation, we write

$$\hat{\psi}(\mathbf{x}) = \sum_{\mathbf{k}\sigma} \phi_{\mathbf{k}\sigma}(\mathbf{x}) c_{\mathbf{k}\sigma} = \begin{cases} \sum_{\mathbf{k}\sigma} \phi_{\mathbf{k}\sigma}(\mathbf{x}) d_{\mathbf{k}\sigma} = \hat{\psi}^{(-)}(\mathbf{x}), & k > k_F \text{ annihilate particles} \\ \sum_{\mathbf{k}\sigma} \phi_{\mathbf{k}\sigma}(\mathbf{x}) h^\dagger_{-\mathbf{k}\sigma} = \hat{\psi}^{(+)}(\mathbf{x}), & k < k_F \text{ create holes} \end{cases}$$

We find that[7]

$$\hat{\psi}^{(-)}(\mathbf{x}) \left| \Phi_0 \right\rangle = 0, \quad \hat{\psi}^{(+)\dagger}(\mathbf{x}) \left| \Phi_0 \right\rangle = 0.$$

Thus, $\hat{\psi}^{(-)}$ and $\hat{\psi}^{(+)\dagger}$ are annihilation parts, while $\hat{\psi}^{(+)}$ and $\hat{\psi}^{(-)\dagger}$ are creation parts.

In order to state Wick's theorem, we have to introduce and elaborate on the concepts of normal ordering and contractions.

Normal Ordering and Its Physical Significance

We will focus here on fermionic operators of a noninteracting system, which are relevant to our current discussion of perturbation theory with N interacting fermions.

We shall generalize the decomposition to other possible bases for the creation and annihilation operators, besides the position representation previously described. To simplify the notation, we will set A as a generic operator, and write

$$A = A^- + A^+.$$

A. Normal Ordering and Contractions

A product of operators $A_i^\pm$ is normally ordered if all factors A_i^- are on the right of the factors A_j^+:

$$A_1^+ \cdots A_k^+ A_{k+1}^- \cdots A_n^-.$$

In particular, a product of operators of the same type, $A_1^+ \cdots A_k^+$ or $A_1^- \ldots A_\ell^-$, and their permutations, is normally ordered.

The very usefulness of the definition is the obvious property that the expectation value on $\left| \Psi_g \right\rangle$ of a normally ordered operator product is always zero:

$$\left\langle \Psi_g \right| A_1^+ \cdots A_n^- \left| \Psi_g \right\rangle = 0.$$

---

[7] The notation $\pm$ carries over from the original application of Wick's theorem in relativistic quantum field theory, where (+) and (−) refer to a Lorentz-invariant decomposition into positive and negative frequency parts.

It is clear that any product of operators $A_1 A_2 \ldots A_n$ can be written as a sum of normally ordered terms. One first writes every factor as $A_i^+ + A_i^-$ and gets 2^n terms. In each term, the components A_i^- are brought to the right by successive anticommutations. After much boring work, the desired expression will be obtained. Wick's theorem is an efficient answer to this particular problem: to write a product $A_1 \ldots A_n$ as a sum of normally ordered terms. The theorem is an extremely useful operator identity, with important corollaries. To state and prove it, we need some technical tools.

The normal-ordering operator brings a generic product into a normal form. If the product contains k factors A_i^+ mixed with $n - k$ factors A_i^-, it is

$$N\left[A_1^\pm \ldots A_n^\pm\right] = (\pm 1)^P \, A_{i_1}^+ \cdots A_{i_k}^+ \cdots A_{i_n}^-, \qquad (15.98)$$

where P is the permutation that brings the sequence $1 \ldots n$ to be normal ordered into the ordered sequence $i_1 \ldots i_n$. It may appear that normal ordering is not unambiguous, since $(+)$ operators and $(-)$ operators can be given separately different orders. However, the different expressions are actually the same operator, because A^+ operators anticommute exactly among themselves, and the same is true for A^- operators. For example, $N\left[A_1^+ A_2^+\right]$ can be written as $A_1^+ A_2^+$, or with the factors exchanged, $-A_2^+ A_1^+$, the two operators coincide.

By linearity, we extend the action of N-ordering from products of components $A_i^\pm$ to products of operators A_i. For example:

$$
\begin{aligned}
N\left[A_1 A_2\right] &= N\left[\left(A_1^+ + A_1^-\right)\left(A_2^+ + A_2^-\right)\right] \\
&= N\left[A_1^+ A_2^+\right] + N\left[A_1^+ A_2^-\right] + N\left[A_1^- A_2^-\right] + N\left[A_1^- A_2^+\right] \\
&= A_1^+ A_2^+ + A_1^+ A_2^- + A_1^- A_2^- - A_2^+ A_1^-. \qquad (15.99)
\end{aligned}
$$

The following property follows from (15.98):

$$N\left[A_1 \ldots A_n\right] = (-1)^P \, N\left[A_{i_1} \ldots A_{i_n}\right].$$

A product $A_1 \ldots A_n$ can be written as a sum of normally ordered terms. For two operators, the process is straightforward:

$$
\begin{aligned}
A_1 A_2 &= \left(A_1^+ + A_1^-\right)\left(A_2^+ + A_2^-\right) \\
&= A_1^+ A_2^+ + A_1^+ A_2^- + A_1^- A_2^- + A_1^- A_2^+ = N\left[A_1 A_2\right] + \left\{A_1^-, A_2^+\right\}.
\end{aligned}
$$

The vanishing of the expectation of a normally ordered product implies

$$\left\{A_1^-, A_2^+\right\} = \left\langle \Psi_g \left| A_1 A_2 \right| \Psi_g \right\rangle.$$

This c-number is called a **contraction** of the two operators, and will be denoted with superscripts that label the pair:

$$A_1^c A_2^c = \left\{A_1^-, A_2^+\right\} = \left\langle \Psi_g \left| A_1 A_2 \right| \Psi_g \right\rangle.$$

We then write the useful formula

$$A_1 A_2 = N\left[A_1 A_2\right] + A_1^c A_2^c.$$

The contraction of two operators can be extended to the case where a product of n' operators is inserted between them:

$$A_1^c \, (A_{1'} \ldots A_{n'}) \, A_2^c = (-1)^{n'} \, A_1^c \, A_2^c \, (A_{1'} \ldots A_{n'}).$$

B. Wick's Theorem

We begin by stating two lemmas; the proof of the second can be done by induction. We wish to set into normal order a single $(-)$ operator that multiplies on the left some $(+)$ operators. We need to bring it to the right by iterated transpositions.

Lemma I:

$$
\begin{aligned}
A_0^- &\left(A_1^+ \cdots A_n^+ \right) \\
&= \left\{ A_0^-, A_1^+ \right\} \left(A_2^+ \cdots A_n^+ \right) - A_1^+ A_0^- \left(A_2^+ \cdots A_n^+ \right) \\
&= A_0^c A_1^c \left(A_2^+ \cdots A_n^+ \right) - A_1^+ \left\{ A_0^-, A_2^+ \right\} \left(A_3^+ \cdots A_n^+ \right) + A_1^+ A_2^+ A_0^- \left(A_3^+ \cdots A_n^+ \right) \\
&= A_0^c A_1^c \left(A_2^+ \cdots A_n^+ \right) + A_0^c A_1^+ A_2^c \left(A_3^+ \cdots A_n^+ \right) + A_1^+ A_2^+ A_0^- \left(A_3^+ \cdots A_n^+ \right) = \cdots \\
&= (-1)^n \left(A_1^+ \cdots A_n^+ \right) A_0^- + \sum_i A_0^c \, (\ldots) \, A_i^c \, (\ldots A_n^+) \\
&= N \left[A_0^- A_1^+ \cdots A_n^+ \right] + \sum_{i=1}^n N \left[A_0^c \ldots A_i^c \ldots A_n^+ \right]
\end{aligned}
$$

Lemma II:

$$A_0^- N \left[A_1 \ldots A_n \right] = N \left[A_0^- A_1 \ldots A_n \right] + \sum_i A_0^c N \left[\ldots A_i^c \ldots A_n \right]$$

This can be proved by induction.

Wick's theorem:

$$
\begin{aligned}
A_1 A_2 \ldots A_n &= N \left[12 \ldots n \right] + \sum_c N \left[\ldots i^c \ldots j^c \ldots n \right] \\
&\quad + \sum_{c,d} N \left[\ldots i^c \ldots r^d \ldots j^c \ldots s^d \ldots n \right] + \cdots
\end{aligned}
$$
(15.100)

The first sum runs over single contractions of pairs, the second sum runs over double contractions, and so on. If n is even, one ends with terms that consist only of products of contractions (c-numbers).

An important consequence of Wick's operator identity is a rule for the expectation value of a product of destruction and creation operators. Particle number conservation requires that the number of destructors equals that of creators.

As a corollary to Wick's theorem, we can write

$$\langle \Psi_g | A_1 \ldots A_{2n} | \Psi_g \rangle = \sum_{\mathcal{P}_2} (-1)^P \langle A_{i_1} A_{j_1} \rangle \ldots \langle A_{i_n} A_{j_n} \rangle.$$
(15.101)

The sum is over the set $\mathcal{P}_2$ of partitions of $1, \ldots, 2n$ into sets of pairs $\{(i_1, j_1) \ldots (i_n, j_n)\}$ $((i, j)$ and (j, i) are the same pair). P is the permutation that takes $1 \ldots 2n$ into the sequence $i_1, j_1, \ldots, i_n, j_n$.

As examples of the corollary,

$$\left\langle \Psi_g \left| 1234 \right| \Psi_g \right\rangle = \langle 12 \rangle \langle 34 \rangle - \langle 13 \rangle \langle 24 \rangle + \langle 14 \rangle \langle 23 \rangle .$$

Wick's Theorem with Time Order

If the time evolution of operators α_a and $\alpha_a^\dagger$ is simply a multiplication by some time-dependent phase factor (c-number), the discussion of normal ordering and contractions of operators $A_i(t_i)$ remains unaltered. An important variant of Wick's theorem deals with the normal ordering of a time-ordered product. Let us begin with two operators, and apply Wick's theorem:

$$
\begin{aligned}
T\, A_1(t_1)\, A_2(t_2) &= \Theta(t_1 - t_2) \left\{ N\left[A_1(t_1)\, A_2(t_2)\right] + A_1^c(t_1)\, A_2^c(t_2) \right\} \\
&\quad - \Theta(t_2 - t_1) \left\{ N\left[A_2(t_2)\, A_1(t_1)\right] + A_2^c(t_2)\, A_1^c(t_1) \right\} \\
&= \Theta(t_1 - t_2)\, N\left[A_1(t_1)\, A_2(t_2)\right] + \Theta(t_2 - t_1)\, N\left[A_1(t_1)\, A_2(t_2)\right] \\
&\quad + \Theta(t_1 - t_2)\, A_1^c(t_1)\, A_2^c(t_2) - \Theta(t_2 - t_1)\, A_2^c(t_2)\, A_1^c(t_1) \\
&= N\left[A_1(t_2)\, A_2(t_1)\right] + \overbrace{A_2(t_2)\, A_1(t_1)} .
\end{aligned}
\tag{15.102}
$$

We introduce the time-ordered contraction (T-contraction)

$$\overbrace{A_1(t_1)\, A_2(t_2)} = \left\langle \Psi_g \left| T\left[A_1(t_1)\, A_2(t_2)\right] \right| \Psi_g \right\rangle .$$

The T-contraction of two operators with a product of k operators in between inherits the property of ordinary contractions

$$\overbrace{A_1(t_1)\, (\ldots)\, A_2(t_2)} = (-1)^k\, \overbrace{A_1(t_1)\, A_2(t_2)}\, (\ldots).$$

T-contractions have a new property, not shared by an ordinary contraction

$$\overbrace{A_1(t_1)\, A_2(t_2)} = -\, \overbrace{A_2(t_2)\, A_1(t_1)} .$$

For field operators, we have the explicit expressions

$$\overbrace{\hat{\psi}(t_1)\hat{\psi}^\dagger(t_2)} = \left\langle T\left[\hat{\psi}(t_1)\hat{\psi}^\dagger(t_2)\right] \right\rangle = i\, G^{(0)}(1, 2)$$

$$\overbrace{\hat{\psi}(t_1)\hat{\psi}(t_2)} = \left\langle T\left[\hat{\psi}(t_1)\hat{\psi}(t_2)\right] \right\rangle = i\, F^{(0)}(1, 2)$$

$$\overbrace{\hat{\psi}^\dagger(t_1)\hat{\psi}^\dagger(t_2)} = \left\langle T\left[\hat{\psi}^\dagger(t_1)\hat{\psi}^\dagger(t_2)\right] \right\rangle = i\, F^{\dagger(0)}(1, 2).$$

In systems where $|\Psi_g\rangle$ has a definite number of particles, the anomalous correlators $F^{(0)}$ and $F^{\dagger(0)}$ are equal to zero. They are nonzero in the Bardeen–Cooper–Schrieffer theory.

For the time ordering of several operators, Wick's theorem retains the same formulation as in (15.102), with T-contractions replacing ordinary ones. Indeed $T\,[A_1(t_1)\dots A_n(t_n)]$ corresponds to a time ordered sequence $(-1)^P\,[A_{i_1}(t_{i_1})\dots A_{i_n}(t_{i_n})]$, to which the previous formulation of Wick's theorem applies. Thus contractions act on time-ordered pairs. Normal ordering of permuted operators can be restored into normal ordering of operators in the primitive sequence $1\dots n$. The permutation sign factors cancel out exactly.

15.4.5 Feynman Diagrams

To prepare for the development of the diagrammatic depiction of the different terms in the perturbation expansion (15.97), we shall consider up to linear terms in $\mathcal{H}_I$, and write

$$iG_{\sigma\sigma'} \;\rightarrow\; iG_{\sigma\sigma'}^{(0)} + iG_{\sigma\sigma'}^{(1)}$$

with

$$i\underline{G}_{\sigma\sigma'}^{(1)} = -\frac{i}{\hbar}\int_{-\infty}^{\infty} dt_1 \left\langle \Phi_0 \left| T\left[\mathcal{H}_I(t_1)\,\hat{\psi}_{I\sigma}(\mathbf{x},t)\,\hat{\psi}_{I\sigma'}^{\dagger}\left(\mathbf{x}',t'\right)\right]\right|\Phi_0\right\rangle$$

$$\mathcal{H}_I = \frac{1}{2}\iint d\mathbf{x}_1\,d\mathbf{x}_2\,\hat{\psi}_{\sigma_1}^{\dagger}(\mathbf{x}_1)\,\hat{\psi}_{\sigma_2}^{\dagger}(\mathbf{x}_2)\,V\left(|\mathbf{x}_1-\mathbf{x}_2|\right)\,\hat{\psi}_{\sigma_2'}(\mathbf{x}_2)\,\hat{\psi}_{\sigma_1'}(\mathbf{x}_1). \qquad (15.103)$$

$\underline{G}$ indicates that we are considering only the numerator in (15.97). Writing

$$V_{\sigma_1\sigma_2\sigma_2'\sigma_1'}(\mathbf{x}_1,t_1;\mathbf{x}_2,t_2) = V\left(|\mathbf{x}_1-\mathbf{x}_2|\right)\,\delta(t_1-t_2)\,\delta_{\sigma_1\sigma_1'}\,\delta_{\sigma_2\sigma_2'}, \qquad (15.104)$$

where σ is the spin quantum number, and using the compact form $x \equiv (\mathbf{x},t)$, we have

$$i\underline{G}_{\sigma\sigma'}^{(1)} = -\frac{i}{\hbar}\frac{1}{2}\sum_{\substack{\sigma_1\sigma_1'\\ \sigma_2\sigma_2'}}\int_{-\infty}^{\infty} d^4x_1\,d^4x_2\,V_{\sigma_1\sigma_2\sigma_2'\sigma_1'}(x_1,x_2)$$

$$\times \left\langle \Phi_0 \left| T\left[\hat{\psi}_{\sigma_1}^{\dagger}(x_1)\,\hat{\psi}_{\sigma_2}^{\dagger}(x_2)\,\hat{\psi}_{\sigma_2'}(x_2)\,\hat{\psi}_{\sigma_1'}(x_1)\,\hat{\psi}_{\sigma}(x)\,\hat{\psi}_{\sigma'}^{\dagger}(x')\right]\right|\Phi_0\right\rangle. \qquad (15.105)$$

The problem of interacting electrons has been reduced to determining expectation values of its noninteracting counterpart. We need to apply Wick's theorem by considering all possible pair contractions. We find that each particle emerging from a creation operator in (15.105) has to be destroyed by an annihilation operator. Thus, we can pair every creation operator with every annihilation operator. However, we should keep in mind the following simplifying points to obtain the final expression for (15.105) in terms of the free Green functions:

- The ground state is that of $\mathcal{H}_0$.
- The operators now have the form

$$\hat{\psi}_{\sigma}^{\dagger}(\mathbf{x},t) = \exp[i\mathcal{H}_0 t/\hbar]\,\hat{\psi}_{S\sigma}^{\dagger}(\mathbf{x})\,\exp[-i\mathcal{H}_0 t/\hbar].$$

For eigenstates of $\mathcal{H}_0$ representing a crystalline solid, the operators can be written as

$$\exp[i\mathcal{H}_0 t/\hbar]\,c_{n\mathbf{k}\sigma}^{\dagger}\,\exp[-i\mathcal{H}_0 t/\hbar] \;\rightarrow\; e^{i\varepsilon_{n\mathbf{k}}t/\hbar}\,c_{n\mathbf{k}\sigma}^{\dagger}.$$

- We note that an injected electron into a state $n\mathbf{k}\sigma$ of a noninteracting system must be removed from the same state, since it evolves in time unperturbed. This allows the insertion of $|\Phi_0\rangle\langle\Phi_0|$ between pairs of creation and annihilation operators to obtain

$$\langle\Phi_0| c_{n\mathbf{k}\sigma}(t)\, c_{n\mathbf{k}\sigma}^\dagger(t_1) |\Phi_0\rangle,$$

which is just the noninteracting single-particle Green function. Inspection of (15.105) shows that we can generalize this procedure to all creation and annihilation operators it contains, since they represent a noninteracting system. We should point out that this is just the outcome of applying Wick's theorem.

This produces six terms in all, which can be written in space-time, after applying fermionic commutations, as

$$i\underline{G}_{\sigma\sigma'}^{(1)} = -\frac{i}{\hbar}\frac{1}{2} \sum_{\sigma_1\sigma_1',\,\sigma_2\sigma_2'} \int_{-\infty}^{\infty} d^4x_1\,d^4x_2\; V_{\sigma_1\sigma_2\sigma_2'\sigma_1'}(x_1,x_2)$$

$$\times\; \Bigg\{ iG_{\sigma\sigma'}^{(0)}(x,x') \left[iG_{\sigma_1'\sigma_1}^{(0)}(x_1,x_1)\, iG_{\sigma_2'\sigma_2}^{(0)}(x_2,x_2) - iG_{\sigma_1'\sigma_2}^{(0)}(x_1,x_2)\, iG_{\sigma_2'\sigma_1}^{(0)}(x_2,x_1) \right]$$

$$+\, iG_{\sigma\sigma_1}^{(0)}(x,x_1) \left[iG_{\sigma_1'\sigma_2}^{(0)}(x_1,x_2)\, iG_{\sigma_2'\sigma'}^{(0)}(x_2,x') - iG_{\sigma_1'\sigma'}^{(0)}(x_1,x')\, iG_{\sigma_2'\sigma_2}^{(0)}(x_2,x_2) \right]$$

$$+\, iG_{\sigma\sigma_2}^{(0)}(x,x_2) \left[iG_{\sigma_2'\sigma_1}^{(0)}(x_2,x_1)\, iG_{\sigma_1'\sigma'}^{(0)}(x_1,x') - iG_{\sigma_2'\sigma'}^{(0)}(x_2,x')\, iG_{\sigma_1'\sigma_2}^{(0)}(x_1,x_1) \right] \Bigg\}.$$

$$(15.106)$$

Thus, using Wick's theorem, we can express each term in the perturbation expansion as product of noninteracting Green functions. This is the general methodology behind perturbation theory in quantum field theory.

Equal-Time Operators

In (15.106), we find Green functions of the form $G^{(0)}(x,x)$. This quantity is ambiguous since it represents a contraction of ψ and $\psi^\dagger$ at equal times, but the time-ordered product is undefined at equal times. Moreover, we surmise from (15.24) that $G^{(0)}$ has a discontinuity at equal times $t-t'=0$. Consequently, taking the limit from either side will yield different results. However, the definitions (15.103) and (15.104) show that the operators of $\mathcal{H}_I$ come in sets of four having identical times, and a term such as $G^{(0)}(x,x)$ arises from a contraction of two fields within $\mathcal{H}_I$ where they appear in the form $\psi^\dagger(x)\psi(x)$ with the adjoint field always occurring to the left of the field

$$\mathcal{H}_I \propto \hat\psi_{\sigma_1}^\dagger(\mathbf{x}_1)\, \hat\psi_{\sigma_2}^\dagger(\mathbf{x}_2)\, \hat\psi_{\sigma_2'}(\mathbf{x}_2)\, \hat\psi_{\sigma_1'}(\mathbf{x}_1) \;\rightarrow\; \hat\psi_{\sigma_1}^\dagger(\mathbf{x}_1)\, \hat\psi_{\sigma_1'}(\mathbf{x}_1)\, \hat\psi_{\sigma_2}^\dagger(\mathbf{x}_2)\, \hat\psi_{\sigma_2'}(\mathbf{x}_2),\; \mathbf{x}_1 \neq \mathbf{x}_2.$$

When we replace this term by propagators, the backward time propagation introduces a factor of -1 for fermions. We therefore should interpret the Green function at equal times as

$$G_{\sigma\sigma'}^{(0)}(x,x) = \lim_{t' \to t} \left\langle \Phi_0 \left| \hat{\psi}_\sigma(\mathbf{x},t) \hat{\psi}_{\sigma'}^\dagger(\mathbf{x},t') \right| \Phi_0 \right\rangle$$

$$= -\left\langle \Phi_0 \left| \hat{\psi}_{\sigma'}^\dagger(\mathbf{x}) \, \hat{\psi}_\sigma(\mathbf{x}) \right| \Phi_0 \right\rangle = -\delta_{\sigma\sigma'} \frac{n^{(0)}(\mathbf{x})}{2s+1}$$

for a fermionic system with spin s. $n^{(0)}(\mathbf{x})$ is the electron density in the noninteracting system, which can be different from $n(\mathbf{x})$ of the interacting system. For a uniform system, such as the jellium, $n^{(0)}(\mathbf{x}) = n(\mathbf{x})$.

In other operator bases, the time-ordering operator still does not know how to place the operators. Realizing that this ambiguity always comes from operators in $\mathcal{H}_I$, which should be normal ordered, we set the following general rule:

> If two operators occur at equal time, they should be *normal ordered*

Operationally, this means that if we encounter a Green function of the form $G^{(0)}(\mathbf{x},\mathbf{x};t=0)$, we should interpret it as

$$G^{(0)}(\mathbf{x},\mathbf{x};t=0^-) \equiv \lim_{\epsilon \to 0} G_{\sigma_1'\sigma_1}^{(0)}(\mathbf{x},t;\mathbf{x},t+\epsilon),$$

where 0^- means the limit as zero is approached from the negative side.

Perturbation Expansion and Feynman Diagrams in Different Representations

At this point, we should note that terms in perturbation theory could become quite long, and that a succinct description of the different contributions would be desirable. This is achieved with the aid of Feynman diagrams,[8] which represent, in a compact form, the different contributions obtained from the Wick's decomposition. From a diagrammatic perspective, a Green function G, of an interacting system, can be represented by an infinite sum of Feynman diagrams, each term of the sum corresponding to a decomposition of G into contracted products. For example, the perturbation expansion (15.106) generates six diagrams, but not all are different!

We start with establishing a connection between perturbation expansion terms and the methodology of constructing equivalent diagrams. Once we become familiar with this connection, we shall reverse our strategy; start with diagram construction and then write the corresponding expression, guided by the diagram structure.

A. Diagrammatic Representations in $(\mathbf{x},t)$-Space

1. The full Green function is represented by a double line, and an arrow that indicates the direction of time propagation

$$\text{Vertex} \qquad\qquad \text{Vertex}$$

$$\underbrace{\Psi(\mathbf{x}_2)\,\Psi^\dagger(\mathbf{x}_2)}\; V(\mathbf{x}_2 - \mathbf{x}_1)\; \underbrace{\Psi(\mathbf{x}_1)\,\Psi^\dagger(\mathbf{x}_1)}$$

---

[8] The article [97] gives an interesting historical perspective to the development of Feynman diagrams.

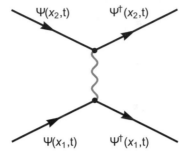

Figure 15.16 Graph representing the interaction vertices.

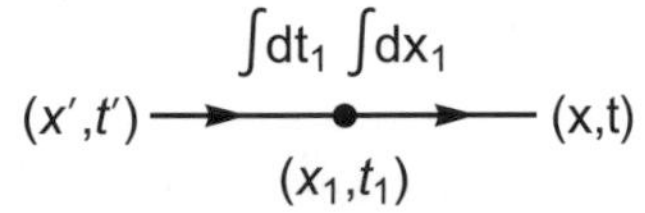

$$\int dt_1 \int dx_1$$

$$(x',t') \longrightarrow \bullet \longrightarrow (x,t)$$
$$(x_1,t_1)$$

Figure 15.17 Integration or summation over internal variables.

2. In Figure 15.16, we introduce the diagram representing the interaction $\mathcal{H}_I(x_1, x_2)$ in (15.105),

a. The wavy line represents the instantaneous Coulomb interaction between two vertices, with two field operators at one vertex and two at the other.

b. Destruction operators are represented by the lines ending at the interaction vertex:

$$\overbrace{\Psi(\mathbf{x}_2)\,\Psi^\dagger(\mathbf{x}_2)}^{\text{Vertex}}\; V(\mathbf{x}_2 - \mathbf{x}_1)\; \overbrace{\Psi(\mathbf{x}_1)\,\Psi^\dagger(\mathbf{x}_1)}^{\text{Vertex}}$$

$$\underbrace{\longrightarrow \bullet \longrightarrow}_{(x_2,t)} \qquad \underbrace{\longrightarrow \bullet \longrightarrow}_{(x_1,t)}$$

c. Creation operators are depicted by lines emanating from the vertex.

3. The internal variables are present in every Feynman diagram, whenever interactions appear. They could be space-time $(\mathbf{x}, t)$ or wavevector energy $(\mathbf{k}, \omega)$, or a combination of them. Because these variables describe events taking place between the initial and final spatiotemporal points of the Green function, which are free of any constraints on the exact time or place, they must be integrated, or summed, over their domains. Figure 15.17 shows the internal event as a black dot, represented by the position and time variables $\mathbf{x}_1$ and t_1.

4. Finally, we have the noninteracting field operators representing the injection and extraction of our test particle, depicted as the extra lines in Figure 15.18.

With these graph elements defined, we can now construct the graphs representing the six different terms of (15.106) by implementing all possible connections of lines starting at $\mathbf{x}'$ in

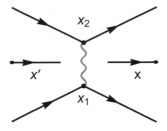

Figure 15.18 Lines to be joined in first order.

Figure 15.18 and joining them with the lines ending at $\mathbf{x}_1$, $\mathbf{x}_2$, or $\mathbf{x}$. The resulting diagrams are shown together in Figure 15.19 with the appropriate Green functions of (15.106).

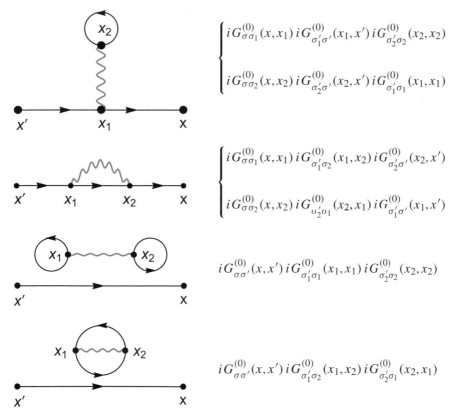

$$\begin{cases} iG^{(0)}_{\sigma\sigma_1}(x,x_1)\,iG^{(0)}_{\sigma'_1\sigma'}(x_1,x')\,iG^{(0)}_{\sigma'_2\sigma_2}(x_2,x_2) \\[2ex] iG^{(0)}_{\sigma\sigma_2}(x,x_2)\,iG^{(0)}_{\sigma'_2\sigma'}(x_2,x')\,iG^{(0)}_{\sigma'_1\sigma_1}(x_1,x_1) \end{cases}$$

$$\begin{cases} iG^{(0)}_{\sigma\sigma_1}(x,x_1)\,iG^{(0)}_{\sigma'_1\sigma_2}(x_1,x_2)\,iG^{(0)}_{\sigma'_2\sigma'}(x_2,x') \\[2ex] iG^{(0)}_{\sigma\sigma_2}(x,x_2)\,iG^{(0)}_{\sigma'_2\sigma_1}(x_2,x_1)\,iG^{(0)}_{\sigma'_1\sigma'}(x_1,x') \end{cases}$$

$$iG^{(0)}_{\sigma\sigma'}(x,x')\,iG^{(0)}_{\sigma'_1\sigma_1}(x_1,x_1)\,iG^{(0)}_{\sigma'_2\sigma_2}(x_2,x_2)$$

$$iG^{(0)}_{\sigma\sigma'}(x,x')\,iG^{(0)}_{\sigma'_1\sigma_2}(x_1,x_2)\,iG^{(0)}_{\sigma'_2\sigma_1}(x_2,x_1)$$

- The sense of time propagation is represented by an arrow along each line. The convention adopted here is that time increases to the right, so that particles propagate from left to right, while holes move from right to left.
- We note that each of the top two diagrams depicts a pair of Green functions. Each pair has identical graphs, but with $\mathbf{x}_1$ and $\mathbf{x}_2$ interchanged. Moreover, we find that in these diagrams all parts are connected.

- In the second diagram, an exchange occurs in the contraction with $\mathbf{x}$, whereas the point $\mathbf{x}'$ has the same contraction. First, the injected electron travels to point $\mathbf{x}_1, t_1$, where it interacts with another electron at point $\mathbf{x}_2$. This electron at $\mathbf{x}_2$ then travels to the finish point, where it is taken out, with another electron at $\mathbf{x}_1$ traveling to $\mathbf{x}_2$ in order to leave the system in the same state it started. Consequently, such diagrams are referred to as exchange diagrams.

- The remaining two diagrams have disconnected parts.

- In the first and third diagrams, we find lines starting and ending at the same vertex, forming a fermion loop. They describe a single scattering event at $(\mathbf{x}_i, t_i)$ represented by $G^{(0)}_{\sigma'_1 \sigma_1}(x_i, x_i)$. It can be thought of as spontaneous creation and reannihilation of a particle. Recall that they arise from $\mathcal{H}_I$, and should be interpreted as

$$\lim_{\epsilon \to 0} G^{(0)}_{\sigma'_1 \sigma_1}(x_1, t_1; x_1, t_1 + \epsilon)$$

and they represent electronic densities.

In this representation, the diagrams have a very nice physical interpretation: the Green function $G(\mathbf{x}, t; \mathbf{x}', t')$ is the probability amplitude for a particle to get from point $\mathbf{x}'$, at time t' to point $\mathbf{x}$ at time t. The Feynman expansion then basically is saying that the electron can get between these two points in all possible ways, weighted by the appropriate amplitude. The zeroth-order Feynman graph is then the direct path.

B. Diagrammatic Representations in $(\mathbf{k}, t)$-Space $\mathcal{H}_I$ can be expressed as

$$\mathcal{H}_I = \frac{1}{\Omega} \sum_{\substack{\mathbf{k}_1, \mathbf{k}'_1 \\ \mathbf{q}}} \frac{V(\mathbf{q})}{2} c^\dagger_{\mathbf{k}_1 - \mathbf{q}} c^\dagger_{\mathbf{k}_2 + \mathbf{q}} c_{\mathbf{k}_2} c_{\mathbf{k}_1}. \tag{15.107}$$

We note that the only difference between momentum space and real space is that we integrate over internal momenta, rather than internal positions.

- Every $G^{(0)}_{\sigma\sigma'}(\mathbf{k}, t - t')$ is depicted as a line with an arrow pointing from t' to t.

- Every $V(\mathbf{q})$ is depicted as a wiggly line without an arrow.

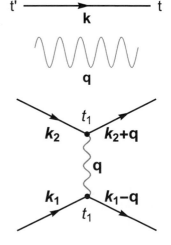

- Momentum is conserved at each vertex, where an interaction links two points at equal time.

Now, with the aid of this basic diagrammatic representation, we shall attempt to construct a depiction of the terms in the perturbation expansion for the interacting Green function. This function is represented diagrammatically by a double line with an arrow, namely,

$$G(\mathbf{k}, t - t') = \quad t' \Longrightarrow_{\mathbf{k}} t \quad = G^{(0)} + G^{(1)} + G^{(2)} + \cdots$$

We must remember that we are still ignoring the denominator, $S(\infty, -\infty)$. We consider perturbation terms up to $O(\mathcal{H}_I^2)$:

1. **Zeroth-order:**

$$t' \Longrightarrow_{\mathbf{k}}^{G(\mathbf{k}, t-t')} t = t' \quad t' \longrightarrow_{\mathbf{k}}^{G^{(0)}(\mathbf{k}, t-t')} t$$

The interpretation of this is that the particle with momentum $\mathbf{k}$ injected at time t_1 travels unperturbed until it is extracted at time t. It does not suffer any scattering off any other particles.

Thus we can interpret the first-order terms as suffering a single scattering event, the second order involves scattering twice, and so on. Quantum mechanically, we sum up the appropriate weight of all of these events of getting from time t_1 to time t, to obtain the total probability amplitude, G.

2. **First-order terms:**
We know from (15.106) that there are six possible pairings of this expectation value – hence there are six first-order terms contributing to $G^{(1)}$, which can be represented by six diagrams. We consider the first pairing,

$$\left\langle T \left[\overbrace{c_{\mathbf{k}}(t)\, c_{\mathbf{k}_1 - \mathbf{q}}^{\dagger}(t_1)}\ \overbrace{c_{\mathbf{k}_2 + \mathbf{q}}^{\dagger}(t_1)\, c_{\mathbf{k}_2}(t_1)}\ \overbrace{c_{\mathbf{k}_1}(t_1)\, c_{\mathbf{k}}^{\dagger}(t')} \right] \right\rangle.$$

This gives

$$G_a^{(1)}\left(\mathbf{k}, t - t'\right) = (-1)\,\frac{1}{\Omega} \int_{-\infty}^{\infty} dt_1$$

$$\times \sum_{\substack{\mathbf{k}, \mathbf{k}' \\ \mathbf{q}}} \frac{V(\mathbf{q})}{2}\left\langle T \left[\overbrace{c_{\mathbf{k}}(t)\, c_{\mathbf{k}_1-\mathbf{q}}^{\dagger}(t_1)}^{\mathbf{k}_1 - \mathbf{q} = \mathbf{k}}\ \overbrace{c_{\mathbf{k}_2}(t_1)\, c_{\mathbf{k}_2+\mathbf{q}}^{\dagger}(t_1)}^{\mathbf{k}_2 = \mathbf{k}_2 + \mathbf{q}}\ \overbrace{c_{\mathbf{k}_1}(t_1)\, c_{\mathbf{k}}^{\dagger}(t')}^{\mathbf{k}_1 = \mathbf{k}} \right] \right\rangle$$

$$= i\,\frac{1}{\Omega} \int_{-\infty}^{\infty} dt_1 \sum_{\mathbf{k}_2} G^{(0)}(\mathbf{k}, t - t_1)\, G^{(0)}(\mathbf{k}_2, 0^-)\, G^{(0)}(\mathbf{k}, t_1 - t').$$

$$(15.108)$$

The corresponding diagram is

$$G_a^{(1)}\left(\mathbf{k}, t - t'\right) =$$

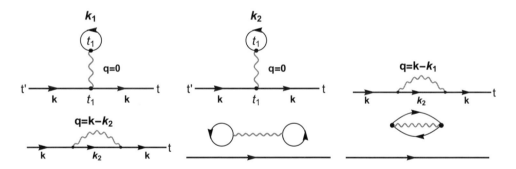

The six first-order diagrams are shown in Figure 15.19. An examination of the diagrams reveals that there are two pairs of identical diagrams, (a,b) and (c,d), and the last two diagrams, (e) and (f), are disconnected. We will return to these features later.

3. **Second-order terms:**

The second-order term involves

$$\left\langle T\left[c_{\mathbf{k}}(t)\, \mathcal{H}_I(t_1)\, \mathcal{H}_I(t_2)\, c_{\mathbf{k}}^\dagger(t_1)\right]\right\rangle. \tag{15.109}$$

When we substitute for $\mathcal{H}_I$ in this expression, we find that each second-order diagram will contain five fermionic lines and two interaction lines, which yield $5! = 120$ Wick contractions, each represented by a diagram. However, as we encountered in the first-order case, some of the diagrams will be identical to other ones. A small sample of second-order diagrams are shown in Figure 15.20.

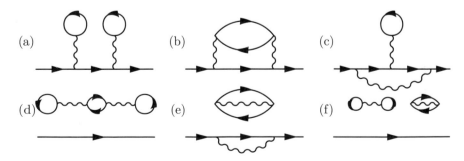

Figure 15.19 The six first-order diagrams, from the six Wick contractions.

Figure 15.20 A small selection of second-order diagrams.

C. Diagrammatic Representations in $(\mathbf{k}, \omega)$-Space Finally, we write the Feynman–Dyson perturbation series in the frequency domain by performing a Fourier transformation of the Green function. We note that the time integrals over the products of the Green function have the form of a convolution, as we find in (15.108), where

$$G^{(1)}\left(\mathbf{k}, t - t'\right) = i \frac{1}{\Omega} \int_{-\infty}^{\infty} dt_1 \sum_{\mathbf{k}_2} G^{(0)}(\mathbf{k}, t - t_1)\, G^{(0)}(\mathbf{k}_2, 0^-)\, G^{(0)}(\mathbf{k}, t_1 - t').$$

The individual Green functions in the integrand depend on time via differences $t_i - t_{i+1}$. Transformation to the frequency domain will yield simple products of Green functions, which is a further simplification in the perturbation series for the Green function.

How this Fourier transformation is performed, and what implications it has, will be explained using the following first-order exchange diagram, but with a time-dependent perturbation $V(\mathbf{q}, t) \to V(\mathbf{q}, \varepsilon_{\mathbf{q}})$:

$$G^{(1)}\left(\mathbf{k}, t - t'\right) =$$

$$\begin{array}{cccccc} t' & \mathbf{k} & t_1 & \mathbf{k{-}q} & t_2 & \mathbf{k} \quad t \end{array}$$

First, we have the single Green function transform

$$G^{(1)}\left(\mathbf{k}, t - t'\right) = \int \frac{d\omega}{2\pi} e^{-i\omega(t-t')} G^{(1)}(\mathbf{k}, \omega), \tag{15.110}$$

then

$$G^{(1)}\left(\mathbf{k}, t - t'\right) = i \frac{1}{\Omega} \int_{-\infty}^{\infty} \int_{-\infty}^{\infty} dt_1\, dt_2 \sum_{\mathbf{q}} \Big[V(\mathbf{q}, t_2 - t_1)\, G^{(0)}(\mathbf{k}, t - t_2)$$

$$\times\, G^{(0)}(\mathbf{k} - \mathbf{q}, t_2 - t_1)\, G^{(0)}(\mathbf{k}, t_1 - t') \Big],$$

which gives rise to the factors

$$\exp[i\varepsilon_{\mathbf{q}}(t_2 - t_1)],\ \exp[-i\omega(t - t_2)],\ \exp[i\omega_2(t_2 - t_1)],\ \exp[i\omega_1(t_1 - t')]$$

for the interaction, and first, second, and third free electron Green functions, respectively. Collecting all these terms together, the time-dependent part becomes

$$\int dt_2 \int dt_1\, e^{-i\omega t}\, e^{-i(\omega_2 + \varepsilon_{\mathbf{q}} - \omega)t_2}\, e^{-i(\omega_1 - \varepsilon_{\mathbf{q}} - \omega_2)t_1}\, e^{i\omega_1 t'},$$

which yields $2\pi\, \delta(\omega_2 + \varepsilon_{\mathbf{q}} - \omega) \times 2\pi\, \delta(\omega_1 - \varepsilon_{\mathbf{q}} - \omega_2)$, guaranteeing the conservation of energy at the vertices and removing the integrals over the internal frequencies ω_1, ω_2 introduced by the Fourier transformation. Then, only the factor $e^{-i\omega(t-t')}$ is left, which resembles the one in the Fourier transformation in (15.110). The corresponding diagram is as follows

$$G^{(1)}(\mathbf{k},\omega) =$$

whereas the corresponding contribution to the frequency-dependent Green function becomes

$$G^{(1)}(\mathbf{k},\omega) = i\,\frac{1}{\Omega}\,G^{(0)}(\mathbf{k},\omega)\left(\sum_{\mathbf{q}} V(\mathbf{q},\varepsilon_{\mathbf{q}})\,G^{(0)}(\mathbf{k}-\mathbf{q},\omega-\varepsilon_{\mathbf{q}})\right)G^{(0)}(\mathbf{k},\omega).$$

D. Redundancies and Simplifications in the Diagrammatic Expansion So far, we have been writing down Wick contraction terms in the perturbation expansion, an d then figuring out how to construct the corresponding diagram with the aid of the basic graph components we established. It would be very desirable to reverse the procedure: draw the diagram and then write down the corresponding expression accordingly. One compelling reason to adopt such recipe is the observation that in each order of perturbation, many contractions appear to correspond to the same graph! To achieve such a goal, we shall consider several features of the perturbation expansion that present redundancies we can eliminate.

1. Redundancies Associated with Interaction Lines We shall use the example of a third-order diagram to illustrate this type of redundancy (see Figure 15.21). The third-order terms arise from contractions of

$$\left\langle T\left[c_{\mathbf{k}}(t)\,\mathcal{H}_I(t_1)\,\mathcal{H}_I(t_2)\,\mathcal{H}_I(t_3)\,c_{\mathbf{k}}^{\dagger}(t')\right]\right\rangle. \tag{15.111}$$

It contains three interaction lines and seven fermionic lines, and therefore leads to $7! = 5,040$ possible Wick contractions. How can we generalize the identification of diagrammatic redundancies among these contractions?

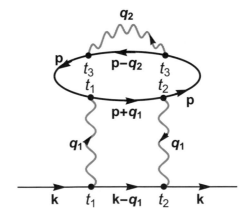

Figure 15.21 A third-order diagram.

(i) Permutations of interaction lines: First, we examine the three interaction lines, which occur at t_1, t_2, and t_3. It is obvious that we have Wick's contractions arising from permutations of the three lines, $3! = 6$. Such contractions will create exactly the same product of Green functions, except with the interaction lines in a different order. Consequently, the corresponding diagrams will look identical, except for the labeling. However, since we integrate over all internal variables anyway, this interchange of interaction lines is irrelevant, and their contribution to $G^{(3)}$ will be identical. Therefore, we only need to draw and evaluate such diagram once, then we multiply it by the factor $3!$, or, in general, $n!$, where n is the number of interaction lines or the order of the perturbation expansion.

We recall that in the S matrix expansion, a factor $1/n!$ is associated with terms of order n, namely,

$$S(\infty, -\infty) = \sum_{n=0}^{\infty} \frac{i^n}{n!} T \int_{-\infty}^{\infty} dt_1 \ldots \int_{-\infty}^{\infty} dt_n \, \mathcal{H}_I(t_1) \ldots \mathcal{H}_I(t_n).$$

This $1/n!$ factor then is canceled by the $n!$ symmetry factor for each diagram.

(ii) Reversal symmetry of an interaction line: If we inspect each individual interaction line further, we find that in the momentum-time representation it has a direction, and we have two possible ways of orienting it. In the position-time representation, we effectively interchange the internal variables, $\mathbf{x}_i$ and $\mathbf{x}_j$, at its vertices. In either case, such operation will yield diagrams having the same topology. Again, integrating over these internal variables will yield the same value for the contribution of each diagram. It will also mean that we need only *topologically distinct* diagrams, and for each interaction line, we have a factor $V(\mathbf{q})$, instead of the $V(\mathbf{q})/2$ that appears in the Hamiltonian.

(iii) Multiplicative factors: The fact that any contraction term of order n has n interaction lines, and $2n + 1$ Green functions, means that the corresponding nth-order diagram must be associated with a factor

$$\underset{G \to -i \langle \ldots \rangle}{(-i)} \times \underset{iG^{(0)}s}{(i)^{2n+1}} \times \underset{S\text{-matrix expansion}}{(-i)^n} = i^n.$$

(iv) Transposition of fermionic operators:

(a) Two transpositions are required (see Figure 15.22). The first disentangles the contractions, and a second one gets the middle contraction into the form of a Green function, giving an overall factor of $(+1)$.

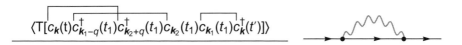

Figure 15.22 Transpositions.

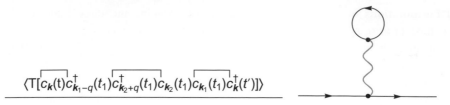

$$\frac{\langle T[\overbrace{c_{\boldsymbol{k}}(t)\,c^{\dagger}_{\boldsymbol{k}_1-q}(t_1)}\,\overbrace{c^{\dagger}_{\boldsymbol{k}_2+q}(t_1)\,c_{\boldsymbol{k}_2}(t_1)}\,\overbrace{c_{\boldsymbol{k}_1}(t_1)\,c^{\dagger}_{\boldsymbol{k}}(t')}]\rangle}{}$$

Figure 15.23 Loops.

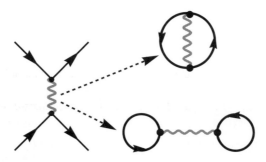

Figure 15.24 Diagrams for the denominator of the Green function in first order.

(b) Correct Green function operator ordering for the middle contraction (equal time) requires operator switching and gives rise to a factor of (-1). Diagrammatically, we associate the (-1) factor with the appearance of a loop (see Figure 15.23). We infer that a factor of $(-1)^L$ will be associated with a diagram containing L closed fermionic loops.

2. *Vacuum Graph Cancellation* So far, we have ignored the denominator of (15.97). Up to first order in $\mathcal{H}_I$, it gives

$$\langle\Phi_0|\,S(\infty,-\infty)\,|\Phi_0\rangle \sim \langle\Phi_0|\,\Phi_0\rangle - \frac{i}{\hbar}\int_{-\infty}^{\infty}dt_1\,\langle\Phi_0|\,T\,\mathcal{H}_I(t_1)\,|\Phi_0\rangle$$

$$= 1 - \frac{i}{\hbar}\int_{-\infty}^{\infty}d^4x_1\,d^4x_2\,V_{\sigma_1\sigma_2\sigma_2'\sigma_1'}\left(x,x'\right)$$

$$\times\,\langle\Phi_0|\,T\,\hat{\psi}^{\dagger}_{\sigma_1}(x_1)\,\hat{\psi}^{\dagger}_{\sigma_2}(x_2)\,\hat{\psi}_{\sigma_2'}(x_2)\,\hat{\psi}_{\sigma_1'}(x_1)\,|\Phi_0\rangle$$

$$= F_0 + F_1. \tag{15.112}$$

Figure 15.24 shows the diagrams obtained from the corresponding Wick contractions. We note that there are no external fermionic lines; consequently, all its diagrams should be closed ones.

We can express $\langle\Phi_0|\,S(\infty,-\infty)\,|\Phi_0\rangle$ as

$$\langle\Phi_0|\,S(\infty,-\infty)\,|\Phi_0\rangle = \sum_i F_i.$$

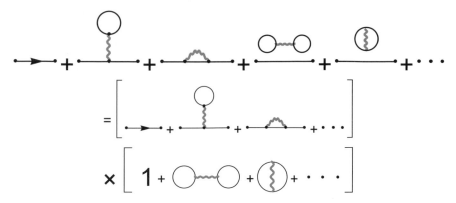

Figure 15.25 A common factor can be extracted from the first-order diagrams.

Terms in the series $\langle\Phi_0|\,S(\infty,-\infty)\,|\Phi_0\rangle$, called *vacuum polarization terms*, have corresponding vacuum polarization diagrams of the closed type. Examination of the numerator of (15.97) shows that some contractions of $\mathcal{H}_I$ in $\langle T\,[c(t)\,\mathcal{H}_I(t_n)\ldots\,\mathcal{H}_I(t_1)\,c^\dagger(t')]\rangle$ do not link to the external femionic points $c(t)$ and $c^\dagger(t')$. Such contractions correspond to diagrams having disconnected closed components. It can be shown, as depicted schematically in Figure 15.25, that there exists a factorization that separates the disconnected closed-type components from connected ones. The cancellation theorem, which we will not prove,[9] demonstrates that in such factorization the closed diagrams in the denominator exactly cancel the disconnected counterpart diagrams of the perturbation expansion in the numerator. This means that diagrams can factorize as shown in Figure 15.25, and that by considering only connected diagrams, we effectively cancel the denominator. The Green function is then given by the sum of all connected diagrams as

$$i\,G_{\sigma\sigma'}\left(x,x'\right) = \sum_{n=0}^{\infty} \frac{1}{n!}\left(\frac{-i}{\hbar}\right)^n \int_{t_0}^{t} dt_1\ldots\int_{t_0}^{t} dt_n$$

$$\times\,\left\langle\Phi_0\left|T\left[\mathcal{H}_I(t_1)\,\ldots\,\mathcal{H}_I(t_n)\,\hat{\psi}_{H\sigma}(\mathbf{x},t)\,\hat{\psi}_{I\sigma'}^\dagger\left(\mathbf{x}',t_1\right)\right]\right|\Phi_0\right\rangle_{\text{connected}}.$$
(15.113)

E. The Feynman Rules We now summarize the Feynman rules for calculating the general nth-order term in the perturbation expansion of G. We give them first in direct space and then in momentum space.

(a) Draw all topologically distinct connected diagrams with n interaction (wavy) lines V and $2n+1$ Green functions $G^{(0)}$ solid lines.
(b) Each wavy line corresponds to the interaction (15.104) (its Fourier transform in $\mathbf{k}-\omega$ basis).

---

[9] For a proof, see [61], sec. 3.8.

(c) Vertex and line labeling:

    (i) **x** $-\, t$ basis: Label each vertex point by space-time x_i; each solid line represents a Green function $G^{(0)}(x, x')$ running from x' to x.

    (ii) **k** $-\, \omega$ basis: Assign a direction to each interaction line; associate directed momentum and frequency to each line (solid and wavy) and conserve energy and momentum at each vertex.

(d) Integrate all internal variables over space-time (or **k** $-\, \omega$). Sum over all internal spin indices.

(e) Each diagram carries a factor $(-1)^L$, where L is the number of closed fermion loops in the diagram.

(f) Each diagram contributes to G a factor $-i(-i/\hbar)^n (i)^{2n+1} = (i/\hbar)^n$.

15.4.6 Infinite Sums: Dyson Series and Self-Energy

Examination of the perturbation series and the corresponding diagrams show that some diagrams can be constructed by linking diagrams that appear in lower-order in the expansion with a fermionic line ($G^{(0)}$). Figure 15.26 shows second-order diagrams comprised of first-order ones connected by a fermionic line. This suggests that we can identify basic diagrammatic *building blocks*, known as *irreducible* or *proper* diagrams, which can be used to generate all diagrams and, hence, the entire perturbation expansion. This is a powerful revelation that can lead to potentially significant simplifications.

We define an irreducible diagram as one that cannot be disconnected in two parts by cutting a single internal electron line. If a diagram is not irreducible, as the ones in figure 15.26, it is referred to as a *reducible* diagram.

Next, we define an irreducible self-energy diagram (ISED) as one obtained by removing the external lines from an irreducible diagram. Equivalently, the mathematical expression for the ISED is obtained by deleting the two factors of $G^{(0)}$ due to the external lines. The total ISED sum can now be represented by Figure 15.27.

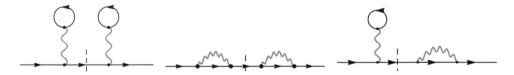

Figure 15.26 Second-order diagrams consisting of linked first-order direct and exchange ones. Each can be cut into two diagrams along the dashed line.

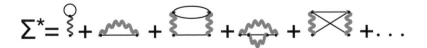

Figure 15.27 The irreducible self-energy is shown as the sum of irreducible diagrams. Shown are first-order ISEDs and a few second-order ISEDs.

The complete one-particle interacting Green function can be recovered by reconstructing the clipped diagrams as

$$\Rrightarrow = \rightarrow + \rightarrow \boxed{\Sigma^*} \Rrightarrow \qquad (15.114)$$

$$G \quad = \quad G^{(0)} \quad + \quad G^{(0)} \quad \Sigma^* \quad G$$

which is the Dyson equation. The appearance of the double line on the right-hand side indicates the self-consistent nature of the equation.

The advantage of this diagrammatic representation is that it can be conveniently terminated at any order, or, depending on the application, we may include only certain classes of diagrams. In many cases, the application of perturbation theory requires summation over whole classes of diagrams in order to obtain meaningful and reliable results. For example, the divergences of perturbation terms in the interacting electron gas that we encountered in Chapter 14 can be removed by performing an infinite sum of the terms arising from the first two diagrams in Figure 15.27.

To illustrate this procedure further, we shall consider the interacting electron gas case in more detail. We recall that after eliminating duplicates and disconnected diagrams, we were left with two first-order diagrams, namely those shown in Figure 15.28.

However, the Hartree term has $\mathbf{q} = 0$ due to momentum conservation and is canceled by the background interaction. We consider the Fock term, which gives a contribution

$$i G^{(0)}(\mathbf{k}, \omega) \underbrace{\int \frac{d\varepsilon}{2\pi} \frac{1}{\Omega} \sum_{\mathbf{q}} V(\mathbf{q}) \, G^{(0)}(\mathbf{k} - \mathbf{q}, \omega - \varepsilon) \, G^{(0)}(\mathbf{k}, \omega)}_{\text{call this } \Sigma_F^*(\mathbf{k}, \omega) \equiv \ \rule{0pt}{0pt}}$$

in first-order. The second-order contribution has the form illustrated in Figure 15.29.

This contribution can be cast in the form

$$G^{(0)}(\mathbf{k}, \omega) \, \Sigma_F^*(\mathbf{k}, \omega) \, G^{(0)}(\mathbf{k}, \omega) \, \Sigma_F^*(\mathbf{k}, \omega) \, G^{(0)}(\mathbf{k}, \omega)$$

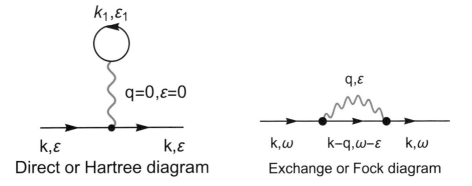

Figure 15.28 First-order terms.

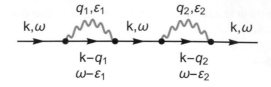

Figure 15.29 Second-order contribution to the Fock term.

having the diagrammatic representation. When we include all higher-order terms,

we generate an infinite series, known as a Dyson series:

$$
= G^{(0)}(\mathbf{k}, \omega) \sum_{n=0}^{\infty} \left(\Sigma^*(\mathbf{k}, \omega)\, G^{(0)}(\mathbf{k}, \omega) \right)^n
$$

$$
= \frac{G^{(0)}(\mathbf{k}, \omega)}{1 - \Sigma_F^*(\mathbf{k}, \omega)\, G^{(0)}(\mathbf{k}, \omega)} = \frac{1}{G^{(0)-1}(\mathbf{k}, \omega) - \Sigma_F^*(\mathbf{k}, \omega)}
$$

Inserting the explicit form of $G^{(0)} = (\omega - \varepsilon(\mathbf{k}))^{-1}$, we get

$$
G(\mathbf{k}, \omega) = \frac{1}{\omega - \varepsilon(\mathbf{k}) - \Sigma_F^*(\mathbf{k}, \omega) \pm i\eta}. \tag{15.115}
$$

$\Sigma_F^*(\mathbf{k}, \omega)$ is the self-energy of the interacting electron gas. The process of carrying out the summation is often referred to as *renormalization*

In general, as shown in Figure 15.27, we identify the self-energy $\Sigma^*(\mathbf{k}, \omega)$ with the sum of all proper diagrams coming from all orders of perturbation, namely, the ones that *cannot be split into parts by breaking fermion lines.*

We note that to obtain the complete interacting Green function, we must augment the self-energy diagrams with external fermion lines at its end vertices; one line will carry momentum $\mathbf{k}$ and energy ω to the first vertex, and the other will carry them away from the last vertex.

The Dyson series allows us to cast the interacting Green function in a form that can be identified with that of the noninteraction counterpart, but with a renormalized energy spectrum

$$
\varepsilon^{(0)}(\mathbf{k}) \;\rightarrow\; \varepsilon(\mathbf{k}) = \varepsilon^{(0)}(\mathbf{k}) + \operatorname{Re} \Sigma^*(\mathbf{k}, \omega).
$$

It also allows us to manage the number of independent diagrams we need to include.

15.4.7 *Recasting the Hartree–Fock Approach*

We are interested here in examining the Hartree–Fock theory from the perspective of the Green function formalism and the Feynman–Dyson perturbative approach. Our approach may be viewed as nonperturbative in the sense that the Green function will contain all powers in the interaction, but will only contain a selected class of diagrams.

Self-Energy in the Hartree–Fock Approximation

Diagrammatically, the one-particle Green function up to first-order is

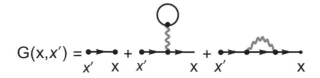

Mathematically, it is given by

$$G = G^{(0)} + G^{(0)} \, \Sigma \, G^{(0)} + \cdots$$

with the corresponding self-energy shown in Figure 15.30.

Terminating the perturbation expansion at this first-order is just the Hartree–Fock (HF) approximation we discussed in Chapter 14. We can go beyond this approximation using Dyson's equation, and we obtain

$$G = G^{(0)} + G^{(0)} \, \Sigma_{\mathrm{HF}} \, G,$$

which is shown diagrammatically in Figure 15.31. By iteration, we can construct a Green function with the interaction still restricted to the HF terms, or the HF self-energy, but extended to all orders. It is instructive to carry out the calculation of the HF self-energy.

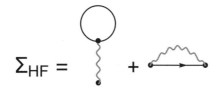

Figure 15.30 Self-energy in first order of the interaction.

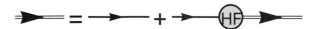

Figure 15.31 Diagrammatic representation of Dyson's equation.

From (15.106), we only consider contributions related to the diagrams in Figure 15.30, namely,

$$iG^{(1)}_{\sigma\sigma'} = \frac{i}{\hbar} \sum_{\substack{\sigma_1\sigma'_1 \\ \sigma_2\sigma'_2}} \int_{-\infty}^{\infty} d^4x_1 \, d^4x_2 \, V_{\sigma_1\sigma_2\sigma'_2\sigma'_1}(x_1,x_2) \, iG^{(0)}_{\sigma\sigma_1}(x,x_1)$$

$$\times \left[iG^{(0)}_{\sigma'_1\sigma_2}(x_1,x_2) \, iG^{(0)}_{\sigma'_2\sigma'}(x_2,x') - iG^{(0)}_{\sigma'_1\sigma'}(x_1,x') \, iG^{(0)}_{\sigma'_2\sigma_2}(x_2,x_2) \right]. \qquad (15.116)$$

We make use of the following properties:

- The noninteracting Green function is diagonal in spin,

$$G^{(0)}_{\sigma\sigma'} = \delta_{\sigma\sigma'} \, G^{(0)}.$$

- We assume the interaction to be spin independent and instantaneous,

$$V_{\sigma_1\sigma_2\sigma'_2\sigma'_1} \sim \delta_{\sigma_1\sigma'_1} \, \delta_{\sigma_2\sigma'_2} \, \delta(t_1 - t_2).$$

- The free fermion Green function in position representation is as follows:

$$\left\langle T\left[\hat{\Psi}(\mathbf{x},t)\,\hat{\Psi}^\dagger(\mathbf{x}_1,t_1)\right]\right\rangle = \sum_{\mathbf{k}\mathbf{k}_1} \phi_{\mathbf{k}}(\mathbf{x})\,\phi^*_{\mathbf{k}_1}(\mathbf{x}_1)\,\left\langle T\left[c_{\mathbf{k}}(t)\,c^\dagger_{\mathbf{k}_1}(t_1)\right]\right\rangle$$

$$= \sum_{\mathbf{k}\mathbf{k}_1} \phi_{\mathbf{k}}(\mathbf{x})\,\phi^*_{\mathbf{k}_1}(\mathbf{x}_1)\,\left\langle T\left[c_{\mathbf{k}}\,c^\dagger_{\mathbf{k}_1}\right]\right\rangle e^{-i(\varepsilon_{\mathbf{k}}t - \varepsilon_{\mathbf{k}_1}t_1)}$$

$$= \sum_{\mathbf{k}} \phi_{\mathbf{k}}(\mathbf{x})\,\phi^*_{\mathbf{k}}(\mathbf{x}_1)\,\delta_{\mathbf{k}\mathbf{k}_1}\,e^{-i\varepsilon_{\mathbf{k}}(t-t_1)} = \delta(\mathbf{x} - \mathbf{x}_1)\,\delta(t - t_1)$$

- Equal space-time Green function

$$i\bar{G}^{(0)}\left(x_2, x_2^+\right) = -n^{(0)}(x_2)$$

is the density per spin of the noninteracting system.

We obtain

$$\Sigma^{*(1)}_{\sigma\sigma'}(x,x') = -\frac{1}{\hbar}\,\delta_{\sigma\sigma'}\delta(t - t_1)\left[\delta\left(\mathbf{x} - \mathbf{x}'\right)(2S+1)\int dx_2\,i\bar{G}^{(0)}\left(x_2, x_2^+\right)\right.$$

$$\left. \times V\left(|\mathbf{x} - \mathbf{x}_2|\right) - V\left(|\mathbf{x} - \mathbf{x}'|\right)\,i\bar{G}^{(0)}\left(\mathbf{x},t;\mathbf{x}',t^+\right)\right], \qquad (15.117)$$

where $(2S+1)$ accounts for the sum over spin. Consequently, the integral in the first self-energy term is just the Coulomb interaction of the fermionic test particle with all the particles of the system. The unphysical self-interaction is canceled by its counterpart in the Fock term.

Figure 15.32 Hartree–Fock irreducible part of the self-energy.

The self-consistency built in the Hartree–Fock theory allows the replacement noninteracting Green functions $G^{(0)}$ with similar diagrams for the interacting Green function G, i.e.,

$$\Sigma^*_{\mathrm{HF}}(x, x') = -\frac{1}{\hbar} \delta(t - t') \left[-\delta \left(\mathbf{x} - \mathbf{x}' \right) (2S + 1) \int d\mathbf{x}_2 \, n \left(x_2 \right) V \left(|\mathbf{x} - \mathbf{x}_2| \right) \right.$$

$$\left. - V \left(|\mathbf{x} - \mathbf{x}'| \right) iG \left(\mathbf{x}, t; \mathbf{x}', t^+ \right) \right],$$ (15.118)

as depicted in Figure 15.32.

We note that Σ^*_{HF} is frequency independent, since it depends on $\delta(t - t_1)$. It is purely real, which manifests the Hartree–Fock approximation as a $\mathbf{k}$-dependent shift in the non-interacting energy poles.

15.5 The Two-Particle Green Function and RPA

As we have shown in Chapter 14, the Hartree–Fock approach, as a first-order approximation, does not give an adequate account of the density density response function of the interacting electron gas. We can attribute these shortfalls to the long interaction range character of the Coulomb potential, which also gives rise to an infrared divergence because of the $1/q^2$ behavior. To make progress, we shall explore the next level of approximations where we consider higher-order diagrams, and attempt to carry out a resummation à la Dyson. Moreover, we shall explore introducing screening effects that would reduce the range of the Coulomb interactions.

We start by inspecting second-order diagrams for self-energy, especially

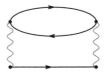

We note the appearance of a new element: a fermionic bubble, also called a polarization loop. Actually, these bubbles are very important in quantum field theory.

15.5.1 A Digression: Diagram Scaling

We consider the self-energy diagrams of order n in the $(\mathbf{k}, \omega)$ representation. They contain n interaction lines, $2n$ vertices and $2n - 1$ fermionic lines. Integration is carried out over n

Table 15.1 *Scaling of different quantities appearing in self-energy diagrams.*

Quantity	Per component	Total
Fermionic lines	k_F^{-2}	$\left(k_F^{-2}\right)^{2n-1}$
Internal variables	k_F^5	$\left(k_F^5\right)^n$
Coulomb interaction	k_F^{-2}	$\left(k_F^{-2}\right)^n$

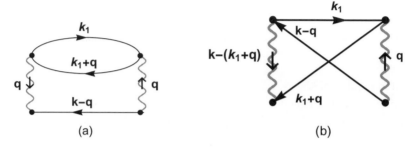

Figure 15.33 Second-order self-energy diagrams.

internal momenta, each with three components, and n internal energies. However, because of conservation of momenta at each vertex, we have $2n$ Kirchhoff-like constraints, reducing the number of independent variables to n. We now scale all energies and momenta in terms of k_F as shown in Table 15.1. Thus, a diagram of order n scales as

$$\left(k_F^5\right)^n \times \left(k_F^{-2}\right)^{2n-1} \times \left(k_F^{-2}\right)^n = k_F^{-(n-2)} \propto r_s^{(n-2)}$$

Implies that for $r_s < 1 \quad \Rightarrow \quad \Sigma^{*(n+1)} \leq \Sigma^{*(n)}$.

As we found out earlier, quasiparticles are only well-defined close to the Fermi surface, and that is precisely where Green functions assume large amplitudes, namely, for $k \simeq k_F$. Thus, we can discern that only configurations where all momenta appearing in the self-energy fermionic lines are close to k_F contribute significantly. First, we consider the second-order diagram (a) in Figure 15.33. For small q, we find that $V^2(q) \propto 1/q^4$, and that the summation over $\mathbf{k}_1$, with $k_1 \simeq k_F$, is unconstrained over a shell of width $2q$. We have learned in the previous chapter that such a second-order term diverges. In contrast, we find in diagram (b) that for small q and $|\mathbf{k} - \mathbf{q} - \mathbf{k}_1|$ in the interaction lines, the phase space for $k_1 \simeq k_F$ becomes severely restricted. We can thus infer that because of the relatively large difference of phase space volume over which $k_1 \simeq k_F$ the contribution of diagram (a) to the self-energy in second order dwarfs that of (b), $\Sigma^*(a)/\Sigma^*(b) \gg 1$.

We can extend this argument to higher-order diagrams and ascertain that the dominant diagram in any order will be the one with maximum phase space volume for all internal momenta to have $k_i \simeq k_F$, and that will be when the internal momenta are least constrained to be in the vicinity of the Fermi surface. We also note that these dominant diagrams have the highest degree of divergence, namely, the highest power of $1/q^2 \to 1/q^{2n}$. These dominant diagrams are called *bubble* or *ring* diagrams. Although the contribution of each individual diagram diverges, their Dyson, infinite sum in the high-density regime, $r_s < 1$, gives a finite result! This inference describes the screening of Coulomb potential by free electrons, and is the basis for the random phase approximation, as we shall show later in this chapter.

15.5.2 Density–Density Correlation Function

We now consider the density–density Green function, or the two-particle Green function, defined as

$$\Pi(\mathbf{q}, t) = i \left\langle GS \left| T \left[\hat{\rho}_\sigma(\mathbf{q}, t) \, \hat{\rho}_{\sigma'}(-\mathbf{q}, 0) \right] \right| GS \right\rangle, \tag{15.119}$$

where $|GS\rangle$ is the ground state, and

$$\hat{\rho}(\mathbf{q}, t) = \sum_{\mathbf{k}\sigma} c_{\mathbf{k},\sigma}^\dagger(t) \, c_{\mathbf{k}+\mathbf{q},\sigma}(t).$$

We obtain

$$\Pi(\mathbf{q}, t) = i \sum_{\substack{\mathbf{k}\mathbf{k}' \\ \sigma\sigma'}} \left\langle GS \left| T \left[c_{\mathbf{k},\sigma}^\dagger(t) \, c_{\mathbf{k}+\mathbf{q},\sigma}(t) \, c_{\mathbf{k}',\sigma'}^\dagger(0) \, c_{\mathbf{k}'-\mathbf{q},\sigma'}(0) \right] \right| GS \right\rangle. \tag{15.120}$$

This quantity describes the propagation of a particle–hole pair created at time 0: the particle is created in state $\mathbf{k}', \sigma'$, and the hole is created by annihilating a particle in state $\mathbf{k}' - \mathbf{q}, \sigma'$. The pair is ultimately annihilated in states $\mathbf{k}, \sigma$ and $\mathbf{k} + \mathbf{q}, \sigma$ at time t. By analogy to the case of the single-particle Green function, this propagator will therefore have poles at energies corresponding to stable particle–hole excitations of the system. The summation over $\mathbf{k}$ and $\mathbf{k}'$ describes the propagation of a superposition of many particle–hole excitations: such superpositions are in fact collective excitations. Actually, we are just setting here a density fluctuation, and its propagator is sometimes called the density-fluctuation propagator.

For an interacting system, we have the perturbative expansion

$$\Pi(\mathbf{q}, t) = i \frac{\left\langle \Phi_0 \left| T \left[\hat{\rho}_\sigma(\mathbf{q}, t) \, \hat{\rho}_{\sigma'}(-\mathbf{q}, 0) S(\infty, -\infty) \right] \right| \Phi_0 \right\rangle}{\left\langle \Phi_0 \left| S(\infty, -\infty) \right| \Phi_0 \right\rangle} \tag{15.121}$$

with all quantities appearing in the interaction representation, and all expectation values with respect to the noninteracting ground state. We can, therefore, follow the same diagrammatic rules developed previously for any order in the perturbation expansion, including vacuum elimination.

15.5.3 Excitation Spectrum of the Noninteracting System

We start by examining the zeroth-order in the perturbation series, $\Pi_0(\mathbf{q}, t)$, which is just the two-particle correlation function of the noninteracting system. $\Pi_0(\mathbf{q}, t)$ provides information about the excitation spectrum of the Sommerfeld Fermi gas, with a fixed number of particles, N. We evaluate this with the aid of Wick's theorem. There are two possible contractions:

$$1. \; \langle T[\overset{\displaystyle\frown}{c^\dagger_{\mathbf{k}\sigma}(t) c_{\mathbf{k}+q,\sigma}(t)} \overset{\displaystyle\frown}{c^\dagger_{\mathbf{k}',\sigma'}(0) c_{\mathbf{k}'-q,\sigma'}(0)}] \rangle$$

$$2. \; \langle T[\overset{\displaystyle\frown}{c^\dagger_{\mathbf{k}\sigma}(t) \; c_{\mathbf{k}+q,\sigma}(t) \; c^\dagger_{\mathbf{k}',\sigma'}(0) \; c_{\mathbf{k}'-q,\sigma'}(0)}] \rangle$$

The first contraction requires $\mathbf{q} = 0$ in order to be nonzero, giving

$$\left(\sum_{\mathbf{k},\sigma} \left\langle c^\dagger_{\mathbf{k},\sigma} \, c_{\mathbf{k},\sigma} \right\rangle \right) \left(\sum_{\mathbf{k}',\sigma'} \left\langle c^\dagger_{\mathbf{k}',\sigma'} \, c_{\mathbf{k}',\sigma'} \right\rangle \right) \propto N^2.$$

It describes the self-correlation of the uniform electron density, and we ignore this contribution. Thus, there is only one possible contraction, which, using the definition $G^{(0)} = -i \left\langle T c \, c^\dagger \right\rangle$ and performing three fermionic swaps, is written as

$$\Pi_0(\mathbf{q}, t) = i \sum_{\mathbf{k}\sigma} G^{(0)}_\sigma(\mathbf{k} + \mathbf{q}, t) \, G^{(0)}_\sigma(\mathbf{k}, -t) \tag{15.122}$$

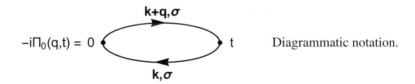

$-i\Pi_0(\text{q,t}) = 0$ Diagrammatic notation.

We note that the momentum $\mathbf{q}$ is an external one, whereas $\mathbf{k}$ is the internal momentum, which together with an internal spin are summed over. We also note that a hole can be considered as an electron propagating backward in time; which is the reason such diagram is known as a particle–hole loop, or a polarization loop.

Next, we perform a Fourier transform with respect to time in order to work in the momentum-energy representation. Since we have a product of two Green functions in time, we obtain a convolution in the energy domain, namely,

$$\Pi_0(\mathbf{q}, \omega) = \int_{-\infty}^{\infty} dt \, e^{i\omega t} \, \Pi_0(\mathbf{q}, t) = i \sum_{\mathbf{k},\sigma} \int \frac{d\varepsilon}{2\pi} G^{(0)}_\sigma(\mathbf{k}, \varepsilon) G^{(0)}_\sigma(\mathbf{k} + \mathbf{q}, \varepsilon + \omega). \tag{15.123}$$

It is diagrammatically depicted as

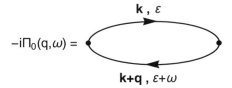

$$-i\Pi_0(q,\omega) =$$

Given that

$$G^{(0)}(\mathbf{k},\varepsilon) = \frac{1}{\varepsilon - \varepsilon(\mathbf{k}) + i\eta\,\text{sign}[\varepsilon(\mathbf{k})]},$$

where $\text{sign}[\varepsilon(\mathbf{k})]$ identifies whether it represents a particle or a hole, we obtain

$$\Pi_0(\mathbf{q},\omega) = \sum_{\mathbf{k},\sigma} \int \frac{d\varepsilon}{2i\pi}\, G_\sigma^{(0)}(\mathbf{k},\varepsilon)\, G_\sigma^{(0)}(\mathbf{k}+\mathbf{q},\varepsilon+\omega)$$

$$= -2 \sum_{\mathbf{k}} \int \frac{d\varepsilon}{2i\pi} \left[\frac{1}{\varepsilon - \varepsilon_{\mathbf{k}} + i\eta\,\text{sign}(\varepsilon_{\mathbf{k}})} \right] \left[\frac{1}{\varepsilon + \omega - \varepsilon_{\mathbf{k}+\mathbf{q}} + i\eta\,\text{sign}(\varepsilon_{\mathbf{k}+\mathbf{q}})} \right].$$

We carry out the integral over ε in the complex plane and make use of the residue theorem. We note that when $\varepsilon_{\mathbf{k}+\mathbf{q}}$ and $\varepsilon_{\mathbf{k}}$ have the same sign, so that both are $\lessgtr 0$, the poles in the complex ε-plane are either in the lower or upper half-plane. Taking the contour in the pole-free half, and noting that the asymptotic value of the integral is of order $1/\varepsilon^2$, yields zero. The integral is nonzero only if $\varepsilon_{\mathbf{k}+\mathbf{q}}$ and $\varepsilon_{\mathbf{k}}$ have opposite signs, in which case evaluating the residue yields

$$\Pi_0(\mathbf{q},\omega) = -2 \sum_{\mathbf{k}} \left[\frac{\Theta(-\varepsilon_{\mathbf{k}+\mathbf{q}})\,\Theta(\varepsilon_{\mathbf{k}})}{\omega - \varepsilon_{\mathbf{k}+\mathbf{q}} + \varepsilon_{\mathbf{k}} + i\eta} - \frac{\Theta(\varepsilon_{\mathbf{k}+\mathbf{q}})\,\Theta(-\varepsilon_{\mathbf{k}})}{\omega - \varepsilon_{\mathbf{k}+\mathbf{q}} + \varepsilon_{\mathbf{k}} - i\eta} \right]. \qquad (15.124)$$

The physical interpretation to the theta functions is as follows:

$$\Theta(-\varepsilon_{\mathbf{k}}) \Rightarrow n_p = n_{\mathbf{k}}^0 \qquad \text{a particle in the Fermi sea}$$

$$\Theta(\varepsilon_{\mathbf{k}}) \Rightarrow n_h = 1 - n_{\mathbf{k}}^0 \qquad \text{a hole outside the Fermi sea.}$$

Consequently, we can regard the first term in (15.124) as the retarded propagator of a particle–hole pair, with $|\mathbf{k}+\mathbf{q}| > k_F$ and $|\mathbf{k}| < k_F$, corresponding to excitation energy $\omega_{\mathbf{q}} = \varepsilon_{\mathbf{k}+\mathbf{q}} - \varepsilon_{\mathbf{k}} > 0$. The second term is just the time-reversed partner, namely the advanced propagator.

The Dynamical Form Factor

The imaginary part of Π_0 is known as the dynamical form factor

$$S_0(\mathbf{q},\omega) = \frac{1}{\pi}\, \text{Im}\, \Pi_0(\mathbf{q},\omega). \qquad (15.125)$$

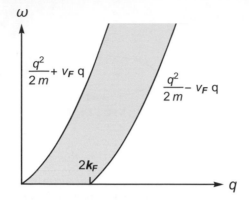

Figure 15.34 Electron–hole excitation range shown as the gray region.

Recalling that $\mathrm{Im}\frac{1}{x-i\eta} = \pi\,\delta(x)$, we obtain

$$S_0(\mathbf{q},\omega) = 2\int\frac{d\mathbf{k}}{(2\pi)^3}\;\underbrace{n^0(\mathbf{k})\left(1 - n^0(\mathbf{k}+\mathbf{q})\right)}\;\underbrace{\delta(\omega - \varepsilon_{\mathbf{k}+\mathbf{q}} + \varepsilon_{\mathbf{k}})}\,. \qquad (15.126)$$

A nonzero $n^0(\mathbf{k})$ implies that the state $\mathbf{k}$ is occupied, and that an excitation of the particle in it will leave behind a hole, while a zero $n^0(\mathbf{k}+\mathbf{q})$ indicates an empty state that can be filled with a particle of momentum $\mathbf{k}+\mathbf{q}$. Conservation of energy implies that the dynamical form factor provides information about *real* or *absorbtive processes*. This is just the manifestation of all possible particle–hole pairs that carry momentum $\mathbf{q}$ and satisfy energy conservation. It defines the spectrum of *real excitations of the Fermi gas*.

To determine the spectrum, we inspect the energy conserving delta function

$$0 \le \omega = \frac{q^2}{2m} + \mathbf{v_k}\cdot\mathbf{q} \;\Rightarrow\; \begin{cases} \omega_{\max} = \dfrac{q^2}{2m} + v_F q & q \le 2k_F \\[2ex] \omega_{\min} = \dfrac{q^2}{2m} - v_F q & q \ge 2k_F. \end{cases}$$

The electron–hole excitation spectrum is shown in Figure 15.35. We find that at a fixed $\mathbf{q}$, $S_0(\mathbf{q},\omega)$ gives a wide range of allowed energies. It is then clear that we have what is known as *incoherent* excitations for all particle–hole events: an excitation in an initial state $(\mathbf{q},\omega)$ can immediately decay into excitations with energies $< \omega$ but with the same $\mathbf{q}$ for the Sommerfeld gas. This is in clear contrast with the coherent single-particle excitation events manifest in the corresponding spectral function

$$A(\mathbf{k},\omega) = \delta[\omega - \varepsilon(\mathbf{k})].$$

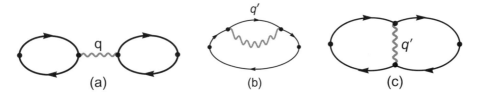

Figure 15.35 First-order diagrams for Π.

Virtual Excitations and the Lindhard Function

If the energy conservation condition is not realized, the corresponding contribution is reactive, not dissipative, and it only appears in the real part of the density–density correlation function. The intermediate state then is only virtual. To see this, we now consider the real part of Π_0 in (15.124), which, after some elementary algebra, can be written as

$$\text{Re}\,\Pi_0(\mathbf{q},\omega) = 2 \int \frac{d\mathbf{k}}{(2\pi)^3} \frac{n^0(\mathbf{k}) - n^0(\mathbf{k}+\mathbf{q})}{\omega - \varepsilon_{\mathbf{k}+\mathbf{q}} + \varepsilon_{\mathbf{k}}}. \tag{15.127}$$

Performing the integral yields

$$\text{Re}\,\Pi_0(\mathbf{q},\omega) = \mathcal{D}(\varepsilon_F)\, F\left(q/2k_F, \omega/4\varepsilon_F\right)$$

$$F(\tilde{q},\tilde{\omega}) = \frac{1}{8\tilde{q}} \left\{ \left[1 - \left(\tilde{q} + \frac{\tilde{\omega}}{\tilde{q}} \right)^2 \right] \ln \left| \frac{\tilde{q} + (\tilde{\omega}/\tilde{q}) + 1}{\tilde{q} + (\tilde{\omega}/\tilde{q}) - 1} \right| \right.$$

$$\left. + \left[1 - \left(\tilde{q} - \frac{\tilde{\omega}}{\tilde{q}} \right)^2 \right] \ln \left| \frac{\tilde{q} - (\tilde{\omega}/\tilde{q}) + 1}{\tilde{q} - (\tilde{\omega}/\tilde{q}) - 1} \right| \right\} + \frac{1}{2}, \tag{15.128}$$

where $\mathcal{D}(\varepsilon_F)$ is the density of states at ε_F. This is just the *Lindhard formula* we derived in Chapter 7. It carries information about the probability and phase of all possible *virtual processes*.

We note that in the limit $\lim_{\omega,\,q\to 0}\,\text{Im}\Pi_0 = 0$, while (15.128) reveals that

$$\text{Re}\left[\Pi_0(\mathbf{q} \to 0, \omega \to 0) \right] = \mathcal{D}(\varepsilon_F) = \frac{m\,k_F}{\pi^2 \hbar^2} = \frac{3}{2} \frac{n}{\varepsilon_F}. \tag{15.129}$$

15.5.4 Interacting Systems and the Random Phase Approximation (RPA)

In the preceding sections, we have attempted to explain the significance of $\Pi(\mathbf{q},\omega)$ and we have examined in detail Π_0, the zeroth-order contribution to Π. Now we shall explore the case when interactions are incorporated, and investigate their ramifications.

Irreducible Polarization Bubble

The first-order diagrams for $\Pi(\mathbf{q}, \omega)$ are shown in Figure 15.35. By analogy to the treatment of the self-energy, we will define the exact irreducible polarization bubble, $\Pi(\mathbf{q}, \omega)$, as the sum over all diagrams that *cannot be cut into two by splitting a single interaction line* (an external line), namely,

$$\Pi(q,\omega) = \bigcirc = \bigcirc + \bigcirc + \bigcirc + \ldots$$

In fact, we can obtain the exact charge susceptibility, which is the response to external Coulomb perturbations, by constructing a Dyson series of the irreducible polarization bubble as

$$\chi(q,\omega) = \bigcirc = \bigcirc + \bigcirc \!\!\!\sim\!\!\! \bigcirc + \ldots$$

or

$$\chi(\mathbf{q},\omega) = \frac{\Pi(\mathbf{q},\omega)}{1 - V_c(\mathbf{q})\,\Pi(\mathbf{q},\omega)}.$$

Screened Interaction

We can also obtain an exact *effective screened Coulomb interaction*, which we represent diagrammatically by a double wiggle line, by attaching external interaction lines, namely,

or

$$V_{\text{eff}}^{\text{exact}}(\mathbf{q},\omega) = \frac{V_c(\mathbf{q})}{1 - V_c(\mathbf{q})\,\Pi(\mathbf{q},\omega)}.$$

The Random Phase Approximation

A phase space analysis of the diagrams in Figure 15.35 shows that diagram (a) has the dominant contribution to the correlation function, where each of the two Π_0 bubbles has an unconstrained sum. Again, by extension, we argue that the dominant contribution at any order consists of interacting Π_0 bubbles. A justifiable approximation can now be made in terms of a Dyson-like sum involving Π_0 bubbles, which is known as the random phase approximation (RPA):

$$\chi^{\text{RPA}}(q,\omega) = \bigcirc = \bigcirc + \bigcirc \!\!\!\sim\!\!\! \bigcirc +$$

$$= \Pi_0(\mathbf{q},\omega) + \Pi_0(\mathbf{q},\omega)\left[V_c(\mathbf{q})\,\Pi_0(\mathbf{q},\omega)\right] + \Pi_0(\mathbf{q},\omega)\left[V_c(\mathbf{q})\,\Pi_0(\mathbf{q},\omega)\right]^2 + \cdots$$

$$= \frac{\Pi_0(\mathbf{q},\omega)}{1 - V_c(\mathbf{q})\,\Pi_0(\mathbf{q},\omega)}. \tag{15.130}$$

Imaginary part of χ^{RPA}

We expand $\chi^{\text{RPA}}(\mathbf{q}, \omega)$ as

$$\chi^{\text{RPA}}(\mathbf{q}, \omega) = \frac{\text{Re}\,[\Pi_0] + i\,\text{Im}\,[\Pi_0]}{(1 - V_c(\mathbf{q})\text{Re}\,[\Pi_0]) - i V_c(\mathbf{q})\text{Im}\,[\Pi_0]},$$

which yields

$$\text{Im}\left[\chi^{\text{RPA}}(\mathbf{q}, \omega)\right] = \frac{\text{Im}\,[\Pi_0]}{(1 - V_c(\mathbf{q})\text{Re}\,[\Pi_0])^2 + V_c^2(\mathbf{q})\,(\text{Im}\,[\Pi_0])^2}.$$

The numerator indicates that the electron–hole excitations spectrum carries over to the interacting case. Moreover, outside the electron–hole continuum, where $\text{Im}\,[\Pi_0] = 0$, new excitations appear when $1 - V_c(\mathbf{q})\text{Re}\,[\Pi_0] = 0$. These excitations are just the plasmons that are manifest in Coulomb interacting systems, which we will describe later in this chapter. Such excitations are also manifest as zero-sound excitations in systems with delta-function, or short-range, interactions, which is the subject of Problem 14.

We can extend our approximation to the effective screened interaction, and write

Physically, we could say that two particles could interact directly, or that one particle interacts with a density fluctuation in the electron sea induced by the other particle, or the interaction of the two particles is mediated by two intermediate density fluctuations, and so on. Mathematically, we see that

$$V^{\text{RPA}}(\mathbf{q}, \omega) = V_c(\mathbf{q}) + V_c(\mathbf{q})\left[V_c(\mathbf{q})\,\Pi_0(\mathbf{q}, \omega)\right] + V_c(\mathbf{q})\left[V_c(\mathbf{q})\,\Pi_0(\mathbf{q}, \omega)\right]^2 + \cdots$$

$$= \frac{V_c(\mathbf{q})}{1 - V_c(\mathbf{q})\,\Pi_0(\mathbf{q}, \omega)} = \frac{V_c(\mathbf{q})}{\epsilon(\mathbf{q}, \omega)} \tag{15.131}$$

$$\epsilon^{-1}(\mathbf{q}, \omega) = 1 + V_c(\mathbf{q})\,\chi^{\text{RPA}}(\mathbf{q}, \omega), \quad \text{dielectric function.}$$

Notice that the effective interaction is ω-dependent, which means that it acquires a time dependence, in contrast to the instantaneous character of the original nonrelativistic Coulomb interaction. This is to be expected, since interactions via density fluctuations certainly involve time delay.

We now consider static screening by examining the effective interaction in the limit $\omega = 0$. Using the expression for $\text{Re}\,\Pi_0(\mathbf{q}, \omega)$ given in (15.129), we obtain

$$1 + V_c(\mathbf{q})\,\Pi_0(\mathbf{q} \to 0, \omega = 0) = 1 + \frac{6\pi n e^2}{q^2 \varepsilon_F}, \tag{15.132}$$

which yields

$$V^{\text{RPA}}(\mathbf{q} \to 0, \omega \to 0) = \frac{4\pi e^2}{q^2 + \kappa^2}, \tag{15.133}$$

where

$$\kappa = \frac{1}{\xi_{TF}} = \left(\frac{6\pi n e^2}{\varepsilon_F} \right)^{1/2} \tag{15.134}$$

is the inverse Thomas–Fermi screening length we derived in Chapter 2. The effective interaction potential becomes a short-range Yukawa type, namely,

$$V_{\text{eff}}(q \to 0, 0) \propto \frac{e^{-\kappa r}}{r}.$$

Consequently, we see that in the presence of the polarizable medium (the electron sea), the long-range Coulomb interaction actually becomes short range. To physically interpret this screening mechanism, we consider the scenario where we introduce a negative point charge. Then the electrons in its vicinity will be repelled from this region, creating a region around this point of lower than average density, namely, a screening region of positive charge. At a long distance, another electron sees not only the initial negative charge, but also the screening region around it, which almost cancel each other, hence making the Coulomb interaction effectively short ranged.

15.5.5 Plasma Oscillations

Next we consider the case of nonzero frequency $\omega \neq 0$. We shall confine the discussion to the domain of high frequencies, long wave lengths, and low temperatures, and in particular outside the electron–hole continuum, where Im $[\Pi_0] = 0$. Writing the dielectric function explicitly as

$$\epsilon^{\text{RPA}}(\mathbf{q}, \omega) = 1 - \frac{4\pi e^2}{q^2} \int \frac{2 d\mathbf{k}}{(2\pi)^3} \frac{n^0(\mathbf{k}) - n^0(\mathbf{k} + \mathbf{q})}{\omega - \varepsilon_{\mathbf{k}+\mathbf{q}} + \varepsilon_{\mathbf{k}}},$$

subject to the conditions $\omega \gg v_F q$, $q \ll k_F$, $k_B T \ll \varepsilon_F$. We note that for long wavelength, $q \to 0$,

$$n^0(\mathbf{k}) - n^0(\mathbf{k} + \mathbf{q}) = \mathbf{q} \cdot \nabla_{\mathbf{k}} \, \varepsilon_{\mathbf{k}} \frac{dn(\varepsilon)}{d\varepsilon} = v_{\mathbf{k}} q \, \cos\theta \left. \frac{dn(\varepsilon)}{d\varepsilon} \right|_{\varepsilon_F}$$

$$\varepsilon_{\mathbf{k}+\mathbf{q}} - \varepsilon_{\mathbf{k}} = \frac{\mathbf{q} \cdot \nabla_{\mathbf{k}} \, \varepsilon_{\mathbf{k}}}{m} = \mathbf{q} \cdot \mathbf{v}_{\mathbf{k}} = v_{\mathbf{k}} q \cos\theta$$

and we write

$$2 \int \frac{d\mathbf{k}}{(2\pi)^3} \frac{n^0(\mathbf{k}) - n^0(\mathbf{k} + \mathbf{q})}{\omega - \varepsilon_{\mathbf{k}+\mathbf{q}} + \varepsilon_{\mathbf{k}}} = \frac{1}{2\pi^2} \int dk \, k^2 \, \delta(\varepsilon_{\mathbf{k}} - \varepsilon_F) \int_{-1}^{1} dx \frac{v_{\mathbf{k}} q x}{\omega - v_{\mathbf{k}} q x}$$

$$= \frac{1}{2\pi^2} \int dk \, k^2 \frac{\delta(k - k_F)}{v_F} \int_{-1}^{1} dx \frac{v_{\mathbf{k}} q x}{\omega - v_{\mathbf{k}} q x}$$

$$= \frac{1}{2\pi^2} k_F^2 \frac{1}{v_F} \frac{\omega}{v_F q} \int_{-q v_F / \omega}^{q v_F / \omega} du \frac{u}{1 - u}$$

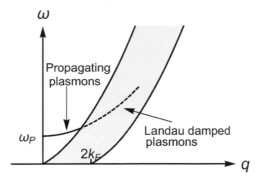

Figure 15.36 Plasmon branch with its propagating and damped parts. Note that Im $\chi(\mathbf{q}, \omega) \neq 0$ only in the shaded area.

Since $qv_F/\omega \ll 1$, we expand

$$\frac{u}{1-u} \simeq u + u^2 + u^3 + u^4$$

and we obtain

$$\text{Re } \Pi_0(\mathbf{q}, \omega) = \frac{1}{2\pi^2} \frac{k_F^2 \omega}{v_F^2 q} \left[\frac{1}{2} u^2 + \frac{1}{3} u^3 + \frac{1}{4} u^4 + \frac{1}{5} u^5 \right]_{-qv_F/\omega}^{qv_F/\omega}$$

$$= \frac{n}{m} \frac{q^2}{\omega^2} \left[1 + \frac{3}{5} \left(\frac{qv_F}{\omega} \right)^2 \right],$$

where we used $v_F = k_F/m$ and $3\pi^2 n = k_F^3$. The RPA dielectric function becomes

$$\epsilon^{\text{RPA}}(\mathbf{q}, \omega) = 1 - \frac{4\pi e^2}{q^2} \Pi_0(\mathbf{q}, \omega) = 1 - \frac{\omega_p^2}{\omega^2} \left[1 + \frac{3}{5} \left(\frac{qv_F}{\omega} \right)^2 \right].$$

The plasmon dispersion, shown in Figure 15.36, is obtained from the zeros of the dielectric function

$$\epsilon^{\text{RPA}}(\mathbf{q}, \omega) = 0 \quad \Rightarrow \quad \omega^2 \simeq \omega_P^2 + \frac{3}{5} (qv_F)^2 \quad \Rightarrow \quad \omega(q) \simeq \omega_P + \frac{3}{10} \frac{v_F^2}{\omega_p} q^2.$$

Once the dispersion curve of plasmons enters in the particle–hole continuum, where Im $[\Pi_0] \neq 0$, plasmons become unstable to decay into an electron–hole pair.

15.6 Finite Temperature Green Functions

In a large number of applications in many-body theory, we need to include the effects of temperature. In such cases, the Green functions take the form

$$G_{AB}(t) = \langle A(t) B(0) \rangle = -i \text{Tr} [\rho \, A(t) \, B(0)],$$

where ρ is the density operator

$$\rho = \frac{1}{Z}\, e^{-\beta\mathcal{H}}, \qquad Z = \mathrm{Tr}\left[e^{-\beta\mathcal{H}}\right].$$

$\mathcal{H} \to \mathcal{H} - \mu N$ and the operators $A(t)$ and $B(t_1)$ are in the Heisenberg picture. If we adopt our earlier approach, we would use perturbation theory, and assume that the Hamiltonian can be split into a solvable part $\mathcal{H}_0$, and a perturbative quantity $\mathcal{H}'$ that contains the remaining terms. We then seek an expansion in powers of $\mathcal{H}'$. We note that at zero temperature, we replace the traces over $\hat{\rho}$ with ground-state averages, which allows us to transform the problem to the interaction picture, thanks to the Gell-Mann–Low theorem and the scattering matrix, and to solve it perturbatively within the noninteraction framework.

15.6.1 *Perturbative Expansion of ρ*

In the present case, the difficulty lies in the expansion of ρ. We note that the density operator ρ satisfies the Bloch equation of motion [30, 63]

$$-\frac{\partial\rho}{\partial\beta} = \mathcal{H}\,\rho,$$

which is similar to the Schrödinger equation of motion of $S(t,t_1)$, but with the time variable it replaced by β:

$$S \leftrightarrow \rho \qquad it \leftrightarrow \beta \qquad \mathcal{H} \leftrightarrow \mathcal{H} - \mu N.$$

Consequently, it is possible to perform a simple expansion of ρ in powers of $\mathcal{H}'$. However, such expansion will imply that the "evolution" of the Hamiltonian will not be along the real-time axis, but along the inverse temperature axis. Merging this expansion with that of $S(t,t_1)$ leads to a dilemma, since we have no clue how to "time-order" product of operators that evolve along two orthogonal axes.

15.6.2 *Analytic Continuation*

From a more mathematical perspective, it would be desirable to include $\rho = e^{-\beta\mathcal{H}}$ in the group of time-evolution operators. This would require that we analytically continue the time-evolution group to complex times, $t \to t + i\tau = z$, and introduce new operators $e^{iz\mathcal{H}}$ that evolve in imaginary rather than real time. Examining the corresponding Green function

$$t > 0,\ G_{AB}(t) = \frac{-i}{Z}\,\mathrm{Tr}\left[e^{-\beta\mathcal{H}}\, e^{i\mathcal{H}t/\hbar}\, A\, e^{-i\mathcal{H}t/\hbar}\, B\right]$$

$$\to\ G_{AB}(z) = \frac{-i}{Z}\,\mathrm{Tr}\left[e^{-(\hbar\beta+\tau)\mathcal{H}/\hbar}\, e^{i\mathcal{H}t/\hbar}\, A\, e^{-i\mathcal{H}t/\hbar}\, B\right]$$

$$t < 0,\ G_{AB}(t) = \frac{\pm i}{Z}\,\mathrm{Tr}\left[e^{-\beta\mathcal{H}}\, e^{-i\mathcal{H}t/\hbar}\, A\, e^{i\mathcal{H}t/\hbar}\, B\right]$$

$$\to\ G_{AB}(z) = \frac{\pm i}{Z}\,\mathrm{Tr}\left[e^{-(\hbar\beta-\tau)\mathcal{H}/\hbar}\, e^{i\mathcal{H}t/\hbar}\, A\, e^{-i\mathcal{H}t/\hbar}\, B\right],$$

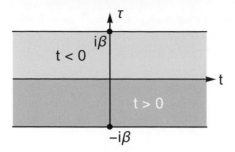

Figure 15.37 Analytic domains in the complex time plane of the Green function: dark gray $t > 0$, light gray $t < 0$.

where, in the second line, we used the cyclic property of the trace, and $+$ corresponds to fermions and $-$ to bosons. We find that e^{izH} is mathematically well defined on the strips, shown in Figure 15.37

$$\begin{cases} -\hbar\beta \leq \tau \leq 0, \ -\infty \leq t \leq \infty, & t > 0 \\ 0 \leq \tau \leq \hbar\beta, \ -\infty \leq t \leq \infty, & t < 0 \end{cases}$$

since $\mathcal{H}$ is bounded from below. This defines the domains of complex time for which the preceding Green function is analytic, and shows that the Green function of complex time is discontinuous on the whole real axis of time [114, 125].

In an informal way, we can associate the boundaries $\pm\hbar\beta$ along the imaginary axis with decoherence induced by thermal fluctuations, namely, thermal fluctuations introduce an uncertainty $k_B T$ in energies, in which case,

$$\tau_T = \frac{\hbar}{k_B T}$$

represents a characteristic time of a thermal fluctuation. Processes of duration longer than τ_T lose their phase coherence, so coherent quantum processes are limited within a world of finite temporal extent, $\hbar\beta$.

15.6.3 Imaginary Time

The solution to this problem, first proposed by Takeo Matsubara [126], relies on taking advantage of the similarity between the operators $e^{-\beta\mathcal{H}}$ and $e^{-i\mathcal{H}t/\hbar} = U(t)$, but limit our domain to the imaginary axis, where

$$\hat{\rho} = \frac{e^{-\beta\mathcal{H}}}{\text{Tr}\,e^{-\beta\mathcal{H}}} = \frac{U(-i\hbar\beta)}{\text{Tr}\,[U(-i\hbar\beta)]}.$$

The Matsubara approach is to treat time as an imaginary temperature. For time $t = -i\hbar\tau$, the Green function is well defined in the interval $-\beta \leq \tau \leq \beta$. We then replace the real-time evolution, carried along t by U (or S), with an imaginary-time evolution

$$-i\tau, \ \tau \in \mathbb{R} : e^{-i\mathcal{H}t/\hbar} \ \rightarrow \ e^{-i\mathcal{H}(-i\tau)} = e^{-\mathcal{H}\tau},$$

which evolve in imaginary rather than real time. This will allow us to write a tractable expansion for the Green function of these new imaginary-time operators. The properties of $e^{-\mathcal{H}\tau}$ allow the imaginary-time functions to be analytically continued to recover the real-time ones.

This choice renders finite temperature many-body physics to be just as manageable as its zero temperature counterpart. Amazingly, in many cases, the imaginary-time, or Matsubara, formulation is easier to handle. We can carry over almost everything we have developed at zero temperature:

- The Schrödinger, Heisenberg, and interaction representations
- Wick's theorem
- Feynman diagram expansions

We shall now recast these concepts and procedures in imaginary time.

15.6.4 *Representations*

To develop the imaginary-time formulation of the Heisenberg and interaction representations in terms of $t \to -i\tau\hbar$, we start with writing the real-time Schrödinger equation

$$i\hbar \frac{\partial}{\partial t} |\psi_S\rangle = \mathcal{H} |\psi_S\rangle,$$

which then becomes

$$-\frac{\partial}{\partial \tau} |\psi_S\rangle = \mathcal{H} |\psi_S\rangle \tag{15.135}$$

so the time-evolved wavefunction is given by

$$|\psi_S(\tau)\rangle = e^{-\mathcal{H}\tau} |\psi_S(0)\rangle. \tag{15.136}$$

The imaginary-time evolution of an operator $\mathcal{O}$ in the Heisenberg representation can then be written as

$$\mathcal{O}_H(\tau) = e^{\mathcal{H}\tau} \mathcal{O}_S e^{-\mathcal{H}\tau}. \tag{15.137}$$

Its Heisenberg equation of motion becomes

$$\frac{\partial \mathcal{O}_H}{\partial \tau} = [\mathcal{H}, \mathcal{O}_H]. \tag{15.138}$$

Example: Case of a Free Fermionic Particle

For the free particle Hamiltonian, we obtain

$$\mathcal{H} = \sum_{\mathbf{k}} \varepsilon_{\mathbf{k}} c_{\mathbf{k}}^{\dagger} c_{\mathbf{k}} \quad \Rightarrow \quad \begin{cases} \dfrac{\partial c_{\mathbf{k}}}{\partial \tau} = [\mathcal{H}, c_{\mathbf{k}}] \\[2mm] \dfrac{\partial c_{\mathbf{k}}^{\dagger}}{\partial \tau} = \left[\mathcal{H}, c_{\mathbf{k}}^{\dagger}\right], \end{cases}$$

which yields

$$c_{\mathbf{k}}(\tau) = e^{-\varepsilon_{\mathbf{k}}\tau}\, c_{\mathbf{k}}$$

$$c_{\mathbf{k}}^{\dagger}(\tau) = e^{\varepsilon_{\mathbf{k}}\tau}\, c_{\mathbf{k}}^{\dagger}.$$

We should caution here about an important difference between real-time and imaginary-time formalism: in the latter, the *Heisenberg creation and annihilation operator are not Hermitian conjugates*:

$$c_{\mathbf{k}}^{\dagger}(\tau) = (c_{\mathbf{k}}(-\tau))^{\dagger} \neq (c_{\mathbf{k}}(\tau))^{\dagger}.$$

Next we consider the modifications to be made to the interaction representation, where the fast time evolution of the state ket, arising from the Hamiltonian $\mathcal{H}_0$, is eliminated. Starting with the state ket

$$|\psi_I(\tau)\rangle = e^{\mathcal{H}_0\tau}\, |\psi_S(\tau)\rangle = e^{\mathcal{H}_0\tau}\, e^{-\mathcal{H}\tau}\, |\psi_H\rangle = U_I(\tau)\, |\psi_H\rangle, \tag{15.139}$$

where $U_I(\tau) = e^{\mathcal{H}_0\tau}\, e^{-\mathcal{H}\tau}$ is the time-evolution operator. The relationship between the Heisenberg and the interaction representation of operators is given by

$$A_H(\tau) = e^{\mathcal{H}\tau}\, A_S\, e^{-\mathcal{H}\tau} = U_I^{-1}(\tau)\, A_I(\tau)\, U_I(\tau). \tag{15.140}$$

The equation of motion for $U_I(\tau)$ is given by

$$-\frac{\partial}{\partial\tau} U_I(\tau) = -\frac{\partial}{\partial\tau}\left[e^{\mathcal{H}_0\tau}\, e^{-\mathcal{H}\tau} \right] = e^{\mathcal{H}_0\tau}\, \mathcal{H}'\, e^{-\mathcal{H}\tau}$$

$$= e^{\mathcal{H}_0\tau}\, \mathcal{H}'\, e^{-\mathcal{H}_0\tau}\, U_I(\tau) = \mathcal{H}_I\, U_I(\tau). \tag{15.141}$$

These equations parallel those in real time, and by following exactly analogous procedures, we find that the imaginary-time-evolution operator in the interaction representation is given by a time-ordered exponential of the form

$$U_I(\tau) = T_\tau\, \exp\left[-\int_0^\tau d\tau'\, \mathcal{H}_I(\tau') \right]. \tag{15.142}$$

15.6.5 Matsubara Functions: Imaginary-Time Green Functions

Green functions for operators with imaginary-time arguments are referred to as *Matsubara* or *temperature* Green functions. The Matsubara function is defined as

$$\mathcal{G}_{\hat{A}\hat{B}}(\tau_1, \tau_2) = -\left\langle T_\tau\, \hat{A}(\tau_1)\, \hat{B}(\tau_2) \right\rangle = -\frac{1}{Z}\, \mathrm{Tr}\left[e^{-\beta\mathcal{H}}\, \hat{A}(\tau_1)\, \hat{B}(\tau_2) \right], \tag{15.143}$$

where $\hat{A}(\tau_1)$ is again in the Heisenberg representation. The brackets denote a thermal average, and the symbol T_τ denotes time ordering. It means that operators are ordered chronologically, and just like the real-time counterpart, the later "times" are placed to the left

$$T_\tau \, \hat{A}(\tau_1) \, \hat{B}(\tau_2) = \Theta(\tau_1 - \tau_2) \, \hat{A}(\tau_1) \, \hat{B}(\tau_2) - \Theta(\tau_2 - \tau_1) \, \hat{B}(\tau_2) \, \hat{A}(\tau_1) \qquad (15.144)$$

for fermions. Matsubara functions are used only for thermal equilibrium calculations at temperature T.

We note that $\mathcal{G}_{AB}(\tau_1, \tau_2)$ is a function of the time difference:

$$\mathcal{G}_{AB}(\tau_1, \tau_2) = \mathcal{G}_{AB}(\tau_1 - \tau_2, 0).$$

This follows from the cyclic properties of the trace. We have for $\tau_1 > \tau_2$

$$\mathcal{G}_{AB}(\tau_1, \tau_2) = -\Theta(\tau_1 - \tau_2) \frac{1}{Z} \, \mathrm{Tr} \left[e^{-\beta \mathcal{H}} \, e^{\mathcal{H}\tau_1} \, \hat{A} e^{-\mathcal{H}\tau_1} \, e^{\mathcal{H}\tau_2} \, \hat{B} \, e^{-\mathcal{H}\tau_2} \right]$$

$$+ \Theta(\tau_2 - \tau_1) \frac{1}{Z} \, \mathrm{Tr} \left[e^{-\beta \mathcal{H}} \, e^{\mathcal{H}\tau_2} \, \hat{B} \, e^{-\mathcal{H}\tau_2} \, e^{\mathcal{H}\tau_1} \, \hat{A} \, e^{-\mathcal{H}\tau_1} \right]$$

$$= -\Theta(\tau_1 - \tau_2) \frac{1}{Z} \, \mathrm{Tr} \left[e^{-\beta \mathcal{H}} \, e^{\mathcal{H}(\tau_1 - \tau_2)} \, \hat{A} e^{-\mathcal{H}(\tau_1 - \tau_2)} \, \hat{B} \right]$$

$$+ \Theta(\tau_2 - \tau_1) \frac{1}{Z} \, \mathrm{Tr} \left[e^{-\beta \mathcal{H}} \, \hat{B} \, e^{\mathcal{H}(\tau_1 - \tau_2)} \, \hat{A} \, e^{-\mathcal{H}(\tau_1 - \tau_2)} \right]$$

$$= \mathcal{G}_{AB}(\tau_1 - \tau_2, 0). \qquad (15.145)$$

Similar results follow for $\tau_2 > \tau_1$. Hence, we can write $\mathcal{G}_{AB}(\tau_1 - \tau_2, 0) = \mathcal{G}_{AB}(\tau)$ instead of $\mathcal{G}_{AB}(\tau_1, \tau_2)$. As we noted earlier, for $\mathcal{G}_{AB}(\tau_1, \tau_2)$ to converge, the condition $-\beta < \tau_1 - \tau_2 < \beta$ must be satisfied.

Periodicities in Imaginary Time

An important property of the Matsubara function is that it is periodic for bosons and antiperiodic for fermions over an interval β (see Figure 15.38):

$$\mathcal{G}_{\hat{A}\hat{B}}(\tau) = \begin{cases} -\mathcal{G}_{\hat{A}\hat{B}}(\tau \pm \beta), & \text{fermions} \\ \\ \mathcal{G}_{\hat{A}\hat{B}}(\tau \pm \beta), & \text{bosons} \end{cases} \qquad \text{for } \tau \lessgtr 0. \qquad (15.146)$$

This, again, follows from the cyclic properties of the trace, since (15.146) for $\tau < 0$ is

$$\mathcal{G}_{\hat{A}\hat{B}}(\tau + \beta) = -\frac{1}{Z} \, \mathrm{Tr} \left[e^{-\beta \mathcal{H}} \, e^{\mathcal{H}(\tau + \beta)} \, \hat{A} \, e^{-\mathcal{H}(\tau + \beta)} \, \hat{B} \right]$$

$$= -\frac{1}{Z} \, \mathrm{Tr} \left[e^{\mathcal{H}\tau} \, \hat{A} \, e^{-\mathcal{H}\tau} \, e^{-\beta \mathcal{H}} \, \hat{B} \right]$$

$$= -\frac{1}{Z} \, \mathrm{Tr} \left[e^{-\beta \mathcal{H}} \, \hat{B} \, e^{\mathcal{H}\tau} \, \hat{A} \, e^{-\mathcal{H}\tau} \right] = -\frac{1}{Z} \, \mathrm{Tr} \left[e^{-\beta \mathcal{H}} \, \hat{B} \, \hat{A}(\tau) \right]$$

$$= \frac{1}{Z} \, \mathrm{Tr} \left[e^{-\beta \mathcal{H}} \, T_\tau \left(\hat{A}(\tau) \, \hat{B} \right) \right] = \mp \mathcal{G}_{\hat{A}\hat{B}}(\tau) \qquad (15.147)$$

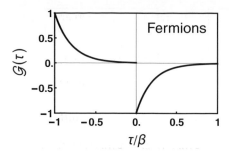

Figure 15.38 Antiperiodicity of a fermionic Green function.

and similarly for $\tau > 0$. The negative sign reveals that the Matsubara Green function is *antiperiodic* for fermions! This symmetry property of periodicity or anti-periodicity will be used to extend the definition of the Matsubara function $\mathcal{G}_{AB}(\tau)$ to the entire imaginary time axis [4].

Example: A noninteracting fermionic system with Hamiltonian

$$\mathcal{H} = \sum_{\mu} \varepsilon_\lambda \, \hat{c}^\dagger_\lambda \, \hat{c}_\lambda,$$

where $\varepsilon_\lambda = E_\lambda - \mu$ is the one-particle energy, shifted by the chemical potential. The corresponding Matsubara function is

$$G_{\lambda\lambda'}(\tau - \tau') = -\frac{1}{Z_0}\left[\Theta(\tau - \tau')\,\hat{c}_\lambda(\tau)\,\hat{c}^\dagger_{\lambda'}(\tau') - \Theta(\tau' - \tau)\,\hat{c}^\dagger_{\lambda'}(\tau')\,\hat{c}_\lambda(\tau)\right]. \quad (15.148)$$

Using the operators' time evolution we derived,

$$\hat{c}_\lambda(\tau) = e^{-\varepsilon_\lambda \tau}\,\hat{c}_\lambda(0)$$

$$\hat{c}^\dagger_\lambda(\tau) = e^{\varepsilon_\lambda \tau}\,\hat{c}^\dagger_\lambda(0). \quad (15.149)$$

Substituting in the Matsubara function, and using the equal-time expectation value of the fields

$$\left\langle \hat{c}^\dagger_{\lambda'}\,\hat{c}_\lambda \right\rangle = \delta_{\lambda\lambda'}\,n_F(\varepsilon_\lambda) = \frac{1}{e^{\beta\varepsilon_\lambda} + 1}$$

$$\left\langle \hat{c}_\lambda\,\hat{c}^\dagger_{\lambda'} \right\rangle = \delta_{\lambda\lambda'}\,(1 - n_F(\varepsilon_\lambda)),$$

we obtain

$$\mathcal{G}_{\lambda\lambda'}(\tau - \tau') = \delta_{\lambda\lambda'}\,\mathcal{G}_\lambda(\tau - \tau')$$

$$\mathcal{G}_\lambda(\tau) = -\frac{1}{Z_0}\,e^{-\varepsilon_\lambda \tau}\left[(1 - n_F(\varepsilon_\lambda))\,\Theta(\tau) - n_F(\varepsilon_\lambda)\Theta(-\tau)\right].$$

$\mathcal{G}_\lambda(\tau)$ is shown in Figure 15.39.

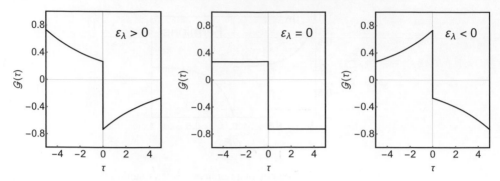

Figure 15.39 Matsubara function $\mathcal{G}^{(0)}(\mathbf{k}, \tau)$ for a noninteracting fermion system.

An important type of Matsubara functions are the single-particle fermionic correlation functions $\mathcal{G}$, which are defined as

$$\mathcal{G}_{\sigma\sigma'}(\mathbf{x}, \tau; \mathbf{x}', \tau') = -\left\langle T_\tau \left[\hat{\psi}_\sigma(\mathbf{x}, \tau)\, \hat{\psi}^\dagger_{\sigma'}(\mathbf{x}', \tau')\right]\right\rangle, \text{ real space}$$

$$\mathcal{G}_{\sigma\sigma'}(\lambda, \tau; \lambda', \tau') = -\left\langle T_\tau \left[\hat{\psi}_\sigma(\lambda, \tau)\, \hat{\psi}^\dagger_{\sigma'}(\lambda', \tau')\right]\right\rangle, [\lambda] - \text{space}. \tag{15.150}$$

15.6.6 Fourier Transform of the Matsubara Functions: Fourier Series in Imaginary Time

Now we will derive the Fourier transform of $\mathcal{G}_{\hat{A}\hat{B}}(\tau)$ to the frequency domain. We use the symmetry property twice

$$\mathcal{G}_{\hat{A}\hat{B}}(-\beta) = \mp \mathcal{G}_{\hat{A}\hat{B}}(0) = (\mp)^2\, \mathcal{G}_{\hat{A}\hat{B}}(\beta) = \mathcal{G}_{\hat{A}\hat{B}}(\beta)$$

to demonstrate that the Matsubara function is periodic with period 2β on the interval $-\beta < \tau < \beta$, as shown in Figure 15.40. This means that we have a discrete Fourier series on that interval given by[10]

$$\mathcal{G}_{\hat{A}\hat{B}}(n) = \frac{1}{2} \int_{-\beta}^{\beta} d\tau\, e^{in\pi\tau/\beta}\, \mathcal{G}_{\hat{A}\hat{B}}(\tau)$$

$$\mathcal{G}_{\hat{A}\hat{B}}(\tau) = \frac{1}{\beta} \sum_{n=-\infty}^{\infty} e^{-in\pi\tau/\beta}\, \mathcal{G}_{\hat{A}\hat{B}}(n). \tag{15.151}$$

Employing the symmetry property (15.147) one more time, we can recast the transform in a simpler form, namely,

---

[10] This transform was first suggested by Abrikosov, Gor'kov, and Dzyaloshinskii [4].

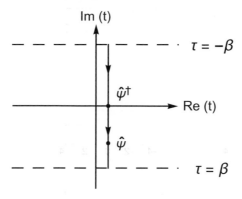

Figure 15.40 Contour for time ordering in imaginary time. Only the time difference is important. The contour is translated slightly along the real-time axis for clarity.

$$\mathcal{G}_{\hat{A}\hat{B}}(n) = \frac{1}{2} \int_0^\beta d\tau \; e^{in\pi\tau/\beta} \, \mathcal{G}_{\hat{A}\hat{B}}(\tau) + \frac{1}{2} \int_{-\beta}^0 d\tau \; e^{in\pi\tau/\beta} \, \mathcal{G}_{\hat{A}\hat{B}}(\tau)$$

$$= \frac{1}{2} \int_0^\beta d\tau \; e^{in\pi\tau/\beta} \, \mathcal{G}_{\hat{A}\hat{B}}(\tau) + e^{-in\pi} \frac{1}{2} \int_0^\beta d\tau \; e^{in\pi\tau/\beta} \, \mathcal{G}_{\hat{A}\hat{B}}(\tau + \beta)$$

$$= \frac{1}{2} \left(1 \mp e^{-in\pi} \right) \int_0^\beta d\tau \; e^{in\pi\tau/\beta} \, \mathcal{G}_{\hat{A}\hat{B}}(\tau). \tag{15.152}$$

The factor $\left(1 \mp e^{-in\pi} \right)$ leads to

$$\mathcal{G}_{\hat{A}\hat{B}}(n) = \int_0^\beta d\tau \; e^{in\pi\tau/\beta} \, \mathcal{G}_{\hat{A}\hat{B}}(\tau), \qquad \begin{cases} n \text{ odd} & \text{fermions} \\ n \text{ even} & \text{bosons.} \end{cases} \tag{15.153}$$

We use the following notation to distinguish the Fourier transforms of the Matsubara functions for bosons and fermions:

$$\mathcal{G}_{\hat{A}\hat{B}}(i\Omega_n) = \int_0^\beta d\tau \; e^{i\Omega_n\tau} \, \mathcal{G}_{\hat{A}\hat{B}}(\tau), \qquad \Omega_n = \frac{(2n+1)\pi}{\beta} \qquad \text{fermions}$$

$$\mathcal{G}_{\hat{A}\hat{B}}(i\omega_n) = \int_0^\beta d\tau \; e^{i\omega_n\tau} \, \mathcal{G}_{\hat{A}\hat{B}}(\tau), \qquad \omega_n = \frac{2n\pi}{\beta} \qquad \text{bosons.} \tag{15.154}$$

Both variables $i\omega_n$ and $i\Omega_n$ are called *Matsubara frequency*. We note, at this point, that the temperature appears in the Matsubara frequencies via β.

15.6.7 Analytic Continuation for the Matsubara and Green Functions

We shall now demonstrate the advantages of using the Matsubara functions. First, we will establish that the Matsubara and the Green functions are manifestations of the same analytic function $\mathcal{F}_{\hat{A}\hat{B}}(z)$ in the complex frequency plane: in the upper half-plane, it is equal to

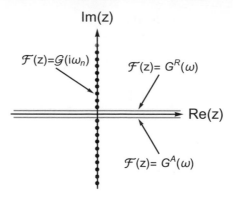

Figure 15.41 Analytical structure of $\mathcal{F}_{\hat{A}\hat{B}}(z)$ in the complex frequency plane. It reduces to either $G^{R}_{\hat{A}\hat{B}}(\omega)$, $G^{A}_{\hat{A}\hat{B}}(\omega)$ or $\mathcal{G}_{\hat{A}\hat{B}}(i\omega_n)$, depending on the value of the complex frequency z. There is a branch cut along the real axis.

$\mathcal{G}_{\hat{A}\hat{B}}(i\omega_n)$ on the imaginary axis and $G^{R}_{\hat{A}\hat{B}}(\omega)$ on the real axis, while in the lower complex plane the advanced Green function emerges on the real axis (see Figure 15.41).

Consequently, determination of one of the two guarantees that the other follows by analytic continuation. As a matter of fact, we will demonstrate that the analytic continuation actually amounts to

$$G^{R}_{\hat{A}\hat{B}}(\omega) = \mathcal{G}_{\hat{A}\hat{B}}(i\omega_n \to \omega + i\eta),$$

where η is positive and infinitesimal constant. However, it is more advantageous to start with obtaining the Matsubara function for the following reasons:

1. $\mathcal{G}_{\hat{A}\hat{B}}(\tau)$ is periodic along the imaginary-time axis.
2. $\mathcal{G}_{\hat{A}\hat{B}}(z)$ does not have statistical factors $n_F(\omega)$ or $n_B(\omega)$ on the imaginary-frequency axis. The information about the statistical probabilities implicitly enters through the Matsubara frequency sums. In contrast, we note that these statistical factors would complicate integrals over $G_{\hat{A}\hat{B}}(\omega)$.
3. The Matsubara function is based on a thermal equilibrium theory. We note that in some cases the limit $T \to 0$ of a thermal equilibrium function can be different from results obtained from a $T = 0$ analysis. We maintain that the $T \to 0$ formalism is designed to get the true ground state, including the prediction of equal thermal occupation for degenerate states, whereas a $T = 0$ calculation may identify only one of the available degenerate states, or because adiabatic turning on of the interaction may miss state crossings or a phase transition and land in an excited state.

We can, therefore, discern that the Matsubara approach provides an expedient recipe for obtaining its Green function counterparts.

We use the Lehmann representation to establish the relation between the two functions $\mathcal{G}_{AB}$ and G^{R}_{AB}. Our earlier results for the zero temperature, retarded, single-particle Green function can now be extended to finite temperatures as

$$G_{\hat{A}\hat{B}}(\omega) = -\frac{i}{Z} \int_{-\infty}^{\infty} dt \, e^{i\omega t} \, G_{\hat{A}\hat{B}}(t)$$

$$= -\frac{i}{Z} \int_{-\infty}^{\infty} dt \, e^{i\omega t} \, \left\langle e^{-\beta\mathcal{H}} \, T \left[\hat{A}(t) \, \hat{B}(0) \right] \right\rangle$$

$$= -\frac{i}{Z} \int_{-\infty}^{\infty} dt \, e^{i\omega t} \, \left\langle e^{-\beta\mathcal{H}} \left[\Theta(t) \, \hat{A}(t) \, \hat{B}(0) - \Theta(-t) \, \hat{B}(0) \, \hat{A}(t) \right] \right\rangle .$$

We calculate the first term as

$$-\frac{i}{Z} \int_{-\infty}^{\infty} dt \, e^{i\omega t} \, \left\langle e^{-\beta\mathcal{H}} \, \Theta(t) \, \hat{A}(t) \, \hat{B}(0) \right\rangle$$

$$= -\frac{i}{Z} \int_{-\infty}^{\infty} dt \, e^{i\omega t} \, \left\langle e^{-\beta\mathcal{H}} \, \Theta(t) \left[e^{i\mathcal{H}t/\hbar} \, \hat{A} \, e^{-i\mathcal{H}t/\hbar} \, \hat{B}(0) \right] \right\rangle$$

$$= -\frac{i}{Z} \int_{0}^{\infty} dt \, e^{i\omega t} \, \mathrm{Tr} \left[e^{-\beta\mathcal{H}} \, e^{i\mathcal{H}t/\hbar} \, \hat{A} \, e^{-i\mathcal{H}t/\hbar} \, \hat{B}(0) \right]$$

$$= -\frac{i}{Z} \int_{0}^{\infty} dt \, e^{i\omega t} \, \sum_{\lambda\lambda'} \left\langle \lambda \left| e^{-\beta\mathcal{H}} \, e^{i\mathcal{H}t/\hbar} \, \hat{A} \right| \lambda' \right\rangle \left\langle \lambda' \left| e^{-i\mathcal{H}t/\hbar} \, \hat{B}(0) \right| \lambda \right\rangle$$

$$= -\frac{i}{Z} \int_{0}^{\infty} dt \, e^{i\omega t} \, \sum_{\lambda\lambda'} e^{-\beta\varepsilon_\lambda} \, e^{i\varepsilon_\lambda t/\hbar} \left\langle \lambda \left| \hat{A} \right| \lambda' \right\rangle e^{-i\varepsilon_{\lambda'} t/\hbar} \left\langle \lambda' \left| \hat{B} \right| \lambda \right\rangle$$

$$= \frac{1}{Z} \sum_{\lambda\lambda'} \frac{\left\langle \lambda \left| \hat{A} \right| \lambda' \right\rangle \left\langle \lambda' \left| \hat{B} \right| \lambda \right\rangle}{\omega + \varepsilon_\lambda - \varepsilon_{\lambda'} + i\eta} \, e^{-\beta\varepsilon_\lambda} .$$

Carrying a similar calculation for the second term, we get

$$G_{\hat{A}\hat{B}}(\omega) = \frac{1}{Z} \sum_{\lambda\lambda'} \frac{\left\langle \lambda \left| \hat{A} \right| \lambda' \right\rangle \left\langle \lambda' \left| \hat{B} \right| \lambda \right\rangle}{\omega + \varepsilon_\lambda - \varepsilon_{\lambda'} + i\eta} \left(e^{-\beta\varepsilon_\lambda} + e^{-\beta\varepsilon_{\lambda'}} \right) . \tag{15.155}$$

As for the Matsubara function, we write for $\tau > 0$,

$$\mathcal{G}_{\hat{A}\hat{B}}(\tau) = -\frac{1}{Z} \, \mathrm{Tr} \left[e^{-\beta\mathcal{H}} \, e^{\mathcal{H}\tau} \, \hat{A} \, e^{-\mathcal{H}\tau} \, \hat{B} \right]$$

$$= -\frac{1}{Z} \sum_{\lambda\lambda'} e^{-\beta\varepsilon_\lambda} \left\langle \lambda \left| \hat{A} \right| \lambda' \right\rangle \left\langle \lambda' \left| \hat{B} \right| \lambda \right\rangle e^{\tau(\varepsilon_\lambda - \varepsilon_{\lambda'})} , \tag{15.156}$$

and transforming to the frequency domain, we get

$$\mathcal{G}_{\hat{A}\hat{B}}(i\omega_m) = \int_{0}^{\beta} d\tau \, e^{i\omega_m \tau} \left(-\frac{1}{Z} \sum_{\lambda\lambda'} e^{-\beta\varepsilon_\lambda} \left\langle \lambda \left| \hat{A} \right| \lambda' \right\rangle \left\langle \lambda' \left| \hat{B} \right| \lambda \right\rangle e^{\tau(\varepsilon_\lambda - \varepsilon_{\lambda'})} \right)$$

$$= -\frac{1}{Z} \sum_{\lambda\lambda'} e^{-\beta\varepsilon_\lambda} \frac{\left\langle \lambda \left| \hat{A} \right| \lambda' \right\rangle \left\langle \lambda' \left| \hat{B} \right| \lambda \right\rangle}{i\omega_m + \varepsilon_\lambda - \varepsilon_{\lambda'}} \left(e^{i\omega_m \beta} \, e^{\beta(\varepsilon_\lambda - E_{\lambda'})} - 1 \right)$$

$$= -\frac{1}{Z} \sum_{\lambda\lambda'} e^{-\beta\varepsilon_\lambda} \frac{\left\langle \lambda \left| \hat{A} \right| \lambda' \right\rangle \left\langle \lambda' \left| \hat{B} \right| \lambda \right\rangle}{i\omega_m + \varepsilon_\lambda - \varepsilon_{\lambda'}} \left(-e^{\beta(\varepsilon_\lambda - \varepsilon_{\lambda'})} - 1 \right)$$

$$= \frac{1}{Z} \sum_{\lambda\lambda'} \frac{\left\langle \lambda \left| \hat{A} \right| \lambda' \right\rangle \left\langle \lambda' \left| \hat{B} \right| \lambda \right\rangle}{i\omega_m + \varepsilon_\lambda - \varepsilon_{\lambda'}} \left(e^{-\beta\varepsilon_\lambda} + e^{-\beta\varepsilon_{\lambda'}} \right). \tag{15.157}$$

Equations (15.155) and (15.157) show that $\mathcal{G}_{\hat{A}\hat{B}}(i\omega_m)$ and $G^R_{\hat{A}\hat{B}}(\omega)$ have identical expressions. We can actually obtain both functions from

$$\mathcal{F}_{\hat{A}\hat{B}}(z) = \frac{1}{Z} \sum_{\lambda\lambda'} \frac{\left\langle \lambda \left| \hat{A} \right| \lambda' \right\rangle \left\langle \lambda' \left| \hat{B} \right| \lambda \right\rangle}{z + \varepsilon_\lambda - \varepsilon_{\lambda'} + i\eta} \left(e^{-\beta\varepsilon_\lambda} + e^{-\beta\varepsilon_{\lambda'}} \right), \tag{15.158}$$

which is analytic in the upper and lower half-plane, but has a series of poles at $\varepsilon_{\lambda'} - \varepsilon_\lambda$ along the real axis.

Knowledge of $\mathcal{G}_{AB}(i\omega_n)$ means that we know the value of the function $\mathcal{F}_{AB}(z)$ of the complex variable z on an infinite set of points on the imaginary axis $z_n = i\Omega_n$. This is sufficient to determine the full $\mathcal{F}_{AB}(z)$ in the complex plane, with the proviso that, since $\mathcal{F}_{AB}(z) = G_{AB}(z)$, it has to be analytic in the upper half-plane:

$$G^R_{\hat{A}\hat{B}}(\omega) = G_{\hat{A}\hat{B}}(i\omega_n \rightarrow \omega + i\eta). \tag{15.159}$$

This is illustrated in Figure 15.42.

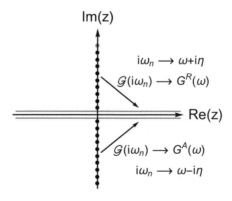

Figure 15.42 The analytic continuation procedure in the complex z-plane where the Matsubara function defined for $z = \omega_n$ goes to the retarded or advanced Green functions defined infinitesimally close to real axis.

Noninteracting Matsubara Function in the Frequency Representation

For the fermionic Hamiltonian

$$\mathcal{H}_0 = \sum_\lambda \varepsilon_\lambda \, c_\lambda^\dagger \, c_\lambda,$$

we found in the previous example that

$$\mathcal{G}_\lambda^{(0)}(\tau) = -\frac{1}{Z_0} \left\langle T \left[c_\lambda(\tau) \, c_\lambda^\dagger(0) \right] \right\rangle$$

$$= -\Theta(\tau) \left\langle c_\lambda(\tau) \, c_\lambda^\dagger(0) \right\rangle + \Theta(-\tau) \left\langle c_\lambda^\dagger(0) \, c_\lambda(-\tau) \right\rangle$$

$$= -\frac{1}{Z_0} \left[\Theta(\tau) \left\langle c_\lambda \, c_\lambda^\dagger \right\rangle - \Theta(-\tau) \left\langle c_\lambda^\dagger \, c_\lambda \right\rangle \right] e^{-\varepsilon_\lambda \tau} \qquad (15.160)$$

$$= -\frac{1}{Z_0} e^{-\varepsilon_\lambda \tau} \left[(1 - n_F(\varepsilon_\lambda)) \, \Theta(\tau) - n_F(\varepsilon_\lambda) \Theta(-\tau) \right] \qquad (15.161)$$

The fermionic Matsubara function in the frequency domain is

$$\mathcal{G}_\lambda^{(0)}(i\Omega_n) = \int_0^\beta d\tau \, e^{i\Omega_n \tau} \, \mathcal{G}_\lambda^{(0)}(\tau), \qquad \Omega_n = \frac{(2n+1)\pi}{\beta}, \ \tau > 0$$

$$= -\left(1 - n(\varepsilon_\lambda) \right) \int_0^\beta d\tau \, e^{i\Omega_n \tau} \, e^{-\varepsilon_\lambda \tau}$$

$$= -\left(1 - n(\varepsilon_\lambda) \right) \frac{1}{i\Omega_n - \varepsilon_\lambda} \left(e^{i\Omega_n \beta} \, e^{-\varepsilon_\lambda \beta} - 1 \right)$$

$$= \frac{1}{i\Omega_n - \varepsilon_\lambda} \qquad (15.162)$$

because $e^{i\Omega_n \tau} = -1$ and $\left(1 - n(\varepsilon_\lambda) \right) = \left(e^{-\varepsilon_\lambda \beta} + 1 \right)^{-1}$.

According to our recipe (15.159), the retarded free particles Green functions are

$$G_\lambda^{(0)}(\omega) = \frac{1}{\omega - \varepsilon_\lambda + i\eta}.$$

15.6.8 Evaluation of Matsubara Sums

In carrying out calculations with Matsubara functions, we often encounter cases where products of Matsubara functions appear, for example

$$\frac{1}{\beta} \sum_{i\Omega_n} \mathcal{G}(\lambda, i\Omega_n) \, \mathcal{G}(\lambda', i\Omega_n + i\omega_m) \, e^{i\Omega_n \tau}, \qquad \tau > 0. \qquad (15.163)$$

In such situations, it is preferable to use partial fractions in such a way that we will basically always have to evaluate sums such as

$$S(\xi, \tau) = \frac{1}{\beta} \sum_n \frac{e^{i\nu_n \tau}}{i\nu_n - \xi} = \sum_n f(i\nu_n), \qquad \nu_n \equiv \Omega_n, \omega_n.$$

Consequently, we have to learn how to handle sums over Matsubara frequencies.

To perform the sum over Matsubara frequencies, the standard trick is to go to the complex plane and to use the theory of residues. The key observation, which allows us to evaluate the Matsubara sums, is that the Fermi and Bose distribution functions $n_\xi(\varepsilon)$, $\xi = F, B$, once analytically continued to complex variables z, have poles on the imaginary axis precisely at the Matsubara frequencies

$$n_F(z) = \frac{1}{e^{\beta z} + 1} = \frac{1}{2}\left[1 - \tanh\left(\frac{\beta \varepsilon}{2}\right)\right], \qquad \begin{cases} \text{poles } z = \dfrac{i(2n+1)\pi}{\beta} = i\Omega_n \\[2mm] \text{Residue } = -\dfrac{1}{\beta} \end{cases}$$

$$n_B(z) = \frac{1}{e^{\beta z} - 1} = \frac{1}{2}\left[\coth\left(\frac{\beta \varepsilon}{2}\right) - 1\right], \qquad \begin{cases} \text{poles } z = \dfrac{i2n\pi}{\beta} = i\omega_n \\[2mm] \text{Residue } = \dfrac{1}{\beta}. \end{cases}$$

The idea, then, is to consider the function $g(z) = n_\xi(z) f(z)$, where $g(|z| \to \infty) = 0$.
$g(z)$ has two types of poles:

1. The poles at the Matsubara frequencies $\nu = \Omega_n, \omega_n$, originating from $n_\xi(z)$, with residues $\pm\frac{1}{\beta}$
2. The poles originating from $f(z)$

Since $n_\xi(z)$ is regular at infinity, we require that $f(|z| \to \infty) = 0$. Assuming that $f(z)$ has no poles on the imaginary axis, we integrate along the following contour

$$\frac{1}{2\pi i} \oint_C dz\, n_\xi(z) f(z) = \sum_n \text{Res}[n_\xi(i\nu_n)]\, f(i\nu_n)$$

$$= \pm\frac{1}{\beta} \sum_n f(i\nu_n) = S(\xi, \tau)$$

$\nu_n \equiv \Omega_n, \omega_n$
$+$ and $-$ signs indicate bosons and fermions, respectively.
This is precisely the sum we want to calculate.

Contour C

This contour can be deformed. It still yields

$$\frac{1}{2\pi i} \oint_C dz\, f(z)\, n_\xi(z) = \pm \frac{1}{\beta} \sum_n f(i\nu_n)$$

$$= S(x, \tau).$$

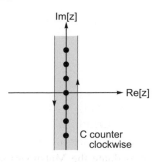

Deformed contour C

Alternative but equivalent contours include the following:

- We choose the *clockwise* contour enveloping the region, which catches all poles of the integrand in the region of the imaginary axis. It gives

$$\frac{1}{2\pi i} \oint_C dz\, f(z)\, n_\xi(z) = \mp \frac{1}{\beta} \sum_n f(i\nu_n).$$

- We can equally use the contours C_1, C_2, which contain all poles of the integrand in the area that excludes the imaginary axis, namely, the white region, which is enclosed counterclockwise, and we obtain

$$\frac{1}{2\pi i} \oint_{C_1, C_2} dz\, f(z)\, n_\xi(z) = \sum_j n_\xi(z_j)\, \mathrm{Res}[f(z_j)].$$

Equivalent contours

Therefore,

$$S(\xi, \tau) = \frac{1}{\beta} \sum_n f(i\nu_n) = \pm \sum_j n_\xi(z_j)\, \mathrm{Res}[f(z_j)].$$

In some cases, $f(z)$ may have branch cuts. This will slightly complicate the calculation. We need to deform the contour in order to avoid the branch cuts.

As an example, we consider the case where $f(z)$ has a branch cut at $z = x + i\omega$, with $x \in [-\infty; \infty]$, as shown in Figure 15.43. Let us consider the white area enclosed inside the contour depicted in Figure 15.43. In this area, there are no poles, hence the contour integral is zero. On the other hand, this contour integral is also equal to the contribution of the poles inside the yellow area, which includes the imaginary axis, plus the integral along the contour that encloses the branch cut. Therefore,

$$\frac{1}{\beta} \sum_n f(i\nu_n) = \pm \int \frac{dx}{2\pi i}\, n_\xi(x + i\omega)\left[f(x + i\omega + i0^+) - f(x + i\omega - i0^+)\right].$$

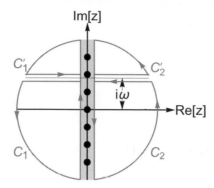

Figure 15.43 Integration contour around a branch cut.

15.6.9 *Matsubara Approach to the Two-Particle Green Function*

As an application of the Matsubara function and imaginary-time formalism, we shall rederive the two-particle correlation function χ at finite temperatures. With the aid of Wick's theorem, we obtain

$$\chi(\mathbf{q}, \tau) = -\sum_{\substack{\mathbf{k}, \mathbf{k}' \\ \sigma, \sigma'}} \left\langle T_\tau \left[c^\dagger_{\mathbf{k}, \sigma}(\tau + \xi)\, c_{\mathbf{k}+\mathbf{q}, \sigma}(\tau)\, c^\dagger_{\mathbf{k}', \sigma'}(\xi)\, c_{\mathbf{k}'-\mathbf{q}, \sigma'}(0) \right] \right\rangle$$

$$= \sum_{\substack{\mathbf{k}, \mathbf{k}' \\ \sigma, \sigma'}} \left\{ \left\langle T_\tau \left[c_{\mathbf{k}+\mathbf{q}, \sigma}(\tau)\, c^\dagger_{\mathbf{k}', \sigma'}(0) \right] \right\rangle \left\langle T_\tau \left[c_{\mathbf{k}'-\mathbf{q}, \sigma'}(0)\, c^\dagger_{\mathbf{k}, \sigma}(\tau) \right] \right\rangle \right.$$

$$\left. - \left\langle T_\tau \left[c^\dagger_{\mathbf{k}, \sigma}(\tau + \xi)\, c_{\mathbf{k}+\mathbf{q}, \sigma}(\tau) \right] \right\rangle \left\langle T_\tau \left[c^\dagger_{\mathbf{k}', \sigma'}(\xi)\, c_{\mathbf{k}'-\mathbf{q}, \sigma'}(0) \right] \right\rangle \right\}$$

$$\chi(\mathbf{q}, \tau) = \sum_{\substack{\mathbf{k}, \mathbf{k}' \\ \sigma, \sigma'}} \left[\mathcal{G}^{(0)}_{\mathbf{k}+\mathbf{q}, \sigma}(\tau)\, \mathcal{G}^{(0)}_{\mathbf{k}, \sigma}(-\tau) - \rho^{(0)}_{\mathbf{q}, \sigma}\, \rho^{(0)}_{-\mathbf{q}, \sigma} \right], \tag{15.164}$$

where ξ is an infinitesimal positive constant that places the creation operators to the left of annihilation ones for equal-time ordering. The second term does not depend on τ, so that its Fourier transform with respect to $\tau \sim \delta_{\omega_n, 0}$. Thus, it will disappear in the process of analytical continuation $i\omega_n \to \omega + i\eta$. The first term comprises products of two Matsubara functions. Before taking its Fourier transform, we should recall an important property of the density–density correlation function. The density operators $\hat{\rho}_{\mathbf{q}}$ are physical observables; they are hermitian and they commute

$$\left[\hat{\rho}_{\mathbf{q}}, \hat{\rho}_{\mathbf{q}'} \right] = 0.$$

Hence, they are bosonic operators, and we expect that their correlation function should reflect bosonic character – it should be a function of a bosonic Matsubara frequency. The Fourier transform of a product in the time domain is a convolution in the frequency

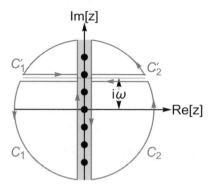

Figure 15.44 Integration contour around a branch cut.

domain. Because one function has argument τ while the other has argument $-\tau$, the internal frequencies in the two have the same sign. Fourier transform of the first term and substitution of the exact results for the electron Matsubara functions gives

$$\chi(\mathbf{q}, i\omega_n) = \frac{1}{\beta} \sum_{i\Omega_m} \sum_{\mathbf{k}\sigma} \mathcal{G}_{\mathbf{k}+\mathbf{q}\sigma}^{(0)}(i\Omega_m + i\omega_n)\, \mathcal{G}_{\mathbf{k}\sigma}^{(0)}(i\Omega_m)$$

$$= \frac{2}{\beta} \sum_{\mathbf{k}, i\Omega_m} \frac{1}{\left[i(\Omega_m + \omega_n) - (\varepsilon_{\mathbf{k}+\mathbf{q}} - \mu)\right]\left[i\Omega_m - (\varepsilon_{\mathbf{k}} - \mu)\right]}. \tag{15.165}$$

ω_n and Ω_m are bosonic and fermionic Matsubara frequencies, respectively.

In order to perform the summation over Ω_m, we write it as an integral over the contour in the complex plane shown in Figure 15.44

$$\chi(\mathbf{q}, i\omega_n) = \frac{2}{\beta} \sum_{\mathbf{k}} \frac{1}{2\pi i} \int_{C_1 + C_2} dz\, \frac{n_F(z)}{\left[z - (\varepsilon_{\mathbf{k}} - \mu)\right]\left[z + i\omega_n - (\varepsilon_{\mathbf{k}+\mathbf{q}} - \mu)\right]}. \tag{15.166}$$

The integrand has poles at the fermionic Matsubara frequencies $z = i\Omega_m$, at $z = \varepsilon_{\mathbf{k}} - \mu$, and at $z = (\varepsilon_{\mathbf{k}+\mathbf{q}} - \mu) - i\omega_n$, with residues

$$\frac{n_F(\varepsilon_{\mathbf{k}} - \mu)}{\varepsilon_{\mathbf{k}} - \varepsilon_{\mathbf{k}+\mathbf{q}} + i\omega_n} \quad \text{and} \quad \frac{n_F(\varepsilon_{\mathbf{k}+\mathbf{q}} - \mu - i\omega_n)}{\varepsilon_{\mathbf{k}+\mathbf{q}} - \varepsilon_{\mathbf{k}} - i\omega_n} = -\frac{n_F(\varepsilon_{\mathbf{k}+\mathbf{q}} - \mu)}{\varepsilon_{\mathbf{k}} - \varepsilon_{\mathbf{k}+\mathbf{q}} + i\omega_n},$$

where we used the periodicity of n_F

$$\left(e^{\beta(x - i2n\pi/\beta)} + 1\right)^{-1} = \left(e^{\beta x} + 1\right)^{-1}.$$

We obtain

$$\chi(\mathbf{q}, i\omega_n) = \frac{2}{\beta} \sum_{\mathbf{k}} \frac{1}{2\pi i} \int_{C_1 + C_2} dz \frac{n_F(z)}{[z - (\varepsilon_{\mathbf{k}} - \mu)][z + i\omega_n - (\varepsilon_{\mathbf{k+q}} - \mu)]}. \quad (15.167)$$

Now the analytical continuation $i\omega_n \to \omega + i\eta$ is easily done and one recovers the result (15.124) and (15.127), but at finite temperature, namely,

$$\chi(\mathbf{q}, \omega) = -2 \sum_{\mathbf{k}} \frac{n_F(\varepsilon_{\mathbf{k+q}} - \mu) - n_F(\varepsilon_{\mathbf{k}} - \mu)}{\omega - \varepsilon_{\mathbf{k+q}} + \varepsilon_{\mathbf{k}} + i\eta}. \quad (15.168)$$

We consider the case of $T \sim 0$ and $q \ll k_F$. We use the approximation

$$n_F(\varepsilon_{\mathbf{k+q}}) - n_F(\varepsilon_{\mathbf{k}}) \simeq \frac{\partial n_F(\varepsilon_{\mathbf{k}})}{\partial \varepsilon_{\mathbf{k}}} \nabla_{\mathbf{k}} \varepsilon_{\mathbf{k}} \cdot \mathbf{q} = \delta(\varepsilon_{\mathbf{k}} - \varepsilon_F) \frac{\mathbf{k} \cdot \mathbf{q}}{m} = \frac{1}{v_F} \delta(k - k_F) \frac{\mathbf{k} \cdot \mathbf{q}}{m}$$

and obtain

$$\chi(\mathbf{q}, i\omega_n) = -2 \int \frac{d\mathbf{k}}{(2\pi)^3} \frac{1}{v_F} \delta(k - k_F) \frac{kq \cos\theta}{m} \frac{1}{i\omega_n - kq\cos\theta/m}$$

$$= -2 \frac{mk_F}{4\pi^2 \hbar} \int_{-1}^{1} dx \frac{v_F qx}{i\omega_n - v_F qx} = -\mathcal{D}(\varepsilon_F) \left[1 + \frac{i\omega_n}{2v_F q} \ln\left[\frac{i\omega_n + v_F q}{i\omega_n - v_F q}\right]\right]$$

$$\int_{-1}^{1} \frac{x \, dx}{ia - x} = -2 + ia \ln\left[\frac{ia + 1}{ia - 1}\right].$$

Analytic continuation to the real-frequency axis yields

$$\chi(\mathbf{q}, \omega + i\eta) = \mathcal{D}(\varepsilon_F) \left[1 + \frac{\omega}{2v_F q} \ln\left[\frac{\omega + v_F q + i\eta}{\omega - v_F q + i\eta}\right]\right].$$

Taking the real and imaginary parts, we obtain the density–density correlation function as

$$\operatorname{Re}\left[\chi(\mathbf{q}, \omega)\right] = \mathcal{D}(\varepsilon_F) \left[1 + \frac{\omega}{2v_F q} \ln\left[\frac{\omega - v_F q}{\omega + v_F q}\right]\right]$$

$$\operatorname{Im}\left[\chi(\mathbf{q}, \omega)\right] = \pi \mathcal{D}(\varepsilon_F) \frac{\omega}{v_F q} \Theta\left(v_F q - |\omega|\right).$$

Table 15.2 lists the Matsubara frequency summations for some simple rational functions $g(z)$

$$S_\xi = \xi \frac{1}{\beta} \sum_n g(i\omega_n),$$

Table 15.2 *Table of Matsubara frequency summations.*

$g(i\omega)$	S_ξ
$(i\omega - \varepsilon)^{-1}$	$-\xi\, n_\xi(\varepsilon) - 1/2$
$(i\omega - \varepsilon)^{-2}$	$-\xi\, n_\xi'(\varepsilon) = \beta\, n_\xi(\varepsilon)(\xi + n_\xi(\varepsilon))$
$(i\omega - \varepsilon)^{-m}$	$-\dfrac{\xi}{(m-1)!}\, \partial_\varepsilon^{m-1}\, n_\xi(\varepsilon)$
$\dfrac{1}{(i\omega - \varepsilon_1)(i\omega - \varepsilon_2)}$	$-\xi\, \dfrac{n_\xi(\varepsilon_1) - n_\xi(\varepsilon_2)}{\varepsilon_1 - \varepsilon_2}$
$\dfrac{1}{(i\omega - \varepsilon_1)^2(i\omega - \varepsilon_2)^2}$	$\dfrac{\xi}{(\varepsilon_1 - \varepsilon_2)^2}\left(\dfrac{2[n_\xi(\varepsilon_1) - n_\xi(\varepsilon_2)]}{\varepsilon_1 - \varepsilon_2} - \left[n_\xi'(\varepsilon_1) + n_\xi'(\varepsilon_2) \right] \right)$
$\dfrac{1}{(i\omega - \varepsilon_1)^2 - \varepsilon_2^2}$	$\eta\, c_\eta(\varepsilon_1, \varepsilon_2)$
$\dfrac{(i\omega)^2}{(i\omega)^2 - \varepsilon^2}$	$-\dfrac{\varepsilon}{2}\left[1 + 2\xi\, n_\xi(\varepsilon) \right]$
$\dfrac{1}{(i\omega)^2 - \varepsilon^2}$	$-\dfrac{1}{2\varepsilon}\left[1 + 2\xi\, n_\xi(\varepsilon) \right] = \eta\, c_\eta(0, \varepsilon)$
$\dfrac{1}{\left[(i\omega)^2 - \varepsilon^2 \right]^2}$	$-\dfrac{\eta}{2\varepsilon^2}\left(c_\eta(0, \varepsilon) + n_\eta'(\varepsilon) \right)$
$\dfrac{(i\omega)^2}{\left[(i\omega)^2 - \varepsilon^2 \right]^2}$	$\dfrac{\eta}{2}\left(c_\eta(0, \varepsilon) - n_\eta'(\varepsilon) \right)$

where $\xi = \pm 1$ denotes the type of statistics. The quantities c_ξ and n_ξ are defined as

$$c_\xi(a, b) = \frac{n_\xi(a + b) - n_\xi(a - b)}{2b} = \begin{cases} \dfrac{1}{4b}\left[\tanh\left(\dfrac{\beta(a + b)}{2} \right) - \tanh\left(\dfrac{\beta(a - b)}{2} \right) \right] \\[2mm] = \dfrac{\sinh(\beta b)}{2b\,[\cosh(\beta a) + \cosh(\beta b)]}, \quad \text{fermions} \\[4mm] \dfrac{1}{4b}\left[\coth\left(\dfrac{\beta(a - b)}{2} \right) - \coth\left(\dfrac{\beta(a + b)}{2} \right) \right] \\[2mm] = \dfrac{\sinh(\beta b)}{2b\,[\cosh(\beta a) - \cosh(\beta b)]}, \quad \text{bosons} \end{cases}$$

$$n_\xi'(\varepsilon) = -\beta\, n_\xi(\varepsilon)\left(1 + \xi\, n_\xi(\varepsilon) \right) = \begin{cases} \dfrac{\beta}{4}\, \mathrm{csch}^2\left(\dfrac{\beta\varepsilon}{2} \right) & \text{fermions} \\[3mm] \dfrac{\beta}{4}\, \mathrm{sech}^2\left(\dfrac{\beta\varepsilon}{2} \right) & \text{bosons.} \end{cases}$$

Exercises

15.1   The Hamiltonian of a bosonic system is

$$\mathcal{H} = \varepsilon_1\, a^\dagger\, a + \varepsilon_2\, b^\dagger\, b + \frac{\Delta}{2}\left(a^\dagger\, b^\dagger + b\, a\right).$$

ε_1, ε_2, and Δ are real and positive, and $\Delta < (\varepsilon_1 + \varepsilon_2)$.

Derive a canonical transformation that diagonalizes this Hamiltonian. Write down explicit expressions for the transformation parameters, the eigenenergies, and eigenvectors.

15.2   Discontinuity in the Green function:

Show that

$$G\left(\mathbf{x}, t + \epsilon; \mathbf{x}', t)\right) - G\left(\mathbf{x}, t; \mathbf{x}', t + \epsilon\right) = -i\delta\left(\mathbf{x} - \mathbf{x}'\right).$$

15.3   Physical quantities from the time-ordered Green function:

The time-ordered Green function is defined as

$$G_{\sigma\sigma'}(x; x') = -i\left\langle T\left[\hat{\Psi}_\sigma(x)\, \hat{\Psi}^\dagger_{\sigma'}(x')\right]\right\rangle, \qquad x = (\mathbf{x}, t). \tag{15.169}$$

(a) Use the expressions for the particle density and particle current of a single-particle wavefunction $\psi(\mathbf{x})$ to write down the second-quantized expressions for particle density and current density operators in position representation.

(b) Show that the expectation values of the particle density and current density of a many-body system can be obtained from the Green function by taking the limits

$$\langle n_\sigma(x)\rangle = \pm i \lim_{\substack{\mathbf{x}' \to \mathbf{x} \\ t' \to t+0}} G_{\sigma\sigma}(x, x')$$

$$\langle j_\sigma\rangle = \pm\frac{\hbar}{2m} \lim_{\substack{\mathbf{x}' \to \mathbf{x} \\ t' \to t+0}} (\nabla_\mathbf{x} - \nabla_{\mathbf{x}'})\, G_{\sigma\sigma}(x, x').$$

In both cases the plus and minus signs refer to bosons and fermions, respectively.

15.4   Density in momentum space:

(a) Express the density of fermionic particles in momentum space $n_\mathbf{p} = \left\langle c^\dagger_\mathbf{p}\, c_\mathbf{p}\right\rangle$ in terms of the Green function $G_{\sigma\sigma'}(\varepsilon, \mathbf{p})$.

(b) Calculate n_p for a noninteracting Fermi gas using the explicit form

$$G_{\sigma\sigma'}(\varepsilon, \mathbf{p}) = \frac{1}{\varepsilon - \varepsilon(\mathbf{p}) + \mu + i0\operatorname{sgn}(\varepsilon)},$$

where $\varepsilon(\mathbf{p}) = p^2/2m$.

15.5   Friedel oscillations in two dimensions (see Figure 15.45):

Consider a two-dimensional noninteracting Fermi gas in the presence of an impurity, represented by a delta-functional potential $U\delta^{(2)}(x)$. The Hamiltonian of the perturbation can be written as

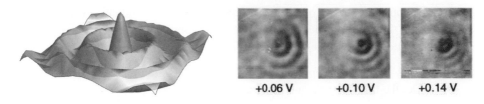

+0.06 V   +0.10 V   +0.14 V

Figure 15.45 Friedel oscillations in two dimensions. STM images from Kanisawa et al. [100] are shown on left.

$$\mathcal{H}_1(t) = \int d\mathbf{x}\, \Psi^\dagger(\mathbf{x},t)\, \Psi(\mathbf{x},t)\, U\, \delta(\mathbf{x}) = U\, \Psi^\dagger(0,t)\, \Psi(0,t).$$

(a) Assume that U is small and calculate the first-order correction to the Green function in momentum space, $G^{(1)}(\mathbf{k},\omega)$.

(b) Express the change in the fermion density in terms of $G^{(1)}(\mathbf{k},\omega)$.

(c) Determine the fermion density change as a function of $r = |\mathbf{x}|$, the distance from the impurity, for $r \gg k_F^{-1}$ in the first order in U.

(d) Show that the density of fermions oscillates as a function of r. What is the period of these oscillations?

15.6 Friedel oscillations in three dimensions:

Repeat the preceding problem for the case of a three dimensional noninteracting Fermi gas in the presence of an impurity – a delta-functional potential $U\delta^{(3)}(x)$.

15.7 Scattering of free electrons from an external potential (impurity):

Consider a free noninteracting electron system with Hamiltonian

$$\mathcal{H}_0 = \sum_{\mathbf{k}} \varepsilon_{\mathbf{k}}\, c_{\mathbf{k}}^\dagger\, c_{\mathbf{k}}.$$

Scattering from an external, or impurity, potential is given by the Hamiltonian

$$\mathcal{H}' = \int d\mathbf{x}\, \Psi^\dagger(\mathbf{x})\, V(\mathbf{x})\, \Psi(\mathbf{x}) \;\rightarrow\; \sum_{\mathbf{k}\mathbf{k}_1} V(\mathbf{k} - \mathbf{k}_1)\, c_{\mathbf{k}}^\dagger\, c_{\mathbf{k}_1}.$$

(a) Draw the Feynman diagram series representing the interaction.

(b) The sum of diagrams involving single- and multiple-scattering processes is known as the T-matrix, $t_{\mathbf{k},\mathbf{k}_1}(\omega)$. Derive a Dyson series expression for $t_{\mathbf{k},\mathbf{k}_1}$.

(c) Redraw the expansion diagrams in terms of $t_{\mathbf{k},\mathbf{k}_1}(\omega)$. Write the corresponding Green function.

(d) Consider the case of a delta-function potential $V(\mathbf{x}) = \delta(\mathbf{x})$:

1. What can you conclude about the momentum dependence of t?

2. Obtain a closed expression for the Dyson series for t.

3. Write t in two dimensions by changing the sum over momenta to an integral over energy. Hint: use the density of states.

4. Evaluate t by extending the integral to the complex energy plane $\omega \rightarrow z$.

5. Consider the case of an attractive potential, $V < 0$, and take the lower bound of the free electron spectrum to be $-W$. Show that it gives rise to a bound state. Hint: assume some high upper bound on the spectrum $\gg |z|$.

15.8 **Frequency-momentum representation of Feynman diagrams:**
Consider the following Feynman diagrams:

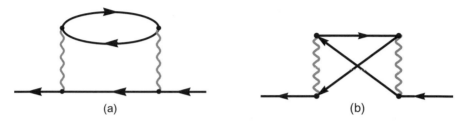

(a)   (b)

(a) Label the components of the diagrams, showing momentum conservation in labeling.
(b) Show that the scattering processes at each interaction are virtual in (a) and (b).

15.9 **Single-particle phonon Green function:**
In the process of phonon second quantization, we defined the Fourier component of the ionic displacement $\mathbf{u}_{\mathbf{q},\lambda}$, of mode $(\mathbf{q}, \lambda)$, in terms of the creation and annihilation operators $b^{\dagger}_{\mathbf{q},\lambda}$ and $b_{\mathbf{q},\lambda}$, as

$$\mathbf{u}_{\mathbf{q},\lambda} = \sqrt{\frac{\hbar}{2M\omega_{\mathbf{q}\lambda}}} (b^{\dagger}_{-\mathbf{q},\lambda} + b_{\mathbf{q},\lambda})\,\hat{\mathbf{e}}_{\lambda},$$

where M is the ionic mass, $\hat{\mathbf{e}}_{\lambda}$ the polarization vector, and $\omega_{\mathbf{q}\lambda}$ the mode frequency.

(a) Write down the noninteracting Hamiltonian $\mathcal{H}_0$ and give the relation between $\omega_{\mathbf{q},\lambda}$ and $\omega_{-\mathbf{q},\lambda}$.
(b) We define

$$\mathbf{A}_{\mathbf{q},\lambda} = \left(b^{\dagger}_{-\mathbf{q},\lambda} + b_{\mathbf{q},\lambda}\right)\hat{\mathbf{e}}_{\lambda}, \qquad \mathbf{A}^{\dagger}_{\mathbf{q},\lambda} = \left(b_{-\mathbf{q},\lambda} + b^{\dagger}_{\mathbf{q},\lambda}\right)\hat{\mathbf{e}}_{\lambda} = \mathbf{A}_{-\mathbf{q},\lambda}.$$

Write $\mathbf{A}_{\mathbf{q},\lambda}$ in the Heisenberg representation, namely $\mathbf{A}_{\mathbf{q},\lambda}(t)$. Hint: use the Heisenberg equations of motion to determine $b^{\dagger}_{\mathbf{q},\lambda}(t)$ and $b_{\mathbf{q},\lambda}(t)$.
(c) We now define the phonon Green function with respect to the displacement operators $\mathbf{u}_{\mathbf{q}\lambda\alpha}$ as

$$\mathcal{D}_{\alpha,\beta}(\mathbf{q},\lambda;t) = -\left\langle T\left[\mathbf{u}_{\mathbf{q}\lambda\alpha}(t)\mathbf{u}^{\dagger}_{\mathbf{q}\lambda\beta}(0)\right]\right\rangle$$

$$\propto -\left\langle T\left[\mathbf{A}_{\mathbf{q}\lambda\alpha}(t)\mathbf{A}^{\dagger}_{\mathbf{q}\lambda\beta}(0)\right]\right\rangle,$$

where α and β are Cartesian components of $\hat{\mathbf{e}}_{\lambda}$. Derive an expression for the noninteracting Green function $\mathcal{D}^0_{\alpha,\beta}(\mathbf{q},\lambda;t)$ at zero temperature.

(d) Derive an expression for $\mathcal{D}^0_{\alpha,\beta}(\mathbf{q}, \lambda; \omega)$.

(e) What is the corresponding spectral function $A^0_{\alpha,\beta}(\mathbf{q}, \lambda; \omega)$?

15.10 Phonon–phonon interaction:

The Hamiltonian of interacting phonons is given by the following:

$$\mathcal{H} = \sum_{\mathbf{q}} \hbar \omega_{\mathbf{q}}\, b^\dagger_{\mathbf{q}} b_{\mathbf{q}} + g \sum_{\mathbf{q}_1 \mathbf{q}} \left(b^\dagger_{\mathbf{q}_1 - \mathbf{q}}\, b^\dagger_{\mathbf{q}} b_{\mathbf{q}_1} + \mathrm{HC} \right).$$

We define the corresponding Green function as

$$D(\mathbf{q}, t) = -i \left\langle T \left[b_{\mathbf{q}}(t)\, b^\dagger_{\mathbf{q}}(0) \right] \right\rangle.$$

(a) Treating the last term in the Hamiltonian as a perturbation, what is the lowest-order nonvanishing term in the Feynman–Dyson expansion? Write down the corresponding Green function. Hint: remember to write the operators in the interaction picture and apply Wick's theorem.

(b) Try to invent Feynman rules for this case. (In $(\mathbf{q}; \omega)$-space).

(c) Draw three topologically different diagrams for the phonon Green function $D(\mathbf{q}; \omega)$.

15.11 Noninteracting Green function for topological insulators:

Consider the surface of a topological insulator as a noninteracting two-dimensional electron system (in the x-y plane) described by the Hamiltonian

$$\mathcal{H}_{\mathrm{TI}} = \hbar\, v_F \sum_{\mathbf{k}, \alpha, \beta} c^\dagger_{\mathbf{k}\alpha}\, \mathbf{k} \cdot \boldsymbol{\sigma}_{\alpha\beta}\, c_{\mathbf{k}\beta},$$

where α, β index the electron spin ($\uparrow$ or $\downarrow$), $\boldsymbol{\sigma} = (\sigma_x; \sigma_y)$ is the Pauli matrix vector, $\mathbf{k} = (k_x; k_y)$ is the electron wave vector, and v_F the Fermi velocity. In this problem, we are interested in calculating the Green function in the ground state at $T = 0$.

(a) Use the Heisenberg equations of motion to determine the time dependence of $c^\dagger_{\mathbf{k}\sigma}(t)$ and $c_{\mathbf{k}\sigma}(t')$ ($\sigma \in \{\uparrow, \downarrow\}$).

(b) Use the results you obtained in part (a) to show that the noninteracting retarded Green function

$$G^R_{\alpha\beta}(\mathbf{k}, t - t') = -i\Theta(t - t') \left\langle \left[c^\dagger_{\mathbf{k}\alpha}(t), c_{\mathbf{k}\beta}(t') \right] \right\rangle$$

at $T = 0$ is given by

$$G^R(\mathbf{k}, t - t') = -i\Theta(t - t')\, \exp\left[-iv_F\, \mathbf{k} \cdot \boldsymbol{\sigma}\, (t - t') \right].$$

Note that in this case G^R is a 2×2 matrix in spin space.

(c) Show that in the frequency domain

$$G^R(\mathbf{k}, \omega) = \int d\tau\, e^{i\omega\tau}\, G^R(\mathbf{k}, \tau)$$
$$= \frac{v_F \mathbf{k} \cdot \boldsymbol{\sigma} - \omega - i\eta}{v_F^2 k^2 - \omega^2}.$$

15.12 Consider electrons interacting via the Coulomb interaction.

(a) Draw all possible topologically distinguishable diagrams for the one-particle Green function in second order in the interaction.

(b) Pick one of the diagrams and write down the corresponding mathematical expression.

15.13 Wick's theorem: Write the following expectation value in terms of noninteracting Green functions:

$$\left\langle T\left[c_{\mathbf{k}_1}^\dagger(t_1)\, c_{\mathbf{k}_2}^\dagger(t_2)\, c_{\mathbf{k}_3}(t_3)\, c_{\mathbf{k}_4}(t_4) \right] \right\rangle.$$

15.14 Plasma oscillations in two dimensions:

Consider the case of a two-dimensional electron gas in a heterostructure quantum well.

(a) Calculate its dielectric response function $\epsilon(\mathbf{q}, \omega)$ in the RPA approximation.

(b) From the zeros of $\epsilon(\mathbf{q}, \omega)$, derive the dispersion relation of plasma oscillations in the two-dimensional electron gas. For what wavevectors are plasma oscillations Landau damped, i.e., by particle–hole pair excitations?

(c) Calculate the screened Coulomb interaction in a two-dimensional electron gas in the RPA approximation.

(d) Show that, at small wavevectors $q \ll k_F$ and low frequencies $\omega \ll \varepsilon_F$, the screened interaction is given by

$$V^{\mathrm{RPA}}(\mathbf{q}) = \frac{e^2}{2(q + k_{\mathrm{TF}})},$$

where k_{TF} is the two-dimensional Thomas–Fermi screening wavevector. Calculate k_{TF} for the case of a GaAs/GaAlAs heterostructure, for which the effective electron mass is $m^* = 0.067m$, relative permitivity $\epsilon_r = 13$, and electron density $n = 10^{15}$ m^2. How does it compare to the Fermi wavelength for this system?

15.15 Density–density correlations in a Sommerfeld gas:

(a) Use the definition of the Green function

$$-i\left\langle T\Psi(\mathbf{x}, 0)\Psi^\dagger(\mathbf{x}', 0) \right\rangle$$

to obtain an expression for $G^{(0)}(x, x')$ for a Sommerfeld fermionic gas.

(b) Use Wick's theorem to obtain an expression for the density–density correlation function

$$\left\langle \Psi^\dagger(x)\Psi(x)\Psi^\dagger(y)\Psi(y) \right\rangle$$

in terms of the free particle Green functions.

15.16 Harmonic oscillator in contact with a reservoir:

Consider a harmonic oscillator of frequency Ω, interacting with a reservoir of oscillators with frequencies ω_i, represented by the Hamiltonian

$$\mathcal{H} = \Omega\, a^\dagger a + \sum_i \omega_i\, b_i^\dagger b_i + \sum_i g_i \left(a^\dagger b_i + a\, b_i^\dagger \right).$$

Two Matsubara functions are defined as

$$\mathcal{D}(\tau) = -\left\langle T_\tau \left(a(\tau)\, a^\dagger(0) \right) \right\rangle$$
$$\mathcal{F}_i(\tau) = -\left\langle T_\tau \left(b_i(\tau)\, a^\dagger(0) \right) \right\rangle.$$

(a) Use the Heisenberg equations of motion to determine the time dependence of $a^\dagger(\tau)$, $a(\tau)$ and $b_i^\dagger(\tau)$, $b_i(\tau)$.

(b) Use the results to derive expressions for the equations of motion of $\mathcal{D}(\tau)$ and $\mathcal{F}_i(\tau)$.

(c) Fourier-transform the propagators' expressions to obtain $\mathcal{D}(i\omega_n)$ and $\mathcal{F}_i(i\omega_n)$.

(d) Use analytic continuation to derive an expression for the retarded Green function $\mathcal{D}^R(\omega)$.

16

Path Integrals

In this chapter, we present an alternative formulation to the operator framework of quantum mechanics and quantum statistics. This equivalent approach is known as the *path integral formalism*. The formalism is based on the functional integral, which extends Feynman's path integral formulation to systems with an infinite number of degrees of freedom. As such, it enables us to derive standard results, such as perturbation expansions, in an economical way, and to set up nonperturbative approaches. The path integral approach also presents a unified view of many concepts and theoretical methods that can be found in various fields of physics: condensed-matter many-body physics, quantum optics, nuclear physics, and quantum field theory, which made it very popular.

Path integral methods involve infinite products of integrals, classical action, and Lagrangians, where the action represents the central quantity that enters into the description of transition amplitudes – the propagators. Historically, the introduction of Lagrangians within a quantum-mechanical setting was suggested by Dirac, but the mathematical foundation and beauty were put forward by Feynman. The path integral description actually enhances our fundamental understanding of quantum mechanics by revealing the underlying similarity and connection between quantum mechanics and classical statistical mechanics.

For the sake of completeness, we start with a brief presentation of Feynman's path integral approach to single-particle quantum mechanics in terms of the classical action. Next, we consider the partition function of statistical physics as an imaginary-time propagator and derive its representation in terms of the Euclidean action. Finally, we derive the functional integral representation of the partition function of a system of interacting quantum particles (bosons or fermions). This representation is based on second-quantized fields and coherent states. We discuss the perturbative calculation of the partition function and the correlation functions (Green functions), and its possible representation in terms of Feynman diagrams.

In comparison with the operator formalism that was widely applied in the many-body theory [60, 143], the coherent state path-integral formalism has a prominent advantage that in the calculations within this formalism, the use of the operator commutators and the Wick theorem is completely avoided.

16.1 Functionals and Variational Principles

The mathematical basis for the discussion of extremum principles is provided by *variational calculus*. In order to prepare the ground for the following sections, but without going into much detail, we discuss here a typical, fundamental problem of variational calculus. It involves finding a real function $y(x)$ of a real variable x such that a given functional $I[y]$ of this function assumes an extreme value

$$I[y] \overset{\text{def}}{=} \int_{x_1}^{x_2} dx \, f(y(x), y'(x), x), \qquad y'(x) \equiv \frac{d}{dx} y(x) \qquad (16.1)$$

is a functional of y, with f a given function of y, y' and the variable x. x_1 and x_2 are arbitrary, but fixed, endpoints. The problem is to determine those functions $y(x)$ that take given values $y_1 = y(x_1)$ and $y_2 = y(x_2)$ at the endpoints and that make the functional $I[y]$ an extremum. In other words, one supposes that all possible functions $y(x)$ that assume the given boundary values are inserted into the integral (16.1) and that its numerical value is calculated. What we are looking for are those functions for which this value assumes an extremal value – a maximum or a minimum, or, possibly, a saddle point. As a first step, we investigate the equality

$$I(\alpha) \overset{\text{def}}{=} \int_{x_1}^{x_2} dx \, f(y(x, \alpha), y'(x, \alpha), x). \qquad (16.2)$$

where $y(x, \alpha) = y(x) + \alpha \xi(x)$ with $\xi(x_1) = \xi(x_2) = 0$. This means that we embed $y(x)$ in a set of comparative curves that fulfill the same boundary conditions as $y(x)$. The next step is to calculate the so-called variation of I, that is, the quantity

$$\delta I \overset{\text{def}}{=} \frac{dI}{d\alpha} \, d\alpha = \int_{x_1}^{x_2} dx \left\{ \frac{\partial f}{\partial y} \frac{dy}{d\alpha} + \frac{\partial f}{\partial y'} \frac{dy'}{d\alpha} \right\} d\alpha. \qquad (16.3)$$

Clearly, $dy'/d\alpha = (d/dx)(dy/d\alpha)$. If the second term is integrated by parts,

$$\int_{x_1}^{x_2} dx \, \frac{\partial f}{\partial y'} \frac{d}{dx} \left(\frac{dy}{d\alpha} \right) = -\int_{x_1}^{x_2} dx \, \frac{dy}{d\alpha} \frac{d}{dx} \left(\frac{\partial f}{\partial y'} \right) + \frac{\partial f}{\partial y'} \frac{dy}{d\alpha} \Big|_{x_1}^{x_2},$$

the boundary terms do not contribute, because $dy/d\alpha = \xi(x)$ vanishes at x_1 and at x_2. Thus

$$\delta I = \int_{x_1}^{x_2} dx \left\{ \frac{\partial f}{\partial y} - \frac{d}{dx} \frac{\partial f}{\partial y'} \right\} \frac{dy}{d\alpha} \, d\alpha. \qquad (16.4)$$

The expression in curly parentheses

$$\frac{\partial f}{\partial y} - \frac{d}{dx} \frac{\partial f}{\partial y'} \overset{\text{def}}{=} \frac{\delta f}{\delta y} \qquad (16.5)$$

is called the *variational derivative* of f by y. It is useful to introduce the notation $(dy/d\alpha) \, d\alpha = \delta y$ and to interpret δy as an infinitesimal variation of the curve $y(x)$. $I(\alpha)$

assumes an extreme value when $\delta I = 0$. As this must hold true for arbitrary variations δy, the integrand in (16.4) must vanish:

$$\frac{\partial f}{\partial y} - \frac{\mathrm{d}}{\mathrm{d}x}\frac{\partial f}{\partial y'} = 0. \tag{16.6}$$

This is *Euler's differential equation of variational calculus*. With a substitution of the Lagrangian $\mathcal{L}(q,\dot{q},t)$ for $f(y,y',x)$, we arrive at Lagrange's equation.

16.2 Quantum Propagators and Path Integrals

The dynamical information in quantum mechanics is contained in the matrix elements of the time-evolution operator $U(t_f;t_i)$. For a time-independent Hamiltonian $\mathcal{H}$, it is given by

$$U(t_f,t_i) = e^{-i(t_f-t_i)\mathcal{H}/\hbar}.$$

We evaluate the probability amplitude for a particle to start at position x_i at time t_i, and end at position x_f at time t_f as[1]

$$U(x_f,t_f;x_i,t_i) = \langle x_f \left| U(t_f,t_i) \right| x_i \rangle,$$

where $|x_f\rangle$, $|x_i\rangle$ are eigenkets of the Schrödinger position operator $\hat{x}$. It is instructive to work in the Heisenberg representation where operators are time dependent, while states are time independent. However, we recall that the basis kets, or eigenkets, become time dependent, namely,

$$|x,t\rangle = e^{i\mathcal{H}t/\hbar} |x\rangle, \text{ and } \hat{x}(t) |x,t\rangle = x |x,t\rangle.$$

We then find that

$$U(x_f,t_f;x_i,t_i) = \langle x_f \left| U(t_f,t_i) \right| x_i \rangle = \langle x_f \left| e^{-i\mathcal{H}t_f/\hbar} e^{i\mathcal{H}t_i/\hbar} \right| x_i \rangle = \langle x_f,t_f | x_i,t_i \rangle. \tag{16.7}$$

Since the Heisenberg eigenkets form a complete orthonormal set at any time t

$$\int dx \, |x,t\rangle \langle x,t| = \mathbb{I},$$

we can write

$$\langle x_f,t_f | x_i,t_i \rangle = \int dx \, \langle x_f,t_f | x,t \rangle \langle x,t | x_i,t_i \rangle, \qquad t_f > t > t_i, \tag{16.8}$$

which shows that the probability amplitude to go from x_i at t_i to x_f at t_f is equal to the superposition of products of probability amplitudes to go first at t to all possible x, $\langle x,t | x_i,t_i \rangle$, then to x_f at t_f, $\langle x_f,t_f | x,t \rangle$, for all possible xs. This is just the *composition property* of the transition amplitude.

We now capitalize on the composition property and divide the interval $t_f - t_i = t$ into $N+1$ subintervals $\delta t = \frac{t_f-t_i}{N+1}$ and write

---

[1] For simplicity and without loss of generality, we shall consider the one-dimensional case.

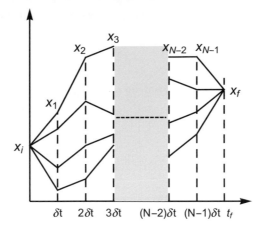

Figure 16.1 Amplitude as a sum over all N-legged paths.

$$\langle x_f, t_f \,|\, x_i, t_i \rangle = \int \left(\prod_{m=1}^{N} dx_m \right) \langle x_f, t_f \,|\, x_N, \delta t \rangle \, \langle x_n, \delta t \,|\, x_{N-1}, \delta t \rangle \dots \langle x_1, \delta t \,|\, x_i, t_i \rangle$$

$$= \int \left(\prod_{m=1}^{N} dx_m \right) \langle x_f \,\big|\, e^{-i\mathcal{H}\,\delta t} \,\big|\, x_N \rangle \langle x_N \,\big|\, e^{-i\mathcal{H}\,\delta t} \,\big|\, x_{N-1} \rangle \dots \langle x_1 \,\big|\, e^{-i\mathcal{H}\,\delta t} \,\big|\, x_i \rangle$$

Giving the amplitude as a sum over all N-legged paths, as shown in Figure 16.1.

We first consider the infinitesimal case $t_f - t_i = \epsilon \to 0$ and a general Hamiltonian of the form

$$\mathcal{H} = \frac{\hat{p}^2}{2m} + V(\hat{x}).$$

We have [2]

$$\Delta U(x_2, x_1; \epsilon) = \langle x_2 \,\big|\, e^{-i\epsilon\,\mathcal{H}} \,\big|\, x_1 \rangle \approx \langle x_2 \,\big|\, e^{-i\epsilon\,\hat{p}^2/2m}\, e^{-i\epsilon\,V(\hat{x})} \,\big|\, x_1 \rangle$$

(we set $\hbar = 1$ for the moment). The propagator is then calculated by inserting the resolution of the identity in the momentum basis, and finally taking the limit $\epsilon \to 0$

$$\Delta U(x_2, x_1; \epsilon) = \int \frac{dp}{2\pi} \, \langle x_2 \,\big|\, e^{-i\epsilon\,\hat{p}^2/2m} \,\big|\, p \rangle \langle p \,\big|\, e^{-i\epsilon\,V(\hat{x})} \,\big|\, x_1 \rangle$$

$$= \int \frac{dp}{2\pi} \, \exp\left[-i\epsilon \left(\frac{p^2}{2m} + V(x_1) \right) \right] \langle x_2 | p \rangle \langle p | x_1 \rangle$$

$$= \int \frac{dp}{2\pi} \, \exp\left[-i\epsilon \left\{ \left(\frac{p^2}{2m} + V(x_1) \right) + ip \left(\frac{x_2 - x_1}{\epsilon} \right) \right\} \right]. \qquad (16.9)$$

---

[2] To first order in ϵ, $e^{\epsilon\,A + \epsilon\,B} = e^{\epsilon\,A}\, e^{\epsilon\,B}\, e^{-[A,B]\epsilon^2/2}$, by the Baker–Hausdorff identity.

In the third line, we used $\langle x | p \rangle = e^{ipx/\hbar}$. Anticipating that $\epsilon \to 0$, we set $\frac{x_2-x_1}{\epsilon} = \dot{x}_1$ and write

$$\Delta U(x_2,x_1;\epsilon) = \int \frac{dp_1}{2\pi} \exp\left[i\epsilon\,(p_1\dot{x}_1 - \mathcal{H}(p_1,x_1))\right]. \tag{16.10}$$

Or, we can carry out the p integral, and obtain

$$\Delta U(x_2,x_1;\epsilon) = \sqrt{\frac{m}{2\pi i\epsilon}}\, \exp\left[i\frac{\epsilon}{2m}\left\{\left(\frac{x_2-x_1}{\epsilon}\right)^2 - V(x_1)\right\}\right]$$

$$= \sqrt{\frac{m}{2\pi i\epsilon}}\, \exp\left[iS(x_2,x_1;\epsilon)\right], \tag{16.11}$$

where we made the substitution $p' = p - \frac{m}{\epsilon}(x_2 - x_1)$ and gave ϵ a very small imaginary part to ensure the p integral convergence. Equation (16.11) shows that the argument of the exponential is just the infinitesimal action $S(x_2,x_1;\epsilon)$. We can now write

$$U(x_f,t_f;x_i,t_i) = \int \prod_{m=1}^{N} dx_m\,\Delta U(x_m,x_{m-1};\delta t). \tag{16.12}$$

Using (16.10), we get

$$U(x_f,t_f;x_i,t_i) = \int \left(\prod_{m=1}^{N} dx_m\right)\left(\prod_{m=1}^{N} \frac{dp_m}{2\pi}\right) \exp\left[i\delta t \sum_{m=0}^{N+1}\left(p_m\dot{x}_m - \mathcal{H}(p_m,x_m)\right)\right]$$

and with the aid of (16.11), we write

$$U(x_f,t_f;x_i,t_i) = \left(\frac{m}{2\pi i\epsilon}\right)^{N/2} \int \left(\prod_{m=1}^{N} dx_m\right) \exp\left[i\sum_{m=1}^{N} S(x_m,x_{m-1};\delta t)\right]$$

$$= \left(\frac{m}{2\pi i\delta t}\right)^{N/2} \int \left(\prod_{m=1}^{N} dx_m\right) \exp\left[i\sum_{m=1}^{N} \frac{\delta t}{2m}\left\{\left(\frac{x_m-x_{m-1}}{\delta t}\right)^2 - V(x_{m-1})\right\}\right]. \tag{16.13}$$

The integral over $x_1 \ldots x_{N-1}$ becomes a functional integral, also known as a path integral, over all paths $x(t)$ that start at x_i at time $t = t_i$ and end at x_N at time $t = t_f$. The prefactor in (16.11) gives rise to an overall (infinite) normalization.

Taking the limit $N \to \infty$, we write

$$\lim_{N\to\infty} \sum_{m=1}^{N} \frac{\delta t}{2m}\left(\frac{x_m-x_{m-1}}{\delta t}\right)^2 = \int_{t_i}^{t_f} dt\,\frac{m}{2}\dot{x}^2$$

$$\lim_{N\to\infty} \delta t \sum_{m=1}^{N} V(x_{m-1}) = \int_{t_i}^{t_f} dt\,V(x)$$

$$\lim_{N\to\infty} \left(\frac{mN}{2\pi it}\right)^{N/2} \prod_{m=1}^{N-1} dx_m = \int \mathcal{D}[x], \qquad \lim_{N\to\infty} \prod_{m=1}^{N-1} \frac{dp_m}{2\pi} = \int \mathcal{D}[p],$$

where $\mathcal{D}[x/p]$ is an integral measure. Since the measure is ill defined when $N \to \infty$, it may be necessary to obtain its normalization through identifying analytically solvable limits. We can then write (16.9) as

$$U(x_N, t_f; x_0, t_i) = \begin{cases} \int_{x_i}^{x_f} \mathcal{D}[x]\mathcal{D}[p] \exp\left[\frac{i}{\hbar}\int_{t_i}^{t_f} dt\left(p\dot{x} - \mathcal{H}(p,x)\right)\right] & \text{Phase space} \\ \int_{x_i}^{x_f} \mathcal{D}[x] \exp\left[\frac{i}{\hbar}\int_{t_i}^{t_f} dt\left(\frac{m}{2}\dot{x}^2 - V(x)\right)\right] & \text{Configuration space} \end{cases}$$

$$= \int_{x_i}^{x_f} \mathcal{D}[x] \exp\left[\frac{i}{\hbar}\int_{t_i}^{t_f} dt \mathcal{L}[\dot{x}, x]\right] = \int_{x_i}^{x_f} \mathcal{D}[x] \exp\left[\frac{i}{\hbar}S[x(t)]\right],$$

(16.14)

where $\mathcal{L}$ is the Lagrangian and S the classical action associated with the trajectory $x(t)$. The integration is to be understood as over all paths that start at $x(t_i) = x_i$ and end at $x(t_f) = x_f$. The path integral representation suggests a beautiful interpretation of the quantum-mechanical transition amplitude: the particle takes all possible trajectories with each trajectory $x(t)$ contributing $\exp[-iS[x(t)]/\hbar]$ to the amplitude. Note that we have momentarily restored $\hbar$ since it plays an essential role in the discussion of the classical limit given later in this chapter.

Alternatively, we can consider the action (16.14) as a functional that takes as argument a function $x(t)$ and returns a number, the action $S[x]$. We can then regard the integral as a functional integral where one has to integrate over all functions satisfying the boundary conditions $x(t_i) = x_i$ and $x(t_f) = x_f$. Since these functions represent paths, as shown in Figure 16.2, one refers to this kind of functional integral also as path integral.

The Feynman path integral is an exact representation of the evolution operator $U(t) = e^{-i\mathcal{H}t/\hbar}$, yet it does not incorporate any-quantum mechanical operators. It provides a fresh starting point for the formulation of quantum mechanics. However, such path integrals do not produce any new results in the quantum mechanics of a single particle.

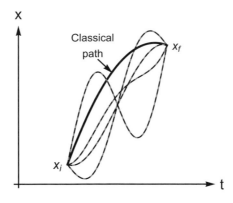

Figure 16.2 Quantum-mechanical propagation of a particle from x_i to x_f.

Moreover, most if not all quantum-mechanical calculations based on path integrals can be done with considerably greater ease using standard formulations of quantum mechanics.

In contrast, we find that path integrals, or more appropriately functional integrals, play a much more important role in quantum field theory, both relativistic and nonrelativistic. For example, they provide a relatively easy way to quantization and to expressions for correlation functions. Also, there are a whole host of nonperturbative phenomena such as solitons and instantons that are most easily viewed via path integrals. Furthermore, the close relation between statistical mechanics and quantum mechanics, or statistical field theory and quantum field theory, is plainly visible via path integrals.

16.2.1 Classical Limit and the Stationary Phase Approximation

As we can see from (16.14), all paths have the same absolute magnitude but with different phase: the phase of every path $x(t)$ is its classical action $S[x(t)]$ divided by the microscopic action $\hbar$. Thus, each path taken in isolation is equally important. The classical path is no more important than any arbitrarily complicated path! However, since the path integral is a coherent summation of paths, different paths interfere with one another. The important trajectories that contribute significantly to the path integral must be those for which the action varies very slowly when the path $x(t)$ is slightly deformed, which means that they leave the action S stationary,

$$\frac{\delta S[x]}{\delta x} = 0.$$

By contrast, the contributions of the rapidly oscillating exponential would add up to zero. Thus, in the classical limit there is only one trajectory $x_c(t)$ that contributes, and it is the solution of the classical equation of motion.

The classical limit means that $\hbar \to 0$. To examine the propagator around that limit, we set $x(t) = x_c(t) + \eta(t)$, and expand the action to second order in $\eta(t)$,

$$U(x_f, x_i; t_f - t_i) \simeq e^{iS_{cl}[x_c]/\hbar} \int_{\eta(t_i)=0}^{\eta(t_f)=0} \mathcal{D}[\eta] \, \exp\left[\frac{i}{2\hbar} \int_{t_i}^{t_f} dt \, \frac{\delta^{(2)} S[x]}{\delta x(t)\delta x(t')} \, \eta(t)\,\eta(t')\right]$$

$$\simeq e^{iS_{cl}[x_c]/\hbar} \, \det\left[\frac{1}{2\pi\hbar} \frac{\delta^{(2)} S[x]}{\delta x(t)\delta x(t')}\right]^{-1/2}, \tag{16.15}$$

where we used Gaussian integration methods.[3] The quantum correction to the classical action measures the strength of the quantum fluctuations. In general, the action for the quantum fluctuations η is not the same as the action for the classical trajectory.

---

[3] The determinant of a matrix is equal to the product of its eigenvalues.

Path Integral of the Harmonic Oscillator

$$U(x_f, x_i; t_f - t_i) = \int_{x(t_i)=x_i}^{x(t_f)=x_f} \mathcal{D}[x(t)] \, \exp\left[i\frac{m}{2\hbar} \int dt \, \left(\dot{x}^2 + \omega^2 x^2\right)\right]$$

The classical trajectory is given by

$$x_{cl}(t) = x_i \frac{\sin(t_f - t)}{\sin(t_f - t_i)} + x_f \frac{\sin(t - t_i)}{\sin(t_f - t_i)}.$$

The action along the classical path is

$$S_{cl} = \int_{t_i}^{t_f} dt \, \left(\frac{m}{2} \dot{x}_{cl}^2 + \frac{m}{2}\omega^2 x_{cl}^2\right) = \frac{m\omega}{2} \frac{\left(x_i^2 + x_f^2\right)\cos\omega(t_f - t_i) - 2x_i x_f}{\sin\omega(t_f - t_i)}.$$

Writing $x(t) = x_{cl}(t) + \eta(t)$, and evaluating the functional derivative $\delta^2 S/\delta x_1 \delta x_2$ in (16.15), we obtain

$$S[x] = S[x_{cl}] + \frac{1}{2} \int_{t_i}^{t_f} dt \, \left(m\dot{\eta}^2 - m\omega^2 \eta^2\right).$$

Since $\eta(t_i) = \eta(t_f) = 0$, we use the expansion

$$\eta(t) = \sum_{n=1}^{\infty} a_n \, \sin\left(\frac{n\pi(t - t_i)}{T}\right), \quad T = t_f - t_i.$$

The orthogonality of the different modes in the time integral leads to the action

$$S = S_{cl} + \sum_{n=1}^{\infty} \frac{m}{2} \left(\frac{(n\pi)^2}{T} - \omega^2 T\right) \frac{a_n^2}{2}.$$

Therefore, the path integral is an infinite number of Fresnel integrals over a_n:

$$\langle x_f, t_f \mid x_i, t_i \rangle$$

$$= e^{-iS_{cl}/\hbar} \sqrt{\frac{m}{2i\pi\hbar T}} \prod_{n=1}^{\infty} \sqrt{\frac{m}{2i\pi\hbar T}} \frac{n\pi}{\sqrt{2}} \int da_n \, \exp\left[\frac{i}{\hbar}\frac{m}{2}\left(\frac{(n\pi)^2}{T} - \omega^2 T\right)\frac{a_n^2}{2}\right]$$

$$= e^{-iS_{cl}/\hbar} \sqrt{\frac{m}{2i\pi\hbar T}} \prod_{n=1}^{\infty} \left[1 - \left(\frac{\omega T}{n\pi}\right)^2\right]^{-1/2}.$$

We use the sine identity

$$\prod_{n=1}^{\infty} \left(1 - \frac{x^2}{n^2}\right) = \frac{\sin \pi x}{\pi x}$$

to obtain

$$\langle x_f, t_f \mid x_i, t_i \rangle = e^{-iS_{cl}/\hbar} \sqrt{\frac{m\omega}{2\pi i\hbar \, \sin(\omega T)}}.$$

16.3 Path Integrals in Statistical Mechanics

Path integral methods offer an elegant and powerful way of doing statistical mechanics. The reason for this is that the central object in statistical mechanics, the partition function, can be written as a path integral. This, in turn, means that the properties of a quantum-mechanical system, at finite temperatures, can also be given a path integral description. Consequently, we can establish a unified view of quantum and statistical mechanics that allows us to apply many powerful methods known in statistical mechanics to calculate correlation functions in quantum mechanics. A statistical ensemble in equilibrium at a finite temperature $1/\beta$ is described in terms of a partition function

$$Z = \mathrm{Tr}\, e^{-\beta \mathcal{H}}.$$

The partition function can be written in a basis of eigenstates

$$Z = \int dx \, \langle x | e^{-\beta \mathcal{H}} | x \rangle = \int dx \int dx_1 \, \langle x | x_1 \rangle \langle x_1 | e^{-\beta \mathcal{H}} | x \rangle,$$

where $\langle x_1 | e^{-\beta \mathcal{H}} | x \rangle$ can now be viewed as analytic continuation of the propagator to imaginary time $\tau = it$, where we consider propagation from $\tau = 0$ to $\tau = \beta\hbar$, a transformation known as a Wick rotation [194] that actually takes us from a Minkowski to a Euclidean space.[4] Using this transformation, we precisely track the same steps followed previously, so that (16.13) becomes

$$\langle x_f | e^{-\beta \mathcal{H}} | x_i \rangle = \lim_{N \to \infty} \left(\frac{m}{2\pi\hbar\delta\tau} \right)^{N/2} \int \prod_{n=1}^{N-1} dx_n$$

$$\times \exp\left[\frac{\delta\tau}{\hbar} \sum_{n=0}^{N-1} \left\{ \frac{m}{2} \left(\frac{x_{n+1} - x_n}{(-i\delta\tau)} \right)^2 - V(x_n) \right\} \right],$$

where $\delta\tau = \beta\hbar/N$. Going over to a continuum description, we arrive at an imaginary-time functional integral

$$\langle x_f | e^{-\beta \mathcal{H}} | x_i \rangle = \mathcal{N} \int \mathcal{D}[x(\tau)] \, e^{-S_E[x]/\hbar}, \qquad (16.16)$$

where now S_E is the *Euclidean action*

$$S_E[x(\tau)] = \int_0^{\hbar\beta} d\tau \left[\frac{m}{2} \left(\frac{dx}{d\tau} \right)^2 + V(x) \right] = \int_0^{\hbar\beta} d\tau \, \mathcal{H}(x), \qquad (16.17)$$

---

[4] Under the transformation $t \to -i\tau$, the Minkowski metric $ds^2 = -(dt^2) + \sum_{i=1}^{d} dx_i^2$ becomes Euclidean

$ds^2 = d\tau^2 + \sum_{i=1}^{d} dx_i^2$, where time is restricted to the imaginary axis in complex time space. A Wick rotation is a rotation from the real-time axis to the imaginary-time axis or from a Minkowski space to a Euclidean one. Taking a problem expressed in Minkowski space with coordinates $(t, \mathbf{x})$ and substituting $t = -i\tau$ sometimes yields a problem in real Euclidean coordinates $(\tau, \mathbf{x})$ that is easier to solve. This solution may then be analytically continued to yield a solution to the original problem.

where the Hamiltonian replaces the Lagrangian in the action, since $t = -i\tau$. The positive form of S_E eliminates convergence problems in the integral – an advantage of imaginary time.

Since the partition function is a trace over states, we must use boundary conditions such that the initial and final states are the same state $x(0) = x(\hbar\beta)$ and sum over all such states. In other words, we must have periodic boundary conditions in imaginary time:

$$x(\tau) = x(\tau + \hbar\beta). \tag{16.18}$$

Thus, we find that the partition function can be expressed as

$$Z = \mathcal{N} \int dx \int_{x(0)=x}^{x(\beta\hbar)=x} \mathcal{D}[x(\tau)]\, e^{-S_E[x]/\hbar} \tag{16.19}$$

and the endpoint x is integrated over.

Real-time dynamics and quantum statistical mechanics are thus related by the transformation $t \to -it$, a Wick rotation of $\pi/2$ in the complex time plane.

Partition Function of the Harmonic Oscillator

The periodicity (16.18) allows us to expand $x(\tau)$ in a Matsubara Fourier series

$$x(\tau) = \frac{1}{\sqrt{\beta}} \sum_{n=-\infty}^{\infty} a_n\, e^{-i\omega_n\tau}, \qquad \omega_n = \frac{2n\pi}{\hbar\beta}$$

with $a_n^* = a_{-n}$ since x is real, and we find

$$\frac{1}{\hbar} \int_0^{\hbar\beta} d\tau\, x(\tau) y(\tau) = \sum_{m,n} a_m a_n \int_0^{\hbar\beta} \frac{d\tau}{\hbar\beta} e^{i(\omega_m + \omega_n)\tau} = \sum_{m,n} a_m a_n\, \delta_{n,-m} = \sum_n a_{-n} a_n.$$

The Euclidean action becomes

$$S_E[x(\tau)] = \int_0^{\hbar\beta} d\tau\, \frac{m}{2} \left[\frac{dx}{d\tau}\frac{dx}{d\tau} + \omega^2 x(\tau)x(\tau) \right]$$

$$= \frac{m}{2} \sum_{-\infty}^{\infty} |a_n|^2 \left(i\omega_n\, i\omega_{-n} + \omega^2 \right) = \frac{m}{2} \sum_{-\infty}^{\infty} \left(\omega_n^2 + \omega^2 \right) |a_n|^2$$

and the partition function becomes

$$Z = \mathcal{N} \int_{-\infty}^{\infty} da_0 \left(\prod_{n\geq 1} da_n\, db_n \right) \exp\left[-\frac{m}{2} \left(\omega^2 a_0^2 - 2\sum_{n\geq 1} \left(\omega_n^2 + \omega^2 \right) \left(a_n^2 + b_n^2 \right) \right) \right]$$

$$= \mathcal{N} \sqrt{\frac{2\pi}{m\omega^2}} \prod_{n=1}^{\infty} \frac{\pi}{m\left(\omega_n^2 + \omega^2 \right)}$$

$$= \mathcal{N} \sqrt{\frac{2\pi}{m\omega^2}} \left(\prod_{n=1}^{\infty} \frac{\hbar^2\beta^2}{\pi m n^2} \right) \left(\prod_{n=1}^{\infty} \left[1 + \left(\frac{\hbar\beta\omega}{2\pi n} \right)^2 \right] \right) = \frac{\mathcal{N}'}{\sinh(\beta\hbar\omega/2)}, \tag{16.20}$$

where we used the identity

$$\prod_{n=1}^{\infty} \left(1 + \frac{x^2}{n^2}\right) = \frac{\sinh(\pi x)}{\pi x}.$$

Recall that $\mathcal{N} \propto \lim_{N \to \infty} (mN/2\pi\hbar\beta)^{N/2}$, so that the term $\propto 1/n^2$ is compensated for by an infinite normalization constant in the integral measure. We can find the normalization constant $\mathcal{N}'$ by taking the limit $\beta \to \infty$, namely,

$$\lim_{\beta \to \infty} Z \simeq 2\mathcal{N}' e^{-\beta\hbar\omega/2} = e^{-\beta\varepsilon_0},$$

where $\varepsilon_0 = \hbar\omega/2$ is the ground-state energy. We finally obtain

$$Z = \frac{1}{2\sinh(\beta\hbar\omega/2)} = \frac{e^{-\beta\hbar\omega/2}}{e^{\beta\hbar\omega} - 1}.$$

16.3.1 Correlation Functions

We are usually interested in measuring thermal averages of local observables or products of local observables. Theoretically, this is obtained from thermal averages of corresponding operators, namely,

$$\left\langle \hat{O}_1 \hat{O}_2 \ldots \hat{O}_n \right\rangle = \frac{1}{Z} \mathrm{Tr} \left(\hat{O}_1 \hat{O}_2 \ldots \hat{O}_n \, e^{-\beta\mathcal{H}} \right).$$

Such averages can readily be obtained from a modified partition function

$$Z(h_1, h_2, \ldots, h_n) = \mathrm{Tr} \, e^{-\beta(\mathcal{H} + \mathcal{H}_{ext})}$$

$$\mathcal{H}_{ext} = \sum_i h_i \, \hat{O}_i, \tag{16.21}$$

where $\mathcal{H}_{ext}$ represents the Hamiltonian of external sources h_i that couple to the relevant operators $\hat{O}_i$.

This allows us to write

$$\left\langle \hat{O}_i \ldots \hat{O}_l \right\rangle = \frac{1}{Z[\{h_i = 0\}]} \frac{\partial^n Z[h_1, h_2, \ldots, h_n]}{\partial h_i \ldots \partial h_l} \Bigg|_{\{h_i = 0\}}. \tag{16.22}$$

Now the question arises as to the possibility of extending this thermal averaging process to correlation functions of the form

$$\left\langle \hat{O}(\tau_1) \hat{O}(\tau_2) \ldots \hat{O}(\tau_n) \right\rangle = \frac{\int_{x(0)=x(\beta)} \mathcal{D}[x(\tau)] \, \mathcal{O}(\tau_1) \mathcal{O}(\tau_2) \ldots \mathcal{O}(\tau_n) \, e^{-S_E/\hbar}}{\int_{x(0)=x(\beta)} \mathcal{D}[x(\tau)] \, e^{-S_E/\hbar}}, \tag{16.23}$$

where $\mathcal{O}(\tau_i) = \mathcal{O}(x(\tau_i))$. To interpret such a correlation function in imaginary time, we start with considering the action of operators on a given state function $|\psi(t)\rangle$ at different times. Operator $\hat{O}_1$ acts at t_1, such that

$$\hat{O}_1 |\psi(t_1)\rangle = \hat{O}_1 U(t_1, 0) |\psi(0)\rangle$$

followed by the action of $\hat{O}_2$ at $t_2 > t_1$:

$$\hat{O}_2 U(t_2, t_1) \hat{O}_1 U(t_1, 0) |\psi(0)\rangle .$$

We now consider the projection of this state onto the unperturbed state $|\psi(t_2)\rangle$, namely,

$$\langle\psi(0)| U(0, t_2) \hat{O}_2 U(t_2, t_1) \hat{O}_1 U(t_1, 0) |\psi(0)\rangle = \langle\psi(0)| \hat{O}_{2H}(t_2) \hat{O}_{1H}(t_1) |\psi(0)\rangle ,$$

where $\hat{O}_{iH}(t_i)$ is in the Heisenberg picture. Now we analytically continue to imaginary time and write

$$\frac{\mathrm{Tr}\left[e^{-\beta\mathcal{H}} \hat{O}_{2H}(\tau_2) \hat{O}_{1H}(\tau_1)\right]}{\mathrm{Tr}\, e^{-\beta\mathcal{H}}},$$

where now $\hat{O}_H^\dagger(\tau) \neq \left[\hat{O}_H(\tau)\right]^\dagger$. Since the operators may not commute, we introduce time ordering and write

$$\frac{\mathrm{Tr}\left[e^{-\beta\mathcal{H}} T_\tau \left(\hat{O}_{2H}(\tau_2) \hat{O}_{1H}(\tau_1)\right)\right]}{\mathrm{Tr}\, e^{-\beta\mathcal{H}}}.$$

Using a complete set of basis functions, we have

$$\mathrm{Tr}\left[e^{-\beta\mathcal{H}}\hat{O}_{2H}(\tau_2)\hat{O}_{1H}(\tau_1)\right]$$

$$= \iiint dx\, dx_1\, dx_2 \left\langle x \left| e^{-(\beta-\tau_2)\mathcal{H}} \right| x_2\right\rangle \times \left\langle x_2 \left| \hat{O}_2\, e^{-(\tau_2-\tau_1)\mathcal{H}} \right| x_1\right\rangle \left\langle x_1 \left| \hat{O}_1\, e^{-\tau_1\mathcal{H}} \right| x\right\rangle$$

$$= \iiint dx\, dx_1\, dx_2\, \mathcal{O}_2(x_2)\, \mathcal{O}_1(x_1) \left\langle x \left| e^{-(\beta-\tau_2)\mathcal{H}} \right| x_2\right\rangle\left\langle x_2 \left| e^{-(\tau_2-\tau_1)\mathcal{H}} \right| x_1\right\rangle\left\langle x_1 \left| e^{-\tau_1\mathcal{H}} \right| x\right\rangle$$

$$= \iiint dx\, dx_1\, dx_2\, \mathcal{O}_2(x_2)\, \mathcal{O}_1(x_1) \int_{x(\tau_2)=x_2}^{x(\beta)=x} \mathcal{D}[x(\tau)] \int_{x(\tau_1)=x_1}^{x(\tau_2)=x_2} \mathcal{D}[x'(\tau)]$$

$$\times \int_{x(0)=x}^{x(\tau_1)=x_1} \mathcal{D}[x''(\tau)]\, e^{-\int_{\tau_2}^{\beta\hbar} d\tau\, \mathcal{L}_E(x(\tau))}\, e^{-\int_{\tau_1}^{\tau_2} d\tau\, \mathcal{L}_E(x(\tau))}\, e^{-\int_0^{\tau_1} d\tau\, \mathcal{L}_E(x(\tau))}$$

$$= \int_{x(0)=x(\beta)} \mathcal{D}[x(\tau)]\, \mathcal{O}_2(x_2)\, \mathcal{O}_1(x_1)\, e^{-\int_0^{\beta\hbar} d\tau\, \mathcal{L}_E(x(\tau))},$$

which is just the numerator in (16.23). Thus, we find that the time-ordered correlation functions are naturally manifest as moments of the path integral. Furthermore, the beauty of the path integrals lies in the fact that all the operators in the correlation function disappear and are replaced by classical c-numbers.

The fact that all paths satisfy the condition $x(0) = x(\beta)$ means that all correlation functions can be expanded in terms of a Matsubara Fourier series. The real-time correlation function is obtained through analytic continuation of Matsubara sums. Thus, path integrals present very convenient tools to obtain time-ordered correlation functions.

Correlation Function in Terms of State Kets

We often deal with transition amplitudes between two state kets, say $|\psi_i\rangle$ and $|\psi_f\rangle$. We can then write the transition amplitude as

$$\langle\psi_f|\,\psi_i\rangle = \int dx_f\,dx_i\,\langle\psi_f|\,x_f,t_f\rangle\langle x_f,t_f|\,x_i,t_i\rangle\langle x_i,t_i|\,\psi_i\rangle$$

$$= \mathcal{N}\int dx_f\,dx_i\,\psi_f^*\left(x_f,t_f\right)\psi_i\left(x_i,t_i\right)\int \mathcal{D}[x]\,e^{\frac{i}{\hbar}S[x]}$$

where we used the fact that $\langle x,t|\,\psi\rangle$ is just the wavefunction $\psi(x,t)$. This immediately shows that we can extend this definition to time-ordered correlation functions and write

$$\overline{\langle\psi_f|\,\psi_i\rangle} = \langle\psi_f|\,T\,[\mathcal{O}(\tau_1)\,\mathcal{O}(\tau_2)\dots\mathcal{O}(\tau_n)]\,|\psi_i\rangle$$

$$= \mathcal{N}\int dx_f\,dx_i\,\psi_f^*\left(x_f,t_f\right)\psi_i\left(x_i,t_i\right)\int \mathcal{D}[x]\,\mathcal{O}(\tau_1)\,\mathcal{O}(\tau_2)\dots\mathcal{O}(\tau_n)\,e^{\frac{i}{\hbar}S[x]}.$$

We can generate the correlation functions if we follow (16.21) and write a modified action in the form

$$S[x,\mathbf{h}] = S[x] + \int_{t_i}^{t_f} dt\,\mathcal{O}(x(t))\cdot\mathbf{h}, \tag{16.24}$$

where $\mathcal{O}(x(t)) = [\mathcal{O}_1(x_1(t_1))\dots\mathcal{O}_n(x_n(t_n))]$ and $\mathbf{h}^T = [h_1\dots h_n]$. We can then define the correlation functions as

$$\langle\psi_i|\,T\,[\mathcal{O}(\tau_1)\,\mathcal{O}(\tau_2)\dots\mathcal{O}(\tau_n)]\,|\psi_i\rangle = \mathcal{N}\,(-i\hbar)^n\int dx_f\,dx_i\,\psi_f^*\left(x_f,t_f\right)\psi_i\left(x_i,t_i\right)$$

$$\times \int \mathcal{D}[x]\,\frac{\delta^{(n)}S[x,\mathbf{h}]}{\delta h_1(t_1)\dots\delta h_n(t_n)}\bigg|_{\mathbf{h}=0} e^{\frac{i}{\hbar}S[x]}. \tag{16.25}$$

16.4 Functional Integral in Many-Particle Systems

In developing the preceding path integrals, we employed single-particle eigenstates, mainly position and momentum eigenkets. However, here we will deal with many-body systems where the most convenient formalism that accounts for many-body symmetries involves canonically second-quantized operators. As we have shown in Chapter 13, the eigenfunctions of such operators are coherent states parameterize by complex scalar variables (Grassmann variables in case of fermions). We thus anticipate that in developing a functional integral approach for many-particle systems, one that parallels the path integral, we will be replacing operators with their corresponding scalar variables.

To develop such a functional integral framework, we will start with deriving a functional integral representation of the partition function of a many-particle system following the general procedure that we have outlined.

16.4.1 Coherent-State Functional Integral

We write the grand canonical partition function as

$$Z = \text{Tr}\left[e^{-\beta(\mathcal{H}-\mu N)}\right] = \sum_n \left\langle n \left| e^{-\beta(\mathcal{H}-\mu N)} \right| n \right\rangle, \tag{16.26}$$

where μ is the chemical potential, N the number operator, and $\{|n\rangle\}$ a basis of Fock space. We start with inserting the coherent states resolution of the identity

$$Z = \sum_n \int d\left(\bar{\boldsymbol{\psi}}, \boldsymbol{\psi}\right) e^{-\sum_\alpha \bar{\psi}_\alpha \psi_\alpha} \left\langle n \middle| \boldsymbol{\psi} \right\rangle \left\langle \boldsymbol{\psi} \left| e^{-\beta(\mathcal{H}-\mu N)} \right| n \right\rangle, \tag{16.27}$$

where $|\boldsymbol{\psi}\rangle$ is a many-body coherent state with ψ_α a c-number for bosons and a Grassmann variable for fermions. The integrating element $d\left(\bar{\boldsymbol{\psi}}, \boldsymbol{\psi}\right)$ is given by

$$d\left(\bar{\boldsymbol{\psi}}, \boldsymbol{\psi}\right) = \begin{cases} \prod_\alpha \dfrac{d\bar{\psi}_\alpha \, d\psi_\alpha}{2\pi i} & \text{(bosons)} \\ \prod_\alpha d\bar{\psi}_\alpha \, d\psi_\alpha & \text{(fermions)} \end{cases}$$

In order to remove the sum over n, we need to bring the matrix elements into the resolution of identity form $\mathbb{I} = \sum_n |n\rangle \langle n|$. This means that we first commute $\langle n| \boldsymbol{\psi} \rangle$ with $\left\langle \boldsymbol{\psi} \left| e^{-\beta(\mathcal{H}-\mu N)} \right| n \right\rangle$. We note, however, that in case of fermions, Grassmann variables anticommute, and this leads to [5]

$$\langle n| \boldsymbol{\psi} \rangle \langle \boldsymbol{\psi}| n \rangle = \langle -\boldsymbol{\psi}| n \rangle \langle n| \boldsymbol{\psi} \rangle,$$

where $\langle -\boldsymbol{\psi}| = \langle \varnothing| \exp\left[-\sum_\alpha \bar{\psi}_\alpha c_\alpha^\dagger\right]$. Making use of the fact that both $\mathcal{H}$ and N contain even numbers of creation and annihilation operators, and thus are not susceptible to sign changes, we write Z as

$$Z = \sum_n \int d\left(\bar{\boldsymbol{\psi}}, \boldsymbol{\psi}\right) e^{-\sum_\alpha \bar{\psi}_\alpha \psi_\alpha} \left\langle \zeta\boldsymbol{\psi} \left| e^{-\beta(\mathcal{H}-\mu N)} \right| n \right\rangle \left\langle n \middle| \boldsymbol{\psi} \right\rangle$$

$$= \int d\left(\bar{\boldsymbol{\psi}}, \boldsymbol{\psi}\right) e^{-\sum_\alpha \bar{\psi}_\alpha \psi_\alpha} \left\langle \zeta\boldsymbol{\psi} \left| e^{-\beta(\mathcal{H}-\mu N)} \right| \boldsymbol{\psi} \right\rangle, \tag{16.28}$$

where $\zeta = -1$ for fermions and $+1$ for bosons.

We start with dividing the interval β into N segments $\delta\beta = \beta/N$, such that

$$\left\langle \zeta\boldsymbol{\psi} \left| e^{-\beta(\mathcal{H}-\mu N)} \right| \boldsymbol{\psi} \right\rangle = \left\langle \zeta\boldsymbol{\psi} \left| e^{-\delta\beta(\mathcal{H}-\mu N)} \wedge e^{-\delta\beta(\mathcal{H}-\mu N)} \wedge \ldots \right| \boldsymbol{\psi} \right\rangle. \tag{16.29}$$

Next we insert the resolution of identity

$$\mathbb{I} = \int d[\bar{\boldsymbol{\psi}}, \boldsymbol{\psi}] \, e^{-\bar{\boldsymbol{\psi}}\cdot\boldsymbol{\psi}} \, |\boldsymbol{\psi}\rangle \langle \boldsymbol{\psi}|$$

---

[5] With $|n\rangle = a_1^\dagger a_2^\dagger \ldots a_n^\dagger |\varnothing\rangle$, we write

$$\langle n| \boldsymbol{\psi} \rangle = \langle \varnothing| a_n \ldots a_2 a_1 |\boldsymbol{\psi}\rangle = \psi_n \ldots \psi_2 \psi_1 \langle \varnothing| \boldsymbol{\psi}\rangle = \psi_n \ldots \psi_2 \psi_1$$

$$\langle \boldsymbol{\psi}| n \rangle = \bar{\psi}_1 \bar{\psi}_2 \ldots \bar{\psi}_n$$

$$\langle n| \boldsymbol{\psi} \rangle \langle \boldsymbol{\psi}| n \rangle = \psi_n \ldots \psi_2 \psi_1 \bar{\psi}_1 \bar{\psi}_2 \ldots \bar{\psi}_n = \psi_1 \bar{\psi}_1 \psi_2 \bar{\psi}_2 \ldots \psi_n \bar{\psi}_n$$

$$= \left(-\bar{\psi}_1 \psi_1\right)\left(-\bar{\psi}_2 \psi_2\right) \ldots \left(-\bar{\psi}_n \psi_n\right) = \langle -\boldsymbol{\psi}| n \rangle \langle n| \boldsymbol{\psi} \rangle.$$

at each wedge location. We expand the incremental exponent

$$
\begin{aligned}
\left\langle \boldsymbol{\psi}' \left| e^{-\delta\beta(\mathcal{H}(a_\alpha^\dagger, a_\alpha) - \mu N)} \right| \boldsymbol{\psi} \right\rangle &= \left\langle \boldsymbol{\psi}' \left| \mathbb{I} - \delta\beta(\mathcal{H}(a_\alpha^\dagger, a_\alpha) - \mu N) \right| \boldsymbol{\psi} \right\rangle + \mathcal{O}(\delta\beta)^2 \\
&= \langle \boldsymbol{\psi}' \mid \boldsymbol{\psi} \rangle - \delta\beta \left\langle \boldsymbol{\psi}' \left| (\mathcal{H}(a_\alpha^\dagger, a_\alpha) - \mu N) \right| \boldsymbol{\psi} \right\rangle + \mathcal{O}(\delta\beta)^2 \\
&= \langle \boldsymbol{\psi}' \mid \boldsymbol{\psi} \rangle \left[\mathbb{I} - \delta\beta \left\{ \mathcal{H}\left(\boldsymbol{\psi}', \boldsymbol{\psi}\right) - \mu N\left(\boldsymbol{\psi}', \boldsymbol{\psi}\right) \right\} \right] + \mathcal{O}(\delta\beta)^2 \\
&\approx e^{\bar{\boldsymbol{\psi}}' \cdot \boldsymbol{\psi}} \, e^{-\delta\beta[\mathcal{H}(\boldsymbol{\psi}', \boldsymbol{\psi}) - \mu N(\boldsymbol{\psi}', \boldsymbol{\psi})]},
\end{aligned}
$$

with $A\left(\boldsymbol{\psi}', \boldsymbol{\psi}\right) = \dfrac{\langle \boldsymbol{\psi}' | \mathcal{H} | \boldsymbol{\psi} \rangle}{\langle \boldsymbol{\psi}' | \boldsymbol{\psi} \rangle}$, $A = \mathcal{H}, N$. Note that $\langle \boldsymbol{\psi}' \mid \boldsymbol{\psi} \rangle$ commutes with everything because it is bilinear in $\boldsymbol{\psi}$. The operators $a^\dagger, a$ can be bosonic ($b^\dagger, b$) or fermionic ($c^\dagger, c$). Substituting back in (16.29), we get

$$
Z = \lim_{N \to \infty} \int d[\bar{\boldsymbol{\psi}}, \boldsymbol{\psi}] \int_{\substack{\bar{\boldsymbol{\psi}}_N = \zeta\bar{\boldsymbol{\psi}}_0 = \zeta\bar{\boldsymbol{\psi}} \\ \boldsymbol{\psi}_N = \zeta\boldsymbol{\psi}_0 = \zeta\boldsymbol{\psi}}} \left(\prod_{n=1}^{N-1} \prod_\alpha d\bar{\psi}_{\alpha,n} d\psi_{\alpha,n} \right) \exp\left[-\delta\beta \sum_{n=1}^{N-1} \sum_\alpha \frac{\bar{\psi}_{\alpha,n} \psi_{\alpha,n}}{\delta\beta} \right]
$$

$$
\times \exp\left[\sum_{n=1}^{N-1} \delta\beta \left(\sum_\alpha \frac{\bar{\psi}_{\alpha,n} \psi_{\alpha,n-1}}{\delta\beta} - \left(\mathcal{H}(\bar{\boldsymbol{\psi}}_n, \boldsymbol{\psi}_{n-1}) - \mu N(\bar{\boldsymbol{\psi}}_n, \boldsymbol{\psi}_{n-1}) \right) \right) \right]. \qquad (16.30)
$$

Taking the continuum limit $N \to \infty$,

$$
\delta\beta \sum_{n=0}^N \to \int_0^\beta d\tau; \quad \frac{\boldsymbol{\psi}_n - \boldsymbol{\psi}_{n-1}}{\delta\beta} \to \partial_\tau \boldsymbol{\psi}\big|_{\tau = n\delta\beta}; \quad \prod_{n=0}^N d[\bar{\boldsymbol{\psi}}_n, \boldsymbol{\psi}_n] \to \mathcal{D}(\bar{\boldsymbol{\psi}}, \boldsymbol{\psi}).
$$

We finally arrive at

$$
Z = \int_{\substack{\bar{\boldsymbol{\psi}}_N = \zeta\bar{\boldsymbol{\psi}}_0 \\ \boldsymbol{\psi}_N = \zeta\boldsymbol{\psi}_0}} \mathcal{D}(\bar{\boldsymbol{\psi}}, \boldsymbol{\psi}) \, e^{-S[\bar{\boldsymbol{\psi}}, \boldsymbol{\psi}]}
$$

$$
S[\bar{\boldsymbol{\psi}}, \boldsymbol{\psi}] = \int_0^\beta d\tau \left(\bar{\boldsymbol{\psi}} \, \partial_\tau \boldsymbol{\psi} + \mathcal{H}\left(\boldsymbol{\psi}, \boldsymbol{\psi}\right) - \mu N\left(\boldsymbol{\psi}, \boldsymbol{\psi}\right) \right). \qquad (16.31)
$$

For a general normal-ordered Hamiltonian with two-particle interactions of the form

$$
\mathcal{H} - \mu N = \sum_{\alpha\beta} \left(h_{\alpha\beta} - \mu\delta_{\alpha\beta} \right) a_\alpha^\dagger a_\beta + \sum_{\alpha\beta\gamma\delta} V_{\alpha\beta\gamma\delta} \, a_\alpha^\dagger a_\beta^\dagger a_\gamma a_\delta,
$$

we get

$$
S[\bar{\boldsymbol{\psi}}, \boldsymbol{\psi}] = \int_0^\beta d\tau \left[\sum_{ij} \bar{\psi}_i(\tau) \left[(\partial_\tau - \mu)\delta_{ij} + h_{ij} \right] \psi_j(\tau) + \sum_{ijkl} V_{ijkl} \bar{\psi}_i(\tau) \bar{\psi}_j(\tau) \psi_k(\tau) \psi_l(\tau) \right].
$$

$$
(16.32)
$$

Since $\psi(\tau) = \zeta \psi(\tau + \beta)$, it is convenient to express S in terms of Matsubara Fourier series, $\psi_i(\tau) = (1/\sqrt{\beta}) \sum_n \psi_{in} e^{-i\nu_n \tau}$:

$$S[\bar{\psi}, \psi] = \sum_{ij,n} \bar{\psi}_{in} \left[(-i\nu_n - \mu) \delta_{ij} + h_{ij} \right] \psi_{jn}$$

$$+ \frac{1}{\beta} \sum_{ijkl} \sum_{n_1, n_2, n_3, n_4} V_{ijkl} \, \bar{\psi}_{in_1} \bar{\psi}_{jn_2} \psi_{kn_3} \psi_{ln_4} \, \delta_{\nu_{n_1} + \nu_{n_2} + \nu_{n_3} + \nu_{n_4}}. \qquad (16.33)$$

Partition Function of Ideal Quantum Gases

We start with diagonalizing h_{ij} in (16.32) with a unitary transformation, so that in the absence of two-body interactions in (16.33), the action reads

$$S_0 = \sum_{\alpha, n} \bar{\eta}_{\alpha, n} \left[-i\nu_n + \mathcal{E}_\alpha \right] \eta_{\alpha, n},$$

where $\mathcal{E}_\alpha = \varepsilon_\alpha - \mu$. The partition function becomes

$$Z_0 = \prod_\alpha Z_\alpha^{(0)} \Rightarrow Z_\alpha^{(0)} = \int \mathcal{D}[\bar{\eta}_\alpha, \eta_\alpha] \, e^{-\beta \sum_n \bar{\eta}_{\alpha, n} \left(-i\nu_n + \mathcal{E}_\alpha \right) \eta_{\alpha, n}}$$

$$= \prod_{n=-\infty}^{\infty} \left[-i\nu_n + \mathcal{E}_\alpha \right]^{-\zeta}$$

yielding a grand potential

$$\Omega_0 = -\frac{1}{\beta} \ln \psi_0 = \frac{\zeta}{\beta} \sum_{\alpha, n} \ln \left(-i\nu_n + \mathcal{E}_\alpha \right).$$

This Matsubara sum is divergent; the continuum time limit of the partition function is therefore ill defined. To circumvent this difficulty, we consider the mean particle number it yields

$$\langle N \rangle = -\frac{\partial \Omega_0}{\partial \mu} = -\frac{\zeta}{\beta} \sum_{\alpha, n} \frac{1}{-i\nu_n + \mathcal{E}_\alpha} = \sum_\alpha \frac{1}{e^{\beta(\varepsilon_\alpha - \mu)} - \zeta}$$

or

$$\Omega_0 = \frac{\zeta}{\beta} \sum_\alpha \ln \left(1 - \zeta \, e^{-\beta(\varepsilon_\alpha - \mu)} \right).$$

If we start with the discretized expression of the partition function (16.30), we obtain the action

$$S = \sum_\alpha \sum_{m=1}^{N} \left[\bar{\eta}_{\alpha, m} \left(\eta_{\alpha, m} - \eta_{\alpha, m-1} \right) - \delta\beta \, \mathcal{E}_\alpha \, \bar{\eta}_{\alpha, m} \, \eta_{\alpha, m-1} \right].$$

Substituting the Matsubara Fourier series

$$\eta_{\alpha,m} = \frac{1}{\sqrt{N}} \sum_{n=0}^{\infty} a_{\alpha,n} \, e^{-i\omega_n \tau_m}$$

for the ηs, where $a_{\alpha,n}$ are complex numbers for bosons and Grassmann variables for fermions. We obtain

$$Z_0 = \lim_{N\to\infty} \int \prod_{n=0}^{N-1} d(\mathbf{a}_n^*, \mathbf{a}_n) \, \exp\left[-\sum_{\alpha} \sum_{n=0}^{N-1} a_{\alpha,n}^* \, a_{\alpha,n} \left[1 + (\delta\beta\mathcal{E}_\alpha - 1) \, e^{i\nu_n \delta\beta} \right] \right]$$

$$= \lim_{N\to\infty} \prod_{\alpha} \prod_{n=0}^{N-1} \left[1 + (\delta\beta\mathcal{E}_\alpha - 1) \, e^{i\nu_n \delta\beta} \right]^{-\zeta}.$$

Writing $1 - \delta\beta \, \mathcal{E}_\alpha \simeq e^{-\phi_\alpha}$ and using the identity

$$\prod_{n=0}^{N-1} \left(1 - e^{i\nu_n \delta\beta - \phi} \right) = 1 - \zeta \, e^{-N\phi},$$

we arrive at

$$Z_0 = \lim_{N\to\infty} \prod_{\alpha} \left(1 - \zeta \, e^{-N\phi_\alpha} \right)^{-\zeta} = \prod_{\alpha} \left(1 - \zeta \, e^{-\beta\mathcal{E}} \right)^{-\zeta}. \tag{16.34}$$

16.4.2 The Hubbard–Stratonovich transformation

The preceding formulation of functional integral furnishes a way of discussing the physics of interacting fermions (bosons), which in many cases turns out to be remarkably transparent and appealing. Particularly when considering cases with a broken symmetry, where the use of the *Hubbard–Stratonovich transformation* often allows the identification of the emerging order parameter and treating its fluctuations in a way that is both physically transparent and systematic.

The Hubbard–Stratonovich transformation (HST) is a general method for replacing a two-body interaction with the interaction of a one-body with external bosonic exchange fields. This transformation is an operator generalization of the Gaussian integral

$$\int_{-\infty}^{\infty} dx \, \exp\left[-\pi x^2 - 2\pi^{1/2} x \, \mathcal{O} \right] = \exp\left[\mathcal{O}^2 \right], \tag{16.35}$$

where $\mathcal{O}$ is an operator. Thus, if a quadratic term like the right side of (16.35) appears in the partition function, representing a two-body interaction, it can be replaced by the left-hand side of (16.35), which contains only one-body interactions. However, these single particles now interact with an external bosonic field that must be integrated.

In order to apply the HST to the case of a bilinear interaction of the form

$$\mathcal{H}_I(\tau_i) = V_{12}\, \mathcal{O}_1(\tau_i)\, \mathcal{O}_2(\tau_i),$$

we rewrite the interaction in the form

$$\mathcal{H}_I = V_{12}\left[\left(\frac{\mathcal{O}_1+\mathcal{O}_2}{2}\right)^2 - \left(\frac{\mathcal{O}_1-\mathcal{O}_2}{2}\right)^2\right],$$

which is valid for commuting $\mathcal{O}_1$ and $\mathcal{O}_2$. We then obtain for the ith slice of a functional integral

$$\lim_{N\to\infty}\exp\left[-\frac{\beta}{N}\mathcal{H}_I(\tau_i)\right] = \exp\left[-\frac{\beta}{N}V_{12}\left(\frac{\mathcal{O}_1+\mathcal{O}_2}{2}\right)^2\right] \times \exp\left[+\frac{\beta}{N}V_{12}\left(\frac{\mathcal{O}_1-\mathcal{O}_2}{2}\right)^2\right].$$

Performing HST, we get

$$\exp\left[+\frac{\beta}{N}V_{12}\left(\frac{\mathcal{O}_1-\mathcal{O}_2}{2}\right)^2\right] = \int_{-\infty}^{\infty}dx_i\,\exp\left[-\pi V_{12}^{-1}x_i^2\right]\exp\left[-\left(\frac{\pi\beta}{N}\right)^{1/2}x_i\left[\mathcal{O}_1-\mathcal{O}_2\right]\right]$$

$$\exp\left[-\frac{\beta}{N}V_{12}\left(\frac{\mathcal{O}_1+\mathcal{O}_2}{2}\right)^2\right] = \int_{-\infty}^{\infty}dy_i\,\exp\left[-\pi V_{12}^{-1}y_i^2\right]\exp\left[-i\left(\frac{\pi\beta}{N}\right)^{1/2}y_i\left[\mathcal{O}_1+\mathcal{O}_2\right]\right].$$

Setting $x_i \to x_i/\sqrt{N}$, $y_i \to y_i/\sqrt{N}$, and introducing the notation

$$\int D[x(y)] \to \lim_{N\to\infty}\int_{-\infty}^{\infty}\prod_{i=1}^{N}\frac{dx_i(y_i)}{\sqrt{N}},$$

leads to the functional integral form

$$\int D[x]\,D[y]\exp\left[-\pi\int_0^\beta d\tau\left[x^2(\tau)+y^2(\tau)\right]\right]T_\tau\left[\exp\left(-\int_0^\beta d\tau\,\sqrt{\pi}\left[\mathcal{O}_1(\tau)-\mathcal{O}_2(\tau)\right]x(\tau)\right)\right.$$

$$\left. \times \exp\left(-i\int_0^\beta d\tau\,\sqrt{\pi}\left[\mathcal{O}_1(\tau)+\mathcal{O}_2(\tau)\right]y(\tau)\right)\right].$$

Introducing the complex field $\phi(\tau) = x(\tau) + iy(\tau)$, we obtain the simpler form

$$\int D[\phi]\exp\left[-\pi\int_0^\beta d\tau\,\phi(\tau)\,V_{12}^{-1}\,\phi^*(\tau)\right]$$

$$\times T_\tau\left[\exp\left(-\int_0^\beta d\tau\,\sqrt{\pi}\left[\mathcal{O}_1(\tau)\phi(\tau)-\mathcal{O}_2(\tau)\phi^*(\tau)\right]\right)\right]. \tag{16.36}$$

In most cases, we will have $\mathcal{O}_2 = \mathcal{O}_1^\dagger$. Choosing the appropriate HS transformation is always an educated guess. To explore the physical ramifications of the HST, we consider some typical two-body fermionic interactions. Actually, we can pair the c-operators in three different but physically meaningful ways:

(i) For Hartree or direct Coulomb interactions, we have

$$\frac{1}{2} \sum_{\substack{\mathbf{kk'},\mathbf{q} \\ \sigma,\sigma'}} \frac{4\pi e^2}{q^2} c^\dagger_{\mathbf{k}+\mathbf{q},\sigma} c^\dagger_{\mathbf{k'}-\mathbf{q},\sigma'} c_{\mathbf{k'},\sigma'} c_{\mathbf{k},\sigma} = \frac{1}{2} \sum_{\substack{\mathbf{q} \\ \sigma,\sigma'}} \frac{4\pi \left(e\rho_\sigma(\mathbf{q})\right) \left(e\rho_{\sigma'}(-\mathbf{q})\right)}{q^2} + \text{const.}$$

With $\mathcal{O}_1 = e \sum_{\sigma=\uparrow\downarrow} \rho_\sigma(\mathbf{q})$, $\mathcal{O}_2 = e \sum_{\sigma=\uparrow\downarrow} \rho_\sigma(-\mathbf{q})$, the HST yields

$$-\frac{1}{8\pi} \sum_{\mathbf{q}} q^2 \phi_{-\mathbf{q}} \phi_{\mathbf{q}} + \sum_{\mathbf{q}} \left[ie\rho(\mathbf{q})\phi_{-\mathbf{q}} - ie\rho(-\mathbf{q})\phi_{\mathbf{q}}\right].$$

The inverse Fourier transform of the first term gives $\frac{1}{8\pi} (\nabla\phi(x))^2 = E^2/8\pi$, which is the bosonic electric field. For a Coulomb electron gas, we have

$$S = \int d\mathbf{x}\, d\tau \left[\bar{\psi}(\mathbf{x}) \left(\partial_\tau - \frac{p^2}{2m} + e\phi(\mathbf{x}) - \mu\right) \psi(\mathbf{x}) - \frac{1}{2} (\nabla\phi)^2\right]$$

(ii) We consider the contact interaction model

$$g \sum_{s,s'} \Psi^\dagger_s(x) \Psi^\dagger_{s'}(x) \Psi_{s'}(x) \Psi_s(x) = -\mathbf{s}(x) \cdot \mathbf{s}(x),$$

where $\mathbf{s}(x) = \Psi^\dagger_s(x)\sigma_{ss'}\Psi_{s'}(x)$. Introducing an exchange field $\mathbf{m}$, HST takes the form

$$\int D[\mathbf{m}] \exp\left[-\int_0^\beta d\tau\, d\mathbf{x} \left(m^2 - 2\mathbf{m} \cdot \mathbf{s}\right)\right] = \mathcal{N} \exp\left[\int_0^\beta d\tau\, d\mathbf{x}\, s^2(\mathbf{x})\right]$$

and yields

$$S = \int_0^\beta d\tau\, d\mathbf{x} \left[\sum_{s=\uparrow\downarrow} \bar{\psi}_s \left(\partial_\tau - \mathcal{H}_0 - \mu\right) \psi_s - 2g\mathbf{m} \cdot \mathbf{s} + gm^2\right].$$

Note that $\mathbf{m}$ here is a field that is allowed to fluctuate, and we integrate over all possible paths of this field.

(iii) We consider the pairing Hamiltonian associated with superconductivity

$$-g \sum_{\mathbf{kk'}} c^\dagger_{\mathbf{k}\uparrow} c^\dagger_{-\mathbf{k}\downarrow} c_{-\mathbf{k'}\downarrow} c_{\mathbf{k'}\uparrow} = -g \left(\sum_{\mathbf{k}} c^\dagger_{\mathbf{k}\uparrow} c^\dagger_{-\mathbf{k}\downarrow}\right) \left(\sum_{\mathbf{k}} c_{-\mathbf{k}\downarrow} c_{\mathbf{k}\uparrow}\right).$$

Applying HST with a complex pairing field Δ, we get

$$\bar{\Delta} \sum_{\mathbf{k}} c^\dagger_{\mathbf{k}\uparrow} c^\dagger_{-\mathbf{k}\downarrow} + \sum_{\mathbf{k}} c_{-\mathbf{k}\downarrow} c_{\mathbf{k}\uparrow} \Delta + \frac{\bar{\Delta}\,\Delta}{g}.$$

Mean Field and the Saddle-Point Approximation

So far, the treatment of two-body interaction and its variants that we have presented is exact; no approximations have been introduced. The Hubbard–Stratonovich transform gives the exact expression

$$Z = \int D[\phi^*, \phi] D[\bar{\psi}, \psi] \exp\left[-\pi \int_0^\beta d\tau \boldsymbol{\phi}(\tau) V^{-1} \boldsymbol{\phi}^*(\tau)\right] \exp\left[-\int_0^\beta d\tau \left(\bar{\boldsymbol{\psi}} \partial_\tau \boldsymbol{\psi} + \mathcal{H}_{\text{eff}}\right)\right]$$

$$= \int D[\phi^*, \phi] D[\bar{\psi}, \psi] e^{-S[\phi^*, \phi, \bar{\psi}, \psi]} = \int D[\phi^*, \phi] e^{-S_{\text{eff}}[\phi^*, \phi]}$$

$$\mathcal{H}_{\text{eff}} = \bar{\boldsymbol{\psi}} \, \mathcal{H}_0 \, \boldsymbol{\psi} + \sum_i \left(\bar{\mathcal{O}}_i \, \phi_i + \mathcal{O}_i \, \phi_i^*\right), \tag{16.37}$$

where the $\mathcal{O}_i$s are bilinear in the ψs. We can integrate over the coherent state variables in (16.37), but we do not know how to handle the functional integral over $\phi(x,t)$. Note that the two-body interactions become encoded in the dynamics of the bosonic field ϕ, and there are various approximations to treat it. The simplest one is to subject the action to a stationary phase treatment, which amounts to performing a saddle-point approximation for the functional integral over ϕ, whereby we extremize the action

$$\frac{\delta S_{\text{eff}}[\phi^*, \phi]}{\delta \phi_i^*} = \frac{\delta \ln Z}{\delta \phi_i^*} = \frac{1}{Z} \int D[\bar{\psi}, \psi] \left(V_i^{-1} \phi_i + \mathcal{O}_i\right) e^{-S[\phi^*, \phi, \bar{\psi}, \psi]} = 0,$$

which is actually equivalent to a classical approximation. It is clear that such a procedure will lead to a fixed configuration of the field ϕ, namely

$$\phi_i^0 = -V_i \, \langle \mathcal{O}_i \rangle,$$

whereby the stationary auxiliary field is the average of the two-body interaction over the associated operator. This is just the mean-field approximation, where for example a quartic term like $c_i^\dagger c_j c_k^\dagger c_l$ becomes $\langle c_i^\dagger c_j \rangle c_k^\dagger c_l + c_i^\dagger c_j \langle c_k^\dagger c_l \rangle$.[6] Spontaneous symmetry breaking occurs when the *order parameter* ϕ_i^0 or $\langle \mathcal{O}_i \rangle$ assumes a nonzero value. In both methods, order parameters can then be solved self-consistently to yield the decoupled Hamiltonian.

In the example of the Coulombic electron gas presented earlier, we get

$$\frac{\delta S_{\text{eff}}[\phi^*, \phi]}{\delta \phi_i^*} = \frac{1}{Z} \int D[\bar{\psi}, \psi] \left(e\rho(\mathbf{x}) + \nabla^2 \phi(\mathbf{x})\right) e^{-S} = 0,$$

and the saddle-point solution satisfies Gauss's law:

$$-\nabla^2 \phi = \nabla \cdot \mathbf{E} = e\rho(\mathbf{x}).$$

Spatial and temporal homogeneity translates to a uniform potential ϕ_0, and charge neutrality sets it to zero.

Effective Action and Fluctuations

It is clear that we can calculate systematic corrections to the lowest-order stationary-phase approximation ϕ^0. After integrating out the ψ's, we can explore the properties of functional

---

[6] Actually, both methods are equivalent. But we find that the Hubbard–Stratanovich transformation is more systematic, as it might be easier to figure out how to go beyond the mean field. It might also be easier to combine different kinds of channels; for instance, in the case of superconductivity, we would select $c_i^\dagger c_j c_k^\dagger c_l \to \langle c_i^\dagger c_k^\dagger \rangle c_j c_l + c_i^\dagger c_k^\dagger \langle c_j c_l \rangle$. But we should keep in mind that HSTs are arbitrary, we can combine an arbitrary number of them, and the different emerging mean-field theories give different results.

fluctuations of the bosonic auxiliary field, $\delta\phi = \phi - \phi^0$. We use the Gaussian form of the ψ's to obtain

$$\int D[\bar\psi,\psi]\exp\left[-\int_0^\beta d\tau\left(\bar\psi\left(\partial_\tau+\mathcal{H}_0-\mu\right)\psi+\mathcal{H}_I^{\text{eff}}\right)\right] = \det^\zeta\left(\partial_\tau+\tilde{\mathcal{H}}_0-\mu+\tilde{\mathcal{H}}_I^{\text{eff}}[\phi^*,\phi]\right),$$

where $\tilde{\mathcal{H}}$ is a matrix representation, and for simplicity of notation we set $\delta\phi \to \phi$. We then get[7]

$$Z = \int D[\phi^*,\phi]\,e^{-S_{\text{eff}}[\phi^*,\phi]}$$

$$S_{\text{eff}}[\phi^*,\phi] = \int_0^\beta d\tau\,\boldsymbol{\phi}(\tau)\,V^{-1}\,\boldsymbol{\phi}^*(\tau) - \zeta\,\text{Tr}\ln\left(\underbrace{\partial_\tau+\tilde{\mathcal{H}}_0-\mu}+\tilde{\mathcal{H}}_I^{\text{eff}}[\phi^*,\phi]\right) \qquad (16.38)$$

$$-G^{(0)-1},$$

where we used the identity $\ln\det = \text{Tr}\ln$. We note that

$$\zeta\,\text{Tr}\ln\left(-G^{(0)-1}+\tilde{\mathcal{H}}_I^{\text{eff}}\right) = \zeta\,\text{Tr}\ln\left(-G^{(0)-1}\right)\left(1-G^{(0)}\tilde{\mathcal{H}}_I^{\text{eff}}\right)$$

$$= \zeta\,\text{Tr}\ln\left(-G^{(0)-1}\right)+\zeta\,\text{Tr}\ln\left(1-G^{(0)}\tilde{\mathcal{H}}_I^{\text{eff}}\right)$$

$$= \zeta\,\text{Tr}\ln\left(-G^{(0)-1}\right)-\zeta\sum_{n=1}^\infty\frac{1}{n!}\text{Tr}\left(G^{(0)}\tilde{\mathcal{H}}_I^{\text{eff}}\right)^n.$$

The first term involves just the noninteracting Green function, and we note that $\exp\left[\text{Tr}\ln\left(-G^{(0)-1}\right)\right] = \exp\left[-\text{Tr}\ln\left(-G^{(0)}\right)\right] = \det G^{(0)-1} = Z_0$, the noninteracting partition function. The last term is a perturbative series that can be treated using Feynman diagram techniques. Putting everything together, we get

$$Z = Z_0\int D[\phi^*,\phi]\exp\left[-\int_0^\beta d\tau\,\boldsymbol{\phi}(\tau)\,V^{-1}\,\boldsymbol{\phi}^*(\tau)-\zeta\sum_{n=1}^\infty\frac{1}{n!}\text{Tr}\left(G^{(0)}\tilde{\mathcal{H}}_I^{\text{eff}}\right)^n\right].$$

$$(16.39)$$

To illustrate this treatment, we shall continue with our Coulombic electron gas example. Recall that ϕ here represents a scalar photon field that mediates Coulomb interaction – it is real and periodic $\phi(\tau+\beta) = \phi(\tau)$. Using a plane-wave basis and the Matsubara Fourier series,[8] we find that

$$Z = Z_0\int D[\phi^*,\phi]\exp\left[-\int_0^\beta d\tau\,\boldsymbol{\phi}(\tau)\,V^{-1}\,\boldsymbol{\phi}^*(\tau)-\zeta\sum_{n=1}^\infty\frac{1}{n!}\text{Tr}\left(G^{(0)}\tilde{\mathcal{H}}_I^{\text{eff}}\right)^n\right],$$

---

[7] $\det A = \prod_i a_i$, where a_i are the eigenvalues of matrix A. Thus, $\ln\det(A) = \sum_i \ln(a_i)$. Note that the eigenvalues of $\ln[A]$ are $\ln(a_i)$.

[8] Ω_n and ω_m are fermionic and bosonic Matsubara frequencies, respectively.

where we used (15.165) for $i\omega_n = 0$. Substitution in (16.38) yields

$$S_{\text{eff}}[\phi] = \frac{1}{2} \sum_{\mathbf{q}} \left(\frac{q^2}{4\pi} - e^2 \Pi_0(\mathbf{q}) \right) |\phi(\mathbf{q})|^2 + \mathcal{O}(e^4)$$

$$\simeq \frac{1}{2} \sum_{\mathbf{q}} \frac{q^2}{4\pi} \left(1 - \frac{4\pi e^2}{q^2} \Pi_0(\mathbf{q}) \right) |\phi(\mathbf{q})|^2 = \sum_{q} \epsilon(q) \frac{q^2}{4\pi},$$

which describes the screening of Coulomb field fluctuations by the electron gas.

Exercises

16.1 Given the action

$$S = \int_{t_a}^{t_b} dt \left(\frac{m}{2} \dot{x}^2 - V(x) \right).$$

(a) Determine $\dfrac{\delta S}{\delta x(u)}$ and $\dfrac{\delta^2 S}{\delta x(u) \delta x(u')}$.

   Hint: use the identity $\dfrac{\delta x(t)}{\delta x(u)} = \delta(x - u)$ and the chain rule.

(b) Substitute in expansion of the action

$$S = \int_{t_i}^{t_f} dt \frac{\delta^{(2)} S[x]}{\delta x(t) \delta x(t')} \eta(t) \eta(t').$$

16.2 The Hamiltonians of both fermionic and bosonic harmonic oscillator can be expressed as $\mathcal{H} = \hbar\omega \left(N + \frac{1}{2} \right)$, where N is the corresponding number operator.

(a) Using their respective commutators, show that the Hamiltonians can be expressed as

$$\mathcal{H}_B = \frac{\hbar\omega}{2} \left\{ b^\dagger, b \right\}$$

$$\mathcal{H}_F = \frac{\hbar\omega}{2} \left[c^\dagger, c \right].$$

(b) Defining the fermionic ground state as $c \, |0\rangle = 0$, show that

$$\mathcal{H}_F \, |0\rangle = -\frac{\hbar\omega}{2} \, |0\rangle.$$

   Now use the raising operator to show that Hilbert space for the fermionic oscillator is of dimension 2, and determine the energy of the upper level.

(c) Derive the coherent state partition function of the fermionic oscillator.

   Hint: use the Matsubara Fourier series and the Euler formula

$$\cosh \left(\frac{x}{2} \right) = \prod_{n=1}^{\infty} \left[1 + \left(\frac{x}{\pi(2n+1)} \right)^2 \right]$$

and fix the normalization by taking the limit $\beta \to \infty$.

16.3 Consider the infinite-range Ising model, where the coupling constant $J_{ij} = J_0$, $\forall\, i, j$ – no restriction to nearest-neighbor interactions. The Hamiltonian is then given by

$$\mathcal{H} = \frac{J_0}{2} \sum_{ij} s_i\, s_j - \mu B \sum_i s_i\,.$$

(a) Explain why this model only makes sense if $J_0 = J/N$, where N is the number of spins in the system.

(b) Use the Hubbard–Stratonovich transformation

$$\exp\left[\frac{u}{2N} x^2\right] = \sqrt{\frac{Nu}{2\pi}} \int_{-\infty}^{\infty} dm\, \exp\left[-\frac{Nu}{2} m^2 + uxm\right]$$

to show that the partition function is given by

$$Z_N = \sqrt{\frac{NJ\beta}{2\pi}} \int dm\, e^{-N\beta S}$$

$$S = \frac{J}{2} m^2 - \frac{1}{\beta} \ln\left[2\cosh\left(\beta(\mu B + Jm)\right)\right].$$

(c) In the thermodynamic limit, $N \to \infty$, use the saddle-point approximation to determine the mean-field value m_0. Determine the average spin $\langle s \rangle$ as

$$\langle s \rangle = \frac{1}{N\beta} \frac{\partial \ln Z_N}{\partial B}$$

and show that it satisfies the self-consistency condition

$$\langle s \rangle = \tanh\left[\beta\,\left(J\,\langle s \rangle + \mu B\right)\right].$$

(d) Derive an expression for the free energy $F = -k_B T\,\ln Z_N$.

16.4 Consider the imaginary-time action of a bosonic system

$$S[\psi] = \int d\mathbf{x} \int_0^\beta d\tau \left[\bar{\psi}\,\partial_\tau\,\psi + \frac{1}{2m}\left|\nabla_{\mathbf{x}}\psi\right|^2 - \mu\,\bar{\psi}\psi + \frac{g}{2}\,\bar{\psi}\bar{\psi}\psi\psi\right],$$

where ψ is a complex boson field, and g is a repulsive contact interaction ($\hbar = 1$).

(a) Show that the homogeneous saddle point of the action is $\psi_0 = \sqrt{\rho_0}\, e^{i\theta}$, where $\rho_0 = \mu/g$ and θ is a fixed arbitrary phase.

(b) Determine the excitation spectrum by expanding the field around the saddle point $\psi(\mathbf{x}, \tau) = \psi_0 + \phi(\mathbf{x}, \tau)$, and terminating at terms quadratic in ϕ. Show that the Gaussian action can be written in the spinor form

$$S[\phi] = \frac{1}{2} \int_0^\beta d\tau \int d\mathbf{x}\,[\bar{\phi}\ \ \phi] \begin{pmatrix} \partial_\tau - \frac{\nabla^2}{2m} + g|\psi_0|^2 & g\psi_0^2 \\ g\bar{\psi}_0^2 & -\partial_\tau - \frac{\nabla^2}{2m} + g|\psi_0|^2 \end{pmatrix} \begin{bmatrix} \bar{\phi} \\ \phi \end{bmatrix}.$$

(c) Show that the quasiparticle spectrum has the form

$$\omega_{\mathbf{q}}^2 = v_p^2 q^2 + \left(\frac{q^2}{2m}\right)^2,$$

where $v_p^2 = \rho_0 g/m$.

17

Boson Systems: Bose–Einstein Condensation and Superfluidity

17.1 Ideal Bose Fluid and Bose–Einstein Condensation

We have obtained in the preceding chapter (16.34), the grand partition function of an ideal Bose gas as $Z_G = \prod_\alpha \left(1 - e^{-\beta(\varepsilon_\alpha - \mu)}\right)^{-1}$, where α labels the quantum states of the gas. Here, we consider free plane-wave states, and write

$$Z_G = \prod_{\mathbf{k}} \left(1 - e^{-\beta(\varepsilon_{\mathbf{k}} - \mu)}\right)^{-1},$$

with the grand potential and particle number given by

$$\Psi = k_B T \sum_{\mathbf{k}} \ln\left(1 - e^{-\beta(\varepsilon_{\mathbf{k}} - \mu)}\right); \quad N = \sum_{\mathbf{k}} \left(\frac{1}{e^{-\beta(\varepsilon_{\mathbf{k}} - \mu)} - 1}\right). \tag{17.1}$$

We notice immediately that at finite temperatures we require that $\mu < \varepsilon_{\mathbf{k}}$ $\forall \mathbf{k}$; otherwise, the occupation number of some state k', with $\varepsilon_{k'} = \mu$, will equal infinity!. Since the lowest state here is $\varepsilon_0 = 0$, μ has to be ≤ 0. However, at $T = 0$ this condition is relaxed. We recall that the quantity $e^{\beta\mu} = \alpha$ is the fugacity, and thus $0 \leq \alpha \leq 1$.

Because bosonic states have no restriction on their occupation numbers, one has to be very careful when making the transition from a summation to an integral over $\mathbf{k}$. To underscore this point, we shall contrast two cases.

17.1.1 Fixed Chemical Potential

First, we consider a 3D system in contact with a particle reservoir that fixes the value of μ. We introduce the integral form of Ψ and N with the aid of the density of states $\mathcal{D}(\varepsilon) = \frac{\Omega}{(2\pi)^2} \left(\frac{2m}{\hbar^2}\right)^{3/2} \varepsilon^{1/2}$, and write

$$\Psi = k_B T \frac{\Omega}{(2\pi)^2} \left(\frac{2m}{\hbar^2}\right)^{3/2} \int_0^\infty \ln\left(1 - e^{-\beta(\varepsilon - \mu)}\right) \varepsilon^{1/2} \, d\varepsilon$$

$$= -\frac{2}{3} \frac{\Omega}{(2\pi)^2} \left(\frac{2m}{\hbar^2}\right)^{3/2} \int_0^\infty \frac{\varepsilon^{3/2}}{e^{\beta(\varepsilon - \mu)} - 1} \, d\varepsilon, \text{ (integration by parts)} \tag{17.2}$$

$$N = \frac{\Omega}{(2\pi)^2} \left(\frac{2m}{\hbar^2}\right)^{3/2} \int_0^\infty \frac{\varepsilon^{1/2} e^{-\beta(\varepsilon-\mu)}}{1 - e^{-\beta(\varepsilon-\mu)}} \, d\varepsilon \qquad (17.3)$$

$$U = -\frac{\partial \ln Z_G}{\partial \beta}\bigg|_{\beta\mu} = \frac{\Omega}{(2\pi)^2} \left(\frac{2m}{\hbar^2}\right)^{3/2} \int_0^\infty \frac{\varepsilon^{3/2}}{e^{\beta(\varepsilon-\mu)} - 1} \, d\varepsilon = \frac{3}{2} \Psi. \qquad (17.4)$$

Since $\Psi = U - TS - \mu N = -PV$, we obtain the equation of state

$$PV = \frac{2}{3} U.$$

17.1.2 Fixed Particle Number: Bose–Einstein Condensation

Now we consider a system of bosonic particles enclosed in an impermeable box and in contact with a temperature reservoir T, with particle number N fixed. Then according to (17.2), and with the fugacity $\alpha = e^{\beta\mu}$, we have

$$\begin{aligned}
N &= \frac{\Omega}{(2\pi)^2} \left(\frac{2m}{\hbar^2}\right)^{3/2} \int_0^\infty \frac{\varepsilon^{1/2} e^{-\beta(\varepsilon-\mu)}}{1 - e^{-\beta(\varepsilon-\mu)}} \, d\varepsilon \\
&= \frac{\Omega}{(2\pi)^2} \left(\frac{2m}{\hbar^2}\right)^{3/2} (k_B T)^{3/2} \int_0^\infty \frac{\alpha \, x^{1/2} e^{-x}}{1 - \alpha \, e^{-x}} \, dx \\
&= \frac{\Omega}{\sqrt{\pi} \lambda_T^3} \sum_{\ell=0}^\infty \alpha^{\ell+1} \int_0^\infty x^{1/2} e^{-(\ell+1)x} \, dx \\
&= \frac{\Omega}{\lambda_T^3} \sum_{\ell=1}^\infty \frac{\alpha^\ell}{\ell^{3/2}} = \frac{\Omega}{\lambda_T^3} F_{3/2}(\alpha),
\end{aligned} \qquad (17.5)$$

where $\lambda_T = \sqrt{\dfrac{2\pi\hbar^2}{mk_B T}}$ is the thermal de Broglie wavelength, and $F_{3/2}$ is known as the *polylogarithmic* function (see Figure 17.1). What happens when we lower the temperature to T'? Will μ remain constant? At $T' < T$, $\lambda_{T'} > \lambda_T$, $\alpha' = e^{\beta'\mu} < \alpha$, since $\mu < 0$. Assuming μ remains constant, then $N' < N$! We have a problem unless μ increases $\Rightarrow \mu \to 0$, $\alpha \to 1$.

We rewrite (17.5) in the form

$$\frac{N\lambda_T^3}{g\Omega} = \frac{n \, \lambda_T^3}{g} = F_{3/2}(\alpha).$$

But how far can μ increase to compensate for all the other terms? That would depend on the maximum value of $F_{3/2}$- 2.612 for $\mu = 0$, $\alpha = 1$!

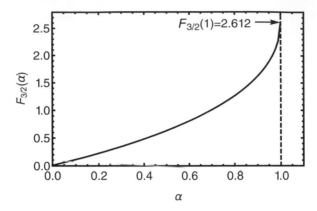

Figure 17.1 The *polylogarithmic* function $F_{3/2}$ as a function of the fugacity α.

A moment's reflection reveals the source of the problem:

We consider the occupation of the lowest energy state for 1 liter ($=10^{-3}\text{m}^3$) of ^{4}He with atomic mass $\simeq 6.6 \times 10^{-27}$ kg enclosed in a cubic box of side L:

$$\varepsilon_0 = \varepsilon_{111} \simeq \frac{\hbar^2 k_{111}^2}{2m} = \frac{3h^2}{8mL^2}, \Rightarrow \mathbf{k}_{111} = \left(\frac{\pi}{L}, \frac{\pi}{L}, \frac{\pi}{L}\right)$$

$$= 2.0 \times 10^{-37}\text{J} \Rightarrow T = 2 \times 10^{-14} \, K.$$

The first excited state is triply degenerate with energy

$$\varepsilon_{211} = 4.0 \times 10^{-37} \, \text{J} \Rightarrow T = 4 \times 10^{-14}\text{K},$$

$$\varepsilon_{211} - \varepsilon_{111} = 2.0 \times 10^{-37} \, \text{J} \Rightarrow T = 2 \times 10^{-14}\text{K}.$$

Thus, the discrete states are very closely spaced in energy – far closer than $k_B T$ at any reasonable temperature. We might well have felt confident in replacing the sum by an integral!

First we note that at $T = 0$, all N particles occupy the ground state, in which case $\mu = \varepsilon_0$! We examine closely the population of the ground state and first-excited state as the chemical potential approaches ε_0 from below. In particular, we inquire as to the value of μ for which the population of the ground state alone is comparable to the entire number of particles in the gas. Let the population of the ground state at a very low temperature be $n_0 \simeq 10^{22}$, then

$$n_0 \sim 10^{22} = \frac{1}{e^{\beta(\varepsilon_{111}-\mu)} - 1} \simeq \frac{k_B T}{\varepsilon_0 - \mu}, \quad \Rightarrow \quad 10^{-22} \simeq \beta\left(\varepsilon_{111} - \mu\right).$$

Notice that because the total number of particles is very large but finite, $\mu < \varepsilon_0$ at finite temperatures.

We find that for $T = 10\text{K}$

$$(\varepsilon_{111} - \mu) = 10^{-22} k_B T = 10^{-21} k_B = 13.8 \times 10^{-44} \text{J},$$

and we obtain for the population of the first excited state

$$\frac{n_1}{n_0} = \frac{\varepsilon_0 - \mu}{(2\varepsilon_0 - \mu)} = \frac{13.8 \times 10^{-44}}{2.0 \times 10^{-37}} \simeq 10^{-7}$$

$$n_1 = 10^{-7} \times 10^{21} = 10^{14}.$$

The population of the higher states continues to fall extremely rapidly. As the temperature decreases further, μ cannot approach closer to the ground state-energy than $\beta(\mu - \epsilon_{111}) \sim 1/N \sim 10^{-23}$ (at which value the ground state would host all N particles in the gas!). Hence, the ground state shields all other states from too close an approach of μ, and each other state individually can host only a relatively small number of particles. Together, of course, the remaining states host all the particles not in the ground state. All states other than the ground state are adequately represented by the integral over the density of orbital states function. *Only the ground state needs special consideration.*

The ground-state energy must be separately and explicitly listed in the sum over states. We identify a demarcation temperature T_{BE} below which the *integral analysis* fails, and the ground state begins to be macroscopically populated, namely,

$$T_{BE} = \frac{2\pi\hbar^2}{mk_B} \left(\frac{N}{2.612 g\Omega}\right)^{2/3} = \frac{1.687 \times 10^{-18}}{m} \left(\frac{n}{g}\right)^{2/3} \tag{17.6}$$

with m in atomic units. Alternatively, we may write

$$\frac{N}{g\Omega} \lambda_{T_{BE}}^3 = 2.612. \tag{17.7}$$

Below T_{BE}, the number of particles in the excited states N_e is given by

$$\frac{N_e}{g\Omega} \lambda_T^3 = 2.612, \tag{17.8}$$

and thus we obtain the following expression for the number of particles in the excited states at $T < T_{BE}$:

$$N_e(T) = \left(\frac{\lambda_{T_{BE}}}{\lambda_T}\right)^3 N = \left(\frac{T}{T_{BE}}\right)^{3/2} N. \tag{17.9}$$

The number of condensed particles is then

$$N_c = N \left[1 - \left(\frac{T}{T_{BE}}\right)^{3/2}\right]. \tag{17.10}$$

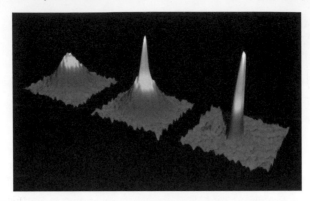

Figure 17.2 Spectroscopic (reciprocal space) images of a gas of atoms at 400 nK (left), 200 nK (middle), and 50 nK (right). The peak in the momentum distribution signals the onset of condensation.

An interesting observation becomes clear if we rewrite (17.7) in terms of the specific volume per particle, $v = \Omega/N = \ell^3$, namely

$$\frac{\lambda_{T_{BE}}^3}{v} = \left(\frac{\lambda_{T_{BE}}}{\ell}\right)^3 = 2.612\,g,$$

This relation reveals that condensation is initiated when the thermal de Broglie wavelength becomes comparable to the mean free path of the bosonic particles, shown in Figure 17.3!

The appearance of a macroscopic population is referred to as a condensate phase in analogy to the gas/liquid transformation.

However, in contrast to the latter, where a clear phase separation with a well-demarcated boundary (interface) between the liquid and gas phases in physical space, no such separation takes place between the condensate and the remaining population of the excited Bose gas states. This is because the ground-state wavefunction permeates the whole volume Ω occupied by the system.

> Yet we should also make the observation that there appears a phase separation in **k**-space, as shown in Figure 17.2.

17.2 Interacting Bosons and Superfluidity

Before the observation of Bose–Einstein condensation (BEC) in optically trapped super-cooled atoms, it was thought to occur in superfluid ^{4}He, which liquifies around 4.2 K and goes over to the superfluid state at $T = 2.18$ K. The latter transition is marked by a second-order transition lambda point. This is similar to a BEC, but we should note that, strictly speaking, BEC deals with bosons in an ideal gas phase. Here, helium is in the liquid phase, and there are significant interactions between He atoms not present in the theory of

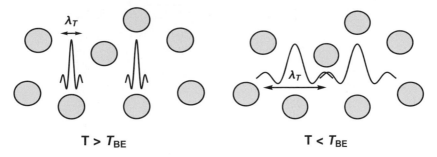

Figure 17.3 Thermal de Broglie wavelength.

a noninteracting gaseous BEC. Therefore, we cannot treat the ^{4}He system using the same approach as for an BEC in an ideal gas. However, the fact that ^{4}He atoms are bosons still remains significant.

17.2.1 An Early Historic perspective

The discovery of superfluidity in liquid ^{4}He was announced to the scientific world on January 8, 1938 [7, 8, 101]. This prompted Fritz London to suggest that the transition at T = 2.18 K in liquid ^{4}He was due to the formation of a BEC of ^{4}He atoms of ideal Bose gas distorted by intermolecular forces [120, 121].[1] This observation is famous because it signals the historical jump of quantum mechanics from the microscopic physics of atoms to macroscopic systems such as a liter of liquid helium. Tisza then followed by suggesting that the spectacular superfluidity effects observed were related to the coherent motion of the Bose condensate. He launched the famous "two fluid theory" [178, 180, 179], where helium can be considered as a mixture of two fluids, "superfluid" and "normal": at $T = 0$, liquid helium is a purely superfluid condensate, whereas with increasing temperature excited atoms are formed constituting the normal component. The most important success of Tisza theory was the prediction of what was later coined as *second sound*[2] by Landau. In 1941, Landau reformulated Tisza's two-fluid model on a more rigorous basis. The principles and consequences were similar, except for the nature of the normal component of the quantum liquid. Landau's intuition was that it consisted of collective excitations, and not of single-atom excitations, as Tisza had assumed. In order to fit the measured specific heat, he had to assume the existence of two different kinds of excitations: a linearly dispersive phonon branch, with a slope corresponding to the velocity of sound, and a quadratic curve with an energy gap for a new class of excitations he called "rotons" (see Figure 17.4).[3] Based on

---

[1] London realized that for a noninteracting gas with the mass and density of ^{4}He, the BEC phenomenon would occur at 3.3 K, which suggested to him that this was exactly what was going on at the observed lambda transition (2.17 K).

[2] First sound is ordinary sound, which consists of fluctuations in total density. Its velocity is only weakly dependent on temperature. Second sound has a velocity that is a strong function of temperature, becoming zero at the λ-point. Second sound consists of fluctuations (propagating waves) in entropy or temperature in helium. See [53] for more detail.

[3] I. E. Tamm suggested to Landau to call these excitations "rotons."

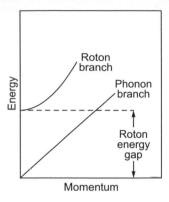

Figure 17.4 Landau's phonon/roton model.

these assumptions, Landau proposed that, for a superfluid flowing at a velocity v at zero temperature, dissipation can only result from the emission of either phonons or rotons, so that, from the conservation of energy and momentum in this process, dissipation is only possible if

$$v \sim 60\,\text{m/s} > \sqrt{\frac{\Delta}{2\mu}} \quad \text{(rotons)},$$

where Δ is the roton energy gap and μ a mass about six to eight times the mass of a ^{4}He atom. He also predicted that there should exist two different types of waves in superfluid helium: ordinary sound and heat waves that he called "second sound." However, Landau's theory is not a microscopic theory. Moreover, the fact that ^{4}He atoms are bosons is never used.

17.2.2 Bogoliubov Theory of Superfluid ^{4}He: Weakly Interacting Bosons

Our current understanding that BEC forms the microscopic basis of Landau's theory of superfluidity was developed from the initial work of Bogoliubov in 1947. Bogoliubov's theory is based on a physical assumption that a weakly nonideal Bose gas can condense à la BEC. The existence of the Bose condensate leads to a unique macroscopic wavefunction of the whole system – a collective effect. This means that the presence of even a weak interaction transforms single-particle excitations into the spectrum of collective excitations.

Following Bogoliubov's recipe, we write the Hamiltonian in the field operators representation as

$$\mathcal{H} = \int d\mathbf{x}\, \frac{\hbar^2}{2m}\, \nabla \psi^\dagger(\mathbf{x}\, \nabla \psi(\mathbf{x} + \frac{1}{2} \int d\mathbf{x}\, V(\mathbf{x}' - \mathbf{x})\, \psi^\dagger(\mathbf{x})\, \psi^\dagger(\mathbf{x}')\, \psi(\mathbf{x}')\, \psi(\mathbf{x}), \quad (17.11)$$

where $V(\mathbf{x})$ is a short-range, two-particle interaction potential. For a uniform gas in volume Ω, we express the field operators as

$$\psi = \sum_{\mathbf{k}} \frac{e^{i\mathbf{k} \cdot \mathbf{x}}}{\sqrt{\Omega}} b_{\mathbf{k}},$$

where $b_{\mathbf{k}}$ are annihilation operators. Substituting in (15.23.32), we get

$$\mathcal{H} = \sum_{\mathbf{k}} \varepsilon_{\mathbf{k}} b_{\mathbf{k}}^{\dagger} b_{\mathbf{k}} + \frac{1}{2\Omega} \sum_{\mathbf{k}, \mathbf{k}', \mathbf{q}} V_{\mathbf{q}} b_{\mathbf{k}-\mathbf{q}}^{\dagger} b_{\mathbf{k}'+\mathbf{q}}^{\dagger} b_{\mathbf{k}'} b_{\mathbf{k}} \tag{17.12}$$

$$\varepsilon_{\mathbf{k}} = \frac{\hbar^2 \mathbf{k}^2}{2M},$$

where $\varepsilon_{\mathbf{k}}$ is the energy of a helium atom of mass M, and $V_{\mathbf{q}}$ is the Fourier transform of $V(\mathbf{x})$.

17.2.3 A Simplified Picture of the Small k Excitation Spectrum

To simplify the derivation while preserving the physics, we shall replace the short-range potential with a contact one, and write

$$\mathcal{H} = \sum_{\mathbf{k}} \varepsilon_{\mathbf{k}} b_{\mathbf{k}}^{\dagger} b_{\mathbf{k}} + \frac{V_0}{2\Omega} \sum_{\mathbf{k}, \mathbf{k}'} b_{\mathbf{k}'}^{\dagger} b_{\mathbf{k}}^{\dagger} b_{\mathbf{k}'} b_{\mathbf{k}}. \tag{17.13}$$

Again, following Bogoliubov's prescription, we argue that the idea of macroscopic condensation of particles into the $k = 0$ state can carry over from noninteracting systems to interacting ones. This macroscopic number of order of Avogadro's number, we call N_0, becomes the expectation value of $b_0^{\dagger} b_0$. In fact, $b_0^{\dagger} b_0^{\dagger}$ operating on a wavefunction with N_0 particles in the $k = 0$ state would give $\sqrt{(N_0 + 1)(N_0 + 2)}$ times the state with $N_0 + 2$ particles, but since $2 \ll N_0$, we ignore this difference. We take $b_0^{\dagger} b_0^{\dagger} = b_0 b_0$ to be equal to N_0, or $b_0^{\dagger} = b_0 = \sqrt{N_0}$.

In a noninteracting gas at $T = 0$, all the atoms are in the condensate and $N_0 = N$. In a weakly interacting gas, the occupation numbers for states with $|\mathbf{k}| \neq 0$ are finite but small. This means that, to first approximation, we can neglect all terms in the Hamiltonian containing operators $b_{\mathbf{k}}$ and $B_{\mathbf{k}}^{\dagger}$, $k \neq 0$. Such an approximation implies that $N_0 \sim N$, and we replace b_0 with $\sqrt{N}$. The ground-state energy becomes

$$\varepsilon_H = N\varepsilon_0 + \frac{V_0}{2\Omega} \left\langle \Phi \left| b_0^{\dagger} b_0^{\dagger} b_0 b_0 \right| \Phi \right\rangle = N\varepsilon_0 + \frac{V_0}{2\Omega} N(N-1) = N\varepsilon_0 + \frac{V_0 N^2}{2\Omega},$$

where Φ is the Hartree wavefunction.

The dependence of the energy on the volume implies that this system will support long wavelength longitudinal sound waves. The sound wave velocity for a classical fluid is given by

$$v = \sqrt{\frac{R}{\rho}}$$

$$R = -\Omega \left(\frac{\partial P}{\partial \Omega}\right)_N, \quad P = -\left(\frac{\partial \varepsilon_H}{\partial \Omega}\right)_N,$$

where ρ is the density, R is the bulk modulus, and P the pressure. We find that

$$v = \sqrt{\frac{\Omega \left(\partial^2 \varepsilon_H / \partial \Omega^2\right)}{\rho}} = \sqrt{\frac{N V_0}{M \Omega}}. \tag{17.14}$$

V_0 has to be positive – a repulsive potential – for the sound velocity to be real.

Thus, for a system of unit volume, we anticipate boson excitations at small $\mathbf{k}$ with dispersion $\omega_{\mathbf{k}}$ given by

$$\omega_{\mathbf{k}}^2 = \frac{N V_0 \mathbf{k}^2}{M}. \tag{17.15}$$

Bogoliubov's method shows how these excitations arise as a modification of the single-particle excitation spectrum.

17.2.4 Interacting Boson System and Bose–Einstein Condensation

We now rewrite the Hamiltonian (17.12) dropping all terms of order less than N_0. Provided that $V_{\mathbf{q}} = V_{-\mathbf{q}}$, we obtain

$$\mathcal{H} \simeq \sideset{}{'}\sum_{\mathbf{k}} \varepsilon_{\mathbf{k}} b_{\mathbf{k}}^\dagger b_{\mathbf{k}} + \frac{1}{2} N_0^2 V_0 + N_0 V_0 \sideset{}{'}\sum_{\mathbf{k}} b_{\mathbf{k}}^\dagger b_{\mathbf{k}} + N_0 \sideset{}{'}\sum_{\mathbf{k}'} V_{\mathbf{k}'} b_{\mathbf{k}'}^\dagger b_{\mathbf{k}'} + \frac{1}{2} N_0 \sideset{}{'}\sum_{\mathbf{q}} V_{\mathbf{q}} \left(b_{\mathbf{q}} b_{-\mathbf{q}} + b_{-\mathbf{q}}^\dagger b_{\mathbf{q}}^\dagger\right)$$

where the prime over sums excludes the zero term. We set

$$N = N_0 + \sum_{\mathbf{k}} b_{\mathbf{k}}^\dagger b_{\mathbf{k}}; \qquad \xi_{\mathbf{k}} = N_0 V_{\mathbf{k}}; \qquad \hbar \Omega_{\mathbf{k}} = \varepsilon_{\mathbf{k}} + \xi_{\mathbf{k}}$$

and assume that $N - N_0 \ll N_0$ [4]. We write

$$\mathcal{H} = \frac{1}{2} N^2 V_0 + \sum_{\mathbf{k}} \hbar \Omega_{\mathbf{k}} b_{\mathbf{k}}^\dagger b_{\mathbf{k}} + \frac{1}{2} \sum_{\mathbf{k}} \xi_{\mathbf{k}} \left(b_{\mathbf{k}} b_{-\mathbf{k}} + b_{\mathbf{k}}^\dagger b_{-\mathbf{k}}^\dagger\right). \tag{17.16}$$

We note that this Hamiltonian is nondiagonal in the operators $b_{\mathbf{k}}$ and $b_{\mathbf{k}}^\dagger$. This is precisely where Bogoliubov introduced his canonical transformation that we presented in Chapter 13. He wrote

$$\alpha_{\mathbf{k}} = (\cosh \theta_{\mathbf{k}}) \, b_{\mathbf{k}} - (\sinh \theta_{\mathbf{k}}) \, b_{-\mathbf{k}}^\dagger, \tag{17.17}$$

where $\theta_{\mathbf{k}}$ is not defined for now. $\alpha_{\mathbf{k}}$ satisfies the bosonic commutator $\left[\alpha_{\mathbf{k}}, \alpha_{\mathbf{k}'}^\dagger\right] = \delta_{\mathbf{k}, \mathbf{k}'}$. Now we require that the Hamiltonian becomes diagonal in the α representation, taking the form

---

[4] This assumption may not hold for real liquid helium.

$$\mathcal{H} = E_0 + \sum_{\mathbf{k}} \hbar\omega_{\mathbf{k}} \, \alpha_{\mathbf{k}}^\dagger \alpha_{\mathbf{k}}. \tag{17.18}$$

We express $\alpha_{\mathbf{k}}^\dagger \alpha_{\mathbf{k}}$ in terms of the bs and set $\omega_{\mathbf{k}}$ and $\theta_{\mathbf{k}}$ such as to match Hamiltonian (17.16):

$$
\begin{aligned}
\alpha_{\mathbf{k}}^\dagger \alpha_{\mathbf{k}} &= \left[(\cosh \theta_{\mathbf{k}}) \, b_{\mathbf{k}}^\dagger - (\sinh \theta_{\mathbf{k}}) \, b_{-\mathbf{k}} \right] \left[(\cosh \theta_{\mathbf{k}}) \, b_{\mathbf{k}} - (\sinh \theta_{\mathbf{k}}) \, b_{-\mathbf{k}}^\dagger \right] \\
&= \left(\cosh^2 \theta_{\mathbf{k}} \right) b_{\mathbf{k}}^\dagger b_{\mathbf{k}} + \left(\sinh^2 \theta_{\mathbf{k}} \right) \left(1 + b_{-\mathbf{k}}^\dagger b_{-\mathbf{k}} \right) \\
&\quad - (\cosh \theta_{\mathbf{k}} \, \sinh \theta_{\mathbf{k}}) \left(b_{\mathbf{k}} b_{-\mathbf{k}} + b_{-\mathbf{k}}^\dagger b_{\mathbf{k}} \right).
\end{aligned} \tag{17.19}
$$

We note that $\omega_{\mathbf{k}} = \omega_{-\mathbf{k}}$ because of time-reversal invariance, and we set $\theta_{\mathbf{k}} = \theta_{-\mathbf{k}}$, to obtain

$$
\begin{aligned}
\sum_{\mathbf{k}} \hbar\omega_{\mathbf{k}} \, \alpha_{\mathbf{k}}^\dagger \alpha_{\mathbf{k}} &= \sum_{\mathbf{k}} \hbar\omega_{\mathbf{k}} \, \cosh(2\theta_{\mathbf{k}}) \, b_{\mathbf{k}}^\dagger b_{\mathbf{k}} + \sum_{\mathbf{k}} \hbar\omega_{\mathbf{k}} \, \sinh^2 \theta_{\mathbf{k}} \\
&\quad - \frac{1}{2} \sum_{\mathbf{k}} \hbar\omega_{\mathbf{k}} \, \sinh(2\theta_{\mathbf{k}}) \left(b_{\mathbf{k}} b_{-\mathbf{k}} + b_{-\mathbf{k}}^\dagger b_{\mathbf{k}}^\dagger \right).
\end{aligned} \tag{17.20}
$$

To reproduce (17.16), apart from a constant, we set

$$\omega_{\mathbf{k}} \cosh(2\theta_{\mathbf{k}}) = \Omega_{\mathbf{k}}; \qquad \hbar\omega_{\mathbf{k}} \sinh(2\theta_{\mathbf{k}}) = -\xi_{\mathbf{k}}$$

and obtain

$$\hbar^2 \omega_{\mathbf{k}}^2 = \hbar^2 \Omega_{\mathbf{k}}^2 - \xi_{\mathbf{k}}^2$$

$$\mathcal{E}(\mathbf{k}) = \hbar\omega_{\mathbf{k}} = \sqrt{(\varepsilon_{\mathbf{k}} + N_0 V_{\mathbf{k}})^2 - (N_0 V_{\mathbf{k}})^2} = \sqrt{\varepsilon_{\mathbf{k}}^2 + 2\varepsilon_{\mathbf{k}} N_0 V_{\mathbf{k}}}. \tag{17.21}$$

Thus, the Bogoliubov approach produced a ground state of energy

$$E_0 = \frac{1}{2} N^2 V_0 + \frac{1}{2} \sum_{\mathbf{k} \neq 0} \left[\mathcal{E}(\mathbf{k}) - N_0 V_{\mathbf{k}} - \varepsilon_{\mathbf{k}} + \frac{(N_0 V_{\mathbf{k}})^2}{2\varepsilon_{\mathbf{k}}} \right]$$

above which new noninteracting elementary collective excitations, or quasiparticles, emerge, with dispersion relation (17.21). These noninteracting quasiparticles are annihilated and created by the operators $\alpha_{\mathbf{k}}$ and $\alpha_{\mathbf{k}}^\dagger$, respectively. The ground state now acts as a vacuum for the new quasiparticle excitations

$$\alpha_{\mathbf{k}} \, |\emptyset\rangle = 0, \qquad \forall \, \mathbf{k} \neq 0. \tag{17.22}$$

We find that the original system of interacting particles can now be described in terms of a Hamiltonian of independent quasiparticles having energy $\hbar\omega_{\mathbf{k}}$.

The interesting thing about these excitations is the way their energy varies with $\mathbf{k}$. When we consider the region where $\varepsilon_{\mathbf{k}} = \hbar^2 k^2 / 2m \ll N_0 V_{\mathbf{k}}$, we find that

$$\omega_{\mathbf{k}} \simeq \sqrt{\frac{N V_{\mathbf{k}}}{M}} \, k. \tag{17.23}$$

This excitation looks more like phonons (*Bogoliubov sound*) than like free particles. However, when $\mathbf{k}$ becomes large, so that $\varepsilon \gg N_0 V_{\mathbf{k}}$, then the excitations will once again be

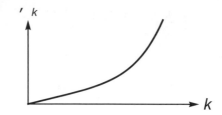

Figure 17.5 Dispersion relation for weak interactions.

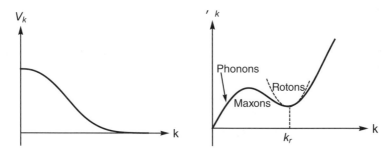

Figure 17.6 Left: a possible form of the effective interaction between helium atoms. It leads to a dispersion curve with a minimum, as shown on the right.

particlelike. The boson dispersion relation will look like the one depicted in Figure 17.5 for contact interaction, where $V_{\mathbf{k}} = V$. The detailed shape of the dispersion curve $\omega(\mathbf{k})$ will depend strongly on the form of $V_{\mathbf{k}}$. In real liquid helium, atoms interact strongly, and we need to introduce modifications to the previous analysis: $V_{\mathbf{k}}$ takes a form similar to that shown in the left graph in Figure 17.6, which leads to the dispersion relation for liquid ^{4}He measured by inelastic neutron scattering, shown on the right. It confirms Landau's model of phonon/roton collective excitations. It is also in accord with the superfluid properties of this system at low temperatures.

A Note about Rotons

The elementary excitations in superfluid liquid ^{4}He named rotons have an unusual dispersion curve. The energy is an approximately quadratic function of $(\mathbf{k} - \mathbf{k}_r)$,

$$\varepsilon(\mathbf{k}) = \Delta + \frac{\hbar^2(\mathbf{k} - \mathbf{k}_r)^2}{2\mu}.$$

At low temperature and pressure, approximate values for the parameters are $\mathbf{k}_r = 1.92\text{Å}^{-1}$, $\Delta/k_B = 8.62$ K, and $\mu = m_{\text{He}}/6$. For $\mathbf{k} > \mathbf{k}_c$, a roton group velocity is parallel to its momentum, while for $\mathbf{k} < \mathbf{k}_c$ the velocity and momentum are antiparallel. When $\mathbf{k} = \mathbf{k}_c$, the roton has nonzero momentum but zero velocity. Consequently, a roton at the minimum has the curious property of remaining stationary in the liquid despite its large momentum. Rotons just to the left of the minimum, which, even more astonishingly, move backward in the opposite direction to their momenta, are known as

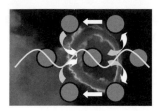

Figure 17.7 Rotons, according to Landau, may be described as microscopic *smoke rings.*

R^- rotons; those on the high-momentum side are known as R^+ rotons. These kinematic properties lead to unusual trajectories when rotons scatter or experience external forces, as rendered in Figure 17.7.

17.2.5 Superfluidity

The linear behavior $\omega(k) \propto k$ at low energies is a characteristic feature of the quasiparticles, which is to be compared with the energy dependence of free particles $\omega(k) \propto k^2$. This result is the core of the remarkable phenomenon of superfluidity in which certain liquids, like ^{4}He, suffer no friction in going through capillaries, i.e., they exhibit no viscosity. To explain the origin of this phenomenon, we start with defining

$$v_c = \min \frac{\omega(k)}{k}, \tag{17.24}$$

where v_c is called the *critical velocity*. Note that from (17.23) $v_c > 0$, and that for a noninteracting system, with

$$\varepsilon = \frac{\hbar^2 k^2}{2m},$$

$v_c = 0$ and hence no excitation occurs.

To understand superfluidity, we follow an argument due to Landau.

Landau's theory of superfluids is based on the *Galilean transformation* of energy and momentum. We consider a fluid of energy E and momentum P in a reference frame K. Expressing its energy and momentum in a frame K', moving with a relative velocity $\mathbf{V}$ with respect to the reference frame K, we obtain the relations

$$\mathbf{P'} = \mathbf{P} - M\mathbf{V}$$

$$E' = \frac{P'^2}{2M} = \frac{|\mathbf{P} - M\mathbf{V}|^2}{2M}$$
$$= E - \mathbf{P} \cdot \mathbf{V} + \frac{1}{2}MV^2,$$

where M is the total mass of the fluid.

Now consider a fluid at zero temperature, in which all particles are in the ground state and flowing through a capillary at constant velocity $\mathbf{v}_s$. In the rest frame K of the fluid, the capillary moves with a velocity $-\mathbf{v}_s$. The friction between the fluid and the capillary can cause quasiparticle excitations in the fluid by transforming kinetic energy to internal energy, or excitation energies. If a generated quasiparticle has momentum $\mathbf{k}$ and energy $\varepsilon(\mathbf{k})$, then the total energy in the rest frame is $E_0 + \varepsilon(\mathbf{k})$, where E_0 is the ground-state energy. Transforming back to the laboratory frame K' where the tube is now at rest, we have $\mathbf{V} = -\mathbf{v}_s$. This fluid will have the energy

$$E' = E + [\varepsilon(\mathbf{k}) + \mathbf{k} \cdot \mathbf{v}_s], \tag{17.25}$$

where $E = E_0 + Mv^2/2$. If no excitation is present, the energy of the fluid is E. The presence of the excitation causes the energy of the fluid to change by the amount $[\varepsilon(\mathbf{k}) + \mathbf{k} \cdot \mathbf{v}_s]$. Since the energy of flowing fluid decreases due to friction, we have

$$\varepsilon(\mathbf{k}) + \mathbf{k} \cdot \mathbf{v}_s \ < \ 0, \tag{17.26}$$

and the condition, $\varepsilon(\mathbf{k}) + \mathbf{k} \cdot \mathbf{v}_s \ \geq \ \varepsilon(\mathbf{k}) - kv_s$, must be satisfied. It follows, therefore, that if quasiparticles satisfy the property

$$v_s \ \geq \ \frac{\varepsilon(k)}{k}, \tag{17.27}$$

then friction will occur. However, the right-hand side of (17.27) is just the v_c of (17.24). Therefore, friction will occur when $v_s \geq v_c$.

If, however, $0 < v < v_c$ then the velocity will be in the gap where it is positive yet below the threshold for creating quasiparticles. The fluid will then move through the capillary tube without dissipation – the fluid will exhibit *superfluidity*.

For helium, Figure 17.8 shows two values of k where $\varepsilon(k)/k$ is a minimum, $k = 0$, and $k = k_c$. The minimum at $k = 0$ corresponds to low temperatures. At such temperatures, superfluidity occurs when

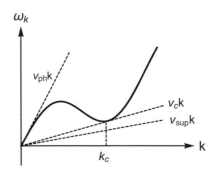

Figure 17.8 ^{4}He dispersion curve with two characteristic velocities: the phonon, or Bogoliubov sound velocity v_{ph}, and the critical velocity v_c. Also shown a typical superfluid velocity v_{sup}.

$$v_s \;\leq\; v_{\text{sound}} = \left.\frac{d\varepsilon(k)}{dk}\right|_{k=0}, \tag{17.28}$$

where v_s is the Bogoliubov sound velocity. For slightly higher temperatures, there is a minimum at K_c, and superfluidity occurs for

$$v \;\leq\; v_c = \left.\frac{d\varepsilon(k)}{dk}\right|_{k=k_c}, \tag{17.29}$$

where the quasiparticles are rotons.

17.3 Ginzburg–Landau Theory of Superfluidity

17.3.1 A Macroscopic Wavefunction!

The motivation behind this approach is the idea of a macroscopically occupied ground state. From the thermodynamic perspective, we can envision the condensate as a reservoir of particles whose number needs not to be specified because it does not contribute to the thermodynamic functions of the system. Thus, it may inject an unlimited number of particles into the system. Such a reservoir presents a macroscopic system that may be described by a macroscopic, classical wavefunction. To develop this approach, we start with defining the field operator $\hat{\Psi}(\mathbf{x})$ as

$$\hat{\Psi}(\mathbf{x}) = \sum_i \psi_i(\mathbf{x})\,\hat{a}_i,$$

where the c-number eigenfunctions $\psi_i(\mathbf{x})$ satisfy the orthonormal condition

$$\int d\mathbf{x}\,\psi_i^*(\mathbf{x})\,\psi_j(\mathbf{x}) = \delta_{ij}.$$

When a lowest-energy single-particle state (ground state) has a macroscopic occupation, we can separate the field operator into the condensate term ($i = 0$, lowest-energy ground state) and the noncondensate components ($i \neq 0$ excited states):

$$\hat{\Psi}(\mathbf{x}) = \psi_0\, a_0 + \sum_{i \neq 0} \psi_i(\mathbf{x})\,\hat{a}_i.$$

This expression of the field operator implicitly introduces the Bogoliubov approximation, where $\hat{a}_0$ is now replaced by the c-number $a_0 = \sqrt{N_0}$; $\left\langle \hat{a}_0^\dagger \hat{a}_0 \right\rangle = N_0$. By defining the classical field $\Psi_0 = \sqrt{N_0}\,\psi_0$ and the quantum field $\delta\hat{\Psi} = \sum_{i \neq 0} \psi_i(\mathbf{x})\,\hat{a}_i$, we can obtain the Bogoliubov ansatz:

$$\hat{\Psi}(\mathbf{x}) = \Psi_0(\mathbf{x}) + \delta\hat{\Psi}(\mathbf{x}).$$

The classical field is interpreted as a kind of *macroscopic wavefunction*, while the quantum field is treated as a small perturbation to the classical field. In this case, the microscopic field operators of the quantum fluid acquire an expectation value

$$\left\langle \hat{\Psi}(\mathbf{x}) \right\rangle_0 = \Psi_0(\mathbf{x}) = |\Psi_0(\mathbf{x})| \, e^{i\phi(\mathbf{x})} \qquad \text{Macroscopic wavefunction} \qquad (17.30)$$

complete with phase. In the language of symmetry, breaking it can be considered as creating a two-component order parameter, the magnitude of which determines the density of particles in the superfluid

$$|\Psi_0(\mathbf{x})|^2 = n_s(\mathbf{x}), \qquad (17.31)$$

while, as we shall see, the phase $\phi(\mathbf{x})$ characterizes the coherence and superfluid phenomena, where the *twist*, or gradient, of the phase determines the superfluid velocity:

$$\mathbf{v}_s(\mathbf{x}) = \frac{\hbar}{m} \, \nabla \, \phi(\mathbf{x}). \qquad (17.32)$$

The idea that the wavefunction can acquire a kind of Newtonian reality in a superfluid or, as we shall see later, in a superconductor goes deeply against our training in quantum physics: at first sight, it appears to defy the Copenhagen interpretation of quantum mechanics, in which $\hat{\Psi}(\mathbf{x})$ is an unobservable variable. The bold idea suggested by Ginzburg and Landau is that $\Psi_0(\mathbf{x})$ is a macroscopic manifestation of quintillions of particles – bosons – all condensed into precisely the same quantum state.

17.3.2 Spontaneous Symmetry Breaking

Spontaneous symmetry breaking (SSB) implies that under certain conditions the symmetry of the ground state is reduced with respect to that of the Hamiltonian. The concept of spontaneous symmetry breaking is subtle. As we will show, it is closely linked to manifest degeneracies of the ground state.

From a conceptual point of view, it is more appealing to associate Bose–Einstein condensation with the phenomenon of spontaneous symmetry breaking of a continuous symmetry than with macroscopic occupation of a single-particle level. The continuous symmetry is the global $U(1)$ gauge symmetry, the freedom in the choice of the global phase of many-particle wavefunctions. This symmetry is responsible for conservation of total particle number.

We shall interpret Bose–Einstein condensation at zero temperature in terms of spontaneous symmetry breaking of a continuous symmetry by explicitly constructing a ground state that breaks the global $U(1)$ gauge symmetry. First, we need to explore how the global $U(1)$ gauge symmetry organizes the Hilbert space spanned by eigenstates of the Hamiltonian. We also need to show how the thermodynamic limit plays an essential role in the construction, especially in the presence of interactions.

Degeneracy and the Breaking of Ground-State Symmetry

We consider any symmetry operator $\hat{\mathcal{O}} \in \mathcal{G}$, where $\mathcal{G}$ is the symmetry group of $\mathcal{H}$. By definition, $\hat{\mathcal{O}}$ must satisfy

$$\left[\hat{\mathcal{O}}, \mathcal{H} \right] = 0$$

which implies that for a ground-state $|\Phi_0\rangle$ of $\mathcal{H}$,

$$\mathcal{H}\,|\Phi_0\rangle = E_0\,|\Phi_0\rangle$$

$$\hat{\mathcal{O}}\,\mathcal{H}\,|\Phi_0\rangle = E_0\,\hat{\mathcal{O}}\,|\Phi_0\rangle = \hat{\mathcal{O}}\mathcal{H}\hat{\mathcal{O}}^{-1}\,\hat{\mathcal{O}}\,|\Phi_0\rangle = \mathcal{H}\,\hat{\mathcal{O}}\,|\Phi_0\rangle$$

- If $|\Phi_0\rangle$ is nondegenerate, i.e., it is the only state with energy E_0, then

$$\mathcal{O}\,|\Phi_0\rangle = e^{i\theta}\,|\Phi_0\rangle\,,$$

which means that a nondegenerate ground state must have the full symmetry of the Hamiltonian. In the language of symmetry, $|\Phi_0\rangle$ must transform according to the *trivial*, or *identity*, *representation* of $\mathcal{G}$. The jargon is that $|\Phi_0\rangle$ transforms as a *singlet*.

- If $|\Phi_0\rangle$ belongs to an ℓ fold degenerate subspace, then we may have

$$\mathcal{O}\,|\Phi_0\rangle = |\Phi_0'\rangle,$$

where $|\Phi_0'\rangle$ must belong to the degenerate ℓ-dimensional subspace that contains $|\Phi_0\rangle$, and it is said that $|\Phi_0\rangle$ transforms as an ℓ-multiplet.

Thus, a nondegenerate ground state $|\Phi_0\rangle$ must transform as a singlet under the operations of the symmetry group of the Hamiltonian. In such a case, *the phenomenon of spontaneous symmetry breaking is precluded, since it requires a change of ground-state symmetry.*

We thus conclude that the many-particle ground-state wavefunction must transform nontrivially under the symmetry group of the Hamiltonian that is to be broken!

Noninteracting Bosons Systems

We start by recalling that the field operators $\hat{\Psi}(\mathbf{x},t)$, $\hat{\Psi}^\dagger(\mathbf{x}',t)$ obey the commutator

$$\left[\hat{\Psi}(\mathbf{x},t),\,\hat{\Psi}^\dagger(\mathbf{x}',t)\right] = \delta\left(\mathbf{x}-\mathbf{x}'\right)$$

with the boson Fock space

$$\mathbb{F} = \bigoplus_{N=0}^{\infty} \mathbb{H}_+^{\otimes N} \quad \Rightarrow \quad \mathbb{H}_+^{\otimes N} = \bigotimes_{\text{symm}}^{N} \mathbb{H}^{(1)},$$

where $\mathbb{H}_+^{\otimes N}$ is the N-particle subspace spanned by the symmetrized many-particle states

$$|n_0,\ldots,n_{i-1},n_i,n_{i+1},\ldots\rangle = \prod_i^{\text{symm}} \frac{\left(a_i^\dagger\right)^{n_i}}{\sqrt{n_i!}}\,|0\rangle \quad \Rightarrow \quad \sum_i n_i = N. \qquad (17.33)$$

The second-quantized total number operator is

$$\widehat{\mathcal{N}} = \int d\mathbf{x}\,\hat{\Psi}^\dagger(\mathbf{x},t)\,\mathbb{I}\,\hat{\Psi}(\mathbf{x}',t) = \sum_n \hat{a}_n^\dagger\,\hat{a}_n.$$

It is explicitly time independent.

1. Global Gauge Symmetry

$U(1)$ Gauge Symmetry

$U(1)$ gauge action transforms each single-particle state as

$$U(\varphi) : |\psi_i\rangle \mapsto e^{i\varphi} |\psi_i\rangle$$

For a system of N indistinguishable particles, each single-particle state in the product state $|\Phi\rangle_N$ transforms under U with the same phase factor. This means that $U(1)$ action on $\mathbb{H}_+^{\otimes N}$ yields

$$U(\varphi) : |\Phi\rangle_N \mapsto e^{iN\varphi} |\Phi\rangle_N ,$$

which leaves the Hamiltonian invariant. For arbitrary particle number

$$U(\varphi) : |\Phi(0)\rangle \mapsto e^{i\varphi\hat{\mathcal{N}}} |\Phi\rangle = |\Phi(\varphi)\rangle , \tag{17.34}$$

$\hat{\mathcal{N}}$ is the generator of $U(1)$.

We consider the action of the infinitesimal operation $U(\delta\varphi)$:

$$U(\delta\varphi) |\Phi(0)\rangle = |\Phi(\delta\varphi)\rangle = \sum_{\ell=0}^{\infty} \frac{1}{\ell!} \frac{d^\ell |\Phi\rangle}{d\varphi^\ell}\Big|_{\varphi=0} \delta\varphi^\ell$$

$$= \exp\left[\delta\varphi \frac{d}{d\varphi} \right] |\Phi(0)\rangle$$

$$= \exp\left[i\delta\varphi \left(-i\frac{d}{d\varphi} \right) \right] |\Phi(0)\rangle . \tag{17.35}$$

Comparing with (17.34), we identify the action of the number operator as

$$\hat{\mathcal{N}} = -i\frac{d}{d\varphi}, \quad \varphi = \text{phase.} \tag{17.36}$$

Thus, we find that phase and particle number are conjugate operators:

$$\left[\hat{\varphi}, \hat{\mathcal{N}} \right] = -i \left(\varphi \frac{d}{d\varphi} - \frac{d}{d\varphi} \varphi \right)$$

$$= -i \left(\varphi \frac{d}{d\varphi} - \frac{d\varphi}{d\varphi} - \varphi \frac{d}{d\varphi} \right)$$

$$= i. \tag{17.37}$$

The number operator $\hat{\mathcal{N}}$ is the infinitesimal generator of global gauge transformations by which all states in $\bigotimes_{\text{symm}}^{N} \mathbb{H}^{(1)}$ are multiplied by the same phase factor. A *global $U(1)$ gauge* transformation on Fock space is implemented by

$$|n_0, \ldots, n_{i-1}, n_i, n_{i+1}, \ldots\rangle \quad \Rightarrow \quad e^{i\varphi\hat{\mathcal{N}}} |n_0, \ldots, n_{i-1}, n_i, n_{i+1}, \ldots\rangle \quad \forall \varphi \in \mathbb{R}$$

or the transformation $\quad \begin{cases} \hat{a}_n \rightarrow e^{i\varphi\hat{\mathcal{N}}} \hat{a}_n e^{-i\varphi\hat{\mathcal{N}}} = e^{-i\varphi} \hat{a}_n \\ \hat{a}_n^\dagger \rightarrow e^{i\varphi\hat{\mathcal{N}}} \hat{a}_n^\dagger e^{-i\varphi\hat{\mathcal{N}}} = e^{+i\varphi} \hat{a}_n^\dagger. \end{cases} \tag{17.38}$

Equation (17.38) tells us that annihilation operators carry particle number -1, while the creation operators carry particle number $+1$. As a consequence, we find that

$$e^{i\varphi\hat{N}}\,\hat{\Psi}(\mathbf{x},t)\,e^{-i\varphi\hat{N}} = e^{-i\varphi}\,\hat{\Psi}(\mathbf{x},t)$$

$$e^{i\varphi\hat{N}}\,\hat{\Psi}^{\dagger}(\mathbf{x},t)\,e^{-i\varphi\hat{N}} = e^{+i\varphi}\,\hat{\Psi}^{\dagger}(\mathbf{x},t). \tag{17.39}$$

Particle Number Conservation

For a Hamiltonian invariant under $U(1)$,

$$\mathcal{H} = e^{i\hat{N}\delta\varphi}\,\mathcal{H}\,e^{-i\hat{N}\delta\varphi}$$

$$= \left(1 + i\delta\varphi\hat{N}\right)\mathcal{H}\left(1 - i\delta\varphi\hat{N}\right) = \mathcal{H} + i\varphi\left[\hat{N}, \mathcal{H}\right]$$

$$\left[\hat{N}, \mathcal{H}\right] = 0. \tag{17.40}$$

Consequently, Heisenberg equation of motion reads

$$i\hbar\frac{d\hat{N}}{dt} = \left[\hat{N}, \mathcal{H}\right] = 0, \qquad N \text{ conserved.}$$

Phase and Particle Number Uncertainty

Since $\hat{\varphi}$ and $\hat{N}$ do not commute, they cannot be measured simultaneously. We have the uncertainty relation

$$\delta\hat{\varphi}\,\delta\hat{N} \leq \frac{1}{2}. \tag{17.41}$$

Thus, the commutator between the total number operator $\hat{N}$ and the single-particle Hamiltonian $\mathcal{H}$ vanishes

$$[\mathcal{H}, \hat{N}] = 0$$

for a system invariant under global gauge transformation, which confirms the stipulation made earlier, that this symmetry is responsible for total particle number conservation. This means that the system becomes restricted to the Fock subspace $\bigotimes_{\text{symm}}^{N}\mathbb{H}^{(1)}$. As a result, the normalized ground-state $|\Phi_0\rangle$ of the many-particle Hamiltonian becomes confined to the $\bigotimes_{\text{symm}}^{N}\mathbb{H}^{(1)}$ space, which guarantees that it is nondegenerate, since it is derived from the corresponding single-particle ground state. Consequently, we infer that spontaneous symmetry breaking can never take place under such constraint. By extension, it is believed that spontaneous symmetry breaking is always ruled out for an interacting Hamiltonian defined on the Hilbert space $\bigotimes_{\text{symm}}^{N}\mathbb{H}^{(1)}$.

Next we shall demonstrate how the *expectation value of $\hat{\Psi}(\mathbf{x},t)$ in the ground-state $|\Phi_0\rangle$ of the many-body system can be used as a signature of spontaneous symmetry breaking of the $U(1)$ symmetry.* This allows us to interpret the quantum statistical average of $\hat{\Psi}(\mathbf{x},t)$ as a temperature-dependent order parameter.

As shown previously, the quantum field $\hat{\Psi}(\mathbf{x},t)$ transforms according to

$$e^{i\varphi\hat{N}}\,\hat{\Psi}(\mathbf{x},t)\,e^{-i\varphi\hat{N}} = e^{-i\varphi}\,\hat{\Psi}(\mathbf{x},t) \qquad \forall\,\mathbf{x}, t$$

under any global gauge transformation. Since the ground-state $|\Phi_0\rangle$ of $\mathcal{H}$ confined to $\bigotimes_{\text{symm}}^{N} \mathbb{H}^{(1)}$ is unique, it transforms like a singlet under $U(1)$:

$$e^{i\varphi\hat{N}} |\Phi_0\rangle = e^{+i\varphi N_0} |\Phi_0\rangle, \qquad \langle\Phi_0| e^{-i\varphi\hat{N}} = \langle\Phi_0| e^{-i\varphi N_0}.$$

Using these two transformations, we find

$$\left\langle \Phi_0 \left| \left[e^{i\varphi\hat{N}} \hat{\Psi}(\mathbf{x},t) e^{-i\varphi\hat{N}} \right] \right| \Phi_0 \right\rangle = e^{-i\varphi} \langle\Phi_0| \hat{\Psi}(\mathbf{x},t) |\Phi_0\rangle$$

$$\left(\left\langle \Phi_0 \left| e^{i\varphi\hat{N}} \right) \hat{\Psi}(\mathbf{x},t) \left(e^{-i\varphi\hat{N}} \right| \Phi_0 \right\rangle \right) = \langle\Phi_0| \hat{\Psi}(\mathbf{x},t) |\Phi_0\rangle,$$

which implies that $\langle\Phi_0| \hat{\Psi}(\mathbf{x},t) |\Phi_0\rangle = 0$. Intuitively, the action of $\hat{\Psi}(\mathbf{x},t)$ on an eigenstate of $\hat{N}$ such as $|\Phi_0\rangle$ is to lower the total number of particle by one, thereby producing a state orthogonal to $|\Phi_0\rangle$.

We conclude that spontaneous symmetry breaking requires quantum degeneracy of the ground state with *orthogonal* ground states that are related by the action of the $U(1)$ symmetry group.

We shall explore how SSB can be implemented through the construction of a ground-state $|\psi\rangle \in \mathbb{F}$ that is *an eigenstate of $\hat{\Psi}(\mathbf{x},t)$ and thus cannot be an eigenstate of $\hat{N}$*.

2. The Thermodynamic Limit and Coherent States

In order to support operations on the Fock space $\mathbb{F}$ with an unrestricted boson number, instead of $\bigotimes_{\text{symm}}^{N} \mathbb{H}^{(1)}$ with a fixed boson number, we have to invoke the thermodynamic limit, namely,

$$\Omega, N \rightarrow \infty, \quad \frac{N}{\Omega} = n.$$

The thermodynamic limit implies the application of the *grand canonical ensemble*.

To construct the sought after ground state in Fock space, we consider a noninteracting system, with single-particle Hamiltonian

$$\mathcal{H}_\mu = \mathcal{H} - \mu\hat{N} \quad \Rightarrow \quad \mathcal{H}_\mu |\psi_n\rangle = (\varepsilon_n - \mu) |\psi_n\rangle, \qquad |\psi_n\rangle = \frac{1}{\sqrt{\Omega}} e^{i\mathbf{k}_n \cdot \mathbf{x}}.$$

Recalling that in the thermodynamic derivation of the BEC the chemical potential acquired the value of the single-particle ground-state energy at the onset of the transition, we set the chemical potential $\mu = \varepsilon_0$, in order to ensure that the single-particle ground-state energy of $\mathcal{H}_\mu$ vanishes, and that the corresponding normalized eigenfunction becomes $\psi_0(\mathbf{x}) = 1/\sqrt{\Omega}$. This choice also guarantees that the states

$$|n_0, 0, \ldots\rangle = \frac{\left(b_0^\dagger\right)^{n_0}}{\sqrt{n_0!}} |\varnothing\rangle, \qquad n_0 = 0, 1, 2, \ldots, \rightarrow \infty \tag{17.42}$$

are orthogonal and degenerate eigenstates of $\mathcal{H}_\mu$ in $\mathbb{F}$. Furthermore, this choice of μ guarantees that $\mathcal{H}_\mu$ has countably many orthogonal ground states provided the volume Ω is finite. The degeneracy implies that any linear combination of states of the form (17.42) is a ground state of $\mathcal{H}_\mu$ with $\mu = \varepsilon_0$. We shall focus on the continuous family of normalized bosonic coherent ground states defined in terms of a complex parameter α,

$$|\alpha\rangle_{gs} = e^{-\Omega|\alpha|^2/2} \sum_{n_0=0}^{\infty} \frac{\left(\sqrt{\Omega}\,\alpha\right)^{n_0}}{\sqrt{n_0!}} |n_0, 0, \ldots\rangle = e^{-\Omega|\alpha|^2/2} e^{\sqrt{\Omega}\,\alpha\,\hat{b}_0^\dagger} |\varnothing\rangle$$

$$= e^{-\Omega|\alpha|^2/2} e^{\sqrt{\Omega}\,\alpha\,\hat{b}_0^\dagger} e^{-\sqrt{\Omega}\,\alpha^*\,\hat{b}_0} |\varnothing\rangle = e^{-\Omega|\alpha|^2/2} e^{\sqrt{\Omega}\left(\alpha\,\hat{b}_0^\dagger - \alpha^*\,\hat{b}_0\right)} |\varnothing\rangle$$

$$= \hat{D}(\sqrt{\Omega}\alpha, 0, \ldots) |\varnothing\rangle, \tag{17.43}$$

where we used $e^{\hat{A}} e^{\hat{B}} = e^{[\hat{A},\hat{B}]/2} e^{\hat{A}+\hat{B}}$. The unitary 99 operator $\hat{D}(\sqrt{\Omega}\alpha, 0, \ldots)$ rotates the vacuum into the boson coherent state

$$\left|\sqrt{\Omega}\alpha, 0, \ldots\right\rangle_{cs} = e^{-\Omega|\alpha|^2/2} e^{\sqrt{\Omega}\,\alpha\,\hat{b}_0^\dagger} |\varnothing\rangle. \tag{17.44}$$

As we showed in Chapter 13, the bosonic coherent states form an overcomplete set of the Fock space, with the overlaps

$$_{cs}\langle \alpha | \varnothing \rangle = e^{-\Omega|\alpha|^2/2}, \quad _{cs}\langle \alpha | \alpha' \rangle_{cs} = e^{-\Omega|\alpha-\alpha'|^2/2}.$$

However, the $\sqrt{\Omega}$ scaling guarantees that all the rotated vacua in (17.44) become orthogonal in the thermodynamic limit. Moreover, we note that the degeneracy in the thermodynamic limit becomes uncountably infinite.

With this construction being established, we need to confirm that each ground-state $|\psi\rangle_{gs}$ in (17.43) is an eigenstate of the quantum fields $\hat{\Psi}(\mathbf{x}, t)$, but not an eigenstate of $\hat{N}$

$$\hat{\Psi}(\mathbf{x}) |\alpha\rangle_{gs} = e^{-\Omega|\alpha|^2/2} \frac{\hat{b}_0}{\sqrt{\Omega}} e^{\sqrt{\Omega}\,\alpha\,\hat{b}_0^\dagger} |\varnothing\rangle = \frac{e^{-\Omega|\alpha|^2/2}}{\sqrt{\Omega}} \left[\hat{b}_0, e^{\sqrt{\Omega}\,\alpha\,\hat{b}_0^\dagger}\right] |\varnothing\rangle$$

$$= \frac{e^{-\Omega|\alpha|^2/2}}{\sqrt{\Omega}} \sum_{n=0}^{\infty} \frac{\left(\sqrt{\Omega}\alpha\right)^n}{n!} \left[\hat{b}_0, \left(\hat{b}_0^\dagger\right)^n\right] |\varnothing\rangle$$

$$= \frac{e^{-\Omega|\alpha|^2/2}}{\sqrt{\Omega}} \sum_{n=1}^{\infty} \frac{n\left(\sqrt{\Omega}\alpha\right)^n}{n!} \left(\hat{b}_0^\dagger\right)^{n-1} |\varnothing\rangle = \alpha |\alpha\rangle_{gs}$$

and

$$e^{-i\varphi\hat{N}} |\alpha\rangle_{gs} = e^{-i\varphi\,\hat{b}_0^\dagger\hat{b}_0} e^{-\Omega|\alpha|^2/2} \sum_{n_0=0}^{\infty} \frac{\left(\sqrt{\Omega}\,\alpha\right)^{n_0}}{\sqrt{n_0!}} |n_0, 0, \ldots\rangle$$

$$= e^{-\Omega|\alpha|^2/2} \sum_{n_0=0}^{\infty} \frac{\left(\sqrt{\Omega}\,\alpha\,e^{-i\varphi}\right)^{n_0}}{\sqrt{n_0!}} |n_0, 0, \ldots\rangle$$

$$= e^{\sqrt{\Omega}\exp(-i\varphi)\,\alpha\,\hat{b}_0^\dagger} |0\rangle = \left|e^{-i\varphi}\,\alpha\right\rangle_{gs},$$

where we used

$$
a\left(a^\dagger\right)^n = aa^\dagger \left(a^\dagger\right)^{n-1} = \left(1 + a^\dagger a\right)\left(a^\dagger\right)^{n-1}
$$
$$
= \left(a^\dagger\right)^{n-1} + a^\dagger a \left(a^\dagger\right)^{n-1} = n\left(a^\dagger\right)^{n-1} + \left(b^\dagger\right)^n a.
$$

Thus, we find that

$$
\hat{\Psi}(\mathbf{x},t)\,|\alpha\rangle_{\text{gs}} = \alpha\,|\alpha\rangle_{\text{gs}}, \qquad e^{-i\varphi\hat{N}}\,|\alpha\rangle_{\text{gs}} = \left|e^{-i\varphi}\,\alpha\right\rangle_{\text{gs}}.
$$

Normalization of the single-particle eigenfunction $\psi_0(\mathbf{x}) = 1/\sqrt{\Omega}$ and the property that coherent states are eigenstates of annihilation operators guarantee that the quantum field $\hat{\Psi}(\mathbf{x},t)$ acquires the expectation value α with the particle density $|\alpha|^2$ in the ground-state manifold

$$
_{\text{gs}}\langle\psi|\,\hat{\Psi}(\mathbf{x},t)\,|\psi\rangle_{\text{gs}} = \alpha, \qquad _{\text{gs}}\langle\psi|\,\hat{\Psi}^\dagger(\mathbf{x},t)\,\hat{\Psi}(\mathbf{x},t)\,|\psi\rangle_{\text{gs}} = |\alpha|^2. \tag{17.45}
$$

Systems of Interacting Bosons

We note that in the presence of *boson pair interaction*, the ground-state energy will depend on N. Thus, the ploy of setting the chemical potential $\mu = \varepsilon_0$ in order to obtain unrestricted many-body ground state breaks down. We then demand instead that the chemical potential be constrained to fix the particle density at

$$
\langle\Phi_0|\,\hat{\Psi}^\dagger(\mathbf{x},t)\,\hat{\Psi}(\mathbf{x},t)\,|\Phi_0\rangle = \frac{N}{\Omega}
$$

in the thermodynamic limit at zero temperature. At finite temperatures, the right-hand side is unchanged whereas the left-hand side becomes a statistical average in the grand canonical ensemble. Since the modulus $|\alpha|^2 = N/\Omega$ is now predetermined, a degenerate manifold of ground states satisfying (17.45) is not anymore parameterized by α but by $\varphi = \arg(\alpha) \in [0, 2\pi]$.

> The mechanics of SSB can be still defined in terms of nonsinglet ground states, which defines a nonzero $\langle\Phi_0|\,\hat{\Psi}(\mathbf{x},t)\,|\Phi_0\rangle$ as an order parameter.

1. Population Fluctuations and Phase Locking

The population and phase of the coherent state have finite fluctuations

$$
\left\langle\Delta\hat{N}^2\right\rangle = \left\langle\hat{N}\right\rangle = |\alpha|^2
$$
$$
\left\langle\Delta\hat{\varphi}^2\right\rangle = \frac{1}{4\left\langle\hat{N}\right\rangle} = \frac{1}{4|\alpha|^2} \tag{17.46}
$$

according to the uncertainty (17.41). Thus, the phase is stabilized, $\left\langle\Delta\hat{\varphi}^2\right\rangle \ll 1$, only at a cost of the increased population fluctuation $\left\langle\Delta\hat{N}^2\right\rangle \gg 1$.

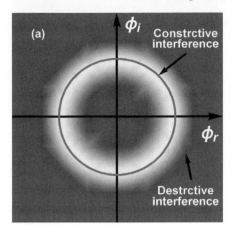
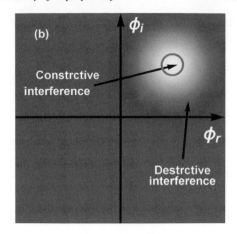

Figure 17.9 Coherent states space defined by the complex parameter $\alpha = \phi_r + i\phi_i$. (a) A particle number eigenstate $|N\rangle_0$ constructed by the coherent superposition of coherent states. (b) A coherent state $|\varphi\rangle_0$ constructed by the coherent superposition of particle number eigenstates.

We compare such a coherent state with the particle number eigenstate

$$|N\rangle = \int_0^{2\pi} d\varphi \, e^{iN\varphi} \, |\alpha(\varphi)\rangle . \tag{17.47}$$

Equation (17.47) shows that the particle number state $|N\rangle$ is constructed as a coherent superposition of coherent states with different eigenvalues as shown in Figure 17.9(a). The constructive and destructive interferences result in the fixed particle number, but the phase is completely spread out. Similarly, the coherent state is expanded by the coherent superposition of the particle number eigenstates

$$|\varphi\rangle = \sum_N e^{-|\alpha|^2/2} \frac{\alpha^N}{\sqrt{N!}} \, |N\rangle . \tag{17.48}$$

As shown in Figure 17.9(b), the constructive and destructive interferences among different particle number eigenstates result in the stabilized phase but the finite particle number noise. The interaction energy among the condensate is the same for the two states, (17.47) and (17.48), since both states have identical average particle number $\langle \widehat{N} \rangle = N$.

2. Bogoliubov Quantum Depletion and Phase Stabilization

We now consider the Bogoliubov quantum depletion term of the Hamiltonian

$$\mathcal{H}_B = \frac{1}{2\Omega} \sum_\mathbf{q} V_\mathbf{q} \, b_0^\dagger b_0^\dagger b_\mathbf{q} b_{-\mathbf{q}} + \text{H.C.},$$

which allows virtual excitations of two particles out of the condensate. In order to take such quantum depletion into account, we will consider a variational state

$$|\psi_0\rangle = e^{\alpha b_0^\dagger + \sum_\mathbf{q} \lambda_\mathbf{q} b_\mathbf{q}^\dagger b_{-\mathbf{q}}^\dagger} \, |\varnothing\rangle = |\alpha\rangle \otimes \sum_\mathbf{q} \left[|0\rangle_\mathbf{q} |0\rangle_{-\mathbf{q}} + \lambda_\mathbf{q} |1\rangle_\mathbf{q} |1\rangle_{-\mathbf{q}} + \cdots \right],$$

where the variational parameter $\lambda_{\mathbf{q}}$ is determined through minimization of the interaction energy.

The Bogoliubov interaction energy is given by

$$E_B = \langle \psi_0 | \mathcal{H}_B | \psi_0 \rangle = \sum_{\mathbf{q}} \frac{V_{\mathbf{q}}}{2\Omega} \left(\alpha_0^{*2} \lambda_{\mathbf{q}} + \text{c.c.} \right), \tag{17.49}$$

where we used $\langle \alpha | b_0^{\dagger} = \langle \alpha | \alpha_0^*$ and $a_{\mathbf{q}} a_{-\mathbf{q}} |1\rangle_{\mathbf{q}} |1\rangle_{-\mathbf{q}} = |0\rangle_{\mathbf{q}} |0\rangle_{-\mathbf{q}}$. The higher-order terms such as $|2\rangle_{\mathbf{q}} |2\rangle_{-\mathbf{q}}$, $|3\rangle_{\mathbf{q}} |3\rangle_{-\mathbf{q}}$, ... are neglected. If we express the complex excitation amplitudes as $\alpha = |\alpha| e^{i\varphi_0}$ and $\lambda_{\mathbf{q}} = |\lambda_{\mathbf{q}}| e^{i\varphi_{\mathbf{q}}}$, (17.49) becomes

$$E_B = \sum_{\mathbf{q}} \frac{V_{\mathbf{q}}}{\Omega} |\alpha|^2 |\lambda_{\mathbf{q}}| \cos\left(2\varphi_0 - \varphi_{\mathbf{q}}\right). \tag{17.50}$$

The Bogoliubov interaction energy is minimal when $2\varphi_0 - \varphi_{\mathbf{q}} = \pi$. It is energetically favorable that the condensate has a well-defined phase and the excitations are phase-locked to the condensate with a 180° phase difference. The reduced energy is macroscopic, $\frac{V_{\mathbf{q}}}{\Omega} |\alpha|^2 |\lambda_{\mathbf{q}}| \sim g n_0 \sqrt{N_{\mathbf{q}}}$, where $n_0 = |\alpha|^2/\Omega$ and $N_{\mathbf{q}}$ is the average population of the excitation modes. In fact, the quantum depletion is bound to occur due to the Bogoliubov Hamiltonian. Then the phase stabilization of the condensate is preferred and the phase locking $2\varphi_0 - \varphi_{\mathbf{q}} = \pi$ is implemented simultaneously. However, this argument is incomplete, in the sense that it appears not to put a lower bound on the extent of depletion: (17.50) is negative and proportional to the $|\lambda_{\mathbf{q}}|$s, pointing to the possibility of continuous growth of the excitations, and a substantial quantum depletion of the condensate. Actually, it turns out that the π-phase difference between the condensate and the excitations guarantees this does not happen and the quantum depletion is kept at a minimum level that quantum mechanics allows.

Population fluctuations and phase stabilization of the condensate, which are caused by phase locking between the condensate and excitations, are genuine signatures of spontaneous symmetry breaking, which is distinct from the standard picture of Bose–Einstein condensation of a noninteracting ideal gas. The stabilized phase is responsible for superfluidity, whereby a superfluid current is generated by the gradient of the phase, as will be presented later in this chapter.

17.3.3 The Gross–Pitaevskii Equation

In second quantization, the interacting boson gas Hamiltonian takes the form

$$\mathcal{H} = \int d\mathbf{x} \, \Psi^{\dagger}(\mathbf{x}, t) \left[-\frac{\hbar^2}{2m} \nabla^2 + V_{\text{ext}}(\mathbf{x}) \right] \Psi(\mathbf{x}, t)$$
$$+ \frac{1}{2} \iint d\mathbf{x} \, d\mathbf{x}' \, V(\mathbf{x}' - \mathbf{x}) \, \Psi^{\dagger}(\mathbf{x}, t) \, \Psi^{\dagger}(\mathbf{x}', t) \, \Psi(\mathbf{x}', t) \, \Psi(\mathbf{x}, t),$$

where, for example, V_{ext} may represent the trapping potential. The quantum-mechanical field operator $\hat{\Psi}(\mathbf{x},t)$ satisfies the Heisenberg equation of motion

$$
\begin{aligned}
i\hbar \frac{\partial}{\partial t} \hat{\Psi}(\mathbf{x},t) &= \left[\hat{\Psi}(\mathbf{x},t), \mathcal{H} \right] \\
&= \left[-\frac{\hbar^2}{2m} \nabla^2 + V_{ext}(\mathbf{x},t) + \int d\mathbf{x}\, \hat{\Psi}^\dagger(\mathbf{x}',t)\, V(\mathbf{x}' - \mathbf{x})\, \hat{\Psi}(\mathbf{x}',t) \right] \hat{\Psi}(\mathbf{x},t).
\end{aligned}
\tag{17.51}
$$

We are interested in the $T < T_{BE}$ regime where the thermal wavelength is much larger than the interaction range; this allows us to set $V(\mathbf{x}' - \mathbf{x}) = g\,\delta(\mathbf{x}' - \mathbf{x})$, and obtain

$$
\mathcal{H} = \int d\mathbf{x} \left[\Psi^\dagger(\mathbf{x},t) \left(-\frac{\hbar^2}{2m} \nabla^2 + V_{ext}(\mathbf{x}) \right) \Psi(\mathbf{x},t) + \frac{g}{2} \left(\Psi^\dagger(\mathbf{x},t) \right)^2 (\Psi(\mathbf{x},t))^2 \right].
$$

Mean-Field Approximation

Next we apply the mean-field approximation by neglecting quantum fluctuations and correlations. More specifically, we replace $\hat{\Psi}(\mathbf{x},t)$ by $\Psi_0(\mathbf{x},t)$, and introduce the factorization ansatz

$$
\left\langle \Psi^\dagger\, \Psi\, \Psi \right\rangle = \left\langle \Psi^\dagger \right\rangle \langle \Psi \rangle \langle \Psi \rangle = \Psi_0^\dagger\, \Psi_0\, \Psi_0.
$$

We will drop the subscript 0 for compactness, and write

$$
i\hbar \frac{\partial}{\partial t} \Psi(\mathbf{x},t) = \left[-\frac{\hbar^2}{2m} \nabla^2 + V_{ext}(\mathbf{x},t) + g\,|\Psi(\mathbf{x},t)|^2 \right] \Psi(\mathbf{x},t),
\tag{17.52}
$$

which is referred to as the time-dependent Gross–Pitaevskii equation. It satisfies the continuity equation

$$
\frac{\partial n(\mathbf{x},t)}{\partial t} + \nabla\cdot\mathbf{j}(\mathbf{x},t) = 0,
$$

where

$$
\Psi(\mathbf{x},t) = |\Psi(\mathbf{x},t)|\, e^{i\varphi(\mathbf{x},t)}, \qquad n(\mathbf{x},t) = |\Psi(\mathbf{x},t)|^2
\tag{17.53}
$$

$$
\mathbf{j}(\mathbf{x},t) = \frac{\hbar}{m} \Im\left(\Psi^* \nabla\Psi \right) = \frac{\hbar}{m} n(\mathbf{x},t)\, \nabla\varphi(\mathbf{x},t).
\tag{17.54}
$$

This result means that the superfluid velocity $\mathbf{v}_s$ of the condensate is related to the gradient of the phase φ as $\mathbf{v}_s = (\hbar/m)\,\nabla\varphi(\mathbf{x},t)$.

The stationary solution of (17.52) has the form of

$$
\Psi(\mathbf{x},t) = \Phi(\mathbf{x})\, e^{-i\mu t/\hbar}.
$$

For simplicity, setting $V_{ext}(\mathbf{x},t) = 0$ we get

$$
\left[-\frac{\hbar^2}{2m} \nabla^2 - \mu + g\,|\Psi(\mathbf{x})|^2 \right] \Psi(\mathbf{x}) = 0.
\tag{17.55}
$$

The corresponding Hamiltonian is given by

$$\mathcal{H} = \int d\mathbf{x} \left[\frac{\hbar^2}{2m} \nabla \Psi^*(\mathbf{x}) \, \nabla \Psi(\mathbf{x}) - \mu \, |\Psi|^2 + \frac{g}{2} \, |\Psi(\mathbf{x})|^4 \right]. \qquad (17.56)$$

For a uniform gas, in the absence of the external potential, $\nabla \Psi = 0$, and we get $E = (g/2)n^2\Omega$. The chemical potential is given by

$$\mu = \frac{\partial E}{\partial N} = \frac{\partial E}{\partial n} \frac{\partial n}{\partial N} = gn.$$

Bogoliubov Treatment of Fluctuations

To account for the quantum fluctuations about the mean field, we decompose the field operator into a mean field and fluctuations components:

$$\Psi = \langle \Psi \rangle + \hat{\psi},$$

where the operator $\hat{\psi}$ represents the fluctuations, and both Ψ and $\hat{\psi}$ obey the boson commutation relations. We substitute this expansion in (17.56) and keep terms up to second order in $\hat{\psi}$ and $\hat{\psi}^\dagger$. The zeroth-order term is nothing but the energy function $\langle \mathcal{H} \rangle$, and the first-order terms vanish identically since Ψ satisfies the Gross–Pitaevskii equation. The second-order terms are

$$\int d\mathbf{x} \, \hat{\psi}^\dagger(\mathbf{x}) \left[-\frac{\hbar^2}{2m} + V_{\text{ext}}(\mathbf{x}) + g \, |\Psi(\mathbf{x})|^2 - \mu \right] \hat{\psi}(\mathbf{x})$$

$$+ \frac{g}{2} \int d\mathbf{x} \left[\Psi^2(\mathbf{x}) \, \hat{\psi}^\dagger(\mathbf{x}) \hat{\psi}^\dagger(\mathbf{x}) + \left(\Psi^*(\mathbf{x}) \right)^2 \hat{\psi}(\mathbf{x}) \, \hat{\psi}(\mathbf{x}) \right]$$

from which we can derive the equation of motion for $\hat{\psi}$.

17.3.4 The Ginzburg-Landau Free Energy

Identifying (17.56) as $\mathcal{H} - \mu N$ with

$$N = \int d\mathbf{x} \, \Psi^\dagger(\mathbf{x}) \, \Psi(\mathbf{x})$$

means that we are working in the grand canonical ensemble, where we define $\langle \mathcal{H} - \mu N \rangle$ as the Ginzburg–Landau (GL) free energy density F_{GL},

$$F_{GL}[\Psi, \nabla \Psi] = -\frac{\hbar^2}{2m} \, |\nabla \Psi|^2 + (V_{\text{ext}}(\mathbf{x}) - \mu) \, |\Psi(\mathbf{x})|^2 + \frac{g}{2} \, |\Psi(\mathbf{x})|^4 \qquad (17.57)$$

and $\Psi(\mathbf{x})$ becomes a macroscopic GL order-parameter field. A few remarks about the GL free energy are appropriate:

- The GL free energy is to be interpreted as the energy density of a condensate of bosons in which the "classicized" field operator $\Psi(\mathbf{x})$ behaves as a complex order parameter, with $|\Psi(\mathbf{x})|^2 = n_s$.

- We may interpret the term containing $|\nabla \Psi|^2 \propto \left\langle \nabla \hat{\psi}^{\dagger}(\mathbf{x}) \nabla \hat{\psi}(\mathbf{x}) \right\rangle$ as a kinetic energy, with the associated constant appropriately set to $\hbar^2/2m$.
- The last term is just the particle–particle interaction.
- The ratio $\dfrac{\hbar^2}{2m\mu} = \dfrac{\alpha}{\mu}$ has the dimension (length)^2.
- Linear static response and healing length for a 1D order parameter:

    The linearized GL free energy is

    $$-\alpha \, |\nabla \Psi|^2 - \mu \, |\Psi(\mathbf{x})|^2$$

    and we write the equation of motion with a delta-function perturbation as

    $$\alpha \, \nabla^2 \Psi + \mu \, \Psi = \delta(\mathbf{x}).$$

    Taking the Fourier transform, we obtain

    $$\Psi(\mathbf{q}) = \frac{1}{\alpha} \frac{1}{q^2 + (1/\xi)^2} \;\Rightarrow\; \Psi(\mathbf{x}) = \frac{\exp[-r/\xi]}{\alpha}, \tag{17.58}$$

    where $\xi = \sqrt{\alpha/\mu}$ is the healing length, or the decay length scale of the system's response. The healing length, in general, parameterizes the range over which superfluid order is affected by a local perturbation.
- Effective potential is as follows:

    $$V_{GL}(\mathbf{x}) = -\mu \, \Psi^* \Psi + \frac{g}{2} \left(\Psi^* \Psi \right)^2. \tag{17.59}$$

    Variation with respect to Ψ^* gives

    $$\Psi \left(-\mu + g \, \Psi^* \Psi \right) = 0.$$

    – For $\mu < 0$, Figure 17.10 (top-left), the potential has a single minimum at $\Psi = 0$, which means that no stable condensate amplitude exists.

    – For $\mu > 0$, Figure 17.10 (top-right and bottom) shows that the potential has the Mexican hat configuration. It has a circle of minima at $|\Psi| = \sqrt{\dfrac{\mu}{g}} = \eta$, and a superfluid density $n_s^{(0)} = \eta^2$. The solution to the stationary phase is continuously degenerate, so that each configuration $\Psi(\mathbf{x}) = \eta \, e^{i\varphi(\mathbf{x})}$, $\varphi = [0, 2\pi)$ is a solution. This raises the question of which of these configurations is the right one.

- Freezing out amplitude fluctuations, so that $\Psi(x) = \sqrt{n_s^{(0)}} \, e^{i\varphi(x)}$, yields

    $$\nabla \Psi = i \, (\nabla \varphi) \, \Psi$$
    $$|\nabla \Psi|^2 = n_s \, |\nabla \varphi|^2,$$

    which leads to a kinetic energy term associated with the phase *twist*:

    $$\frac{\hbar^2}{2m} n_s \, |\nabla \varphi|^2 = \frac{mn_s}{2} \left(\frac{\hbar}{m} \nabla \varphi \right)^2 = \frac{mn_s}{2} v_s^2. \tag{17.60}$$

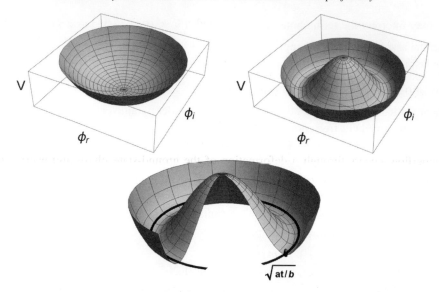

Figure 17.10   Effective Ginzburg–Landau potential for $\mu < 0$ (left) and $\mu > 0$ (right).

Since mn_s is the mass density, we see that a twist of the phase results in an increase in the kinetic energy that we may associate with a *superfluid* velocity:

$$\mathbf{v}_s = \frac{\hbar}{m} \nabla \varphi. \tag{17.61}$$

The equilibrium state of uniform phase is the state when no currents are present.

17.3.5 Superflow: Phase Symmetry, Gradient, and Rigidity

The GL free energy has global phase symmetry

$$\Psi(\mathbf{x}, t) \;\rightarrow\; \Psi(\mathbf{x}, t)\, e^{i \Delta \varphi}, \tag{17.62}$$

which is spontaneously broken with the choice of an arbitrary phase angle φ.

For $\Psi = |\Psi(\mathbf{x})|\, e^{i\varphi(\mathbf{x})}$, the gradient $\nabla = (\nabla |\Psi| + i \nabla \varphi\, |\Psi|)\, e^{i\varphi}$ leads to

$$F_{GL} = \frac{\hbar^2}{2m} |\Psi|^2 (\nabla \varphi)^2 + \left[\frac{\hbar^2}{2m} (\nabla |\Psi|)^2 + \mu |\Psi|^2 + \frac{g}{2} |\Psi|^4 \right]. \tag{17.63}$$

The term in square brackets describes the energy cost of variations in the magnitude of the order parameter. As we have shown, order-parameter amplitude fluctuations are confined to scales shorter than the healing length ξ. The first term actually describes *phase rigidity*. On longer length scales, the physics is entirely controlled by the phase degrees of freedom, so that

$$F_{GL} \sim \frac{\rho_\varphi}{2} (\nabla \varphi)^2 + \text{constant}. \tag{17.64}$$

The quantity $\rho_\varphi = \rho_s = \frac{\hbar^2}{m} n_s$ is often called the *superfluid phase stiffness*. From a microscopic point of view, the phase rigidity term is simply the kinetic energy of particles in the condensate, but from a macroscopic view, it is an elastic energy associated with the twisted phase. The only way to reconcile these two viewpoints is if a twist of the condensate wavefunction results in a coherent flow of particles.

Superfluid Vortices

Conventional particle flow is achieved by the addition of excitations above the ground state, but superflow occurs through a deformation of the ground-state phase and every single particle moves in perfect synchrony. The phase of the order parameter plays the role of a velocity potential and $\mathbf{v}_s$ is referred to as a velocity field. Because the superfluid velocity is the gradient of a scalar function, its rotation vanishes identically:

$$\mathbf{v}_s = \frac{\hbar}{m} \nabla\varphi \quad \Rightarrow \quad \text{Curl } \mathbf{v} = 0. \tag{17.65}$$

Thus, a superfluid is irrotational. This means that there can be no local rotational motion of the superfluid component. This is really a consequence of the quantization of angular momentum, as we will see more clearly in a moment. Yet, it is possible to have a finite hydrodynamic circulation around any loop that cannot shrink to nothing while remaining in the fluid, namely,

$$\kappa = \oint_C \mathbf{v}_s \cdot d\boldsymbol{\ell}.$$

For example, a circuit around a solid cylinder passing through the fluid, as shown in Figure 17.11. However, the circulation cannot take any value. If we substitute (17.65) into this integral, we obtain

$$\kappa = \frac{\hbar}{m} \oint_C \nabla\varphi \cdot d\boldsymbol{\ell} = n\,\frac{2\pi\hbar}{m}, \tag{17.66}$$

where n must be an integer in order to satisfy the condition that the condensate wavefunction be single valued. This means that the superfluid circulation must be quantized in units of $2\pi\hbar/m$. This circulation is macroscopically large, in the sense that it can be measured in a macroscopic mechanical experiment, and this fact provides the clearest evidence that superfluidity is indeed a "quantum mechanism on a macroscopic scale." It arises from the quantization of angular momentum, combined with the fact that all the particles in the condensate must have the same angular momentum. In the absence of any quantized circulation, there can be no local angular momentum, as we have seen in connection with (17.65).

Figure 17.11   A loop around which there can be a finite superfluid circulation.

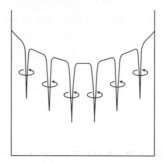

Figure 17.12 Vortex lines. A rotating superfluid becomes normal within narrow regions where vortex lines penetrate.

When a container of a superfluid is rotated, the surface of the superfluid shows a parabolic meniscus on the periphery, as in the case of a normal fluid. This shows that the value of the surface integral of vorticity, $\int \text{Curl } \mathbf{v}_s \cdot d\mathbf{S}$, is nonzero in apparent contradiction with (17.65). Onsager resolved this paradox by assuming that Curl $\mathbf{v}_s$ is nonzero only within microscopic regions where the liquid is not a superfluid. These singular regions are where vortex lines penetrate, as shown in Figure 17.12. The size of each vortex line is on the order of the healing length. Integration of $\mathbf{v}_s$ along a closed vortex contour gives

$$\oint \mathbf{v}_s \cdot d\boldsymbol{\ell} = \frac{\hbar}{m} \oint \nabla\varphi(\mathbf{x},t) \cdot d\boldsymbol{\ell}, \qquad (17.67)$$

reproducing the quantization of circulation.

As examples of superfluid quantized circulation units, we have

$$\kappa_0 = \frac{\hbar}{m} \simeq \begin{cases} 9.97 \times 10^{-4} \text{ cm}^2/\text{s for } ^4\text{He}, \\ 4.59 \times 10^{-5} \text{ cm}^2/\text{s for } ^{87}Rb. \end{cases}$$

For a container radius R and angular frequency of rotation ω, we get

$$\oint \mathbf{v}_s \cdot d\boldsymbol{\ell} = \omega R \times 2\pi R = 2\pi R^2 \omega = n\kappa_0.$$

The observed meniscus on the periphery can be explained if the vortex lines are distributed with density $n/(\pi R^2) = 2\omega/\kappa_0$. For the case of $\omega/2\pi = 100$ Hz, we have $2\omega/\kappa_0 \simeq 0.013/\mu\text{m}^2$ for ^{4}He and $0.3/\mu\text{m}^2$ for ^{87}Rb.

17.3.6 Time-Dependent Ginzburg–Landau Equation and the Goldstone Mode

In order to describe the dynamics of the order-parameter fields, we obtain a time-dependent GL free energy for the superfluid from the Gross–Pitaevskii equation

$$i\hbar \frac{d}{dt}\Psi = -\frac{\hbar^2}{2m}\nabla^2\Psi + \left(-\mu + g\,|\Psi|^2\right)\Psi \qquad (17.68)$$

ignoring the external potential. This allows us to write

$$F_{GL}\left[\Psi(\mathbf{x},t),\Psi^*(\mathbf{x},t)\right] = \frac{\hbar^2}{2m}\nabla\Psi^*(\mathbf{x},t)\cdot\nabla\Psi(\mathbf{x},t) - \mu\,\Psi^*(\mathbf{x},t)\Psi(\mathbf{x},t)$$
$$+\frac{g}{2}\,\Psi^*(\mathbf{x},t)\Psi^*(\mathbf{x},t)\Psi(\mathbf{x},t)\Psi(\mathbf{x},t)$$

and thus reproduce (17.68) as

$$i\hbar\frac{d}{dt}\,\Psi = \frac{\delta}{\delta\Psi^*}\,F_{GL}\left[\Psi(\mathbf{x},t),\Psi^*(\mathbf{x},t)\right] = -\frac{\hbar^2}{2m}\nabla^2\Psi - \mu\,\Psi + g\,|\Psi|^2\,\Psi \qquad (17.69)$$

and

$$-i\hbar\frac{d}{dt}\,\Psi^* = \frac{\delta}{\delta\Psi}\,F_{GL}\left[\Psi(\mathbf{x},t),\Psi^*(\mathbf{x},t)\right]. \qquad (17.70)$$

These are just *Hamilton's equations from classical mechanics*, where F_{GL} plays the role of the classical Hamiltonian, and $\Psi(\mathbf{x},t)$ and $\Psi^*(\mathbf{x},t)$ are canonically conjugate. Expressing the order parameter as

$$\Psi(\mathbf{x},t) = \sqrt{n_s(\mathbf{x},t)}\,e^{i\varphi(\mathbf{x},t)}$$

and substituting in (17.69) and (17.70), using the chain rule, we obtain

$$\hbar\frac{d}{dt}\,\varphi(\mathbf{x},t) = -\frac{\delta F_{GL}}{\delta n_s(\mathbf{x},t)} \qquad (17.71)$$

$$\hbar\frac{d}{dt}\,n_s(\mathbf{x},t) = +\frac{\delta F_{GL}}{\delta\varphi(\mathbf{x},t)}. \qquad (17.72)$$

Equations (17.71) and (17.72) are also Hamilton's equations of motion, and furthermore the phase φ and the density n_s are canonically conjugate, as expected.

Bogoliubov Meets GL

An important solution to (17.68) is given by

$$\Psi(\mathbf{x},t) = \Psi_0\,e^{i(\mathbf{k}\cdot\mathbf{x}-\omega t)} = \sqrt{\frac{\mu}{g}}\,e^{i(\mathbf{k}\cdot\mathbf{x}-\omega t)} \qquad (17.73)$$

and yields the dispersion relation

$$\hbar\omega = \frac{\hbar^2 k^2}{2m^*}.$$

This solution represents the superflow of the Bose–Einstein condensate. In order to see that it describes the superfluid, namely there is no viscosity, we study microscopic excitations above the condensate. Recall that the viscosity arises when the overall fluid flow induces energy/momentum transferred to microscopic excitations, namely heat. To conform with Landau's superfluid criterion, we would like to see that the bulk flow of Bose–Einstein cannot create such microscopic excitations below a critical velocity.

We consider fluctuation about the static expectation value $\Psi_0 = \sqrt{\mu/g}$, in both density and phase, namely,

$$\Psi = (\Psi_0 + \chi) \, e^{i\varphi},$$

where both χ and φ are real-valued fields. Substituting in (17.68),

$$i\hbar \, \dot{\chi} - \hbar \, \Psi_0 \, \dot{\varphi} + \frac{\hbar^2}{2m} \left(\nabla^2 \chi + i2 \, (\nabla \chi) \cdot \nabla \, \varphi + (\Psi_0 + \chi) \left(- (\nabla \varphi)^2 + i \nabla^2 \varphi \right) \right)$$
$$+ \mu \, (\Psi_0 + \chi) - g \, (\Psi_0 + \chi)^3 = 0. \tag{17.74}$$

Since we are interested in small fluctuations, we linearize the equation and obtain

$$i\hbar \, \dot{\chi} - \hbar \, \Psi_0 \, \dot{\varphi} + \frac{\hbar^2}{2m} \left(\nabla^2 \chi + i \Psi_0 \nabla^2 \varphi \right) - 2\mu \, \chi = 0. \tag{17.75}$$

Both the real and imaginary parts of this equation must both be satisfied, which leads to the two coupled equations

$$\hbar \, \dot{\chi} + \Psi_0 \, \frac{\hbar^2}{2m} \, \nabla^2 \varphi = 0 \tag{17.76}$$

$$\left(2\mu - \frac{\hbar^2}{2m} \nabla^2 \right) \chi + \hbar \, \Psi_0 \, \dot{\varphi} = 0. \tag{17.77}$$

Operating on (17.76) by $\left(2\mu - \frac{\hbar^2}{2m} \nabla^2 \right)$, and on (17.77) by $\hbar \frac{\partial}{\partial t}$, we obtain

$$-\hbar^2 \, \ddot{\varphi} + \left(2\mu - \frac{\hbar^2}{2m} \nabla^2 \right) \frac{\hbar^2}{2m} \nabla^2 \varphi = 0. \tag{17.78}$$

A plane-wave solution $\varphi \sim \exp \left[\frac{\varepsilon t - \mathbf{p} \cdot \mathbf{x}}{\hbar} \right]$ yields the dispersion relation

$$\varepsilon^2(\mathbf{k}) = \left(2\mu + \frac{p^2}{2m} \right) \frac{p^2}{2m}, \tag{17.79}$$

which is just that of the Bogoliubov model.

Figure 17.13 shows a measurement of the excitation spectrum in a trapped BEC. The dispersion shows good agreement with (17.79), indicating that a delta-function approximation to the interaction potential is a good one. In contrast, the excitation spectrum of liquid ^{4}He, shown in Figure 17.14, does not support a delta-function approximation to the interaction.

The Goldstone Mode and Zero Sound in a Neutral Superfluid

A continuous symmetry like (17.62) normally implies the existence of a gapless Goldstone mode: if we make phase gradients at arbitrarily large wavelengths, they will have arbitrarily small restoring forces leading to arbitrarily slow motions, hence the dispersion should satisfy $\omega(\mathbf{q}) \to 0$ as $\mathbf{q} \to 0$. Such a mode exists in a neutral superfluid. The time-dependent

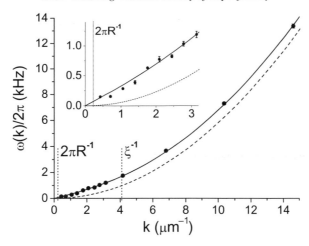

Figure 17.13 Excitation spectrum $\omega(k)$ of a trapped Bose–Einstein condensate. The solid line is the Bogoliubov spectrum in the local density approximation (LDA). The dashed line is the parabolic free-particle spectrum. For most points, the error bars are not visible on the scale of the figure. The inset shows the linear phonon regime. Taken from [171].

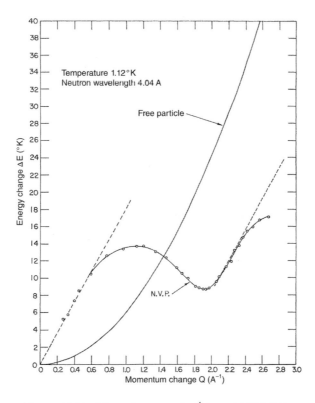

Figure 17.14 Dispersion curve for ^{4}He at 1.12°K [91].

GL equations are all the machinery needed to find the small oscillation modes, analogous to lattice waves, or spin waves. They consist of oscillations of phase and superfluid density (90° out of phase). We obtain from (17.70) and (17.71)

$$\hbar \frac{d}{dt} \delta\phi(\mathbf{x}, t) = -g\, \delta n_s(\mathbf{x}, t) \tag{17.80}$$

$$\hbar \frac{d}{dt} \delta n_s(\mathbf{x}, t) = +\frac{\hbar^2 n_0}{m^*} \nabla^2 \delta\phi(\mathbf{x}, t) \tag{17.81}$$

with $n_0 = |\Psi_0|^2$. (Actually, in (17.80), a $\nabla^2 \delta n_s$ term also appears on the right-hand side, but it is negligible in the long-wavelength limit.) It is because of the continuous phase symmetry that only gradient terms can appear on the right-hand side of (17.81). When we combine (17.80) and (17.81), we obtain

$$\frac{d^2}{dt^2} \delta n_s = v_0^2 \nabla^2 \delta n_s \tag{17.82}$$

$$v_0^2 = \frac{n_0 g}{m^*}.$$

Actually, we can identify (17.82) with the equation of ordinary sound waves in a fluid, by setting the bulk modulus $B = n_0^2 g = n_0^2 \partial^2 F_{GL}/\partial n_s^2$ and the density $\rho = m^* n_s$. So our gapless Goldstone mode can simply be identified with the zero-frequency mode of the superfluid phonon excitations.

Alternatively, we can consider amplitude and phase fluctuations in the order parameter, and we write

$$\Psi = (\Psi_0 + \delta\psi)\, e^{i\varphi(\mathbf{x})}$$

The GL functional becomes

$$F[\Psi] \simeq \int d\mathbf{x} \left[\frac{1}{2m^*} \left(\frac{\hbar}{i} (\nabla \delta\psi)\, e^{i\varphi(\mathbf{x})} + \hbar\, (\Psi_0 + \delta\psi)\, (\nabla\varphi)\, e^{i\varphi(\mathbf{x})} \right)^* \right.$$
$$\left. \times \left(\frac{\hbar}{i} (\nabla \delta\psi)\, e^{i\varphi(\mathbf{x})} + \hbar (\Psi_0 + \delta\psi)(\nabla\varphi)\, e^{i\varphi(\mathbf{x})} \right) - \mu\, (\Psi_0 + \delta\psi)^2 + \frac{g}{2} (\Psi_0 + \delta\psi)^4 \right].$$

Keeping terms up to second order in the fluctuations,

$$F[\Psi] \simeq \int d\mathbf{x} \left[\frac{1}{2m^*} \left\{ \left(\frac{\hbar}{i} (\nabla \delta\psi) \right)^* \cdot \left(\frac{\hbar}{i} (\nabla \delta\psi) \right) + \left(\frac{\hbar}{i} (\nabla \delta\psi) \right)^* \cdot \left(\hbar \Psi_0 \nabla\varphi \right) \right. \right.$$
$$\left. + \left(\frac{\hbar}{i} (\nabla \delta\psi) \right) \cdot \left(\hbar \Psi_0 \nabla\varphi \right)^* \right\} + \frac{1}{2m^*} \left(\hbar \Psi_0 \nabla\varphi \right) \cdot \left(\hbar \Psi_0 \nabla\varphi \right)^*$$
$$\left. - 2\mu\, \Psi_0 \delta\psi + 2g\, \Psi_0^3 \delta\psi - \mu\, \delta\psi^2 + 3g\, \Psi_0^2 \delta\psi^2 \right].$$

Note that the first-order terms cancel since we are expanding about a minimum of the potential. Furthermore, up to second order there are no terms mixing amplitude fluctuations $\delta\psi$ and phase fluctuations φ. We can simplify the expression

$$F[\Psi] \simeq \int dx \left[\left(\frac{\hbar}{i} (\nabla \delta \psi) \right)^* \cdot \left(\frac{\hbar}{i} (\nabla \delta \psi) \right) + \frac{1}{2m^*} \left(\hbar \Psi_0 \nabla \varphi \right) \cdot \left(\hbar \Psi_0 \nabla \varphi \right)^* + 2\mu \delta \psi^2 \right].$$

Taking a Fourier transform, we obtain

$$F[\psi] \simeq \sum_{\mathbf{k}} \left[\left(2\mu + \frac{\hbar^2 k^2}{2m^*} \right) \delta \psi_{\mathbf{k}}^* \, \delta \psi_{\mathbf{k}} + \frac{\mu}{g} \frac{\hbar^2 k^2}{2m^*} \varphi_{\mathbf{k}}^* \, \varphi_{\mathbf{k}} \right].$$

We see that amplitude fluctuations are gapped (massive) with an energy gap 2μ. They are not degenerate. In contrast, phase fluctuations are *ungapped* (*massless*) with quadratic dispersion:

$$\varepsilon_{\mathbf{k}}^\varphi = -\frac{\mu}{g} \frac{\hbar^2 k^2}{2m^*} = n_s \frac{\hbar^2 k^2}{2m^*}.$$

This ungapped mode is the Goldstone mode, characteristic of systems with spontaneously broken continuous symmetries.

Exercises

17.1 (a) Given the grand canonical partition function

$$Z = \mathrm{Tr} \left[e^{-\beta \left(\mathcal{H} - \mu \hat{N} \right)} \right],$$

show that for a general system of conserved particles at chemical potential μ, the total particle number in thermal equilibrium can be written as

$$N = -\frac{\partial F}{\partial \mu} \Rightarrow F = -k_B T \ln Z.$$

(b) Apply this to a single energy level, where

$$\mathcal{H} - \mu \hat{N} = (\varepsilon - \mu) \, a^\dagger a$$

and $a^\dagger$ creates either a Fermion, or a boson, to show that

$$F = \pm k_B T \ln \left[1 \mp e^{-\beta(\varepsilon - \mu)} \right]$$

$$\langle n \rangle = \frac{1}{e^{\beta(\varepsilon - \mu)} \mp 1},$$

where the upper sign refers to bosons, the lower to fermions. Sketch the occupancy as a function of ε for the case of fermions and bosons. Why does μ have to be negative for bosons?

17.2 Bose–Einstein condensates created inside optical atom traps contain alkali atoms at densities of about $10^{14} - 10^{15}$ cm^3.

(a) What is the Bose–Einstein transition temperature of a gas of sodium atoms at a density 10^{15} cm^3? (Give your answer in microkelvin.) How are such temperatures attained in practice?

(b) Liquid helium has a density of 122 g/liter at its boiling point. Compare its theoretical Bose–Einstein condensation temperature with its superfluid transition temperature (2.21 K). Why are the two numbers not the same?

17.3 A three-dimensional gas of noninteracting bosonic particles obeys the dispersion relation

$$\varepsilon(\mathbf{k}) = A \, |\mathbf{k}|^{1/2} .$$

(a) Obtain an expression for the density $n(T, \alpha)$, where $\alpha = e^{\mu/k_B T}$ is the fugacity. Simplify your expression as best you can, adimensionalizing any integral or infinite sum that may appear. You may find it convenient to define

$$\mathrm{Li}_\nu(\alpha) = \frac{1}{\Gamma(\nu)} \int_0^\infty dt \, \frac{t^{\nu-1}}{\alpha^{-1} e^t - 1} = \sum_\ell \frac{\alpha^\ell}{\ell^\nu} .$$

Note that $\mathrm{Li}_\nu(1) = \zeta(\nu)$, the Riemann zeta function.
(b) Find the critical temperature for Bose condensation, $T_{BE}(n)$, where n is the gas density. Your expression should only include the density n, the constant A, physical constants, and numerical factors (which may be expressed in terms of integrals or infinite sums).
(c) What is the condensate density n_0 when $T = T_{BE}/2$?

17.4 Consider a free Bose gas in which the particle spectrum is given by

$$\varepsilon(\mathbf{k}) = a \, |\mathbf{k}|^s .$$

Determine for which values of s the system undergoes a Bose–Einstein condensation in three, two, and four dimensions.

17.5 Consider a system of bosons created by the field $\Psi^\dagger(\mathbf{x})$, and having a pair interaction potential of the form

$$V(r) = \begin{cases} U & r < R \\ 0 & r > R. \end{cases}$$

(a) Write the interaction terms in second quantized form, as a function of field operators.
(b) Switch to the momentum basis, where

$$\Psi(\mathbf{x}) = \int \frac{d\mathbf{k}}{(2\pi)3} \, c_\mathbf{k} \, e^{i\mathbf{k}\cdot\mathbf{x}} .$$

Verify that

$$\left[c_\mathbf{k}, c_{\mathbf{k}'}^\dagger \right] = (2\pi)^3 \, \delta^{(3)} \left(\mathbf{k} - \mathbf{k}' \right)$$

and write the interaction in this basis. Sketch the form of the interaction in momentum space.

17.6 Start from the definition of the one-body density matrix for the case of an ideal gas in a 3D box,

$$n^{(1)}(\mathbf{x}, \mathbf{x}') = \sum_{\mathbf{k}} f_{\mathbf{k}} \, \varphi^*(\mathbf{x}) \, \varphi(\mathbf{x}'),$$

where the eigenfunctions of $\mathcal{H} = \frac{\mathbf{p}^2}{2m}$ are plane waves

$$\varphi(\mathbf{x}) = \frac{1}{\sqrt{\Omega}} \, e^{i\mathbf{k}\cdot\mathbf{x}}$$

and

$$f_{\mathbf{k}} = \left(e^{\beta(\varepsilon_{\mathbf{p}} - \mu)} - 1 \right)^{-1}.$$

(a) Show that $\varphi_i^*(\mathbf{x})$ are also the eigenfunctions of the one-body density matrix with the eigenvalues given by the average occupation number f_i, i.e.,

$$\int d\mathbf{x}' \, n^{(1)}(\mathbf{x}; \mathbf{x}') \, \varphi_i^*(\mathbf{x}') = f_i \, \varphi(\mathbf{x}).$$

(b) Derive an analytic form for $n^{(1)}(\mathbf{x}, \mathbf{x}')$, and show that it depends on $|\mathbf{x} - \mathbf{x}'| = s$.
(c) Show that the behavior of $n^{(1)}(s)$ at large distances ($s = |\mathbf{x} - \mathbf{x}'| \gg \lambda_T$) for $T < T_{BE}$ follows

$$n^{(1)}(s) \simeq n_0 + \frac{1}{\lambda_T^2 s},$$

where n_0 is the condensate density.
(d) Repeat the same calculation for $T > T_{BE}$ and show that now the decay to zero is of the type of Yukawa law

$$n^{(1)}(s) \simeq \frac{e^{\beta\mu}}{\lambda_T^2 s} \exp\left[-\sqrt{4\pi \left(1 - e^{\beta\mu}\right)} \, \frac{s}{\lambda_T} \right].$$

(e) Consider the case of a classical gas, where the distribution function is the Maxwell–Boltzmann one, $f_{\mathbf{k}} = e^{-\beta\varepsilon_{\mathbf{p}}}$, and evaluate the one-body density matrix $n^{(1)}(s)$. Show that if you consider the short-distance behavior $s \ll \lambda_T$ of the general quantum case, you get exactly the same result.

17.7 If a free bosonic particle Hamiltonian is given by

$$\mathcal{H} = \hbar\omega \, a^\dagger a,$$

what is the time dependence of the unitary evolution operator

$$U = \exp\left[-\frac{i}{\hbar} \mathcal{H}t \right]$$

for a particle number eigenstate $|n\rangle$ and coherent state $|\alpha\rangle$?
    Explain why the preceding two states have distinct time dependence.

17.8　(a) Show that for an arbitrary analytic operator function $f(\hat{A})$ it holds that

$$e^S f(\hat{A}) e^{-S} = f\left(e^S \hat{A} e^{-S}\right),$$

a result that is easily generalized to the case where instead of $\hat{A}$ we have several operators.

Hint: as a first step, consider the case where $f(\hat{A}) = \hat{A}^n$ with integer number n.

(b) As an application of the general result in part (a), show that for the Glauber displacement operator

$$D(\alpha) \equiv \exp\left(\alpha\, a^\dagger - \alpha^*\, a\right), \alpha \in \mathbb{C}$$

generating coherent states when acting on the vacuum $|\varnothing\rangle$ of the bosonic mode a satisfies the action

$$D^\dagger(\alpha)\, f\left(a, a^\dagger\right) D(\alpha) = f\left(a + \alpha, a^\dagger + \alpha^*\right),$$

where $f\left(a, a^\dagger\right)$ is an arbitrary analytic function.

17.9　Show that the action of the creation operator on the coherent state $|\alpha\rangle$ is

$$a^\dagger\, |\alpha\rangle = \frac{d}{d\alpha}\, |\alpha\rangle.$$

17.10　Weisskopf's rendition of a roton is shown in Figure 17.15.

Inelastic neutron scattering is used to measure the ^{4}He dispersion. Consider the case where a neutron scatters off a He atom in the liquid, where the He atom acquires a velocity $\mathbf{v}$. Surrounded by other atoms, the atom is forced to rotate about an adjacent atom.

(a) Calculate the kinetic energy in the center of mass (COM) frame of the two He atoms and compare it to that in the lab frame.

(b) Quantize the rotational energy and determine the moment of inertia in the COM frame. Assume an interatomic distance of d.

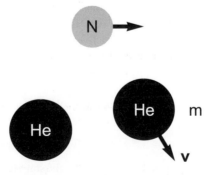

Figure 17.15　Development of a roton.

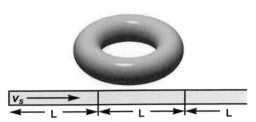

Figure 17.16 Circulation of a superfluid.

(c) Assume that the kinetic energy in the COM frame corresponds to the lowest angular momentum excitation. Relate this energy to the energy imparted by the neutron to the He atom. Denote the latter by Δ, and note that this is the energy loss recorded in the experiment.

(d) Determine the momentum transfer p_0 to the He atom in terms of Δ.

(e) Obtain numerical values for Δ and p_0 and compare them to those determined experimentally: $\Delta / k_B = 10\text{K}$ and $p_0/\hbar = 20\ \text{nm}^{-1}$.

Although Weisskopf's model is very simple, it is able to give order of magnitude agreement. The real situation is likely to be more complex.

17.11 Circulation, vortices, and Landau's two-fluid model:

Consider the case of a circulating flow of a superfluid in a torus that continues indefinitely, without dissipation (see Figure 17.16). We want to understand this, and the connection with Bose–Einstein condensation.

(a) Consider first flow states in the ideal Bose gas at zero temperature when all the particles are condensed into a single state. We model the torus as a length L with periodic boundary conditions. A flowing state is given by Bose condensing into the lowest single-particle state with nonvanishing momentum. Start with this case of an ideal "noninteracting" Bose gas, with all N particles in that state.

   (i) What is the single-particle ground state as set by the boundary conditions?
   (ii) Write the many-body wavefunction and its total momentum $\mathbf{P}$.
   (iii) Use the result for $\mathbf{P}$ to derive an expression for the total momentum density in terms of the superfluid mass density and superflow velocity.
   (iv) What is the superflow velocity? If we express the single particle wavefunction in the form $|\psi|\, e^{i\varphi}$, it allows the results to be generalized to more complicated flow fields that correspond to a $\psi(\mathbf{x})$ that is not simply a plane wave. Write down the expression for the superflow in this latter case.
   (v) Given the quantization of $\mathbf{k}$ you obtained in part (i), determine the *circulation*, namely, the line integral of the velocity around a closed loop:

$$\oint \mathbf{v}_s \cdot d\boldsymbol{\ell}.$$

Realizing that this arises because of the single-valuedness of the wavefunction, write down the general expression of this integral in terms of $\nabla \varphi$. You have just obtained the quantization condition for superflow vortices!

Onsager suggested the following Gedanken experiment in 1949 [147]:

Now we observe that a torus can be converted into a simply-connected space by shrinking the hole. If a circulating superfluid is subjected to such a deformation of its container, it must retain a quantized vortex in its interior.

(b) With this result in hand, consider the adiabatic continuity of turning on the interparticle pair interactions at zero temperature.

   (i)   Write down the interacting boson Hamiltonian in second-quantized form.
   (ii) Interactions may reduce the condensate density. Describe the scattering mechanism that may reduce the condensate density.
   (iii) By examining the Hamiltonian, what kind of momentum conservation do you infer?
   (iv) If the adiabatic turnon starts with the circulating superflow state described earlier for the noninteracting system, what does this imply for the superflow in the interacting system?

(c) Next, consider interactions with the containing walls. Assume a superflow momentum density

$$\mathbf{g} = \rho_s \, \mathbf{v}_s.$$

There are two ways for $\mathbf{g}$ to decay:

   (i) $\mathbf{v}_s$ may decrease. Discuss the conditions constraining this decay.
   (ii) ρ_s may decrease. Analyze this process in terms of Landau's superfluid argument involving Galilean transformation. (Consider the simple case of delta-function interactions.)

(d) At nonzero temperatures, thermal excitations may redistribute and change the total momentum. Consider the situation where the condensed state flows at the superfluid velocity $\mathbf{v}_s$, and the walls of the container are moving with velocity $\mathbf{v}_n$.

   (i) The walls provide a *momentum bath*. Equilibrium in contact with a momentum bath moving at velocity $\mathbf{v}$ is given by the Boltzmann factor $e^{-\beta (E - \mathbf{p} \cdot \mathbf{v})}$, where $\beta = 1/k_B T$.

        Determine the number of excitations, $n_B(\varepsilon(\mathbf{k})) = \left(e^{\beta \varepsilon} - 1\right)^{-1}$, with momentum $\hbar \mathbf{k}$ and excitation energy $\varepsilon(\mathbf{k}) + \hbar \mathbf{k} \cdot \mathbf{v}_s$ at temperature T.
   (ii) What is the total momentum $\mathbf{P}(T)$ of the flowing ground state plus the momentum of the excitations?
   (iii) For small $\mathbf{v}_s - \mathbf{v}_n$, write a Taylor expansion of the n_B term. What are the zeroth-order and first-order terms?

(iv) Show that the momentum density can now be written as

$$\mathbf{g}(T) = \frac{\mathbf{P}(T)}{V} = \rho \, \mathbf{v}_s + \rho_n \, (\mathbf{v}_n - \mathbf{v}_s)$$

$$= \rho_s \, \mathbf{v}_s + \rho_n \, \mathbf{v}_n,$$

where

$$\rho_n = \frac{1}{3V} \sum_{\mathbf{k}} \hbar^2 k^2 \left(-\frac{\partial n_B}{\partial \varepsilon(\mathbf{k})} \right)$$

and $\rho_s = \rho - \rho_n$. This definition corresponds to Landau's *two fluid model*.

17.12 Nature of ground state:

The Hamiltonian of the interacting Bose system is given by

$$\mathcal{H} = \sum_{\mathbf{k}} \varepsilon_{\mathbf{k}} \, a_{\mathbf{k}}^\dagger a_{\mathbf{k}} + \frac{1}{2} \sum_{\mathbf{k}, \mathbf{k}', \mathbf{q}} V_{\mathbf{q}} \, a_{\mathbf{k}-\mathbf{q}}^\dagger \, a_{\mathbf{k}'+\mathbf{q}}^\dagger \, a_{\mathbf{k}'} \, a_{\mathbf{k}}$$

$$= \mathcal{H}_0 + \mathcal{H}_{\text{int}}. \tag{17.83}$$

(a) Show that the unitary transformation

$$\mathcal{T}_\varphi = \exp\left[-i\varphi \sum_{\mathbf{k}} a_{\mathbf{k}}^\dagger a_{\mathbf{k}} \right] = \exp\left[-i\varphi \, \hat{N} \right],$$

where $\hat{N}$ is the total number operator, leads to

$$a_{\mathbf{k}} \quad \rightarrow \quad \mathcal{T}_\varphi^\dagger \, a_{\mathbf{k}} \, \mathcal{T}_\varphi = e^{-i\varphi} \, a_{\mathbf{k}}$$

$$a_{\mathbf{k}}^\dagger \quad \rightarrow \quad \mathcal{T}_\varphi^\dagger \, a_{\mathbf{k}}^\dagger \, \mathcal{T}_\varphi = e^{i\varphi} \, a_{\mathbf{k}}^\dagger$$

for all φ. Note that the continuous variable φ defines the continuous group $U(1)$. (Make use of one of the Baker–Housdorff formulas.)

(b) How does this transformation affect the Hamiltonian?

(c) What implications does it have on the total number of particles?

(d) Using your conclusions from the previous parts, describe the possible outcomes when $\mathcal{T}_\varphi$ acts on the normalized eigenfunctions of the system, $|\psi_n\rangle$. Compare the symmetries of the possible outcomes with that of the Hamiltonian.

(e) If the ground-state wavefunction $|\psi_0\rangle$ is nondegenerate, what is the expectation value $\langle \psi_0 | a_0 | \psi_0 \rangle$? Base your argument on the action of $\mathcal{T}_\varphi$ on $|\psi_0\rangle$ and on a_0.

(f) How about if there is a continuum of degenerate ground states?

(g) Based on your knowledge of the quantum harmonic oscillator, what type of wavefunction would you expect to yield a nonzero value of $\langle \psi_0 | a_0 | \psi_0 \rangle$?

17.13 Condensation in multiple single-particle states:

Consider the possibility that, for an interacting boson system, condensation may occur in more than one low-lying single-particle state. The interaction Hamiltonian takes the usual form

$$\mathcal{H}_I = \frac{1}{2V} \sum_{\mathbf{k},\mathbf{k}',\mathbf{q}} V(\mathbf{q}) \, \hat{a}^\dagger_{\mathbf{k}-\mathbf{q}} \, \hat{a}^\dagger_{\mathbf{k}'+\mathbf{q}} \, \hat{a}_{\mathbf{k}'} \, \hat{a}_{\mathbf{k}}.$$

(a) If all N particles are condensed in the lowest energy state $|\Phi_0\rangle$, calculate the interaction energy.

(b) Now suppose the condensate is fragmented into two single-particle states $|\Phi_0\rangle$ and $|\Phi_1\rangle$ with population N_0 and N_1, $N_0 + N_1 = N$, where the kinetic energies of both states are infinitesimally close. Construct the many-particle wavefunction and determine the interaction energy. Remember that the interaction energy involves all possible contractions of operators.

   Based on your findings, do you think that fragmentation is favored? Explain why.

17.14 Phase and density as conjugate operators:

Consider bosonic particles in a periodic box of volume Ω. The field operator can be expanded as

$$\hat{\Psi}(\mathbf{x}) = \frac{1}{\sqrt{\Omega}} \sum_{\mathbf{k}} e^{i\mathbf{k}\cdot\mathbf{x}} \, \hat{a}_{\mathbf{k}}.$$

To study condensation, we separate out the lowest energy state and write

$$\hat{\Psi}(\mathbf{x}) = \hat{A} + \frac{1}{\sqrt{\Omega}} \sum_{\mathbf{k}}' e^{i\mathbf{k}\cdot\mathbf{x}} \, \hat{a}_{\mathbf{k}},$$

where

$$\hat{A} = \frac{1}{\sqrt{\Omega}} \, \hat{a}_0, \quad \text{and} \quad \left[\hat{A}, \hat{A}^\dagger \right] = \frac{1}{\Omega}.$$

(In the limit of large N, the operators $\hat{A}$ and $\hat{A}^\dagger$ almost commute and can with a very good approximation be considered as classical, i.e., as c-numbers.)

   Suppose that we express $\hat{A}$ and $\hat{A}^\dagger$ in terms of the density and phase operators as

$$\hat{A} = e^{i\hat{\varphi}} \sqrt{\hat{n}}$$
$$\hat{A}^\dagger = \sqrt{\hat{n}} \, e^{-i\hat{\varphi}}.$$

(a) Use the commutator of $\hat{A}$ and $\hat{A}^\dagger$ together with one of the Baker–Housdorff identities to determine the commutator

$$\left[\hat{\varphi}, \hat{n} \right].$$

(b) Show that

$$e^{i\hat{\varphi}} \, \hat{n} = \left(\hat{n} + \frac{1}{\Omega} \right) e^{i\hat{\varphi}}.$$

(c) Determine the commutator

$$\left[e^{ia\hat{p}/\hbar}, x\right],$$

where $\hat{x}$ and $\hat{p}$ are conjugate position and momentum, respectively. What does this commutator imply? Based on your answer, how would you describe the relation in part (b)?

17.15 Coherent states:

(a) Show that the coherent state

$$|\alpha\rangle = e^{\sqrt{\Omega}\alpha\, a^{\dagger}} |\varnothing\rangle$$

is an eigenstate of the annihilation operator a.

(b) Derive an expression for the action of the creation operator $a^{\dagger}$ on the coherent state.

(c) Determine the overlap of two coherent states $|\alpha\rangle$ and $|\alpha'\rangle$ and thus normalize the preceding coherent state.

(d) Determine the overlap after normalization.

(e) Show that the probability of being in a state with n particles is a Poisson distribution:

$$p(n) = \frac{\lambda^{n}}{n!} e^{-|\lambda|}, \qquad \lambda = |\alpha|^{2}.$$

17.16 Quantum and thermal depletion:
We discussed quantum depletion in the context of phase stabilization and phase locking. In this problem, we are going to explore obtaining some quantitative estimates for both quantum and thermal depletions in a weakly interacting boson gas.

(a) Calculate the average number of quasiparticles $N_{\mathbf{k}}^{\text{qp}} = \left\langle \alpha_{\mathbf{k}}^{\dagger} \alpha_{\mathbf{k}} \right\rangle$ with momentum $\mathbf{k}$. Recall that it must obey the Bose–Einstein distribution with the chemical potential $\mu = 0$.

(b) Express the average number of real particles $N_{\mathbf{k}}^{\text{rp}} = \left\langle a_{\mathbf{k}}^{\dagger} a_{\mathbf{k}} \right\rangle$ in terms of $N_{\mathbf{k}}^{\text{qp}}$, $u_{\mathbf{k}}$ and $v_{\mathbf{k}}$. In the process, determine the number of particles in the condensate as a function of temperature.

(c) Can you separate the resultant depletion into quantum and thermal components? What is the depletion mechanism at $T = 0$?

18

Landau Fermi Liquid Theory

18.1 Introduction

In describing most metals and insulators, as for example in calculating band structures and other properties, one starts from a picture of noninteracting electrons. However, as we showed in Chapter 1, the Coulomb interaction energy among electrons is actually very large, and one might wonder why it is appropriate to assume that noninteracting electrons make a sensible starting point. Yet, we find that the noninteracting model of the Fermi gas reproduces many qualitative features of metallic behavior, such as a well-defined Fermi surface, a linear specific heat capacity, and a temperature-independent paramagnetic susceptibility – evidence of remarkably strong robustness against perturbation.

The underlying idea, first phrased in these terms by the great physicist Lev Landau, is that electrons in a real metal form a **Fermi liquid**, which bears the same relation to the Fermi gas of free electrons that a normal liquid bears to a normal gas: *the interactions in the liquid are much stronger than in the gas, but there is no change in symmetry or in the fundamental nature of the state.* In particular, the elementary excitations of the ground state bear the same quantum numbers as ordinary electrons. Landau's theory, which we will justify in this chapter, explains how these excitations can wind up as *electrons dressed by particle–hole pairs*, which renormalize the mass (enhancing it by a factor up to 103 in so-called heavy-fermion compounds) and some other properties but not the charge e and fermionic statistics. Such *Landau Fermi liquid behavior* appears in many contexts – in metals at low temperatures, in the core of neutron stars, and in liquid ^{3}He, and most recently, it has become possible to create Fermi liquids with tunable interactions in atom traps. We should note that our understanding of Landau Fermi liquids is intimately linked with the idea of adiabaticity, which was introduced earlier.

18.1.1 Preliminaries: Noninteracting Free-Particle Fermion Gas

As we have shown in Chapter 1, the noninteracting free-particle gas at $T = 0$ is characterized by

$$\text{Dispersion:} \quad \varepsilon_{\mathbf{k}}^{\text{free}} = \frac{\hbar^2 k^2}{2m}, \quad k_F = \left(3\pi^2 n\right)^{1/3}$$

$$\varepsilon_F = \left(3\pi^2\right)^{2/3} \frac{\hbar^2}{2m} n^{2/3}, \quad \text{Total energy:} = \frac{3}{5} N \varepsilon_F.$$

We can express these quantities in terms of r_s as

$$v_F = \frac{\hbar k_F}{m} = \frac{4.2}{r_s} \times 10^8 \text{ cm/sec}, \quad \varepsilon_F = \frac{50.1}{r_s^2} \text{ eV}, \quad T_F = \frac{58.2}{r_s^2} \times 10^4 \text{ K}.$$

Finite Temperature Properties

At finite temperatures, the Fermi–Dirac distribution is

$$f(\varepsilon) = \left(e^{\beta(\varepsilon - \mu)} + 1\right)^{-1},$$

and the total energy per unit volume can be expressed as

$$\mathcal{E} = \int_0^\infty d\varepsilon \, \varepsilon \, \mathcal{D}(\varepsilon) \, f(\varepsilon) = \frac{\sqrt{2m^3}}{\pi^2 \hbar^3} \int_0^\infty d\varepsilon \, \frac{\varepsilon^{3/2}}{e^{\beta(\varepsilon - \mu)} + 1},$$

where $\mathcal{D}(\varepsilon)$ is the density of states. The specific heat is given by

$$C_v = \frac{\partial \mathcal{E}}{\partial T} = \int_0^\infty d\varepsilon \, \varepsilon \, \mathcal{D}(\varepsilon) \, \frac{\partial f(\varepsilon)}{\partial T}$$

subject to the constraint

$$n = \int_0^\infty d\varepsilon \, \mathcal{D}(\varepsilon) \, f(\varepsilon),$$

which determines μ and its temperature dependence. Using the relation

$$\frac{dn}{dT} = 0 \implies \varepsilon_F \frac{dn}{dT} = 0 = \varepsilon_F \int_0^\infty d\varepsilon \, \mathcal{D}(\varepsilon) \, \frac{\partial f(\varepsilon)}{\partial T},$$

we write

$$C_v = \int_0^\infty d\varepsilon \, (\varepsilon - \varepsilon_F) \, \mathcal{D}(\varepsilon) \, \frac{\partial f(\varepsilon)}{\partial T}$$

$$= \int_0^\infty d\varepsilon \, (\varepsilon - \varepsilon_F) \, \mathcal{D}(\varepsilon) \, \frac{\beta \, e^{\beta(\varepsilon - \mu)}}{\left(e^{\beta(\varepsilon - \mu)} + 1\right)^2} \left[\frac{\varepsilon - \mu}{T} - \frac{\partial \mu}{\partial T} \right].$$

At finite but low temperatures, the Fermi function only changes in a regime $\pm k_B T$ around the Fermi energy. In metals, $E_F \sim 10$ eV, and $E_F/k_B \sim 10^5$ K, which is huge compared to room temperature. Thus, the change in internal energy, at a small but finite temperature, will involve electrons close to the Fermi level. Their excitation energy is $\sim k_B T$, so that the relative number of excited states is only $\mathcal{D}(E_F) k_B T$. In that range, we neglect $\partial \mu / \partial T \ll 1$ and set $\mu \sim \varepsilon_F$; we have

$$C_v \simeq \frac{1}{k_B T^2} \int_0^\infty d\varepsilon \, (\varepsilon - \varepsilon_F)^2 \, \mathcal{D}(\varepsilon) \, \frac{e^{\beta(\varepsilon - \varepsilon_F)}}{\left(e^{\beta(\varepsilon - \varepsilon_F)} + 1\right)^2}$$

$$= k_B^2 T \int_{-\beta \varepsilon_F}^\infty dx \, \mathcal{D}\left(\frac{x}{\beta} + \varepsilon_F\right) \frac{x^2 \, e^x}{(e^x + 1)^2}.$$

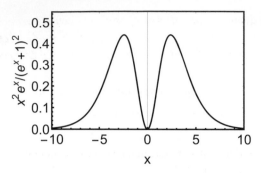

Figure 18.1 $\frac{x^2 e^x}{(e^x+1)^2}$ versus x.

But, as shown in Figure 18.1, $\frac{x^2 e^x}{(e^x+1)^2}$ is negligible outside the range $-10 < x > 10$, and for $T \ll T_F$ we can set $\mathcal{D}((x/\beta) + \varepsilon_F) \sim \mathcal{D}(\varepsilon_F)$. Thus,

$$C_v \sim k_B^2 T \, \mathcal{D}(\varepsilon_F) \int_{-\beta \varepsilon_F}^{\infty} dx \, \frac{x^2 e^x}{(e^x + 1)^2} \sim \frac{\pi^2}{3} \mathcal{D}(\varepsilon_F) \, k_B^2 T = \frac{\pi^2 k_B}{2} \, n \, \frac{T}{T_F}. \qquad (18.1)$$

Similarly, we can derive an expression for the magnetic susceptibility as follows. The magnetization is given by

$$M = g\mu_B \left(n_\uparrow - n_\downarrow \right) = g\mu_B \frac{n}{2} \int_0^{\varepsilon_F} d\varepsilon \left[\mathcal{D}(\varepsilon + g\mu_B B) - \mathcal{D}(\varepsilon - g\mu_B B) \right].$$

For small fields, $\mathcal{D}(\varepsilon \pm g\mu_B B) \sim \mathcal{D}(\varepsilon) \pm \frac{\partial \mathcal{D}}{\partial \varepsilon} g\mu_B B$, and we get

$$M = 2g^2 \mu_B^2 \frac{n}{2} B \int d\varepsilon \, \frac{\partial \mathcal{D}}{\partial \varepsilon} = 2g^2 \mu_B^2 \, n \, \mathcal{D}(\varepsilon_F) \, B,$$

and the susceptibility is

$$\chi = \frac{\partial M}{\partial B} = g^2 \mu_B^2 \, n \, \mathcal{D}(\varepsilon_F).$$

Since both the specific heat and the magnetic susceptibility are proportional to the density of states, the ratio of these two quantities $W = \chi/C_v$, often called the *Wilson ratio*, or *Stoner enhancement factor* in the context of ferromagnetism. It is set purely by the size of the magnetic moment:

$$W = 3 \left(\frac{g^2 \mu_B^2}{\pi k_B} \right)^2.$$

The measured Wilson ratio in ^{3}He is about 10 times larger than predicted by the noninteracting free-fermion model.

Quantitative discrepancies between measured physical properties of liquid ^{3}He and corresponding ideal Fermi gas predictions, motivated Landau to develop his theory. He used the adiabatic idea, in a brilliantly qualitative fashion, to formulate his theory of interacting Fermi liquids that produced agreement with experiment [115, 116]. Landau's revolutionary hypothesis that the adiabatic evolution of low-energy excited states occurs in a bulk system, and that by justifying this simple assumption we can predict a lot of physical properties is one of the major achievements of contemporary condensed matter theory.

18.2 Landau's Concept of Quasiparticles

The key concept underlying Fermi liquid theory is adiabacity: as the interactions are slowly turned on, it is assumed that the noninteracting states smoothly and continuously evolve into interacting states, and that this evolution takes place without encountering any singular behavior. Such a singularity would signal an instability of the ground state and should be viewed as a phase transition. Thus, if there are no phase transitions, there should be a smooth connection between noninteracting and interacting states. In particular, the quantum numbers used to label the noninteracting states should also be good quantum numbers in the presence of interactions – the low-energy excitations of an interacting Fermi system are in one-to-one correspondence to the excitations of a noninteracting Fermi gas. As was mentioned previously, the theory was originally developed for ^{3}He. One simplifying aspect of ^{3}He is the absence of an underlying crystalline lattice.

18.2.1 Particles and Holes

Because of the Pauli principle, the ground state of a Sommerfeld gas is the Fermi sphere. The quantum numbers of its excited states are the occupations $n_{\mathbf{k}\sigma}$ of single-particle states, which are characterized by momentum and spin: $|\mathbf{k}\sigma\rangle$.

We excite the system by promoting a certain number of fermionic particles across the Fermi surface yielding particles above and an equal number of holes below the Fermi surface. These "elementary excitations" are quantified by $\delta n_{\mathbf{k}} = n_{\mathbf{k}} - n_{\mathbf{k}}^{(0)}$,"

$$
\delta n_{\mathbf{k}} =
\begin{cases}
\delta_{\mathbf{k},\mathbf{k}'}, & \text{for particles} \quad |\mathbf{k}'| > k_F \\
-\delta_{\mathbf{k},\mathbf{k}'}, & \text{for holes} \quad\;\; |\mathbf{k}'| < k_F.
\end{cases}
\tag{18.2}
$$

For excitations induced by thermal fluctuations, $|\delta n| \sim 1$ only for excitation energies within $k_B T$ of ε_F, as shown in Figure 18.2. The energy of the noninteracting system can be defined as a functional of the occupation

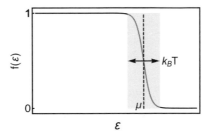

Figure 18.2 Fermi–Dirac distribution at temperature T.

$$E - E_0 = \sum_{\mathbf{k}} \frac{\hbar^2 k^2}{2m} \left(n_{\mathbf{k}} - n_{\mathbf{k}}^{(0)} \right) = \sum_{\mathbf{k}} \frac{\hbar^2 k^2}{2m} \delta n_{\mathbf{k}}. \tag{18.3}$$

When the system is placed in contact with a particle reservoir, the appropriate potential is the free energy, which for $T = 0$ is $F = E - \mu N$, and

$$F - F_0 = \sum_{\mathbf{k}} \left(\frac{\hbar^2 k^2}{2m} - \mu \right) \delta n_{\mathbf{k}}. \tag{18.4}$$

The free energy of a particle, with momentum $\hbar \mathbf{k}$ and $\delta n_{\mathbf{k'}} = \delta_{\mathbf{k}, \mathbf{k'}}$, is

$$\frac{\hbar^2 k^2}{2m} - \mu$$

and it corresponds to an excitation outside the Fermi sphere. The free energy of a hole $\delta n_{\mathbf{k'}} = -\delta_{\mathbf{k}, \mathbf{k'}}$, an excitation within the Fermi sphere, is

$$\mu - \frac{\hbar^2 k^2}{2m}.$$

Now, since $\mu \sim \hbar^2 k_F^2 / 2m$ at low temperatures, the free energy of a particle or hole at $|\mathbf{k}| = k_F$ is zero, and we can write the free energy of an excitation as

$$\left| \frac{\hbar^2 k^2}{2m} - \mu \right|. \tag{18.5}$$

Always a positive number that indicates the system is stable to excitations!

18.2.2 Quasiparticles and Quasiholes at $T = 0$

We consider a system of interacting electrons, and assume that this interaction grows adiabatically slowly, so that the system remains in the ground state as it evolves from a Sommerfeld gas and emerges into a fully interacting system. While the Fermi gas eigenstate is indexed by $n_{\mathbf{k}}^{(0)}$, the interacting system eigenstate will evolve quasistatically from $n_{\mathbf{k}}^{(0)}$ to $n_{\mathbf{k}}$. However, we note that for an isotropic system, $n_{\mathbf{k}}^{(0)} = n_{\mathbf{k}}$.

Now we consider the scenario where we add an electron of momentum $\mathbf{p} = \hbar \mathbf{k}$ to the initial Sommerfeld gas, and then slowly turn on the interaction (see Figure 18.3). As the interaction is switched on, the electrons in the vicinity of our injected one are slowly perturbed, and the injected particle becomes *dressed* by these interactions. Yet, since momentum is conserved, we have created an excitation (the electron plus its cloud) of momentum $\hbar \mathbf{k}$. We call this particle plus cloud a *quasiparticle*. This scenario carries over to the case of injecting a hole of momentum $\mathbf{p} = \hbar \mathbf{k}$ below the Fermi surface.

18.2.3 Quasiparticle Lifetime

Momentum and spin of excitations can only serve as quantum numbers if an excitation with $|\mathbf{k}\sigma\rangle$ does not immediately decay into other states. Since the eigenstate basis vectors of the noninteracting system, in general, become unstable and damp out after a certain time

 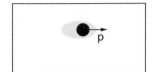 

Figure 18.3 We inject an electron of momentum $\hbar\mathbf{k}$ into a Sommerfeld gas, and slowly increase the interaction to its full value. As the combined system evolves, the electron becomes dressed by interactions with electrons in its vicinity (shown as a gray ellipse) which changes the effective mass but not the momentum of this single-particle excitation, designated a quasiparticle.

period τ in the real system, the adiabatic switching on of interaction should require a time much shorter than τ. However, if the interaction is turned on too fast, the final state is no longer an eigenstate of the system. It is therefore necessary that the time period τ should be large, which implies that the excited states have a long lifetime.

As we have demonstrated in Chapter 16, phase space constraints limit the excited states lifetime to $\tau \propto (\varepsilon - \varepsilon_F)^{-2}$ at $T = 0$. In other words, the Landau quasiparticles become *better and better defined* as one gets closer to the Fermi surface. This is a remarkable result since it confirms that we can view the system as having single-particle excitations that resemble those of the original fermions, but with renormalized parameters. Note that this does not mean that close to the Fermi surface the interactions are disappearing from the system. They are present and can be extremely strong and can affect the dynamics and other properties of the system. Moreover, the quasiparticle excitations are not eigenstates of the interacting system, since they have a finite lifetime!

At finite temperatures, T plays the role of energy, setting the characteristic spread of the thermal distribution about the Fermi surface. The Fermi liquid decay rate is then

$$\frac{1}{\tau} = a\ (\varepsilon - \varepsilon_F)^2 + b\ (k_B T)^2,$$

where a and b are constants. The larger of temperature or energy terms dominates. The kinematic constraints on the decay rate lead to a characteristic form for the electrical resistivity

$$\rho(T) \propto \frac{1}{\tau} \propto T^2.$$

This is often taken as an identifying signature of Fermi liquid behavior.

A quasiparticle is then a manifestation of the adiabatic evolution of the noninteracting fermion into an interacting environment. The conserved quantum numbers of this excitation, its spin and its "charge" and momentum, are unchanged, but Landau reasoned that its dynamical properties, for example the effective magnetic moment and mass of the quasiparticle, would be *renormalized* to new values μ^* and m^* respectively. These *renormalizations* of the quasiparticle mass and magnetic moment are elegantly accounted for in Landau Fermi liquid theory in terms of a small set of *Landau parameters* that characterize the interaction.

18.2.4 Quasiparticle Energy

As in the noninteracting system, excitations will be described by deviations of occupation numbers from their ground-state values $n_{\mathbf{k}}^{(0)}$

$$\delta n_{\mathbf{k}} = n_{\mathbf{k}} - n_{\mathbf{k}}^{(0)}$$

At low temperatures, $\delta n_{\mathbf{k}} \sim 1$ only for $k \sim k_F$, where the particles are sufficiently long lived. We should underscore here that only $\delta n_{\mathbf{k}}$ will be physically relevant, and not $n_{\mathbf{k}}^{(0)}$ or $n_{\mathbf{k}}$.

For the Sommerfeld gas,

$$E\left[\delta n_{\mathbf{k}}\right] - E_0 = \sum_{\mathbf{k}} \frac{\hbar^2 k^2}{2m} \delta n_{\mathbf{k}}. \tag{18.6}$$

For the interacting system, $E\left[\delta n_{\mathbf{k}}\right]$ becomes a much more complicated functional. If, however, $\delta n_{\mathbf{k}}$ is small (so that the system is close to its ground state), then we may expand:

$$E\left[\delta n_{\mathbf{k}}\right] = E_0 + \sum_{\mathbf{k}} \varepsilon_{\mathbf{k}} \, \delta n_{\mathbf{k}} + \mathcal{O}\left(\delta n_{\mathbf{k}}^2\right), \tag{18.7}$$

where $\varepsilon_{\mathbf{k}} = \delta E / \delta n_{\mathbf{k}}$. Note that $\varepsilon_{\mathbf{k}}$ is intensive (independent of the system volume). If $\delta n_{\mathbf{k}} = \delta_{\mathbf{k},\mathbf{k}'}$, then $E \sim E_0 + \varepsilon_{\mathbf{k}'}$ and the energy of the quasiparticle of momentum $\hbar \mathbf{k}'$ is $\varepsilon_{\mathbf{k}'}$.

Actually, we only need define $\varepsilon_{\mathbf{k}}$ near the Fermi surface, where $\delta n_{\mathbf{k}}$ is finite. So we may approximate

$$\varepsilon_{\mathbf{k}} \simeq \mu + \hbar \left(\mathbf{k} - \mathbf{k}_F\right) \cdot \frac{\nabla_{\mathbf{k}}}{\hbar} \varepsilon_{\mathbf{k}} \bigg|_{k=k_F},$$

where $\left(\nabla_{\mathbf{k}} \, \varepsilon_{\mathbf{k}}\right) / \hbar = \mathbf{v}_{\mathbf{k}}$ is the group velocity of the quasiparticle.

We may learn more about $\varepsilon_{\mathbf{k}}$ by employing the symmetries of our system. In the absence of a magnetic field, we have

$$\begin{cases} \varepsilon_{\mathbf{k},\sigma} = \varepsilon_{-\mathbf{k},-\sigma} & \text{under time reversal} \\ \varepsilon_{\mathbf{k},\sigma} = \varepsilon_{-\mathbf{k},\sigma} & \text{under inversion} \end{cases}$$

and we find that $\varepsilon_{\mathbf{k},\sigma} = \varepsilon_{-\mathbf{k},\sigma} = \varepsilon_{\mathbf{k},-\sigma}$, does not depend on σ. Furthermore, for an isotropic system $\varepsilon_{\mathbf{k}}$ depends only upon the magnitude of $\mathbf{k}$, with $\mathbf{k}$ and $\mathbf{v}_{\mathbf{k}}$ parallel. We define m^* as the constant of proportionality at the Fermi surface:

$$v_F = \frac{\hbar k_F}{m^*}.$$

Thus, for an interacting Fermi liquid at the Fermi surface, the density of states is

$$\mathcal{D}_{\text{interacting}}(\varepsilon_F) = \frac{m^* k_F}{\pi^2 \hbar^2}.$$

m^* is usually $> m$, accounting for the fact that the quasiparticle may be viewed as a dressed particle, and must *drag* this dressing along with it. In this sense, the effective mass to some extent accounts for the interaction between the particles.

18.3 Landau Fermi Liquid

The power of the Landau Fermi liquid theory lies in its ability to parameterize the interactions in terms of a small number of multipole parameters called *Landau parameters*. These parameters describe how the original noninteracting Fermi liquid theory is renormalized by the feedback effect of interactions on quasiparticle energies.

18.3.1 Free Energy and Interparticle Interactions

The thermodynamics of the system depends upon the free energy F, which at zero temperature is

$$F - F_0 = E - E_0 - \mu (N - N_0).$$

Since our quasiparticles are formed by adiabatically switching on the interaction in the $N + 1$ particle ideal system, adding one quasiparticle to the system adds one real particle. Thus,

$$N - N_0 = \sum_{\mathbf{k}} \delta n_{\mathbf{k}}$$

and since

$$E - E_0 = \sum_{\mathbf{k}} \varepsilon_{\mathbf{k}} \, \delta n_{\mathbf{k}},$$

we obtain

$$F - F_0 \simeq \sum_{\mathbf{k}} (\varepsilon_{\mathbf{k}} - \mu) \, \delta n_{\mathbf{k}}$$

As shown in Figure 18.4, we will be interested in excitations of the system that distort the Fermi surface by an amount proportional to δ. The validity of theory/expansion requires that

$$\frac{1}{N} \sum_{\mathbf{k}} |\delta n_{\mathbf{k}}| \ll 1,$$

where $\delta n_{\mathbf{k}} \neq 0$, $\varepsilon_{\mathbf{k}} - \mu$ will also be of order δ. Therefore,

$$\sum_{\mathbf{k}} (\varepsilon_{\mathbf{k}} - \mu) \, \delta n_{\mathbf{k}} \sim \mathcal{O}(\delta^2).$$

Thus, for consistency we must add the next term in the Taylor series expansion of the energy to the expression for the free energy, and we write

$$F - F_0 = \sum_{\mathbf{k}} (\varepsilon_{\mathbf{k}} - \mu) \, \delta n_{\mathbf{k}} + \frac{1}{2} \sum_{\mathbf{k}, \mathbf{k}'} f_{\mathbf{k}, \mathbf{k}'} \, \delta n_{\mathbf{k}} \, \delta n_{\mathbf{k}'} + \mathcal{O}(\delta^3). \qquad (18.8)$$

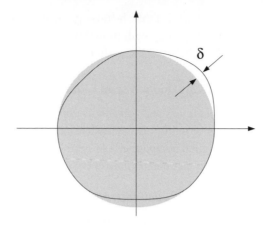

Figure 18.4 We consider small distortions of the Fermi surface, proportional to δ, so that $\frac{1}{N} \sum_{\mathbf{k}} |\delta n_{\mathbf{k}}| \ll 1$.

The second-order coefficients

$$f_{\mathbf{k},\mathbf{k}'} = \frac{\delta E}{\delta n_{\mathbf{k}}\, \delta n_{\mathbf{k}'}}$$

describe the interactions between quasiparticles at the Fermi surface. These partial derivatives are evaluated in the presence of an otherwise *frozen* Fermi sea, where all other quasiparticle occupancies are fixed. Landau was able to show that in an isotropic Fermi liquid, the quasiparticle mass m^* is related to the dipolar component of these interactions, as we shall shortly demonstrate. The Landau interaction can be regarded as an interaction operator that acts on a thin shell of quasiparticle states near the Fermi surface. Writing the quasiparticle occupancy as $n_{\mathbf{k}\sigma} = c^{\dagger}_{\mathbf{k}\sigma} c_{\mathbf{k}\sigma}$, where $c^{\dagger}_{\mathbf{k}\sigma}$ is the quasiparticle creation operator, then we can compare

$$\mathcal{H}_I \sim \frac{1}{2} \sum_{\mathbf{k}\sigma,\mathbf{k}'\sigma'} f_{\mathbf{k}\sigma,\mathbf{k}'\sigma'}\, \delta n_{\mathbf{k}\sigma}\, \delta n_{\mathbf{k}'\sigma'}$$

with

$$\mathcal{H}_I \sim \frac{1}{2} \sum_{\substack{\mathbf{k}\sigma,\mathbf{k}'\sigma' \\ \mathbf{q}}} V(\mathbf{q})\, c^{\dagger}_{\mathbf{k}+\mathbf{q}\sigma}\, c^{\dagger}_{\mathbf{k}'-\mathbf{q}\sigma'}\, c_{\mathbf{k}'\sigma'}\, c_{\mathbf{k}\sigma},$$

which shows that the Landau interaction term is a *forward scattering amplitude* between quasiparticles whose initial and final momenta are unchanged.

Volume dependence of $f_{\mathbf{k},\mathbf{k}'}$: Each sum over $\mathbf{k}$ is proportional to the volume

$$\sum_{\mathbf{k}} \rightarrow \frac{V}{(2\pi)^3},$$

and since $F \propto V$, it must be that $f_{\mathbf{k},\mathbf{k}'} \sim 1/V$. However, it is also clear that $f_{\mathbf{k},\mathbf{k}'}$ is an interaction between quasiparticles, each of which is spread out over the whole volume V, so the probability that they will interact is $\sim \lambda_{TF}^3/V$, thus

$$f_{\mathbf{k},\mathbf{k}'} \sim \frac{\lambda_{TF}^3}{V^2}. \tag{18.9}$$

Spin dependence of $f_{\mathbf{k},\mathbf{k}'}$: We can also reduce the spin dependence of $f_{\mathbf{k},\mathbf{k}'}$ to a symmetric and antisymmetric part. In the absence of an external field, the system should be invariant under time reversal, and hence the interaction is invariant under spin rotations. In addition, it has the following symmetries

$$f_{\mathbf{k}\sigma,\mathbf{k}'\sigma'} = f_{-\mathbf{k}-\sigma,-\mathbf{k}'-\sigma'} \quad \text{time reversal}$$

$$f_{\mathbf{k}\sigma,\mathbf{k}'\sigma'} = f_{-\mathbf{k}\sigma,-\mathbf{k}'\sigma'} \quad \text{inversion,}$$

Therefore,

$$f_{\mathbf{k}\sigma,\mathbf{k}'\sigma'} = f_{\mathbf{k}-\sigma,\mathbf{k}'-\sigma'}. \tag{18.10}$$

Thus f depends only on the relative orientations of the spins σ and σ', and there are only two independent components $f_{\mathbf{k}\uparrow,\mathbf{k}'\uparrow}$ and $f_{\mathbf{k}\uparrow,\mathbf{k}'\downarrow}$, which allows us to split f into symmetric and antisymmetric parts:

$$f_{\mathbf{k},\mathbf{k}'}^s = \frac{1}{2}\left(f_{\mathbf{k}\uparrow,\mathbf{k}'\uparrow} + f_{\mathbf{k}\uparrow,\mathbf{k}'\downarrow}\right), \qquad f_{\mathbf{k},\mathbf{k}'}^a = \frac{1}{2}\left(f_{\mathbf{k}\uparrow,\mathbf{k}'\uparrow} - f_{\mathbf{k}\uparrow,\mathbf{k}'\downarrow}\right). \tag{18.11}$$

$f_{\mathbf{k},\mathbf{k}'}^a$ may be interpreted as an exchange interaction, or

$$f_{\mathbf{k}\sigma,\mathbf{k}'\sigma'} = f_{\mathbf{k},\mathbf{k}'}^s + \boldsymbol{\sigma} \cdot \boldsymbol{\sigma}' f_{\mathbf{k},\mathbf{k}'}^a, \tag{18.12}$$

where $\boldsymbol{\sigma}$ and $\boldsymbol{\sigma}'$ are the Pauli matrices.

Since $\delta n_{\mathbf{k}}$ is only of order one near the Fermi surface, we will only care about $f_{\mathbf{k},\mathbf{k}'}$ on the Fermi surface, where it should be continuous and changes slowly as it crosses the Fermi surface. In an isotropic Landau Fermi liquid, the physics is invariant under spatial rotations. In this case, $\mathbf{k} = k_F \hat{\mathbf{e}}_{\mathbf{k}}$ and $\mathbf{k}' = k_F \hat{\mathbf{e}}_{\mathbf{k}'}$, and $f_{\mathbf{k},\mathbf{k}'}$ only depends on the angle between $\mathbf{k}$ and $\mathbf{k}'$, namely $\cos\theta = \hat{\mathbf{e}}_{\mathbf{k}} \cdot \hat{\mathbf{e}}_{\mathbf{k}'}$, where the $\hat{\mathbf{e}}$s are unit vectors. In turn, $f_{\mathbf{k},\mathbf{k}'}^s$ and $f_{\mathbf{k},\mathbf{k}'}^a$ will also depend only on θ. We can express either $f_{\mathbf{k},\mathbf{k}'}^a$ or $f_{\mathbf{k},\mathbf{k}'}^s$ in a multipole expansions in terms of Legendre polynomials as

$$f_{\mathbf{k},\mathbf{k}'}^\alpha = \sum_\ell f_\ell^\alpha P_\ell(\cos\theta). \tag{18.13}$$

We can then speak of interactions as monopolar, dipolar, quadrupolar, etc.

The f parameters have the dimension of energy; conventionally they are defined as dimensionless quantities via

$$V \mathcal{D}(\varepsilon_F) f_\ell^\alpha = V \frac{m^* k}{\pi^2 \hbar^2} f_\ell^\alpha = F_\ell^\alpha. \tag{18.14}$$

The coefficients F_ℓ^α are the Landau parameters.

18.3.2 Local Energy of a Quasiparticle

We consider an interacting system with a certain distribution of excited quasiparticles $\delta n_{\mathbf{k}'}$. To this, we add another quasiparticle of momentum $\hbar \mathbf{k}$, so that $\delta n_{\mathbf{k}'} \rightarrow \delta n_{\mathbf{k}'} + \delta_{\mathbf{k}, \mathbf{k}'}$.

From (18.8), the free energy of the additional quasiparticle is

$$\tilde{\varepsilon}_{\mathbf{k}} - \mu = \varepsilon_{\mathbf{k}} - \mu + \sum_{\mathbf{k}'} f_{\mathbf{k}, \mathbf{k}'} \, \delta n_{\mathbf{k}'}. \tag{18.15}$$

Both terms here are $\mathcal{O}(\delta)$. The second term describes the free energy of a quasiparticle due its interaction with other quasiparticles in the system (a Hartree-like term): an energy change induced by the polarization of the Fermi sea. $\tilde{\varepsilon}_{\mathbf{k}}$ plays the part of the local energy of a quasiparticle, in the sense that $\nabla_{\mathbf{x}} \tilde{\varepsilon}_{\mathbf{k}}$ is the force the system exerts on the additional quasiparticle. When the quasiparticle is added to the system, the system develops inhomogeneity so that $\delta n_{\mathbf{k}'} = \delta n_{\mathbf{k}'}(\mathbf{x})$. The system will react to this inhomogeneity by minimizing its free energy so that

$$\nabla_{\mathbf{x}} F = 0.$$

However, only the additional free energy due to the added particle (18.15) is inhomogeneous, as shown in Figure 18.5, and has a nonzero gradient. Thus, the system will exert a force

$$-\nabla_{\mathbf{x}} \tilde{\varepsilon}_{\mathbf{k}} = -\nabla_{\mathbf{x}} \sum_{\mathbf{k}'} f_{\mathbf{k}, \mathbf{k}'} \, \delta n_{\mathbf{k}'}$$

on the added quasiparticle resulting from interactions with other quasiparticles.

Equilibrium Distribution of Quasiparticles at Finite T

$\tilde{\varepsilon}_{\mathbf{k}}$ also plays an important role in the finite temperature properties of the system. For $\sum \delta n_{\mathbf{k}} \ll N$, we can write

$$\delta n_{\mathbf{k}} = \langle \delta n_{\mathbf{k}} \rangle + (\delta n_{\mathbf{k}} - \langle \delta n_{\mathbf{k}} \rangle),$$

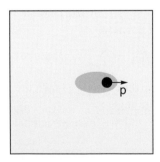

Figure 18.5 The addition of an extra particle to a homogeneous system will induce forces on the quasiparticle that tend to restore equilibrium.

where the first term is $O(\delta)$, and the second $O(\delta^2)$. Thus,

$$\delta n_{\mathbf{k}} \, \delta n_{\mathbf{k}'} \simeq - \langle \delta n_{\mathbf{k}} \rangle \, \langle \delta n_{\mathbf{k}'} \rangle + \langle \delta n_{\mathbf{k}} \rangle \, \delta n_{\mathbf{k}'} + \langle \delta n_{\mathbf{k}'} \rangle \, \delta n_{\mathbf{k}}.$$

We may use this to rewrite the energy of our interacting system:

$$
\begin{aligned}
E - E_0 &\simeq \sum_{\mathbf{k}} \varepsilon_{\mathbf{k}} \, \delta n_{\mathbf{k}} - \frac{1}{2} \sum_{\mathbf{k}, \mathbf{k}'} f_{\mathbf{k}, \mathbf{k}'} \langle \delta n_{\mathbf{k}} \rangle \langle \delta n_{\mathbf{k}'} \rangle + \sum_{\mathbf{k}, \mathbf{k}'} f_{\mathbf{k}, \mathbf{k}'} \langle \delta n_{\mathbf{k}'} \rangle \, \delta n_{\mathbf{k}} \\
&= \sum_{\mathbf{k}} \left(\varepsilon_{\mathbf{k}} + \sum_{\mathbf{k}'} f_{\mathbf{k}, \mathbf{k}'} \langle \delta n_{\mathbf{k}'} \rangle \right) \delta n_{\mathbf{k}} - \frac{1}{2} \sum_{\mathbf{k}, \mathbf{k}'} f_{\mathbf{k}, \mathbf{k}'} \langle \delta n_{\mathbf{k}} \rangle \langle \delta n_{\mathbf{k}'} \rangle \\
&= \sum_{\mathbf{k}} \langle \tilde{\varepsilon}_{\mathbf{k}} \rangle \, \delta n_{\mathbf{k}} - \frac{1}{2} \sum_{\mathbf{k}, \mathbf{k}'} f_{\mathbf{k}, \mathbf{k}'} \langle \delta n_{\mathbf{k}} \rangle \langle \delta n_{\mathbf{k}'} \rangle \, .
\end{aligned}
\tag{18.16}
$$

We now use (18.16) to determine the fermion occupation probability. Noting that the second term on the right-hand side is a constant and has no effect, we obtain

$$n_{\mathbf{k}}(T, \mu) = \frac{1}{\exp\left[\beta \left(\langle \tilde{\varepsilon}_{\mathbf{k}} \rangle - \mu\right)\right] + 1}$$

or

$$\delta n_{\mathbf{k}}(T, \mu) = \frac{1}{\exp\left[\beta \left(\langle \tilde{\varepsilon}_{\mathbf{k}} \rangle - \mu\right)\right] + 1} - \Theta(k_F - k).$$

However, for an isotropic system,

$$\langle \tilde{\varepsilon}_{\mathbf{k}} - \varepsilon_{\mathbf{k}} \rangle = \sum_{\mathbf{k}'} f_{\mathbf{k}, \mathbf{k}'} \langle \delta n_{\mathbf{k}'} \rangle$$

must be *independent of the location of* $\mathbf{k}$ *on the Fermi surface* (and, of course, spin), and is thus constant. To see this, we reconsider the Legendre polynomial expansion discussed earlier:

$$
\begin{aligned}
\langle \tilde{\varepsilon}_{\mathbf{k}} - \varepsilon_{\mathbf{k}} \rangle &= \sum_{\mathbf{k}'} f_{\mathbf{k}, \mathbf{k}'} \langle \delta n_{\mathbf{k}'} \rangle \\
&\propto \sum_{\ell} \int d\mathbf{k} \, f_\ell \, P_\ell(\cos\theta) \, \langle \delta n_{\mathbf{k}'} \rangle \\
&\propto f_0 \int d\mathbf{k} \, \langle \delta n_{\mathbf{k}'} \rangle = 0.
\end{aligned}
$$

- In going from the second to the third line, we made use of the isotropy of the system, so that $\langle \delta n_{\mathbf{k}'} \rangle$ is independent of the angle θ.
- The evaluation in the third line follows from particle number conservation.

Thus, to lowest order in δ,

$$n_{\mathbf{k}}(T, \mu) = \frac{1}{\exp\left[\beta \left(\varepsilon_{\mathbf{k}} - \mu\right)\right] + 1} + O(\delta^4).$$

Local Equilibrium Distribution

Now suppose we introduce a local weak perturbation, such as δ discussed in the preceding subsection, to an isotropic system at zero temperature. For example, such a perturbation could be caused by a sound wave or a weak magnetic field. This perturbation will cause a small deviation of the equilibrium distribution function, leading to a new *local equilibrium distribution*

$$\bar{n}_{\mathbf{k}} = n(\tilde{\varepsilon}_{\mathbf{k}} - \mu),$$

where the argument of the rhs indicates that this is the distribution corresponding to the local energy discussed in the preceding subsection. The gradient of the local energy yields a force that tries to restore the equilibrium distribution $n(\varepsilon_{\mathbf{k}} - \mu)$ derived. The deviation from true equilibrium is

$$\delta n_{\mathbf{k}} = n_{\mathbf{k}} - \bar{n}_{\mathbf{k}} = \delta \bar{n}_{\mathbf{k}} + \frac{\partial n(\varepsilon_{\mathbf{k}} - \mu)}{\partial \varepsilon_{\mathbf{k}}} (\tilde{\varepsilon}_{\mathbf{k}} - \varepsilon_{\mathbf{k}})$$

$$= \delta \bar{n}_{\mathbf{k}} + \frac{\partial n(\varepsilon_{\mathbf{k}} - \mu)}{\partial \varepsilon_{\mathbf{k}}} \sum_{\mathbf{k}'} f_{\mathbf{k},\mathbf{k}'} \, \delta n_{\mathbf{k}'}. \tag{18.17}$$

At zero temperature, the factor

$$\frac{\partial n(\varepsilon_{\mathbf{k}} - \mu)}{\partial \varepsilon_{\mathbf{k}}} = -\delta(\varepsilon_{\mathbf{k}} - \mu), \tag{18.18}$$

so both $\delta n_{\mathbf{k}}$ and $\delta \bar{n}_{\mathbf{k}}$ are restricted to the Fermi surface. Since the perturbation of interest is small, we may expand both $\delta n_{\mathbf{k}}$ and $\delta \bar{n}_{\mathbf{k}}$ in a series of Legendre polynomials, and we will also split them into symmetric and antisymmetric parts (as we did with $f_{\mathbf{k},\mathbf{k}'}$ previously). For example,

$$\delta n_{\mathbf{k}}^{s} = \sum_{\ell} \delta(\varepsilon_{\mathbf{k}} - \mu) \, \delta n_{\ell}^{s} \, P_{\ell}. \tag{18.19}$$

If we make a similar expansion for the antisymmetric and symmetric parts of $\delta \bar{n}_{\mathbf{k}}$, and substitute this back into (18.17), then

$$\delta \bar{n}_{\ell}^{a} = \left(1 + \frac{F_{\ell}^{a}}{2\ell + 1}\right) \delta n_{\ell}^{a}$$

$$\delta \bar{n}_{\ell}^{s} = \left(1 + \frac{F_{\ell}^{a}}{2\ell + 1}\right) \delta n_{\ell}^{s}. \tag{18.20}$$

18.3.3 Effective Mass m^* of Quasiparticles

Using this formulation of the interacting Fermi gas, Landau was able to link the renormalization of quasiparticle mass to the dipole component of the interactions F_1^s. As the fermion moves through the medium, the backflow of the surrounding fluid enhances its effective mass according to the relation

$$m^* = \left(1 + \frac{F_1^s}{3}\right) m.$$

The relation between effective mass m^* and the interaction term f will now be derived. The sum of momenta in a unit volume is equal to a flow of mass. The momentum of a unit volume of the Fermi liquid is the same as the momentum of the quasiparticle in this volume. The current of the particles in the Fermi liquid is equal to the current of quasiparticles:

$$\int \frac{d\mathbf{k}}{(2\pi)^3} \, \mathbf{k} \, n_\mathbf{k} = m \int \frac{d\mathbf{k}}{(2\pi)^3} \, \mathbf{v} \, n_\mathbf{k} = m \int \frac{d\mathbf{k}}{(2\pi)^3} \, \nabla_\mathbf{k} \, \tilde{\varepsilon}_\mathbf{k} \, n_\mathbf{k}. \qquad (18.21)$$

We apply variation with respect to $n_\mathbf{k}$ and remember that

$$\delta \int \frac{d\mathbf{k}}{(2\pi)^3} \, \nabla_\mathbf{k} \, \tilde{\varepsilon}_\mathbf{k} \, n_\mathbf{k} = \int \frac{d\mathbf{k}}{(2\pi)^3} \, \nabla_\mathbf{k} \, \tilde{\varepsilon}_\mathbf{k} \, \delta \, n_\mathbf{k} + \int \frac{d\mathbf{k}}{(2\pi)^3} \, \nabla_\mathbf{k} \, \delta \, \tilde{\varepsilon}_\mathbf{k} \, n_\mathbf{k}$$

and that (18.15) gives

$$\delta \tilde{\varepsilon}_\mathbf{k} = \int \frac{d\mathbf{k}'}{(2\pi)^3} \, \left(f_{\mathbf{k},\mathbf{k}'} \, \delta n_{\mathbf{k}'} \right)$$

$$\int \frac{d\mathbf{k}}{(2\pi)^3} \, \frac{\mathbf{k}}{m} \, \delta n_\mathbf{k} = \int \frac{d\mathbf{k}}{(2\pi)^3} \, \nabla_\mathbf{k} \, \tilde{\varepsilon}_\mathbf{k} \, \delta n_\mathbf{k} + \int \frac{d\mathbf{k} d\mathbf{k}'}{(2\pi)^6} \, \left(\nabla_\mathbf{k} \, f_{\mathbf{k},\mathbf{k}'} \, \delta n_{\mathbf{k}'} \right) \, n_\mathbf{k}$$

$$= \int \frac{d\mathbf{k}}{(2\pi)^3} \, \nabla_\mathbf{k} \, \tilde{\varepsilon}_\mathbf{k} \, \delta n_\mathbf{k} - \int \frac{d\mathbf{k} d\mathbf{k}'}{(2\pi)^6} \, f_{\mathbf{k},\mathbf{k}'} \, \left(\nabla_{\mathbf{k}'} \, n_{\mathbf{k}'} \right) \, \delta n_\mathbf{k}, \qquad (18.22)$$

where in the last line we interchanged the dummy variables $\mathbf{k}$ and $\mathbf{k}'$ and integrated by parts. The average over spin indices is taken since n and ε do not depend upon spin here. Note that only f^s survives. Since $\delta n_\mathbf{k}$ is arbitrary, it follows that

$$\frac{\mathbf{k}}{m} = \nabla_\mathbf{k} \, \tilde{\varepsilon}_\mathbf{k} - \int \frac{d\mathbf{k}'}{(2\pi)^3} \, f_{\mathbf{k},\mathbf{k}'}^s \, \left(\nabla_{\mathbf{k}'} \, n_{\mathbf{k}'} \right). \qquad (18.23)$$

With $n_\mathbf{k} = \Theta(k_F - k)$ at $T = 0$, we obtain

$$\nabla_{\mathbf{k}'} \, n_{\mathbf{k}'} = -\frac{\mathbf{k}'}{k'} \, \delta(k' - k_F).$$

However, since both k and k' are restricted to the Fermi surface, we can orient k along the z-axis and write

$$\frac{\mathbf{k}'}{k'} = \cos \theta.$$

Moreover, since f depends only upon the angle θ between $\mathbf{k}$ and $\mathbf{k}'$, we obtain the following relation between the effective mass m^* of the quasiparticles and the mass m of the fermion

$$\frac{1}{m} = \frac{1}{m^*} + \frac{k_F}{(2\pi)^2} \int d\cos\theta \, f^s(\theta) \, \cos\theta. \qquad (18.24)$$

The translation invariance property of an isotropic liquid was also used here.

Recalling that

$$f^s(\theta) = \sum_\ell f_\ell^s \, P_\ell(\theta)$$

and that

$$F_\ell^s = \mathcal{D}(\tilde{\varepsilon}_F) \, f_\ell^s = \frac{m^* k_F}{\pi^2} f_\ell^s$$

we find that

$$\frac{1}{m} = \frac{1}{m^*} + \frac{F_1^s}{3m^*} \qquad \Rightarrow \qquad m^* = \left(1 + \frac{F_1^s}{3}\right) m. \qquad (18.25)$$

Another way to understand quasiparticle mass renormalization is to consider the current carried by a quasiparticle. The total number of particles is conserved and we can ascribe a particle current $\mathbf{v}_F = \mathbf{p}_F/m^*$ to each quasiparticle. We can rewrite this current in the form

$$\mathbf{v}_F = \frac{\mathbf{p}_F}{m^*} = \underbrace{\frac{\mathbf{p}_F}{m}}_{\text{bare current}} - \overbrace{\frac{\mathbf{p}_F}{m} \frac{F_1^s}{1 + F_1^s}}^{\text{backflow}}$$

18.3.4 Equilibrium Properties

Specific Heat

We apply the definition of the fermionic specific heat derived in (18.1), but with the modified density of state, namely

$$C_v = \frac{\pi^2}{3} \mathcal{D}(\tilde{\varepsilon}_F) \, k_B^2 T = \frac{m^* k_F}{3} k_B^2 T. \qquad (18.26)$$

Thus, measuring the electronic contribution to the specific heat C_v yields information about the effective mass m^*, and hence F_1^s.

Compressibility

We now turn to the compressibility. The pressure of the system is defined as

$$P = -\frac{\partial E}{\partial V} = -\frac{\partial E}{\partial n} \frac{\partial n}{\partial V}.$$

With $n = \frac{N}{V}$,

$$P = \frac{n}{V} \frac{\partial E}{\partial n} = n \frac{\partial \varepsilon}{\partial n}.$$

But

$$\kappa^{-1} = -V \frac{\partial P}{\partial V} = n \frac{\partial P}{\partial n} = n^2 \frac{\partial^2 \varepsilon}{\partial n^2}.$$

But the chemical potential is the derivative of energy with respect to the number of particles

$$\mu = \frac{\partial E}{\partial N} = \frac{\partial \varepsilon}{\partial n}$$

so that

$$\kappa^{-1} = n^2 \frac{\partial \mu}{\partial n}.$$

We will now calculate $\frac{d\mu}{dn}$. If the particle density of the system is changed by δn, then

$$\delta n = \sum_{\mathbf{k}} \delta n_{\mathbf{k}}$$

$$\delta n_{\mathbf{k}} = \frac{\partial n_{\mathbf{k}}}{\partial \tilde{\varepsilon}_{\mathbf{k}}} (\delta \tilde{\varepsilon}_{\mathbf{k}} - \delta \mu)$$

$$\delta \tilde{\varepsilon}_{\mathbf{k}} = \frac{2}{(2\pi)^3} \int d\mathbf{k}' \, f_{\mathbf{k},\mathbf{k}'} \, \delta n_{\mathbf{k}'}. \tag{18.27}$$

The quasiparticle energy $\delta \tilde{\varepsilon}_{\mathbf{k}}$ depends on μ only through its dependence on $\delta n_{\mathbf{k}}$

As we vary the chemical potential, the resulting variations are isotropic and spin independent. We therefore can conclude that from the Landau parameters, only F_0^s can play a role, and we write

$$\delta \tilde{\varepsilon}_{\mathbf{k}} = \frac{F_0^s}{\mathcal{D}(\tilde{\varepsilon}_F)} \sum_{\mathbf{k}'} \delta n_{\mathbf{k}'} = \frac{F_0^s}{\mathcal{D}(\tilde{\varepsilon}_F)} \delta n$$

and we get

$$\delta n = \sum_{\mathbf{k}} \delta n_{\mathbf{k}} = \sum_{\mathbf{k}} \frac{\partial n_{\mathbf{k}}}{\partial \tilde{\varepsilon}_{\mathbf{k}}} \left(\frac{F_0^s}{\mathcal{D}(\tilde{\varepsilon}_F)} \delta n - \delta \mu \right). \tag{18.28}$$

As $T \to 0$

$$\sum_{\mathbf{k}} \frac{\partial n_{\mathbf{k}}}{\partial \tilde{\varepsilon}_{\mathbf{k}}} = \int d\varepsilon \, \mathcal{D}(\varepsilon) \frac{\partial n}{\partial \varepsilon} = -\mathcal{D}(\tilde{\varepsilon}_F),$$

which yields

$$\delta n = \mathcal{D}(\tilde{\varepsilon}_F) \delta \mu - F_0^s \delta n$$

and we get

$$\frac{\delta n}{\delta \mu} = \frac{\mathcal{D}(\tilde{\varepsilon}_F)}{1 + F_0^s}, \tag{18.29}$$

which leads to an expression for the compressibility κ

$$\kappa = \frac{1}{n^2} \frac{\mathcal{D}(\tilde{\varepsilon}_F)}{1 + F_0^s} = \frac{m^*/m}{1 + F_0^s} \kappa^{(0)}, \qquad (18.30)$$

where $\kappa^{(0)}$ is the compressibility of the noninteracting system.

The important things are that we again find a renormalization m^*/m with respect to the noninteracting compressibility, as for the specific heat. The novel aspect, however, is that a further renormalization occurs due to the quasiparticle interactions. In fact, depending on the sign of F_0^s, this can lead to a sizable change in κ. Moreover, if $F_0^s \le -1$, the (18.30) leads to a divergence of κ or a negative sign. This immediately tells us that the Fermi liquid is unstable and the whole concept of quasiparticles breaks down.

Magnetic Susceptibility

We will now determine the (spin) magnetic susceptibility of a Fermi liquid. Thus we need to consider its response to an external magnetic field. Here we will be interested in the effect of the Zeeman coupling, which causes the quasiparticle energy to change by an amount that depends on the spin polarization

$$-\frac{1}{2} \hbar \, \mathrm{g} \, \sigma_z \, B,$$

where g is the gyromagnetic ratio, σ_z is the diagonal Pauli matrix, and B is the external (uniform) magnetic field. By taking into account also the change caused to the distribution functions, we find

$$\delta\tilde{\varepsilon}_{\mathbf{k}} = -\frac{1}{2} \hbar \, \mathrm{g} \, \sigma_z \, B + \frac{2}{(2\pi)^3} \int d\mathbf{k}' \, f_{\mathbf{k}\sigma, \mathbf{k}'\sigma'} \, \delta n_{\mathbf{k}'\sigma'}$$

with $\delta n_{\mathbf{k}'}$ given (18.28).

The chemical potential is a scalar (and time-reversal invariant) quantity, and as such it cannot have a linear variation with the magnetic field. Hence the only possible dependence of μ with B must be an even power and (at least) of order B^2. Thus, it does not contribute to the magnetic susceptibility (within linear response), and we will neglect this contribution. Hence, $\delta n_{\mathbf{k}} \propto \delta\tilde{\varepsilon}_{\mathbf{k}}$, are independent of the direction of the momentum $\mathbf{k}$, and have opposite sign for $\uparrow$ and $\downarrow$ quasiparticles. Since $\delta n_{\mathbf{k}} \ne 0$ only for $\mathbf{k}$ on the Fermi surface (which we will assume to be isotropic), we find

$$\frac{2}{(2\pi)^3} \int d\mathbf{k}' \, f_{\mathbf{k}\sigma, \mathbf{k}'\sigma'} \, \delta n_{\mathbf{k}'\sigma'} = 2 f_0^a \, \delta n_\sigma = \sigma_z \, f_0^a \left(\delta n_\uparrow - \delta n_\downarrow \right),$$

where δn_σ is the change in the total number of particles (per unit volume) with spin σ. Hence,

$$\delta n_\sigma = \frac{\mathcal{D}(\tilde{\varepsilon}_F)}{2} \left(\frac{1}{2} \hbar \, \mathrm{g} \, \sigma_z \, B - 2 f_0^a \, \delta n_\sigma \right).$$

The net spin polarization is

$$\delta n_\uparrow - \delta n_\downarrow = \frac{\hbar}{2} g \frac{\mathcal{D}(\tilde{\varepsilon}_F)}{1 + F_0^a} B \qquad (18.31)$$

and the total magnetization M is

$$M = \frac{\hbar^2}{4} g^2 \frac{\mathcal{D}(\tilde{\varepsilon}_F)}{1 + F_0^a} B.$$

We can thus identify the spin susceptibility χ with

$$\chi = \frac{\hbar^2}{4} g^2 \frac{\mathcal{D}(\tilde{\varepsilon}_F)}{1 + F_0^a} = \frac{m^*/m}{1 + F_0^a} \chi^{(0)}, \qquad (18.32)$$

which is the (Pauli) spin susceptibility of a free Fermi gas with mass m^*, with the Fermi liquid correction.

We again have two distinct contributions to the renormalization with respect to the noninteracting electron gas: one from the effective mass and a second from the quasiparticle interactions. If we calculate now the Wilson ratio, we find

$$W = \frac{1}{1 + F_0^a} W^{(0)}.$$

It is thus important to note that the Fermi gas value $W^{(0)}$ can be easily changed to values of the order $1 \dots 10$ by the quasiparticle interactions. For Fermi liquids with strong ferromagnetic exchange interactions between fermions, a negative F_0^a enhances the Pauli susceptibility. This accounts for the enhanced Pauli susceptibility in liquid ^{3}He, with $W \sim 4$, and for the much enhanced $W = 10$ in Pd metal. Compelling evidence for Landau Fermi liquid theory comes from heavy-fermion materials, for example in UPt$_3$, $\frac{m^*}{m} \sim 17$ as measured by de Haas–van Alphen and confirmed by specific heat. Typical data about Landau parameters in ^{3}He, effective masses, and susceptibilities are given in Table 18.1.

We again have to require that $F_0^a > -1$, in order for the Fermi liquid concept to be valid. Otherwise, we will in general observe a ferromagnetic instability, called a *Stoner instability*: it is an example of a ferromagnetic quantum critical point – a point where quantum zero-point fluctuations of the magnetization develop an infinite range of correlations in space and time. At such a point, the Wilson ratio will diverge.

Table 18.1 *Landau parameters for ^{3}He and selected m^* and χ^*.*

Parameter	0 bar	26 bar	System	m^*/m	$\chi/\chi^{(0)}$
m^*/m	2.8	5.26	Nb	2	1
F_0^s	9.28	67.17	^{3}He	6	20
F_1^s	5.39	12.79	Heavy	100	100
F_0^a	−0.696	−0.76	fermions		

18.4 Microscopic Verification of Landau's Phenomenological Fermi Liquid Theory

The microscopic verification of Landau's Fermi liquid theory (FLT) is actually derived from the one-particle Green function and its spectral function developed in Chapter 15. We recall that the Green function characterizes the response of an interacting system to injection of an additional particle or hole. We start with the analytically continued Green function

$$G_{\mathbf{k}\sigma}(z) = \frac{1}{z + \mu - \varepsilon_{\mathbf{k}} - \Sigma_{\mathbf{k}\sigma}(z)}$$

with the self-energy function $\Sigma_{\mathbf{k}\sigma}(z)$ being unknown. As we have made clear, this function contains all information about interactions, but, in general, is hard to calculate. However, in the vicinity of the Fermi energy, $z = 0$ and $|\mathbf{k}| = k_F$, $\Sigma_{\mathbf{k}\sigma}(z)$ can in many cases be expanded in a Taylor series:

$$\Sigma_{\mathbf{k}\sigma}(z) = \Sigma_{\mathbf{k}\sigma}(0) + \left.\frac{\partial \Sigma_{\mathbf{k}\sigma}(z)}{\partial z}\right|_{z=0} z + \left.\frac{\partial^2 \Sigma_{\mathbf{k}\sigma}(z)}{\partial z^2}\right|_{z=0} z^2 + \cdots$$

For a Fermi liquid, the derivatives have the following properties: assuming $z = \omega + i0^+$, and $\omega \in \mathbb{R}$,

$$\Sigma_{\mathbf{k}\sigma}(0) \in \mathbb{R}$$

$$\left.\frac{\partial \Sigma_{\mathbf{k}\sigma}(z)}{\partial z}\right|_{z=0} \leq 0$$

$$\left.\frac{\partial^2 \Sigma_{\mathbf{k}\sigma}(z)}{\partial z^2}\right|_{z=0} = -i\eta, \qquad \eta > 0. \tag{18.33}$$

Inserting the self-energy expansion into the Green function, and introducing the abbreviations

$$Z_{k_F}^{-1} = 1 - \left.\frac{\partial \Sigma_{\mathbf{k}\sigma}(z)}{\partial z}\right|_{z=0} \geq 1$$

$$\tilde{\varepsilon}_{\mathbf{k}} = Z_{k_F}\,\varepsilon_{\mathbf{k}}$$

$$\tilde{\mu} = Z_{k_F}\,(\mu - \Sigma_{\mathbf{k}\sigma}(0)), \tag{18.34}$$

we obtain

$$G_{\mathbf{k}\sigma}(z) = \frac{Z_{k_F}}{\omega + \tilde{\mu} - \tilde{\varepsilon}_{\mathbf{k}} + i\eta\omega^2}. \tag{18.35}$$

Note that the Fermi wave vector k_F is implicitly defined via

$$0 = \mu - \varepsilon_{k_F} - \Sigma_{\mathbf{k}\sigma}(0).$$

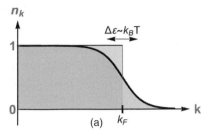
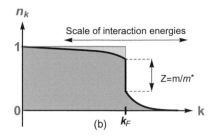

Figure 18.6 (a) In a noninteracting Fermi liquid, a temperature $T \ll E_F/k_B$ slightly blurs the Fermi surface. (b) In a Landau Fermi liquid, the exclusion principle stabilizes the jump in occupancy at the Fermi surface, even though the bare interaction energy is far greater than the Fermi energy.

Close to the Fermi energy $\omega = 0$, the imaginary part becomes small and hence the Green function has a simple pole with a weight Z_{k_F}. For the noninteracting electron gas, the Green function has the form

$$G^{(0)}_{\mathbf{k}\sigma}(z) = \frac{1}{\omega + \mu - \varepsilon_{\mathbf{k}} + i0^+}$$

and the interacting Green function is of similar structure. The weight of the pole is, however, $Z_{k_F} < 1$, because the interacting Green function does not describe a real particle, but a quasiparticle with the factor Z_{k_F} as its weight. Finally, close to the Fermi wavevector, we may expand $\varepsilon_{\mathbf{k}}$ in a Taylor series as

$$\varepsilon_{\mathbf{k}} = \mu + \frac{\hbar k_F}{m} |\mathbf{k} - \mathbf{k}_F|$$

and hence

$$\tilde{\varepsilon}_{\mathbf{k}} = \tilde{\mu} + \frac{\hbar k_F Z_{k_F}}{m^*} |\mathbf{k} - \mathbf{k}_F|.$$

However, Landau's phenomenological FLT asserts that $m^* = m/Z_{k_F}$ must hold, hence sometimes Z_{k_F} is referred to as the mass renormalization. Yet, we should note that we introduced a generalization of Landau's concept: the renormalizations can be $\mathbf{k}$-dependent.

The approximate form (18.35) for the Green function shows that the momentum distribution function $\tilde{n}_{\mathbf{k}\sigma}$ at $T = 0$ has a jump at $\mathbf{k}_F$, namely

$$\tilde{n}_{\mathbf{k}\to\mathbf{k}_F^+,\sigma} - \tilde{n}_{\mathbf{k}\to\mathbf{k}_F^-,\sigma} = Z_{k_F},$$

as shown in Figure 18.6(b).

Another cornerstone of FLT is the phase space argument[1] that close to the Fermi energy at zero temperature, the width $1/\tau_{\mathbf{k}}$ of the coherent quasiparticle peak is proportional to $(\varepsilon_{\mathbf{k}} - \mu)^2$ so that near the Fermi energy the lifetime is long and quasiparticles are well defined.

---

[1] See Section 15.2.6

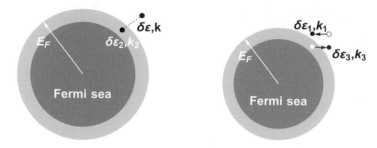

Figure 18.7 Left: initial state with an injected single quasiparticle above a filled Fermi sea. Right: final state with two quasiparticles and a quasihole. Annular light gray areas represent quasiparticles.

Furthermore, we note that the repulsive interaction between the added particle and those already in the Fermi sea catapults particles from below to above the Fermi surface, creating electron–hole pairs as shown in Figure 18.7. The possible terms in a perturbative description of this process are constrained by the conservation laws of charge, particle number, energy, momentum, and spin, and lead to an expression of $\left|\Psi_{\mathbf{k}\sigma}^{N+1}\right\rangle$ of the type

$$\left|\Psi_{\mathbf{k}\sigma}^{N+1}\right\rangle = Z_{\mathbf{k}}^{1/2}\, c_{\mathbf{k}\sigma}^{\dagger}\left|\Psi^{N}\right\rangle + \frac{1}{\Omega^{3/2}} \sum_{\substack{\mathbf{k}_1,\mathbf{k}_2,\mathbf{k}_3 \\ \sigma_1,\sigma_2,\sigma_3}} \alpha(\mathbf{k}_1\sigma_1,\mathbf{k}_2\sigma_2,\mathbf{k}_3\sigma_3)$$

$$\times\, \delta_{\mathbf{k},\,\mathbf{k}_1-\mathbf{k}_2+\mathbf{k}_3}\, \delta_{\sigma;\sigma_1,\sigma_2,\sigma_3}\, c_{\mathbf{k}_3}^{\dagger}\, c_{\mathbf{k}_2}\, c_{\mathbf{k}_1}^{\dagger}\left|\Psi^{N}\right\rangle + \cdots \tag{18.36}$$

The dots indicate higher-order terms, for which two or more particle–hole pairs are created, and $\delta_{\sigma;\sigma_1,\sigma_2,\sigma_3}$ expresses conservation of spin under vector addition. The multiple particle–hole pairs for a fixed total momentum can be created with a continuum of momenta of the individual bare particles and holes. Therefore, an added particle with fixed momentum has a wide distribution of energies.[2] However, if $Z_{\mathbf{k}}$ defined by (18.36) is finite, there is a well-defined feature in this distribution at $\tilde{\varepsilon}_{\mathbf{k}}$, which is in general different from the noninteracting value $\varepsilon_{\mathbf{k}} = \hbar^2 k^2/2m$.

From (18.36), we have a more physical definition of $Z_{\mathbf{k}}$: $Z_{\mathbf{k}}$ is the projection amplitude of $\left|\Psi_{\mathbf{k}}^{N+1}\right\rangle$ onto the state with one bare particle added to the ground state, since the projection of all other terms in the expansion vanish in the thermodynamic limit in the perturbative expression manifest in (18.36):

$$Z_{\mathbf{k}} = \left\langle \Psi_{\mathbf{k}}^{N+1}\middle| c_{\mathbf{k}}^{\dagger}\middle|\Psi^{N}\right\rangle.$$

In other words, $Z_{\mathbf{k}}$ is the overlap of the ground-state wavefunction of a system of interacting $N+1$ fermions of total momentum $\mathbf{k}$ with the wavefunction of N interacting particles and a bare particle of momentum $\mathbf{k}$. Accordingly, $Z_{\mathbf{k}}$ is called the *quasiparticle amplitude*. The Landau theory implicitly assumes that $Z_{\mathbf{k}}$ is finite. Furthermore, it asserts that for small ω

---

[2] See section 15.5.3

and $|\mathbf{k}|$ close to k_F, the physical properties can be calculated from quasiparticles that carry the same quantum numbers as the particles, namely, charge, spin, and momentum. These quasiparticles may be simply defined by a creation operator $\gamma_{\mathbf{k}\sigma}^{\dagger}$:

$$\left|\Psi_{\mathbf{k}\sigma}^{N+1}\right\rangle = \gamma_{\mathbf{k}\sigma}^{\dagger}\left|\Psi^{N}\right\rangle.$$

In other words, in FLT the single-particle character still lives.

Equation (18.36) allows us to consider a possible scenario where FLT breaks down: the quasiparticle amplitude $Z_{\mathbf{k}}$ becomes zero when the states $\left|\Psi_{\mathbf{k}}^{N+1}\right\rangle$ and $c_{\mathbf{k}}^{\dagger}\left|\Psi^{N}\right\rangle$ are orthogonal. This can happen if terms involving the number of particle–hole pairs become divergent. In other words, the addition of a particle or a hole to the system creates a divergent number of particle–hole pairs, so that the leading term does not have a finite weight in the thermodynamic limit. From (18.34), which links the Zs to Σs, we find that the single-particle self-energy must be singular as a function of ω at $k \simeq k_F$. This in turn means that, unlike Landau Fermi liquids, the Green functions contain branch cuts rather than the poles. Moreover, if a divergent number of low-energy particle–hole pairs is created upon the addition of a bare particle, it means that the low-energy response functions (which all involve creating particle–hole pairs) are also divergent.

18.4.1 The Meaning of the Fermi Surface and the Fermi Sphere

Some final remarks are in order. The picture of the Fermi liquid ground state as a filled Fermi sphere of quasiparticles is misleading in the sense that the states deep inside the sphere cannot be thought of as quasiparticles in momentum eigenstates; to highlight this point, Figure 18.7 depicts the nonquasiparticle states as dark gray areas. Close to $\mathbf{k}_F$, and for T small compared to the Fermi energy, the distribution of the quasiparticles in terms of the renormalized quasiparticle energies, $\tilde{\varepsilon}_{\mathbf{k}}$ is taken to be the Fermi–Dirac distribution.

Another possible misconception relates to the momentum distribution $\tilde{n}(\mathbf{k})$, which for free fermions at zero T is just a step function at the Fermi momentum k_F. In the interacting case, the quasiparticle distribution function $n_{\mathbf{k}}$ has the same behavior and, as discussed previously, k_F is related to the density just as in the free case. The meaning of $n_{\mathbf{k}}$ is, however, not that of the distribution of physical momentum in the ground state. Obviously, this distribution does not vanish for $k > k_F$ since scattering processes will populate the ground state with electron–hole pairs. This is not in contradiction with the notion that such a state does not have any quasielectron or quasiholes in it. As is illustrated in Figure 18.6, it is depleted below k_F and augmented above k_F, with a discontinuity at $T = 0$, whose value is shown in microscopic theory to be Z_{k_F}. The qualitative feature of the distribution is shown in Figure 18.6. $\tilde{n}(\mathbf{k})$ is no longer just a step function; it does have a discontinuity at the Fermi momentum. It is this discontinuity that is characteristic for a Fermi liquid and that guarantees that there are low-lying quasiparticle-quasihole excitations just as in the free Fermi gas.

Exercises

18.1 Uniaxial compressibility:

We consider a system of electrons upon which an uniaxial pressure in z-direction acts. Assume that this pressure causes a deformation of the Fermi surface $k = k_F^0$ of the form

$$k_F(\phi;\theta) = k_F^0 + \gamma \frac{1}{k_F^0}\left[3k_z^2 - \left(k_F^0\right)^2\right] = k_F^0 + \gamma\, k_F^0\left[3\cos^2\theta - 1\right], \quad (18.37)$$

where $\gamma = (P_z - P_0)/P_0$ is the anisotropy of the applied pressure.

(a) Can you explain why this deformation of the Fermi surface is adequate? How will this assumption be modified if we had a uniaxial tensile stress?

(b) Show that for $\gamma \ll 1$, the deformed Fermi surface $k_F(\phi;\theta)$ encloses the same volume as the nondeformed one, k_F^0, where terms of order $O\left(\gamma^2\right)$ can be neglected.

(c) The deformation of the Fermi surface effects a change in the distribution function of the electrons. Using Landau's Fermi liquid theory, calculate the uniaxial compressibility

$$\kappa_u = \frac{1}{\Omega}\frac{\partial^2 E}{\partial P_z^2},$$

which is caused by the deformation given in (18.37) (E denotes the Landau energy functional).

(d) What is the stability condition of the Fermi liquid against the deformation given in (18.37)?

18.2 Zero sound in Fermi liquid:

The transport equation for the distribution function $\delta n_\mathbf{k}(\mathbf{x},t)$ in a Fermi liquid like ^{3}He looks quite similar to the Boltzmann equation

$$\frac{\partial \delta n_\mathbf{k}}{\partial t} + \mathbf{v}_\mathbf{k}\cdot\nabla_\mathbf{x}\left(\delta n_\mathbf{k} - \frac{\partial \delta n_\mathbf{k}^{(0)}}{\partial \varepsilon_\mathbf{k}}\delta\varepsilon_\mathbf{k}\right) = \frac{dn}{dt}\bigg|_{\text{collision}},$$

where $\delta\varepsilon_\mathbf{k}$ is the shift of the quasiparticle energy due to both external fields and the Landau interaction, and $n^{(0)}$ is the Fermi function.

(a) Assume that the system is in the collisionless regime for frequencies of interest, i.e., neglect the right-hand side of the equation. Fourier-transform this equation with respect to time and space to obtain

$$(\omega - \mathbf{q}\cdot\mathbf{v}_\mathbf{k})\,\delta n_\mathbf{k} - \delta(\varepsilon_\mathbf{k} - \varepsilon_F)\,\mathbf{q}\cdot\mathbf{v}_\mathbf{k}\left(\sum_{\mathbf{k}'} f_{\mathbf{k}\mathbf{k}'}\,\delta n_{\mathbf{k}'}\right) = 0.$$

(b) Assume that $f_{\mathbf{kk'}} = F_0^s / N_0$, i.e., the Landau interaction is isotropic and nonmagnetic. Show that the restriction of $\delta n_{\mathbf{k}}$ to the Fermi sphere $\delta n_{\hat{\mathbf{k}}}$ obeys

$$\delta n_{\hat{\mathbf{k}}} = \frac{\mathbf{q} \cdot \mathbf{v_k}}{\omega - \mathbf{q} \cdot \mathbf{v_k}} \, F_0^s \int \frac{d\Omega'}{4\pi} \, \delta n_{\hat{\mathbf{k}'}}.$$

(c) Expand $\delta n_{\hat{\mathbf{k}}}$ in Legendre polynomials on the Fermi sphere, and show there is a solution $\delta n_{\hat{\mathbf{k}}}$ with isotropic symmetry only if

$$1 + F_0^s \left(1 + \frac{1}{2} \frac{\omega}{q v_F} \ln \left(\frac{\omega - q v_F}{\omega + q v_F} \right) \right) = 0. \tag{18.38}$$

(d) Plot graphically the solution of (18.38). This is the dispersion for collisionless zero sound, a sound mode in a degenerate Fermi liquid that has no analog in the noninteracting system.

18.3 First-sound velocity:

Derive an expression for the first-sound velocity in terms of Landau parameters.

19

Non-Fermi Liquids, the Luttinger Liquid, and Bosonization

19.1 Introduction

The Fermi liquid description of metals outlined in the previous chapter is one of the most successful theories in condensed matter physics. It can be applied to describe vastly different systems, ranging from liquid ^{3}He to metals such as copper or gold to complicated compounds such as the heavy fermion intermetallic compound $CeCu_6$, where the Coulomb interaction in strongly localized f-electron shells leads to gigantic interaction effects and a hundredfold increase of the effective masses. The specific heat behavior of the heavy fermion compound $CeAl_3$ is shown in Figure 19.1. As in the Sommerfeld gas, the specific heat is linear in temperature at low T, $C_v \simeq \gamma T$, but the value of the Sommerfeld constant γ is about a thousand times as large as one would estimate from the density of states of a typical metal. Such observations dramatically illustrate the power and range of validity of the Fermi liquid ideas.

19.1.1 Routes to Breakdown of Landau Theory

However, since the early 1980s a variety of metallic compounds have been discovered that, at low temperatures, display fundamentally different thermodynamic and transport properties from those of the usual Fermi liquid metallic systems. They have often been referred to as *non-Fermi liquids*. The most prominently discussed materials that exhibit properties qualitatively different from FLT predictions are the normal phase of high-temperature superconductors for a range of compositions near their highest T_c[1]. At very low levels of doping, they are insulating antiferromagnets. A T versus doping phase diagram is shown in Figure 19.2. At increased doping levels, they become conducting, and the exact temperature and doping level determine which phase they will be in. The dome shape of the superconducting region peaks at optimal doping. In the underdoped pseudo-gap region, the single-particle density of states is suppressed without the onset of global phase coherence indicative of superconductivity. In the strange metal, the resistivity $\rho \propto T$. The pseudogap

---

[1] More than 100,000 papers have been written on the subject of high-temperature cuprate superconductors.

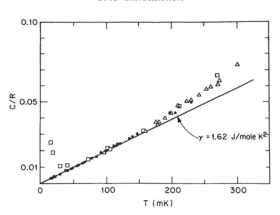

Figure 19.1 Specific heat of CeAl$_3$ at low temperatures: zero field (•, $\triangle$) and in 10 kOe ($\square$), from [18].

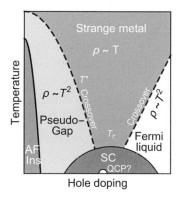

Figure 19.2 Schematic T-vs.-doping (x) phase diagram for cuprate superconductors showing location of possible quantum critical points (QCPs). One of these QCPs may be responsible for the anomalous normal state that develops above the pseudogap scale.

and strange metal regimes are separated by a line of crossover[2] temperatures T*. The boundary of the pseudogap region at low doping levels is unknown. The transition between the Fermi liquid phase and the strange-metal phase occurs gradually (by crossover). The white half-disc indicates possible quantum critical point at which the temperature T* goes to absolute zero. A fundamental characteristic of such systems is that the low-energy properties in a wide range of their phase diagram are dominated by singularities as a function of energy and temperature. Deviations from Fermi liquid behavior have become a central topic in the experimental and theoretical studies of correlated electronic systems, triggered by the discovery of high-temperature superconductivity, the success in synthesizing effectively

---

[2] The term *crossover* is typically used when a system changes from one type of critical behavior to another. It is driven by changes in some relevant variables such as anisotropy, magnetic field, or finite size. The crossover is typically smooth, taking place in a region rather than a precise point. It can be characterized by a crossover exponent.

low-dimensional materials, and the study of compounds that can be tuned through zero-temperature phase transitions, such as heavy fermions. Consequently, our perspective on Fermi liquids has changed significantly due both to such experimental developments, and to accompanying changes in our theoretical outlook.

19.1.2 Thermally Induced Classical Phase Transitions

It has of course been known for a long time that FLT breaks down in the fluctuation regime of classical, finite temperature, continuous phase transitions. The order parameter (magnetization, staggered magnetization, etc.) fluctuates coherently over increasing distances and time scales as the transition is approached. The spatial correlations of the order parameter fluctuations become long ranged. Close to the critical point T_c, their typical length scale, the correlation length, ξ, diverges as

$$\xi \propto \left| \frac{T - T_c}{T_c} \right|^{-\nu},$$

where ν is the correlation length critical exponent. In addition to the long-range correlations in space, there are analogous long-range correlations of the order-parameter fluctuations in time. The characteristic time scale of the fluctuations is the correlation time, τ_c. As the critical point is approached, the correlation time diverges as

$$\tau_c \propto \xi^z,$$

where z is the dynamic critical exponent.

At a finite temperature critical point, the critical long-wavelength fluctuations of the order parameter do not involve quantum mechanics. This is because thermal fluctuations destroy the coherence of quantum fluctuations on time scales longer than

$$\tau = \frac{\hbar}{k_B T}.$$

Any continuous finite-temperature phase transition is classical in the following sense:

There exists a characteristic frequency ω_c for order-parameter fluctuations, which tends to zero at the transition. *The system will behave classically if the transition temperature T_c satisfies $k_B T_c \gg \hbar \omega_c$,* even when quantum effects are important at short length scales.

It is not that quantum mechanics is unimportant in these cases, for in its absence there would not be an ordered state, such as a superfluid or a superconductor. However, sufficiently close to the critical point, quantum fluctuations are important at the microscopic scale but not at the longer length scales that control the critical behavior. In the jargon of statistical mechanics: *quantum mechanics is needed for the existence of an order parameter, but it is classical thermal fluctuations that govern it at long wavelengths.* We recall that near the superfluid transition in ^{4}He, the order parameter is a complex-valued field that is related to the underlying condensate wavefunction. However, its critical fluctuations can be captured exactly by doing classical statistical mechanics with an effective Hamiltonian for the order-parameter field – as we have encountered in the phenomenological Ginsburg–Landau free energy functional.

Thus, when a system approaches a second-order classical phase transition where fluctuations of the order parameter slow down and occur over increasingly long wavelengths, we can envision that a moving quasiparticle can easily generate a large disturbance in the system. This disturbance can, in turn, affect other quasiparticles in the vicinity, dramatically enhancing the scattering cross section in such a way that the quasiparticle lifetime vanishes. Eventually, this process is terminated at the ordering temperature, where the fluctuations become locked into a long-range ordered state. Below this temperature, our initial assumptions again remain valid and the Landau quasiparticle is saved.

19.2 Quantum Criticality and Quantum Critical Points

The breakdown of FLT occurs in a more substantial region of the phase diagram around the quantum critical point (QCP) where the transition temperature tends to zero as a function of some external nonthermal control parameter r; see Figure 19.3.

A. Quantum critical point A *quantum critical point* arises when a system undergoes a continuous transition from one phase to another at *zero temperature*. Actually the transition is between two ground states, since it is at $T = 0$. The distinct ground states themselves occur due to competing interactions of a many-body system that can be tuned by an external nonthermal parameter, such as pressure or doping. At the critical point, the system is in a quantum coherent superposition of the degenerate "ordered" and "disordered" states. Away from the QCP, the nonthermal control parameter r tunes the amount of zero-point motion of the constituent "particles." We can say that such a parameter controls quantum-mechanical tunneling dictated by Heisenberg's uncertainty principle, changing the degree of quantum fluctuations. This is the analog of varying the thermal fluctuations in the case of temperature-driven classical phase transitions. Phase transitions at $T = 0$ are referred to

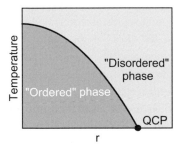

Figure 19.3 Schematic phase diagram near a quantum critical point. The parameter r along the x-axis may represent pressure, doping, a ratio of coupling constants, etc. Whenever the critical temperature vanishes, a QCP, indicated by a dot, is encountered. In the vicinity of such a point, quantum-mechanical zero-point fluctuations dominate.

as *quantum phase transitions*, because they are dominated by quantum effects – quantum mechanics determines the fluctuation of the order parameter. Consequently, quantum phase transitions require quantum statistical description of their critical fluctuations.

B. Quantum critical regime The excitation spectrum immediately above this quantum critical state, $T \neq 0$, may be distinctly different from the excitations of either phase, the disordered and the ordered one. The physics of this *quantum critical regime*, which corresponds to a wide parameter range at nonzero temperature, is controlled by collective fluctuations of the QCP. However, in addition to quantum fluctuations, the physical properties in this regime show also unusual temperature dependence essentially due to thermal excitation of the anomalous spectrum, so that the quantum critical behavior extends up to elevated temperatures. Quantum criticality plays a special role in strongly correlated electron systems; it provides a mechanism for both the non-Fermi liquid behavior and emergent phases such as unconventional superconductivity.

The interplay of classical and quantum fluctuations leads to an interesting phase diagram in the vicinity of the QCP. Two cases need to be distinguished, depending on whether long-range order can exist at finite temperatures:

(i) Figure 19.4(a) describes the situation where order only exists at $T = 0$.[3] In this case, there will be no true phase transition in any real experiment carried out at finite temperature. However, the finite-T behavior is characterized by three very different regimes, separated by *crossovers*, depending on whether the behavior is dominated by thermal or quantum fluctuations of the order parameter. In the thermally disordered region, the long-range order is destroyed mainly by thermal-order parameter fluctuations, while in the quantum-disordered region the physics is dominated by quantum fluctuations. In the latter, the system essentially resembles that in its quantum-disordered ground state at $r > r_c$. In between is the so-called quantum critical region, where both types of fluctuations are important. It is located near the critical parameter value $r = r_c$ at *comparatively high temperatures*. Its boundaries are determined by the condition $k_B T > \hbar\omega_c \propto |r - r_c|^{\nu z}$: the system "looks critical" with respect to the tuning parameter r, but is driven away from criticality by thermal fluctuations. Thus, the physics in the quantum critical region is controlled by the thermal excitations of the quantum critical ground state, whose main characteristic is the absence of conventional quasiparticle-like excitations. This causes unusual finite-temperature properties in the quantum critical region, such as unconventional power laws, non-Fermi liquid behavior, etc. Universal behavior is only observable in the vicinity of the QCP, i.e., when the correlation length is much larger than microscopic length scales. Quantum critical behavior is thus cut off at high temperatures when $k_B T$ exceeds characteristic microscopic energy scales of the problem; in magnets this cutoff is set by the typical exchange energy.

---

[3] A typical example is the case of two-dimensional magnets with continuous $SU(2)$ symmetry, where order at finite T is forbidden by the Mermin–Wagner theorem.

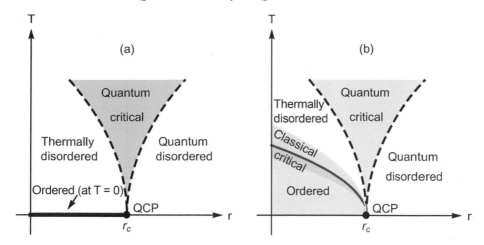

Figure 19.4 Schematic phase diagrams in the vicinity of a QCP. The horizontal axis represents the control parameter r used to tune the system through the quantum phase transition, and the vertical axis is the temperature, T. (a) Order is only present at zero temperature. The dashed lines indicate the boundaries of the quantum critical region where the leading critical singularities can be observed; these crossover lines are given by $k_B T \propto |r - r_c|^{\nu z}$. (b) Order can also exist at finite temperature. The solid line marks the finite-temperature boundary between the ordered and disordered phases. Close to this line, the critical behavior is classical.

(ii) If order also exists at finite temperatures (Figure 19.4(b)), the phase diagram is even richer. Here, a real phase transition is encountered upon variation of r at low T; the QCP can be viewed as the endpoint of a line of finite-temperature transitions. As discussed earlier, classical fluctuations will dominate in the vicinity of the finite-T phase boundary, but this region becomes narrower with decreasing temperature, such that it might even be unobservable in a low-T experiment. The fascinating quantum critical region is again at finite temperatures above the QCP.

C. Quantum-classical mapping To gain a deeper understanding of the relation between classical and quantum behavior, and the possible quantum-classical crossover, we have to recall general features from quantum statistical mechanics.

The starting point for the derivation of thermodynamic properties is the partition function

$$Z = \text{Tr } e^{-\beta \mathcal{H}}, \qquad \beta = \frac{1}{k_B T},$$

where $\mathcal{H} = \mathcal{H}_{\text{kin}} + \mathcal{H}_{\text{pot}}$ is the Hamiltonian characterizing a d-dimensional system.

In a classical system, the kinetic and potential part of $\mathcal{H}$ commute; thus Z factorizes

$$Z = Z_{\text{kin}} \times Z_{\text{pot}},$$

indicating that in a classical system, statics and dynamics decouple. The kinetic contribution to the free energy usually does not display any singularities since it derives from the product of simple Gaussian integrals. Therefore, one can study classical phase transitions using effective time-independent theories, which naturally live in d dimensions.

In contrast, in a quantum problem the kinetic and potential parts of $\mathcal{H}$ in general do not commute, and the quantum-mechanical partition function does not *factorize*, which implies that statics and dynamics are always coupled. An order-parameter field theory needs to be formulated in terms of space- and time-dependent fields.

As we discussed in Section 15.6, the canonical density operator $e^{-\beta\mathcal{H}}$ looks exactly like a time-evolution operator in imaginary time if we identify $\hbar\beta = -i\tau$. Therefore, it proves convenient to introduce an imaginary time direction into the system. At zero temperature, the imaginary time acts similar to an additional space dimension since the extension of the system in this direction is infinite.

D. Partition functions and path integrals We now focus on the expression for Z. Upon writing the trace in terms of a complete set of states,

$$Z(\beta) = \sum_n \left\langle n \left| e^{-\beta\mathcal{H}} \right| n \right\rangle. \tag{19.1}$$

Z takes the form of a sum of imaginary-time transition amplitudes for the system to start in some state $|n\rangle$ and return to the same state after an imaginary time interval $-i\hbar\beta$. Thus we see that calculating the thermodynamics of a quantum system is the same as calculating transition amplitudes for its evolution in imaginary time, with the total time interval fixed by the temperature of interest. The fact that the time interval happens to be imaginary is not central. The key idea we hope to relay is that (19.1) should evoke an image of quantum dynamics and temporal propagation. This way of looking at things can be given a particularly beautiful and practical implementation in the language of Feynman's path-integral formulation of quantum mechanics. Feynman's prescription is that the net transition amplitude between two states of the system can be calculated by summing amplitudes for all possible paths between them. The path taken by the system is defined by specifying the state of the system at a sequence of finely spaced intermediate time steps. Formally, we write

$$e^{-\beta\mathcal{H}} = \left[e^{-\mathcal{H}\delta\tau/\hbar} \right]^N,$$

where $\delta\tau$ is a time interval that is small on the time scales of interest, and N is a large integer chosen so that $N\delta\tau = \hbar\beta$. We then insert a sequence of sums over complete sets of intermediate states into the expression for $Z(\beta)$:

$$Z(\beta) = \sum_n \sum_{m_1, m_2, \ldots, m_N} \left\langle n \left| e^{-\mathcal{H}\delta\tau/\hbar} \right| m_1 \right\rangle$$
$$\times \left\langle m_1 \left| e^{-\mathcal{H}\delta\tau/\hbar} \right| m_2 \right\rangle \left\langle m_2 \right| \ldots \left| m_N \right\rangle \left\langle m_N \left| e^{-\mathcal{H}\delta\tau/\hbar} \right| n \right\rangle. \tag{19.2}$$

This rather messy expression actually has a rather simple physical interpretation: Equation (19.2) has the form of a classical partition function – if we think of imaginary time as an additional spatial dimension, then we can regard (19.2) as a sum over configurations

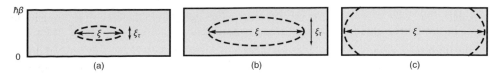

Figure 19.5 Illustration of growing correlation volume as the $(T = 0)$ critical coupling r_c is approached in a system with finite extent in the temporal direction. (a) The correlation time is much shorter than $\hbar\beta$. (b) It is comparable. (c) The system is very close to the critical point, and the correlation time (that the system would have had at zero temperature) greatly exceeds $\hbar\beta$.

expressed in terms of a transfer matrix. In particular, if our quantum system lives in d dimensions, the expression for its partition function looks like a classical partition function for a system with $(d + 1)$ dimensions, except that the extra dimension is finite in extent $-\hbar\beta$ in units of time. As $T \to 0$, the system size in this extra "time" direction diverges, and we get a truly $(d + 1)$-dimensional, effective classical system.

E. Quantum-classical crossover

Once $\hbar\beta < \xi_\tau \sim \xi^z$, the system realizes that it is effectively d-dimensional and not $(d+1)$-dimensional. The actual correlation time saturates at $\hbar\beta$, and the corresponding $T = 0$ correlation length at which this occurs is the quantum-to-classical crossover length.

Since the physics has to be continuous in temperature, the question arises of how large the temperature has to be before the system "knows" that its dimension has been reduced. The answer to this is illustrated in Figure 19.5. When the coupling r is far away from the zero temperature critical coupling r_c, the correlation length ξ is not large, and the corresponding correlation time $\xi_\tau \propto \xi^z$ is small. As long as the correlation time is smaller than the system "thickness" $\hbar\beta$, the system does not realize that the temperature is finite. That is, the characteristic fluctuation frequencies obey $\hbar\omega \gg k_B T$, and so are quantum mechanical in nature. However, as the critical coupling is approached, the correlation time grows and eventually exceeds $\hbar\beta$. (More precisely, the correlation time that the system would have had at zero temperature exceeds $\hbar\beta$; the actual fluctuation correlation time is thus limited by temperature.) At this point, the system "knows" that the temperature is finite and realizes that it is now effectively a d-dimensional classical system rather than a $(d + 1)$-dimensional system. The jargon is the system will crossover from a $(d + 1)$-dimension critical behavior to a d-dimension one.

19.2.1 Quantum Criticality and Heavy Fermions

Though high T_c cuprates triggered the exploration of non-Fermi liquid system and quantum criticality, it was heavy-fermion compounds that provided the experimental means to carry out the research, because of the ability to tune their transition temperature to $T = 0$ via pressure, doping, and magnetic fields. Experimental measurements on three-dimensional heavy-fermion compounds suggest that the quasiparticle mass also diverges in the approach to an antiferromagnetic QCP. The divergence of the electron mass at an antiferromagnetic

QCP has important consequences, for it indicates that antiferromagnetism causes a breakdown in the Fermi liquid concept. In these materials, quasiparticle masses of order 100 are recorded, but sometimes in excess of 1,000 bare electron masses have been recorded, a significant fraction of which is thought to derive from their close vicinity to an antiferromagnetic QCP.

19.3 Interacting 1D Electron Gas, Tomonaga–Luttinger Liquid, and Bosonization

It has been known since the 1960s that an interacting one-dimensional electron gas clearly breaks FLT, and that fermion physics in one dimension is distinctively different from the physics in higher dimensions. Without any calculations, it is easy to see that interactions have drastic effects compared to higher dimensions.

As shown in Figure 19.6(a), nearly free quasiparticle excitations exist in high dimension. By contrast, Figure 19.6(b) shows that an electron in one dimension that tries to propagate has to push its neighbors because of electron–electron interactions. No individual motion is possible. Any individual excitation has to become a collective one. This *collectivization* of excitations is obviously a major difference between the one-dimensional world and worlds of higher dimensions. The quasiparticle excitations are replaced in 1D by the Tomonaga–Luttinger densitylike excitations [4] having a completely different nature. It clearly invalidates any possibility to have a Fermi liquid theory work. This demonstrates that the physical properties of the one-dimensional electron gas are drastically different from the ones of a corresponding free electron gas.

For fermions with spin, this is even worse. Only collective excitations can exist, which implies that a single fermionic excitation has to split into a collective excitation carrying charge (like a sound wave) and a collective excitation carrying spin (like a spin wave).

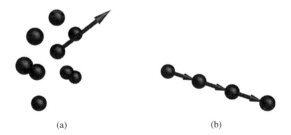

(a)   (b)

Figure 19.6 (a) In high dimensions, nearly free quasiparticle excitations that look nearly like individual particles are possible. (b) In a one-dimensional interacting system, an individual electron cannot move without pushing all the electrons. Thus, only collective excitation can exist.

---

[4] The typical properties of a one-dimensional fermionic quantum liquid were first found in the models proposed by Tomonaga [181], and Luttinger [123]. Tomonaga's important step was to realize and use the fact that the low-energy spectrum of noninteracting fermions in one dimension is identical to that of a harmonic chain. This allows us to describe the interacting fermions as a system of coupled oscillators. Luttinger's calculation of the momentum distribution in the ground state marks the appearance of power laws for interacting fermions in one dimension.

These excitations have in general different velocities, so the electron has to break into two elementary excitations. These properties, quite different from the ones of a Fermi liquid, will be the essence of the Luttinger liquid.[5]

As we have learned earlier, the onset of a Peierls transition and of charge-density waves in 2D and 3D is strongly dependent on the degree of nesting in a system. By contrast, $2k_F$ nesting in one dimension is the rule rather than the exception, regardless of the precise dispersion relation. Since the susceptibility diverges at $2k_F$, we found that any perturbation theory in the interaction to be singular at this wavevector. Alternatively, the fact that a perturbation theory diverges is an indication that the ground state of the interacting system is quite different from the one before the onset of perturbing interactions. We thus recover from this argument that the physical properties of interacting electrons in one dimension, however weak the interaction, are drastically different from the free electron ones.

19.3.1 A Heuristic Primer to Bosonization

To develop an intuitive picture of the meaning of bosonization, we shall examine the extreme limits of strong and weak interactions in the 1D electron gas.

Peculiarity of One Dimension: Noninteracting Electrons in 1D

As we have seen in Section 15.5, the signature excitations of the electron gas are of the particle–hole type that form a continuum as a function of their momentum q. In other words, for $d > 1$ and $q < 2k_F$ we can create particle–hole pairs of arbitrarily low energy by killing a particle just below the Fermi surface at one point and recreating the particle just above the Fermi surface at another point, as shown in Figure 19.7. The particle–hole excitations thus lead to a continuum extending to zero energy for all $|\mathbf{q}|$ vectors smaller than $2k_F$.

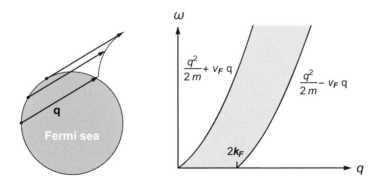

Figure 19.7 Particle–hole spectrum for two- or three-dimensional systems.

---

[5] The name Luttinger liquid was termed for this behavior by Haldane [81].

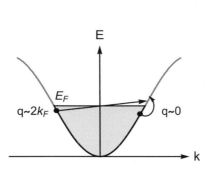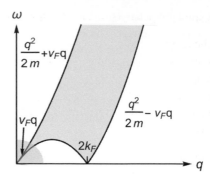

Figure 19.8 Particle–hole spectrum for one-dimensional systems. In one dimension, contrary to higher dimensions, particle–hole excitations have both a well-defined momentum and energy, for small momentum q, indicated by the gray-shaded region.

In one dimension, the Fermi surface is reduced to two points, and one cannot play with angles to increase the momentum q without moving away from the Fermi surface in energy. Since the only way to get a low-energy excitation is to destroy and recreate pairs close to the *Fermi surface*, the only places where the particle–hole energy can reach zero are for $q = 0$ and for $q = 2k_F$. In order to explore the behavior of the particle–hole spectrum in one dimension, shown in Figure 19.8, we start by writing the excitation energy $E_{\mathbf{k}}(\mathbf{q})$ as

$$E_{\mathbf{k}}(\mathbf{q}) = \varepsilon(\mathbf{k} + \mathbf{q}) - \varepsilon(\mathbf{k}),$$

where $\varepsilon(\mathbf{k})$ should be occupied and $\varepsilon(\mathbf{k} + \mathbf{q})$ empty. We examine the possible values of $E_{\mathbf{k}}(\mathbf{q})$ for the standard quadratic dispersion

$$\varepsilon(\mathbf{k}) = \frac{k^2 - k_F^2}{2m}.$$

It is easy to check that for $k \in [k_F - q, k_F]$ the average value $E(\mathbf{q})$ of $E_{\mathbf{k}}(\mathbf{q})$ and the dispersion $\delta E(\mathbf{q}) = \max(E(\mathbf{q})) - \min(E(\mathbf{q}))$ are

$$E(\mathbf{q}) = \frac{k_F q}{m} = v_F q$$

$$\delta E(\mathbf{q}) = \frac{q^2}{2m} = \frac{E^2(\mathbf{q})}{m v_F^2}. \tag{19.3}$$

A similar calculation can be made by expanding the energy around k_F. If one writes

$$\varepsilon(\mathbf{k}) = v_F (k - k_F) + \frac{\lambda}{2} (k - k_F)^2,$$

then it is obvious that

$$E(\mathbf{q}) = v_F q$$

$$\delta E(\mathbf{q}) = \lambda q^2 = \frac{\lambda}{v_F^2} E^2(\mathbf{q}). \tag{19.4}$$

The results (19.3) or (19.4) clearly show the following:

(i) The average energy of a particle–hole excitation is only dependent on its momentum q, endowing the particle–hole excitations with well-defined momentum q and energy $E(q)$

(ii) The dispersion in energy $\delta E(q)$ goes to zero much faster than the average energy, regardless of the dispersion relation $\varepsilon(\mathbf{k})$, provided it has a finite slope at the Fermi level.

This means that in one dimension the particle–hole excitations are well-defined "particles" – objects with well-defined momentum and energy – which become longer and longer lived when the energy approaches zero. It is reminiscent of Fermi liquid quasiparticles. Because these excitations consist of particle–hole pairs, they are bosonic in nature. These bosonic quasiparticles will just be the key to solving our one-dimensional problem – an observation that is at the root of the bosonization method.

From an alternatively perspective, we can surmise that for small q and energy $E(q)$, the particle–hole spectrum resembles a sound mode with dispersion $\omega = v_F q$. This suggests that the low-energy excitations of the 1D electron gas are similar to that of a 1D elastic medium, a picture that remains valid even in the presence of interactions. To see this, we will now consider the opposite extreme of strongly interacting electrons.

Strong Repulsive Interactions: The 1D Wigner Crystal

In the opposite limit, the potential energy will dominate. In this regime, the lowest-energy electronic configuration is the Wigner crystal, shown in Figure 19.9, with a lattice constant $a = 1/n_0$ – n_0 is the linear electron density – with *phononlike* low-energy excitations.

We define the phononlike electron displacements as

$$x_i = x_i^0 + \frac{a}{\pi} \theta_i,$$

where x_i^0 is the equilibrium position of the ith electron, and the angular displacement variable θ_i advances by π when the crystal is displaced by one lattice constant. The phononlike excitations are represented by the Lagrangian:

$$
\begin{aligned}
\mathcal{L} &= \int dx \, \frac{ma}{2\pi^2} \, \dot\theta^2(x) - \frac{1}{2} \iint dx\, dx' \, V(x - x') \, \delta n(x) \, \delta n(x') \\
&= \int dx \, \frac{ma}{2\pi^2} \, \dot\theta^2(x) - \frac{V_0}{2} \int dx \, \delta n^2(x).
\end{aligned}
\tag{19.5}
$$

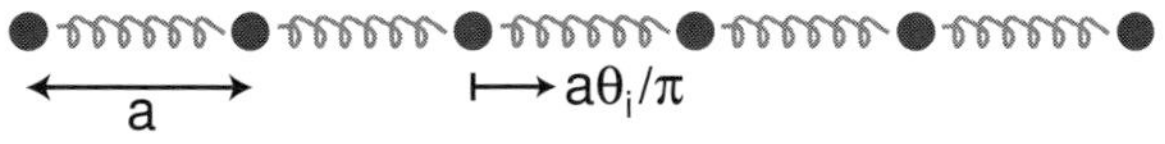

Figure 19.9 A Wigner crystal and the phonon displacement coordinate θ_i.

$\delta n(x)$ is the deviation of the density from its average value n_0. We have introduced the following assumptions in deriving (19.5):

- θ_i is assumed to vary slowly and is treated as a continuous variable.
- The long wavelength assumption, together with a short-ranged screened Coulomb interaction, allows for the effective potential to be represented by a delta function with $V_0 = \int dx\, V(x)$.[6]

Noting that for $\theta(L) - \theta(0) = -\pi$, exactly one extra electron resides between 0 and L, we express δn as

$$\delta n(x) = -\frac{\partial_x \theta(x)}{\pi},$$

and we arrive at the Lagrangian form

$$\mathcal{L} = \int dx \left[\frac{ma}{2\pi^2} (\partial_t \theta)^2 - \frac{V_0}{2\pi^2} (\partial_x \theta)^2 \right].$$

It can be recast as

$$\mathcal{L} = \frac{\hbar}{2\pi g} \int dx \left[\frac{1}{v_\rho} (\partial_t \theta)^2 - v_\rho (\partial_x \theta)^2 \right]. \qquad (19.6)$$

where the interaction parameter g is given by

$$g = \sqrt{\frac{\pi \hbar v_F}{V_0}} \qquad (19.7)$$

and the phonon velocity defined as

$$v_\rho = \sqrt{\frac{V_0}{ma}} = \frac{v_F}{g}.$$

The similarity in behavior of the low-energy Wigner crystal elastic phonon theory, representing strong interaction $g \ll 1$, with the "sound mode" for noninteracting electrons, suggests that (19.6) may remain correct for weaker interactions. However, (19.7) may need to be modified in such a case.

Quantum Fluctuations in 1D

As we already know, the Mermin–Wagner theorem stipulates that in one and two dimensions, long-range order is destroyed by thermal or quantum fluctuations. Here, we shall explore how quantum fluctuations in 1D destroy the long-range Wigner crystalline order at zero temperature. We represent the crystalline order by the Fourier series of the electron density

$$n(x) = \sum_{q \sim 0} n_q\, e^{iqx} + \sum_{q \sim 2k_F} n_q\, e^{iqx} + \cdots, \qquad (19.8)$$

---

[6] For a Coulomb interaction $V(x) = e^2/x$ screened at large distances by a ground plane at distance R_s, we have $V_0 = 2e^2 \ln(R_s/a)$.

where the main contributions will come from the long wavelength fluctuations of the density

$$\sum_{q\sim 0} n_q \, e^{iqx} = n_0 - \frac{\partial_x \theta(x)}{\pi} \tag{19.9}$$

and from fluctuations at the wavelength of the Wigner crystal $a = 1/n_0 = 2\pi/2k_F$, represented by the second term. The latter may be characterized by a slowly varying complex function $n_{2k_F}(x)$ delineating the amplitude and the phase of the oscillation at $2k_F$,

$$\sum_{q\sim 2k_F} n_q \, e^{iqx} = n_{2k_F}(x) \, e^{i2k_Fx} + c.c. \tag{19.10}$$

Since $2k_F = 2\pi n_0$, and an electron is added when θ changes by $-\pi$, we may write

$$n_{2k_F}(x) \sim e^{2i\theta(x)}. \tag{19.11}$$

For a perfect crystal, a $2k_F$ Bragg peak corresponds to

$$\langle n_{2k_F} \rangle \neq 0. \tag{19.12}$$

As we have shown in Chapter 6, its amplitude is attenuated by the Debye–Waller factor, arising from phonon thermal fluctuations. We will now show that quantum fluctuations give rise to a logarithmically divergent Debye–Waller factor that may destroy the crystalline order. Recall from Section 6.8 that

$$\langle n_{2k_F} \rangle = \left\langle e^{2i\theta(x,\tau)} \right\rangle = \exp\left[-\frac{1}{2}\left\langle (2\theta)^2 \right\rangle \right]$$

and we write

$$\left\langle \theta^2 \right\rangle = \sum_{q,\omega} \left\langle |\theta(q,\omega)|^2 \right\rangle = \int \frac{dq\, d\omega}{(2\pi)^2} \frac{\pi g v_\rho}{\omega^2 + v_\rho^2 q^2},$$

where the integrand is obtained from the Euler–Lagrange equation associated with Lagrangian (19.6). We use the relation $\omega = v_\rho q$ to get

$$\int \frac{2\pi q\, dq}{(2\pi)^2} \frac{\pi g}{q^2} = \frac{g}{2} \ln\left[\frac{L}{a} \right].$$

We find that the Debye–Waller factor arising from 1D quantum fluctuations logarithmically diverges as $L \to \infty$. We arrive at

$$\langle n_{2k_F} \rangle = e^{-2\langle \theta^2 \rangle} \sim \left(\frac{a}{L} \right)^g \to 0.$$

A similar calculation for the correlation function

$$\lim_{x\to\infty} \left\langle n_{2k_F}(x)\, n_{2k_F}(0) \right\rangle \neq 0 \tag{19.13}$$

gives

$$\left\langle n_{2k_F}(x)\, n_{2k_F}(0) \right\rangle = \exp\left[\frac{1}{2}\left\langle (2\theta(x) - 2\theta(0))^2 \right\rangle \right] = \left(\frac{a}{x} \right)^{2g}.$$

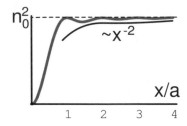

Figure 19.10 Pair correlation function for noninteracting electrons. The oscillations indicate power law crystalline order.

Thus, the crystalline correlations decay as a power law at long distances. For strong interactions $g \ll 1$, the exponent is very close to zero, and a Wigner crystal precariously persists. By comparison, long-range correlations for noninteracting electrons can be extracted from the pair correlation function, shown in Figure 19.10,

$$\left\langle \Psi^\dagger(x)\, \Psi^\dagger(0)\, \Psi(0)\, \Psi(x) \right\rangle = n_0^2 \left[1 - \frac{\sin^2(k_F x)}{x^2} \right],$$

which shows that

$$\left\langle n_{2k_F}(x)\, n_{2k_F}(0) \right\rangle \sim \frac{1}{x^2}.$$

The power law decay of the oscillations suggests that noninteracting electrons are also described by (19.6) with $g = 1$.

We surmise that (19.6) is indeed more general than the Wigner crystal limit; however, the form of g should be system dependent.

19.3.2 Spinless 1D Fermion System: The Tomonaga–Luttinger Model

As we have learned, the low-energy long wavelength particle–hole excitations of the 1D system exhibit nearly linear dispersion and thus well-defined excitations. The linearity of low-energy excitations are a manifestation of the near-linear electronic band structure dispersion in the vicinity of the Fermi points of the 1D Fermi gas, as shown in Figure 19.11(a). The Tomonaga–Luttinger (TL) model [74] extends the linearity to the entire electron–hole excitation spectrum by replacing the original parabolic dispersion with the purely linear spectrum of Figure 19.11.(a). It should be clear that the low-lying excitations in both models are the same. In order to completely eliminate the particle–hole pair energy dependence on the initial state momentum k, for all q, the TL model extends the electronic energy dispersion down to $-\infty$, $k \in [-\infty, \infty]$, as shown in Figure 19.11(b). This modification makes the model exactly solvable even in the presence of nontrivial and possibly strong interactions. It is the simplest model, and perhaps the most studied, which describes the low-lying excitations around the two Fermi points in a 1D Fermi gas with density–density interactions.

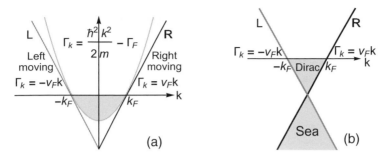

Figure 19.11 (a) Parabolic energy band of noninteracting electrons in 1D, shown in light gray. The gray area is the Fermi sea. Linearized energy bands in the vicinity of the Fermi level are extended along the black and gray lines, which introduces two species of fermions (right (R) and left (L) going fermions). (b) The linear dispersion is extended down to $-\infty$, thus replacing the Fermi sea with the Dirac sea.

Since the states deep below the Fermi level are inert, the TL model is expected to have the same low-energy behavior as the real electron gas: replacing the original parabolic dispersion, which has a finite number of electrons filling the band, with an infinite *Dirac sea* in the TL model should not change the physics close to the Fermi points. However, the TL model requires the introduction of *chirality*, namely, two species of fermions: right (*R*) and left (*L*) going fermions. We write the Hamiltonian of this system as

$$\mathcal{H} = \sum_{\eta=\pm 1} v_F \int dx \, \Psi_\eta^\dagger(x) \, (i\eta \, \partial_x - k_F) \, \Psi_\eta(x) - \frac{1}{2} \int dx \, dx' \rho(x) \, V(x - x')\rho(x').$$
(19.14)

v_F is the Fermi velocity, and $\eta = \pm$ labels the two species: $R \to +$, $L \to -$. The field operator $\Psi(x)$ and the total density operator ρ are defined as

$$\Psi_\eta(x) = \left(\frac{2\pi}{\mathfrak{L}}\right)^{1/2} \sum_k e^{iqx} \, c_{\eta(k_F+k)}$$

$$\rho(x) = \Psi_+^\dagger(x) \, \Psi_+(x) + \Psi_-^\dagger(x) \, \Psi_-(x)$$
(19.15)

where we introduced periodic boundary conditions with period $\mathfrak{L}$, so that $k = 2\pi n/\mathfrak{L}$ and $q = 2\pi n_q/\mathfrak{L}$.[7] The vacuum state, shown in Figure 19.12, is defined as

$$|\emptyset\rangle : \begin{cases} c_{\eta k} |\emptyset\rangle &= 0, \, k, n > 0 \\ c_{\eta k}^\dagger |\emptyset\rangle &= 0, \, k, n \leq 0. \end{cases}$$
(19.16)

Now, we shall use the method of *bosonization* [58, 74, 76, 163, 188], specific to 1D, to obtain an exact solution to the TL model. The bosonization procedure can be formulated

---

[7] Discreteness and unboundedness of k are essential prerequisites for a systematic derivation of the bosonization scheme and its identities. Discreteness provides systematic bookkeeping of states, and unboundedness allows for the definition of proper bosonic operators.

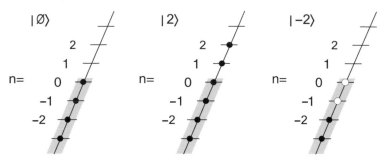

Figure 19.12 The vacuum state $|\emptyset\rangle$: $c_{\eta k} |\emptyset\rangle = 0$, $k(n) > 0$ and $c_{\eta k}^\dagger |\emptyset\rangle = 0$, $k(n) \leq 0$. The states containing two extra particles, $c_{\eta,1}^\dagger c_{\eta,2}^\dagger |\emptyset\rangle$, and two fewer particles, $c_{\eta,0} c_{\eta,-1} |\emptyset\rangle$, are also shown.

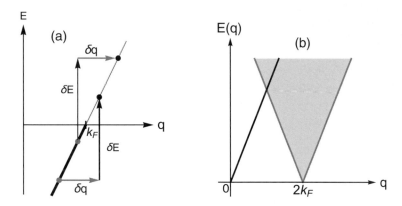

Figure 19.13 (a) In a 1D linearized regime, the excitation energy has a single value δE for a given δq. (b) Particle–hole excitation spectrum for the linearized model (19.14). For small q, particle–hole excitations are sharp excitations with well-defined energy and momentum.

precisely for fermions with a linear energy-momentum relation in the form of operator identities. The underlying physics is that in 1D we cannot perform particle–particle exchange – there is no real difference between fermions and bosons.

19.3.3 Bosonization of the Noninteracting Tomonaga–Luttinger Model

The noninteracting Hamiltonian is expressed in the momentum representation as

$$\mathcal{H}_0 = \sum_{k;\eta=\pm} v_F \, (\eta \, k - k_F) \, c_{\eta,k}^\dagger \, c_{\eta,k}, \tag{19.17}$$

having particle–hole excitations with well-defined momentum q and energy

$$E_{\eta,k}(q) = \eta \left[v_F \, (k + q) - v_F \, k \right] = \eta \, v_F \, q \tag{19.18}$$

totally independent of k, as shown in Figure 19.13(a). The particle–hole excitation spectrum is shown in Figure 19.13(b).

The fact that the particle–hole excitations of this model are well defined suggests that we can try to rewrite (19.17) in a basis representing such excitations. This is also motivated by our earlier argument that only collective excitations can remain in one dimension. Density fluctuations,

$$
\begin{cases}
\rho^\dagger(q) = \sum_k c^\dagger_{k+q} c_k, & |q| > 0 \\
\rho(q) = \sum_k c^\dagger_k c_{k+q} = \sum_k c^\dagger_{k-q} c_k = \rho^\dagger(-q),
\end{cases}
\tag{19.19}
$$

which consist of superpositions of electron–hole excitations, are obvious natural candidates for such representation, and as we recall, the ρ s are bosonic operators.

If we can demonstrate that a density fluctuation of momentum q has a well-defined energy, we can proceed to second-quantize it: we can identify $\rho(q)$ or $\rho^\dagger(q)$ with bosonic operators b_q and $b^\dagger_q$, or a linear combination thereof. The advantage is that the interaction Hamiltonian becomes quadratic when expressed in terms of density operators, or their bosonic equivalent

$$
\mathcal{H}_{int} = \frac{1}{2\mathfrak{L}} \sum_q v(q)\, \rho(q)\, \rho(-q) = \frac{1}{2\mathfrak{L}} \sum_q v(q) \left(\ldots b_q + \cdots b^\dagger_q \right)^2,
\tag{19.20}
$$

and is easy to diagonalize.

We start by constructing the noninteracting Hamiltonian in terms of these operators, using the following steps.

Normal Ordering

In the TL model, the number of electrons in the Dirac sea – the vacuum state $|\emptyset\rangle$ with respect to particle–hole excitations – is infinite, since all the negative energy states are occupied. Therefore, it only makes sense to talk about deviations in the density from the density of the vacuum state. It turns out that the process of normal-ordering produces the desired deviations: We recall that normal ordering of operators is equivalent to subtracting the average value in the vacuum, namely,

$$
N[AB] = AB - \langle \emptyset | AB | \emptyset \rangle .
$$

The normal-ordered density is defined as

$$
N\left[\rho_\eta(x)\right] = N\left[\Psi^\dagger_\eta(x)\, \Psi_\eta(x) \right]
$$

and the Fourier component $\rho_\eta(q)$ of the density is given by

$$
N\left[\rho_\eta(x)\right] = \frac{1}{\mathfrak{L}} \sum_q N\left[\rho_\eta(q)\right] e^{iqx}
$$

$$
N\left[\rho_\eta(q)\right] =
\begin{cases}
\sum_k c^\dagger_{\eta,k+q} c_{\eta,k} & (q \neq 0) \\
\sum_k \left[c^\dagger_{\eta,k} c_{\eta,k} - \left\langle \emptyset \left| c^\dagger_{\eta,k} c_{\eta,k} \right| \emptyset \right\rangle \right] = \mathcal{N}_\eta & (q = 0).
\end{cases}
\tag{19.21}
$$

$\mathcal{N}_\eta$ is the operator that counts the number of η-electrons relative to $|\emptyset\rangle$, for example, the states shown in Figure 19.12 give 0, 2, and -2. From now on, we shall identify ρ_η as $N[\rho_\eta]$, and write

$$\rho_\eta(x) = \frac{2\pi}{\mathcal{L}} \sum_{|q|>0} \rho_\eta(q)\, e^{iqx} + \frac{2\pi}{\mathcal{L}} \mathcal{N}_\eta. \tag{19.22}$$

Commutators

The commutator of the density operators for two different species is obviously zero

$$\left[\rho_+^\dagger(q),\, \rho_-^\dagger(q') \right] = 0,$$

while for same species, we write

$$\left[\rho_\eta^\dagger(q),\, \rho_\eta^\dagger(-q') \right] = \sum_{k_1,k_2} \left[c_{\eta,k_1+q}^\dagger c_{\eta,k_1},\, c_{\eta,k_2-q'}^\dagger c_{\eta,k_2} \right]$$

$$= \sum_{k_1,k_2} \left(c_{\eta,k_1+q}^\dagger c_{\eta,k_2}\, \delta_{k_1,k_2-q'} - c_{\eta,k_2-q'}^\dagger c_{\eta,k_1}\, \delta_{k_1+q,k_2} \right)$$

$$= \sum_{k_2} \left(N\left[c_{\eta,k_2+q-q'}^\dagger c_{\eta,k_2} \right] - N\left[c_{\eta,k_2-q'}^\dagger c_{\eta,k_2-q} \right] \right)$$

$$+ \sum_{k_2} \left(\left\langle \emptyset \middle| c_{\eta,k_2+q-q'}^\dagger c_{\eta,k_2} \middle| \emptyset \right\rangle - \left\langle \emptyset \middle| c_{\eta,k_2-q'}^\dagger c_{\eta,k_2-q} \middle| \emptyset \right\rangle \right)$$

$$= \sum_{k_2} \left(\left\langle \emptyset \middle| c_{\eta,k_2+q-q'}^\dagger c_{\eta,k_2} \middle| \emptyset \right\rangle - \left\langle \emptyset \middle| c_{\eta,k_2-q'}^\dagger c_{\eta,k_2-q} \middle| \emptyset \right\rangle \right), \tag{19.23}$$

where we avoided pending infinities in the sums via normal ordering.

Since $\left\langle \emptyset \middle| c_{\eta,k_2}^\dagger c_{\eta,k_1} \middle| \emptyset \right\rangle = \delta_{k_1,k_2}$, we arrive at

$$\left[\rho_\eta^\dagger(q),\, \rho_\eta^\dagger(-q') \right] = \delta_{q,q'} \sum_{k_2} \left[\left\langle \emptyset \middle| c_{\eta,k_2}^\dagger c_{\eta,k_2} \middle| \emptyset \right\rangle - \left\langle \emptyset \middle| c_{\eta,k_2-q}^\dagger c_{\eta,k_2-q} \middle| \emptyset \right\rangle \right] \tag{19.24}$$

and we obtain

$$\left[\rho_\eta^\dagger(q),\, \rho_{\eta'}^\dagger(-q') \right] = -\delta_{\eta\eta'}\, \delta_{q,q'}\, \eta\, \frac{\mathcal{L}}{2\pi} \int_0^q dk = -\delta_{\eta\eta'}\, \delta_{q,q'}\, \frac{\eta\, \mathcal{L}}{2\pi}\, q. \tag{19.25}$$

An alternative perspective, which reveals the physical meaning of the density operators, is to consider how the operators act on the ground state. For $q > 0$, $\rho_+^\dagger(q) |\emptyset\rangle$ is a superposition of states with a single particle–hole excitation at k and $k+q$, as shown in Figure 19.14, where the hole has momentum $-q < k < 0$. But for $q < 0$, $\rho_+^\dagger(-q) |\emptyset\rangle = 0$ because in $|\emptyset\rangle$ there are no empty states with momentum less than an occupied state. Moreover, when

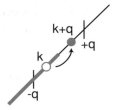

Figure 19.14 A particle–hole excitation created by the chiral density operator.

$\rho_\eta^\dagger(-q)$ acts on $\rho_\eta^\dagger(q) |\emptyset\rangle$, the only thing it can do is put the excited particle back into the hole. Thus we conclude that

$$\left[\rho_R^\dagger(q), \rho_R^\dagger(-q) \right] |\emptyset\rangle = \frac{\mathfrak{L}}{2\pi} \int_{-q}^{0} dq \, |\emptyset\rangle = \frac{\mathfrak{L}}{2\pi} q \, |\emptyset\rangle .$$

Equation (19.25) underscores the bosonic nature of the density operator. Moreover, we note that for the TL model (19.17) this is an exact result.

Equivalence of Fermion and Boson Hamiltonians

We note that

$$\rho_-^\dagger(q > 0) \, |\emptyset\rangle = 0$$
$$\rho_+^\dagger(q < 0) \, |\emptyset\rangle = 0 \qquad (19.26)$$

so that these density operators can be identified with destruction operators for bosons, as can be surmised from Figure 19.15. This allows us to define the boson creation and destruction operators, for $q \neq 0$, as

$$\begin{cases} b_q^\dagger = i\sqrt{\frac{2\pi}{\mathfrak{L}|q|}} \sum_\eta \Theta(\eta q) \rho_\eta^\dagger(q), \\[2mm] b_q = -i\sqrt{\frac{2\pi}{\mathfrak{L}|q|}} \sum_\eta \Theta(\eta q) \rho_\eta^\dagger(-q), \end{cases} \Rightarrow \left\{ \rho_{\eta,q} = \sqrt{\frac{\mathfrak{L}|q|}{2\pi}} \left(\Theta(\eta q) \, b_q + \Theta(-\eta q) \, b_{-q}^\dagger \right), \right.$$

$$(19.27)$$

where Θ is the Heaviside step function, and $q = (2\pi n_q)/\mathfrak{L}$, $n_q \in \mathbb{Z}$. We note from (19.25) that the commutators of the b operators become

$$\left[b_{\eta,q}, b_{\eta',q'}^\dagger \right] = \delta_{\eta\eta'} \, \delta_{qq'} . \qquad (19.28)$$

Or we can define species bosonic operators as

$$\begin{cases} b_{\eta,q}^\dagger = i\sqrt{\frac{2\pi}{\mathfrak{L}|q|}} \, \rho_\eta^\dagger(\eta q), \\[2mm] b_{\eta,q} = -i\sqrt{\frac{2\pi}{\mathfrak{L}|q|}} \, \rho_\eta(\eta q), \end{cases} \qquad q > 0. \qquad (19.29)$$

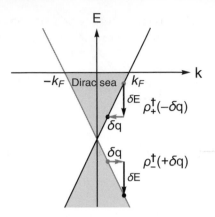

Figure 19.15 Depiction of $\rho_+^\dagger(\delta q < 0)\,|\emptyset\rangle$ and $\rho_-^\dagger(\delta q > 0)\,|\emptyset\rangle$.

Here we consider the commutators of the Hamiltonian with the boson operators. For $q > 0$, $b_q^\dagger = \sqrt{2\pi/\mathcal{L}|q|}\,\rho_+(q)$; we obtain

$$\left[b_q^\dagger, \mathcal{H}_0\right] = i\left(\frac{2\pi}{\mathcal{L}|q|}\right)^{1/2}\sum_{\eta,k}\left[\rho_+^\dagger(q),\, v_F\,(\eta\,k - k_F)\,c_{\eta,k}^\dagger c_{\eta,k}\right]$$

$$= i\left(\frac{2\pi}{\mathcal{L}|q|}\right)^{1/2}\sum_{k,k_1} v_F\,(k - k_F)$$

$$\times\left(c_{+,k_1+q}^\dagger\left\{c_{+,k_1}c_{+,k}^\dagger\right\}c_{+,k} - c_{+,k}^\dagger\left\{c_{+,k}c_{+,k_1+q}^\dagger\right\}c_{+,k_1}\right)$$

$$= i\left(\frac{2\pi}{\mathcal{L}|q|}\right)^{1/2}\sum_{k,k_1} v_F\,(k - k_F)\left(c_{+,k_1+q}^\dagger c_{+,k}\,\delta_{k_1,k} - c_{+,k}^\dagger c_{+,k_1}\,\delta_{k_1+q,k}\right)$$

$$= i\left(\frac{2\pi}{\mathcal{L}|q|}\right)^{1/2} v_F\left[\sum_k k\,c_{+,k-q}^\dagger c_{+,k} - \sum_{k_1}(k_1 + q)\,c_{+,k_1-q}^\dagger c_{+,k_1}\right]$$

$$= -\left(\frac{2\pi}{\mathcal{L}|q|}\right)^{1/2}\sum_k v_F\,q\,c_{+,k-q}^\dagger c_{+,k} = -v_F\,q\,b_q, \tag{19.30}$$

which shows that states created by $b_{+,q}^\dagger$ are eigenstates of $\mathcal{H}_0$, with energy $v_F q$: $b_{+,q}^\dagger$ creates particle–hole pairs, each having total momentum q and energy $\varepsilon_{k+q} - \varepsilon_k = v_F q$, independent of k because of the linearity of the spectrum. Thus, *states created by $b_{+,q}^\dagger$ are linear combinations of individual electron–hole excitations all with the same energy, and therefore are also eigenstates of* (19.17). So, if we assume for the moment that the basis generated by the operators b is complete, then the *results* (19.27) *define completely the*

Hamiltonian in the boson basis. It is easy to check, from the boson commutation relations, that the operator that would satisfy the commutation rules (19.27) is

$$
\mathcal{H} = \frac{2\pi v_F}{\mathfrak{L}} \sum_{\eta} \left[\sum_{q \neq 0} \rho_\eta(q)\, \rho_\eta(-q) + \frac{\mathcal{N}_\eta(\mathcal{N}_\eta + 1)}{2} \right]
$$

$$
= \left[\sum_{q \neq 0} v_F\, |q|\, b_q^\dagger\, b_q + \frac{\pi v_F}{\mathfrak{L}} \mathcal{N}_\eta\, (\mathcal{N}_\eta + 1) \right], \tag{19.31}
$$

where $\pi v_F \left(\mathcal{N}_+ (\mathcal{N}_+ + 1) + \mathcal{N}_- (\mathcal{N}_- + 1) \right) / \mathfrak{L}$ is the energy of the particles added/ removed from $|\emptyset\rangle$ and counted for by the operators $\mathcal{N}_\eta$ – the so-called *zero modes*. Remarkably, (19.31) shows that, contrary to naive expectations, the kinetic energy, which is normally quadratic in fermions operators, can also be expressed as quadratic terms in boson operators. Thus, adding the interaction will keep the Hamiltonian quadratic and allows for solving the problem in a remarkably simple way.

Single Fermion Operators

Up to this point, the construction does not allow for a direct calculation of correlation functions involving individual creation or destruction operators, like the one-particle Green function. Such correlation functions become tractable by representing single-particle operators in terms of the boson operators [188].

(a) Bosonic Restructuring of Fock Space As depicted in Figure 19.12, the highest filled level of the vacuum state $|\emptyset\rangle$ is labeled by $n = 0$ and the lowest empty level by $n = 1$, for all η. We shall use this $|\emptyset\rangle$ as reference state relative to which the occupations of all other states in Fock space $\mathbb{F}$ are specified. The set of all states with the same $\mathcal{N}_\eta$-eigenvalues, N_+, N_-, will be referred to as $|N\rangle \equiv |N_+, N_-\rangle$-particle and defines the Hilbert space $\mathbb{H}_N$. It contains infinitely many states, corresponding to different configurations of particle–hole excitations, all of which will generically be denoted by $|N\rangle$. Furthermore, for given N, we set $|N\rangle_0$ as the N-particle state that has no particle–hole excitations. Since it is the lowest-energy state in $\mathbb{H}_N$, we shall call it the N-particle ground state, $b_{\eta, q} |N\rangle_0 = 0 \,\forall \eta, q$. Any other state in $\mathbb{H}_N$ is obtained as $|N\rangle = f(b^\dagger) |N\rangle_0$.

It is then possible to restructure the Fock space $\mathbb{F}$ of states spanned by the $c_{\eta k}$ operators as a direct sum, $\mathbb{F} = \bigoplus_N \mathbb{H}_N$ over Hilbert spaces $\mathbb{H}_N$ characterized by fixed particle numbers $N = N_+, N_-$, within each of which all excitations are particle–hole-like and hence have bosonic character.

(b) Klein Factors $F_\eta^\dagger$, F_η To connect the various $\mathbb{H}_N$'s, we need to define "ladder operators" that raise or lower the total η-fermion number by one – an action that bosonic operators are unable to do. In addition, they also ensure that fermion fields of different species anticommute. Such operators are known as Klein factors, and denoted by $F_\eta^\dagger$ and F_η. They need to satisfy the following commutators $\left[b_{\eta,q}, F_{\eta'}^\dagger\right] = \left[b_{\eta,q}, F_{\eta'}\right] = \left[b_{\eta,q}^\dagger, F_{\eta'}\right] = \left[b_{\eta,q}^\dagger, F_{\eta'}\right] = 0,$ $\forall\, \eta, \eta', q.$ Moreover, their action on a generic state $|N\rangle$ is defined as

$$F_+^\dagger\, |N\rangle = f(b^\dagger)\, C_+\, |N_+ + 1, N_-\rangle_0 = f(b^\dagger)\, (-1)^{N_-/2}\, |N_+ + 1, N_-\rangle_0,$$

$$F_+\, |N\rangle = f(b^\dagger)\, C_+\, |N_+ - 1, N_-\rangle_0 \qquad (19.32)$$

similarly for $F_-^\dagger$, F_-. C_η is the fermion *phase-counting* operator defined in Section 13.3.2 Since the spectrum of $\mathcal{N}_\eta$ is unbounded from above or below, F_η is unitary: $F_\eta^{-1} = F_\eta^\dagger$, and can be written as $F_\eta = e^{i\theta_\eta}$, $\theta^\dagger = \theta$. The Klein factors obey the commutations

$$\left\{F_\eta^\dagger, F_{\eta'}\right\} = 2\delta_{\eta\eta'}\ \forall\, \eta, \eta', (F_\eta^\dagger F_\eta = F_\eta F_\eta^\dagger = 1) \qquad (19.33)$$

$$\left\{F_\eta^\dagger, F_{\eta'}^\dagger\right\} = \{F_\eta, F_{\eta'}\} = 0\ \eta \neq \eta', \qquad (19.34)$$

$$\left[\mathcal{N}_\eta, F_{\eta'}^\dagger\right] = \delta_{\eta\eta'}\, F_\eta^\dagger, \quad \left[\mathcal{N}_\eta, F_{\eta'}\right] = -\delta_{\eta\eta'}\, F_\eta. \qquad (19.35)$$

(c) Real Space Representations

1. Bosonic Field Operators We now define the hermitian boson field $\phi_\eta(x)$ in terms of the b operators as

$$\phi_\eta(x) = -\eta\, \left(\frac{2\pi}{\mathfrak{L}}\right)^{1/2} \sum_{\eta q > 0} \frac{1}{\sqrt{\eta q}} \left(e^{-iqx}\, b_{\eta,q} + e^{iqx}\, b_{\eta,q}^\dagger\right) e^{-aq/2} = \varphi_\eta(x) + \varphi_\eta^\dagger(x).$$

$a > 0$ is an infinitesimal parameter needed to regularize ultraviolet ($q \to \infty$) divergent momentum sums – it will be set to zero at the end of the calculations. We will calculate the commutators $\left[\varphi_\eta(x), \varphi_\eta^\dagger(x')\right]$ and $\left[\phi_\eta(x), \partial_{x'}\, \phi_\eta^\dagger(x')\right]$, which we will need later in deriving relevant correlation functions,

$$\left[\varphi_\eta(x), \varphi_\eta^\dagger(x')\right] = \frac{2\pi}{\mathfrak{L}} \sum_{\eta q > 0} \frac{e^{q[i(x'-x)-a]}}{|q|} \qquad \left[b_{\eta,q}, b_{\eta,q}^\dagger\right] = \sum_{n_q} \frac{e^{2\pi n_q [i(x'-x)-a]/\mathfrak{L}}}{n_q}$$

$$= -\ln\left[1 - \exp\left(\frac{2\pi\left[i(x-x') - a\right]}{\mathfrak{L}}\right)\right]$$

$$\simeq -\ln\left[\frac{2\pi\left[i(x-x') - a\right]}{\mathfrak{L}}\right], \quad \mathfrak{L} \to \infty. \qquad (19.36)$$

We used the series $\ln(1 - y) = -\sum_{n=1}^\infty y^n/n$ in the second line. The second commutator gives

$$\left[\phi_\eta(x), \partial_{x'}\, \phi_\eta^\dagger(x')\right] = \partial_{x'}\left(\left[\varphi_\eta(x), \varphi_\eta^\dagger(x')\right] + \left[\varphi_\eta^\dagger(x), \varphi_\eta(x')\right]\right)$$

$$= \partial_{x'}\left(\left[\varphi_\eta(x), \varphi_\eta^\dagger(x')\right] - \left[\varphi_\eta(x), \varphi_\eta^\dagger(x')\right]^\dagger\right)$$

$$= -\partial_{x'}\left(\ln\left[1 - \exp\left([2\pi i(x - x') - a]/\mathfrak{L}\right)\right] - H.C.\right)$$

$$= -\frac{2\pi i}{\mathfrak{L}}\,\frac{\exp\left([2\pi i(x - x') - a]/\mathfrak{L}\right)}{1 - \exp\left([2\pi i(x - x') - a]/\mathfrak{L}\right)} - H.C.$$

$$= -\frac{2\pi i}{\mathfrak{L}}\,\frac{1}{\exp\left([2\pi i(x' - x) + a]/\mathfrak{L}\right)} - H.C. \tag{19.37}$$

For $\mathfrak{L} \gg 1$, we expand $(\exp(y) - 1)^{-1} = 1/y - 1/2 + \mathcal{O}(y)$, and write

$$\left[\phi_\eta(x), \partial_{x'}\phi_\eta^\dagger(x')\right] = \left(\frac{1}{x' - x + ia} - H.C.\right) + \frac{2\pi i}{\mathfrak{L}} + \mathcal{O}\left(\frac{1}{\mathfrak{L}^2}\right)$$

$$= -2\pi i\left(\frac{a/\pi}{(x' - x)^2 + a^2} - \frac{1}{\mathfrak{L}}\right) + \mathcal{O}\left(\frac{1}{\mathfrak{L}^2}\right)$$

$$= -2\pi i\left(\delta(x' - x)^2 - \frac{1}{\mathfrak{L}}\right) + \mathcal{O}\left(\frac{1}{\mathfrak{L}^2}\right). \tag{19.38}$$

where we have taken the limit $a \to 0$. Moreover, in the limit $\mathfrak{L} \to \infty$, we find that $\partial_x\phi/2\pi$ acts like the conjugate momentum to the field ϕ, a property that we shall shortly exploit.

We also find that $\rho_\eta(x)$ depends linearly on $\partial_x\phi(x)$, as

$$\rho_\eta(x) = \frac{2\pi}{\mathfrak{L}}\sum_{\eta q > 0} \rho_\eta(q)\,e^{iqx} + \frac{2\pi}{\mathfrak{L}}\,N_\eta$$

$$= \left(\frac{2\pi}{\mathfrak{L}}\right)^{1/2}\sum_{\eta q > 0} i\sqrt{\eta q}\left(e^{-iqx}b_{\eta,q} - e^{iqx}b_{\eta,q}^\dagger\right) + \frac{2\pi}{\mathfrak{L}}\,N_\eta$$

$$= \partial_x\phi_\eta(x) + \frac{2\pi}{\mathfrak{L}}\,N_\eta \qquad (a \to 0), \tag{19.39}$$

showing that in the limit $\mathfrak{L} \to \infty$, $\rho_\eta(x)$ and $\phi(x)$ become conjugate variables. Since $\int \rho\,dx$ is a particle number, we discern that ϕ must be a phase.

2. Anomalous Commutators The commutation relations of the density operator $\rho_\eta(x)$, as defined in (19.22), are given by

$$\left[\rho_\eta(x), \rho_\eta(x')\right] = \frac{1}{\mathfrak{L}^2}\sum_{qq'}{}' e^{iqx - iq'x'}\,\left[\rho_\eta(q), \rho_\eta(q')\right] = \frac{1}{2\pi\mathfrak{L}}\sum_q{}' q\,e^{iq(x - x')}$$

$$= \frac{1}{2\pi\mathfrak{L}}\,\frac{\partial}{i\partial x}\sum_q{}' e^{iq(x - x')} = \pm\frac{1}{2\pi i}\,\frac{\partial}{\partial x}\,\delta(x - x'), \tag{19.40}$$

where the $+$ sign is for right movers. These commutators involve space derivatives of delta functions instead of canonical delta functions. It is an example of a commutator that is nonzero only for a system of infinite number of particles, but, as we know, vanishes for any finite system. Such commutators are called *anomalous* commutators, and such relations are known as the *Kac–Moody algebra*. They are a consequence of the separation of the electrons into two different species with infinite bands.

3. Elastic String Representation We now introduce two new fields $\Phi(x)$ and $\Theta(x)$ as

$$\Phi(x) = \frac{1}{2} \sum_\eta \left[\phi_\eta(x) - \frac{\pi x}{\mathcal{L}} \, \mathcal{N}_\eta \right] \tag{19.41}$$

$$\Theta(x) = -\frac{1}{2} \sum_\eta \eta \left[\phi_\eta(x) + \frac{\pi x}{\mathcal{L}} \, \mathcal{N}_\eta \right]. \tag{19.42}$$

In the limit $\mathcal{L} \to \infty$, the canonically conjugate momentum to Φ, $\Pi(x)$, is defined as

$$\Pi(x) = \frac{1}{\pi} \, \partial_x \, \Theta(x) \tag{19.43}$$

$$\left[\Phi(x), \Pi(x') \right] = -\frac{1}{4\pi} \left(\left[\phi_+(x), \partial_{x'} \, \phi_+^\dagger(x') \right] - \left[\phi_-(x), \partial_{x'} \, \phi_-^\dagger(x') \right] \right)$$

$$= i\delta(x - x') - \frac{i}{\mathcal{L}}. \tag{19.44}$$

It is then physically appealing to rewrite the Hamiltonian (19.31) as

$$\mathcal{H}_0 = \frac{v_F}{4\pi} \int_{-\mathcal{L}/2}^{\mathcal{L}/2} dx \left[(\partial_x \phi_+(x))^2 + (\partial_x \phi_+(x))^2 \right] + \frac{\pi v_F}{2\mathcal{L}} \left[((\mathcal{N}_+ + \mathcal{N}_-)^2 + (\mathcal{N}_+ - \mathcal{N}_n)^2 \right]$$

$$= \frac{v_F}{2\pi} \int_{-\mathcal{L}/2}^{\mathcal{L}/2} dx \left[(\pi \, \Pi(x))^2(x) + \left(\partial_x \Phi(x) \right)^2 \right], \tag{19.45}$$

which reveals that the Hamiltonian is just that of an elastic string, with Φ the local displacement field. The corresponding Hamilton equations of motion are

$$\dot{\Pi}(x) = -\frac{\delta \mathcal{H}_0}{\delta \Phi(x)} = v_F \, \partial_x^2 \, \Phi(x), \qquad \dot{\Phi}(x) = \frac{\delta \mathcal{H}_0}{\delta \Pi(x)} = v_F \, \Pi(x)$$

$$\partial_x \, \dot{\Phi}(x) = v_F \, \partial_x \, \Pi(x). \tag{19.46}$$

The last line has the form of the continuity equation

$$\dot{\rho} + \partial_x J = 0 \quad \Rightarrow \quad J = v_F \, (\rho_+ - \rho_-), \ \rho = \rho_+ + \rho_-.$$

Moreover, Hamilton's equations yield the wave equation

$$\ddot{\Phi} - v_F^2 \, \partial_x^2 \, \Phi = 0$$

with the dispersion relation $\omega = v_F \, q$.

(d) The Bosonization Identity We consider the single-particle operator $\Psi_\eta(x)$ that destroys a right/left going fermion at point x, namely,

$$\Psi_\eta(x) = \frac{1}{\sqrt{\mathcal{L}}} \sum_k e^{ikx} \, c_{k\eta}.$$

Using (19.27), it is easy to show that

$$\left[b_{\eta',q}, \psi_\eta(x)\right] = \delta_{\eta\eta'}\, \alpha_q(x)\, \psi_\eta(x) \tag{19.47}$$

$$\left[b_{\eta',q}^\dagger, \psi_\eta(x)\right] = \delta_{\eta\eta'}\, \alpha_q^*(x)\, \psi_\eta(x), \tag{19.48}$$

where $\alpha_q(x) = \sqrt{2\pi/\mathfrak{L}|q|}\, e^{iqx}$. Equation (19.47) indicates that $\psi_\eta(x)\,|N\rangle_0$ is an eigenstate of the bosonic annihilation operator $b_{\eta,q}$, with eigenvalue $\alpha_q(x)$:

$$b_{\eta',q}\, \psi_\eta(x)\,|N\rangle_0 = \delta_{\eta\eta'}\, \alpha_q(x)\, \psi_\eta(x)\,|N\rangle_0\,.$$

Hence, it must have a coherent-state representation of the form

$$\psi_\eta(x)\,|N\rangle_0 = \exp\left[\sum_q \alpha_q(x) b_{\eta,q}^\dagger\right] F_\eta \lambda_\eta(x)\,|N\rangle_0 = e^{-i\varphi_\eta^\dagger(x)} F_\eta \lambda_\eta(x)\,|N\rangle_0, \tag{19.49}$$

where $\lambda_\eta(x)$ is a phase factor to be determined, and its inclusion is required since no action has been performed on $|N\rangle_0$. We can say that the Fock–number states in fermionic Hilbert space correspond to coherent states in the bosonic Hilbert space.

Multiplying both sides of (19.49) by $_0\langle N|\, F^\dagger$, we obtain

$$_0\langle N|\, F^\dagger \psi_\eta(x)\,|N\rangle_0 =\,_0\langle N|\, e^{-i\varphi_\eta^\dagger(x)} \lambda_\eta(x)\,|N\rangle_0 = \lambda_\eta(x)\,_0\langle N|\, e^{-i\varphi_\eta^\dagger(x)}\,|N\rangle_0 = \lambda_\eta(x)$$

since $_0\langle N|\, b_q^\dagger = 0$. But

$$_0\langle N|\, F^\dagger \psi_\eta(x)\,|N\rangle_0 = \sqrt{\frac{2\pi}{\mathfrak{L}}}\,_0\langle N|\, F^\dagger \sum_k e^{-ikx} c_{\eta,k_F,N_\eta}\,|N\rangle_0\,.$$

Since $_0\langle N|\, F^\dagger$ does not contain any particle–hole pairs, the only nonzero term in the summation comes from $c_{\eta,k}\,|N\rangle_0 = F_\eta\,|N\rangle_0$, that is, the annihilation of the highest η-momentum level. Thus we obtain

$$\lambda_\eta(x) = \sqrt{\frac{2\pi}{\mathfrak{L}}}\, \exp\left[-ik_{F,N_\eta}x\right]. \tag{19.50}$$

Alternatively, the commutator of $\Psi_\eta^\dagger(x)$ with the density operator

$$\left[\psi_\eta^\dagger(x), \rho_\eta^\dagger(x')\right] = -\delta(x-x')\, \psi_\eta^\dagger(x) \tag{19.51}$$

shows that $\psi_\eta^\dagger(x)$ increases the density by $\delta(x'-x)$ – the density of the additional particle. Since the density operator is bosonic, we find that if we express the single electron operator in the form of a displacement operator $\psi \sim e^B$ (see Exercise 19.5), where B is a linear function of the boson operators, its commutator with the density operator yields

$$\left[e^B, \rho_\eta(x')\right] = \left[B, \rho_\eta(x')\right] e^B, \tag{19.52}$$

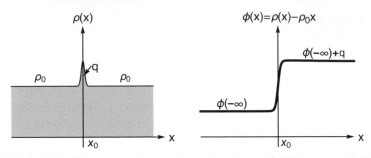

Figure 19.16 A kink in ϕ at point x_0 corresponds to the creation of a particle at this point. The amplitude of the kink gives the charge of the particle.

which has the same form as the commutator of $\psi_\eta^\dagger(x)$ with $b_{\eta,q}^\dagger$. Comparing (19.40), (19.51), and (19.52), we arrive at

$$B = \mp 2\pi i \int_{-\infty}^x dx' \left[\rho_\eta(x') - \rho_{0\eta} \right], \tag{19.53}$$

where the upper sign goes with $\eta = +$. Thus the action of $\psi_\eta^\dagger(x)$ is to add an electron and to displace the boson modes because of the sudden appearance of one electron.

To visualize this operation, we consider the following scenario. First, the field ϕ is obviously related to the density of particles. Remember that the long wavelength part of the density is simply $\rho(x) = \partial_x \phi(x)/\pi$. If we add a single particle via $\Psi_+^\dagger(x_0)$, we see immediately the two facts illustrated in Figure 19.16. A particle at point x_0 would manifest itself as $\rho(x_0) = \delta(x - x_0) = \partial_x \phi(x = x_0)$; this would appear as a kink in ϕ. One sees that the step in ϕ is a measure of the total charge added in the system. Of course, in the case of removal of a particle the converse is true.

For a general state in $\mathbb{H}_N$, we have $|N\rangle = f(\{b_q^\dagger\}) |N\rangle_0$, and we write

$$\psi_\eta(x) |N\rangle = \psi_\eta(x) f\left(\{b_{\eta',q}^\dagger\}\right) |N\rangle_0 = f\left(\{b_{\eta',q}^\dagger - \delta\eta\eta' \alpha_q^*(x)\}\right) \psi_\eta(x) |N\rangle_0$$

$$= F_\eta \, \lambda_\eta(x) \, e^{-i\varphi_\eta^\dagger(x)} \, f\left(\{b_{\eta',q}^\dagger - \delta\eta\eta' \alpha_q^*(x)\}\right) |N\rangle_0 \,.$$

Using the operator identity

$$f(A + C) = e^{-B} f(A) e^B; \qquad C = [A, B], \ [A, C] = [B, C] = 0$$

with $A = b_{\eta,q}^\dagger$, $B = i\varphi_\eta(x)$ and $C = -\delta\eta\eta' \alpha_q^*(x)$, we obtain

$$\psi_\eta(x) |N\rangle = F_\eta \, \lambda_\eta(x) \, e^{-i\varphi_\eta^\dagger(x)} \, e^{-i\varphi_\eta(x)} \, f\left(\{b_{\eta',q}^\dagger\}\right) e^{i\varphi_\eta(x)} |N\rangle_0$$

$$= F_\eta \, \lambda_\eta(x) \, e^{-i\varphi_\eta^\dagger(x)} \, e^{-i\varphi_\eta(x)} \, f\left(\{b_{\eta',q}^\dagger\}\right) |N\rangle_0$$

$$= \sqrt{\frac{2\pi}{\mathfrak{L}}} \, F_\eta \, \exp\left[-ik_{F,N_\eta} x\right] e^{-i\varphi_\eta^\dagger(x)} \, e^{-i\varphi_\eta(x)} |N\rangle, \tag{19.54}$$

where we used $b_{\eta,q} |N\rangle_0 = 0$ and the Baker–Housdorff identity.

What (19.54) expresses is that creation or annihilation of a localized charge involves physics on two length scales. The usual de Broglie wavelength of a particle moving with the Fermi momentum gives the $e^{\pm i k_F x}$ factor, while the exponent with the boson fields causes a shake-up of a lot of low-energy collective modes of the electron gas. The length scale for the latter depends on the energy involved, and is given by $q^{-1} \sim v_F/\varepsilon$, where ε is the available energy. The decomposition of the single-electron operator into an infinite number of low-energy states is what gives rise to many of the peculiar properties of 1D metals.

This completes our derivation of the bosonization formulas. Since the derivations were based on elementary operator identities, we can safely proceed to calculate, for example, the anticommutator $\left\{ \psi_\eta, \psi_\eta^\dagger \right\}$ or the correlator $\left\langle \psi_\eta \psi_\eta^\dagger \right\rangle$.

19.3.4 Interacting Tomonaga–Luttinger Model

Having established the fermion–boson mapping, we first turn to examine interaction effects. The interaction has the form

$$\mathcal{H}_{\text{int}} = \int dx \, \rho(x) \, V(x - x') \, \rho(x'). \tag{19.55}$$

We note that the interaction couples to the total density. Thus, to correctly consider interactions involving left and right movers, we need first to refine our definition of the fermionic field operator:

$$\Psi(x) = \frac{1}{\mathcal{L}} \sum_k e^{ikx} \, c_k.$$

Since we are actually interested in low-energy excitations close to the Fermi surface, we confine the sum to momenta close to $+k_F$ and momenta close to $-k_F$:

$$\Psi(x) \simeq \frac{1}{\mathcal{L}} \left[\sum_{-\Lambda < k - k_F < \Lambda} e^{ikx} \, c_k + \sum_{-\Lambda < k + k_F < \Lambda} e^{ikx} \, c_k \right],$$

where Λ is a momentum cutoff such that $v_F \Lambda \sim W$, the electronic bandwidth. A physical electron is neither a right mover nor a left mover; it is actually a right mover for $k > 0$ and a left mover for $k < 0$, so that the destruction operator becomes

$$c_k = \Theta(k) \, c_{+,k} + \Theta(-k) \, c_{-,k}.$$

The density operator can then be written as

$$\rho(q) = \sum_k c_{k-q}^\dagger c_k = \sum_k \left[\Theta(k - q) \, \Theta(k) \, c_{+,k-q}^\dagger c_{+,k} + \Theta(-k + q) \, \Theta(-k) \, c_{-,k-q}^\dagger c_{-,k} \right.$$

$$\left. + \, \Theta(k - q) \, \Theta(-k) \, c_{+,k-q}^\dagger c_{-,k} + \Theta(-k + q) \, \Theta(k) \, c_{-,k-q}^\dagger c_{+,k} \right]. \tag{19.56}$$

Since both k and $k - q$ are restricted to be in the vicinity of $\pm k_F$, the only two possibilities for q then become (i) $|q|/k_F \sim 0$ (k and $k - q$ have the same sign), or (ii) $|q| \sim 2k_F$ (k and $k - q$ have opposite signs). The first case is associated with the top line of (19.56), where the electron density fluctuation is just the sum of right and left movers' components. The second case corresponds to the terms on the second line of (19.56), which involves electron transfer from one side of the Fermi surface to the other. We can also write

$$\Psi(x) = \Psi_+(x) + \Psi_-(x)$$

so that the density is expressed as

$$\Psi^\dagger(x)\, \Psi(x) = \Psi_+^\dagger(x)\, \Psi_+(x) + \Psi_-^\dagger(x)\, \Psi_-(x) + \Psi_+^\dagger(x)\, \Psi_-(x) + \Psi_-^\dagger(x)\, \Psi_+(x)$$

$$= \rho_+(x) + \rho_-(x) + S_\pm + S_\pm^\dagger. \tag{19.57}$$

Substituting the density expression (19.57) in the interaction Hamiltonian (19.55), we obtain different interaction processes, the *g-ology*, that specify the type of incoming or outgoing fermions – right or left movers. The *g-ology* classification of these processes is shown in Figure 19.17. The g_4 process only couples fermions on the same side of the Fermi surface. The g_2 process couples fermions from one side of the Fermi surface with fermions on the other side, yet each species stays on its side of the Fermi surface after the interaction – *forward scattering*. In contrast, the g_1 process corresponds to a $2k_F$ *backscattering* where fermions exchange sides. Note that for spinless fermions g_2 and g_1 processes are identical since one can exchange the indistinguishable outgoing particles. This is not the case of spinful fermions, since the interaction has to conserve spin, and processes g_2 and g_1 are different.

Spinless Case

We write the g_4 process for right movers in bosonic representation as

$$g_4\, \Psi_\pm^\dagger(x)\, \Psi_\pm(x)\, \Psi_\pm^\dagger(x)\, \Psi_\pm(x) = g_4\, \rho_\pm(x)\, \rho_\pm(x). \tag{19.58}$$

The sum of the two processes leads for the g_4 interaction to

$$g_4 \left[\rho_+(x)\, \rho_+(x) + \rho_-(x)\, \rho_-(x) \right] = g_4 \left[\left(\pi\, \Pi(x) \right)^2 + \left(\partial_x\, \Phi(x) \right)^2 \right].$$

The contribution of g_2 processes can be expressed as

$$g_2\, \Psi_+^\dagger(x)\, \Psi_+(x)\, \Psi_-^\dagger(x)\, \Psi_-(x) = g_2\, \rho_+(x)\, \rho_-(x) = -g_2 \left[\left(\pi\, \Pi(x) \right)^2(x) - \left(\partial_x\, \Phi(x) \right)^2 \right].$$

The g_2 interaction does not commute with the kinetic energy $[\mathcal{H}_2, \mathcal{H}_0] \neq 0$. It can therefore modify the ground state by exciting particle–hole pairs out of the Fermi sea. On the other hand, g_4 commutes $[\mathcal{H}_4, \mathcal{H}_0] = 0$, and the Fermi sea remains the ground state in the presence of g_4 alone.

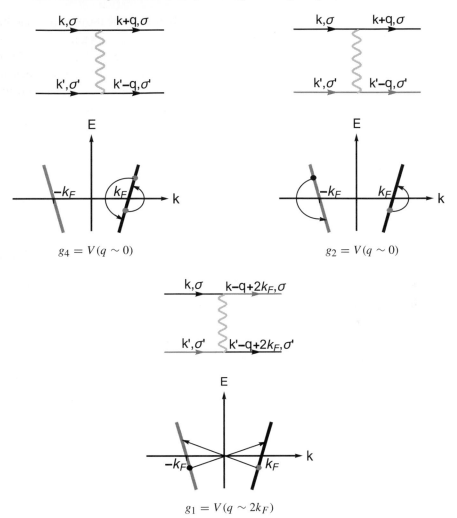

Figure 19.17 The important low-energy interaction processes are classified in three types. A black line represents a right-moving fermion with $k \sim +k_F$ and a gray line a left-moving fermion with $k \sim -k_F$.

The total interaction Hamiltonian is given by[8]

$$\mathcal{H}_{\text{int}} = \frac{1}{2\pi} \int_{-\mathcal{L}/2}^{\mathcal{L}/2} dx \left[(g_4 - g_2) \left(\pi \, \Pi(x) \right)^2 + (g_4 + g_2) \left(\partial_x \, \phi(x) \right)^2 \right]. \qquad (19.59)$$

---

[8] Notice that we considered q-independent interaction processes – a δ-function interaction in real space. Taking q-dependent interaction is not much more complicated. Actually, as long as the limit $g(q \to 0)$ is finite, the physics we describe asymptotically holds.

Adding the kinetic energy term, we write the full Hamiltonian in the form

$$\mathcal{H} = \frac{1}{2\pi} \int_{-\mathcal{L}/2}^{\mathcal{L}/2} dx \left[u\, g \left(\pi \Pi(x) \right)^2 + \frac{u}{g} \left(\partial_x \phi(x) \right)^2 \right], \tag{19.60}$$

where

$$\begin{cases} ug = v_F \left[1 + \dfrac{g_4}{v_F} - \dfrac{g_2}{v_F} \right] \\[2mm] \dfrac{u}{g} = v_F \left[1 + \dfrac{g_4}{v_F} + \dfrac{g_2}{v_F} \right] \end{cases} \Rightarrow \begin{cases} u = v_F \sqrt{(1 + v_4)^2 - v_2^2}, \\[2mm] g = \sqrt{\dfrac{1 + v_4 - v_2}{1 + v_4 + v_2}}, \end{cases} \tag{19.61}$$

g is a dimensionless parameter, and u has the dimensions of a velocity. $v_i = g_i/v_F$ is a dimensionless coupling constant. It is clear that $g < 1$ for repulsive interactions ($v_2 > 0$) and $g > 1$ for attractive ones ($v_2 < 0$). We see that the Hamiltonian remains *quadratic*, even in the presence of interactions. As discussed before, this is one of the main attractions of the bosonization method. To diagonalize the Hamiltonian, we Fourier-transform (19.60)

$$\mathcal{H} = \frac{u}{2\mathcal{L}} \sum_q \left[g\, \Pi_{-q}\, \Pi_q + \frac{q^2}{g}\, \Phi_{-q}\, \Phi_q \right] \tag{19.62}$$

together with

$$\left[\Phi_q, \Pi_{-q'} \right] = i \mathcal{L}\, \delta qq'.$$

Next we define the boson operators that diagonalize (19.62)

$$a_q = \frac{1}{\sqrt{\mathcal{L}}} \left(\sqrt{\frac{|q|}{g}}\, \Phi_q + i\sqrt{\frac{g}{|q|}}\, \Pi_q \right), \quad a_q^\dagger = \frac{1}{\sqrt{\mathcal{L}}} \left(\sqrt{\frac{|q|}{g}}\, \Phi_{-q} - i\sqrt{\frac{g}{|q|}}\, \Pi_{-q} \right),$$

which leads to the diagonalized form

$$\mathcal{H} = \sum_q u\, |q|\, a_q^\dagger a_q. \tag{19.63}$$

Equation (19.63) shows that the low-energy physics of the interacting TL model is described by *free bosonic excitations*. It is also quite remarkable to find how simple it is to solve the interacting 1D Hamiltonian of spinless fermions!

Alternatively, we can express the interacting Hamiltonian in terms of the b_q operators and obtain

$$\mathcal{H} = \mathcal{H}_b + \mathcal{H}_N$$

$$\mathcal{H}_b = v_F (1 + v_4) \sum_{q>0} q \left[\sum_{\eta=\pm} b_{\eta,q}^\dagger b_{\eta,q} + \lambda \left(b_{\eta,q}^\dagger b_{-\eta,q}^\dagger + b_{\eta,q} b_{-\eta,q} \right) \right],$$

$$\mathcal{H}_N = \frac{\pi v_F}{\mathcal{L}} (1 + v_4) \left[\sum_\eta \mathcal{N}_\eta^2 + 2\lambda\, \mathcal{N}_+ \mathcal{N}_- \right] \tag{19.64}$$

with $\lambda = v_2/(1+v_4)$. $\mathcal{H}_b$ is easily diagonalized with the aid of a Bogoliubov transformation of the form

$$
\begin{cases}
\beta_{1,q} = \cosh\theta\, b_{+,q} + \sinh\theta\, b^\dagger_{-,q}, \\
\beta^\dagger_{2,q} = \sinh\theta\, b_{+,q} + \cosh\theta\, b^\dagger_{-,q}
\end{cases}
\Rightarrow
\begin{cases}
b_{+,q} = \cosh\theta\, \beta_{1,q} - \sinh\theta\, \beta^\dagger_{2,q}, \\
b^\dagger_{-,q} = -\sinh\theta\, \beta_{1,q} + \cosh\theta\, \beta^\dagger_{2,q}
\end{cases}
$$

$$
\tanh(2\theta) = \lambda \;\rightarrow\; \theta = \frac{1}{4}\ln\left[\frac{1+\lambda}{1-\lambda}\right] = \frac{1}{2}\ln(g). \tag{19.65}
$$

The diagonalized Hamiltonian reads

$$
\mathcal{H}_b = u \sum_{q>0,\, v=1,2} q\, \beta^\dagger_{v,q}\, \beta_{v,q} \tag{19.66}
$$

and

$$
\beta_{1,q} = \frac{1}{2}\left[\left(\frac{1}{\sqrt{g}} + \sqrt{g}\right) b_{+,q} + \left(\frac{1}{\sqrt{g}} - \sqrt{g}\right) b^\dagger_{-,q}\right],
$$

$$
\beta^\dagger_{2,q} = \frac{1}{2}\left[\left(\frac{1}{\sqrt{g}} - \sqrt{g}\right) b_{+,q} + \left(\frac{1}{\sqrt{g}} + \sqrt{g}\right) b^\dagger_{-,q}\right]. \tag{19.67}
$$

From (19.65), we obtain

$$
\varphi_{1,2}(x) = \sqrt{\frac{2\pi}{\mathfrak{L}}} \sum_{q>0} \frac{1}{\sqrt{q}} \left(e^{\mp iqx}\, \beta_{(1,2),q} + e^{\pm iqx}\, \beta^\dagger_{(1,2),q}\right) e^{-aq/2}
$$

$$
= \cosh\gamma\, \varphi_{+,-}(x) - \sinh\gamma\, \varphi_{-,+}(x). \tag{19.68}
$$

The Spinful Case

For fermions with spins, each interaction can take two values $(g_\parallel, g_\perp)$ depending on whether the spins σ and σ' of the two interacting fermions are parallel $g_\parallel$ or opposite $g_\perp$. Adding the spin degrees of freedom modifies $\mathcal{H}_b$ in (19.64) to read

$$
\mathcal{H}_b = v_F \sum_{q>0,\, \eta=\pm} q\, b^\dagger_{\eta,q,\sigma}\, b_{\eta,q,\sigma}
$$

$$
+ \sum_{\substack{q>0\eta=\pm \\ \sigma\sigma'}} q\left[\left(g_{4\parallel}\delta_{\sigma,\sigma'} + g_{4\perp}\delta_{\sigma,-\sigma'}\right) b^\dagger_{\eta,q,\sigma}\, b_{\eta,q,\sigma'}\right.
$$

$$
\left. + \left(g_{2\parallel}\delta_{\sigma,\sigma'} + g_{2\perp}\delta_{\sigma,-\sigma'}\right)\left(b^\dagger_{\eta,q,\sigma}\, b^\dagger_{-\eta,q,\sigma'} + b_{\eta,q,\sigma}\, b_{-\eta,q,\sigma'}\right)\right]. \tag{19.69}
$$

Naturally, the interaction couples ↑-spin with ↓-spin. To remove such spin couplings, we define total bosonic charge and spin operators as

$$
\begin{cases}
b_{\eta,q,c} = \dfrac{1}{\sqrt{2}}\left[b_{\eta,q,\uparrow} + b_{\eta,q,\downarrow}\right], \\[2mm]
b_{\eta,q,s} = \dfrac{1}{\sqrt{2}}\left[b_{\eta,q,\uparrow} - b_{\eta,q,\downarrow}\right],
\end{cases}
\qquad
\left[b^\dagger_{\eta,q,c},\, b_{\eta,q,s}\right] = 0.
$$

We find that $\mathcal{H}_b$ separates into two mutually commuting Hamiltonians, one for the charge, and one for the spin degree of freedom:

$$\mathcal{H}_b = \mathcal{H}_c + \mathcal{H}_s$$

$$\mathcal{H}_c = \sum_{\substack{q>0 \\ \eta=\pm}} q \left[(v_F + g_{4c}) \, b^\dagger_{\eta,q,c} \, b_{\eta,q,c} + g_{2c} \left(b^\dagger_{\eta,q,c} \, b^\dagger_{-\eta,q,c} + b_{\eta,q,c} \, b_{-\eta,q,c} \right) \right]$$

$$\mathcal{H}_s = \sum_{\substack{q>0 \\ \eta=\pm}} q \left[(v_F + g_{4s}) \, b^\dagger_{\eta,q,s} \, b_{\eta,q,s} + g_{2s} \left(b^\dagger_{\eta,q,s} \, b^\dagger_{-\eta,q,s} + b_{\eta,q,s} \, b_{-\eta,q,s} \right) \right], \quad (19.70)$$

where $g_{i,\frac{c}{s}} = g_{i\parallel} \pm g_{i\perp}$. It is then straightforward to find a Bogoliubov transformation that diagonalizes each of the two independent Hamiltonians. It is also clear that the spinful system supports two independent branches of excitations: density waves and spin waves. There will now be a velocity u_κ and a coupling constant g_κ for each excitation sector, $\kappa = c, s$, given by

$$u_c = v_F \sqrt{\left(1 + \frac{g_{4\kappa}}{v_F} \right)^2 - \left(\frac{g_{2\kappa}}{v_F} \right)^2},$$

$$g_\kappa = \sqrt{\frac{v_F + g_{4\kappa} - g_{2\kappa}}{v_F + g_{4\kappa} + g_{2\kappa}}}. \quad (19.71)$$

The dynamics of the Luttinger model, as written in terms of decoupled charge and spin excitations, is a manifestation of the phenomenon of spin-charge separation, an important feature of fermionic systems in one spatial dimension. An electron that is introduced into an interacting system will rapidly decay into its constituent elementary excitations: charge and spin density modes that propagate at different velocities and that will spatially separate with time. The full Hilbert space can be completely represented as a product of charge excitations and spin excitations. This phenomenon is completely absent in higher dimensions, especially in systems described by Landaus Fermi liquid theory.

19.3.5 Green Functions

In this chapter, we were able to reduce the interacting fermion system to a system of non-interacting bosons. It should straightforward to compute all correlations of interest for such a noninteracting system. As a simple example we will calculate the equal time Green function

$$G^>(x) = -i \left\langle \psi(x) \psi^\dagger(0) \right\rangle$$

$$\psi(x) = \psi_+(x) + \psi_-(x)$$

at zero temperature. To simplify the calculations further, we shall ignore any processes that convert a fermion on one of the dispersion branches to one on the other dispersion branch's nondiagonal contributions such as $\langle \psi_- \psi^\dagger_+ \rangle$. We need to compute only the diagonal ones $\langle \psi_+ \psi_+^\dagger \rangle$ and $\langle \psi_- \psi_-^\dagger \rangle$. By reflection symmetry, they are equal except for $x \to -x$, so we focus on one of them, namely on the right branch, and we compute $G_+^>(x) = -i \langle \psi_+(x) \psi_+^\dagger(0) \rangle$.

From (19.49), we write

$$\psi^\dagger(0) \, |N\rangle_0 = \frac{1}{\mathfrak{L}} \, F_+^\dagger \, e^{i\varphi_+^\dagger(0)} \, e^{i\varphi_+(0)} \, |N\rangle_0$$

$$_0\langle N| \, \psi_+(x) =_0 \langle N| \, \frac{1}{\mathfrak{L}} \, e^{-i\varphi_+^\dagger(x)} \, e^{-i\varphi_+(x)} \, F_+.$$

Since the Hamiltonian is diagonal in the β operators, we can express φ_+ as

$$\varphi_+(x) = -\sum_{q>0} \frac{1}{n_q} \, e^{iqx - aq/2} \left[\cosh(\theta) \, \beta_{1q} - \sinh(\theta) \, \beta_{2,q}^\dagger \right] \tag{19.72}$$

$$= C \, \varphi_1(x) - S \, \varphi_2^\dagger(x), \tag{19.73}$$

where $C \equiv \cosh(\theta)$ and $S \equiv \sinh(\theta)$. Using (19.72), we get

$$e^{i\varphi_+^\dagger(x)} \, e^{i\varphi_+(x)} = e^{i\left(C \, \varphi_1^\dagger(x) - S \, \varphi_2(x)\right)} \, e^{i\left(C \, \varphi_1(x) - S \, \varphi_2^\dagger(x)\right)}$$

$$= e^{iC \, \varphi_1^\dagger(x)} \, e^{-iS \, \varphi_2(x)} \, e^{-iS \, \varphi_2^\dagger(x)} \, e^{iC \, \varphi_1(x)}$$

$$= e^{i\left(C \, \varphi_1^\dagger(x) - S \, \varphi_2^\dagger(x)\right)} \, e^{i(C \, \varphi_1(x) - S \, \varphi_2(x))} \, e^{-S^2\left[\varphi_2, \varphi_2^\dagger\right]}.$$

From (19.36), we have $\left[\varphi_2, \varphi_2^\dagger\right] = -\ln(2\pi a/\mathfrak{L})$, and we obtain

$$e^{-S^2\left[\varphi_2, \varphi_2^\dagger\right]} = \left(\frac{2\pi a}{\mathfrak{L}}\right)^{S^2}.$$

The correlation function is then given by

$$\langle \psi(x) \psi^\dagger(0) \rangle = \frac{1}{\mathfrak{L}} \left(\frac{2\pi a}{\mathfrak{L}}\right)^{2S^2} e^{ik_{F,N_+}x} \left\langle e^{i\left(C \, \varphi_1^\dagger(x) - S \, \varphi_2^\dagger(x)\right)} \, e^{i\left(C \, \varphi_1^\dagger(0) - S \, \varphi_2^\dagger(0)\right)} \right\rangle$$

$$= \frac{1}{\mathfrak{L}} \left(\frac{2\pi a}{\mathfrak{L}}\right)^{2S^2} e^{ik_{F,N_+}x} \, e^{C^2[\varphi_1(x), \varphi_1(0)] + S^2[\varphi_2(x), \varphi_2(0)]}.$$

For $\mathfrak{L} \gg |x|$, we use the approximation in (19.36) and we obtain

$$G_+^>(x) = \frac{-i}{2\pi(a-ix)} \left(\frac{\mathfrak{L}}{2\pi}\right)^{2S^2} \left(\frac{1}{(a+ix)(a-ix)}\right)^{S^2} \left(\frac{2\pi a}{\mathfrak{L}}\right)^{2S^2} e^{ik_{F,N_+}x}.$$

Finally, we arrive at

$$G_+^>(x) = \frac{1}{2\pi(x+ia)} \left(\frac{a^2}{a^2+x^2}\right)^{S^2} e^{ik_{F,N_+}x}. \tag{19.74}$$

We note that in the absence of interactions, $S = 0$, we have

$$G >_+ (x) = \frac{-i e^{i k_F, \mathcal{N}_+ x}}{2\pi (a - ix)} \xrightarrow{x \gg a} \frac{e^{i k_F, \mathcal{N}_+ x}}{2\pi x},$$

whereas we find that

$$\int_{-\infty}^{\infty} \Theta(k - a) \exp(ikx)\, dk = -\frac{i e^{iax}}{x},$$

If we identify a with k_F, then we obtain the step function

$$1 - n_F = \Theta(k - k_F),$$

which is just the distribution of holes in a Fermi gas. If we now turn on the interactions, we find that for $x \to \infty$ $G >_+ (x) \propto a^{2S^2}/x^{1+2S^2}$, where a^{2S^2} is kept for ensuring the correct dimension. But

$$\int_{-\infty}^{\infty} \frac{e^{-i(k - k_F)x}}{x^{1+2g}}\, dx \propto |k - k_F|^{2g}$$

and we obtain the a power law jump for the momentum occupation number

$$n(k) \propto \frac{1}{2} - \text{sign}(k - k_F)\, a^{2S^2}\, |k - k_F|^{S^2}. \tag{19.75}$$

Instead of the discontinuity at k_F that signals in a Fermi liquid that fermionic quasiparticles are sharp excitations, one finds in one dimension an essential power law singularity. Formally, this corresponds to $Z = 0$. This is the signature that individual fermionic excitations cannot survive in one dimension. As we already discussed, they are converted into collective ones. As a result, the decay of a single-particle Green function is always faster than in a free electron gas. Note that the position of the singularity is still at k_F.

Exercises

19.1 The RPA susceptibility of an interacting 1D electron system is given by

$$\chi(q, \omega) = \frac{\Pi_0(q, \omega)}{1 - V_c(q) \Pi_0(q, \omega)},$$

where

$$\Pi_0(q, \omega) = \sum_k \frac{\tanh[\beta(\varepsilon_{k+q} - \mu)/1] - \tanh[\beta(\varepsilon_k - \mu)/2]}{\omega + \varepsilon_{k+q} - \varepsilon_k + i\eta}.$$

(a) Determine $\Pi_0(q, \omega)$ for $q \ll k_F$ by expressing as $\varepsilon_{k+q} = \varepsilon_k + q v_F \,\text{sign}(k)$ and integrating over ε_k:

$$\int_0^{\infty} dx\, [\tanh(x + \alpha) - \tanh(x)] = \ln[1 + \tanh(\alpha)].$$

(b) The Coulomb potential in 1D is given by

$$V_c(q) \approx -e^2 \ln[qa], \quad qa \ll 1$$

for a wire of diameter a. Calculate $\chi(q,\omega)$.

(c) Determine the imaginary part of $\chi(q,\omega)$. What type and character of the excitations do you discern?

19.2 Derive the commutators (19.47) and (19.48).

19.3 Consider the displacement operator $D(z) = \exp\left[zb^\dagger - z^*b\right]$. Show that

$$D^\dagger(z)\,b\,D(z) = b + z$$
$$D^\dagger(z)\,b^\dagger\,D(z) = b + z^*.$$

The name displacement operator comes precisely from the fact that it *displaces* the operator b by a constant z.

20

Electron–Phonon Interactions

20.1 Introduction

The formulation of Coulomb interactions and electron–phonon interactions can be considered as the most successful achievements in many-body condensed matter physics. The effects of electron–phonon interactions proliferate many aspects of materials physical properties, including transport and thermodynamic ones. The most spectacular and profound manifestation is found in the effective mutual attraction of electron pairs in conventional superconductivity. In the present chapter, we will develop the basic formulation, and apply it in following chapters to superconductivity.

Herbert Fröhlich, a German-born British physicist, derived the key Hamiltonian that describes the electron–phonon interaction and that bears his name. Fröhlich realized the similarity of electron–phonon interaction to the electron Coulomb interactions of quantum electrodynamics (QED). He, together with John Bardeen, was the first to perceive the possibility that such interactions can produce an effective attraction between electrons.

20.2 The Phonon Hamiltonian in Second Quantization

We will first consider the case of a monatomic one-dimensional solid to describe the process of second quantization and diagonalization. The classical Hamiltonian may be written as

$$\mathcal{H} = \sum_j \left[\frac{p_j^2}{2m} + \frac{m\omega_0^2}{2} \left(u_{j+1} - u_j \right)^2 \right]. \tag{20.1}$$

We first-quantize the problem by setting

$$\left[u_j, p_k \right] = i\hbar \, \delta_{j,k}.$$

Next, because of translational invariance, we transform u_j to momentum space by setting

$$u_j = N^{-1/2} \sum_q u_q \, e^{iqja}; \qquad u_q = N^{-1/2} \sum_q u_j \, e^{-iqka}. \tag{20.2}$$

Because of the hermiticity of u_j, we require

$$u_j = N^{-1/2} \sum_q u_q \, e^{iqja} = u_j^\dagger = N^{-1/2} \sum_q u_q^\dagger \, e^{-iqja}. \tag{20.3}$$

This relation is satisfied if

$$u_q = u_{-q}^\dagger \tag{20.4}$$

and we may write

$$u_j = \frac{1}{2} N^{-1/2} \sum_q \left(u_q \, e^{iqja} + u_q^\dagger \, e^{-iqja} \right). \tag{20.5}$$

To determine the form of the momentum P conjugate to u_q, we construct the Lagrangian, noting the following transformations

$$\sum_j \left(\dot{u}_j \right)^2 = \frac{1}{N} \sum_{j;q,q'} \dot{u}_q \dot{u}_{q'} \, e^{(q+q')ja} = \sum_q \dot{u}_q \dot{u}_{-q}$$

$$\sum_j \left(u_{j+1} - u_j \right)^2 = \frac{1}{N} \sum_{j;q,q'} u_q u_{q'} \, e^{iqja} \left(e^{iqa} - 1 \right) e^{iq'ja} \left(e^{iq'a} - 1 \right)$$

$$= 2 \sum_q u_q u_{-q} \left(1 - \cos(qa) \right) \tag{20.6}$$

and the Lagrangian is written as

$$\mathcal{L} = \frac{m}{2} \sum_q \dot{u}_q \dot{u}_{-q} - \sum_q u_q u_{-q} \left(1 - \cos(qa) \right). \tag{20.7}$$

The conjugate momentum is then

$$P_{-q} = \frac{\partial \mathcal{L}}{\partial \dot{u}_q} = \frac{m}{2} \dot{u}_{-q} = P^\dagger_{-q}$$

and

$$\mathcal{H} = \sum_q \left[\frac{1}{2m} P_q P_{-q} + m\omega_0^2 \left(1 - \cos(qa) \right) u_q u_{-q} \right]$$

$$= \sum_q \left[\frac{1}{2m} P_q P_{-q} + \frac{m\omega_q^2}{2} u_q u_{-q} \right] \tag{20.8}$$

with

$$\omega_q^2 = 2\omega_0^2 \left(1 - \cos(qa) \right).$$

To diagonalize the Hamiltonian, we use the following transformation

$$
\begin{cases}
b_q^\dagger = \sqrt{\dfrac{1}{2m\hbar\omega_q}} \left(m\omega_q\, u_{-q} - i\, P_q \right) \\[3mm]
b_q = \sqrt{\dfrac{1}{2m\hbar\omega_q}} \left(m\omega_q\, u_q + i\, P_{-q} \right)
\end{cases}
\Rightarrow
\begin{cases}
u_q = \sqrt{\dfrac{\hbar}{2m\omega_q}} \left(b_q^\dagger + b_{-q} \right) \\[3mm]
P_q = \sqrt{\dfrac{\hbar}{2m\omega_q}} \left(b_{-q}^\dagger - b_q \right)
\end{cases}
\tag{20.9}
$$

and we obtain

$$
\mathcal{H} = \sum_q \hbar\omega_q \left(b_q^\dagger b_q + \frac{1}{2} \right).
\tag{20.10}
$$

20.2.1 Matsubara Green Function for Free Phonons

Extending the definition of u_q in terms of the b_q operators in (20.9) to three dimensions, we write

$$
\mathbf{u}_{\mathbf{q},\lambda} = \sqrt{\frac{\hbar}{2M\omega_{\mathbf{q}\lambda}}} \, (b_{-\mathbf{q},\lambda}^\dagger + b_{\mathbf{q},\lambda}) \, \hat{\mathbf{e}}_{\mathbf{q}\lambda} = \sqrt{\frac{\hbar}{2M\omega_{\mathbf{q}\lambda}}} \, \mathbf{B}_{\mathbf{q},\lambda},
$$

where $\hat{\mathbf{e}}_{\mathbf{q}\lambda}$ is the polarization vector of mode $\mathbf{q}\lambda$, and we introduced the new operators $\mathbf{B}_{\mathbf{q},\lambda}$ as

$$
\mathbf{B}_{\mathbf{q},\lambda} = \left(b_{-\mathbf{q},\lambda}^\dagger + b_{\mathbf{q},\lambda} \right) \hat{\mathbf{e}}_{\mathbf{q}\lambda}, \qquad
\mathbf{B}_{\mathbf{q},\lambda}^\dagger = \left(b_{-\mathbf{q},\lambda} + b_{\mathbf{q},\lambda}^\dagger \right) \hat{\mathbf{e}}_{\mathbf{q}\lambda} = \mathbf{B}_{-\mathbf{q},\lambda}.
$$

In the interaction (or Heisenberg) picture, we have

$$
\mathbf{B}_{\mathbf{q},\lambda}(\tau) = e^{\mathcal{H}_0 \tau}\, \mathbf{B}_{\mathbf{q},\lambda}\, e^{-\mathcal{H}_0 \tau} = \left(b_{-\mathbf{q},\lambda}^\dagger\, e^{-\omega_{\mathbf{q},\lambda}\tau} + b_{\mathbf{q},\lambda}\, e^{\omega_{\mathbf{q},\lambda}\tau} \right) \hat{\mathbf{e}}_\lambda,
$$

where $\mathcal{H}_0$ is the noninteracting phonon Hamiltonian.

We now define the noninteracting phonon Matsubara function with respect to the correlation function of the displacement operators $\mathbf{u}_{\mathbf{q}\lambda\alpha}$ as

$$
C_{\mathbf{u}}(\mathbf{q},\tau) = -\left\langle T_r \left[\mathbf{u}_{\mathbf{q}\lambda\alpha}(\tau)\mathbf{u}_{\mathbf{q}\lambda\beta}(0) \right] \right\rangle = \frac{\hbar}{2M\omega_{\mathbf{q}\lambda}}\, \mathcal{D}_{\alpha,\beta}^\lambda(\mathbf{q},\tau)
$$

$$
\mathcal{D}_{\alpha,\beta}^\lambda(\mathbf{q},\tau) = -\left\langle T_\tau \left[\mathbf{B}_{\mathbf{q}\lambda\alpha}(\tau)\mathbf{B}_{\mathbf{q}\lambda\beta}^\dagger(0) \right] \right\rangle.
\tag{20.11}
$$

For phonons, $\mathcal{D}$ is real since $\mathcal{H}$ and $\mathbf{u}$ are real, which requires

$$
\mathcal{D}_{\alpha,\beta}^\lambda(\mathbf{q},\tau) = \mathcal{D}_{i\alpha, j\beta}(\mathbf{q}, -\tau).
$$

Thus, the Matsubara function is given by

$$
\mathcal{D}_{\alpha,\beta}^\lambda(\mathbf{q},\tau) =
\begin{cases}
-\left\{ \left[n(\omega_{\mathbf{q},\lambda}) + 1 \right] e^{-\omega_{\mathbf{q},\lambda}\tau} + n(\omega_{\mathbf{q},\lambda}) e^{\omega_{\mathbf{q},\lambda}\tau} \right\} e_{\mathbf{q}\alpha}^\lambda e_{\mathbf{q}\beta}^\lambda & \text{for } \tau > 0 \\[3mm]
-\left\{ n(\omega_{\mathbf{q},\lambda}) e^{-\omega_{\mathbf{q},\lambda}\tau} + \left[n(\omega_{\mathbf{q},\lambda}) + 1 \right] e^{\omega_{\mathbf{q},\lambda}\tau} \right\} e_{\mathbf{q}\alpha}^\lambda e_{\mathbf{q}\beta}^\lambda & \text{for } \tau < 0.
\end{cases}
\tag{20.12}
$$

In the frequency domain, it becomes

$$\mathcal{D}_{\alpha,\beta}^{\lambda}(\mathbf{q}, i\omega_n) = \int_0^{\beta} d\tau \, e^{i\omega_n \tau} \, \mathcal{D}_{\alpha,\beta}^{\lambda}(\tau)$$

$$= -\left\{ [n(\omega_{\mathbf{q},\lambda}) + 1] \int_0^{\beta} d\tau \, e^{(i\omega_n - \omega_{\mathbf{q},\lambda})\tau} + n(\omega_{\mathbf{q},\lambda}) \int_0^{\beta} d\tau \, e^{(i\omega_n + \omega_{\mathbf{q},\lambda})\tau} \right\} e_{\mathbf{q}\alpha}^{\lambda} e_{\mathbf{q}\beta}^{\lambda}$$

$$= -\left\{ [n(\omega_{\mathbf{q},\lambda}) + 1] \frac{e^{-\beta\omega_{\mathbf{q},\lambda}} - 1}{i\omega_n - \omega_{\mathbf{q},\lambda}} + n(\omega_{\mathbf{q},\lambda}) \frac{e^{\beta\omega_{\mathbf{q},\lambda}} - 1}{i\omega_n + \omega_{\mathbf{q},\lambda}} \right\} e_{\mathbf{q}\alpha}^{\lambda} e_{\mathbf{q}\beta}^{\lambda}$$

$$= \left(\frac{1}{i\omega_n - \omega_{\mathbf{q}\lambda}} - \frac{1}{i\omega_n + \omega_{\mathbf{q}\lambda}} \right) e_{\mathbf{q}\alpha}^{\lambda} e_{\mathbf{q}\beta}^{\lambda}$$

$$= \frac{2\omega_{\mathbf{q}\lambda}}{(i\omega_n)^2 - \omega_{\mathbf{q}\lambda}^2} \, e_{\mathbf{q}\alpha}^{\lambda} e_{\mathbf{q}\beta}^{\lambda}. \tag{20.13}$$

By analytic continuation, we obtain

$$\mathcal{D}_{\alpha,\beta}^{R\lambda}(\mathbf{q}, \omega) = \left(\frac{1}{\omega - \omega_{\mathbf{q}\lambda} + i\eta} - \frac{1}{\omega + \omega_{\mathbf{q}\lambda} - i\eta} \right) e_{\mathbf{q}\alpha}^{\lambda} e_{\mathbf{q}\beta}^{\lambda}.$$

The spectral function can be obtained by taking the imaginary part of $\mathcal{D}^R(\omega)$:

$$\mathcal{A}_{\alpha,\beta}(\omega) = 2\pi \sum_{\mathbf{q}\lambda} \left(\delta(\omega - \omega_{\mathbf{q}\lambda}) + \delta(\omega + \omega_{\mathbf{q}\lambda}) \right) e_{\mathbf{q}\alpha}^{\lambda} e_{\mathbf{q}\beta}^{\lambda}. \tag{20.14}$$

20.3 Electron–Phonon Interactions: The Fröhlich Hamiltonian

We shall consider here some of the consequences of the interaction of phonons with electrons, and in particular with the electrons in a simple metal. The subject is a complicated and difficult one, in that we need to call on most of the knowledge that we have of the behavior of the electron gas and of lattice vibrations. A complete calculation should really start with the Hamiltonian of a lattice of bare ions, whose mutual interaction would include the long-range Coulomb potential. One would then add the electron gas, which would shield the potential due to the ions in the manner discussed earlier in the one-electron picture. It is, however, possible to explore many of the consequences of the electron–phonon interaction by use of a simpler model. In this model, we take for granted the concept of screening, and assume that the ions interact with each other and with the electrons only through a short-range screened potential, and we treat the electrons themselves as independent fermions. For a monatomic crystal, our unperturbed Hamiltonian is then simply

$$\mathcal{H}_0 = \sum_{\mathbf{k}} \mathcal{E}_{\mathbf{k}} c_{\mathbf{k}}^{\dagger} c_{\mathbf{k}} + \sum_{\mathbf{q}s} \hbar\omega_{\mathbf{q}s} b_{\mathbf{q}s}^{\dagger} b_{\mathbf{q}s}, \tag{20.15}$$

the phonon frequencies $\omega_{\mathbf{q}s}$ being proportional to $\mathbf{q}$ as $\mathbf{q} \to 0$; in other words, we are dealing only with acoustic modes. To this we add the interaction, $\mathcal{H}_I$, of the electrons with the screened ions. We assume that at any point the potential due to a particular ion depends only on the distance from the center of the ion – an assumption known as the *rigid-ion approximation* – so that in the second-quantized notation,

$$\mathcal{H}_I = \sum_{\mathbf{k},\mathbf{k}',\mathbf{l}} \langle \mathbf{k}| V(\mathbf{r} - \mathbf{l} - \mathbf{u_l})|\mathbf{k}'\rangle \, c_{\mathbf{k}}^\dagger c_{\mathbf{k}'}$$

$$= \sum_{\mathbf{k},\mathbf{k}',\mathbf{l}} e^{i(\mathbf{k}'-\mathbf{k})\cdot(\mathbf{l}+\mathbf{u_l})} \, V_{\mathbf{k}-\mathbf{k}'} \, c_{\mathbf{k}}^\dagger c_{\mathbf{k}'}. \tag{20.16}$$

Here $V(\mathbf{r})$ is the potential due to a single ion at the origin, and $V_{\mathbf{k}-\mathbf{k}'}$ its Fourier transform. With the assumption that the displacement $\mathbf{u_l}$ of the ion whose equilibrium position is $\mathbf{l}$ is sufficiently small that $(\mathbf{k}' - \mathbf{k}) \cdot \mathbf{u_l} \ll 1$, we can write

$$e^{i(\mathbf{k}'-\mathbf{k})\cdot\mathbf{u_l}} \simeq 1 + i(\mathbf{k}' - \mathbf{k}) \cdot \mathbf{u_l}$$

$$= 1 + iN^{-1/2}(\mathbf{k}' - \mathbf{k}) \cdot \sum_{\mathbf{q}} e^{i\mathbf{q}\cdot\mathbf{l}} \, \mathbf{u_q}, \tag{20.17}$$

where we expanded $\mathbf{u_l}$ in terms of its Fourier components. Then $\mathcal{H}_I$ can be split into two parts,

$$\mathcal{H}_I = \mathcal{H}_{\text{Bloch}} + \mathcal{H}_{\text{e-p}}. \tag{20.18}$$

The first term, $\mathcal{H}_{\text{Bloch}}$, is the usual lattice potential encountered in electronic structure calculations, and is independent of lattice displacements. We have

$$\mathcal{H}_{\text{Bloch}} = \sum_{\mathbf{k},\mathbf{k}',\mathbf{l}} e^{i(\mathbf{k}'-\mathbf{k})\cdot\mathbf{l}} \, V_{\mathbf{k}-\mathbf{k}'} \, c_{\mathbf{k}}^\dagger c_{\mathbf{k}'}$$

$$= N \sum_{\mathbf{k},\mathbf{G}} V_{-\mathbf{G}} \, c_{\mathbf{k}-\mathbf{G}}^\dagger c_{\mathbf{k}}, \tag{20.19}$$

where the $\mathbf{G}$s are reciprocal lattice vectors, and

$$\mathcal{H}_{\text{e-p}} = iN^{-1/2} \sum_{\mathbf{k},\mathbf{k}',\mathbf{q};\mathbf{l}} e^{i(\mathbf{k}'-\mathbf{k}+\mathbf{q})\cdot\mathbf{l}} \, (\mathbf{k}' - \mathbf{k}) \cdot \mathbf{u_q} \, V_{\mathbf{k}-\mathbf{k}'} \, c_{\mathbf{k}}^\dagger c_{\mathbf{k}'}$$

$$= iN^{1/2} \sum_{\mathbf{k},\mathbf{k}'} (\mathbf{k}' - \mathbf{k}) \cdot \mathbf{u}_{\mathbf{k}-\mathbf{k}'} \, V_{\mathbf{k}-\mathbf{k}'} \, c_{\mathbf{k}}^\dagger c_{\mathbf{k}'}. \tag{20.20}$$

Expressing $\mathbf{u_q}$ in terms of phonon annihilation and creation operators,

$$\mathbf{u}_{\mathbf{q},\lambda} = \sqrt{\frac{\hbar}{2M\omega_{\mathbf{q}\lambda}}} (b_{-\mathbf{q},\lambda}^\dagger + b_{\mathbf{q},\lambda}) \, \hat{\mathbf{e}}_\lambda, \tag{20.21}$$

where $\mathbf{e}_\lambda$ is the mode polarization vector, and substituting in (23.8), we get

$$\mathcal{H}_{\text{e-p}} = i \sum_{\mathbf{k},\mathbf{k}',\lambda} \sqrt{\frac{N\hbar}{2M\omega_{\mathbf{k}'-\mathbf{k},\lambda}}} (\mathbf{k}' - \mathbf{k}) \cdot \hat{\mathbf{e}}_\lambda \, V_{\mathbf{k}-\mathbf{k}'} \, (b_{\mathbf{k}'-\mathbf{k},\lambda}^\dagger + b_{\mathbf{k}'-\mathbf{k},\lambda}) \, c_{\mathbf{k}}^\dagger c_{\mathbf{k}'}, \tag{20.22}$$

where the summation now also includes the three polarization vectors, $\mathbf{e}_\lambda$, of the phonons. For simplicity, we shall assume the phonon spectrum to be isotropic, so that the phonons

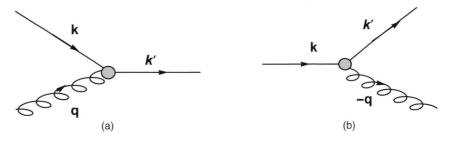

Figure 20.1 The Fröhlich Hamiltonian includes an interaction term in which an electron is scattered from $\mathbf{k}'$ to $\mathbf{k}$ with either emission (a) or absorption (b) of a phonon. In each case, the total wavevector is conserved.

will either be longitudinally or transversely polarized. Only the longitudinal modes, for which $\mathbf{s}$ is parallel to $\mathbf{k}' - \mathbf{k}$, then enter $\mathcal{H}_{\text{e-p}}$. We shall also neglect the effects of $\mathcal{H}_{\text{Bloch}}$, the periodic potential of the stationary lattice. With these simplifications, we arrive at the *Fröhlich Hamiltonian* (see Figure 20.1)

$$\mathcal{H} = \sum_{\mathbf{k}} \mathcal{E}_{\mathbf{k}}\, c_{\mathbf{k}}^{\dagger} c_{\mathbf{k}} + \sum_{\mathbf{q}\lambda} \hbar\omega_{\mathbf{q}\lambda}\, b_{\mathbf{q}\lambda}^{\dagger} b_{\mathbf{q}\lambda} + \sum_{\mathbf{k},\mathbf{k}'} g_{\mathbf{k}\mathbf{k}'}\, (b_{\mathbf{k}'-\mathbf{k},\lambda}^{\dagger} + b_{\mathbf{k}-\mathbf{k}',\lambda})\, c_{\mathbf{k}}^{\dagger} c_{\mathbf{k}'}, \quad (20.23)$$

where the electron–phonon matrix element is defined by

$$g_{\mathbf{k}\mathbf{k}'} = i\sqrt{\frac{N\hbar}{2M\omega_{\mathbf{k}'-\mathbf{k}}}}\,|\mathbf{k}' - \mathbf{k}|\, V_{\mathbf{k}-\mathbf{k}'}, \quad (20.24)$$

with the phonon wavevector $\mathbf{q} = \mathbf{k} - \mathbf{k}'$, reduced to the first Brillouin zone (BZ) if necessary.

The interaction $\mathcal{H}_{\text{e-p}}$ can be considered as being composed of two parts: terms involving $b_{-\mathbf{q}}^{\dagger} c_{\mathbf{k}}^{\dagger} c_{\mathbf{k}'}$ and terms involving $b_{\mathbf{q}} c_{\mathbf{k}}^{\dagger} c_{\mathbf{k}'}$. These may be represented by the diagrams shown in Figures 20.1(a) and 20.1(b), respectively. In the first diagram, an electron is scattered from $\mathbf{k}'$ to $\mathbf{k}$ with the emission of a phonon of wavevector $\mathbf{k}' - \mathbf{k}$. The total wavevector is conserved, as is always the case in a periodic system, unless the vector $\mathbf{k}' - \mathbf{k}$ lies outside the first BZ, in which case $\mathbf{q} = \mathbf{k}' - \mathbf{k} + \mathbf{G}$ for some nonzero $\mathbf{G}$. Such electron–phonon Umklapp processes do not conserve wavevector, and are important in contributing to the electrical and thermal resistivity of metals.

20.4 Matsubara Approach to Electron–Phonon Interactions

20.4.1 Electron Self-Energy: Leading Correction

The leading correction to the electron spectrum, representing virtual phonon exchange, is given by the diagram in Figure 20.2. The corresponding self-energy is given by

$$\Sigma_{ep}(\mathbf{k}, i\Omega_n) = -\frac{1}{\mathcal{V}\beta} \sum_{\mathbf{q}, m} |g_{\mathbf{q}}|^2\, \mathcal{G}^{(0)}(\mathbf{k} + \mathbf{q}, i\Omega_n + \omega_m)\, \mathcal{D}^{(0)}(\mathbf{q}, i\omega_m). \quad (20.25)$$

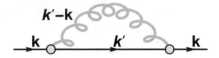

Figure 20.2 The lowest-order diagram describing the renormalization of the electron spectrum due to phonons.

Performing Matsubara sums by contour integration (problem), as well as analytic continuation, we arrive at

$$\Sigma_{ep}(\mathbf{k}, \varepsilon) = \frac{1}{\mathcal{V}} \sum_{\mathbf{q}} |g_{\mathbf{q}}|^2 \left[\frac{n_B(\omega_{\mathbf{q}}) + n_F(\xi_{\mathbf{k}+\mathbf{q}})}{\varepsilon - \xi_{\mathbf{k}+\mathbf{q}} + \omega_{\mathbf{q}} + i\eta} - \frac{n_B(\omega_{\mathbf{q}}) + 1 - n_F(\xi_{\mathbf{k}+\mathbf{q}})}{\varepsilon - \xi_{\mathbf{k}+\mathbf{q}} - \omega_{\mathbf{q}} - i\eta} \right] \quad (20.26)$$

Since $\omega_{\mathbf{q}} \leq \omega_D$, and $\omega_D \ll E_F$, the major contribution will come from $\varepsilon \sim E_F$.

The electron quasiparticle acquires a finite lifetime leading to a linewidth

$$\Gamma_{\mathbf{k}} = -2 \text{Im} \, \Sigma_{ep}(\mathbf{k}, \varepsilon). \quad (20.27)$$

Substituting $\mathbf{k}' = \mathbf{k} + \mathbf{q}$, we write

$$\text{Im} \, \Sigma_{ep}(\mathbf{k}, \varepsilon) = -\frac{\pi}{\mathcal{V}} \sum_{\mathbf{k}'} |g_{\mathbf{k}'-\mathbf{k}}|^2 \Big[\delta \left(\varepsilon - \xi_{\mathbf{k}'} + \omega_{\mathbf{k}'-\mathbf{k}} \right) \left(n_B(\omega_{\mathbf{k}'-\mathbf{k}}) + n_F(\xi_{\mathbf{k}'}) \right)$$

$$+ \delta \left(\varepsilon - \xi_{\mathbf{k}'} - \omega_{\mathbf{k}'-\mathbf{k}} \right) \left(n_B(\omega_{\mathbf{k}'-\mathbf{k}}) + 1 - n_F(\xi_{\mathbf{k}'}) \right) \Big]. \quad (20.28)$$

Setting $\varepsilon = E_F = 0$, we get

$$\text{Im} \, \Sigma_{ep}(\mathbf{k}) = -\frac{\pi}{\mathcal{V}} \int d\omega \sum_{\mathbf{k}'} |g_{\mathbf{k}'-\mathbf{k}}|^2 \, \delta(\omega - \omega_{\mathbf{q}}) \Big[\delta \left(\omega - \xi_{\mathbf{k}'} \right) \left(n_B(\omega) + n_F(\xi_{\mathbf{k}'}) \right)$$

$$+ \delta \left(\omega + \xi_{\mathbf{k}'} \right) \left(n_B(\omega) + 1 - n_F(\xi_{\mathbf{k}'}) \right) \Big]$$

$$\simeq -\frac{\pi}{\mathcal{V}} \int d\omega \int \frac{d\mathbf{k}'}{(2\pi)^3} \, |g_{\mathbf{k}'-\mathbf{k}}|^2 \, \delta(\omega - \omega_{\mathbf{k}'-\mathbf{k}}) \delta(\xi_{\mathbf{k}'})$$

$$\times \Big[2 n_B(\omega) + n_F(\omega + \varepsilon) + n_F(\omega - \varepsilon) \Big]. \quad (20.29)$$

But $d\mathbf{k}' = 2\pi k'^2 dk' d\cos\theta$, and $\int d\cos\theta \, \delta(\xi_{\mathbf{k}'}) = (k' v_F)^{-1}$, which leaves $d^2 k' = 2\pi k' dk'$. We now write $\text{Im} \, \Sigma_{ep}(\mathbf{k})$ in terms of the *spectral function* $\alpha^2 F(\omega)$[1] introduced by McMillan [127], as

$$\text{Im} \, \Sigma_{ep}(\mathbf{k}) = -\frac{\pi}{\mathcal{V}} \int d\omega \, \alpha_{\mathbf{k}}^2 F(\omega) \Big[2 n_B(\omega) + n_F(\omega + \varepsilon) + n_F(\omega - \varepsilon) \Big]. \quad (20.30)$$

$\alpha^2 F(\omega)$ is defined as

$$\alpha_{\mathbf{k}}^2 F(\omega) = \frac{1}{(2\pi)^3} \int \frac{d^2 k'}{v_F} \, |g_{\mathbf{k}'-\mathbf{k}}|^2 \, \delta \left[\omega - \omega_{\mathbf{k}'-\mathbf{k}} \right].$$

---

[1] $\alpha^2 F(\omega)$ is actually a single function, and not a product of two functions.

$\alpha_{\mathbf{k}}^2 F(\omega)$ is the frequency spectrum one obtains by starting at a point on the Fermi surface $\mathbf{k}$ and integrating over all other points on the Fermi surface $\mathbf{k}'$. It will vary from point to point on the Fermi surface of a metal. The spectral function contains the essential information related to the electron–phonon coupling of the specific electronic state $\mathbf{k}$. A convenient measure for the strength of the electron–phonon coupling is the dimensionless coupling parameter

$$\lambda(\hat{\mathbf{k}}) = 2 \int_0^{\omega_D} \frac{d\omega}{\omega} \, \alpha_{\mathbf{k}}^2 F(\omega). \tag{20.31}$$

20.4.2 Interacting Electronic Matsubara Function

The full interacting electronic Matsubara function is given by

$$\mathcal{G}_\sigma(\mathbf{k}, \tau) = - \sum_{m=0}^\infty \frac{1}{m!} \left(\frac{-1}{\hbar}\right)^m \int_0^\beta d\tau_1 \dots \int_0^\beta d\tau_m \frac{\left\langle T_\tau \left[\mathcal{H}_I(\tau_1) \dots \mathcal{H}_I(\tau_m) c_{\mathbf{k}\sigma}(\tau) c_{\mathbf{k}\sigma}^\dagger(0) \right] \right\rangle_0}{\langle S_I(\beta, 0) \rangle_0},$$

where $\langle \dots \rangle_0$ indicates averaging with respect to the noninteracting Hamiltonian, and

$$\mathcal{H}_I(\tau_j) = \sum_{\substack{\mathbf{k}\sigma \\ \mathbf{q}\lambda}} g_{\mathbf{q}\lambda} \, c_{\mathbf{k}+\mathbf{q},\sigma}^\dagger(\tau_j) \, c_{\mathbf{k},\sigma}(\tau_j) \, \mathbf{B}_{\mathbf{q},\lambda}(\tau_j), \tag{20.32}$$

where, for now, the Coulomb interaction has been omitted. Because the averaging is done with respect to the noninteracting Hamiltonian, the electron and phonon degrees of freedom decouple, and the thermal averages becomes

$$\left\langle T_\tau \left[\mathbf{B}_{\mathbf{q}_1,\lambda_1}(\tau_1) \dots \mathbf{B}_{\mathbf{q}_n,\lambda_n}(\tau_n) c_{\mathbf{k}+\mathbf{q}_1,\sigma}^\dagger(\tau_1) c_{\mathbf{k},\sigma}(\tau_1) \dots c_{\mathbf{k}+\mathbf{q}_n,\sigma}^\dagger(\tau_n) c_{\mathbf{k},\sigma}(\tau_n) c_{\mathbf{k}\sigma}(\tau) c_{\mathbf{k}\sigma}^\dagger \right] \right\rangle_0$$

$$= \left\langle T_\tau \left[\mathbf{B}_{\mathbf{q}_1,\lambda_1}(\tau_1) \dots \mathbf{B}_{\mathbf{q}_n,\lambda_n}(\tau_n) \right] \right\rangle_0 \left\langle c_{\mathbf{k}+\mathbf{q}_1,\sigma}^\dagger(\tau_1) c_{\mathbf{k},\sigma}(\tau_1) \dots c_{\mathbf{k}+\mathbf{q}_n,\sigma}^\dagger(\tau_n) c_{\mathbf{k},\sigma}(\tau_n) c_{\mathbf{k}\sigma}(\tau) c_{\mathbf{k}\sigma}^\dagger \right]_0.$$

It is clear that only an even number of phonon operators will give nonvanishing thermal averages, thus $m \to 2n$.

Upon performing bosonic Wick's contractions, the electronic Matsubara function expansion will contain products of single-particle Green functions of the form

$$g_{\mathbf{q}_i \lambda_i} \, g_{\mathbf{q}_j \lambda_j} \left\langle T_\tau \left[\mathbf{B}_{\mathbf{q}_i,\lambda_i}(\tau_i) \, \mathbf{B}_{\mathbf{q}_j,\lambda_j}(\tau_j) \right] \right\rangle_0$$

$$= \left| g_{\mathbf{q}_i \lambda_i} \right|^2 \left\langle T_\tau \left[\mathbf{B}_{\mathbf{q}_i,\lambda_i}(\tau_i) \, \mathbf{B}_{-\mathbf{q}_i,\lambda_i}(\tau_j) \right] \right\rangle_0 \delta_{\mathbf{q}_j,-\mathbf{q}_i} \delta_{\lambda_i,\lambda_j}$$

$$= - \left| g_{\mathbf{q}_i \lambda_i} \right|^2 \mathcal{D}_\lambda^{(0)}(\mathbf{q}_i, \tau_i - \tau_j) \, \delta_{\mathbf{q}_j,-\mathbf{q}_i} \delta_{\lambda_i,\lambda_j}. \tag{20.33}$$

We should be aware that the process of changing $m \to 2n$ involves some combinatoric steps: (i) the prefactor $(-1)^m/m!$ becomes $1/(2n)!$; (ii) each of the n factors of the form (20.33) contributes a minus sign, leading to $(-1)^n$; (iii) a factor $(2n)!/(n!\,n!)$ arises from the choice of n independent momenta $\mathbf{q}_j$ among the $2n$; and (iv) another factor $n!/2^n$

comes from the fact that all choices of possible ways to combine the remaining n momenta to the chosen ones and to symmetrize the pairs, leading to the same result.

Hence the net prefactor is $(-1))^n/2^n n!$. We can then write the one-electron Matsubara function as

$$\mathcal{G}_\sigma(\mathbf{k},\tau) = \sum_{n=0}^{\infty} \frac{1}{n!} \left(\frac{-1}{\hbar}\right)^n \int_0^\beta d\tau_1 \dots \int_0^\beta d\tau_n \frac{\left\langle T_\tau \left[\tilde{\mathcal{H}}_I(\tau_1)\dots\tilde{\mathcal{H}}_I(\tau_n) c_{\mathbf{k}\sigma}(\tau) c^\dagger_{\mathbf{k}\sigma}(0)\right]\right\rangle_0}{\langle U_I(\beta,0)\rangle_0},$$

where the phonon-mediated electron–electron interaction becomes

$$\tilde{\mathcal{H}}_I(\tau_1) = \frac{1}{2} \sum_{\substack{\mathbf{k}_1\sigma_1 \\ \mathbf{k}_2\sigma_2}} \sum_{\mathbf{q}\lambda} |g_{\mathbf{q}\lambda}|^2 \, \mathcal{D}^{(0)}_\lambda(\mathbf{q}_i,\tau_i - \tau_j) c^\dagger_{\mathbf{k}_1+\mathbf{q},\sigma_1}(\tau_j) c^\dagger_{\mathbf{k}_2-\mathbf{q},\sigma_2}(\tau_i) c_{\mathbf{k}_2,\sigma_2}(\tau_i) c_{\mathbf{k}_1,\sigma_1}(\tau_j).$$

This interaction is reminiscent of the nonrelativistic Coulomb interaction, but it is retarded, namely nonlocal in time: $|g_{\mathbf{q}\lambda}|^2 \, \mathcal{D}^{(0)}_\lambda(\mathbf{q}_i,\tau_i - \tau_j)$. The Feynman rules remain the same as for Coulomb interactions.

20.5 Electron–Phonon Interactions in the Jellium Model

20.5.1 Jellium Model for Phonons and Einstein Oscillations

As we have seen earlier, in the jellium model we replaced the lattice with a rigid (static), smooth, positively charged ion density, while we treated the electrons as particles. Here, we shall treat both ions and electrons as *fluids*, so that the only difference between the ions and the electrons is the different ratios of mass density to charge density: both fluids have the same average charge density, whereas the mass density $\rho^m_{ion}/\rho^m_{electron} \sim 10^5$.

For pedagogical reasons, we shall start with treating the electron jellium as a negatively charged rigid background, while allowing the ion fluid to be a deformable fluid, described as a continuous charge density function $\rho_{ion}(\mathbf{x})$. We then try to explore what the ionic normal modes are like in such a jellium model. Of course, we understand that this is not an appropriate description of the electrons, which are actually much more mobile.

In such a system, any deviation from equilibrium $\delta\rho_{ion}(\mathbf{x}) = \rho_{ion} - \rho^0_{ion}$ will give rise to an electric field $\mathbf{E}$, such that

$$\nabla\cdot\mathbf{E} = \frac{Ze}{\epsilon_0} \delta n_{ion}(\mathbf{x}),$$

where ϵ_0 is the electric permittivity of vacuum, and Ze is the ionic charge. The electric field, in turn, exerts a force field on the ions,

$$\nabla\cdot\mathbf{f} = \frac{Z^2 e^2 n^0_{ion}}{\epsilon_0} \delta\rho_{ion}(\mathbf{x}).$$

We write the continuity equation as

$$\partial_t \rho_{ion} + \nabla\cdot(\rho_{ion}\mathbf{v}) = \partial_t \delta\rho_{ion} + \rho^0_{ion} \, \nabla\cdot\mathbf{v} = 0, \qquad (20.34)$$

where we assumed that $\delta\rho_{\text{ion}} \ll \rho_{\text{ion}}^0$. Differentiating (20.34) with respect to time, and using $\mathbf{f} = M\,\partial_t \mathbf{v}$, where M is the ionic mass, we get

$$\partial_t^2 \delta\rho_{\text{ion}} + \frac{1}{M}\,\nabla\cdot\mathbf{f} = 0 \Rightarrow \Omega_p^2\,\delta\rho_{\text{ion}} - \frac{Z^2 e^2 n_{\text{ion}}^0}{M\epsilon_0}\,\delta\rho_{\text{ion}} = 0$$

$$\Rightarrow \Omega_p = \sqrt{\frac{Z^2 e^2 n_{\text{ion}}^0}{M\epsilon_0}} = \sqrt{\frac{Z e^2 n_{\text{el}}^0}{M\epsilon_0}}, \tag{20.35}$$

where we used the time-dependence $\delta\rho_{\text{ion}}(\mathbf{x}, t) = \delta\rho_{\text{ion}}(\mathbf{x})\, e^{i\Omega_p t}$. Equation (20.35) describes an Einstein-mode-like (wavevector-independent) optical phonon. This is actually the result of long-range Coulomb interaction, which gives rise to ionic *plasmon* excitations, an artifact of the rigidity of the electronic background.

Actually, recalling the Born–Oppenheimer approximation, the electrons should follow the ion density deformations instantaneously, maintaining charge neutrality throughout the material at all times, thus effectively screening the long-range Coulomb interaction between the ions.[2] Our bare-phonon Hamiltonian gives the wrong answer because it did not let the electrons follow the ion motion. Permitting the electrons to follow the ion motion is another set of words to describe the screening by the electrons of the motion of the ions. This means that a deformation $\delta\rho_{\text{ion}}$ of the ion charge density will induce an instantaneous deformation $-\delta\rho_{\text{ion}}$ in the electron charge density, leading to a local increase in the electronic kinetic energy-à la Thomas–Fermi.

To estimate the resulting restoring force, we simply look for changes in *electron pressure* associated with changes in electronic densities. Given that the ground-state energy for N electrons confined to a volume V is

$$E_{\text{el}} = \frac{\hbar^2}{10\pi^2 m_e}\frac{\left(3\pi^2 N\right)^{5/3}}{V^{2/3}},$$

the corresponding electronic pressure equals

$$P_e = \frac{\hbar^2}{15\pi^2 m_2}\left(3\pi^2 n_{\text{el}}\right)^{5/3}.$$

The ionic motion is driven by the gradient of the electronic pressure P_e:

$$\mathbf{f} = \nabla P_e = \frac{\partial P_e}{\partial n_{\text{el}}}\nabla n_{\text{el}} = -\frac{\hbar^2 k_F^2}{3 m_e}\nabla\,\delta\rho_{\text{ion}}.$$

Substituting for $\mathbf{f}$ in (20.35), we obtain

$$\partial_t^2 \delta\rho_{\text{ion}} - \frac{2Z}{3M}\varepsilon_F \nabla^2 \delta\rho_{\text{ion}} = 0. \tag{20.36}$$

---

[2] When the electrons oscillate at their natural frequency coP, the ions are too heavy to follow, so the electron motion is not screened. The electrons can freely oscillate at their plasma frequency, while the ions cannot.

Solutions of (20.36) have a linear dispersion relation

$$\omega_q = \sqrt{\frac{Zm_e}{3M}}\, v_F\, q, \tag{20.37}$$

which is known as the *Bohm–Staver formula*. It is a longitudinal mode associated with changes in the charge density; no transverse modes emerge.

20.5.2 Microscopic Theory of Electron–Phonon Interaction in the Jellium Model: Effective Interaction

We now present a more microscopic picture of how the conduction electrons determine the phonon dispersion relation. We still work within the framework of the jellium model, and determine the modification to the bare ion Einstein dispersion relation by electron-phonon interactions, which will be treated within perturbation theory.

Bare Phonon Matsubara Function for Jellium

As we have shown, in the absence of electron–phonon interaction, the phonons are the quantized plasma oscillations of the ion jellium, having an Einstein dispersion at a frequency Ω_p. Thus, we define the bare electron–phonon coupling $g_{\mathbf{q}}$ with the aid of (20.24) as

$$\frac{1}{V}\left|g_{\mathbf{q}}\right|^2 = \frac{1}{V}\left|\sqrt{\frac{N\hbar}{2M\Omega_p}}\,|\mathbf{q}|\,V_{\mathbf{q}}\right|^2$$

$$= \frac{1}{V}\left(\frac{Ze^2}{\epsilon_0 q^2}\right)^2 \frac{N\hbar}{2M\Omega_p} \tag{20.38}$$

$$= \frac{e^2}{\epsilon_0 q^2}\,\frac{Z^2 e^2 N}{\epsilon_0 M V}\,\frac{1}{2\Omega_p} = \frac{1}{2}\,V_c(\mathbf{q})\,\Omega_p, \tag{20.39}$$

where V_c is the Coulomb potential. Thus the resulting, bare, phonon-mediated electron–electron interaction is

$$\frac{1}{V}\left|g_{\mathbf{q}}\right|^2 \mathcal{D}^{(0)}(\mathbf{q}, i\omega_n) = V_c(\mathbf{q})\,\frac{\Omega_p^2}{(i\omega_n)^2 - \Omega_p^2}. \tag{20.40}$$

Effective Electron–Electron Interaction in the Jellium Model

The total interaction between electrons is the sum of the pure electronic Coulomb interaction and the bare interaction due to phonon exchange. In the Feynman diagramatic representation, the total bare, effective electron–electron interaction line is given by

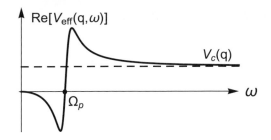

Figure 20.3 Frequency dependence of the real part of the bare, effective electron–electron interaction $V_{\text{eff}}(\mathbf{q}, \omega)$ at fixed $\mathbf{q}$. The interaction is attractive for $\omega < \Omega_p$, and $V_{\text{eff}}^{(0)}(\mathbf{q}, \omega) \to V_c(\mathbf{q})$ for $\omega \to \infty$.

$$V_{\text{eff}}(\mathbf{q}, i\omega_n) = V_c(\mathbf{q}) + \frac{1}{V}|g_{\mathbf{q}}|^2 \mathcal{D}^{(0)}(\mathbf{q}, i\omega_n)$$

$$= V_c(\mathbf{q}) + V_c(\mathbf{q})\frac{\Omega_p^2}{(i\omega_n)^2 - \Omega_p^2} = V_c(\mathbf{q})\frac{(i\omega_n)^2}{(i\omega_n)^2 - \Omega_p^2}.$$

$$(20.41)$$

The real frequency effective interaction $(i\omega_n \to \omega + i\eta)$ is

$$V_{\text{eff}}(\mathbf{q}, \omega) = V_c(\mathbf{q})\frac{\omega^2}{\omega^2 - \Omega_p^2 + i\eta}. \qquad (20.42)$$

The real part of $V_{\text{eff}}(\mathbf{q}, \omega)$ is shown in Figure 20.3. It is seen that the bare, effective electron–electron interaction becomes negative for $\omega < \Omega_p$, i.e., at low frequencies the electron–phonon interaction combined with the originally fully repulsive Coulomb interaction results in an attractive effective electron–electron interaction. At high frequencies, the normal Coulomb interaction is recovered.

20.5.3 RPA Screening of Effective Electron–Electron Interaction

The effective electronic Coulomb interaction between two electrons in a metal is strongly screened at all except very short distances. As we have shown in Chapter 15, this screening is well described within RPA. We shall now treat the effective interaction derived previously within the RPA framework. Expressing $V_{\text{eff}}^{\text{RPA}}(\mathbf{q})$ as a Dyson equation involving the simple bubble $-\Pi_0$, we write

$$-V_{\text{eff}}^{\text{RPA}}(\mathbf{q}, i\omega_n) = \qquad = \qquad + \qquad , \qquad (20.43)$$

which can be rewritten in the form

$$-V_{\text{eff}}^{\text{RPA}}(\mathbf{q}, i\omega_n) = \text{\Large MMMM} = \frac{\text{\Large MMM}}{1 - \text{\Large (——)}} = \frac{-V_{\text{eff}}(\mathbf{q})}{1 - V_{\text{eff}}(\mathbf{q})\, \Pi_0(\mathbf{q}, i\omega_n)}.$$

$$(20.44)$$

Expanding (20.44), we obtain

$$
-V_{\text{eff}}^{\text{RPA}}(\mathbf{q}, i\omega_n) = \frac{V_c(\mathbf{q})\, \frac{(i\omega_n)^2}{(i\omega_n)^2 - \Omega_p^2}}{1 - V_c(\mathbf{q})\, \frac{(i\omega_n)^2}{(i\omega_n)^2 - \Omega_p^2}\, \Pi_0(\mathbf{q}, i\omega_n)}
$$

$$
= V_c(\mathbf{q})\, \frac{(i\omega_n)^2}{(i\omega_n)^2 - \Omega_p^2 - V_c(\mathbf{q})\, (i\omega_n)^2\, \Pi_0(\mathbf{q}, i\omega_n)}
$$

$$
= \frac{V_c(\mathbf{q})}{1 - V_c(\mathbf{q})\, \Pi_0(\mathbf{q}, i\omega_n)}\, \frac{(i\omega_n)^2\, (1 - V_c(\mathbf{q})\, \Pi_0(\mathbf{q}, i\omega_n))}{(i\omega_n)^2\, (1 - V_c(\mathbf{q})\, \Pi_0(\mathbf{q}, i\omega_n)) - \Omega_p^2}
$$

$$
= V_c^{\text{RPA}}(\mathbf{q})\, \frac{(i\omega_n)^2}{(i\omega_n)^2 - \frac{\Omega_p^2}{1 - V_c(\mathbf{q})\, \Pi_0(\mathbf{q}, i\omega_n)}}
$$

$$
= V_c^{\text{RPA}}(\mathbf{q})\, \frac{(i\omega_n)^2}{(i\omega_n)^2 - \omega_{\mathbf{q}}^2},
$$

$$(20.45)$$

where $\omega_{\mathbf{q}}$ is the renormalized phonon frequency

$$\omega_{\mathbf{q}} = \frac{\Omega_p}{\sqrt{1 - V_c(\mathbf{q})\, \Pi_0(\mathbf{q}, i\omega_n)}}.$$

20.5.4 *Phonon Matsubara Function in RPA: Bohm–Staver Formula and the Kohn Anomaly*

Although we arrived at an expression for the RPA-screened phonon frequencies, it is instructive to derive the phonon Matsubara function in RPA. We write the corresponding Dyson equation as

$$\mathcal{D}^{\text{RPA}}(\mathbf{q}, i\omega_n) = \text{\Large OOOOOO} = \text{\Large OOOO} + \text{\Large OOO}\boxed{\text{RPA}}\text{\Large OOO}.$$

The solution for the renormalized phonon line is

$$\mathcal{D}^{\text{RPA}}(\mathbf{q}, i\omega_n) = \frac{\text{\Large OOO}}{1 - \text{\Large (}\boxed{\text{RPA}}\text{)}\,\text{\Large OOO}} = \frac{-\mathcal{D}^{(0)}(\mathbf{q}, i\omega_n)}{1 - \chi^{\text{RPA}}(\mathbf{q}, i\omega_n)\, |g_{\mathbf{q}}|^2\, \mathcal{D}^{(0)}(\mathbf{q}, i\omega_n)}.$$

Using (20.13) and (20.40), we obtain

$$\mathcal{D}^{\text{RPA}}(\mathbf{q}, i\omega_n) = \frac{2\Omega}{\left[(i\omega_n)^2 - \Omega^2\right] - \Omega^2\, V_c(\mathbf{q})\, \chi^{\text{RPA}}(\mathbf{q}, i\omega_n)} = \frac{2\Omega}{(i\omega_n)^2 - \omega_{\mathbf{q}}^2}, \qquad (20.46)$$

where

$$\omega_{\mathbf{q}} = \Omega\sqrt{1 + V_c(\mathbf{q})\,\chi^{\text{RPA}}(\mathbf{q}, i\omega_n)} = \frac{\Omega}{\sqrt{\epsilon^{\text{RPA}}(\mathbf{q}, i\omega_n)}} = \sqrt{\frac{Ze^2 n^{(0)}}{\epsilon^{\text{RPA}}\epsilon_0 M}} \qquad (20.47)$$

is the renormalized phonon frequency due to electronic RPA screening. If we substitute the Thomas–Fermi approximation of the dielectric response function, $\epsilon(\mathbf{q}, 0) = 1 + \frac{\kappa_{\text{TF}}^2}{q^2} \to \frac{\kappa_{\text{TF}}^2}{q^2}$ as $q \to 0$, we find

$$\omega_{\mathbf{q}} = v_F q \sqrt{\frac{Zm_e}{3M}},$$

which is the Bohm–Staver expression for the phonon dispersion. We thus arrive at the satisfactory conclusion that, once the electron–phonon interaction is taken into account, the phonon frequencies are renormalized from Ω to $\omega_{\mathbf{q}}$, where $\omega_{\mathbf{q}}$ is proportional to q for long wavelengths. Of course, we already reached the same conclusion based on macroscopic arguments.

More Useful Form for $V_{\text{eff}}^{\text{RPA}}$

While (20.44) is correct, a physically more transparent form of $V_{\text{eff}}^{\text{RPA}}$ is obtained by expanding the infinite series (20.43), and then collecting all the diagrams containing only Coulomb interaction lines into one sum (this simply yields the RPA-screened Coulomb interaction $W^{\text{RPA}}(\mathbf{q}) = V_c(\mathbf{q})/\epsilon^{\text{RPA}}$), while collecting the remaining diagrams containing a mix of Coulomb and phonon interaction lines into another sum:

The renormalized coupling $g_{\mathbf{q}}^{\text{RPA}}$

$$g_{\mathbf{q}}^{\text{RPA}} = \cdots = \cdots + \cdots = \left(1 + V_c\,\chi^{\text{RPA}}\right) g_{\mathbf{q}} = \frac{g_{\mathbf{q}}}{\epsilon^{\text{RPA}}} \qquad (20.48)$$

is the sum of all diagrams between the outgoing left (incoming right) vertex and the first (last) phonon line, while the renormalized phonon line is given by (20.46).

The final form of the RPA-screened phonon-mediated electron–electron interaction is now obtained by combining (20.48) and (20.46)

$$\frac{1}{V}\left|g_{\mathbf{q}}^{\text{RPA}}\right|^2 \mathcal{D}^{\text{RPA}}(\mathbf{q}, i\omega_n) = \text{(image)} = \frac{|g_{\mathbf{q}}|^2}{\left(\epsilon^{\text{RPA}}\right)^2} \frac{2\Omega}{(i\omega_n)^2 - \omega_{\mathbf{q}}^2}$$

$$\rightarrow \frac{V_c(\mathbf{q})}{\epsilon^{\text{RPA}}} \frac{\omega_{\mathbf{q}}^2}{(i\omega_n)^2 - \omega_{\mathbf{q}}^2},$$

where the unscreened phonon frequency Ω and the unscreened Coulomb interaction $V_c(\mathbf{q})$ have been replaced by their RPA-screened counterparts $\omega_{\mathbf{q}}$ and $V_c(\mathbf{q})\epsilon^{\text{RPA}}$, respectively. The form of the effective electron–electron interaction then becomes

$$V_{\text{eff}}^{\text{RPA}}(\mathbf{q}, i\omega_n) = \frac{V_c(\mathbf{q})}{\epsilon^{\text{RPA}}} + \left|g_{\mathbf{q}}^{\text{RPA}}\right|^2 \mathcal{D}^{\text{RPA}}(\mathbf{q}, i\omega_n) = -\frac{V_c(\mathbf{q})}{\epsilon^{\text{RPA}}} \frac{(i\omega_n)^2}{(i\omega_n)^2 - \omega_{\mathbf{q}}^2}$$

This form for $V_{\text{eff}}^{\text{RPA}}$ will be used in the following derivation of the pair-scattering vertex.

20.6 Phonon Frequencies and the Kohn Effect

The effect of the electron–phonon interaction on the phonon spectrum may be seen by using perturbation theory to calculate the total energy of the system described by the Fröhlich Hamiltonian to second order in $\mathcal{H}_{\text{e-p}}$. We have

$$\mathcal{E} = \mathcal{E}_0 + \left\langle \Phi \left| \mathcal{H}_{\text{e-p}} \right| \Phi \right\rangle + \left\langle \Phi \left| \mathcal{H}_{\text{e-p}} (\mathcal{E}_0 - \mathcal{H}_0)^{-1} \mathcal{H}_{\text{e-p}} \right| \Phi \right\rangle, \tag{20.49}$$

with $\mathcal{E}_0$ the unperturbed energy of the state Φ having $n_{\mathbf{q}}$ phonons in the longitudinal mode $\mathbf{q}$ and $n_{\mathbf{k}}$ electrons in the state $\mathbf{k}$. The first-order term vanishes from this expression, since the components of $\mathcal{H}_{\text{e-p}}$ act on Φ either to destroy or to create one phonon, and the resulting wavefunction must be orthogonal to Φ. In second order, there is a set of nonvanishing terms, as the phonon destroyed by the first factor of $\mathcal{H}_{\text{e-p}}$ to act on Φ can be replaced by the second factor $\mathcal{H}_{\text{e-p}}$, and vice versa.

We then find the contribution, $\mathcal{E}_2$, of the second-order terms to be

$$\mathcal{E}_2 = \left\langle \Phi \left| \sum_{\mathbf{k},\mathbf{k}',\mathbf{q}} M_{\mathbf{k},\mathbf{k}'} \left(b_{-\mathbf{q}}^\dagger + b_{\mathbf{q}} \right) c_{\mathbf{k}}^\dagger c_{\mathbf{k}'} (\mathcal{E}_0 - \mathcal{H}_0)^{-1} \right. \right.$$

$$\left. \times \sum_{\mathbf{k}'',\mathbf{k}''',\mathbf{q}'} M_{\mathbf{k}'',\mathbf{k}'''} \left(b_{-\mathbf{q}'}^\dagger + b_{\mathbf{q}'} \right) c_{\mathbf{k}''}^\dagger c_{\mathbf{k}'''} \left| \Phi \right\rangle \right.$$

$$= \left\langle \Phi \left| \sum_{\mathbf{k},\mathbf{k}',\mathbf{q}} |M_{\mathbf{k},\mathbf{k}'}|^2 \left[b_{-\mathbf{q}}^\dagger c_{\mathbf{k}}^\dagger c_{\mathbf{k}'} (\mathcal{E}_0 - \mathcal{H}_0)^{-1} b_{-\mathbf{q}} c_{\mathbf{k}'}^\dagger c_{\mathbf{k}} \right. \right. \right.$$

$$\left. \left. + b_{\mathbf{q}} c_{\mathbf{k}}^\dagger c_{\mathbf{k}'} (\mathcal{E}_0 - \mathcal{H}_0)^{-1} b_{\mathbf{q}}^\dagger c_{\mathbf{k}'}^\dagger c_{\mathbf{k}} \right] \right| \Phi \right\rangle. \tag{20.50}$$

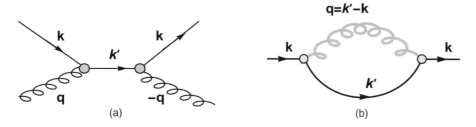

Figure 20.4 These two processes contribute to the energy of the electron–phonon system in second-order perturbation theory.

All other terms have zero matrix element. The first term in brackets in (20.50) can be represented as in Figure 20.4(a).

An electron is first scattered from $\mathbf{k}$ to $\mathbf{k}'$ with the absorption of a phonon of wavevector $-\mathbf{q} = \mathbf{k}' - \mathbf{k}$. The factor $(\mathcal{E}_0 - \mathcal{H}_0)^{-1}$ then measures the amount of time the electron is allowed by the Uncertainty Principle to stay in the intermediate state $\mathbf{k}'$. In this case, the energy difference between the initial and intermediate states is $\mathcal{E}_\mathbf{k} + \hbar\omega_{-\mathbf{q}} - \mathcal{E}_{\mathbf{k}'}$, and so a factor of $(\mathcal{E}_\mathbf{k} + \hbar\omega_{-\mathbf{q}} - \mathcal{E}_{\mathbf{k}'})^{-1}$ is contributed. The electron is then scattered back into its original state with the reemission of the phonon. We can represent the second term in (20.50) by Figure 20.4(b), and there find an energy denominator of $\mathcal{E}_\mathbf{k} - \hbar\omega_\mathbf{q} - \mathcal{E}_{\mathbf{k}'}$.

A rearrangement of the as and the cs into the form of number operators then gives

$$\mathcal{E}_2 = \sum_{\mathbf{k},\mathbf{k}'} \left| M_{\mathbf{k},\mathbf{k}'} \right|^2 \langle n_\mathbf{k}\,(1 - n_{\mathbf{k}'}) \rangle \left(\frac{\langle n_{-\mathbf{q}} \rangle}{\mathcal{E}_\mathbf{k} - \mathcal{E}_{\mathbf{k}'} + \hbar\omega_{-\mathbf{q}}} + \frac{\langle n_\mathbf{q} + 1 \rangle}{\mathcal{E}_\mathbf{k} - \mathcal{E}_{\mathbf{k}'} - \hbar\omega_\mathbf{q}} \right). \tag{20.51}$$

Here $\langle n_\mathbf{k} \rangle$ and $\langle n_{\mathbf{k}'} \rangle$ are electron occupation numbers while $\langle n_{-\mathbf{q}} \rangle$ and $\langle n_\mathbf{q} \rangle$ refer to phonon states. It may be assumed that $\omega_\mathbf{q} = \omega_{-\mathbf{q}}$, and hence that in equilibrium $\langle n_{-\mathbf{q}} \rangle = \langle n_\mathbf{q} \rangle$. One may then rearrange (20.51), to find

$$\mathcal{E} = \mathcal{E}_0 + \sum_{\mathbf{k},\mathbf{k}'} \left| M_{\mathbf{k},\mathbf{k}'} \right|^2 \langle n_\mathbf{k} \rangle \left[\frac{2(\mathcal{E}_\mathbf{k} - \mathcal{E}_{\mathbf{k}'})\langle n_\mathbf{q} \rangle}{(\mathcal{E}_\mathbf{k} - \mathcal{E}_{\mathbf{k}'})^2 - (\hbar\omega_\mathbf{q})^2} + \frac{1 - \langle n_{\mathbf{k}'} \rangle}{\mathcal{E}_\mathbf{k} - \mathcal{E}_{\mathbf{k}'} - \hbar\omega_\mathbf{q}} \right], \tag{20.52}$$

the terms in $\langle n_\mathbf{k} n_{\mathbf{k}'} n_\mathbf{q} \rangle$ canceling by symmetry.

The effect of the electron–phonon interaction on the phonon spectrum is contained in the term proportional to $\langle n_\mathbf{q} \rangle$ in (20.52). We identify the perturbed phonon energy, $\hbar\omega_\mathbf{q}^{(p)}$, with the energy required to increase $\langle n_\mathbf{q} \rangle$ by unity, and so find

$$\hbar\omega_\mathbf{q}^{(p)} = \frac{\partial \mathcal{E}}{\partial \langle n_\mathbf{q} \rangle}$$

$$= \hbar\omega_\mathbf{q} + \sum_\mathbf{k} \left| M_{\mathbf{k},\mathbf{k}'} \right|^2 \frac{2\,(\mathcal{E}_\mathbf{k} - \mathcal{E}_{\mathbf{k}'})\,\langle n_\mathbf{k} \rangle}{(\mathcal{E}_\mathbf{k} - \mathcal{E}_{\mathbf{k}'})^2 - (\hbar\omega_\mathbf{q})^2}. \tag{20.53}$$

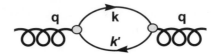

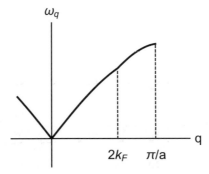

Figure 20.5 This alternative way of considering the process of Figure 20.4(a) suggests that a phonon spends part of its time as a virtual electron–hole pair.

Figure 20.6 The Kohn effect causes a kink to appear in the phonon dispersion curve when the phonon wavevector **q** is equal to the diameter of the Fermi surface.

If we neglect the phonon energy in the denominator in comparison with the electron energies, we have

$$\hbar\omega_{\mathbf{q}}^{(p)} = \frac{\partial \mathcal{E}}{\partial \langle n_{\mathbf{q}} \rangle} = \hbar\omega_{\mathbf{q}} - \sum_{\mathbf{k}} 2 \left| M_{\mathbf{k},\mathbf{k}'} \right|^2 (\mathcal{E}_{\mathbf{k}'} - \mathcal{E}_{\mathbf{k}})^{-1} \langle n_{\mathbf{k}} \rangle, \tag{20.54}$$

where, as before $\mathbf{k}' = \mathbf{k} - \mathbf{q}$.

One may picture the origin of this change in phonon frequency by redrawing Figure 20.4(a) in the form of Figure 20.5, in which the first interaction is represented, not as the scattering of an electron, but as the creation of an electron–hole pair. One can then say that it is the fact that the phonon spends part of its time in the form of an electron–hole pair that modifies its energy.

One interesting consequence of (20.54) occurs when **q** has a value close to the diameter, $2k_F$, of the Fermi surface. Let us suppose **q** to be in the x-direction and of magnitude $2k_F$, and evaluate $\hbar \partial \omega_{\mathbf{q}}^{(p)} / \partial q_x$. If we neglect the variation of $g_{\mathbf{k}\mathbf{k}'}$ with **q** the electron–phonon interaction contributes an amount

$$\sum_{\mathbf{k}} 2 \left| M_{\mathbf{k},\mathbf{k}'} \right|^2 (\mathcal{E}_{\mathbf{k}-\mathbf{q}} - \mathcal{E}_{\mathbf{k}})^{-2} \langle n_{\mathbf{k}} \rangle \frac{\partial \mathcal{E}_{\mathbf{k}-\mathbf{q}}}{\partial q_x}. \tag{20.55}$$

On substituting for $\mathcal{E}_{\mathbf{k}-\mathbf{q}}$ one finds the summation to contain the factor $\langle n_{\mathbf{k}} \rangle$ $(k_x - k_F)^{-2}$. These cause a logarithmic divergence when the summation is performed, and thus indicate that the phonon spectrum has the form indicated in Figure 20.6. The kink in the spectrum when $q = 2k_F$ reflects the infinite group velocity of the phonons at that point, and constitutes the *Kohn effect*. Its importance lies in the fact that even for very complex metals there should always be such an image of the Fermi surface in the phonon spectrum.

20.7 Polarons and Mass Enhancement

Just as the energy of the phonons in a crystal is altered by interaction with the electrons, so also does the converse process occur. We examine (20.52) in the limit of low temperatures, when $\langle n_q \rangle$ vanishes for all $\mathbf{q}$. The perturbed energy of an electron – once again the energy needed to fill an initially empty unperturbed state – is given by

$$\frac{\partial \mathcal{E}}{\partial \langle n_{\mathbf{k}} \rangle} = \mathcal{E}_{\mathbf{k}} + \sum_{\mathbf{k}'} \left| M_{\mathbf{k},\mathbf{k}'} \right|^2 \left[\frac{1 - \langle n_{\mathbf{k}'} \rangle}{\mathcal{E}_{\mathbf{k}} - \mathcal{E}_{\mathbf{k}'} - \hbar\omega_q} - \frac{\langle n_{\mathbf{k}'} \rangle}{\mathcal{E}_{\mathbf{k}'} - \mathcal{E}_{\mathbf{k}} - \hbar\omega_q} \right]$$

$$= \mathcal{E}_{\mathbf{k}} + \sum_{\mathbf{k}'} \left| M_{\mathbf{k},\mathbf{k}'} \right|^2 \left[\frac{1}{\mathcal{E}_{\mathbf{k}} - \mathcal{E}_{\mathbf{k}'} - \hbar\omega_q} - \frac{2\hbar\omega_q \langle n_{\mathbf{k}'} \rangle}{(\mathcal{E}_{\mathbf{k}} - \mathcal{E}_{\mathbf{k}'})^2 - (\hbar\omega_q)^2} \right]. \tag{20.56}$$

The first term in the brackets is independent of $n_{\mathbf{k}'}$, and is thus a correction to the electron energy that would be present for a single electron in an insulating crystal. Indeed, in an ionic crystal the effect of this term may be so great as to change markedly the effective mass of an electron at the bottom of the conduction band.

It then becomes reasonable to use the term *polaron* to describe the composite particle shown in Figure 20.4(b) that is the electron with its attendant cloud of virtual phonons. The name arises because one considers the positive ions to be attracted toward the electron, and thus to polarize the lattice. If this polarization is too great, then second-order perturbation theory is inadequate, and different methods must be used.

The second term in the brackets in (20.56) expresses the dependence of the electron energy on the occupancy of the other k-states. It has the effect of causing a kink in the electronic energy dispersion curves, as shown in Figure 20.7. This kink occurs at the Fermi wavevector k_F, and leads to a change in the group velocity $\mathbf{v}_{\mathbf{k}}$ of the electron.

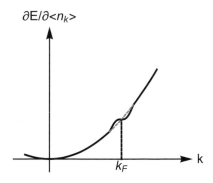

Figure 20.7 The electron–phonon interaction changes the effective electron energy in such a way that the velocity is lowered in the vicinity of the Fermi surface. This gives rise to an increase in the observed electronic specific heat.

We find an expression for $\hbar \mathbf{v_k}$ by differentiating (20.56) with respect to $\mathbf{k}$. Instead of the free electron expression, we find

$$\hbar \mathbf{v_k} = \frac{\partial \mathcal{E}_\mathbf{k}}{\partial \mathbf{k}} \left[1 - \frac{d}{d\mathcal{E}_\mathbf{k}} \sum_{\mathbf{k}'} |M_{\mathbf{k},\mathbf{k}'}|^2 \frac{2\hbar\omega_\mathbf{q} < n_{\mathbf{k}'} >}{(\mathcal{E}_\mathbf{k} - \mathcal{E}_{\mathbf{k}'})^2 - (\hbar\omega_\mathbf{q})^2} \right]. \tag{20.57}$$

We then argue that the major contribution to the derivative of the summation comes from the rapid variation of $< n_{\mathbf{k}'} >$ at the Fermi surface. We change the summation over $\mathbf{k}'$ to an integral over $\mathcal{E}_{\mathbf{k}'}$ and make the approximations of replacing $g_{\mathbf{kk}'}$ and $\omega_\mathbf{q}$ by their average values $\bar{M}$ and $\bar{\omega}$, and $\mathcal{D}(\mathcal{E}_{\mathbf{k}'})$ by its value at the Fermi energy ζ. With $\mathcal{E}_{\mathbf{k}'} - \mathcal{E}_\mathbf{k}$ written as η, we then have

$$\hbar \mathbf{v_k} \simeq \frac{\partial \mathcal{E}_\mathbf{k}}{\partial \mathbf{k}} \left[1 - 2\hbar\bar{\omega}\mathcal{D}(\zeta)|\bar{M}|^2 \frac{d}{d\mathcal{E}_\mathbf{k}} \int \frac{< n(\mathcal{E}_\mathbf{k} + \eta) >}{\eta^2 - (\hbar\bar{\omega})^2} d\eta \right]. \tag{20.58}$$

Since

$$\frac{d < n(\mathcal{E}_\mathbf{k} + \eta) >}{d\mathcal{E}_\mathbf{k}} = -\delta(\mathcal{E}_\mathbf{k} + \eta - \zeta), \tag{20.59}$$

we find

$$\hbar \mathbf{v_k} \simeq \frac{\partial \mathcal{E}_\mathbf{k}}{\partial \mathbf{k}} \left[1 + \frac{2\hbar\bar{\omega}\mathcal{D}(\zeta)|\bar{M}|^2}{(\mathcal{E}_\mathbf{k} - \zeta)^2 - (\hbar\bar{\omega})^2} \right]. \tag{20.60}$$

The infinities that this expression predicts when $\mathcal{E}_\mathbf{k} = \zeta \pm \hbar\bar{\omega}$ are a spurious consequence of our averaging procedure. The value predicted when $\mathbf{k}$ lies in the Fermi surface, when $\mathcal{E}_\mathbf{k} = \zeta$, is more plausible, and gives us the result

$$\mathbf{v_k} \simeq \mathbf{v_k^0}(1 - \alpha) \tag{20.61}$$

where $\mathbf{v_k^0}$ is the unperturbed velocity and

$$\alpha = \frac{2\mathcal{D}(\zeta)|\bar{M}|^2}{(\hbar\bar{\omega})}. \tag{20.62}$$

This decrease in the electron velocity is equivalent to an increase in the density of states by the factor $(1 - \alpha)^{-1}$. Because $\mathbf{v_k^0}$ is inversely proportional to the electron mass, m, it is common to discuss the increase in the density of states in terms of an increase in the effective mass of the electron. One refers to $(1 - \alpha)^{-1}$ as the *mass enhancement* factor due to electron-phonon interactions.

Exercises

20.1 Momentum dependence of electron–phonon coupling:

Following are three different types of short-range coupling of a single electron with Einstein phonons (or, more generally, dispersionless bosons) of frequency ω, on a one-dimensional lattice (lattice constant $= 1$) with N sites. The coupling constants g, ϕ, and ϕ_b are dimensionless, and ω has units of energy. For each of these couplings, find the equivalent momentum-space representation, by Fourier-transforming the corresponding electron–phonon coupling Hamiltonian in real space to the form

$$\mathcal{H}_{\text{e-ph}} = \frac{1}{\sqrt{N}} \sum_{k,q} \gamma(k,q) \, c^\dagger_{k+q} \, c_k \left(b^\dagger_{-q} + b_q \right).$$

(a) Holstein-type (purely local) coupling

$$\mathcal{H}_{\text{e-ph}} = g \omega \sum_i c^\dagger_i c_i \left(b^\dagger_i + b_i \right).$$

(b) Su–Schrieffer–Heeger (Peierls-type) coupling

$$\mathcal{H}_{\text{e-ph}} = g \omega \sum_i \left(c^\dagger_{i+1} c_i + hc \right) \left(b^\dagger_{i+1} + b_{i+1} - b^\dagger_i - b_i \right).$$

(c) "Breathing" coupling (relevant in cuprate high-Tc superconductors)

$$\mathcal{H}_{\text{e-ph}} = g \omega \sum_i c^\dagger_i c_i \left(b^\dagger_{i-1/2} + b_{i-1/2} - b^\dagger_{i+1/2} - b_{i+1/2} \right).$$

Here, $i \pm 1/2$ refers to the fact that the Einstein oscillators are placed in the middle between two sites.

Then comment on the differences between these three types of electron–phonon interaction as far as the momentum dependence of the vertex function $(\mathbf{k}, \mathbf{q})$ is concerned.

20.2 Polarons at weak coupling:

Electrons in a conduction band of a semiconductor form a very dilute gas and as a result, they are not "aware" of the presence of each other. Consequently, their Green function is of noninteracting type, given by

$$G(\mathbf{p}, E) = \frac{1}{E - \dfrac{p^2}{2m} - i0}.$$

However, the interaction of electrons with phonons gives rise to the distortion of the crystalline lattice, which leads to the formation of a new particle, called a polaron (an electron with a deformed lattice cloud around it).

The interaction between the electrons and the phonons is given by

$$\mathcal{H}_{\text{int}} = g \int d\mathbf{x} \, \Psi^\dagger(\mathbf{x}) \Psi(\mathbf{x}) \hat{A}(\mathbf{x}),$$

where $\hat{A}(\mathbf{x})$ is the phonon field operator. The Green function of the phonons, defined as

$$\mathcal{D} = -i \left\langle T \hat{A}(\mathbf{x}_f, t_f) \hat{A}(\mathbf{x}_i, t_i) \right\rangle,$$

Figure 20.8 The simplest electronic self-energy diagram. The phonons are represented by a wavy line, while the electrons are represented by straight lines.

is given by

$$\mathcal{D}(\mathbf{q}, E) = \frac{c^2 q^2}{E^2 - c^2 q^2 + i0} \, \Theta(q_D - q),$$

where c is the speed of sound, q_D is the Debye wavevector.

The simplest electronic self-energy diagram is shown in Figure 20.8.

(a) Write down the expression corresponding to $\Sigma(\mathbf{p}, E)$.

(b) Show that if $p > mc$, $\Sigma(\mathbf{p}, E)$ is not a real function, but rather a complex function. Calculate Im $\Sigma(\mathbf{p}, E)$ and figure out the polaron lifetime.

(c) Show that for $E \sim p^2/2m$ and for $p \ll mc$, Σ is real and takes the form

$$\Sigma(p, E) = \varepsilon_0 - \alpha_1 \left(E - \frac{p^2}{2m} \right) - \alpha_2 \frac{p^2}{2m}.$$

Show that ε_0 is the binding energy of this composite particle, while α_2 gives the effective mass of a polaron m^*. Find m^*.

20.3 The partition function of an electron–phonon system can be written as the path integral

$$Z = \int D(\bar{\psi}, \psi) \, D(\bar{\phi}, \phi) \, \exp\left[-S_e \left[\bar{\psi}, \psi \right] - S_p \left[\bar{\phi}, \phi \right] - S_{ep} \left[\bar{\psi}, \psi; \bar{\phi}, \phi \right] \right]$$

with fermion Grassmann fields $\bar{\psi}, \psi$ representing the electrons and phonon complex fields $\bar{\phi}, \phi$, with action for the phonons S_p and for the electron–phonon interaction S_{ep} given by

$$S_p \left[\bar{\phi}, \phi \right] = \sum_q \bar{\phi}_q \left(-i\omega_n + \omega_{\mathbf{q}} \right) \phi_q$$

$$S_{ep} \left[\bar{\psi}, \psi; \bar{\phi}, \phi \right] = \frac{1}{\sqrt{\Omega}} \sum_q g_{\mathbf{q}} \, \rho_q \left(\phi_q + \bar{\phi}_{-q} \right)$$

with the electronic density operator $\rho_q = \sum_{\mathbf{k}} \bar{\psi}_{k+q} \psi_k$, electron–phonon coupling $g_{\mathbf{q}}$ and the kinetic term $S_e[\bar{\psi}, \psi]$ for the electrons. The four-vector $q = (i\omega_n, \mathbf{q})$ denotes frequency and momentum, and the sum $\sum_q$ includes a Matsubara and momentum sums.

Perform a Gaussian integration over the phonon fields $\bar{\phi}, \phi$ in Z and derive an effective action $S_{\text{eff}}[\bar{\psi}, \psi]$ for the electronic system (ignore terms that do not depend on $\bar{\psi}, \psi$):

$$Z = \int D(\bar{\psi}, \psi) \; \exp\left[-S_{\text{eff}}\left[\bar{\psi}, \psi\right]\right]$$

$$\exp\left[-S_{\text{eff}}\left[\bar{\psi}, \psi\right]\right] = \int D(\bar{\phi}, \phi) \; \exp\left[-S_e\left[\bar{\psi}, \psi\right] - S_p\left[\bar{\phi}, \phi\right] - S_{ep}\left[\bar{\psi}, \psi; \bar{\phi}, \phi\right]\right].$$

Analytically continuing $\omega_n \to -i\omega$, show that the effective action contains an interaction term between the electrons that becomes attractive for $\omega < \omega_{\mathbf{q}}$.

21

Microscopic Theory of Conventional Superconductivity

21.1 Introduction

In 1911, three years after he liquified helium, Heike Kamerlingh-Onnes, in his quest to study materials at ever lower temperatures, happened to find that the electrical resistance of some metallic materials suddenly vanished at temperatures near absolute zero. He called the phenomenon superconductivity, and scientists soon found additional materials that exhibited this property. But no one could completely explain how it worked. For the next few decades, many prominent physicists worked to develop a theory of the mechanism underlying superconductivity, but no one had much success, and some despaired of figuring it out. One such physicist, Felix Bloch, was quoted as proposing *Bloch's theorem: Superconductivity is impossible.*

In 1957, John Bardeen, Leon Cooper, and Robert Schrieffer presented their complete theory of superconductivity, finally explaining a phenomenon that had been a mystery to physicists since its discovery in 1911. Richard Feynman later recalled that he had "spent an awful lot of time in trying to understand it and doing everything by means of which I could approach it I developed an emotional block against the problem of superconductivity, so that when I learned about the BCS paper I could not bring myself to read it for a long time."

Properties Exhibited by Superconducting Materials

Figure 21.1 shows the properties exihibited by superconducting materials. The properties include the following:

1. **Zero resistance.** Below a material T_c, the DC electrical resistivity ρ is zero, not just very small. This leads to the possibility of a related effect,
2. **Persistent currents.** If a current is set up in a superconductor with multiply connected topology, e.g., a torus, it will flow forever without any driving voltage. (In practice, experiments have been performed in which persistent currents flow for several years without signs of degrading).
3. **Perfect diamagnetism.** A superconductor expels a weak magnetic field nearly completely from its interior (screening currents flow to compensate the field within a surface layer of a few 100 or 1,000 Å, and the field at the sample surface drops to zero over this layer). This phenomenon of perfect diamagnetism is known as the **Meissner effect**.

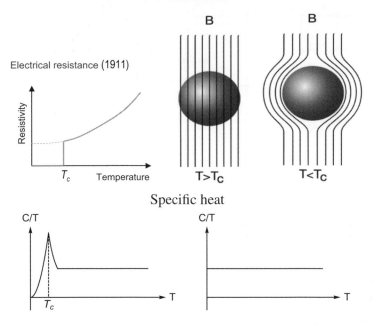

Figure 21.1 Properties of superconductors.

4. **Energy gap.** Most thermodynamic properties of a superconductor are found to vary as $e^{-\Delta/(k_B T)}$, indicating the existence of a gap, or energy interval with no allowed eigenenergies, in the energy spectrum. Idea: when there is a gap, only an exponentially small number of particles have enough thermal energy to be promoted to the available unoccupied states above the gap. In addition, this gap is visible in electromagnetic absorption: send in a photon at low temperatures (strictly speaking, $T = 0$), and no absorption is possible until the photon energy reaches 2Δ, i.e., until the energy required to break a pair is available.

Imagine a snapshot of a single electron entering a region of the metal; it will create a net positive charge density near itself by attracting the oppositely charged ions (see Figure 21.2). Crucial here is that a typical electron close to the Fermi surface moves with velocity $v_F = \hbar k_F/m_e$ which is much larger than the velocity of the ions, $v_I = v_F m_e/M$. So by the time ($\tau \sim 2\pi/\omega_D \sim 10^{-13}$ s), the ions have polarized themselves, the first electron is long gone (it has moved a distance v_F ($\sim 10^8$ cm/s) $\tau \sim 1000$ Å, and the second electron can happen to be passing by and to lower its energy through interaction with the concentration of positive charge before the ionic fluctuation relaxes away. This gives rise to an effective attraction between the two electrons as shown in Figure 21.2, which may be large enough to overcome the repulsive Coulomb interaction. As we described here, the attraction of the second electron to the first does not happen instantaneously, it is delayed through the reaction of the slow ions; it is referred to as a *retarded phonon-mediated interaction*. If it were instantaneous, namely, it has a $\delta(t)$ form,

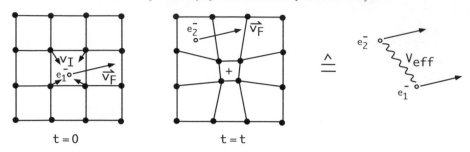

$$t = 0 \qquad\qquad t = t$$

Figure 21.2 Effective attraction of two electrons due to phonon exchange.

it would have been frequency independent; however, its retarded nature endows it with a frequency dependence.

21.2 Electronic Instability against Electron–Phonon Interaction

21.2.1 *Leon Cooper and the Pairing Hypothesis*

The existence of such an obvious phase transition as that involved in superconductivity led to a long search for a mechanism that would lead to an attractive interaction between electrons. A great deal became known about the phenomenology of superconductivity in the 1950s, and it was already suspected that the electron–phonon interaction was responsible. The discovery of the *isotope effect* provided corroborating evidence that the electron–phonon interaction was indeed the mechanism responsible: it was found that the transition temperature T_c for some metals depended on their atomic mass M_A. Early on, it appeared that T_c was proportional to $M_A^{-1/2}$, and hence to the Debye temperature T_D. But more recent measurements have shown a wider variety of power laws; for instance, an isotope effect was completely absent in osmium, while a dependence of approximately $T_c \simeq M_A^2$ was found in α-uranium. Yet the microscopic form of the wavefunction was unknown. A clue was provided by Leon Cooper [47], who showed that the noninteracting Fermi sea is unstable toward the addition of a single pair of electrons with attractive interactions. To follow Cooper's argument, we introduce two assumptions:

(i) We suppose that the mechanism responsible for superconductivity is an attractive interaction between pairs of particles, each occupying states having energy within an energy shell of width 2δ at the Fermi surface.

(ii) We should focus on pairs of particles with total momentum zero, as will be justified later.

We now explore how these assumptions can be cast into a theory of the superconducting state. What we need to establish is that such pairs with zero total momentum will bind together to produce the desired energy gap, given an attractive force that mirrors the Fröhlich interaction previously derived.

Binding Energy of a Cooper Pair

We consider a pair of particles at the Fermi level that in the absence of the interaction have a total energy of $2\mathcal{E}_F$. To explore whether a binding can arise, we need to derive the ground state of the Schrödinger equation

$$\left[-\frac{\hbar^2}{2m}\left\{\nabla_1^2 + \nabla_2^2\right\} + v(\mathbf{x}_1, \mathbf{x}_2)\right]\psi = \mathcal{E}\,\psi, \tag{21.1}$$

where $v(\mathbf{x}_1, \mathbf{x}_2)$ represents the attractive interaction. For binding to occur, we need to show that the ground-state eigenvalue is reduced below $2\mathcal{E}_F$.

The noninteracting states have the form

$$\psi(\mathbf{k}_1, \mathbf{k}_2; \mathbf{x}_1, \mathbf{x}_2) = \Omega^{-1}\,e^{i(\mathbf{k}_1 \cdot \mathbf{x}_1 + \mathbf{k}_2 \cdot \mathbf{x}_2)}\,\chi$$

$$= \Omega^{-1}\,e^{i[\mathbf{K} \cdot (\mathbf{x}_1 + \mathbf{x}_2) + \mathbf{k} \cdot (\mathbf{x}_1 - \mathbf{x}_2)/2]}\,\chi = \Omega^{-1}\,e^{i(\mathbf{K} \cdot \mathbf{R} + \mathbf{k} \cdot \mathbf{r})}\,\chi, \tag{21.2}$$

where $\mathbf{R}$, $\mathbf{K}$ and $\mathbf{r}$, $\mathbf{k}$ are the center of mass and relative coordinate and conjugate momenta, respectively; χ is the spin wavefunction. It is more favorable to have a singlet spin wavefunction to take maximum advantage of the attractive potential with a symmetric spatial wavefunction.

Moreover, the total momentum

$$\mathbf{k}_1 + \mathbf{k}_2 = \mathbf{K}$$

is conserved, as shown in Figure 21.3. We should note, as depicted in Figure 21.4, that if $K \neq 0$, the phase space for the attractive scattering is dramatically reduced. So the system can always lower its energy by creating $\mathbf{K} = 0$ pairs. Henceforth, we will make this assumption, as Cooper did.

If we choose states with zero total momentum, i.e., $\mathbf{K} = 0$, then

$$\psi(\mathbf{k}; \mathbf{x}_1, \mathbf{x}_2) = \Omega^{-1}\,e^{i\mathbf{k} \cdot (\mathbf{x}_1 - \mathbf{x}_2)}\,\chi_{\text{singlet}}. \tag{21.3}$$

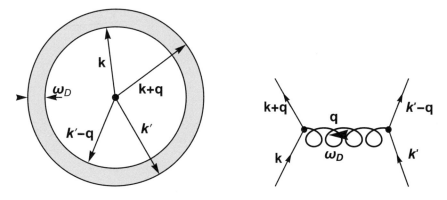

Figure 21.3 Electrons scattered by phonon exchange are confined to a shell of thickness ω_D about the Fermi surface.

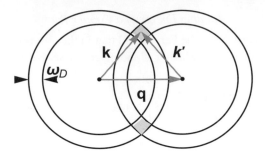

Figure 21.4 To get (attractive) scattering with finite COM momentum **K**, we need both electron energies to be within ω_D of the Fermi level with very little phase space.

To solve (21.1), we expand ψ in terms of the set of noninteracting states, with the constraint that the Fermi sea states are occupied. Thus we write the perturbed wavefunction as

$$\psi(\mathbf{x}_1, \mathbf{x}_2) = \Omega^{-1} \chi_{\text{singlet}} \sum_{|\mathbf{k}'|>k_F} \alpha_{\mathbf{k}'} \, e^{i\mathbf{k}'\cdot(\mathbf{x}_1-\mathbf{x}_2)}, \tag{21.4}$$

where we have implicitly formed two-particle unperturbed states with individual momenta $\mathbf{k}'$ and $-\mathbf{k}'$.

At this stage, to keep the details at a minimum, we will implicitly deal with a singlet spin state and drop χ_{singlet}. Then the wavefunction (21.4) may be substituted into (21.1), and we obtain

$$\Omega^{-1} \sum_{|\mathbf{k}'|>k_F} \alpha_{\mathbf{k}'} \left[\mathcal{E}_{\mathbf{k}'} - \mathcal{E} + v(\mathbf{x}_1, \mathbf{x}_2) \right] e^{i\mathbf{k}'\cdot(\mathbf{x}_1-\mathbf{x}_2)} = 0, \tag{21.5}$$

where $\mathcal{E}_{\mathbf{k}'} = \hbar^2 k'^2/m$. If we now multiply both sides by $e^{-i\mathbf{k}\cdot(\mathbf{x}_1-\mathbf{x}_2)}$ and integrate over $\mathbf{x}_1$ and $\mathbf{x}_2$, we find

$$\alpha_{\mathbf{k}} \left[\mathcal{E}_{\mathbf{k}} - \mathcal{E} \right] = - \sum_{|\mathbf{k}'|>k_F} \alpha_{\mathbf{k}'} \, v(\mathbf{k}, -\mathbf{k}, \mathbf{k}', -\mathbf{k}'), \tag{21.6}$$

where $v(\mathbf{k}, -\mathbf{k}, \mathbf{k}', -\mathbf{k}')$ represents the matrix elements of the interaction. We choose the simplest possible Hermitian form

$$v(\mathbf{k}, -\mathbf{k}, \mathbf{k}', -\mathbf{k}') = \begin{cases} -V/\Omega; & \mathcal{E}_{\mathbf{k}} > \mathcal{E}_F, \ \mathcal{E}_{\mathbf{k}'} > \mathcal{E}_F + \delta, \\[2mm] 0; & \text{otherwise,} \end{cases} \tag{21.7}$$

thus restricting the scattering states to the energy range $\mathcal{E}_F, \mathcal{E}_F + \delta$.

The choice of the negative sign is, of course, to get an attractive interaction, the constancy of the matrix element is really reflecting some average over the energy shell, and the factor Ω^{-1} is to make v of the order of unity.

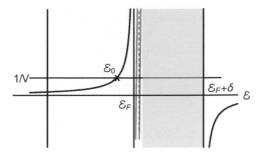

Figure 21.5 Graphic display of the solutions of (21.9). Isolated cross shows bound state with volume-independent energy.

If we substitute these matrix elements in (21.6) for $\alpha_{\mathbf{k}}$, we find

$$\alpha_{\mathbf{k}} = \frac{V}{\mathcal{E}_{\mathbf{k}} - \mathcal{E}} \frac{1}{\Omega} \sum_{\mathcal{E}_F < \mathcal{E}_{\mathbf{k}'} < \mathcal{E}_F + \delta} \alpha_{\mathbf{k}'}. \tag{21.8}$$

Though the solution of such an equation for the $\alpha_{\mathbf{k}}$'s is, in general, difficult, we can simply determine the eigenvalues in this case by summing each side of (21.8) over $\mathbf{k}$ values in the energy range between $\mathcal{E}_F$ and $\mathcal{E}_F + \delta$, and then a common factor $\sum_{\mathcal{E}_F < \mathcal{E}_{\mathbf{k}} < \mathcal{E}_F + \delta} \alpha_{\mathbf{k}}$ appears. The resulting equation for the eigenvalue $\mathcal{E}$ takes the form

$$\frac{1}{V} = \frac{1}{\Omega} \sum_{\mathcal{E}_F < \mathcal{E}_{\mathbf{k}} < \mathcal{E}_F + \delta} \frac{1}{\mathcal{E}_{\mathbf{k}} - \mathcal{E}}. \tag{21.9}$$

We apply a standard graphic technique to find the roots of (21.9): we simply plot both sides as a function of $\mathcal{E}$ and find where the resulting curves intersect, as shown in Figure 21.5. While the left-hand side is simply a positive constant, it will be seen that as $\mathcal{E}$ goes from $\mathcal{E}_F$ to $\mathcal{E}_F + \delta$, the right-hand side varies very rapidly between plus and minus infinity. During each such fluctuation, it takes on the value $1/V$ once. Thus, in this energy range δ above the Fermi level, we find a continuum of solutions.

But the basic point is that there is a portion of the curve below $\mathcal{E}_F$ that separates off from the continuum and decreases monotonically to zero as $\mathcal{E} \to -\infty$. This branch of the curve takes on the value $1/V$ at some volume-independent energy below the Fermi surface. This is the bound state we have been searching for.

Now that the presence of this bound-state level, with energy $\mathcal{E}_0$, has been established, we can perform a simple calculation to determine the binding energy of the pair – electrons in such bound states are called Cooper pairs. We consider (21.9) specifically to the ground state, and transform the sum to an integral via $\frac{1}{\Omega} \sum \to \int d\varepsilon \mathcal{D}(\epsilon)$, where $\mathcal{D}$ is the density of states, and we find

$$\frac{1}{V} = \int_{\mathcal{E}_F}^{\mathcal{E}_F + \delta} \frac{\mathcal{D}(\varepsilon) d\varepsilon}{\varepsilon - \mathcal{E}_0}. \tag{21.10}$$

We note that in the range of integration we expect $(\varepsilon - \mathcal{E}_0)^{-1}$ to vary much more rapidly than the density of states. This allows us to set $\mathcal{D}(\varepsilon) \sim \mathcal{D}(\varepsilon_F)$, since the energy width δ is known to be small. The resulting integral is trivial, and we find a binding energy given by

$$\mathcal{E}_F - \mathcal{E}_0 = \frac{\delta}{\left(\exp\left[\frac{1}{\mathcal{D}V}\right] - 1\right)} \simeq \delta \exp\left[\frac{-1}{\mathcal{D}V}\right], \tag{21.11}$$

where we have assumed in the second step that $e^{\frac{1}{\mathcal{D}V}} \gg 1$, which is justified, a posteriori, by the fact that in metals experiment indicates that $\mathcal{E}_F - \mathcal{E}_0 \simeq 10^{-4}$ eV, while $\delta \simeq 10^{-1}$ eV.

The interesting point about (21.11) is that, although the binding energy is small, we *cannot* expand it by perturbation theory in powers of V. The dependence on V is that of an essential singularity, i.e., a nonanalytic function of the parameter. Thus we may expect never to arrive at this result at any order in perturbation theory, an unexpected problem that hindered theoretical progress for a long time. Thus in the subsequent generalization of this two-particle theory to the many-body problem, we must avoid perturbation theory. In fact, we shall employ the variational method.

Size of Cooper Pair

The extent of the Cooper pair wavefunction in space (see Figure 21.6) may be estimated as follows: the orbital wavefunction for a pair may be written as

$$\psi(\mathbf{x}_1, \mathbf{x}_2) = \phi_{\mathbf{K}}(\mathbf{x}_1 - \mathbf{x}_2) \exp\left[i\mathbf{K} \cdot \left(\frac{\mathbf{x}_1 + \mathbf{x}_2}{2}\right)\right], \tag{21.12}$$

where $\hbar\mathbf{K}$ is the momentum of the center of mass. For the singlet spin case, $\phi_{\mathbf{K}}$ is symmetric, and considering the center of mass at rest, $\mathbf{K} = 0$, we get

$$\psi(\mathbf{x}_1, \mathbf{x}_2) = \phi_0(\mathbf{r}) = \sum_{\mathbf{k}} \alpha(\mathbf{k}) e^{i\mathbf{k} \cdot (\mathbf{x}_1 - \mathbf{x}_2)}, \quad \mathbf{r} = \mathbf{x}_1 - \mathbf{x}_2, \tag{21.13}$$

which is also an eigenfunction of angular momentum. The $\mathbf{k}$ sum is over states near the Fermi surface.

The mean square radius of a Cooper pair is

$$\langle r^2 \rangle = \frac{\int |\phi_0(\mathbf{r})|^2 r^2 \, d\mathbf{r}}{\int |\phi_0(\mathbf{r})|^2 \, d\mathbf{r}} = \frac{\sum_{\mathbf{k}} |\nabla_{\mathbf{k}} \alpha(\mathbf{k})|^2}{\sum_{\mathbf{k}} |\alpha(\mathbf{k})|^2}. \tag{21.14}$$

With the potential (21.7), the two-body Schrödinger equation leads to the solution for the momentum eigenfunctions $\alpha(\mathbf{k})$ as

$$\alpha(\mathbf{k}) = \frac{1}{\mathcal{E}_{\mathbf{k}} - \mathcal{E}} \frac{V}{\Omega} \sum_{\mathcal{E}_F < \mathcal{E}_{\mathbf{k}'} < \mathcal{E}_F + \delta} \alpha_{\mathbf{k}'} = \frac{\beta}{\mathcal{E}_{\mathbf{k}} + \Delta - 2\mathcal{E}_F}, \tag{21.15}$$

where Δ is the binding energy for the pair relative to the Fermi level, and β is a constant independent of $\mathbf{k}$. We find that

$$\sum_{\mathbf{k}} |\alpha(\mathbf{k})|^2 = \beta^2 \int_{\mathcal{E}_F}^{\mathcal{E}_F + \varepsilon_D} d\varepsilon \, \mathcal{D}(\varepsilon) \frac{1}{(2\mathcal{E}_{\mathbf{k}} + \Delta - 2\mathcal{E}_F)^2} \simeq \frac{\mathcal{D}(\mathcal{E}_F)}{2\Delta}$$

$$\nabla_{\mathbf{k}} \alpha(\mathbf{k}) = \frac{-\beta}{(\mathcal{E}_{\mathbf{k}} + \Delta - 2\mathcal{E}_F)^2} \nabla_{\mathbf{k}} \mathcal{E}_{\mathbf{k}}$$

$$\sum_{\mathbf{k}} |\nabla_{\mathbf{k}} \alpha(\mathbf{k})|^2 = \beta^2 \int_{\mathcal{E}_F}^{\mathcal{E}_F + \varepsilon_D} d\varepsilon \, \mathcal{D}(\varepsilon) \frac{\hbar^2 v_F^2}{(2\mathcal{E}_{\mathbf{k}} + \Delta - 2\mathcal{E}_F)^4} \simeq \frac{2\mathcal{D}(\mathcal{E}_F)\hbar^2 v_F^2}{3\Delta^3},$$

which yields

$$\left(\langle r^2 \rangle\right)^{1/2} = \frac{2}{\sqrt{3}} \frac{\hbar v_F}{\Delta} \simeq 10^{-4} \, \text{cm}, \tag{21.16}$$

where v_F is the velocity at the Fermi surface. This quantity is essentially the coherence length and represents the distance over which the superconducting electrons are correlated, or equivalently the range of local order. The rough magnitude of a Cooper pair binding energy obtained above conforms with the order of magnitude provided by experiment, while the estimated size relates to the corresponding coherence length. Thus, any theory developed to describe superconductivity should consistently comply with such values. However, before carrying out this task, we find that a closer examination of the Cooper hypothesis raises several questions:

- Why not turn on the attractive interaction for other electrons in the Fermi sea, so that they too form pairs and lower the energy of the system still further?
- If we do so, do all the electrons in the Fermi sea participate?
- Will the pairing of many other electrons essentially modify the attractive interaction responsible for the pairing?
- Does the pair binding energy depend on the number of emerging pairs?

Such questions lead us to the realization that to explain superconductivity, we need to take into account the fact that such interactions simultaneously occur between many electrons,

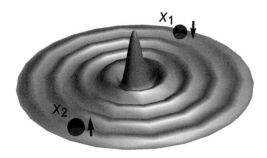

Figure 21.6 Illustrative rendition of a Cooper pair wavefunction.

and that Cooper pairs do interact with one another. Thus, while Cooper's idea was crucial to the development of the theory, it is only a starting point toward the construction of a many-body theory.

21.2.2 Diagramatic Derivation of the Cooper Instability

We note that the participation of many electrons in the pairing process may induce an *instability* of the Fermi liquid state of the normal metal, giving rise to a ground state of completely different character and symmetry than the Fermi liquid state. This instability should be manifest in some diverging susceptibility we can calculate. Guided by the Cooper recipe, we focus on the pair interaction correlation function

$$\mathcal{G}(\mathbf{q}, \tau) = \frac{1}{\Omega} \sum_{\mathbf{k}\mathbf{k}'} \left\langle a_{-\mathbf{k}'+\mathbf{q}\downarrow}(\tau)\, a_{\mathbf{k}'\uparrow}(\tau)\, a^{\dagger}_{-\mathbf{k}+\mathbf{q}\downarrow}(0)\, a^{\dagger}_{\mathbf{k}\uparrow}(0) \right\rangle.$$

Here we introduce two electron quasiparticles occupying time-reversed states of the form $|\mathbf{k} \uparrow\rangle$ and $|-\mathbf{k} \downarrow\rangle$ into the system, subject to an effective electron–electron interaction potential, of the form prescribed in Section 20.5, namely,

$$V_{\text{eff}}(\mathbf{q}, i\omega_n) = \frac{V_c(\mathbf{q})}{\epsilon^{\text{RPA}}} + \left| g_{\mathbf{q}}^{\text{RPA}} \right|^2 \mathcal{D}^{\text{RPA}}(\mathbf{q}, i\omega_n)$$

$$= -\frac{V_c(\mathbf{q})}{\epsilon^{\text{RPA}}} \frac{(i\omega_n)^2}{(i\omega_n)^2 - \omega_{\mathbf{q}}^2}$$

and allow them to evolve through multiple scattering before being extracted.

Construction of the Effective Pair-Scattering Vertex

Fourier-transforming $\mathcal{G}$ and using the four-momentum notation $\bar{k} = (\mathbf{k}; i\Omega_n)$ we represent the multiple electron–electron scattering process by the infinite diagramatic sum. It is sufficient to consider the dominant diagrams at each order. These are the diagrams that do not contain crossing interaction lines, namely,

Such diagrams are called *ladder diagrams*, because all Fermion lines run forward in time, indicating exclusively particlelike character.

We now define the pair scattering vertex $\Gamma(\tilde{k};\tilde{k}') = \boxed{{}^{\text{r}}}$ for the infinite ladder-diagram sum according to the Dyson diagramatic equation

$$\tag{21.17}$$

which can be expressed as the integral equation

$$\Gamma\left(\bar{k};\bar{k}'\right) = -V_{\text{eff}}(\bar{k} - \bar{k}') + \frac{1}{\beta} \sum_{\bar{k}_1} \left[-V_{\text{eff}}(\bar{k} - \tilde{k}_1) \right] \mathcal{G}_\uparrow^{(0)}(\bar{k}_1)\, \mathcal{G}_\downarrow^{(0)}(-\bar{k}_1)\, \Gamma\left(\bar{k}_1;\bar{k}'\right).$$

At this point, we need to simplify the form of the scattering potential, to obtain a manageable solution while maintaining the physics we are seeking. We recall that we are actually interested in attractive interactions, which appear in the low-energy domain of $V_{\text{eff}}^{\text{RPA}}$, bounded by $\omega_{\mathbf{q}} \leq \omega_D$, the Debye frequency. Thus, we introduce the approximation

$$V_{\text{eff}}(\mathbf{q}; i\omega_n) = \begin{cases} -V_0, & |i\omega_n| \leq \omega_D \\ 0, & |i\omega_n| > \omega_D. \end{cases} \tag{21.18}$$

The equation for $\Gamma(\bar{k};\bar{k}_1)$ reduces to

$$\Gamma(\bar{k};\bar{k}') = V_0 + \frac{V_0}{\beta} \sum_{\substack{i\omega_n \\ |\omega_n| < \omega_D}} \frac{1}{V} \sum_{\mathbf{k}_1} \mathcal{G}_\uparrow^{(0)}(\mathbf{k}_1, i\omega_n)\, \mathcal{G}_\downarrow^{(0)}(-\mathbf{k}_1, i\omega_n)\, \Gamma\left(\bar{k}_1;\bar{k}'\right). \tag{21.19}$$

The external momentum $\bar{k}$ does not appear in the summation on the right-hand side, which justifies setting $\Gamma(\bar{k};\bar{k}') = \Gamma(\bar{k}')$, $\Gamma(\bar{k}_1;\bar{k}') = \Gamma(\bar{k}')$. As a matter of fact, it becomes clear that $\Gamma(\bar{k};\bar{k}')$ is a constant, Γ. Rearranging (21.19), we arrive at

$$\Gamma = \frac{V_0}{1 - \frac{V_0}{\beta} \sum_{\substack{i\omega_n \\ |i\omega_n| < \omega_D}} \frac{1}{V} \sum_{\mathbf{k}_1} \mathcal{G}_\uparrow^{(0)}(\bar{k}_1)\, \mathcal{G}_\downarrow^{(0)}(-\bar{k}_1)}. \tag{21.20}$$

With

$$\mathcal{G}_\sigma^{(0)}(\mathbf{k}_1, i\Omega_n) = \frac{1}{i\Omega_n - \varepsilon_{\mathbf{k}_1}}, \qquad i\Omega_n = \frac{(2n+1)\pi}{\beta},$$

we evaluate explicitly the quantity

$$\frac{V_0}{\beta} \sum_{\substack{i\omega_n \\ |i\omega_n| < \omega_D}} \frac{1}{V} \sum_{\mathbf{k}_1} \mathcal{G}_\uparrow^{(0)}(\bar{k}_1) \, \mathcal{G}_\downarrow^{(0)}(-\bar{k}_1) = \frac{V_0}{\beta} \sum_{|i\omega_n|}^{\omega_D} \frac{1}{V} \sum_{\mathbf{k}_1} \left(\frac{1}{i\omega_n - \varepsilon_{\mathbf{k}_1}} \right) \left(\frac{1}{-i\omega_n - \varepsilon_{\mathbf{k}_1}} \right)$$

$$= \frac{V_0}{\beta} \sum_{|i\omega_n|}^{\omega_D} \int_{-\infty}^{\infty} d\varepsilon \, \frac{\mathcal{D}(\varepsilon_F)/2}{\omega_n^2 + \varepsilon^2} = \frac{V_0 \, \mathcal{D}(\varepsilon_F)}{2\beta}$$

$$\times \sum_{|i\omega_n| < \omega_D} \frac{\pi}{|\omega_n|}$$

$$= V_0 \, \mathcal{D}(\varepsilon_F) \sum_{n=0}^{\beta\omega_D/2\pi} \frac{1}{2n+1},$$

where $\mathcal{D}(\varepsilon_F)/2$ is the density of states per spin. Using the relation

$$\sum_{k=0}^{n} \frac{1}{2k+1} = \frac{1}{2} \left[C + \ln(n) \right] + \ln(2) + \frac{B_2}{8n^2} + \cdots,$$

where $C \simeq 0.577216$ is the Euler constant and B_n are Bernoulli's numbers, we get

$$\frac{V_0}{\beta} \sum_{i\omega_n| < |i\omega_n\omega_D} \frac{1}{V} \sum_{\mathbf{k}_1} \mathcal{G}_\uparrow^{(0)}(\bar{k}_1) \, \mathcal{G}_\downarrow^{(0)}(-\bar{k}_1) \simeq \frac{V_0 \, \mathcal{D}(\varepsilon_F)}{2} \left[C + \ln\left(4\frac{\beta\omega_D}{2\pi} \right) \right]. \quad \beta\omega_D \gg 1.$$

Altogether, we find

$$\Gamma = \frac{V_0}{1 - \frac{V_0 \, \mathcal{D}(\varepsilon_F)}{2} \left[C + \ln\left(\frac{2\omega_D}{\pi k_B T} \right) \right]}. \tag{21.21}$$

We note that $\ln\left(\frac{2\omega_D}{\pi k_B T} \right)$ increases with decreasing T, indicating that Γ will diverge at a temperature $T = T_c$, satisfying

$$\frac{V \, \mathcal{D}(\varepsilon_F)}{2} \left[C + \ln\left(\frac{2\omega_D}{\pi k_B T_c} \right) \right] = 1 \;\Rightarrow\; \ln\left[\frac{2\omega_D e^C}{\pi k_B T_c} \right] = \frac{2}{V \, \mathcal{D}(\varepsilon_F)}$$

$$\text{yielding} \quad k_B T_c = \underbrace{\frac{2e^C}{\pi}}_{1.13387 \,\sim\, 1.} k_B T_D \, \exp\left[\frac{-2}{V \, \mathcal{D}(\varepsilon_F)} \right]$$

From this equation, T_c is found to be

$$T_c = T_D \, \exp\left[\frac{-2}{V \, \mathcal{D}(\varepsilon_F)} \right]. \tag{21.22}$$

This is the *Cooper instability*. Its characteristic temperature scale is smaller than the Debye temperature by a factor of about $1/100$ due to the exponential factor, typically yielding a T_c

of a few Kelvin. It is important to note that T_c is not analytic in V_0 at $V_0 = 0$ (the function has an essential singularity there). Thus T_c cannot be expanded into a Taylor series around the noninteracting limit. This means that we cannot obtain T_c in perturbation theory in V_0 to any finite order. BCS theory is indeed nonperturbative.

21.3 Superconductivity and the BCS Hamiltonian

We note that V_{eff} is a retarded potential, and thus is frequency dependent. However, as demonstrated by BCS, the interesting physics can be extracted from a model employing an attractive instantaneous effective potential. We can then write the effective electronic Hamiltonian in the form

$$\mathcal{H} = \sum_{\mathbf{k}\sigma} \varepsilon_{\mathbf{k}} \, c_{\mathbf{k}\sigma}^\dagger \, c_{\mathbf{k}\sigma} + \frac{1}{2} \sum_{\substack{\mathbf{k}, \mathbf{k}_1, \mathbf{q} \\ \sigma, \sigma'}} V_{\mathbf{q}} \, c_{\mathbf{k}+\mathbf{q},\sigma}^\dagger \, c_{\mathbf{k}_1-\mathbf{q},\sigma'}^\dagger \, c_{\mathbf{k}_1,\sigma'} \, c_{\mathbf{k},\sigma}. \tag{21.23}$$

Following Cooper's prescription, we shall retain only time-reversed states, and consider the reduced electronic Hamiltonian

$$\mathcal{H}_{\text{BCS}} = \sum_{\mathbf{k}\sigma} \varepsilon_{\mathbf{k}} \, c_{\mathbf{k}\sigma}^\dagger \, c_{\mathbf{k}\sigma} + \frac{1}{2} \sum_{\mathbf{k}, \mathbf{k}'} V_{\mathbf{k}\mathbf{k}'} \left(c_{\mathbf{k}'\downarrow}^\dagger \, c_{-\mathbf{k}'\uparrow}^\dagger \, c_{-\mathbf{k}\uparrow} \, c_{\mathbf{k}\downarrow} + c_{\mathbf{k}'\uparrow}^\dagger \, c_{-\mathbf{k}'\downarrow}^\dagger \, c_{-\mathbf{k}\downarrow} \, c_{\mathbf{k}\uparrow} \right)$$

$$= \sum_{\mathbf{k}} \varepsilon_{\mathbf{k}} \left(c_{\mathbf{k}}^\dagger c_{\mathbf{k}} + c_{-\mathbf{k}}^\dagger c_{-\mathbf{k}} \right) + \sum_{\mathbf{k}, \mathbf{k}'} V_{\mathbf{k}\mathbf{k}'} \, c_{\mathbf{k}'}^\dagger \, c_{-\mathbf{k}'}^\dagger \, c_{-\mathbf{k}} \, c_{\mathbf{k}}, \tag{21.24}$$

where we used the relation $V_{\mathbf{k}\mathbf{k}'} = V_{-\mathbf{k},-\mathbf{k}'}$, and the notation $c_{\mathbf{k}'}^\dagger \equiv c_{\mathbf{k}'\uparrow}^\dagger$; $c_{-\mathbf{k}'}^\dagger \equiv c_{-\mathbf{k}'\downarrow}^\dagger$, etc. This is the model Hamiltonian of Bardeen, Cooper, and Schrieffer [24].

As we have noted previously, we do not expect a perturbation expansion in terms of V_{eff} to be useful in finding the eigenstates of a Hamiltonian such as (21.24). A variational approach in which a brilliant guess at the form of the trial wavefunction $|\Psi\rangle$ was used by Bardeen, Cooper, and Schrieffer in their original 1957 seminal work. For pedagogical reasons, we shall take a slightly different route to obtain the same results, and turn for inspiration to the only problem that we have yet attempted without using perturbation theory – the Bogoliubov theory of helium.

21.3.1 The Bogoliubov–Valatin Transformation

Mean-Field Approximation

$\mathcal{H}_{\text{BCS}}$ is still an interacting Hamiltonian with a quartic term, and hence in general difficult to solve. However, we will resort to a mean-field approach, and introduce the following quantities:

$$c_{-\mathbf{k}} c_{\mathbf{k}} = \left\langle c_{-\mathbf{k}} c_{\mathbf{k}} \right\rangle + \left(c_{-\mathbf{k}} c_{\mathbf{k}} - \left\langle c_{-\mathbf{k}} c_{\mathbf{k}} \right\rangle \right)$$

$$c_{\mathbf{k}}^\dagger c_{-\mathbf{k}}^\dagger = \left\langle c_{\mathbf{k}}^\dagger c_{-\mathbf{k}}^\dagger \right\rangle + \left(c_{\mathbf{k}}^\dagger c_{-\mathbf{k}}^\dagger - \left\langle c_{\mathbf{k}}^\dagger c_{-\mathbf{k}}^\dagger \right\rangle \right). \tag{21.25}$$

In the normal state, $\langle c_{\mathbf{k}}^{\dagger} c_{-\mathbf{k}}^{\dagger} \rangle = \langle c_{-\mathbf{k}} c_{\mathbf{k}} \rangle = 0$, since they are expectation numbers of operators that change the number of particles in the system. Yet, as we will see, the complex field

$$\phi_{\mathbf{k}} = \langle c_{-\mathbf{k}} c_{\mathbf{k}} \rangle = \langle c_{\mathbf{k}}^{\dagger} c_{-\mathbf{k}}^{\dagger} \rangle^{*}$$

will furnish a nonzero order parameter in the superconducting phase. We note that $\langle c_{-\mathbf{k}} c_{\mathbf{k}} \rangle \neq 0$ means that the new state is one without a definite number of particles. This is directly related to the fact that gauge invariance symmetry, which conserves particle number, is broken.

The mean-field approximation amounts to neglecting terms quadratic in fluctuations (second terms in (21.25)). We then write the following:

$$c_{\mathbf{k}'}^{\dagger} c_{-\mathbf{k}'}^{\dagger} c_{-\mathbf{k}} c_{\mathbf{k}} \simeq \phi_{\mathbf{k}'}^{*} \phi_{\mathbf{k}} + \phi_{\mathbf{k}'}^{*} \left(c_{-\mathbf{k}} c_{\mathbf{k}} - \phi_{\mathbf{k}} \right) + \left(c_{\mathbf{k}'}^{\dagger} c_{-\mathbf{k}'}^{\dagger} - \phi_{\mathbf{k}'}^{*} \right) \phi_{\mathbf{k}}$$

$$= c_{\mathbf{k}'}^{\dagger} c_{-\mathbf{k}'}^{\dagger} \phi_{\mathbf{k}} + \phi_{\mathbf{k}'}^{*} c_{-\mathbf{k}} c_{\mathbf{k}} - \phi_{\mathbf{k}'}^{*} \phi_{\mathbf{k}}. \qquad (21.26)$$

To satisfy nonconservation of particles, we need to introduce the chemical potential, so that the BCS-Hamiltonian becomes

$$\mathcal{H}_{\text{BCS}} \rightarrow \mathcal{H}_{\text{BCS}} - \mu N \simeq \sum_{\mathbf{k}} (\varepsilon_{\mathbf{k}} - \mu) \left(c_{\mathbf{k}}^{\dagger} c_{\mathbf{k}} + c_{-\mathbf{k}}^{\dagger} c_{-\mathbf{k}} \right)$$

$$+ \sum_{\mathbf{k}, \mathbf{k}'} V_{\mathbf{k}\mathbf{k}'} \left(c_{\mathbf{k}'}^{\dagger} c_{-\mathbf{k}'}^{\dagger} \phi_{\mathbf{k}} + \phi_{\mathbf{k}'}^{*} c_{-\mathbf{k}} c_{\mathbf{k}} - \phi_{\mathbf{k}'}^{*} \phi_{\mathbf{k}} \right). \qquad (21.27)$$

Within this mean-field approximation, we replaced the interacting fermionic system with a noninteracting one subject to a field that has to be determined self-consistently. Since the Hamiltonian is now bilinear in the fermionic operators, it is possible to diagonalize it by means of the Bogoliubov transformation:

$$\gamma_{\mathbf{k}} = u_{\mathbf{k}} c_{\mathbf{k}} - v_{\mathbf{k}} c_{-\mathbf{k}}^{\dagger}; \qquad \gamma_{-\mathbf{k}} = u_{\mathbf{k}} c_{-\mathbf{k}} + v_{\mathbf{k}} c_{\mathbf{k}}^{\dagger},$$

$$\gamma_{\mathbf{k}}^{\dagger} = u_{\mathbf{k}} c_{\mathbf{k}}^{\dagger} - v_{\mathbf{k}} c_{-\mathbf{k}}; \qquad \gamma_{-\mathbf{k}}^{\dagger} = u_{\mathbf{k}} c_{-\mathbf{k}}^{\dagger} + v_{\mathbf{k}} c_{\mathbf{k}}, \qquad (21.28)$$

$u_{\mathbf{k}}$ and $v_{\mathbf{k}}$ are chosen to be real and positive and to obey the conditions

$$u_{\mathbf{k}} = u_{-\mathbf{k}}, \quad v_{\mathbf{k}} = -v_{-\mathbf{k}}, \quad u_{\mathbf{k}}^{2} + v_{\mathbf{k}}^{2} = 1$$

that guarantee the *fermion anticommutation* relations

$$\{\gamma_{\mathbf{k}}^{\dagger}, \gamma_{-\mathbf{k}'}^{\dagger}\} = \{\gamma_{\mathbf{k}}, \gamma_{-\mathbf{k}'}\} = \{\gamma_{\mathbf{k}}, \gamma_{\mathbf{k}'}\} = 0$$

$$\{\gamma_{\mathbf{k}}^{\dagger}, \gamma_{\mathbf{k}'}\} = \{\gamma_{-\mathbf{k}}^{\dagger}, \gamma_{-\mathbf{k}'}\} = \delta_{\mathbf{k}\mathbf{k}'}.$$

Inverting the transformations (21.28)

$$c_{\mathbf{k}} = u_{\mathbf{k}} \gamma_{\mathbf{k}} + v_{\mathbf{k}} \gamma_{-\mathbf{k}}^{\dagger}; \qquad c_{-\mathbf{k}} = u_{\mathbf{k}} \gamma_{-\mathbf{k}} - v_{\mathbf{k}} \gamma_{\mathbf{k}}^{\dagger},$$

$$c_{\mathbf{k}}^{\dagger} = u_{\mathbf{k}} \gamma_{\mathbf{k}}^{\dagger} + v_{\mathbf{k}} \gamma_{-\mathbf{k}}; \qquad c_{-\mathbf{k}}^{\dagger} = u_{\mathbf{k}} \gamma_{-\mathbf{k}}^{\dagger} - v_{\mathbf{k}} \gamma_{\mathbf{k}}, \qquad (21.29)$$

substituting for the c operators in $\mathcal{H}_{\text{BCS}}$, and setting ϕ to be real, we get

$$
\mathcal{H}_{\text{BCS}} = \sum_{\mathbf{k}} \left[(\varepsilon_{\mathbf{k}} - \mu) \left(u_{\mathbf{k}}^2 - v_{\mathbf{k}}^2 \right) + 2 \sum_{\mathbf{k}'} V_{\mathbf{k}\mathbf{k}'} \, \phi_{\mathbf{k}'} \, u_{\mathbf{k}} v_{\mathbf{k}} \right] \gamma_{\mathbf{k}}^{\dagger} \gamma_{\mathbf{k}}
$$

$$
+ \sum_{\mathbf{k}} \left[(\varepsilon_{\mathbf{k}} - \mu) \, u_{\mathbf{k}} v_{\mathbf{k}} + \frac{1}{2} \sum_{\mathbf{k}'} V_{\mathbf{k}\mathbf{k}'} \, \phi_{\mathbf{k}'} \left(u_{\mathbf{k}}^2 - v_{\mathbf{k}}^2 \right) \right] \left(\gamma_{\mathbf{k}}^{\dagger} \gamma_{-\mathbf{k}}^{\dagger} + \gamma_{-\mathbf{k}} \gamma_{\mathbf{k}} \right).
$$

$$(21.30)$$

We diagonalize $\mathcal{H}_{\text{BCS}}$ by requiring

$$
2 \, (\varepsilon_{\mathbf{k}} - \mu) \, u_{\mathbf{k}} v_{\mathbf{k}} = - \left(u_{\mathbf{k}}^2 - v_{\mathbf{k}}^2 \right) \sum_{\mathbf{k}'} V_{\mathbf{k}\mathbf{k}'} \, \phi_{\mathbf{k}'} = \left(u_{\mathbf{k}}^2 - v_{\mathbf{k}}^2 \right) \Delta_{\mathbf{k}}, \tag{21.31}
$$

where we defined

$$
\Delta_{\mathbf{k}} = - \sum_{\mathbf{k}'} V_{\mathbf{k}\mathbf{k}'} \, \phi_{\mathbf{k}'}, \tag{21.32}
$$

The minus sign was taken into account due to the fact that $V_{\mathbf{k}\mathbf{k}'} < 0$. Equation (21.31) together with the condition $u_{\mathbf{k}}^2 + v_{\mathbf{k}}^2 = 1$ constitutes the equations that determine the coefficients $u_{\mathbf{k}}$ and $v_{\mathbf{k}}$ of the canonical transformation. Setting

$$
u_{\mathbf{k}} = \cos(\theta_{\mathbf{k}}), \qquad v_{\mathbf{k}} = \sin(\theta_{\mathbf{k}}),
$$

we express (21.31) as

$$
(\varepsilon_{\mathbf{k}} - \mu) \sin(2\theta_{\mathbf{k}}) = \Delta_{\mathbf{k}} \cos(2\theta_{\mathbf{k}}) \implies \tan(2\theta_{\mathbf{k}}) = \frac{\Delta_{\mathbf{k}}}{\varepsilon_{\mathbf{k}} - \mu},
$$

which yields

$$
\cos(2\theta_{\mathbf{k}}) = u_{\mathbf{k}}^2 - v_{\mathbf{k}}^2 = \pm \frac{\varepsilon_{\mathbf{k}} - \mu}{E_{\mathbf{k}}}
$$

$$
\sin(2\theta_{\mathbf{k}}) = \pm \frac{\Delta_{\mathbf{k}}}{E_{\mathbf{k}}}
$$

$$
E_{\mathbf{k}} = \sqrt{(\varepsilon_{\mathbf{k}} - \mu)^2 + \Delta_{\mathbf{k}}^2} = \sqrt{\mathcal{E}_{\mathbf{k}}^2 + \Delta_{\mathbf{k}}^2}. \tag{21.33}
$$

Substituting the preceding results into (21.30), we arrive at

$$
\mathcal{H}_{\text{BCS}}^{D} = \pm \sum_{\mathbf{k}} E_{\mathbf{k}} \gamma_{\mathbf{k}}^{\dagger} \gamma_{\mathbf{k}}. \tag{21.34}
$$

As shown in Figure 21.7, the quasiparticles spectrum has now a gap of 2Δ. The quasiparticles represented by (21.34) are referred to as Bogoliubov quasiparticles.

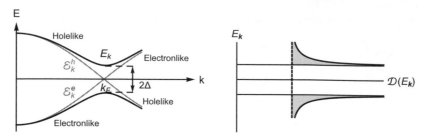

Figure 21.7 Left: dispersion curve of the Bogoliubov quasiparticles $E_{\mathbf{k}}$ (solid black line). Right: density of states.

Physical Picture of BCS Theory: Electron–Hole Mixing and Andreev Reflection

We can consider the annihilation of an electron in state $-\mathbf{k} \downarrow$ as the creation of a hole $\mathbf{k} \uparrow$, and we write

$$\gamma_{\mathbf{k}}^{\dagger} = u_{\mathbf{k}} c_{\mathbf{k}}^{\dagger} - v_{\mathbf{k}} c_{-\mathbf{k}} = u_{\mathbf{k}} c_{\mathbf{k}}^{\dagger} - v_{\mathbf{k}} h_{\mathbf{k}}^{\dagger}.$$

Thus we find that such excitation involves coherent superposition of electron and hole excitations, which is depicted in Figure 21.7 by the gray dispersion curves. When the two dispersion curves cross at the Fermi surface, mixing between electrons and holes destroys the Fermi surface and gives rise to the gap in the dispersion (21.34). This picture can be extended to the interpretation of the pairing terms in the BCS mean-field Hamiltonian

$$\mathcal{H}_{\text{int}}(\mathbf{k}) = \Delta^{*} c_{-\mathbf{k}} c_{\mathbf{k}} + \Delta c_{-\mathbf{k}}^{\dagger} c_{\mathbf{k}}^{\dagger} = \Delta^{*} h_{\mathbf{k}}^{\dagger} c_{\mathbf{k}} + \Delta h_{\mathbf{k}} c_{\mathbf{k}}^{\dagger}$$

whereby the Δ^{*} term now represents the simultaneous creation of a condensate pair and a hole. This process is represented by the Feynman diagram

Electron   Hole

——————→   ✗   ←——————

k, ω Δ^{*} $-k, -\omega$

It is reminiscent of the *Andreev reflection* at the interface of a normal metal and a superconductor, as shown in Figure 21.8.

$$\text{Andreev reflection}: \quad e^{-} \;\rightleftharpoons\; \text{Pair}^{2-} + h^{+}.$$

It conserves spin, momentum, and current, for a hole in the state $(-\mathbf{k}, \downarrow)$ has spin up, momentum $+\mathbf{k}$, and carries a current $I = (-e) \times (-\nabla \varepsilon_{\mathbf{k}}) = e \nabla \varepsilon_{\mathbf{k}}$.

Superconducting Ground State

The ground state of the Bogoliubov quasiparticles $|\Psi_0\rangle$ is just the vacuum state for the new operators

$$\gamma_{\mathbf{k}} |\Psi_0\rangle = \gamma_{-\mathbf{k}} |\Psi_0\rangle = 0, \qquad \langle \Psi_0| \gamma_{-\mathbf{k}}^{\dagger} = \langle \Psi_0| \gamma_{\mathbf{k}}^{\dagger} = 0. \tag{21.35}$$

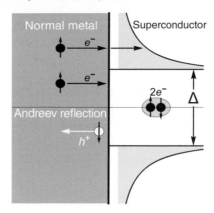

Figure 21.8 Andreev reflection: an incident electron with energy $< \Delta$ scatters back as a hole, leaving a condensate pair in the superconductor.

To derive this ground state, we consider the property that fermionic operators, such as the γs, obey the identity

$$\gamma_{\mathbf{k}}\gamma_{\mathbf{k}} = \gamma_{-\mathbf{k}}\gamma_{-\mathbf{k}} = 0$$

and explore the construction

$$\gamma_{\mathbf{k}} \left(\gamma_{\mathbf{k}}\gamma_{-\mathbf{k}} |\emptyset\rangle \right) = 0,$$

where $|\emptyset\rangle$ is the vacuum state of the free electron system. This construction suggests that $|\Psi_0\rangle$ that satisfies (21.35) is simply obtained by operating on $|\emptyset\rangle$ with all the $\gamma_{\mathbf{k}}$ and $\gamma_{-\mathbf{k}}$, and we obtain

$$\left(\prod_{\mathbf{k}} \gamma_{\mathbf{k}}\gamma_{-\mathbf{k}} \right) |\emptyset\rangle = \left[\prod_{\mathbf{k}} \left(u_{\mathbf{k}}c_{\mathbf{k}} - v_{\mathbf{k}}c^{\dagger}_{-\mathbf{k}} \right) \left(u_{\mathbf{k}}c_{-\mathbf{k}} + v_{\mathbf{k}}c^{\dagger}_{\mathbf{k}} \right) \right] |\emptyset\rangle$$

$$= \left[\prod_{\mathbf{k}} \left(u_{\mathbf{k}}v_{\mathbf{k}} + v_{\mathbf{k}}^2 \, c^{\dagger}_{\mathbf{k}}c^{\dagger}_{-\mathbf{k}} \right) \right] |\emptyset\rangle = \left[\prod_{\mathbf{k}} v_{\mathbf{k}} \left(u_{\mathbf{k}} + v_{\mathbf{k}} \, c^{\dagger}_{\mathbf{k}}c^{\dagger}_{-\mathbf{k}} \right) \right] |\emptyset\rangle .$$

Here $u_{\mathbf{k}}$ is the amplitude for a pair of orbitals to be empty, and $v_{\mathbf{k}}$ is the amplitude for them to contain a Cooper pair, so we require $|u_{\mathbf{k}}|^2 + |v_{\mathbf{k}}|^2 = 1$, and $\sum_{\mathbf{k}} |v_{\mathbf{k}}|^2 = N$ in a system containing $2N$ electrons on average.

To normalize this we divide by the product of all the $u_{\mathbf{k}}v_{\mathbf{k}}$s to obtain

$$|\Psi_0\rangle = \left[\prod_{\mathbf{k}} (1 + z_{\mathbf{k}}c^{\dagger}_{\mathbf{k}}c^{\dagger}_{-\mathbf{k}}) \right] |\emptyset\rangle = \exp\left[\sum_{\mathbf{k}} z_{\mathbf{k}} \, c^{\dagger}_{\mathbf{k}}c^{\dagger}_{-\mathbf{k}} \right] |\emptyset\rangle = \exp\left[\hat{\Lambda}^{\dagger} \right] |\emptyset\rangle , \quad (21.36)$$

which is just a coherent state. This result derives from the fact that

$$\exp\left[z_{\mathbf{k}}c^{\dagger}_{\mathbf{k}}c^{\dagger}_{-\mathbf{k}} \right] |\emptyset\rangle = (1 + z_{\mathbf{k}}c^{\dagger}_{\mathbf{k}}c^{\dagger}_{-\mathbf{k}}) |\emptyset\rangle$$

because of the identity

$$c_{\mathbf{k}}^{\dagger} c_{\mathbf{k}}^{\dagger} = 0.$$

This wavefunction is a linear combination of simpler wavefunctions containing different numbers of particles, which means that it is not an eigenstate of the total number operator $\hat{N}$. Our familiarity with the concept of the chemical potential μ teaches us not to be too concerned about this fact, however, as long as we make sure that the average value is kept constant. We can expand $|\Psi_0\rangle$ as a coherent sum of pair states:

$$|\Psi_0\rangle = \sum_n \frac{1}{n!} \left(\hat{\Lambda}^{\dagger}\right)^n |\emptyset\rangle = \sum_n \frac{1}{n!} |n\rangle.$$

The Gap Parameter Δ

In the ground state,

$$\phi_{\mathbf{k}} = \left\langle c_{-\mathbf{k}} c_{\mathbf{k}} \right\rangle = \left\langle (u_{\mathbf{k}} \gamma_{-\mathbf{k}} - v_{\mathbf{k}} \gamma_{\mathbf{k}}^{\dagger})(u_{\mathbf{k}} \gamma_{\mathbf{k}} + v_{\mathbf{k}} \gamma_{-\mathbf{k}}^{\dagger}) \right\rangle = u_{\mathbf{k}} v_{\mathbf{k}} = \frac{\Delta_{\mathbf{k}}}{2E_{\mathbf{k}}}. \tag{21.37}$$

Substituting for $\phi_{\mathbf{k}}$ in (21.32), we obtain the homogeneous equation

$$\Delta_{\mathbf{k}} = -\frac{1}{2} \sum_{\mathbf{k}'} V_{\mathbf{k}\mathbf{k}'} \frac{\Delta_{\mathbf{k}'}}{E_{\mathbf{k}'}}, \tag{21.38}$$

which supports the trivial solution $\Delta_{\mathbf{k}} = 0$.

In order to solve this equation for $\Delta_{\mathbf{k}}$, we resort to the simple contact potential model

$$V_{\mathbf{k}\mathbf{k}'} = \begin{cases} -\dfrac{V}{\Omega} & \mathcal{E} \leq \hbar\omega_D \\[2mm] 0 & \text{Otherwise} \end{cases} \tag{21.39}$$

we used in deriving the binding energy of the Cooper pair. Here Ω is the volume of the system. Substituting for $V_{\mathbf{k}\mathbf{k}'}$ in (21.32), we find that Δ becomes momentum independent, namely,

$$\Delta_{\mathbf{k}} = -\sum_{\mathbf{k}'} V_{\mathbf{k}\mathbf{k}'} \phi_{\mathbf{k}'} = \frac{V}{\Omega} \sum_{\mathbf{k}'} \phi_{\mathbf{k}'} \Rightarrow \Delta \quad \text{(momentum independent)}$$

so that (21.38) gives

$$\Delta = \frac{V}{2\Omega} \sum_{\mathbf{k}'} \frac{\Delta}{E_{\mathbf{k}'}}. \tag{21.40}$$

Then the gap equation reduces to

$$
\begin{aligned}
1 &= \frac{V}{2\Omega} \sum_{\substack{\mathbf{k} \\ |\varepsilon_{\mathbf{k}}-\mu|<\hbar\omega_D}} \frac{1}{\sqrt{(\mathcal{E}_{\mathbf{k}}-\mu)^2 + \Delta^2}} \\
&= \frac{V}{2} \int_{-\hbar\omega_D}^{\hbar\omega_D} \frac{d\mathcal{E}\, \mathcal{D}(\mathcal{E})}{\sqrt{\mathcal{E}^2 + \Delta^2}} = V\,\mathcal{D}(0) \int_0^{\hbar\omega_D} \frac{d\mathcal{E}}{\sqrt{\mathcal{E}^2 + \Delta^2}} \\
&= V\,\mathcal{D}(0) \ln\left[\frac{\hbar\omega_D + \sqrt{(\hbar\omega_D)^2 + \Delta^2}}{\Delta} \right] \sim V\,\mathcal{D}(0) \ln\left(\frac{2\hbar\omega_D}{\Delta} \right),
\end{aligned}
\tag{21.41}
$$

leading finally to

$$
\Delta = 2\hbar\omega_D \exp\left[-\frac{1}{V\,\mathcal{D}(0)} \right],
\tag{21.42}
$$

where the gap depends in a nonanalytic fashion on the coupling constant, underscoring that such a result cannot be obtained in the frame of a perturbation theory.

The constancy of the gap allows us to obtain closed form expressions for the coefficients $u_{\mathbf{k}}$ and $v_{\mathbf{k}}$. Using the condition $u_{\mathbf{k}}^2 + v_{\mathbf{k}}^2 = 1$, and (21.33), we get

$$
u_{\mathbf{k}}^2 = \frac{1}{2}\left[1 + \frac{\mathcal{E}_{\mathbf{k}}}{\sqrt{\mathcal{E}_{\mathbf{k}}^2 + \Delta^2}} \right], \qquad v_{\mathbf{k}}^2 = \frac{1}{2}\left[1 - \frac{\mathcal{E}_{\mathbf{k}}}{\sqrt{\mathcal{E}_{\mathbf{k}}^2 + \Delta^2}} \right].
\tag{21.43}
$$

We can also determine the ground-state expectation value

$$
\left\langle c_{\mathbf{k}}^\dagger c_{\mathbf{k}} \right\rangle = \left\langle (u_{\mathbf{k}}\gamma_{\mathbf{k}}^\dagger + v_{\mathbf{k}}\gamma_{-\mathbf{k}})(u_{\mathbf{k}}\gamma_{\mathbf{k}} + v_{\mathbf{k}}\gamma_{-\mathbf{k}}^\dagger) \right\rangle = v_{\mathbf{k}}^2,
$$

which is shown in Figure 21.9. It represents the occupation number of the original fermions in the new ground state.

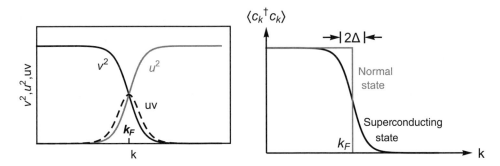

Figure 21.9 Left: distribution of v^2, u^2 and uv. Right: the occupation number distribution of the original fermions in the new ground state.

In fact, if $\Delta = 0$, then we have

$$v_{\mathbf{k}}^2 = \frac{1}{2}\left[1 - \frac{\mathcal{E}_{\mathbf{k}}}{|\mathcal{E}_{\mathbf{k}}|}\right] = \Theta\left[\mu - \varepsilon_{\mathbf{k}}\right],$$

which is the expected distribution in the normal metallic phase.

For the superconducting state with $\Delta \neq 0$, a gap opens destroying the Fermi surface, and the momentum distribution function shows a smooth decrease across the Fermi energy.

Gauge Symmetry Breaking and Complex Order Parameter

The BCS Hamiltonian is invariant under the $U(1)$ transformation $c_{\mathbf{k}\sigma} \rightarrow e^{i\alpha} c_{\mathbf{k}\sigma}$. This gauge symmetry is broken in the BCS coherent ground-state $|\Psi_0\rangle$

$$|\Psi_0\rangle \rightarrow |\alpha\rangle = \left[\prod_{\mathbf{k}}(1 + e^{2i\alpha}\,z_{\mathbf{k}}c_{\mathbf{k}}^{\dagger}c_{-\mathbf{k}}^{\dagger})\right]|0\rangle = \sum_{n}\frac{e^{2in\alpha}}{n!}\,|n\rangle\,.$$

Moreover, under such transformation, the order parameter

$$\Delta = -\frac{V}{\Omega}\sum_{\mathbf{k}}\langle\alpha\,|c_{-\mathbf{k}\downarrow}\,c_{\mathbf{k}\uparrow}|\,\alpha\rangle$$

acquires a phase $\Delta \rightarrow e^{2i\alpha}\,\Delta$. However, the energy of the BCS state is unchanged by the gauge transformation, so the states $|\alpha\rangle$ actually form a family of degenerate broken symmetry states.

Again, as in the case of superfluid coherent state, the number and phase operators form conjugate variables satisfying the commutator $\left[\alpha, \hat{N}\right] = i$.

Transition Temperature

In the superconducting state, the Bogoliubov quasiparticles behave like ordinary fermions, and satisfy the fermionic properties

$$\left\langle\gamma_{\mathbf{k}'}^{\dagger}\,\gamma_{-\mathbf{k}'}^{\dagger}\right\rangle = \left\langle\gamma_{-\mathbf{k}'}\,\gamma_{\mathbf{k}'}\right\rangle = 0$$

$$\left\langle\gamma_{-\mathbf{k}'}\,\gamma_{-\mathbf{k}'}^{\dagger}\right\rangle = 1 - n_F(E_{\mathbf{k}'}), \quad \left\langle\gamma_{\mathbf{k}'}^{\dagger}\,\gamma_{\mathbf{k}'}\right\rangle = n_F(E_{\mathbf{k}'}).$$

We make use of these properties by substituting for the c-annihilation operators that appear in the gap equation in terms of the γ-creation and annihilation operators:

$$\Delta_{\mathbf{k}} = -\sum_{\mathbf{k}'}V_{\mathbf{k}\mathbf{k}'}\left\langle c_{-\mathbf{k}'}c_{\mathbf{k}'}\right\rangle_T$$

$$= -\sum_{\mathbf{k}'}V_{\mathbf{k}\mathbf{k}'}\left\langle\left(-v_{\mathbf{k}'}\,\gamma_{\mathbf{k}'}^{\dagger} + u_{\mathbf{k}'}\,\gamma_{-\mathbf{k}'}\right)\left(u_{\mathbf{k}'}\,\gamma_{\mathbf{k}'} + v_{\mathbf{k}'}\,\gamma_{-\mathbf{k}'}^{\dagger}\right)\right\rangle$$

$$= \sum_{\mathbf{k}'}V_{\mathbf{k}\mathbf{k}'}\left[u_{\mathbf{k}'}\,v_{\mathbf{k}'}\left\langle\gamma_{-\mathbf{k}'}\,\gamma_{-\mathbf{k}'}^{\dagger} - \gamma_{\mathbf{k}'}^{\dagger}\,\gamma_{\mathbf{k}'}\right\rangle + v_{\mathbf{k}'}^2\left\langle\gamma_{\mathbf{k}'}^{\dagger}\,\gamma_{-\mathbf{k}'}^{\dagger}\right\rangle - u_{\mathbf{k}'}^2\left\langle\gamma_{-\mathbf{k}'}\,\gamma_{\mathbf{k}'}\right\rangle\right].$$

The last two terms vanish. Thus, for $T > 0$ the gap parameter becomes temperature dependent. Equation (21.38) is replaced by

$$\Delta_{\mathbf{k}} = -\frac{1}{2} \sum_{\mathbf{k}'} V_{\mathbf{k}\mathbf{k}'} \frac{\Delta_{\mathbf{k}'}}{E_{\mathbf{k}'}} \left[1 - 2n_F(E_{\mathbf{k}'})\right]$$

$$= -\frac{1}{2} \sum_{\mathbf{k}'} V_{\mathbf{k}\mathbf{k}'} \frac{\Delta_{\mathbf{k}'}}{E_{\mathbf{k}'}} \tanh\left[\frac{\beta E_{\mathbf{k}'}}{2}\right], \qquad (21.44)$$

which goes over into (21.38) for $T \to 0$. Proceeding as before, we obtain the finite temperature gap equation

$$1 = \frac{V}{2} \int_{-\hbar\omega_D}^{\hbar\omega_D} \frac{d\mathcal{E}\, \mathcal{D}(\mathcal{E})}{\sqrt{\mathcal{E}^2 + \Delta^2}} = V\, \mathcal{D}(0) \int_0^{\hbar\omega_D} \frac{d\mathcal{E}}{\sqrt{\mathcal{E}^2 + \Delta^2}} \tanh\left[\frac{\beta}{2}\sqrt{\mathcal{E}^2 + \Delta^2}\right]. \qquad (21.45)$$

The integrand is a decreasing function of both T and Δ. It follows that the solution of (21.45), $\Delta(T)$, decreases with increasing temperature until it vanishes at a critical temperature T_c. For weak coupling ($k_B T_c \ll \hbar\omega_D$), the function $\Delta(T)$, depicted in Figure 21.10, is universal. For $T = 0$, the solution of the gap equation is given by (21.42). The magnitude of T_c in the BCS model is found from (21.45) by setting $\Delta(T_c) = 0$:

$$V\, \mathcal{D}(\zeta) \int_0^{\hbar\omega_D/k_B T_c} dx\, x^{-1} \tanh(x/2) = 1. \qquad (21.46)$$

Integrating by parts

$$\ln\left(\frac{\hbar\omega_D}{k_B T_c}\right) \tanh\left(\frac{\hbar\omega_D}{2k_B T_c}\right) - \int_0^{\hbar\omega_D/k_B T_c} dx\, \ln(x) \frac{d}{dx} \tanh(x/2) = \frac{1}{V\, \mathcal{D}(\zeta)}. \qquad (21.47)$$

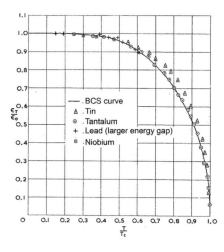

Figure 21.10 Reduced values of the energy gaps for lead, tin, tantalum, and niobium as a function of reduced temperature, compared to the theoretical curve (solid line) of BCS. From [182]

For weak-coupling superconductors, $\hbar\omega_D/2k_BT_c \gg 1$, we set $\tanh(\hbar\omega_D/2k_BT_c) = 1$ and extend the upper limit of the integral to infinity, and obtain

$$\ln\left(\frac{\hbar\omega_D}{2k_BT_c}\right) - \int_0^\infty dx \, \ln(x)\frac{d}{dx}\tanh(x/2) = \ln\left(\frac{\hbar\omega_D}{2k_BT_c}\right) - \ln(0.44) = \frac{1}{V\,\mathcal{D}(\zeta)}, \tag{21.48}$$

which leads to

$$k_BT_c = 1.14\hbar\omega_D \, e^{-1/V\,\mathcal{D}(\zeta)}. \tag{21.49}$$

We also arrive at the relation

$$\frac{2\Delta(0)}{k_BT_c} = 3.5,$$

a result in adequate agreement with experimentally observes values. This relation, often used to check whether a superconducting material is BCS-like, is well satisfied in elemental superconductors:

Al	Cd	Ga	Hg	In	La	Nb	Pb	Sn	Ta	Tl	V	Zn
3.53	3.44	3.5	3.95	3.65	3.72	3.65	3.95	3.6	3.63	3.63	3.50	3.44

We will waive the discussion of thermodynamic properties and the Josephson effect associated with the phenomenon of superconductivity, and refer the reader to the book by Tinkham [177], for example.

21.3.2 Gor'kov Equations and the Anomalous Green Function

Gor'kov [1] developed a powerful method for understanding superconductivity through a set of coupled equations for the time evolution of Green functions [77]. The equations couple the normal Matsubara functions, $\mathcal{G}(\tau)$, and the anomalous Matsubara functions, $\mathcal{F}(\tau)$, relevant to Cooper pairing. The pairing is described by introducing the new correlation function, $\mathcal{F}$, for particles of opposite spin:

$$\mathcal{G}(\mathbf{k},\tau) = -\left\langle T_\tau \, c_{\mathbf{k}\sigma}(\tau) \, c_{\mathbf{k}\sigma}^\dagger(0)\right\rangle$$

$$\mathcal{F}(\mathbf{k},\tau) = \left\langle T_\tau \, c_{-\mathbf{k}\downarrow}(\tau) \, c_{\mathbf{k}\uparrow}(0)\right\rangle, \quad \mathcal{F}^\dagger(\mathbf{k},\tau) = \left\langle T_\tau \, c_{\mathbf{k}\uparrow}^\dagger(\tau) \, c_{-\mathbf{k}\downarrow}^\dagger(0)\right\rangle. \tag{21.50}$$

The equations of motion follow from the time evolution of the fermion field operators, in which the interaction terms are again treated in a mean-field manner. From such equations, essentially all interesting physical quantities can be obtained.

---

[1] Gor'kov first heard of the microscopic BCS theory of superconductivity during a seminar given in late 1957 by Nikolai Bogoliubov at the Landau Institute. Within a few weeks, and before he had access to the BCS papers, Gor'kov derived what are now known as the Gor'kov equations, which established the field-theoretical formulation of superconductivity. In 1959, he used them to derive the Ginzburg–Landau equations and resolved the conceptual meaning of the superconducting order parameter. His theory opened the way to describe inhomogeneous superconducting states in the presence of magnetic fields or electrical currents.

We shall use the equation of motion approach to construct the Matsubara functions for the Hamiltonian defined in (21.23):

$$\mathcal{H} = \sum_{k\sigma} \varepsilon_k c^\dagger_{k\sigma} c_{k\sigma} + \frac{1}{2} \sum_{\substack{k,k',q \\ \sigma,\sigma'}} V_q c^\dagger_{k+q,\sigma} c^\dagger_{k'-q,\sigma'} c_{k',\sigma'} c_{k,\sigma}.$$

We shall set $V_{q=0} = 0$, since the pair coupling is through the longitudinal acoustic branch.

We start with

$$\frac{dc_{k\sigma}}{d\tau} = \left[\mathcal{H}, c_k\right] = -\varepsilon_k c_k - \sum_{\substack{k'q \\ \sigma'}} V_q c^\dagger_{k'-q\sigma'} c_{k'\sigma'} c_{k-q\sigma}, \qquad (21.51)$$

together with the equation of motion for $\mathcal{G}$, namely,

$$\frac{\partial \mathcal{G}(k,\tau)}{\partial \tau} = -\delta(\tau) - \left\langle T_\tau \left[\frac{dc_{k\sigma}(\tau)}{d\tau}\right] c^\dagger_{k\sigma}(0) \right\rangle$$

$$= -\delta(\tau) + \varepsilon_k \mathcal{G}(k,\tau) + \sum_{\substack{k'q \\ \sigma'}} V_q \left\langle T_\tau c^\dagger_{k'-q\sigma'}(\tau) c_{k'\sigma'}(\tau) c_{k-q\sigma}(\tau) c^\dagger_{k\sigma}(0) \right\rangle.$$

$$(21.52)$$

At this point, we need to carry out Wick's contractions. But before we proceed, we need to recall that the superconducting transition is characterized by spontaneous breaking of gauge symmetry associated with particle number, or charge conservation. Consequently, we find that electron pairs may *disappear* into the condensate, or *emerge* from the condensate, without changing the macroscopic state of the system, with number of pairs $\mathcal{N}_{pair} \to \infty$! This scenario allows us to introduce some modified contractions:

$$\left\langle T_\tau c^\dagger_{k'-q\sigma'}(\tau) c_{k'\sigma'}(\tau) c_{k-q\sigma}(\tau) c^\dagger_{k\sigma}(0) \right\rangle = \underbrace{\left\langle T_\tau c^\dagger_{k'-q\sigma'}(\tau) c_{k'\sigma'}(\tau) \right\rangle \left\langle T_\tau c_{k-q\sigma}(\tau) c^\dagger_{k\sigma}(0) \right\rangle}$$

$$q = 0$$

$$+ \left\langle T_\tau c^\dagger_{k'-q\sigma'}(\tau) c_{k-q\sigma}(\tau) \right\rangle \left\langle T_\tau c_{k'\sigma'}(\tau) c^\dagger_{k\sigma}(0) \right\rangle$$

$$+ \langle N| T_\tau \underbrace{c_{k'\sigma'}(\tau) c_{k-q\sigma}(\tau)} |N+2\rangle \langle N+2| T_\tau \underbrace{c^\dagger_{k'-q\sigma'}(\tau) c^\dagger_{k\sigma}(0)} |N\rangle$$

$$k' = -k + q \qquad\qquad k' = -k + q$$

$$= n_{k-q}\, \mathcal{G}(k,\tau) - \mathcal{F}(-k+q,0)\, \mathcal{F}^\dagger(k,\tau). \qquad (21.53)$$

The equation for $\mathcal{G}$ becomes

$$\left(\frac{\partial}{\partial \tau} - \varepsilon_k - \sum_q V_q n_{k-q}\right) \mathcal{G}(k,\tau) + \sum_q V_q \mathcal{F}(-k+q,0)\, \mathcal{F}^\dagger(k,\tau) = \delta(\tau). \quad (21.54)$$

The self-energy term $\Sigma = \sum_q V_q n_{k-q}$ is just the exchange energy arising from the effective electron–phonon interaction. It is usually small for weak coupling and is not very interesting to us, since it just leads to some irrelevant renormalization of the energy spectrum. Thus, in the following we shall drop it. We shall also introduce the quantity

$$\Delta_{\mathbf{k}} = -\sum_{\mathbf{q}} V_{\mathbf{q}} \mathcal{F}(-\mathbf{k}+\mathbf{q}, 0), \tag{21.55}$$

which is the equivalent of the gap function of the BCS theory. Equation (21.54) becomes

$$\left(\frac{\partial}{\partial \tau} - \mathcal{E}_{\mathbf{k}}\right) \mathcal{G}(\mathbf{k}, \tau) - \Delta_{\mathbf{k}} \mathcal{F}^{\dagger}(\mathbf{k}, \tau) = \delta(\tau). \tag{21.56}$$

We now need to derive the equation of motion for $\mathcal{F}^{\dagger}$

$$\frac{\partial \mathcal{F}^{\dagger}(\mathbf{k}, \tau)}{\partial \tau} = -\delta(\tau) \left\langle \left\{ c_{\mathbf{k}\uparrow}^{\dagger}, c_{-\mathbf{k}\downarrow}^{\dagger} \right\} \right\rangle + \left\langle T_{\tau} \frac{\partial c_{\mathbf{k}\uparrow}^{\dagger}(\tau)}{\partial \tau} c_{-\mathbf{k}\downarrow}^{\dagger}(0) \right\rangle = \left\langle T_{\tau} \frac{\partial c_{\mathbf{k}\uparrow}^{\dagger}(\tau)}{\partial \tau} c_{-\mathbf{k}\downarrow}^{\dagger}(0) \right\rangle. \tag{21.57}$$

With

$$\frac{dc_{\mathbf{k}\sigma}^{\dagger}}{d\tau} = \left[\mathcal{H}, c_{\mathbf{k}\sigma}^{\dagger} \right] = \mathcal{E}_{\mathbf{k}} c_{\mathbf{k}\sigma}^{\dagger} + \sum_{\substack{\mathbf{k}'\mathbf{q} \\ \sigma'}} V_{\mathbf{q}} c_{\mathbf{k}+\mathbf{q}}^{\dagger} c_{\mathbf{k}'+\mathbf{q}\sigma'}^{\dagger} c_{\mathbf{k}'\sigma'} \tag{21.58}$$

we obtain

$$\frac{\partial \mathcal{F}^{\dagger}(\mathbf{k}, \tau)}{\partial \tau} = \mathcal{E}_{\mathbf{k}} \mathcal{F}^{\dagger}(\mathbf{k}, \tau) + \sum_{\substack{\mathbf{k}'\mathbf{q} \\ \sigma'}} V_{\mathbf{q}} \left\langle T_{\tau} c_{\mathbf{k}-\mathbf{q}\uparrow}^{\dagger}(\tau) c_{\mathbf{k}'+\mathbf{q}\sigma'}^{\dagger}(\tau) c_{\mathbf{k}'\sigma'}(\tau) c_{-\mathbf{k}\downarrow}^{\dagger}(0) \right\rangle. \tag{21.59}$$

Again, we write the contractions

$$\left\langle T_{\tau} c_{\mathbf{k}-\mathbf{q}\uparrow}^{\dagger}(\tau) c_{\mathbf{k}'+\mathbf{q}\sigma'}^{\dagger}(\tau) c_{\mathbf{k}'\sigma'}(\tau) c_{-\mathbf{k}\downarrow}^{\dagger}(0) \right\rangle$$

$$= \left\langle T_{\tau} c_{\mathbf{k}-\mathbf{q}\uparrow}^{\dagger}(\tau) c_{\mathbf{k}'+\mathbf{q}\sigma'}^{\dagger}(\tau) \right\rangle \left\langle T_{\tau} c_{\mathbf{k}'\sigma'}(\tau) c_{-\mathbf{k}\downarrow}^{\dagger}(0) \right\rangle$$

$$+ \left\langle T_{\tau} c_{\mathbf{k}-\mathbf{q}\uparrow}^{\dagger}(\tau) c_{\mathbf{k}'\sigma'}(\tau) \right\rangle \left\langle T_{\tau} c_{\mathbf{k}'+\mathbf{q}\sigma'}^{\dagger}(\tau) c_{-\mathbf{k}\downarrow}^{\dagger}(0) \right\rangle$$

$$+ \left\langle T_{\tau} c_{\mathbf{k}'+\mathbf{q}\sigma'}^{\dagger}(\tau) c_{\mathbf{k}'\sigma'}(\tau) \right\rangle \left\langle T_{\tau} c_{\mathbf{k}-\mathbf{q}\uparrow}^{\dagger}(\tau) c_{-\mathbf{k}\downarrow}^{\dagger}(0) \right\rangle$$

$$= \mathcal{F}^{\dagger}(\mathbf{k} - \mathbf{q}, 0) \mathcal{G}(-\mathbf{k}, \tau) + n_{\mathbf{k}-\mathbf{q}} \mathcal{F}^{\dagger}(\mathbf{k}, \tau) + n_{\mathbf{k}} \mathcal{F}^{\dagger}(\mathbf{k}, \tau).$$

Again, the $\mathbf{q} = 0$ term vanishes, and we neglect the contribution to the self-energy term. We arrive at

$$\frac{\partial \mathcal{F}^{\dagger}(\mathbf{k}, \tau)}{\partial \tau} = \mathcal{E}_{\mathbf{k}} \mathcal{F}^{\dagger}(\mathbf{k}, \tau) + \Delta_{\mathbf{k}}^{\dagger} \mathcal{G}(-\mathbf{k}, \tau). \tag{21.60}$$

Equations (21.56) and (21.60) are known as the Gor'kov equations. Fourier transformation of the Matsubara functions yields

$$(i\Omega_n - \mathcal{E}_{\mathbf{k}}) \mathcal{G}(\mathbf{k}, i\Omega_n) + \Delta_{\mathbf{k}} \mathcal{F}^{\dagger}(\mathbf{k}, i\Omega_n) = 1$$

$$(i\Omega_n + \mathcal{E}_{\mathbf{k}}) \mathcal{F}^{\dagger}(\mathbf{k}, i\Omega_n) + \Delta_{\mathbf{k}}^{\dagger} \mathcal{G}(\mathbf{k}, i\Omega_n) = 0. \tag{21.61}$$

We obtain the solutions

$$\mathcal{G}(\mathbf{k}, i\Omega_n) = -\frac{i\Omega_n + \mathcal{E}_{\mathbf{k}}}{\Omega_n^2 + E_{\mathbf{k}}^2}$$

$$\mathcal{F}^{\dagger}(\mathbf{k}, i\Omega_n) = \mathcal{F}(\mathbf{k}, i\Omega_n) = \frac{\Delta_{\mathbf{k}}}{\Omega_n^2 + E_{\mathbf{k}}^2}, \tag{21.62}$$

where $E_{\mathbf{k}}^2 = \mathcal{E}_{\mathbf{k}}^2 + \Delta_{\mathbf{k}}^2$. Analytic continuation of (21.62) shows that the corresponding Green functions have poles at $\varepsilon = \pm E_{\mathbf{k}}$.

To complete the solution, we need to determine the gap function. First, we note that

$$\mathcal{F}(\mathbf{k}, \tau) = \frac{1}{\beta} \sum_{i\Omega_n} e^{-i\Omega_n \tau} \, \mathcal{F}(\mathbf{k}, i\Omega_n).$$

Thus,

$$\mathcal{F}(\mathbf{k}, 0) = \frac{1}{\beta} \sum_{i\Omega_n} \mathcal{F}(\mathbf{k}, i\Omega_n) = \frac{1}{\beta} \sum_{i\Omega_n} \frac{\Delta_{\mathbf{k}}}{\Omega_n^2 + E_{\mathbf{k}}^2} = \frac{\Delta_{\mathbf{k}}}{2 E_{\mathbf{k}}} \tanh \frac{\beta E_{\mathbf{k}}}{2}.$$

Consequently, the gap equation (21.55) becomes

$$\Delta_{\mathbf{k}} = -\sum_{\mathbf{q}} V_{\mathbf{q}} \frac{\Delta_{\mathbf{k}-\mathbf{q}}}{2 E_{\mathbf{k}-\mathbf{q}}} \tanh \frac{\beta E_{\mathbf{k}-\mathbf{q}}}{2}.$$

Again, using the contact potential approximation (21.39), we obtain the gap equation (21.45).

Nambu Spinors, Bogoliubov–de Gennes Hamiltonian, and the Gor'kov Green Functions

Now we present Yoichiro Nambu's pedagogical approach using the spinor formulation of BCS theory. As we have seen, the BCS quasiparticles consist of a superposition of electrons and holes. Nambu introduced an *isospin* construct that describes orientations in charge (electron–hole) space, defining the quasiparticles composition. The *Nambu spinor* is defined as

$$\psi_{\mathbf{k}} = \begin{pmatrix} c_{\mathbf{k}} \\ c_{-\mathbf{k}}^{\dagger} \end{pmatrix} \quad \begin{matrix} \text{electron annihilation} \\ \text{hole annihilation} \end{matrix} \qquad \psi_{\mathbf{k}}^{\dagger} = \begin{pmatrix} c_{\mathbf{k}}^{\dagger} \\ c_{-\mathbf{k}} \end{pmatrix}, \quad \begin{matrix} \text{electron creation} \\ \text{hole creation} \end{matrix} \tag{21.63}$$

Nambu spinors satisfy conventional fermionic field anticommutator

$$\{\psi_{\mathbf{k}}, \psi_{\mathbf{k}'}\} = \delta_{\mathbf{k}\mathbf{k}'} \, \mathbb{I},$$

they describe electrons and holes and not up and down electron spins. These spinors allow us to combine the kinetic and pairing energy terms in the BCS Hamiltonian into a single-vector field, a magnetic field–like quantity that acts in isospin space. The kinetic energy can be written as

$$\sum_{\mathbf{k}} \mathcal{E}_{\mathbf{k}} \left(c_{\mathbf{k}}^{\dagger} c_{\mathbf{k}} + c_{-\mathbf{k}}^{\dagger} c_{-\mathbf{k}} \right) = \sum_{\mathbf{k}} \mathcal{E}_{\mathbf{k}} \left(c_{\mathbf{k}}^{\dagger} c_{\mathbf{k}} - \underbrace{c_{-\mathbf{k}} c_{-\mathbf{k}}^{\dagger}}_{\text{hole}} + 1 \right)$$

$$= \sum_{\mathbf{k}} \begin{pmatrix} c_{\mathbf{k}}^{\dagger} & c_{-\mathbf{k}} \end{pmatrix} \begin{bmatrix} \mathcal{E}_{\mathbf{k}} & 0 \\ 0 & -\mathcal{E}_{\mathbf{k}} \end{bmatrix} \begin{pmatrix} c_{\mathbf{k}} \\ c_{-\mathbf{k}}^{\dagger} \end{pmatrix} + \underbrace{\sum_{\mathbf{k}} \mathcal{E}_{\mathbf{k}}}_{\text{constant}}$$

where $\mathcal{E}_{\mathbf{k}} = \varepsilon_{\mathbf{k}} - \mu$. The energy $-\mathcal{E}_{\mathbf{k}}$ is the energy to create a hole. We will drop the constant remainder term $\sum_{\mathbf{k}} \mathcal{E}_{\mathbf{k}}$. Similarly, we obtain

$$\frac{V}{2\Omega} \sum_{\mathbf{k}, \mathbf{k}'} \left(c_{\mathbf{k}'}^{\dagger} c_{-\mathbf{k}'}^{\dagger} \phi_{\mathbf{k}} + \phi_{\mathbf{k}'}^{*} c_{-\mathbf{k}} c_{\mathbf{k}} - \phi_{\mathbf{k}'}^{*} \phi_{\mathbf{k}} \right)$$

$$= \frac{V}{2\Omega} \left[\left(\sum_{\mathbf{k}'} c_{\mathbf{k}'}^{\dagger} c_{-\mathbf{k}'}^{\dagger} \right) \left(\sum_{\mathbf{k}} \phi_{\mathbf{k}} \right) + \left(\sum_{\mathbf{k}'} \phi_{\mathbf{k}'}^{*} \right) \left(\sum_{\mathbf{k}} c_{-\mathbf{k}} c_{\mathbf{k}} \right) - \left(\sum_{\mathbf{k}'} \phi_{\mathbf{k}'}^{*} \right) \left(\sum_{\mathbf{k}} \phi_{\mathbf{k}} \right) \right]$$

$$= \frac{1}{2} \sum_{\mathbf{k}} \left[\Delta^{*} c_{-\mathbf{k}} c_{\mathbf{k}} + c_{\mathbf{k}}^{\dagger} c_{-\mathbf{k}}^{\dagger} \Delta - \frac{\Delta^{*} \Delta}{V/\Omega} \right] = \sum_{\mathbf{k}} \begin{pmatrix} c_{\mathbf{k}}^{\dagger} & c_{-\mathbf{k}} \end{pmatrix} \begin{bmatrix} 0 & \Delta \\ \Delta^{*} & 0 \end{bmatrix} \begin{pmatrix} c_{\mathbf{k}} \\ c_{-\mathbf{k}}^{\dagger} \end{pmatrix} - \frac{\Delta^{*} \Delta}{V/\Omega}$$

$$= \sum_{\mathbf{k}} \psi_{\mathbf{k}}^{\dagger} \begin{bmatrix} 0 & \Delta_1 - i\Delta_2 \\ \Delta_1 + i\Delta_2 & 0 \end{bmatrix} \psi_{\mathbf{k}} - \frac{\Delta^{*} \Delta}{V/\Omega}. \tag{21.64}$$

We can now combine the kinetic and pairing terms into a single matrix:

$$\mathcal{H}_{\text{BCS}} = \sum_{\mathbf{k}} \psi_{\mathbf{k}}^{\dagger} \begin{bmatrix} \mathcal{E}_{\mathbf{k}} & \Delta_1 - i\Delta_2 \\ \Delta_1 + i\Delta_2 & -\mathcal{E}_{\mathbf{k}} \end{bmatrix} \psi_{\mathbf{k}} = \sum_{\mathbf{k}} \psi_{\mathbf{k}}^{\dagger} \left[\mathcal{E}_{\mathbf{k}} \tau_3 + \Delta_1 \tau_1 + \Delta_2 \tau_2 \right] \psi_{\mathbf{k}}, \tag{21.65}$$

where we used the Pauli isospin matrices τ, and the definition of Δ given in (21.40). The generalized complex pairing field Δ can be regarded as a transverse field in isospin space. Equation (21.65) is also referred to as the Bogoliubov–de Gennes Hamiltonian.

We now recast (21.65) in two forms:

- The first form is

$$\mathcal{H} = \sum_{\mathbf{k}} \left[\Delta_1 \psi_{\mathbf{k}}^{\dagger} \tau_1 \psi_{\mathbf{k}} + \Delta_2 \psi_{\mathbf{k}}^{\dagger} \tau_2 \psi_{\mathbf{k}} + \mathcal{E}_{\mathbf{k}} \psi_{\mathbf{k}}^{\dagger} \tau_3 \psi_{\mathbf{k}} \right]$$

$$= \sum_{\mathbf{k}} \left[\Delta_1 \hat{\tau}_1 + \Delta_2 \hat{\tau}_2 + \mathcal{E}_{\mathbf{k}} \hat{\tau}_3 \right]$$

$$\hat{\tau}_{1\mathbf{k}} = \left(c_{\mathbf{k}\uparrow}^{\dagger} c_{-\mathbf{k}\downarrow}^{\dagger} + c_{-\mathbf{k}\uparrow} c_{\mathbf{k}\downarrow} \right)$$

$$\hat{\tau}_{2\mathbf{k}} = \left(c_{\mathbf{k}\uparrow}^{\dagger} c_{-\mathbf{k}\downarrow}^{\dagger} - c_{-\mathbf{k}\uparrow} c_{\mathbf{k}\downarrow} \right)$$

$$\hat{\tau}_{3\mathbf{k}} = \left(c_{\mathbf{k}\uparrow}^{\dagger} c_{\mathbf{k}\uparrow} - c_{-\mathbf{k}\downarrow} c_{-\mathbf{k}\downarrow}^{\dagger} \right) = \left(n_{\mathbf{k}\uparrow} + n_{-\mathbf{k}\downarrow} - 1 \right).$$

In a normal metal, $|\Delta| = 0$, $\tau_{3\mathbf{k}} = (n_{\mathbf{k}\uparrow} + n_{-\mathbf{k}\downarrow} - 1) = \mathrm{sgn}(k_F - k)$, the isospin points "up" in the doubly occupied states below the Fermi surface, and "down" in the empty states above the Fermi surface.

- The second form is

$$\mathcal{H}_{\mathrm{BCS}} = \sum_{\mathbf{k}} \psi_{\mathbf{k}}^{\dagger} \left[\mathbf{h}_{\mathbf{k}} \cdot \boldsymbol{\tau} \right] \psi_{\mathbf{k}} - \Delta^* \Delta \tag{21.66}$$

$$\mathbf{h}_{\mathbf{k}} \equiv \left(\Delta_1, \Delta_2, \mathcal{E}_{\mathbf{k}} \right), \qquad |\mathbf{h}_{\mathbf{k}}| = E_{\mathbf{k}}.$$

$\mathbf{h}_{\mathbf{k}}$ plays the role of a *field* acting in isospin space. We define the unit vector

$$\hat{\mathbf{n}}_{\mathbf{k}} = \left(\frac{\Delta_1}{E_{\mathbf{k}}}, \frac{\Delta_2}{E_{\mathbf{k}}}, \frac{\mathcal{E}_{\mathbf{k}}}{E_{\mathbf{k}}} \right).$$

For Δ real, we find that

$$\hat{\mathbf{n}}_{\mathbf{k}} = \left(\frac{\Delta}{E_{\mathbf{k}}}, 0, \frac{\mathcal{E}_{\mathbf{k}}}{E_{\mathbf{k}}} \right) = \left(u_{\mathbf{k}}^2 - v_{\mathbf{k}}^2, 0, u_{\mathbf{k}} v_{\mathbf{k}} \right).$$

Gor'kov Green Functions in the Nambu Representation

The Gor'kov Green functions can be conveniently expressed in isospin space as the outer product of the Nambu spinors:

$$\tilde{\mathcal{G}}(\mathbf{k}, \tau) = -\left\langle T_{\tau} \, \psi_{\mathbf{k}}(\tau) \otimes \psi_{\mathbf{k}}^{\dagger}(0) \right\rangle = \begin{pmatrix} -\left\langle T_{\tau} \, c_{\mathbf{k}\uparrow}(\tau) \, c_{\mathbf{k}\uparrow}^{\dagger}(0) \right\rangle & -\left\langle T_{\tau} \, c_{-\mathbf{k}\downarrow}(\tau) \, c_{\mathbf{k}\uparrow}(0) \right\rangle \\ -\left\langle T_{\tau} \, c_{\mathbf{k}\uparrow}^{\dagger}(\tau) \, c_{-\mathbf{k}\downarrow}^{\dagger}(0) \right\rangle & -\left\langle T_{\tau} \, c_{-\mathbf{k}\downarrow}(\tau) \, c_{-\mathbf{k}\downarrow}^{\dagger}(0) \right\rangle \end{pmatrix}$$

$$= \begin{pmatrix} \mathcal{G}_{\uparrow\uparrow}(\mathbf{k}, \tau) & \mathcal{F}_{\downarrow\uparrow}(\mathbf{k}, \tau) \\ \mathcal{F}_{\downarrow\uparrow}^{\dagger}(\mathbf{k}, \tau) & \mathcal{G}_{\downarrow\downarrow}(\mathbf{k}, \tau) \end{pmatrix} \tag{21.67}$$

This compact representation is known as the Nambu–Gor'kov Green functions.

Using (21.62), we write

$$\tilde{\mathcal{G}}(\mathbf{k}, i\Omega_n) = \frac{1}{(i\Omega_n)^2 - \mathcal{E}_{\mathbf{k}}^2} \begin{bmatrix} i\Omega_n + \mathcal{E}_{\mathbf{k}} & \Delta^* \\ \Delta & i\Omega_n - \mathcal{E}_{\mathbf{k}} \end{bmatrix}. \tag{21.68}$$

To bring out the physical content of (21.68), we shall reproduce it diagrammatically. The noninteracting electron and hole propagators are the diagonal components of the Nambu–Gor'kov propagator:

$$\tilde{\mathcal{G}}_0(\mathbf{k}, i\Omega_n) = \frac{1}{i\Omega_n - \varepsilon_{\mathbf{k}} \tau_3} = \begin{bmatrix} \dfrac{1}{i\Omega_n - \varepsilon_{\mathbf{k}}} & \\ & \dfrac{1}{i\Omega_n + \varepsilon_{\mathbf{k}}} \end{bmatrix}.$$

These two components are represented by the diagrams

$$\longrightarrow^{\mathbf{k}} \equiv \mathcal{G}_0(\mathbf{k}) = \frac{1}{i\Omega_n - \varepsilon_{\mathbf{k}}}$$

$$\longleftarrow^{-\mathbf{k}} \equiv -\mathcal{G}_0(-\mathbf{k}) = \frac{1}{i\Omega_n + \varepsilon_{\mathbf{k}}}.$$

Anticommutation of creation and annihilation operators produced the minus sign in the hole propagator.

The off-diagonal scattering terms convert an electron into a hole plus a condensate pair and vice versa: à la Andreev scattering processes,

$$\Delta^* c_{-\mathbf{k}\downarrow} c_{\mathbf{k}\uparrow} \equiv \underset{\mathbf{k} \qquad \omega^* \qquad -\mathbf{k}}{\longrightarrow \times \longleftarrow}$$

$$\Delta^* c_{\mathbf{k}\uparrow}^{\dagger} c_{-\mathbf{k}\downarrow}^{\dagger} \equiv \underset{-\mathbf{k} \qquad \omega \qquad \mathbf{k}}{\longleftarrow \times \longrightarrow}$$

The Feynman diagrams for $\mathcal{G}(\mathbf{k})$, electron or hole propagators, must involve an even number of Andreev scattering events, and are given by

$$\Longrightarrow = \xrightarrow{\mathbf{k}} + \xrightarrow{\mathbf{k}} \times \xleftarrow{-\mathbf{k}} \times \xrightarrow{\mathbf{k}} + \xrightarrow{\mathbf{k}} \times \xleftarrow{-\mathbf{k}} \times \xrightarrow{\mathbf{k}} \times \xleftarrow{-\mathbf{k}} \times \xrightarrow{\mathbf{k}} + \cdots$$

The corresponding self-energy diagram is given by

$$\Sigma(\mathbf{k}) = \bullet\!\!\Sigma\!\!\bullet = \times\!\!\xleftarrow{-\mathbf{k}}\!\!\times = \frac{|\Delta|^2}{i\Omega_n + \varepsilon_{\mathbf{k}}}.$$

Then $\mathcal{G}(\mathbf{k})$ is represented as

$$\mathcal{G}(\mathbf{k}) = \longrightarrow + \xrightarrow{\Sigma} + \xrightarrow{\Sigma}\xrightarrow{\Sigma} + \cdots$$

$$= \frac{1}{i\Omega_n - \varepsilon_{\mathbf{k}} - \Sigma(i\Omega_n)} = \frac{1}{i\Omega_n - \varepsilon_{\mathbf{k}} - \dfrac{|\Delta|^2}{i\Omega_n + \varepsilon_{\mathbf{k}}}} = \frac{i\Omega_n + \varepsilon_{\mathbf{k}}}{(i\Omega_n)^2 - \mathcal{E}_{\mathbf{k}}^2}. \qquad (21.69)$$

The anomalous propagator is also given by

$$\xLeftarrow{} = \xrightarrow{-\mathbf{k}} \times \xrightarrow{\mathbf{k}} + \xleftarrow{-\mathbf{k}} \times \xrightarrow{\mathbf{k}} \times \xleftarrow{-\mathbf{k}} \times \xrightarrow{\mathbf{k}} + \cdots$$

$$= \xleftarrow{-\mathbf{k}} \times \xrightarrow{\mathbf{k}} \qquad (21.70)$$

so that

$$\mathcal{F}(\mathbf{k}, i\Omega_n) = \frac{\Delta}{i\Omega_n + \varepsilon_{\mathbf{k}}} \frac{1}{i\Omega_n - \varepsilon_{\mathbf{k}} - \dfrac{|\Delta|^2}{i\Omega_n + \varepsilon_{\mathbf{k}}}} = \frac{\Delta}{(i\Omega_n)^2 - \mathcal{E}_{\mathbf{k}}^2}. \qquad (21.71)$$

21.4 Ginzburg–Landau Theory of Superconductivity

From a fundamental point of view, the main triumph of BCS theory was its derivation of the gapped energy spectrum from whence superconductivity follows. However, the BCS theory describes homogeneous, or clean, superconductors. Actually, there are many cases where spatial inhomogeneity is pervasive, for example in the case of a normal metal/superconductor interface, or in the case of superconducting alloys. In such situations, the microscopic theory becomes difficult to apply, and the availability of a more macroscopic theory would be desirable. Moreover, neither the BCS variational approach nor the Valatin–Bogoliubov formulation allow for the definition of an order parameter and its symmetry in the superconducting state. The Ginsburg–Landau (GL) theory was one of the most important phenomenological breakthroughs prior to the development of the microscopic theory of superconductivity [75]. GL introduced a pseudowavefunction $\phi(\mathbf{x})$ as a macroscopic complex order parameter, where $|\phi|^2$ serves as the density of superconducting electrons. The GL theory was based on a free energy expansion using a spatially inhomogeneous order parameter.

21.4.1 Pair Wavefunction as the Order Parameter

We may identify the pair wavefunction, shown in Figure 21.11, with the quantity

$$\phi_{\mathbf{k}} = \langle c_{-\mathbf{k}}\, c_{\mathbf{k}} \rangle = u_{\mathbf{k}}\, v_{\mathbf{k}} = \frac{\Delta}{2 E_{\mathbf{k}}} = \frac{\Delta}{2\sqrt{\mathcal{E}_{\mathbf{k}}^2 + \Delta^2}}$$

since $c_{-\mathbf{k}}\, c_{\mathbf{k}}$ annihilate two electrons from the Fermi sea and create a singlet pair. We write the corresponding real space wavefunction as

$$\psi(\mathbf{x}) = \sum_{|\mathbf{k}|} \phi_{\mathbf{k}}\, e^{i\mathbf{k}\cdot\mathbf{x}}.$$

Since both Δ and $\mathcal{E}_{\mathbf{k}}$ do not depend on the direction of $\mathbf{k}$, $\psi(\mathbf{x})$ represents *s-wave pairing*, and we get

$$\psi(\mathbf{x}) = \frac{\Omega}{(2\pi)^3} \int d\mathbf{k}\, \frac{\Delta}{2 E_{\mathbf{k}}}\, e^{i\mathbf{k}\cdot\mathbf{x}},$$

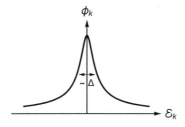

Figure 21.11 The pair function $\phi_{\mathbf{k}}$ as a function of $\mathcal{E}_{\mathbf{k}}$.

where Ω is the volume of the system. Since $\hbar\omega_D \ll E_F$, we use the approximation $\mathcal{E}_{\mathbf{k}} = \varepsilon_{\mathbf{k}} - E_F \sim \hbar v_F (k - k_F)$, and we define $a_F = \hbar v_F k_F / \Delta$

$$
\psi(\mathbf{x}) = \frac{\Omega}{(2\pi)^3} \int d\mathbf{k} \, \frac{e^{i\mathbf{k}\cdot\mathbf{x}}}{2\sqrt{\left(\frac{\hbar v_F}{\Delta}k - a_F\right)^2 + 1}}
$$

$$
= \frac{\Omega}{(2\pi)^3} \int dk \, k^2 \, \frac{1}{2\sqrt{\left(\frac{\hbar v_F}{\Delta}k - a_F\right)^2 + 1}} \int d\Omega_k \, e^{i\mathbf{k}\cdot\mathbf{x}}
$$

$$
= \frac{\Omega}{4\pi^2} \int dk \, k \, \frac{\sin(kr)}{r} \, \frac{1}{\sqrt{\left(\frac{\hbar v_F}{\Delta}k - a_F\right)^2 + 1}}
$$

$$
= \frac{\Omega}{4\pi^2} \left(\frac{\Delta}{\hbar v_F}\right)^3 \frac{1}{(r/\xi)} \int dy \, y \, \frac{\sin(yr/\xi)}{\sqrt{(y - a_F)^2 + 1}}
$$

$$
= \text{dimensionless constant} \times f(r/\xi)
$$

where $\xi = \hbar v_F / \Delta \simeq 10^{-4}$ cm, is the defining length scale of the pair wavefunction. We find that the pair wavefunction varies on a length scale of a micron in real space, which is very huge compared to the microscopic length scales. Such a macroscopic character of the pair wavefunction justifies it to be considered as an order parameter.

21.4.2 *The Ginzburg–Landau Free Energy Expansion*

In addition to its success in handling systems with spatial inhomogeneities, the GL theory can determine the nonlinear response to fields strong enough to change the superconducting density, producing results in good agreement with experimental findings. Yet initial interest was limited because of its phenomenological foundation. It was therefore important to demonstrate that it can be corroborated on microscopic grounds. Gor'kov showed, in 1959, that the GL theory was, in fact, derivable as a rigorous limiting case of the microscopic theory. Gor'kov's formulation was based on Green functions, which allowed him to extend the formulation to systems with spatial inhomogeneity [78, 79].[2] Moreover, the GL order parameter was identified with the pair wavefunction and is proportional to the energy gap.

Derivation of the Ginzburg–Landau Free Energy Expansion from Microscopics

Gor'kov realized that close to the transition temperature $|T - T_c| \ll 1$, where the GL theory is valid, the gap is small, $\Delta \ll k_B T_c$, which allows perturbative expansions in terms of Δ. Here we will depart from Gor'kov's derivation and treat this expansion in the path

---

[2]　Gor'kov published two papers in early 1959; the first treated clean superconductors, and the second considered superconducting alloys.

integral formalism. We start with the BCS Lagrangian expressed in terms of field operators
$\psi_\sigma(\mathbf{x}) = \frac{1}{\Omega} \sum_\mathbf{k} e^{i\mathbf{k}\cdot\mathbf{x}} c_{\mathbf{k}\sigma}$

$$\mathcal{H} = \sum_\sigma \int d\mathbf{x}\, \psi_\sigma^\dagger(\mathbf{x}) \left(\partial_\tau - \frac{\nabla^2}{2m} - \mu\right) \psi_\sigma(\mathbf{x}) - g \int d\mathbf{x}\, \psi_\uparrow^\dagger(\mathbf{x})\, \psi_\downarrow^\dagger(\mathbf{x})\, \psi_\downarrow(\mathbf{x})\, \psi_\uparrow(\mathbf{x})$$

and with the partition function expressed in terms of coherent state path integral

$$Z = \int D[\phi,\bar\phi]\, e^{-S[\phi,\bar\phi]}$$

$$S[\bar\phi,\phi] = \int_0^\beta d\tau \int d\mathbf{x} \left[\bar\phi_\sigma \left(\partial_\tau - \frac{\nabla^2}{2m} - \mu\right) \phi_\sigma + g\, \bar\phi_\uparrow\, \bar\phi_\downarrow\, \phi_\downarrow\, \phi_\uparrow \right], \qquad (21.72)$$

where $S[\bar\phi,\phi]$ is the action in imaginary time, and $\phi(\mathbf{x},\tau)$ are Grassmann field variables. Applying the Hubbard–Stratonovich transformation, we get

$$\exp\left[g \int_0^\beta d\tau \int d\mathbf{x}\, \bar\phi_\uparrow\, \bar\phi_\downarrow\, \phi_\downarrow\, \phi_\uparrow \right]$$

$$\Rightarrow \int D[\Delta,\bar\Delta]\, \exp\left[-\int_0^\beta d\tau \int d\mathbf{x} \left[-\frac{|\Delta|^2}{g} - \bar\Delta\, \phi_\downarrow\, \phi_\uparrow + \Delta\, \bar\phi_\uparrow\, \bar\phi_\downarrow \right] \right],$$

where $\Delta(\mathbf{x},\tau)$ and $\bar\Delta(\mathbf{x},\tau)$ are dynamically fluctuating bosonic auxiliary fields satisfying $\Delta(\mathbf{x},\tau+\beta) = \Delta(\mathbf{x},\tau)$, while the fermionic Grassmann variables satisfy $\phi(\mathbf{x},\tau+\beta) = -\phi(\mathbf{x},\tau)$.

Treating Δ as a constant field yields an action identical to the mean-field action obtained from BCS, and renders Δ a superconducting order parameter. However, now it can be treated as a dynamical field, and in particular, it has a phase that can fluctuate.

Using the Nambu spinor representation,

$$\overline\Phi = [\bar\phi_\uparrow \quad \phi_\downarrow], \qquad \Phi = \begin{bmatrix} \phi_\uparrow \\ \bar\phi_\downarrow \end{bmatrix},$$

we write the partition function as

$$Z = \int D[\bar\phi,\phi] \int D[\bar\Delta,\Delta]\, \exp\left[\int_0^\beta d\tau \int d\mathbf{x} \left[-\frac{|\Delta|^2}{g} - \overline\Phi\, \mathcal{G}^{-1}\, \Phi \right] \right] \qquad (21.73)$$

$$\mathcal{G}^{-1} = \left(\left[\mathcal{G}^{(0)}\right]^{-1} + \boldsymbol\Delta \right) = \begin{pmatrix} \left[\mathcal{G}_p^{(0)}\right]^{-1} & 0 \\ 0 & \left[\mathcal{G}_h^{(0)}\right]^{-1} \end{pmatrix} + \begin{pmatrix} 0 & \Delta \\ \Delta^* & 0 \end{pmatrix}.$$

$\mathcal{G}^{-1}$ is inverse Nambu–Gor'kov Green function, with

$$\left[\mathcal{G}_p^{(0)}\right]^{-1} = -\partial_\tau + \frac{\nabla^2}{2m} + \mu, \quad \text{noninteracting particle}$$

$$\left[\mathcal{G}_h^{(0)}\right]^{-1} = -\partial_\tau - \frac{\nabla^2}{2m} - \mu, \quad \text{noninteracting hole}.$$

Notice that the bosonic field $\Delta(\mathbf{x}, \tau)$ is transverse, since it is proportional to τ_1.

Integrating out the fermion fields, we obtain [3]

$$Z = \int D[\bar{\Delta}, \Delta] \, e^{-\tilde{S}[\bar{\Delta}, \Delta]}$$

$$\tilde{S}[\bar{\Delta}, \Delta] = \int_0^\beta d\tau \int d\mathbf{x} \, \frac{|\Delta|^2}{g} - \ln \det\left(\mathcal{G}^{-1}\right)$$

$$= \int_0^\beta d\tau \int d\mathbf{x} \, \frac{|\Delta|^2}{g} - \operatorname{Tr} \ln\left(\mathcal{G}^{-1}\right). \tag{21.74}$$

We can now derive the GL functional by expanding the logarithm in powers of the symmetry-breaking self-energy Δ and keeping terms up to fourth order,

$$\ln \mathcal{G}^{-1} = \ln\left(\left[\mathcal{G}^{(0)}\right]^{-1} + \Delta\right) = \ln\left[\mathcal{G}^{(0)}\right]^{-1} + \ln\left(\mathbb{I} + \mathcal{G}_0\,\Delta\right)$$

$$= \ln\left[\mathcal{G}^{(0)}\right]^{-1} + \sum_{n=1}^\infty \frac{(-1)^{n+1}}{n}\,(\mathcal{G}_0\,\Delta)^n.$$

The odd-n terms have vanishing trace over the Nambu space, and $\operatorname{Tr} \ln \mathcal{G}_0^{-1}/\beta$ is just the free energy of the normal phase, F_n. We thus obtain

$$F - F_n = \frac{1}{\Omega}\left\{\int_0^\beta d\tau \int d\mathbf{x} \, \frac{|\Delta|^2}{g} - \frac{1}{\beta} \operatorname{Tr}\left[\frac{1}{2}\,(\mathcal{G}_0\,\Delta)^2 + \frac{1}{4}\,(\mathcal{G}_0\,\Delta)^4\right]\right\}. \tag{21.75}$$

We first calculate the fourth-order contribution, where we can assume that the pairing field Δ is uniform. The trace over the Nambu space yields merely a factor of 2 and the space-time trace may be simply performed in the momentum-frequency representation, using the free propagator

$$\left[\mathcal{G}^{(0)}\right]^{-1}(\mathbf{k}) = \begin{pmatrix} i\Omega_n + \mathcal{E}_{\mathbf{k}} & 0 \\ 0 & i\Omega_n - \mathcal{E}_{\mathbf{k}} \end{pmatrix}.$$

The fourth-degree term becomes

$$\frac{1}{4\Omega\beta} \operatorname{Tr} (\mathcal{G}_0\,\Delta)^4 = \frac{|\Delta|^4}{2\beta} \sum_n \int \frac{d\mathbf{k}}{(2\pi)^3} \left[\mathcal{G}_p^{(0)}(i\Omega_n, \mathbf{k})\,\mathcal{G}_h^{(0)}(i\omega_n, \mathbf{k})\right]^2$$

$$= \frac{|\Delta|^4}{2\beta} \sum_n \int \frac{d\mathbf{k}}{(2\pi)^3} \frac{1}{\left(\omega_n^2 + \mathcal{E}_{\mathbf{k}}^2\right)^2}$$

---

[3] $\det A = \prod_i a_i$ where a_i are the eigenvalues of matrix A. Thus, $\ln \det(A) = \sum_i \ln(a_i)$. Note that the eigenvalues of $\ln[A]$ are $\ln(a_i)$.

$$= \frac{|\Delta|^4 \mathcal{D}}{2\beta} \sum_n \int \frac{d\mathcal{E}}{(\omega_n^2 + \mathcal{E}^2)^2} = \frac{|\Delta|^4 \mathcal{D}}{2\beta} \sum_n \frac{\pi}{4|\omega_n|^3}$$

$$= \frac{|\Delta|^4 \mathcal{D} \beta^2}{8\pi^2} \sum_{n=0}^{\infty} \frac{1}{(2n+1)^3}$$

$$= \frac{7\zeta(3)\mathcal{D}}{64\pi^2 T_c^2} |\Delta|^4,$$

where we used the identity

$$\sum_{n=0}^{\infty} \frac{1}{(2n+1)^\ell} = \frac{2^\ell - 1}{2^\ell} \zeta(\ell), \quad \zeta \text{ is the Riemann zeta function.}$$

Next, we consider the quadratic contribution to the free energy (21.75), and allow the pairing field Δ to vary in space. Expanding the trace, we obtain

$$\frac{1}{2\Omega\beta} \operatorname{Tr} \left(\mathcal{G}^{(0)} \mathbf{\Delta} \right)^2 = \frac{1}{\beta} \sum_{\mathbf{k}} \langle \mathbf{k} | \mathcal{G}_p^{(0)} \Delta \, \mathcal{G}_h^{(0)} \bar{\Delta} | \mathbf{k} \rangle$$

$$= \frac{1}{\beta} \sum_{\mathbf{q}} \Delta_{\mathbf{q}} \bar{\Delta}_{-\mathbf{q}} \underbrace{\sum_{\mathbf{k}} \mathcal{G}_p^{(0)}(\mathbf{k}) \, \mathcal{G}_h^{(0)}(\mathbf{k}+\mathbf{q})}_{\Pi_0(\mathbf{q},0).}$$

The sum over $\mathbf{k}$ is just the polarization function $\Pi_0(\mathbf{q},0)$ of (17.170). In order to allow slow variations of Δ, we perform a Taylor expansion of $\Pi_0(\mathbf{q},0)$ in $\mathbf{q}$ up to second order[4]

$$\Pi_0(\mathbf{q},0) = \Pi_0(0,0) + \frac{\mathbf{q}^2}{2} \partial_q^2 \Pi_0(0,0) + \cdots$$

The term involving $\Pi_0(0,0)$ is

$$\left(\sum_{\mathbf{q}} \Delta_{\mathbf{q}} \bar{\Delta}_{-\mathbf{q}} \right) \Pi_0(0,0) = -|\Delta|^2 \frac{1}{\beta} \sum_n \int \frac{d\mathbf{k}}{(2\pi)^3} \frac{1}{\omega_n^2 + \mathcal{E}_{\mathbf{k}}^2}$$

$$= |\Delta|^2 \mathcal{D} \int_0^{\hbar\omega_D} d\mathcal{E} \frac{1}{\beta} \sum_n \frac{1}{\omega_n^2 + \mathcal{E}^2}$$

$$= -|\Delta|^2 \mathcal{D} \int_0^{\hbar\omega_D} \frac{d\mathcal{E}}{\mathcal{E}} \tanh \frac{\beta\mathcal{E}}{2} = -|\Delta|^2 \mathcal{D} \ln \frac{\hbar\omega_D}{k_B T},$$

where we followed the integration procedure in (21.46) and (21.47). Combining quadratic terms in Δ independent of q, we get

$$\left[\frac{1}{g} - \mathcal{D} \ln \frac{\hbar\omega_D}{k_B T} \right] |\Delta|^2 = \mathcal{D} \ln \frac{T}{T_c} |\Delta|^2 \sim \mathcal{D} \frac{T_c - T}{T_c} |\Delta|^2. \tag{21.76}$$

---

[4] $\partial_q \Pi_0(0,0)$ vanishes because it is linear in $\mathbf{k}$.

To evaluate the $q^2 \partial_q^2 \Pi_0(0,0)$ term, we note that

$$\partial_q^2 \Pi_0(0,0) \rightarrow \frac{1}{3m^2\beta} \sum_n \int \frac{d\mathbf{k}}{(2\pi)^3} k^2 \frac{1}{(i\omega_n + \mathcal{E}_\mathbf{k})^3 (i\omega_n - \mathcal{E}_\mathbf{k})} = \frac{7\zeta(3)\mathcal{D}k_F^2}{48m^2\pi^2 T_c^2},$$

which leads to

$$\frac{7\zeta(3)\mathcal{D}k_F^2}{48m^2\pi^2 T_c^2} \sum_\mathbf{q} \mathbf{q}^2 \Delta_\mathbf{q} \bar{\Delta}_{-\mathbf{q}} = \frac{7\zeta(3)\mathcal{D}k_F^2}{48m^2\pi^2 T_c^2} \frac{1}{\Omega} \int d\mathbf{x} |\partial_\mathbf{x}\Delta|^2. \tag{21.77}$$

Collecting all terms in the free energy expansion, we finally arrive at

$$\Delta F_{\mathrm{GL}} = F - F_n = \int d\mathbf{x} \left[\frac{7\zeta(3)\mathcal{D}k_F^2}{48m^2\pi^2 T_c^2} |\partial_\mathbf{x}\Delta|^2 + \mathcal{D} \frac{T - T_c}{T_c} |\Delta|^2 + \frac{7\zeta(3)\mathcal{D}}{64\pi^2 T_c^2} |\Delta|^4 + \cdots \right]$$

$$= \int d\mathbf{x} \left[\frac{7\zeta(3)n}{32m\pi^2 T_c^2} |\partial_\mathbf{x}\Delta|^2 + \mathcal{D} \frac{T - T_c}{T_c} |\Delta|^2 + \frac{7\zeta(3)\mathcal{D}}{64\pi^2 T_c^2} |\Delta|^4 + \cdots \right]. \tag{21.78}$$

In the last line, we used $\mathcal{D} = mk_F/2\pi^2$ and $n = k_F^3/3\pi^2$. Setting $\psi = \frac{\sqrt{7\zeta(3)n}}{4\pi T_c}\Delta$, we rewrite (20.21.78) as

$$\Delta F_{\mathrm{GL}} = \int d\mathbf{x} \left[\frac{1}{2m} |\partial_\mathbf{x}\psi|^2 + a \frac{T - T_c}{T_c} |\psi|^2 + \frac{b}{2} |\psi|^4 + \cdots \right], \tag{21.79}$$

where ψ has the character of a macroscopic wavefunction. We note that ΔF_{GL} has a global $U(1)$ gauge symmetry.

We consider the special case of a uniform $\psi = |\psi| e^{i\alpha}$, such that

$$\Delta F_{\mathrm{GL}} = \int d\mathbf{x} \left[a t \psi^* \psi + \frac{b}{2} \psi^* \psi |\psi|^2 \right], \tag{21.80}$$

where $t = \left(\frac{T - T_c}{T_c} \right)$. We note that ΔF is independent of the phase α. Extremizing the free energy with respect to $|\psi|$ yields

$$\begin{cases} |\psi| = \qquad\quad 0, & t > 0, \;\Rightarrow\; \Delta F = \qquad 0 \\[2mm] |\psi| = \sqrt{\dfrac{a(T_c - T)/T_c}{b}}, & t < 0 \;\Rightarrow\; \Delta F = -\dfrac{a^2}{b} \left(\dfrac{T - T_c}{T_c} \right)^2, \end{cases} \tag{21.81}$$

leaving a continuous degeneracy in the phase $\alpha = [0, 2\pi]$.

21.4.3 Electromagnetic Fields and Gauge Invariance

A superconductor can be viewed as a charged condensate fluid, which naturally couples to electromagnetic fields and exhibits a charged superflow. A key feature of a superconductor is the manifestation of the phenomenon called the *Meissner effect*, its capacity to expel magnetic fields from its bulk. Although in principle it is possible to derive the Meissner effect and other electrodynamic properties of a superconductor starting from the microscopic formulation, the GL theory provides an elegant and more physically transparent way to derive such macroscopic properties of superconductors.

In the spirit of treating the order parameter as a macroscopic wavefunction, we note that in the presence of a vector potential, the kinetic energy term in the Hamiltonian becomes

$$-\frac{1}{2m} \int d\mathbf{x} \; \psi^* \; (\nabla + ie\mathbf{A})^2 \; \psi = \frac{1}{2m} \int d\mathbf{x} \; |(\nabla + ie\mathbf{A}) \; \psi|^2 \qquad (21.82)$$

upon integration by parts. Thus the introduction of electromagnetic fields modifies the GL free energy to read

$$\Delta F_{\text{GL}} = \int d\mathbf{x} \left[\underbrace{\frac{1}{2m^*} \left|(\partial_{\mathbf{x}} + ie^*\mathbf{A})\psi\right|^2 + a\,t\,|\psi|^2 + \frac{b}{2}\,|\psi|^4}_{\Delta F_\psi} + \underbrace{\frac{1}{2}\,(\nabla\times\mathbf{A})^2}_{\Delta F_A} \right], \quad (21.83)$$

where we set the scalar potential $\varphi = 0$, and we added the electromagnetic field free energy.

The appearance of the vector potential $\mathbf{A}$, a gauge field, evokes the process of local $U(1)$ gauge transformation:

$$\mathbf{A} \;\rightarrow\; \mathbf{A} + \nabla\alpha(\mathbf{x}).$$

However, as we know from quantum mechanics, such transformations require that we transform the wavefunction according to

$$\psi(\mathbf{x}) \;\rightarrow\; e^{-ie^*\alpha(\mathbf{x})} \, \psi(\mathbf{x}),$$

which clearly leaves the free energy ΔF invariant. Thus we find that presence of the gauge field promotes the global $U(1)$ symmetry to a local one!

We note that we can rewrite the first term in ΔF as

$$\int d\mathbf{x} \; \frac{1}{2m^*} \left|(\partial_{\mathbf{x}} + ie^*\mathbf{A})\psi\right|^2 = \int d\mathbf{x} \; \frac{1}{2m^*} \left[|\nabla\psi|^2 + \left(\nabla\alpha + e^*\mathbf{A}\right)^2 |\psi|^2 \right].$$

The energy of the superconducting state below T_c is lower than that of the normal state by the *condensation energy*, given in (21.81). Thus we realize that spatial variations of $|\psi|$ will cost a significant fraction of the condensation energy in the region of space where it occurs. In contrast, the zero-field free energy is actually invariant with respect to changes in α, so fluctuations of α alone will essentially cost no energy. Consequently, if we apply a weak magnetic field described by $\mathbf{A}$ to the system, we do not expect it to couple to $|\psi|$ but rather to the phase α, since it is a small perturbation. The superconducting kinetic energy density should then reduce to the second term.

21.4.4 Ginzburg–Landau Equations

The GL equations of motion are obtained through varying ΔF_{GL} with respect to the order-parameter ψ and the vector potential $\mathbf{A}$.

Variation with respect to **A** yields

$$\delta \Delta F_{GL} = - \int_{\mathbf{x}} \delta \mathbf{A} \cdot \left[-\frac{ie^*}{2m^*} \left(\psi^* \nabla \psi - \nabla \psi^* \psi \right) - \frac{e^{*2}}{m^*} |\psi|^2 \, \mathbf{A} \right] + \int (\nabla \times \delta \mathbf{A}) \cdot \mathbf{B}$$

$$= - \int_{\mathbf{x}} \left[\delta \mathbf{A} \cdot \mathbf{J}_s(\mathbf{x}) - \delta \mathbf{A} \cdot \nabla \times \mathbf{B} \right], \tag{21.84}$$

where $\mathbf{J}_s(\mathbf{x})$ is the supercurrent density – the probability current. We used the vector identity

$$(\nabla \times \delta \mathbf{A}) \cdot \mathbf{B} = \underbrace{\nabla \cdot (\delta \mathbf{A} \times \mathbf{B})}_{0} + \delta \mathbf{A} \cdot (\nabla \times \mathbf{B}) .$$

Extremizing the total variation, we obtain

$$\frac{\delta F}{\delta \mathbf{A}} = -\mathbf{J}_s(\mathbf{x}) + \frac{\nabla \times \mathbf{B}}{\mu_0} = 0, \tag{21.85}$$

the first GL equation of motion, which is just Ampère's equation.

To vary with respect to ψ, we will use (21.82) and write

$$\Delta F_{GL} = \int_{\mathbf{x}} \left[\frac{1}{2m^*} \psi^* \left(\nabla + ie^* \mathbf{A} \right)^2 \psi + a \, t \, \psi^* \psi + \frac{b}{2} |\psi^* \psi|^2 + \frac{1}{2} (\nabla \times \mathbf{A})^2 \right].$$

Varying with respect to ψ^*, we obtain

$$\delta \Delta F_{GL} = \int dx \left(\delta \psi^* \left[\frac{1}{2m^*} \left(\nabla + ie^* \mathbf{A} \right)^2 \psi + a \, t \, \psi + b \, |\psi|^2 \, \psi \right] \right). \tag{21.86}$$

Extremization then yields GL's second equation of motion

$$\frac{1}{2m^*} \left(\nabla + ie^* \mathbf{A} \right)^2 \psi + a \, t \, \psi + b \, |\psi|^2 \, \psi = 0, \tag{21.87}$$

which respects gauge invariance.

Ginzburg–Landau Coherence Length

In the absence of electromagnetic fields, and interpreting ψ as a macroscopic wavefunction, we write

$$F[\psi] = \int dx \left[\frac{1}{2m^*} |\nabla \psi|^2 + a \, t \, |\psi|^2 + \frac{b}{2} |\psi|^4 \right].$$

The corresponding Euler–Lagrange equation as

$$\frac{1}{2m^*} \nabla^2 \psi + a \, \psi + b \, |\psi|^2 \, \psi = 0.$$

For pedagogical reasons, we consider the case of a superfluid filling the half-space $z > 0$, for $t < 0$. We impose the boundary condition $\psi(z = 0) = 0$ and assume that the solution only depends on z. Then

$$\frac{1}{2m^*} \psi''(z) - a \, |t| \psi(z) + b \, |\psi(z)|^2 \, \psi(z) = 0.$$

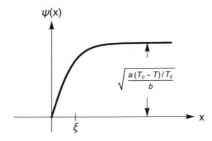

Figure 21.12 Solution to the nonlinear Schrödinger equation.

For $z \to \infty$, $|\psi|^2$ should approach the uniform superconductor density n_s, namely

$$\lim_{z \to \infty} \psi(z) = \sqrt{\frac{a|t|}{b}} = \sqrt{\frac{a(T_c - T)/T_c}{b}} = \sqrt{n_s}.$$

Writing

$$\psi(z) = \sqrt{\frac{a(T_c - T)/T_c}{b}} \, f(z) = \sqrt{n_s} \, f(z),$$

we obtain

$$-\frac{1}{2m^*a|t|} \, f''(z) + f(z) - f^3(z) = 0 \tag{21.88}$$

with

$$\xi^2 = \frac{1}{2m^*a|t|} = \frac{1}{2m^*a(T_c - T)/T_c} > 0,$$

the *Ginzburg–Landau coherence length*. As a correlation length in a second-order phase transition, the Ginzburg–Landau ξ actually diverges at $T = T_c$.

Equation (21.88) is the nonlinear Schrödinger equation, which has the solution

$$f(z) = \tanh\left[\frac{z}{\sqrt{2}\xi}\right]$$

shown in Figure 21.12. It satisfies the boundary conditions at $z = 0$ and $z \to \infty$.

21.4.5 Meissner Effect

We note that when we use $\psi = |\psi| e^{i\alpha}$ in the expression for the supercurrent $\mathbf{J}_s$, we find that

$$\psi^* \nabla \psi = \left(|\psi| \, e^{-i\alpha}\right) \nabla \left(|\psi| \, e^{i\alpha}\right) = i \, |\psi|^2 \, \nabla\alpha + |\psi| \, \nabla |\psi|$$

$$\left(\nabla \psi^*\right) \psi = \nabla \left(|\psi| \, e^{-i\alpha}\right) \left(|\psi| \, e^{i\alpha}\right) = -i \, |\psi|^2 \, \nabla\alpha + |\psi| \, \nabla |\psi|$$

and we obtain

$$\mathbf{J}_s(\mathbf{x}) = \frac{e^*\hbar}{m^*} |\psi|^2 \, \nabla\alpha - \frac{e^{*2}}{m^*} |\psi|^2 \, \mathbf{A}$$

$$= \frac{e^* n_s}{m^*} \left(\nabla\alpha - \frac{e^*}{\hbar} \mathbf{A} \right) = e^* n_s \, \mathbf{v}_s, \tag{21.89}$$

where we have identified

$$\mathbf{v}_s = \frac{\hbar}{m^*} \left(\nabla\alpha - \frac{e^*}{\hbar} \mathbf{A} \right)$$

as the superfluid velocity, which is invariant under a local gauge transformation. We find that the onset of a supercurrent can be effected by either a twist in the phase or by an external vector potential. Substituting for $\mathbf{J}_s$ in (21.85), we get

$$\nabla\times\mathbf{B} = -\frac{n_s e^{*2}}{m^*} \left(\mathbf{A} - \frac{\hbar}{e^*} \nabla\alpha \right).$$

For a uniform n_s, we obtain

$$\nabla\times\nabla\times\mathbf{B} = \nabla\left(\nabla\cdot\mathbf{B}\right) - \nabla^2\mathbf{B} = -\nabla^2\mathbf{B} = \nabla\times\mathbf{J} = -\frac{n_s e^{*2}}{m^*}\mathbf{B},$$

where we have used the identity $\nabla\times\nabla\psi = 0$ to eliminate $\nabla\alpha$. This leads to

Meissner effect $\begin{cases} \nabla^2\mathbf{B} = \dfrac{1}{\lambda_L^2}\mathbf{B} & \text{London equation} \\[2ex] \dfrac{1}{\lambda_L^2} = \dfrac{n_s e^{*2}}{m^*} & \text{London penetration depth.} \end{cases}$

This equation[5] demonstrates the remarkable phenomenon of complete diamagnetism: the magnetic fields are completely expelled from superconductors. For the case of a superfluid filling the half-space $z > 0$, (21.90) reduces to the 1D equation

$$\lambda_L^2 \frac{d^2 B}{dz^2} = B$$

with solutions of the form $B(z) \sim B_0 \, e^{-z/\lambda_L}$. Near the surface of a superconductor, magnetic fields only penetrate a distance depth λ_L into the condensate, because persistent supercurrents screen the field out of the superconductor. Figure 21.13 shows a magnetically levitated magnet in the vicinity of a superconductor – a manifestation of the Meissner effect.

Another interesting observation can be seen if we assume that α is uniform, so that (21.89) becomes

$$\mathbf{J}_s(\mathbf{x}) = -\frac{e^{*2} n_s}{m^*} \mathbf{A}.$$

---

[5] It was first derived by Fritz and Heinz London in 1935 [122] The London equation can also be derived directly from BCS theory.

Figure 21.13 Levitation due to Meissner effect: a magnet levitating above a high-temperature superconductor, cooled with liquid oxygen.

Taking the time derivative, and working in a gauge where $\varphi = 0$, so $E = -i\partial_t \mathbf{A}$, we find that

$$\partial_t \mathbf{J}_s = \frac{e^{*2} n_s}{m^*} \mathbf{E}.$$

This shows that for a constant current, there is no electric field – a system with a finite current and zero electric field must have zero resistivity!

21.4.6 Fluctuations of the Order Parameter and the Anderson–Higgs Mechanism

It is instructive at this point to explore the nature and characteristics of long wavelength fluctuations in the order parameter for many reasons. To start with, we find that the London equation in the form

$$\mathbf{J}_s = \rho_s \mathbf{A}$$

violates local gauge invariance. Three papers published independently by Anderson [11, 12], Bogoliubov [32]), and Nambu [139] showed that the violation of local gauge symmetry can be resolved by introducing order-parameter fluctuations into the theory. Anderson, Bogoliubov, and Nambu independently came up with the so-called Nambu–Goldstone phase mode that corresponds to phase fluctuations of the order parameter of neutral superfluids.[6] The possibility of an additional fluctuation in the order-parameter amplitude was not given special attention at that time, probably since it was not needed to get a consistent theory of superconductivity.

Fluctuations in Superconducting Fluids and the Anderson–Higgs Mechanism:
Eating Up the Goldstone Mode

When discussing fluctuations about the mean-field state of a superconductor, we have to take the coupling to the electromagnetic field into account. We proceed analogously to the case of a neutral superfluid but with the kinetic term

---

[6] As we discussed in Section 17.3.6, the massless Nambu–Goldstone mode arises from the spontaneous breaking of a continuous symmetry of the order-parameter phase.

$$\frac{1}{2m^*} \left| \left(\frac{\hbar}{i} \nabla - \frac{e^*}{c} \mathbf{A} \right) \psi \right|^2 .$$

Setting

$$\psi = (\psi_0 + \delta\psi)\, e^{i\alpha(\mathbf{x})}$$

and following the steps of Section 17.3.6, we obtain

$$F[\delta\psi, \alpha, \mathbf{A}] \simeq \int d\mathbf{x} \left[\left(\frac{\hbar}{i}(\nabla\delta\psi) \right)^* \cdot \left(\frac{\hbar}{i}(\nabla\delta\psi) \right) - 2a\, \delta\psi^2 \right.$$

$$\left. - \frac{\hbar^2}{2m^*} \frac{a}{b} \left(\nabla\alpha - \frac{e^*}{\hbar c}\mathbf{A} \right)^* \left(\nabla\alpha - \frac{e^*}{\hbar c}\mathbf{A} \right) + \frac{1}{8\pi}(\nabla\times\mathbf{A})^* \cdot (\nabla\times\mathbf{A}) \right].$$

Note that the phase α of the macroscopic wavefunction and the vector potential appear in the combination $\nabla\alpha - \frac{e^*}{\hbar c}\mathbf{A}$ [7]. The physics should be invariant under the gauge transformation

$$\mathbf{A} \to \mathbf{A} + \nabla\chi, \quad \phi \to \phi - \frac{1}{c}\dot{\chi}, \quad \psi \to e^{ie^*\chi/\hbar c}\,\psi,$$

where ϕ is the scalar electric potential and $\chi(\mathbf{x}, t)$ is an arbitrary scalar field. We make use of this gauge invariance by choosing

$$\chi = -\frac{\hbar c}{e^*}\alpha.$$

Under this transformation, we get

$$\mathbf{A} \to \mathbf{A} - \frac{\hbar c}{e^*}\nabla\alpha = \mathbf{A}'$$

$$\psi = (\psi_0 + \delta\psi)\, e^{i\alpha} \to (\psi_0 + \delta\psi)\, e^{i(\alpha - \alpha)} = \psi_0 + \delta\psi.$$

The macroscopic wavefunction becomes purely real and positive. The Landau functional thus transforms into

$$F[\delta\psi, \mathbf{A}'] \to \int d\mathbf{x} \left[\left(\frac{\hbar}{i}(\nabla\delta\psi) \right)^* \cdot \frac{\hbar}{i}(\nabla\delta\psi) - 2a\, \delta\psi^2 \right.$$

$$\left. - \frac{a}{b}\frac{e^{*2}}{2m^*c^2}\mathbf{A}'^* \cdot \mathbf{A}' + \frac{1}{8\pi}(\nabla\times\mathbf{A}')^* \cdot (\nabla\times\mathbf{A}') \right]$$

(note that $\nabla\times\mathbf{A}' = \nabla\times\mathbf{A}$). Thus the phase no longer appears in F; it has been *absorbed into the vector potential*! The sum of the phase gradient and the vector potential creates a field with both longitudinal and transverse character.

---

[7] Note that the vector potential, which gives rise to transverse electromagnetic waves, becomes coupled to gradients of the phase, which are longitudinal in character.

Furthermore, dropping the prime and taking the Fourier transform, we arrive at

$$
F[\delta\psi, \mathbf{A}] \simeq \sum_{\mathbf{k}} \left[\left(-2a + \frac{\hbar^2 k^2}{2m^*} \right) \delta\psi_{\mathbf{k}}^* \delta\psi_{\mathbf{k}} - \frac{a}{b} \frac{e^{*2}}{2m^* c^2} \mathbf{A}_{\mathbf{k}}^* \cdot \mathbf{A}_{\mathbf{k}} \right.
$$

$$
\left. + \frac{1}{8\pi} (\mathbf{k} \times \mathbf{A}_{\mathbf{k}})^* \cdot (\mathbf{k} \times \mathbf{A}_{\mathbf{k}}) \right]
$$

$$
= \sum_{\mathbf{k}} \left[\left(-2a + \frac{\hbar^2 k^2}{2m^*} \right) \delta\psi_{\mathbf{k}}^* \delta\psi_{\mathbf{k}} \right.
$$

$$
\left. + n_s \frac{e^{*2}}{2m^* c^2} \left[\mathbf{A}_{\mathbf{k}}^{\parallel *} \cdot \mathbf{A}_{\mathbf{k}}^{\parallel} + \mathbf{A}_{\mathbf{k}}^{\perp *} \cdot \mathbf{A}_{\mathbf{k}}^{\perp} \right] + \frac{1}{8\pi} k^2 \mathbf{A}_{\mathbf{k}}^{\perp *} \mathbf{A}_{\mathbf{k}}^{\perp} \right].
$$

Obviously, amplitude fluctuations decouple from electromagnetic fluctuations and behave like those of a neutral superfluid. It is interesting to discuss the electromagnetic fluctuations further. The term proportional to $-a/b = n_s$ is due to superconductivity. Without it, we would have the free-field functional

$$
F[\mathbf{A}] = \frac{1}{8\pi} (\mathbf{k} \times \mathbf{A}_{\mathbf{k}})^* \cdot (\mathbf{k} \times \mathbf{A}_{\mathbf{k}}) = \frac{1}{8\pi} k^2 \mathbf{A}_{\mathbf{k}}^{\perp *} \mathbf{A}_{\mathbf{k}}^{\perp},
$$

which represents purely gapless transverse modes.

All three components of $\mathbf{A}$ appear now; the longitudinal one has been introduced by absorbing the phase $\alpha(\mathbf{x})$. Even more importantly, all components obtain a term with a constant coefficient – a mass term. Thus the electromagnetic field inside a superconductor becomes massive. This is the famous Anderson–Higgs mechanism. The same general idea is also thought to explain the masses of elementary particles, although in a more complicated way. The *Higgs bosons* in our case are the amplitude-fluctuation modes described by $\delta\psi$ [8]. Contrary to what is said in popular discussions, they are not responsible for giving mass to the field $\mathbf{A}$. Rather, they are left over when the phase fluctuations are eaten by the field $\mathbf{A}$.

Amazingly, by absorbing the phase of the order parameter, we arrive at a purely electromagnetic action, but one in which the phase stiffness of the condensate $\nabla\alpha$ imparts a new quadratic term in the action of the electromagnetic field – a "mass term." Like a python that has swallowed its prey whole, the new gauge field is transformed into a much more sluggish object: it is heavy and weak.

---

[8] As it does not carry any spin or charge, in principle, the amplitude mode of the superconducting order parameter, or the Higgs mode, does not couple directly to any external probe. However, when superconductivity (SC) coexists with a charge density wave order, the amplitude mode of the CDW order couples to the Higgs mode by modulating the density of states at the Fermi level, thus shaking the SC condensate by modulating the amplitude of the superconducting order parameter. This allows the indirect detection of the Higgs mode by spectroscopic probes. Experimentally, the Higgs mode becomes active by removing spectral weight from the CDW amplitude mode upon entering the SC state. The requisite of a coexisting CDW mode and the observation of a transfer of spectral weight from the CDW amplitude mode to the Higgs mode in the SC state can thus be considered key predictions of the Higgs mode scenario.

The mass term in the superconducting case can be thought of as leading to the Meissner effect (finite penetration depth λ). Indeed, we can write

$$F[\delta\psi, \mathbf{A}] \simeq \sum_{\mathbf{k}} \left[\left(-2a + \frac{\hbar^2 k^2}{2m^*} \right) \delta\psi_{\mathbf{k}}^* \, \delta\psi_{\mathbf{k}} + \frac{1}{8\pi} \sum_{\mathbf{k}} \left[\frac{1}{\lambda_L^2} \mathbf{A}_{\mathbf{k}}^* \cdot \mathbf{A}_{\mathbf{k}} + k^2 \, \mathbf{A}_{\mathbf{k}}^{\perp *} \mathbf{A}_{\mathbf{k}}^{\perp} \right] \right],$$

where λ_L is just the London penetration depth we encountered in the Meissner effect. Moreover, we find that the last two terms lead to exactly the $\mathbf{B}$ equation

$$\frac{1}{\lambda_L^2} \mathbf{A}^{\perp *} \cdot \mathbf{A}^{\perp} + k^2 \, \mathbf{A}_{\mathbf{k}}^{\perp *} \mathbf{A}_{\mathbf{k}}^{\perp} \;\to\; \nabla^2 \mathbf{A}^{\perp} = \frac{1}{\lambda_L^2} \mathbf{A}^{\perp}.$$

The photon mass is proportional to $1/\lambda_L$.

In concluding, we should note the following observation: in a neutral superfluid, spontaneous symmetry breaking is associated with a global $U(1)$ gauge symmetry and leads to the emergence of a massless Goldstone bosonic mode. By contrast, in a charged superfluid – a superconductor – spontaneous symmetry breaking is associated with a local gauge symmetry, due to the presence of the gauge field $\mathbf{A}$. In this case, spontaneous symmetry breaking leads to a massive photon instead of a massless Goldstone boson.

Exercises

21.1 Finite temperature gap equation:
      The BCS gap equation is expressed as

$$\Delta = -V \sum_{\mathbf{k}} \langle c_{-\mathbf{k}\uparrow} c_{\mathbf{k}\downarrow} \rangle.$$

(a) Given that the anomalous Green's function is defined as

$$F_{\downarrow\uparrow}(\mathbf{k}, \tau) = -\left\langle T_\tau c_{-\mathbf{k}\downarrow}^\dagger(\tau) c_{\mathbf{k}\uparrow}^\dagger(0) \right\rangle$$

$$\mathcal{F}(\mathbf{k}, i\omega_n) = \frac{\Delta}{(i\omega_n)^2 - \mathcal{E}_{\mathbf{k}}^2},$$

derive an expression for the gap equation in terms of a Matsubara sum.

(b) Show that by performing the sum over Matsubara frequencies you obtain the temperature dependence of the gap as

$$\frac{1}{V} = \int_0^\infty d\mathcal{E} \, \frac{\tanh[\beta E/2]}{E},$$

where $E^2 = \mathcal{E}^2 + \Delta^2(T)$.

21.2 Peierls transition in 1D à la superconductivity: electron perspective:
      Consider the Frölich Hamiltonian

$$\mathcal{H}_F = \sum_{k,\sigma} \varepsilon(k) \, c_{k\sigma}^\dagger c_{k,\sigma} + \sum_{k,q,\sigma} g(q) \, c_{k-q,\sigma}^\dagger c_{k,\sigma} \left(b_{-q}^\dagger + b_q \right) + \sum_q \omega(q) \, b_q^\dagger b_q.$$

We expect that at low temperatures $< T_{\text{Peierls}}$ the phonon mode $q = 2k_F$ will be macroscopically occupied. Accordingly, we replace operators $b_{q=\pm 2k_F}$ and $b^\dagger_{q=\pm 2k_F}$ by c-numbers $\tilde{b} = \langle b_{q=\pm 2k_F} \rangle$.

(a) How will this modify the Frölich Hamiltonian?
(b) Using the modified Hamiltonian, obtain an expression for $\langle \mathcal{H} \rangle$.
(c) Derive an expression for $\tilde{b}$ in terms of average values of electron operators by minimizing $\langle \mathcal{H} \rangle$ with respect to $\tilde{b}$. What is the physical interpretation of the expression you obtained?
(d) By substituting it back in the modified Hamiltonian, you should obtain a familiar form reminiscent of superconductivity.
(e) Proceed with the aid of the machinery developed for the superconductivity problem to obtain an expression for T_{Peierls} and the gap temperature dependence.

21.3 **Peierls instability: phonon perspective**

Find the renormalized acoustic phonon spectra $\bar{\omega}_{\mathbf{q}}$ in one dimension. The renormalization is due to interaction with the one-dimensional Fermi gas (Fermi surface consists of two points: k_F and $-k_F$).

(a) Consider only the lowest-order diagram in $\Pi(\mathbf{q}, \omega)$.
(b) Both integrals, over $!$ and over k, can be taken without any approximation.
(c) Assume that $c \ll v_F = k_F/m$. Then the spectrum renormalization is proportional to Re $\Pi(\mathbf{q}, \omega) \simeq \Pi(\mathbf{q}, 0)$. Find $\bar{\omega}_{\mathbf{q}}$ for the following cases:

(i) $q \to 0$.
(ii) $q = 2k_F + \delta q$. In the vicinity of $2k_F$, the renormalized spectrum becomes negative: $\bar{\omega}_{\mathbf{q}} < 0$. This signifies an instability (Peierls instability). Any idea why and what this instability leads to?
Sketch $\bar{\omega}_{\mathbf{q}}$ as a function of q.

21.4 **Electron–phonon coupling:**

The lowest-order correction to the fermionic self-energy due to electron–phonon coupling takes the form

$$\Sigma(i\Omega_n, \mathbf{k}) = \frac{1}{\beta \mathcal{V}} \sum_{m, \mathbf{q}} \mathcal{D}^{(0)}(i\omega_m, \mathbf{q}) \, \mathcal{G}^{(0)}(i\Omega_n + i\omega_m, \mathbf{k} + \mathbf{q}),$$

where the phonon Green's function is given by

$$\mathcal{D}^{(0)}(i\omega_m; \mathbf{q}) = \frac{-2\omega_{\mathbf{q}}}{(i\omega_m)^2 - \omega_{\mathbf{q}}^2}.$$

Perform the Matsubara frequency summation.

21.5 d-wave superconductivity (high Tc):

Consider the Nambu-type Hamiltonian

$$\mathcal{H} = \sum_{\mathbf{k}} \begin{bmatrix} c_{\mathbf{k}\uparrow}^{\dagger} & c_{-\mathbf{k}\downarrow} \end{bmatrix} \begin{pmatrix} \varepsilon_{\mathbf{k}} - \mu & \Delta_{\mathbf{k}} \\ \Delta_{\mathbf{k}} & \mu - \varepsilon_{\mathbf{k}} \end{pmatrix} \begin{bmatrix} c_{\mathbf{k}\uparrow} \\ c_{-\mathbf{k}\downarrow}^{\dagger} \end{bmatrix}.$$

Obtain the spectrum by diagonalizing the Hamiltonian. The spectrum should contain two bands with energy $E_{+}(\mathbf{k})$ and $E_{-}(\mathbf{k})$, respectively.

Show that $E_{+}(\mathbf{k}) \geq 0$ and $E_{-}(\mathbf{k}) \leq 0$. In addition, show that $E_{+}(\mathbf{k}) = -E_{-}(\mathbf{k})$. Compute the superconducting gap $\Delta_{\mathrm{sc}} = E_{+}(\mathbf{k}) - E_{-}(\mathbf{k})$.

(a) s-wave superconductor:

Here, we consider an s-wave superconductor with $\Delta_{\mathbf{k}}$ being a constant (independent of $\mathbf{k}$) $\Delta_{\mathbf{k}} = \Delta$. Find the minimum value of $\Delta_{\mathrm{sc}}(\mathbf{k})$. Show that the minimum is reached at the Fermi surface $\varepsilon_{\mathbf{k}} = \mu$.

Prove that the minimum value of $\Delta_{\mathrm{sc}}(\mathbf{k})$ only relies on the value of Δ. As long as $\Delta > 0$, the two bands never cross, so the system is always gapped.

(b) d-wave superconductor:

Now, we consider a d-wave superconductor with $\Delta_{\mathbf{k}} = \Delta(k_x^2 - k_y^2)$, and $\varepsilon_{\mathbf{k}} = \frac{k^2}{2m}$. Find the minimum value of $\Delta_{\mathrm{sc}}(\mathbf{k}) = E_{+}(\mathbf{k}) - E_{-}(\mathbf{k})$. Prove that the minimum value is reached when two conditions are satisfied:

1. At the Fermi surface, $\varepsilon_{\mathbf{k}} = \mu$.
2. Along the diagonal directions, $k_x = \pm k_y$.

Show that there are four $\mathbf{k}$-points at which $\Delta_{\mathbf{k}}$ reaches its minimum value. These points are known as nodes or nodal points.

Expand $E_{+}(\mathbf{k})$ and $E_{-}(\mathbf{k})$ near one of nodal points and show that the dispersion is linear near this point (i.e., this is a Dirac point).

21.6 BCS variational method to superconductivity:

Consider the interacting Hamiltonian

$$\mathcal{H} = \mathcal{H}_0 + \mathcal{H}_1$$

$$\mathcal{H}_0 = \sum_{\mathbf{k}\sigma} \xi_{\mathbf{k}} \, c_{\mathbf{k}\sigma}^{\dagger} \, c_{\mathbf{k}\sigma}$$

$$\mathcal{H}_1 = \sum_{\mathbf{k}\mathbf{k}'} V_{\mathbf{k}\mathbf{k}'} \, c_{+\mathbf{k}\uparrow}^{\dagger} \, c_{-\mathbf{k}\downarrow}^{\dagger} \, c_{-\mathbf{k}'\downarrow} \, c_{+\mathbf{k}'\uparrow}$$

$$\xi_{\mathbf{k}} = \varepsilon_{\mathbf{k}} - \mu.$$

The interaction matrix elements obey $V_{\mathbf{k}\mathbf{k}'} = V_{\mathbf{k}'\mathbf{k}}^{*}$. The BCS variational wavefunction is defined by

$$|\Psi\rangle = \prod_{\mathbf{k}} \left(u_{\mathbf{k}} + v_{\mathbf{k}} \, e^{i\varphi} \, c_{+\mathbf{k}\uparrow}^{\dagger} \, c_{-\mathbf{k}\downarrow}^{\dagger} \right) |\varnothing\rangle$$

in terms of the electron creation and annihilation operators. This BCS wavefunction can be thought of as a coherent state for Cooper pairs with $v_{\mathbf{k}}$ and $u_{\mathbf{k}}$ being the amplitudes to have or not to have, a Cooper pair with relative momentum $\mathbf{k}$, respectively. The c-numbers $v_{\mathbf{k}}$ and $u_{\mathbf{k}}$ obey the normalization conditions

$$v_{\mathbf{k}}^2 + u_{\mathbf{k}}^2 = 1.$$

$\varphi \in [0, 2\pi]$ is a global phase.

(a) Express the expectation value in the variational state $|\Psi\rangle$ of the kinetic energy $\langle\Psi|\mathcal{H}_0|\Psi\rangle$ terms of the parameters $u_{\mathbf{k}}$, $v_{\mathbf{k}}$, and φ.

(b) Express the expectation value in the variational state $|\Psi\rangle$ of the interacting energy $\langle\Psi|\mathcal{H}_1|\Psi\rangle$ terms of the parameters $u_{\mathbf{k}}$, $v_{\mathbf{k}}$, and φ.

(c) Does $\langle\Psi|\mathcal{H}|\Psi\rangle$ depend on the global phase φ?

From now on, we assume that the matrix elements of the interaction potential take the reduced form

$$V_{\mathbf{k}\mathbf{k}'} = -V.$$

Define the complex-valued parameter

$$\Delta := V \sum_{\mathbf{k}} u_{\mathbf{k}} v_{\mathbf{k}}.$$

(d) Express $\langle\Psi|\mathcal{H}|\Psi\rangle$ in terms of Δ and $v_{\mathbf{k}}$ only.

(e) Minimalize $\langle\Psi|\mathcal{H}|\Psi\rangle$ with respect to $v_{\mathbf{k}}$ and show that

$$\left\{ \begin{aligned} u_{\mathbf{k}}^2 &= \frac{1}{2}\left(1 + \frac{\xi_{\mathbf{k}}}{E_{\mathbf{k}}}\right) \\ v_{\mathbf{k}}^2 &= \frac{1}{2}\left(1 - \frac{\xi_{\mathbf{k}}}{E_{\mathbf{k}}}\right) \end{aligned} \right\} \qquad E_{\mathbf{k}} = \sqrt{\xi_{\mathbf{k}}^2 + \Delta^2}.$$

(f) Express $\langle\Psi|\mathcal{H}|\Psi\rangle$ and Δ in terms of $\xi_{\mathbf{k}}$ and $E_{\mathbf{k}}$.

(g) Consider the subspaces for $\mathbf{k}|$, $\uparrow$ and $-\mathbf{k}$ $\downarrow$. Show that the states

$$\left(u_{\mathbf{k}} + v_{\mathbf{k}} c_{+\mathbf{k}\uparrow}^\dagger c_{-\mathbf{k}\downarrow}^\dagger\right)|\varnothing\rangle, \quad c_{\mathbf{k}\uparrow}^\dagger|\varnothing\rangle, \quad c_{-\mathbf{k}\downarrow}^\dagger|\varnothing\rangle, \quad \left(u_{\mathbf{k}} + v_{\mathbf{k}} c_{+\mathbf{k}\uparrow}^\dagger c_{-\mathbf{k}\downarrow}^\dagger\right)|\varnothing\rangle$$

are orthogonal to each other and normalized to one.

(h) Consider the state $|2, \mathbf{k}\rangle$, which is defined as

$$|2, \mathbf{k}\rangle = \left(u_{\mathbf{k}}^* + v_{\mathbf{k}}^* c_{+\mathbf{k}\uparrow}^\dagger c_{-\mathbf{k}\downarrow}^\dagger\right) \prod_{\mathbf{k}'}{}' \left(u_{\mathbf{k}'} + v_{\mathbf{k}'} c_{+\mathbf{k}'\uparrow}^\dagger c_{-\mathbf{k}'\downarrow}^\dagger\right)|\varnothing\rangle.$$

Show that

$$\langle 2, \mathbf{k}|\mathcal{H}|2, \mathbf{k}\rangle - \langle\Psi|\mathcal{H}|\Psi\rangle \simeq 2E_{\mathbf{k}}.$$

21.7 Anomalous Green function and coherence length:

(a) In superconductors, there is a characteristic length scale ξ called the coherence length. One of its possible definitions is the extent of the pair correlations. Consider the anomalous correlations in real space:

$$\mathcal{F}(\mathbf{x} - \mathbf{y}) = \langle \psi_\downarrow(x)\psi_\uparrow(y) \rangle.$$

It decays at a certain length scale ξ. Calculate this length in the BCS ground state.

Hint 1: in the BCS ground state, different wavevectors k are decoupled, so it is convenient to do a calculation at a given k vector, and then Fourier-transform.

Hint 2: only a vicinity of the Fermi surface contributes to this anomalous correlator, so you may linearize the electron spectrum near the Fermi surface.

Hint 3: you will find $\xi = v_F/\Delta$.

(b) For aluminum, find in the literature the value of the gap Δ and estimate the superconducting coherence length ξ.

22

Quantum Theory of Magnetism: Exchange Coupling Mechanisms

22.1 Introduction

One of the profound and insightful surprises in physics is that magnetism is an inherently quantum mechanical effect. In that sense, the title of "quantum magnetism" may be as redundant. In the classical viewpoint, magnetic moments arise from electric currents. As we know from classical electromagnetism, a current density $\mathbf{j}(\mathbf{x})$ produces a magnetic moment

$$\boldsymbol{\mu} = \frac{1}{2} \int d\mathbf{x} \ (\mathbf{x} \times \mathbf{j})$$

and the moments interact via the dipole–dipole interaction. However, the Bohr–van Leeuwen theorem[1] showed that this cannot be the origin of the magnetism found in magnetic materials. In a classical system, charges cannot flow in thermodynamic equilibrium, and hence there are no magnetic moments at the outset. The Bohr–van Leeuwen theorem was discovered by Niels Bohr in 1911 in his doctoral dissertation and was later rediscovered by Hendrika Johanna van Leeuwen in her doctoral thesis in 1919.

In contrast, quantum mechanics allows nonvanishing charge currents in the ground state: the current density of an electron in state $\psi(\mathbf{x})$ is given by

$$\mathbf{j}(\mathbf{x}) = -\frac{e\hbar}{2im_e} \left[\psi^*(\mathbf{x}) \, \nabla \psi(\mathbf{x}) - \left(\nabla \psi^*(\mathbf{x})\right) \, \psi(\mathbf{x}) \right],$$

which can be nonvanishing for a complex wavefunction $\psi(\mathbf{x})$. For a state of angular momentum $\mathbf{L}$, the electron has a magnetic moment

$$\boldsymbol{\mu} = -\frac{e\hbar}{2m_e} \, \langle \mathbf{L} \rangle = \mu_B \, \langle \mathbf{L} \rangle \,.$$

Also the electron carries spin $\mathbf{S}$, which produces a magnetic moment

$$\boldsymbol{\mu}_S = -g_e \, \mu_B \, \langle \mathbf{S} \rangle \,,$$

where $g_e = 2$ is the gyromagnetic ratio for an electron.

---

[1] In 1919, van Leeuwen demonstrated that the classical Boltzmann statistics applied rigorously to any dynamical system must lead to a zero susceptibility. John van Vleck stated the Bohr–van Leeuwen theorem as "At any finite temperature, and in all finite applied electric or magnetic fields, the net magnetization of a collection of electrons in thermal equilibrium vanishes identically." [184].

Atomic moments are thus of the scale of μ_B. This leads to dipolar interaction energy of the order of 0.05 meV for two such moments, a distance of 1Å apart, which corresponds to a temperature < 1 K. However, we know that magnetic ordering persists at much higher temperatures.[2] Consequently, we surmise that such ordering must be derived from interactions other than the dipolar type. We now know that manifest magnetic properties arise from the combined interplay of the Pauli principle, the Coulomb repulsion (Coulomb exchange), and electron hopping (kinetic exchange) – electronic properties that, naively, seem to be unrelated to magnetism. The interplay gives rise to effective couplings between magnetic moments in solids. This makes magnetism in solids solely a quantum-mechanical effect and means that classical physics cannot account for diamagnetism, paramagnetism, ferromagnetism, or any other magnetic-ism! In this chapter, we shall introduce and develop the mechanisms of exchange interactions between paramagnetic ions that lead to spin-dependent coupling of their magnetic moments. Itinerant electron magnetism will be covered in Chapter 23. First, it is appropriate to mention that the basic concept of the quantum-mechanical exchange interaction was developed by Heisenberg (1928) [90] and Dirac (1929) [51]. The introduction of such concepts was the basis of subsequent theoretical work on ferromagnetism, ferrimagnetism, and antiferromagnetism: Van Vleck (1937) [185] and Néel (1932), and [140], (1948) [142].

22.2 Heisenberg/Dirac Exchange Hamiltonian

22.2.1 Heitler–London Model: Two-Center, Two-Electron System

We consider a system of two identical paramagnetic ions having one unpaired electron each, in addition to the closed-shell core whose role is neglected. The ions are assumed fixed, with interatomic spacing R_{ab}. The Hamiltonian of the system can be expressed as

$$\mathcal{H} = \mathcal{H}_a + \mathcal{H}_b + \mathcal{H}_{ab} \tag{22.1}$$

$$\mathcal{H}_a = \frac{p_1^2}{2m} - \frac{Ze^2}{|\mathbf{r}_1 - \mathbf{r}_a|}, \quad \mathcal{H}_b = \frac{p_2^2}{2m} - \frac{Ze^2}{|\mathbf{r}_2 - \mathbf{r}_b|}.$$

The configuration is illustrated in Figure 22.1. $\mathcal{H}_a$, $\mathcal{H}_b$ are the Hamiltonians of electrons 1, 2 in the field of ions core a, b, respectively. The atomic orbitals that are solutions of the corresponding Schrödinger equation are denoted by ϕ_a and ϕ_b:

$$\mathcal{H}_a \phi_a(\mathbf{x}_1) = E_a \phi_a(\mathbf{x}_1), \quad \mathcal{H}_b \phi_b(\mathbf{x}_2) = E_b \phi_b(\mathbf{x}_2). \tag{22.2}$$

$\mathcal{H}_{ab}$ is the interaction Hamiltonian and has the form

$$\mathcal{H}_{ab} = \frac{Z^2 e^2}{R_{ab}} - \frac{Ze^2}{|\mathbf{r}_1 - \mathbf{r}_b|} - \frac{Ze^2}{|\mathbf{r}_2 - \mathbf{r}_a|} + \frac{e^2}{|\mathbf{r}_1 - \mathbf{r}_2|}. \tag{22.3}$$

---

[2] For a system like magnetite (Fe_3O_4), magnetic order persists until about 860°K.

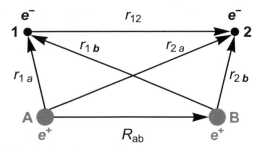

Figure 22.1 Two-center, two-electron system.

Here $Z^2 e^2 / R_{ab}$ is the mutual repulsion of the two ion cores, each with a charge Ze. The second and third terms in (22.3) represent the attractive potentials between electron 1 and ion b and electron 2 and ion a, respectively; e^2 / r_{12} is the mutual Coulomb repulsion between the two electrons. Hamiltonian (22.1) has no explicit dependence on the electron spin variables. In general, spin dependence emerges when we invoke the exclusion principle via the antisymmetrization of the many-electron state function, here, the two electrons in question. Defining the one-electron spin orbitals as

$$\psi_i(\xi) \equiv \psi_i(\mathbf{x}, \sigma) = \phi_i(\mathbf{x}) \, \chi(\sigma), \tag{22.4}$$

we write the eigenstates of the S^2 and S^z operators for the two electron system as the triplets

$$^3|1,1\rangle_{ab} = \frac{[\psi_a \psi_b]}{(1 - S_{ab}^2)^{1/2}}, \quad ^3|1,0\rangle_{ab} = \frac{[\psi_a \bar{\psi}_b] + [\bar{\psi}_a \psi_b]}{\{2(1 - S_{ab}^2)\}^{1/2}}$$

$$^3|1,-1\rangle_{ab} = \frac{[\bar{\psi}_a \bar{\psi}_b]}{(1 - S_{ab}^2)^{1/2}}. \tag{22.5}$$

and the singlets

$$^1|0,0\rangle_{ab} = \frac{[\psi_a \bar{\psi}_b] - [\bar{\psi}_a \psi_b]}{\{2(1 + S_{ab}^2)\}^{1/2}} \tag{22.6}$$

$$\begin{cases} ^1|0,0\rangle_{aa} = [\psi_a \bar{\psi}_a] \\ ^1|0,0\rangle_{bb} = [\psi_b \bar{\psi}_b]. \end{cases} \tag{22.7}$$

The determinantal states are indicated by square brackets, namely,

$$[\psi_a \psi_b] = \frac{1}{\sqrt{2}} \begin{vmatrix} \phi_{a\uparrow}(1) & \phi_{b\uparrow}(1) \\ \phi_{a\uparrow}(2) & \phi_{b\uparrow}(2) \end{vmatrix}, \quad [\psi_a \bar{\psi}_b] = \frac{1}{\sqrt{2}} \begin{vmatrix} \phi_{a\uparrow}(1) & \phi_{b\downarrow}(1) \\ \phi_{a\uparrow}(2) & \phi_{b\downarrow}(2) \end{vmatrix}.$$

The overlap integral for the orbitals $\phi_a(\mathbf{r})$ and $\phi_b(\mathbf{r})$ is

$$S_{ab} = \langle \phi_a(\mathbf{x}) | \phi_b(\mathbf{x}) \rangle = \int d\mathbf{x} \, \phi_a^*(\mathbf{x}) \, \phi_b(\mathbf{x}). \tag{22.8}$$

The triplet states in (22.5) and the singlet state in (22.6) correspond to the states considered by Heitler and London for the hydrogen molecule H-H. The singlet states in (22.7) represent ionic configurations such as H^+H^- and H^-H^+ in which one hydrogen has two electrons and the other is a bare proton. The singlet configuration (22.6) – the ground state in the hydrogen molecule – will interact with the excited-state configurations (22.7), giving rise to a second-order correction to its energy.[3]

We shall use ^3E to denote the energy of the triplet states, and ^1E to denote that of the singlet state including the *second-order corrections due to the interaction with excited states*. We express these energies in the compact form

$$^{(2S+1)}E = K - \{S(S+1) - 1\}J, \tag{22.9}$$

where

$$\begin{cases} K \equiv \dfrac{^1E + {}^3E}{2} = \left\langle \phi_a(\mathbf{x}_1)\,\phi_b(\mathbf{x}_2) \left| \dfrac{e^2}{r_{12}} \right| \phi_a(\mathbf{x}_1)\,\phi_b(\mathbf{x}_2) \right\rangle & \text{Coulomb energy} \\[4mm] J \equiv \dfrac{^1E - {}^3E}{2} = \left\langle \phi_a(\mathbf{x}_1)\,\phi_b(\mathbf{x}_2) \left| \dfrac{e^2}{r_{12}} \right| \phi_a(\mathbf{x}_2)\,\phi_b(\mathbf{x}_1) \right\rangle & \text{Exchange energy.} \end{cases}$$

We now use the relation

$$\mathbf{S} \cdot \mathbf{S} = s_1^2 + s_2^2 + 2\mathbf{s}_1 \cdot \mathbf{s}_2 = \frac{3}{2} + 2\mathbf{s}_1 \cdot \mathbf{s}_2$$

to write

$$S(S+1) - 1 = \frac{1}{2}\,(1 + 4\mathbf{s}_1 \cdot \mathbf{s}_2), \tag{22.10}$$

which has the value 1 for the triplet state and -1 for the singlet. The relation in (22.9) can then be recast as

$$^{(2S+1)}E = K - \frac{1}{2}\,(1 + 4\mathbf{s}_1 \cdot \mathbf{s}_2)\,J. \tag{22.11}$$

Examining the definition of the exchange energy

$$J = \frac{1}{2}\,(^1E - {}^3E),$$

we find that, for J positive, the triplet state has lower energy, and parallel spin alignment (ferromagnetism) is energetically favored. Conversely, for negative J, the singlet state is more stable, and antiferromagnetic ordering is favored.

We may express J in terms of explicit forms of ^1E and ^3E as[4]

---

[3] We note that such corrections do not exist for the triplet states within the two-orbital manifold ϕ_a and ϕ_b, since excited states having two electrons with parallel spins either in ϕ_a or ϕ_b are precluded by the exclusion principle. Moreover, since the Hamiltonian is spin independent, states with different S quantum number cannot interact.

[4] The derivation is left as an exercise.

$$J = \frac{1}{1 - S_{ab}^4} \left\{ \left(\left\langle ab \left| \frac{e^2}{r_{12}} \right| ba \right\rangle + S_{ab} \left\langle a \left| \frac{-Ze^2}{r_{2a}} \right| b \right\rangle + S_{ab} \left\langle b \left| \frac{-Ze^2}{r_{1b}} \right| a \right\rangle \right) \right.$$

$$\left. - S_{ab}^2 \left(\left\langle ab \left| \frac{e^2}{r_{12}} \right| ab \right\rangle + \left\langle b \left| \frac{-Ze^2}{r_{1a}} \right| b \right\rangle + \left\langle a \left| \frac{-Ze^2}{r_{2b}} \right| a \right\rangle \right) \right\}$$

$$- \frac{1}{1 + S_{ab}^2} \left\{ \frac{\left| \langle a | -Ze^2/r_{1a} | b \rangle \right|^2}{\Delta E(b \to a)} + \frac{\left| \langle b | -Ze^2/r_{2b} | a \rangle \right|^2}{\Delta E(a \to b)} \right\}. \tag{22.12}$$

The constant inter-ion core term ($Z^2 e^2 / R_{ab}$) is omitted.

We note that the last two terms in (22.12) arise from the interaction of the ground singlet with the two excited singlets representing the ionic configurations. The energy corrections appear in second order. $\Delta E(a \to b)$ in the denominator actually represents the repulsive energy when two electrons are in the same orbital state. The explicit forms of the quantities occurring in (22.12) are as follows:

$$\langle a | V(\mathbf{x}) | b \rangle = \int d\mathbf{x} \, \phi_a^*(\mathbf{x}) \, V(\mathbf{x}) \, \phi_b(\mathbf{x})$$

is the matrix element of the one-electron operator $V(\mathbf{r})$ connecting the orbitals ϕ_a and ϕ_b, and

$$\left\langle ab \left| \frac{e^2}{r_{12}} \right| cd \right\rangle \equiv \int d\mathbf{x}_1 d\mathbf{x}_2 \, \phi_a^*(\mathbf{x}_1) \, \phi_b^*(\mathbf{x}_2) \left(\frac{e^2}{r_{12}} \right) \phi_c(\mathbf{x}_1) \, \phi_d(\mathbf{x}_2).$$

Expression (22.12) is a more exact representation of the exchange energy than that considered by Heisenberg (1928). If we neglect quadratic and higher powers of the overlap integral S_{ab}, we obtain a simpler form of the Heisenberg exchange energy that includes electron transfer effects

$$J_H = \left\langle ab \left| \frac{e^2}{r_{12}} \right| ba \right\rangle - 2 \left(S_{ab} + \Gamma_{ab} \right) \langle b | V | a \rangle, \tag{22.13}$$

where

$$\begin{cases} \Gamma_{ab} = \dfrac{\langle a | V | b \rangle}{\Delta E(a \to b)} \\ V = \dfrac{Ze^2}{r_{1b}}. \end{cases} \tag{22.14}$$

The spin-dependent part of the energy (22.11) can thus be expressed as

$$\mathcal{H}(\text{Heisenberg}) = -2 J_H \, s_1 \cdot s_2 \tag{22.15}$$

with J_H given by (22.13).

The exchange integral $\langle ab | e^2/r_{12} | ba \rangle$ is always positive definite

$$
\begin{aligned}
\left\langle ab \left| \frac{e^2}{r_{12}} \right| ba \right\rangle &= \int d\mathbf{x}_1 d\mathbf{x}_2 \, \phi_a^*(\mathbf{x}_1) \phi_b^*(\mathbf{x}_2) \left(\frac{e^2}{r_{12}} \right) \phi_b(\mathbf{x}_1) \phi_a(\mathbf{x}_2) \\
&= \frac{1}{\Omega} \sum_{\mathbf{q}} \frac{4\pi e^2}{q^2} \int d\mathbf{x}_1 d\mathbf{x}_2 \, \phi_a^*(\mathbf{x}_1) \phi_b^*(\mathbf{x}_2) \, e^{i\mathbf{q}\cdot(\mathbf{r}_1-\mathbf{r}_2)} \, \phi_b(\mathbf{x}_1) \phi_a(\mathbf{x}_2) \\
&= \frac{4\pi e^2}{\Omega} \sum_{\mathbf{q}} \frac{1}{q^2} \int d\mathbf{x}_1 \phi_a^*(\mathbf{x}_1) \phi_b(\mathbf{x}_1) e^{i\mathbf{q}\cdot\mathbf{r}_1} \int d\mathbf{x}_2 \phi_b^*(\mathbf{x}_2) \phi_a(\mathbf{x}_2) e^{-i\mathbf{q}\cdot\mathbf{r}_2} \\
&= \frac{4\pi e^2}{\Omega} \sum_{\mathbf{q}} \frac{1}{q^2} \left| \int d\mathbf{x} \, \phi_a^*(\mathbf{x}) \, \phi_b(\mathbf{x}) \, e^{i\mathbf{q}\cdot\mathbf{r}} \right|^2 > 0.
\end{aligned}
$$

It can be considered as the self-energy of the complex overlap charge $e\phi_a^*(\mathbf{x}_1)\phi_b(\mathbf{x}_1)$. If this term dominates, then J_H would be positive, and the ferromagnetic (triplet) state would be favored. If, however, the second term, containing the overlap integral S_{ab} and the transfer amplitude Γ_{ab}, dominates, J_H would be negative and the antiferromagnetic (singlet) state is favored.

22.2.2 Exchange Hamiltonian for N Localized Spin

We now develop the formulation of the exchange Hamiltonian for a system consisting of N magnetic ions, each having one localized unpaired electron in addition to its core electrons. We shall assume that the core electrons are not involved in the interaction process. The Hamiltonian of the system is given by

$$
\mathcal{H} = \sum_i \frac{p_i^2}{2m} + \sum_{n,l} V(\mathbf{x}_i - \mathbf{R}_n) + \sum_{i<j} \frac{e^2}{r_{ij}}, \tag{22.16}
$$

where the subscript i index the ith electron, and $V(\mathbf{r_i} - \mathbf{R_n})$ is its potential energy operator in the field of the nth ion core. e^2/r_{ij} is the two-body Coulomb interaction between electrons i and j.

Hamiltonian (22.16) can be expressed in terms fermionic creation and annihilation operators as

$$
\mathcal{H} = \sum_{lm\sigma} V_{lm} c_{l\sigma}^\dagger c_{m\sigma} + \frac{1}{2} \sum_{\substack{jlmn \\ \sigma\sigma'}} U_{mn}^{jl} c_{l\sigma'}^\dagger c_{j\sigma}^\dagger c_{m\sigma} c_{n\sigma'}, \tag{22.17}
$$

where V_{lm} is the matrix element of the one-body Hamiltonian connecting the orbital states ϕ_l and ϕ_m, namely,

$$
V_{lm} = \left\langle \phi_l \left| \frac{p^2}{2m} + V(\mathbf{x}) \right| \phi_m \right\rangle \tag{22.18}
$$

and

$$U_{mn}^{jl} = \left\langle \phi_j(\mathbf{x}_1)\phi_l(\mathbf{x}_2) \left| \frac{e^2}{r_{12}} \right| \phi_m(\mathbf{x}_1)\phi_n(\mathbf{x}_2) \right\rangle. \tag{22.19}$$

In this formulation, it is assumed that all unoccupied states are energetically far removed from the singly occupied states. We simplify the analysis further by assuming that orbitals ϕ_l and ϕ_m are orthogonal with an overlap parameter $S_{lm} = 0$.

Connect spin operators to fermionic creation and annihilation operators as follows:

$$c_{i\uparrow}^\dagger c_{i\uparrow} + c_{i\downarrow}^\dagger c_{i\downarrow} \equiv 1 \equiv N_{i\uparrow} + N_{i\downarrow} \quad \text{(singly occupied orbitals)}$$

$$c_{i\uparrow}^\dagger c_{i\uparrow} - c_{i\downarrow}^\dagger c_{i\downarrow} = 2s_i^z$$

$$c_{i\uparrow}^\dagger c_{i\downarrow} = s_i^+ \equiv s_i^x + is_i^y$$

$$c_{i\downarrow}^\dagger c_{i\uparrow} = s_i^- \equiv s_i^x - is_i^y. \tag{22.20}$$

Potential and Kinetic Exchange

For the sake of pedagogy, we separate the Hamiltonian (22.17) into diagonal and off-diagonal terms

$$\mathcal{H} = \mathcal{H}_0 + \mathcal{H}_{ex} + \mathcal{H}_{corr} + \mathcal{H}_{tr} \tag{22.21}$$

with

$$\mathcal{H}_0 = \sum_{l\sigma} \varepsilon_l\, c_{l\sigma}^\dagger c_{l\sigma} + \frac{1}{2} \sum_{\sigma\sigma'} K_{lm}\, c_{l\sigma}^\dagger c_{l\sigma}\, c_{m\sigma'}^\dagger c_{m\sigma'} \tag{22.22}$$

$$K_{lm} = \left\langle \phi_l(\mathbf{x}_1)\phi_m(\mathbf{x}_2) \left| \frac{e^2}{r_{12}} \right| \phi_l(\mathbf{x}_1)\phi_m(\mathbf{x}_2) \right\rangle,$$

where ε_l is the one-electron orbital energy. $\mathcal{H}_{ex}$ is the spin-dependent exchange term, called *potential exchange*, expressed as

$$\mathcal{H}_{ex} = \frac{1}{2} \sum_{\substack{l\neq m \\ \sigma\sigma'}} J_{lm}\, c_{l\sigma'}^\dagger c_{m\sigma}^\dagger c_{l\sigma} c_{m\sigma'} \tag{22.23}$$

$$J_{lm} = \left\langle \phi_l(\mathbf{x}_1)\phi_m(\mathbf{x}_2) \left| \frac{e^2}{r_{12}} \right| \phi_m(\mathbf{x}_1)\phi_l(\mathbf{x}_2) \right\rangle.$$

The summation over the spin variables σ

$$\sum_{\sigma\sigma'} c_{l\sigma'}^\dagger c_{m\sigma}^\dagger c_{l\sigma} c_{m\sigma'} = c_{l\uparrow}^\dagger c_{m\uparrow}^\dagger c_{l\uparrow} c_{m\uparrow} + c_{l\downarrow}^\dagger c_{m\downarrow}^\dagger c_{l\downarrow} c_{m\downarrow}$$

$$+ c_{l\uparrow}^\dagger c_{m\downarrow}^\dagger c_{l\downarrow} c_{m\uparrow} + c_{l\downarrow}^\dagger c_{m\uparrow}^\dagger c_{l\uparrow} c_{m\downarrow} \tag{22.24}$$

can be expressed in terms of the spin operators using the relations (22.20) as

$$-\left(\frac{1}{2} + 2\mathbf{s_l} \cdot \mathbf{s_m}\right), \tag{22.25}$$

which allows us to rewrite the expression for $\mathcal{H}_{ex}$ as

$$\mathcal{H}_{ex} = -\frac{1}{4} \sum_{l \neq m} J_{lm} \left(1 + 4\,\mathbf{s_l} \cdot \mathbf{s_m}\right). \tag{22.26}$$

To this point, we have considered terms of (22.21) in which one-electron transfer processes were not included. We now take into account the other two terms of (22.21), namely

$$\mathcal{H}_{corr} = \sum_{m\sigma} U_{mm}\, c^\dagger_{m\sigma} c_{m\sigma} c^\dagger_{m\bar\sigma} c_{m\bar\sigma}, \tag{22.27}$$

where U_{mm} is the Coulomb repulsion between two electrons with antiparallel spins residing in the same orbital ϕ_m. Since the singly occupied orbitals on different ions are considered equivalent, we write $U_{mm} = U$. The last term

$$\mathcal{H}_{tr} = \sum_{lm\sigma} V_{lm}\, c^\dagger_{l\sigma} c_{m\sigma} \tag{22.28}$$

involves electron transfer from ion m (in orbital ϕ_m) to ion l (in orbital ϕ_l).

We write the Hamiltonian as

$$\mathcal{H} = \mathcal{H}_d + \mathcal{H}_{tr},$$

where $\mathcal{H}_d$ does not contain off-diagonal electron transfer terms, and is given by

$$\mathcal{H}_d = \sum_{l\sigma} \mathcal{E}_l\, c^\dagger_{l\sigma} c_{l\sigma} + U \sum_{m\sigma} c^\dagger_{m\sigma} c_{m\sigma} c^\dagger_{m\bar\sigma} c_{m\bar\sigma} \tag{22.29}$$

with $\mathcal{E}_l$ the spin-independent Hartree–Fock single-particle energy

$$\mathcal{E}_l = \varepsilon_l + \frac{1}{2} \sum_{m\sigma} \left(K_{lm} - \frac{1}{2} J_{lm}\right) \langle n_{m\sigma} \rangle.$$

We apply the perturbative canonical transformation

$$\mathcal{H}_T = e^{-i\mathcal{S}}\, \mathcal{H}\, e^{i\mathcal{S}}$$

$$= \mathcal{H} + i[\mathcal{H}, \mathcal{S}] + \frac{i^2}{2}\left[[\mathcal{H}, \mathcal{S}], \mathcal{S}\right] + \cdots$$

$$\equiv \mathcal{H}_d + \mathcal{H}_{tr} + i[\mathcal{H}_d, \mathcal{S}] + i[\mathcal{H}_{tr}, \mathcal{S}] + \frac{i^2}{2}\left[[\mathcal{H}, \mathcal{S}], \mathcal{S}\right] + \cdots \tag{22.30}$$

in order to eliminate the off-diagonal transfer terms in (22.28) in first order. The Hermitian generator of the canonical transformation, $\mathcal{S}$, is obtained from the condition

$$\mathcal{H}_{tr} + i[\mathcal{H}_d, \mathcal{S}] = 0. \tag{22.31}$$

We express $\mathcal{S}$ as

$$\mathcal{S} = \sum_{lm\sigma} A_{lm} \, c_{l\sigma}^{\dagger} c_{m\bar{\sigma}}^{\dagger} c_{m\sigma} c_{l\bar{\sigma}} + \text{h.c.,} \tag{22.32}$$

where h.c. denotes the Hermitian conjugate, and the A_{lm} are coefficients to be determined with the aid of (22.31). After some algebraic manipulations, making use of the fermion commutation relations, we obtain

$$A_{lm} = \frac{i V_{lm}}{\mathcal{E}_l + U - \mathcal{E}_m}. \tag{22.33}$$

Since all magnetic ions are identical, we have $\mathcal{E}_l = \mathcal{E}_m$, and we obtain, to second order in the perturbation from the fourth and fifth terms of (22.30), the spin-dependent interaction term

$$\mathcal{H}_{\text{KE}} = \sum_{l \neq m, \sigma} \frac{V_{lm}^2}{U} \, c_{l\sigma}^{\dagger} c_{m\bar{\sigma}}^{\dagger} c_{l\bar{\sigma}} c_{m,\sigma}. \tag{22.34}$$

Carrying out the summation over the spin variables and making use of (22.20), $\mathcal{H}_{\text{KE}}$ becomes

$$\mathcal{H}_{\text{KE}} = -\sum_{l \neq m} \frac{V_{lm}^2}{U} \frac{1}{2} \left(1 - 4\mathbf{s}_l \cdot \mathbf{s}_m \right)$$

$$= \text{constant} + 2 \sum_{l \neq m} \frac{V_{lm}^2}{U} \, \mathbf{s}_l \cdot \mathbf{s}_m, \tag{22.35}$$

which is identified as the *kinetic exchange* term (see Figure 22.2). It clearly *stabilizes the antiferromagnetic (AFM) state*. We note that the hopping V_{lm} is a one-body process and is allowed by the Pauli principle only when the spins at neighboring sites are antiparallel. This is also reflected in the presence of the projection operator $\frac{1}{2}(1 - 4\mathbf{s}_l \cdot \mathbf{s}_m)$ in (22.35), which annihilates the triplet state for which $\mathbf{s}_l \cdot \mathbf{s}_m = \frac{1}{4}$.

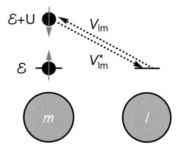

Figure 22.2 Kinetic exchange involves a second-order virtual process, whereby an electron hops to doubly occupy it neighboring orbital.

Combining (22.26) and (22.35), we obtain the generalized effective Hamiltonian

$$\mathcal{H}_{H\text{-}D} = C - \sum_{l \neq m} \left(J_{lm} - \frac{2|V_{lm}|^2}{U} \right) \mathbf{s}_l \cdot \mathbf{s}_m, \tag{22.36}$$

which we consider as stemming from the work of Heisenberg and Dirac.

While the *kinetic exchange* term always supports the AFM state, the *potential exchange*, coupling the spins of two different ions is positive definite, and hence favors ferromagnetic coupling, as we have seen earlier. The overall interaction depends on the difference between these two contributions.

22.3 Indirect Exchange Mechanisms

The exchange processes presented in the preceding section involved localized electrons and gave rise to ionic magnetic moments. However, a different scenario may arise where the magnetic nature of a system is associated with itinerant conduction electrons or with electrons of diamagnetic ions. These will lead to indirect exchange processes via polarization of the electron system.

We shall focus here on indirect exchange mechanism arising from s–d or s–f interactions, shown in Figure 22.3, and ignore possible direct exchange effects that we have treated in the previous section.

We consider the Hamiltonian

$$\mathcal{H} = \mathcal{H}_s + \mathcal{H}_d + \mathcal{H}_{ex}(s\text{-}d), \tag{22.37}$$

where the conduction electrons are represented by

$$\mathcal{H}_s = \sum_{k\sigma} \varepsilon_{\mathbf{k}} \, c_{\mathbf{k}\sigma}^\dagger c_{\mathbf{k}\sigma}. \tag{22.38}$$

$\varepsilon_{\mathbf{k}}$ is the single-particle unperturbed energy of an electron in the conduction state

$$|\mathbf{k}\sigma\rangle = \phi_{\mathbf{k}}(\mathbf{x}) \, \chi(\sigma) = |\mathbf{k}\rangle = \frac{1}{\sqrt{\Omega}} \, \exp(i\mathbf{k} \cdot \mathbf{x}) \, u_{\mathbf{k}}(\mathbf{x}) \, \chi(\sigma),$$

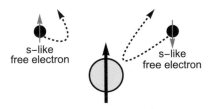

Paramagnetic metal ion
with 3d or 4f electrons

Figure 22.3 Scattering of a free electron by a paramagnetic ion. Electrons with spin parallel to the local spin of the ion scatter differently from those with antiparallel spin.

where Ω is the volume. The localized d (or f) Hamiltonian is given by

$$\mathcal{H}_d = \sum_{l\sigma} \mathcal{E}_l \, c_{l\sigma}^\dagger c_{l\sigma}, \tag{22.39}$$

where $\mathcal{E}_l$ is the one-electron energy of a nondegenerate orbital ϕ_l at site $\mathbf{R}_l$; $c_{l\sigma}^\dagger$, $c_{l\sigma}$ are the corresponding electron creation and annihilation operators.

The s–d exchange interaction Hamiltonian is modeled as

$$\mathcal{H}_{ex}(s\text{-}d) = -\frac{1}{N} \sum_{\mathbf{k},\mathbf{k}',l} J\big(|\mathbf{k} - \mathbf{k}'|\big) \, \exp\{i(\mathbf{k} - \mathbf{k}') \cdot \mathbf{R}_l\}$$

$$\times \left\{ \big(c_{\mathbf{k}'\uparrow}^\dagger c_{\mathbf{k}\uparrow} + c_{\mathbf{k}'\downarrow}^\dagger c_{\mathbf{k}\downarrow}\big) S_l^z + c_{\mathbf{k}'\uparrow}^\dagger c_{\mathbf{k}\downarrow} \, S_l^- + c_{\mathbf{k}'\downarrow}^\dagger c_{\mathbf{k}\uparrow} \, S_l^+ \right\}, \tag{22.40}$$

where N is the number of lattice sites. The localized d/f electrons are represented by spin operators, obtained after summation over fermionic operators using (22.20). The effective exchange coefficient is given by

$$J\big(|\mathbf{k} - \mathbf{k}'|\big) = N \left\langle \phi_{\mathbf{k}'}(\mathbf{x}_1)\phi_l(\mathbf{x}_2 - \mathbf{R}_l) \left| \exp\{-i(\mathbf{k} - \mathbf{k}') \cdot \mathbf{R}_l\} \frac{e^2}{r_{12}} \right| \phi_l(\mathbf{x}_1 - \mathbf{R}_l)\phi_{\mathbf{k}}(\mathbf{x}_2) \right\rangle$$

$$\tag{22.41}$$

For simplicity, we assumed that the exchange integral depends only on $|\mathbf{k} - \mathbf{k}'|$, which is true for spherically symmetric orbitals.

22.3.1 Mean-Field Spin-Dependent Energy Shift

Next we include the diagonal part of $\mathcal{H}_{ex}(s\text{-}d)$ $(\mathbf{k} = \mathbf{k}')$ in $\mathcal{H}_s$, and write

$$\mathcal{H}_s = \sum_{\mathbf{k}\sigma} (\varepsilon_\mathbf{k} - \bar{S}J(0)) \, c_{\mathbf{k}\sigma}^\dagger c_{\mathbf{k}\sigma}, \tag{22.42}$$

where $\bar{S} = 1/N \sum_l \langle S_l^z \rangle$, which is proportional to the net magnetization of the localized spins. By doing so, we effectively extracted the *mean-field* terms from the s–d exchange Hamiltonian. We note that it gives rise to an equal and opposite energy shift of the $\uparrow$ (+) and $\downarrow$ (−) conduction electron states, and leads to unequal conduction electron occupation numbers of the two spin orientations – majority and minority orientations.

22.3.2 Spin Polarization of Conduction Electrons

We now consider the effect of off-diagonal terms $\mathcal{H}'_{ex}(s\text{-}d)$ of $\mathcal{H}_{ex}(s\text{-}d)$, which we consider as perturbations on the conduction electrons. To first order, the perturbed wavefunction is given by

$$\psi_{\mathbf{k}\sigma} = \phi_{\mathbf{k}\sigma} + \sum_{\mathbf{k}'\sigma'} \frac{\langle \mathbf{k}'\sigma' | \, \mathcal{H}'_{ex}(s\text{-}d) \, | \mathbf{k}\sigma \rangle}{\varepsilon_{\mathbf{k}\sigma} - \varepsilon_{\mathbf{k}'\sigma'}} \phi_{\mathbf{k}'\sigma'}. \tag{22.43}$$

Substituting from (22.40), we obtain

$$\psi_{\mathbf{k}(\pm)} = \phi_{\mathbf{k}(\pm)} - \frac{1}{N} \sum_{\mathbf{k}',l}' J(|\mathbf{k} - \mathbf{k}'|)\, e^{i(\mathbf{k}-\mathbf{k}')\cdot\mathbf{R}_l}$$

$$\times \left\{ \frac{S_l^{\pm}\, \phi_{\mathbf{k}'(\mp)}}{\varepsilon_{\mathbf{k}(\pm)} - \varepsilon_{\mathbf{k}'(\mp)}} \pm \frac{S_l^z\, \phi_{\mathbf{k}'(\pm)}}{\varepsilon_{\mathbf{k}(\pm)} - \varepsilon_{\mathbf{k}'(\pm)}} \right\}. \qquad (22.44)$$

The prime over the summation sign excludes $\mathbf{k} = \mathbf{k}'$.

The polarizing effect of the exchange mechanism on the conduction electron with spin ↑ (+) relative to that with spin ↓ (−) is clear in (22.44). The change in the conduction electron state in (22.44) comprises two distinct terms appearing in the braces. The first term involves a spin-flip of the d spin, accompanied by a reversal of the conduction electron spin. The second term involves interactions that do not change the spins of either the d or conduction electrons.

The modified densities of up and down spins are

$$\rho_{\pm}(\mathbf{x}) = \sum_{\mathbf{k}}^{\mathbf{k}_{F(\pm)}} \psi_{\mathbf{k}(\pm)}^{*}\, \psi_{\mathbf{k}(\pm)} = \left(\frac{1}{\Omega} \sum_{\mathbf{k}}^{\mathbf{k}_{F(\pm)}} \right) \mp \frac{1}{N\Omega} \sum_{\mathbf{k},\mathbf{k}'} \frac{J(\mathbf{k} - \mathbf{k}')}{\varepsilon_{\mathbf{k}(\pm)} - \varepsilon_{\mathbf{k}'(\pm)}}$$

$$\times \sum_{l} \left[\exp\{i(\mathbf{k} - \mathbf{k}') \cdot (\mathbf{x} - \mathbf{R}_l)\} + \text{h.c.} \right] \langle S_l^z \rangle.$$

As shown in Figure 22.4, k_{F+} and k_{F-} are the electron wavevector magnitude for up and down spins at the new Fermi energy E_F, respectively, satisfying

$$\frac{1}{\Omega} \sum_{\mathbf{k}}^{\mathbf{k}_{F(\pm)}} = n_{(\pm)} = \frac{n_c}{2} \pm \left(\frac{3n_c}{4E_F} \right) \frac{J(0)}{N} \sum_{l} \langle S_l^z \rangle. \qquad (22.45)$$

$n_c = n_+ + n_-$ is the total density of conduction electrons, and E_F is the unperturbed Fermi energy. We note that $(3n_c/4E_F)$ is the density of states of free electrons at the Fermi energy, and $\pm(J(0)/N) \sum_l \langle S_l^z \rangle$ is just the energy shift of ↑ and ↓ electrons. The expression for the spin density becomes

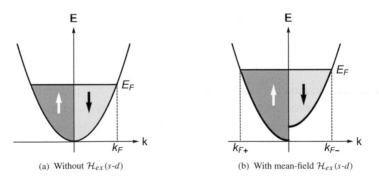

(a) Without $\mathcal{H}_{ex}(s\text{-}d)$ (b) With mean-field $\mathcal{H}_{ex}(s\text{-}d)$

Figure 22.4 Effect of the mean field.

$$\rho_\pm(\mathbf{x}) = \frac{n_c}{2\Omega} \pm \left(\frac{3n_c}{4E_F}\right) \frac{J(0)}{N\Omega} \sum_l \langle S_l^z \rangle$$

$$\mp \frac{1}{N\Omega} \sum_{\substack{\mathbf{k},\mathbf{q} \\ q>0}} \frac{J(\mathbf{q})}{\varepsilon_{\mathbf{k}-\mathbf{q}(\pm)} - \varepsilon_{\mathbf{k}(\pm)}} \sum_l [\exp\{i\mathbf{q}\cdot(\mathbf{x}-\mathbf{R}_l)\} + \text{h.c.}]\langle S_l^z \rangle, \quad (22.46)$$

where we set $\mathbf{k} - \mathbf{k}' = \mathbf{q}$. In carrying out the summation over $\mathbf{k}$ in (22.46), we shall ignore $\pm \bar{S} J(0)$ when choosing the upper limit of the wavevector, since they lead to second-order terms in $J(q)/E_F$. We use parabolic dispersion relation

$$\varepsilon_{\mathbf{k}(\pm)} = \frac{\hbar^2 k^2}{2m^*} \mp \bar{S} J(0)$$

for the conduction electrons, with an effective mass m^*, to obtain

$$\sum_{\mathbf{k}}^{k_F} \left(\frac{1}{\varepsilon_{\mathbf{k}-\mathbf{q}} - \varepsilon_{\mathbf{k}}}\right) = \left(\frac{3}{16}\frac{n_c}{E_F}\right) f(x), \quad x = \frac{q}{2k_F}$$

$$f(x) = 1 + \left(\frac{1-x^2}{x}\right) \ln\left|\frac{1+x}{1-x}\right| \qquad \text{Lindhard function.} \qquad (22.47)$$

To carry out the summation over $\mathbf{q}$, we shall assume that $J(\mathbf{q})$ is a very slowly varying function of $\mathbf{q}$ in the vicinity of the Fermi surface ($|\mathbf{k}|, |\mathbf{k}'| \simeq k_F$), where we expect to have the maximum contribution. Converting the summation into an integral, and using the integral representation

$$\ln\left|\frac{2k_F+q}{2k_F-q}\right| = 2\int_0^\infty dy\, \frac{\sin(2k_F y)\,\sin(qy)}{y},$$

we obtain [157]

$$\frac{1}{N}\sum_{\mathbf{q}} f(q/2k_F)\,\exp(i\mathbf{q}\cdot\mathbf{u}) = -24\pi\frac{n_c}{N} F(2k_F u) \qquad (22.48)$$

$$\text{Ruderman–Kittel function} \quad F(x) = \frac{x\cos x - \sin x}{x^4}.$$

Approximating $J(q) = J_0$ and substituting back in (22.46), we find

$$\rho_\pm(\mathbf{x}) = \frac{n_c}{2} \pm \frac{9\pi}{2}\left(\frac{n_c}{\Omega}\right)\frac{J_0}{E_F}\frac{n_c}{N} \sum_l F(2k_F|\mathbf{x}-\mathbf{R}_l|)\langle S_l^z \rangle. \qquad (22.49)$$

The physical interpretation of (22.49) is as follows: the first term is just the spin density of the conduction electrons in the absence of the s–d interaction. The effective spin polarization induced by $\mathcal{H}_{ex}(s\text{-}d)$ is accounted for in the second term.

We find that the net spin polarization, shown in Figure 22.5, is

$$\rho_+(\mathbf{x}) - \rho_-(\mathbf{x}) = \frac{9\pi}{E_F}\frac{n_c^2 J_0}{N\Omega} \sum_l \langle S_l^z \rangle \frac{\cos(2k_F|\mathbf{x}-\mathbf{R}_l|)}{(k_F|\mathbf{x}-\mathbf{R}_l|)^3}. \qquad (22.50)$$

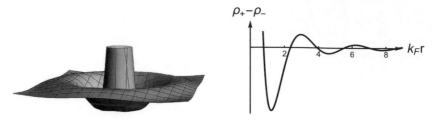

Figure 22.5 Spin polarization.

We note the oscillatory behavior of this inhomogeneous spin density polarization; it has a maximum value at the magnetic ion and decreases asymptotically as $1/R^3$.

22.3.3 Indirect Exchange Coupling and RKKY

We now explore the nature of the indirect exchange coupling between localized $d(f)$ spins engendered by the induced polarization of the conduction electrons. In second-order perturbation, we obtain

$$
\mathcal{H}_{\text{eff}}^{ex} = \sum_{\substack{l \neq m \\ \sigma}} \sum_{\substack{\mathbf{k}, \mathbf{k}' \\ \sigma'}} \frac{\langle \mathbf{k}\sigma | \mathcal{H}_{ex}(s\text{-}d) | \mathbf{k}'\sigma' \rangle \langle \mathbf{k}'\sigma' | \mathcal{H}_{ex}(s\text{-}d) | \mathbf{k}\sigma \rangle}{\varepsilon_{\mathbf{k}\sigma} - \varepsilon_{\mathbf{k}'\sigma'}}. \tag{22.51}
$$

Substituting for $\mathcal{H}_{ex}(s\text{-}d)$ from (22.40), and determining the matrix element of the fermion operators with respect to the Bloch states, we obtain

$$
\mathcal{H}_{\text{eff}}^{ex} = \frac{1}{N^2} \sum_{l \neq m} \sum_{\mathbf{k}, \mathbf{k}'} \frac{J^2(|\mathbf{k} - \mathbf{k}'|)}{\varepsilon_{\mathbf{k}} - \varepsilon_{\mathbf{k}}'} \, \exp\{i\,(\mathbf{k} - \mathbf{k}') \cdot \mathbf{R}_{lm}\}
$$
$$
\times \left[\left\{ n_{\mathbf{k}(+)}\big(1 - n_{\mathbf{k}'(+)}\big) + n_{\mathbf{k}(-)}\big(1 - n_{\mathbf{k}'(-)}\big) \right\} S_l^z S_m^z \right.
$$
$$
\left. + \, n_{\mathbf{k}(-)}\big(1 - n_{\mathbf{k}'(+)}\big) S_l^+ S_m^- + n_{\mathbf{k}(+)}\big(1 - n_{\mathbf{k}'(-)}\big) S_l^- S_m^+ \right] \tag{22.52}
$$

with $\mathbf{R}_{lm} = \mathbf{R}_l - \mathbf{R}_m$, and $n_{\mathbf{k}(\pm)}$ is the fermion occupation number for the states $|\mathbf{k}\pm\rangle$. Applying these approximations, together with $n_{\mathbf{k}(+)} = n_{\mathbf{k}(-)} = n_{\mathbf{k}}$, we find that at absolute zero, $\mathcal{H}_{\text{eff}}^{ex}$ reduces to

$$
\mathcal{H}_{\text{eff}}^{ex} = -\frac{3}{8} \frac{n_c J_0^2}{N^2 E_F} \sum_{\mathbf{q}} \sum_{l \neq m} f(\mathbf{q}) \, \exp(i\mathbf{q}.\mathbf{R}_{lm}) \, (\mathbf{S}_l \cdot \mathbf{S}_m). \tag{22.53}
$$

Using (22.48), we obtain the final expression

$$
\mathcal{H}_{\text{eff}}^{ex} = -\frac{9\pi}{2} \frac{J_0^2}{E_F} \left(\frac{n_c}{N}\right)^2 \sum_{l \neq m} F(2k_F R_{lm}) \, (\mathbf{S}_l \cdot \mathbf{S}_m). \tag{22.54}
$$

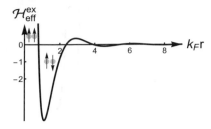

Figure 22.6 Effective exchange Hamiltonian for d magnetic ions, showing regions of ferromagnetic and antiferromagnetic interactions.

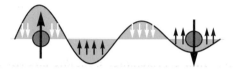

Figure 22.7 Scattering of a free electron by a paramagnetic ion with electron spin parallel and antiparallel to the local spin of the ion.

The coupling magnitude falls off as $1/R_{lm}^3$, where R_{lm} is the distance between magnetic ions l and m. However, the sign of the effective interaction depends on the oscillating function $F(x)$.

This type of exchange coupling, depicted in Figures 22.6 and 22.7, is referred to as the RKKY exchange, after Ruderman–Kittel[157], Kasuya[102], and Yosida[204]. The importance of s–d exchange for spin coupling in magnetic metals and alloys was first suggested by Vonsovskii[190] and then by Zener[206]. This inter-ion interaction can be either ferromagnetic or antiferromagnetic, depending on their separation. We also note that the indirect coupling (22.54) has a long-range character, compared with the Heisenberg-type exchange interaction. Such long-range oscillatory exchange is behind the occurrence of spin-glass phases in dilute metallic alloys such as $Cu_{1-x}Mn_x$, $x < 1\%$, where the Mn ions with localized magnetic moments are randomly dispersed through the nonmagnetic host Cu matrix.

22.4 Exchange Interactions in Magnetic Insulators

In this section, we study the origin, diversity, and nature of exchange interactions in magnetic compounds, which, in contrast to metals and alloys, are actually insulators. These insulating magnetic compounds contain paramagnetic cations and diamagnetic anions (ligands, in chemical jargon); the latter constitute the main matrix of the crystalline lattice. Within this matrix, the paramagnetic ions occupy sites with well-defined symmetry – tetrahedral, octahedral, cubic, etc. – and, accordingly, well-defined coordination.

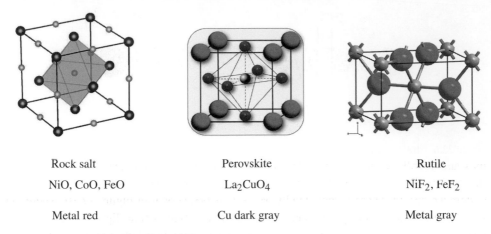

Rock salt	Perovskite	Rutile
NiO, CoO, FeO	La$_2$CuO$_4$	NiF$_2$, FeF$_2$
Metal red	Cu dark gray	Metal gray

Figure 22.8 Examples of magnetic insulators and their structure.

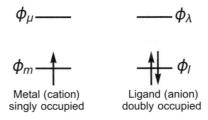

Figure 22.9 Energy levels and occupancy of the metal–ligand system.

The diamagnetic matrix essentially determines the electronic state of the paramagnetic ions, in particular the character of the valence shell, which contains unpaired electrons. Furthermore, the diamagnetic ions have a mediating influence in regard to the character of the ensuing coupling between the spins of the paramagnetic ions.

It is important to stress here that the paramagnetic ions in these compounds exhibit strong magnetic coupling despite their large separation due to the intervening diamagnetic anions (O^{2-}, F^-, S^{2-}, etc.). Thus, we infer that the large magnitude of the magnetic coupling in such systems cannot be attributed to direct exchange–type interactions. We argue that indirect exchange mechanisms, involving spin-dependent excited configurations of the intermediate anions, or of the *cation–anion–cation* unit as a whole, must play a dominant role. This type of spin coupling was first introduced by Kramers [113], and subsequently several distinct mechanisms have been proposed by others. We shall adopt a unified approach to treat the different mechanisms that appear in these insulating compounds.

In the following analysis, we still consider the magnetic-electron orbitals as fairly localized, but with some covalent mixing with the diamagnetic ligand ion orbitals. Essentially, the basis orbitals will be assumed to be localized and orthogonal Hartree–Fock orbitals of the entire crystal. The zeroth-order ground-state configuration of the system, shown in Figure 22.9, consists of singly occupied (SO) orbitals, ϕ_m, of the paramagnetic metallic

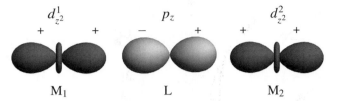

Figure 22.10 A collinear unit of metal-ligand-metal ($M_1 - L - M_2$).

ions and doubly occupied (DO) s/p valence orbitals of the diamagnetic, ligand ions, ϕ_l. In addition, the basis set includes completely empty orbitals ϕ_μ and ϕ_λ, respectively. The system we consider here is modeled by $M_1 - L - M_2$, shown in Figure 22.10, where the Ms represent the paramagnetic metal ions and L the diamagnetic ligand ion. With this prescription, we set out to formulate the exchange interaction between electrons of the singly occupied orbitals of the paramagnetic ions.

22.4.1 Derivation of the Effective Exchange Hamiltonian

The aforementioned orbitals furnish the basis set for the fermion creation and annihilation operators, and we write

$$\mathcal{H} = \mathcal{H}_0 + \mathcal{H}_{dir} + \mathcal{H}_{tr} + \mathcal{H}_{pol} + \mathcal{H}_{sup} \tag{22.55}$$

$$\mathcal{H}_0 = \text{const.} + \sum_{m\sigma} \varepsilon_m\, c_{m\sigma}^\dagger\, c_{m\sigma} + \sum_{l\sigma} \varepsilon_l\, c_{l\sigma}^\dagger\, c_{l\sigma}$$

$$+ \sum_{\mu\sigma} \varepsilon_\mu\, c_{\mu\sigma}^\dagger\, c_{\mu\sigma} + U \sum_{m\sigma} c_{m\sigma}^\dagger\, c_{m\sigma}\, c_{m\bar\sigma}^\dagger\, c_{m\bar\sigma}, \tag{22.56}$$

where ε_m, ε_l, and ε_μ represent the one-electron energies of the SO, DO, and empty Hartree–Fock orbitals, respectively. The last term is the onsite Coulomb repulsion. The remaining terms in (22.55) are

$$\mathcal{H}_{dir} = \frac{1}{2} \sum_{\substack{m_1 m_2 \\ \sigma_1 \sigma_2}} J_{m_1 m_2}^{m_2 m_1} c_{m_1\sigma_1}^\dagger c_{m_2\sigma_2}^\dagger c_{m_1\sigma_2} c_{m_2\sigma_1} = \text{const.} - \sum_{m_1 < m_2} 2 J_{m_1 m_2}^{m_2 m_1}\, \mathbf{s}_{m_1} \cdot \mathbf{s}_{m_2}, \tag{22.57}$$

the direct Heisenberg-type exchange interaction involving SO orbitals, and

$$\mathcal{H}_{tr} = \sum_{m_1, m_2, \sigma} t_{m_1 m_2}\, c_{m_1\sigma}^\dagger\, c_{m_2\sigma} \tag{22.58}$$

the spin-independent electron transfer process. The last two terms in $\mathcal{H}$ are new:

$$\mathcal{H}_{pol} = \sum_{\substack{l m_1 \mu \\ \sigma\sigma'}} J_{l m_1}^{m_1 \mu}\, c_{m_1\sigma'}^\dagger\, c_{\mu\sigma}^\dagger\, c_{m_1\sigma}\, c_{l\sigma'} + \text{h.c.} \tag{22.59}$$

$$J_{l m_1}^{m_1 \mu} = \left\langle \phi_{m_1} \phi_\mu \middle| \mathcal{H}_{12}(12) \middle| \phi_l \phi_{m_1} \right\rangle$$

represents the exchange polarization interactions, and involves one-orbital transition $(l \to \mu)$ along with spin exchange. Finally, we have the two-electron excitation terms given by

$$\mathcal{H}_{sup} = \sum G^{m_1 m_2}_{l_1 l_2} c^\dagger_{m_1 \sigma_1} c^\dagger_{m_2 \sigma_2} c_{l_2 \sigma_2} c_{l_1 \sigma_1} + \text{h.c.}$$

$$+ \sum G^{\mu_1 \mu_2}_{m_1 m_2} c^\dagger_{\mu_1 \sigma_1} c^\dagger_{\mu_2 \sigma_2} c_{m_2 \sigma_2} c_{m_1 \sigma_1} + \text{h.c.} \tag{22.60}$$

$$G^{\mu_1 \mu_2}_{m_1 m_2} = \langle \phi_{\mu_1} \phi_{\mu_2} | \mathcal{H}_{12}(12) | \phi_{m_1} \phi_{m_2} \rangle.$$

The effective exchange coupling contributions of $\mathcal{H}_{tr} + \mathcal{H}_{pol} + \mathcal{H}_{sup}$ is determined in second order with the aid of the perturbative canonical transformation

$$\mathcal{H}_{eff} = e^{-iS} \mathcal{H} e^{iS} = \mathcal{H} + i[\mathcal{H}, S] + \frac{i^2}{2} \big[[\mathcal{H}, S], S\big] + \cdots \tag{22.61}$$

The generator S is determined from the condition

$$\mathcal{H}_{tr} + \mathcal{H}_{pol} + \mathcal{H}_{sup} + i[\mathcal{H}_0, S] = 0, \tag{22.62}$$

and the effective interactions are derived, to second order, from

$$\mathcal{H}^{ex}_{eff} = \frac{i}{2} [\mathcal{H}_{tr} + \mathcal{H}_{pol} + \mathcal{H}_{sup}, S]. \tag{22.63}$$

This procedure yields several terms, including irrelevant scattering processes whose contribution to exchange arise in higher order. We will only consider terms that produce effective exchange coupling in second order. Such interaction terms for magnetic compounds were first formulated elegantly by Anderson [13,16].

Direct kinetic exchange: For interactions involving single-electron transfer between SO orbitals, we obtain the familiar term

$$\mathcal{H}^{ex}_{eff}(tr) = - \sum_{\substack{m_1 m_2 \\ \sigma}} \frac{|t_{m_1 m_2}|^2}{\Delta E(m_1 \to m_2)} c^\dagger_{m_1 \sigma} c^\dagger_{m_2 \bar\sigma} c_{m_1 \bar\sigma} c_{m_2 \sigma}, \tag{22.64}$$

where $t_{m_1 m_2}$ is the hopping parameter between m_1 and m_2 (see Figure 22.11). $\Delta E(m_1 \to m_2)$ is the energy involved in the transfer of an electron from the orbital ϕ_{m_1}, to the orbital ϕ_{m_2}. In the present case, it is nearly equal to U, the onsite Coulomb repulsion. On carrying out the spin summation (22.64) becomes

$$\mathcal{H}^{ex}_{eff}(tr) = - \sum_{m_1 m_2} \frac{|t_{m_1 m_2}|^2}{U} \frac{1}{2} \left(1 - 4\, \mathbf{s}_{m_1} \cdot \mathbf{s}_{m_2}\right)$$

$$= \text{constant} - \sum_{m_1 m_2} 2\left(-\frac{|t_{m_1 m_2}|^2}{U}\right) \mathbf{s}_{m_1} \cdot \mathbf{s}_{m_2}. \tag{22.65}$$

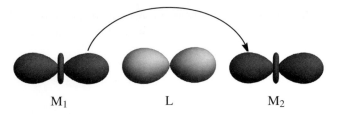

Figure 22.11 Schematic representation of the kinetic exchange interaction involving one-electron transfer.

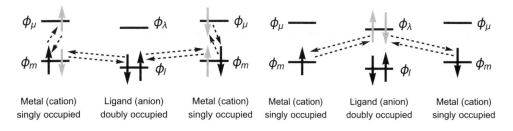

| Metal (cation) | Ligand (anion) | Metal (cation) | Metal (cation) | Ligand (anion) | Metal (cation) |
| singly occupied | doubly occupied | singly occupied | singly occupied | doubly occupied | singly occupied |

Figure 22.12 Schematic representation of the polarization exchange interaction involving two-electron virtual transfer.

Polarization exchange: The spin-polarization contribution involves a virtual excitation of an electron from an SO orbital (say ϕ_{m_1}) to an empty orbital (ϕ_μ) along with the excitation of an electron from DO orbital (say ϕ_l) to ϕ_{m_1} (see Figure 22.12).

The effective interaction turns out to be

$$\mathcal{H}_{\text{eff}}^{ex}(pol) = \sum_{\substack{l\mu m_1 m_2 \\ \sigma_1 \sigma_2}} \frac{J_{lm_1}^{m_1\mu} (J_{lm_2}^{m_2\mu})^*}{\Delta E(l \to \mu)} c_{m_1\sigma_1}^\dagger c_{m_2\sigma_2}^\dagger c_{m_1\sigma_2} c_{m_2\sigma_1},$$

which on spin summation becomes

$$\mathcal{H}_{\text{eff}}^{ex}(pol) = -\sum_{\substack{m_1 m_2 \\ l\mu}} \frac{J_{lm_1}^{m_1\mu} (J_{lm_2}^{m_2\mu})^*}{\Delta E(l \to \mu)} \frac{1}{2} \left(1 + 4\,\mathbf{s}_{m_1} \cdot \mathbf{s}_{m_2}\right)$$

$$= \text{constant} - 2 \sum_{m_1 m_2} J_{\text{eff}}(pol)\, \mathbf{s}_{m_1} \cdot \mathbf{s}_{m_2}. \tag{22.66}$$

Here $\Delta E(l \to \mu) \simeq (\varepsilon_\mu - \varepsilon_l)$- the energy difference corresponding to the orbitals ϕ_μ and ϕ_l; $\Delta E(l \to \mu)$ is positive here. However, the sign of $J_{\text{eff}}(pol)$ is determined by the product of hybrid integrals $J_{lm_1}^{m_1\mu}$ and $(J_{lm_2}^{m_2\mu})^*$. It is not possible to predict the sign without making a careful analysis of the relative symmetry of the orbitals involved.

It may be noted that the spin-polarization mechanism is analogous to the RKKY mechanism of exchange in dilute alloys and rare-earth metals discussed earlier. In magnetic insulators, the electrons of the intervening diamagnetic ligand ion (e.g., O^{2-}, S^{2-}) play the same role as the conduction electrons in metals.

Superexchange: Finally, there are contributions that arise from two-electron excitations. The effective interaction turns out to be

$$\mathcal{H}_{\text{eff}}^{ex}(\text{super}) = - \left[\sum_{\substack{lm_1m_2 \\ \sigma}} \frac{G_{m_1m_2}^{ll} G_{ll}^{m_1m_2}}{\Delta E(m_1m_2 \leftarrow ll)} + \sum_{\substack{m_1m_2\mu \\ \sigma}} \frac{G_{\mu\mu}^{m_1m_2} G_{m_1m_2}^{\mu\mu}}{\Delta E(\mu\mu \leftarrow m_1m_2)} \right]$$

$$\times c_{m_1\sigma}^{\dagger} c_{m_2,-\sigma}^{\dagger} c_{m_1,-\sigma} c_{m_2\sigma},$$

Upon spin summation, we get

$$\mathcal{H}_{\text{eff}}^{ex}(\text{super}) = \text{const.} + 2 \sum_{m_1m_2} \left\{ \sum_l \frac{|G_{ll}^{m_1m_2}|^2}{\Delta E(m_1m_2 \leftarrow ll)} + \sum_\mu \frac{|G_{m_1m_2}^{\mu\mu}|^2}{\Delta E(\mu\mu \leftarrow m_1m_2)} \right\}$$

$$\times \left(\mathbf{s}_{m_1} \cdot \mathbf{s}_{m_2} \right). \qquad (22.67)$$

This mechanism stabilizes the AF state. The first term of (22.67) represents the process in which two electrons from the DO orbital (say ϕ_l) make a virtual transition to the adjoining SO orbitals, one each to ϕ_{m_1} and ϕ_{m_2}, as shown in Figure 22.13. The second term denotes two-electron virtual transitions, one each from ϕ_{m_1} and ϕ_{m_2}, to a nondegenerate empty orbital ϕ_μ.

Taking all the processes previously noted, the effective exchange Hamiltonian can be written as

$$\mathcal{H}_{\text{eff}}^{ex}(\text{total}) = -2 \sum_{m_1 < m_2} J_{m_1m_2}(\text{total}) \, \mathbf{s}_{m_1} \cdot \mathbf{s}_{m_2}, \qquad (22.68)$$

where $J_{m_1m_2}(\text{total})$ is the algebraic sum of the coefficients occurring in (22.57), (22.65), (22.66), and (22.67).

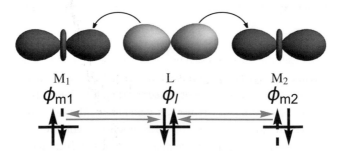

Figure 22.13 Schematic representation of the superexchange interaction involving two-electron transfer.

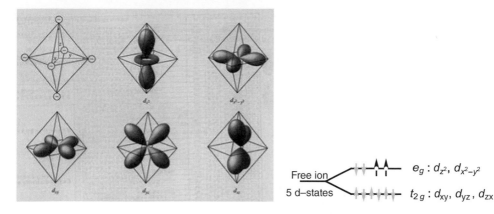

Figure 22.14 d-Orbitals splitting in octahedral environment. Electron configuration for Ni^{++} is in the lower figure (hole orbitals are indicated by white discs and hole spin by black arrows).

22.4.2 Spin–Orbit Coupling: Single-Ion Magnetic Anisotropy

When a paramagnetic ion is placed in a crystalline environment, it experiences a reduction in symmetry. For example, in NiO with rock salt structure, the Ni^{++} ion has an octahedral environment. Accordingly, the $3d$-states of the Ni^{++} ion are split into two manifolds, t_{2g} and e_g, as shown in Figure 22.14. The e_g manifold orbitals point directly toward the negative O^{--} ions, while the t_{2g} manifold orbitals are directed between the O^{--} ions; consequently, the electrons in the e_g manifold will suffer higher repulsion energies, and the e_g energy is found to be higher than that of the t_{2g} manifold. The typical scale of this splitting is $\sim$1–2 eV. This splitting can be represented by an anisotropic electrostatic potential, coined *crystal field potential*, due to the neighboring ions.

Crystal Field and Quenching of Angular Momentum

We discern that the crystal field is a real potential that favors particular orientations of the orbital wavefunctions. As a consequence of being a real and time reversal invariant perturbation, we find that a nondegenerate ground state $|\psi_0\rangle$ has to be real.[5] The reality of $|\psi_0\rangle$ implies that the expectation value of the angular momentum must vanish

$$\langle\psi_0|\,\mathbf{L}\,|\psi_0\rangle = 0$$

since the angular momentum operator is a purely imaginary operator, while the expectation value has to be a real quantity. The jargon used in such a case is that the angular momentum is *quenched*, and only the spin degree of freedom remains. There is however a caveat: if the crystal field does not entirely remove a degeneracy, then one may construct from degenerate real wavefunctions combinations for which this equality does not hold. It is, however, usually the case that such degeneracies do not survive, since, according to the

---

[5] See Section 6.6.

Jahn–Teller theorem, it becomes energetically favorable to pay the cost of elastic energy of deforming a lattice to lower symmetry in order to gain a larger reduction in electronic energy associated with the splitting of the degenerate state.

Single-Site Magnetic Anisotropy

When complete quenching of angular momentum occurs, namely, $\langle \psi_0 | \mathbf{L} | \psi_0 \rangle = 0$, it is clear that the spin–orbit perturbation $\mathcal{H}_{SO} = \lambda \, \mathbf{L} \cdot \mathbf{S}$ has no effect at first order in perturbation theory, since there is no remaining degeneracy to break. However, this term may still have an effect in second-order perturbation theory. Irrespective of the spin state, we integrate out the orbital excitations perturbatively, and write

$$\mathcal{H}_S^{AN} = |\lambda|^2 \sum_n \frac{\langle 0 | L_\mu | n \rangle \langle n | L_\nu | 0 \rangle}{E_0 - E_n} \, S_\mu \, S_\nu = S_\mu \, \Lambda_{\mu\nu} \, S_\nu,$$

where $\Lambda_{\mu\nu}$ is a symmetric second-order tensor that depends on the local crystalline symmetry and on the ordering of the excited orbitals. When $\Lambda_{\mu\nu}$ is diagonalized through spin-space rotation that brings the principal axes, it takes the general form

$$\mathcal{H}_S^{AN} = -\frac{1}{2} D_\| \, S_z^2 + \frac{1}{2} D_\perp \left(S_x^2 - S_y^2 \right).$$

Since a symmetric second-rank tensor is decomposable into a scalar and a traceless symmetric second-rank tensor, we find that one combination of the three independent eigenvalues of $\Lambda_{\mu\nu}$ corresponds to a trivial constant $\propto S(S + 1)$. For tetrahedral or cubic local symmetry, the tensor is proportional to the identity matrix, and there is no nontrivial quadratic term; the first anisotropic term is

$$\mathcal{E}_{\text{cubic}}^{An}(\mathbf{S}) = C \left(S_x^4 + S_y^4 + S_z^4 \right).$$

Finally, a symmetry rule requires that when the local environment has mirror planes (symmetry under reflection in those planes), the principal axis directions must lie in them or perpendicular to them.

22.4.3 Dzialoshinskii–Moriya Anisotropic Exchange Interaction

The Dzialoshinskii–Moriya interaction (DMI) was first proposed by I. Dzialoshinskii [57] to describe manifest weak ferromagnetism in some antiferromagnets; the analysis was based on symmetry considerations. DMI is an antisymmetric form of exchange interaction that is only allowed when inversion symmetry is broken. Subsequently, T. Moriya proposed the first microscopic model, based on the Anderson model of superexchange [131]. He also wrote down the symmetry rules that constrain the orientation of the DMI vector, depending on the atomic arrangement in the crystal. The presence of DMI was confirmed by the analysis of the magnetization curves of various crystals.

Microscopically, DMI exchange between spins is the result of interplay of onsite spin–orbit coupling and intersite scalar exchange interaction, usually of the superexchange type.

To explore this type of interaction, we consider two magnetic ions with no orbital degeneracy and with quenched angular momentum. We study the simultaneous action of spin–orbit and scalar exchange couplings as expressed in the Hamiltonian [203]

$$\mathcal{H}_S = \lambda \ (\mathbf{L}_1 \cdot \mathbf{S}_1 + \mathbf{L}_2 \cdot \mathbf{S}_2) + \hat{\mathbf{J}} \mathbf{S}_1 \cdot \mathbf{S}_2.$$

We consider the third-order perturbation process where, for example,

- Ion 1 is excited to orbital state n_1 by $\mathbf{L}_1 \cdot \mathbf{S}_1$.
- Exchange interaction takes place between ion 1 in $|n_1\rangle$ and ion 2 in its ground state.
- Finally, $\mathbf{L}_1 \cdot \mathbf{S}_1$ returns ion 1 to its ground state $|0_1\rangle$.

We find that $\mathbf{J}$ would depend on what orbital state the electron is in; however, normally this fact is irrelevant because the orbital states are split, and so one may consider $\mathbf{J} \rightarrow \mathbf{J} = \langle 0 | \hat{\mathbf{J}} | 0 \rangle$ but combined with the spin-orbit coupling. We can then express the third-order perturbation as

$$\mathcal{H}_S^{(3)} = - \sum_{\mu\nu} \left[S_{1\mu} \, \Gamma_{\mu\nu}^{(1)} (\mathbf{S}_1 \cdot \mathbf{S}_2) \, S_{1\nu} + S_{2\mu} \, \Gamma_{\mu\nu}^{(2)} (\mathbf{S}_1 \cdot \mathbf{S}_2) \, S_{2\nu} \right] \tag{22.69}$$

$$[2pt]\Gamma_{\mu\nu}^{(1)} = 2\lambda^2 \sum_{n_1, n_1'} \frac{\langle 0_1 | L_\mu | n_1 \rangle \, J\left(n_1 0_2, n_1' 0_2\right) \langle n_1' | L_\nu | 0_1 \rangle}{(\varepsilon_{n_1} - \varepsilon_{0_1})(\varepsilon_{n_1'} - \varepsilon_{0_1})}. \tag{22.70}$$

$J\left(n_1 0_2, n_1 0_2\right)$ is the exchange integral between $|n_1\rangle$ and the ground state of ion 2, however, we also include the possibility of other excited ion present in $J\left(n_1 0_2, n_1' 0_2\right)$. Taking into account all the off-diagonal terms of the exchange interaction with respect to the orbital states, reduces the perturbation to an effective second order of the form

$$\mathcal{H}_S^{(2)} = - \lambda \sum_{n_1} \left[\frac{\langle 0_1 0_2 | \mathbf{L}_1 \cdot \mathbf{S}_1 | n_1 0_2 \rangle \langle n_1 0_2 | \mathbf{J} \mathbf{S}_1 \cdot \mathbf{S}_2 | 0_1 0_2 \rangle}{E_{0_1 0_2} - E_{n_1 0_2}} \right.$$
$$\left. + \frac{\langle 0_1 0_2 | \hat{\mathbf{J}} \mathbf{S}_1 \cdot \mathbf{S}_2 | n_1 0_2 \rangle \langle n_1 0_2 | \mathbf{L}_1 \cdot \mathbf{S}_1 | 0_1 0_2 \rangle}{E_{0_1 0_2} - E_{n_1 0_2}} \right] + \left(1 \leftrightarrow 2 \right).$$

This may be simplified because for real wavefunctions, we have

$$\langle \psi \, |\mathbf{L}| \, \phi \rangle = - \langle \phi \, |\mathbf{L}| \, \psi \rangle \quad \text{and} \quad \left\langle \psi \left| \hat{\mathbf{J}} \right| \phi \right\rangle = \left\langle \phi \left| \hat{\mathbf{J}} \right| \psi \right\rangle,$$

hence

$$\mathcal{H}_S^{(2)} = -\lambda \sum_{n_1, \mu} \left[\frac{\langle 0_1 0_2 | L_{1\mu} | n_1 0_2 \rangle \left\langle n_1 0_2 \left| \hat{\mathbf{J}} \mathbf{S}_1 \cdot \mathbf{S}_2 \right| 0_1 0_2 \right\rangle}{E_{0_1 0_2} - E_{n_1 0_2}} \right] [S_{1\mu}, \mathbf{S}_1 \cdot \mathbf{S}_2] + \left(1 \leftrightarrow 2 \right).$$

Figure 22.15 Canted antiferromagnet, producing ferrimagnetism.

Using the commutation relations

$$[S_{1x}, S_{1x}S_{2x} + S_{1y}S_{2y} + S_{1z}S_{2z}] = i\left(0 + S_{1z}S_{2y} - S_{1y}S_{2z}\right) = -i\left(\mathbf{S}_1 \times \mathbf{S}_2\right)_x$$

and writing the sum over n_1 as $\boldsymbol{\xi}_1$, we obtain

$$\mathcal{H}_S^{(2)} = \left(\boldsymbol{\xi}_1 - \boldsymbol{\xi}_2\right) \cdot \left(\mathbf{S}_1 \times \mathbf{S}_2\right), \quad i\boldsymbol{\xi}_1 = -\lambda \sum_{n_1} \left[\frac{\langle 0_1 0_2 \left|\mathbf{L}_1\right| n_1 0_2\rangle \left\langle n_1 0_2 \left|\hat{\mathbf{J}}\mathbf{S}_1 \cdot \mathbf{S}_2\right| 0_1 0_2\right\rangle}{E_{0_1 0_2} - E_{n_1 0_2}}\right].$$

This interaction will vanish if $\boldsymbol{\xi}_1 = \boldsymbol{\xi}_2$. Thus, the existence of this exchange mechanism will require that the two ions have different chemical environments, in other words, different site symmetries where the orbital splitting will be different.

We define a vector exchange constant $\mathbf{D} = \boldsymbol{\xi}_1 - \boldsymbol{\xi}_2$, and express the DMI exchange as[6]

$$\mathcal{H}_{\mathrm{DM}} = -\mathbf{D} \cdot \mathbf{S}_1 \times \mathbf{S}_2.$$

This interaction tends to align the spins perpendicular to each other and to $\mathbf{D}$ which lies along the symmetry axis. When coexisting with an antiferromagnetic interaction, typically $|\mathbf{D}| \ll |\mathbf{J}|$, it will induce a canting of the spins away from their antiferromagnetic configuration, as illustrated in Figure 22.15, and gives rise to a nonvanishing expectation of the net magnetization. This ferromagnetic parasitic effect leads to the formation of weak ferrimagnetism.

Symmetry Considerations for Dzialoshinskii–Moriya Interaction

We now explore when such a term is allowed by symmetry. We have to consider all symmetry operations of the crystal that leave the center point C on the bond between the two spins fixed – its Wyckoff symmetry designation. The configurations are depicted in Figure 22.16.

---

[6] A general bilinear interaction between two spins can be written as $\mathbf{S}_1^T \underline{\mathbf{V}} \mathbf{S}_2$, where $\underline{\mathbf{V}}$ is a general rank 2 tensor, which can be decomposed into

$$J = \frac{1}{3}\mathrm{Tr}[\underline{\mathbf{V}}], \qquad \text{scalar exchange constant}$$

$$\underline{\mathbf{V}}^s = \frac{1}{2}\left[\underline{\mathbf{V}} + \underline{\mathbf{V}}^T\right] - J\,\mathbb{I}, \qquad \text{traceless symmetric exchange tensor (anisotropic exchange)}$$

$$\underline{\mathbf{V}}^a = \frac{1}{2}\left[\underline{\mathbf{V}} - \underline{\mathbf{V}}^T\right], \qquad \text{antisymmetric exchange tensor}$$

We can write

$$\mathbf{S}_1^T \underline{\mathbf{V}}^a \mathbf{S}_2 = \mathbf{D} \cdot (\mathbf{S}_1 \times \mathbf{S}_2).$$

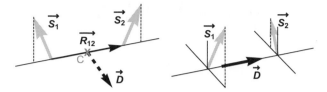

Figure 22.16 Dzialoshinskii–Moriya interactions for different cantings. C is bond center.

1. C is an inversion center of the crystal. The inversion operator $\mathcal{I}$ interchanges the two spins but does not otherwise change them (spins are pseudovectors), and we obtain

$$\mathbf{S}_1 \times \mathbf{S}_2 \quad \overset{\mathcal{I}}{\rightarrow} \quad \mathbf{S}_2 \times \mathbf{S}_1 = -\mathbf{S}_1 \times \mathbf{S}_2.$$

In order to preserve Hamiltonian invariance, $\mathbf{D}$ must vanish. In this case, there cannot be a Dzialoshinsky–Moriya term.

2. We consider the case of a twofold rotation axis C_2 through C perpendicular to the bond. The mapping is

$$S_{1x} \overset{C_2}{\leftrightarrow} -S_{2x}$$

$$S_{1y} \overset{C_2}{\leftrightarrow} -S_{2y}$$

$$S_{1z} \overset{C_2}{\leftrightarrow} +S_{2z}, \qquad C_2 \text{ along } z$$

and we obtain

$$\mathcal{H}_{\mathrm{DM}} = D_x \left(S_{1y}S_{2z} - S_{1z}S_{2y}\right) + D_y \left(S_{1z}S_{2x} - S_{1x}S_{2z}\right) + D_z \underline{\left(S_{1x}S_{2y} - S_{1y}S_{2x}\right)}.$$

The underlined term changes sign under C_2, whereas the others do not. We find that $\mathcal{H}_{\mathrm{DM}}$ is invariant only if $D_z = 0$, which is along the twofold axis: $\mathbf{D}$ must be perpendicular to the C_2 symmetry axis.

Moriya has given the following rules for the allowed directions of $\mathbf{D}$:

(i) Mirror plane through C perpendicular to $\mathbf{R}_{12}$: $\mathbf{D} \perp \mathbf{R}_{12}$
(ii) Mirror plane containing $\mathbf{R}_{12}$: $\mathbf{D}$ perpendicular to mirror plane
(iii) Twofold rotation axis perpendicular to $\mathbf{R}_{12}$ through C: $\mathbf{D} \perp$ twofold rotation axis
(iv) n-fold rotation ($n > 2$) along $\mathbf{R}_{12}$: $\mathbf{D} \parallel \mathbf{R}_{12}$

22.4.4 Double Exchange (DE) Interaction: Mixed-Valence Systems

In mixed-valence conductors, which are doped magnetic insulators, a special exchange interaction takes place. Instead of the s–d interaction that led to the RKKY mechanism, here, it involves d–d electron interaction: The d orbitals are split by the crystal field. The high-lying orbitals contain the conducting d electrons and the low-lying orbitals present

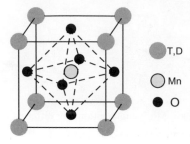

Figure 22.17 Manganite with perovskite structure.

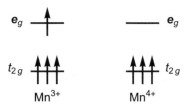

Figure 22.18 The two valence states of the Mn ion found in doped manganites, leading to a noninteger electron occupation of the e_g derived conduction band. Notice that in Mn^{3+}, the exchange coupling between the e_g and t_{2g} electrons is dictated by Hund's first rule and is ferromagnetic.

localized states. The interaction between the two components is intratomic and arises from Hund's first rule: maximizing spin. Consequently, it is ferromagnetic.

The most widely known systems that exhibit such behavior are the manganites, which have the perovskite structure (see Figure 22.17). The undoped insulator has chemical structure like $La^{3+}Mn^{3+}O_3^{2-}$.

The conducting doped manganites have the general chemical formula $T_{(1-x)}D_xMnO_3$, where T is a trivalent rare-earth ion (T = La, Pr, Nd,...) and D is a divalent alkali ion (D = Ca, Sr, . . .). In the manganites, the crystal field is octahedral, and gives rise to e_g-t_{2g} splitting with $E_{e_g} > E_{t_{2g}}$. Divalent doping results in depleting some of the e_g electrons, giving rise to Mn^{4+} ions, and hence mixed valence.

The doping gives rise to two states, shown in Figure 22.18, $\begin{cases} \psi_1: & Mn^{3+} O^{2-} Mn^{4+} \\ \psi_2: & Mn^{4+} O^{2-} Mn^{3+} \end{cases}$,

which are degenerate in energy.[7] However, they are not eigenstates of the system since the e_g electron can effectively hop from the Mn^{3+} ion to the adjacent Mn^{4+} ion through the intermediate O^{2-} ion, as is visualized in Figure 22.19. As two simultaneous processes are involved, this mechanism was coined *double exchange* (DE) by Zener [206].

---

[7] A necessary condition for this degeneracy, and, hence, metallic conductivity, is that the spins of their respective d-shells point in the same direction because the spin of the e_g electron does not change in the hopping process and Hund's coupling punishes antialignment of unpaired electrons. This establishes the correlation between ferromagnetism and metallic conductivity in doped manganites.

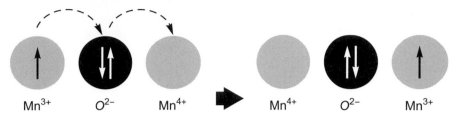

Figure 22.19 Double exchange process with one electron hopping from the intermediate O^{2-} ion to the right Mn^{4+} ion and simultaneously one electron from the left Mn^{3+} ion to the O^{2-} ion.

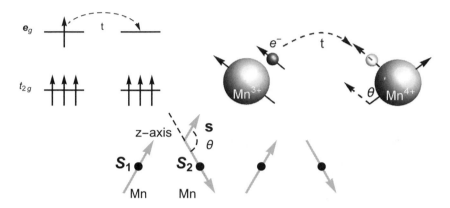

Figure 22.20 Left: the effective hopping for a fully polarized case. Right: spin-canted state where the spin $\mathbf{s}$ of the transferred e_g electron has to be projected onto the onsite spin $\mathbf{S}_2$.

An alternative way to describe DE processes [15] is to replace the double hopping mechanism with a second-order transfer process, represented by an effective electron hopping between nearest-neighbor Mn sites. It is proportional to the square of the hopping involving the p-oxygen and d-manganese orbitals (t_{pd}).

Within this framework, we consider the total spin of the localized t_{2g} to be classical (not quite justifiable for $S = 3/2$), and generalize the DE mechanism to the case of relative nearest-neighbor canted orientation with an angle θ, shown in Figure 22.20.[8] The effective hopping is obtained by projecting the spin $\mathbf{s}$ of the transferred e_g electron onto the onsite spin $\mathbf{S}_2$, which is proportional to $\cos(\theta/2)$. If $\theta = 0$, the hopping is t, the largest, while if $\theta = \pi$, corresponding to an antiferromagnetic background, then the hopping cancels. It follows that there is a direct connection between conductivity and ferromagnetism.

To justify $\cos(\theta/2)$ dependence, and to deduce the double exchange Hamiltonian, we consider two Mn^{4+} ions and a single e_g electron. We set J_H as Hund's onsite coupling energy constant between either of two ion cores, having t_{2g}-spin $S = 3/2$, and a single e_g mobile electron, with $S = 1/2$. The Hamiltonian matrix $\mathcal{H}$ for the system of two ions and the electron is written in block form as

---

[8] Anticipating the presence of both AFM, together with the FM arising from the DE mechanism.

$$\mathcal{H} = \begin{pmatrix} -J_H \, (\mathbf{S}_{t_{2g}}^1 \cdot \mathbf{S}_{e_g}) & t\mathbb{I} \\ t\mathbb{I} & -J_H \, (\mathbf{S}_{t_{2g}}^2 \cdot \mathbf{S}_{e_g}) \end{pmatrix}. \tag{22.71}$$

Setting $S_{t_{2g}} = S$, we have

$$\mathbf{S}_{t_{2g}} \cdot \mathbf{S}_{e_g} = \begin{cases} \dfrac{-J_H \, S}{2} & S_{\text{total}} = S + \tfrac{1}{2} \\[2mm] \dfrac{J_H \, (S+1)}{2} & S_{\text{total}} = S - \tfrac{1}{2} \end{cases}$$

so that in the diagonal representation, we write

$$-J_H \, \mathbf{S}_{t_{2g}} \cdot \mathbf{S}_{e_g} = \begin{pmatrix} \dfrac{-J_H \, S}{2} & 0 \\[2mm] 0 & \dfrac{J_H \, (S+1)}{2} \end{pmatrix}.$$

However, the two diagonal representations of the diagonal blocks in (22.71) are inclined at an angle θ, corresponding to the classical angle between the two Mn^{4+} core spins. Thus, we need to rotate the spin-1/2 off-diagonal hopping blocks, and write

$$t\,\mathbb{I} \rightarrow \begin{pmatrix} t \cos\left(\dfrac{\theta}{2}\right) & -t \sin\left(\dfrac{\theta}{2}\right) \\[3mm] t \sin\left(\dfrac{\theta}{2}\right) & t \cos\left(\dfrac{\theta}{2}\right) \end{pmatrix}.$$

The total Hamiltonian expressed in terms of the interacting basis states, and after rearrangement is given by

$$\mathcal{H} = \begin{pmatrix} \dfrac{-J_H \, S}{2} & t \cos\left(\dfrac{\theta}{2}\right) & t \sin\left(\dfrac{\theta}{2}\right) & 0 \\[3mm] t \cos\left(\dfrac{\theta}{2}\right) & \dfrac{-J_H \, S}{2} & 0 & -t \sin\left(\dfrac{\theta}{2}\right) \\[3mm] t \sin\left(\dfrac{\theta}{2}\right) & 0 & \dfrac{J_H \, (S+1)}{2} & t \cos\left(\dfrac{\theta}{2}\right) \\[3mm] 0 & -t \sin\left(\dfrac{\theta}{2}\right) & t \cos\left(\dfrac{\theta}{2}\right) & \dfrac{J_H \, (S+1)}{2} \end{pmatrix}.$$

In many cases, the intratomic Hund exchange $J_H \gg t$, and we may approximately write

$$\varepsilon \simeq \frac{-J_H \, S}{2} \pm t \cos\left(\frac{\theta}{2}\right).$$

We can then represent the actual crystalline system by the Hamiltonian

$$\mathcal{H} = -t \cos(\theta/2) \sum_{\langle ij \rangle} c_{i\sigma}^\dagger c_{j\sigma} - J_H \sum_i c_{i\sigma}^\dagger \, \boldsymbol{\sigma} \, c_{j\sigma} \cdot \mathbf{S}_i. \tag{22.72}$$

We may also include another intersite exchange coupling, usually antiferromagnetic superexchange, and write

$$\mathcal{H} = -t\cos(\theta/2) \sum_{\langle ij \rangle} c_{i\sigma}^{\dagger} c_{j\sigma} - J_H \sum_i c_{i\sigma}^{\dagger} \sigma\, c_{j\sigma} \cdot \mathbf{S}_i + J \sum_{\langle ij \rangle} \mathbf{S}_i \cdot \mathbf{S}_j. \tag{22.73}$$

However, usually $J_H \gg J$. Since the fraction of Mn^{4+} is equal to the doping concentration x, we can write the average energy per site as

$$\frac{E}{N} = zJS^2 \cos\theta - zxt \cos\frac{\theta}{2}, \tag{22.74}$$

where N is total number of sites, and z the number of nearest-neighbor Mn. Minimizing (22.74), we obtain

$$\cos\frac{\theta}{2} = \frac{xt}{4JS^2},$$

which shows that even at small doping x, the antiferromagnetic lattice is already canted. Moreover, at $x = x_c = 4JS^2/t$ the system becomes completely ferromagnetic!

Exercises

22.1 A toy model: kinetic intersite exchange:

As a toy model, we consider the minimal model of an H_2 molecule with strong onsite repulsion. We consider a system with two (orthogonal) orbitals, 1 and 2, separated by a small distance. There are two electrons in the system, one electron per orbital. The Hamiltonian in this case is the two-site Hubbard model

$$\mathcal{H} = -t\left(c_{1\uparrow}^{\dagger} c_{2\uparrow} - c_{1\uparrow} c_{2\uparrow}^{\dagger} + c_{1\downarrow}^{\dagger} c_{2\downarrow} - c_{1\downarrow} c_{2\downarrow}^{\dagger} \right) + U\left(n_{1\uparrow} n_{1\downarrow} + n_{2\uparrow} n_{2\downarrow} \right).$$

(a) Use the Pauli principle to specify the allowed states of the system, and, hence, its Hilbert space basis.

(b) Use the basis states of part (a) to construct the two-site Hubbard Hamiltonian matrix.

(c) Solve the eigenvalue problem, and obtain the eigenvalues ε and the eigenfunctions in terms of the parameters U and t.

(d) Plot ε/t versus U/t.

(e) Show that for the case $U \gg t$, the singlet state's energy is lowered by $2t^2/U$, which is just the magnitude of the kinetic exchange.

(f) Alternatively, use the resolvent method to show the result of part (d).

22.2 Rudermann–Kittel–Kasuya–Yosida interaction:

In this problem, we will approach the RKKY interaction in a reverse way. We start with the Hamiltonian

$$\mathcal{H} = \mathcal{H}_s + \mathcal{H}_{sd} = \sum_{\mathbf{k}\sigma} \varepsilon_{\mathbf{k}\sigma} c_{\mathbf{k}\sigma}^{\dagger} c_{\mathbf{k}\sigma} - J \sum_i \mathbf{S}_i \cdot \mathbf{s}_i,$$

where $\mathcal{H}_s$ is the conduction electrons, Hamiltonian, and $\mathbf{s}_i$ and $\mathbf{S}_i$ are the spin operators for the electrons and magnetic ions at lattice site i, respectively.

(a) Using (22.20) and the relation

$$\mathbf{S}_i \cdot \mathbf{s} = S^z s^z + \frac{1}{2}\left(S^+ s^- + S^- s^+\right),$$

and Fourier transforming the creation and annihilation operators into $\mathbf{k}$-space, show that $\mathcal{H}_{sd}$ can be written as

$$\mathcal{H}_{sd} = \frac{J}{2N}\sum_i \sum_{\mathbf{kq}} e^{i\mathbf{q}\cdot\mathbf{R}_i}\left[S_i^z\left(c_{\mathbf{k}+\mathbf{q}\uparrow}^\dagger c_{\mathbf{k}\uparrow} + c_{\mathbf{k}-\mathbf{q}\downarrow}^\dagger c_{\mathbf{k}\downarrow}\right) + S_i^+ c_{\mathbf{k}+\mathbf{q}\downarrow}^\dagger c_{\mathbf{k}\uparrow} + S_i^- c_{\mathbf{k}+\mathbf{q}\uparrow}^\dagger c_{\mathbf{k}\downarrow}\right],$$

where N is number of lattice sites, and $\mathbf{R}_i$ are the lattice vectors.

(b) Now we treat $\mathcal{H}_{sd}$ as a perturbation. In the absence of $\mathcal{H}_{sd}$, the electron and ion systems are independent, and the combined unperturbed wavefunction becomes $|\{\mathbf{k}\}, d\rangle = |\{\mathbf{k}\}\rangle \, |d\rangle$. Defining the ground state as $|k_F\rangle |d\rangle$, where $|k_F\rangle$ stands for the Fermi sea, show that the first-order perturbation energy vanishes.

(c) Obtain expressions for the matrix elements

$$\left\langle \mathbf{k}', m_s' \left| c_{\mathbf{k}+\mathbf{q}\uparrow}^\dagger c_{\mathbf{k}\uparrow} + c_{\mathbf{k}-\mathbf{q}\downarrow}^\dagger c_{\mathbf{k}\downarrow} \right| \mathbf{k}, m_s \right\rangle, \quad \left\langle \mathbf{k}', m_s' \left| c_{\mathbf{k}+\mathbf{q}\downarrow}^\dagger c_{\mathbf{k}\uparrow} \right| \mathbf{k}, m_s \right\rangle$$

$$\left\langle \mathbf{k}', m_s' \left| c_{\mathbf{k}+\mathbf{q}\uparrow}^\dagger c_{\mathbf{k}\downarrow} \right| \mathbf{k}, m_s \right\rangle.$$

(d) Show that the energy correction in second order

$$E^{(2)} = \sum_{\substack{|\{\mathbf{k}\}, d'\rangle \\ \neq |k_F, d\rangle}} \frac{\left|\langle k_F, d| \, \mathcal{H}_{sd} \, |\{\mathbf{k}\}, d'\rangle\right|^2}{E_{k_F, d}^{(0)} - E_{\{\mathbf{k}\}, d'}^{(0)}}$$

becomes

$$E^{(2)} = \frac{J^2}{4N^2}\sum_{ij}\sum_{\mathbf{kq}}\sum_{m_s} \frac{\Theta \, e^{\mathbf{q}\cdot(\mathbf{R}_i - \mathbf{R}_j)}}{\varepsilon(\mathbf{k}+\mathbf{q}) - \varepsilon(\mathbf{k})}$$

$$\times \langle d| \langle m_s| \left[S_i^z \left\{ 4S_j^z \left(s^z\right)^2 + 2S_j^+ \left(s^z s^-\right) + 2S_j^- \left(s^z s^+\right) \right\} \right.$$

$$+ S_i^+ \left\{ 2S_j^z \left(s^z s^-\right) + S_j^+ \left(s^-\right)^2 + S_j^- \left(s^- s^+\right) \right\}$$

$$\left. + S_i^- \left\{ 2S_j^z \left(s^z s^+\right) + S_j^+ \left(s^+ s^-\right) + S_j^- \left(s^+\right)^2 \right\} \right] |m_s\rangle |d\rangle$$

after substituting the matrix elements you obtained in part (c) and using the completeness of $|d\rangle$ and $|m_s\rangle$. Θ represents

$$\Theta = \Theta(k_F - |\mathbf{k}|)\,\Theta(|\mathbf{k}+\mathbf{q}| - k_F).$$

(e) Substituting for the spin operators (s^+, s^-, s^z) in terms of the Pauli matrices, show that the second-order energy reduces to

$$E^{(2)} = \frac{J^2\hbar^2}{2N^2} \sum_{ij} \sum_{\mathbf{kq}} \Theta \, e^{\mathbf{q}\cdot(\mathbf{R}_i-\mathbf{R}_j)} \frac{\langle d|\, \mathbf{S}_i \cdot \mathbf{S}_j \,|d\rangle}{\varepsilon(\mathbf{k}+\mathbf{q}) - \varepsilon(\mathbf{k})} = -\sum_{ij} J_{ij}^{\text{RKKY}} \, \mathbf{S}_i \cdot \mathbf{S}_j.$$

(f) Using a parabolic dispersion for the unperturbed conduction electrons, with an effective mass m^*, and replacing the summations over $\mathbf{k}$ and $\mathbf{q}$ with integrals, show that

$$J_{ij}^{\text{RKKY}} = \frac{J^2 m^* \Omega^2}{4\pi^2 N^2 R_{ij}^2} \int_0^{k_F} dk'k' \int_{k_F}^{\infty} dk\, k \, \frac{\sin(kR_{ij}) \sin(k'R_{ij})}{k^2 - k'^2},$$

where m^* is the effective mass and Ω the volume.

(g) Obtain an expression for J_{ij} by setting the lower limit on the second integral to zero, and using the integral

$$\int_0^{\infty} dk\, k \, \frac{\sin(kR_{ij})}{k^2 - k'^2} = \frac{\pi}{2} \cos(k'R_{ij}).$$

22.3 **A more complicated toy model: superexchange**

The model is based on the configuration of two metallic ions (with a singly occupied d orbital each) and the ligand ion (with doubly occupied p orbital), shown in Figure 22.21. The Hamiltonian is given by

$$\mathcal{H} = \sum_{\sigma} \left[\varepsilon_d \sum_i n_{i\sigma} + \varepsilon_p \, n_{p\sigma} + t_{pd} \sum_i \left(c_{i\sigma}^\dagger c_{p\sigma} + c_{p\sigma}^\dagger c_{i\sigma} \right) \right] + U_d \sum_i n_{i\uparrow} n_{i\downarrow},$$

where ε_d, ε_p are the orbital energies, U_d is the onsite repulsion for a d orbital, and t_{pd} is the hopping parameter.

(a) The Hilbert space basis states can be classified according to the relative orientation of the spins in the two d orbitals, as shown in Figure 22.22.

(i) Specify the states that can be derived from the state with parallel spin orientation in the two d orbitals.
(ii) Specify the states that can be derived from the state with antiparallel spin orientation in the two d orbitals.

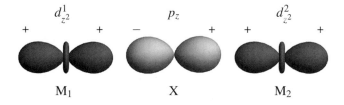

Figure 22.21 Superexchange model.

Figure 22.22 Hilbert space basis states.

Hint: A d orbital can be doubly occupied.

(b) Since the Hamiltonian does not contain spin-flip operations, the corresponding matrix is block-diagonalized. Use the states you have obtained to write down the blocks. Set the zero of energy at $2(\varepsilon_d + \varepsilon_p)$, which is the energy of the states shown in Figure 22.22. Note the t_{pd} sign difference for hoppings to left and right metal sites.

(c) Use the resolvent method to show that only the kinetic exchange is active in the parallel spin case.

(d) Apply the resolvent method to the antiparallel case. In this case, however, the Hamiltonian block matrix may be triply partitioned:

$$\begin{pmatrix} H_{00} & H_{01} & 0 \\ H_{10} & H_{11} & H_{12} \\ 0 & H_{21} & H_{22} \end{pmatrix}$$

H_{00} is a low-energy 2×2 submatrix, H_{11} a 4×4 block, and H_{22} a 3×3 block. The last two blocks contain high-energy states with at least one doubly occupied d-orbital. Set $E = 0$ and apply the method of projected resolvent twice.

(e) Extract the effective Hamiltonian of the low-energy sector. Then simplify the expression using the approximation $(A + B)^{-1} \approx A^{-1}(\mathbb{I} - BA^{-1})$.

(f) Explain the nature of the terms in your final expression of the effective Hamiltonian.

22.4 Orbitals with cubic symmetry:

Consider the effect of cubic crystal fields on the fivefold degenerate d orbitals. The single-particle potential on an electron, projected into this quintuplet, can in general be expressed as a function of the three orbital angular momentum operators, L_x, L_y, L_z, which are 5×5 matrices, since $\mathbf{L} \cdot \mathbf{L} = \ell(\ell + 1)$ with $\ell = 2$.

(a) Find the general form of the Hamiltonian as a function of $\mathbf{L}$, assuming cubic symmetry – that is, the symmetries are those of a cube with corners at $(\pm 1; \pm 1; \pm 1)$ and the atomic nucleus at its center. Apart from a trivial constant, there should be only one free parameter not fixed by symmetry.

(b) Show that the five levels split into a triplet and a doublet. Find a basis for each that is real, by using, instead of spherical harmonic functions of angle, second-order polynomials in x, y, z. The triplet and doublet states are called t_{2g} and e_g orbitals, respectively.

22.5 Spin-state transition:

Consider the cubic situation of the previous problem. Let us denote by $a = 1; 2; 3$ the t_{2g} orbitals and $a = 4; 5$ the e_g orbitals. Consider the single-ion Hamiltonian

$$\mathcal{H} = \frac{\Delta}{2} \left[\sum_{i=4}^{5} n_i - \sum_{i=1}^{3} n_i \right] + U \sum_i n_{i\uparrow} n_{i\downarrow} - J \sum_{i<j} \mathbf{S}_i \cdot \mathbf{S}_j,$$

where n_i is the number of electrons in orbital i, $n_{i\alpha}$ is the number with spin $\alpha = \uparrow, \downarrow$, and the sums, unless otherwise specified, are over all five orbitals. Assuming U, Δ, J are all positive, find the ground-state spin as a function of these parameters, for the Co^{3+} ion, which has 6 d electrons.

22.6 Crystal field potential:

Consider a tetragonal lattice with lattice parameters $a = b \neq c$. The crystal field potential can be approximated by placing equal point charges at the six nearest neighbors of a lattice point, representing the ligands occupying these sites.

(a) Use the identity

$$\frac{1}{|\mathbf{x} - \mathbf{x}'|} = 4\pi \sum_{\ell} \sum_{m=-\ell}^{\ell} \frac{1}{2\ell + 1} \frac{r_<^\ell}{r_>^{\ell+1}} Y_{\ell m}^*(\theta', \varphi') Y_{\ell m}(\theta, \varphi)$$

to obtain an expansion of the crystal potential at the lattice point, up to $\ell = 4$. Use your results to obtain a similar expression for the octahedral environment.

(b) In the case of tetragonal crystal field, use degenerate perturbation theory to determine the energy splittings in the d orbitals. Determine the splitting in the limit $V_{\text{tetragonal}} \to V_{\text{octahedral}}$.

(c) In the octahedral case, determine the ground-state degeneracy for electron fillings d^n, $n = 1 \to 9$ using Hund's rules.

(d) In the case of a ground-state degeneracy, how can such degeneracy be lifted?

$\lambda_{Ni} = -315/8.06 \, \text{meV}$

22.7 In Section 22.4 we discussed how the crystal field together with spin–orbit coupling lead to single-ion anisotropy contribution to the spin Hamiltonian. Here, we consider the case where in addition we have an applied magnetic field $\mathbf{H}$. Show that, to second order, the anisotropy contribution to the Hamiltonian can be written as

$$\mathcal{H} = \lambda^2 \left[\Lambda_{zz} - \frac{1}{2}(\Lambda_{xx} + \Lambda_{yy}) \right] \left[S_z^2 - \frac{1}{3}S(S+1) \right] - \frac{\lambda^2}{2}(\Lambda_{xx} - \Lambda_{yy})\left(S_x^2 - S_y^2 \right)$$

$$+ \mu_B \left(g_{xx} S_x H_x + g_{yy} S_y H_y + g_{zz} S_z H_z \right),$$

where λ is the spin–orbit coupling parameter, Λ is the second-order single-site anisotropy tensor along the principal axes, and

$$g_{ii} = (g_e - 2\lambda \Lambda_{ii}) = 2(1 - \lambda \Lambda_{ii}),$$

where $g_e = 2$ is the electron gyromagnetic ratio.

22.8 Consider the case of Ni^{2+} ($(3d)^8$), with orbital singlet and spin triplet ground state. Given the angular momentum operators for $S = 1$

$$S_x = \begin{pmatrix} 0 & \frac{1}{\sqrt{2}} & 0 \\ \frac{1}{\sqrt{2}} & 0 & \frac{1}{\sqrt{2}} \\ 0 & \frac{1}{\sqrt{2}} & 0 \end{pmatrix}, \quad S_y = \begin{pmatrix} 0 & \frac{-i}{\sqrt{2}} & 0 \\ \frac{i}{\sqrt{2}} & 0 & \frac{-i}{\sqrt{2}} \\ 0 & \frac{i}{\sqrt{2}} & 0 \end{pmatrix}, \quad S_z = \begin{pmatrix} 1 & 0 & 0 \\ 0 & 0 & 0 \\ 0 & 0 & -1 \end{pmatrix},$$

Use the spin Hamiltonian you obtained in Exercise 22.7 to determine the ensuing energy splitting in the absence of a magnetic field. What is the corresponding spin eigenfunction? What is the expectation value of S_z in the ground state?

23

Quantum Theory of Magnetism: Magnetic Insulator Ground States and Spin-Wave Excitations

23.1 Introduction

In the previous chapters, we discussed the various mechanisms of exchange interactions in magnetic solids. The effective exchange interaction may have a positive or a negative sign depending on the relative strengths of the various competing processes. This eventually determines the type of manifest magnetic ordering. Thus, at absolute zero temperature, we can picture spins of the paramagnetic ions aligned in a definite pattern (ferromagnetic, ferrimagnetic, antiferromagnet, or helimagnet) apart from the possible zero-point fluctuations. This state is considered to be the magnetic ground state of the solid.

We note that the formation of ordinary band insulators, where Coulomb interactions are neglected, requires completely filled bands with equal number of both spin orientations, resulting in an electronic configuration that does not allow spontaneous magnetism. Consequently, we surmise that for an insulator to acquire spontaneous magnetism, it cannot be a band insulator, and we cannot exclude interelectron Coulomb interactions.

We shall now consider the nature of the magnetic ground states of such insulators and characterize the elementary excitations above these ground states. We note that any local deviation from perfect alignment of the spin system will not in general remain locally confined, but, owing to exchange coupling, will propagate like a wave. These low-lying excitations are called *spin waves* and were first introduced by Bloch in 1930. The spin wave can be thought of as one spin reversal spread coherently over the entire crystal. When spin waves are quantized, we refer to the state of excitation in terms of a certain number of *magnons* in a particular mode. Spin waves are manifestations of the Goldstone mode corresponding to continuous spin rotation symmetry. They acquire a gap in cases where such symmetry is absent because of single-site magnetic anisotropy. However, if a continuous symmetry still persists, as in the case of an easy plane anisotropy ferromagnet, one expects a gapless mode to still be manifest.

23.2 Possible Ground States of the Classical Heisenberg Hamiltonian

Although electrons have spin $1/2$, many atoms exhibit quite large spin magnitude S. This can be attributed to the presence of a large number of electrons in an atomic shell, together with the action of spin–orbit coupling. The classical limit is achieved when $S \to \infty$, so that

$$\frac{1}{S^2}\left[S_\alpha, S_\beta\right] = i\epsilon_{\alpha\beta\gamma}\frac{S_\gamma}{S} \to 0$$

The Heisenberg model then reduces to its classical Hamiltonian form, where the spin **S** is described by a three-component vector of magnitude S, as

$$\mathcal{H}_H = -J\sum_{<ij>} \mathbf{S}_i \cdot \mathbf{S}_j,$$

where $\langle ij \rangle$ indicates nearest neighbors. To determine the ground state, we need to minimize this Hamiltonian, subject to the constraint

$$\mathbf{S}_i \cdot \mathbf{S}_i = S^2, \qquad \mathbf{S}_i^* = \mathbf{S}_i, \quad \mathbf{S}_i \ \text{real}.$$

We expand $\mathbf{S}_i$ in terms of its lattice Fourier components as

$$\mathbf{S}_i = \frac{1}{\sqrt{N}}\sum_{\mathbf{q}} e^{i\mathbf{q}\cdot\mathbf{R}_i}\,\mathbf{S}_{\mathbf{q}}, \qquad \mathbf{S}_{-\mathbf{q}} = \mathbf{S}_{\mathbf{q}}^*,$$

where N is the number of lattice sites. Substituting in $\mathcal{H}_H$, and setting $\mathbf{R}_j = \mathbf{R}_i + \Delta\mathbf{R}_s$, $s = 1,\ldots, \mathcal{Z}$, where $\mathcal{Z}$ is the coordination number, we obtain

$$\mathcal{H}_H = -\frac{J}{N}\sum_{\mathbf{q},\mathbf{q}'}\sum_{s=1}^{\mathcal{Z}} e^{i\mathbf{q}'\cdot\Delta\mathbf{R}_s}\sum_{\mathbf{R}_i} e^{i(\mathbf{q}+\mathbf{q}')\cdot\mathbf{R}_i}\,\mathbf{S}_{\mathbf{q}}\cdot\mathbf{S}_{\mathbf{q}'}$$

$$= -J\mathcal{Z}\sum_{\mathbf{q}}\gamma_{\mathbf{q}}\,\mathbf{S}_{\mathbf{q}}\cdot\mathbf{S}_{-\mathbf{q}} = -J\mathcal{Z}\sum_{\mathbf{q}}\gamma_{\mathbf{q}}\,\mathbf{S}_{\mathbf{q}}\cdot\mathbf{S}_{-\mathbf{q}}$$

$$\gamma_{\mathbf{q}} = \sum_{s=1}^{\mathcal{Z}}\frac{1}{\mathcal{Z}}e^{i\mathbf{q}\cdot\Delta\mathbf{R}_s}, \tag{23.1}$$

where $\gamma_{\mathbf{q}}$ is the structure factor, which determines the energy. The constraint becomes

$$\mathbf{S}_i \cdot \mathbf{S}_i = \frac{1}{N}\sum_{\mathbf{q},-\mathbf{q}'} e^{i(\mathbf{q}-\mathbf{q}')\cdot\mathbf{R}_i}\,\mathbf{S}_{\mathbf{q}}\cdot\mathbf{S}_{-\mathbf{q}'} = S^2.$$

We stipulate that $\mathcal{H}_H$ has a minimum at some $\gamma_{\mathbf{Q}}$, with a degeneracy depending on the action of the crystallographic point group on $\mathbf{Q}$ in the Brillouin zone. However, we are still left with satisfying the constraint.

23.2.1 Ferromagnetism

For the case $\mathbf{Q} = 0$, we obtain

$$\mathbf{S}_i = \mathbf{S}_{\mathbf{Q}=0}, \qquad \forall i,$$

which is just uniform spin polarization, namely, a ferromagnetic state.

23.2.2 Magnetic Phases with Nonzero Q

For a spatial modulation of wavevector $\mathbf{Q}$, we consider a modulation of $2\mathbf{Q}$ in the constraint, which yields

$$S^2 \sum_{\mathbf{R}_i} e^{i2\mathbf{Q}\cdot\mathbf{R}_i} = \frac{1}{N} \sum_{\substack{\mathbf{q},-\mathbf{q}' \\ \mathbf{R}_i}} e^{i(2\mathbf{Q}-\mathbf{q}-\mathbf{q}')\cdot\mathbf{R}_i} \mathbf{S}_\mathbf{q} \cdot \mathbf{S}_{-\mathbf{q}'}$$

$$NS^2 \sum_{\mathbf{G}} \delta_{2\mathbf{Q}-\mathbf{G}} = \sum_{\mathbf{q},\mathbf{G}} \mathbf{S}_\mathbf{q} \cdot \mathbf{S}_{2\mathbf{Q}-\mathbf{q}+\mathbf{G}}.$$

Setting, appropriately, $\mathbf{q} = \mathbf{Q}$ in the first Brillouin zone, we obtain

$$\mathbf{S}_\mathbf{Q} \cdot \mathbf{S}_\mathbf{Q} = 0.$$

Recalling that $\mathbf{S}_\mathbf{Q}$ can, in general, be a complex quantity

$$\begin{cases} \mathbf{S}_\mathbf{Q} = \mathbf{S}_\mathbf{Q}^r + i\mathbf{S}_\mathbf{Q}^i \\ \mathbf{S}_{-\mathbf{Q}} = \mathbf{S}_\mathbf{Q}^r - i\mathbf{S}_\mathbf{Q}^i \end{cases} \Rightarrow \begin{cases} \mathbf{S}_\mathbf{Q} \cdot \mathbf{S}_{-\mathbf{Q}} = \left(\mathbf{S}_\mathbf{Q}^r\right)^2 + \left(\mathbf{S}_\mathbf{Q}^i\right)^2 \\ \mathbf{S}_\mathbf{Q} \cdot \mathbf{S}_\mathbf{Q} = \left(\mathbf{S}_\mathbf{Q}^r\right)^2 - \left(\mathbf{S}_\mathbf{Q}^i\right)^2 + 2i\mathbf{S}_\mathbf{Q}^r \cdot \mathbf{S}_\mathbf{Q}^i \end{cases}$$

For $\mathbf{S}_\mathbf{Q} \neq 0$, we obtain the solution

$$\begin{cases} \left|\mathbf{S}_\mathbf{Q}^r\right| = \left|\mathbf{S}_\mathbf{Q}^i\right| \\ \mathbf{S}_\mathbf{Q}^r \perp \mathbf{S}_\mathbf{Q}^i \end{cases}$$

where $\mathbf{S}_\mathbf{Q}^r$ and $\mathbf{S}_\mathbf{Q}^i$ are mutually perpendicular vectors of equal lengths. To determine the real space modulation, we take the inverse Fourier transform, and remembering that $\mathbf{S}_i$ is real

$$\begin{aligned} \mathbf{S}_i &= \frac{1}{\sqrt{N}} \left[e^{i\mathbf{Q}\cdot\mathbf{R}_i} \mathbf{S}_\mathbf{Q} + e^{-i\mathbf{Q}\cdot\mathbf{R}_i} \mathbf{S}_{-\mathbf{Q}} \right] \\ &= \frac{1}{\sqrt{N}} \left[e^{i\mathbf{Q}\cdot\mathbf{R}_i} \left(\mathbf{S}_\mathbf{Q}^r + i\mathbf{S}_\mathbf{Q}^i \right) + e^{-i\mathbf{Q}\cdot\mathbf{R}_i} \left(\mathbf{S}_\mathbf{Q}^r - i\mathbf{S}_\mathbf{Q}^i \right) \right] \\ &= \frac{2}{\sqrt{N}} \left[\mathbf{S}_\mathbf{Q}^r \cos\left(\mathbf{Q}\cdot\mathbf{R}_i\right) - \mathbf{S}_\mathbf{Q}^i \sin\left(\mathbf{Q}\cdot\mathbf{R}_i\right) \right] \\ &= S \left[\cos\left(\mathbf{Q}\cdot\mathbf{R}_i\right) \hat{\mathbf{n}}_r - \sin\left(\mathbf{Q}\cdot\mathbf{R}_i\right) \hat{\mathbf{n}}_i \right], \end{aligned} \tag{23.2}$$

where $\hat{\mathbf{n}}_s = \mathbf{S}_\mathbf{Q}^s / |\mathbf{S}_\mathbf{Q}^r|$ are unit vectors. The directions of the mutually perpendicular vectors $\hat{\mathbf{n}}_r$ and $\hat{\mathbf{n}}_i$ remain arbitrary with respect to $\mathbf{Q}$, leaving a substantial degeneracy. However, this degeneracy can be eliminated by the presence of anisotropic terms in the Hamiltonian.

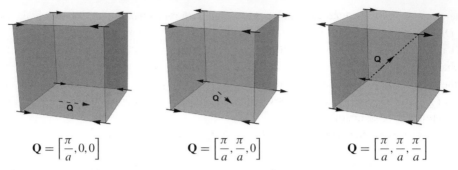

$$\mathbf{Q} = \left[\frac{\pi}{a}, 0, 0\right] \qquad \mathbf{Q} = \left[\frac{\pi}{a}, \frac{\pi}{a}, 0\right] \qquad \mathbf{Q} = \left[\frac{\pi}{a}, \frac{\pi}{a}, \frac{\pi}{a}\right]$$

Figure 23.1 A simple cubic system.

We find from (23.2) that for $\mathbf{Q} \neq 0$ we have helical or antiferromagnetic (AFM) configurations, depending on the magnitude of $\mathbf{Q}$ and its relative orientation with $\mathbf{S_Q}$. For example if we set $\mathbf{Q}$ to be inside the BZ, and set $\hat{\mathbf{n}}_r \equiv \hat{\mathbf{x}}$ and $\hat{\mathbf{n}}_i \equiv -\hat{\mathbf{y}}$, we obtain

$$S_i^x = S \cos\left(\mathbf{Q} \cdot \mathbf{R}_i\right)$$
$$S_i^y = S \sin\left(\mathbf{Q} \cdot \mathbf{R}_i\right)$$
$$S_i^z = 0,$$

which displays a helical phase with fixed S in any plane normal to $\mathbf{Q}$. Moreover, an incommensurate helical phase is manifest for $\mathbf{Q}$ not a high-symmetry point in the BZ, when $m\mathbf{Q} \neq \mathbf{G} \ \forall m$.

However, when $\mathbf{Q}$ is a high-symmetry point on the BZ boundary, we obtain AFM phases, such as the ones shown in Figure 23.1 for a simple cubic system as shown in Figure 23.1.

23.3 Ferromagnetic Insulators

We consider a ferromagnet with an arbitrary spin S, described by the Hamiltonian

$$H_F = \text{constant} - \sum_{l<m} 2J\left(\mathbf{R}_{lm}\right) \mathbf{S}_l \cdot \mathbf{S}_m - \mu_B B \sum_l S_l^z$$

$$+ \frac{1}{2} \sum_{lm} D_{lm} \left[\mathbf{S}_l \cdot \mathbf{S}_m - \frac{3(\mathbf{S}_l \cdot \mathbf{R}_{lm})(\mathbf{S}_m \cdot \mathbf{R}_{lm})}{R_{lm}^2} \right] + \mathcal{H}_{an}, \qquad (23.3)$$

where $\mathbf{R}_{lm} = \mathbf{R}_l - \mathbf{R}_m$. The first term, after the constant, is the isotropic Heisenberg exchange Hamiltonian and for a ferromagnetic ordering $J(\mathbf{R}_{lm}) > 0$. The second term is the Zeeman energy of the spin system in the external magnetic field $\mathbf{B}$, which is directed along the z-axis. The third term is the Hamiltonian for the magnetic dipole–dipole interaction. The last term, $\mathcal{H}_{an}$, is the single-ion anisotropic energy.

For the classical magnetic interaction

$$D_{lm} = \frac{g^2 \mu_B^2}{R_{lm}^3}, \qquad (23.4)$$

where μ_B is the Bohr magneton and g is the spectroscopic splitting factor. A typical value of D_{lm} is of the order of $10\,\mu$ eV, which is quite small when compared to $J \sim 1 - 100$ me V. It should be noted that D_{lm} will have a complicated form for anisotropic exchange interaction. The last term, $\mathcal{H}_{an}$, is the single-ion anisotropic energy. The presence of the dipolar term complicates matters considerably. We find that neither $\mathbf{S}^z = \sum_l \hat{S}_l^z$ nor $\mathbf{S}^2 = (\sum_l \mathbf{S}_l)^2$ commutes with the second part of the dipolar interaction term. Hence these are no longer constants of motion.

We next express the scalar product $\mathbf{S}_l \cdot \mathbf{S}_m$ in terms of $\hat{S}^+$, $\hat{S}^-$, and $\hat{S}^z$

$$\mathbf{S}_l \cdot \mathbf{S}_m = \hat{S}_l^x \hat{S}_m^x + \hat{S}_l^y \hat{S}_m^y + \hat{S}_l^z \hat{S}_m^z = \frac{1}{2} \left(\hat{S}_l^+ \hat{S}_m^- + \hat{S}_l^- \hat{S}_m^+ \right) + \hat{S}_l^z \hat{S}_m^z, \tag{23.5}$$

so that (23.3) becomes

$$\mathcal{H}_F = - \sum_{l<m} 2J'(\mathbf{R}_{lm}) \left[\frac{1}{2} \left(\hat{S}_l^+ \hat{S}_m^- + \hat{S}_l^- \hat{S}_m^+ \right) + \hat{S}_l^z \hat{S}_m^z \right]$$

$$- g\mu_B B \sum_l \hat{S}_l^z - \frac{3}{2} \sum_{lm} D_{lm} \frac{(\mathbf{S}_l \cdot \mathbf{R}_{lm})(\mathbf{S}_m \cdot \mathbf{R}_{lm})}{R_{lm}^2} + \mathcal{H}_{an}, \tag{23.6}$$

where we defined $J'(\mathbf{R}_{lm}) = J(\mathbf{R}_{lm}) - D_{lm}$.

The operator $\hat{S}_l^-$ ($\hat{S}_l^+$) creates (destroys) spin deviations on a specific site. Furthermore, the product of operators such as $\hat{S}_l^+ \hat{S}_m^-$ exchange spin deviations between two sites.

23.3.1 Ground State of Ferromagnetic Insulators

Henceforth, we shall consider the simple case of a nearest-neighbor ferromagnetically coupled insulator. This system can be represented by the Heisenberg Hamiltonian

$$H_F = -2J \sum_{<lm>} \mathbf{S}_l \cdot \mathbf{S}_m$$

$$= -2J \sum_{lm} \left[\left(\hat{S}_l^z \hat{S}_m^z - \frac{1}{4} \right) + \frac{1}{2} \left(\hat{S}_l^+ \hat{S}_m^- + \hat{S}_l^- \hat{S}_m^+ \right) \right], \tag{23.7}$$

with $J > 0$, and $\langle lm \rangle$ indicates nearest neighbor (nn). Both dipolar and other anisotropic interactions are neglected. We find that both $\mathbf{S}^z$ and $\mathbf{S}^2$ commute with this Hamiltonian, and, therefore, M_s and S, S being the maximal spin per site, are good quantum numbers, with $M_s = NS, (N-1)S, \ldots, -NS$.

We consider the state of maximal spin alignment at every site, i.e., $\hat{S}_l^+ |S, S\rangle_l = 0$,

$$|\Psi_0\rangle \equiv |\emptyset\rangle = \prod_l |S, S\rangle_l. \tag{23.8}$$

We note that

$$\hat{S}_l^+ \hat{S}_m^- |\Psi_0\rangle = \hat{S}_l^- \hat{S}_m^+ \prod_l |S, S\rangle_l = 0,$$

and the ground-state energy is then

$$\langle \Psi_0 | H_F | \Psi_0 \rangle = -2 J S^2 N.$$

Now we consider the expectation value $\langle \psi | \mathcal{H}_H | \psi \rangle$ in an arbitrary product state

$$|\psi\rangle = \prod_l |S, m_l\rangle,$$

which is generally not an eigenstate of $\mathcal{H}_H$. The expectation value is

$$\langle \psi | \mathcal{H}_H | \psi \rangle = -\frac{1}{2} J \sum_{lm} \left[\underbrace{\frac{\langle \psi | \hat{S}_l^+ \hat{S}_m^- | \psi \rangle}{2}}_{= 0} + \underbrace{\frac{\langle \psi | \hat{S}_l^+ \hat{S}_m^- | \psi \rangle}{2}}_{= 0} + \langle \psi | \hat{S}_l^z \hat{S}_m^- | \psi \rangle \right]$$

$$= -\frac{1}{2} J \sum_{lm} m_l m_m > -\frac{1}{2} J S^2 N.$$

Since the $|\psi\rangle$s form a basis, we conclude that all eigenenergies, in particular the ground-state energy, are larger than or equal to $-\frac{1}{2} J S^2 N$. Since $|\Psi_0\rangle$ is an eigenstate with eigenenergy $-\frac{1}{2} J S^2 N$, $|\Psi_0\rangle$ must be a ground state. Moreover, $|\Psi_0\rangle$ is an eigenstate of the Hamiltonian

$$H_F | \Psi_0 \rangle = -2 J S^2 N | \Psi_0 \rangle.$$

We have found an example of spontaneous symmetry breaking, where the ground-state manifold of the completely ferromagnetic Heisenberg model consists of fully polarized states (maximum S_{tot}). Typical ground states thus have $\langle S \rangle \neq 0$, so that they distinguish a certain direction in spin space. Yet the Hamiltonian $\mathcal{H}_H$ does not. Thus the symmetry of a typical ground state is lower than that of the Hamiltonian. We have a case of spontaneously broken symmetry!

23.3.2 *Spin Waves in Ferromagnetic Insulators*

The manifold of lowest excited states consists of configurations in which a single *spin deviation* from the alignment of the completely ferromagnetic Ψ_0 state takes place (see Figure 23.2). It is represented by

$$\Psi_1(l) = \left[\frac{1}{2S} \right]^{1/2} \hat{S}_l^- | \emptyset \rangle = |S, S, \ldots, S - 1(l), S, \ldots, S\rangle. \qquad (23.9)$$

Figure 23.2 A single local spin-flip excitation of a ferromagnetic system. The resulting state is not an eigenstate of the Heisenberg Hamiltonian.

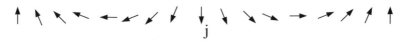

Figure 23.3 If we spread out the spin-flip over a wider region, then we can create a lower-energy excitation. A spin-wave is the completely delocalized analog of this with one net spin-flip.

These spin-deviation states are orthonormal:

$$\langle \Psi_1(m)|\Psi_1(l)\rangle = \frac{1}{2S}\langle \emptyset|\hat{S}_m^+ \hat{S}_l^-|\emptyset\rangle$$

$$= \frac{1}{2S}\langle \emptyset|2\delta_{lm}\,\hat{S}_l^z + \hat{S}_m^-\hat{S}_l^+|\emptyset\rangle$$

$$= \frac{1}{2S}\langle \emptyset|2\delta_{lm}\,\hat{S}_l^z|\emptyset\rangle = \delta_{lm}.$$

The states $\Psi_1(l)$ are localized, they are not eigenfunctions of the Hamiltonian, but do furnish a convenient basis for the solution of the problem of determining the eigenstates. If we spread out the spin-flip over a wider region then we can create a spin wave, as shown in Figure 23.3.

Low-Energy Excitation Spectrum: Holstein–Primakoff (HP) Operators

In 1940, Holstein and Primakoff showed how spin operators could be expressed in terms of true bosonic-fields.[1] The HP representation is best understood as a special case of the Schwinger coupled-Boson representation.

Coupled-Boson Representation; Schwinger Bosons

The general matrix structure of the angular momentum operators

$$\langle j'm'|\hat{\mathbf{J}}^{\pm}|jm\rangle = \delta_{jj'}\delta_{m\pm 1,m'}\,\hbar\,\sqrt{(j\mp m)(j\pm m+1)}$$

$$\langle j'm'|\hat{J}^z|jm\rangle = \delta_{jj'}\delta_{m,m'}\,\hbar m$$

$$\langle j'm'|\hat{J}^2|jm\rangle = \delta_{jj'}\delta_{m,m'}\,\hbar^2 j(j+1) \tag{23.10}$$

resembles in many respects the matrix structure of harmonic oscillator operators

$$\hat{a}^{\dagger} = \frac{1}{\sqrt{2\hbar}}\left(\frac{\hbar}{i}\frac{\partial}{\partial x}+ix\right); \quad \hat{a} = \frac{1}{\sqrt{2\hbar}}\left(\frac{\hbar}{i}\frac{\partial}{\partial x}-ix\right) \tag{23.11}$$

with commutation relations

$$\left[\hat{a}_l,\hat{a}_m^{\dagger}\right] = \delta_{lm}, \quad \left[\hat{a}_l,\hat{a}_m\right] = \left[\hat{a}_l^{\dagger},\hat{a}_m^{\dagger}\right] = 0$$

and

$$\left[\hat{n}_l,\hat{a}_m^{\dagger}\right] = \delta_{lm}\,\hat{a}_m^{\dagger}; \quad \left[\hat{n}_l,\hat{a}_m\right] = -\delta_{lm}\,\hat{a}_m.$$

---

[1] While still in graduate school, Primakoff and Holstein developed the theory of spin waves.

The normalized two-particle states are $\hat{a}_1^\dagger \hat{a}_2^\dagger |\emptyset\rangle$ and $(\hat{a}_1^{\dagger 2}/\sqrt{2})|\emptyset\rangle$. The general many-particle state is

$$\frac{\left(\hat{a}_1^\dagger\right)^{n_1}\left(\hat{a}_2^\dagger\right)^{n_2}\cdots}{\sqrt{(n_1!\,n_2!\ldots}}|\emptyset\rangle$$

with $n_i > 0$, and the total occupation is given by $\sum_i n_i$.

Schwinger showed that with the aid of only two harmonic oscillators, the entire matrix structure of a single angular momentum could be exactly reproduced. Labeling the two oscillators by 1 and 2, we introduce the *spinor operators*

$$\hat{\mathbf{a}}^\dagger = \left(\hat{a}_1^\dagger, \hat{a}_2^\dagger\right) \qquad \text{and} \qquad \hat{\mathbf{a}} = \begin{pmatrix} \hat{a}_1 \\ \hat{a}_2 \end{pmatrix}, \tag{23.12}$$

which are merely two-component vectors with operator components. Moreover, if we contract these operators with the Pauli spin matrices, we obtain the desired representation. That is, let

$$J^z = \frac{\hbar}{2}\,\hat{\mathbf{a}}^\dagger \cdot \hat{\sigma}_z \cdot \hat{\mathbf{a}} = \frac{\hbar}{2}\,\hat{a}^\dagger \begin{bmatrix} 1 & 0 \\ 0 & -1 \end{bmatrix} \hat{a} = \frac{\hbar}{2}\,(\hat{n}_1 - \hat{n}_2),$$

and similarly for the other components, with the following compact result:

$$\hat{\mathbf{J}} = \frac{\hbar}{2}\,\hat{\mathbf{a}}^\dagger \cdot \hat{\sigma} \cdot \hat{\mathbf{a}}. \tag{23.13}$$

We next define

$$\hat{\mathbf{j}} = \frac{1}{2}\,\hat{\mathbf{a}}^\dagger \cdot \hat{\mathbf{I}} \cdot \hat{\mathbf{a}} = \frac{1}{2}\,(\hat{n}_1 + \hat{n}_2) \tag{23.14}$$

so that we write the state $|jm\rangle$ as

$$|jm\rangle = \frac{(\hat{a}_1^\dagger)^{j+m}(\hat{a}_2^\dagger)^{j-m}}{\sqrt{(j+m)!\,(j-m)!}}|\emptyset\rangle.$$

It may be verified that the eigenvalues of $\hat{\mathbf{j}}$ are indeed $j = 0,\ 1/2,\ 1,\ 3/2,\ldots$, where $\mathbf{J}^2$ has eigenvalue $\hbar^2\,j(j+1)$. The coupled-Boson operators have more flexibility than the original angular momentum operators. For example, we can make use of the extra degrees of freedom to construct the so-called hyperbolic operators, which conserve m, but raise or lower j.

Holstein and Primakoff proposed an irreducible representation of the Schwinger coupled-Boson in a subspace of fixed j according to the following argument: since from (23.14) we have

$$\hat{\mathbf{j}} = \frac{1}{2}\left(\hat{a}_1^\dagger \hat{a}_1 + \hat{a}_2^\dagger \hat{a}_2\right),$$

we find that

$$\hat{a}_2^\dagger \hat{a}_2 = 2\hat{j} - \hat{n}_1 = \sqrt{(2\hat{j} - \hat{n}_1)}\sqrt{(2\hat{j} - \hat{n}_1)}.$$

In a subspace of fixed j, this equation is solved by treating $\hat{a}_2$ and its conjugate as two diagonal operators

$$\hat{a}_2 = \hat{a}_2^\dagger = (2j)^{1/2}\sqrt{1 - \frac{\hat{\mathbf{n}}}{2j}}, \tag{23.15}$$

and hence we find

$$\mathbf{J}^+ = \hbar\, \hat{a}_1^\dagger \hat{a}_2 = \hbar\,(2j)^{1/2}\,\hat{\mathbf{a}}^\dagger \sqrt{1 - \frac{\hat{\mathbf{n}}}{2j}},$$

$$\mathbf{J}^- = \hbar\, \hat{a}_2^\dagger \hat{a}_1 = \hbar\,(2j)^{1/2}\sqrt{1 - \frac{\hat{\mathbf{n}}}{2j}}\,\hat{\mathbf{a}},$$

$$\mathbf{J}^{\mathbf{z}} = \hbar(\hat{\mathbf{n}} - j). \tag{23.16}$$

If $n > 2j$, the formalism is incorrect; however, within that range the commutation relations

$$\hat{J} \times \hat{J} = i\hbar\hat{J}$$
$$\left[\hat{J}^z, \hat{J}^\pm\right] = \pm\hbar\hat{J}^\pm,$$
$$\left[\hat{J}^+, \hat{J}^-\right] = , 2\hbar\hat{J}^z,$$
$$\left[\hat{J}^2, \hat{J}\ \right] = \left[\hat{J}^2, \hat{J}^\pm\right] = 0 \tag{23.17}$$

are satisfied. Remembering that

$$\mathbf{J}^{\mathbf{z}}\left|jm\right\rangle = \hbar m\left|jm\right\rangle = \hbar(n - j)\left|jm\right\rangle,$$

we find that $n = j - m$, and hence we obtain

$$\left\langle m\right|\hat{J}^+\left|m - 1\right\rangle = \left\langle n\right|\hat{J}^+\left|n + 1\right\rangle = \sqrt{(j - m + 1)(j + m)}$$
$$= \sqrt{(n + 1)(2j - n)}. \tag{23.18}$$

n can thus be identified as the *angular-momentum-deviation number operator* from the state with maximal m, i.e., $m = j$. Consequently, Holstein and Primakoff (HP) introduced the operator $\hat{\mathbf{n}}_l$ as the spin-deviation number operator from the ground-state value S at site l, namely,

$$\hat{\mathbf{n}}_l = \mathbf{a}_l^\dagger \mathbf{a}_l = S - \hat{\mathbf{S}}_l^z, \tag{23.19}$$

and thus were able to write the equations for $\hat{\mathbf{S}}_l^-$ and $\hat{\mathbf{S}}_l^+$ as

$$\hat{\mathbf{S}}_l^+\left|n_l\right\rangle = (2S)^{1/2}\left[1 - \frac{n_l - 1}{2S}\right]^{1/2}\sqrt{n_l}\left|n_l - 1\right\rangle,$$

$$\hat{\mathbf{S}}_l^-\left|n_l\right\rangle = (2S)^{1/2}\sqrt{n_l + 1}\left[1 - \frac{n_l}{2S}\right]^{1/2}\left|n_l + 1\right\rangle, \tag{23.20}$$

which can be recast in the form of (23.16) as

$$\hat{S}_l^- = (2S)^{1/2}\, \hat{a}_l^\dagger \sqrt{1 - \frac{\hat{a}_l^\dagger \hat{a}_l}{2S}}$$

$$\hat{S}_l^+ = (2S)^{1/2} \sqrt{1 - \frac{\hat{a}_l^\dagger \hat{a}_l}{2S}}\, \hat{a}_l$$

$$S_l^z = S - \hat{a}_l^\dagger \hat{a}_l. \tag{23.21}$$

Moreover, substituting (23.21) into the third equation in (23.17) yields

$$\left[\hat{a}_l, \hat{a}_m^\dagger\right] = \delta_{lm}, \tag{23.22}$$

all other commutators of $\hat{a}_l^\dagger$ and $\hat{a}_l$ being zero. Thus we interpret $\hat{a}_l^\dagger$ and $\hat{a}_l$ as spin-deviation creation and annihilation operators that obey Bose commutation relations.

A few words about the HP radical

$$\sqrt{1 - \frac{\hat{a}_l^\dagger \hat{a}_l}{2S}}$$

are in order. For any given value of S (integer or half-integer), it is always possible to construct a polynomial of order $2S$ that has the same structure as the HP radical in the physical domain, i.e., $0 \le \hat{a}_l^\dagger \hat{a}_l \le 2S$. Outside this physical domain, of course, the correspondence is lost. For large values of S, however, it is not easy to construct the preceding polynomial. On the other hand, for small values of $n_l/2S$ the Taylor expansion, namely

$$\sqrt{1 - \frac{\hat{a}_l^\dagger \hat{a}_l}{2S}} = 1 - \frac{1}{2}\left(\frac{\mathbf{n}_l}{2S}\right) + \frac{1}{8}\left(\frac{\mathbf{n}_l}{2S}\right)^2 + \cdots, \tag{23.23}$$

becomes asymptotically exact for $n_l/2S \rightarrow 0$. Accordingly, for small fractional spin deviation $(n_l/2S) \ll 1$, we can truncate the preceding expansion at an appropriate stage. This is evidently valid in the low-temperature regime.

We shall derive the excitation spectrum for the isotropic Heisenberg Hamiltonian, (23.7), plus the Zeeman interaction

$$\mathcal{H}_F = -J \sum_{lm} \left[\frac{1}{2}\left(S_l^+ S_m^- + S_l^- S_m^+\right) + S_l^z S_m^z\right] - g\mu_B \mathbf{B} \sum_l S_l^z \tag{23.24}$$

and making use of the relations (23.21), we obtain, in the site representation:

$$\mathcal{H}_F = -J \sum_{<lm>} \left[S\hat{a}_l^\dagger \sqrt{1 - \frac{\hat{a}_l^\dagger \hat{a}_l}{2S}} \sqrt{1 - \frac{\hat{a}_m^\dagger \hat{a}_m}{2S}}\, \hat{a}_m + S\sqrt{1 - \frac{\hat{a}_l^\dagger \hat{a}_l}{2S}}\, \hat{a}_l\, \hat{a}_m^\dagger \sqrt{1 - \frac{\hat{a}_m^\dagger \hat{a}_m}{2S}} \right.$$

$$\left. + \left(S - \hat{a}_l^\dagger \hat{a}_l\right)\left(S - \hat{a}_m^\dagger \hat{a}_m\right) \right] + g\mu_B B \sum_l \hat{a}_l^\dagger \hat{a}_l$$

$$\approx \mathcal{H}_{bl} + \mathcal{H}_q + O\left(\frac{1}{S}\right)$$

$$\mathcal{H}_{bl} = -N\mathcal{Z}S^2 J - SJ \sum_{<lm>} \left[\hat{a}_l^\dagger \hat{a}_m + \hat{a}_l \hat{a}_m^\dagger - \hat{a}_l^\dagger \hat{a}_l - \hat{a}_m^\dagger \hat{a}_m \right] + g\mu_B B \sum_l \hat{a}_l^\dagger \hat{a}_l$$

$$\mathcal{H}_q = -J \sum_{<lm>} \left[\hat{a}_l^\dagger \hat{a}_l \hat{a}_m^\dagger \hat{a}_m - \frac{1}{4}\left(\hat{a}_l^\dagger \hat{a}_l^\dagger \hat{a}_l \hat{a}_m + \hat{a}_l^\dagger \hat{a}_m^\dagger \hat{a}_m \hat{a}_m \right. \right.$$

$$\left. \left. + \hat{a}_l^\dagger \hat{a}_l \hat{a}_l \hat{a}_m^\dagger + \hat{a}_l \hat{a}_m^\dagger \hat{a}_m^\dagger \hat{a}_m \right) \right]. \tag{23.25}$$

We shall confine our study to very low temperatures, where $(\langle n_l \rangle /4S) \ll 1$ and hence keep terms to order S in (23.25), which means we neglect $\mathcal{H}_q$, which contains quartic terms. In order to diagonalize $\mathcal{H}_{bl}$, we perform the following canonical transformation

$$\left. \begin{array}{l} \hat{a}_l = \dfrac{1}{\sqrt{N}} \displaystyle\sum_q \exp(-i q \cdot R_l)\hat{a}_q \\[2mm] \hat{a}_l^\dagger = \dfrac{1}{\sqrt{N}} \displaystyle\sum_q \exp(i q \cdot R_l) \hat{a}_q^\dagger \end{array} \right\} \Rightarrow \begin{array}{l} \hat{a}_q = \dfrac{1}{\sqrt{N}} \displaystyle\sum_q \exp(i q \cdot R_l)\hat{a}_l \\[2mm] \hat{a}_q^\dagger = \dfrac{1}{\sqrt{N}} \displaystyle\sum_q \exp(-i q \cdot R_l) \hat{a}_l^\dagger . \end{array} \tag{23.26}$$

After substituting (23.26) in (23.25), the Hamiltonian in (23.25) becomes

$$\mathcal{H}_F = -N\mathcal{Z}S^2 J + \sum_q \hbar\omega(q) \left(\hat{a}_q^\dagger \hat{a}_q + \frac{1}{2} \right)$$

$$= -N\mathcal{Z}S^2 J + \sum_q \hbar\omega(q) \left(\hat{n}_q + \frac{1}{2} \right), \tag{23.27}$$

where

$$\hbar\omega(q) = 2JS \mathcal{Z} \left(1 - \gamma_q\right) + g\mu_B B$$

$$\gamma_q = \frac{1}{\mathcal{Z}} \sum_{s=1}^{\mathcal{Z}} \exp(i q \cdot \Delta R_s). \tag{23.28}$$

This Hamiltonian has the well-known harmonic oscillator form. Here $\hat{a}_q^\dagger \hat{a}_q$ is the number operator $\hat{n}_q$, $\hat{a}_q^\dagger$ creates a spin-wave quantum and $\hat{a}_q$ annihilates a spin-wave quantum. These quanta are referred to as magnons. As can be seen from (23.22), magnons obey Bose statistics. The dispersion relation takes a simple form in the long-wavelength limit

$$\hbar\omega(q) = JS \left(\sum_s \Delta R_s^2 \cos^2 \theta_{q,\Delta R_s} \right) q^2 + g\mu_B B, \tag{23.29}$$

where $\theta_{q,\Delta R_s}$ is the angle between the vectors q and ΔR_s. For a simple cubic system, the mean value of $\cos^2 \theta_{q,\Delta R_s}$ is $1/3$ and the mean value of $\mathcal{Z}\Delta R_s^2$ is $6a^2$, where a is the lattice constant. This yields the dispersion relation

$$\hbar\omega(q) = 2JS a^2 q^2 + g\mu_B B, \tag{23.30}$$

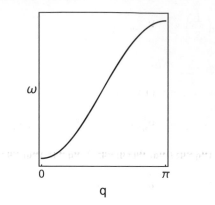

Figure 23.4 Magnon, or spin-wave, dispersion for a simple cubic ferromagnet (lattice constant $a = 1$) along the [1 0 0]-direction. The dispersion is parabolic at small q.

which has a parabolic dependence on the wavevector, as shown in Figure 23.4. The eigenfunctions of the magnon Hamiltonian in the occupation number representation have the well-known form

$$|\ldots n_{\mathbf{q}}\ldots\rangle = \prod_{\mathbf{q}} \frac{1}{\sqrt{n_{\mathbf{q}}!}} (\hat{a}_{\mathbf{q}}^{\dagger})^{n_{\mathbf{q}}} |0\rangle, \qquad (23.31)$$

where $|0\rangle$ is the magnon vacuum state, which corresponds to complete ferromagnetic alignment.

23.3.3 Magnon–Magnon Interactions in Mean Field

The quartic terms in $\mathcal{H}_{\mathbf{q}}$, which we neglected, represent magnon–magnon (particle–particle) interactions. We shall study here the effect of such interactions on the magnon dispersion. We are now familiar with the Hartree–Fock-like approximation; here it is essentially equivalent to RPA. We find

$$\mathcal{H}_{\mathrm{q}} = \frac{-J}{N^2} \sum_{\substack{\mathbf{q},\mathbf{q}_1 \\ \mathbf{q}_2,\mathbf{q}_3}} \sum_{<lm>} \left[e^{-i(\mathbf{q}\cdot\mathbf{R}_l + \mathbf{q}_1\cdot\mathbf{R}_m - \mathbf{q}_2\cdot\mathbf{R}_m - \mathbf{q}_3\cdot\mathbf{R}_l)} - \frac{1}{4}\left(e^{-i(\mathbf{q}\cdot\mathbf{R}_l + \mathbf{q}_1\cdot\mathbf{R}_l - \mathbf{q}_2\cdot\mathbf{R}_l - \mathbf{q}_3\cdot\mathbf{R}_m)} \right. \right.$$

$$+ e^{-i(\mathbf{q}\cdot\mathbf{R}_l + \mathbf{q}_1\cdot\mathbf{R}_m - \mathbf{q}_2\cdot\mathbf{R}_m - \mathbf{q}_3\cdot\mathbf{R}_m)} + e^{-i(\mathbf{q}\cdot\mathbf{R}_m + \mathbf{q}_1\cdot\mathbf{R}_l - \mathbf{q}_2\cdot\mathbf{R}_l - \mathbf{q}_3\cdot\mathbf{R}_l)}$$

$$\left. \left. + e^{-i(\mathbf{q}\cdot\mathbf{R}_m + \mathbf{q}_1\cdot\mathbf{R}_m - \mathbf{q}_2\cdot\mathbf{R}_m - \mathbf{q}_3\cdot\mathbf{R}_l)} \right) \right] \hat{a}_{\mathbf{q}}^{\dagger} \hat{a}_{\mathbf{q}_1}^{\dagger} \hat{a}_{\mathbf{q}_2} \hat{a}_{\mathbf{q}_3}.$$

With $\mathbf{R}_m = \mathbf{R}_l + \Delta\mathbf{R}_m^s$, we obtain

$$\mathcal{H}_{\mathrm{q}} = \frac{-J}{N} \sum_{\mathbf{q},\mathbf{q}_1,\mathbf{q}_2} \sum_{s} \left[e^{-i(\mathbf{q}-\mathbf{q}_2)\cdot\Delta\mathbf{R}_s} - \frac{1}{4}\left(e^{i(\mathbf{q}+\mathbf{q}_1-\mathbf{q}_2)\cdot\Delta\mathbf{R}_s} + e^{i\mathbf{q}\cdot\Delta\mathbf{R}_s} + e^{-i\mathbf{q}\cdot\Delta\mathbf{R}_s} \right. \right.$$

$$\left. \left. + e^{-i(\mathbf{q}+\mathbf{q}_1-\mathbf{q}_2)\cdot\Delta\mathbf{R}_s} \right) \right] \hat{a}_{\mathbf{q}}^{\dagger} \hat{a}_{\mathbf{q}_1}^{\dagger} \hat{a}_{\mathbf{q}_2} \hat{a}_{\mathbf{q}+\mathbf{q}_1-\mathbf{q}_2}$$

$$= \frac{-J}{N} \sum_{\mathbf{q},\mathbf{q}_1,\mathbf{q}_2} \sum_s \left[e^{-i(\mathbf{q}-\mathbf{q}_2)\cdot\Delta\mathbf{R}_s} - \frac{1}{2}\left(\cos\left[\mathbf{q}\cdot\Delta\mathbf{R}_s\right] + \cos\left[(\mathbf{q}+\mathbf{q}_1-\mathbf{q}_2)\cdot\Delta\mathbf{R}_s\right] \right) \right]$$
$$\times \hat{a}_{\mathbf{q}}^\dagger \hat{a}_{\mathbf{q}_1}^\dagger \, \hat{a}_{\mathbf{q}_2} \, \hat{a}_{\mathbf{q}+\mathbf{q}_1-\mathbf{q}_2}.$$

Now we use the HF approximation for the quartic operator product

$$\hat{a}_{\mathbf{q}}^\dagger\hat{a}_{\mathbf{q}_1}^\dagger\hat{a}_{\mathbf{q}_2}\hat{a}_{\mathbf{q}+\mathbf{q}_1-\mathbf{q}_2} \approx \hat{a}_{\mathbf{q}}^\dagger\hat{a}_{\mathbf{q}+\mathbf{q}_1-\mathbf{q}_2}\left\langle \hat{a}_{\mathbf{q}_1}^\dagger \hat{a}_{\mathbf{q}_2}\right\rangle + \left\langle \hat{a}_{\mathbf{q}}^\dagger\hat{a}_{\mathbf{q}+\mathbf{q}_1-\mathbf{q}_2}\right\rangle\hat{a}_{\mathbf{q}_1}^\dagger \hat{a}_{\mathbf{q}_2} + \left\langle \hat{a}_{\mathbf{q}}^\dagger\hat{a}_{\mathbf{q}+\mathbf{q}_1-\mathbf{q}_2}\right\rangle\left\langle \hat{a}_{\mathbf{q}_1}^\dagger \hat{a}_{\mathbf{q}_2}\right\rangle$$
$$+ \left\langle \hat{a}_{\mathbf{q}}^\dagger \hat{a}_{\mathbf{q}_2}\right\rangle\hat{a}_{\mathbf{q}_1}^\dagger \, \hat{a}_{\mathbf{q}+\mathbf{q}_1-\mathbf{q}_2} + \hat{a}_{\mathbf{q}}^\dagger \, \hat{a}_{\mathbf{q}_2} \left\langle \hat{a}_{\mathbf{q}_1}^\dagger \, \hat{a}_{\mathbf{q}+\mathbf{q}_1-\mathbf{q}_2}\right\rangle$$
$$+ \left\langle \hat{a}_{\mathbf{q}}^\dagger \hat{a}_{\mathbf{q}_2}\right\rangle\left\langle \hat{a}_{\mathbf{q}_1}^\dagger \, \hat{a}_{\mathbf{q}+\mathbf{q}_1-\mathbf{q}_2}\right\rangle$$
$$= \begin{cases} \left[n_{\mathbf{q}_1} \hat{a}_{\mathbf{q}}^\dagger \hat{a}_{\mathbf{q}} + n_{\mathbf{q}} \hat{a}_{\mathbf{q}_1}^\dagger \hat{a}_{\mathbf{q}_1} - n_{\mathbf{q}_1} n_{\mathbf{q}} \right] & \to \quad \mathbf{q}_1 = \mathbf{q}_2 \\ \left[n_{\mathbf{q}_1} \hat{a}_{\mathbf{q}}^\dagger \hat{a}_{\mathbf{q}} + n_{\mathbf{q}} \hat{a}_{\mathbf{q}_1}^\dagger \hat{a}_{\mathbf{q}_1} - n_{\mathbf{q}_1} n_{\mathbf{q}} \right] & \to \quad \mathbf{q} = \mathbf{q}_2 \end{cases}.$$

Substituting in $\mathcal{H}_{\mathbf{q}}$, we get

$$\mathcal{H}_{\mathbf{q}} = -J \sum_{\mathbf{q},\mathbf{q}_1;s} \left[1 + e^{-i(\mathbf{q}-\mathbf{q}_1)\cdot\Delta\mathbf{R}_s} - \cos\left[\mathbf{q}\cdot\Delta\mathbf{R}_s\right] \right.$$
$$\left. - \frac{1}{2}\left(\cos\left[\mathbf{q}\cdot\Delta\mathbf{R}_s\right] + \cos\left[\mathbf{q}_1\cdot\Delta\mathbf{R}_s\right] \right) \right] \left[n_{\mathbf{q}_1} \hat{a}_{\mathbf{q}}^\dagger \hat{a}_{\mathbf{q}} + n_{\mathbf{q}} \hat{a}_{\mathbf{q}_1}^\dagger \, \hat{a}_{\mathbf{q}_1} \right]$$
$$= -J \sum_{\mathbf{q},\mathbf{q}_1;s} \left[2 + 2\cos\left[(\mathbf{q}-\mathbf{q}_1)\cdot\Delta\mathbf{R}_s\right] - 2\cos\left[\mathbf{q}\cdot\Delta\mathbf{R}_s\right] - 2\cos\left[\mathbf{q}_1\cdot\Delta\mathbf{R}_s\right] \right] n_{\mathbf{q}_1}\hat{a}_{\mathbf{q}}^\dagger \hat{a}_{\mathbf{q}}$$
$$= \sum_{\mathbf{q}} \hbar\, \Delta\omega_{\mathbf{q}} \, \hat{a}_{\mathbf{q}}^\dagger \hat{a}_{\mathbf{q}},$$

where

$$\hbar\,\Delta\omega_{\mathbf{q}} = -2J \sum_{\mathbf{q}_1;s} \left[1 + \cos\left[(\mathbf{q}-\mathbf{q}_1)\cdot\Delta\mathbf{R}_s\right] - \cos\left[\mathbf{q}\cdot\Delta\mathbf{R}_s\right] - \cos\left[\mathbf{q}_1\cdot\Delta\mathbf{R}_s\right] \right] n_{\mathbf{q}_1}$$
$$= -2\mathcal{Z}SJ \left(\frac{1}{\mathcal{Z}}\sum_s \left(1 - \cos\left[\mathbf{q}\cdot\Delta\mathbf{R}_s\right]\right) \right) \frac{1}{NS}\sum_{\mathbf{q}_1}\left(1 - \cos\left[\mathbf{q}\cdot\Delta\mathbf{R}_s\right]\right) n_{\mathbf{q}_1}$$
$$= -2\mathcal{Z}SJ \left(1 - \gamma_{\mathbf{q}}\right) \frac{1}{NS}\sum_{\mathbf{q}_1}\left(1 - \cos\left[\mathbf{q}\cdot\Delta\mathbf{R}_s\right]\right) n_{\mathbf{q}_1}.$$

In the second line, we used the property

$$\sum_{\mathbf{q}_1}\cos\left[(\mathbf{q}-\mathbf{q}_1)\cdot\Delta\mathbf{R}_s\right] = \sum_{\mathbf{q}_1}\left(\cos\left[\mathbf{q}\cdot\Delta\mathbf{R}_s\right]\cos\left[\mathbf{q}_1\cdot\Delta\mathbf{R}_s\right] + \sin\left[\mathbf{q}\cdot\Delta\mathbf{R}_s\right]\sin\left[\mathbf{q}_1\cdot\Delta\mathbf{R}_s\right] \right)$$
$$= \sum_{\mathbf{q}_1} 2\cos\left[\mathbf{q}\cdot\Delta\mathbf{R}_s\right]\cos\left[\mathbf{q}_1\cdot\Delta\mathbf{R}_s\right].$$

The mean-field renormalization of the ferromagnetic spin-wave dispersion due to magnon–magnon interaction is

$$\hbar\omega_{\mathbf{q}}^{MF} = 2\mathcal{Z}SJ\left(1 - \gamma_{\mathbf{q}}\right)\left[1 - \frac{1}{NS}\sum_{\mathbf{q}_1}\left(1 - \cos\left[\mathbf{q}_1 \cdot \Delta\mathbf{R}_s\right]\right)n_{\mathbf{q}_1}\right].$$

To assess the magnitude of this correction, we consider a simple cubic lattice with the dispersion relation (23.30), valid at low temperatures, we write

$$\frac{1}{N}\sum_{\mathbf{q}_1}\left(1 - \cos\left[\mathbf{q}\cdot\Delta\mathbf{R}_s\right]\right)n_{\mathbf{q}} = \frac{\Omega}{N}\int\frac{d\mathbf{q}}{(2\pi)^3}\frac{1 - \cos\left[\mathbf{q}\cdot\Delta\mathbf{R}_s\right]}{e^{\beta\hbar\omega_{\mathbf{q}}} - 1}$$

$$\simeq \frac{\Omega}{(2\pi)^2 N}\int_0^\infty dq\, q^2\int_{-1}^1 dx\,\frac{1 - \cos(qax)}{e^{2\beta JSa^2q^2} - 1} = \frac{\Omega}{2\pi^2 N}\int_0^\infty dq\, q^2\,\frac{1 - \sin(qa)/qa}{e^{2\beta JSa^2q^2} - 1}$$

$$\simeq \frac{\Omega}{(2\pi)^2 N}\int_0^\infty 2dq\, q^2\,\frac{q^2a^2/6}{e^{2\beta JSa^2q^2} - 1} = \frac{a^5}{(2\pi)^2(2\beta JSa^2)^{5/2}}\int_0^\infty dt\,\frac{t^{3/2}}{e^t - 1}$$

$$= \frac{\zeta(5/2)}{32\pi^{3/2}(2\beta JS)^{5/2}} = 5.3 \times 10^{-4}\left(\frac{k_B}{SJ}\right)^{5/2}T^{5/2}.$$

23.3.4 Thermal Fluctuations in a Ferromagnet

The magnetization (per site) is

$$M = \frac{1}{N}\sum_l\langle S_l^z\rangle.$$

Using the Holstein–Primakoff transformation at leading order, we have

$$M = S - \frac{1}{N}\sum_{\mathbf{k}}\left\langle a_{\mathbf{k}}^\dagger a_{\mathbf{k}}\right\rangle.$$

Now, since the excitations are bosons, the thermal average is given by the Bose distribution, and we obtain

$$\frac{1}{N}\sum_{\mathbf{k}}\left\langle a_{\mathbf{k}}^\dagger a_{\mathbf{k}}\right\rangle \rightarrow \frac{1}{\Omega}\int_{BZ}d\mathbf{k}^d\,\frac{1}{e^{\beta\varepsilon(\mathbf{k})} - 1}.$$

The most interesting aspects of this result are the generic ones, which emerge at low temperature. In that regime, only low-energy magnons are excited, and for these we can take the small wavevector form for their energy. We find, for example from (23.30), that

$$\frac{1}{N}\sum_{\mathbf{k}}\left\langle a_{\mathbf{k}}^\dagger a_{\mathbf{k}}\right\rangle = \frac{k_B T}{J}\int_0^{\sqrt{k_B T/J}}dk\, k^{d-1}\frac{1}{k^2}.$$

The integral is divergent for $T > 0$ in $d = 1$ and $d = 2$, showing that long-range order is not possible in the Heisenberg ferromagnet in low dimensions (an illustration of the Mermin–Wagner theorem, which says that a continuous symmetry cannot be broken spontaneously at finite temperature for $d \leq 2$. In $d = 3$, the integral is $\propto T^{3/2}$.

23.4 Antiferromagnetic Insulators

23.4.1 The Hubbard Model: Kinetic Exchange Revisited

The Hubbard model has served as a kind of Rosetta stone for interacting electron systems. It is defined by the Hamiltonian

$$\mathcal{H} = -\frac{1}{2} \sum_{ij,\sigma} t_{ij} \left(c^{\dagger}_{i\sigma} c_{j\sigma} + c^{\dagger}_{j\sigma} c_{i\sigma} \right) + U \sum_i n_{i\uparrow} n_{i\downarrow} + \mu_B \mathbf{B} \cdot \sum_I c^{\dagger}_{i\alpha} \boldsymbol{\sigma}_{\alpha\beta} c_{i\beta}, \quad (23.32)$$

where the first term describes hopping of electrons between ionic sites in the solid, with $\langle ij \rangle$ denoting sites i and j. The second term describes the local (onsite) energy cost of Coulomb repulsion between two electrons of opposite spins in a single orbital. The last term is the Zeeman interaction of the electron spins with an external magnetic field. Typically, the Hubbard U parameter is on the order of electron volts. The short-ranged character of this interaction is justified on the basis of effective screening of the Coulomb potential in metals. It is a good approximation when the screening length is of the order of the lattice constant. We note that the local nature of the interaction term $U \, n_{i\uparrow} \, n_{i\downarrow}$ favors a local onsite moment. We argue that the chemical potential fixes the mean value of the total occupancy $n_{i\uparrow} + n_{i\downarrow}$, and that it is preferable to maximize the difference $\left| n_{i\uparrow} - n_{i\downarrow} \right|$ in order to minimize the contribution from this interaction. We will focus here on the case of half-filling, namely, $n = 1$, one electron per site on average, and limit the hopping term to nearest neighbors (NN). There are two obvious limiting cases:

- *Insulating atomic limit*:

  We set $t_{ij} = 0$ and $\mathbf{B} = 0$. The ground state has exactly one electron on each lattice site. This state is, however, highly degenerate. In fact, the degeneracy for N sites is 2^N, since each electron has spin-1/2, we have

$$\left| \Phi_0^A \{\sigma_i\} \right\rangle = \prod_i c^{\dagger}_{i\sigma_i} \left| 0 \right\rangle,$$

where the spin configuration $\{\sigma_i\}$ can be chosen arbitrarily. We will deal with the lifting of this degeneracy later. The first excited states feature one lattice site without electron and one doubly occupied site. This state has energy U, and its degeneracy is even higher, i.e., $2^{N-2} N(N-1)$. Even higher excited states correspond to more empty and doubly occupied sites. The system is an insulator, and the density of states is shown in Figure 23.5

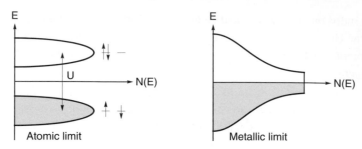

Figure 23.5 Density of states of the Hubbard model in the atomic limit (left) and in the free limit (right).

- *Metallic band limit*:

  We set $U = 0$ and $\mathbf{B} = 0$. The electrons are independent and move freely via hopping processes. The band energy is found through a Fourier transform of the Hamiltonian. With $c_{i\sigma} = \frac{1}{\sqrt{N}} \sum_{\mathbf{k}} c_{\mathbf{k}\sigma} e^{i\mathbf{k}\cdot\mathbf{r}_i}$, we write

  $$-t \sum_{\langle ij \rangle, \sigma} \left(c_{i\sigma}^{\dagger} c_{j\sigma} + h.c. \right) = \sum_{\mathbf{k}\sigma} \varepsilon_{\mathbf{k}} \, c_{\mathbf{k}\sigma}^{\dagger} c_{\mathbf{k}\sigma},$$

  where, for a simple cubic lattice,

  $$\varepsilon_{\mathbf{k}} = -t \sum_{\mathbf{a}} e^{i\mathbf{k}\cdot\mathbf{a}} = -2t \left(\cos(k_x a) + \cos(k_y a) + \cos(k_z a) \right)$$

  and the sum runs over all vectors $\mathbf{a}$ connecting nearest neighbors. The density of states is also shown in Figure 23.5. Obviously, this system is metallic, with a unique ground state:

  $$\left| \Phi_0^M \right| = \prod_{\mathbf{k}} \Theta(-\varepsilon_{\mathbf{k}}) \, c_{\mathbf{k}\uparrow}^{\dagger} c_{\mathbf{k}\downarrow}^{\dagger} \, |0\rangle \, .$$

  Note that $\varepsilon_F = 0$ at half-filling, whereas the bandwidth $2D = 12t$.

Kinetic Exchange Revisited

The Hubbard model provides a pedagogical approach to the kinetic exchange. We shall still focus on the NN hopping Hamiltonian

$$\mathcal{H} = -t \sum_{<ij>, \sigma} c_{i\sigma}^{\dagger} c_{j\sigma} + U \sum_i n_{i\uparrow} n_{i\downarrow}.$$

We emphasize that the form of the interaction makes use of the fact that Pauli exclusion prevents double occupancy of the same site with the same spin, so interactions only arise when there are two particles, one of each spin, on a given site. Also, we still consider the case of half-filling, and consider the limit $U \gg t$. In this case, the model is a *Mott insulator*, since it costs an energy U to move an electron onto an already singly occupied site. One may then ask what the magnetic state of this insulating phase is.

We now consider a pair of sites and compare the energies of the states:

$$|1,1\rangle = \left|\uparrow,\uparrow\right\rangle, \quad |1,0\rangle = \frac{1}{\sqrt{2}}\left(\left|\uparrow,\downarrow\right\rangle + \left|\downarrow,\uparrow\right\rangle\right), \quad |0,0\rangle = \frac{1}{\sqrt{2}}\left(\left|\uparrow,\downarrow\right\rangle - \left|\downarrow,\uparrow\right\rangle\right)$$

(the state $|1,-1\rangle$ is omitted since its behavior is clearly identical to $|1,1\rangle$). Since all of these states have the same occupation of each site, and since the hopping term will produce something orthogonal to each state, it is clear that at leading order,

$$\langle 1,1|\,\mathcal{H}\,|1,1\rangle = \langle 1,0|\,\mathcal{H}\,|1,0\rangle = \langle 0,0|\,\mathcal{H}\,|0,0\rangle.$$

However, since these states are eigenstates of the interaction part, let us now consider the hopping as a perturbation at second order in perturbation theory:

$$\delta E_n^{(2)} = \sum_m{}' \frac{\langle n|\,\delta\mathcal{H}\,|m\rangle \langle m|\,\delta\mathcal{H}\,|n\rangle}{E_n - E_m}.$$

One can immediately see that $\delta\mathcal{H}\,|1,1\rangle = 0$, since the hopping cannot move a spin-up particle to a site already occupied by a spin-up particle. However, for the other states, one has

$$\delta\mathcal{H}\left|\uparrow,\downarrow\right\rangle = -t\left(+\left|\uparrow\downarrow,\_\right\rangle + \left|\_,\uparrow\downarrow\right\rangle\right) \tag{23.33}$$

$$\delta\mathcal{H}\left|\downarrow,\uparrow\right\rangle = -t\left(-\left|\uparrow\downarrow,\_\right\rangle - \left|\_,\uparrow\downarrow\right\rangle\right). \tag{23.34}$$

In order to relate the preceding two terms, it is necessary to choose a sign convention for how the diagramatic configurations relate to ordering of operators. The choice that has been used in the preceding case is

$$\begin{cases} \left|\uparrow,\downarrow\right\rangle = c_{1\uparrow}^\dagger c_{2\downarrow}^\dagger |\varnothing\rangle, & \text{operators appear in the same order as sites} \\[2ex] \left|\uparrow\downarrow,\_\right\rangle = c_{1\uparrow}^\dagger c_{1\downarrow}^\dagger |\varnothing\rangle, & \text{spin-up operators appear first.} \end{cases}$$

Thus, (23.33) can be written as

$$-t\left(c_{1\downarrow}^\dagger c_{2\downarrow} + c_{2\uparrow}^\dagger c_{1\uparrow}\right) c_{1\uparrow}^\dagger c_{2\downarrow}^\dagger = -t\left(-c_{1\downarrow}^\dagger c_{1\uparrow}^\dagger + c_{2\uparrow}^\dagger c_{1\uparrow}^\dagger\right)$$

$$= -t\left(c_{1\uparrow}^\dagger c_{1\downarrow}^\dagger + c_{2\uparrow}^\dagger c_{1\uparrow}^\dagger\right)$$

It is thus clear that one has $\delta\mathcal{H}\,|1,0\rangle = 0$ due to cancellation of the two terms, but $\delta\mathcal{H}\,|0,0\rangle \neq 0$, and so

$$\delta E_{00}^{(2)} = \frac{2t \times 2t}{0 - U} = -\frac{4t^2}{U}.$$

Thus there is an antiferromagnetic interaction, strength $J = 4t^2/U$ favoring projection onto the singlet state. For a pair of spin one-half particles, the operator

$$\mathbf{S}_l \cdot \mathbf{S}_m = \left[(\mathbf{S}_l + \mathbf{S}_m)^2 - \mathbf{S}_l^2 - \mathbf{S}_m^2\right] = \begin{cases} \dfrac{1}{2}\left(0 - \dfrac{3}{2}\right) & \text{Singlet} \\[2mm] \dfrac{1}{2}\left(2 - \dfrac{3}{2}\right) & \text{Triplet} \end{cases}$$

$$= \frac{1}{4} - \mathcal{P}_{\text{singlet}},$$

where $\mathcal{P}_{\text{singlet}}$ projects onto the singlet state, thus this effective hopping term can again be written as $J\,\mathbf{S}_1 \cdot \mathbf{S}_2$, another isotropic magnetic interaction.

23.4.2 Quantum Ground State of Antiferromagnetic Insulators

Thus, at half-filling antiferromagnetic coupling arises in the low-energy Hamiltonian due to virtual hopping processes reducing the energy for antiparallel spin configurations, giving rise to the isotropic Heisenberg model. We will now turn to discuss what states this model supports, but will generalize somewhat to discuss the case where $|S|$ is not necessarily $1/2$, but can be larger – physically, this corresponds to multiple electrons per atom, or, in rare-earth materials, to effects of spin–orbit coupling replacing spin by total angular momentum.

Inadequacy of the Classical Ground State: Bounds on Ground-State Energy

The classical spin antiferromagnetic ground state, known as the Néel state [141] (see Figure 23.6), was described earlier. It has the form ... ↑↓↑↓↑↓ ....

Whereas the ferromagnetic quantum ground state is exactly known (complete maximal spin alignment), the exact ground state of the one-dimensional antiferromagnet has been solved by Bethe in 1931; however, in the case of the two- and three-dimensional antiferromagnet, it is only approximately known. In fact, not only is the Néel state not the ground state in all these cases, it is not even an eigenstate of the Heisenberg Hamiltonian

$$\mathcal{H}_H = J \sum_{<lm>} \mathbf{S}_l \cdot \mathbf{S}_m = J \sum_{<lm>} \left[S_l^z S_m^z + \frac{1}{2}\left(S_l^+ S_m^- + S_l^- S_m^+\right)\right]$$

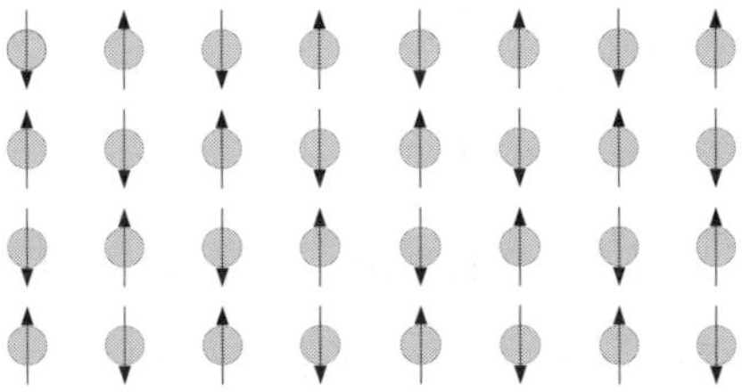

Figure 23.6 The classical Néel state of an antiferromagnet.

since

$$\hat{S}_l^+ \hat{S}_{l+1}^- |\ldots, \uparrow (l-1), \downarrow (l), \uparrow (l+1), \downarrow (l+2), \uparrow (l+3), \ldots\rangle$$
$$= |\ldots, \uparrow (l-1), \uparrow (l), \downarrow (l+1), \downarrow (l+2), \uparrow (l+3), \ldots\rangle, \qquad (23.35)$$

which is not the same as the initial state. Furthermore, there is an additional complication owing to the orientational degeneracy. This can be removed for ferromagnets by an external magnetic field, but for an antiferromagnet, it can only be removed by assuming the presence of an alternating anisotropy field B_{an} that fixes the spin alignment with respect to the crystal axes.

However, we may consider the Néel state as a provisional variational platform, with no parameters to be varied. We shall use it to provide an upper bound on the ground-state energy. The expectation value of the Heisenberg Hamiltonian in the Néel ground state is given by

$$\langle \Psi_{\text{Néel}}| \mathcal{H}_H |\Psi_{\text{Néel}}\rangle = \left\langle \Psi_{\text{Néel}} \left| J \sum_{<lm>} S_l^z S_m^z \right| \Psi_{\text{Néel}} \right\rangle = -\frac{1}{2} J S^2 \mathcal{Z} N$$

for a lattice with coordination number $\mathcal{Z}$. S is the magnitude of the spin, and N is the number of lattice sites, which yields $N\mathcal{Z}/2$ as the number of nearest-neighbor pairs. Alternatively, we may determine a lower bound on the ground-state energy, by considering formation of singlet pairs. For AFM coupling, the ground state is a singlet, $|\Psi_0\rangle = (\uparrow\downarrow - \downarrow\uparrow)/2$, since

$$\mathbf{S}_1 \cdot \mathbf{S}_2 = \frac{1}{2} \left[(\mathbf{S}_1 + \mathbf{S}_2)^2 - \mathbf{S}_1^2 - \mathbf{S}_2^2 \right]$$

and

$$E = \frac{J}{2} [S_{\text{tot}} (S_{\text{tot}} + 1) - S(S+1)] = -JS(S+1),$$

which has the lowest energy, $-JS(S+1)$ for $S_{\text{tot}} = 0$. This is lower than the Néel state, which would have had $-JS^2$; it is lower because it respects the rotation symmetry of the Hamiltonian. We can then define an absolute lowest bound on the ground-state energy via a fictitious system where every possible pair bond forms a singlet. This is not feasible in reality because the same spin cannot form a singlet bond with more than one other spin. The energy of this configuration is

$$-\frac{JS(S+1)N\mathcal{Z}}{2} < E_{\text{gs}} < -\frac{JS^2 N\mathcal{Z}}{2}.$$

23.4.3 Spin Waves in Antiferromagnetic Insulators

Although the state with maximal sublattice magnetization is not actually the ground state of an antiferromagnetic system, we may entertain the idea that the actual ground state admits some spin deviations in the Néel state. Such spin deviations can be accounted for by zero-point fluctuations of the z-component of the spin. For a three-dimensional antiferromagnetic system, the spin deviations were estimated to be of the order of 7% from the maximal sublattice magnetization. So we shall adopt the assumption that though a two-sublattice picture, or a bipartite lattice, is not exact, it may be sufficiently close to the truth to serve as a base on which a more correct theory can be built. The simplest model consists of two interpenetrating cubic sublattices that together form a bcc lattice. The nearest neighbors belong to the two different sublattices. Further, the effective AFM exchange is taken only between nearest neighbors. Thus in the assumed Néel state, we have

$$\mathbf{S}_a^+ |0\rangle - \mathbf{S}_b^- |0\rangle = 0, \tag{23.36}$$

where $\mathbf{S}_a$ and $\mathbf{S}_b$ represent the atomic spin on sublattice A and B, respectively. All of the A spins are chosen to be $\uparrow$, and all of the B spins are $\downarrow$. The Hamiltonian for such a system can be written as

$$\mathcal{H}_{AF} = \mathcal{H}_{\text{ex}} + \mathcal{H}_Z + \mathcal{H}_{\text{an}}$$

$$= 2J \sum_{lm} \mathbf{S}_l \cdot \mathbf{S}_m - g\mu_B B \left[\sum_l \hat{S}_l^z + \sum_m \hat{S}_m^z \right]$$

$$- g\mu_B B_{\text{an}} \left[\sum_l \hat{S}_l^z - \sum_m \hat{S}_m^z \right], \tag{23.37}$$

where l spans sublattice A and m spans sublattice B; B_{an} is a fictitious alternating anisotropy field. The magnitude of the spin at each sublattice is the same, i.e., $S_a = S_b = S$. The spin-deviation operators can be defined in a slightly different manner from the case of a ferromagnet. For the sake of simplicity, we shall retain terms linear in the spin-deviation operators. Thus, for sublattice A and B, we have the following:

Sublattice A	Sublattice B
$\mathbf{S}_l^+ \approx (2S)^{1/2} \hat{\mathbf{a}}_l$	$\mathbf{S}_m^+ \approx (2S)^{1/2} \hat{\mathbf{b}}_m^\dagger$
$\mathbf{S}_l^- \approx (2S)^{1/2} \hat{\mathbf{a}}_l^\dagger$	$\mathbf{S}_m^- \approx (2S)^{1/2} \hat{\mathbf{b}}_m$
$\hat{S}_l^z = S - \hat{\mathbf{a}}_l^\dagger \hat{\mathbf{a}}_l$	$\hat{S}_m^z = \hat{\mathbf{b}}_m^\dagger \hat{\mathbf{b}}_m - S$

$$\tag{23.38}$$

We introduce the spin-wave transformations

$$\hat{\mathbf{a}}_l = \left(\frac{1}{N} \right)^{1/2} \sum_{\mathbf{q}} \exp\left(i\mathbf{q} \cdot \mathbf{R}_L \right) \hat{\mathbf{a}}_{\mathbf{q}}$$

$$\hat{\mathbf{b}}_m = \left(\frac{1}{N} \right)^{1/2} \sum_{\mathbf{q}} \exp\left(-i\mathbf{q} \cdot \mathbf{R}_m \right) \hat{\mathbf{b}}_{\mathbf{q}}, \tag{23.39}$$

where $\hat{\mathbf{a}}_{\mathbf{q}}^\dagger$, $\hat{\mathbf{a}}_{\mathbf{q}}$ and $\hat{\mathbf{b}}_{\mathbf{q}}^\dagger$, $\hat{\mathbf{b}}_{\mathbf{q}}$ are the spin-wave creation and annihilation operators for the two sublattices in question. The wavevector $\mathbf{q}$ spans the N points of the Brillouin zone and corresponds to N spins of each sublattice. The spin-wave Hamiltonian is

$$\mathcal{H}_{AF} = -JS^2 N\mathcal{Z} + 2JS\mathcal{Z} \sum_{\mathbf{q}} \left[\hat{a}_{\mathbf{q}}^\dagger \hat{a}_{\mathbf{q}} + \hat{b}_{\mathbf{q}}^\dagger \hat{b}_{\mathbf{q}} + \gamma_{\mathbf{q}} \left(\hat{a}_{\mathbf{q}}^\dagger \hat{b}_{\mathbf{q}}^\dagger + \hat{a}_{\mathbf{q}} \hat{b}_{\mathbf{q}} \right) \right]$$

$$+ g\mu_B \sum_{\mathbf{q}} \left[(B_{an} + B) \hat{a}_{\mathbf{q}}^\dagger \hat{a}_{\mathbf{q}} + (B_{an} - B) \hat{b}_{\mathbf{q}}^\dagger \hat{b}_{\mathbf{q}} \right] \tag{23.40}$$

$$\gamma_{\mathbf{q}} = \frac{1}{\mathcal{Z}} \sum_j \exp(i\mathbf{q} \cdot \mathbf{d}_j).$$

Unlike the Hamiltonian for the ferromagnet, we here find terms that do not conserve the total number of bosons. We have encountered terms of this form in the Bogoliubov theory of superfluidity, where diagonalization is obtained with the aid of the Bogoliubov canonical transformation

$$\hat{a}_{\mathbf{q}} = \hat{\alpha}_{\mathbf{q}} \cosh \theta_{\mathbf{q}} + \hat{\beta}_{\mathbf{q}}^\dagger \sinh \theta_{\mathbf{q}}$$

$$\hat{b}_{\mathbf{q}} = \hat{\alpha}_{\mathbf{q}}^\dagger \sinh \theta_{\mathbf{q}} + \hat{\beta}_{\mathbf{q}} \cosh \theta_{\mathbf{q}} \tag{23.41}$$

together with their hermitian conjugates. The transformed Hamiltonian reads

$$\mathcal{H}_{AF} = -JS^2 N\mathcal{Z} + JS\mathcal{Z} \sum_{\mathbf{q}} \left(2 \sinh^2 \theta_{\mathbf{q}} - 2\gamma_{\mathbf{q}} \cosh \theta_{\mathbf{q}} \sinh \theta_{\mathbf{q}} \right)$$

$$+ \sum_{\mathbf{q}} \left\{ \left[(\cosh^2 \theta_{\mathbf{q}} + \sinh^2 \theta_{\mathbf{q}})(g\mu_B B_{an} - 2JS\mathcal{Z}) + g\mu_B B - 4\cosh \theta_{\mathbf{q}} \sinh \theta_{\mathbf{q}} JS\mathcal{Z}\gamma_{\mathbf{q}} \right] \hat{\alpha}_{\mathbf{q}}^\dagger \hat{\alpha}_{\mathbf{q}} \right.$$

$$+ \left[(\cosh^2 \theta_{\mathbf{q}} + \sinh^2 \theta_{\mathbf{q}})(g\mu_B B_{an} - 2JS\mathcal{Z}) - g\mu_B B - 4\cosh \theta_{\mathbf{q}} \sinh \theta_{\mathbf{q}} JS\mathcal{Z}\gamma_{\mathbf{q}} \right] \hat{\beta}_{\mathbf{q}}^\dagger \hat{\beta}_{\mathbf{q}}$$

$$\left. + \left[2\cosh \theta_{\mathbf{q}} \sinh \theta_{\mathbf{q}} (g\mu_B B_{an} - 2JS\mathcal{Z}) - 2(\cosh^2 \theta_{\mathbf{q}} + \sinh^2 \theta_{\mathbf{q}}) JS\mathcal{Z}\gamma_{\mathbf{q}} \right] \left[\hat{\alpha}_{\mathbf{q}}^\dagger \hat{\beta}_{\mathbf{q}}^\dagger + \hat{\alpha}_{\mathbf{q}} \hat{\beta}_{\mathbf{q}} \right] \right\}.$$

In order to diagonalize the Hamiltonian, we require that

$$2\cosh \theta_{\mathbf{q}} \sinh \theta_{\mathbf{q}} (g\mu_B B_{an} - 2JS\mathcal{Z}) - (\cosh^2 \theta_{\mathbf{q}} + \sinh^2 \theta_{\mathbf{q}}) 2JS\mathcal{Z}\gamma_{\mathbf{q}} = 0$$

or

$$\tanh 2\theta_{\mathbf{q}} = \frac{\omega_e \gamma_{\mathbf{q}}}{\omega_A - \omega_e} \qquad \begin{cases} \hbar\omega_e = 2\mathcal{Z}SJ \\ \hbar\omega_A = g\mu_B B_{an}. \end{cases} \tag{23.42}$$

Thus the Hamiltonian becomes

$$\mathcal{H}_{AF} = -JS^2 N\mathcal{Z} + JS\mathcal{Z} \sum_{\mathbf{q}} \left(\sqrt{1 - \gamma_{\mathbf{q}}^2} - 1 \right)$$

$$+ \sum_{\mathbf{q}} \left[\hbar\omega_{\mathbf{q}}^{(+)} \left(\hat{\alpha}_{\mathbf{q}}^\dagger \hat{\alpha}_{\mathbf{q}} + \frac{1}{2} \right) + \hbar\omega_{\mathbf{q}}^{(-)} \left(\hat{\beta}_{\mathbf{q}}^\dagger \hat{\beta}_{\mathbf{q}} + \frac{1}{2} \right) \right] \tag{23.43}$$

$$\omega_{\mathbf{q}}^{(\pm)} = \left[(\omega_e + \omega_A)^2 - \omega_e^2 \gamma_{\mathbf{q}}^2 \right]^{1/2} \pm \omega_H, \qquad \hbar\omega_H = g\mu_B B.$$

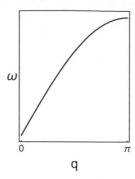

Figure 23.7 Magnon, or spin-wave, dispersion for a body-centered cubic antiferromagnet (lattice constant $a = 1$) along the $[1\,0\,0]$-direction. The dispersion is linear at small q.

We see that there are two magnon branches, with Bose creation and annihilation operators $\hat{\alpha}_{\mathbf{q}}^{\dagger}$, $\hat{\alpha}_{\mathbf{q}}$ and $\hat{\beta}_{\mathbf{q}}^{\dagger}$, $\hat{\beta}_{\mathbf{q}}$. The two branches are degenerate in the absence of the Zeeman term. It should be noted that the anisotropy has been incorporated in the energy. It is readily seen from (23.43) that the lowest excitation energy goes negative when $\omega_H^2 > \omega_A^2 + 2\omega_e\omega_a$, implying instability. This is just the spin-flip transition induced by a magnetic field.

We now consider the dispersion relation in the long-wavelength limit. Setting $B, B_{\text{an}} = 0$, the dispersion relation is considerably simplified. We obtain for the b.c.c. system

$$\hbar\omega_{\mathbf{q}}^{(\pm)} = 2(2\mathcal{Z})^{1/2}\, SJa\, q. \tag{23.44}$$

Thus, unlike the ferromagnetic case, we obtain a linear dispersion relation, as shown in Figure 23.7. In this respect, magnons in antiferromagnets behave like phonons.

Correction to the Antiferromagnetic Ground State

The resulting effective Hamiltonian is approximate because we have neglected magnon interactions. This was the only approximation we have made. The ground state $|0\rangle$ in this noninteracting-magnon approximation satisfies $\hat{\alpha}_{\mathbf{q}}|0\rangle = 0$, $\hat{\beta}_{\mathbf{q}}|0\rangle = 0$. It is the vacuum of the new bosons. In the absence of external fields, the ground-state energy is

$$E_0 = -JS(S+1)N\mathcal{Z} + \sum_{\mathbf{q}} \hbar\omega_{\mathbf{q}} = -JS(S+1)N\mathcal{Z} + JS\mathcal{Z} \sum_{\mathbf{q}} \sqrt{1 - \gamma_{\mathbf{q}}^2}$$

$$= -JS(S+1)N\mathcal{Z} + JS\mathcal{Z}\frac{N^d}{(2\pi)^d}\int d\mathbf{q}^d \sqrt{1 - \left(\frac{1}{d}\sum_{j=1}^{d}\cos(q_j a)\right)^2},$$

where d is the space dimension. The integral can be evaluated numerically (and analytically for $d = 1$). The results are

$$E_0 = dNJS^2 \times \begin{cases} \left(1 + \dfrac{0.363}{S}\right) & \text{For } d = 1 \\[2mm] \left(1 + \dfrac{0.158}{S}\right) & \text{For } d = 2 \\[2mm] \left(1 + \dfrac{0.097}{S}\right) & \text{For } d = 3 \end{cases}$$

The correction is larger for decreasing d. For $d = 1$, the exact ground-state energy is known from the Bethe ansatz. It is very close to E_0.

It is also instructive to derive the sublattice polarization or staggered magnetization

$$M = \langle S_l^z \rangle \Big| l \in A = \langle S_m^z \rangle \Big| m \in B.$$

In the ground state $|0\rangle$,

$$M_0 = \langle 0| S_l^z |0\rangle = \langle 0| S - a_l^\dagger a_l |0\rangle = S - \frac{1}{N} \sum_{\mathbf{q}} \langle 0| a_{\mathbf{q}}^\dagger a_{\mathbf{q}} |0\rangle$$

$$= S - \frac{1}{N} \sum_{\mathbf{q}} \Big\langle 0 \Big| \Big[\cosh^2 \theta_{\mathbf{q}} \, \alpha_{\mathbf{q}}^\dagger \alpha_{\mathbf{q}} + \sinh^2 \theta_{\mathbf{q}} \, \beta_{\mathbf{q}}^\dagger \beta_{\mathbf{q}} + \sinh^2 \theta_{\mathbf{q}}$$

$$- \cosh \theta_{\mathbf{q}} \sinh \theta_{\mathbf{q}} \left(\alpha_{\mathbf{q}} \beta_{\mathbf{q}} + \alpha_{\mathbf{q}}^\dagger \beta_{\mathbf{q}}^\dagger \right) \Big] \Big| 0 \Big\rangle.$$

Since $|0\rangle$ is the vacuum state of $\alpha_{\mathbf{q}}$ and $\beta_{\mathbf{q}}$, we get

$$M_0 = S - \frac{1}{N} \sum_{\mathbf{q}} \sinh^2 \theta_{\mathbf{q}} = S - \frac{1}{N} \sum_{\mathbf{q}} \frac{1}{2} \left(\frac{1}{\sqrt{1 - \gamma_{\mathbf{q}}^2}} - 1 \right) = S + \frac{1}{2} - \frac{1}{2} \frac{a^d}{(2\pi)^d} \int \frac{dq^d}{\sqrt{1 - \gamma_{\mathbf{q}}^2}}.$$

For $d = 1$, the integral is of the form

$$\int \frac{dq}{\sqrt{1 - (1 - a^2 q^2 / 2)^2}} \simeq \frac{1}{a} \int \frac{dq}{q}$$

at small q and thus diverges logarithmically. This indicates that for the $1D$ Heisenberg antiferromagnet even the ground state does not show long-range order. For the $1D$ ferromagnet, we know that the ground state does show long-range order but that the order is destroyed by thermal fluctuations for any $T > 0$. Since thermal fluctuations cannot play a role for the ground state, one says that the magnetic order in the $1D$ antiferromagnet is destroyed by quantum fluctuations.

For $d > 1$, the integral converges. For the models considered earlier, we get

$$M_0 = \begin{cases} S\left(1 - \dfrac{0.197}{S}\right) & \text{For } d = 2 \\[2mm] S\left(1 - \dfrac{0.078}{S}\right) & \text{For } d = 3 \end{cases}$$

Note that for $d = 2$ and $S = 1/2$ we obtain a roughly 40% reduction compared to the Néel state due to quantum fluctuations.

Exercises

23.1 Anisotropic Heisenberg ferromagnet

Consider a 1D Heisenberg-type model with nearest-neighbor-only interactions

$$\mathcal{H} = J_z \sum_i S_i^z S_{i+1}^z + \frac{J_\perp}{2} \sum_i \left(S_i^+ S_{i+1}^- + S_i^- S_{i+1}^+ \right).$$

(a) Show that for $J_z > J_\perp$ the magnon dispersion $E(\mathbf{q})$ is similar to the isotropic case, but with nonzero energy as $q \to 0$.

(b) Show for the opposite case $J_z < J_\perp$ that the ferromagnetically ordered state with magnetization along z is not the ground state!

24

Quantum Theory of Magnetism: Itinerant-Electron Systems and the Kondo Effect

When we think about magnetism in an itinerant electron system, a Sommerfeld model immediately comes to mind. However, we also immediately realize that, depending on the sign of the susceptibility χ, a noninteracting electron gas can exhibit paramagnetism or diamagnetism, but it can never develop a spontaneous magnetic moment: $\mathbf{M}\big|_{\mathbf{B}=0} = 0$.

What then gives rise to magnetism in itinerant electron systems? Again, it must be that Coulomb repulsion between electrons is responsible for magnetism whenever it arises in itinerant-electron systems. At the outset, this might seem odd, since the Coulomb interaction is spin independent. How then can it lead to a spontaneous magnetic moment? To arrive at a reasonable explanation, we shall introduce the Stoner model.

24.1 Stoner Mean-Field Theory: Ferromagnetic Case

Stoner developed a very simple picture of ferromagnetism based on the competition between the kinetic energy cost of making the $\uparrow$ and $\downarrow$ spin electron numbers different and the corresponding gain in exchange energy. The basic idea can be explained as follows.

In the absence of Coulomb interaction, the Pauli principle allows double occupancy of each energy level with electrons of opposite spins, up to the Fermi energy. If, instead, we make the number of $\uparrow$ and $\downarrow$ electrons unequal, we will have to occupy levels above the designated Fermi energy. However, in the presence of Coulomb interaction, such an unequal configuration results in a decrease of exchange energy. In the extreme case of complete polarization, shown in Figure 24.1, the exchange energy cost becomes zero. To make the argument more precise, we consider a system with equal $\uparrow$ and $\downarrow$ filling up to the same E_F. The density of $\uparrow$ or $\downarrow$ electrons is equal to n. We compute the change in energy that results from a reduction in the density of $\downarrow$ spin electrons by δn and at the same time an increase the number of $\uparrow$ spin electrons by δn. The potential energy changes by

$$\Delta V = U\,(n + \delta n)\,(n - \delta n) - U\,n^2 = -U\,(\delta n)^2.$$

Placing an extra δn electrons into the $\uparrow$ group requires occupying energy levels above E_F. With the density of states $\mathcal{D}(E) = dN/dE$, we have $\delta n = \mathcal{D}(E)\,\delta E$. This delineates the range of energies above E_F filled by δn. It also gives the energy range of emptied levels

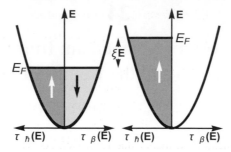

Figure 24.1 Increase in kinetic energy.

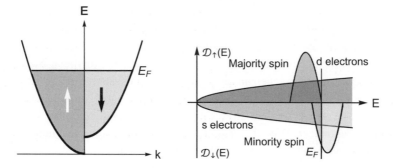

Figure 24.2 Left: Stoner criterion. Right: a more typical rendering of an itinerant ferromagnet is given on the right, showing the high density of states (DOS) of a $3d$ system.

below E_F that used to be occupied by down-spin electrons. The net result of this process is to shift δn electrons up in energy by an amount δE. The change in the kinetic energy is then

$$\delta T = \delta n \, \delta E = \frac{1}{\mathcal{D}(E)} \, (\delta n)^2 .$$

Combining these two contributions,

$$\delta E = \delta T + \delta V = \left(\frac{1}{\mathcal{D}(E)} - U \right) (\delta n)^2 = (1 - U \, \mathcal{D}(E)) \, \frac{(\delta n)^2}{\mathcal{D}(E)} .$$

It is clear that for $U \, \mathcal{D}(E) > 1$ the total energy change $\delta E < 0$, so it is favorable to have different $\uparrow$ and $\downarrow$ electron densities, which favors ferromagnetism. This is called the Stoner criterion (see Figure 24.2). It tells us that magnetism is favored by large electron interactions. As we shall see, this simple calculation yields results in precise agreement with mean-field theory.

24.1.1 Mean-Field Hubbard Model

The simplest model that can describe itinerant magnetism is the Hubbard model:

$$\mathcal{H} = -t \sum_{<ij>,\sigma} \left(c_{i\sigma}^{\dagger} c_{j\sigma} + c_{j\sigma}^{\dagger} c_{i\sigma} \right) + U \sum_{i} n_{i\uparrow} n_{i\downarrow} - \mu_B B \sum_{i} \left(n_{i\uparrow} - n_{i\downarrow} \right).$$

As is quite clear by now, we cannot devise general methods to solve for the exact ground state of such an interacting many-body Hamiltonian. However, we can solve this problem, approximately, using a mean-field theory due to Stoner. As we have stated, the Hubbard model accounts for onsite Coulomb interactions only, and hence it has a constant value in reciprocal space. Thus, we write the Hubbard Hamiltonian in k-space as

$$\mathcal{H} = \sum_{\mathbf{k}\sigma} \varepsilon_{\mathbf{k}} c_{\mathbf{k}\sigma}^{\dagger} c_{\mathbf{k}\sigma} + \frac{U}{2\Omega} \sum_{\substack{\mathbf{k}\mathbf{k}'\mathbf{q} \\ \sigma\sigma'}} c_{\mathbf{k}+\mathbf{q},\sigma}^{\dagger} c_{\mathbf{k}'-\mathbf{q},\sigma'}^{\dagger} c_{\mathbf{k}'\sigma'} c_{\mathbf{k}\sigma}. \tag{24.1}$$

We apply the Hartree–Fock approximation to this model, but search for a ferromagnetic solution by allowing for the expectation values to depend on the direction of the spin. We substitute

$$c_{\mathbf{k}+\mathbf{q},\sigma}^{\dagger} c_{\mathbf{k}'-\mathbf{q},\sigma'}^{\dagger} c_{\mathbf{k}'\sigma'} c_{\mathbf{k}\sigma} = 2 \left\langle c_{\mathbf{k}+\mathbf{q},\sigma}^{\dagger} c_{\mathbf{k}\sigma} \right\rangle c_{\mathbf{k}'-\mathbf{q},\sigma'}^{\dagger} c_{\mathbf{k}'\sigma'} - 2 \left\langle c_{\mathbf{k}+\mathbf{q},\sigma}^{\dagger} c_{\mathbf{k}'\sigma'} \right\rangle c_{\mathbf{k}'-\mathbf{q},\sigma'}^{\dagger} c_{\mathbf{k}\sigma}$$

$$- \left\langle c_{\mathbf{k}+\mathbf{q},\sigma}^{\dagger} c_{\mathbf{k}\sigma} \right\rangle \left\langle c_{\mathbf{k}'-\mathbf{q},\sigma'}^{\dagger} c_{\mathbf{k}'\sigma'} \right\rangle + \left\langle c_{\mathbf{k}+\mathbf{q},\sigma}^{\dagger} c_{\mathbf{k}'\sigma'} \right\rangle \left\langle c_{\mathbf{k}'-\mathbf{q},\sigma'}^{\dagger} c_{\mathbf{k}\sigma} \right\rangle$$

in the interaction component of (24.1), and we obtain the mean-field interaction Hamiltonian

$$V_{\text{int}}^{\text{MF}} = \frac{U}{\Omega} \sum_{\substack{\mathbf{k}\mathbf{k}'\mathbf{q} \\ \sigma\sigma'}} c_{\mathbf{k}+\mathbf{q},\sigma}^{\dagger} \left\langle c_{\mathbf{k}'-\mathbf{q},\sigma'}^{\dagger} c_{\mathbf{k}'\sigma'} \right\rangle c_{\mathbf{k}\sigma} - \frac{U}{\Omega} \sum_{\substack{\mathbf{k}\mathbf{k}'\mathbf{q} \\ \sigma\sigma'}} \left\langle c_{\mathbf{k}+\mathbf{q},\sigma}^{\dagger} c_{\mathbf{k}'\sigma'} \right\rangle c_{\mathbf{k}'-\mathbf{q},\sigma'}^{\dagger} c_{\mathbf{k}\sigma}$$

$$- \frac{U}{2\Omega} \sum_{\substack{\mathbf{k}\mathbf{k}'\mathbf{q} \\ \sigma\sigma'}} \left[\left\langle c_{\mathbf{k}+\mathbf{q},\sigma}^{\dagger} c_{\mathbf{k}\sigma} \right\rangle \left\langle c_{\mathbf{k}'-\mathbf{q},\sigma'}^{\dagger} c_{\mathbf{k}'\sigma'} \right\rangle - \left\langle c_{\mathbf{k}+\mathbf{q},\sigma}^{\dagger} c_{\mathbf{k}'\sigma'} \right\rangle \left\langle c_{\mathbf{k}'-\mathbf{q},\sigma'}^{\dagger} c_{\mathbf{k}\sigma} \right\rangle \right].$$

With the mean-field parameters defined as

$$\left\langle c_{\mathbf{k}\uparrow}^{\dagger} c_{\mathbf{k}'\uparrow} \right\rangle = \delta_{\mathbf{k}\mathbf{k}'} n_{\mathbf{k}\uparrow}, \qquad \left\langle c_{\mathbf{k}\downarrow}^{\dagger} c_{\mathbf{k}'\downarrow} \right\rangle = \delta_{\mathbf{k}\mathbf{k}'} n_{\mathbf{k}\downarrow} \tag{24.2}$$

and the spin densities

$$\bar{n}_{\sigma} = \frac{1}{\Omega} \sum_{\mathbf{k}} \left\langle c_{\mathbf{k}\sigma}^{\dagger} c_{\mathbf{k}\sigma} \right\rangle,$$

we get

$$V_{\text{int}}^{\text{MF}} = U \sum_{\mathbf{k}\sigma\sigma'} c_{\mathbf{k}\sigma}^{\dagger} c_{\mathbf{k}\sigma} \left[\bar{n}_{\sigma'} - \bar{n}_{\sigma} \delta_{\sigma\sigma'} \right] - U\Omega \sum_{\sigma\sigma'} \bar{n}_{\sigma'} \bar{n}_{\sigma} + U\Omega \sum_{\sigma} \bar{n}_{\sigma}^2.$$

The full Stoner mean-field Hamiltonian is now given by

$$\mathcal{H}_{MF} = \sum_{\mathbf{k}\sigma} \mathcal{E}_{\mathbf{k}\sigma}^{MF} c_{\mathbf{k}\sigma}^{\dagger} c_{\mathbf{k}\sigma} - U\Omega \sum_{\sigma\sigma'} \bar{n}_{\sigma'} \bar{n}_{\sigma} + U\Omega \sum_{\sigma} \bar{n}_{\sigma}^2$$

$$\mathcal{E}_{\mathbf{k}\sigma}^{MF} = \varepsilon_{\mathbf{k}} + U\left(\bar{n}_{\uparrow} + \bar{n}_{\downarrow} - \bar{n}_{\sigma}\right) = \varepsilon_{\mathbf{k}} + U\,\bar{n}_{\bar{\sigma}}. \tag{24.3}$$

We obtain the $T = 0$ mean-field solution via the self-consistency equations

$$\bar{n}_{\sigma} = \frac{1}{\Omega} \sum_{\mathbf{k}} \left\langle c_{\mathbf{k}\sigma}^{\dagger} c_{\mathbf{k}\sigma} \right\rangle = \int \frac{d\mathbf{k}}{(2\pi)^3} \, \Theta\left(\mu - \frac{\hbar^2 k^2}{2m} - U\,\bar{n}_{\bar{\sigma}}\right) = \frac{1}{6\pi^2} k_{F\sigma}^3,$$

where $\frac{\hbar^2 k_{F\sigma}^2}{2m} + U\,\bar{n}_{\bar{\sigma}} = \mu$, leading to

$$\frac{\hbar^2}{2m} (6\pi^2)^{2/3} \bar{n}_{\uparrow}^{2/3} + U\,\bar{n}_{\downarrow} = \frac{\hbar^2}{2m} (6\pi^2)^{2/3} \bar{n}_{\downarrow}^{2/3} + U\,\bar{n}_{\uparrow} = \mu. \tag{24.4}$$

Defining the quantities

$$\bar{n} = \bar{n}_{\uparrow} + \bar{n}_{\downarrow}, \quad \zeta = \frac{\bar{n}_{\uparrow} - \bar{n}_{\downarrow}}{\bar{n}}, \quad \gamma = \frac{2mU\,\bar{n}^{1/3}}{(6\pi^2)^{2/3}\,\hbar^2}$$

and rearranging the self-consistency conditions (24.4), we get

$$\bar{n}_{\uparrow}^{2/3} - \bar{n}^{2/3} = \frac{2mU}{(6\pi^2)^{2/3}\,\hbar^2} \left(\bar{n}_{\uparrow} - \bar{n}_{\downarrow}\right) \quad \Rightarrow \quad (1+\zeta)^{2/3} - (1-\zeta)^{2/3} = \gamma\,\zeta. \tag{24.5}$$

A graphic solution of (24.5) is depicted in Figure 24.3. It shows three solution regimes:

$$\gamma < \frac{4}{3}: \qquad \text{Isotropic solution (normal state)} \qquad\qquad \zeta = 0$$

$$\frac{4}{3} < \gamma < 2^{2/3}: \qquad \text{Partial polarization (weak ferromagnet)} \qquad 0 < \zeta < 1$$

$$\gamma > 2^{2/3}: \qquad \text{Full polarization (strong ferromagnet)} \qquad\qquad \zeta = 1$$

We note that the initial slope of the lhs is 4/3, and its value at $\zeta = 1$ is $2^{2/3}$, which signals complete polarization. The different solutions are sketched in Figure 24.4.

24.1.2 *Temperature Dependence of Magnetization in Mean Field*

We write the site occupancy $n_{i\sigma}$ as

$$n_{i\sigma} = \langle n_{\sigma} \rangle + \delta n_{i\sigma}, \tag{24.6}$$

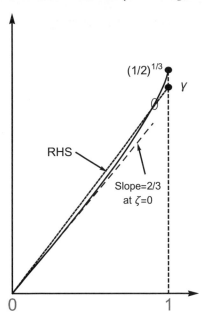

Figure 24.3 Plot of the two sides of (24.5).

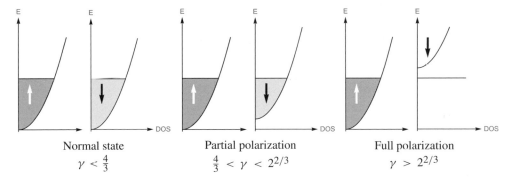

Normal state	Partial polarization	Full polarization
$\gamma < \frac{4}{3}$	$\frac{4}{3} < \gamma < 2^{2/3}$	$\gamma > 2^{2/3}$

Figure 24.4 The three possible solutions of the Stoner model. The polarization is thus a function of the interaction strength; the stronger the interaction, the larger the polarization. The Stoner model provides a clear physical picture for how the exchange interactions induce a ferromagnetic phase transition in a metal with strong onsite interactions.

where $\langle n_{i\sigma} \rangle$ is the thermodynamic average, and $\delta n_{i\sigma}$ the fluctuating component, such that $|\delta n_{i\sigma}|^2 \ll \langle n_\sigma \rangle^2$. Substituting in the Hubbard Hamiltonian, we get

$$\mathcal{H}^{\text{MF}} = \sum_{\mathbf{k}\sigma} \varepsilon_{\mathbf{k}} \, c_{\mathbf{k}\sigma}^\dagger \, c_{\mathbf{k}\sigma} + U \sum_i \left[n_{i\uparrow} \langle n_\downarrow \rangle + n_{i\downarrow} \langle n_\uparrow \rangle + \langle n_\uparrow \rangle \langle n_\downarrow \rangle \right]$$

$$= \sum_{\mathbf{k}\sigma} \left(\varepsilon_{\mathbf{k}} + U \, \langle n_{\bar{\sigma}} \rangle \right) c_{\mathbf{k}\sigma}^\dagger \, c_{\mathbf{k}\sigma} + U N \, \langle n_\uparrow \rangle \langle n_\downarrow \rangle, \tag{24.7}$$

The advantage of this approximation is that the many-body problem is now reduced to an effective one-particle problem, where only the mean electron interaction is taken into account. We express $\langle n_\sigma \rangle$ as

$$\langle n_\sigma \rangle = \frac{1}{N} \sum_{\mathbf{k}} \left\langle c_{\mathbf{k}\sigma}^\dagger c_{\mathbf{k}\sigma} \right\rangle = \int d\varepsilon \, \frac{1}{N} \sum_{\mathbf{k}} \delta \left(\varepsilon - \varepsilon_{\mathbf{k}} - U \left\langle n_{\bar{\sigma}} \right\rangle \right) n_F(\varepsilon)$$

$$= \int d\varepsilon \, \frac{1}{2} \mathcal{D} \left(\varepsilon - U \left\langle n_{\bar{\sigma}} \right\rangle \right) n_F(\varepsilon). \tag{24.8}$$

To solve for the $\langle n_\sigma \rangle$s self-consistently, we introduce the average occupancy, $\bar{n}$, and average spin polarization, m

$$\left. \begin{aligned} \bar{n} &= \langle n_\downarrow \rangle + \langle n_\uparrow \rangle \\ m &= \langle n_\uparrow \rangle - \langle n_\downarrow \rangle \end{aligned} \right\} \quad \Rightarrow \quad \langle n_\sigma \rangle = \frac{1}{2} \left(\bar{n} + \sigma m \right) \tag{24.9}$$

and obtain

$$\bar{n} = \frac{1}{2} \int d\varepsilon \left[\mathcal{D} \left(\varepsilon - U \langle n_\downarrow \rangle \right) + \mathcal{D} \left(\varepsilon - U \langle n_\uparrow \rangle \right) \right] n_F(\varepsilon)$$

$$= \frac{1}{2} \sum_\sigma \int d\varepsilon \, \mathcal{D} \left(\varepsilon - \frac{U\bar{n}}{2} - \sigma \frac{Um}{2} \right) n_F(\varepsilon) \tag{24.10}$$

$$m = \frac{1}{2} \int d\varepsilon \left[\mathcal{D} \left(\varepsilon - U \langle n_\downarrow \rangle \right) - \mathcal{D} \left(\varepsilon - U \langle n_\uparrow \rangle \right) \right] n_F(\varepsilon)$$

$$= -\frac{1}{2} \sum_\sigma \sigma \int d\varepsilon \, \mathcal{D} \left(\varepsilon - \frac{U\bar{n}}{2} - \sigma \frac{Um}{2} \right) n_F(\varepsilon), \tag{24.11}$$

which presents two coupled equations that can be solved approximately for $m \ll \bar{n}$, by allowing a change in the chemical potential μ that depends on temperature and polarization:

$$\mu(m, T) = \varepsilon_F + \frac{U\bar{n}}{2} + \Delta\mu(m, T) = \bar{\varepsilon}_F + \Delta\mu(m, T).$$

For small m, T, and

$$n_F(\varepsilon, T) = \frac{1}{e^{\beta(\varepsilon - \mu(m,T))} + 1} = n_F(\varepsilon, 0) + \left. \frac{\partial n_F}{\partial \mu} \right|_{\mu = \bar{\varepsilon}_F} \Delta\mu(m, T) + \frac{\partial n_F}{\partial T} k_B T, \quad k_B T \ll \varepsilon_F,$$

we expand (24.10) as

$$\bar{n} \simeq \int d\varepsilon \left[\mathcal{D}(\varepsilon) + \frac{1}{2} \left(\frac{Um}{2} \right)^2 \frac{d^2 \mathcal{D}(\varepsilon)}{d\varepsilon^2} + \cdots \right] n_F(\varepsilon)$$

$$\approx \int_0^{\bar{\varepsilon}_F} d\varepsilon \, \mathcal{D}(\varepsilon) + \mathcal{D}(\bar{\varepsilon}_F) \, \Delta\mu + \left[\frac{\pi^2}{6} (k_B T)^2 \frac{d\mathcal{D}(\varepsilon)}{d\varepsilon} + \frac{1}{2} \left(\frac{Um}{2} \right)^2 \frac{d\mathcal{D}(\varepsilon)}{d\varepsilon} \right]_{\bar{\varepsilon}_F},$$

where we used

$$\int_{-\infty}^{\infty} dx \, x^2 \frac{e^x}{(e^x + 1)^2} = \frac{\pi^2}{3}$$

and we expanded $\mathcal{D}(\varepsilon)$ around $\bar{\varepsilon}_F$. We note that

$$\int_0^{\bar{\varepsilon}_F} d\varepsilon \, \mathcal{D}(\varepsilon) = \bar{n}$$

and we obtain

$$\Delta\mu(m, T) = -\frac{\mathcal{D}'(\bar{\varepsilon}_F)}{\mathcal{D}(\bar{\varepsilon}_F)} \left[\frac{\pi^2}{6} (k_B T)^2 + \frac{1}{2} \left(\frac{Um}{2} \right)^2 \right].$$

Next we carry out a similar expansion for m and substitute for $\Delta\mu(m, T)$

$$m \simeq \int d\varepsilon \left[\frac{d\mathcal{D}(\varepsilon)}{d\varepsilon} \frac{Um}{2} + \frac{1}{3!} \left(\frac{Um}{2} \right)^3 \frac{d^3 \mathcal{D}(\varepsilon)}{d\varepsilon^3} + \cdots \right] n_F(\varepsilon)$$

$$\simeq \left[\mathcal{D}(\bar{\varepsilon}_F) + \mathcal{D}'(\bar{\varepsilon}_F) \, \Delta\mu + \frac{\pi^2}{6} (k_B T)^2 \, \mathcal{D}''(\bar{\varepsilon}_F) + \frac{1}{3!} \left(\frac{Um}{2} \right)^2 \mathcal{D}''(\bar{\varepsilon}_F) \right] \frac{Um}{2}$$

$$= A(\bar{\varepsilon}_F, T) \, m - B(\bar{\varepsilon}_F) \, m^3,$$

where

$$A(\bar{\varepsilon}_F, T) = \frac{\mathcal{D}(\bar{\varepsilon}_F) U}{2} \left\{ 1 - \left[\left(\frac{\mathcal{D}'(\bar{\varepsilon}_F)}{\mathcal{D}(\bar{\varepsilon}_F)} \right)^2 - \frac{\mathcal{D}''(\bar{\varepsilon}_F)}{\mathcal{D}(\bar{\varepsilon}_F)} \right] \frac{\pi^2}{6} (k_B T)^2 \right\} = \frac{\mathcal{D}(\bar{\varepsilon}_F) U}{2} \left(1 - \alpha T^2 \right)$$

$$B(\bar{\varepsilon}_F) = \mathcal{D}(\bar{\varepsilon}_F) \left(\frac{U}{2} \right)^3 \left[\frac{1}{2} \left(\frac{\mathcal{D}'(\bar{\varepsilon}_F)}{\mathcal{D}(\bar{\varepsilon}_F)} \right)^2 - \frac{\mathcal{D}''(\bar{\varepsilon}_F)}{3! \, \mathcal{D}(\bar{\varepsilon}_F)} \right].$$

We find that

$$m^2 = \frac{A - 1}{B}, \qquad A, B > 0.$$

Thus a nonzero real root exists for $A > 1$, which yields two conditions for a Fermi liquid instability and the emergence of magnetism

$$\begin{cases} U &> \dfrac{2}{\mathcal{D}(\bar{\varepsilon}_F)} = U_c \\[2ex] T &\leq \dfrac{1}{\alpha} \sqrt{1 - \dfrac{U_c}{U}} = T_c, \end{cases}$$

where T_c is the Curie temperature. The emerging magnetization M exhibits a temperature dependence

$$M \propto \sqrt{T_c - T}.$$

24.2 RPA Susceptibility: Stoner Excitations and Spin Waves

We now turn to explore the nature of excitations above the Stoner ground state. Such excitations are described by the dynamical magnetic response contained in the magnetic susceptibility, which we now consider.

24.2.1 The RPA Magnetic Susceptibility

The magnetic susceptibility, $\chi(\mathbf{x}t; \mathbf{x}', t')$, gives the change in spin density in response to an external applied magnetic field:

$$\delta s_\alpha(\mathbf{x}, t) = \sum_\beta \int d\mathbf{x}' \int dt' \, \chi_{\alpha\beta}(\mathbf{x}t; \mathbf{x}', t') \, B_\beta(\mathbf{x}', t').$$

The spin density $\mathbf{s}(\mathbf{x})$ is defined as

$$s_\alpha(\mathbf{x}) = \sum_{\sigma, \sigma'} \Psi_\sigma^\dagger(\mathbf{x}) \left(\sigma^\alpha\right)_{\sigma\sigma'} \Psi_{\sigma'}(\mathbf{x}),$$

where σ^α are the Pauli matrices. The magnetic field $\mathbf{B}$ enters into the Hamiltonian through the Zeeman coupling,

$$\mathcal{H}_I = -\frac{\hbar}{2} \mu_B g \int d\mathbf{x} \, B(\mathbf{x}, t) \cdot \mathbf{s}(\mathbf{x}).$$

We can express the spin susceptibility in terms of the retarded spin–spin correlation, or Green function as

$$\chi_{\alpha\beta}(\mathbf{x}, t; \mathbf{x}', t') = i \frac{\mu_B g}{2} \, \Theta(t - t') \left\langle \left[s_\alpha(\mathbf{x}, t), s_\beta(\mathbf{x}', t')\right]\right\rangle,$$

where the minus sign characteristic of the Green function is removed by that appearing in $\mathcal{H}_I$. Transformation to frequency space yields

$$\delta s_\alpha(\mathbf{x}, \omega) = \sum_\beta \int d\mathbf{x}' \, \chi_{\alpha\beta}(\mathbf{x}, \mathbf{x}', \omega) \, B_\beta(\mathbf{x}', \omega)$$

$$\chi_{\alpha\beta}(\mathbf{x}, \mathbf{x}', \omega) = i \frac{\mu_B g}{2} \int dt \, e^{i\omega t} \left\langle \left[s_\alpha(\mathbf{x}, t), s_\beta(\mathbf{x}', 0)\right]\right\rangle.$$

Moreover, for spatially homogeneous systems, we may carry out a spatial Fourier transform, to obtain

$$\chi_{\alpha\beta}(\mathbf{q}, \omega) = \frac{1}{\Omega} \int d\mathbf{x} \, d\mathbf{x}' \, e^{i\mathbf{q}\cdot(\mathbf{x}-\mathbf{x}')} \, \chi_{\alpha\beta}(\mathbf{x} - \mathbf{x}', \omega).$$

In the plane wave representation, we write

$$\mathbf{s}(\mathbf{x}, t) = \sum_\mathbf{q} e^{i\mathbf{q}\cdot\mathbf{x}} \, \mathbf{s}(\mathbf{q}, t), \qquad \Rightarrow \qquad s_\alpha(\mathbf{q}, t) = \sum_\mathbf{k} c_{\mathbf{k}+\mathbf{q}, s}^\dagger(t) \, \sigma_{ss'}^\alpha \, c_{\mathbf{k}, s'}(t),$$

$$s^+(\mathbf{q}, t) = \sum_\mathbf{k} c_{\mathbf{k}+\mathbf{q}, \uparrow}^\dagger(t) \, c_{\mathbf{k}\downarrow}(t), \qquad\qquad s^-(\mathbf{q}, t) = \sum_\mathbf{k} c_{\mathbf{k}+\mathbf{q}, \downarrow}^\dagger(t) \, c_{\mathbf{k}\uparrow}(t),$$

$$s^z(\mathbf{q}, t) = \frac{1}{2} \sum_\mathbf{k} \left(c_{\mathbf{k}+\mathbf{q}, \uparrow}^\dagger(t) \, c_{\mathbf{k}\uparrow}(t) - c_{\mathbf{k}+\mathbf{q}, \downarrow}^\dagger(t) \, c_{\mathbf{k}\downarrow}(t)\right).$$

We identify two types of spin susceptibility functions:

- The *transverse* spin susceptibility function, χ^{-+}, which represents the response to a field that couples to the spin-flip (deviation) component of the spin
- The *longitudinal* susceptibility χ^{zz} containing the response to a field that couples to the S_z component

It is particularly interesting to examine the transverse spin fluctuations, or deviations. A uniform transverse spin fluctuation corresponds to a rotation of the magnetization, which costs no energy due to the rotational invariance of the system. If we carry out a slow twist of the magnetization, this costs an energy that goes to zero as the pitch of the twist goes to infinity. The corresponding normal mode is the "Goldstone mode" of the magnet.

For the case we are interested in, namely, ferromagnetic excitations, we shall consider the retarded transverse susceptibility

$$\chi^{-+}(\mathbf{x},t;0,0) = i\,\Theta(t)\,\langle[s^{-}(\mathbf{x},t),s^{+}(0,0)]\rangle = \sum_{\mathbf{q}} \chi^{-+}(\mathbf{q},t).$$

We find that

$$\chi^{-+}(\mathbf{q},t) = i\frac{\mu_B g}{2\Omega}\,\Theta(t)\sum_{\mathbf{k},\mathbf{k}'}\left\langle\left[c^{\dagger}_{\mathbf{k}-\mathbf{q}\downarrow}(t)\,c_{\mathbf{k}\uparrow}(t),\,c^{\dagger}_{\mathbf{k}'+\mathbf{q}\uparrow}(0)\,c_{\mathbf{k}'\downarrow}(0)\right]\right\rangle. \qquad (24.12)$$

Diagrammatically, the susceptibility for noninteracting electrons can be represented as shown in Figure 24.5. The gray circles are spin vertices that require a spin-flip between incoming and outgoing spins, in the case of the transverse susceptibility χ^{-+}.

In the case of noninteracting electron systems, the definition of χ^{-+} in (24.12) clearly resembles Π_0, the polarizability of the noninteracting electron gas, and we write

$$\chi_0^{-+}(\mathbf{q},\omega) = -\frac{\mu_B g}{2\Omega}\Pi_0^{-+}(\mathbf{q},\omega) = -\frac{\mu_B g}{2\Omega}\sum_{\mathbf{k}}\frac{n_{F\uparrow}(\varepsilon_{\mathbf{k}}-\mu) - n_{F\downarrow}(\varepsilon_{\mathbf{k}+\mathbf{q}}-\mu)}{\omega + \varepsilon_{\mathbf{k}} - \varepsilon_{\mathbf{k}+\mathbf{q}} + i\eta}. \qquad (24.13)$$

In a manner similar to our treatment of the interacting electron gas, we use diagrammatic RPA techniques to derive expressions for the susceptibility in the present case. Typical RPA diagrams are shown in Figure 24.6: the RPA amounts to adding to the diagram 24.6(a) all possible interaction vertices of the form 24.6(b) that connect the two arms of the bubble to

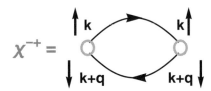

Figure 24.5 Diagrammatic representation of the transverse spin susceptibility χ^{-+} for noninteracting electrons.

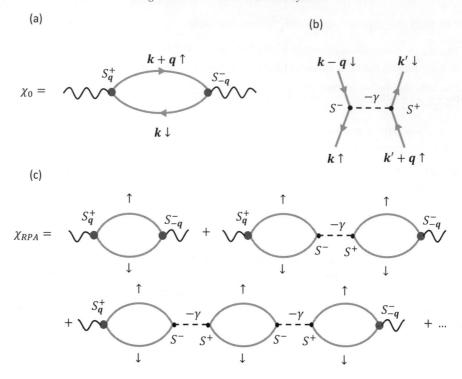

Figure 24.6 (a) Diagrammatic representation of the susceptibility χ_0^{-+}. (b) Diagrammatic representation of the interaction vertex, with strength γ. (c) The sum of bubble diagrams that gives the transverse spin susceptibility up to the RPA. Notice that the only way to connect the bare bubble is through an exchange vertex of the form presented in (b).

a consecutive bubble (as shown in 24.6(c)). If we now consider a local interaction, like the Hubbard model, $U n_{i\uparrow} n_{i\downarrow}$, $\gamma = U$, summing up this series, we get

$$\chi^{-+}(\mathbf{q}, \omega) = \frac{\chi_0^{-+}(\mathbf{q}, \omega)}{1 - U \chi_0^{-+}(\mathbf{q}, \omega)}.$$

Again, it is clear that an instability will occur if the condition $U \Pi_0^{-+}(\mathbf{q}, \omega) \geq 1$ is satisfied. This more general condition is satisfied for the ferromagnetic instability, when we take the limit $\Pi_0^{-+}(\mathbf{q} \to 0, 0) \to \mathcal{D}_\uparrow(\varepsilon_F)$, yielding the Stoner criterion $U \mathcal{D}(\varepsilon_F) \geq 1$.

24.2.2 *Mean-Field Stoner Excitations*

The mean-field approximation amounts to replacing the interacting system with an effective noninteracting system where the original noninteracting particle spectrum gets renormalized. In that spirit, we need to replace the denominator of (24.13) with corresponding terms from the Stoner mean-field spectrum of (24.3), namely,

$$-\varepsilon_{\mathbf{k}} + \varepsilon_{\mathbf{k+q}} \rightarrow -\mathcal{E}_{\mathbf{k}\uparrow}^{\mathrm{MF}} + \mathcal{E}_{\mathbf{k+q}\downarrow}^{\mathrm{MF}} = -\varepsilon_{\mathbf{k}} + \varepsilon_{\mathbf{k+q}} - U\bar{n}\left(1 - \zeta\right) + U\bar{n}\left(1 + \zeta\right)$$
$$[2pt] = \varepsilon_{\mathbf{k+q}} - \varepsilon_{\mathbf{k}} + 2U\bar{n}\zeta$$
$$[2pt] = \varepsilon_{\mathbf{k+q}} - \varepsilon_{\mathbf{k}} + \Delta. \tag{24.14}$$

And we write the noninteracting Stoner magnetic susceptibility as

$$\chi_{0,\,\mathrm{Stoner}}^{-+}(\mathbf{q},\omega) = \frac{\mu_B g}{2\Omega} \sum_{\mathbf{k}} \frac{n_{F\uparrow}(\varepsilon_{\mathbf{k}} - \mu) - n_{F\downarrow}(\varepsilon_{\mathbf{k+q}} - \mu)}{\omega - \varepsilon_{\mathbf{k+q}} + \varepsilon_{\mathbf{k}} - \Delta + i\eta}. \tag{24.15}$$

The excitation energy of the noninteracting Stoner model is given by the poles of the Stoner susceptibility, as

$$\varepsilon_{\mathbf{k+q}} - \varepsilon_{\mathbf{k}} + \Delta = \frac{\hbar^2}{m}\mathbf{k}\cdot\mathbf{q} + \frac{\hbar^2}{2m}q^2 + \Delta$$

depends on $\mathbf{k}$ for a fixed $\mathbf{q} \neq 0$. Therefore, for any $\mathbf{q} \neq 0$ we find a continuum of electron–hole excitations, coined *Stoner particle–hole continuum* (see Figure 24.7). Only for $\mathbf{q} = 0$, the excitation energy is sharp, $\Delta = 2U\bar{n}\,\zeta > 0$. Excitations with vanishing energy exist when the two Fermi spheres above cross, as shown in Figure 24.8 since then we have at the crossing points

$$\varepsilon_{\mathbf{k}} - U\bar{n}\,\zeta = \mathcal{E}_F, \text{and} \varepsilon_{\mathbf{k+q}} + U\bar{n}\,\zeta = \mathcal{E}_F$$

and thus

$$\varepsilon_{\mathbf{k+q}} - \varepsilon_{\mathbf{k}} + 2U\bar{n}\,\zeta = \varepsilon_{\mathbf{k+q}} + U\bar{n}\,\zeta - (\varepsilon_{\mathbf{k}} - U\bar{n}\,\zeta) = \mathcal{E}_F - \mathcal{E}_F.$$

Zero-energy excitations are dangerous, since they suggest an instability of the mean-field ground state. One has to go beyond Stoner theory to see that they do not destroy the ferromagnetic ground state in this case.

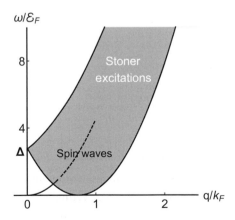

Figure 24.7 Stoner excitations spectrum.

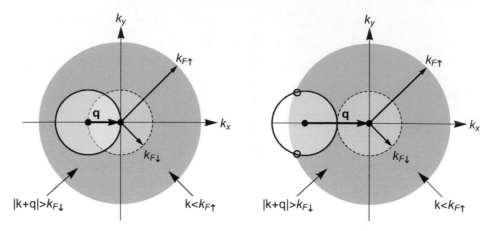

Figure 24.8 Relative disposition of the ↑ and ↓ Fermi seas at two different **q** vectors. The black dots indicate zero energy excitations.

24.2.3 Susceptibility of Interacting System and Spin Waves

Following the RPA procedure, we can write the susceptibility of the interacting ferromagnetic system as

$$\chi^{-+}(\mathbf{q},\omega) = \frac{\chi_{0,\,\text{Stoner}}^{-+}(\mathbf{q},\omega)}{1 - U\chi_{0,\,\text{Stoner}}^{-+}(\mathbf{q},\omega)}.$$

The poles of χ^{-+} determine the excitation spectrum of the itinerant ferromagnet with spin flipping. The poles equation becomes

$$1 - U\chi_{0,\,\text{Stoner}}(\mathbf{q},\omega) = 0 \quad \Rightarrow \quad \chi_{0,\,\text{Stoner}}(\mathbf{q},\omega) = \frac{1}{U}.$$

At $T = 0$, we get

$$\sum_{\mathbf{k}} \frac{\Theta(\mu - \varepsilon_{\mathbf{k}} + \gamma\,\zeta)\Big[1 - \Theta(\varepsilon_{\mathbf{k}+\mathbf{q}} + \gamma\,\zeta - \mu)\Big]}{\omega - \frac{\hbar^2}{m}\mathbf{k}\cdot\mathbf{q} - \frac{\hbar^2}{2m}q^2 - \Delta} = \frac{2\Omega}{\mu_B g U} = \frac{1}{V}. \tag{24.16}$$

We note that there are two types of singularities:

- For fixed **q**, a series of poles fill the branch cut along the real ω-axis, between $\omega_{\min}$ and $\omega_{\max}$. We note that there are as many intersections of the curves, associated with these poles, with the horizontal line $\chi_{0,\,Stoner}^{-+} = 1/U$. These give rise to the Stoner excitations.
- There is one intersection that lies at some ω below the quasicontinuum. At long wavelengths and $\omega \ll \Delta$, we have a new spin-wave excitation branch, reminiscent of plasmons, as indicated by the root $\omega_{\mathbf{q}}$ in Figure 24.9. Clearly, when $U \to 0$, $\omega_{\mathbf{q}} \to \omega_{min}$ and the discrete state merges into the quasicontinuum. However, unlike plasmons, when

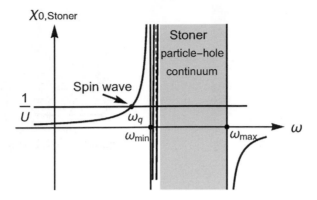

Figure 24.9 Extraction of spin-wave dispersion.

$\mathbf{q} = 0$, (24.16) shows that $\chi_{0,\,Stoner}^{-+} = 1/U$ has one root $\omega = \Delta + \frac{\mu_B g U}{2\Omega}$, so that the spin-wave frequency

$$\lim_{\mathbf{q} \to 0} \omega_{\mathbf{q}} = 0.$$

24.3 Nesting and Spin-Density Waves

We now explore instabilities that may possibly lead to magnetic orderings other than a uniform ferromagnetic phase in itinerant electronic systems. Such instabilities will be manifest as $\omega = 0$ singularities in the response function

$$\chi^{-+}(\mathbf{q}, 0) = \frac{\chi_{0,\,\text{Stoner}}^{-+}(\mathbf{q}, 0)}{1 - U\chi_{0,\,\text{Stoner}}^{-+}(\mathbf{q}, 0)} \quad \Rightarrow \quad 1 - U\chi_{0,\,\text{Stoner}}(\mathbf{q}, 0) = 0$$

at some finite q, indicating a spatially varying magnetization. However, a $\mathbf{q}$ wavevector instability is likely to develop if there exists a large region of the Fermi surface where $\varepsilon_{\mathbf{k}} = \varepsilon_{\mathbf{k}+\mathbf{q}}$ for some fixed vector $\mathbf{q}$, in a nesting fashion similar to the case of charge-density waves described in Chapter 7. As we have seen, in one dimension, the problem is particularly simple because we have natural nesting; and the instability occurs at $q = 2k_F$. In dimensions > 1, such an instability relies on there being a part of the Fermi surface that is parallel to another part, so that a constant wavevector can connect particles and holes across the Fermi sea. Examples of how this may occur are shown in Figure 24.10; note that at half-filling, tight-binding band structures show the presence of nesting across the entire Fermi surface.

Spin-Density Waves versus AFM

The AFM behavior of a half-filled insulator can be regarded as a spatially varying magnetization. However, we should distinguish between such a phase and the presence of *spin-density waves*: the AFM phase displays strong sublattice polarization, in the sense

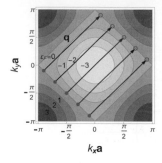 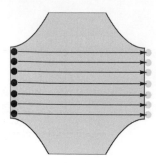

Figure 24.10 Right: half-filling, so $(\pi, \pi)/a_{\text{lattice}}$ wavevector generically leads to nesting. Left: Fermi surface in which a region is nested, so that the same spin-density-wave wavevector couples many points on the Fermi surface.

that each site had its entire magnetic moment points in some direction. By contrast, the *spin-density wave* scenario, which is the subject of this section, describes an instability that grows continuously from zero, and represents a small, partial magnetization at each site. It is however worth noticing that half-filling is a state particularly susceptible to spin-density wave ordering, and that with increasing interaction strength, there is a crossover from the spin-density wave ordering of a weakly interacting half-filled itinerant electron system to the AFM of a half-filled Mott insulator.

24.4 Anderson Model of Magnetic Impurities

The magnetic instabilities previously discussed mainly occur in metals with narrow bands, such as $3d$-bands in the first row transition metals or f-bands in rare-earth metals and metal actinides.

Here we shall study the process of formation of local magnetic moments and the effect of substitutional magnetic impurities in a nonmagnetic host metal. When atoms of these magnetic metals are disolved in nonmagnetic metallic hosts, they sometimes retain many features of magnetism, such as temperature-dependent susceptibility following a Curie-like behavior. Since such behavior cannot be explained by a one-electron model, it is obvious that interelectron interactions must play an essential role. We have studied the s–d exchange interaction, and derived the form of the long-range RKKY interaction among spins localized on different sites in a nonmagnetic metal, such as Mn in Cu. However, when elements of the iron group are introduced into a nonmagnetic metal, it is not always the case that they display a permanent magnetic moment. Fe and Mn in Cu maintain their spins, while Mn in Al does not. For the first case, we may use the s–d exchange interaction model. In the second case, one must understand why there is no magnetic moment on the transition ion impurity.

Another role of magnetic impurity atoms is revealed in the historic origin of the Kondo problem. This problem is interesting in showing a large separation of energy scales between

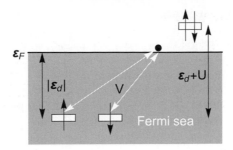

Figure 24.11 P. W. Anderson. Schematic of the Anderson magnetic impurity model.

the microscopic Hamiltonian and the effective binding energy of the ground state, hence producing a wide range of temperatures in which response properties are dominated by excited, rather than ground-state, configurations.

In order to study the conditions for an impurity to sustain a permanent magnetic moment, we require an approach where both cases can be explained with the same model. Such an approach is based on the Anderson model.

24.4.1 Anderson Hamiltonian and Local Moment Formation

Anderson realized that the moment formation has origins in strong correlations, which implies strong Coulomb interactions. Thus, in order to study the formation of localized magnetic states, Anderson proposed a model Hamiltonian [14], which turns out to be very closely related to the Hubbard Hamiltonian discussed earlier. As we have seen so far, Coulomb repulsion, specially onsite between antiparallel spins, tends to localize electrons. In contrast, Anderson also recognized that localized atomic orbitals are amenable to tunneling into free electronic states, as shown in Figure 24.11, when their atomic energy level overlaps with the free electron band.

The Anderson Hamiltonian depicts a model of localized states, represented by the operators[1] $d^\dagger$, d, coupled to delocalized conduction electrons. Because of their spatial confinement, the localized states have a strong Hubbard U, which justifies the neglect of the much weaker repulsive interaction between conduction electrons. The Hamiltonian is written as

$$\mathcal{H}_A = \sum_{\mathbf{k}\sigma} \varepsilon_{\mathbf{k}\sigma}\, c^\dagger_{\mathbf{k}\sigma}\, c_{\mathbf{k}\sigma} + \varepsilon_d \sum_{\sigma} d^\dagger_\sigma d_\sigma + U\, d^\dagger_\uparrow d_\uparrow d^\dagger_\downarrow d_\downarrow + \frac{V}{\sqrt{\Omega}} \sum_{\mathbf{k}\sigma} \left[d^\dagger_\sigma c_{\mathbf{k}\sigma} + c^\dagger_{\mathbf{k}\sigma} d_\sigma \right].$$

The impurity spin operator can be written as

$$\mathbf{S} = \frac{1}{2} \sum_{\sigma\sigma'} d^\dagger_\sigma\, \boldsymbol{\tau}_{\sigma\sigma'}\, d_{\sigma'},$$

---

[1] Although we are using the symbol d to denote creation and annihilation operators associated with the localized orbital, the orbital may be of d or f type. Similarly, we shall use ε_d to denote the orbital energy.

where $\boldsymbol{\tau}$ is the Pauli matrix. We write

$$\mathbf{S} \cdot \mathbf{S} = \frac{5}{4} \left(n_{d\uparrow} + n_{d\downarrow} \right) + \frac{3}{4} n_{d\uparrow} n_{d\downarrow}$$

so that the impurity Hamiltonian can be written as

$$\mathcal{H}_d = \sum_{\mathbf{k}\sigma} \left(\varepsilon_{\mathbf{k}\sigma} + \frac{5}{3} U \right) c_{\mathbf{k}\sigma}^\dagger c_{\mathbf{k}\sigma} - \frac{4}{3} U \mathbf{S} \cdot \mathbf{S}.$$

Thus, if U is "large enough," and if we can arrange things such that $\langle n_d \rangle \neq 0$, the impurity will maximize $\mathbf{S} \cdot \mathbf{S}$, and will acquire a moment!

First, we shall determine the conditions required for the single occupation of the localized level, namely, one electron, rather than zero or two. Second, we shall investigate when this localized magnetic moment remains free, rather than being screened by spin fluctuations of the surrounding Fermi sea.

It is important to note the energy scales of these two different processes; the interesting physics is that two very different energy scales arise from these two problems.

Virtual Bound-State Formation and Hybridization Resonance

In order to separate the different competing mechanisms contained in the Anderson Hamiltonian, we will turn off the interaction U and consider hybridization effects only. The effect is manifest in a narrow resonance induced by the impurity, which is essentially an atomic level broadened by the hybridization with the conduction electrons of the host metal.

We shall examine here the ensuing resonant scattering off a noninteracting d/f-level, using the Green function approach.

Impurity Green Function

The localized retarded Green function G_d is given by

$$G_d(t) = -i\Theta(t) \left\langle \left[d_\sigma(t), d_\sigma^\dagger(0) \right] \right\rangle.$$

The equation of motion is

$$\frac{\partial G_d}{\partial t} = -i\delta(t) \left\langle \left[d_\sigma(t), d_\sigma^\dagger(0) \right] \right\rangle - i\Theta(t) \left\langle \left[\frac{\partial d_\sigma(t)}{\partial t}, d_\sigma^\dagger(0) \right] \right\rangle$$

$$i\frac{\partial G_d}{\partial t} = \delta(t) - i\Theta(t) \left\langle \left[[d_\sigma(t), \mathcal{H}], d_\sigma^\dagger(0) \right] \right\rangle. \tag{24.17}$$

The commutator couples G_d to a mixed propagator

$$G_{\mathbf{k}d}(t) = -i\Theta(t) \left\langle \left[c_{\mathbf{k}\sigma}^\dagger(t), d_\sigma^\dagger(0) \right] \right\rangle$$

according to the equation

$$i\frac{\partial G_d}{\partial t} = \delta(t) + \varepsilon_d G_d + \sum_{\mathbf{k}} V G_{\mathbf{k}d}(t) \quad \Rightarrow \quad (\omega - \varepsilon_d) G_d(\omega) = 1 + \sum_{\mathbf{k}} V G_{\mathbf{k}d}(\omega).$$

We find in the same way that

$$(\omega - \varepsilon_{\mathbf{k}}) \; G_{\mathbf{k}d}(\omega) - V \, G_d(\omega) = 0.$$

Eliminating $G_{\mathbf{k}d}$, we finally have

$$G_d(\omega) = \frac{1}{\omega - \varepsilon_d - \Sigma(d,\omega)} = \frac{1}{G_d^{(0)-1} - \Sigma(d,\omega)}$$

$$\Sigma(d,\omega + i\eta) = \sum_{\mathbf{k}} \frac{V^2}{\omega - \varepsilon_{\mathbf{k}} + i\eta}.$$

We represent the propagator of the bare d/f-electron by a black line, and that of the conduction electron by a gray line:

$$G_d^{(0)}(\omega) = \frac{1}{\omega - \varepsilon_d}, \qquad \underset{d,\,\omega}{\underline{\qquad\longrightarrow\qquad}}$$

$$G_{\mathbf{k}}^{(0)}(\omega) = \frac{1}{\omega - \varepsilon_{\mathbf{k}}}, \qquad \underset{k,\,\omega}{\underline{\qquad\longrightarrow\qquad}}$$

The hybridization enables the d/f-electron to tunnel back and forth into the continuum, with a manifest *self-energy* diagram:

$$\overset{V \qquad\qquad V}{\underset{k,\,\omega}{\underline{\quad\bullet\longrightarrow\bullet\quad}}} \;= \Sigma(d,\omega) = \sum_k \frac{V^2}{\omega - \varepsilon_k + i\eta},$$

where we have assumed a $\mathbf{k}$-independent (contact scattering) hybridization interaction. This diagram can also be considered as an effective retarded (frequency/time-dependent) scattering potential for d/f-electrons, since the electron can spend significant time in the conduction band. The multiple scattering processes of the d/f-electron are represented by

$$\underset{d}{\Longrightarrow} \;=\; \underset{}{\longrightarrow} + \underset{k'}{\longrightarrow\!\bullet\!\triangleright\!\bullet} + \underset{k' \qquad k''}{\longrightarrow\!\bullet\!\triangleright\!\bullet\!\bullet\!\triangleright\!\bullet\longrightarrow}$$

We note that the electron assumes a different momentum each time it tunnels into the conduction band, and we have to sum over all possible values of these intermediate state momenta available in the conduction band.

We assume a broad conduction band of width $[-W, W]$, and we calculate $\Sigma(d, \omega \pm i\eta)$ as

$$\Sigma(d, \omega + i\eta) = \int \frac{d\varepsilon}{\pi} \, \mathcal{D}(\varepsilon) \frac{\pi V^2}{\omega - \varepsilon + i\eta} = \int \frac{d\varepsilon}{\pi} \frac{\Delta(\varepsilon)}{\omega - \varepsilon + i\eta},$$

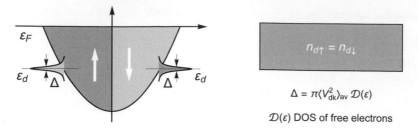

$$\Delta = \pi \langle V_{dk}^2 \rangle_{av} \, \mathcal{D}(\varepsilon)$$

$\mathcal{D}(\varepsilon)$ DOS of free electrons

Figure 24.12 Resonance in the absence of Coulomb interactions. No localized moments.

where $\mathcal{D}(\varepsilon)$ is the DOS of the conduction band, and $\Delta(\varepsilon) = \pi \mathcal{D}_c(\varepsilon) V^2$. We further simplify the problem by assuming a uniform DOS, such that $\Delta(\varepsilon) = \Delta$, and obtain

$$\Sigma(\omega + i\eta) = \frac{\Delta}{\pi} \int_{-W}^{W} \frac{d\varepsilon}{\omega - \varepsilon + i\eta} = \frac{\Delta}{\pi} \left(P \int_{-W}^{W} \frac{d\varepsilon}{\omega - \varepsilon} - i\pi \, \Theta(W - |\omega|) \right)$$

$$= -\frac{\Delta}{\pi} \ln \left[\frac{\omega + W}{\omega - W} \right] - i\Delta \, \Theta(W - |\omega|).$$

Re Σ_c is of order $O(\omega / W)$, which for $\Delta \ll W$ is negligible. Consequently, we obtain the simple form of the d, f-propagator

$$G_d(\omega) = \frac{1}{\omega - \varepsilon_d - i\Delta}.$$

It describes a resonance of width Δ, centered around ε_d, as shown in Figure 24.12, with a density of states

$$\mathcal{D}_d(\omega) = \frac{1}{\pi} \operatorname{Im} G_d(\omega - \varepsilon_d - i\eta) = \frac{\Delta}{(\omega - \varepsilon_d)^2 + \Delta^2}.$$

Conduction Electrons Green Function

Next we consider this multiple scattering process from the perspective of the conduction electrons, where we reverse the roles of the propagators, and write

$$\underset{k \quad k'}{\Longrightarrow} = \underset{k \quad k}{\longrightarrow} + \underset{k \quad k' \quad k}{\bullet\!\!\Rightarrow\!\!\bullet} + \underset{k \quad k'' \quad k'}{\bullet\!\!\Rightarrow\!\!\bullet\cdots\bullet\!\!\Rightarrow\!\!\bullet} \qquad (24.18)$$

First, note that the presence of the impurity breaks the translation symmetry of the Hamiltonian, and $G(\mathbf{k}', \mathbf{k}; \omega)$ is no longer diagonal in the momentum variables. Moreover, we note that the multiple scattering from the impurity can be cast in terms of a $\mathcal{T}$-matrix, which contains the shaded areas in (24.18), and similar diagrams in higher terms – this yields precisely $V^2 G_d$. Consequently, we represent the full (broadened) d, f-propagator by

$$G(\mathbf{k}', \mathbf{k}; \omega) = \delta_{\mathbf{k}\mathbf{k}'} G^{(0)}(\mathbf{k}, \omega) + G^{(0)}(\mathbf{k}, \omega) V^2 G_d(\omega) G^{(0)}(\mathbf{k}', \omega)$$

$$= \delta_{\mathbf{k}\mathbf{k}'} G^{(0)}(\mathbf{k}, \omega) + G^{(0)}(\mathbf{k}, \omega) \mathfrak{T}(\omega) G^{(0)}(\mathbf{k}', \omega). \qquad (24.19)$$

We write the $\mathfrak{T}$-matrix as

$$\mathfrak{T}(\omega) \equiv V^2 G_d(\omega) = \frac{1}{\pi \mathcal{D}(\omega)} \frac{\Delta}{\omega - \varepsilon_d + i\Delta} = -\frac{1}{\pi \mathcal{D}(\omega)} \frac{1}{\frac{\varepsilon_d - \omega}{\Delta} - i}. \qquad (24.20)$$

Scattering Phase Shift and Friedel Sum Rule

We recall from scattering theory that the $\mathcal{S}$-matrix can be expressed as

$$\mathcal{S}(\omega) = e^{2i\delta(\omega)} = 1 - 2\pi i \mathcal{D}(\omega) \mathfrak{T}(\omega + i\eta),$$

where $\delta(\omega)$ is the scattering phase shift, so that

$$\mathfrak{T}(\omega + i\eta) = \frac{1}{2i\pi \mathcal{D}(\omega)} \left[\mathcal{S}(\omega) - 1 \right] = -\frac{1}{\pi \mathcal{D}(\omega)} \frac{1}{\cot[\delta(\omega)] - i}. \qquad (24.21)$$

From (24.20) and (24.21), we obtain the scattering phase shift

$$\delta_d(\omega) = \cot^{-1}\left(\frac{\varepsilon_d - \omega}{\Delta} \right).$$

$\delta_d(\omega)$ increases from $\delta_d(\omega \ll \varepsilon_d) \sim 0$ to $\delta_d(\omega \gg \varepsilon_d) = \pi$; at resonance, $\delta(\varepsilon_d) = \pi/2$.

The Anderson model provides a microscopic manifestation of the Friedel sum rule, which relates the phase shifts of the conduction electrons scattered on the impurity to the number of displaced electrons.

Friedel Sum Rule

Friedel considered the presence of a localized impurity in a host metal [68]. We recall from quantum mechanics that the scattering wavefunction can be written as

$$\Psi(\mathbf{x}) = \frac{1}{k} \sum_l (2l + 1) i^l e^{i\delta_l} P_l(\cos\theta) \psi_l(r).$$

Setting $\psi_l = \varphi_l/r$, and taking spin degeneracy into account, we obtain the total electron number within a sphere of radius R as

$$4\pi \int_0^R dr\, r^2\, n(r) = 4\pi \int_0^R dr\, r^2\, 2 \int_0^{k_F} 4\pi\, dk\, k^2 \frac{1}{(2\pi)^3} \sum_l (2l + 1) \frac{\psi_l^2(r)}{k^2}$$

$$= \frac{4}{\pi} \sum_l (2l + 1) \int_0^{k_F} dk \int_0^R dr\, \varphi_l^2(r).$$

The Schrödinger equation for φ_l is given by

$$
\begin{cases}
\dfrac{d^2\varphi_l}{dr^2} + \left[k^2 - \dfrac{l(l+1)}{r^2} - \dfrac{2mV(r)}{\hbar^2} \right] \varphi_l = 0 \\[2ex]
\dfrac{d^2\bar\varphi_l}{dr^2} + \left[\bar k^2 - \dfrac{l(l+1)}{r^2} - \dfrac{2mV(r)}{\hbar^2} \right] \bar\varphi_l = 0
\end{cases}
$$

From these two equations, we obtain

$$
(\bar k^2 - k^2) \int_0^R dr\, \bar\varphi\, \varphi = \int_0^R dr \left[\bar\varphi \dfrac{d^2\varphi}{dr^2} - \varphi \dfrac{d^2\bar\varphi}{dr^2} \right] = \left[\bar\varphi \dfrac{d\varphi}{dr} - \varphi \dfrac{d\bar\varphi}{dr} \right]_0^R .
$$

which in the limit $\bar k \to k$ yields

$$
\int_0^R dr\, \varphi_l^2(r) = \dfrac{1}{2k} \left[\dfrac{d\varphi}{dk} \dfrac{d\varphi}{dr} - \varphi \dfrac{d^2\varphi}{drdk} \right]_{r=R} .
$$

Using the asymptotic form of φ

$$
\varphi(R) \simeq \sin\left(kR + \delta_l(k) - \dfrac{l\pi}{2} \right),
$$

we obtain

$$
\int_0^R dr\, r^2 n(r) = \dfrac{2}{\pi} \sum_l (2l+1) \int_0^{k_F} dk \left[\left(R + \dfrac{d\delta_l}{dk} \right) - \dfrac{1}{2k} \sin\left(2kR + 2\delta_l(k) - l\pi \right) \right].
$$

The change in electron number is given by

$$
\begin{aligned}
\Delta N &= \int_0^R \left(n(r) - n_0(r) \right] 4\pi\, r^2\, dr \\[1ex]
&= \dfrac{2}{\pi} \sum_l (2l+1) \int_0^{k_F} dk \left[\left(\dfrac{d\delta_l}{dk} \right) - \dfrac{1}{k} \sin\delta_l(k)\, \cos\left(2kR + \delta_l(k) - l\pi \right) \right].
\end{aligned}
$$

Considering the weak k-dependence of $\delta_l(k)$ compared to $2kR$, and partially integrating, we obtain

$$
\Delta N = \dfrac{2}{\pi} \sum_l (2l+1) \left[\delta_l(k_F) - \dfrac{1}{2k_F R} \sin\delta_l(k_F) \sin\left(2k_F R + \delta_l(k_F) - l\pi \right) \right].
$$

In the limit $R \to \infty$, we should obtain the density difference between the impurity and the host, namely

$$
\Delta n = \dfrac{2}{\pi} \sum_l (2l+1)\, \delta_l(k_F).
$$

Thus we surmise that the phase shift $\delta_d = \delta_d(\varepsilon_F = 0)$ at the Fermi surface determines the amount of charge bound inside the resonance. We can determine this charge by using the d/f-spectral function to calculate the ground-state occupancy:

$$n_d = \int_{-\infty}^{0} d\omega \mathcal{D}(\omega) = 2 \int_{-\infty}^{0} \frac{d\omega}{\pi} \frac{\Delta}{(\omega - \varepsilon_d)^2 + \Delta^2} = \frac{2}{\pi} \cot^{-1}\left(\frac{\varepsilon_d}{\Delta}\right) = \frac{\delta_d}{\pi/2}. \quad (24.22)$$

Note that when $\delta_d(0) = \pi/2$, $n_d = 1$. This is a particular example of the *Friedel sum rule*.

24.4.2 Hartree–Fock Physics of the Anderson Model

The Anderson model presents a competition between the Coulomb interaction and hybridization. We would expect that local moments will develop when the Coulomb interaction exceeds the hybridization – the question is how can we quantify this. For $U \neq 0$, we rewrite (24.17) in the form

$$i\frac{\partial G_{d\sigma}}{\partial t} = \delta(t) - i\Theta(t) \left\langle \left[[d_\sigma(t), \mathcal{H}], d_\sigma^\dagger(0) \right] \right\rangle$$

$$= \delta(t) + \varepsilon_d \, G_{d\sigma}(t) + V \sum_{\mathbf{k}} G_{\mathbf{k}d\sigma}(t) - i\Theta(t) U \left\langle \left[d_\sigma(t) n_{\bar{\sigma}}(t), d_\sigma^\dagger(0) \right] \right\rangle$$

and we find that the right-hand side contains a higher-order Green function that eventually leads to an infinite hierarchy of equations. We can avoid this complication by applying the Hartree–Fock approximation to terminate the hierarchy and simplify the problem. This amounts to neglecting the correlations between the "up" and "down" electron in the impurity orbital, namely,

$$n_{\bar{\sigma}}(t) = (n_{\bar{\sigma}}(t) - \langle n_{\bar{\sigma}} \rangle) + \langle n_{\bar{\sigma}} \rangle \simeq \langle n_{\bar{\sigma}} \rangle.$$

The validity of the Hartree–Fock solution would require that the correlation time scale $1/U$ be much larger than the lifetime of the localized state $1/\Delta$. This translates to $\Delta \gg U$, which may not apply in the magnetic regime!

However, to gain an initial insight into the effect of hybridization on local moment formation, Anderson originally developed following Hartree–Fock mean-field treatment: first, we modify the impurity Green function, to read

$$G_{d\sigma}(\omega) = \frac{1}{\omega - \varepsilon_d - U \langle n_{\bar{\sigma}} \rangle - i\Delta},$$

which means that we set

$$\varepsilon_d \rightarrow E_{d\sigma} = \varepsilon_d + U \langle n_{\bar{\sigma}} \rangle$$

$$\langle n_{d\sigma} \rangle = \frac{\delta_{d\sigma}}{\pi} = \frac{1}{\pi} \cot^{-1}\left(\frac{\varepsilon_d + U \langle n_{\bar{\sigma}} \rangle}{\Delta}\right).$$

Anticipating that the development of a magnetic moment amounts to $\langle n_{d\uparrow} \rangle \neq \langle n_{d\downarrow} \rangle$, we introduce the following definitions:

Total occupancy $\qquad n_d = \sum_\sigma \langle n_{d\sigma} \rangle$

Magnetization $\qquad \mu = \langle n_\uparrow \rangle - \langle n_\downarrow \rangle$

$$\langle n_{d\sigma} \rangle = \frac{1}{2} \left(n_d + \sigma \mu \right), \quad \sigma = \pm 1$$

The self-consistent mean-field equation for occupancy and magnetization become

$$n_d = \frac{1}{\pi} \sum_{\sigma = \pm 1} \cot^{-1} \left(\frac{\varepsilon_d + (U/2)\,(n_d - \sigma\,\mu)}{\Delta} \right) \tag{24.23}$$

$$\mu = \frac{1}{\pi} \sum_{\sigma = \pm 1} \sigma \, \cot^{-1} \left(\frac{\varepsilon_d + (U/2)\,(n_d - \sigma\,\mu)}{\Delta} \right). \tag{24.24}$$

To obtain the critical interaction strength U_c, which defines the threshold for local moment formation, we set $\mu = 0$ in (24.23), yielding

$$\frac{\varepsilon_d + (U_c/2)\,n_d}{\Delta} = \cot \left(\frac{\pi n_d}{2} \right).$$

However, to ensure that we are taking the limit $\mu \to 0$, we take the derivative of (24.24) with respect to μ and then set $\mu = 0$; we obtain[2]

$$1 = \frac{U_c}{\pi \Delta} \frac{1}{1 + \left(\frac{\varepsilon_d + (U_c/2)\,n_d}{\Delta} \right)^2} = \frac{U_c}{\pi \Delta} \sin^2 \left(\frac{\pi n_d}{2} \right)$$

so that for a local moment to exist, namely, $n_d = 1$, we obtain

$$U_c = \pi \Delta.$$

For $U > U_c$, there are two solutions, corresponding to an "up" or "down" spin polarization of the d/f-state. We will see that this is an oversimplified description of the local moment, but it gives us an approximate picture of the physics. The total density of states now contains two Lorentzian peaks, located at $\varepsilon_d \pm M$:

$$\mathcal{D}_d(\omega) = \frac{1}{\pi} \left[\frac{\Delta}{(\omega - \varepsilon_d - UM)^2 + \Delta^2} + \frac{\Delta}{(\omega - \varepsilon_d + UM)^2 + \Delta^2} \right].$$

The critical curve obtained by plotting U_c and ε_d as a parametric function of n_d is shown in Figure 24.13.

The Anderson mean-field theory allows a qualitative understanding of the experimentally observed formation of local moments. When dilute magnetic ions are dissolved in a

---

2 $\frac{d\cot^{-1}(x)}{dx} = \frac{1}{1+x^2}.$

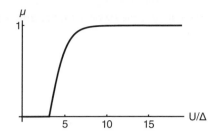

Figure 24.13 Mean-field magnetic moment.

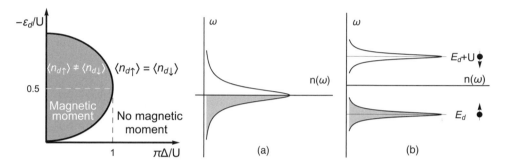

Figure 24.14 Mean-field phase diagram of the Anderson model, illustrating how the d, f-electron resonance splits to form a local moment. (a) $U < \pi \Delta$, single half-filled resonance. (b) $U > \pi \Delta$, up and down components of the resonance are split by an energy U.

metal to form an alloy, the formation of a local moment is dependent on whether the ratio $U/\pi \Delta$ is larger than or smaller than unity. When iron is dissolved in pure niobium, the failure of the moment to form reflects the higher density of states and larger value of Δ in this alloy. When iron is dissolved in molybdenum, the lower density of states causes $U > U_c$, and local moments form.

The case in which a magnetic moment can arise, as illustrated in Figure 24.14, is when the localized level has energy less than the Fermi level, so it will be occupied by at least one electron, but the interaction strength is large enough that the energy to have two electrons is greater than the Fermi level, i.e., $\varepsilon_d < \varepsilon_F < \varepsilon_d + U$.

The singly occupied impurity ground state is given by

$$\left| d_\sigma^1 \right\rangle = d_\sigma^\dagger \prod_{\mathbf{k}}^{\mathbf{k}_F} c_{\mathbf{k}\uparrow}^\dagger c_{\mathbf{k}\downarrow}^\dagger \left| 0 \right\rangle .$$

For simplicity, we also assume that excitations to

$$\left| d^2 \right\rangle = d_{\bar{\sigma}}^\dagger c_{\mathbf{k}\bar{\sigma}} \left| d_\sigma^1 \right\rangle, \quad \text{or} \quad \left| d^0 \right\rangle = c_{\mathbf{k}\sigma}^\dagger d_\sigma \left| d_\sigma^1 \right\rangle,$$

i.e., respectively adding an electron to or removing an electron from the impurity cost the same amount of energy $U/2$. As a result, the situation becomes electron–hole symmetric: there are two Hubbard bands at the same energy that are each other's mirror image through electron–hole inversion. Effectively, the impurity is now a localized $S = 1/2$ system that can switch its magnetization through either of two virtual processes, each with a transition rate $2V^2/U$:

$$\left|d_{\bar{\sigma}}^1\right\rangle = c_{\mathbf{k}'\sigma}^\dagger \, d_\sigma \, d_{\bar{\sigma}}^\dagger \, c_{\mathbf{k}\bar{\sigma}} \left|d_\sigma^1\right\rangle \qquad \text{via } \left|d^2\right\rangle$$

$$\left|d_{\bar{\sigma}}^1\right\rangle = d_{\bar{\sigma}}^\dagger \, c_{\mathbf{k}'\bar{\sigma}}^\dagger \, c_{\mathbf{k}\sigma}^\dagger \, d_\sigma \left|d_\sigma^1\right\rangle \qquad \text{via } \left|d^0\right\rangle.$$

We note that in this simple Hartree–Fock approach, we have a state consisting of a singly occupied d/f-state plus a filled Fermi sea: as shown in Figure 24.15, the occupancy is either $\langle n_{d\uparrow}\rangle = 1$, $\langle n_{d\downarrow}\rangle = 0$, or $\langle n_{d\downarrow}\rangle = 1$, $\langle n_{d\uparrow}\rangle = 0$, accounting for the two degenerate situations. We recall from the Friedel sum rule that the phase shift $\delta(\varepsilon_F)$ divided by π gives the total number of occupying electrons. Thus for this doublet case, we have $\delta_\uparrow = \pi, \delta_\downarrow = 0$ or $\delta_\uparrow = 0, \delta_\downarrow = \pi$, as shown in Figure 24.16.

The question we turn to now it that of the ultimate fate of a local moment immersed in a sea of conduction electrons. In the large U limit, that is the limit in which the onsite repulsion energy U is much greater than the local moment resonance width Δ, we will examine the residual interaction between the local moment and the conduction electrons.

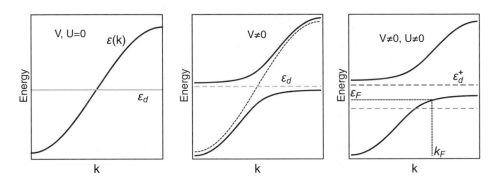

Figure 24.15 Electron dispersions: left $U = V = 0$; middle $V \neq 0$, $U = 0$; right $U, V \neq 0$. ϵ_d is the singly occupied d/f energy level, while $\epsilon_d + U$ is the doubly occupied level.

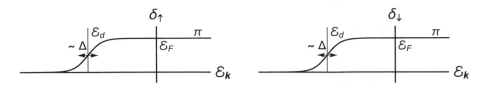

Figure 24.16 The impurity scattering phase shift for the conduction band states, reflecting single-occupancy in the mean-field state.

24.4.3 Relating the Anderson and Kondo Models: Schrieffer–Wolff Transformation

We understand now that in the regime $\Delta \ll |\varepsilon_d|$, $\varepsilon_d \ll \varepsilon_F \equiv 0$, and $\varepsilon_d + U > 0$, the energies of the empty and double occupied state, $|d^0\rangle$ and $|d^2\rangle$, lie above the $|d^1\rangle$ states, and they can be neglected at low temperatures – a local moment represented by a spin-1/2 is formed. This implies that local charge fluctuations on the d-level are suppressed at low temperatures, and only virtual exchange with conduction band electrons survive, leading to spin-flip processes. In other words, in systems that display an impurity moment, we surmise that the prerequisite of large $V \ll U$ means suppressing charge (or occupancy) fluctuations on the impurity, thus effectively quenching the impurity *charge degree of freedom*. The only degree of freedom that remains at the impurity is its spin, which carries low-energy spin excitations. For such systems, we write the Anderson Hamiltonian as

$$\mathcal{H}_A = \mathcal{H}_0 + \mathcal{H}_h$$

$$\mathcal{H}_0 = \sum_{\mathbf{k}\sigma} \varepsilon_{\mathbf{k}\sigma}\, c^\dagger_{\mathbf{k}\sigma}\, c_{\mathbf{k}\sigma} + \varepsilon_d \sum_\sigma d^\dagger_\sigma d_\sigma + U\, d^\dagger_\uparrow d_\uparrow d^\dagger_\downarrow d_\downarrow$$

$$\mathcal{H}_h = \frac{V}{\sqrt{\Omega}} \left(d^\dagger_\sigma c_{\mathbf{k}\sigma} + \text{H.c.} \right),$$

where we treat $\mathcal{H}_h$ as a perturbation. We conjure the scenario where an electron that is strongly localized at the impurity may occasionally hop into the band, or a band electron hops on to the impurity site to gain kinetic energy. From our previous discussions, we infer that this leads to an antiferromagnetic exchange interaction between the local impurity spin and the conduction electron spin at the impurity site.

In 1966, Schrieffer and Wolff [162] proposed a scheme whereby the Anderson Hamiltonian is transformed into an effective one that supports the prescribed low-energy scenario. Following their approach, we consider an effective Hamiltonian restricted to the subspace in which the localized level is singly occupied.

We now follow the prescription proposed by Schrieffer and Wolff that involves the application of a canonical transformation of the form $e^{iS}\mathcal{H}e^{-iS} = \tilde{\mathcal{H}}$. In the general scheme of the Schrieffer–Wolff transformation and within the restricted Hilbert space, we begin by partitioning $\mathcal{H}_h$ into components that effect transitions between Hilbert subspaces with different occupancies, as shown in Figure 24.17,

$$\mathcal{H}_h = \frac{V}{\sqrt{\Omega}} \sum_{\mathbf{k}} [\mathcal{H}_{21}(\mathbf{k}) + \mathcal{H}_{12}(\mathbf{k}) + \mathcal{H}_{01}(\mathbf{k}) + \mathcal{H}_{10}(\mathbf{k})]$$

$$\mathcal{H}_{12}(\mathbf{k}) = \sum_\sigma n_{\bar{\sigma}}\, d^\dagger_\sigma c_{\mathbf{k}\sigma}, \qquad \mathcal{H}_{10}(\mathbf{k}) = \sum_\sigma (1 - n_{\bar{\sigma}})\, c^\dagger_{\mathbf{k}\sigma} d_\sigma,$$

where $n \equiv n_d$ is the population of the d level. We note that $\mathcal{H}_{12}$ takes the localized state from singly occupied to doubly occupied, $n_d = 1 \to 2$, and $\mathcal{H}_{10}$ takes it from singly occupied to empty. The presence of $n_{\bar{\sigma}}$ in $\mathcal{H}_{12}$ and $1 - n_{\bar{\sigma}}$ in $\mathcal{H}_{12}$ ensures that the impurity level is already singly occupied. $\mathcal{H}_{21}$, $\mathcal{H}_{01}$ are the appropriate Hermitian conjugates.

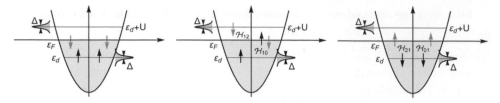

Figure 24.17 Virtual spin-flip transitions under impurity single-occupancy constraint.

At this point, we should be familiar with the application of this type of canonical transformation. We expand to second order

$$\tilde{\mathcal{H}} = \mathcal{H}_0 + i\,[S, \mathcal{H}_0] + \mathcal{H}_{\mathrm{h}} + i\,[S, \mathcal{H}_{\mathrm{h}}] - \frac{1}{2}\,[S, [S, \mathcal{H}_0]] + \cdots$$

and require that $[S, \mathcal{H}_0] = i\mathcal{H}_{\mathrm{h}}$. We then arrive at

$$\tilde{\mathcal{H}} = \mathcal{H}_0 + \frac{i}{2}\,[S, \mathcal{H}_{\mathrm{h}}].$$

To determine S, we note that

$$[\mathcal{H}_{12}(\mathbf{k}), \mathcal{H}_0] = (\varepsilon_{\mathbf{k}} - \varepsilon_d - U)\sum_{\sigma} n_{\bar{\sigma}}\, d^{\dagger}\, c_{\mathbf{k}\sigma}$$

$$[\mathcal{H}_{10}(\mathbf{k}), \mathcal{H}_0] = -(\varepsilon_{\mathbf{k}} - \varepsilon_d)\sum_{\sigma} (1 - n_{\bar{\sigma}})\, c^{\dagger}_{\mathbf{k}\sigma}\, d_{\sigma}.$$

The energy prefactors represent the difference between before and after the application of the corresponding $\mathcal{H}_{\mathrm{h}}$ operator. The Hermitian conjugate terms should have opposite signs. We obtain

$$S = i\,\frac{V}{\sqrt{\Omega}}\sum_{\mathbf{k}}\left[\frac{\mathcal{H}_{21}(\mathbf{k}) - \mathcal{H}_{12}(\mathbf{k})}{\varepsilon_{\mathbf{k}} - \varepsilon_d - U} - \frac{\mathcal{H}_{10}(\mathbf{k}) - \mathcal{H}_{01}(\mathbf{k})}{\varepsilon_{\mathbf{k}} - \varepsilon_d}\right]. \qquad (24.25)$$

In the expansion of $[S, \mathcal{H}_{\mathrm{h}}]$, we only retain terms pertinent to the Hilbert subspace of single occupancy, namely terms like $\mathcal{H}_{12}(\mathbf{k})\mathcal{H}_{21}(\mathbf{k})$ and $\mathcal{H}_{10}(\mathbf{k})\mathcal{H}_{01}(\mathbf{k})$, which represent virtual double or single occupancy. We thus obtain

$$\frac{i}{2}\,[S, \mathcal{H}_{\mathrm{h}}] \approx -\frac{V^2}{2\Omega}\sum_{kk'}\left[-\frac{\mathcal{H}_{12}(\mathbf{k})\mathcal{H}_{21}(\mathbf{k}') + \mathcal{H}_{12}(\mathbf{k}')\mathcal{H}_{21}(\mathbf{k})}{\varepsilon_{\mathbf{k}} - \varepsilon_d - U}\right.$$

$$\left.+\frac{\mathcal{H}_{10}(\mathbf{k})\mathcal{H}_{01}(\mathbf{k}') + \mathcal{H}_{10}(\mathbf{k})\mathcal{H}_{01}(\mathbf{k}')}{\varepsilon_{\mathbf{k}} - \varepsilon_d}\right]. \qquad (24.26)$$

Note the opposite signs coming from the opposite signs of $\mathcal{H}_{21}$ and $\mathcal{H}_{10}$ in (24.25).

To simplify the procedure, we invoke the assumption that the **k**-dependence of the denominators is slow, which is justified when ε_d is far below the Fermi surface, and $\varepsilon_d + U$ far above. This allows us to introduce the quantities

$$J_{12} = \frac{2V^2}{\Omega} \sum_{\mathbf{k}} \frac{1}{\varepsilon_{\mathbf{k}} - \varepsilon_d - U}, \qquad J_{10} = \frac{2V^2}{\Omega} \sum_{\mathbf{k}} \frac{1}{\varepsilon_{\mathbf{k}} - \varepsilon_d}$$

and write

$$\frac{i}{2} [S, \mathcal{H}_{\mathrm{h}}] = -\frac{1}{2\Omega} \sum_{kk'} \left[J_{12} \, \mathcal{H}_{12}(\mathbf{k}) \mathcal{H}_{21}(\mathbf{k}') - J_{10} \, \mathcal{H}_{10}(\mathbf{k}) \mathcal{H}_{01}(\mathbf{k}') \right].$$

Next we consider how products of operators affect the states of the impurity spin and of the conduction electrons, namely,

$$\mathcal{H}_{12}(\mathbf{k}') \mathcal{H}_{21}(\mathbf{k}) = \sum_{\sigma} \left[n_{\bar{\sigma}} \, c_{\mathbf{k}'\sigma}^{\dagger} \, d_{\sigma} \, n_{\bar{\sigma}} \, d_{\sigma}^{\dagger} \, c_{\mathbf{k}\sigma} + n_{\sigma} \, c_{\mathbf{k}'\bar{\sigma}}^{\dagger} \, d_{\bar{\sigma}} \, n_{\bar{\sigma}} \, d_{\sigma}^{\dagger} \, c_{\mathbf{k}\sigma} \right]$$

$$= \sum_{\sigma} \left[n_{\bar{\sigma}} \, (1 - n_{\sigma}) \, c_{\mathbf{k}'\sigma}^{\dagger} \, c_{\mathbf{k}\sigma} - d_{\sigma}^{\dagger} \, d_{\bar{\sigma}} \, c_{\mathbf{k}'\bar{\sigma}}^{\dagger} \, c_{\mathbf{k}\sigma} \right] = (2 \leftrightarrow 1)$$

$$\mathcal{H}_{10}(\mathbf{k}) \mathcal{H}_{01}(\mathbf{k}') = \sum_{\sigma} \left[(1 - n_{\bar{\sigma}}) \, d_{\sigma}^{\dagger} \, c_{\mathbf{k}'\sigma} (1 - n_{\bar{\sigma}}) \, c_{\mathbf{k}\sigma}^{\dagger} \, d_{\sigma} + (1 - n_{\sigma}) \, d_{\bar{\sigma}}^{\dagger} \, c_{\mathbf{k}'\bar{\sigma}} \, (1 - n_{\bar{\sigma}}) \, c_{\mathbf{k}\sigma}^{\dagger} \, d_{\sigma} \right]$$

$$= \sum_{\sigma} \left[-n_{\sigma} \, (1 - n_{\bar{\sigma}}) \, c_{\mathbf{k}\sigma}^{\dagger} \, c_{\mathbf{k}'\sigma} - d_{\bar{\sigma}}^{\dagger} \, d_{\sigma} \, c_{\mathbf{k}\sigma}^{\dagger} \, c_{\mathbf{k}'\bar{\sigma}} \right] = (1 \leftrightarrow 0). \qquad (24.27)$$

Recalling that $n_{\uparrow} (1 - n_{\downarrow}) = 1/2 + S^z$ and $n_{\downarrow} (1 - n_{\uparrow}) = 1/2 - S^z$, we can write

$$(2 \leftrightarrow 1) = n_{\downarrow} (1 - n_{\uparrow}) \, c_{\mathbf{k}'\uparrow}^{\dagger} \, c_{\mathbf{k}\uparrow} - d_{\uparrow}^{\dagger} \, d_{\downarrow} \, c_{\mathbf{k}'\downarrow}^{\dagger} \, c_{\mathbf{k}\uparrow} + n_{\uparrow} (1 - n_{\downarrow}) \, c_{\mathbf{k}'\downarrow}^{\dagger} \, c_{\mathbf{k}\downarrow} - d_{\downarrow}^{\dagger} \, d_{\uparrow} \, c_{\mathbf{k}'\uparrow}^{\dagger} \, c_{\mathbf{k}\downarrow}$$

$$= \left(\frac{1}{2} - S^z \right) c_{\mathbf{k}'\uparrow}^{\dagger} \, c_{\mathbf{k}\uparrow} - S^+ \, c_{\mathbf{k}'\downarrow}^{\dagger} \, c_{\mathbf{k}\uparrow} + \left(\frac{1}{2} + S^z \right) c_{\mathbf{k}'\downarrow}^{\dagger} \, c_{\mathbf{k}\downarrow} - S^- \, c_{\mathbf{k}'\uparrow}^{\dagger} \, c_{\mathbf{k}\downarrow}$$

$$= \frac{1}{2} \left(c_{\mathbf{k}'\uparrow}^{\dagger} \, c_{\mathbf{k}\uparrow} + c_{\mathbf{k}'\downarrow}^{\dagger} \, c_{\mathbf{k}\downarrow} \right) - 2S \cdot c_{\mathbf{k}'\sigma'}^{\dagger} \, \mathbf{s}^{\sigma'\sigma} \, c_{\mathbf{k}\sigma} \qquad (24.28)$$

and

$$(1 \leftrightarrow 0) = -\frac{1}{2} \left(c_{\mathbf{k}'\uparrow}^{\dagger} \, c_{\mathbf{k}\uparrow} + c_{\mathbf{k}'\downarrow}^{\dagger} \, c_{\mathbf{k}\downarrow} \right) - 2S \cdot c_{\mathbf{k}'\sigma'}^{\dagger} \, \mathbf{s}^{\sigma'\sigma} \, c_{\mathbf{k}\sigma},$$

where $\mathbf{s} = \frac{\hbar}{2} \boldsymbol{\sigma}$, with $\boldsymbol{\sigma}$ being the Pauli matrix.

The sum of these terms produces two distinct scattering channels:

- A spin-independent scattering channel, carrying regular impurity disorder scattering, with a coefficient $J_{10} + J_{12}$

- A second channel containing the exchange interaction of the impurity spin with the conduction electrons with a coefficient $J_{10} - J_{12}$

Here we are interested in the second channel, which yields the Kondo Hamiltonian

$$\mathcal{H}_{Kondo} = \sum_{\mathbf{k}\sigma} \varepsilon_{\mathbf{k}\sigma}\, c_{\mathbf{k}\sigma}^\dagger\, c_{\mathbf{k}\sigma} + \frac{J}{\Omega}\, \mathbf{S} \cdot \sum_{\substack{\mathbf{k}\mathbf{k}' \\ \sigma\sigma'}} c_{\mathbf{k}'\sigma'}^\dagger\, \mathbf{s}^{\sigma'\sigma}\, c_{\mathbf{k}\sigma}, \qquad (24.29)$$

where

$$J = J_{10} - J_{12} = \frac{2V^2}{\Omega} \sum_{\mathbf{k}} \frac{-U}{(\varepsilon_{\mathbf{k}} - \varepsilon_d - U)(\varepsilon_{\mathbf{k}} - \varepsilon_d)} \approx -\frac{2V^2}{\Omega}\frac{U}{\varepsilon_d(\varepsilon_d + U)}$$

$$J \approx -\frac{2V^2}{\Omega}\frac{1}{\varepsilon_d}, \qquad U \gg \varepsilon_d \qquad \text{Strong repulsion limit.}$$

J is clearly positive, since $\varepsilon_d < 0$ and $\varepsilon_d + U > 0$. Thus we find that the localized spin couples to the spins of the conduction electrons that overlap it. Such coupling can cause spin-flips and may enhance the scattering. We also note that the coupling is antiferromagnetic, which favors the formation of a singlet; however, this competes with the kinetic energy cost of single occupation of conduction electron levels.

24.5 The Kondo Effect

Normally the resistance of pure metals monotonically decreases as the temperature is lowered, as electronic inelastic scattering processes, mainly due to phonons, are suppressed. Adding impurities is expected to gives rise to a constant offset that does not change the monotonicity. But, the story of the Kondo effect starts in the 1930s, when de Haas and coworkers [48] in Leiden measured the resistance of some metals at very low temperatures. Surprisingly, they found that gold samples' resistance increased rather than decreased upon cooling down below $\sim$8K, as shown in Figure 24.18. Subsequently, it was realized that alloys containing dilute magnetic moments, such as a low concentration of Mn or Fe in Cu, sometimes show similar trends. An explanation of the minimum was given by Kondo in 1964, but it took a further decade before the nature of the ground state was properly understood, and it was only in 1980 that a model for the phenomenon was solved exactly.

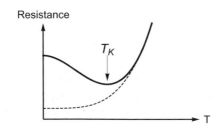

Figure 24.18 Electrical resistance of an alloy exhibiting the Kondo effect.

24.5.1 The Kondo Model

The mean field (Hartree–Fock) approximation to the Anderson model provided an explanation for the local moment formation at magnetic impurity sites in nonmagnetic metallic hosts. However, it failed to predict the strange behavior at low temperatures and low energies that should be manifest in the model. A primary example is the peculiar disappearance of the moments at low temperatures, where they become effectively screened by the conduction electrons. The disappearance is found to be associated with one quasielectron effectively binding with the impurity moment to create a local singlet state that behaves like a nonmagnetic scatterer, at $T = 0$. This physical picture was proposed by Kondo, in his attempt to explain the existence of the resistance minimum described previously. Kondo treated the problem perturbatively in terms of an exchange coupling between the magnetic impurity and the conduction band. The Hamiltonian constructed by Kondo is

$$\mathcal{H}_K = \sum_{k\sigma} \varepsilon_{k\sigma} \, c^\dagger_{k\sigma} \, c_{k\sigma} + \mathcal{H}_s = \sum_{k\sigma} \varepsilon_{k\sigma} \, c^\dagger_{k\sigma} \, c_{k\sigma} + J \, \mathbf{S} \cdot \mathbf{s}(0), \qquad (24.30)$$

where $\mathbf{S}$ represents the local impurity moment and $\mathbf{s}(0)$ is the spin density of the conduction electrons at the impurity site. It is in fact identical to (24.29), when we can write the exchange interaction as

$$\mathcal{H}_s = J \, \mathbf{S} \cdot \mathbf{s}(0) = J \left[S^z \, s^z(0) + \frac{1}{2} \left(S^+ s^-(0) + S^- s^+(0) \right) \right]$$

$$= \frac{J}{\Omega} \sum_{kk'} \left[S^z \left(c^\dagger_{k'\uparrow} c_{k\uparrow} - c^\dagger_{k'\downarrow} c_{k\downarrow} \right) + S^+ c^\dagger_{k'\downarrow} c_{k\uparrow} + S^- c^\dagger_{k'\uparrow} c_{k\downarrow} \right]. \qquad (24.31)$$

We now follow Kondo's perturbative scheme by adiabatically turning on $\mathcal{H}_s$ in a system that consists of a Fermi sea $|FS\rangle$ and an extra electron at the Fermi level, together with the impurity level. The initial state is then $|i\rangle = c^\dagger_{k_i\sigma} |FS\rangle$. We determine the scattering amplitude, with the aid of the $\mathcal{T}$-matrix, between $|i\rangle$ and $|f\rangle = c^\dagger_{k_f\sigma'} |FS\rangle$,

$$\langle f| \, \mathcal{T} \, |i\rangle = 2\pi i \left\langle f \left| \mathcal{H}_s + \mathcal{H}_s \frac{1}{\varepsilon_k - \mathcal{H}_0} \mathcal{H}_s + \cdots \right| i \right\rangle.$$

Since we are interested in scattering events that contribute to the resistivity, we have to consider on-the-energy-shell scattering, where $\varepsilon_{k_i} = \varepsilon_{k_f}$.

Scattering Amplitude

We find that in the case of large U, the Anderson model has a built-in local moment. This local moment provides a scattering potential for the conduction electrons that is quite different from a conventional scalar potential: it endows the impurity scatterer with internal degrees of freedom. We will now explore the ramifications of this difference.

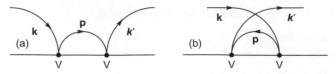

Figure 24.19 Second-order spin-independent scattering processes: (a) direct, (b) exchange.

A. Nonmagnetic Impurity Scattering Pedagogically, it would be instructive to compare the spin-dependent scattering outcome with that of a regular nonmagnetic scattering. We shall first consider the case of a nonmagnetic impurity, with a scattering potential

$$\mathcal{H}_s = \sum_{\mathbf{kk'},\sigma} c_{\mathbf{k'}\sigma}^{\dagger} \, V_{\mathbf{kk'}} \, c_{\mathbf{k}\sigma}.$$

We have to consider the direct and exchange channels in second-order processes, as shown in Figure 24.19:

$$\left\langle 0 \left| c_{\mathbf{k}_f} c_{\mathbf{k}_f}^{\dagger} V_{\mathbf{k}_f \mathbf{p}} c_{\mathbf{p}} \frac{1}{\varepsilon_{\mathbf{k}_i} - \mathcal{H}_0} c_{\mathbf{p}}^{\dagger} V_{\mathbf{pk}_i} c_{\mathbf{k}_i} c_{\mathbf{k}_i}^{\dagger} \right| 0 \right\rangle = V_{\mathbf{k}_f \mathbf{p}} \frac{1 - n_F(\varepsilon_{\mathbf{p}})}{\varepsilon_{\mathbf{k}_i} - \varepsilon_{\mathbf{p}}} V_{\mathbf{pk}_i}$$

$$\left\langle 0 \left| c_{\mathbf{k}_f} c_{\mathbf{p}}^{\dagger} V_{\mathbf{pk}_i} c_{\mathbf{k}_i} \frac{1}{\varepsilon_{\mathbf{k}_i} - \mathcal{H}_0} c_{\mathbf{k}_f}^{\dagger} V_{\mathbf{k}_f \mathbf{p}} c_{\mathbf{p}} c_{\mathbf{k}_i}^{\dagger} \right| 0 \right\rangle = -V_{\mathbf{pk}_i} \frac{n_F(\varepsilon_{\mathbf{p}})}{\varepsilon_{\mathbf{k}_i} - (\varepsilon_{\mathbf{k}_i} - \varepsilon_{\mathbf{p}} + \varepsilon_{\mathbf{k}_f})} V_{\mathbf{k}_f \mathbf{p}}$$

$$= V_{\mathbf{pk}_i} \frac{n_F(\varepsilon_{\mathbf{p}})}{\varepsilon_{\mathbf{k}_f} - \varepsilon_{\mathbf{p}}} V_{\mathbf{k}_f \mathbf{p}}.$$

We note that the first process only takes place if the intermediate state $\mathbf{p}$ is unoccupied, while the second only if it is occupied. Recalling that $\varepsilon_{\mathbf{k}_f} = \varepsilon_{\mathbf{k}_i}$, it is clear that the Fermi function cancels when we take the sum of the two processes. This indicates that there is no significant T dependence to this order in ordinary potential scattering.

B. Magnetic Impurity Scattering Next, we consider the corresponding processes for spin-dependent scattering, where the perturbation is given by $\mathcal{H}_s$ in (24.31). Actually, the exchange interactions of the impurity spin with the conduction electrons produce a correlated choreography, where the impurity spin state at a given time depends on previous scattering events.[3] To demonstrate the ensuing effects, we will calculate the scattering amplitude, to second order in J, which will give the scattering rate to $O(J^3)$.

The first-order contribution to the amplitude is

$$\langle \mathbf{k}_f \uparrow | \mathcal{H}_s | \mathbf{k}_i \uparrow \rangle = -\langle \mathbf{k}_f \downarrow | \mathcal{H}_s | \mathbf{k}_i \downarrow \rangle = \frac{J S^z}{\Omega}$$

$$\langle \mathbf{k}_f \downarrow | \mathcal{H}_s | \mathbf{k}_i \uparrow \rangle = \frac{J S^+}{\Omega}; \qquad \langle \mathbf{k}_f \uparrow | \mathcal{H}_s | \mathbf{k}_i \downarrow \rangle = \frac{J S^-}{\Omega}$$

---

[3] For example, we consider two electrons, both with spin-up, trying to spin-flip scatter from a spin-down impurity. The first electron can exchange its spin with the impurity and leave it spin-up. The second electron therefore cannot spin-flip scatter because of spin conservation. Thus the electrons of the conduction band cannot be treated as independent objects.

giving a temperature-independent scattering rate of

$$\Gamma \propto \frac{N_{\text{imp}}}{\Omega} \frac{J^2}{\omega} \left[S^{z2} + \frac{1}{2} \left(S^+ S^- + S^- S^+ \right) \right] \quad \Rightarrow \quad \frac{N_{\text{imp}}}{\Omega} \frac{J^2}{\omega} S(S+1).$$

The second-order contribution is of the form

$$\sum_p \langle \mathbf{k}_f \uparrow | \mathcal{H}_s | p \rangle \frac{1}{\varepsilon - \varepsilon_p} \langle p | \mathcal{H}_s | \mathbf{k}_i \uparrow \rangle,$$

where, again, $\varepsilon = \varepsilon_{\mathbf{k}_i} = \varepsilon_{\mathbf{k}_f}$, and $|p\rangle$ is an intermediate state with energy ε_p. The direct, or non-spin-flip, processes are given by

$$\sum_p \langle \mathbf{k}_f \uparrow | S^z c^\dagger_{\mathbf{k}_f \uparrow} c_{\mathbf{p}\uparrow} \frac{1}{\varepsilon - \varepsilon_p} S^z c^\dagger_{\mathbf{p}\uparrow} c_{\mathbf{k}_i \uparrow} | \mathbf{k}_i \uparrow \rangle = J^2 \sum_{\mathbf{p}} S^z S^z \frac{1 - n_F(\varepsilon_p)}{\varepsilon_{\mathbf{k}_i} - \varepsilon_{\mathbf{p}}}$$

$$\sum_p \langle \mathbf{k}_f \uparrow | S^z c^\dagger_{\mathbf{p}\uparrow} c_{\mathbf{k}_i \uparrow} \frac{1}{\varepsilon - \varepsilon_p} S^z c^\dagger_{\mathbf{k}_f \uparrow} c_{\mathbf{p}\uparrow} | \mathbf{k}_i \uparrow \rangle = J^2 \sum_{\mathbf{p}} S^z S^z \frac{n_F(\varepsilon_p)}{\varepsilon_{\mathbf{k}_f} - \varepsilon_{\mathbf{p}}}.$$

We thermal-averaged over the intermediate state, by setting

$$c^\dagger_{\mathbf{p}\downarrow} c_{\mathbf{p}\downarrow} \rightarrow \left\langle c^\dagger_{\mathbf{p}\downarrow} c_{\mathbf{p}\downarrow} \right\rangle = n_F(\varepsilon_\mathbf{p}), \qquad \text{and} \qquad c_{\mathbf{p}\downarrow} c^\dagger_{\mathbf{p}\downarrow} \rightarrow \left\langle c_{\mathbf{p}\downarrow} c^\dagger_{\mathbf{p}\downarrow} \right\rangle = 1 - n_F(\varepsilon_\mathbf{p}).$$

Upon adding the two longitudinal contributions, as shown in Figure 24.20, the Fermi distribution cancels out: the final probability does not depend on the occupation of intermediate states and hence on temperature.

The interesting contributions involving spin-flips, shown in Figure 24.21, arise from the processes

$$\sum_p \langle \mathbf{k}_f \uparrow | S^- c^\dagger_{\mathbf{k}_f \uparrow} c_{\mathbf{p}\downarrow} \frac{1}{\varepsilon - \varepsilon_p} S^+ c^\dagger_{\mathbf{p}\downarrow} c_{\mathbf{k}_i \uparrow} | \mathbf{k}_i \uparrow \rangle = J^2 \sum_{\mathbf{p}} S^- S^+ \frac{1 - n_F(\varepsilon_p)}{\varepsilon_{\mathbf{k}_i} - \varepsilon_{\mathbf{p}}}$$

$$\sum_p \langle \mathbf{k}_f \uparrow | S^+ c^\dagger_{\mathbf{p}\downarrow} c_{\mathbf{k}_i \uparrow} \frac{1}{\varepsilon - \varepsilon_p} S^- c^\dagger_{\mathbf{k}_f \uparrow} c_{\mathbf{p}\downarrow} | \mathbf{k}_i \uparrow \rangle = J^2 \sum_{\mathbf{p}} S^+ S^- \frac{n_F(\varepsilon_p)}{\varepsilon_{\mathbf{k}_f} - \varepsilon_{\mathbf{p}}}.$$

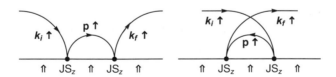

Figure 24.20 Longitudinal terms.

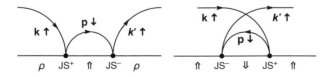

Figure 24.21 Transverse terms.

Combining terms, we then have contributions to the amplitude of

$$\frac{J^2}{\Omega} \sum_{\mathbf{p}} \frac{1}{\varepsilon_{\mathbf{k}} - \varepsilon_{\mathbf{p}}} \left[S^+ S^- n_F(\varepsilon_{\mathbf{p}}) + S^- S^+ \left(1 - n_F(\varepsilon_{\mathbf{p}}) \right) \right]$$

$$= S^- S^+ \frac{J^2}{\Omega} \sum_{\mathbf{p}} \frac{1}{\varepsilon_{\mathbf{k}} - \varepsilon_{\mathbf{p}}} + \left[S^+, S^- \right] \frac{J^2}{\Omega} \sum_{\mathbf{p}} \frac{n_F(\varepsilon_{\mathbf{p}})}{\varepsilon_{\mathbf{k}} - \varepsilon_{\mathbf{p}}}$$

$$= S^- S^+ \frac{J^2}{\Omega} \sum_{\mathbf{p}} \frac{1}{\varepsilon_{\mathbf{k}} - \varepsilon_{\mathbf{p}}} + \frac{J^2 S^z}{\Omega} \sum_{\mathbf{p}} \frac{n_F(\varepsilon_{\mathbf{p}})}{\varepsilon_{\mathbf{k}} - \varepsilon_{\mathbf{p}}}.$$

The factor $[\varepsilon_{\mathbf{k}} - \varepsilon_{\mathbf{p}}]^{-1}$ in the sums on the right-hand side is divergent at the energy $\varepsilon_{\mathbf{k}}$ of the particle whose scattering amplitude we are calculating, which in the case of most interest is the Fermi energy. For the first sum, this divergence is not important, since contributions from above and below the Fermi energy cancel to leave a finite result. But in the second term, the Fermi function limits this cancellation.

In evaluating the sum over the intermediate states $\mathbf{p}$ in the second term, we assume a conduction bandwidth $[-W, W]$ and a flat density of states $\mathcal{D}$, and we obtain

$$\sum_{\mathbf{p}} \frac{n_F(\varepsilon_{\mathbf{p}})}{\varepsilon_{\mathbf{k}} - \varepsilon_{\mathbf{p}}} \simeq \mathcal{D} \int_{-W}^{W} d\varepsilon_{\mathbf{p}} \frac{n_F(\varepsilon_{\mathbf{p}})}{\varepsilon_{\mathbf{k}} - \varepsilon_{\mathbf{p}}}$$

$$= \mathcal{D} \left(\ln |W + \varepsilon_{\mathbf{k}}| + \int_{-W}^{W} d\varepsilon_{\mathbf{p}} \left(-\frac{\partial n_F(\varepsilon_{\mathbf{p}})}{\partial \varepsilon_{\mathbf{p}}} \right) \ln \left[\varepsilon_{\mathbf{k}} - \varepsilon_{\mathbf{p}} \right] \right). \quad (24.32)$$

In the last integral, we can approximate the factor of order $\partial n_F / \partial \varepsilon_{\mathbf{p}}$ as finite only in the range $k_B T$, as shown in Figure 24.22, giving

$$\begin{cases} \ln \varepsilon_{\mathbf{k}} & \text{for } \varepsilon_{\mathbf{k}} > k_B T \\ \ln \left[\frac{W}{k_B T} \right] + \text{const.} & \text{for } \varepsilon_{\mathbf{k}} < k_B T. \end{cases}$$

We obtain

$$\frac{1}{\Omega} \sum_{\mathbf{q}} \frac{n_F(\varepsilon_{\mathbf{q}})}{\varepsilon_{\mathbf{k}} - \varepsilon_{\mathbf{q}}} = \mathcal{D}(\varepsilon_F) \ln \left[\frac{W}{k_B T} \right].$$

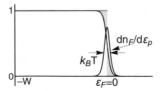

Figure 24.22 Fermi function and its derivative.

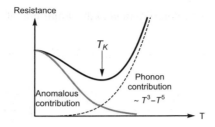

Figure 24.23 Anomalous scattering contribution to the resistivity due to magnetic impurities.

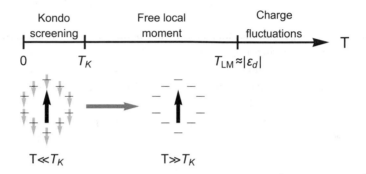

Figure 24.24 Magnetic impurity moment screening above and below the Kondo temperature.

To calculate the scattering rate, we square the combination of the first- and second-order contributions to the amplitude. We find a result proportional to

$$J^2 + 3J^3 \, \mathcal{D}(\varepsilon_F) \, \ln \left[\frac{W}{k_B T} \right].$$

which grows with decreasing temperature. Thus, the spin part of the interaction, to a first approximation, makes a contribution to the resistance of order $J^3 \ln(1/T)$. Combined with the phonon contribution to the electron scattering rate, this yields a minimum in the resistance as a function of temperature, as shown in Figure 24.23, for alloys containing magnetic impurities, which is the quest of the Kondo effect.

The temperature scale at which the logarithmic term becomes important is the Kondo temperature $T_K = W \, e^{-1/2\mathcal{D}J}$, as shown in Figure 24.24. For the physically relevant regime in which $J\mathcal{D}(\varepsilon_F)$ is small, $T_K \ll W$.

We might speculate that when Kondo found this result he might have said "*eureka*," I found it: a contribution to the resistivity that grows as T gets small. Yet, he might also have realized that at temperatures $T \ll T_K$, the divergence of this contribution signals that perturbation theory is breaking down. Thus we arrive at the realization that even though we are dealing with a single impurity, the Kondo problem is actually a complicated many-body case. Once more we encounter a very singular problem where we have to find a way to sum all the processes.

24.5.2 *Variational Approach to the Kondo Problem*

Due to the breakdown of perturbation theory with the appearance of a logarithmic singularity in the electron local moment scattering amplitude, we must seek a different approach to analyzing the Kondo S-(d/f) model. In our pursuit to reveal and understand the nature of the Kondo ground state, we should contemplate the case of strong coupling limit in which $J\mathcal{D}(\varepsilon_F)$ is large, namely, an antiferromagnetic J where the impurity spin binds a conduction electron into a singlet state.

To study this scenario, we shall employ a variational method, starting with the trial wave function [187, 205]

$$|\psi_0\rangle = \left[\alpha_0 + \sum_{k<k_F,\sigma} \alpha_{\mathbf{k}}\, d_\sigma^\dagger c_{\mathbf{k}\sigma}\right] \prod_{\mathbf{k}\leq k_F} \left(c_{\mathbf{k}\uparrow}^\dagger c_{\mathbf{k}\downarrow}^\dagger\right) |0\rangle, \qquad (24.33)$$

where $\prod_{\mathbf{k}\leq k_F} \left(c_{\mathbf{k}\uparrow}^\dagger c_{\mathbf{k}\downarrow}^\dagger\right) |0\rangle$ represents the filled Fermi sea ground state of the pure system. The first term is just the amplitude for a filled Fermi sea and an empty d/f-state, while the second term is a superposition of states with a filled impurity d/f-level and a hole in the Fermi sea at momentum $\mathbf{k}$. Actually, the amplitude α_0 is found to be very small, but it does signal a nonzero probability of finding the d/f-level empty, which leaves a striking footprint in the resultant spectrum. The more obvious feature of this trial state wavefunction is that it represents a spin singlet state, shown in Figure 24.25, rather than a single-spin doublet that we would have naively expected.

To calculate the variational energy of the trial wavefunction, we use the variational energy functional

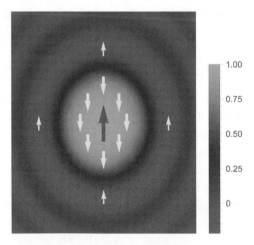

Figure 24.25 Singlet state in the Kondo regime, showing the Kondo cloud.

$$\tilde{E}[|\psi_0\rangle] = \langle\psi_0| \,\mathcal{H}_A \,|\psi_0\rangle - \varepsilon \,\langle\psi_0 \,|\psi_0\rangle$$

$$\langle\psi_0 \,|\psi_0\rangle = |\alpha_0|^2 + 2\sum_{\mathbf{k}} |\alpha_{\mathbf{k}}|^2$$

$$\langle\psi_0| \,\mathcal{H}_A \,|\psi_0\rangle = |\alpha_0|^2 \,\mathcal{E}_0 + 2\sum_{\mathbf{k}} |\alpha_{\mathbf{k}}|^2 \,(\mathcal{E}_0 + \varepsilon_d - \varepsilon_{\mathbf{k}}) + 2V \sum_{\mathbf{k}} \left[\alpha_0^* \,\alpha_{\mathbf{k}\sigma} + \alpha_{\mathbf{k}\sigma}^* \,\alpha_0\right],$$

where ε is a Lagrange multiplier for the normalization condition. We shall measure all energies in the Hamiltonian relative to the Fermi sphere and set $\mathcal{E}_0 = 0$ and take the coefficients to be real. The extremum conditions become

$$\frac{\partial}{\partial\alpha_0} \quad \Rightarrow \quad 2V \sum_{\mathbf{k}} \alpha_{\mathbf{k}} = \varepsilon \,\alpha_0$$

$$\frac{\partial}{\partial\alpha_{\mathbf{k}}} \quad \Rightarrow \quad (\varepsilon_d - \varepsilon_{\mathbf{k}}) \,\alpha_{\mathbf{k}} + V \,\alpha_0 = \varepsilon \,\alpha_{\mathbf{k}}. \tag{24.34}$$

The solution of (24.34) yields the self-consistent equation

$$\varepsilon = 2 \sum_{k<k_F} \frac{V^2}{\varepsilon - \varepsilon_d + \varepsilon_{\mathbf{k}}}. \tag{24.35}$$

We introduce the binding energy $\Delta_K = \varepsilon - \varepsilon_d < 0$, which provides a measure of the energy of the Kondo singlet trial wavefunction to be compared with that of the singly occupied d/f-state plus the filled Fermi sea configuration($\mathcal{E}_0 + \varepsilon_d = \varepsilon_d$). This allows us to write (24.35) as

$$\varepsilon_d + \Delta_K = 2 \sum_{k<k_F} \frac{V^2}{\Delta_K + \varepsilon_{\mathbf{k}}} = 2 \sum_{k<k_F} \frac{V^2}{\Delta_K - |\varepsilon_{\mathbf{k}}|}, \tag{24.36}$$

where in the last line we used the fact that $\varepsilon_{\mathbf{k}} < 0$ since the only available states are those below the Fermi surface. Next we change the sum to an integral, setting $|\varepsilon_{\mathbf{k}}| = \mathcal{E}$, and we write

$$\varepsilon_d + \Delta_K = 2\,\mathcal{D}(0) \int_0^{\varepsilon_F} d\mathcal{E} \,\frac{-V^2}{\mathcal{E} - \Delta_K} = 2\,\mathcal{D}(0)\, V^2 \ln\left[\frac{\varepsilon_F}{|\Delta_K|}\right]. \tag{24.37}$$

We note that this result resembles the BCS self-consistent problem, showing a logarithmically divergent behavior at low energies. When the binding energy $|\Delta_K| \ll \varepsilon_d$, we can neglect Δ_K on the left-hand side and rearrange to get

$$\Delta_K = -\varepsilon_F \exp\left[-\frac{1}{2\,\mathcal{D}(0)\, V^2/\varepsilon_d}\right]. \tag{24.38}$$

But V^2/ε_d is just the value of the coupling constant J in the Kondo Hamiltonian, and we write

$$\Delta_K = -\varepsilon_F \exp\left[-\frac{1}{2\,\mathcal{D}(0)\, J}\right] < 0. \tag{24.39}$$

This result supports the validity of our singlet ground state represented by our trial wave-function. We can attribute the emergence of a logarithmic singularity in our results to the possibility of making hole excitations with arbitrarily low energy. Such low-energy excitations are available because of the presence of a sharp Fermi surface and the absence of an energy gap. Thus it becomes easy to induce low-energy excitations, in order to gain the hybridization energy.

There are thus two very different energy scales associated with the original Anderson problem ($\sim \varepsilon_d$) and with the Kondo problem $k_B T_K \sim \varepsilon_d \exp[-1/2\mathcal{D}(0)J]$. The Kondo temperature can become small, particularly when J is small, and a typical scale is around 10°K. Since this is relatively low, it becomes relevant to ask what occurs at temperatures where a localized moment exists, but the Kondo singlet is thermally dissociated, which we will address next.

d/f-Level Occupation

Within the scenario of large U and $\varepsilon_d < 0$, the occupation of the impurity state $\langle n_d \rangle \sim 1$. Thus, it is more instructive to compute

$$1 - \langle n_d \rangle = \frac{\alpha_0^2}{\alpha_0^2 + 2\sum_{k<k_F} \alpha_\mathbf{k}^2} = \frac{1}{1 + 2\sum_{k<k_F} \left(\frac{\alpha_\mathbf{k}}{\alpha_0}\right)^2}. \tag{24.40}$$

We find from (24.34) that

$$\left(\frac{\alpha_\mathbf{k}}{\alpha_0}\right)^2 = \frac{V^2}{(\Delta_K + \varepsilon_\mathbf{k})^2}.$$

Recalling that $\Delta_K, \varepsilon_\mathbf{k} < 0$, we write

$$2 \sum_{k<k_F} \left(\frac{\alpha_\mathbf{k}}{\alpha_0}\right)^2 = 2\mathcal{D}(0) \int_0^{\varepsilon_F} d\mathcal{E} \frac{V^2}{(|\Delta_K| + |\mathcal{E}|)^2} = 2\mathcal{D}(0) \frac{V^2}{\Delta_K} = \frac{2}{\pi} \frac{\Delta}{\Delta_K}, \tag{24.41}$$

where Δ is the resonance width of the impurity level. Substitution in (24.40) yields

$$1 - \langle n_d \rangle = \frac{1}{1 + \frac{2\Delta}{\pi \Delta_K}} \simeq \frac{\pi \Delta_K}{\Delta} \ll 1,$$

where the used the experimental quantities $\Delta \sim 0.5$ eV, and $\Delta_K \sim 10^{-3}$ eV.

From this result, we see that the deviation of the d/f-level occupation from unity is very small; in the variational Kondo wavefunction, the d/f-level is nearly singly occupied, but not quite. In the preceding example, we obtained the Hartree–Fock solution to the Anderson model. One key feature of this solution was that it predicted a doublet of degenerate ground states for the system. In contrast, the Kondo variational ground state is a singlet and does not distinguish up- from down-spins.

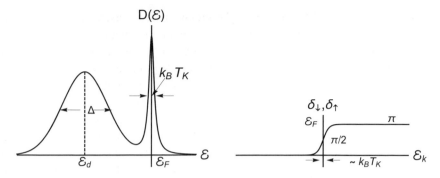

Figure 24.26 For the singlet ground-state function, the excess density of states in the conduction band appears as a Kondo resonance at ε_F, satisfying $\delta_{d\uparrow}(\varepsilon_F) = \delta_{d\downarrow}(\varepsilon_F) \simeq \pi/2$.

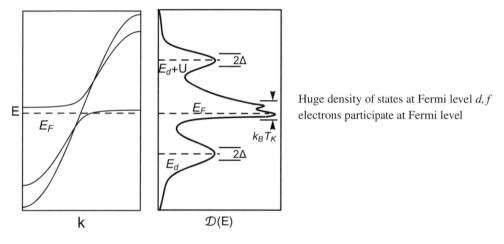

Huge density of states at Fermi level d, f electrons participate at Fermi level

Figure 24.27 Electron dispersions and density of states with $U, V \neq 0$. $k_B T_K$ is the Kondo linewidth, Δ the hybridization parameter, and E_d the ionic d/f energy level.

Kondo Resonance

The result that the d/f-level occupancy is less than unity leads to the presence of extra electron density of states in the conduction band.

Because of the Pauli principle, this excess density of states appears at the Fermi energy and is manifest as a *Kondo resonance peak* of width Δ_K and integrated magnitude of $1 - \langle n_d \rangle = \frac{\pi \Delta_K}{\Delta}$, as shown in Figure 24.26.

We can obtain an alternative physical perspective of this resonance if we consider it through the lens of the Friedel sum rule. We note that the Kondo variational singlet state contains equal probability of up- and down-spin occupation. Moreover, we found that the effective occupation of the d/f-state $\langle n_d \rangle \approx 1$. Consequently, $\langle n_{d\uparrow} \rangle = \langle n_{d\downarrow} \rangle \simeq 1/2$, which, according to the Friedel sum rule, means that at the Fermi energy we have resonances, since $\delta_{d\uparrow} = \delta_{d\downarrow} \simeq \pi/2$. Figure 24.27 shows a typical band dispersion and DOS of a Kondo system."

Exercises

24.1 Spin susceptibility of a superconductor:

Using the Kubo formula, calculate the magnetization $\langle \mu_z \rangle$ of a free Fermi gas associated with an external Zeeman term in the Hamiltonian

$$\mathcal{H}_I = -\int d\mathbf{x}' \, \mu_z(\mathbf{x}') \, B_z(\mathbf{x}', t)$$

$$\mu_z(\mathbf{x}) = \mu_B \left(n_\uparrow(\mathbf{x}) - n_\downarrow(\mathbf{x}) \right).$$

Calculate the spin susceptibility $\chi^{zz}(\mathbf{q}, \omega)$ (follow the Lindhard function derivation in Chapter 7), and show that it is similar to that of the charge susceptibility, up to an overall factor. Show the susceptibility reduces to the Pauli susceptibility, $\chi_{zz}(q \to 0, \omega = 0, T \to 0) = \mu_B^2 N_0$.

Repeat the calculation for the superconducting state, and predict the temperature dependence of the Pauli susceptibility for $T \ll T_c$ in that state.

24.2 Frustration:

On a bipartite lattice (i.e., one in which the neighbors of one sublattice belong to the other sublattice), the ground state (known as a Néel state) of a classical antiferromagnet can adopt a staggered spin configuration in which the exchange energy is maximized. Lattices that cannot be classified in this way are said to be frustrated, and the maximal exchange energy associated with each bond cannot be recovered. Using only symmetry arguments, specify one of the possible ground states of a classical three-site triangular lattice antiferromagnet. (Note that the invariance of the Hamiltonian under a global rotation of the spins means that there is manifold continuous degeneracy in the ground state.) Using this result, construct one of the classical ground states of the infinite triangular lattice.

24.3 Commutators in the Anderson model:

Show that the Anderson model Hamiltonian gives

$$[d_\sigma, \mathcal{H}_A] = \varepsilon_d \, d_\sigma + \sum_{\mathbf{k}} V \, c_{\mathbf{k}\sigma} + U \, d_\sigma \, n_{\bar{\sigma}}.$$

24.4 Ruderman–Kittel–Kasuya–Yosida (RKKY) oscillations:

As we have learned, the RKKY interaction is a mechanism of coupling of localized magnetic moments (nuclear magnetic moments or spins of localized electrons in inner shells) through conduction electrons. The physics of the mechanism is as follows: if we have a localized magnetic moment S, it couples locally to electrons via

$$\mathcal{H}_{\text{int}} = J \, \mathbf{S} \, \Psi_\alpha(\mathbf{x}) \, \sigma_{\alpha\beta} \, \Psi_\beta(\mathbf{x}).$$

Without loss of generality, assume that $\mathbf{S} \| \mathbf{z}$. Then it is equivalent to the potential $\pm JS$ at the position x for up- and down-spins, respectively. This potential leads to the modulation of density (the actual density, not the density of states!)

$$\delta n_\uparrow(\mathbf{y}) = U(\mathbf{x} - \mathbf{y}) \, JS, \qquad \delta n_\downarrow(\mathbf{y}) = -U(\mathbf{x} - \mathbf{y}) \, JS$$

with some function $U(\mathbf{x} - \mathbf{y})$. If there is another local spin at the position $\mathbf{y}$, this modulation of electronic density leads to the interaction between the two magnetic moments given by

$$E_{\text{RKKY}} = 2J^2 \, \mathbf{S} \cdot \mathbf{S}_2 \, U(\mathbf{x} - \mathbf{y}).$$

Our goal is to calculate the function $U(\mathbf{x} - \mathbf{y})$. It is a linear response of the density δn at the position $\mathbf{y}$ to the potential at the position $\mathbf{x}$. Show that this response function is given by the same diagram in Figure 24.9, with the only difference being that now we integrate over the time difference at the positions $\mathbf{x}$ and $\mathbf{y}$ and use the time-ordered Green function. In the frequency representation, it is given by

$$U(R) = \int \frac{d\omega}{2\pi} \, [G(R, \omega)]^2,$$

where $R = |\mathbf{x} - \mathbf{y}|$. Hint: consider a noninteracting electron gas in the presence of a localized spin S, interacting with the local spin density of electrons. The interaction can be written as

$$\mathcal{H}_1 = J \, S \, \delta(\mathbf{x}).$$

Assume that J is small and calculate the spin density of electrons as a function of R the distance to the spin for $R \ll k_F^{-1}$ in the first order in J.

References

[1] Abarenkov, I. V., and Heine, V. 1965. The Model Potential for Positive Ions. *Philos. Mag.*, **12**(117), 529–537.

[2] Abergel, D.S.L., Apalkov, V., Berashevich, J., Ziegler, K., and Chakraborty, T. 2010. Properties of Graphene: A Theoretical Perspective. *Adv. Phys.*, **59**, 261–482.

[3] Abrahams, E., Anderson, P. W., Licciardello, D. C., and Ramakrishnan, T. V. 1979. Scaling Theory of Localization: Absence of Quantum Diffusion in Two Dimensions. *Phys. Rev. Lett.*, **42**, 673–676.

[4] Abrikosov, A. A., Gor'kov, L. P., and Dzyaloshinskii, I. E. 1959. On the Application of Quantum-Field-Theory Methods to Problems of Quantum Statistics at Finite Temperatures. *Soviet Physics JETP*, **36**, 636–641.

[5] Abrikosov, A. A., Gor'kov, L. P., and Dzyaloshinskii, I. E. 1975. *Methods of Quantum Field Theory in Statistical Physics*. Dover.

[6] Aharonov, Y., and Bohm, D. 1959. Significance of Electromagnetic Potentials in the Quantum Theory. *Phys. Rev.*, **115**, 485–491.

[7] Allen, J. F., and Misener, A. D. 1938. Flow of Liquid Helium II. *Nature*, **141**, 75.

[8] Allen, J. F., Peierls, R., and Uddin, M. Z. 1937. Heat Conduction in Liquid Helium. *Nature*, **140**, 62.

[9] Altland, A., and Zirnbauer, M. R. 1997. Nonstandard Symmetry Classes in Mesoscopic Normal-Superconducting Hybrid Structures. *Phys. Rev. B*, **55**, 1142–1161.

[10] Anderson, P. W. 1958a. Absence of Diffusion in Certain Random Lattices. *Phys. Rev.*, **109**, 1492–1505.

[11] Anderson, P. W. 1958b. New Method in the Theory of Superconductivity. *Phys. Rev.*, **110**, 985–986.

[12] Anderson, P. W. 1958c. Random-Phase Approximation in the Theory of Superconductivity. *Phys. Rev.*, **112**, 1900–1916.

[13] Anderson, P. W. 1959. New Approach to the Theory of Superexchange Interactions. *Phys. Rev.*, **115**, 2–13.

[14] Anderson, P. W. 1961. Localized Magnetic States in Metals. *Phys. Rev.*, **124**, 41–53.

[15] Anderson, P. W., and Hasegawa, H. 1955. Considerations on Double Exchange. *Phys. Rev.*, **100**, 675–681.

[16] Anderson, Philip W. 1963. *Theory of Magnetic Exchange Interactions:Exchange in Insulators and Semiconductors*. Solid State Physics, vol. 14. Academic Press. Pages 99–214.

[17] Ando, Tsuneya. 2005. Theory of Electronic States and Transport in Carbon Nanotubes. *J. Phys. Soc. Jpn.*, **74**, 777–817.

[18] Andres, K., Graebner, J. E., and Ott, H. R. 1975. $4f$-Virtual-Bound-State Formation in CeAl$_3$ at Low Temperatures. *Phys. Rev. Lett.*, **35**, 1779–1782.

[19] Armitage, N. P., Mele, E. J., and Vishwanath, A. 2018. Weyl and Dirac Semimetals in Three-Dimensional Solids. *Rev. Mod. Phys.*, **90**, 015001.

[20] Ashcroft, N. W. 1966. Electron–Ion pseudopotentials in metals. *Phys. Lett.*, **23**, 48–50.

[21] Avron, J. E., Osadshy, D., and Seiler, R. 2003. A Topological Look at the Quantum Hall Effect. *Phys. Today*, **56**(8), 38.

[22] Avron, J. E., Seiler, R., and Simon, B. 1983. Homotopy and Quantization in Condensed Matter Physics. *Phys. Rev. Lett.*, **51**, 51–53.

[23] Balachandran, A. P. 1994. *Topology in Physics – A Perspective*. Found. Phys., vol. 24. Kluwer Academic Publishers–Plenum Publishers. Page 455.

[24] Bardeen, J., Cooper, L. N., and Schrieffer, J. R. 1957. Theory of Superconductivity. *Phys. Rev.*, **108**, 1175–1204.

[25] Bernevig, B. Andrei, and Hughes, Taylor. 2013. *Topological Insulators and Topological Superconductors*. Princeton University Press.

[26] Bernevig, B. Andrei, Hughes, Taylor L., and Zhang, Shou-Cheng. 2006. Quantum Spin Hall Effect and Topological Phase Transition in HgTe Quantum Wells. *Science*, **314**(5806), 1757–1761.

[27] Berry, M. V. 1984. Quantal Phase Factors Accompanying Adiabatic Changes. *Proc. R. Soc. Lond. A: Math. Phys. Eng. Sci.*, **392**(1802), 45–57.

[28] Berry, M. V. 1985. Aspects of Degeneracy. In: Casati, Giulio (ed), *Chaotic Behaviour of Quantum Systems, Theory and Applications*. NATO ASI Series.

[29] Bihlmayer, G. 2014. Relativistic Effects in Solids. In: Blügel, Stefan (ed), *Computing Solids: Models, Ab Initio Methods, and Supercomputing*. Lecture Notes 45th IFF Spring School.

[30] Bloch, F. 1932. Zur Theorie des Austauschproblems und der Remanenzerscheinung der Ferromagnetika. *Zeitschrift für Physik*, **74**, 295–335.

[31] Bloch, F. Z. 1929. ber die Quantenmechanik der Elektronen in Kristallgittern. *Physik*, **52**, 555.

[32] Bogolyubov, N. N., Tolmachev, V. V., and Shirkov, D. V. 1958. A New Method in the Theory of Superconductivity. *Fortsch. Phys.*, **6**, 605–682.

[33] Bohm, A., Mostafazadeh, A., Koizumi, H., Niu, Q., and Zwanziger, J. 2010. *The Geometric Phase in Quantum Systems: Foundations, Mathematical Concepts, and Applications in Molecular and Condensed Matter Physics*. Springer Verlag.

[34] Bohm, Arno, Boya, Luis J., and Kendrick, Brian. 1991. Derivation of the Geometrical Phase. *Phys. Rev. A*, **43**, 1206–1210.

[35] Bohm, David, and Pines, David. 1953. A Collective Description of Electron Interactions: III. Coulomb Interactions in a Degenerate Electron Gas. *Phys. Rev.*, **92**, 609–625.

[36] Born, M., and Oppenheimer, R. 1927. Zur Quantentheorie der Molekeln. *Annalen der Physik*, **389**, 457–484.

[37] Boya, Louis J. 1997. Rays and Phases: A Paradox? *Z. Naturforsch.*, **52a**, 63–65.

[38] Brown, Laurie M. 1978. The Idea of the Neutrino. *Phys. Today*, **31**(9), 23–28.

[39] Burdick, Glenn A. 1963. Energy Band Structure of Copper. *Phys. Rev.*, **129**, 138–150.

[40] Callaway, Joseph. 1955. Orthogonalized Plane Wave Method. *Phys. Rev.*, **97**, 933–936.

[41] Castro Neto, A. H., Guinea, F., Peres, N. M. R., Novoselov, K. S., and Geim, A. K. 2009. The Electronic Properties of Graphene. *Rev. Mod. Phys.*, **81**, 109–162.

[42] Cayssol, Jrme. 2013. Various Probes of Dirac Matter: From Graphene to Topological Insulators. *arXiv, [cond-mat.mes-hall]*, 1303.5902.

[43] Chelikowsky, James R., and Cohen, Marvin L. 1976. Nonlocal Pseudopotential Calculations for the Electronic Structure of Eleven Diamond and Zinc-Blende Semiconductors. *Phys. Rev. B*, **14**, 556–582.

[44] Cohen, Marvin L., and Bergstresser, T. K. 1966. Band Structures and Pseudopotential Form Factors for Fourteen Semiconductors of the Diamond and Zinc-Blende Structures. *Phys. Rev.*, **141**, 789–796.

[45] Cohen, Marvin L., and Heine, Volker. 1970. *The Fitting of Pseudopotentials to Experimental Data and Their Subsequent Application.* Solid State Physics, vol. 24. Academic Press. Pages 37–248.

[46] Coker, Ayodele, Lee, Taesul, and Das, T. P. 1980. Investigation of the Electronic Properties of Tellurium-Energy-Band Structure. *Phys. Rev. B*, **22**, 2968–2975.

[47] Cooper, Leon N. 1956. Bound Electron Pairs in a Degenerate Fermi Gas. *Phys. Rev.*, **104**, 1189–1190.

[48] de Haas, W. J., de Boer, J., and van dn Berg, G. J. 1934. The Electrical Resistance of Gold, Copper and Lead at Low Temperatures. *Physica*, **1**, 1115–1124.

[49] del Castillo, Gerardo F. Torres. 2012. *Differentiable Manifolds: A Theoretical Physics Approach.* Birkhuser Basel.

[50] Dirac, P. A. M. 1928. The Quantum Theory of the Electron. *Proc. R. Soc. Lond. A: Math. Phys. Eng. Sci.*, **117**(778), 610–624.

[51] Dirac, P. A. M. 1929. Quantum Mechanics of Many-Electron Systems. *Proc. R. Soc. Lond. A: Math. Phys. Eng. Sci.*, **123**, 714–733.

[52] Dirac, P. A. M. 1931. Quantised Singularities in the Electromagnetic Field,. *Proc. R. Soc. Lond. A: Math. Phys. Eng. Sci.*, **133**(821), 60–72.

[53] Donnelly, R. 2009. The Two-Fluid Theory and Second Sound in Liquid Helium. *Phys. Today*, **38**, 34.

[54] Dyson, F. J. 1949a. The Radiation Theories of Tomonaga, Schwinger, and Feynman. *Phys. Rev.*, **75**, 486–502.

[55] Dyson, F. J. 1949b. The *S* Matrix in Quantum Electrodynamics. *Phys. Rev.*, **75**, 1736–1755.

[56] Dyson, Freeman J. 1962. The Threefold Way. Algebraic Structure of Symmetry Groups and Ensembles in Quantum Mechanics. *J. Math. Phys.*, **3**, 1199–1215.

[57] Dzyaloshinskii, I. 1958. A Thermodynamic Theory of Weak Ferromagnetism of Antiferromagnetics. *J. Phys. Chem. Solids*, **4**, 241–255.

[58] Eggert, Sebastian. 2007. One-Dimensional Quantum Wires: A Pedestrian Approach to Bosonization. In: Kuk, Y. (ed), *Theoretical Survey of One Dimensional Wire Systems.* Sowha Publishing.

[59] Eschrig, Helmut. 2011. *Topology and Geometry for Physics.* Lecture Notes in Physics, vol. 822. Springer-Verlag.

[60] Fetter, A. L., and Walecka, J. D. 1971. *Quantum Theory of Many Particle System.* McGraw-Hill Book Company.

[61] Feynman, R. P. 1949a. Space-Time Approach to Quantum Electrodynamics. *Phys. Rev.*, **76**, 769–789.

[62] Feynman, R. P. 1949b. The Theory of Positrons. *Phys. Rev.*, **76**, 749–759.

[63] Feynman, R. P. 1972. *Statistical Mechanics.* Frontiers in Physics. W. A. Benjamin Inc. (1972), CRC Press (1998).

[64] Fock, V. A. 1930. Näherungsmethode zur Losung des quantenmechanischen Mehrkörperproblem (Approximation method for the solution of the quantum mechanical many-body problem). *Z. Phys.*, **61**, 126–148.

[65] Fradkin, Eduardo. 2013. *Field Theories of Condensed Matter Physics*. second edn. Cambridge University Press.

[66] Franz, Marcel, and Molenkamp, Laurens (eds). 2013. *Topological Insulators*. Contemporary Concepts of Condensed Matter Science, vol. 6. Elsevier.

[67] Fré, Pietro Giuseppe. 2013. *Gravity, a Geometrical Course: Development of the Theory and Basic Physical Applications*. Vol. 1. Springer.

[68] Friedel, J. 1958. Metallic Alloys. *Il Nuovo Cimento (1955-1965)*, **7**, 287–311.

[69] Fruchart, Michel, and Carpentier, David. 2013. An Introduction to Topological Insulators. *Comptes Rendus Physique*, **14**, 779–815.

[70] Fu, Liang. 2011. Topological Crystalline Insulators. *Phys. Rev. Lett.*, **106**, 106802.

[71] Fu, Liang, and Kane, C. L. 2006. Time Reversal Polarization and a $\mathbb{Z}_2$ Adiabatic Spin Pump. *Phys. Rev. B*, **74**, 195312.

[72] Gell-Mann, Murray, and Brueckner, Keith A. 1957. Correlation Energy of an Electron Gas at High Density. *Phys. Rev.* **106**, 364–368.

[73] Gell-Mann, Murray, and Low, Francis. 1951. Bound States in Quantum Field Theory. *Phys. Rev.*, **84**, 350–354.

[74] Giamarchi, Thierry. 2004. *Quantum Physics in One Dimension*. International Series of Monographs on Physics. Oxford University Press.

[75] Ginzburg, V. L., and Landau, L. D. 1950. On the Theory of Superconductivity. *Zh. Eksp. Teor. Fiz.*, **20**, 1064–1082.

[76] Giuliani, Gabriele, and Vignale, Giovanni. 2005. *Quantum Theory of the Electron Liquid*. Cambridge University Press.

[77] Gor'kov, L. P. 1958. On the Energy Spectrum of Superconductors. *Soviet Phys. JETP*, **7**, 505.

[78] Gor'kov, L. P. 1959. Microscopic Derivation of the Ginzburg–Landau Equations in the Theory of Superconductivity. *Soviet Phys. JETP*, **9**, 1364.

[79] Gor'kov, L. P. 1960. Theory of Superconducting Alloys in a Strong Magnetic Field near the Critical Temperature. *Soviet Phys. JETP*, **10**, 998.

[80] Gundmann, Marius. 2016. *The Physics of Semiconductors: An Introduction*. third edn. Graduate Texts in Physics. Springer International Publishing.

[81] Haldane, F. D. M. 1981. "Luttinger Liquid Theory" of One-Dimensional Quantum Fluids. I. Properties of the Luttinger Model and Their Extension to the General 1D Interacting Spinless Fermi Gas. *J. Phys. C: Solid State Phys.*, **14**, 2585.

[82] Haldane, F. D. M. 1988. Model for a Quantum Hall Effect without Landau Levels: Condensed-Matter Realization of the "Parity Anomaly." *Phys. Rev. Lett.*, **61**, 2015–2018.

[83] Haldane, F. D. M. 2014. Attachment of Surface Fermi Arcs to the Bulk Fermi Surface: Fermi-Level Plumbing in Topological Metals. *arXiv[cond-mat.str-el]*, 1401.0529.

[84] Hamann, D. R., Schlüter, M., and Chiang, C. 1979. Norm-Conserving Pseudopotentials. *Phys. Rev. Lett.*, **43**, 1494–1497.

[85] Harrison, W. A. 2012. *Electronic Structure and the Properties of Solids: The Physics of the Chemical Bond*. Dover Books on Physics. Dover Publications.

[86] Hartree, D. R. 1928. The Wave Mechanics of an Atom with a Non-Coulomb Central Field. Part I, II and III. *Math. Proc. Cambridge Philos. Soc.*, **24**, 89, 111, 426.

[87] Hartree, D. R., and Hartree, W. 1935. Self-Consistent Field, with Exchange, for Beryllium. *Proc. R. Soc. Lond. A: Math Phys. Eng. Sci.*, **150**(869), 9–33.

[88] Hasan, M. Z., and Kane, C. L. 2010. Colloquium: Topological Insulators. *Rev. Mod. Phys.*, **82**, 3045–3067.

[89] Heine, Volker. 1970. *The Pseudopotential Concept*. Solid State Physics, vol. 24. Academic Press. Pages 1–36.

[90] Heisenberg, W. 1928. Zur Theorie des Ferromagnetismus. *Zeitschrift für Physik*, **49**, 619–636.

[91] Henshaw, D. G., and Woods, A. D. B. 1961. Modes of Atomic Motions in Liquid Helium by Inelastic Scattering of Neutrons. *Phys. Rev.*, **121**, 1266–1274.

[92] Herring, Conyers. 1937. Accidental Degeneracy in the Energy Bands of Crystals. *Phys. Rev.*, **52**, 365–373.

[93] Hohenberg, P., and Kohn, W. 1964. Inhomogeneous Electron Gas. *Phys. Rev.*, **136**, B864–B871.

[94] Hubbard, J. 1963. Electron Correlations in Narrow Energy Bands. *Proc. R. Soc. Lond.*, **276**, 238–257.

[95] Ivancevic, Vladimir G, and Ivancevic, Tijana T. 2007. *Applied Differential Geometry*. World Scientific.

[96] Jones, R. O. 2015. Density Functional Theory: Its Origins, Rise to Prominence, and Future. *Rev. Mod. Phys.*, **87**, 897–923.

[97] Kaiser, David. 2005. Physics and Feynman's Diagrams. *American Scientist*, **93**(2), 156–165.

[98] Kane, C. L., and Mele, E. J. 2005a. $\mathbb{Z}_2$ Topological Order and the Quantum Spin Hall Effect. *Phys. Rev. Lett.*, **95**, 146802.

[99] Kane, C. L., and Mele, E. J. 2005b. Quantum Spin Hall Effect in Graphene. *Phys. Rev. Lett.*, **95**, 226801.

[100] Kanisawa, K., Butcher, M. J., Yamaguchi, H., and Hirayama, Y. 2001. Imaging of Friedel Oscillation Patterns of Two-Dimensionally Accumulated Electrons at Epitaxially Grown InAs(111) *A* Surfaces. *Phys. Rev. Lett.*, **86**, 3384–3387.

[101] Kapitza, P. 1938. Viscosity of Liquid Helium below the λ-Point. *Nature*, **141**, 74.

[102] Kasuya, Tadao. 1956. A Theory of Metallic Ferro- and Antiferromagnetism on Zener's Model. *Prog. Theor. Phys.*, **16**, 45–57.

[103] Kellermann, E. W. 1940. Theory of the Vibrations of the Sodium Chloride Lattice. *Philos. Trans. R Soc. Lond. Series A: Math. Phys. Eng. Sci.*, **238**(798), 513–548.

[104] King-Smith, R. D., and Vanderbilt, David. 1993. Theory of Polarization of Crystalline Solids. *Phys. Rev. B*, **47**, 1651–1654.

[105] Klitzing, K. v., Dorda, G., and Pepper, M. 1980. New Method for High-Accuracy Determination of the Fine-Structure Constant Based on Quantized Hall Resistance. *Phys. Rev. Lett.*, **45**, 494–497.

[106] Kohmoto, Mahito. 1985. Topological Invariant and the Quantization of the Hall Conductance. *Ann. Phys.*, **160**, 343–354.

[107] Kohn, W., and Rostoker, N. 1954. Solution of the Schrödinger Equation in Periodic Lattices with an Application to Metallic Lithium. *Phys. Rev.*, **94**(Jun), 1111–1120.

[108] Kohn, W., and Sham, L. J. 1965. Self-Consistent Equations Including Exchange and Correlation Effects. *Phys. Rev.*, **140**, A1133–A1138.

[109] König, Markus, Wiedmann, Steffen, and Brüne, Christoph et. al. 2007. Quantum Spin Hall Insulator State in HgTe Quantum Wells. *Science*, **318**(5851), 766–770.

[110] Konschuh, Sergej, Gmitra, Martin, and Fabian, Jaroslav. 2010. Tight-Binding Theory of the Spin-Orbit Coupling in Graphene. *Phys. Rev. B*, **82**, 245412.

[111] Koopmans, T. 1934. ber die Zuordnung von Wellenfunktionen und Eigenwerten zu den Einzelnen Elektronen Eines Atoms. *Physica*, **1**, 104–113.

[112] Korringa, J. 1947. On the Calculation of the Energy of a Bloch Wave in a Metal. *Physica*, **13**, 392–400.

[113] Kramers, H. A. 1934. L'interaction Entre les Atomes Magnétogénes dans un Cristal Paramagnétique. *Physica*, **1**, 182–192.

[114] Kubo, R. 1957. Statistical-Mechanical Theory of Irreversible Processes. I. General Theory and Simple Applications to Magnetic and Conduction Problems. *J. Phys. Soc. of Japan*, **12**, 570–586.

[115] Landau, L. 1957. The Theory of a Fermi Liquid. *Soviet Phys. JETP*, **30**, 1058–1064.

[116] Landau, L. 1959. On the Theory of the Fermi Liquid. *Soviet Phys. JETP*, **35**, 70–74.

[117] LaShell, S., McDougall, B. A., and Jensen, E. 1996. Spin Splitting of an Au(111) Surface State Band Observed with Angle Resolved Photoelectron Spectroscopy. *Phys. Rev. Lett.*, **77**, 3419–3422.

[118] Laughlin, R. B. 1981. Quantized Hall Conductivity in Two Dimensions. *Phys. Rev. B*, **23**, 5632–5633.

[119] Lee, W.S., Vishik, I.M., Lu, D.H., and Shen, Z-X. 2009. A Brief Update of Angle-Resolved Photoemission Spectroscopy on a Correlated Electron System. *J. Phys.: Condens. Matter*, **21**, 164217.

[120] London, F. 1938a. The λ-Phenomenon of Liquid Helium and the Bose–Einstein Degeneracy. *Nature*, **141**, 643.

[121] London, F. 1938b. On the Bose–Einstein Condensation. *Phys. Rev.*, **54**, 947–954.

[122] London, F., and London, H. 1935. The Electromagnetic Equations of the Supracon-ductor. *Proc. R. Soc. Lond. A: Math. Phys. Eng. Sci.*, **149**(866), 71–88.

[123] Luttinger, J. M. 1963. An Exactly Soluble Model of a Many-Fermion System. *J. Math. Phys.*, **4**, 1154–1162.

[124] Majorana, E. 1937. Teoria Simmetrica dell'Elettrone e del Positrone (Theory of the Symmetry of Electrons and Positrons). *Il Nuovo Cimento*, **14**, 171.

[125] Martin, P. C., and Schwinger, Julian. 1959. Theory of Many-Particle Systems. I. *Phys. Rev.*, **115**, 1342–1373.

[126] Matsubara, T. 1955. A New Approach to Quantum-Statistical Mechanics. *Prog. Theor. Phys.*, **14**, 351–378.

[127] McMillan, W. L. 1968. Transition Temperature of Strong-Coupled Superconductors. *Phys. Rev.*, **167**, 331–344.

[128] Mead, C. A. 1980. The Molecular Aharonov–Bohm Effect in Bound States. *Chem. Phys.*, **49**, 23–32.

[129] Mehta, M. L. 2004. *Random Matrices*. Pure and Applied Mathematics, vol. 142. Elsevier.

[130] Moore, J. E., and Balents, L. 2007. Topological Invariants of Time-Reversal-Invariant Band Structures. *Phys. Rev. B*, **75**, 121306.

[131] Moriya, T. 1960. Anisotropic Superexchange Interaction and Weak Ferromagnetism. *Phys. Rev.*, **120**, 91–98.

[132] Mott, N. F. 1949. The Basis of the Electron Theory of Metals, with Special Reference to the Transition Metals. *Proc. Phys. Soci., Section A*, **62**, 416.

[133] Mott, N. F., and Peierls, R. 1937. Discussion of the Paper by de Boer and Verwey. *Proc. Phys. Soci.*, **49**, 72.

[134] Mueller, F. M. 1967. Combined Interpolation Scheme for Transition and Noble Metals. *Phys. Rev.*, **153**, 659–669.

[135] Murakami, S. 2007. Phase Transition between the Quantum Spin Hall and Insulator Phases in 3D: Emergence of a Topological Gapless Phase. *New J. Phys.*, **9**, 356.

[136] Murakami, S., Nagaosa, N., and Zhang, S-C. 2004. Spin-Hall Insulator. *Phys. Rev. Lett.*, **93**, 156804.

[137] Myron, H. W., and Freeman, A. J. 1975. Electronic Structure and Fermi-Surface-Related Instabilities in $1T - TaS_2$ and $1T - TaSe_2$. *Phys. Rev. B*, **11**, 2735–2739.

[138] Nakahara, M. 2003. *Geometry, Topology and Physics*. second edn. Graduate Student Series in Physics. CRC Press; Taylor & Francis.

[139] Nambu, Y. 1960. Quasi-Particles and Gauge Invariance in the Theory of Superconductivity. *Phys. Rev.*, **117**, 648–663.

[140] Néel, L. 1932. *Influence des fluctuations des champs moléculaires sur les propriétés magnétiques des corps*. Theses, Universite de Strasbourg.

[141] Néel, L. 1952. Antiferromagnetism and Ferrimagnetism. *Proc. Phys. Soc., Section A*, **65**(11), 869.

[142] Néel, M. L. 1948. Propriétés magnétiques des ferrites, ferrimagnétisme et antiferromagnétisme. *Ann. Phys.*, **12**, 137–198.

[143] Negele, J. W., and Orland, H. 1998. *Quantum Many-Particle Systems*. Advanced Books Classics. Perseus Books.

[144] Nielsen, H. B., and Ninomiya, M. 1981a. Absence of Neutrinos on a Lattice: (I). Proof by Homotopy Theory. *Nuclear Physics B*, **185**, 20–40.

[145] Nielsen, H. B., and Ninomiya, M. 1981b. Absence of Neutrinos on a Lattice: (II). Intuitive Topological Proof. *Nuclear Physics B*, **193**, 173–194.

[146] Nielsen, H. B., and Ninomiya, M. 1983. The Adler–Bell–Jackiw Anomaly and Weyl Fermions in a Crystal. *Physics Letters B*, **130**, 389–396.

[147] Onsager, L. 1949. Statistical Hydrodynamics. *Il Nuovo Cimento*, **6**, 279–287.

[148] Ortmann, F., Roche, S., and Valenzuela, S. O. (eds). 2015. *Topological Insulators: Fundamentals and Perspectives*. Wiley.

[149] Panati, G. 2007. Triviality of Bloch and Bloch–Dirac Bundles. *Annales Henri Poincaré*, **8**, 995–1011.

[150] Paulatto, L. 2008. *Pseudopotential Methods for DFT Calculations*. Lecture.

[151] Peierls, R. E. 1955. *Quantum Theory of Solids*. Oxford University Press.

[152] Peierls, R. E. 1985. *Bird of Passage: Recollections of a Physicist*. Princeton University Press.

[153] Petersen, L., and Hedegard, P. 2000. A Simple Tight-Binding Model of Spin–Orbit Splitting of sp–Derived Surface States. *Surf. Sci.*, **459**, 49–56.

[154] Resta, R. 1992. Theory of the Electric Polarization in Crystals. *Ferroelectrics*, **136**, 51–55.

[155] Resta, R. 2012. *Geometry and Topology in Electronic Structure Theory*. University of Trieste.

[156] Richard, S., Aniel, F., and Fishman, G. 2004. Energy-Band Structure of Ge, Si, and GaAs: A Thirty-Band $\mathbf{k} \cdot \mathbf{p}$ Method. *Phys. Rev. B*, **70**, 235204.

[157] Ruderman, M. A., and Kittel, C. 1954. Indirect Exchange Coupling of Nuclear Magnetic Moments by Conduction Electrons. *Phys. Rev.*, **96**, 99–102.

[158] Ryu, S., Schnyder, A. P., Furusaki, A., and Ludwig, A. W. W. 2010. Topological Insulators and Superconductors: Tenfold Way and Dimensional Hierarchy. *New J. Phys.*, **12**, 065010.

[159] Sakurai, J. J., and Napolitano, J. 2017. *Modern Quantum Mechanics.* second edn. Cambridge University Press.

[160] Schnyder, A. P., Ryu, S., Furusaki, A., and Ludwig, A. W. W. 2008. Classification of Topological Insulators and Superconductors in Three Spatial Dimensions. *Phys. Rev. B*, **78**, 195125.

[161] Schnyder, A. P., Ryu, S., Furusaki, A., and Ludwig, A. W. W. 2009. Classification of Topological Insulators and Superconductors. *AIP Conf. Proc.*, **1134**, 10–21.

[162] Schrieffer, J. R., and Wolff, P. A. 1966. Relation between the Anderson and Kondo Hamiltonians. *Phys. Rev.*, **149**, 491–492.

[163] Schulz, H. J., Cuniberti, G., and Pieri, P. 2000. Fermi Liquids and Luttinger Liquids. In: Morandi, G., Sodano, P., Tagliacozzo, A., and Tognetti, V. (eds), *Field Theories for Low-Dimensional Condensed Matter Systems.* Lecture Notes of the Les Houches Summer School, vol. 131. Springer.

[164] Semenoff, Gordon W. 1984. Condensed-Matter Simulation of a Three-Dimensional Anomaly. *Phys. Rev. Lett.*, **53**, 2449–2452.

[165] Setyawan, W., and Curtarolo, S. 2010. High-Throughput Electronic Band Structure Calculations: Challenges and Tools. *Comput. Mater. Sci.*, **49**, 299–312.

[166] Shen, S-Q. 2012. *Topological Insulators: Dirac Equation in Condensed Matter.* Springer Series in Solid-State Sciences. Springer-Verlag.

[167] Simon, B. 1983. Holonomy, the Quantum Adiabatic Theorem, and Berry's Phase. *Phys. Rev. Lett.*, **51**, 2167–2170.

[168] Slater, J. C. 1937. Wave Functions in a Periodic Potential. *Phys. Rev.*, **51**(May), 846–851.

[169] Slater, J. C. 1972. *Symmetry and Energy Bands in Crystals.* Dover Publications.

[170] Soluyanov, A. A., Gresch, D., and Wang, Z. et al. 2015. Type-II Weyl Semimetals. *Nature*, **527**, 495.

[171] Steinhauer, J., Ozeri, R., Katz, N., and Davidson, N. 2002. Excitation Spectrum of a Bose–Einstein Condensate. *Phys. Rev. Lett.*, **88**, 120407.

[172] Stone, M., and Goldbart, P. 2009. *Mathematics for Physics: A Guided Tour for Graduate Students.* first edn. Cambridge University Press.

[173] Takahashi, R. 2015. *Topological States on Interfaces Protected by Symmetry.* Springer Theses.

[174] Tamtögl, A. 2012. *Surface Dynamics and Structure of Bi(111) from Helium Atom Scattering.* Ph.D. thesis, Graz University of Technology.

[175] Tanaka, Y., Sato, T., Nakayama, K., et al. 2013. Tunability of the k-Space Location of the Dirac Cones in the Topological Crystalline Insulator $Pb_{1-x}Sn_xTe$. *Phys. Rev. B*, **87**, 155105.

[176] Thouless, D. J., Kohmoto, M., Nightingale, M. P., and den Nijs, M. 1982. Quantized Hall Conductance in a Two-Dimensional Periodic Potential. *Phys. Rev. Lett.*, **49**, 405–408.

[177] Tinkham, M. 1996. *Introduction to Superconductivity.* Dover.

[178] Tisza, L. 1938. Transport Phenomena in Helium II. *Nature*, **141**, 913.

[179] Tisza, L. 1940a. About the Theory of Quantum Liquids. Application to Liquid Helium. II. *J. Phys. Radium*, **1**, 350–358.

[180] Tisza, L. 1940b. On the Theory of Quantum Liquids. Application to Liquid Helium. *J. Phys. Radium*, **1**, 164–172.

[181] Tomonaga, S-i. 1950. Remarks on Bloch's Method of Sound Waves Applied to Many-Fermion Problems. *Prog. Theor. Phys.*, **5**, 544–569.

[182] Townsend, P., and Sutton, J. 1962. Investigation by Electron Tunneling of the Superconducting Energy Gaps in Nb, Ta, Sn, and Pb. *Phys. Rev.*, **128**, 591–595.

[183] Toya, T. 1958. Normal Vibrations of Sodium. *J. Res. Inst. Catal. Hokkaido Univ.*, **6**, 183–195.

[184] Vafek, O., and Vishwanath, A. 2014. Dirac Fermions in Solids: From High-T_c Cuprates and Graphene to Topological Insulators and Weyl Semimetals. *Annual Review of Condensed Matter Physics*, **5**, 83–112.

[185] van Vleck, J. H. 1937. On the Anisotropy of Cubic Ferromagnetic Crystals. *Phys. Rev.*, **52**, 1178–1198.

[186] van Vleck, J. H. 1992. Quantum Mechanics: The Key to Understanding Magnetism. Pages 353–369 of: Lundqvist, S. (ed), *Nobel Lectures, Physics 1971-1980*. World Scientific, Singapore.

[187] Varma, C. M., and Yafet, Y. 1976. Magnetic Susceptibility of Mixed-Valence Rare-Earth Compounds. *Phys. Rev. B*, **13**, 2950–2954.

[188] von Delft, J., and Schoeller, H. 1998. Bosonization for Beginners Refermionization for Experts. *Ann. Phys.*, **7**, 225–305.

[189] von Neuman, J., and Wigner, E. P. 1929. Uber merkwürdige diskrete Eigenwerte. Uber das Verhalten von Eigenwerten bei adiabatischen Prozessen. *Phys. Z.*, **30**, 467–470.

[190] Vonsovskii, S. V. 1946. *Zh. Eksp. Teor. Fiz.*, **16**, 981.

[191] Wallace, P. R. 1947. The Band Theory of Graphite. *Phys. Rev.*, **71**, 622–634.

[192] Wan, X., Turner, A. M., Vishwanath, A., and Savrasov, S. Y. 2011. Topological Semimetal and Fermi-Arc Surface States in the Electronic Structure of Pyrochlore Iridates. *Phys. Rev. B*, **83**, 205101.

[193] Weyl, H. 1929. Elektron und Gravitation I. *Zeitschrift für Physik*, **56**, 330–352.

[194] Wick, G. C. 1954. Properties of Bethe–Salpeter Wave Functions. *Phys. Rev.*, **96**, 1124–1134.

[195] Wigner, E. P. 1932. Ueber die Operation der Zeitumkehr in der Quantenmechanik, Nachrichten von der Gesellschaft der Wissenschaften zu Gttingen. *Math.-Phys. K.*, 546–559.

[196] Wigner, E. P. 1959. *Group Theory and Its Application to the Quantum Mechanics of Atomic Spectra*. Academic Press. Chaps 24 and 26.

[197] Wilczek, F., and Zee, A. 1984. Appearance of Gauge Structure in Simple Dynamical Systems. *Phys. Rev. Lett.*, **52**, 2111–2114.

[198] Wilson, A. 1931a. The Theory of Electronic Semi-Conductors. *Proc. R. Soc. Lond. A: Math. Phys. Eng. Sci.*, **133**(822), 458–491.

[199] Wilson, A. 1931b. The Theory of Electronic Semi-Conductors. II. *Proc. R. Soc. Lond. A: Math. Phys. Eng. Sci.*, **134**(823), 277–287.

[200] Witten, E. 1988. Topological Quantum Field Theory. *Commun. Math. Phys.*, **117**, 353–386.

[201] Woods, A. D. B., Brockhouse, B. N., March, R. H., Stewart, A. T., and Bowers, R. 1962. Crystal Dynamics of Sodium at 90° K. *Phys. Rev.*, **128**, 1112–1120.

[202] Yao, Y., Ye, F., Qi, X.-L., Zhang, S-C., and Fang, Z. 2007. Spin–Orbit Gap of Graphene: First Principles Calculations. *Phys. Rev. B*, **75**, 041401.

[203] Yosida, K. 1996. *Theory of Magnetism*. Springer Series in Solid State Sciences, vol. 122. Springer.

[204] Yosida, Kei. 1957. Magnetic Properties of Cu-Mn Alloys. *Phys. Rev.*, **106**, 893–898.

[205] Yosida, Kei. 1966. Bound State Due to the $s - d$ Exchange Interaction. *Phys. Rev.*, **147**, 223–227.

[206] Zener, C. 1951. Interaction between the d Shells in the Transition Metals. *Phys. Rev.*, **81**, 440–444.

[207] Ziman, J. M. 1965. Transport Phenomena in Helium II. *Proc. Phys. SOC.*, **86**, 337.

Index